KT-376-170

Christ Church
T011183

PHYSICS
Foundations and Applications

volume I

PHYSICS
Foundations and Applications

volume I

ROBERT M. EISBERG
Professor of Physics
University of California, Santa Barbara

LAWRENCE S. LERNER
Professor of Physics
California State University, Long Beach

McGraw-Hill Book Company
New York St. Louis San Francisco Auckland Bogotá Hamburg
Johannesburg London Madrid Mexico Montreal New Delhi
Panama Paris São Paulo Singapore Sydney Tokyo Toronto

PHYSICS: Foundations and Applications, ***volume I***

3 4 5 6 7 8 9 0 RMRM 8 9 8 7 6 5 4 3 2

This book was set in Baskerville by Progressive Typographers.
The editor was John J. Corrigan;
the designer was Merrill Haber;
the production supervisor was Dominick Petrellese.
The photo researcher was Mira Schachne.
The drawings were done by J & R Services, Inc.
Rand McNally & Company was printer and binder.

Library of Congress Cataloging in Publication Data

Eisberg, Robert Martin.
Physics, foundations and applications.

Includes index.
1. Physics. I. Lerner, Lawrence S., date
joint author. II. Title.
QC21.2.E4 530 80-24417
ISBN 0-07-019091-7 (v. 1)

Cover: "Vega-Nor" by Victor de Vasarely, reproduced by courtesy of the Albright-Knox Art Gallery, Buffalo, New York, and the Vasarely Center, New York, New York.

Contents

Preface

Science is constructed of facts, as a house is of stones. But a collection of facts is no more a science than a heap of stones is a house.

HENRI POINCARÉ
SCIENCE AND HYPOTHESIS

In this book we present the science of physics in a carefully structured manner which emphasizes its foundations as well as its applications. The structure is flexible enough, however, for there to be paths through it compatible with the various presentations encountered in introductory physics courses having calculus as a corequisite or prerequisite.

We have always kept in view the idea that a textbook should be a complete study aid. Thus we have started each topic at the beginning and have included everything that a student needs to know. This feature is central to the senior author's successful textbooks on modern physics and on quantum physics.

The book is written in an expansive style. Attention paid to motivating the introduction of new topics is one aspect of this style. Another is the space devoted to showing that physics is an experimentally based science. In Volume I direct experimental evidence is repeatedly brought into the developments by the use of photographs. And although the experiments underlying the topics considered in Volume II generally do not lend themselves to photographic presentation, at least the flavor of the laboratory work is given by including careful descriptions of the experiments. Still another aspect of the expansive style is found in the frequent discussions of the microscopic basis of macroscopic phenomena.

Developments are often presented in "spiral" fashion. That is, a qualitative discussion is followed by a more rigorous treatment. An example is found in the development of Newton's second law. Chapter 1 introduces its most important features in a purely qualitative way. When the second law is treated systematically in Chap. 4, Newton's approach, using intuitive notions of mass and force, is followed by Mach's approach, where mass and force are defined logically in terms of momentum in a manner suggested by the analysis of a set of collision experiments.

The book contains many features designed to help the student. For instance, when a term is defined formally or by implication, or is redefined in a broader way, it is emphasized with boldface letters. And all such items in boldface are listed in the index to make it easy to locate definitions which a student may have forgotten.

It is not intended that course lectures cover every point made in the book. The book can be relied upon to do many of the straightforward things that need to be done, thereby freeing the instructor to concentrate on the things that cause students the most trouble. Instructors interested in teaching a self-paced course will find that the completeness of this book makes it well adapted to use in such a course.

A novel feature of this book is the use of numerical procedures employing programmable calculating devices. At the risk of giving them more emphasis than is warranted by their importance to the book, we describe in the following paragraphs what these procedures make possible, and how they can be implemented. Numerical procedures are used for:

1. Numerical differentiation and integration. For students concurrently studying calculus this drives home the fundamental concepts of a limit, a derivative, and an integral.

2. Assistance in curve plotting. This is put to good use in studying ballistic trajectories, electric field lines and equipotentials, and wave groups.

3. Numerical solution of differential equations. This procedure permits the use of Newton's second law in a variety of cases involving varying forces. It also is applied to the vibration of a circular drumhead, *LRC* circuits, and Schrödinger's equation.

4. Simulation of statistical experiments. The procedure allows fundamental topics of statistical mechanics to be introduced in an elementary way.

5. Multiplication of several 2 by 2 matrices. This makes practical the introduction of a very simple yet very powerful method of doing ray optics.

The principal advantage of using numerical procedures in the introductory course is that it frees the *physics* content of the course from the limitations normally imposed by the students' inability to manipulate differential equations analytically or to handle certain other analytical techniques. To give just one example of the many embodied in this book, we have found that students are quite interested in the numerical work on celestial mechanics and are well able to understand the physics involved. It is the *mathematical* difficulty of the traditionally used analytical techniques that normally mandate the deferral of this material to advanced courses.

The advantages of the numerical procedures go the other way as well —they open up mathematical horizons not usually accessible to the introductory-level student. The analytical solution of a differential equation generally requires an educated guess at the form of the solution. It is precisely such a guess that the student is not prepared to make, or to accept from others. But the numerical solution suggests the correct guess strongly and directly. Our experience is that students armed with such insight can go through the analytical solution confidently. The book exploits this advantage on several occasions.

The numerical work can be presented in the lecture part of a course in various ways. One which has proven to be successful is to demonstrate to the students the first numerical procedure that is emphasized by using a closed-circuit TV system to provide an enlarged view of the display of a programmable calculator or small computer running through the procedure. (Programs for every numerical procedure used in the book, and step-by-step operating instructions, are given in the accompanying pamphlet, the *Numerical Calculation Supplement.*) After the demonstration, a graph of the results obtained is shown to the students by projecting a transparency made from the appropriate figure in the book, and the significance of the results is explained. In subsequent lectures involving numerical procedures, all that need be done is to graph their results and then discuss the meaning of the results. An instructor who is more inclined to numerical procedures may want to give more demonstrations; one who is less convinced of their worth need not give any. The essential point is that explanations of the physics emerging from the numerical work can be well understood by students who do no more than look carefully at graphs of the results obtained.

But it goes without saying that students will get more out of an active involvement with the numerical procedures than a passive one. The most active approach is to ask the students to do several of the homework exercises labeled Numerical in each of the fourteen chapters where some use is made of numerical procedures. But the instructor should not assign too many numerical exercises, particularly at first, because some are rather time-consuming. A good way to start is to make the numerical exercises optional or to give extra credit for them. Instruction in operating a programmable calculator or small computer can be given in a laboratory period or in one or two discussion periods.

We now describe paths which may be taken through this book, other than the one going continuously from the beginning to the end.

1. Several entire topics can be deleted without difficulty. These are: relativity, Chaps. 14 and 15; fluid dynamics, Secs. 16-6 and 16-7; thermal physics, Chaps. 17 through 19; changing electric currents, Chap. 26; electromagnetic waves, Chap. 27; optics, Chaps. 28 and 29; and quantum physics, Chaps. 30 and 31.

2. We believe the book contains as much modern physics as should be in the introductory course. This material is distributed throughout the book, but it has been written in such a way that there will be no problem in presenting it all in the final term. To do so, the following material should be skipped in proceeding through the book, and presented at the end: Chaps. 14 and 15; Secs. 20-1, 20-3, 22-4, 22-5, 23-3, 24-2, 24-4, and 24-5. Then close with Chaps. 30 and 31.

3. In some schools the study of thermal physics is undertaken before that of wave motion. For such a purpose Chaps. 16 through 19 can be treated before Chaps. 12 and 13.

4. If it is desired to present a shorter course in which no major topics are to be deleted, the sections in the following list can be dropped without significantly interrupting the flow of the argument and without passing over material essential to subsequent subject matter. (In some cases it will be necessary to substitute a very brief qualitative summary of the ideas not treated formally when the need for these ideas arises. Sections marked with an asterisk are those to be deleted if it is desired to avoid entirely the wave equation in its various forms. If this is done, electromagnetic radiation may still be treated on a semiquantitative basis.) The sections which can be dropped are: 2-5, 2-8, 3-7, 4-2 (if some of the examples are used later), 5-4, 5-5, 6-1, 6-6, 7-1, 7-3, 8-2, 8-5, 9-7, 10-2, 10-3, 11-2, 11-4, 11-7, 12-3*, 12-4*, 12-5, 12-6, 13-4*, 13-5*, 13-6, 13-7, 13-8, 15-5, 15-6, 16-5, 16-6, 17-5, 18-6, 19-6, 19-7, 20-1, 20-3, 21-5, 21-8, 22-4, 22-5, 23-3, 24-2, 24-4, 24-5, 25-4, 26-6, 26-7, 26-8, 26-9, 27-3*, 27-4*, 27-6, 28-5, 28-7, 29-2, 29-6,

29-7, 30-3, 30-4, 31-3, and 31-4. In addition, any material in small print can be dropped.

Many persons have assisted us in writing this book. In particular, advice on presentation or on technical points and/or aid in producing many of the photographs was given by R. Dean Ayers, Alfred Bork, John Clauser, Roger H. Hildebrand, Daniel Hone, Anthony Korda, Jill H. Larken, Isidor Lerner, Narcinda R. Lerner, Ralph K. Myers, Roger Osborne, and Abel Rosales. The manuscript was reviewed at various stages, in part or in whole, by Raymond L. Askew, R. Dean Ayers, Carol Bartnick, George H. Bowen, Sumner P. Davis, Joann Eisberg, Lila Eisberg, Austin Gleeson, Russell K. Hobbie, William H. Ingham, Isidor Lerner, Ralph K. Myers, Herbert D. Peckham, Earl R. Pinkston, James Smith, Jacqueline D. Spears, Edwin F. Taylor, Gordon G. Wiseman, Mason Yearian, Arthur M. Yelon, and Dean Zollman. Isidor Lerner contributed many of the exercises; others were written by Van Bluemel, Don Chodrow, Eugene Godfredsen, John Hutcherson, William Ingham, Daniel Schechter, and Mark F. Taylor. Dean Zollman assisted greatly in selecting and editing exercises. Don Chodrow and William Ingham checked all solutions, compiled the short answers that appear in the back of the book, and prepared the *Solutions Manual.* Herbert D. Peckham wrote the original versions of the computer programs for the *Numerical Calculation Supplement.* Lila Eisberg played a major role in reading proof and prepared the index. Important contributions to the development of the manuscript and its transformation into a book were made by John J. Corrigan, Mel Haber, Annette Hall, Alice Macnow, Peter Nalle, Janice Rogers, Jo Satloff, and Robert Zappa at McGraw-Hill, and by our photo researcher, Mira Schachne. Many students at the University of California, Santa Barbara, and at California State University, Long Beach, had a real impact on the manuscript by asking just the right questions in class. To all these persons we express our warmest thanks.

Robert M. Eisberg
Lawrence S. Lerner

PHYSICS
Foundations and Applications

volume I

1

An Introduction to Physics

1-1 WHAT IS PHYSICS?

Physics is the systematic study of the basic properties of the universe. Each of these properties is related to interactions among the objects found in the universe. In the branch of physics known as cosmology, the overall structure of the universe itself is analyzed by taking into account interactions between every part of the universe and every other part. But in more typical branches of physics any one object can be assumed to interact in a significant way only with other objects that are not too distant. Thus a property usually can be studied by considering the interactions among a limited set of objects, as well as the interactions this set may have with a few other objects in its vicinity. A set of objects on which attention is focused is called a **system.** The smallest physical systems are those investigated in elementary-particle physics; they are the tiny building blocks from which everything else is constructed. The largest physical system is the entire universe, as it is treated in cosmology. Between these extremes lies a tremendous variety of systems that are studied in physics.

The motivations for studying physical systems are almost as varied as the systems themselves. A container of the superfluid liquid helium is an example of a system that may be studied because of its inherent interest. Often, however, the investigation of a system is directed toward a commercial or social end. This happens in semiconductor physics and medical physics. Sometimes the motivation lies in the light the work will shed on a field of science other than physics; an example is the application of physics to molecular biology. In such a situation, the line separating that field and physics is sure to become blurred and eventually obliterated. This has already occurred in astronomy, chemistry, and several branches of engineering, and is currently occurring in molecular biology.

The strategy used in the study of physical systems is one of the most powerful inventions of the human mind. Its fruits have completely transformed the way the human race lives, the way its members think, and the world they inhabit. The strategy of physics has three characteristic features.

The first feature is that the analysis of a physical system tends to be carried out in terms of the properties of simpler systems. In investigating a system, a physicist seeks to treat separately each factor influencing its behavior. This usually involves switching attention to a collection of simpler systems. Each of these is related in some important way to the original system, but has fewer factors that are vital to its behavior. Being simpler, these systems can be investigated to the extent that their properties are well understood. When this has been done, the information obtained can be mentally reassembled into an understanding of the properties of the original system. But sometimes this last step is not taken, or only partly taken, because it is found that the information of most general significance to physics is obtained in the analysis of one of the simpler systems. You will see an example of this feature in Sec. 1-3.

The second feature of the strategy is that physics is uncompromisingly based on experiment. You will see an example of this also in Sec. 1-3. Sometimes theory suggests experiment. Much more frequently, though, an experimentalist does the pioneering work in a particular area of physics, and the theoretician then synthesizes the results of the experiments and deepens the understanding of their significance. A really good theory suggests new and interesting experiments which can be used to confirm it. But beyond this, such a theory goes on to suggest new areas of interest.

Whether he or she is experimentally or theoretically inclined, the physicist acknowledges that ideas must be tested by experiments. No matter how beautiful an idea may seem, no matter how attached he or she may have become to it, the physicist agrees in advance to abandon the idea if it conflicts with the evidence furnished by a crucial experiment. But experience repeated innumerable times has shown that the abandoned idea is never as beautiful as the one which ultimately emerges to provide an accurate description of nature.

A third feature of the strategy of physics is the frequent use of mathematics. You may have heard the saying that mathematics is the language of physics. It is worth exploring what the saying means. Physics is concerned with the interactions among objects. We believe that objects interact according to certain laws, whether we know those laws or not. Since physical laws are almost always quantitative, it becomes essential to be able to trace quantitative logical connections in studying physical systems. The rules governing all such connections (whether they have anything to do with the physical universe or not) are the subject of mathematics. Thus most of the rules and procedures of mathematics are directly applicable to the understanding of physics. Mathematics is used in physics because in all except the simplest situations it provides by far the most *convenient* way to trace the logical relationships that arise in the analysis of physical systems.

Granted that mathematics is the language of physics, we must stress that this does not mean that mathematics *is* physics, or vice versa. Consequently, when we obtain a result from a mathematical argument, we will be interested principally in both the physical meaning of the steps used to obtain it and the experimental verifiability of the result.

The part of the mathematical language which you will need to use most frequently at first relates certain quantities to the rate of change of other quantities, that is, differential calculus. We assume that you come to the study of physics without having previously studied calculus. All aspects of calculus that you will need for using this book are developed from scratch, as they are needed. But this is not a calculus book, and you certainly should be working through such a book as you work through this one.

In the many years since it evolved in the sixteenth century, the use of the strategy of physics has spread to all fields of science. Indeed, some fields, such as psychology and economics, are considered "scientific" to the extent that they make use of the strategy or parts of it. The strategy is most successfully applied in physics, however, because it is especially suitable for the relatively simple systems which are the main concern of physics. It may be far from easy for a physicist to analyze one by one the factors affecting the behavior of a simple atom, using the language of mathematics, and then to verify in the laboratory the results obtained. But it is still much less difficult than it is for a chemist to carry out the analogous task for a complex molecule. More difficult yet is the task of a biologist in carrying out a similar procedure on the enormously complex molecule called a gene. To make full application of the strategy of physics in studying a living animal is a practical impossibility. The point is that the number of significant factors increases rapidly with the complexity of a system, and these factors become so intimately interrelated that they cannot be separated. Put briefly, physics is the simplest science because it studies the simplest systems. For this reason physics forms the foundation of all other sciences.

The connection between physics and engineering is even more direct than that between physics and any of the other sciences. An engineer deals with systems to which the principles of physics are immediately applicable. Electrical engineering, as an example, is in large part a matter of making highly sophisticated practical applications of the work of nineteenth- and twentieth-century physicists. No matter what field of engineering or science you are planning to enter, you will find continued use for what you learn in the study of physics. You will find use for the specific facts of physics, for the techniques used in solving physics problems, and for the frame of mind you inevitably acquire through the study of physics.

It is thus that physics assumes a central role. Its facts, its procedures, and its view of the world find vital applications everywhere. In addition—and this is a point of great significance to the physicist—the edifice of physics is a beautiful and fascinating work of the intellect. Albert Einstein called it a miracle that the human mind is so constructed as to have the ability to comprehend the universe of which it is a part. Whether you wish to call it a miracle or not, you will come, as your understanding of the subject develops, to agree with the gist of his statement.

1-2 THE DOMAINS OF PHYSICS

It is always useful to divide a complex field into domains. For physics one such division can be made according to the sizes of the objects studied. Specifically, objects are considered small if their sizes are comparable to or smaller than the size of an atom. Objects are considered large if they are larger than an atom. This division into the domains of small and large objects is shown schematically in Fig. 1-1*a*. The domain of small objects is called the **quantum domain,** and the domain of large objects is called the **nonquantum domain.** (This division according to size is not a completely rigid one since in certain circumstances phenomena of the quantum domain can be seen in large objects. The superfluid properties of a container of liquid helium provide an example.)

Another division of physics into domains can be made according to the speeds of the objects being studied, as shown in Fig. 1-1*b*. If the speeds are small compared to the speed of light, they are considered low. If they are comparable to the speed of light, they are considered high. The domain of high speeds is known as the **relativistic domain,** and the domain of low speeds is known as the **nonrelativistic domain.** (There is flexibility in the division according to speed, just as there is in the division according to size. An example is found in the magnetic effects produced when electrons move through a wire. Although the speeds of the electrons are small compared to the speed of light, the effects are, basically, caused by a phenomenon of the relativistic domain.)

A combination of the divisions according to size and speed is indicated in Fig. 1-1*c*. Emphasis has been put on the region which lies in both the domain of large objects and the domain of low speeds. This region has a special significance—it is the one we deal with in our everyday lives—and is given a special name. It is called the **newtonian domain,** in honor of Isaac Newton, the seventeenth-century physicist who played the key role in developing the physics of large objects moving at low speeds.

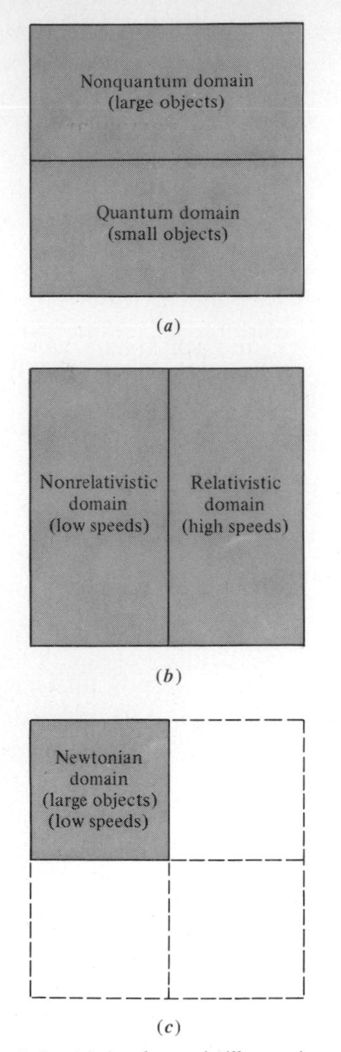

Fig. 1-1 (*a*) A schematic illustration of the division of physics into the nonquantum domain, in which the objects studied are large compared to an atom, and the quantum domain, in which the objects studied are comparable in size to or smaller than an atom. (*b*) The division of physics into the nonrelativistic domain, in which the speeds of the objects studied are low compared to the speed of light, and the relativistic domain, in which the speeds are comparable to the speed of light. (*c*) The domain in which the objects studied both are large compared to an atom and have speeds low compared to the speed of light is called the newtonian domain. The name honors Isaac Newton, who developed the form of mechanics that is used to obtain accurate predictions of the motion of objects in this domain.

When objects interact, they exert forces on one another. The force acting on an object determines how the object moves. Mechanics is the study of the relation between the force and resulting motion. That is, **mechanics** seeks to account quantitatively for the motions of objects having given properties in terms of the forces acting on them. The mechanics of the newtonian domain is known as **newtonian mechanics.** It is the mechanics of systems containing objects which are large compared to an atom and which move at speeds that are low compared to the speed of light. An example is our planetary system. Planets are certainly very large compared to an atom, and they move with speeds quite low in comparison to the speed of light. Near the end of the seventeenth century Newton took a giant step in explaining the astronomical observations that had been made on the planets over the preceding centuries. Out of Newton's work flowed a powerful science capable of explaining an immense variety of familiar natural phenomena.

Until shortly after the turn of the present century it was thought that there were no limits to the applicability of newtonian mechanics. But in 1905 Einstein showed that a quite different (though not unrelated) approach was necessary for the study of objects moving with speeds so high as to be comparable to the speed of light. These objects are in the relativistic domain and must be treated by **relativistic mechanics.**

At about the same time Max Planck, Louis de Broglie, Erwin Schrödinger, and others found that newtonian mechanics could not explain the motion of objects whose size is on the atomic scale or smaller. These objects require the use of **quantum mechanics** because they are in the quantum domain.

Some of the most exciting work in contemporary physics lies in the quantum domain, or in the relativistic domain, or in both. Nevertheless, the logical structure of physics demands that this book begin in the newtonian domain with newtonian mechanics. Furthermore, newtonian mechanics is very important in itself. The fact that it is the oldest form of mechanics by no means renders it obsolete or inactive. In particular, many contemporary engineering and biological applications of physics are based completely on newtonian mechanics. However, two chapters in the middle of the book are devoted to relativistic mechanics, and two chapters at the end of the book to quantum mechanics.

You may have expected mention, in this categorization of physics, of topics such as electromagnetism, heat, acoustics, and solid-state physics. How do these fit into the structure we have set forth?

Solid-state physics is an example of a branch of physics defined by its subject matter—the properties of solids—rather than by a certain procedure used to study the subject matter. A particular problem in solid-state physics, for instance the properties of materials used in transistors, is attacked by employing the mechanics of whichever domain is most appropriate. In most cases, this is quantum mechanics or newtonian mechanics.

Acoustics and heat concern particular phenomena in various kinds of matter, for instance the creation of sonic booms by a supersonic airplane. Very often they can be understood completely in terms of systems of particles acting in conformity with the rules of newtonian mechanics, though some cases require quantum mechanics.

Electromagnetism is a study of the properties and consequences of the **electromagnetic force,** which is one of the four fundamental forces in nature. The electromagnetic force plays an important role in all domains of physics. For instance, it is the force of overwhelming importance in solid-state physics, the study of which involves several domains.

Of the three fundamental forces other than the electromagnetic one, the **gravitational force** is the most familiar. The remaining two, called the **strong nuclear force** and the **weak nuclear force,** are unfamiliar because they operate over only very small distances. Thus we do not experience them directly with our senses. Nevertheless, the strong and weak nuclear forces are the dominant forces holding together the smallest parts of the universe. Since the larger parts are built up of the smaller ones, many large-scale properties of the universe do depend ultimately on the nuclear forces. We encounter all four fundamental forces in our study of physics, but electromagnetism and gravitation will occupy most of our attention.

In the next section we begin to look at the newtonian domain through the eyes of the physicist by asking the basic question of newtonian mechanics: What is the relation between force and motion?

1-3 FORCE AND MOTION IN NEWTONIAN MECHANICS

Whether or not you have previously taken a physics course, you come to the study of physics in this book having more experience with the subject than you may think. You could not have survived in the physical world without a considerable grasp of how it operates, particularly in its mechanical aspects. In this section we make a first approach to newtonian mechanics by investigating the relation between force and motion. This approach takes the form of qualitative discussions based partly on the mechanical intuition you have gained from everyday experience and partly on experimental evidence provided by a set of photographs. Your intuitive understanding of many of the topics considered here makes it possible for the discussions to be quite brief. But each topic is treated in a much more thorough and systematic way in subsequent chapters.

One purpose of presenting such material at this point is to give you a preliminary example of the strategy of physics. The arguments must be qualitative at this stage because we have not established the physical foundations for quantitative ones. Thus mathematics is not applicable, and so the third feature of the strategy is not used. Otherwise, the approach is representative of that actually employed in physics because the first and second features are used. In particular, this section should give you some idea of (1) the way a physicist's investigation of a complicated system leads to the investigation of a set of related but progressively simpler systems and (2) the way experiment is used in this process.

As was said earlier, newtonian mechanics is concerned with familiar objects (objects not very small) moving in familiar ways (at speeds not very high). The aim is to explain the motion of an object in terms of the force acting on it. Motion occurs when an object changes its position. Here we consider the simplest kind of motion, in which the object moves always in the same direction along a straight line. One of the terms used to describe the motion is commonplace. It is *speed.* (In this case of motion along a straight line in only one direction, it is not necessary to make the technical distinction between speed and velocity because the two quantities are equivalent.) Speed is a measure of how much the position of the object changes—in other words, how far it moves—in a certain small interval of time. In fact, the speed of an object moving in a particular direction along a straight line is just the change in its position divided by the time interval in which the change occurs.

The other term used to describe the motion we are considering is almost as widely used in everyday language as speed. It is *acceleration.* The term acceleration is useful if the speed of the object is not constant. Acceleration is a measure of how much the speed of the object changes in a certain small interval of time. Specifically, the acceleration of an object moving in a particular direction along a straight line is just the change in speed divided by the time interval in which the change occurs.

If an object moving in only one direction along a straight line has a *constant speed,* then it has *no acceleration.* If the object has an *increasing speed,* then in common language it is said to be *accelerating,* and in technical language it is said to have an acceleration in the *same direction* as the direction of its motion. If the object has a *decreasing speed,* then commonly it is said to be *decelerating,* and technically it is said to have an acceleration in the *opposite direction* to its direction of motion.

Speed and acceleration are related but distinctly different quantities. Consider the following examples. A car which is stopped at a red light has no speed and no acceleration. A short time after the driver sees the light turn green and steps hard on the gas pedal, the car has a small speed and a large acceleration in the direction of its motion. When the car is moving along a straight stretch of highway and its speed has nearly reached the legal limit, it has a large speed and a small acceleration in the direction of its motion. Shortly after the driver sees a traffic jam ahead and presses hard on the brake pedal, the car has a large speed and a large acceleration in the direction opposite to its motion. The direction of its acceleration is opposite to the direction of its motion because its speed is decreasing.

Force is another term commonly used in everyday language, although in some of these uses the meaning of the word has little relation to its technical meaning. You certainly know that you can exert a force on an object by pushing or pulling on it. The force acts in the direction in which you push or pull, and its strength is a measure of how hard you do it. You probably also know that the earth exerts a force on an object near its surface by means of gravitational attraction, pulling the object in the downward direction (toward the center of the earth). The strength of the force is a measure of the weight of the object; in fact, it *is* the weight.

Familiar but less well understood than the two forces just discussed is the frictional force. This force is produced by the frictional effects which in many situations are experienced by objects. If an object is moving, the frictional force exerted on it by another object is in whatever direction is opposite to its direction of motion past that other object. The strength of the frictional force depends on circumstances that will be mentioned shortly.

With these ideas in mind about the quantities used to describe force and motion, let us return to the central question: What is the relation between force and motion? Three possible answers are (1) there is no relation, (2) the force applied to an object is related to its speed, and (3) the force applied to an object is related to its acceleration. We will find the correct answer by studying a set of simple experiments.

Figure 1-2 is a stroboscopic photograph of the first experiment. A squat cylindrical object, called a puck, was initially motionless on the horizontal top of a table. Then the experimenter pushed the puck in a particular direction so that it moved uniformly in a straight line across the unlubricated tabletop. He reported that in doing this he felt the sensation that the force applied to the puck by his hand was of constant strength. Stroboscopic lights were used to illuminate the puck while it was moving. These lights flashed very briefly at regular intervals of time. A camera viewed the puck from above with its shutter remaining open. With each light flash the position of the puck at that instant was recorded on the film. So the strobe

Fig. 1-2 A strobe photo looking down on a puck that is being continually pushed at constant speed across the top of a table.

photograph is, in effect, a sequence of "snapshots" showing the positions of the puck at equally separated times.

You can see from the strobe photo that the puck moved very nearly equal distances in each of the equal time intervals. Thus its speed was approximately constant while the hand was exerting an approximately constant force on the puck.

From this experiment you can conclude that possible answer 1 to the question posed is wrong. It is evident that there *is* a relation between force and motion. Before the hand pushed the puck, it was stationary. After the hand began exerting the force, the puck moved.

Another conclusion you might draw from the experiment is that answer 2 is right. That is, you might conclude that force is related to speed, since an approximately constant force applied by the hand led to an approximately constant speed of the puck. However, this is *not* a correct conclusion.

If you came to that conclusion, take comfort, because you are in good company. Up to the time of Galileo Galilei (1564–1642) almost all scholars who were concerned about the question believed that a body acted on by a constant force moves with a constant speed. Consider a modern example. You are driving a car along a level, straight road at a certain constant speed by keeping the gas pedal depressed a certain amount, thereby making the engine produce a constant force of propulsion. If you want to travel at a higher constant speed, you must depress the gas pedal more, so as to increase the force produced by the engine of the car. These considerations seem to imply, again, that force is related to speed. But this is not so.

What was not recognized by the ancients, and is still overlooked by some moderns who have not studied physics, is that *two* significant forces act on the car, or on the puck. Each is a bona fide force, although one is more obvious than the other. The obvious force is the one that the engine causes to be applied to the car, or the even more obvious one that is applied by the hand to the puck. The other force is a less apparent—but equally important—force applied by frictional effects. In the case of the car, this force arises primarily from air resistance. For the puck moving at rather low speed across the table, the frictional force acting on it is almost entirely due to the contact between its slightly rough bottom surface and the slightly rough surface of the tabletop. But in both cases the frictional force under consideration always acts in the direction opposite to the direction of motion of the object on which it is exerted.

The frictional force of air resistance is rather complicated in that its strength increases rapidly with the speed of the car. This is why the engine must produce a stronger propulsive force to move the car at a higher fixed speed. The engine must cause a harder push to be applied to the car to compensate for the stronger frictional retarding force that is experienced when the car is moving faster. A contact friction force has the simple property that its strength is almost independent of the speed of the moving object.

The strobe photo in Fig. 1-3 illustrates the effect of contact friction on the motion of the puck in the more elementary situation where it is the only force acting along the *line of motion* (in other words, it is the only force acting either in the direction of motion or in the opposite direction). In the experiment recorded in this photograph, the propelling hand stopped short after the puck had been set into motion. The puck then moved across

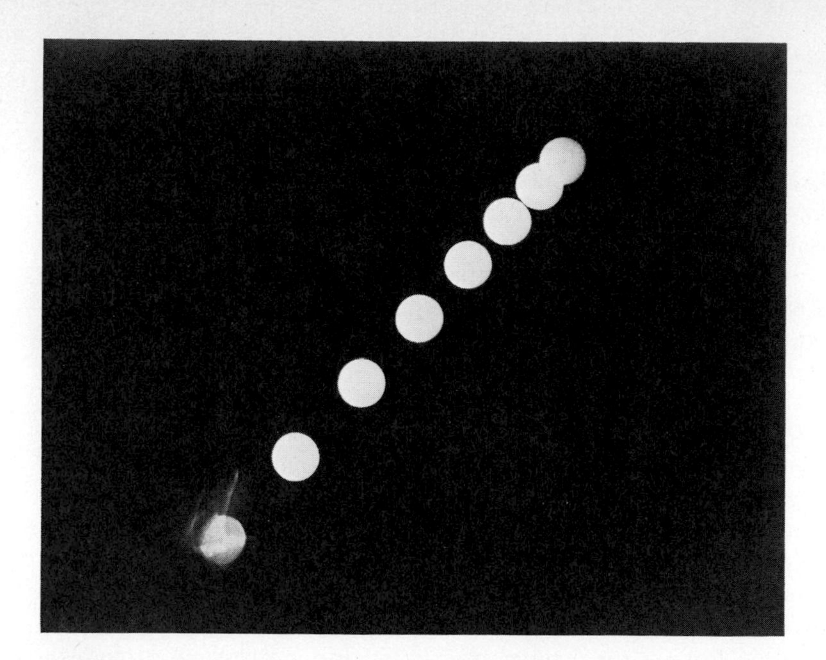

Fig. 1-3 A strobe photo looking down on a puck that is given an initial push across the top of a table. It goes slower and slower, until it comes to rest, because there is appreciable friction. Later in the argument you will be asked to show that the acceleration of the puck is nearly constant while it is slowing down. When it is time to do this, read the following instructions. Begin by labeling as 1 the point at the center of the first image of the puck that is well away from the hand, labeling the next center point as 2, and so on. Then lay a ruler having decimal subdivisions along the line connecting these points, with its left end at some convenient location near the image of the hand. Use the ruler to determine carefully the numbers specifying the locations of the points labeled 1, 2, and so on. These are the "coordinates" of the points. Find the difference between the first pair of coordinates by subtracting coordinate 1 from coordinate 2. This gives the change in the puck's position in the time interval between the corresponding pair of stroboscopic light flashes. So it is a measure of the puck's speed in that time interval. Do the same for coordinates 2 and 3 to measure the speed of the puck in the next time interval. Then subtract the 1,2 difference from the 2,3 difference. The result is a measure of the change in speed between the consecutive time intervals 1,2 and 2,3. Therefore it is a measure of the puck's acceleration in the first part of the motion. The value will be negative because the puck is slowing down. Now repeat the entire procedure using the coordinates of the next three points to obtain a measure of the puck's acceleration in the next part of its motion. You will find that the two values are equal, within the limited accuracy that can be expected from this procedure.

the table under the influence of the only significant force acting on it. This force was the contact friction force acting in the direction opposite to the direction of the puck's motion. It is not surprising that the puck slowed down and soon came to rest. You can see that this is so by noting that in the equal time intervals between consecutive stroboscopic light flashes, the puck moved an ever-decreasing distance. So its speed decreased throughout its motion, until it stopped.

While the puck was slowing down, its acceleration was in the direction opposite to the direction of its motion. And while this was happening, the only significant force acting on the puck was that of contact friction, which also acted in the direction opposite to the direction of its motion. Thus the experiment shows that *a force exerted on the puck in a certain direction leads to an acceleration in that direction.* The experimental result is suggestive.

If you believe that the strength of a contact friction force acting on an object is almost constant as long as the object is moving, then the experiment can be even more suggestive. Use the procedure explained in the caption to Fig. 1-3, and you will find that the amount of the puck's acceleration was nearly constant while it was coming to rest. Thus the experiment shows that an approximately constant force produced an approximately constant acceleration in the direction of the force. It strongly suggests that the correct answer to the question posed is answer 3: *The force applied to an object is related to its acceleration.*

There is additional support for this statement in that it allows you to understand the *unaccelerated* motion of the puck traveling at constant speed across the table in the experiment shown in Fig. 1-2. In that experiment the constant force applied to the puck by the hand acted in the direction of the puck's motion, while the constant force applied to the puck by friction acted in the direction opposite to its motion. In order to make the puck move uniformly, the experimenter adjusted the strength of the hand-applied force to equal that of the oppositely directed friction-applied force. The moving puck was acted on by two forces of equal strength whose effects canceled each other because they were applied in opposite directions. In other words, the puck experienced no *net force.* If the force applied to an object is related to its acceleration, as suggested in the preceding paragraph, and if the word "force" is interpreted to mean "net force," then the unaccelerated motion of the puck in Fig. 1-2 can be understood.

It certainly would be worthwhile to test these ideas by studying the motion of the puck when there really is *no* force at all acting on it. Unfortunately, this cannot be done. But it is possible to eliminate almost completely those forces which act along the puck's line of motion, and these are the important forces for our considerations. This is accomplished in an experiment using a table with an almost friction-free top, called an **air table.**

A picture of an air table is shown in Fig. 1-4. It has a very flat top with a closely spaced gridwork of small holes through which air is forced by a blower. The air coming up through the holes under pressure provides a horizontal cushion on which a puck rides with almost no friction. This is because the puck does not touch the tabletop itself, but literally rides on a thin layer of air. If a puck is set into motion by a short initial push, it will then continue to move freely across the table. For all practical purposes, after the puck has started, no force acts on it in any horizontal direction until the puck hits a barrier at the edge of the table.

However, two forces always act on the puck in vertical directions. One is the downward force of gravity, and the other is the exactly canceling upward supporting force of the air cushion. But these forces are not significant in the air table experiment, since they do not affect motion in a horizontal direction. Downward and upward forces which cancel one another were also acting on the puck in the earlier experiments.

Figure 1-5 shows a strobe photo of a puck moving across the air table after being initially set into motion. It is apparent that the puck does indeed move in a straight line with constant speed when no force at all acts on it along its line of motion. Another way of stating this result is to say that *when there is no net force, there is no acceleration.*

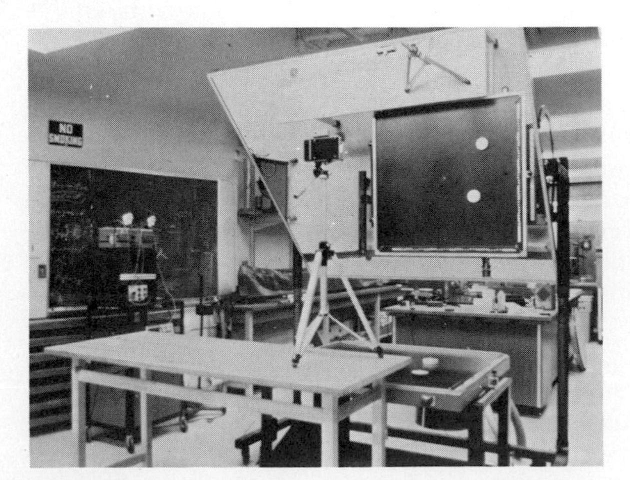

Fig. 1-4 The apparatus used to obtain the air table strobe photos in this book. A camera mounted on a lab bench records a view from above of the air table by means of an inclined mirror. The air table has a black surface to contrast with the white pucks and is illuminated by a pair of strobe lights located to the left of the table. The flexible pipe delivering compressed air to the table can be seen extending from its right-rear corner to the air pump behind it. For the sake of clarity, it was necessary to retouch some of the strobe photos obtained with this apparatus (as well as some of the other strobe photos in the book). But when this was done, care was taken not to change the positions of the pucks (or other objects) recorded in the photos.

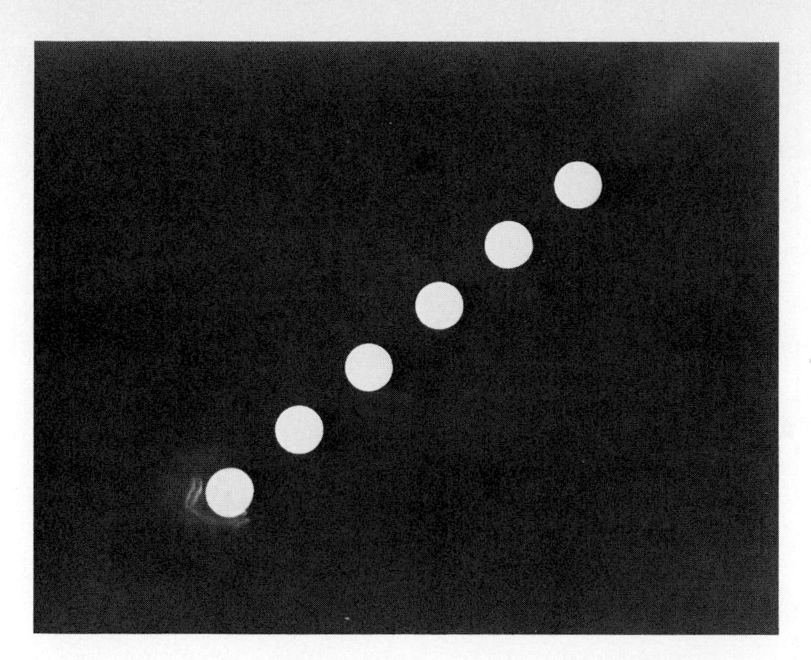

Fig. 1-5 A strobe photo looking down on a puck that is given a short initial push across the top of an air table. It continues to move at constant speed because friction is negligible.

The final experiment to be considered in this section is the simplest of all because it studies the motion of the puck when only one single force acts on it. Furthermore, you will find it easy to agree that the strength of the force is essentially constant throughout the motion of the puck. The experiment is performed by removing the support from beneath the puck, thereby allowing it to fall a short distance near the surface of the earth. Since it does not have a chance to pick up much speed, it never experiences a significant amount of air resistance. Consequently, only one force acts on the puck while it is falling. This net force is the force of gravity exerted on the puck by the earth. The direction of the force is downward. Its strength is what is called the *weight* of the puck. Since the weight is constant throughout the fall, the strength of the net force acting on the puck is constant.

The experiment is illustrated in the strobe photo of Fig. 1-6. You can see that the speed of the puck is increasing throughout its motion. It therefore has an acceleration in the downward direction of its motion. If you carry out the analysis suggested in the caption to Fig. 1-6, you will see that the amount of acceleration is constant while the puck is falling.

Thus the very simple experiment of Fig. 1-6 shows that a constant net force leads to a constant acceleration in the direction of the force. The not quite so simple experiment of Fig. 1-5 showed that when there is no net force, there is no acceleration. These two statements are in agreement with the even more complicated experiments of Figs. 1-3 and 1-2. The experiments on the set of progressively simpler systems are summarized in Fig.

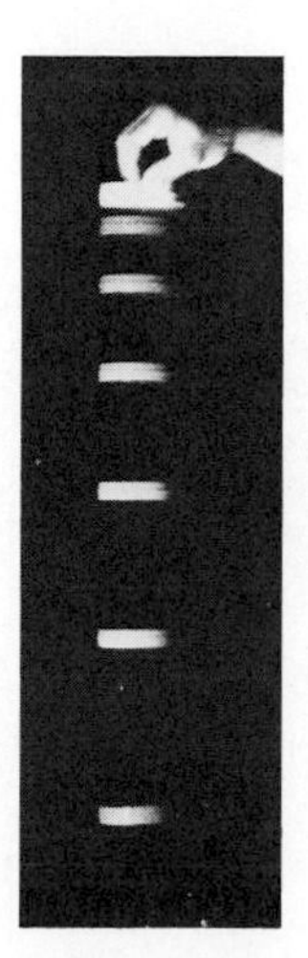

Fig. 1-6 A strobe photo looking at a falling puck in side view. Its speed increases continually because gravity continually pulls it toward the earth. To show that during its fall the puck's acceleration is constant, within the accuracy of the analysis, you can carry out the same procedure as that described in the caption to Fig. 1-3. However, in this case the value you obtain will be positive because the puck is speeding up.

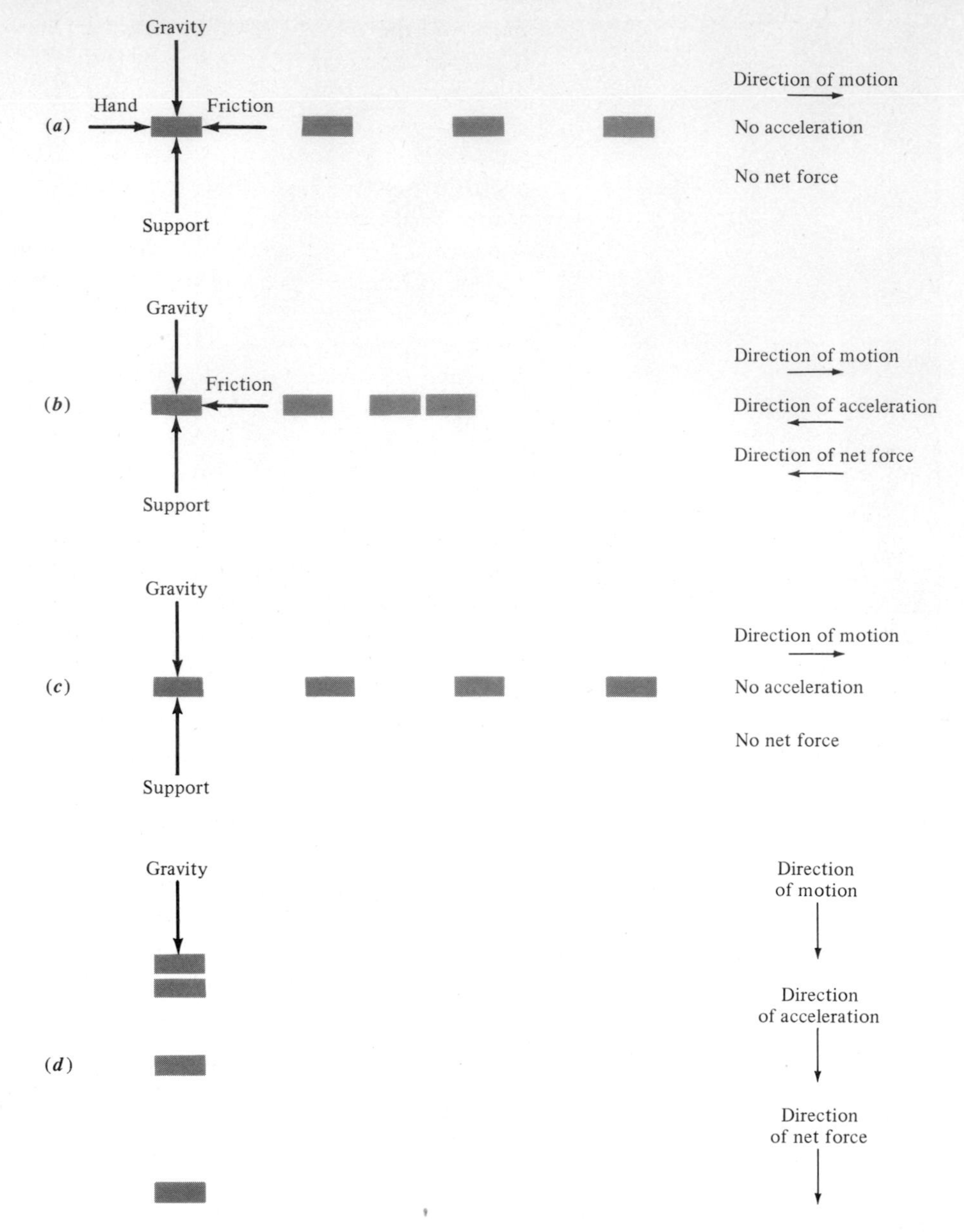

Fig. 1-7 A summary of experiments leading to the conclusion that the net force applied to an object is related to its acceleration. Parts *a*, *b*, *c*, and *d* are schematic views from the side of the puck motions illustrated in the strobe photos of Figs. 1-2, 1-3, 1-5, and 1-6, respectively. All the forces acting on a puck are indicated by arrows on the first puck image. The direction of each of these arrows shows the direction of the force, and its length indicates qualitatively the strength of the force. To the right of each part of the figure arrows show the direction of motion of the puck and, if applicable, the direction of its acceleration and of the net force acting on it. The lengths of these arrows have no significance. Note how the experiments become progressively simpler in that four forces act on the puck in the first experiment, three in the next, two in the next, and only one in the last. This is an example of the way that physicists try to focus their attention on simpler systems that have fewer factors to influence the behavior.

1-7. All the experiments support the conclusion that *the net force applied to an object is related to its acceleration.*

We are not yet in a position to establish the precise nature of the relation between the net force acting on an object and its acceleration. The most rudimentary possibility would be for the two quantities which are in the same direction to have values which are in direct proportion. In Chap. 4 we find that this is, in fact, the case. We also find there that the proportionality constant in the relation between their values is the mass of the object. To be specific, we will learn that when a net force of strength F acts on a body of mass m, it experiences an acceleration in the direction of the force whose value a satisfies the relation $F = ma$. This is the most important equation of newtonian mechanics; it is one of Newton's laws of motion. These laws make it possible to predict accurately the behavior of any system of objects in the newtonian domain if their masses, and the forces acting on them, are known. The detailed development of the laws of motion in Chap. 4 is founded on a quantitative analysis of experiments in which strobe photos record the motion of pucks on an air table. In this section you have been given a qualitative introduction to the experimental technique.

Before we embark on the development of Newton's laws of motion, we must first find out how to specify precisely what is meant by acceleration in situations which are more general than those considered in this section. That is, we must set up a description of motion that will provide an exact connection between acceleration and the directly measurable quantities position and time, for any situation of interest. We begin this task in Chap. 2.

2

Kinematics in One Dimension

2-1 ONE-DIMENSIONAL MOTION

Kinematics (from the Greek word *kinematos,* meaning motion) is the description of motion, without regard to its cause. Kinematics provides the language needed to describe *how* objects move. But it says nothing about *why* they do so; that topic starts in Chap. 4.

An object can move in two basically different ways: it can change only its orientation, or it can change only its location. Figure 2-1*a* illustrates the first type of motion. The brick shown in the figure changes its **orientation** by rotating about its center, while keeping the position of that point fixed. Figure 2-1*b* shows the brick undergoing the second type of motion. It changes its **location** by changing the position of its center, while not rotating about that point. Of course, an object can also experience both types of motion at the same time. But in the first part of this book we are concerned only with *changes in location.* The study of changes in orientation is deferred until Chap. 9.

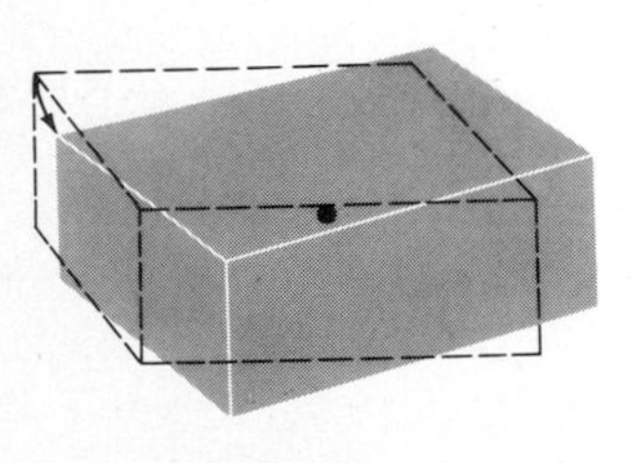

(*a*)

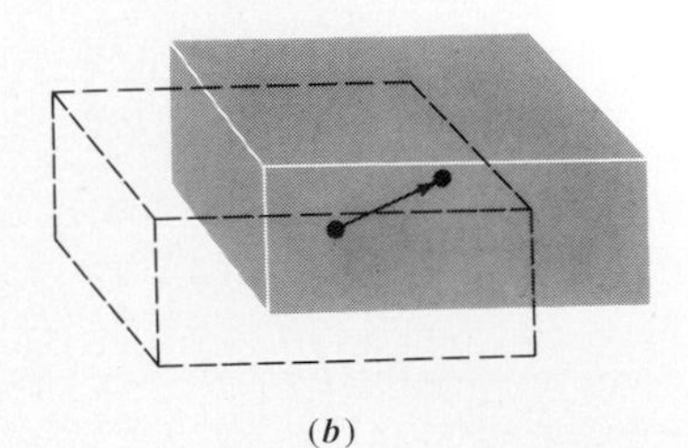

(*b*)

Fig. 2-1 (*a*) An object changing its orientation without changing the position of its center. (*b*) An object changing the position of its center without changing its orientation.

Thus we are concerned at first with an object whose behavior of interest can be described in terms of the motion of its center, or of some other point conveniently used to specify its location. As an example, if you are studying the annual motion of the earth about the sun, you can ignore the fact that it spins daily about its center and focus your attention on the changes in position of the center.

The most general way for a point locating an object to move is in three dimensions. An example of such motion is Fig. 2-2, which shows the changing location of an airplane gaining altitude by flying in a helical path. A less general, but simpler, motion is that confined to two dimensions. An airplane flying in a circle at a constant altitude provides the example illustrated in Fig. 2-3. The most restricted, but simplest, case is motion along a straight line, that is, in **one dimension.** See Fig. 2-4, which represents an airplane flying a fixed course at a constant altitude. Since we always try to start a subject by treating the simplest case, we begin the development of kinematics by considering only motion in *one dimension.* When the physics to be studied in the next chapter leads us to two- or three-dimensional motion, you will find that the extension can be made without too much difficulty.

The location of the airplane in Fig. 2-4 can be specified concisely at any instant by using its line of motion to define an x **axis.** (For the horizontal motion considered here it is natural to call the axis the x axis. But we will call the single axis that suffices to describe one-dimensional motion the x axis, no matter how it is oriented in space.) Some conveniently located fixed reference point, labeled O in the figure, is chosen as the **origin** of the x axis. Then the location of the airplane is given by stating the *distance* from O to the airplane and specifying whether the airplane is on *one* side of O or on the *other* side. This is done by giving the value of a quantity x, the x **coordinate** of the airplane. The *magnitude* of x (that is, its absolute value $|x|$) equals the distance from O to the airplane. The sign of x is *positive* if the direction from O to the airplane is in a direction that has been chosen to be the positive direction, and *negative* otherwise. (For the case illustrated in Fig. 2-4 it would be convenient to choose the positive x direction to be to the right. But the choice used does not really matter, as long as it is used consistently.)

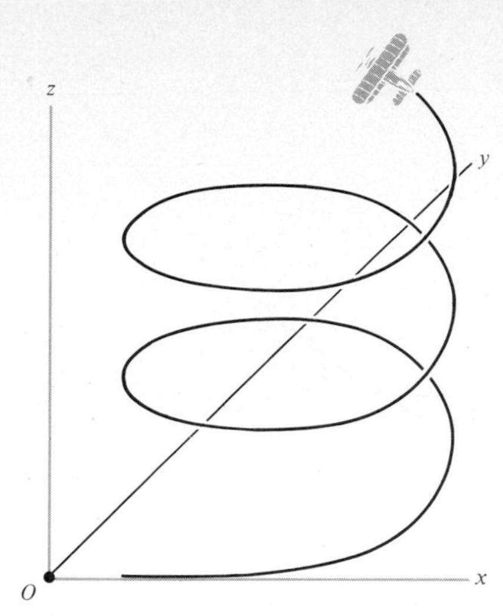

Fig. 2-2 An airplane flying in a helical climb.

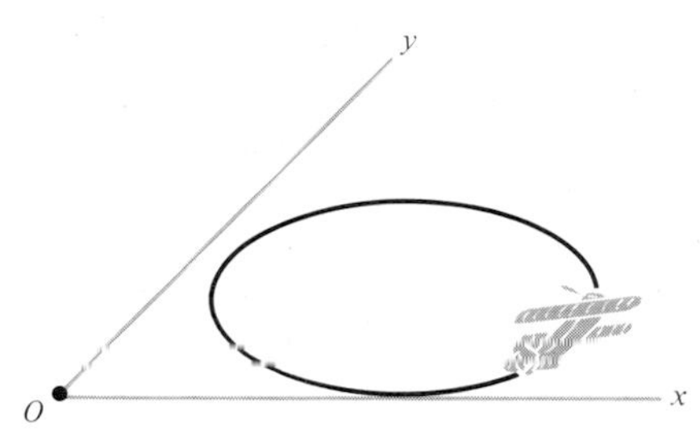

Fig. 2-3 An airplane flying in a circle at fixed altitude.

Fig. 2-4 An airplane flying straight and level.

Consider another example of one-dimensional motion. A car is moving in a certain direction on a straight road. The most rudimentary way of depicting the nature of its motion is on a figure which shows the location of the car at a succession of equally separated times. Figure 2-5 is such a strobe-photo-like illustration showing a car moving uniformly on a straight road past a green traffic light. It is said to have **uniform motion** because it moves the same distance in each of the equal time intervals between the instants at which its location is depicted. In contrast, Fig. 2-6 shows a car moving nonuniformly as it races away from a traffic light when the light turns green. These figures allow a distinction to be made between uniform and nonuniform motion. Such a distinction cannot be made in Fig. 2-4.

Also indicated in Figs. 2-5 and 2-6 is an x axis laid out along the road with its positive direction to the right and its origin located at the traffic light. The location of the car at a certain time is given by the value of x. That time is specified by the quantity t, which is the amount of time that has passed since the instant when the car goes by the light. The subscripts 1, 2, and so on are used on x and t to associate corresponding values. That is, x_1 is the coordinate corresponding to time t_1, x_2 corresponds to t_2, and so

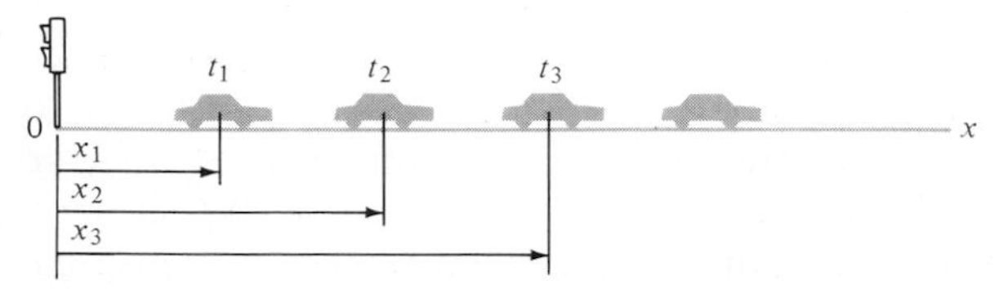

Fig. 2-5 An automobile moving uniformly along a straight road. The first three views show its locations at time t_1, t_2, and t_3 when its distances from a traffic light are, respectively, x_1, x_2, and x_3. All times are measured from when the car passes the light, and all the intervals between successive times are equal.

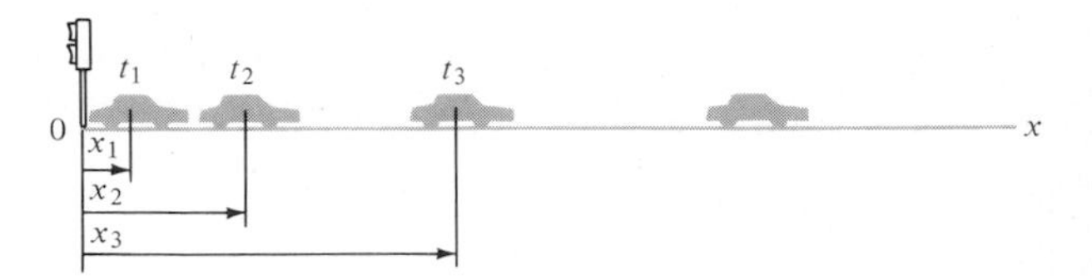

Fig. 2-6 An automobile moving nonuniformly along a straight road.

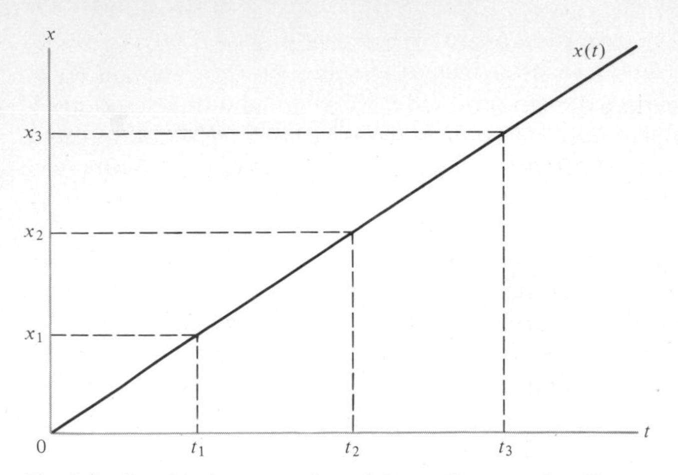

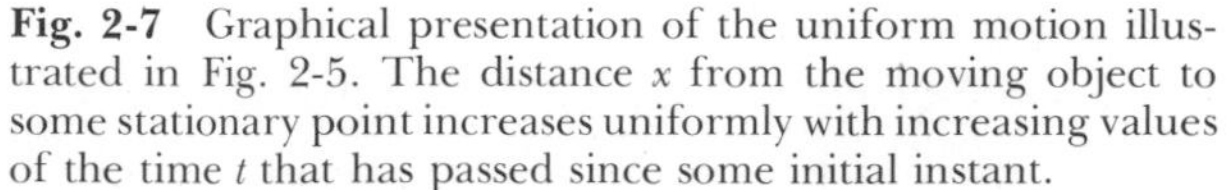

Fig. 2-7 Graphical presentation of the uniform motion illustrated in Fig. 2-5. The distance x from the moving object to some stationary point increases uniformly with increasing values of the time t that has passed since some initial instant.

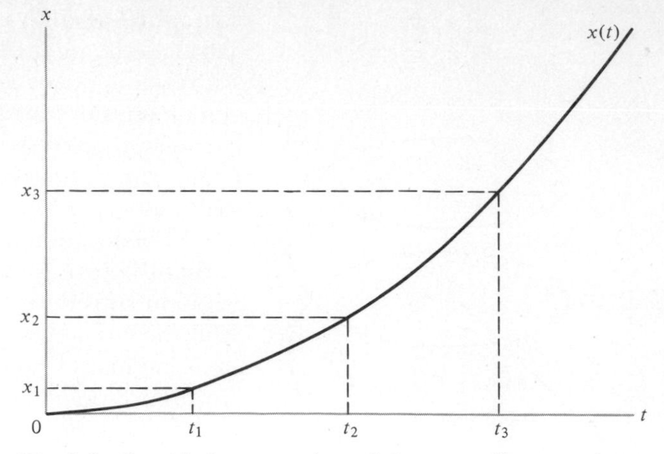

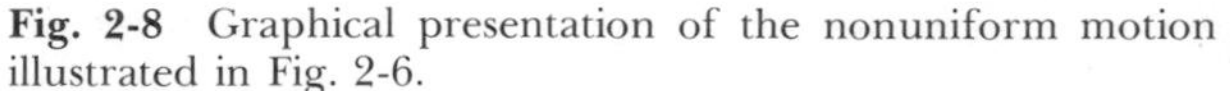

Fig. 2-8 Graphical presentation of the nonuniform motion illustrated in Fig. 2-6.

forth. Actually, there is a value of x corresponding to every possible value of t. The relation between the quantities x and t completely describes the motion of the car.

In Fig. 2-7 this relation is plotted for the motion illustrated in Fig. 2-5, and the same is done in Fig. 2-8 for the motion in Fig. 2-6. Each plot shows the values of x for all values of t that are of interest. So each tells everything there is to know about the relation between x and t, and therefore about the motion of the car. But the x versus t plots present this information in a very much more useful form than the strobe-photo-like figures do. Furthermore, they lead naturally to the fruitful idea of using mathematical functions to describe the motion. That is, the same information which is expressed graphically by plotting x versus t can usually be expressed by a **mathematical function** giving the form of the relation between x and t. The symbolism $x(t)$, seen in Figs. 2-7 and 2-8, is used to indicate that x is a function of t.

We will soon explore the properties of the x versus t plots. But first we must find clear definitions of the quantities x and t that specify position and time. In the next two sections we do this by considering procedures that can be used to *measure* these two quantities.

2-2 POSITION AND UNITS OF LENGTH

If a point locating an object moves in a certain direction along a straight line, its position at any instant can be given by stating its distance from some previously chosen reference point. But how can this distance be measured? The answer is: Construct a measuring stick of a certain length. Then agree that this will be the standard **unit** of length. Finally, measure the distance by counting the number of times the length of the stick can be fitted into the distance. The number obtained (not necessarily an integer) specifies the distance *in terms of* the length unit used.

In physics the agreed-upon unit of length is the meter, with its decimal multiples and submultiples. The meter is used universally in the other sciences too. It is also used in engineering, technology, manufacturing, and everyday affairs throughout the world—with the principal exception of the United States. It is the unit of length used in this book. Indeed, the contemporary version of the metric system, called the Système Internationale (SI), is used almost exclusively. However, we do make occasional comparisons between SI units and those familiar but cumbersome ones still used for nonscientific purposes in the United States. At a later stage we also discuss some units closely related to SI units that are commonly employed in certain branches of physics.

After much study, the metric system was legally adopted in France in 1799, during the later phases of the French Revolution. Its use was compulsory there after 1820. It turned out to be a highly successful attempt to make measurement simpler and more rational. As the metric system spread rapidly through Europe on the winds of revolution, it largely replaced a confusing muddle of local units, which often employed the same name for varying magnitudes. The metric system was perhaps the most popular legacy of the French Revolution, and even in the subsequent political reaction no serious attempt was made to abolish it.

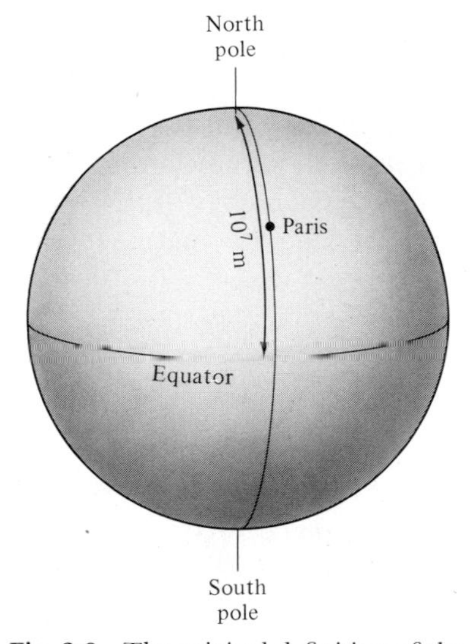

Fig. 2-9 The original definition of the metric unit of length was based on the dimensions of the earth. A meter was defined as 1/10,000,000 of the distance from the equator to the North pole, measured along the circle (called a meridian circle) passing through both poles and Paris.

The original intent of the metric system was to define the **meter** (abbreviated m) in terms of the earth itself. The distance from the equator to the North pole, along the meridian (longitude) circle passing through Paris, was to be precisely 10^7 m, by definition. This definition is illustrated in Fig. 2-9. It soon became clear, however, that it was not possible to measure the dimensions of the earth with anywhere near the same accuracy that could be achieved in the laboratory for a much shorter length. Consequently, the meter was actually defined to be the distance between two marks on a standard meter bar. Considerable pains were taken to make that distance as close as practicable to the best available value for the "terrestrial" meter. But once the meter was defined in 1795, terrestrial distance was measured in terms of the standard meter bar, not vice versa. The bar that was used as the standard meter is located in a vault of the International Bureau of Weights and Measures in a suburb of Paris. It was very carefully constructed and supported and was maintained at a fixed temperature. In due course, copies were sent to other bureaus throughout the world.

A significant change in the definition of the meter was made in 1960. By that time it had become possible to measure a wavelength of certain kinds of light much more accurately than the standard meter bar itself could be measured. So just as the meter had been redefined in terms of the standard meter bar instead of the earth, the meter was newly redefined in terms of the wavelength of the orange light emitted by atoms of krypton-86 in an electric discharge tube. Specifically, the **meter** is now defined to be *exactly* 1,650,763.73 wavelengths of this light. To make the new standard compatible with the old, the value was obtained from a set of careful measurements of the length of the previous standard meter bar, using krypton-86 light and an optical instrument called an interferometer, which is described later in this book. Aside from precision, the advantages of the atomic wavelength definition of the meter include indestructibility and the ability to reproduce it in any properly equipped laboratory without physically transporting a copy of the meter bar from Paris. A particular advantage is stability: We have every reason to believe that the atomic emission wavelength is more stable than the length of a metal bar.

One convenience of the metric system is that it is a decimal system. In other words, the multiples or submultiples of each basic unit are larger or smaller than the basic unit by *exact* factors of 10. For instance, a commonly used multiple of the meter is the **kilometer** (km), which is 10^3 m:

$$1 \text{ km} = 10^3 \text{ m}$$

Some common submultiples are the **centimeter** (cm), which has the value

$$1 \text{ cm} = 10^{-2} \text{ m}$$

and the **millimeter** (mm), which has the value

$$1 \text{ mm} = 10^{-3} \text{ m}$$

One of the tables inside the covers of this book lists the names and values of the prefixes used in the metric system.

The relation between the meter and the unit of length still frequently used in the United States is *precisely*

$$1 \text{ yard} = 0.9144 \text{ m}$$

This is equivalent to the *precise* relation

$$1 \text{ inch} = 2.54 \text{ cm}$$

There is no longer a standard yardstick. The yard (yd) and the inch (in) were redefined in terms of the standard meter in 1959. Other length units and conversion factors can be found in tables inside the covers of the book. Examples 2-1 and 2-2 will give you some experience with metric length units and conversion factors.

EXAMPLE 2-1

Evaluate the radius of the earth in meters by using the relation between the original terrestrial definition of the meter and the dimensions of the earth. Bearing in mind the limited accuracy of that relation (about 1 part in 10^4), quote the value you obtain to only three **significant figures.** That is, express the value in power-of-ten notation, with three digits appearing in the factor multiplying the power of 10.

■ Since $2\pi r$ is the circumference of a meridian circle of radius r passing through the poles and Paris, the quantity $2\pi r/4$ is very close to the distance from the equator to the North pole measured along that meridian. (The value $2\pi r/4$ is not exact because the earth is not exactly spherical.) Invoking the definition, you set $2\pi r/4$ equal to precisely 10^7 m:

$$\frac{2\pi r}{4} = 10^7 \text{ m}$$

Solving for r, you obtain

$$r = \frac{4 \times 10^7 \text{ m}}{2\pi}$$

Carrying out the division and expressing the result to three significant figures, you have

$$r = 6.37 \times 10^6 \text{ m}$$

It is said that there are three significant figures in this result. They are the digits in the factor 6.37. This means that the value quoted for r is 6.37×10^6 m, in contrast to 6.38×10^6 m or 6.36×10^6 m. But no attempt is being made to distinguish between 6.370×10^6 m and 6.371×10^6 m or 6.369×10^6 m.

EXAMPLE 2-2

Express the value of the earth radius r, obtained in Example 2-1, in inches.

■ You have

$$r = 6.37 \times 10^6 \text{ m}$$

To convert r to inches, first multiply the right side of this expression by the fraction 100 cm/1 m. Since 100 cm = 1 m, the value of the fraction is 1, and so the multiplication does not change the value of r. Thus you have

$$r = 6.37 \times 10^6 \text{ m} \times \frac{100 \text{ cm}}{1 \text{ m}}$$

Next multiply by the fraction 1 in/2.54 cm. Again, the value of the fraction is 1, and you have

$$r = 6.37 \times 10^6 \text{ m} \times \frac{100 \text{ cm}}{1 \text{ m}} \times \frac{1 \text{ in}}{2.54 \text{ cm}}$$

Now write the expression in the expanded form

$$r = 6.37 \times 10^6 \times 1 \text{ m} \times \frac{100 \times 1 \text{ cm}}{1 \text{ m}} \times \frac{1 \text{ in}}{2.54 \times 1 \text{ cm}}$$

Then simplify it by canceling the quantity 1 cm appearing in a numerator against the 1 cm appearing in a denominator, and do the same for the quantity 1 m which also appears in both a numerator and a denominator:

$$r = 6.37 \times 10^6 \times \cancel{1 \text{ m}} \times \frac{100 \times \cancel{1 \text{ cm}}}{\cancel{1 \text{ m}}} \times \frac{1 \text{ in}}{2.54 \times \cancel{1 \text{ cm}}}$$

You now have

$$r = \frac{6.37 \times 10^6 \times 100 \times 1 \text{ in}}{2.54}$$

or

$$r = 2.51 \times 10^8 \text{ in}$$

An abbreviated procedure is to write

$$r = 6.37 \times 10^6 \text{ m} \times \frac{100 \text{ cm}}{1 \text{ m}} \times \frac{1 \text{ in}}{2.54 \text{ cm}}$$

Then treating the symbols m and cm themselves as algebraic quantities which can be canceled, make two cancellations, to give

$$r = 6.37 \times 10^6 \times 100 \times \frac{1 \text{ in}}{2.54} = 2.51 \times 10^8 \text{ in}$$

Note that the result is quoted to only three significant figures since it was obtained from the value of r in meters, which was quoted to only three significant figures. To give more significant figures in the result would be to give a misleading impression of its accuracy.

2-3 TIME AND UNITS OF TIME

With the aim of continuing the development of a precise description of motion, we turn now to the definition and measurement of time. Any physical system which has a repetitive behavior can be used to measure the passage of time, if there is reason to believe that each cycle of its behavior accurately reproduces each preceding cycle. The time for one repetition of a cycle of motion is called a **period.** Thus, any system which has a constant period is capable of being used as a clock to measure time. Examples of systems which might be expected to have this property, to a greater or lesser degree, are the pulse beat, the oscillation of a pendulum, the rotation of the earth, and the vibration of the electrons in an atom. In fact, each of these systems has been used to measure time. No matter which system is used, the procedure is to count the number of repetitions, that is, periods, which occur during the interval of time to be measured.

For instance, some time intervals can be measured conveniently by recording how many earth rotations occur during the interval. The time lapse at a particular location from one noon (the instant when the sun appears to reach its highest point in the sky and pass over the meridian circle of the observer) to the next noon, averaged over one year, is used to define the unit of time called the **mean solar day.** Until recently the **second** (s) was defined as $(1/60)(1/60)(1/24) = 1/86{,}400$ of a mean solar day. The first of these factors arises because a second is one-sixtieth of a minute, the next because a minute is one-sixtieth of an hour, and the last because an hour is one twenty-fourth of a day.

When the meter was first introduced, an attempt was also made to introduce a time unit that was a purely decimal subdivision of the day. But few people were willing to accept it since relatively precise timekeeping was already a well-established custom. And so the second was retained as the basic unit of time, in spite of the nondecimal factors relating it to other common units of time. In most branches of physics, however, the second is used almost exclusively as the unit of time. Thus the awkward factors do not cause much trouble.

How is it determined that a repetitive phenomenon *used to define* the passage of time actually continues to repeat itself uniformly? It can be by comparison of the phenomenon with some other repetitive phenomenon. For example, the rotational motion of the earth can be compared to the oscillatory motion of a pendulum by coupling the pendulum to a mechanical counting device to form a pendulum clock. If the oscillations of the pendulum are very carefully stabilized, it is found that the number occurring during one rotation of the earth is reproducible and constant, within the accuracy of the measurement.

An atomic clock can be made by coupling the vibrations of the electrons in cesium atoms to an electronic counter. Figure 2-10 is a photograph of one such clock used by the U.S. Bureau of Standards. When extremely careful comparisons are made among several different atomic clocks, it is found that they agree with one another to better than 1 part in 10^{11}. But when the atomic clocks are compared with the rotation period of the earth, it is found that the earth does not keep exactly in step with the atomic clocks. Instead there are seasonal fluctuations in the earth's rotation period, as judged by the atomic clocks, of about 1 part in 10^8 (about 1000 times greater than the disagreement among the various atomic clocks). There is also an average annual increase of the earth's rotation period of about 2 parts in 10^9. Seasonal and annual changes in the rotation period of the earth can be plausibly attributed to known effects (for example, seasonal variations in wind patterns and continual drag from ocean tides). There are no known reasons why there could be such variations in the period of vibration of atomic electrons. It is therefore believed that atomic clocks provide a true measure of the changes in the rotation period of the earth since they depend on what is thought to be an accurately repetitive phenomenon.

Fig. 2-10 The most recent definition of the metric unit of time is based on the period elapsing between successive vibrations of electrons in cesium-133 atoms. The atomic clock shown in the photograph is an electronic device that counts these vibrations. (*Courtesy of the National Bureau of Standards.*)

As a consequence, the second was redefined in 1967 in terms of the vibration period of the electrons in cesium-133 atoms, when the electrons are making transitions between their so-called hyperfine ground-state levels. The **second** is now defined as *exactly* 9,192,631,770 of these periods. Other time units, used in certain circumstances, can be found in the tables inside the book covers.

EXAMPLE 2-3

Comparison of several atomic clocks indicates that a cesium-133 atomic clock will maintain a constant rate to about 1 part in 10^{11}. Get a feeling for this accuracy by estimating how many years it would take an improperly maintained clock whose rate is consistently slow (or fast) by 1 part in 10^{11}, compared to a properly maintained one, to lose (or gain) 1 s.

■ Since it is a convenient conversion factor to know, you might evaluate first the approximate number of seconds in 1 year (yr):

$$1 \text{ yr} = 365 \text{ days} \times \frac{24 \text{ h}}{1 \text{ day}} \times \frac{60 \text{ min}}{1 \text{ h}} \times \frac{60 \text{ s}}{1 \text{ min}}$$
$$\simeq 3 \times 10^7 \text{ s}$$

(The symbol $\simeq$ means "approximately equal to." The letter h stands for hour, and min abbreviates minute.) Then use this factor to convert 10^{11} s to years:

$$10^{11} \text{ s} \simeq 10^{11} \text{ s} \times \frac{1 \text{ yr}}{3 \times 10^7 \text{ s}}$$
$$\simeq 3 \times 10^3 \text{ yr}$$

From this you can conclude that in about 3000 yr a clock running slow (or fast) by 1 part in 10^{11} would lose (or gain) 1 s.

But it is extremely unlikely that an actual atomic clock would accumulate anything like a 1-s error in 3000 yr (if it was maintained in good operating order). The reason is that its rate will not be consistently slow, or fast, but instead will fluctuate between slow and fast. Thus there would be a very high degree of error compensation over the 3000-yr period, and the accumulated error would be a very small fraction of 1 s.

2-4 VELOCITY

Having considered the procedures used to measure position and time, we turn our attention to a fundamentally important quantity that involves both position and time. This is velocity, the quantity specifying how rapidly the position of an object changes with the passage of time. As a simple specific example, the velocity of a car moving uniformly along a straight road is the change in its position in a certain time, divided by that time. The *magnitude* of the velocity of the car can be expressed in meters per second or in other units of length per time (such as miles per hour in the United States). These are also the units for speed. Indeed, the words "velocity" and "speed" are used interchangeably in everyday language. But a significant distinction between the two is made in technical language. Velocity has a *direction* as well as a *magnitude,* whereas speed has *only magnitude.* The direction of a velocity specifies the direction of the motion it describes. A car moving in one direction along a straight road at 100 kilometers per hour (km/h) does not have the same velocity as one moving at 100 km/h in the opposite direction. The velocities of the two cars have opposite directions. But the cars have the same speeds. In this section you will see how these ideas are extended to handle the very important case of one-dimensional motion which is not uniform. You will also see how the mathematical concept of a derivative arises naturally when such a case is considered.

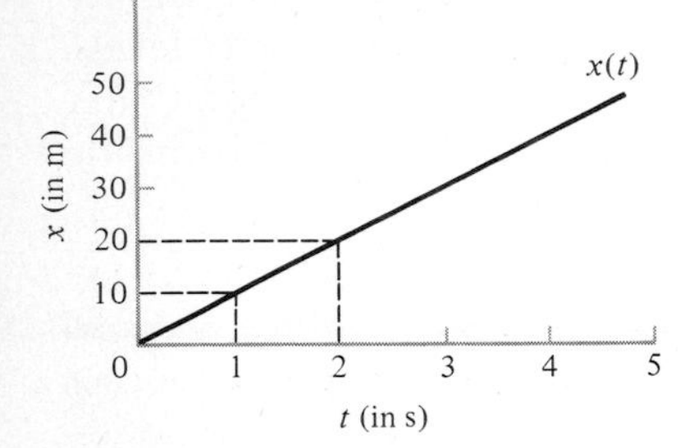

Fig. 2-11 The position x versus the time t for an object moving with constant velocity of magnitude 10 m/s in the direction of increasing x. The velocity has the value $v = +10$ m/s.

Imagine that the car, having *uniform* motion in a certain direction along a straight road, is traveling a distance of 10 m each second. Figure 2-11 describes this motion by plotting the straight line giving the position x of the uniformly moving car versus the time t. The clock used to measure t was zeroed at the instant the car passed the mark on the road from which x is measured. That is why the straight line showing the dependence of x on t passes through the point at $x = 0$ m and $t = 0$ s, the origin of both the x and t axes. The time t is plotted so that it increases to the right, and the position x is plotted so that positive values of x are plotted upward. Positive values of x are those in an agreed-upon direction from the mark on the road. As the values of x become more positive with increasing values of t, the car is moving in the direction of positive x.

That the car is moving uniformly is shown in the way x increases by the same amount in equal intervals of t. The figure indicates that the car is 10 m past the mark after 1 s has elapsed, 20 m past after 2 s, and so on. The car is said to have a constant velocity of 10 meters per second (m/s) in the positive x direction, because x increases by 10 m every time t increases by 1 s. Velocity is assigned the symbol v. *The constant velocity v of an object is defined to have a value equal to the change in its position in a certain time interval, divided by the duration of that time interval.*

We write the values of the pair of quantities specifying position and time, at the first instant considered, as x_i and t_i. These are called **initial values,** and the subscript i stands for initial. A subsequent pair of position and time values is written as x and t. In particular, we let x_i and t_i be the values at the *beginning* of the time interval used in the definition of constant velocity. And we let x and t be the pair at the *end* of the time interval. Then the definition of **constant velocity** v can be written

$$v = \frac{x - x_i}{t - t_i} \quad \text{for constant } v \tag{2-1a}$$

We were using this definition implicitly in the paragraph before the last when we evaluated v for the motion described in Fig. 2-11. Now let us

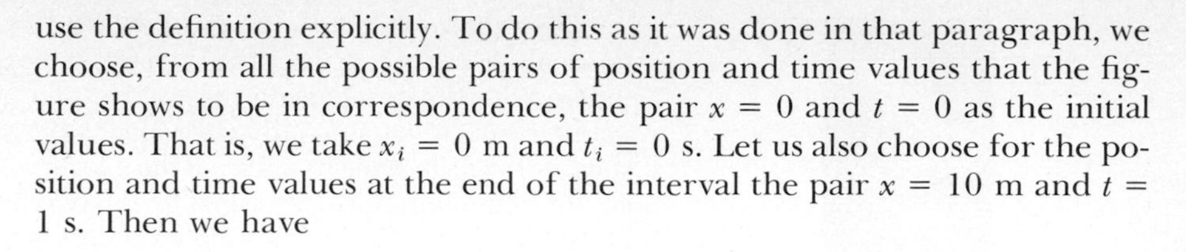

use the definition explicitly. To do this as it was done in that paragraph, we choose, from all the possible pairs of position and time values that the figure shows to be in correspondence, the pair $x = 0$ and $t = 0$ as the initial values. That is, we take $x_i = 0$ m and $t_i = 0$ s. Let us also choose for the position and time values at the end of the interval the pair $x = 10$ m and $t = 1$ s. Then we have

$$v = \frac{10 \text{ m} - 0 \text{ m}}{1 \text{ s} - 0 \text{ s}} = \frac{+10 \text{ m}}{+1 \text{ s}} = +10 \text{ m/s}$$

We can also choose any other pair of x and t values to be those at the end of the interval, but we will obtain the same result for v. Try it with $x = 20$ m and $t = 2$ s. Furthermore, we can alter our choice for the pair of initial values without affecting the result obtained. For example, we can take $x_i = 10$ m and $t_i = 1$ s and, say, $x = 20$ m and $t = 2$ s, to obtain again

$$v = \frac{20 \text{ m} - 10 \text{ m}}{2 \text{ s} - 1 \text{ s}} = \frac{+10 \text{ m}}{+1 \text{ s}} = +10 \text{ m/s}$$

No matter how the definition of Eq. (2-1a) is used on the x versus t curve of Fig. 2-11, it consistently yields the same result for v. That is what we mean when we say that the curve describes motion with a constant velocity.

The definition of Eq. (2-1a) gives both the magnitude of the velocity and its direction. For one-dimensional motion with constant velocity, the **magnitude of a velocity** is just the *magnitude* (in other words, the absolute value) of the fraction $(x - x_i)/(t - t_i)$. The **direction of a velocity** is given by the *sign* of this fraction for motion in one dimension. If an object is moving in the direction chosen to be the direction of positive x, as in Fig. 2-11, then x is more positive than x_i. So the numerator of the fraction, $x - x_i$, is positive. Since time always increases, the denominator of the fraction, $t - t_i$, is positive in all circumstances. Thus the value of v is positive.

But if an object is moving in the direction *opposite* to the direction of positive x, as in Fig. 2-12, then x is less positive than x_i. This causes the value of v calculated from Eq. (2-1a) to be *negative*. You should use that equation to show that the value of v for Fig. 2-12 is $v = -10$ m/s.

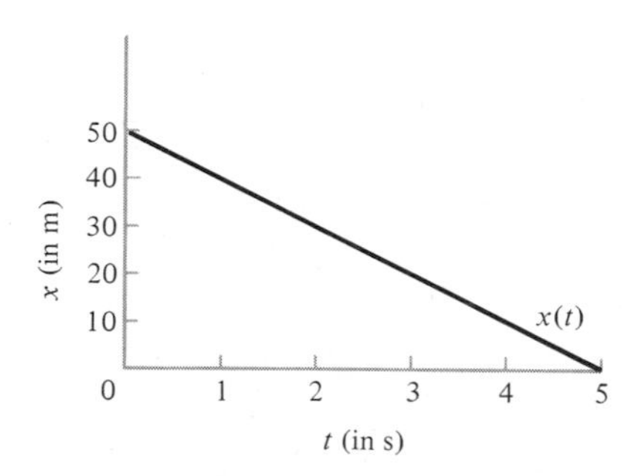

Fig. 2-12 The position x versus the time t for an object moving with constant velocity of magnitude 10 m/s in the direction of decreasing x. The velocity has the value $v = -10$ m/s.

In summary, for one-dimensional motion the direction of a velocity is given by its algebraic sign. Positive v means motion in the agreed-upon direction of positive x. Negative v means motion in the opposite direction. Later in this section, we discuss a specific mechanical system in which negative velocities must be used.

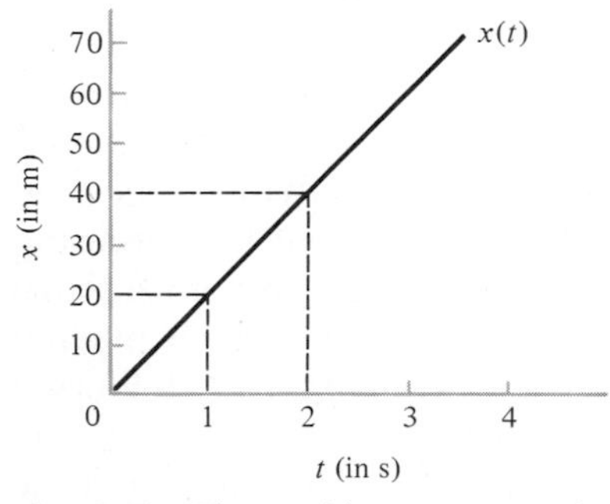

Fig. 2-13 The position x versus the time t for an object moving with constant velocity $v = +20$ m/s.

Figure 2-13 plots x versus t for a car moving with the constant velocity $v = +20$ m/s past the location $x = 0$ at time $t = 0$. You should verify the value of v by applying Eq. (2-1a) to data obtained from the figure. Then compare Fig. 2-13 to Fig. 2-11. You will note that the characteristic which distinguishes them is that the straight line produced by plotting x versus t has a steeper **positive slope** for the figure corresponding to the greater positive velocity. So, the *slope* of the line describing the function $x(t)$ on an x versus t plot is a measure of the velocity of the uniformly moving object. In fact, the slope is numerically *equal* to the velocity since the definition of Eq. (2-1a) is just a way of calculating the slope. The magnitude of the slope gives the magnitude of the velocity, and its sign gives the sign of the velocity. An x versus t plot with a **negative slope** was shown in Fig. 2-12.

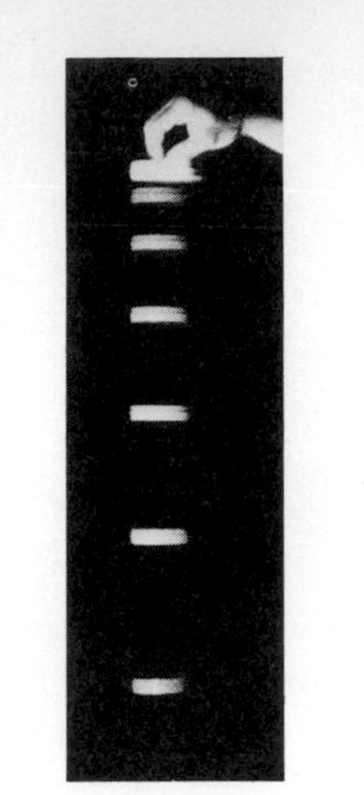

Fig. 2-14 A falling puck.

It might seem that the slope of the line describing a function $x(t)$ depends on the scale used to construct the x versus t plot. For instance, if Fig. 2-13 were redrawn with the numbers on the x axis twice as close together, its slope could be made to appear superficially the same as the slope of the line in Fig. 2-11. But as soon as the difference in scales of the x axes was noted, the actual difference in the slopes of the functions described by the two figures would be apparent. The point is that the slope of a line plotting x versus t is measured by a change in x *in meters*, divided by the corresponding change in t *in seconds*. This quantity is independent of the scales used for the x and t axes.

Before continuing the development of one-dimensional kinematics, we need to introduce a more convenient way of expressing the content of Eq. (2-1*a*). Instead of writing it as

$$v = \frac{x - x_i}{t - t_i} \qquad \text{for constant } v$$

we write the equation as

$$v = \frac{\Delta x}{\Delta t} \qquad \text{for constant } v \tag{2-1b}$$

That is, we express $x - x_i$ as Δx and $t - t_i$ as Δt. We do this by using the uppercase Greek letter Δ (delta) to mean "change in," a usage that is universal in physics and mathematics. Thus Δx means "change in x" and Δt means "change in t." Specifically,

$$\Delta x = x - x_i \tag{2-2a}$$

and

$$\Delta t = t - t_i \tag{2-2b}$$

Remember that the symbol Δ is not itself an algebraic quantity; Δx is *not* "Δ times x," and Δt is *not* "Δ times t."

Armed with this compact notation, we begin an investigation of *nonuniform* motion in one dimension. An example is depicted in Fig 2-14, which reproduces the strobe photo of a falling puck considered in Chap. 1. As we will demonstrate by direct experiment at the end of this chapter, Fig. 2-15 gives quantitatively the position versus time plot for the motion of the puck, or any other object falling freely (that is, with negligible air resistance) near the surface of the earth. Positive values of the position x are measured *vertically downward* from the point at which the object was released from rest. The values of the time t are measured from the instant of release.

The motion of the falling object is nonuniform because its velocity is not constant. So the plot of the function $x(t)$ in Fig. 2-15 is not a line of constant slope. Your intuition may be able to make an immediate connection between the ever-increasing slope of the $x(t)$ curve and the ever-increasing velocity of the falling object. But intuition is not enough. We must learn how to determine precisely from $x(t)$ the velocity of a nonuniformly moving object at any instant t. This will be done by applying the definition $v = \Delta x/\Delta t$ while letting Δt become smaller and smaller.

Suppose that we want to determine the value of the velocity when the value of the time is 1 s. We express this time as $t_i = 1$, simplifying the symbolism to be used in the following discussion by agreeing that in the discus-

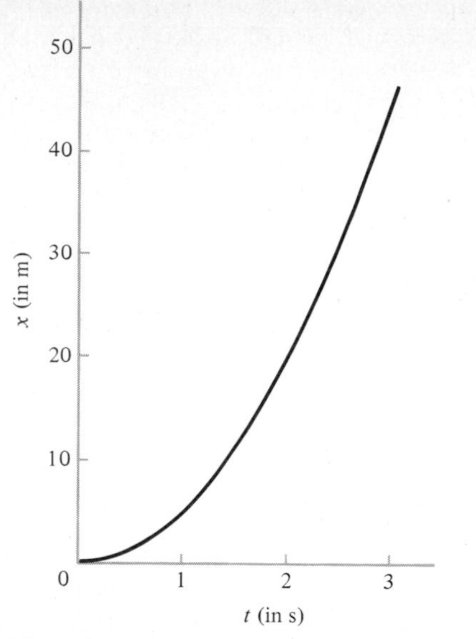

Fig. 2-15 The distance fallen x versus the time t elapsed since the beginning of the fall, for an object in free fall near the surface of the earth. Free fall means that air resistance is negligible. This will be the case for a dense object like a steel ball in the early part of its fall where its speed is low.

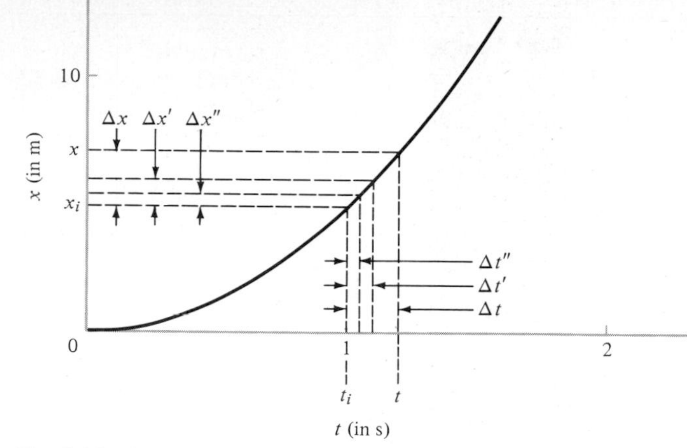

Fig. 2-16 An expanded view of the first part of the $x(t)$ curve in Fig. 2-15.

sion we will measure time in seconds. In Fig. 2-16 (which is Fig. 2-15 drawn to an expanded scale) we illustrate how the value of the velocity v can be determined. Take a small time interval beginning at $t_i = 1$ and ending at $t = 1 + \Delta t$. Then the corresponding values of x_i and x at the beginning and end of the interval are found from the $x(t)$ curve. Their difference, $\Delta x = x - x_i$, is the actual change in position in the particular time interval of duration Δt which begins at $t_i = 1$. Next, evaluate the fraction $\Delta x/\Delta t$. This fraction is clearly related to the velocity of the body at $t_i = 1$. But it is not the velocity at the *precise instant* $t_i = 1$. For one thing, it has to do with the total motion over the total time interval from $t_i = 1$ to $t = 1 + \Delta t$, not just with the motion at the instant $t_i = 1$. Also, the definition $v = \Delta x/\Delta t$ of Eq. (2-1*b*) was restricted to the case where $x(t)$ plots as a straight line. Here the plot is not a straight line.

But is it really necessary that $x(t)$ yield a straight line for *all* values of time in order for the definition $v = \Delta x/\Delta t$ to be applied in the immediate vicinity of $t_i = 1$? No. It is necessary only that the part of the $x(t)$ curve which is *actually used* to evaluate $\Delta x/\Delta t$ be a *sufficiently good approximation* to a straight line.

If you look at the entire $x(t)$ curve in Fig. 2-16, it is obviously not straight. However, if you restrict your attention to the part struck off by Δt, you must inspect it quite carefully to discern that it is not straight. The point is made in Fig. 2-17. The actual motion of the object is *almost* uniform over the time interval Δt. So it should be possible to obtain a good approximation to its velocity v at the instant occurring at the beginning of the interval by evaluating $\Delta x/\Delta t$. In other words, we can use the approximate equality

$$v \simeq \frac{\Delta x}{\Delta t} \quad \text{for nonuniform motion when } \Delta t \text{ is small} \tag{2-3}$$

Is the approximation good enough? If it is not, we can always improve it by simply reducing the size of the time interval Δt. Referring to Fig. 2-16, we can obtain an even better approximation to the value of v at the instant $t_i = 1$ by taking the smaller time interval $\Delta t'$ and then evaluating the ratio $\Delta x'/\Delta t'$. Over this shorter interval the actual $x(t)$ curve is even more closely

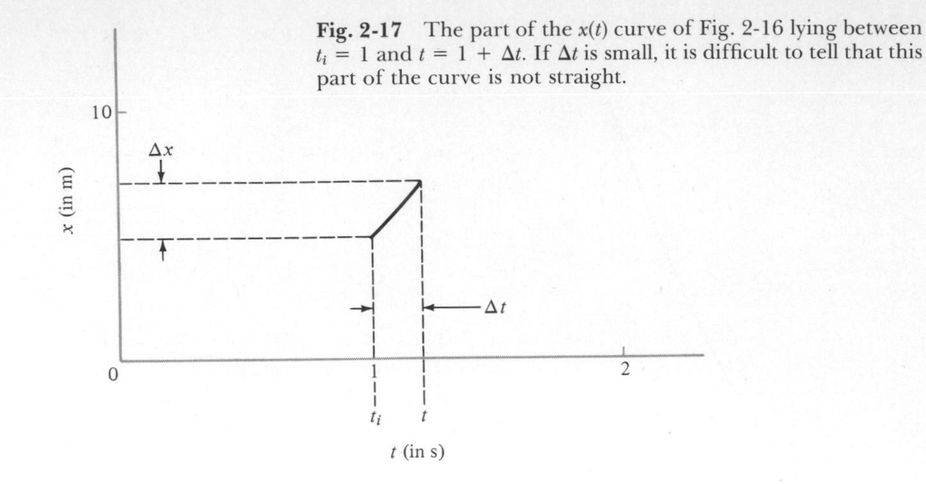

Fig. 2-17 The part of the $x(t)$ curve of Fig. 2-16 lying between $t_i = 1$ and $t = 1 + \Delta t$. If Δt is small, it is difficult to tell that this part of the curve is not straight.

approximated by a straight line. If still greater precision is required, the value of v at $t_i = 1$ can be approximated by using an even smaller time interval $\Delta t''$ and evaluating $\Delta x''/\Delta t''$. And so on.

Numerical examples of this procedure are given in Sec. 2-5. When you inspect them, you will see that there is a substantial difference between the values of $\Delta x/\Delta t$ and $\Delta x'/\Delta t'$. But the difference between $\Delta x'/\Delta t'$ and $\Delta x''/\Delta t''$ is appreciably smaller. Continuing the process, it is found that further reduction in the time interval no longer produces a discernible change in the approximation to the velocity at the instant when the time interval begins. The reason is that the time interval has become so short that the part of the $x(t)$ curve lying within it is essentially a straight line. When this happens, the sequence of approximations to the velocity is said to have **converged** to its **limit.** The velocity v is defined to equal this limit. In other words, by definition v equals the limit of $\Delta x/\Delta t$ as Δt approaches zero. In mathematical notation the definition is written

$$v = \underset{\Delta t \to 0}{\text{limit}} \frac{\Delta x}{\Delta t} \tag{2-4}$$

We have been led by the need to evaluate the velocity for a nonuniformly moving body to the most basic concept of differential calculus. In fact, the development of differential calculus—the work of Isaac Newton and, independently, of Gottfried Wilhelm von Leibniz around 1666—was motivated by the same need. The quantity on the right side of Eq. (2-4) is the *instantaneous rate of change of x with respect to t.* In differential calculus it is called the **derivative of** x **with respect to** t and is commonly writted as dx/dt. That is,

$$\frac{dx}{dt} \equiv \underset{\Delta t \to 0}{\text{limit}} \frac{\Delta x}{\Delta t} \tag{2-5}$$

(The symbol $\equiv$ means "identical to." But dx does not mean "d times x," and dt does not mean "d times t.") With this symbolism, the definition of the **velocity** v can be expressed by

$$v \equiv \frac{dx}{dt} \tag{2-6}$$

The velocity is the derivative of position with respect to time. To be more specific, at any instant the velocity v of an object is obtained by evaluating at that instant the derivative of its position x with respect to time t.

Note that there are no restrictions on the definition of velocity in one-dimensional motion given by Eq. (2-6), as there were on earlier definitions. It applies in all circumstances. Can you see how each of the earlier definitions is contained in Eq. (2-6) as a special case? Specific examples of the evaluation of derivatives are given in Sec. 2-5, using both numerical and analytical methods.

The convergence of $\Delta x/\Delta t$ to its limit, as Δt approaches zero, is illustrated and interpreted from geometric considerations in Fig. 2-18. [This figure is like an expanded view of the important part of Fig. 2-16. But for the sake of clarity it plots a function $x(t)$ with considerably greater curvature than the one plotted in Fig. 2-16.] You can see that the value of $\Delta x/\Delta t$ gives the slope of the straight line connecting the point on the curve at $t_i = 1$ to the farthest point on the curve. The value of $\Delta x'/\Delta t'$ gives the slope of the straight line connecting the $t_i = 1$ point to the intermediate point on the curve. And the value of $\Delta x''/\Delta t''$ gives the slope of the straight line connecting the $t_i = 1$ point to the point nearest to it on the curve. As the time interval is reduced, the slopes of the sequence of straight lines generated approach a value which is the slope of the tangent to the curve of $x(t)$ at $t_i = 1$. Stated in geometric terms, the chords drawn to the curve for successively smaller intervals approximate more and more closely the tangent to the curve, and their slopes approximate more and more closely the slope of the tangent. The slope of the tangent to a curve at some point is commonly called just the "slope of the curve" at that point. So we can say that the slope of the x versus t curve at some instant equals the limit as Δt approaches zero of $\Delta x/\Delta t$ at that instant. Since the limit is the derivative and the derivative is the velocity, we can also say that *the velocity at a particular instant is given by the slope of the position versus time curve at that instant.*

The magnitude of a velocity is called a **speed.** Thus speed is always a positive quantity. It is written symbolically as $|v|$. (The bars are the mathe-

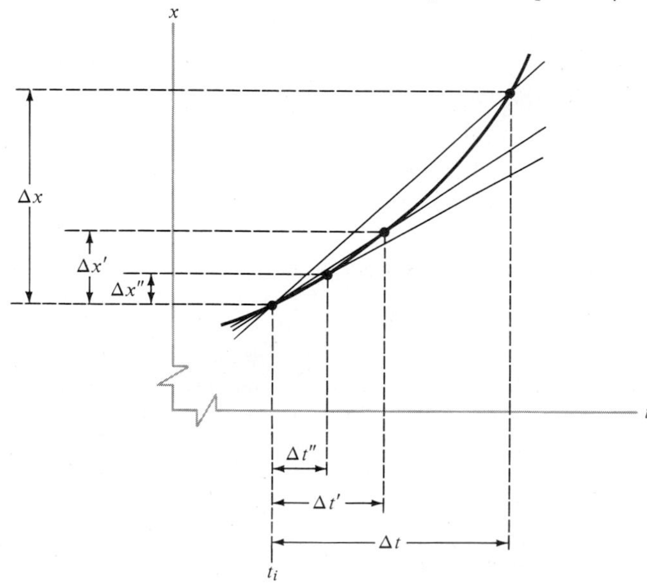

Fig. 2-18 A geometrical representation of the derivative of x with respect to t. The derivative at the value t_i is the slope of the line that is tangent to the x versus t curve at the point corresponding to t_i.

matical symbols for taking the absolute value of a quantity; that is, for deleting the minus sign in the numerical value if the numerical value of the quantity is negative.) The speed $|v|$ provides a useful description of the motion of an object in circumstances where the rapidity of the motion is important but the direction of the motion is not. For instance, the condition that motion be governed by the laws of newtonian mechanics, instead of relativistic mechanics, is that the speed be small compared to the speed of light. The direction of the motion is of no consequence, as far as this distinction is concerned. But in most circumstances the direction in which an object moves *is* important, and so the velocity, not the speed, must be used to describe the motion.

For nonuniform motion it is sometimes useful to speak of the **average velocity,** symbolized by $\langle v \rangle$. (The brackets are the mathematical symbols for taking the average value of a quantity.) If in a time interval of duration Δt the change in the x coordinate of an object is Δx, then its average velocity for the time interval is defined as

$$\langle v \rangle \equiv \frac{\Delta x}{\Delta t} \tag{2-7}$$

As the lack of restrictions implies, the time interval is *not* necessarily small. The average velocity $\langle v \rangle$ can be of greater practical interest than the instantaneous velocity v. For instance, in planning a long automobile trip (on a straight road since we are working in one dimension) you know that there will be variations in v because of traffic, rest stops, and so forth. But to predict the distance you will travel in a day, you can often make a reasonably accurate estimate of $\langle v \rangle$, even though v is unpredictable from minute to minute during the trip. Then you can solve Eq. (2-7) for Δx, to obtain

$$\Delta x = \langle v \rangle \Delta t \tag{2-8}$$

Inserting the estimated value of $\langle v \rangle$ and the time Δt that you plan to spend in traveling, you obtain immediately an estimate for the distance Δx that you will travel.

For uniform motion the velocity v is constant, and so the average velocity $\langle v \rangle$ has the same value as v. The more nonuniform the motion and the larger the time interval Δt, the greater can be the difference between $\langle v \rangle$ and v. Example 2-4 illustrates the relation between these two quantities, as well as the other relations developed in this section.

EXAMPLE 2-4

Figure 2-19 shows a strobe photo of an oscillating pendulum. It comprises an object called a bob tied to one end of a cord of moderate length, whose other end is tied to a fixed support. Figures 2-20*a* and *b* indicate the position at successive times through one oscillation cycle of the bob in a pendulum with a very long cord. Since the distance between the extreme positions of the bob is small compared to the length of the cord, the motion is very nearly one-dimensional. The caption to Fig. 2-20*a* describes the motion in detail. Use Fig. 2-20*b* to answer the following questions concerning the motion.

a. When does the velocity have the largest positive value?

b. When does the velocity have the largest negative value?

c. Is the velocity ever zero? If so, when?

d. What are the answers to the three preceding questions if they are asked about the speed?

Fig. 2-19 A strobe photo of a pendulum with a cord of moderate length.

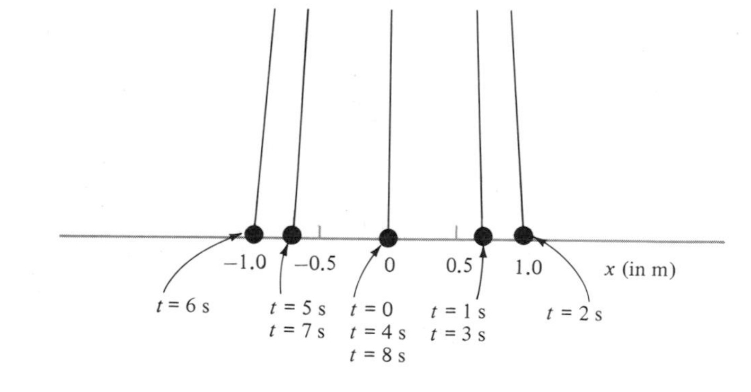

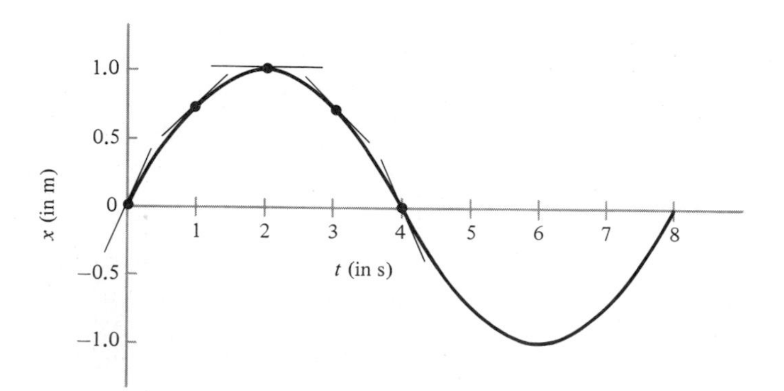

Fig. 2-20 (*a*) A strobe-photo-like representation of one oscillation cycle of a pendulum with a very long cord. The pendulum bob supported by the cord actually travels along a circle centered on the fixed end of the cord (the end being too distant to show on the figure). But since the maximum change in the position of the bob is small compared to the length of the cord, its path will lie very close to the straight line defining the x axis. When $t = 0$ s, the bob swings past $x = 0$ m, going to the right. When $t = 1$ s, the bob passes $x = 0.7$ m. When $t = 2$ s, the bob comes instantaneously to rest at $x = 1.0$ m. Then it reverses its direction of motion, passing $x = 0.7$ m again at $t = 3$ s. The bob continues past $x = 0$ m at $t = 4$ s, and $x = -0.7$ m at $t = 5$ s, coming instantaneously to rest at $x = -1.0$ m when $t = 6$ s. Then it reverses its direction of motion once more, passing $x = -0.7$ m at $t = 7$ s and $x = 0$ m at $t = 8$ s. Having completed the first cycle of its oscillation, the bob continues moving into the next cycle, but this is not shown in the figure. (*b*) An x versus t plot for the motion depicted in part a of this figure. At several values of t the slope of the curve is shown by drawing its tangent.

e. What is the average velocity for the time interval covered by the first half-cycle?

■ **a.** Since the value of the velocity v is the slope of the position versus time curve, and since the slope has its largest positive value at $t = 0$ s and again at $t = 8$ s, the velocity has its largest positive value at these times.

b. The velocity has its largest negative value at $t = 4$ s because at that time the slope of the curve is most negative.

c. The velocity is zero at $t = 2$ s, as well as at $t = 6$ s, since the slope is zero at these instants. If this bothers you, consider the fact that just before $t = 2$ s the bob is moving slowly in the positive x direction, while just after $t = 2$ s it is moving slowly in the negative x direction. Since it switches the direction of its motion, surely there will be an instant when it is not moving.

d. Since speed $|v|$ is the magnitude of velocity, the speed has the same value at $t = 0$ s, $t = 4$ s, and $t = 8$ s. These are the three times when the speed has the largest positive value. The speed never has a negative value. Its value is zero when the value of the velocity is zero, that is, at $t = 2$ s and $t = 6$ s.

e. The half-cycle begins at $t = 0$ s and ends at $t = 4$ s. At both times the bob is at $x = 0$ m. Using these values in the definition of the average velocity for the time interval, you find

$$\langle v \rangle = \frac{0 \text{ m} - 0 \text{ m}}{4 \text{ s} - 0 \text{ s}} = 0 \text{ m/s}$$

You can understand the meaning of this result by noting the symmetry of the x versus t curve about $t = 2$ s. Because of this symmetry the set of values taken on by the velocity from $t = 0$ s to $t = 2$ s is exactly the same as the set of values taken on from $t = 2$ s to $t = 4$ s, except that the signs are reversed. Since $\langle v \rangle$ represents the average of the values of v from $t = 0$ s to $t = 4$ s, and since the contributions to this average from the positive values will just cancel the contributions from the negative values, it is not surprising that the direct application just made of the definition of $\langle v \rangle$ yields the result $\langle v \rangle = 0$ m/s. But such a result is not always obtained. For instance, $\langle v \rangle$ is certainly greater than zero for the interval from $t = 0$ s to $t = 2$ s.

2-5 DIFFERENTIATION

Unless you are already familiar with differential calculus, it will be necessary to take a short break from physics in this section to study differentiation, the mathematical process of evaluating a derivative.

The definition of a derivative given in Eq. (2-5),

$$\frac{dx}{dt} \equiv \lim_{\Delta t \to 0} \frac{\Delta x}{\Delta t}$$

is not very explicit. A more detailed form of the definition provides more information about the procedure actually used in evaluating a derivative. Consider a varying quantity x whose value depends on the value of a varying quantity t. The quantity x is called the **dependent variable,** and the quantity t is called the **independent variable.** The nature of the dependence of x on t is usually given in the form of a mathematical function $x(t)$. By definition, **the derivative of x with respect to t, evaluated at t_i,** is

$$\left(\frac{dx}{dt}\right)_{t_i} = \lim_{\Delta t \to 0} \left(\frac{\Delta x}{\Delta t}\right)_{t_i} = \lim_{\Delta t \to 0} \left[\frac{x(t_i + \Delta t) - x(t_i)}{\Delta t}\right] \tag{2-9}$$

This says that the derivative, at some value of t designated as t_i, is obtained by first finding the difference between the value of the function $x(t)$ when t is $t_i + \Delta t$ and its value when t is t_i, then dividing that difference by Δt, and finally taking the limit of the sequence of results obtained by using successively smaller values of Δt.

Example 2-5 demonstrates the process of taking the limit and thereby illustrates the most important idea of differential calculus. The demonstration involves going through a sequence of numerical calculations to evaluate the derivative, at a particular value of t_i, of a function $x(t)$ that has a particular mathematical form. You will find it worthwhile to employ this numerical method a few times. You can use it to evaluate derivatives for other values of t_i as well as for functions $x(t)$ having other forms. It is feasible to perform the numerical calculations on a manually operated pocket calculator. But it is much easier to use a programmable pocket calculator or a small computer. A suitable program and operating instructions are found in the pamphlet called the Numerical Calculation Supplement that is provided with this book. The program is identified as the numerical differentiation program.

EXAMPLE 2-5

The motion of an object falling freely near the surface of the earth was plotted in Fig. 2-15. The information given by the plot can also be given by the function

$$x(t) = (4.90 \text{ m/s}^2)t^2 \tag{2-10}$$

To verify this, let t have the precise values $t = 0$ s, 1 s, and 2 s in the equation to obtain $x = 0$ m, 4.90 m, and 19.60 m. Then look at the plot. Use the definition of Eq. (2-9) to make a direct numerical evaluation of $(dx/dt)_{t_i}$ for $t_i = 1$ s. Let the first value of Δt be precisely $\Delta t = 0.1$ s and make each subsequent value smaller by a factor of $\frac{1}{2}$ (that is, use $\Delta t = 0.1$ s, 0.05 s, 0.025 s, and so on). Obtain results to an accuracy of two decimal places.

■ The first thing to do is to reexpress Eq. (2-10) in the completely equivalent form

$$x(t) = 4.90t^2 \qquad (x \text{ in m, } t \text{ in s}) \tag{2-11}$$

Since a calculating device deals only with numbers and not with units, you must keep track of the units because it does not. The units are specified in Eq. (2-11) in a way that makes this fact more apparent. Similarly, reexpress t_i and the first value of Δt as $t_i = 1$ and $\Delta t = 0.1$ (t in s). Then perform the calculations, either manually by carrying out the procedure called for in Eq. (2-9) or by using the program which will make a programmable calculator or computer carry out the procedure for you.

First you calculate the number

$$\frac{4.90(1.1)^2 - 4.90(1)^2}{0.1} = 4.90\left[\frac{(1.1)^2 - 1}{0.1}\right] = 4.90\left(\frac{1.21 - 1}{0.1}\right) = 10.29$$

The next number you calculate is

$$\frac{4.90(1.05)^2 - 4.90(1)^2}{0.05} = 4.90\left[\frac{(1.05)^2 - 1}{0.05}\right] = 4.90\left(\frac{1.1025 - 1}{0.05}\right) = 10.05$$

You continue with these calculations, with each successive value of Δt half as large as the one before.

You will obtain the following sequence.

$$\left(\frac{\Delta x}{\Delta t}\right)_{t_i=1} = 10.29, 10.05, 9.92, 9.86, 9.83, 9.82, 9.81, 9.80, 9.80, 9.80 \quad (x \text{ in m, } t \text{ in s})$$

Inspection shows that the sequence of numbers gradually approaches the number

9.80. In other words, the sequence converges to the limit 9.80. Consequently, the derivative has the value

$$\left(\frac{dx}{dt}\right)_{t_i=1} = \underset{\Delta t \to 0}{\text{limit}} \left(\frac{\Delta x}{\Delta t}\right)_{t_i=1} = 9.80 \qquad (x \text{ in m, } t \text{ in s})$$

Inserting the units for x and t to express the result in the manner of Eq. (2-10), you have

$$\left(\frac{dx}{dt}\right)_{ti=1\text{ s}} = 9.80 \text{ m/s}$$

The units in the result are meters per second since a velocity is a length divided by a time, and lengths are measured in meters while times are measured in seconds.

The numerical method of evaluating a derivative that was used in Example 2-5 has the pedagogical advantage of providing a real feeling for the limiting process. But it has the practical disadvantage of being extremely specific. To evaluate $(dx/dt)_{t_i}$ for other values of t_i, you must perform the entire sequence of calculations at each value. Once you understand the concept of a limit and the meaning of a derivative, it is much more convenient to evaluate $(dx/dt)_{t_i}$ from the functional form of $x(t)$ by using the analytical method indicated in Example 2-6.

EXAMPLE 2-6

Make an analytical evaluation of $(dx/dt)_{t_i}$ for $x(t) = ct^2$, where c is a constant, by employing the definition of Eq. (2-9).

■ You must first evaluate the quantity

$$\left(\frac{\Delta x}{\Delta t}\right)_{t_i} = \frac{x(t_i + \Delta t) - x(t_i)}{\Delta t}$$

Setting t equal to $t_i + \Delta t$ in the expression $x(t) = ct^2$, you have $x(t_i + \Delta t) = c(t_i + \Delta t)^2$. Doing the same with t set equal to t_i gives you $x(t_i) = ct_i^2$. So you have

$$\left(\frac{\Delta x}{\Delta t}\right)_{t_i} = \frac{c(t_i + \Delta t)^2 - ct_i^2}{\Delta t}$$

$$= c\left[\frac{t_i^2 + 2t_i\,\Delta t + (\Delta t)^2 - t_i^2}{\Delta t}\right] = c(2t_i + \Delta t)$$

Then take the limit as $\Delta t \to 0$. You get

$$\underset{\Delta t \to 0}{\text{limit}} \left(\frac{\Delta x}{\Delta t}\right)_{t_i} = \underset{\Delta t \to 0}{\text{limit}}\; c(2t_i + \Delta t) = 2ct_i$$

Thus

$$\left(\frac{dx}{dt}\right)_{t_i} = 2ct_i \qquad \text{for } x(t) = ct^2 \text{ and constant } c \tag{2-12a}$$

This result is very useful because it applies to any value of c and any value of t_i. To make a comparison with the specific result of Example 2-5, set $c = 4.90\text{ m/s}^2$ and $t_i = 1$ s. You find

$$\left(\frac{dx}{dt}\right)_{t_i=1\text{ s}} = 2 \times 4.90 \text{ m/s}^2 \times 1 \text{ s} = 9.80 \text{ m/s}$$

in agreement with what was found from using the numerical method for evaluating the derivative.

Since Eq. (2-12a) is valid for any time t_i, the specific time t_i at which dx/dt is evaluated can be dropped from the left side of the equation if the subscript i is deleted from the quantity t on the right side. By doing this, and also substituting ct^2 for x on the left side, the equation assumes the more compact and useful form

$$\frac{d(ct^2)}{dt} = 2ct \qquad \text{for constant } c \tag{2-12b}$$

Equation (2-12*b*) leads to a particular example of an important general rule for the derivative of an expression containing a constant factor such as c. Since the equation is valid for any value of c, it must be valid for the special case $c = 1$. In this case we can write

$$\frac{d(t^2)}{dt} = 2t \tag{2-12c}$$

Using Eq. (2-12*c*) to substitute $d(t^2)/dt$ for $2t$ in Eq. (2-12*b*), we have

$$\frac{d(ct^2)}{dt} = c\,\frac{d(t^2)}{dt} \qquad \text{for constant } c$$

This is the particular example of the general rule.

The rule itself applies to any function of t, not just t^2. That is for *any* $f(t)$

$$\frac{d[cf(t)]}{dt} = c\,\frac{d[f(t)]}{dt} \qquad \text{for constant } c \tag{2-13}$$

The validity of this rule can be seen in an intuitive way by using the relation between derivatives and slopes to interpret its meaning in terms of slopes. In these terms the rule says simply that for any value of the independent variable t, the slope of the curve plotted for $cf(t)$ is c times as steep as the slope of the curve plotted for $f(t)$.

You can make a formal proof of Eq. (2-13) by following the procedure of Example 2-6. In the same way you can prove two other important general rules:

$$\frac{d[f(t) + g(t)]}{dt} = \frac{d[f(t)]}{dt} + \frac{d[g(t)]}{dt} \tag{2-14}$$

and

$$\frac{d[f(t)g(t)]}{dt} = f(t)\,\frac{d[g(t)]}{dt} + g(t)\,\frac{d[f(t)]}{dt} \tag{2-15}$$

where $f(t)$ and $g(t)$ are *any* two functions of t.

Following Example 2-6, you can prove each of the relations

$$\frac{d(t^2)}{dt} = 2t^1$$

$$\frac{d(t^1)}{dt} = 1t^0$$

$$\frac{d(t^0)}{dt} = 0$$

$$\frac{d(t^{-1})}{dt} = -1t^{-2}$$

$$\frac{d(t^{-2})}{dt} = -2t^{-3}$$

The first relation is just Eq. (2-12*c*). In the second and third relations $t^0 = 1$. Written in these consistent forms, the general pattern is easy to discern:

$$\frac{d(t^n)}{dt} = nt^{n-1} \tag{2-16}$$

Any calculus text will prove that this is valid for all values of n, integral or nonintegral.

Two other derivatives, which we will soon find useful, are

$$\frac{d[\sin(\omega t)]}{dt} = \omega \cos(\omega t) \qquad \text{for constant } \omega \tag{2-17}$$

and

$$\frac{d[\cos(\omega t)]}{dt} = -\omega \sin(\omega t) \qquad \text{for constant } \omega \tag{2-18}$$

(The symbol ω is the Greek letter omega.) The quantity ωt, whose sine or cosine is being taken, is an angle expressed in **radians.** The radian measure of an angle is the arc length intercepted by the angle on a circle whose center is at the apex of the angle, divided by the length of the circle's radius. Being a length divided by a length, the angle is unitless. Since ωt must be unitless, and since the units of t are seconds, the units of ω must be reciprocal seconds (s^{-1}).

EXAMPLE 2-7

Derive Eq. (2-17).

■ Applying the definition of Eq. (2-9), you obtain

$$\left\{\frac{d[\sin(\omega t)]}{dt}\right\}_{t_i} = \lim_{\Delta t \to 0} \left[\frac{\sin(\omega t_i + \omega \Delta t) - \sin(\omega t_i)}{\Delta t}\right]$$

Using the trigonometric identity for the sine of the sum of two angles, you get

$$\left\{\frac{d[\sin(\omega t)]}{dt}\right\}_{t_i} = \lim_{\Delta t \to 0} \left[\frac{\sin(\omega t_i)\cos(\omega \Delta t) + \cos(\omega t_i)\sin(\omega \Delta t) - \sin(\omega t_i)}{\Delta t}\right]$$

For the very small values of Δt that you will have when taking the limit as $\Delta t \to 0$, the values of $\omega \Delta t$ will be very small compared to 1. This suggests that you make use of the relations

$$\cos(\omega \Delta t) \simeq 1 \qquad \text{for } \omega \Delta t \ll 1$$

and

$$\sin(\omega \Delta t) \simeq \omega \Delta t \qquad \text{for } \omega \Delta t \ll 1$$

[The first of these makes sense since $\cos 0 = 1$. The second may be new to you. If so, verify it by setting a calculator to the radian mode and evaluating the sine of 1 radian (rad), 0.1 rad, 0.01 rad, and 0.001 rad. Then for each case compare the sine of the angle with the angle itself. Can you use the definitions of the sine and of the radian to explain your observations?] These relations allow you to express the quantity whose limit must be calculated by the approximation

$$\frac{\sin(\omega t_i) + \cos(\omega t_i)(\omega \Delta t) - \sin(\omega t_i)}{\Delta t} = \omega \cos(\omega t_i)$$

As Δt (and therefore $\omega \Delta t$) becomes smaller, the approximation becomes more accurate. In the limit $\Delta t \to 0$ it is exact. Thus taking the limit is a matter of replacing the quantity by $\omega \cos(\omega t_i)$. Doing this, you have

$$\left\{\frac{d[\sin(\omega t)]}{dt}\right\}_{t_i} = \omega \cos(\omega t_i)$$

This equation can be written as

$$\left\{\frac{d[\sin(\omega t)]}{dt}\right\}_{t_i} = [\omega \cos(\omega t)]_{t_i}$$

Since the equation is valid for any value of time t_i, the t_i can be deleted from both sides. Then it becomes Eq. (2-17). A procedure similar to the one used here will yield Eq. (2-18).

The formulas displayed in this section, and a very few more that will be introduced as needed, allow you to evaluate analytically the derivative of any of the functions dealt with in this book.

2-6 ACCELERATION

Let us return to physics and to the task of describing motion in one dimension. When the velocity of a particle is changing, it is said to be accelerating. In Sec. 1-3 we considered qualitatively several examples of acceleration in one-dimensional motion, for instance the motion of a falling object. There we also saw that acceleration plays an extremely important role in newtonian mechanics. Now we consider a *quantitative* way of evaluating acceleration for one-dimensional motion.

By definition, *the acceleration is the derivative of velocity with respect to time.* Using the symbol a for **acceleration,** we can write the definition

$$a \equiv \frac{dv}{dt} \tag{2-19}$$

Since the velocity v is related to the position x and time t by the definition

$$v \equiv \frac{dx}{dt}$$

an expression for a, equivalent to Eq. (2-19), is

$$a = \frac{d}{dt}\left(\frac{dx}{dt}\right) \tag{2-20}$$

The right side of Eq. (2-20) is the derivative with respect to t of the derivative of x with respect to t—in other words, the instantaneous rate of change with respect to t of the instantaneous rate of change of x with respect to t. It is called a **second derivative,** and in the notation of differential calculus it is written in the abbreviated form

$$\frac{d}{dt}\left(\frac{dx}{dt}\right) \equiv \frac{d^2x}{dt^2} \tag{2-21}$$

Thus we can also say that *the acceleration is the second derivative of position with respect to time:*

$$a \equiv \frac{d^2x}{dt^2} \tag{2-22}$$

(Note that just as dx does not mean "d times x," the symbol d^2x does not mean "d^2 times x." Nor does dt^2 mean "d times t^2.")

The definition of Eq. (2-19) can be expressed in geometrical terms by saying *the acceleration at a particular instant is given by the slope of the velocity versus time curve at that instant.* In similar terms, Eq. (2-20) can be stated: *the acceleration at a particular instant is given by the rate of change of the slope of the position versus time curve at that instant.*

In the metric system the unit for measuring the magnitude of acceleration is meters per second per second, since it measures the change per second of the velocity, which is measured in meters per second. A shorter way of expressing the unit is meters per second squared (m/s^2).

Just as a velocity has both a magnitude and a direction, so does an acceleration. And just as is the case for velocity, the direction of an acceleration in one-dimensional motion is specified by its sign. A positive value of a means that the direction of the acceleration is the same as that chosen to be the direction from the coordinate origin to positive values of x. In particular, if an object is moving in the positive x direction so that the sign of v is positive, and if the magnitude of v is increasing, then its acceleration is in the positive x direction and a is positive. If the sign of v is positive but its magnitude is decreasing, then the acceleration of the object is in the negative x direction and a is negative. What would you expect for the sign of a if v is negative and its magnitude is increasing? What is it if v is negative and its magnitude is decreasing? (These questions are answered later in this section.)

Acceleration is a quantity of fundamental interest in mechanics because it is the effect that is directly related to the cause—force. Example 2-8 will give you some experience in calculating acceleration.

EXAMPLE 2-8

In Examples 2-5 and 2-6 the expression

$$x(t) = (4.90\ \text{m/s}^2)t^2 \tag{2-23}$$

was used to represent the motion of an object falling freely near the surface of the earth after it is released from rest at $x = 0$ m and $t = 0$ s, with positive values of x measured downward. Calculate the acceleration of the object.

■ First, evaluate the velocity $v = dx/dt$. You have

$$v = \frac{dx}{dt} = \frac{d[(4.90\ \text{m/s}^2)t^2]}{dt}$$

You can use Eq. (2-13) to remove the constant factor, 4.90 m/s^2, from the derivative. This gives you

$$v = (4.90\ \text{m/s}^2)\frac{d(t^2)}{dt}$$

You can now use Eq. (2-16), with n set equal to 2, to evaluate the derivative of t^2. You obtain

$$v = (4.90\ \text{m/s}^2)2t$$

Simplifying, and indicating explicitly that v depends on t, you obtain

$$v(t) = (9.80\ \text{m/s}^2)t \tag{2-24}$$

Then you evaluate the acceleration:

$$a = \frac{dv}{dt} = \frac{d[(9.80\ \text{m/s}^2)t]}{dt}$$

Using Eq. (2-13) again, you have

$$a = (9.80\ \text{m/s}^2)\frac{dt}{dt}$$

Setting dt/dt equal to 1, in agreement both with Eq. (2-16) for $n = 1$ and with common sense, you obtain

$$a = 9.80 \text{ m/s}^2 \tag{2-25}$$

Thus, you can see from this result that if the falling object obeys the relation between the distance of fall x and time of fall t given by Eq. (2-23), it must be falling with the *constant* acceleration $a = 9.80 \text{ m/s}^2$. The value of a is positive because the acceleration is in the direction that has been chosen to be the positive x direction, namely, downward.

The strobe photo shown in Fig. 2-14 and the analysis of the identical photograph outlined in the caption to Fig. 1-6 provide an experimental demonstration that the motion of an object falling with negligible air resistance actually is a case of motion with constant acceleration. Thus the proportionality between x and t^2 of Eq. (2-23) is physically correct—that is, it conforms to what actually happens—since this is what leads to the constant acceleration of Eq. (2-25). Whether or not the numerical value of the factor 4.90 m/s^2 connecting x and t^2 is correct must also be settled by experiment. Later in this chapter you will see experimental evidence demonstrating that the value of the acceleration actually is $a = 9.80 \text{ m/s}^2$, and consequently that the numerical factor quoted in Eq. (2-23) is correct.

Additional experience in calculating acceleration will be obtained by considering again the motion of a pendulum bob oscillating between positions whose separation is small compared to the length of the cord from which the bob is suspended. Figure 2-21 plots the position, velocity, and acceleration of the bob. The position curve is just like the one in Fig. 2-20*b*, except that the numerical values are different. An equation giving the position of a pendulum bob executing small oscillations is

$$x(t) = c \sin(\omega t) \tag{2-26}$$

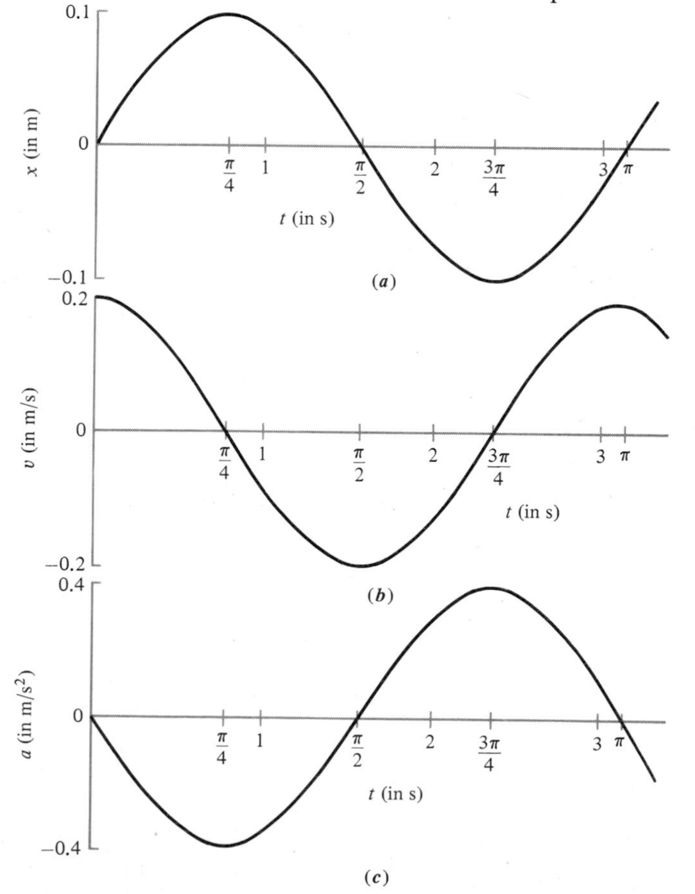

Fig. 2-21 The position x, velocity v, and acceleration a, plotted versus the time t, for one cycle of the motion of a bob at the end of a long pendulum cord.

For the particular case illustrated in Fig. 2-21, the constant c has the numerical value 0.1 m and the constant ω has the numerical value 2 s^{-1}. The top part of the figure is just a plot of Eq. (2-26) for these values. {To verify this, check a few points. For $t = 0$ s the equation gives $x = (0.1 \text{ m})(\sin 0) = 0$ m. For $t = \pi/4$ s, it gives $x = (0.1 \text{ m})[\sin(2 \text{ s}^{-1} \times \pi/4 \text{ s})] = (0.1 \text{ m}) \times (\sin \pi/2) = 0.1$ m. For $t = \pi/2$ s, it gives $x = (0.1 \text{ m})[\sin(2 \text{ s}^{-1} \times \pi/2 \text{ s})] = (0.1 \text{ m})(\sin \pi) = 0$ m.} The middle and lower parts of Fig. 2-20 are plots of equations that will be obtained in Example 2-9 from Eq. (2-26) by differentiating $x(t)$ to obtain $v(t)$ and then by differentiating $v(t)$ to obtain $a(t)$.

EXAMPLE 2-9

Evaluate the velocity, and then the acceleration, for the pendulum bob whose position is given by Eq. (2-26).

■ To determine the velocity, use its definition and Eq. (2-26) to obtain

$$v = \frac{dx}{dt} = \frac{d[c \sin(\omega t)]}{dt}$$

Remembering that c is a constant, you next use Eq. (2-13) and get

$$v = c \frac{d[\sin(\omega t)]}{dt}$$

Since ω is also a constant, Eq. (2-17) is applicable, and it gives you

$$v(t) = c\omega \cos(\omega t) \tag{2-27}$$

The notation makes explicit the fact that v depends on t. For $c = 0.1$ m and $\omega = 2 \text{ s}^{-1}$, this result is in agreement with the curve plotted in Fig. 2-21.

You determine the acceleration by using the expression for $v(t)$ just obtained in the definition:

$$a = \frac{dv}{dt} = \frac{d[c\omega \cos(\omega t)]}{dt}$$

Again employing Eq. (2-13), you have

$$a = c\omega \frac{d[\cos(\omega t)]}{dt}$$

Then Eq. (2-18) is used to yield

$$a(t) = -c\omega^2 \sin(\omega t) \tag{2-28}$$

For $c = 0.1$ m and $\omega = 2 \text{ s}^{-1}$, this result is also in agreement with the curve plotted in Fig. 2-21.

The relation between the $x(t)$ and $v(t)$ curves in Fig. 2-21 was explained qualitatively in Example 2-4. The explanation used the fact, emphasized in Sec. 2-4, that at any t the *slope* of the $x(t)$ curve equals the *value* of the $v(t)$ curve, because $dx/dt = v$. Earlier in this section we pointed out a similar explanation for the relation between the $v(t)$ and $a(t)$ curves. The relation depends on the fact that at any t the *slope* of the $v(t)$ curve equals the *value* of the $a(t)$ curve, since $dv/dt = a$. For examples of both relations, consider Fig. 2-21 at $t = 0$ s. At this instant $x(t)$ has a maximum positive slope and $v(t)$ has a maximum positive value. Also $v(t)$ has zero slope, and $a(t)$ has zero value. At $t = \pi/4$ s, $x(t)$ has zero slope and $v(t)$ has zero value. Also, $v(t)$ has its most negative slope and $a(t)$ has its most negative value. You can continue the analysis yourself.

Now we will consider a direct relation between the curves for $x(t)$ and $a(t)$. At any t the rate of change of slope of the $x(t)$ curve equals the value of the $a(t)$ curve, since $d(dx/dt)/dt = a$. For an example, look at the $x(t)$ curve of Fig. 2-21 in the interval from $t = 0$ s to $t = \pi/2$ s. The slope of $x(t)$ is always becoming less positive, or more negative. The rate of change of slope is therefore negative. This is why $a(t)$ is negative in the interval. For the interval from $t = \pi/2$ s to $t = \pi$ s, the rate of change of slope of $x(t)$ is positive because the slope is always becoming less negative, or more positive. As a consequence, $a(t)$ is positive in the interval.

In the interval from $t = 0$ s to $t = \pi/2$ s, the $x(t)$ curve is said to be **concave downward.** From $t = \pi/2$ s to $t = \pi$ s, it is said to be **concave upward.** In these terms, the relation between $x(t)$ and $a(t)$ can be expressed by saying *the direction of curvature of the* $x(t)$ *plot determines the sign of* $a(t)$. A concave upward $x(t)$ means a positive $a(t)$, and a concave downward $x(t)$ means a negative $a(t)$.

Furthermore, *the magnitude of curvature of the* $x(t)$ *plot determines the magnitude of* $a(t)$. That is, the more sharply the $x(t)$ plot curves, the larger the magnitude of $a(t)$. You can see the truth of this statement in Fig. 2-21. Note that in the vicinity of $t = \pi/4$ s the $x(t)$ plot curves most sharply and $a(t)$ has a maximum magnitude. Near $t = \pi/2$ s, $x(t)$ has minimum curvature and $a(t)$ has minimum magnitude.

In subsequent chapters you will see that these geometric relations between the $x(t)$, $v(t)$, and $a(t)$ curves can help you obtain an intuitive understanding of the behavior of systems governed by newtonian mechanics. You will also see that very similar analyses can fruitfully be employed on other systems too.

2-7 VELOCITY AND POSITION FOR CONSTANT ACCELERATION

In Examples 2-8 and 2-9 the position x of an object was expressed as a function of the time t by quoting the mathematical function relating x to t. Then its velocity v was evaluated by calculating the derivative of x with respect to t. Finally, the acceleration a of the object was evaluated by calculating the derivative of v with respect to t. No matter how an object moves in one dimension, if x is known as a function of t, it is always possible to go from x to v to a by two consecutive differentiations with respect to t.

But the conclusion of Sec. 1-3 implies that in newtonian mechanics the process must generally be carried out in the inverse order, that is, from a to v to x. The point is that in newtonian mechanics force is related to acceleration. In analyzing a mechanical system, the net force acting on an object will usually be determined first. Then Newton's laws of motion will be applied to evaluate the acceleration of the object from the force. To use this information in making a prediction of the object's motion that can be compared with experiment, v and then x must be determined from a.

Is there a way to go from a to v to x? Yes, by a process called *integration*. Integration is the mathematical inverse of differentiation. It is generally a more difficult task than differentiation. In fact, there are cases of great physical interest where integration cannot be performed by an analytical method and a more cumbersome—though conceptually simple—numerical method must be used. (Some idea of what is meant by the distinction between numerical and analytical methods of integration can be obtained by referring to Examples 2-5 and 2-6. Although they involved differentiation, not integration, the first one used a numerical method and the second used an analytical method.) A simple numerical method, carrying out a process amounting to integration because it goes from a to v to x, is introduced in Chap. 5. Considerable use will be made of this method, and variations of it, throughout the book. Analytical integration is deferred as long as possible, specifically until Chap. 7. The purpose is to allow you more time to reach the topic of integration if you are concurrently studying calculus. And in Chap. 7 we present a self-contained, albeit concise, treatment of integration, just as we have done for differentiation in this chapter.

Fortunately, if an object moving in one dimension has *constant acceleration*, as in free fall and many other important situations, there happens to be a very easy way to go from a to v to x. In fact, it is not necessary to use explicitly any of the methods of calculus in the argument.

Here is the argument. Consider an object that is released from an initial position where $x = 0$ at an initial time when $t = 0$ with an initial velocity $v = 0$. The object released from rest then falls freely downward with a constant acceleration a. Positive values of x are in the downward direction, so the same is true of v and a. Since a is constant, with the passage of time v increases steadily from its initial value zero at a constant rate a. Therefore at a subsequent time t the value of v is

$$v = at$$

Furthermore, since v does increase at a constant rate from zero as t increases, the average velocity $\langle v \rangle$ of the body over the interval beginning at

Fig. 2-22 (*a*) Velocity v versus time t for an object starting from rest and moving with constant acceleration. Because the value of the velocity increases uniformly from zero as time passes, the average velocity $\langle v \rangle$ over a time interval zero to t is exactly one-half its instantaneous value v at the end of the interval. That is, $\langle v \rangle = v/2$. A more detailed explanation of why this is so follows. If the velocity is sampled at each of a set of times distributed uniformly over the interval zero to t, the average of these velocities (that is, the sum of their values divided by the number of values being summed) will equal $\langle v \rangle$. The uniformly distributed times are indicated by the ticks on the time axis. For each of these the corresponding value of velocity is indicated by a tick on the velocity axis. Since the relation between velocity and time is given by the straight line plotting $v(t)$ for the case of *constant acceleration*, the ticks on the velocity axis are also uniformly distributed. As a consequence, when the average velocity $\langle v \rangle$ is computed, its value will be exactly at the center of the range zero to v. Thus $\langle v \rangle$ equals $v/2$, the value at the center of this range. The argument can be further clarified by considering the counterexample illustrated in the other part of this figure. (*b*) Here $\langle v \rangle$ is *not* equal to $v/2$ because the acceleration is *not* constant. The example illustrated is one in which the acceleration gradually becomes more positive through the time interval since the slope of $v(t)$ is gradually becoming more positive. (This is the case for a jet plane preparing to take off.) In these circumstances the ticks along the velocity axis are concentrated in the lower part of the range zero to v, and so $\langle v \rangle$ is less than $v/2$. You should sketch a figure illustrating an example in which the acceleration is gradually becoming less positive.

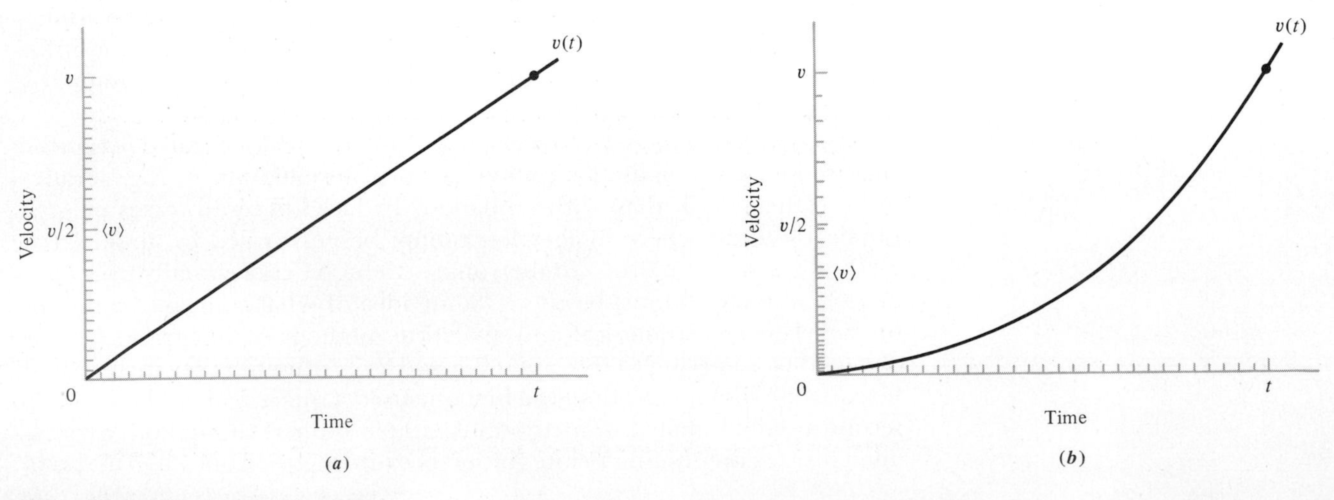

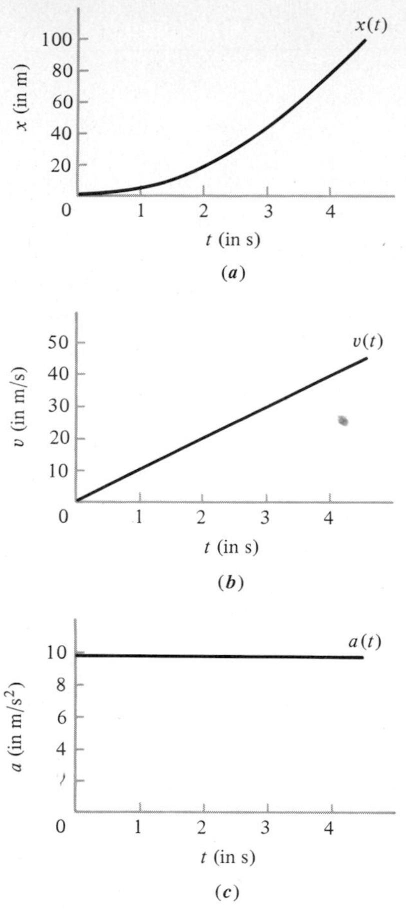

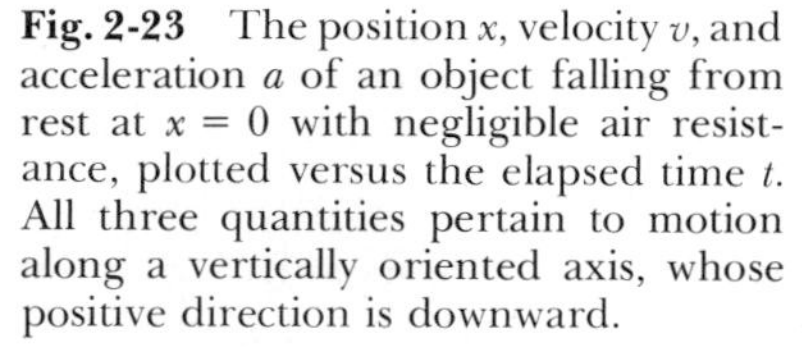

Fig. 2-23 The position x, velocity v, and acceleration a of an object falling from rest at $x = 0$ with negligible air resistance, plotted versus the elapsed time t. All three quantities pertain to motion along a vertically oriented axis, whose positive direction is downward.

time zero and ending at time t will equal *one-half* the final velocity v. (If this is not apparent, look at Fig. 2-22a, which is a plot of $v = at$, and read the explanatory caption.) Thus, for constant acceleration

$$\langle v \rangle = \frac{v}{2}$$

Using $v = at$, we obtain

$$\langle v \rangle = \frac{at}{2}$$

But the total change in position is always the average velocity multiplied by the time interval, as stated in Eq. (2-8):

$$\Delta x = \langle v \rangle \, \Delta t$$

Since the values of both position and time are zero at the beginning of the time interval, their values x and t at the end of the interval give the *changes* Δx and Δt in the values. Thus $\Delta x = x$ and $\Delta t = t$, and we have from Eq. (2-8) the result

$$x = \langle v \rangle t = \left(\frac{at}{2}\right) t$$

or

$$x = \frac{at^2}{2}$$

After specifying the initial values of a body's position and velocity ($x = 0$ and $v = 0$) and the initial value of the time ($t = 0$), we have used a statement about its acceleration (a = constant) to find an expression for its velocity ($v = at$) and then an expression for its position ($x = at^2/2$). Thus we have gone from a to v to x, for this case of constant a. The time dependences of the quantities a, v, and x are shown in Fig. 2-23. A numerical example of the argument used to relate these quantities is given in Table 2-1.

Now we modify the argument to treat a more general case in which the body experiencing constant acceleration has nonzero initial values of x and v, which we designate as x_i and v_i. But since we can usually zero the clock used to measure time at the initial instant, the initial time is again taken to be when $t = 0$. The subsequent values of position, velocity, and time are x, v, and t, as before. The modified argument is as follows. Since a is constant, v increases steadily from its initial value v_i at the constant rate a. Therefore at the time t the instantaneous value of v is given by the important equation

$$v = v_i + at \qquad \text{for constant } a \text{ and } v = v_i \text{ at } t = 0 \qquad (2\text{-}29)$$

Table 2-1

Numerical Example with $a = 9.8$ m/s² of Argument Leading to the Equation $x = at^2/2$, for $x = 0$ and $v = 0$ at $t = 0$

t (in s)	$v = at$ (in m/s)	$\langle v \rangle = v/2$ (in m/s)	$\langle v \rangle t = x$ (in m)	$9.8t^2/2 = at^2/2$ (in m)
0	0	0	0	0
1	9.8	4.9	4.9	4.9
2	19.6	9.8	19.6	19.6
3	29.4	14.7	44.1	44.1
4	39.2	19.6	78.4	78.4
5	49.0	24.5	122.5	122.5

Furthermore, since v does increase at a constant rate from the initial value v_i to the value v in the time interval from zero to t, the average velocity $\langle v \rangle$ is

$$\langle v \rangle = \frac{v_i + v}{2}$$

(How would you modify Fig. 2-21, and its caption, to prove this?) Using Eq. (2-29) to evaluate v, we have

$$\langle v \rangle = \frac{v_i + v_i + at}{2} = v_i + \frac{at}{2}$$

It is still true that $\langle v \rangle$ relates the change in position Δx to the time interval Δt by the equation $\Delta x = \langle v \rangle \Delta t$, and that $\Delta t = t$. But in the present case the initial value of the position is x_i, not zero. So Δx, the difference between the subsequent and initial positions, has the value $\Delta x = x - x_i$. Therefore we have

$$x - x_i = \langle v \rangle t = \left(v_i + \frac{at}{2}\right) t = v_i t + \frac{at^2}{2}$$

or

$$x = x_i + v_i t + \frac{at^2}{2} \qquad \text{for constant } a \text{ and } x = x_i,\ v = v_i \text{ at } t = 0 \qquad (2\text{-}30)$$

Equation (2-30) is important because it is the general expression for the position of an object in the frequently studied case of one-dimensional motion with constant acceleration. Let us interpret each term on the right side of the equation. The third term is the same as the single term $at^2/2$, obtained in the original argument, which gives the value of the position x at time t for motion with constant acceleration a in a case when both the initial position x_i and the initial velocity v_i are zero. The second term takes into account the fact that if v_i is not zero, this velocity acting over a time interval of duration t will make an *additional* contribution $v_i t$ to the value of x. The first term is present to account for the fact that if x_i is not zero, its value must be added into the value of x. Example 2-10 demonstrates the use of Eq. (2-30), and also of Eq. (2-29), in a common situation.

EXAMPLE 2-10

a. A car is traveling along a straight road at a speed of 71 km/h. Seeing a traffic jam ahead, the driver applies the brakes for 2.3 s and reduces the speed to 47 km/h. Assuming the acceleration is constant during the braking period, calculate its value.

b. If the driver continued to apply the brakes so as to maintain this acceleration, what distance would be required to bring the car to a halt from the speed of 47 km/h?

■ **a.** First you should convert the speeds from kilometers per hour to meters per second. You have for the initial speed

$$71 \text{ km/h} = 71 \frac{\text{km}}{\text{h}} \times \frac{10^3 \text{ m}}{1 \text{ km}} \times \frac{1 \text{ h}}{60 \text{ min}} \times \frac{1 \text{ min}}{60 \text{ s}} = 19.7 \text{ m/s}$$

In power-of-ten notation, 71 km/h is 7.1×10^1 km/h and 19.7 m/s is 1.97×10^1 m/s. So the velocity is expressed in meters per second to three significant figures even though it comes from a value expressed in kilometers per hour to only two significant figures. This means that the accuracy of the third significant figure is doubtful. However, it is better to retain it at this stage than to round off and write 19.7 m/s as 20 m/s. The reason is that the value will be used in subsequent calculations, and rounding off could impair their accuracy needlessly. In general, an extra significant figure should be carried in all calculated numbers that will be used in subsequent calculations. But when final results are obtained, they should be rounded off to no more significant figures than are in the least accurate value used to begin the calculations. (The number of significant figures in a final result can be appreciably fewer than in any value entering a calculation if at some point two nearly equal numbers are subtracted. For an example, consider 1.23 − 1.22 = 0.01.)

Converting the final speed from kilometers per hour to meters per second, as above, you find

$$47 \text{ km/h} = 13.1 \text{ m/s}$$

Next choose the direction of motion of the car to be the positive direction, and let time have the value zero at the instant that the braking period begins. The initial and final velocities are then $v_i = 19.7$ m/s, $v = 13.1$ m/s, and the final time is $t = 2.3$ s. These quantities are related to the acceleration a through Eq. (2-29):

$$v = v_i + at$$

Solving for a, you obtain

$$a = \frac{v - v_i}{t} = \frac{13.1 \text{ m/s} - 19.7 \text{ m/s}}{2.3 \text{ s}}$$

or

$$a = -2.87 \text{ m/s}^2$$

The acceleration is negative since the magnitude of the velocity is decreasing. That is, the acceleration is in the direction opposite to the direction of motion. As far as the answer to this part of the example is concerned, the value of a should be rounded off to two significant figures and quoted as

$$a = -2.9 \text{ m/s}^2$$

But when a is used in the next part, all three significant figures should be retained.

b. You can calculate the distance required to bring the car to a stop from a velocity of 47 km/h = 13.1 m/s, at an acceleration of -2.87 m/s², by first using Eq. (2-29) again to calculate the time required and then calculating the distance traveled in this time from Eq. (2-30). In the first step, you use a new choice for time zero, taking it to be the instant when the velocity is 13.1 m/s. Equating the final velocity in Eq. (2-29) to zero, you have

$$0 = v_i + at$$

or

$$t = -\frac{v_i}{a}$$

Setting $v_i = 13.1$ m/s and $a = -2.87$ m/s², you find the value of time when the final velocity is zero to be

$$t = -\frac{13.1 \text{ m/s}}{-2.87 \text{ m/s}^2} = 4.56 \text{ s}$$

In the next step you use Eq. (2-30):

$$x = x_i + v_i t + \frac{at^2}{2}$$

Measuring x from the point where the velocity is $v_i = 13.1$ m/s, you have $x_i = 0$. Setting $a = -2.87$ m/s² and $t = 4.56$ s, you find

$$x = 13.1 \text{ m/s} \times 4.56 \text{ s} + \frac{-2.87 \text{ m/s}^2 \times (4.56 \text{ s})^2}{2} \tag{2-31}$$

or

$$x = 30 \text{ m}$$

This is the distance traveled by the car in coming to rest from a velocity of 47 km/h, quoted to two significant figures.

Calculate the distance the car would travel in the time required for it to come to rest, if it continued to move at 47 km/h instead of having a negative acceleration. You will find that it is just twice as large as the distance traveled in coming to rest. Can you explain why?

The calculation in Example 2-10 suggests that a useful general relation can be obtained by solving Eq. (2-29) for t and then substituting the expression obtained for t into Eq. (2-30). This is exactly what was done in the example to evaluate x. But there it was done for a specific value of t, whereas the general result will apply to any value of t. We have from Eq. (2-29)

$$t = \frac{v - v_i}{a}$$

Inserting this in Eq. (2-30), we obtain

$$x = x_i + v_i \frac{v - v_i}{a} + \frac{a}{2}\frac{(v - v_i)^2}{a^2}$$

$$= x_i + \frac{2v_i v - 2v_i^2}{2a} + \frac{v^2 - 2v_i v + v_i^2}{2a}$$

or

$$x = x_i + \frac{v^2 - v_i^2}{2a} \qquad \text{for constant } a \text{ and } x = x_i,\ v = v_i \text{ at } t = 0 \tag{2-32}$$

This relation is used to find directly the change in position of an object moving with a known constant acceleration, while its velocity changes from one known value to another. Use it to recalculate the distance required for the car in Example 2-10 to stop.

Another convenient relation can be obtained by writing Eq. (2-29) as

$$a = \frac{v - v_i}{t}$$

and then substituting this expression for the a in Eq. (2-30). The result is

$$x = x_i + v_i t + \frac{(v - v_i)t^2}{2t}$$

which simplifies to

$$x = x_i + \frac{(v + v_i)t}{2} \qquad \text{for constant } a \text{ and } x = x_i,\ v = v_i \text{ at } t = 0 \tag{2-33}$$

This relation is employed to find the change in position when the velocity changes from one known value to another, if the value of the constant acceleration is not known but the time during which the velocity changes is known.

The calculation in Example 2-10 provides an excellent illustration of the way each term in a *correct* physical equation has the proper units. Look at Eq. (2-31):

$$x = 13.1 \text{ m/s} \times 4.56 \text{ s} + \frac{-2.87 \text{ m/s}^2 \times (4.56 \text{ s})^2}{2}$$

In the first term on the right side the seconds cancel, leaving the units for the term to be meters. A similar cancellation occurs in the second term, so that its units are also meters. Since the proper units for the term on the left side are meters too, the equation is consistent as far as units are concerned. If somehow an error had been made in deriving Eq. (2-30),

$$x = x_i + v_i t + \frac{at^2}{2}$$

so that the last term on the right side was mistakenly thought to be $at/2$, Example 2-10 would have made the mistake very apparent. In such a situation the corresponding term in Eq. (2-31) would have been found to have the units meters per second, instead of the required meters.

But it is not necessary to work through a specific numerical calculation [such as in Eq. (2-31)] to search for errors in a newly obtained equation [like Eq. (2-30)] by checking the consistency of the units. Just inspect the factors in each term of the equation from the point of view of their units. Any system of units can be used for this purpose. What is really important in the analysis of Eq. (2-30) is that x is a length (not that it is a length measured in the particular units called meters), that v is a length divided by a time, that a is a length divided by the square of a time, and that t is a time. It is said that x has the **dimensions** of length. The dimensions of v are length divided by time, that is, (length)(time)$^{-1}$. The dimensions of a are length divided by time squared, that is, (length)(time)$^{-2}$. And the dimensions of t are time. A **dimensional analysis** demonstrating the consistency of Eq. (2-30) is carried out by writing it and then writing beneath it an equation showing the dimensions of each term:

$$\begin{array}{ccccccc} x & = & x_i & + & v_i t & + & \dfrac{at^2}{2} \\ \text{length} & = & \text{length} & + & (\text{length})(\text{time})^{-1}(\text{time}) & + & (\text{length})(\text{time})^{-2}(\text{time})^2 \end{array}$$

(No dimensions are indicated for the factor $\frac{1}{2}$ because it is a pure number.) By treating the words "length" and "time" as quantities that can be manipulated according to the rules of algebra, it is seen that each of the three terms on the right side has the dimensions of length. Since the dimensions of the term on the left side are also length, the equation is dimensionally consistent.

Dimensional analysis is a very useful tool for finding errors. Carry out a dimensional analysis showing the inconsistency of an equation which is like Eq. (2-30) except that the third term on the right side has the form $a^2t^2/2$, instead of $at^2/2$. It would be a very good idea for you to get into the habit of doing a dimensional analysis on any equation you develop in the process of working through the exercises in this book. If you find an equation has inconsistent dimensions, you know that an error has been made. Of course, you cannot use dimensional analysis to check the consistency of numerical factors in equations, such as the factor $\frac{1}{2}$ in the third term of Eq. (2-30), since pure numbers are dimensionless and thus play no role in the analysis.

You can give Eq. (2-30), and also Eq. (2-29), a *complete* check by differentiating. This has been done already in Example 2-8 for the special case $x_i = 0$, $v_i = 0$, and $a = 9.80 \text{ m/s}^2$. Repeat the calculation of Example 2-8 without specifying the values of x_i, v_i, and a. By differentiating Eq. (2-30) with respect to t, you will show that the velocity is given by Eq. (2-29). Differentiating again, you will show that the acceleration has the value a, where a is a constant. This successfully finishes the verification, since the assumption of a constant acceleration a was the basis used to obtain Eqs. (2-29) and (2-30).

Section 2-8 closes the chapter by presenting additional applications of Eqs. (2-29) and (2-30). It will also serve to remind you that even though physics makes much use of mathematics, it is a science based on experiment.

2-8 VERTICAL FREE FALL

In Fig. 2-14 we presented a strobe photo of an object falling with negligible air resistance very near the surface of the earth. It showed qualitatively that the object experiences a constant downward acceleration. Quantitative results were quoted in Example 2-8 which led to the numerical value of the gravitational acceleration. But no experimental basis for the quantitative results was given. Here we rectify this omission by considering a measurement of the value of the very important quantity, the acceleration due to gravity.

One way to make such a measurement would be to obtain a strobe photo much like the one in Fig. 2-14, but with a meter stick and a clock included. It could then be analyzed quantitatively in the manner indicated in the caption to Fig. 1-6, the one from which Fig. 2-14 is reproduced. If you look back, you will see how this could lead to a numerical value of the acceleration of the falling object. However, you could not expect the value obtained to be very accurate because the analysis involves measuring first the difference between two pairs of positions, to obtain two velocities, and then the difference between these two velocities. Precision is lost in the succession of subtractions.

A reasonably accurate value of the gravitational acceleration can be obtained by measuring the *total time* required for an object to fall freely from rest over a certain *total distance.* Then the measurement can be analyzed by applying Eq. (2-30), with $x_i = 0$ and $v_i = 0$:

$$x = \frac{at^2}{2}$$

Fig. 2-24 An experiment measuring the gravitational acceleration g.

Figure 2-24 is a photograph of such a measurement. A steel ball is initially held at the top of a meter stick by an electromagnet. When a switch is thrown, a relay interrupts the current to the magnet. This releases the ball, allowing it to fall from rest. The relay simultaneously sends current to start the clock. After falling some distance, the ball knocks a switch open, thereby interrupting the current to the clock and stopping it.

The photograph shows that 0.45 s of time is required for the ball to fall from rest through a distance of 100 cm. Solving the preceding equation for a and substituting in these values for t and x, we obtain

$$a = \frac{2x}{t^2} = \frac{2(100 \text{ cm})}{(0.45 \text{ s})^2} = 9.9 \times 10^2 \text{ cm/s}^2 = 9.9 \text{ m/s}^2$$

This result is within 1 percent of the value that is obtained from a series of more accurate measurements of the magnitude of the **gravitational acceleration** near the earth's surface. That magnitude is

$$g = 9.80 \text{ m/s}^2 \tag{2-34}$$

Following common convention, the symbol g is used for the magnitude of the gravitational acceleration near the surface of the earth. For reasons that are explained in Sec. 5-4, the value of g differs from place to place by as much as several parts in 1000. The value tends to be smaller near the equator and larger near the poles, but there are also minute local variations. The value quoted in Eq. (2-34) has been averaged over various locations in the United States and then rounded off to three significant figures. This value of g is generally used throughout the book.

It is possible to verify that the gravitational acceleration is essentially constant throughout the motion of an object falling freely near the surface of the earth, and to do so with more accuracy than is possible from the experiment and analysis of Fig. 1-6. This can be done by repeating the experiment and analysis of Fig. 2-24 for several different values of x, showing that the same value of g is always obtained.

It is also possible to show experimentally that g does not depend on the nature of the freely falling object. Take two different objects which are both compact enough to fall a short distance without air resistance playing a significant role, say a small coin and a large coin. Hold them at the same height above the floor, and then release them at the same time. You will see and hear them hit the floor at very nearly the same time, showing that they have traveled with very nearly the same acceleration. In Sec. 4-2 we discuss the reason for this.

The experiment just suggested is sometimes called the "Leaning Tower of Pisa" experiment. According to legend, Galileo dropped a large cannonball and a small musket ball from the Leaning Tower and proved to a large audience of dumbstruck university professors that the two balls hit the ground simultaneously, in contrast to their expectation that the heavier ball would hit much sooner. In fact, Galileo probably never dropped anything from the Leaning Tower, though he almost certainly did the experiment we suggest that you do. In any case, no public demonstration of this sort was recorded.

However, in his famous *Dialogues Concerning the Two Great World Systems*, Galileo describes such an experiment in a hypothetical way and states clearly that the two balls will not *quite* strike the ground simultaneously. He then attributes the small difference in time of fall to air resistance. He bases his argument for an equal gravitational acceleration of all objects, in the absence of air resistance, on the following subtle, somewhat negative grounds: Suppose that a heavy object does fall with greater acceleration than a light one. Tie the two together with a string. Will the heavy object then pull the light one down faster than it would go by itself, and will the light object retard the heavy one, so that the acceleration of the combination is intermediate between those of the two separate objects? Or does the combination constitute an object heavier than either alone, which therefore falls faster than either? The only way to escape this contradiction is to agree that there is no difference in the motions in the first place.

Example 2-11 uses the equations relating position, velocity, and acceleration in one-dimensional motion with constant acceleration, and the measured value of the gravitational acceleration, to solve a rather complicated problem involving vertical free fall.

EXAMPLE 2-11

A child leans out of the window of a building at a height 10.0 m above the ground and throws a ball vertically upward with velocity 12.0 m/s. Neglecting air resistance, predict the maximum height above ground attained by the ball and also the total elapsed time at the moment it hits the ground.

■ Call the initial height above ground of the ball h and its maximum height H, the initial vertical velocity v_i, and the total elapsed time T. Your task is to find H and T

in terms of the given values of h, v_i and the known value of the gravitational acceleration g. It is best not to insert the actual numerical values of h, v_i, and g until the end of the analysis. You should take the vertical line on which the ball moves as the x axis. Also, you can choose the origin of that axis at the initial location of the ball, and choose its positive direction to be *upward.*

The two useful relations are those describing one-dimensional motion with constant acceleration, Eq. (2-29):

$$v = v_i + at$$

and Eq. (2-30):

$$x = x_i + v_i t + \frac{at^2}{2}$$

Choosing the positive x direction to be upward means that you *must* set $a = -g$. The negative sign expresses the fact that the gravitational acceleration, being always downward, is in the *negative* x direction. Also, you will have $x_i = 0$, since the origin of the x axis has been fixed at the initial location of the ball. So you have

$$v = v_i - gt \tag{2-35}$$

and

$$x = v_i t - \frac{gt^2}{2} \tag{2-36}$$

At the top of the ball's path its velocity will instantaneously be $v = 0$. It has finished going up, and it has not yet started going down. Using this condition in Eq. (2-35) yields

$$0 = v_i - gt$$

or

$$t = \frac{v_i}{g}$$

for the time when the ball reaches the top of the path. When you substitute this value of t into Eq. (2-36), you obtain

$$x = v_i \frac{v_i}{g} - \frac{g}{2}\left(\frac{v_i}{g}\right)^2$$

or

$$x = \frac{v_i^2}{2g}$$

for the maximum x coordinate of the ball. To find its maximum height H above the ground, you add the height of the window above the ground to the quantity x and obtain

$$H = \frac{v_i^2}{2g} + h \tag{2-37}$$

You can find the time at which the ball hits the ground from Eq. (2-36). Set $x = -h$, and thus equate the position of the ball with ground level. This gives you

$$-h = v_i t - \frac{gt^2}{2}$$

or

$$\frac{gt^2}{2} - v_i t - h = 0$$

Now solve this equation for the value of t corresponding to $x = -h$ by applying the standard expression for the solution to a general quadratic equation. (Recall that if $at^2 + bt + c = 0$, then $t = [-b \pm (b^2 - 4ac)^{1/2}]/2a$.) You get

$$t = \frac{v_i \pm \sqrt{v_i^2 + 2gh}}{g}$$

Since $(v_i^2 + 2gh)^{1/2}$ is greater than v_i, and since you should have t greater than zero, you will want the positive root. Thus the total elapsed time is

$$T = \frac{v_i + \sqrt{v_i^2 + 2gh}}{g} \tag{2-38}$$

Determining numerical values of H and T, in terms of the given numerical values of h, v_i, and g, is now simply an exercise in "plugging in" values and doing arithmetic. From Eq. (2-37) you find

$$H = \frac{(12.0\ \text{m/s})^2}{2 \times 9.80\ \text{m/s}^2} + 10.0\ \text{m} = 17.4\ \text{m}$$

And Eq. (2-38) gives you

$$T = \frac{1}{9.80\ \text{m/s}^2}\left[12.0\ \text{m/s} + \sqrt{(12.0\ \text{m/s})^2 + 2 \times 9.80\ \text{m/s}^2 \times 10.0\ \text{m}}\right]$$
$$= 3.11\ \text{s}$$

Note the way that values expressed to three significant figures are used consistently throughout these two independent calculations. In contrast to the situation in Example 2-10, here no calculated numbers are used in subsequent calculations. So it is not necessary to bother with carrying an extra significant figure on numbers produced at intermediate stages. This is one of the advantages of not inserting actual numerical values until the end of an analysis.

You could have chosen $x = 0$ to be at ground level and chosen the positive direction of the x axis to be downward. It will be worthwhile to repeat the analysis, making these choices, and show that the same final results are obtained. Still another approach to the problem is possible, if you wish to avoid the general quadratic equation that arises in determining the total time for the ball to hit the ground. You can break the analysis into two parts: (1) the trip up to the maximum height, and (2) the trip from that height to ground. Do this, and compare the results with those obtained here.

EXERCISES

Group A

2-1. *Speeds of various objects.* Give the approximate value of the speed, in meters per second (m/s), for each object listed. Also express each speed as a fraction of the speed of light, which is 3.00×10^8 m/s. In cases where full information is not provided, make reasoned estimates in order to obtain your results. (Note the table of conversion factors inside one of the book covers.)

a. An ant crawling

b. A person walking at a comfortable pace

c. A track star running the mile

d. An automobile on a superhighway

e. A cruising jet airliner (650 mi/h, or about 90 percent of the speed of sound)

f. A near-earth artificial satellite (orbital radius of 7000 km; orbital period of approximately 90 min)

g. The moon in its orbit around the earth (orbital radius of approximately 380,000 km; orbital period of 27.3 days)

h. The earth in its orbit around the sun (orbital radius of approximately 93,000,000 mi; orbital period of 365.3 days)

2-2. *Travel time in radio communication.* The speed of radio waves is the same as the speed of light waves, namely 3.00×10^8 m/s. Calculate the time required for radio waves to make each of the trips listed.

a. From a seacoast radio transmitter to a ship located 100 km offshore

b. From a ground station to a synchronous communications satellite orbiting at an altitude of 36,000 km

c. From an astronaut on the lunar surface to a control center on Earth (a distance of approximately 380,000 km)

d. Back to Earth from the Viking space probes upon their arrival at Mars (a distance of approximately 380,000,000 km)

2-3. *Debunking a rumor.* Suppose that a person claiming to possess psychic powers announces to you that the sun has just exploded. How long will you have to wait in order to be sure that the claim is untrue? (The distance from the sun to earth is 1.50×10^8 km, and the speed of light is given in Exercise 2-1.)

2-4. *The definition of velocity.* Explain how the general definition of velocity for one-dimensional motion, Eq. (2-6), contains in it the definitions given by Eqs. (2-1) and (2-3).

2-5. *The loneliness of planet dwellers.* The tremendous distances encountered in astronomy have led to the use of the travel time of light over a fixed distance to define a suitably large distance unit. A **light-year** is the distance traveled by light during one year.

a. Find the number of kilometers in one light-year.

b. Proxima Centauri, the nearest known star beyond the sun, is located 4.1×10^{13} km away. Express its distance in light-years.

c. Suppose an earthling decides to attempt to establish radio contact with inhabitants of a (hypothetical) planet orbiting Proxima Centauri. How long after initiating transmission should the earthling begin to listen for a reply?

2-6. *No house calls.* An unstaffed space probe passing near the planet Saturn is sending a continuous radio transmission back toward Earth. Suddenly one of the probe's instruments malfunctions in a way that cannot be handled by the probe's onboard computer. However, the malfunction affects the radio transmission in a recognizable manner. If the control personnel on Earth send corrective commands as soon as they receive the first sign of trouble, how much time elapses between the malfunction and the arrival at the probe of the corrective commands? Saturn is about 1.4×10^9 km from Earth.

2-7. *Strike up the band.* A marching band is performing on a football field during half time. The band members are standing in a rectangular formation, with the back row located 30 m behind the front row. They are playing a musical composition whose tempo is 150 beats per minute.

a. After the sound from the first rank passes a listener on the sideline, how much time elapses before the sound from the back row reaches the listener? Assume that the band members are perfectly synchronized. Sound waves travel at 340 m/s.

b. What fraction of a beat is your result for part *a*?

2-8. *Velocity from a graph of position versus time.*

a. Calculate the velocity for the motion depicted in Fig. 2-11, using values for a time interval starting at $t_i = 0$ s and ending at $t = 2$ s.

b. Repeat the procedure of part *a* for the motion depicted in Fig. 2-12.

c. Repeat the procedure of part *a* for the motion depicted in Fig. 2-13.

2-9. *Jumping the gun?* A person watching a track meet is sitting 150 m from the starting line.

a. After each race starts, how much time elapses before the spectator hears the starter's pistol? The speed of sound is 340 m/s.

b. Assuming that the runners' reaction time is 0.2 s, approximately how far from the starting line will the runners be at this instant?

2-10. *Analytical differentiation.* Use the analytical method of Example 2-6 to evaluate $(dx/dt)_{t_i}$, where $x(t) = ct^3$. Here c is a constant. Determine the value of the derivative for $c = 2$ m/s³ at $t_i = 1$ s and at $t_i = 2$ s. (The numerical method is treated in Exercise 2-42.)

2-11. *Rules for derivatives.*

a. Following Example 2-6, verify Eq. (2-16) for the case $n = -1$.

b. Prove Eq. (2-14).

c. Prove Eq. (2-15).

d. Following Example 2-7, verify Eq. (2-18).

2-12. *Sandy landing.* A steel ball falls from a height of 8.0 m onto smooth sand. It makes a depression in the sand 0.40 cm deep.

a. What is the average acceleration during the time it takes to stop the ball?

b. How long does it take to stop the ball?

2-13. *Flipping a coin to determine g.* In a crude experiment to determine the magnitude g of the acceleration of gravity, a coin is tossed vertically upward and its height of rise is measured. If it rises 1.0 m and the time between leaving and returning is 1.0 s as nearly as can be determined, what value of g results from these measurements?

2-14. *Looking for constant acceleration.* Which equation(s) among the following represent(s) motion in which the acceleration is constant? The symbols t, x, and v represent time, position coordinate, and velocity for one-dimensional motion. In each case, the symbol k represents a constant with the appropriate physical dimensions.

a. $v = kt$

b. $x = kt$

c. $v = kx$

d. $x = kt^2$

e. $v^2 = kx$

2-15. *Deer on the road.* A motorist is traveling along a straight highway at an initial speed of 20 m/s. A deer ambles onto the road 50 m ahead and stops.

a. What is the minimum deceleration which will bring the car to a halt before it strikes the deer?

b. Reevaluate your result for part *a* to take into account that the motorist has a reaction time of 0.30 s.

2-16. *Faster and faster.* Starting at $t = 0$ s, a body accelerates from rest, moving in a straight line with acceleration numerically equal to g. Find the body's final speed, distance traveled, and average speed for the time interval starting at $t = 0$ s and ending at

a. $t = 1.0$ s

b. $t = 1.0$ min

c. $t = 1.0$ h

d. $t = 1.0$ day

e. $t = 1.0$ month

Express the speeds in meters per second and also as a fraction of the speed of light. Express distances traveled in meters and, for part *e*, in light-years. (The light-year is defined in Exercise 2-5.)

2-17. *On the open road.*

a. Estimate the average speed in miles per hour (mi/h) of an automobile that is used primarily for highway driving. (In estimating the average speed, count only the time that the car is on the road!)

b. How many hours are spend in driving 100,000 mi?

c. If in one year a woman drives 15,000 mi, how many hours per day does she spend driving? What fraction of a typical 16-h waking period is this?

2-18. *In city traffic.*

a. Estimate the average speed in miles per hour (mi/h) of an automobile which is used primarily in city traffic. (In estimating the average speed, count only the time that the car is on the road!)

b. How many hours are spent in driving 100,000 mi?

c. If in one year a commuter drives 8000 mi, how many hours per day does he spend driving? What fraction of a typical 16-h waking period is this?

2-19. *When do we start?* Modify Eqs. (2-29), (2-30), (2-32), and (2-33) so that they apply to a situation in which the value of time at the initial instant is $t = t_i \neq 0$, rather than $t = 0$.

2-20. *What if?* Calculate the distance the car in Example 2-10*b* would travel in 4.56 s if it continued to move at 47 km/h. Explain why this distance is twice the distance it travels in coming to rest in 4.56 s with a constant negative acceleration.

2-21. *Average velocity.* Evaluate the average velocity in the motion depicted in Fig. 2-20 for the time interval

a. Starting at $t = 0$ s and ending at $t = 2$ s.

b. Starting at $t = 2$ s and ending at $t = 6$ s.

c. Starting at $t = 0$ s and ending at $t = 6$ s.

d. Starting at $t = 0$ s and ending at $t = 8$ s.

2-22. *Basic juggling.* A juggler wishes to have exactly three balls in the air at all times. He wants to throw a ball every 0.50 s.

a. With what initial speed must he throw each ball?

b. How high will each ball rise above his hands?

2-23. *Well, well.* A boy drops a stone into a deep water well and hears the splash 3.0 s later. How far did the stone fall before striking the surface of the water? The speed of sound is 340 m/s.

Group B

2-24. *Up and back.* A rocket rises vertically with constant acceleration $2.0g$ ($g = 9.8$ m/s^2). The acceleration lasts for 1.0 min, after which the engine is cut off.

a. How fast is the rocket going at cutoff?

b. How high is the rocket at cutoff?

c. How much higher will the rocket rise after cutoff, assuming that the value of g due to the attraction of the earth is substantially constant, and neglecting air resistance?

d. How long does it take to achieve this additional height?

e. Neglecting air resistance, how long will it take the rocket to return to the ground from its highest point?

f. With what speed does the rocket hit the ground?

2-25. *Back and forth.* Beginning at time $t = 0$, a pendulum bob executes small oscillations, so that its position x is given by $x = 0.030 \sin(2t)$, where x is measured in meters and t in seconds.

a. During the first cycle of oscillation, for what value of t does the position have its maximum positive value?

b. What is the velocity of the pendulum bob at this instant?

c. When during the first cycle does the acceleration attain its maximum positive value? What is that maximum value?

2-26. *A confirmation.* Verify that Eqs. (2-29) and (2-30) describe motion with constant acceleration. That is, differentiate Eq. (2-30) with respect to time to show that the velocity is given by Eq. (2-29). Then differentiate Eq. (2-29) with respect to time to show that the acceleration is indeed equal to the constant denoted by the symbol a.

2-27. *Average velocity for motion under constant acceleration.* Figure 2-22 and its caption describe the motion of an object starting from rest at time zero and undergoing constant acceleration. The description establishes that the average velocity $\langle v \rangle$ over a time interval beginning at time $t_i = 0$ and ending at time t is given by $\langle v \rangle = v/2$, where v is the velocity at time t. Construct an appropriate graph and accompanying discussion to prove that, for motion under constant acceleration, the average velocity $\langle v \rangle$ over any time interval is given by $\langle v \rangle = (v_i + v)/2$, where v_i is the velocity at the beginning of the interval and v is the velocity at the end of the time interval.

2-28. *Carry on.* The text presents on analysis of Fig. 2-21, describing the relations between the $x(t)$ and $v(t)$ curves and between the $v(t)$ and $a(t)$ curves. The analysis covers the time interval starting at $t = 0$ s and ending at $t = \pi/2$ s. Extend the analysis to cover the time interval starting at $t = \pi/2$ s and ending at $t = \pi$ s.

2-29. *Gasoline alley.* A car built for drag racing is able to accelerate from rest to 60 mi/h in 3.5 s.

a. What is its average acceleration during this time? Express your result in miles per hour per second, in meters per second squared, and as a multiple of g, the acceleration due to gravity.

b. Assuming that the acceleration is actually constant, what distance does the car travel during this time?

c. Suppose that throughout a $\frac{1}{4}$-mi race this car maintains an acceleration equal to that found in part a. What would its final speed be? What would its average speed be? How long would it take to complete the race?

2-30. *Intermediate juggling.* If a juggler knows that the shortest time interval she needs between tossing successive balls into the air is 0.30 s, what is the maximum number of balls she can keep in the air in a room in which the ceiling is 3.0 m above her hands?

2-31. *Roger the scientific detective.* Roger, the staff resident in a college dormitory, looks out his window and sees that water balloons are falling past his window. He is unable to lean out far enough to see the culprit, but he notices that each water balloon strikes the sidewalk 0.80 s after passing his window. Roger's room is on the fifth floor, 15.0 m above the sidewalk. Assuming that the balloons are being released from rest, how far above Roger is the release point?

2-32. *Matters of choice.*

a. Repeat the calculation of Example 2-11, choosing $x = 0$ to be at ground level.

b. Repeat the calculation of Example 2-11, choosing the positive direction of the x axis to be downward.

c. Repeat the calculation of Example 2-11, breaking the motion into two parts: the trip up to maximum height and the descent.

2-33. *Major league pop-up.* Professional baseball players sometimes hit "pop-ups" that go straight up, so that they can be caught right at home plate. Such pop-ups can remain airborne for several seconds.

a. Neglecting air resistance, what is the initial speed of a ball popped straight up if 6.0 s elapses before it is caught?

b. What is the maximum height of the pop-up described in part *a*?

Group C

2-34. *Same time, same station.* A homebound commuter usually arrives at her hometown train station at 5 p.m., just as her husband is arriving to meet her in the family car. One day she leaves work early and catches a train that arrives at the station at 4 p.m. She decides to walk home and immediately starts out along the same route that her husband uses. Her husband leaves home at the usual time, driving at his usual speed of 50 km/h. They meet on the way and drive home at the same speed, arriving 15 min earlier than usual.

a. How far did the commuter walk?

b. When did husband and wife meet?

c. What was the commuter's walking speed?

d. Can you determine the total distance between the commuter's home and the train station? Explain your answer.

2-35. *One after another.* At $t = 0$ s, a child throws a ball upward with an initial speed of 15.0 m/s. At $t = 0.50$ s, he tosses a second ball upward at the same speed.

a. Construct graphs showing how the acceleration, velocity, and position of the first ball depend on time, from $t = 0$ s until the ball returns to the ground.

b. Describe how to obtain the corresponding graphs for the second ball.

c. At what instant do the two balls have the same vertical position? What is their common position at that instant? Obtain your results using both an analytical method and a graphical method.

2-36. *Safe passage.* Mary Smith is driving along a two-lane highway, following another motorist, who is traveling at 80 km/h. Mary is keeping a safe distance, with (the center of) her car 30 m behind (the center of) the car ahead of her. However, the speed limit is 90 km/h, and Mary wishes to pass.

a. When she finds a clear opportunity, Mary begins to accelerate and changes to the other lane. She accelerates uniformly from 80 km/h to 90 km/h in 5.0 s and then maintains a constant speed. At the end of the acceleration, how far behind the other car is Mary?

b. From the time she begins to accelerate, how long is it before she pulls even with the other car? How far does Mary travel during this time?

c. When she is 30 m ahead of the other car, Mary changes back to the travel lane. How long does the entire passing procedure require? How far does Mary travel during this time?

d. If another car is approaching at a speed of 90 km/h, how far away must it be from Mary's car at the beginning of the procedure, in order for Mary to be able to vacate the passing lane by the time the approaching car is 150 m away?

2-37. *Catching up.* A college student drops a ball out a window on the top floor of the science building. She throws another ball straight down after the first one 1.0 s later. The second ball leaves her hand with a speed of 20 m/s.

a. Neglecting air resistance, and assuming that the balls continue to fall freely, how long after the first ball is dropped will the second ball overtake it?

b. How far above the ground must the release point be located in order for the two balls to hit the ground at the same time?

2-38. *A daredevil parachutist.* An aerial stunt man plans a spectacular jump from the World Trade Center in New York City. His assistant will drop a (packed) parachute from the top of one of the towers, 411 m above the street. The stunt man will then drop from a launch point 50 m below his assistant, timing his jump so that he can grab the parachute as it falls by him, strap it on, open the chute, and float safely to the street.

a. The ripcord must be pulled when the parachutist is at least 250 m above the street in order for him to land safely. The stunt man has found from experience in high-altitude skydiving that he can strap on a parachute and pull the ripcord in 3.0 s flat. After his assistant drops the chute, what is the earliest time at which the stunt man can jump? The latest time? (Neglect any change in the stunt man's motion as he catches the chute.)

b. If things were to go badly and the stunt man fell freely all the way to the street, how long would it take? What speed would he have on impact? (*Note:* Air resistance, which would need to be taken into account in order to obtain accurate results, is analyzed in Chap. 5.)

c. At the insistence of city officials, a layer of foam 10 m thick is placed over the street during the jump. If the stunt man hit the foam traveling at the speed found in part *b*, what constant upward acceleration would be just sufficient to bring him to rest within a distance of 10 m? Give your result in meters per second squared and also as a multiple of g.

2-39. *Upstairs, downstairs.* Two dormitory roommates, Hugh and Lou, decide to play an unusual game of catch. Hugh stands on a balcony, and Lou stands on the ground directly below. The balcony is 10 m above the ground. Hugh and Lou each throw the ball directly toward the other with an initial speed of 15 m/s.

a. How long does it take for the ball to travel from Lou up to Hugh? How fast is the ball traveling when Hugh catches it?

b. How long does it take for the ball to descend from Hugh to Lou? How fast is the ball traveling when Lou catches it?

c. Compare the round-trip travel time with the time that would be required if the ball traveled with a constant speed of 15 m/s in both directions.

d. Alice, another resident of the dormitory, lives 15 m above the two roommates. She decides to douse Hugh and Lou with two water-filled balloons. She watches them toss the ball back and forth several times and learns to anticipate when Hugh will release the ball. Alice wants to drop the balloons so that Hugh and Lou will be hit simultaneously, just as the ball is reaching Lou. When should Alice drop each balloon? Must she drop both balloons before Hugh throws the ball downward? How long after Hugh throws the ball will the first balloon fall past him? How long after that will the second balloon strike him?

e. Alice's diabolical plot works perfectly. However, Hugh and Lou are accustomed to Alice's pranks, and each of them has a rotten tomato handy. They grab the tomatoes, count to three, and simultaneously hurl the tomatoes at Alice. Each tomato has an initial speed of 30 m/s. How soon after the tomatoes are thrown does Alice need to be out of the way?

f. Both tomatoes miss on the way up, but then Alice makes a mistake. She leans out to gloat at her wet victims. Both tomatoes strike her on the back of the head. Whose tomato hits Alice first? When does it hit her, and how fast is it traveling? When does the other tomato hit, and how fast is it traveling?

Numerical

2-40. *Numerical evaluation of a derivative: I.* Use the numerical method of Example 2-5 to evaluate $(dx/dt)_{t_i}$, where $x(t) = (4.90\ \mathrm{m/s^2})t^2$, for $t_i = 2$ s. Stop the calculation after several consecutive numbers in the sequence have the same value to two decimal places. This value is the two-decimal-place limit.

2-41. *Numerical evaluation of a derivative: II.* Continue the calculation of Exercise 2-40 to obtain numbers in the sequence of results beyond the two-decimal-place limit. Show that the numbers in the sequence eventually begin to fluctuate about the limit. These fluctuations are a result of calculation round-off error. Use the definition of a derivative to explain this phenomenon in detail. Will it be a practical limitation to the utility of the numerical method for evaluating derivatives?

2-42. *Numerical method in action: I.* Use the numerical method of Example 2-5 to evaluate $(dx/dt)_{t_i}$ where $x = (2\ \mathrm{m/s^3})t^3$ for $t_i = 1$ s and for $t_i = 2$ s. Compare your answers with the result of Exercise 2-10.

2-43. *Numerical method in action: II.* Use the numerical method of Example 2-5 to evaluate $(dx/dt)_{t_i}$, where $x(t) = \sin[(2\ \mathrm{s^{-1}})t]$, for $t_i = 0$ s, $\pi/8$ s, $\pi/4$ s, $3\pi/8$ s, and $\pi/2$ s. Plot x versus t and also plot dx/dt versus t for these values of t_i. Comment on the apparent relation between the two plots.

2-44. *Midpoint versus endpoint.* The definition of a derivative given in Eq. (2-9) is called the endpoint definition. An alternative definition, called the midpoint definition, is

$$\left(\frac{dx}{dt}\right)_{t_i} = \operatorname*{limit}_{\Delta t \to 0} \left[\frac{x(t_i + \Delta t/2) - x(t_i - \Delta t/2)}{\Delta t}\right]$$

These two definitions are equivalent since they lead to the same limiting value as $\Delta t \to 0$. However, the midpoint definition forms a basis of a numerical method for evaluating derivatives superior to the method based on the endpoint definition. That is, the sequence of numbers obtained using the midpoint method converges more rapidly to the limit than does the sequence of numbers obtained using the endpoint method.

a. Convert the endpoint program for numerical differentiation, given in the Numerical Calculation Supplement, into a midpoint program.

b. Use this midpoint program to repeat the calculations of Exercise 2-42. Compare the rates of convergence of the midpoint and endpoint methods. How many steps are required to reach the two-decimal-place limit in each case?

c. Use the midpoint program to repeat the calculation of Example 2-5. Explain why the midpoint results converge immediately in this particular case.

3 Kinematics in Two and Three Dimensions

3-1 PROJECTILE MOTION

If the universe were one-dimensional, physics would be much simpler. But that would hardly compensate for the loss of richness of phenomena which makes the physical world so fascinating. Very many of the most important phenomena of physics simply could not take place in a one-dimensional world.

Now that you have begun to feel at home with the kinematics used to describe motion in one dimension, it is time to see what generalizations are required to extend kinematics so that it applies to motion in two dimensions, and ultimately in three dimensions as well. This extension not only introduces additional physics, but also calls for—and leads naturally to—a convenient and powerful mathematical tool known as the *vector*.

Let us begin our study of physics in two dimensions by considering a straightforward question: After an object is projected in a direction parallel to the earth's surface, it has a horizontal motion. But it also has a downward vertical motion resulting from the influence of gravity. Is the vertical motion affected by the horizontal motion, and vice versa?

Every child is aware that there are two conceivable answers to this question. The first one, which has strong appeal in spite of its implausibility, appears regularly in the cartoons of Saturday morning television. See Fig. 3-1. The villain, in pursuit of the hero, runs over the edge of a cliff. He continues to move horizontally for some distance until he comes to a stop with feet windmilling, and only then does he plunge vertically downward.

Children never fail to laugh at this cliché. It is based on a notion that things should not be able to move simultaneously in both the horizontal and vertical directions, that somehow they should have to lose their horizontal motion before they can begin vertical motion. This view is not

(a) (b) (c) (d)

Fig. 3-1 Motion in the "cartoon universe," in which horizontal and vertical motion cannot take place simultaneously. In this sequence of cartoon stills, the villain runs off the edge of the cliff, but does not begin to fall until his horizontal motion has completely ceased. Only then does he plunge straight downward to disaster.

merely childish; it was shared in at least some degree by most physicists until near the end of the Renaissance. And yet children *know* that real objects do not behave in "cartoon" fashion. A child who really thought they did could never learn to catch a ball.

You might guess that the analysis of motion would be very simple in a universe where things could not move horizontally and vertically at the same time, since the motions could be handled one at a time. But in fact such a universe would be more complicated than the one in which we actually live. The "cartoon law of motion" implies a very strong *interaction* between horizontal and vertical motion, which would have to be taken into account in a complicated way.

Fortunately, the way things actually behave is much simpler. The rule underlying that behavior is *exactly the opposite* of what happens in the cartoon: Under the influence of gravity, objects change their vertical positions by amounts which have *absolutely nothing to do* with changes in their horizontal positions, even though the vertical and horizontal motions take place simultaneously. The point is made clear by the experiment shown in the strobe photo of Fig. 3-2. In this experiment, two steel balls are initially at the same height. Ball 1 is loaded into a spring gun which can shoot it as a **projectile,** and ball 2 is held by an electromagnet. A single switch simultaneously triggers the gun and turns off the electromagnet, and both balls start to move simultaneously. Ball 2 is simply dropped and has no horizon-

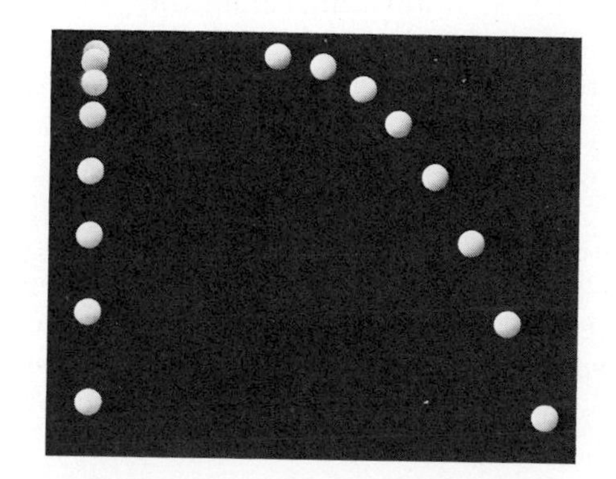

Fig. 3-2 Strobe photograph of two balls which are set into motion simultaneously, but in different ways, and which then continue to move under the influence of gravity. Ball 1, on the right, is shot horizontally from a spring gun. Ball 2, on the left, is simply released at the same moment by turning off the electromagnet which holds it at the same initial height as ball 1.

tal velocity. The spring gun has been aimed in the horizontal plane, so that ball 1 leaves the gun with a horizontal velocity. So far as *horizontal* motion is concerned, ball 1 acts like the air table puck in Fig. 1-5, which moved equal distances in the equal time intervals from one light flash to the next. That is, the horizontal velocity of the ball remains constant. This can be verified by measuring the distances through which the ball travels horizontally between its successive positions. The *vertical* motion of ball 1 is the same as that of ball 2, which falls straight down with constant acceleration. This statement is based directly on the evidence of Fig. 3-2, since both balls travel the same distances vertically between their successive positions.

Thus the ball shot from the gun does not "know" it is moving vertically, as far as its horizontal motion is concerned. Likewise, it does not "know" it is moving horizontally, as far as its vertical motion is concerned. This is fortunate, since we already understand how to analyze both motion with constant velocity and motion with constant acceleration. The problem, then, is to combine these motions in two dimensions and thus to determine the path of the ball (or projectile), which is called its **trajectory.**

In order to describe the trajectory in a specific fashion, it is necessary first to choose an origin, positive directions, and distance scales for the coordinate axes. In making these choices, we specify what mathematicians call a *coordinate system.* There is a further requirement, however, if a coordinate system is to be useful in making physical measurements. There must be an *observer,* real or imagined, who makes the necessary measurements and specifies them in terms of the coordinates. The observer has a fixed location with respect to the coordinate axes being used. Taken together, the observer and the coordinate axes constitute a **frame of reference.**

In Fig. 3-2 it is convenient to measure both the horizontal coordinate x and the vertical coordinate y of ball 1 from its starting point in the spring gun. So we fix the origin of these coordinates at the gun. Also, we choose the positive direction of the x axis toward the right and the positive direction of the y axis upward. Since the ball always moves downward, this choice means that the values of y will all be negative. (While it might seem "natural" to avoid negative quantities by choosing the downward direction as that of positive y, this is not done because it is conventional to draw the positive y axis of a coordinate system in the direction 90° counterclockwise from that of the positive x axis.) We denote the initial velocity of ball 1 as it is shot from the gun by v_i, the velocity of its motion in the horizontal direction at any time by v_x, and the velocity of its motion in the vertical direction at any time by v_y. Since the horizontal and vertical motions are *mutually independent,* the equations describing the motion of the ball after it leaves the gun can be written *separately.* We use g to represent the magnitude of the gravitational acceleration. These equations are then written

Horizontal	**Vertical**
$v_x = v_i = \text{constant}$ (3-1*a*)	$v_y = -gt$ (3-1*b*)
which leads to the relation	which leads to the relation
$x = v_i t$ (3-2*a*)	$y = -\dfrac{gt^2}{2}$ (3-2*b*)

Equations (3-2*a*) and (3-2*b*) are independent descriptions of motion, one involving the coordinate x and the other the coordinate y. *But x and y both depend on a common variable, the time t.* Such equations are called **parametric equations,** and the common variable (here t) is called the **parameter.**

The algebraic operation of eliminating t from the pair of equations yields a single equation expressing a relation between the value of x and that of y for the position of the ball at *any* value of t. To find this equation, square Eq. (3-2*a*) to obtain $x^2 = v_i^2 t^2$, and solve for t^2:

$$t^2 = \frac{x^2}{v_i^2}$$

Insert this value of t^2 into Eq. (3-2*b*) to obtain

$$y = -\frac{g}{2v_i^2}x^2 \tag{3-3}$$

Equation (3-3) describes all the points through which the ball passes. It is the equation of a parabola. Thus *the trajectory is parabolic.* This result was first derived by Galileo, probably quite early in the seventeenth century. Specific cases of Eqs. (3-2*a*) and (3-2*b*), and of the connection between them given by Eq. (3-3), are worked out in Example 3-1.

EXAMPLE 3-1

Measurements made on the strobe photo of Fig. 3-2 show that ball 1 leaves the gun with a horizontal velocity $v_x = v_i = 1.2$ m/s.

a. Calling the time when the ball leaves the gun $t = 0$, find the quantities x and y, which describe the horizontal and vertical positions of the ball with respect to the gun, at the times $t = 0.25$ s and $t = 0.50$ s.

b. Give the equation for the trajectory of the ball, and then use it to check the relation between the values of x and y found for $t = 0.50$ s.

■ **a.** When you insert the numerical values into Eq. (3-2*a*), you obtain for

$$t = 0.25 \text{ s}$$

the value

$$x = v_i t = 1.2 \text{ m/s} \times 0.25 \text{ s} = 0.30 \text{ m}$$

From Eq. (3-2*b*) you obtain

$$y = -\frac{gt^2}{2} = -\frac{9.8 \text{ m/s}^2 \times (0.25 \text{ s})^2}{2} = -0.31 \text{ m}$$

For

$$t = 0.50 \text{ s}$$

you have

$$x = 1.2 \text{ m/s} \times 0.50 \text{ s} = 0.60 \text{ m}$$

and

$$y = -\frac{9.8 \text{ m/s}^2 \times (0.50 \text{ s})^2}{2} = -1.2 \text{ m}$$

b. Substituting the numerical values of g and v_i into Eq. (3-3), you can express the trajectory in the form

$$y = -\frac{9.8 \text{ m/s}^2}{2 \times (1.2 \text{ m/s})^2} x^2 = (-3.4 \text{ m}^{-1})x^2$$

For the particular value $x = 0.60$ m, which you found for $t = 0.50$ s, this equation predicts

$$y = (-3.4 \text{ m}^{-1}) \times (0.60 \text{ m})^2 = -1.2 \text{ m}$$

in agreement with the corresponding value of y found in part *a*.

Now we will try out all the ideas we have developed on a more general case of projectile motion. The experiment illustrated by the strobe photo of Fig. 3-3 demonstrates what happens if the spring gun used to project the ball is not aimed horizontally. (For practical reasons, it is necessary to compress the spring more tightly in this experiment than in the experiment of Fig. 3-2.)

Inspection of Fig. 3-3 shows immediately that the horizontal distance between successive positions of the ball is constant, just as was the case in Fig. 3-2. If we call the horizontal velocity v_x, the horizontal distance covered is given by the relation

$$x = v_x t \tag{3-4}$$

which is valid for constant v_x if we set $t = 0$ at the moment of launching.

Is it still true that the vertical motion is one of constant downward acceleration g, in spite of the fact that the ball has an initial upward vertical velocity v_{yi}? It is not easy to answer this question by looking at the strobe photo. But let us assume that again the horizontal motion is irrelevant, as far as the vertical motion is concerned. If this is so, the vertical motion of the ball will be just the same as if it were moving in *only* the vertical direction. We know from the study of one-dimensional vertical motion in Chap. 2 that the acceleration will then indeed be downward with constant magnitude g.

If the assumption is correct, it also follows that Eq. (2-30), $x = x_i + v_i t + at^2/2$, can be applied to the vertical motion with appropriate changes

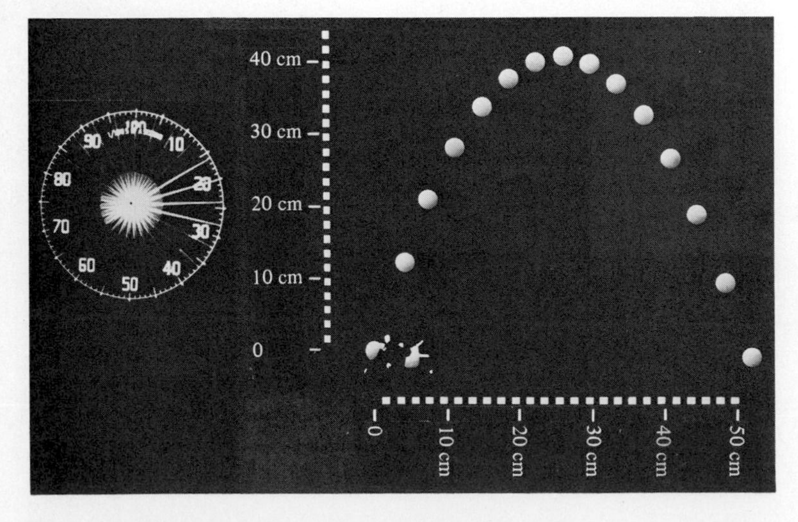

Fig. 3-3 Strobe photograph of a ball shot from a spring gun in a nonhorizontal direction. The clock shows that 0.040 s passes between each flash of the strobe light and the next one. Only part of the spring gun can be seen.

in notation. Since we are calling the vertical coordinate y, with the positive direction upward and the gun located at the origin where $y_i = 0$, Eq. (2-30) becomes

$$y = v_{yi}t - \frac{gt^2}{2} \qquad (3\text{-}5)$$

Equations (3-4) and (3-5) have the common variable time, just like the simpler Eqs. (3-2*a*) and (3-2*b*) which apply to the case of the horizontally aimed gun. Solving Eq. (3-4) for t and substituting into Eq. (3-5) gives

$$y = \frac{v_{yi}}{v_x} x - \frac{g}{2v_x^2} x^2 \qquad (3\text{-}6)$$

Like Eq. (3-3), this is the equation of a parabola, although this time the turning point (the point where the slope of the parabola is zero) is not at the origin. Example 3-2 will give you a direct experimental verification of Eq. (3-6), and therefore also of Eqs. (3-4) and (3-5).

EXAMPLE 3-2

Using measurements made on the photograph of Fig. 3-3, find the position coordinates x and y for one of the images of the ball. Measure also the (constant) horizontal velocity v_x and the initial vertical velocity v_{yi}. Use Eq. (3-6) and the measured values of x, v_x, and v_{yi} to calculate y, and compare the calculated value of y with the measured value.

■ Choose any image of the ball, say the eleventh, counting the image of the ball in the gun as zero. (This is the fourth image of the ball past the top of the trajectory.) Choose the positive x direction to the right and the positive y direction upward. Take the starting position of the ball (in the gun) to be the origin, so that $x_i = 0$ and $y_i = 0$. When you scale off horizontal and vertical distances, you find that the coordinates of the eleventh image of the ball are

$$x_{11} = 41.5 \text{ cm} \qquad \text{and} \qquad y_{11} = 25.7 \text{ cm}$$

A digression concerning significant figures is appropriate here. You found in making measurements on Fig. 3-3 that it is not possible to measure the location of an image of the ball within 0.1 cm. That is, the last digit in the value of x or y immediately above is not a significant figure in the sense of the discussion in Example 2-1. For example, $x = 41.5$ cm cannot mean *less than* $x = 41.6$ cm or *more than* $x = 41.4$ cm. Nevertheless, it is not desirable to drop the last digit completely, because you *can* make the measurements—or at least estimate them reliably—within a few tenths of a centimeter. Put another way, the possible precision of the measurement is such that you would not be exploiting it fully if you discarded the digits to the right of the decimal point. In cases like this, you should retain the last digit even though it is not *fully* significant. However, you must bear in mind that the significance of the last digit in the final result will be the same (at best) as that of the corresponding digit in the least precise measurement.

To find an accurate value of v_x, you choose two well-separated images of the ball, say image 3 and image 11. Call the time interval between successive flashes of the strobe light a flash interval. Since there are eight flashes between the two images of interest, the time elapsed as the ball passes from image 3 to image 11 is $\Delta t = 8$ flash intervals. Scaling off the horizontal distance, you obtain $x_{11} - x_3 = 29.7$ cm. Thus you have for the horizontal velocity

$$v_x = \frac{x_{11} - x_3}{\Delta t} = \frac{29.7 \text{ cm}}{8 \text{ flash intervals}} = 3.71 \text{ cm/flash interval}$$

Using the timer in the figure to measure the number of flashes occurring in 1 s, you find that 1 flash interval = 0.0400 s. You thus have

$$v_x = \frac{3.71 \text{ cm}}{1 \text{ flash interval}} \times \frac{1 \text{ flash interval}}{0.0400 \text{ s}} = 92.8 \text{ cm/s}$$

It is not possible to obtain a precise value of v_{yi} from the strobe photo by direct measurement, since the ball has that initial vertical velocity only at the instant of departure from the gun. Nevertheless, you can make an approximation by working backward. Call the measured heights of images 1 and 2 of the moving ball y_1 and y_2. The *average* vertical velocity of the ball between flashes 1 and 2 is

$$\langle v_{y\,1\to 2} \rangle = \frac{y_2 - y_1}{\Delta t} = \frac{19.2 \text{ cm} - 10.5 \text{ cm}}{0.0400 \text{ s}} = 218 \text{ cm/s}$$

According to the argument made in Sec. 2-7, this is also the *instantaneous* velocity at the *midpoint* of the time interval between flashes 1 and 2. That is, it is the instantaneous velocity in the y direction at approximately $1\frac{1}{2}$ flash intervals = 0.0600 s after the ball is launched. This is an approximation, since the ball was probably not launched simultaneously with a flash. Nevertheless, you can see from the figure that the vertical distance between images 0 and 1 is greater than that between images 1 and 2, so that the chances are that the firing of the gun nearly coincided with a flash. Thus you can take the value of the average velocity $\langle v_{y\,1\to 2} \rangle$ to be the value of the instantaneous velocity $v_{y\,1\,1/2}$.

Now that you know $v_{y\,1\,1/2} = 218$ cm/s (to a good approximation), you can apply Eq. (2-29) to evaluate the ball's initial vertical velocity v_{yi}. In the present notation, it is

$$v_{y\,1\,1/2} = v_{yi} - gt$$

where $t = 0.0600$ s is the time $1\frac{1}{2}$ flash intervals after the initial time $t = 0$. Solving for v_{yi}, you obtain

$$v_{yi} = v_{y\,1\,1/2} + gt$$

Since you are measuring distances in centimeters, you must express the value of g in centimeters per second squared. You have

$$g = 9.80 \text{ m/s}^2 \times \frac{100 \text{ cm}}{1 \text{ m}} = 980 \text{ cm/s}^2$$

You insert this value, and the other numerical values, into the equation immediately above to obtain, within the accuracy of the approximation and the measurements themselves,

$$v_{yi} = 218 \text{ cm/s} + 980 \text{ cm/s}^2 \times 0.0600 \text{ s} = 277 \text{ cm/s}$$

Using this value in Eq. (3-6), you obtain the result

$$y = \frac{277 \text{ cm/s}}{92.8 \text{ cm/s}} \times 41.5 \text{ cm} - \frac{980 \text{ cm/s}^2}{2 \times (92.8 \text{ cm/s})^2} \times (41.5 \text{ cm})^2$$

or

$$y = 25.9 \text{ cm}$$

This compares well with the directly measured value, which is $y = 25.7$ cm. Indeed, you could not expect a better correspondence, given the limits on the accuracy of the measurements used.

3-2 PROPERTIES OF VECTORS

In the preceding section we learned how to find the quantities x and y that specify, at any time t, the horizontally- and vertically-measured positions of a ball with respect to a gun from which it it shot. But there are other quantities which are useful in providing the same kind of information. One of these is the "slant distance" from the gun to the ball. This distance r is the length of a straight line extending from the gun to the ball. Refer to Fig. 3-4. In this figure, the pythagorean theorem tells us that

$$r^2 = x^2 + y^2$$

Thus the distance from the gun to the ball is

$$r = \sqrt{x^2 + y^2} \tag{3-7}$$

The positive root is taken because a *distance* is always positive. The value of r provides a partial specification of the position of the ball relative to the gun. To complete the specification it is necessary to give also the direction from the gun to the ball. The angle ϕ in Fig. 3-4 does this. It is measured counterclockwise from the positive x axis to the line of length r and, by the definition of the tangent function, satisfies the relation

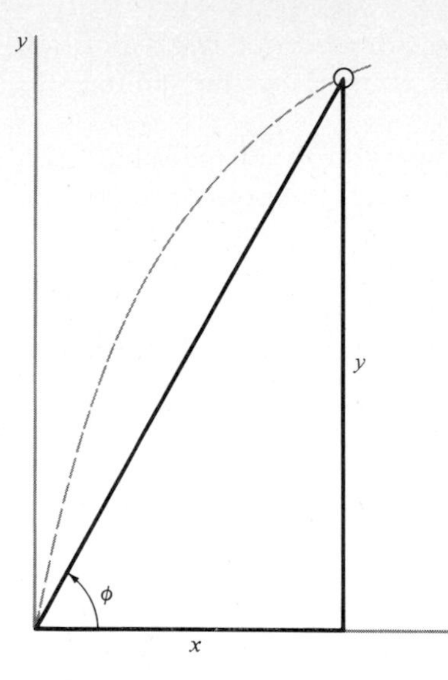

Fig. 3-4 The slant distance r from the gun to the ball at any moment is the hypotenuse of the right triangle of which the horizontal distance x and the vertical distance y are the sides. The direction from the gun to the ball is specified by the angle ϕ. The dashed gray curve represents the trajectory.

$$\tan \phi = \frac{y}{x}$$

Thus the angle ϕ is given by

$$\phi = \tan^{-1} \frac{y}{x} \tag{3-8}$$

(The symbol "$\tan^{-1}$" means that ϕ is the angle whose tangent is y/x.) For any value of t, the values of x and y can be found from Eqs. (3-4) and (3-5), and then used in Eqs. (3-7) and (3-8) to evaluate the quantities r and ϕ.

There is a striking contrast between the way the horizontally- and vertically-measured relative positions are added in Eq. (3-7) and the familiar way, called **algebraic addition,** that quantities such as time and volume are added. For example, a 2-s time interval followed by a 3-s time interval constitutes a total time interval of 2 s + 3 s = 5 s. For another example, if you fill a 5-liter (L) bottle with water and then fill a 1-L bottle from the 5-L bottle, the water remaining in the latter amounts to 5 L + (−1 L) = 4 L. A quantity which adds to a like quantity in this manner can be completely specified, in terms of an agreed-upon unit, by a single number preceded by a positive sign (usually unwritten) or by a negative sign. Such a quantity is called a **scalar.**

A quantity which adds to a like quantity in the *different* way that relative positions add requires more than a single number to specify it completely. Such a quantity is called a **vector.** Relative position is not the only vector quantity. For example, velocity and acceleration are vectors, as we will see in the next section. In this section we develop the properties of vectors. Our development will be guided by considering the mathematical properties of position vectors. But any other quantity which partakes of the same *mathematical* properties is a vector, no matter what *physical* attribute it describes.

We have learned already that there are two different ways to specify a vector:

1. In a situation confined to two dimensions, a *vector* can be specified in terms of two *scalars.* These scalars are the two **components** of the vector. For example, a vector describing the position of an object with respect to the origin can be specified, using a set of mutually perpendicular axes, in terms of its components. These are the pair of scalars designated as x and y in Fig. 3-4. Because the properties of vectors are general, we now introduce

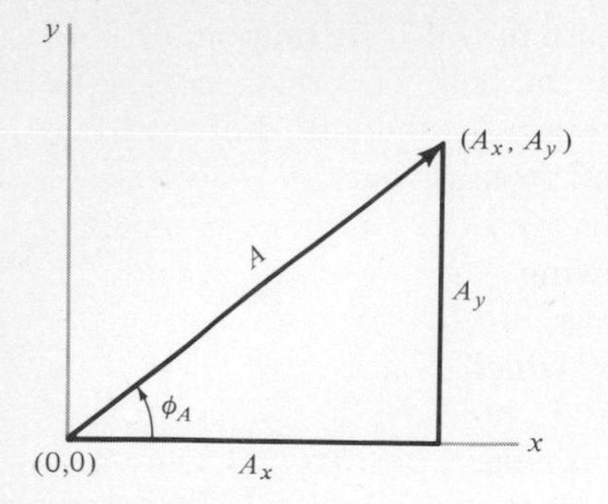

Fig. 3-5 Illustrating two ways of describing the position of the point (A_x, A_y) with respect to the origin (0, 0), as discussed in the text. The figure assumes that both A_x and A_y have positive values.

a notation that more clearly applies to *any* vector and not just a position vector. We do this by replacing the symbols x and y by the symbols A_x and A_y.

In Fig. 3-5, a pair of mutually perpendicular coordinate axes has been drawn on a plane. The origin of these coordinates is the point in the plane specified by the pair of values ($A_x = 0$, $A_y = 0$), or simply (0, 0) for short. Now, *any other point in the plane can be uniquely specified by an ordered pair of scalars (numbers)* (A_x, A_y). [It is purely conventional that A_x is written first, but it is important to stick to the convention once it is established; (5, 3) is not the same point as (3, 5).]

In essence, this ordered pair prescribes the position of the point (A_x, A_y) with respect to the origin. It does so by specifying a pathway for reaching it from the origin. The prescription reads: *Beginning at the origin, measure off a length equal to the magnitude of* A_x*, extending in the positive direction along the x axis if* A_x *is positive and in the negative direction if* A_x *is negative. Next turn your ruler by 90°, so that it is parallel to the y axis. Then measure off a length equal to the magnitude of* A_y*, extending in the positive direction along the y axis if* A_y *is positive and in the negative direction if* A_y *is negative. You have now located the point* (A_x, A_y) *in a completely unambiguous way.*

2. In two dimensions, a different pair of scalars can also be used to specify a vector. One of these scalars gives the **magnitude** of the vector (that is, its absolute numerical value), and the other gives its **direction.** Examples are the pair of scalars (r, ϕ) in Fig. 3-4. This second way of specifying a vector is the algebraic equivalent of the "natural" way to depict a vector geometrically, as an arrow whose length and orientation represent the magnitude and direction of the vector. Such an arrow, of length A directed at angle ϕ_A, can be seen in Fig. 3-5.

The ordered pair of scalars (A, ϕ_A) can be used in a prescription for locating a point on the plane by following a different pathway: *Begin at the origin, with your ruler along the positive x axis. Then turn the ruler counterclockwise through an angle* ϕ_A*. Next measure off a length A. You are now at the same point on the plane as prescribed in 1, provided that*

$$A = \sqrt{A_x^2 + A_y^2} \qquad \text{and} \qquad \phi_A = \tan^{-1}\frac{A_y}{A_x} \tag{3-9}$$

The rules quoted in Eqs. (3-9) for obtaining the values of the ordered pair (A, ϕ_A) from those of the ordered pair (A_x, A_y) are identical to the rules set forth in Eqs. (3-7) and (3-8), except for the change to the more general notation. There are also rules for obtaining (A_x, A_y) from (A, ϕ_A). These rules are found by applying the definition of the sine and cosine functions to the triangle with sides A_x, A_y, and A in Fig. 3-5. These definitions are

$$\cos\phi_A = \frac{A_x}{A} \qquad \text{and} \qquad \sin\phi_A = \frac{A_y}{A}$$

Multiplication of both sides of each of these equalities by A yields

$$A_x = A\cos\phi_A \qquad \text{and} \qquad A_y = A\sin\phi_A \tag{3-10}$$

We now make an important observation. Given the ordered pair (A_x, A_y), we can find the ordered pair (A, ϕ_A), and vice versa. Both ordered pairs uniquely specify the same point in space, which can therefore be labeled equally well with either pair. To put it another way, *the vector* **A** *(which* *determines a certain position with respect to the origin) is uniquely specified by an ordered pair of numbers, each of which is a scalar.*

It is conventional in printed material to distinguish vectors from scalars by using **boldface** type for all letters designating vector quantities, as has been done immediately above for the vector **A.** (In handwritten material, vectors are usually denoted by drawing small arrows above the symbols: $\vec{A}$. This is a more evocative notation, but it is awkward to set the arrows in type.) In dealing with situations which are definitely two- or three-dimensional, we will always use a letter in boldface type to represent a vector and the *same* letter in italic type to represent the magnitude of that vector. Thus the magnitude of the vector **A** is the scalar A.

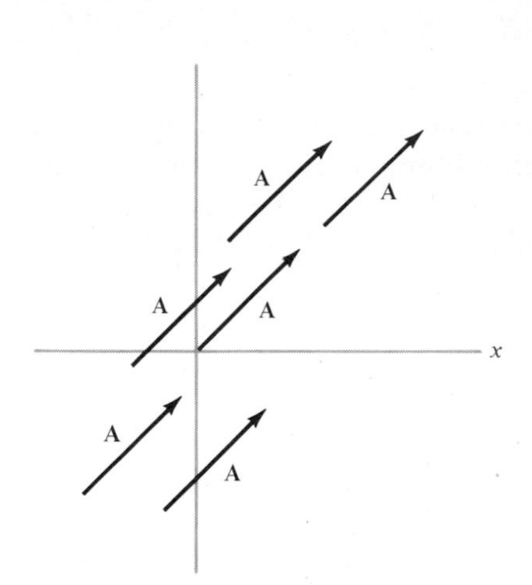

Fig. 3-6 All the vectors shown are the same vector, since they have identical magnitudes and directions. The location of a vector is not one of its intrinsic properties.

Now that we have said what a vector is, it is equally important to make clear what it is *not.* In the (A, ϕ_A) representation (which corresponds most closely to intuition) a vector is *completely specified* by a *magnitude* and a *direction.* Thus a vector has *no other* properties. In particular—and this sometimes strikes people as surprising at first—a vector has no particular location. For example, all the vectors **A** shown in Fig. 3-6 are the same vector. Thus a vector may be moved without changing it in any way, provided no change is made in its magnitude or its direction. In other words, *a vector can be moved parallel to itself without changing it.*

If the vectors in Fig. 3-6 are position vectors, then each shows the position of the point at its head with respect to that of the point at its tail. For one of these vectors the tail happens to be at the origin and so it gives the position of a point with respect to the origin. For the other vectors this is not the case. But in all cases the vector depicts exactly the same position of the point at its head *relative* to that of the point at its tail. This is *all* the information that can be provided by a position vector since, like any other vector, it is completely specified by its magnitude and its direction.

Sometimes the location of a vector must be specified for *other* reasons. For example, the effect produced by a force exerted on a lever depends on the point of application of the force. So even though a force can be represented completely by a magnitude and a direction, which are not altered by changing its location—that is, force is a vector—its physical effect depends on the point at which it is applied. But note that the significance of the point of application of the force is a consequence of the overall physical situation and is not a property of the vector itself. You will study cases of this sort in Chap. 9.

How are vectors "added"? We have already seen how to construct a vector out of its components. This is what is done in Fig. 3-5. The operation already contains implicitly the essentials of vector addition. We now make the vector addition process explicit.

Consider a pair of components (A_x, A_y), assuming for simplicity that both have positive values. Figure 3-7*a* shows a vector $\mathbf{A}_x$ of length A_x extending in the positive direction along the x axis. Also shown is a vector $\mathbf{A}_y$ of length A_y extending in the positive direction along the y axis.

The vector $\mathbf{A}_y$ can be moved parallel to itself so that its tail coincides with the head of the vector $\mathbf{A}_x$, as shown in Fig. 3-7*b*. For the reasons just discussed, this has no effect on the vector. However, the vectors shown in Fig. 3-7*b* display pictorially the prescription of Fig. 3-5 for constructing the vector **A,** which has the components (A_x, A_y). That is, $\mathbf{A}_x$ represents the process of measuring off a length A_x in the positive direction along the x

axis, while $\mathbf{A}_y$ represents the subsequent process of measuring off a length A_y parallel to the y axis in the positive direction of that axis. *Thus the sequence of the two processes is equivalent to specifying the vector* **A.** This fact is represented in mathematical notation in the form

$$\mathbf{A} = \mathbf{A}_x + \mathbf{A}_y \tag{3-11}$$

We therefore say that **A** is the **vector sum** of the vectors $\mathbf{A}_x$ and $\mathbf{A}_y$. The use of the symbol "+" and of the term "sum" does *not* imply that a vector sum is the *same* as an ordinary algebraic sum. Rather it implies an analogy between the two distinct mathematical operations.

While it is not a universal convention, we will call $\mathbf{A}_x$ and $\mathbf{A}_y$ the **constituent vectors** of **A.** This is to distinguish them from the *components* A_x and A_y, which are *scalars.*

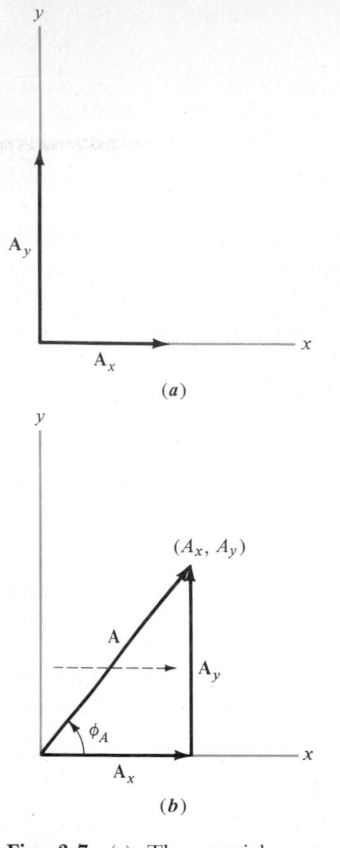

Fig. 3-7 (*a*) The special vector $\mathbf{A}_x$, whose direction is that of the x axis, has magnitude A_x. Similarly, the special vector $\mathbf{A}_y$, whose direction is that of the y axis, has magnitude A_y. (*b*) The vector $\mathbf{A}_y$ is unchanged by moving it parallel to itself until its tail coincides with the head of the vector $\mathbf{A}_x$. A vector **A** is constructed with its tail at the origin and its head coincident with the head of $\mathbf{A}_y$. This vector, whose magnitude is $A = (A_x^2 + A_y^2)^{1/2}$ and whose direction is $\phi = \tan^{-1}(A_y/A_x)$, is the vector sum of the vectors $\mathbf{A}_x$ and $\mathbf{A}_y$; that is, $\mathbf{A} = \mathbf{A}_x + \mathbf{A}_y$. The components of **A** are (A_x, A_y).

Although the components A_x and A_y are not the same as the constituent vectors $\mathbf{A}_x$ and $\mathbf{A}_y$, there is a close connection between them. The connection can be made explicit by defining a quantity called the unit vector. We *define* the **unit vector** $\hat{\mathbf{x}}$ (spoken as "x hat") to be the vector in the positive x direction having magnitude 1, that is, *unit magnitude. The vector* $\mathbf{A}_x$, *which has magnitude* A_x, *is just* A_x *times as "long" as the unit vector* $\hat{\mathbf{x}}$ *and has the same direction as* $\hat{\mathbf{x}}$. This is represented mathematically by the identity

$$\mathbf{A}_x \equiv A_x\hat{\mathbf{x}} \tag{3-12}$$

(This "product" is a special case of multiplication of a vector by a scalar, an operation which is defined and discussed more generally later in this section.) Equation (3-12) is a very convenient way of expressing individually the two essential properties of the vector $\mathbf{A}_x$: its magnitude A_x and its direction $\hat{\mathbf{x}}$. The idea is easily extended to any other vector. For the vector $\mathbf{A}_y$ we have

$$\mathbf{A}_y \equiv A_y\hat{\mathbf{y}} \tag{3-13}$$

And for *any* vector **A,** having *any* direction $\hat{\mathbf{A}}$, whether it is a position vector or some other type of vector, we can write

$$\mathbf{A} \equiv A\hat{\mathbf{A}} \tag{3-14}$$

If **A** is a position vector it is not correct to say that the unit vector $\hat{\mathbf{A}}$ has magnitude 1 m. The unit of measurement is associated with the *magnitude* of a vector, not with its *direction.* This point is demonstrated in Example 3-3.

EXAMPLE 3-3

If the direction of the unit vector $\hat{\mathbf{A}}_x$ denotes eastward, and the direction of the unit vector $\hat{\mathbf{A}}_y$ denotes northward, find the components of the position vector **A** when $A = 50.0$ m and its direction is 30° north of east. Then write expressions for the constituent vectors $\mathbf{A}_x$ and $\mathbf{A}_y$ of **A.**

■ You begin by drawing a sketch of the situation, as in Fig. 3-8*a.* The orientation of the vector **A** is shown relative to the directions of the unit vectors $\hat{\mathbf{A}}_x$ and $\hat{\mathbf{A}}_y$.

You now use Eqs. (3-10) to find the components A_x and A_y of the vector **A.** The first of these equations gives you

$$A_x = A\cos\phi_A = 50.0 \text{ m} \times \cos 30°$$

or

$$A_x = 43.3 \text{ m}$$

The second gives you

$$A_y = A \sin \phi_A = 50.0 \text{ m} \times \sin 30°$$

or

$$A_y = 25.0 \text{ m}$$

The corresponding constituent vector $\mathbf{A}_x$ is the product of the component A_x and the unit vector $\hat{\mathbf{A}}_x$, and similarly for the constituent vector $\mathbf{A}_y$. Thus

$$\mathbf{A}_x = A_x\hat{\mathbf{A}}_x = (43.3 \text{ m})(1 \text{ eastward})$$

and

$$\mathbf{A}_y = A_y\hat{\mathbf{A}}_y = (25.0 \text{ m})(1 \text{ northward})$$

The constituent vectors are shown in Fig. 3-8*b*. The vector sum of its two constituent vectors is equivalent to the vector **A**; that is,

$$\mathbf{A} = \mathbf{A}_x + \mathbf{A}_y$$

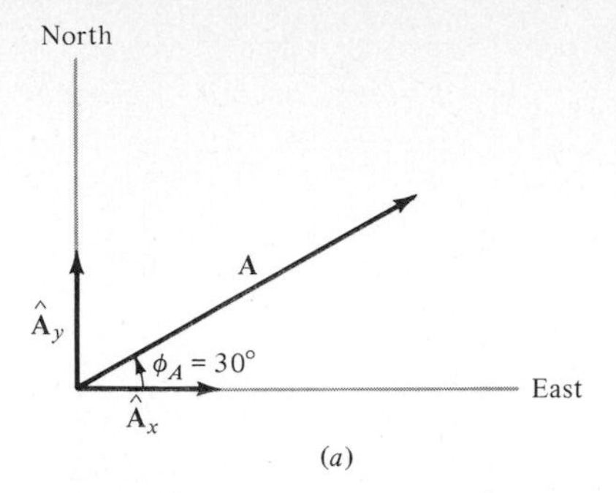

Fig. 3-8*a* The vector **A** oriented with respect to the unit vector $\hat{\mathbf{A}}_x$, representing "eastward," and the unit vector $\hat{\mathbf{A}}_y$, representing "northward."

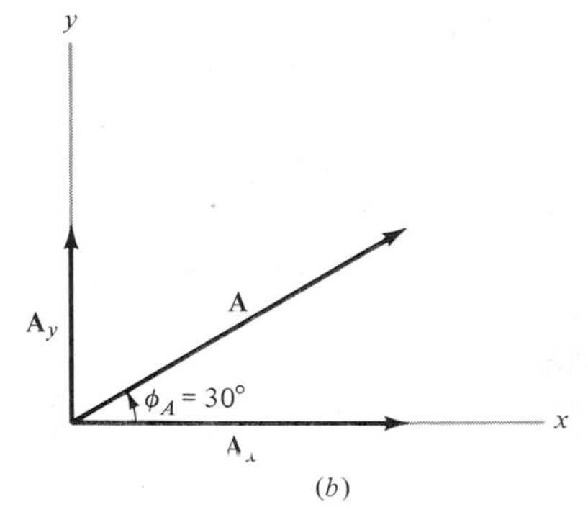

Fig. 3-8*b* Illustrating the constituent vectors $\mathbf{A}_x$ and $\mathbf{A}_y$ of the vector **A**.

We now generalize the idea of vector addition so as to give meaning to the addition of two vectors having *any* orientation. Earlier in this section we specified the position of a point with respect to the origin by prescribing two pathways for reaching it. One was the "direct route" along the vector **A**. The other followed the constituent vectors $\mathbf{A}_x$ and $\mathbf{A}_y$ in succession. In Fig. 3-9, two pathways are shown from (0, 0) to the arbitrary point (C_x, C_y). One is again the "direct route." The other follows the vectors **A** and **B** in succession. While **A** and **B** are not constituent vectors, they certainly do comprise a pathway from the origin to the point (C_x, C_y). It is entirely reasonable to state the equivalence of the two pathways in the figure by means of the equation.

$$\mathbf{C} = \mathbf{A} + \mathbf{B} \tag{3-15}$$

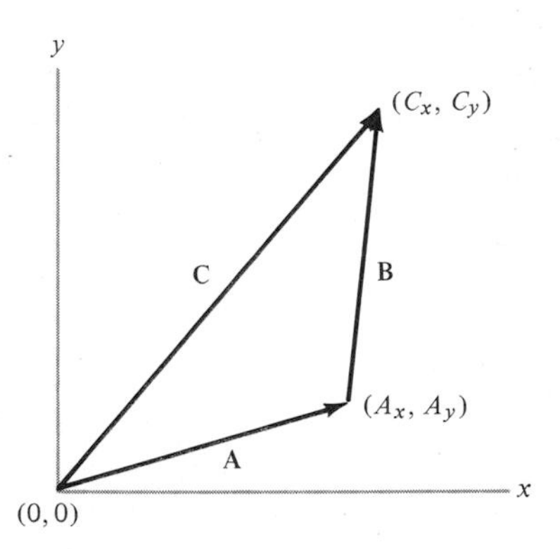

Fig. 3-9 The point (C_x, C_y) can be reached from the origin either via the straight pathway along the vector **C** or by first following the pathway from the origin to the head of the vector **A** at (A_x, A_y), and then following the vector **B** to (C_x, C_y). The equivalence of the two processes is denoted by the vector sum **C** = **A** + **B**.

The sum of two vectors is found geometrically by making the tail of the second coincide with the head of the first and then drawing a vector from the tail of the first to the head of the second. To put it another way, the vector **A** locates the point at its head having coordinates (A_x, A_y) with respect to the origin (0, 0). The vector **B** locates, with respect to the point (A_x, A_y), the point at *its* head having coordinates (C_x, C_y). The vector **C** locates the point having coordinates (C_x, C_y) directly with respect to the origin. Thus **C**, whose tail is at (0, 0) and whose head is at (C_x, C_y), accomplishes in a single step the same process of location which is accomplished *sequentially* by **A**, with its tail at (0, 0) and its head at (A_x, A_y), and by **B**, with its tail at (A_x, A_y) and its head at (C_x, C_y). This process is called **vector addition.**

What is the justification for giving the name "addition" to the vector process sketched in Fig. 3-9, as is done in Eq. (3-15)? The justification is easiest to see if the process is described in equivalent algebraic terms. In Fig. 3-10*a* and *b,* the vector **A** has *x* and *y* components (A_x, A_y), and the vector **B** has components (B_x, B_y). (Note that the same vector **B** is shown in both figures.) The vector **C**, which is the vector sum of **A** and **B**, has components (C_x, C_y). From inspection of the figures you can see that if **C** = **A** + **B**, then $C_x = A_x + B_x$ and $C_y = A_y + B_y$. That is, the sum of two vectors is

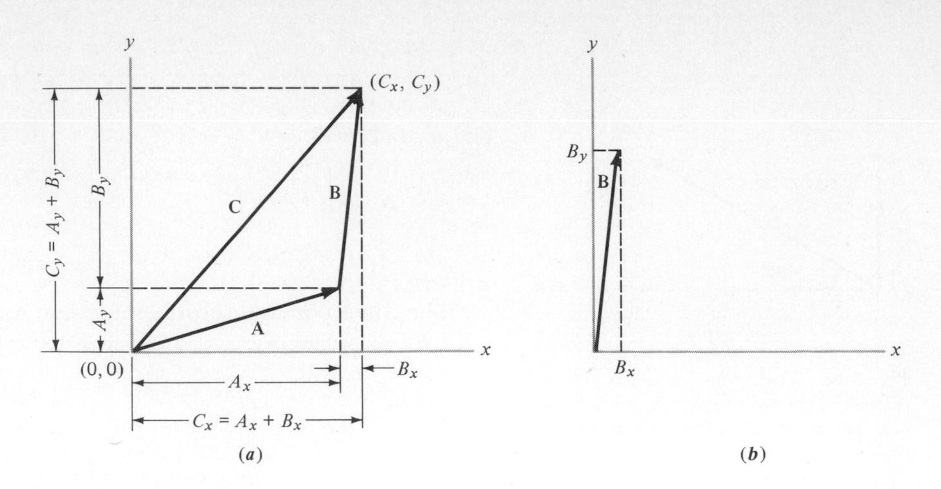

Fig. 3-10 (*a*) The *x* and *y* components of **A** and **C** are shown. (*b*) The vector **B** is shown separately, moved so that its tail coincides with the origin. Its *x* and *y* components are shown. Compare both parts of this figure with Fig. 3-9 to see that B_x is the difference between C_x and A_x, and that B_y is the difference between C_y and A_y. Thus $C_x = A_x + B_x$ and $C_y = A_y + B_y$. Consequently, the vector from the origin to $(A_x + B_x, A_y + B_y)$ is the vector **C**.

a vector whose components are the sums of the corresponding components. We are thus justified in writing

$$\mathbf{C} = \mathbf{A} + \mathbf{B}$$

as the complete equivalent of the pair of equations

$$C_x = A_x + B_x \tag{3-16a}$$

and

$$C_y = A_y + B_y \tag{3-16b}$$

Equations (3-16) amount to an algebraic method of finding the sum of any two vectors whose components are known. The name "vector addition" is thus justified both in the sense that the algebraic process for carrying it out involves algebraic additions of the scalar components, and in the sense that there is a strong analogy between the process of vector addition (whether it is carried out algebraically or geometrically) and the process of algebraic addition.

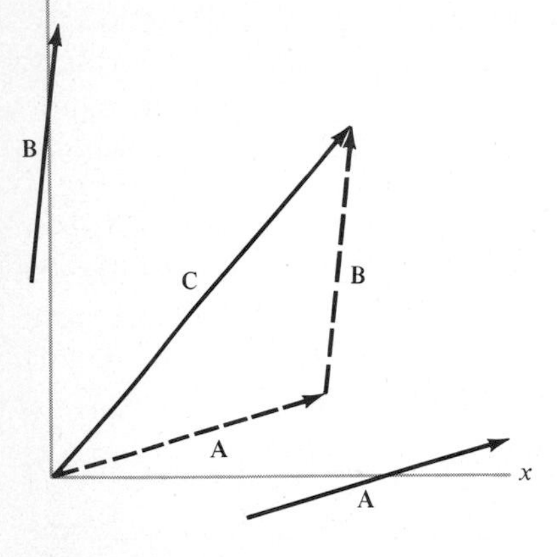

Fig. 3-11 In this case, the vectors **A** and **B** (shown by solid lines) are not conveniently located with the tail of **A** at the origin and the tail of **B** coinciding with the head of **A**. But since a vector is not changed by moving it parallel to itself, **A** and **B** can always be moved to the locations shown by dashed lines. The vector summation can then be carried out as described in Fig. 3-9 or Fig. 3-10.

Figure 3-11 depicts two vectors **A** and **B,** neither of which lies with its tail at the origin. Nor do the two vectors describe, as they are shown in the figure, a single pathway from one point to another. Nevertheless, the two vectors can be added geometrically. This is done by reducing the situation to the simpler one shown in Fig. 3-9, exploiting the fact that the significance of a vector is independent of its location. It is necessary only to move **A**, without changing its magnitude or its direction, until its tail lies at the origin and then to move **B** until its tail coincides with the head of **A**. Thus Eq. (3-15), **C** = **A** + **B,** holds as well for Fig. 3-11 as it does for Fig. 3-9.

EXAMPLE 3-4

Vector **A** has magnitude $A = 4.00$ m and is directed at an angle $\phi_A = -45.0°$ from the positive *x* axis of a particular reference frame, with positive angles measured counterclockwise. The magnitude of vector **B** is $B = 2.00$ m, and its direction from the *x* axis is given by the angle $\phi_B = +120.0°$.

a. Evaluate the magnitude and direction of their vector sum, **C** = **A** + **B**, by using a geometrical method based directly on the definition of vector addition specified by Eq. (3-15) and Figs. 3-9 and 3-11.

b. Then evaluate the magnitude and direction of **C** using an algebraic method based on the component addition process specified by Eqs. (3-16) and Fig. 3-10*a*. Compare your results with those obtained in part *a*.

■ **a.** Working as carefully as you can, you use a compass, protractor, and graph paper to lay off the magnitudes and directions of vectors **A** and **B** to a convenient scale on a set of *xy* coordinates. In Fig. 3-12*a* this is done with the tails of both vectors at the origin of coordinates. Then use the drawing instruments to move vector **B**, without changing its length or direction, so that its tail is at the head of vector **A,** as in Fig. 3-12*b*. Now you can connect the tail of vector **A** to the head of vector **B** with a vector labeled **C**. By the definition of vector addition, **C** = **A** + **B**. Measuring its length and direction, you obtain $C = 2.13$ m and $\phi_C = -31.1°$. However, the accuracy of the last digit in both numbers is doubtful.

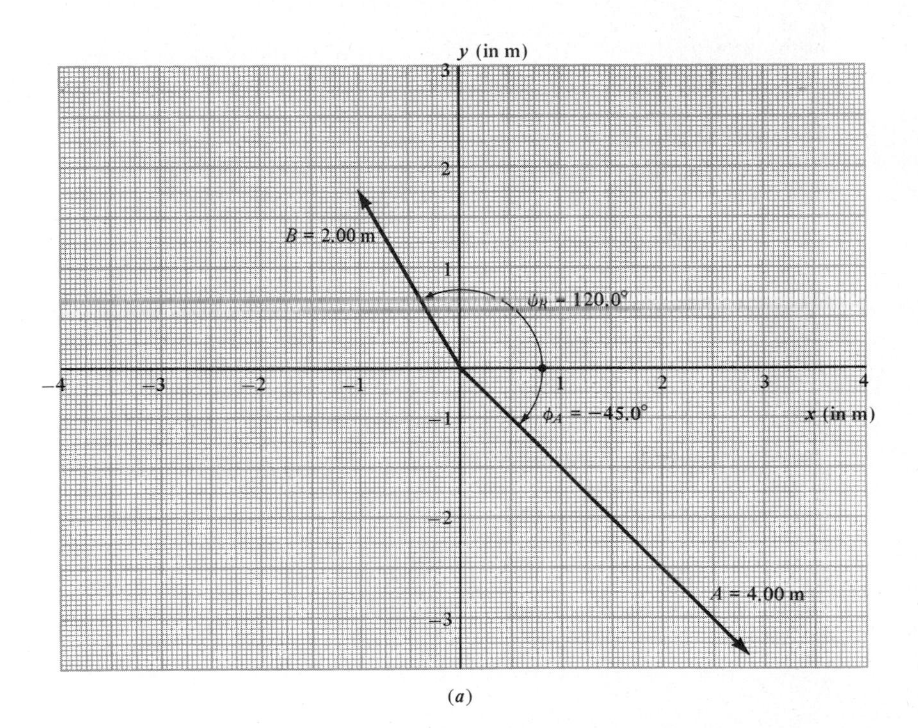

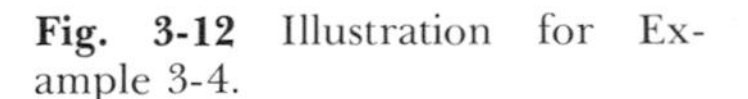
Fig. 3-12 Illustration for Example 3-4.

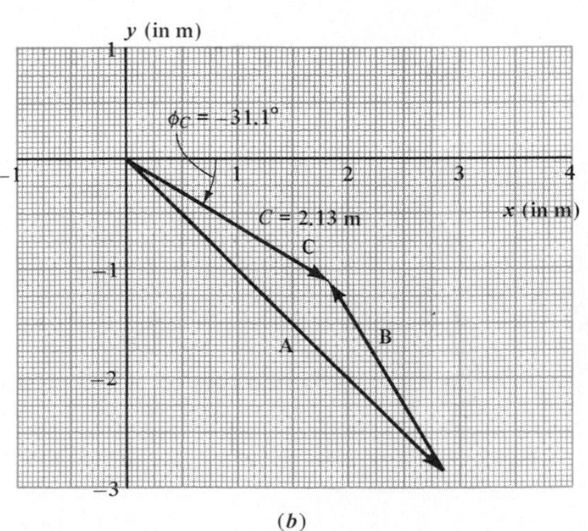

If you use this method again, you will likely want to skip making a construction similar to Fig. 3-12*a* and go instead directly to a construction similar to Fig. 3-12*b*. An alternative procedure begins with a construction like that of Fig. 3-12*a*. You then draw a line parallel to **B** that goes through the head of **A**, and also a line parallel to **A** that goes through the head of **B**. You will now have a parallelogram. Next draw a vector whose tail is at the origin and whose head is at the intersection of the two parallel lines you have constructed. This vector is $\mathbf{C} = \mathbf{A} + \mathbf{B}$. Carry out this procedure by drawing the parallel lines on Fig. 3-12*a*, and then explain why it works.

b. The algebraic method requires that you first use Eqs. (3-10) to determine the x and y components of vectors **A** and **B**. For **A** you have

$$A_x = A \cos \phi_A = 4.00 \text{ m} \times \cos(-45.0°) = +2.83 \text{ m} \quad \text{and} \quad A_y = A \sin \phi_A = 4.00 \text{ m} \times \sin(-45.0°) = -2.83 \text{ m}$$

For **B** you have

$$B_x = B \cos \phi_B = 2.00 \text{ m} \times \cos(120.0°) = -1.00 \text{ m} \quad \text{and} \quad B_y = B \sin \phi_B = 2.00 \text{ m} \times \sin(120.0°) = +1.73 \text{ m}$$

You sum these components algebraically to find the components C_x and C_y, using Eqs. (3-16*a*) and (3-16*b*). You obtain

$$C_x = A_x + B_x = +2.83 \text{ m} - 1.00 \text{ m} = +1.83 \text{ m} \quad \text{and} \quad C_y = A_y + B_y = -2.83 \text{ m} + 1.73 \text{ m} = -1.10 \text{ m}$$

To find the magnitude C and direction ϕ_C of the vector **C**, you employ Eqs. (3-9). The magnitude is

$$C = \sqrt{C_x^2 + C_y^2} = \sqrt{(1.83 \text{ m})^2 + (-1.10 \text{ m})^2} = 2.14 \text{ m}$$

The direction is

$$\phi_C = \tan^{-1}\frac{C_y}{C_x} = \tan^{-1}\left(\frac{-1.10 \text{ m}}{+1.83 \text{ m}}\right) = -31.0°$$

These results compare well with those found by the geometrical method. Of the two methods, the algebraic one is probably faster, and certainly more accurate—particularly if you use a calculator intended for scientific work. If you use a programmable calculator, you can program it to use the algebraic method and add any number of vectors automatically.

What has been said above for the summation of two vectors applies to any number of vectors. The vectors must be placed head to tail. Their sum is defined as the vector joining the tail of the first with the head of the last. See Fig. 3-13. Can you write a set of equations to describe the equivalent algebraic addition process?

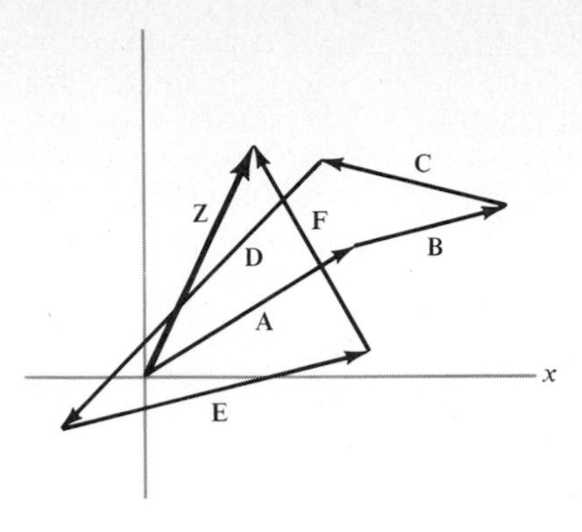

Fig. 3-13 Any number of vectors can be added by placing them head to tail, and then joining the tail of the first to the head of the last. In the figure, $\mathbf{Z} = \mathbf{A} + \mathbf{B} + \mathbf{C} + \mathbf{D} + \mathbf{E} + \mathbf{F}$.

You are familiar with the fact that the algebraic sum of two numbers does not depend on the order of addition. That is, $a + b = b + a$ is always true. This property of addition is called **commutativity.** Vector sums are also commutative. To see this, note that the components of a vector, being themselves scalars, add commutatively:

$$C_x = A_x + B_x = B_x + A_x \qquad \text{and} \qquad C_y = A_y + B_y = B_y + A_y$$

Therefore the coordinates $(B_x + A_x, B_y + A_y)$, which locate the head of the vector $\mathbf{B} + \mathbf{A}$ when its tail is at the origin, are identical with the coordinates $(A_x + B_x, A_y + B_y)$, which locate the head of the vector $\mathbf{A} + \mathbf{B}$ when its tail is at the origin. That is, the two vectors are the same vector:

$$\mathbf{A} + \mathbf{B} = \mathbf{B} + \mathbf{A} \tag{3-17}$$

This rule applies as well to the sum of any number of vectors, which can be added in any order. Figure 3-14 justifies the commutative rule from a geometrical point of view.

The idea of vector addition leads to a definition for the negative of a vector. For any vector $\mathbf{A}$ there is always a vector $\mathbf{A}'$ which has the property that

$$\mathbf{A} + \mathbf{A}' = \mathbf{0} \tag{3-18}$$

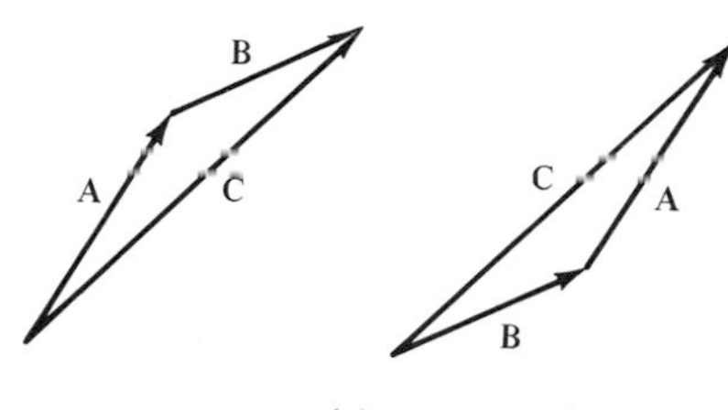

Such a pair of vectors is illustrated in Fig. 3-15. You can see from a geometrical point of view that $\mathbf{A}'$ satisfies Eq. (3-18) by imagining $\mathbf{A}'$ to be moved so that its tail coincides with the head of $\mathbf{A}$. Then observe that the two vectors add to "nothing." The symbol $\mathbf{0}$ denotes the **zero, or null, vector.** Since the null vector has no particular direction, we will often ignore its vectorial nature and write it as the scalar 0.

The vector $\mathbf{A}'$ is called the **negative of A.** That is, by definition

$$\mathbf{A}' \equiv -\mathbf{A}$$

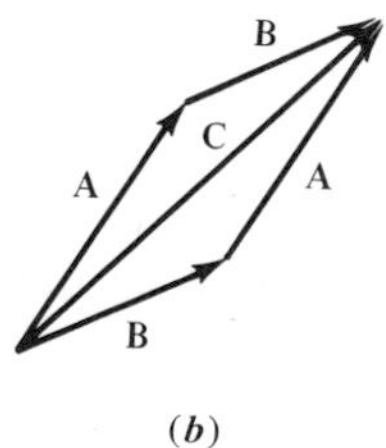

Fig. 3-14 Geometric justification for the commutative rule $\mathbf{A} + \mathbf{B} = \mathbf{B} + \mathbf{A}$ for vector addition. In (*a*), the vector $\mathbf{C}$ is constructed by adding $\mathbf{A}$ and $\mathbf{B}$ in both possible orders. Note that the addition process does not depend upon the choice of any particular frame of reference, as long as the scale of length is understood. (*b*) The vector $\mathbf{C}$ is uniquely specified as the directed diagonal of the parallelogram of directed sides $\mathbf{A}$ and $\mathbf{B}$. One pair of sides represents the sum $\mathbf{A} + \mathbf{B}$, and the other pair the sum $\mathbf{B} + \mathbf{A}$.

Stated in words, *the negative of any vector is a vector of equal magnitude and opposite direction.* In view of this definition, the negative of a vector can always be constructed by "turning the vector around," that is, by moving its head to the opposite end.

The operation of vector subtraction is defined in a way which follows directly from the definition of the negative of a vector. In complete analogy with ordinary algebraic subtraction, we define

$$\mathbf{A} - \mathbf{B} \equiv \mathbf{A} + (-\mathbf{B}) \tag{3-19}$$

Fig. 3-15 The vector $\mathbf{A}'$ has the same magnitude as the vector $\mathbf{A}$, but its direction is opposite that of $\mathbf{A}$. As explained in the text, the vector sum $\mathbf{A} + \mathbf{A}' = 0$, and $\mathbf{A}'$ is defined to be the negative of $\mathbf{A}$; $\mathbf{A}' \equiv -\mathbf{A}$.

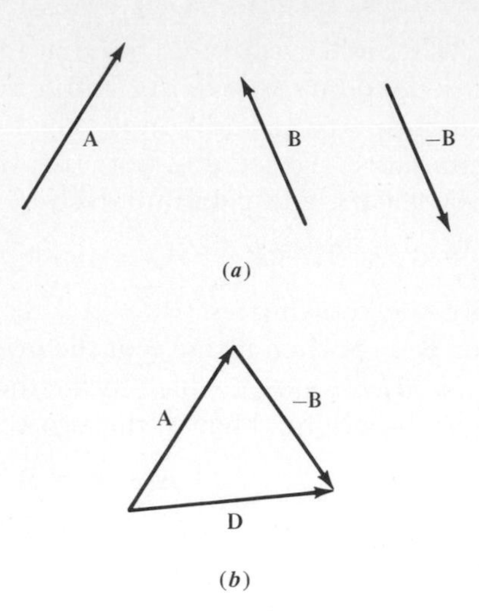

Fig. 3-16 Vector subtraction by the geometric method. (*a*) The vector $-\mathbf{B}$ is constructed by "turning **B** around." (*b*) The vector sum $\mathbf{D} = \mathbf{A} + (-\mathbf{B})$ is the desired vector difference $\mathbf{A} - \mathbf{B}$.

Figure 3-16 depicts this operation. Given the two vectors **A** and **B,** the negative $-\mathbf{B}$ of **B** is first constructed. Then $-\mathbf{B}$ is *added* to **A** to obtain the vector sum $\mathbf{D} = \mathbf{A} + (-\mathbf{B})$. According to Eq. (3-19), this is identical to the **vector difference**

$$\mathbf{D} = \mathbf{A} - \mathbf{B}$$

Can you show geometrically that $\mathbf{B} - \mathbf{A} = -(\mathbf{A} - \mathbf{B}) = -\mathbf{D}$?

EXAMPLE 3-5

Use the algebraic method to find the difference $\mathbf{D} = \mathbf{A} - \mathbf{B}$ of the two vectors whose sum $\mathbf{C} = \mathbf{A} + \mathbf{B}$ you found in Example 3-4.

▪ First you must extend the algebraic method of summing the components of two vectors to obtain the components of their vector sum, so that it can be used to obtain the components of their vector difference. This is easy to do. Employing the unit vectors $\hat{\mathbf{x}}$ and $\hat{\mathbf{y}}$, you write

$$\mathbf{A} = A_x\hat{\mathbf{x}} + A_y\hat{\mathbf{y}}$$

and

$$\mathbf{B} = B_x\hat{\mathbf{x}} + B_y\hat{\mathbf{y}}$$

Then subtract corresponding sides of the second equation from those of the first. You obtain

$$\mathbf{A} - \mathbf{B} = (A_x - B_x)\hat{\mathbf{x}} + (A_y - B_y)\hat{\mathbf{y}}$$

Now write the vector difference $\mathbf{D} = \mathbf{A} - \mathbf{B}$ in terms of its components and the unit vectors. You have

$$\mathbf{A} - \mathbf{B} = \mathbf{D} = D_x\hat{\mathbf{x}} + D_y\hat{\mathbf{y}}$$

Comparison with the equation displayed immediately above shows that

$$D_x = A_x - B_x \tag{3-20a}$$

and

$$D_y = A_y - B_y \tag{3-20b}$$

You can apply this method immediately to the problem at hand since A_x, A_y, B_x, and B_y were evaluated in Example 3-4. Using these values, you have

$$D_x = A_x - B_x \qquad \text{and} \qquad D_y = A_y - B_y$$
$$= +2.83 \text{ m} - (-1.00 \text{ m}) \qquad = -2.83 \text{ m} - (+1.73 \text{ m})$$
$$= +3.83 \text{ m} \qquad = -4.56 \text{ m}$$

The magnitude D and direction ϕ_D of the vector **D** are given by

$$D = \sqrt{D_x^2 + D_y^2}$$
$$= \sqrt{(+3.83 \text{ m})^2 + (-4.56 \text{ m})^2}$$
$$= 5.96 \text{ m}$$

and

$$\phi_D = \tan^{-1} \frac{D_y}{D_x}$$
$$= \tan^{-1} \left(\frac{-4.56 \text{ m}}{3.83 \text{ m}} \right)$$
$$= -50.0°$$

You should check these results by using the geometrical method for obtaining a vector difference.

An important conclusion obtained in Example 3-5 is that *the components of the negative of a vector are equal to the negatives of the corresponding components of the vector.*

The **multiplication of a vector by a scalar** is an extension of the idea of ordinary scalar multiplication. Consider the vector **P** given by

$$\mathbf{P} = c\mathbf{A} \tag{3-21a}$$

By definition, *the magnitude of* **P** *is the magnitude of the scalar c multiplied by the magnitude of* **A**. That is,

$$P = |c|A \tag{3-21b}$$

where $|c|$ is the absolute value of the scalar c. Also by definition, *the direction of* **P** *is the same as that of* **A** *if c is a positive number; it is the same as that of* $-$**A** *if c is a negative number.* We can express both the magnitude and the direction of **P** by means of the vector equation

$$\mathbf{P} = cA\hat{\mathbf{A}} \tag{3-22}$$

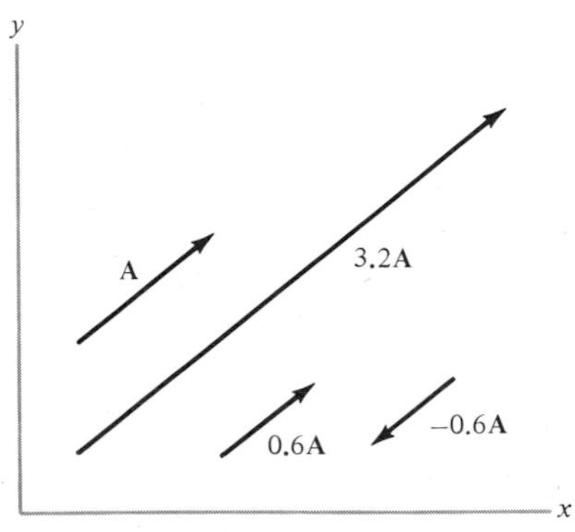

Fig. 3-17 Multiplication of a vector **A** by the scalars 3.2, 0.6, and −0.6.

In the case $c > 0$, the quantity cA is positive since A is always positive; thus **P** has the same direction as **A.** In the case $c < 0$, the quantity cA is negative; thus the direction of **P** is opposite that of **A.** If $|c| > 1$, then **P** is longer than **A.** If $|c| < 1$, **P** is shorter than **A.** This definition is illustrated in Fig. 3-17. Can you see a connection between the definition of the negative of a vector and the definition of multiplication of a vector by a negative scalar? Are these independent definitions?

The definition of multiplication of a vector by a scalar can be extended so as to define the operation of **division of a vector by a scalar.** We define

$$\frac{\mathbf{A}}{c} \equiv \frac{1}{c}\mathbf{A} \tag{3-23}$$

That is, *the vector* **A** *divided by the scalar c is defined to be the product of the vector* **A** *with the scalar 1/c.* Division of a vector or a scalar *by* a vector has no meaning.

The ideas just developed for vectors in two dimensions can be readily extended into three-dimensional space. Figure 3-18 is a perspective representation of a frame of reference in three-dimensional space. It is specified by the mutually perpendicular x, y, and z axes with their (identical) scales. By using this frame of reference, any point in space can be uniquely specified by an **ordered triplet** of scalars (A_x, A_y, A_z). (Just as in two-dimensional space, the number of separate scalars required to do this is equal to the number of dimensions.) Such a point—the point (7.0 m, 4.0 m, 5.0 m)—is depicted in Fig. 3-18.

In complete parallelism to the geometrical representation of a two-dimensional vector, a three-dimensional vector can be visualized as a directed line (or arrow) extending from the origin (0, 0, 0) to the point (A_x, A_y, A_z), as shown in the figure. This vector **A** has the *three* constituent vectors $\mathbf{A}_x$, $\mathbf{A}_y$, and $\mathbf{A}_z$, whose magnitudes A_x, A_y, and A_z are the three components of **A**. In mathematical language, we can express this statement in the form

$$\mathbf{A} = \mathbf{A}_x + \mathbf{A}_y + \mathbf{A}_z \tag{3-24}$$

and

$$\mathbf{A} = A_x\hat{\mathbf{x}} + A_y\hat{\mathbf{y}} + A_z\hat{\mathbf{z}} \tag{3-25}$$

where $\hat{\mathbf{x}}$, $\hat{\mathbf{y}}$, and $\hat{\mathbf{z}}$ are the unit vectors in the directions of the positive x, y, and z axes, respectively.

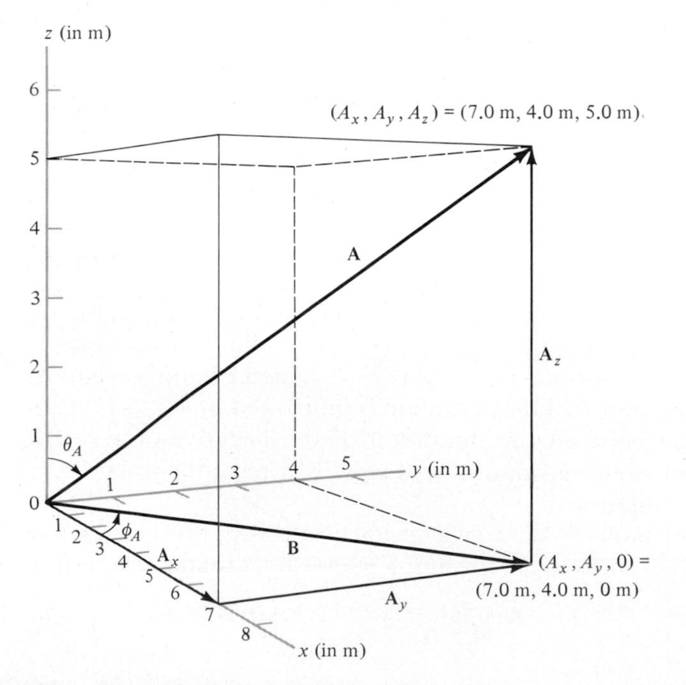

Fig. 3-18 A perspective view of the three-dimensional vector **A**, whose components are $A_x = 7$, $A_y = 4$, and $A_z = 5$. The constituent vectors $\mathbf{A}_x$, $\mathbf{A}_y$, and $\mathbf{A}_z$ are shown, as is the vector **B** which is the projection of **A** on the xy plane. The magnitude A of **A** is evaluated in Example 3-6. The direction of **A** can be specified by means of the *two* angles θ_A and ϕ_A shown in the figure. The angle θ_A is that between **A** and the z axis, while ϕ_A is the angle between the x axis and **B**, the base of the vertical triangle of sides **B**, $\mathbf{A}_z$, and **A**.

EXAMPLE 3-6

Find the magnitude of the vector **A** in Fig. 3-18, whose tail lies at the origin and whose head lies at the point (7.0 m, 4.0 m, 5.0 m).

■ Just as in two dimensions, this example requires the application of the pythagorean theorem. However, it must be applied twice.

First, note that the vector **A** and its constituent vector $\mathbf{A}_z$ are the hypotenuse and one side of a right triangle whose plane is perpendicular to the *xy* plane (the plane defined by the *x* and *y* axes). As shown in Fig. 3-18, the other side of this triangle is the vector **B,** which lies along the line of intersection of the two planes. This vector is the **projection** of **A** on the *xy* plane; you may imagine it as the "shadow" cast by **A** on the *xy* plane when the "sun" is located very far away on the *z* axis.

Using the pythagorean theorem, you have

$$A^2 = B^2 + A_z^2$$

But the vector **B** is itself the hypotenuse of the right triangle in the *xy* plane whose sides are the constituent vectors $\mathbf{A}_x$ and $\mathbf{A}_y$. Thus you can use the pythagorean theorem again to obtain

$$B^2 = A_x^2 + A_y^2$$

Combining the two equations immediately above gives you

$$A^2 = A_x^2 + A_y^2 + A_z^2$$

or

$$A = \sqrt{A_x^2 + A_y^2 + A_z^2} \tag{3-26}$$

This is the form taken by the pythagorean theorem in three dimensions.

You can now use Eq. (3-26) to find the magnitude of the vector **A** whose head lies at (7.0 m, 4.0 m, 5.0 m). You have

$$A = \sqrt{(7.0\text{ m})^2 + (4.0\text{ m})^2 + (5.0\text{ m})^2} = 9.5\text{ m}$$

Like a two-dimensional vector, a three-dimensional vector is completely specified by its magnitude and direction. It can therefore be moved parallel to itself without being changed.

The geometrical and algebraic rules for addition and subtraction of three-dimensional vectors, and for multiplication and division of a three-dimensional vector by a scalar, are completely analogous to the corresponding operations in two dimensions. Because of the difficulty of drawing accurate three-dimensional representations, however, the geometrical method is rarely used. The algebraic method of vector addition is a direct extension of that given by Eqs. (3-16) for two-dimensional vectors. There are now three components instead of two, and the vector sum **C** = **A** + **B** is equivalent to the component equations

$$C_x = A_x + B_x \tag{3-27a}$$

$$C_y = A_y + B_y \tag{3-27b}$$

$$C_z = A_z + B_z \tag{3-27c}$$

The other vector operations already discussed for two-dimensional vectors are extended similarly.

3-3 POSITION, VELOCITY, AND ACCELERATION VECTORS

At any instant of time the position of a moving body relative to the origin of a coordinate system can be described by the position vector extending from the origin to the body. (**Body** is a technical term for an object whose motion is being considered.) Since the velocity of the body is defined as the rate of change of its position, the fact that its position is described by a vector suggests that its velocity also may be described by a vector. And since the acceleration of the body is defined as the rate of change of its velocity, the same line of thought suggests that the acceleration may be a vector quantity, too. In this section we show that both suggestions are correct by making use of the concept called a *differential.* Furthermore, we develop some very useful relations among the position, velocity, and acceleration vectors.

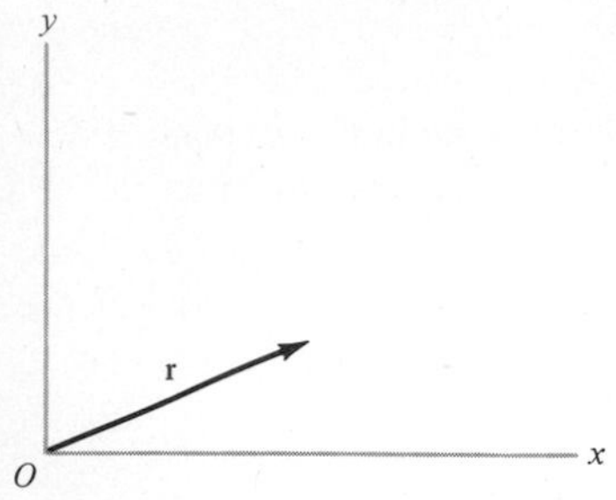

In Fig. 3-19, the vector **r** specifies the position of a moving body with respect to the origin of a particular frame of reference at a certain instant of time t. The velocity of the body is defined to be the instantaneous rate of change of **r** with respect to t. Stated mathematically, we define the velocity to be

$$\frac{d\mathbf{r}}{dt}$$

Fig. 3-19 The vector **r** specifies the position of a point with respect to the origin of a particular frame of reference.

Except for the fact that in two or three dimensions position must be treated as a vector, this quantity is exactly the same in concept as the quantity

$$\frac{dx}{dt}$$

which is used to define velocity when the motion is one-dimensional.

In both cases the derivative with respect to time is defined to be the *limiting value* of a sequence of fractions

$$\frac{\text{(change of position)}}{\text{(time interval over which change of position occurs)}}$$

The sequence of fractions is obtained by making the time interval smaller and smaller. In the multidimensional case, each fraction in this sequence consists of a numerator $\Delta\mathbf{r}$ and a corresponding denominator Δt, both of which have certain particular values.

The symbol $d\mathbf{r}$ denotes an *infinitesimal change* in **r.** It is called the **differential** of **r.** Physicists conventionally use $d\mathbf{r}$ to represent a value of $\Delta\mathbf{r}$ which is the numerator of a fraction belonging to the sequence when the value of the fraction is *very close* to the limiting value of the sequence. In other words, physicists use the symbol $d\mathbf{r}$ to represent an extremely small—or infinitesimally small—$\Delta\mathbf{r}$. In like manner, they use the symbol dt to represent the *corresponding* infinitesimally small value of the denominator Δt. The word "corresponding" means that $d\mathbf{r}$ and dt represent the $\Delta\mathbf{r}$ and the Δt of the *same* fraction. While both $d\mathbf{r}$ and dt are infinitesimally small, the value of the fraction $d\mathbf{r}/dt$ formed by division of $d\mathbf{r}$ by dt is just a value of $\Delta\mathbf{r}/\Delta t$ (not necessarily small) which is very near the limiting value. Hence this *fraction* is essentially the same as the *derivative* of **r** with respect to t. In other words, the derivative of **r** with respect to t can be treated as a fraction whose numerator is $d\mathbf{r}$ and whose denominator is dt. In fact, treating the derivative of *any* function of *any* independent variable as a fraction is legitimate for all the cases normally encountered in describing physical measurements.

It is possible to concoct mathematical functions which behave in peculiar fashion as the limit is approached, and there can then be difficulties with treating the derivative of the function as a fraction. But these difficulties arise with discontinuous functions not normally used in describing the continuous behavior with which this book is concerned.

Mathematicians can legitimately take exception to this approach, since the nature of their work leads them to take particular interest in the abnormal cases. Physicists, on the other hand, are interested in useful calculational tools and are not so concerned with what they call "pathological" functions. In any case, the disharmony between the intuitive notion of the physicist and the rigorous approach of the mathematician has come to be reconciled in recent years through work in the field of mathematics called nonstandard analysis, in which the intuitive idea of differentials is reformulated on a rigorous basis.

The fact that the velocity $d\mathbf{r}/dt$ is a vector quantity follows immediately from two things. One is that we can treat $d\mathbf{r}/dt$ as the vector quantity $d\mathbf{r}$ divided by the scalar quantity dt. The other is that a vector divided by a scalar is a vector; see Eq. (3-23). Hence we are justified in using vector notation to designate the velocity $\mathbf{v}$, and in writing the equation defining it as

$$\mathbf{v} \equiv \frac{d\mathbf{r}}{dt} \tag{3-28}$$

The **velocity** *vector* **v** *is the derivative of the position vector* **r** *with respect to the time t.*

The definition of the differential, taken together with treatment of the derivative as a fraction, makes possible some useful algebraic manipulations which have direct physical meaning. Multiplying Eq. (3-28) on both sides by the differential quantity dt gives

$$d\mathbf{r} = \mathbf{v}\, dt \tag{3-29a}$$

This equation tells us that a body moving with velocity $\mathbf{v}$ for the infinitesimal time interval dt will change its position by the infinitesimal amount $d\mathbf{r}$ given by the product of $\mathbf{v}$ and dt. The vector $d\mathbf{r}$ can also be expressed as

$$d\mathbf{r} = \mathbf{r} - \mathbf{r}_i$$

where $\mathbf{r}_i$ is the initial position vector of the body—at the beginning of the infinitesimal time interval—and $\mathbf{r}$ is its position vector at the end of the interval. The three vectors are shown in Fig. 3-20. The vectors $\mathbf{r}$ and $\mathbf{r}_i$ are supposed to be only slightly different in magnitude and direction.

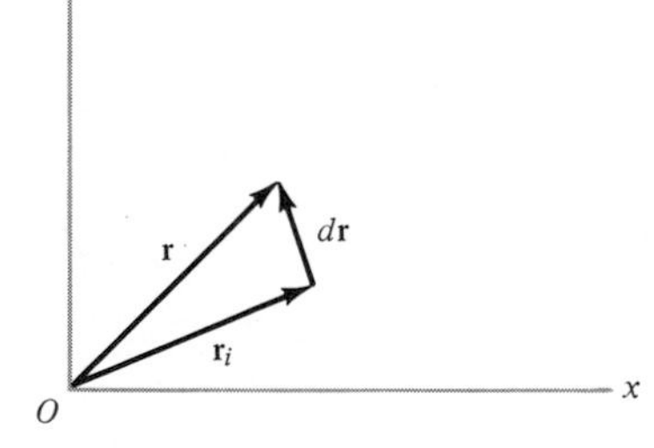

Fig. 3-20 If the point located in Fig. 3-19 moves with respect to the origin, the vector $\mathbf{r}$ changes with time. Over an infinitesimal time interval dt, the change in position is given by the vector $d\mathbf{r} = \mathbf{r} - \mathbf{r}_i$, where $\mathbf{r}$ and $\mathbf{r}_i$ are respectively the final and initial position vectors. The length of the infinitesimal vector $d\mathbf{r}$ must of course be exaggerated in the figure. The easiest way to verify that the three vectors in the figure are arranged in such a way as to satisfy the equation $d\mathbf{r} = \mathbf{r} - \mathbf{r}_i$ is to rewrite it in the form $\mathbf{r} = \mathbf{r}_i + d\mathbf{r}$. Then note that $\mathbf{r}$ extends from the tail of $\mathbf{r}_i$ to the head of $d\mathbf{r}$, in agreement with the rule for vector addition.

The utility of Eq. (3-29*a*) comes from the fact that even if $\mathbf{v}$ varies with time, each time interval dt is so short that $\mathbf{v}$ may be considered constant over any particular time interval. Equation (3-29*a*) is very similar to the relation $\Delta\mathbf{r} = \mathbf{v}\Delta t$ which involves finite quantities. But the latter is valid only for *constant* **v**. The possibility of treating a variable as an "instantaneous constant" is essential to the development of integral calculus, as you will see in Chap. 7.

The vectors $\mathbf{r}_i$ and $\mathbf{r}$ depend on the frame of reference from which they are measured. In Fig. 3-21, for instance, a second ("primed") frame of reference is chosen with axes x' and y'. These axes have *fixed* locations with respect to the axes x and y in the first ("unprimed") frame of reference. The initial and final position vectors $\mathbf{r}'_i$ and $\mathbf{r}'$ in the primed frame of reference

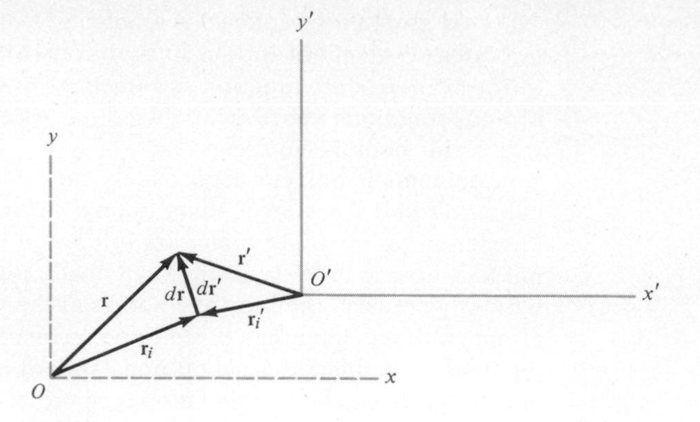

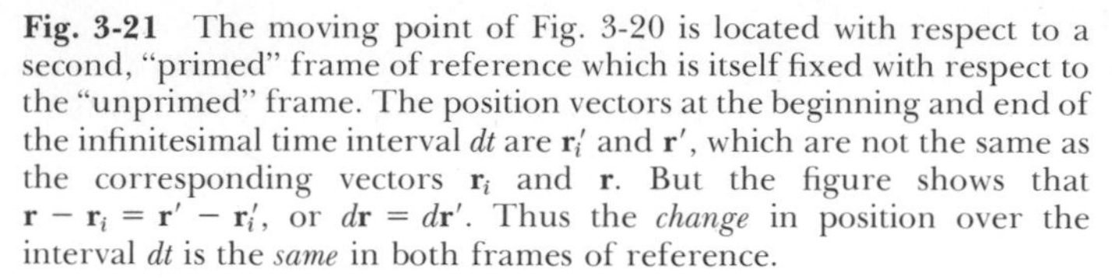

Fig. 3-21 The moving point of Fig. 3-20 is located with respect to a second, "primed" frame of reference which is itself fixed with respect to the "unprimed" frame. The position vectors at the beginning and end of the infinitesimal time interval dt are $\mathbf{r}_i'$ and $\mathbf{r}'$, which are not the same as the corresponding vectors $\mathbf{r}_i$ and $\mathbf{r}$. But the figure shows that $\mathbf{r} - \mathbf{r}_i = \mathbf{r}' - \mathbf{r}_i'$, or $d\mathbf{r} = d\mathbf{r}'$. Thus the *change* in position over the interval dt is the *same* in both frames of reference.

Fig. 3-22 The displacement vector $d\mathbf{s}$ is shown in an arbitrary frame of reference to emphasize its independence of the choice of the frame in which the position vectors $\mathbf{r}$ and $\mathbf{r}_i$ are specified. The vector $d\mathbf{s}$ is identical to the vector $d\mathbf{r}$ in Fig. 3-21. (In actuality, both $d\mathbf{r}$ and $d\mathbf{s}$ are infinitesimal. For purposes of illustration, they must necessarily be shown as finite.)

are not the same as the corresponding position vectors $\mathbf{r}_i$ and $\mathbf{r}$ in the unprimed frame of reference. But the vector $d\mathbf{r}$ denoting the *change in position* with respect to the unprimed frame is *identical* to the vector $d\mathbf{r}'$ denoting the change in position with respect to the primed frame.

In order to make this important fact clear, we introduce a new name and a new symbol for the infinitesimal "change in position $d\mathbf{r}$." We call it the infinitesimal **displacement** $d\mathbf{s}$. Hence displacement means "change in position," and

$$d\mathbf{s} \equiv d\mathbf{r}$$

In terms of the displacement, we can describe what happens to the moving body in the infinitesimal time interval dt *without* having to specify the position vectors $\mathbf{r}_i$ and $\mathbf{r}$ (or $\mathbf{r}_i'$ and $\mathbf{r}'$). In Fig. 3-22, the same displacement vector $d\mathbf{s}$ is shown in a frame of reference whose origin is chosen quite arbitrarily. Indeed, it is often possible to draw a displacement vector without bothering to draw the frame of reference at all.

Since the displacement $d\mathbf{s}$ is another name for the change in position $d\mathbf{r}$, Eq. (3-29*a*), $d\mathbf{r} = \mathbf{v}\,dt$, can be written equally well

$$d\mathbf{s} = \mathbf{v}\,dt \tag{3-29b}$$

Similarly, the definition of velocity given by Eq. (3-28), $\mathbf{v} \equiv d\mathbf{r}/dt$, can be written equally well

$$\mathbf{v} \equiv \frac{d\mathbf{s}}{dt} \tag{3-30}$$

Thus *velocity is the infinitesimal displacement vector* $d\mathbf{s}$ *divided by the corresponding infinitesimal time interval* dt. Whether $\mathbf{v}$ is related to an infinitesimal displacement vector $d\mathbf{s}$, as in Eqs. (3-29*b*) and (3-30), or to an infinitesimal change $d\mathbf{r}$ in a position vector $\mathbf{r}$, as in Eqs. (3-28) and (3-29*a*), depends only on the emphasis desired in a particular situation.

The **acceleration** *vector* **a** *is the derivative of the velocity vector* **v** *with respect to the time t.* Expressed in mathematical notation, this definition is

$$\mathbf{a} \equiv \frac{d\mathbf{v}}{dt} \tag{3-31}$$

Acceleration is a vector because an infinitesimal change $d\mathbf{v}$ in a velocity is a vector, because the corresponding infinitesimal interval dt of time is a scalar, and because a vector divided by a scalar is a vector. Except for the fact that it takes into account the vectorial nature of acceleration in two or three dimensions, Eq. (3-31) does not differ from Eq. (2-19),

$$a \equiv \frac{dv}{dt}$$

which defined acceleration for one-dimensional motion.

We can substitute the definition $\mathbf{v} = d\mathbf{r}/dt$ into Eq. (3-31) to obtain

$$\mathbf{a} \equiv \frac{d}{dt}\left(\frac{d\mathbf{r}}{dt}\right) \equiv \frac{d^2\mathbf{r}}{dt^2} \tag{3-32}$$

That is, *the* **acceleration** *vector* **a** *is the second derivative of the position vector* **r** *with respect to time.*

The identity $d\mathbf{s} \equiv d\mathbf{r}$ can be used to write Eq. (3-32) in the equivalent form

$$\mathbf{a} \equiv \frac{d^2\mathbf{s}}{dt^2} \tag{3-33}$$

Again, it is a matter of the emphasis desired as to whether Eq. (3-32) or Eq. (3-33) is used.

Now we will make use of the vector definition of acceleration, Eq. (3-31), to gain new insight into the relation between acceleration and velocity. Multiplying both sides of the equation by dt yields the relation

$$d\mathbf{v} = \mathbf{a}\, dt \tag{3-34}$$

That is, an object having acceleration **a** during the infinitesimal time interval dt will change its velocity by the infinitesimal amount $d\mathbf{v}$.

The simplest case to which the differential relation of Eq. (3-34) can be applied is the one in which an object accelerates in the direction in which it is already moving. This case is illustrated in Fig. 3-23. The object, whose initial velocity is $\mathbf{v}_i$, experiences an acceleration **a** in the same direction as $\mathbf{v}_i$. During the infinitesimal time interval dt, the acceleration is essentially constant, and according to Eq. (3-34) there is an infinitesimal change $d\mathbf{v} = \mathbf{a}\, dt$ in the velocity. The velocity at the end of the interval is thus

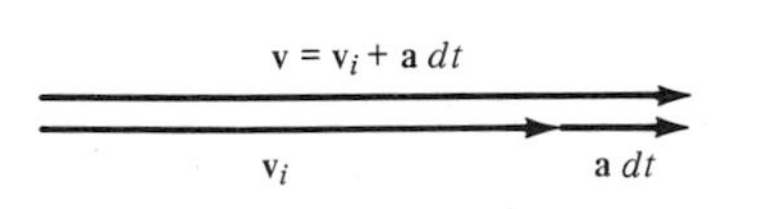

Fig. 3-23 At the beginning of the infinitesimal time interval dt, an object moves with instantaneous velocity $\mathbf{v}_i$. During the time interval, it experiences an acceleration **a** whose direction is the same as that of $\mathbf{v}_i$. At the end of the time interval, its instantaneous velocity is $\mathbf{v} = \mathbf{v}_i + \mathbf{a}\, dt$. The magnitude of the velocity is changed, but the direction of the velocity is unchanged.

$$\mathbf{v} = \mathbf{v}_i + d\mathbf{v}$$

or

$$\mathbf{v} = \mathbf{v}_i + \mathbf{a}\, dt \tag{3-35}$$

As you can see from Fig. 3-23, the acceleration has changed the magnitude of the velocity, but not its direction, in this simple case.

But Eq. (3-35) is not restricted to cases where the acceleration is parallel to the initial velocity. Consider the situation represented in Fig. 3-24. Here again an acceleration **a** is applied for an infinitesimal time dt. But this

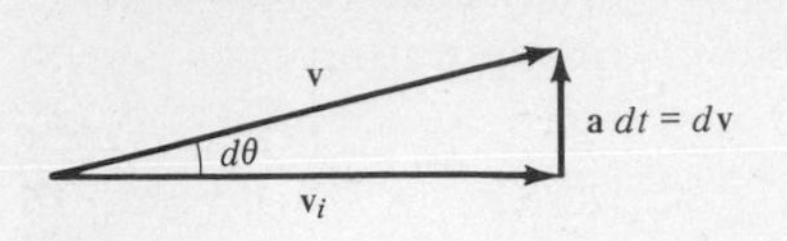

Fig. 3-24 At the beginning of the infinitesimal time interval dt, an object moves with instantaneous velocity $\mathbf{v}_i$. During the time interval, it experiences an acceleration **a** whose direction is perpendicular to that of $\mathbf{v}_i$. At the end of the time interval, its instantaneous velocity is $\mathbf{v} = \mathbf{v}_i + \mathbf{a}\,dt$, just as in the parallel case illustrated in Fig. 3-23. In the present case, however, the magnitude of the velocity is unchanged, as explained in the text. But the direction of the velocity is changed.

time **a** is perpendicular to the initial velocity $\mathbf{v}_i$. While Eq. (3-35) still applies, we must now consider explicitly the fact that the addition on the right side is vectorial. At first glance, the result of the acceleration may *appear* to be a change in both the magnitude and the direction of the velocity. But the appearance is misleading, since the figure necessarily exaggerates the length of the infinitesimal vector $d\mathbf{v}$. It is shown in the small-print section immediately below that when **a** is perpendicular to $\mathbf{v}_i$, the result is actually a change in the direction but *not* in the magnitude of the velocity, if the time interval dt is infinitesimal.

In Fig. 3-24, the infinitesimal angle $d\theta$ between $\mathbf{v}_i$ and $\mathbf{v}$ is given by the expression

$$\tan d\theta = \frac{dv}{v_i}$$

where $d\theta$ is expressed in radians and $dv = a\,dt$. Since $d\theta$ is small and is expressed in radians, we can use the approximation $\tan d\theta = d\theta$, where $d\theta \ll 1$ rad. (You can check the validity of this approximation on your pocket calculator, using successively smaller values for the angle such as 0.1 rad, 0.01 rad, 0.001 rad, and so forth.) The equation displayed above then simplifies to the form

$$d\theta = \frac{dv}{v_i}$$

Thus $d\theta$, the infinitesimal change in the direction of $\mathbf{v}_i$ over the infinitesimal time interval dt, is proportional to dv/v_i.

Now let us consider the change in the magnitude of $\mathbf{v}_i$ produced by the same $d\mathbf{v}$. By the pythagorean theorem we have

$$v^2 = v_i^2 + (dv)^2$$

Dividing both sides of this equation by v_i^2 and taking the square root, we obtain

$$\frac{v}{v_i} = \left[1 + \left(\frac{dv}{v_i}\right)^2\right]^{1/2}$$

We now use the approximation

$$(1 + z)^{1/2} \simeq 1 + \tfrac{1}{2}z \qquad \text{where } z \ll 1 \tag{3-36}$$

which is valid for any number z. (Again, you can evaluate the accuracy of this approximation on your pocket calculator by trying successively smaller values of z, for example, 0.1, 0.01, 0.001, and so forth.) The approximation can be applied to the last equation by writing the dimensionless quantity $(dv/v_i)^2$ for the number z. So doing yields

$$\frac{v}{v_i} \simeq 1 + \frac{1}{2}\left(\frac{dv}{v_i}\right)^2$$

or

$$\frac{v - v_i}{v_i} \simeq \frac{1}{2}\left(\frac{dv}{v_i}\right)^2$$

This tells us that the fractional change in the magnitude of **v** is proportional to the *square* of the infinitesimally small quantity dv/v_i. But if a quantity is small, its square is very much smaller. Thus the fractional change in *magnitude* of **v** resulting from the perpendicular acceleration **a**, being equal to $\frac{1}{2}$ times the square of an infinitesimal, is negligible compared to the change in *direction*, which is equal to the first power of the same infinitesimal. This proves the assertion that an acceleration **a** acting perpendicular to the velocity **v** of a moving object changes the direction, but not the magnitude, of the velocity.

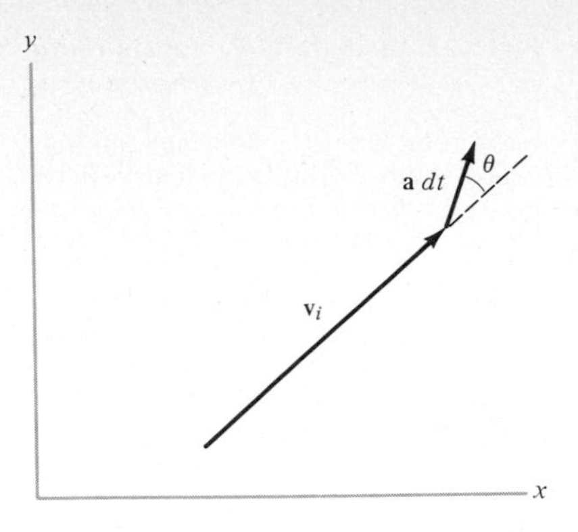

Fig. 3-25 At the beginning of an infinitesimal time interval dt, an object moves with instantaneous velocity $\mathbf{v}_i$. During the time interval, it experiences an acceleration **a** whose direction is arbitrary.

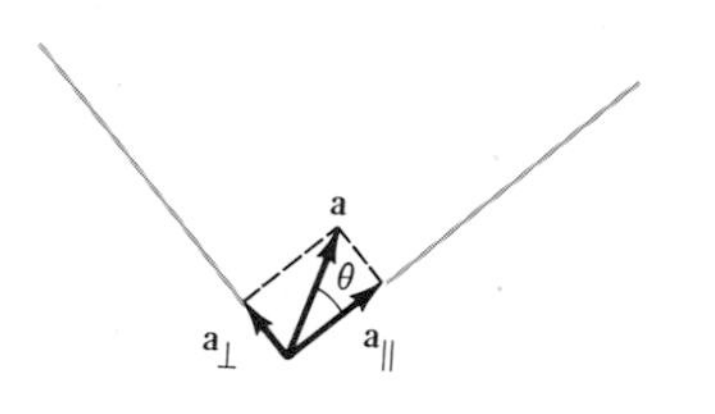

Fig. 3-26 Directions are chosen parallel and perpendicular to the direction of motion in Fig. 3-25. The acceleration vector **a** is drawn, and the constituent vectors $\mathbf{a}_{\parallel}$ and $\mathbf{a}_{\perp}$ constructed. Each of these constituent vectors can be treated as one of the special cases of Figs. 3-23 and 3-24.

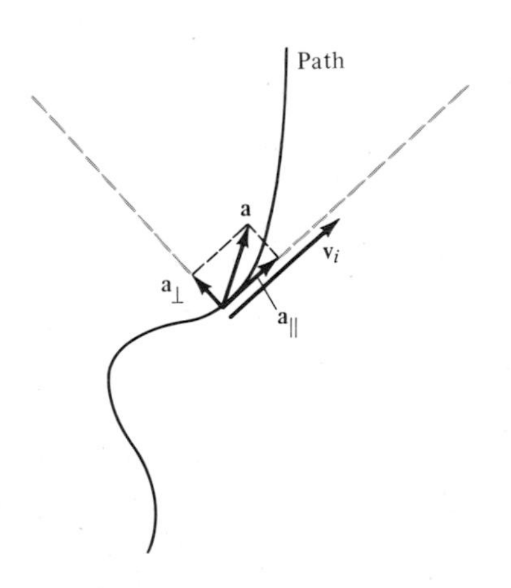

Fig. 3-27 The vectors $\mathbf{a}$, $\mathbf{a}_{\parallel}$, $\mathbf{a}_{\perp}$, and $\mathbf{v}_i$ of Figs. 3-25 and 3-26 indicated in relation to the path of motion. The vectors $\mathbf{v}_i$ and $\mathbf{a}_{\parallel}$ are tangent to the path, while $\mathbf{a}_{\perp}$ is perpendicular to the path. In Sec. 3-5 it will be shown that $\mathbf{a}_{\perp}$ "points to the inside of the curve."

We have now discussed two special cases: (1) where **a** is parallel to **v** and (2) where **a** is perpendicular to **v**. An example of the first case is vertical fall. In vertical fall there is an acceleration $\mathbf{a} = d\mathbf{v}/dt$ because the magnitude v of **v** changes. But its direction $\hat{\mathbf{v}}$ remains constant. The second case is exemplified by a satellite in a circular orbit, discussed in Sec. 3-5. For such a case, an acceleration $\mathbf{a} = d\mathbf{v}/dt$ exists because the direction $\hat{\mathbf{v}}$ of the velocity **v** changes, while its magnitude v remains constant.

What about the general case where the angle between **a** and **v** is arbitrary, as in Fig. 3-25? In this case, of which projectile motion is an example, we can always pick a set of axes so that one axis is parallel to, and the other perpendicular to, the instantaneous direction of motion, that is, the direction of $\mathbf{v}_i$. This has been done in Fig. 3-26. We can then "resolve" the vector **a** into two constituent vectors $\mathbf{a}_{\parallel}$ and $\mathbf{a}_{\perp}$. In other words, we consider **a** to be the sum of $\mathbf{a}_{\parallel}$ and $\mathbf{a}_{\perp}$:

$$\mathbf{a} = \mathbf{a}_{\parallel} + \mathbf{a}_{\perp} \tag{3-37}$$

Each of the constituent vectors $\mathbf{a}_{\parallel}$ and $\mathbf{a}_{\perp}$ can be dealt with as one of the two special cases discussed above.

The vector $\mathbf{a}_{\parallel}$ is called the **tangential acceleration,** since it is directed along the tangent of the curve describing the path of the object. For the vector $\mathbf{a}_{\perp}$ Newton coined the name **centripetal** (that is, center-seeking) **acceleration** since, as you will see in Sec. 3-5, it is directed inward along the local radius of curvature of the path. The relation of the tangential and centripetal accelerations to the path of motion is shown in Fig. 3-27.

To summarize, velocity is a *vector quantity* and is therefore specified by *two* attributes, *magnitude* and *direction.* Whenever *either* (*or both*) of these attributes is *changing,* the velocity is changing and there will be an *acceleration* equal to the rate of change of that velocity.

Before ending this discussion of directed quantities, it is important to explain the relation between the notation introduced in the present chapter for such quantities and the notation introduced in the preceding chapter. In two- or three-dimensional situations the vector symbolism of this chapter must be used for a directed quantity, such as a velocity. But in one-dimensional situations either the vector symbolism can be used, or the symbolism of Chap. 2 can be used. These two symbolisms arise from the fact that in one dimension there are two ways to treat a quantity which has both magnitude and direction. Both ways are indicated in Fig. 3-28. The first way, shown in Fig. 3-28*a* and *b*, is to regard the quantity as a vector whose possible directions are restricted to just two opposite directions. The second way, shown in Fig. 3-28*c* and *d*, is to regard the quantity as a scalar, and hence a quantity whose numerical value can be either positive or negative. The direction of the quantity is then designated by the sign of the numerical value. This second procedure is the one used in Chap. 2. (It cannot be applied to situations involving more than one dimension, simply because two signs, + and −, are not adequate to describe all possible directions.) When the second procedure is used, the italic symbol v represents a velocity (whose direction may be positive or negative), and the italic symbols in absolute value signs, $|v|$, represents its magnitude, a speed. When using this symbolism, the value of the quantity v can be either positive or negative, according to its direction. In contrast, the value of the quantity v can be *only* positive when it represents the magnitude of the vector **v**. Despite the pos-

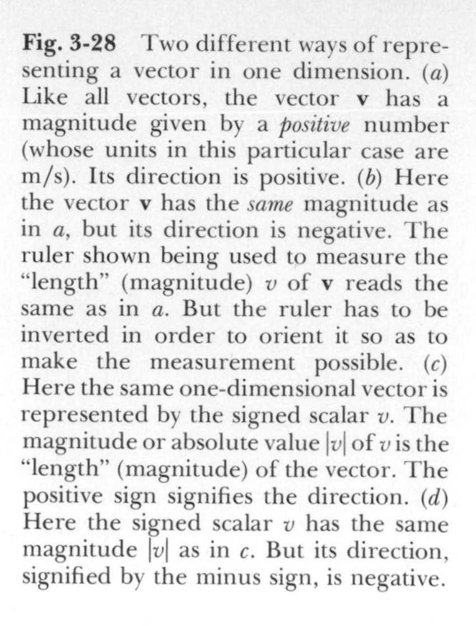

Fig. 3-28 Two different ways of representing a vector in one dimension. (*a*) Like all vectors, the vector **v** has a magnitude given by a *positive* number (whose units in this particular case are m/s). Its direction is positive. (*b*) Here the vector **v** has the *same* magnitude as in *a*, but its direction is negative. The ruler shown being used to measure the "length" (magnitude) v of **v** reads the same as in *a*. But the ruler has to be inverted in order to orient it so as to make the measurement possible. (*c*) Here the same one-dimensional vector is represented by the signed scalar v. The magnitude or absolute value $|v|$ of v is the "length" (magnitude) of the vector. The positive sign signifies the direction. (*d*) Here the signed scalar v has the same magnitude $|v|$ as in *c*. But its direction, signified by the minus sign, is negative.

sibility of confusion arising from the two possible meanings of an italic symbol, we will use both symbolisms because one is more convenient in some one-dimensional situations and the other is more convenient in others. But we will always specify clearly which symbolism is intended if there is any ambiguity. This will be done by stating whether we treat a directed quantity as a vector or as a signed scalar. We use the term **signed scalar** to refer to a quantity whose value can be either positive or negative. (The term is redundant since by definition a scalar can be of either sign. But it helps emphasize the distinction being made between such a quantity and a quantity which is a magnitude—that is, one which is necessarily positive.)

In two-dimensional situations a vector can be represented either by two components or by a length and an angle. Both components are always signed scalars. The same is true of the angle. But the length is always a magnitude. How would you extend these statements to three-dimensional situations?

3-4 THE PARABOLIC TRAJECTORY

We will now rephrase the conclusion of Sec. 3-1 concerning the motion of a projectile in precise vectorial terms and then continue with the discussion of projectile motion in those terms.

How does a projectile move, in the absence of air resistance, when it is given an arbitrary initial velocity $\mathbf{v}_i$? We already have a sort of answer in Eq. (3-6). At any moment, the relation between the x and y coordinates of the projectile is

$$y = \frac{v_{yi}}{v_x} x - \frac{g}{2v_x^2} x^2$$

The quantity v_x is the horizontal component of the projectile velocity. Its value is a constant equal to its initial value, since the horizontal motion is one of constant velocity. The quantity v_{yi} is the initial value of the vertical component of the projectile velocity. And g is the magnitude of the gravitational acceleration. Thus v_x and v_{yi}, the two components of $\mathbf{v}_i$, specify the value of the constant factor in each of the terms on the right side of the equation. But this equation is not in the most convenient form, since we usually know the magnitude and direction v_i and θ of the initial projectile velocity, and not its x and y components v_x and v_{yi}. However, we can use Eqs. (3-10), in the form

$$v_x = v \cos \theta \qquad \text{and} \qquad v_y = v \sin \theta$$

to rewrite Eq. (3-6) in terms of v_i and θ (we simplify the notation by using θ instead of ϕ_v). Setting $v = v_i$ and $v_y = v_{yi}$, and substituting the values of v_x and v_{yi} thus obtained into Eq. (3-6), we have the following relation between r_y and r_x, the components of the instantaneous position vector $\mathbf{r}$ of the projectile:

$$r_y = \frac{v_i \sin \theta}{v_i \cos \theta} r_x - \frac{g}{2v_i^2 \cos^2 \theta} r_x^2$$

or

$$r_y = (\tan \theta) r_x - \frac{g}{2(v_i \cos \theta)^2} r_x^2 \tag{3-38}$$

This equation is less complicated than it looks. Suppose, for the moment, that we could "switch off" gravity—that we could set $g = 0$. Equation (3-38) would then become simply $r_y = (\tan \theta) r_x$. But $\tan \theta = v_{yi}/v_x$, and so we would have $r_y/r_x = v_{yi}/v_x$. Thus the ratio of the distance traveled in the y direction to that traveled in the x direction would simply be the ratio of the components v_{yi} and v_x of the velocity in those directions. The reason is that both components of the velocity (and not just v_x) would remain constant at their initial values if g were equal to zero.

Now let us "turn on" gravity again and consider the second term on the right side of the equation. It is of the same general form as the right side of Eq. (3-3),

$$r_y = -\frac{g}{2v_i^2} r_x^2$$

which describes the trajectory of a horizontally launched projectile. Indeed, the second term of Eq. (3-38) reduces to $-(g/2v_i^2)r_x^2$ when $\theta = 0$. In Eq. (3-38), however, $v_i \cos \theta$, the horizontal component of the initial velocity, takes the place of v_i in the simple case of horizontal launching.

With this general qualitative picture in mind, let us proceed to a quantitative picture of the parabolic trajectory. In hitting a baseball or firing a gun, we usually have control over the initial elevation angle θ and the magnitude v_i of the initial velocity. Examples 3-7 and 3-8 are studies of what happens when θ is varied while v_i is kept constant and when v_i is varied while θ is kept constant.

The calculations necessary for Examples 3-7 and 3-8 can be performed (given some patience) on any pocket calculator. However, they are more conveniently done on a programmable pocket calculator or a small computer. A trajectory plotting program is listed in the Numerical Calculation Supplement.

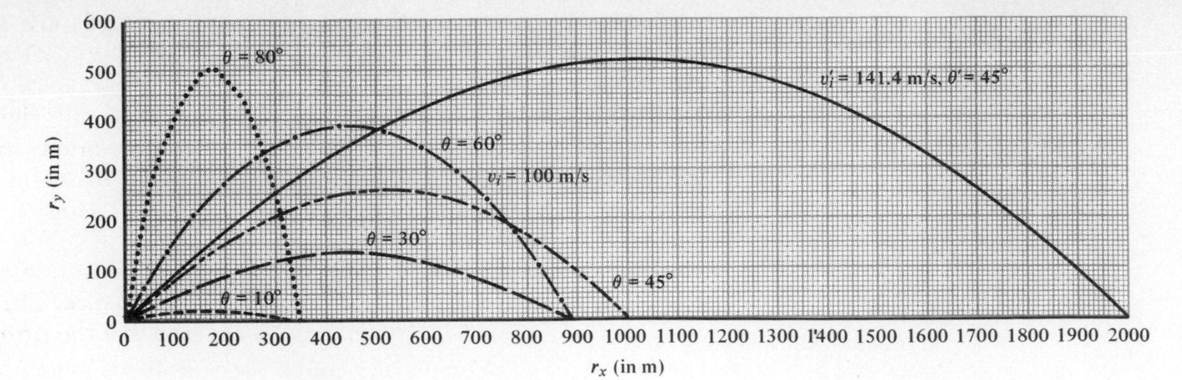

Fig. 3-29 Plots of the trajectories calculated for projectiles whose initial speed is $v_i = 100$ m/s, shot from a gun whose elevation angle is successively $\theta = 10°, 30°, 45°, 60°$, and $80°$. Note that maximum range is attained when $\theta = 45°$. Also shown is the trajectory for a projectile whose initial speed is $v_i' = 141.4$ m/s, or $\sqrt{2}$ greater than v_i, with elevation angle $\theta' = 45°$. The range is twice the maximum attained by the slower projectile.

EXAMPLE 3-7

Use Eq. (3-38) to plot the trajectories of a projectile having an initial velocity of magnitude $v_i = 100$ m/s, for elevation angles $\theta = 10°, 30°, 45°, 60°$, and $80°$.

■ As in Example 2-5, you must keep track of the proper units, since a calculating device deals with numbers only. You therefore rewrite Eq. (3-38) in the completely equivalent form

$$r_y = (\tan\theta)r_x - \frac{9.8}{2 \times (100)^2 \cos^2\theta} r_y^2 \qquad (r_x \text{ and } r_y \text{ in meters})$$

The family of plots shown in Fig. 3-29 was obtained by drawing smooth curves through points gotten by calculating values of r_y for successive values of r_x which were taken 50 m apart. You will find it worthwhile to calculate and plot one such trajectory yourself, choosing any values you like for v_i and θ and thus specifying $\mathbf{v}_i$.

As you would expect, increasing θ increases the maximum height y_{max} to which the projectile rises. At first the **range** R (the horizontal distance from the origin to the point where the projectile returns to $y = 0$) increases with increasing θ, but then it decreases. The maximum range appears to be attained with $\theta = 45°$. You can check this point by running some more calculations yourself, choosing angles close to 45° and keeping v_i constant.

All the trajectories of Fig. 3-29 other than the 45° trajectory fall into pairs of equal range R. The members of each pair have elevation angles lying at equal angles above and below 45°, as can be seen from Fig. 3-29. This remarkable fact leads us to look for a qualitative reason why it should be so. With fixed v_i, a large angle θ means that the projectile will rise high and hence will take a relatively long time to reach the ground again. However, the component $v_x = v_i \cos\theta$ will be small, since $\cos\theta$ is small for large angles. If T is the total time the projectile spends aloft, $R = v_x T$ will not be large. On the other hand, if θ is small, v_x will be large but the projectile will not remain above the ground very long. Somewhere in the middle (the calculation suggests $\theta = 45°$) the range is maximized.

EXAMPLE 3-8

Repeat the calculation of Example 3-7 with elevation angle $\theta' = 45°$, but increase v_i by a factor $\sqrt{2}$ so that $v_i' = 141.4$ m/s.

■ When you repeat the calculation with these new values, you obtain a plot like the longest trajectory in Fig. 3-29. Comparing the range with that obtained for a 45° elevation angle in Example 3-7, you see that the range has doubled. Since the initial speed is increased by $\sqrt{2}$, this suggests that the range is proportional to v_i^2.

We will now generalize the ideas arising from the numerical calculations of Examples 3-7 and 3-8, and derive an analytical expression for the range R in terms of v_i and θ, the magnitude and direction of the initial velocity vector $\mathbf{v}_i$. We have set up the problem so that the projectile starts at the coordinates $x = 0$, $y = 0$. When it returns to the ground (which is assumed to be level), it does so at the specific coordinates $x = R$, $y = 0$. We can therefore rewrite Eq. (3-38) in this special case by substituting the value R for x and the value 0 for y. We find

$$0 = (\tan \theta)R - \frac{g}{2v_i^2 \cos^2 \theta} R^2$$

Since $R \neq 0$, we can divide this quadratic equation through by R. Doing so reduces it immediately to a linear equation which can be solved for R to yield

$$R = \frac{v_i^2}{g} 2 \tan \theta \cos^2 \theta \tag{3-39}$$

This expression can be simplified by noting that

$$2 \tan \theta \cos^2 \theta = 2 \frac{\sin \theta}{\cos \theta} \cos^2 \theta = 2 \sin \theta \cos \theta$$

Using the standard trigonometric identity $2 \sin \theta \cos \theta = \sin(2\theta)$, and substituting into Eq. (3-39), we have

$$R = \frac{v_i^2}{g} \sin(2\theta) \tag{3-40}$$

As suggested by the numerical solutions, the range depends on the *square* of the initial speed v_i.

The range R depends on the sine of *twice* the elevation angle θ. Furthermore, R is a maximum when $\sin(2\theta)$ has its maximum value of 1. This occurs when $\theta = 45°$, a result which is probably in rough but reasonable agreement with your experience in throwing balls, in spite of the fact that we have ignored air resistance. In any case, the analytical result certainly agrees with the numerical solutions plotted in Fig. 3-29.

The symmetry observed in Fig. 3-29 for elevation angles symmetric about 45° finds its analytical expression in the symmetry of the function $\sin(2\theta)$ about the angle $\theta = 45°$. That is, since $\sin[2(45° - \alpha)] = \sin[2(45° + \alpha)]$, the range will be the same for any two elevation angles $\theta = 45° \pm \alpha$ which are equal amounts greater than and less than 45°.

The small angle of each pair gives a flat trajectory, and the large angle a high trajectory. Air resistance tends to affect the high trajectory more, since it is longer. In baseball, the high trajectory is called a pop fly. Such a fly is easy to catch, since the time of flight is so long that the fielder has plenty of time to get into position. The low trajectory is called a line drive. It is much harder to catch because it takes much less time than the pop fly to reach the same point in the field.

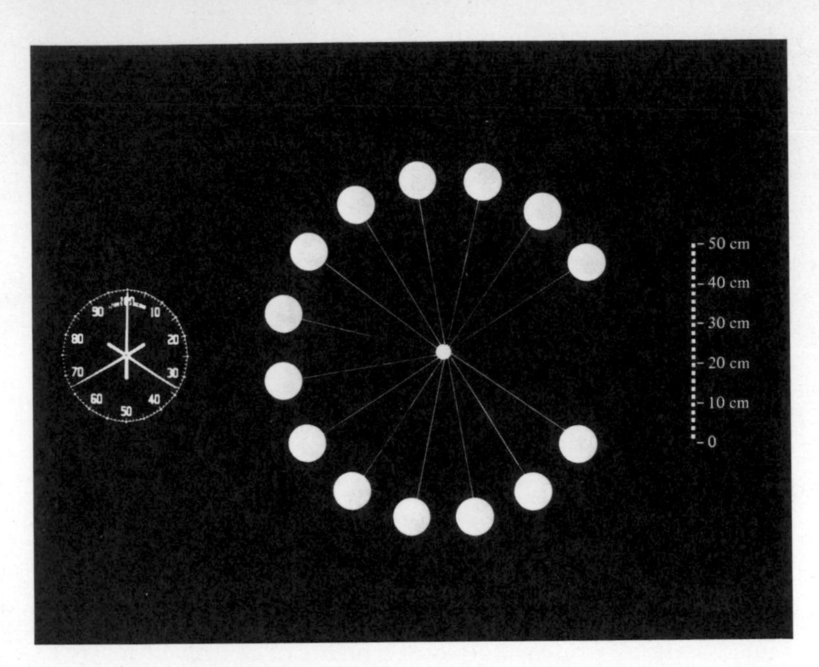

Fig. 3-30 Strobe photo of an air-table puck in uniform circular motion. A string attached to the puck is threaded over a low-friction pulley through a hole in the table. A weight hanging at the other end of the string (not seen) maintains a constant tension in the string. By means of repeated trials, the puck is set into motion in such a way that it moves in a circle. The speed of the puck is constant, as can be seen from the uniform spacing between successive puck images. The clock hand makes one revolution per second. It is illuminated by the strobe light to show the strobe flash interval.

3-5 UNIFORM CIRCULAR MOTION AND CENTRIPETAL ACCELERATION

The case of uniform circular motion is the most important application of the general idea of centripetal acceleration, which we developed in Sec. 3-3. It is also the simplest. Out of this application will come a quantitative result which has direct bearing on the motion of such things as artificial and natural satellites, of nuclear particles in accelerators, of bodies whirling at the ends of strings, and of flywheels spinning on shafts.

The air table experiment illustrated in the strobe photo of Fig. 3-30 provides an example of uniform circular motion. A puck is connected to a string. The string is threaded through a hole in the center of the air table, over a pulley which can swivel freely with very little friction. A small weight hanging from the end of the string beneath the air table maintains a constant tension in the string. If you look carefully at the photograph, you can see the string stretching between the pulley and the puck.

To begin the experiment, the puck is carefully set into circular motion by projecting it at a certain speed and distance from the center, with a direction of motion perpendicular to the string. The proper speed for producing circular motion at that distance is found by trial and error. The fact that the orbit is circular ensures that the velocity of the puck always remains perpendicular to the string, since the tangent to a circle is perpendicular to the radius at the point of tangency. Inspection of the figure will satisfy you that the orbit is indeed circular and that the speed is constant.

The magnitude of the velocity vector—the speed of the puck—is not changing. The puck is nevertheless accelerating, since the direction of the velocity vector is continually changing. We will determine the acceleration of the puck by measurements on the photograph. Then we will use the ideas suggested by this experiment to derive a mathematical expression for the centripetal acceleration.

Figure 3-31 is a copy of Fig. 3-30, in which vectors are added to represent the change in position of the puck between flash 1 and flash 2 of the strobe light, and its change in position between flash 2 and flash 3. Each vector is drawn with its tail at the center of a puck image and its head at the center of the next image. Thus each vector shows the change in the posi-

Fig. 3-31 The vectors, drawn in over a copy of Fig. 3-30, represent the displacements of the puck during the time intervals between successive flashes of the strobe light. If we choose the unit of time to be one flash interval, the vectors represent the displacements per unit time interval, which are the average velocities during those intervals.

tion of the puck during a time interval between consecutive flashes of the strobe light. It shows this *directly,* without referring to the change in a vector describing the position of the puck with respect to some particular reference frame. Hence the vector is a displacement vector. The word is used in complete analogy to the way it is used in the preceding section. The only difference is that here the displacement vectors are not of infinitesimal magnitude, because here the time intervals during which the displacements occur are not of infinitesimal duration. The *average velocity* of the puck in the time interval between any consecutive pair of flashes is found by dividing the vector depicting its displacement from one flash to the next by the duration of the time interval between the flashes. This is analogous to the statement, made in Sec. 3-4, that instantaneous velocity is found by dividing infinitesimal displacement by the corresponding infinitesimal time interval. Since the time interval between any two successive flashes of the strobe light is always the same, the displacement vectors are proportional to the average velocities during the corresponding time intervals. In fact, we will use this fixed time interval as the *unit* of time. If we do so, the numerical value of the time interval is 1 unit, and so the vectors shown in Fig. 3-31 *are* the average velocities. In other words, we will measure velocity in terms of displacement per strobe flash interval.

The puck experiences an *average acceleration* as its average velocity changes from the value it has over the time interval between flash 1 and flash 2 to the value it has over the time interval between flash 2 and flash 3. This average acceleration can be constructed by using the fact that the final average velocity vector is the sum of the initial average velocity vector and the vector specifying the change in velocity. With this in mind, we move the final velocity vector so that its tail coincides with the tail of the initial velocity vector, as in Fig. 3-32. In the process, the direction and length of the final vector are not changed. As you can see from Fig. 3-32, the solid vector (in white) is the *change* in average velocity over one flash interval, because the initial average velocity vector (dashed white) plus the change in velocity (solid white) equals the final average velocity (dashed gray). Since we are

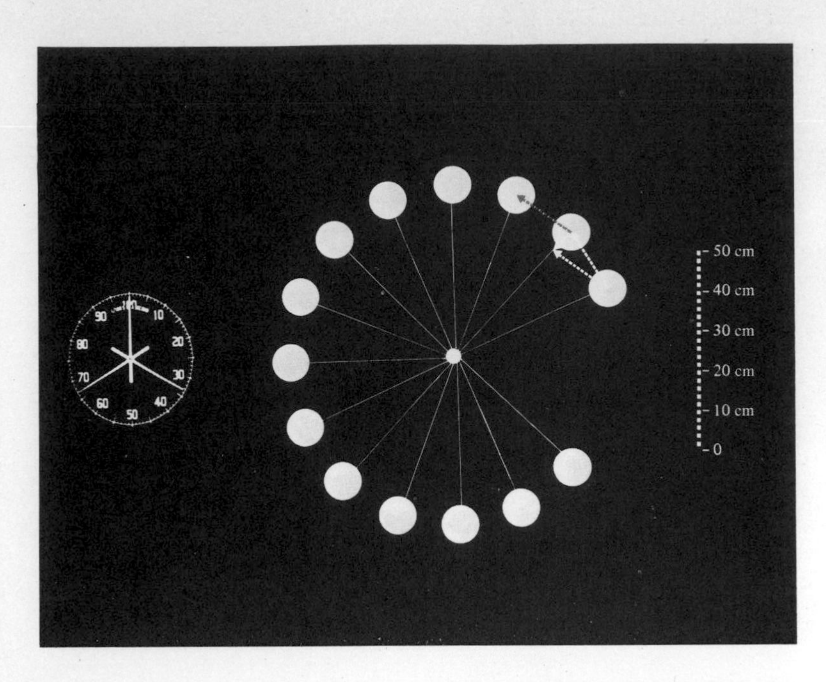

Fig. 3-32 Measurement of the average acceleration between flash intervals. The second velocity vector (dashed gray) is moved parallel to itself so that its tail coincides with the tail of the first velocity vector (dashed white). The geometric procedure for subtracting vectors, described in Sec. 3-2, is then carried out. The solid white vector is the difference between the second average velocity vector and the first, and is thus the change in average velocity between the first flash interval and the second. This change in average velocity per flash interval is the average acceleration.

taking the strobe flash interval to be the unit of time, the solid white vector is the change in average velocity per unit time; that is, it is the average acceleration. You should note that the acceleration vector points toward the center of the circular path of the puck.

As was implied at the beginning of the discussion, the puck can be started in different ways, and will then move in different paths. In the experiment depicted by the strobe photo of Fig. 3-33, the puck is simply released from rest at some distance from the center. Not surprisingly, the puck starts to move on a straight line toward the center. In fact, it accelerates toward the center under the influence of the tension in the string, as you can see from the increasing magnitudes of the displacements of the puck between successive flashes. For this second experiment, the tension in

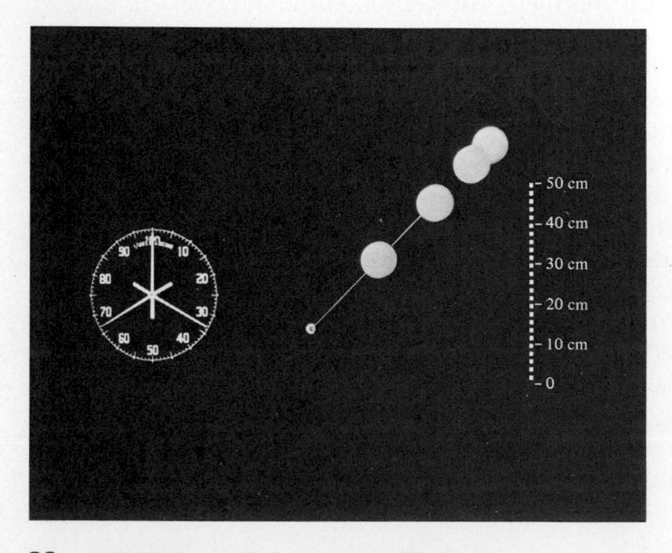

Fig. 3-33 In this strobe photograph, the experimental setup is identical with that of Fig. 3-30. Here, however, the air puck has been released from rest.

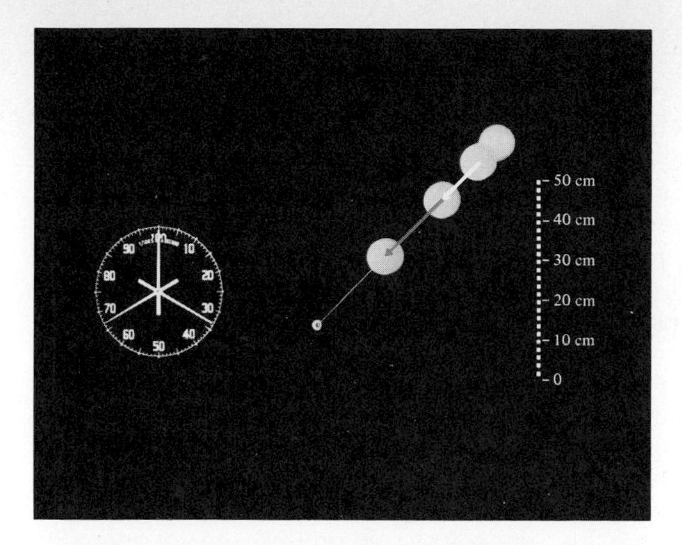

Fig. 3-34 The vectors, drawn in over a copy of Fig. 3-33, represent the displacements of the puck during the time intervals between successive flashes of the strobe light. As in Fig. 3-31, the vectors denote the average velocities during these time intervals, if the unit of time is chosen to be one flash interval.

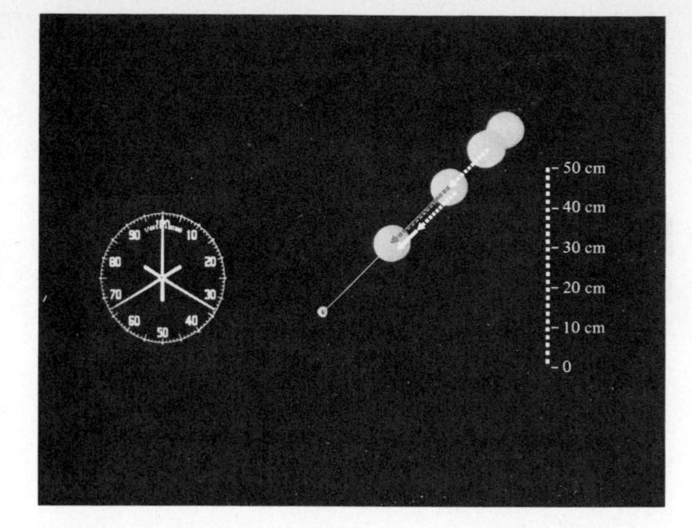

Fig. 3-35 Measurement of the average acceleration between two successive flash intervals. The procedure followed is the same as that described in the caption to Fig. 3-32.

the string has been adjusted to the same value as in the first experiment with the puck in a circular orbit. The same flash interval (or time unit) and the same distance unit are used in the second experiment, as shown by the clock and the distance scale.

We want to measure the average acceleration in this second experiment. In Fig. 3-34 the average velocity vectors are constructed for two successive time intervals. In Fig. 3-35, the difference of these two velocities is used to determine the average acceleration. As before, the dashed white vector is moved without altering its length or direction, so that the tails of the two velocity vectors coincide. The difference, which is the average acceleration, is the solid white vector. Note its magnitude and direction and then compare those values with the magnitude and direction of the average acceleration found for the circular-motion experiment in Fig. 3-32.

The comparison shows that the accelerations are of the same magnitude, within experimental accuracy, and that both point toward the center. This is true even though the different initial conditions have led to quite different motions.

In view of the connection between force and acceleration established in Sec. 1-3, this result should not be a mystery. If the tension in the string is the same in both experiments, then in both the force acting on the puck is the same—the force has a certain fixed magnitude and is always directed inward toward the center of the air table. Thus in both experiments the puck should have an acceleration which is the same in magnitude and is always directed toward the center of the air table. In the second experiment (Figs. 3-33 through 3-35) the acceleration is of the more obvious kind illustrated in Fig. 3-28*a,* where **a** is parallel to $\mathbf{v}_i$ so that the magnitude of the velocity changes but its direction remains constant. In the first experiment (Figs. 3-30 through 3-32) **a** is perpendicular to $\mathbf{v}_i$, as in Fig. 3-28*b*. Consequently, the direction of the velocity changes, but its magnitude remains constant. But in both cases there is an acceleration resulting from a change in a property of the velocity vector.

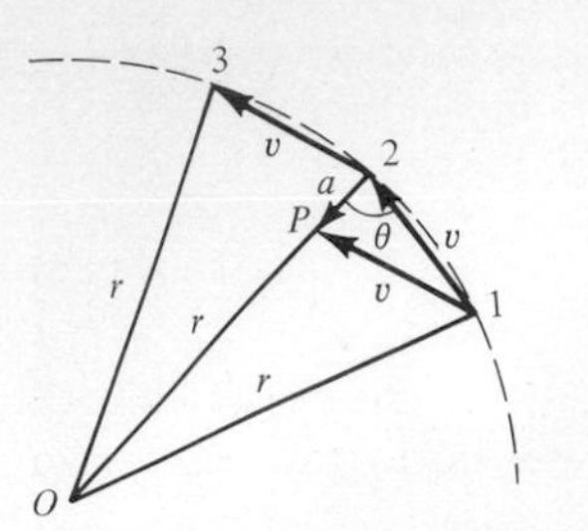

Fig. 3-36 Idealized representation of the experimental analysis of Fig. 3-32. A body moves in a circular path of radius r, shown by the dashed curve. The center of the circle is at O. In an unspecified unit time interval, the body passes from position 1 to position 2, and in an equal time interval it passes from position 2 to position 3. The vectors of magnitude v from 1 to 2 and from 2 to 3 are the average velocities (displacements per unit time interval) over those intervals. The vector of magnitude a (drawn from 2 to P) represents the average acceleration of the body from the first time interval to the second. The angle θ is a base angle both of the isosceles triangle $12P$ and of the isosceles triangle $O12$. The triangles are thus similar, and $a/v = v/r$, or $a = v^2/r$.

So far we have been concerned with experimental measurements on a body moving uniformly in a circle. Now we will devise a theoretical account of this motion. The diagram in Fig. 3-36 follows very closely the construction in Fig. 3-32. It shows successive positions, labeled 1, 2, 3, of a body moving uniformly in a circular path centered on the point O. The time interval between positions 1 and 2 is the same as that between positions 2 and 3. These positions are connected to each other, and to the center of the circle, by straight lines. As before, the length of the line from 1 to 2 denotes the magnitude v of the average velocity vector over the first time interval. (We continue to use the duration of the time interval as a convenient unit to define velocities and accelerations.) The magnitude of the average velocity over the second time interval has the same value v, although the direction of this average velocity is different. The second average velocity vector is also shown after having been moved parallel to itself so as to bring its tail to point 1. The figure shows that its head is then at point P, a point located on the radial line $O2$. (It can be shown on the basis of symmetry considerations that P actually lies on that line, and you should do this yourself later.)

The average acceleration between the two successive time intervals is the vector from point 2 to point P, and its magnitude a is the length of that vector. We want to determine how a is related to v and to the radius r of the circle. This relation depends on the fact that the triangle $12P$ is *similar* to the triangle $O12$. [That is, both triangles are of the same shape and differ only in size. This is true because both triangles are *isosceles* (each triangle has two sides of equal length) and they have a common base angle—the angle $12P$ or $12O$. Two isosceles triangles of the same base angle are similar.]

The ratios of corresponding sides of similar triangles are equal. Thus a/v for triangle $12P$ is equal to v/r for triangle $O12$. That is,

$$\frac{a}{v} = \frac{v}{r}$$

or

$$a = \frac{v^2}{r} \tag{3-41a}$$

The symbols a and v in this relation represent the *average* values of the magnitudes of the acceleration and the velocity. However, the argument leading to Eq. (3-41a) imposes no special conditions on the (equal) time intervals required for the body of Fig. 3-36 to pass from position 1 to position 2 and from position 2 to position 3. While practical difficulties would prevent the reduction of the strobe flash interval in the experiment of Figs. 3-30 through 3-32 to an arbitrarily small value, no such difficulty stands in the way of the theoretical analysis of Fig. 3-36. Reducing the time interval to an infinitesimally small value leads to an infinitesimal separation of points 1, 2, and 3 in the figure, so that there is no longer a distinction to be made between the average and instantaneous magnitudes of the velocity and of the acceleration. But reducing the time interval makes no change at all in the argument leading to Eq. (3-41a). Thus Eq. (3-41a) is an *exact* relation between the *instantaneous* magnitudes a and v, and the value of r. Furthermore, for any time interval the direction of the average acceleration is always toward the center of the circle. Hence the same is true of the direction of the instantaneous acceleration obtained from the argument when the time interval is infinitesimal. Thus we can write a vector expression for the **centripetal acceleration $\mathbf{a}_c$** in the form

$$\mathbf{a}_c = -\frac{v^2}{r}\hat{\mathbf{r}} \tag{3-41b}$$

where $\hat{\mathbf{r}}$ is a unit vector pointing from the center of the circle to the instantaneous location of the moving body. We rederive this important equation in Chap. 9, using a very different and more formal argument.

The derivation given above is only slightly modified from the original form in which it was given by Christian Huygens (1629–1695). You will see in Chap. 11 the important role which it played in Newton's development of mechanics. Huygens' life overlapped those of both Galileo (1565–1642) and Newton (1642–1727), and his contributions to physics and technology were prolific and wide-ranging. Although he was Dutch, he spent much of his working life in Paris, and he was a principal contributor to the establishment of modern physics in France.

Example 3-9 is an application of Eq. (3-41*b*).

EXAMPLE 3-9

A child whirls a stone, tied to a string, around her head at a speed of 6.0 m/s. The stone describes a horizontal circle of radius 0.75 m. What is the centripetal acceleration of the stone?

■ Use Eq. (3-41*b*) to obtain

$$\mathbf{a}_c = -\frac{(6.0\ \text{m/s})^2}{0.75\ \text{m}}\hat{\mathbf{r}} = -48\ \text{m/s}^2 \times \hat{\mathbf{r}}$$

That is, the magnitude of the acceleration is 48 m/s², and its direction is inward. The magnitude a_c is relatively large, being 4.9 times the magnitude of the gravitational acceleration $g = 9.8\ \text{m/s}^2$.

3-6 THE MINIMUM-ORBIT EARTH SATELLITE

The idea of launching a **satellite** into orbit around the earth originated with Newton, although nearly three centuries had to elapse before the actuality came within range of available technology. Newton's illustration of satellite motion is reproduced in Fig. 3-37. The trajectories of cannonballs, launched parallel to the surface of the earth from a mountaintop, are shown for a set of initial velocities of increasing magnitude. The illustration speaks for itself!

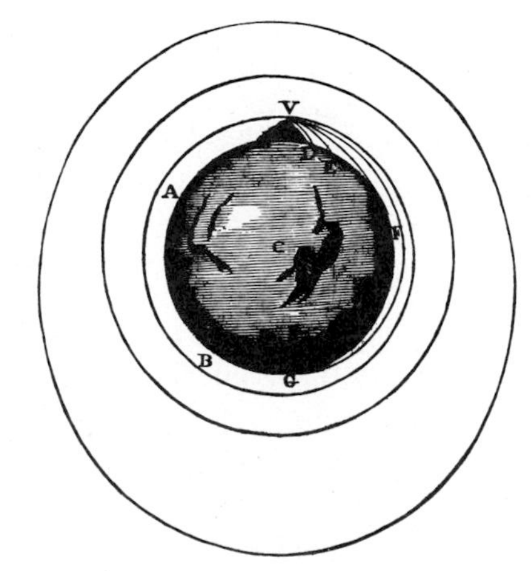

Fig. 3-37 A hypothetical method for launching an earth satellite, as illustrated by a figure in Newton's *Principia Mathematica*. A cannon on a mountaintop fires a series of cannonballs horizontally, with successively increasing muzzle speeds. The shortest trajectory is close to a parabola (it would be exactly a parabola if the earth were flat or if the range were short enough that the curvature of the earth did not need to be taken into consideration). If the muzzle speed is great enough, the cannonball clears the earth and returns to its starting point. Two orbits are shown for satellites launched from points higher above the earth than the mountaintop. (*Courtesy of the New York Public Library.*)

An earthbound object, like the puck on an air table in Fig. 1-5, maintains a constant velocity if no net force acts on it. The same is true of a satellite. It would move in a straight line with constant speed—that is, it would move with constant velocity—*if* there were no net force acting on it. But there *is* a net force—the gravitational force—exerted on the satellite by the earth in the direction toward the center of the earth. So the satellite must accelerate toward the center of the earth. If it did not have a tangential velocity, the satellite would do this by falling directly toward the center of the earth, in analogy to the motion of the air table puck in Fig. 3-33. The magnitude of its velocity would thus change, but not the direction. Instead, the satellite accelerates toward the center of the earth (despite the fact that its distance from the center of the earth is constant) because it has a tangential velocity which is continually changing in direction, but not in magnitude. The satellite's motion is analogous to that of the air table puck in Fig. 3-30.

The easiest satellite orbit to achieve is an approximately circular orbit just barely high enough above the earth's surface to avoid excessive friction from air resistance. In this **minimum earth orbit,** at an altitude of about 160 km, air resistance is sufficiently small to allow the satellite a lifetime of a few weeks.

What is the speed of such a satellite? What is its period of revolution (the time required for it to make one trip around its orbit)? In the minimum-orbit case, the satellite is so close to the surface of the earth, in comparison to the earth's radius, that it is a good approximation to equate the magnitude a_c of its acceleration to g, the magnitude of the gravitational acceleration it would have immediately above the earth's surface. Thus we can set $a_c = g$ in Eq. (3-41*a*), $a_c = v^2/r$, and write the magnitude of the acceleration as

$$g = \frac{v^2}{r}$$

where r is the radius of the satellite's circular orbit. Solving for its speed v, we obtain

$$v = \sqrt{rg} \tag{3-42}$$

To the same degree of approximation, we can set r equal to the earth's radius. The original definition of the meter tells us immediately that $2\pi r/4 = 10^7$ m, so that $r = 6.4 \times 10^6$ m. We thus have

$$v = (6.4 \times 10^6 \text{ m} \times 9.8 \text{ m/s}^2)^{1/2}$$

or

$$v = 7.9 \times 10^3 \text{ m/s} = 7.9 \text{ km/s}$$

In order to find the period of revolution T, note that it is just the distance once around the orbit divided by the speed of the satellite. Thus we have

$$T = \frac{2\pi r}{v} \tag{3-43}$$

Substituting numerical values into this equation gives

$$T = \frac{2\pi \times 6.4 \times 10^6 \text{ m}}{7.9 \times 10^3 \text{ m/s}}$$

$$= 5.1 \times 10^3 \text{ s} = 85 \text{ min}$$

or just a little under an hour and a half. The actual initial period of Sputnik 1 (launched October 4, 1957) was 96 min. The discrepancy is due to the fact that the orbit was not exactly circular.

Sputnik 10, launched about $3\frac{1}{2}$ years later, had a much more closely circular orbit. Its maximum and minimum altitudes were 247 km and 175 km, so that the departure from circularity was only about 1 percent. Its orbit period was 5142 s (85 min 42 s), in close agreement with the prediction of Eq. (3-43).

The period T can be expressed directly in terms of r and g by substituting Eq. (3-42) into Eq. (3-43). This gives

$$T = \frac{2\pi r}{\sqrt{rg}}$$

or

$$T = 2\pi \sqrt{\frac{r}{g}} \tag{3-44}$$

The magnitude a_c of the centripetal acceleration of an earth satellite in a circular orbit of *any* radius can be expressed directly in terms of r and T. From Eq. (3-43) we have

$$v^2 = \frac{4\pi^2 r^2}{T^2} \tag{3-45}$$

Substituting this value into Eq. (3-41a), $a_c = v^2/r$, we obtain

$$a_c = \frac{4\pi^2 r}{T^2} \tag{3-46}$$

Example 3-10 applies Eq. (3-46) to the very important practical case of a communications satellite.

EXAMPLE 3-10

The period of the synchronous satellites used for long-distance communication is 23 h 56 min (h stands for hours). This is the time required for the earth to make one rotation with respect to the "fixed stars." The orbit of the satellite is circular. If the orbit lies in the same plane as the earth's equator, and the direction of revolution of the satellite around the earth is the same as that of the rotation of the earth about its own axis, the satellite appears to be permanently suspended above some point on the equator. (This greatly facilitates the aiming of antennas on the earth at the satellite and of antennas on the satellite at fixed earth stations.) The altitude of such a satellite is about 35,800 km. Find the acceleration of gravity at this altitude.

■ Equation (3-46) will yield the desired magnitude a_c of the acceleration due to gravity. The period T is given, but you must determine the value of the orbit radius r. You add the distance from the center of the earth to its surface, 6.4×10^6 m, to the altitude, or distance from the earth's surface to the satellite orbit, 35.8×10^6 m. This gives you $r = 42.2 \times 10^6$ m. You then convert 23 h 56 min to seconds, and obtain the value $T = 8.62 \times 10^4$ s for the period of the satellite. Using these figures in Eq. (3-46) gives you the acceleration

$$a_c = \frac{4\pi^2 \times 42.2 \times 10^6 \text{ m}}{(8.62 \times 10^4 \text{ s})^2} = 0.224 \text{ m/s}^2$$

This is smaller than the acceleration of gravity near the earth's surface, $g = 9.80$ m/s^2, by a factor of 2.29×10^{-2}.

You will see in Chap. 11 how Newton used a natural earth satellite (the moon) in an approximately circular orbit of radius 4×10^8 m to evaluate a_c at that distance from the earth, and the crucial role this evaluation played in Newton's development of the "law of gravity."

3-7 THE CONICAL PENDULUM AND THE BANKING OF CURVES

In Example 3-9 we considered the motion of a stone on the end of a string, being whirled in a circular orbit. Let us now consider the entire system—the string as well as the stone.

Imagine what happens as the child puts her stone into motion. At first the stone hangs vertically. As she begins to swing it, it describes a circular path, so that the string itself describes a cone. As the stone moves faster and faster, it describes a wider and wider circle, so that the string describes a broader and broader cone.

Let us consider the stone when it has been spun up to a particular speed. If the child then holds her end of the string still, the stone will continue indefinitely along the same circular pathway, neglecting friction.

This system is called a **conical pendulum.** It is shown in perspective in Fig. 3-38*a* and in side view at a particular instant of time in Fig. 3-38*b*. The length of the string is l, and it makes an angle θ with the vertical. The vector $\mathbf{r}$ locates the stone, which we call the pendulum bob, with respect to the center of its circular path.

If the string were suddenly cut, the bob would instantaneously acquire a downward acceleration $\mathbf{g}$. If, on the other hand, gravity could suddenly be "turned off," the tension in the string at that instant would result in an instantaneous acceleration $\mathbf{a}_s$ of the bob in the direction along the string; $\mathbf{a}_s$ is shown in Fig. 3-38*b*.

Since the bob neither ascends nor descends, its actual vertical acceleration must be zero. Thus $\mathbf{a}_{sy}$, the vertical constituent vector of $\mathbf{a}_s$, must be

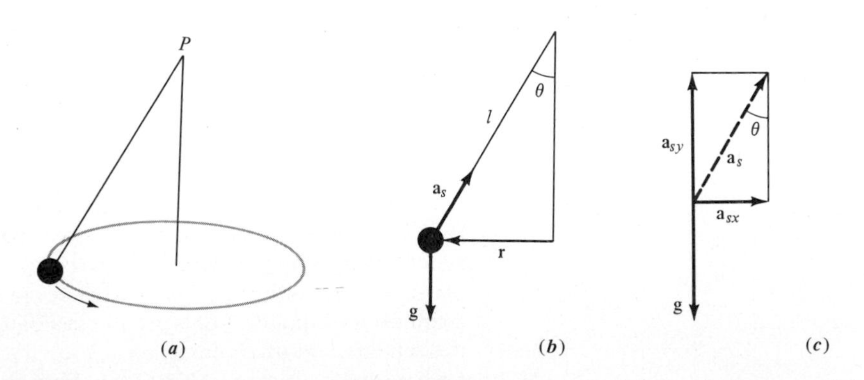

Fig. 3-38 (*a*) A conical pendulum. The bob whirls in a horizontal circle. The string, which is supported at point P, describes a cone. (*b*) A "side view" of the system at a particular instant. The horizontal circular orbit of the bob has radius r; its instantaneous location with respect to the center of the orbit is specified by the vector $\mathbf{r}$. The string, of length l, makes an angle θ with the vertical. In the absence of the string tension, the bob would fall with vertical acceleration $\mathbf{g}$. If gravity were suddenly to disappear, the tension in the string would result in an instantaneous acceleration of the bob given by the vector $\mathbf{a}_s$, whose direction is along the string. (*c*) The vectors $\mathbf{a}_s$ and $\mathbf{g}$ are shown, together with the horizontal and vertical constituent vectors of $\mathbf{a}_s$, which are, respectively, $\mathbf{a}_{sx}$ and $\mathbf{a}_{sy}$.

equal in magnitude to $\mathbf{g}$ and oppositely directed, so that $\mathbf{a}_{sy} + \mathbf{g} = 0$. See Fig. 3-38$c$, where we have constructed the horizontal and vertical constituent vectors $\mathbf{a}_{sx}$ and $\mathbf{a}_{sy}$ of $\mathbf{a}_s$. The condition that the two vertical accelerations add to zero gives the following relation between a_{sy}, the magnitude of $\mathbf{a}_{sy}$, and g, the magnitude of $\mathbf{g}$:

$$a_{sy} = g \tag{3-47}$$

But the figure shows that $a_{sy} = a_s \cos \theta$, where a_s is the magnitude of $\mathbf{a}_s$. Thus we have

$$a_s \cos \theta = g \tag{3-48}$$

The horizontal constituent vector $\mathbf{a}_{sx}$ of the vector $\mathbf{a}_s$ is the centripetal acceleration associated with the circular motion of the pendulum bob in the horizontal plane. The vector $\mathbf{a}_{sx}$ has the direction $-\hat{\mathbf{r}}$, and its magnitude is $a_{sx} = a_s \sin \theta$. If the speed of the bob is v and r is the radius of its circular orbit, then Eq. (3-41a) requires that

$$a_{sx} = \frac{v^2}{r} \tag{3-49}$$

or

$$a_s \sin \theta = \frac{v^2}{r} \tag{3-50}$$

If we divide Eq. (3-50) by Eq. (3-48), we obtain

$$\frac{a_s \sin \theta}{a_s \cos \theta} = \frac{v^2}{rg}$$

or

$$\tan \theta = \frac{v^2}{rg} \tag{3-51}$$

This relation will appear again shortly in what will seem at first to be a totally different situation.

Equation (3-51) can be inconvenient because it contains three independent variables, v, r, and θ. Note, however, that $r = l \sin \theta$. Consequently, we can write Eq. (3-51) in the form

$$\tan \theta \sin \theta = \frac{v^2}{gl} \tag{3-52}$$

which contains only two variables, θ and v.

A still more convenient expression of this relation can be written in terms of the period T (the time required by the bob to complete one orbit) rather than the speed of the bob v. Substituting $v^2 = 4\pi^2 r^2/T^2$ [see Eq. (3-45)] into Eq. (3-51), we obtain

$$\tan \theta = \frac{4\pi^2 r}{gT^2}$$

or

$$T = 2\pi \sqrt{\frac{r}{g \tan \theta}}$$

Again making the substitution $r = l \sin \theta$, we have

$$T = 2\pi \sqrt{\frac{l}{g} \cos \theta} \tag{3-53}$$

This equation tells us that the period of the pendulum depends on the angle θ. The dependence is weak, however, since T depends on the square root of the cosine of θ, which does not change very rapidly as θ changes. You will find it worthwhile to check on this behavior qualitatively with a makeshift pendulum and a watch.

The design of highway curves may seem a far cry from a child swinging a stone on a string. However, there is a close connection between the two, as far as the analysis of the motion is concerned.

If a car is driven around an unbanked curve too fast, it will skid outward. The reason for this is easily understood in terms of the centripetal acceleration present when the car (which we assume to have a constant speed) follows a curved path. If the radius of the curve is r, there is a centripetal acceleration of magnitude v^2/r.

But an acceleration implies the existence of a *net force acting in the direction of the acceleration,* as we learned in Sec. 1-3. What can supply this force to the car, in a direction perpendicular to that of its forward motion? It must be the roadway, which exerts an inward radial force on the tires (in addition to the vertical force it always exerts on the tires in order to hold the car up). If the car is going too fast for the curve, the necessary centripetal acceleration will require an inward radial force in excess of the maximum frictional force which can be exerted by the road on the tires. The car then skids. That is, the curve it follows is gentler (has a larger radius) than the curve of the road. Clearly, the result can be disastrous.

Even if the car does not skid, going around unbanked curves can be an uncomfortable process for passengers (and a potentially damaging one for cargo). Just as the roadway must exert a force on the car to accelerate it inward, the car must apply a force to the passengers in order to accelerate them inward, mostly through the seats of their pants. If various parts of a passenger's body are to follow, a similar force must be applied to them by means of appropriate muscle tensions. But the muscles of the back are not very well adapted to pulling sideways on the shoulders and head. There is a distinct improvement in comfort (to say nothing of safety and economy) if the direction of the force can be altered to lie parallel to the spine.

The general method for producing this alteration is to tilt the road inward, so that the road presses against the car in a nonvertical direction. This inclination of the roadway, called **banking,** is shown in Fig. 3-39a.

We could analyze the situation from scratch, but it is possible to save repetitive work by making a simple observation. The road surface serves the same purpose here as does the string of the conical pendulum—it prevents the object in question (in this case the car and its contents) from moving in the vertical direction. Indeed, let us perform the following operation in our mind's eye: From a point P directly above the center of curvature of the road, extend a cable downward to the body of the car, adjusting the height of the cable at point P so that it is exactly perpendicular to the properly banked road, as in Fig. 3-39b. The cable then makes an angle θ with the vertical, just as the road does with the horizontal. Increase

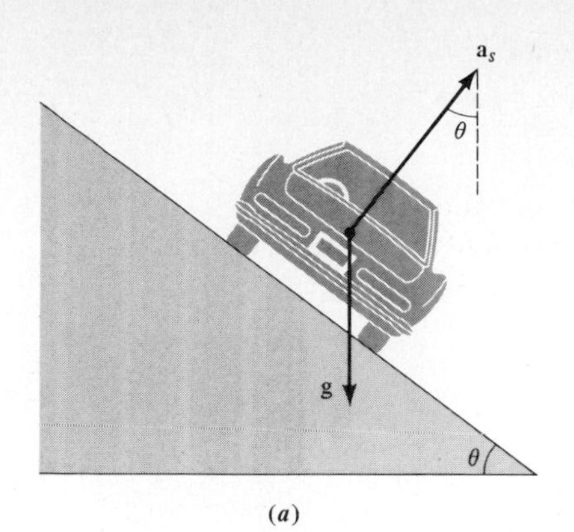

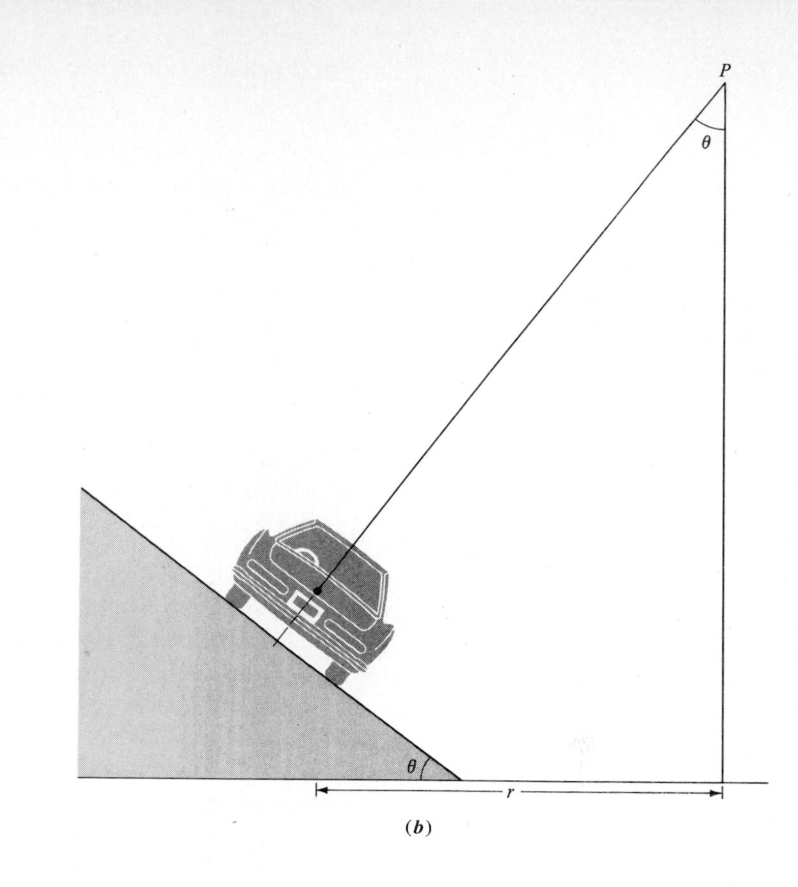

Fig. 3-39 (*a*) An automobile on a banked roadway. In the absence of the support of the roadway, the automobile would fall with vertical acceleration **g**. If gravity were suddenly to disappear, the pressure exerted by the roadway on the automobile would result in an acceleration of the automobile given by the vector $\mathbf{a}_s$, whose direction is in the plane of the page and perpendicular to that of the roadway. (*b*) In imagination, a cable is attached to the automobile and supported from a point *P* located above the center of curvature of the roadway, at such a height that the cable is perpendicular to the roadway. The tension in the cable is increased until it entirely supports the automobile, and the roadway is removed. The automobile thus becomes the bob of a conical pendulum.

the tension in the cable until the road exerts zero force on the car; then remove the road. This thought process converts the car into a conical pendulum, and the applicable equation is Eq. (3-51),

$$\tan\theta = \frac{v^2}{rg} \tag{3-54}$$

At this inclination angle θ, the passengers are tilted so that the forces exerted on them by the seats of the car are parallel to their spines.

EXAMPLE 3-11

Find the magnitude of the acceleration $\mathbf{a}_s$ in Fig. 3-39*a*. What will be the magnitude and direction of $\mathbf{a}_s$ if $r = 100$ m and $v = 15.0$ m/s (about 34 mi/h)?

■ Since a car moving on a properly banked road is equivalent to the bob of a conical pendulum, the conditions governing the vector $\mathbf{a}_s$ in Fig. 3-39*a* are the same as those governing the equivalent vector $\mathbf{a}_s$ for the conical pendulum, shown in Fig. 3-38*b*. That is, the horizontal and vertical components of $\mathbf{a}_s$ must be those given by Eqs. (3-49) and (3-47),

$$a_{sx} = \frac{v^2}{r} \qquad \text{and} \qquad a_{sy} = g$$

To find the magnitude of the vector $\mathbf{a}_s$, you use the pythagorean theorem:

$$a_s = (a_{sx}^2 + a_{sy}^2)^{1/2}$$
$$= \left[\left(\frac{v^2}{r}\right)^2 + g^2\right]^{1/2} = \left(\frac{v^4}{r^2} + g^2\right)^{1/2} \tag{3-55}$$

For the numbers given, you have the magnitude

$$a_s = \left[\frac{(15.0 \text{ m/s})^4}{(100 \text{ m})^2} + (9.80 \text{ m/s}^2)^2\right]^{1/2} = 10.1 \text{ m/s}^2$$

or about 3 percent more than the ordinary acceleration of gravity.

You can solve Eq. (3-54) for θ and use the result to calculate the ideal banking angle. You have

$$\theta = \tan^{-1}\frac{v^2}{rg}$$

or

$$\theta = \tan^{-1}\frac{(15.0 \text{ m/s})^2}{100 \text{ m} \times 9.80 \text{ m/s}^2} = 12.9°$$

Thus the direction of $\mathbf{a}_s$ is about 13° from the vertical. This is a relatively steep angle of bank under ordinary roadway conditions.

3-8 THE GALILEAN TRANSFORMATIONS

There are many occasions when two persons who are moving with respect to each other observe the same phenomenon and then use the laws of physics to analyze their observations. In order to compare their results, the two observers must be able to describe their observations in mutually intelligible terms. The rules which they must employ in doing so depend on how they are moving with respect to each other. In this section we consider the simplest and most important case, in which one observer moves at *constant velocity* with respect to the other. In such a case these rules are called the **Galilean transformations.**

For a specific example, let observer O be stationed at the side of a straight road. Observer O' is in a car moving at constant speed along the road. The phenomenon they both observe is the motion of a car C, which is traveling along the road in the same direction as O' but at a higher speed. Both O' and C pass O at the same instant. Figure 3-40a shows the situation, *as seen by* O. She measures the position of O' with respect to the origin of her reference frame at instants separated by equal time intervals. Since the situation is one-dimensional, she is free to describe the results by using either signed scalar quantities or vector quantities. She chooses signed scalars. Thus she describes the position of O' by means of the coordinate x, taking the values of x to be positive when the direction from her coordinate origin to O' is the same as the direction in which O' is moving. She finds that the displacement of O' during each of the equal time intervals Δt has the same value ΔX. So she concludes that the velocity of O' with respect to herself has the constant value $V = \Delta X/\Delta t$.

While this is happening, O also observes car C. She finds that the displacement of C has the same value Δx for each time interval Δt. Thus she concludes that the velocity of C with respect to herself has the constant value $v = \Delta x/\Delta t$.

What result does O' obtain when he observes the velocity of C with respect to himself? Figure 3-41 depicts the motion of C, *as seen by* O'. He describes the position of C in terms of its coordinate x' measured from his coordinate origin, taking the positive direction to be the same as that chosen by O. Using the same time interval Δt which O uses, he finds the displacement of car C to be $\Delta x'$ during each time interval. He therefore finds the velocity of C with respect to himself to have the constant value $v' = \Delta x'/\Delta t$.

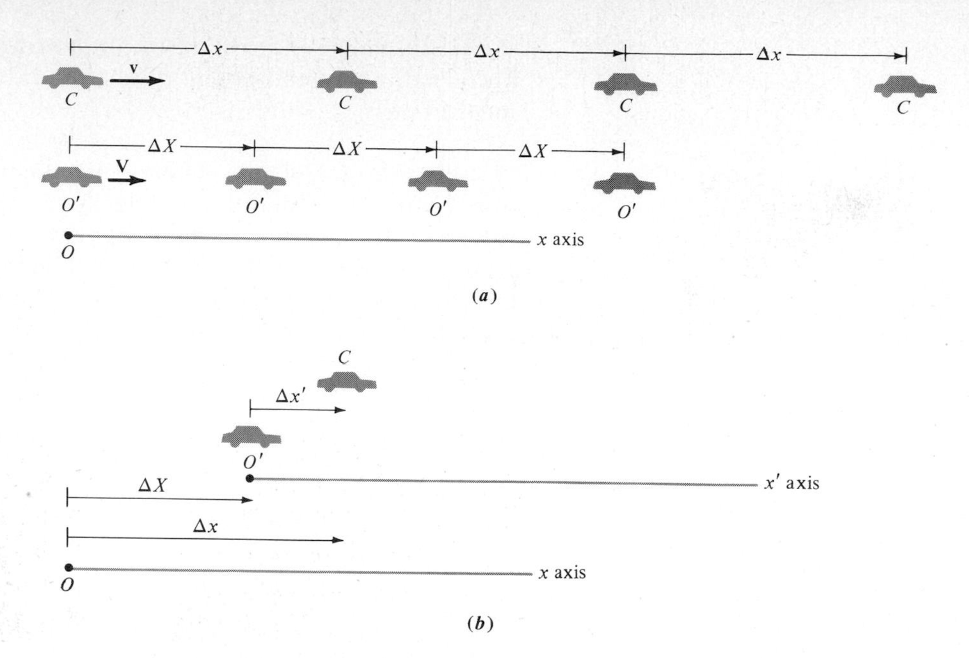

Fig. 3-40 (*a*) Two cars, one containing observer O' and the other known as car C, move parallel to the positive x axis of the reference frame of observer O. They pass O at the same instant. She measures their positions at the ends of equal time intervals Δt after they pass her. She finds that O' has a constant velocity $V = \Delta X/\Delta t$ and that C has a constant velocity $v = \Delta x/\Delta t$. (*b*) The positions of O' and C at the end of the first time interval Δt, showing that $\Delta x = \Delta X + \Delta x'$ so that $\Delta x' = \Delta x - \Delta X$.

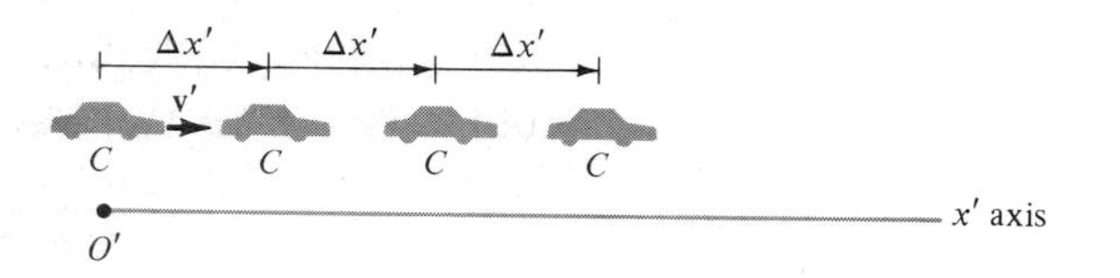

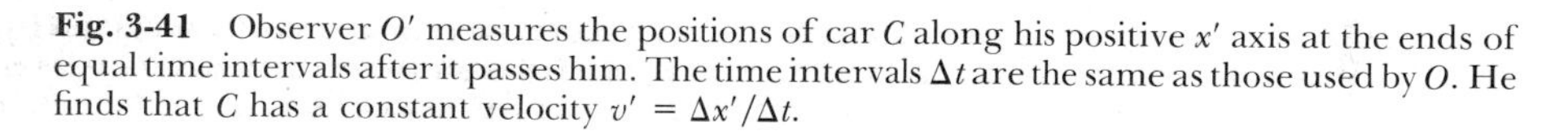

Fig. 3-41 Observer O' measures the positions of car C along his positive x' axis at the ends of equal time intervals after it passes him. The time intervals Δt are the same as those used by O. He finds that C has a constant velocity $v' = \Delta x'/\Delta t$.

Figure 3-40*b* shows how O can deduce from her own measurements what value of v' will be measured by O'. She notes that

$$\Delta x' = \Delta x - \Delta X$$

Dividing both sides of this equation by Δt, she has

$$\frac{\Delta x'}{\Delta t} = \frac{\Delta x}{\Delta t} - \frac{\Delta X}{\Delta t}$$

Since all the velocities involved are constant, she can write the last equation in the form

$$v' = v - V$$

Thus she concludes that v' (the velocity of C as measured by O') is the difference between v (the velocity of C as she measures it) and V (the velocity of O' as she measures it). This relation is a special case of the Galilean velocity transformation. She uses it to predict the value of v' and then communicates the result to O'. He compares the prediction with the result of his own direct measurement of v' and finds the prediction agrees with the measurement.

We will now obtain general expressions for the Galilean transformations from a consideration of Fig. 3-42. This figure shows the unprimed reference frame whose origin is O. At an arbitrary fixed location in this frame is an observer O (not shown). The figure also shows the primed reference frame whose origin is O'. Fixed at an arbitrary location in this frame is another observer O' (also not shown). (It is customary to denote the origin of a frame of reference by the same symbol used for the observer fixed to that frame, even though the observer is not necessarily stationed at the origin.)

As long as two conditions are satisfied, observer O' will move at constant velocity $\mathbf{V}$ with respect to observer O. The first of these conditions is that the origin O' of the primed reference frame must move at constant velocity $\mathbf{V}$ with respect to the origin O of the unprimed frame. The second condition is that neither of the reference frames may rotate about its own origin, as seen from the other. (If there were such rotation, the observers could not be moving at constant velocity with respect to each other unless they happened to be at the origins of their respective frames.) The figure is drawn for a case in which the axes of the primed frame are parallel to the corresponding axes of the unprimed frame. However, this is not essential to the argument which follows. Indeed, it is not even necessary to define specific axes. We assume, for convenience, that the origins of the two reference frames coincide precisely at a certain instant, which we call $t = 0$. That is, at the time $t = 0$ (but *only* then) any vector $\mathbf{r}$ which locates a body B with respect to the origin O is identical with the vector $\mathbf{r}'$ which locates B with respect to the origin O'.

At any other time t, the position of origin O' relative to origin O is given by the vector $\mathbf{V}t$ shown in the figure. This is so since O' is moving with respect to O at the velocity $\mathbf{V}$. At time t the observers stationed in their ref-

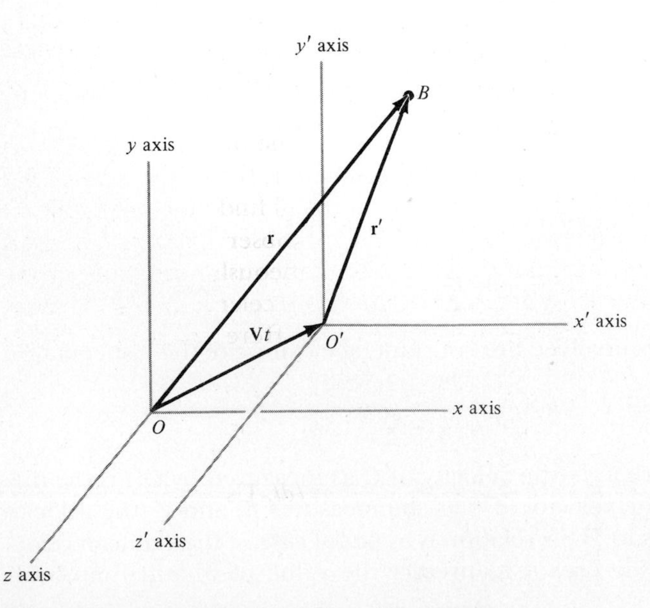

Fig. 3-42 A pair of three-dimensional frames of reference moving relative to each other at constant velocity. An observer in the unprimed frame measures the velocity of the primed frame and finds it to have the arbitrary but constant magnitude and direction which are expressed by the velocity vector $\mathbf{V}$ (not shown). If the two reference frames coincide at the instant $t = 0$, the position of the origin O' is given relative to that of origin O at any time t by means of the position vector $\mathbf{V}t$. At that time the observers in the two frames simultaneously measure the position of body B relative to the origins of their frames, expressing the results by the position vectors $\mathbf{r}$ and $\mathbf{r}'$. The figure shows that $\mathbf{r} = \mathbf{r}' + \mathbf{V}t$, so that $\mathbf{r}' = \mathbf{r} - \mathbf{V}t$.

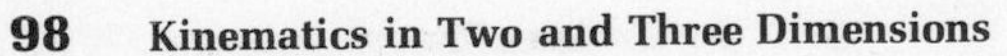

erence frames simultaneously measure the position of body B. The observer in the unprimed frame will represent that position by the vector $\mathbf{r}$ extending from O to B, while the one in the primed frame will represent it by the vector $\mathbf{r}'$ extending from O' to B. Inspection of the figure (which is just a three-dimensional generalization of Fig. 3-40*b*) shows that the relation between the two position vectors is

$$\mathbf{r}' = \mathbf{r} - \mathbf{V}t \tag{3-56}$$

The comparison of position made possible by this equation is called the **Galilean position transformation,** a name often given to the equation itself.

The rules for differentiating vector quantities are the same as those for differentiating scalar quantities. We can differentiate each term in Eq. (3-56) with respect to time to obtain

$$\frac{d\mathbf{r}'}{dt} = \frac{d\mathbf{r}}{dt} - \frac{d(\mathbf{V}t)}{dt}$$

But we have stipulated that the velocity $\mathbf{V}$ of O' with respect to O be a constant. Hence we can use the rules for differentiation given by Eq. (2-13), and by Eq. (2-16), to obtain

$$\frac{d\mathbf{r}'}{dt} = \frac{d\mathbf{r}}{dt} - \mathbf{V}\frac{dt}{dt}$$

and then

$$\frac{d\mathbf{r}'}{dt} = \frac{d\mathbf{r}}{dt} - \mathbf{V}$$

Employing the definitions $\mathbf{v} = d\mathbf{r}/dt$ and $\mathbf{v}' = d\mathbf{r}'/dt$, we have

$$\mathbf{v}' = \mathbf{v} - \mathbf{V} \tag{3-57}$$

The comparison of velocity made possible by this equation is called the **Galilean velocity transformation.** Stated in words, it says *the velocity of a body with respect to a primed frame is the vector difference between its velocity with respect to an unprimed frame and the constant velocity of the primed frame with respect to the unprimed frame.*

An important special consequence of Eq. (3-57) is that when $\mathbf{v}$ is constant, then $\mathbf{v}'$ is also constant. That is, *if a primed reference frame moves at constant velocity with respect to an unprimed frame, then a body observed to move relative to the unprimed frame at constant velocity will be observed to move relative to the primed frame with a different, but still constant, velocity.*

In general, the observer in the unprimed frame of reference will not find the velocity $\mathbf{v}$ of a body B to be constant. Rather, the value of $\mathbf{v}$ will depend on when it is measured. Thus the observer will find body B in general to have a nonzero acceleration $\mathbf{a}$. Similarly, the observer in the primed frame will find from measurements, made simultaneously with those of the observer in the unprimed frame, that body B has acceleration $\mathbf{a}'$. The connection between $\mathbf{a}$ and $\mathbf{a}'$ can be determined by differentiating Eq. (3-57) with respect to time. This gives

$$\frac{d\mathbf{v}'}{dt} = \frac{d\mathbf{v}}{dt} - \frac{d\mathbf{V}}{dt}$$

Since $\mathbf{V}$ is constant, $d\mathbf{V}/dt = 0$. Thus $d\mathbf{v}'/dt = d\mathbf{v}/dt$. Using the definitions $\mathbf{a} = d\mathbf{v}/dt$ and $\mathbf{a}' = d\mathbf{v}'/dt$, we obtain

$$\mathbf{a}' = \mathbf{a} \tag{3-58}$$

In words, *if a primed reference frame moves at constant velocity with respect to an unprimed frame, then the acceleration of a body will be observed to have the same value with respect to either frame.* This equality of accelerations is called the **Galilean acceleration transformation.**

It was Einstein who first named the transformations given by Eqs. (3-56), (3-57), and (3-58) in honor of Galileo. They have an intimate connection with Newton's laws of motion. This connection is discussed in Chap. 4.

Examples 3-12 and 3-13 illustrate the application of the Galilean transformation equations.

EXAMPLE 3-12

Observer O drops a stone from the top of a skyscraper. Observer O', riding in an elevator, starts down from the top of the skyscraper at the instant when the stone is dropped. The elevator accelerates very quickly to a downward velocity of magnitude $V = 5.0$ m/s and then maintains that velocity steadily. At the time $t = 3.0$ s after the stone is dropped, find the position, the velocity, and the acceleration of the stone relative to O. Then find the position, the velocity, and the acceleration of the stone relative to O'.

■ Since the problem is essentially one-dimensional, you can use the signed scalar representation of one-dimensional vectors, as discussed in Sec. 3-2. You then represent the vectors **r**, **v**, and **a** by the scalars x, v, and a, and use Eq. (2-30),

$$x = x_i + v_i t + \frac{at^2}{2}$$

to find the position x. Taking $x = 0$ at the top of the skyscraper and taking the downward direction as the positive x direction, you have $x_i = 0$, $v_i = 0$, $a = +g = +9.8\ \text{m/s}^2$, and thus

$$\begin{aligned} x &= 0 + 0 + \frac{9.8\ \text{m/s}^2 \times (3.0\ \text{s})^2}{2} \\ &= +44\ \text{m} \end{aligned}$$

The positive value of x denotes the downward direction.

Using Eq. (2-29),

$$v = v_i + at$$

you find the velocity of the stone with respect to O to be

$$v = 0 + 9.8\ \text{m/s}^2 \times 3.0\ \text{s} = +29\ \text{m/s}$$

Again, the positive value indicates a downward direction.

The acceleration of a freely falling body, as seen by the observer O who is stationary with respect to the earth, is known to be the constant gravitational acceleration. (Indeed, this underlies the validity of the two calculations immediately above.) You thus have

$$a = +g = +9.8\ \text{m/s}^2$$

Here the positive sign means the acceleration is downward.

You can now find x' by using Eq. (3-56) and the value of x just calculated. Substituting the signed scalar V for the vector **V**, you have

$$x' = x - Vt$$

or

$$x' = 44\ \text{m} - 5.0\ \text{m/s} \times 3.0\ \text{s} = +29\ \text{m}$$

That is, the stone is located 29 m below observer O' at the end of 3.0 s.

Substituting signed scalars for vectors in Eq. (3-57), you have

$$v' = v - V$$

Inserting the numerical values gives you

$$v' = 29 \text{ m/s} - 5.0 \text{ m/s} = +24 \text{ m/s}$$

Thus O' sees the velocity of the stone to be 24 m/s downward.

From Eq. (3-58), $\mathbf{a}' = \mathbf{a}$, you have

$$a' = +g = +9.8 \text{ m/s}^2$$

Observer O' sees the stone to have the same downward acceleration as that seen by O.

EXAMPLE 3-13

In order to reach his destination on schedule, an airline pilot wishes to fly over the ground at a "ground speed" of 250 m/s along a "course made good" which is in a northeasterly direction (in other words, 45.0° north of east). The weather bureau tells him that the wind is blowing due eastward with a speed of 30 m/s. Find the required "airspeed" of the airplane (its speed relative to the air through which it flies, read by the pilot on his airspeed indicator) and the required "heading" (the geographical direction in which the nose of the plane should be pointed, read by the pilot on his compass), if the plane is to reach the right place at the right time.

■ Although it is possible to think of the pilot as both the moving observer and the observed moving body, you may find it easier to analyze the problem if you make a separation as follows. Imagine an observer O' in a balloon which floats along with the wind. His velocity with respect to an observer O on the ground is $\mathbf{V}$ = 30 m/s (eastward). Since the airplane is also carried along by the wind as it moves through the air, O' will observe the proper airspeed and heading which comprise the velocity $\mathbf{v}'$. Can you give a more detailed explanation of why this is true? At the same time O observes the desired ground speed and course made good, $\mathbf{v}$ = 250 m/s (northeastward. You should make a sketch like that of Fig. 3-43 to depict the relationship among the velocities $\mathbf{v}$, $\mathbf{v}'$, and $\mathbf{V}$. The resulting vector diagram is the pictorial equivalent of Eq. (3-57), $\mathbf{v}' = \mathbf{v} - \mathbf{V}$. If you draw the diagram carefully to scale, you can measure the desired value of $\mathbf{v}'$ with a ruler and a protractor.

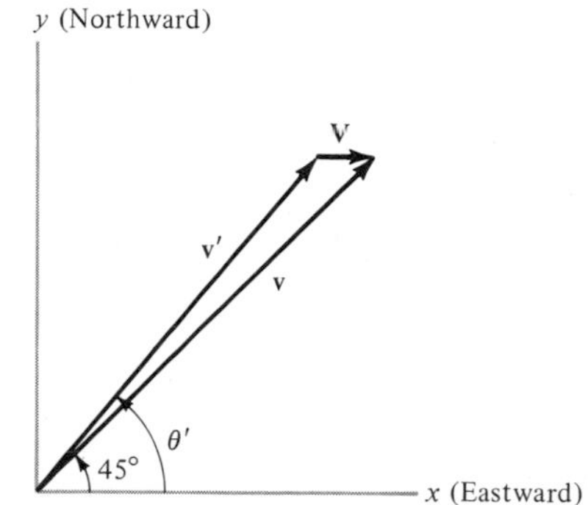
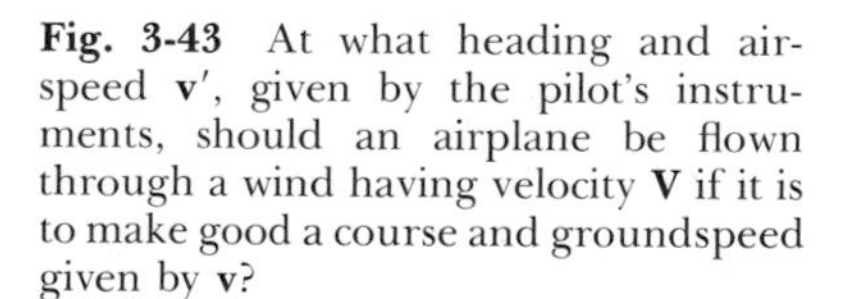

Fig. 3-43 At what heading and airspeed $\mathbf{v}'$, given by the pilot's instruments, should an airplane be flown through a wind having velocity $\mathbf{V}$ if it is to make good a course and groundspeed given by $\mathbf{v}$?

However, it is usually both easier and more accurate to obtain the desired result by algebraic calculation. As you saw in Sec. 3-2, the sum of the vectors $\mathbf{v}$ and $-\mathbf{V}$ required to solve Eq. (3-57) for $\mathbf{v}'$ can be found by adding the vectors component by component. Since the problem is essentially two-dimensional, you need only the x and y components, and Eq. (3-57) can be written in the equivalent component form as

$$v'_x = v_x - V_x \tag{3-59a}$$

and

$$v'_y = v_y - V_y \tag{3-59b}$$

To save some work, take the positive x direction to be eastward, that is, the direction in which O sees O' to be moving. The components of $\mathbf{V}$ then become

$$V_x = V = 30 \text{ m/s} \qquad \text{and} \qquad V_y = 0$$

Now calculate the x and y components of the desired ground velocity $\mathbf{v}$. You have

$$v_x = v \cos 45.0° = 250 \text{ m/s} \times \cos 45.0° = 177 \text{ m/s}$$

and

$$v_y = v \sin 45.0° = 250 \text{ m/s} \times \sin 45.0° = 177 \text{ m/s}$$

Inserting the numerical values of v_x and V_x into Eq. (3-59a), you obtain

$$v'_x = 177 \text{ m/s} - 30 \text{ m/s} = 147 \text{ m/s}$$

And the numerical values of v_y and V_y inserted into Eq. (3-59b) give you

$$v'_y = 177 \text{ m/s} - 0 = 177 \text{ m/s}$$

You can now calculate the magnitude and direction of the velocity $\mathbf{v}'$ of the plane as seen by O' in the balloon (and also of the plane with respect to the air as measured by the pilot's instruments). You have for the magnitude

$$v' = [(v'_x)^2 + (v'_y)^2]^{1/2} = [(147 \text{ m/s})^2 + (177 \text{ m/s})^2]^{1/2} = 230 \text{ m/s}$$

For the direction you have

$$\theta' = \tan^{-1}\frac{v'_y}{v'_x} = \tan^{-1}\frac{177 \text{ m/s}}{147 \text{ m/s}} = 50.3°$$

Since the positive x direction has been taken to be eastward, this angle signifies a direction 50.3° north of east, and the pilot should fly the plane with a velocity

$$\mathbf{v}' = v'\hat{\mathbf{v}}' = (230 \text{ m/s})(50.3° \text{ north of east})$$

through the air.

EXERCISES

Group A

3-1. *Determining g.* A simple laboratory apparatus for determining g is shown in Fig. 3E-1. A ball is traveling horizontally with a known speed as it leaves the end of a curved incline at the edge of a table. The table height AB is measured, and so is the distance BC from the base of the table to point C, where the ball strikes the floor.

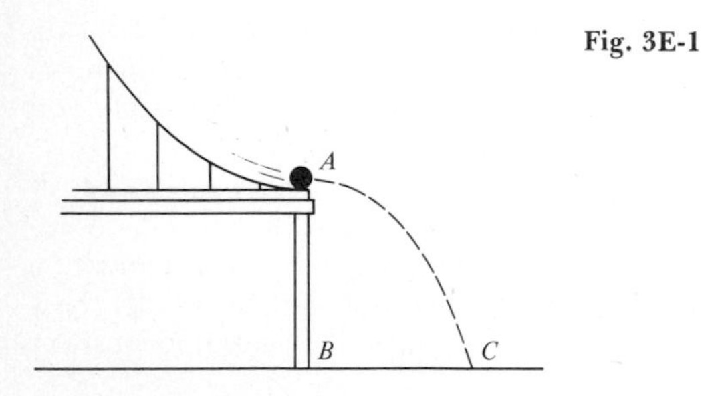

Fig. 3E-1

a. If the ball's speed as it passes point A is 3.0 m/s and the distance BC is found to be 1.5 m, how long was the ball in flight?

b. Let the table height AB be 1.0 m. Use the information in part a to find the value of g.

3-2. *Marking the spot.* An aerial search party locates the spot where a ship has sunk. The fliers plan to mark the spot by dropping a buoy. They are flying horizontally at speed v and altitude h.

a. Neglecting air resistance, how far will the buoy travel horizontally before splash-down?

b. As the plane makes a direct approach over the site, the crew member in charge of releasing the buoy monitors the line of sight to the "target." What angle should the line of sight make with the vertical at the proper time to drop the buoy?

c. Evaluate your results for v = 150 m/s and h = 3000 m.

d. Where will the plane be when the buoy strikes the water?

3-3. *A game of catch.* Two youngsters are playing catch. They stand 2.0 m apart. On each throw, the ball rises and falls 1.5 m.

a. What is the time of rise? The time of fall? The time of flight?

b. What is the horizontal component of the initial velocity? The vertical component?

c. What angle does the initial velocity make with the horizontal? What is the initial speed of the ball?

d. What is the speed of the ball when it is caught?

3-4. *Water fun.* Three tilted hose nozzles A, B, and C are fixed on the ground. The nozzles are inclined to the horizontal at 30°, 45°, and 60°, respectively. Streams of water issue from the three hoses at identical speeds.

a. What is the ratio of the maximum heights of rise?

b. What is the ratio of the ranges?

3-5. *Out in left field.* A baseball outfielder has a throwing range of 80 m when he throws the ball at 30° above the horizontal. Assuming the same initial speed, how high would the ball rise if the player threw it vertically upward?

3-6. *Grasshopping.* A typical adult grasshopper has a jumping range of 0.75 m. Assume that the launch angle is 45°.

a. What is the horizontal component of the grasshopper's velocity?

b. How long is the grasshopper in flight? Neglect air resistance and aerodynamic lift.

3-7. *Pigskin's progress.* A football player is carrying the ball at a speed of 7.5 m/s. His path makes an angle of 35° with the sidelines. At what rate is he approaching the goal line?

3-8. *Add 'em up.* Figure 3E-8 shows six vectors of the indicated magnitudes, each at an angle of 60° with the adjacent vectors. What is the magnitude of the resultant vector? What is its direction with respect to vector **A**?

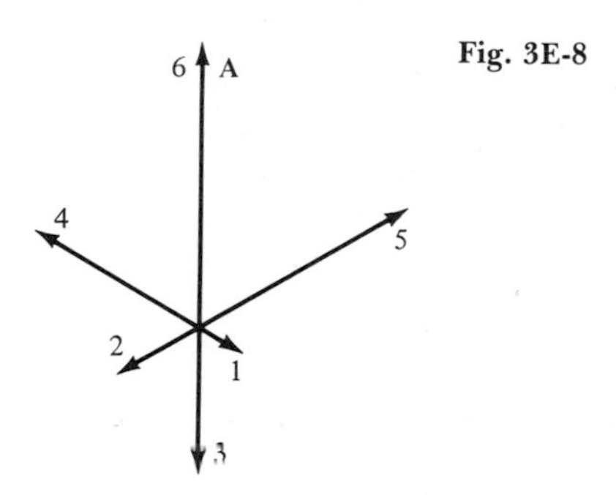

Fig. 3E-8

3-9. *Graphical representation of vectors.*

a. Construct a pair of mutually perpendicular coordinate axes. Draw a vector, **A**, from the origin to the point (3, 4). Draw a vector, **B**, from the origin to the point (4, 3).

b. Are A and B equal?

c. Are **A** and **B** equal?

3-10. *Vector addition.* Figure 3E-10 shows a square whose sides are each 1 unit in length. Consider each side to be a vector, as indicated. Keeping in mind that a vector

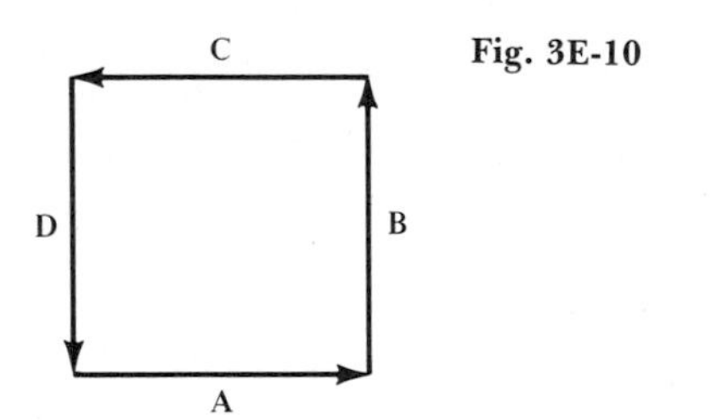

Fig. 3E-10

can be moved parallel to itself without changing it, find the magnitudes and directions of the following vectors.

a. **A** + **B**
b. **A** + **C**
c. **A** + **D**
d. **A** − **B**
e. **A** − **C**
f. **A** − **D**
g. **A** + **B** + **C** + **D**
h. **A** + **B** − **C** − **D**
i. **A** + 2**B**

3-11. *Vector practice.* Vectors **A** and **B** lie in the xy plane. As is customary, the positive y direction makes an angle of +90° with the positive x direction. Vector **A** is 5.0 cm long and makes an angle of +60° with the x axis. Vector **B** is 5.0 cm long and makes an angle of −60° with the x axis.

a. Construct a diagram in which **A** and **B** are represented.

b. Find the magnitude and direction of **A** + **B**; **A** − **B**; **B** − **A**.

3-12. *Net displacement.* The following five vectors represent displacements on the earth's surface:

(1) 1 m north
(2) 2 m 30° east of north
(3) 3 m 60° east of north
(4) 4 m east
(5) 5 m southeast

a. Construct an xy diagram in which the positive x direction represents east and the positive y direction represents north. Choose an appropriate scale and carefully represent each vector. Position the five vectors in such a way that you can immediately find their vector sum from the diagram. Use a ruler and protractor to determine the magnitude and direction of their vector sum.

b. Using the x and y axes defined in part *a*, find the x and y components of each of the five given vectors. Use these to find the components of their vector sum.

c. Verify that the results of parts *a* and *b* are consistent.

3-13. *Unscheduled stop.* On a calm day, an airplane flying at 400 km/h flies east for 1 h, then south for $\frac{1}{2}$ h. It then flies toward its starting point for $\frac{1}{2}$ h before making a forced landing. How far and in what direction from its starting point does the plane land?

3-14. *Trading far for high.* Sally's maximum range in throwing a baseball is R_{max}.

a. Show that if she throws the baseball vertically upward with the same initial speed, it will attain a height equal to $\frac{1}{2}R_{max}$.

b. How does the height found in part *a* compare with the height attained when Sally is throwing for maximum range?

3-15. *How much string?* A hose nozzle N is strapped at point A to a fixed rod, $ABCD$. See Fig. 3E-15. Strings are tied to the rod at equal distances so that $AB = BC = CD$. The lengths of the strings are adjusted so that the lower

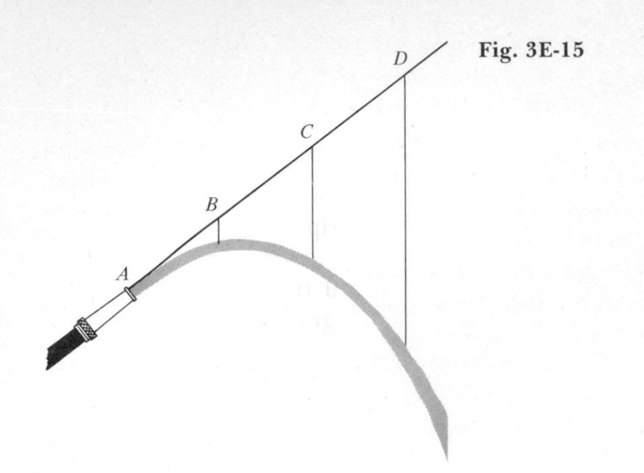

Fig. 3E-15

ends just touch the curved stream of water issuing from the nozzle. If the length of the string at B is 5.0 cm, what must be the string lengths at C and D?

3-16. *Equalizing height and range.* What is the angle of elevation of a launcher which throws a projectile to a height equal to the range?

3-17. *Interpreting a trajectory diagram.* Figure 3E-17 shows a portion of the path of a body moving in the xy plane. If the body moves with constant speed, at which point along its trajectory is the magnitude of its acceleration the greatest? Explain your choice.

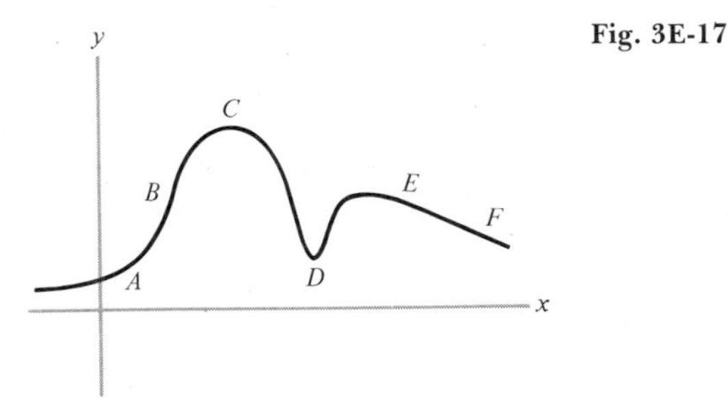

Fig. 3E-17

3-18. *High-speed centrifuges.* High-speed centrifuges have been successfully operated at 60,000 revolutions per minute.

a. If the radius of the centrifuge is 20 cm, what is the magnitude of the acceleration at the circumference? Express your answer in meters per second per second.

b. What is the ratio of this value to g, the acceleration due to gravity?

3-19. *Uniform circular motion.* A body moves at constant speed in a circular path whose circumference is 60 m. It completes one revolution every 12 s.

a. What is the body's speed?

b. What is its average velocity over one complete revolution?

c. At any given instant, what is the magnitude of the body's acceleration? What is the direction of its acceleration?

3-20. *Windy day.* A steamer is sailing west at 25 knots (12.9 m/s). A steady wind is blowing over the ocean from the south at 10 knots (5.1 m/s). What wind speed and wind direction are indicated by an anemometer and a wind vane mounted on the ship?

3-21. *Airspeed versus ground speed.* An airplane is headed 20° east of north at an airspeed of 200 km/h. A wind is blowing from the east at 50 km/h.

a. What is the ground speed of the plane?

b. What is the plane's course made good? (That is, in what direction is the plane moving relative to the ground?)

Group B

3-22. *Muzzle velocity.* The term muzzle "velocity" frequently is used for the speed at which a projectile leaves a gun. In order to determine the muzzle velocity of a particular type of bullet fired from a rifle, the rifle is mounted and carefully leveled. Then it is fired at a target at a known distance L. The vertical distance d from the aim point to the actual impact point is measured.

a. Derive an expression for the muzzle velocity v in terms of L, d, and the acceleration due to gravity g.

b. For a target distance $L = 3.00 \times 10^2$ m, the measured drop is $d = 1.30$ m. What is the muzzle velocity?

c. If the rifle is fired horizontally from a height of 1.70 m, where does the bullet strike the ground if it misses the target?

3-23. *Raindrops.* During a rainstorm, raindrops are observed to be striking the ground at an angle of 35° with the vertical. The wind speed is 4.5 m/s (10 mi/h). Assuming that the horizontal velocity component of the raindrops is the same as the speed of the air, what is the vertical velocity component of the raindrops? What is their speed? (As is discussed in Chap. 4, air resistance operates in such a way that the raindrops do not accelerate, but fall at a constant speed called the terminal speed.)

3-24. *How far away and how high?* An explorer on a plain carefully sights on a distant mountain. He finds that the mountain is located 20° east of north. After traveling 10 km due northward, he makes new measurements. He finds that the mountain now lies 25° east of north.

a. What is the distance from the second sighting point to the mountain?

b. As viewed from the second sighting point, the peak of the mountain is elevated 8.0° above the horizon. How high above the plain does the mountain extend?

3-25. *An acute case of constituent vectors.* A vector can be resolved into two constituent vectors not necessarily at right angles. It is useful to do this for the conical pendulum treated in the text (Sec. 3-7). Resolve the vector $\mathbf{g}$ into constituents along the string and in the horizontal direction. Then obtain Eq. (3-51), $\tan \theta = v^2/rg$.

3-26. *What's up, Doc?* The motion picture comedy *What's Up, Doc?* features a wild automobile chase which ends with several cars hurtling into San Francisco Bay from the end of a dock. Photographically speeding up the action is a common technique in slapstick cinema, and anyone who watches the film must hope (for the sake of the stunt drivers) that the action did not really happen that fast. The cars appear to be traveling at least 22 m/s (50 mi/h) as they leave the dock.

Based on several viewings of the sequence, reasonable estimates for three properties of the trajectories are:

Vertical drop: 6 m

Horizontal distance traveled before splash-down: 14 m

Angle between velocity vector and water surface at splash-down: 35°

a. Use the vertical drop and horizontal distance estimates to calculate the initial speed of the automobiles. Express your result in meters per second.

b. Use the vertical drop and impact angle estimates to calculate the initial speed.

c. By how much do the results of parts *a* and *b* differ? By how much does the *average* of your two calculated speeds differ from 22 m/s?

d. Suppose an object in free fall were photographed and then the action were speeded up greatly. In what way would the film sequence appear "unnatural"?

(*Note:* The data given above are estimates based on a visual recollection. The authors have no documentary evidence that the film employed speeded-up sequences.)

3-27. *All the way or by relay?* A baseball outfielder wishes to get the ball from his position to home plate as soon as possible. An infielder is ready to act as relay man if the outfielder decides not to throw the ball all the way to the plate. Both the relay man and the outfielder have the same maximum throwing speed v_0. The relay man requires a time interval Δt to catch the ball, turn, and throw it again. The field is soggy, so using bounces is not a sensible strategy.

a. If the outfielder is at a distance R_{max} so far from home plate that he can just barely get the ball there on the fly, how long will the ball be in flight if he throws it all the way?

b. If the outfielder throws instead to the relay man who is standing halfway to home plate, how long will it take the ball to arrive at home?

c. Evaluate your results for $v_0 = 35$ m/s and $\Delta t = 0.5$ s. Which method is quicker?

3-28. *Snow fun.* Hugh and Lou are having a snowball fight. They are standing 40 m apart, and Hugh decides to throw two snowballs at the same initial speed of 30 m/s, but at different times and elevation angles, so that they will hit Lou simultaneously.

a. What are the two elevation angles that Hugh must use?

b. How long after the first snowball is thrown must Hugh throw the second one? How long after that will both snowballs land?

3-29. *Effective gravity on a rotating earth.* At the equator, the effective value of g is smaller than at the poles. One reason for this is the centripetal acceleration due to the earth's rotation. The magnitude of the centripetal acceleration must be subtracted from the magnitude of the acceleration due purely to gravity in order to obtain the effective value of g.

a. Calculate the fractional diminution of g at the equator as a result of the earth's rotation. Express your result as a percentage.

b. How short would the earth's period of rotation have to be in order for objects at the equator to be "weightless" (that is, in order for the effective value of g to be zero)?

c. How would the period found in part *b* compare with that of a satellite skimming the surface of an airless earth?

3-30. *Flying in circles.* As indicated in Fig. 3E-30, a plane flying at constant speed is banked at angle θ in order to fly in a horizontal circle of radius r. Its motion can be analyzed by analogy with the conical pendulum of the text. The aerodynamic lift force acts generally upward at right angles to the plane's wings and fuselage. This lift force corresponds to the tension provided by the string in the conical pendulum.

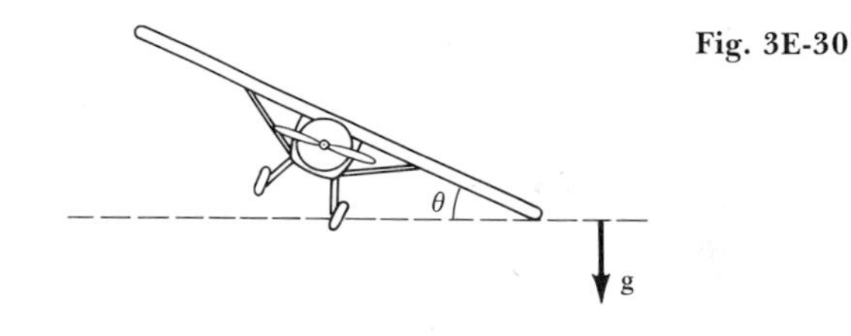

Fig. 3E-30

a. Obtain the equation for the required banking angle θ in terms of v, r, and g.

b. What is the required angle for $v = 60$ m/s (216 km/h) and $r = 1.0$ km?

c. When a plane that has been flying straight enters a turn, it is necessary to increase the engine power to maintain constant speed and altitude. Can you explain why?

3-31. *Row, row, row.* A rowboat is pointing perpendicularly to the bank of a river. The rower can propel the boat with a speed of 3.0 m/s *with respect to the water.* The river has a current of 4.0 m/s.

a. Construct a diagram in which the two velocities are represented as vectors.

b. Find the vector which represents the boat's velocity with respect to the shore.

c. At what angle is this vector inclined to the direction in which the boat is pointing? What is the boat's speed with respect to the launch point?

d. If the river is 100 m wide, how far downstream of the launch point is the rowboat when it reaches the opposite bank?

3-32. *The open road.* A truck is traveling due north and descending a 10 percent grade (angle of slope = $\tan^{-1} 0.10 = 5.7°$) at a constant speed of 90 km/h. At the base of the hill there is a gentle curve, and beyond that the road is level and heads 30° east of north. A southbound police car with a radar unit is traveling at 80 km/h along the level road at the base of the hill, approaching the truck. What is the velocity vector of the truck with respect to the police car?

Group C

3-33. *Out of the park.* Strongarm Sam, a baseball player, decides to throw a ball out of the stadium. The stadium seats slope upward at angle α, as shown in Fig. 3E-33.

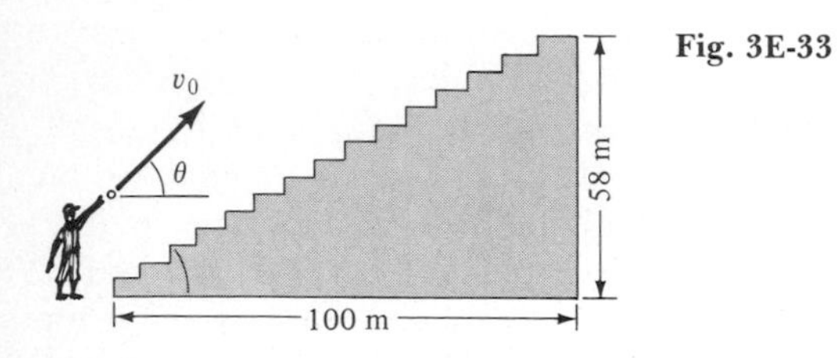

Fig. 3E-33

a. Find the horizontal distance R_x a baseball will travel before landing in the seats if the ball is thrown with speed v_0 at an elevation angle θ above the horizontal.

b. Assuming that Sam's maximum throwing speed is independent of direction, what elevation angle θ_{max} should he use to maximize R_x?

c. What is the value of θ_{max} if $\alpha = 30°$?

d. Suppose the stands are 100 m deep and 58 m high (corresponding to $\alpha = 30°$). What minimum throwing speed is needed to get the ball out of the park? *Note:* A speed of 40 m/s corresponds to a very respectable major league fastball.

3-34. *Throwing down a mountain.* A mountain climber stands on top of a peak whose straight sides slope down at an angle β with the horizontal. The climber wishes to throw a rock as far as possible down the slope. Adapt the result of Exercise 3-33*a* to determine the elevation angle for maximum range.

3-35. *Music on the subway.* A subway rider is playing a record on a battery-powered record player. The record has radius r and rim speed v_0. The train is traveling due north on level tracks at speed u_0. Use the coordinate directions indicated in Fig. 3E-35; where the positive x direction is east and the positive y direction is north. As shown in the figure, the position of any point on the rim of the record can be specified by the angle θ, whose vertex is at the center of the record. The angle θ is measured *clockwise* from north to the rim point. (Records turn clockwise as they are played.)

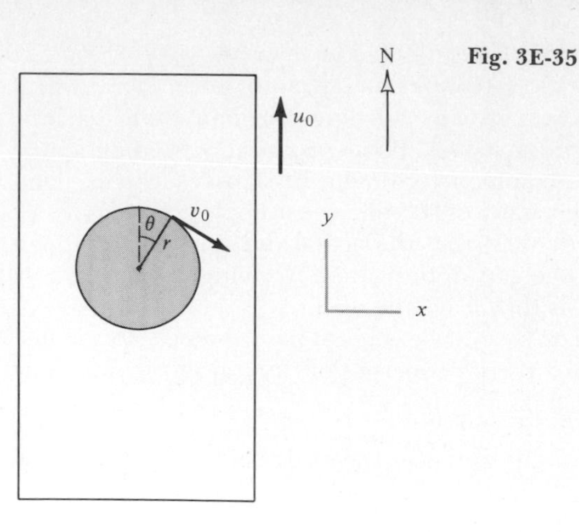

Fig. 3E-35

a. What is the velocity vector **w** of some point on the rim of the record *with respect to the tracks?* Express **w** in terms of u_0, v_0, and the angle θ.

b. Evaluate your result for a 45-rpm record if u_0 = 5.0 m/s (18 km/h). The radius of a 45-rpm record is 0.087 m. What is the maximum speed of any point along the rim, with respect to the tracks? What is the maximum angle between **w** and the positive y direction (due north)? For what value of θ does this angle occur?

3-36. *Swimming across.* A swimmer can swim at a speed of 0.70 m/s with respect to the water. She wants to cross a river which is 50 m wide and has a current of 0.50 m/s.

a. If she wishes to land on the other bank at a point directly across the river from her starting point, in what direction will she need to swim? How rapidly will she increase her distance from the near bank? How long will it take her to cross?

b. If, rather, she decides to cross in the shortest possible time, in what direction must she swim? How rapidly will she increase her distance from the near bank? How long will it take her to cross? How far downstream will she be when she lands?

Numerical

3-37. *Parabolic trajectory I.* Use the trajectory plotting program in the Numerical Calculation Supplement to plot the trajectory of a projectile having an initial velocity of magnitude v_i = 70.7 m/s (this is 100 m/s divided by $\sqrt{2}$) for an elevation angle of $\theta = 45°$. Compare the range of the projectile with the range obtained in Example 3-7 for v_i = 100 m/s and $\theta = 45°$.

3-38. *Parabolic trajectory II.* Using v_i = 100 m/s and several values of θ in the neighborhood of 45°, show that the maximum range of the projectile considered in Example 3-7 occurs when $\theta = 45°$. Do this by running the trajectory plotting program in the Numerical Calculation Supplement.

3-39. *Parabolic trajectory III.* When the x and y values provided by the trajectory plotting program in the Numerical Calculation Supplement are plotted, the graph resembles a strobe photo of a projectile. It is not very difficult to identify the reason for the resemblance: The program evaluates y for uniformly spaced values of x. Since the x component of the projectile's velocity is constant, the projectile's x coordinate increases uniformly with time. Thus the plotted x and y values are those occurring at equal intervals of time, just as in a strobe photo.

a. Run the program to obtain a set of points showing the positions of the projectile on a trajectory for $v_i = 100$ m/s and $\theta = 70°$, with successive values of x taken to be 50 m apart. Construct a plot of these points, but do *not* connect the points with a smooth curve; that obscures the points, as can be seen in Fig. 3-29.

b. Making use of the fact that the speed is proportional to the separation between adjacent points on the plot, determine the relative values of the speed of the projectile in various parts of its trajectory. Explain why the speed behaves as it does.

c. Explain why it is *not* possible to use this procedure to compare the speeds of projectiles in two different trajectories.

3-40. *A strobed trajectory program.*

a. Overcome the limitation described in Exercise 3-39*c*. That is, write a calculator or computer program that will produce a strobe-photo-like set of points on any projectile trajectory in such a way that it *is* possible to compare the speeds of projectiles on different trajectories. Do this by having the program use Eqs. (3-4) and (3-5) to evaluate x and y at the *same* set of uniformly separated values of t for all trajectories.

b. Use your program to obtain sets of points on the trajectories for $v_i = 100$ m/s, $\theta = 35°$ and $v_i = 100$ m/s, $\theta = 55°$, with successive values of t taken 1.0 s apart. Compare the speeds of the projectile in the two different trajectories.

c. Compare the total times of flight of the projectile in the two different trajectories, by simply counting the number of points on each trajectory. Explain why the times of flight are different although the ranges are the same.

3-41. *Adding vectors by machine.*

a. Write a program that instructs a calculator or computer to evaluate the sum of any number of vectors and to present the result in component form and as a magnitude and direction. Use the algebraic procedure of Example 3-4*b*. Test the program by repeating the calculation done there.

b. Use your program to evaluate the sum of the following set of vectors: (1) $r_1 = 1.57$ m, $\phi_1 = 28.3°$; (2) $r_2 = 6.03$ m, $\phi_2 = 258.6°$; (3) $r_3 = 4.67$ m, $\phi_3 = -105.6°$; (4) $r_4 = 3.71$ m, $\phi_4 = 96.2°$.

4

Newton's Laws of Motion

4-1 NEWTON'S FIRST LAW AND INERTIAL REFERENCE FRAMES

In Chap. 1 we began our discussion of newtonian mechanics with the basic question: What is the relation between force and motion? The experiments this question led us to analyze are summarized in Fig. 4-1, which is a reproduction of Fig. 1-7. From the analysis we concluded that an answer is: The net force acting on an object is related to its acceleration. But since we had not even defined acceleration precisely in Chap. 1, the relation had to remain qualitative.

Our task in this chapter is to obtain a quantitative relation between force and motion. In Sec. 4-2 we go through preliminary considerations that use little more than an intuitive understanding of the crucial quantities force and mass. The relation among force, mass, and acceleration that will emerge from this treatment is a form of *Newton's second law of motion.* The form is applicable to most of the systems studied in newtonian mechanics, but not to them all. In order to get the general form of the law needed in some of our work with newtonian mechanics, and to develop a thorough understanding of force and mass, we reconsider Newton's second law in Secs. 4-3 and 4-4 from a rigorous and quite different approach. This involves analyzing strobe photos of collisions between pucks on an air table. As an additional advantage, the rigorous approach will make it possible for us to derive in Sec. 4-5 *Newton's third law of motion,* which relates the forces that objects exert on each other when they interact. Newton's third law also enters into the preliminary considerations of Sec. 4-2, but there it is given only an intuitive justification.

In this section we are concerned with *Newton's first law of motion.* It has to do with the motion of a single body on which *no* net force acts. Figure 4-1 will remind you that as long as there is no net force applied to an air table puck, there is no change in its velocity, as observed by a stroboscopic

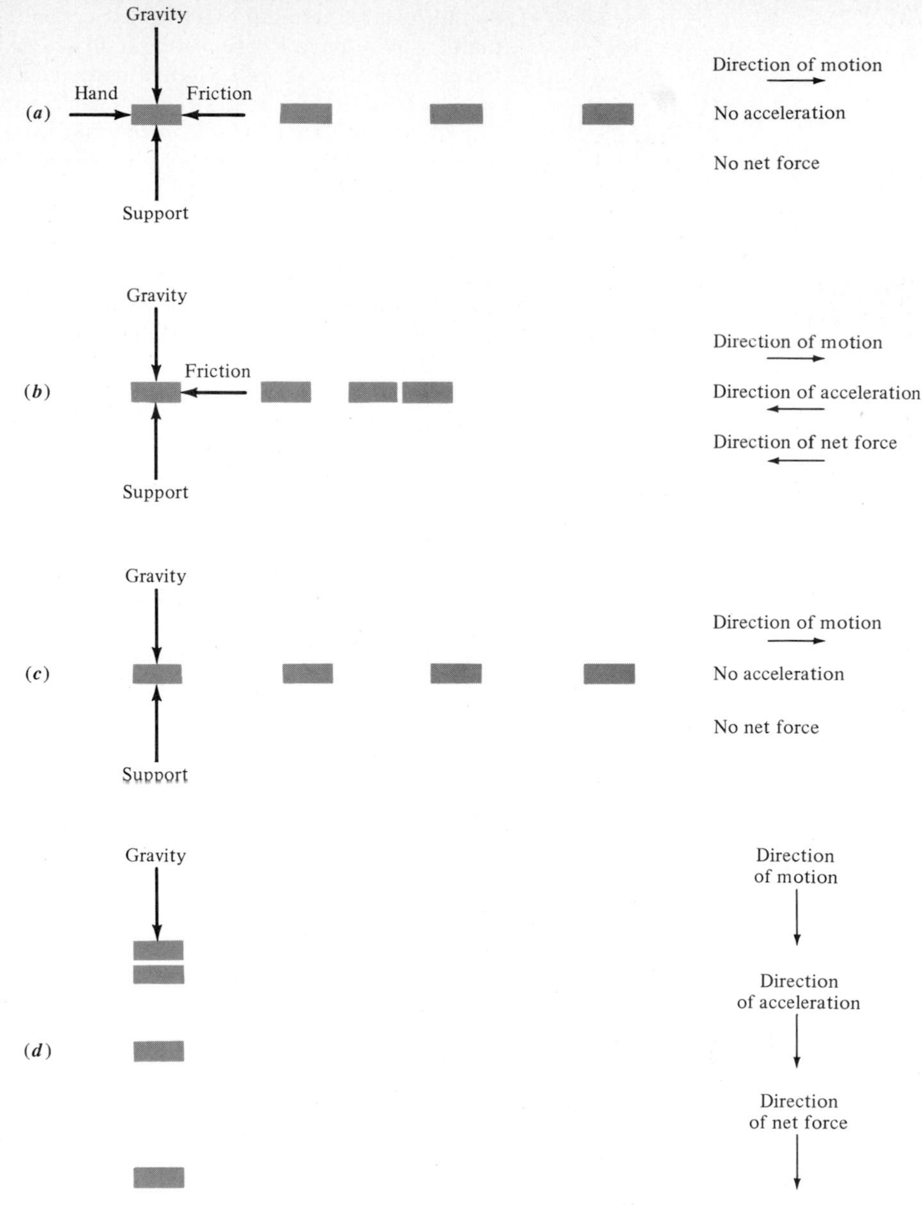

Fig. 4-1 A summary of experiments considered in Chap. 1, which led to the conclusion that the net force applied to an object is related to its acceleration.

camera system supported from the ground. Since the camera is performing the function of an observer, the situation can be described as follows: *If no net force is applied to a body, it maintains a constant velocity with respect to an observer fixed to the earth's surface.* This statement is Galileo's form of the *law of inertia.* The name is appropriate, since the word "inertia" means the tendency to avoid changing a state of motion. The law of inertia was modified by Newton, in a way that is explained later, to become his first law of motion.

The law of inertia played a key role in Galileo's 1632 *Dialogue Concerning the Two Chief World Systems,* his magnificent argument in defense of the Copernican view that the earth revolves around the sun, not vice versa. In the Aristotelian view almost universally held at the time, a fundamental distinction was made between celestial matter, which moved of its own accord, and terrestrial matter, which moved only under the influence of a continually applied force.

In a series of arguments, Galileo destroyed the distinction between celestial and terrestrial matter. It was thus important for him to demonstrate that if celestial matter could move forever without any force being applied, the same was true of terrestrial matter as well. Galileo perceived that it was possible and desirable to neglect friction, which had always been regarded as an essential aspect of the motion of bodies. This was far from obvious in a day when the common experience of motion was with such things as oxcarts on rough roads.

Nevertheless, Galileo was able to *imagine* how bodies would move in the absence of friction centuries before the invention of the air table. His argument goes as follows: Imagine a ball rolling with negligible friction. If it rolls down an inclined plane, it will accelerate. If it rolls up an inclined plane, it will decelerate. Now imagine the ball rolling on a *level* plane. It will neither accelerate nor decelerate—neglecting friction—and so it must move at constant velocity.

Isaac Newton (1642–1727) was born the year after Galileo's death. In 1666 Cambridge University was closed because of a plague epidemic, and Newton spent the time at home in what must be the most productive single year of scientific endeavor in history. During 1666 Newton developed the major part of his mechanics and (among other things) invented calculus.

Much later, under pressure from friends, Newton prepared a systematic, formal account of his work in mechanics. It was published in Latin in 1686 under the title *Philosophiae Naturalis Principia Mathematica,* or *Mathematical Principles of Natural Philosophy*. The organization of the work is that of a classical euclidean geometry text, with definitions, axioms or laws, and theorems. Newton adapted Galileo's law of inertia into his first axiom, that is, his first law of motion. After his book had won wide acclaim, Newton acknowledged his debts to Galileo and other predecessors in the charming statement "If I have seen farther than others, it is because I have stood on the shoulders of giants."

In Chap. 1 we replaced Galileo's imaginary experiment justifying the law of inertia with a real experiment made possible by the air table and stroboscopic photography. Figure 4-2 repeats the experiment. It depicts a strobe photo of a puck set into motion across the nearly frictionless horizontal top of an air table by a launching apparatus at the upper right of the photograph. Figure 4-3 demonstrates the constancy of the puck's subsequent velocity. We use the duration of the *strobe flash interval* as the *time unit,* just as in Sec. 3-5. Then the dashed white vector is not only the displacement of the puck during a time interval at the beginning of its motion but also its average velocity during that time interval. Similarly, the dashed gray vector is the average velocity of the puck during a time interval at the end of its motion. These initial and final average velocity vectors are moved, without changing their directions or lengths, to form the solid white and gray vectors. This is done to make it easier to compare the initial and final average velocity vectors. Such a comparison shows that they are essentially the same, so that the average velocity is constant. In fact, it is not necessary to make a distinction between average and instantaneous quantities when we speak of the initial and final velocities. Thus we can say that the velocity of the puck is observed to be essentially constant during its motion.

The downward gravitational force exerted on the puck by the earth is exactly compensated by the upward force exerted on it by the air film under the puck. And the puck does not experience significant frictional force acting in the horizontal direction. So after it leaves the launcher, the puck is a body experiencing essentially no *net* applied force. Furthermore, it is viewed by an observer (the camera) fixed to the surface of the earth. Thus Galileo's law of inertia should describe the behavior of the puck, and it does.

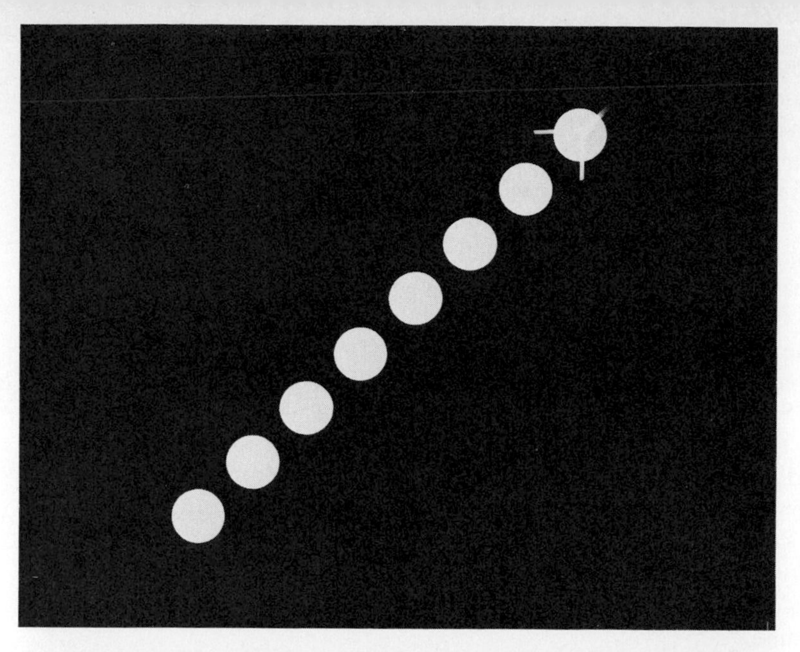

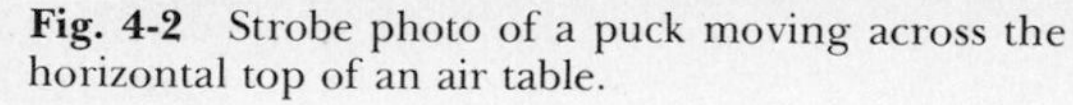

Fig. 4-2 Strobe photo of a puck moving across the horizontal top of an air table.

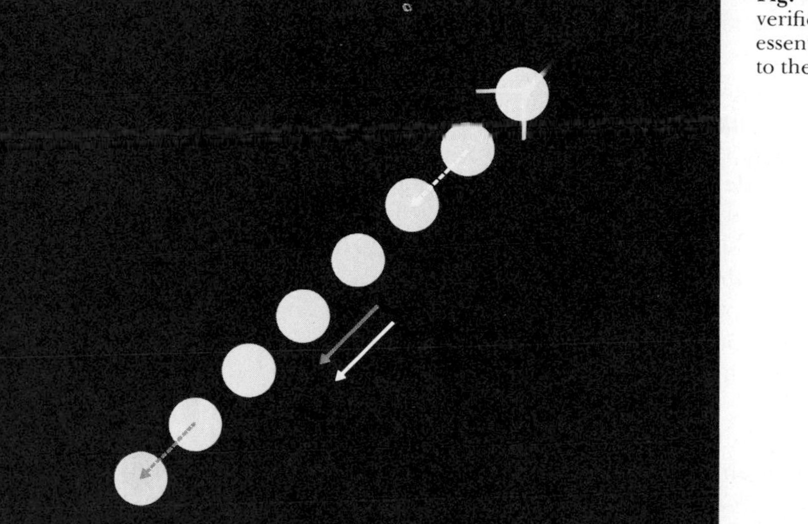

Fig. 4-3 An analysis, explained in the text, which verifies the conclusion that the velocity of the puck is essentially constant, as observed by the camera, fixed to the surface of the earth, that took the photograph.

Very accurate measurements would show that the velocity of the puck in Fig. 4-2 is not quite constant. Part of the change in its velocity is due to the fact that a small amount of friction still acts on the puck, even though it is on an air table. More interesting for our purpose here is the part that is due to the small accelerations of the observer fixed to the surface of the earth. If a puck were launched with a velocity of 1.0 m/s due north across a completely frictionless air table located at a latitude of 45° North, an observer standing next to the table would see that after 1.0 s the puck had gained a velocity of 1.0×10^{-4} m/s to the east—providing the observer had equipment sensitive enough to measure such a velocity gain. The reason for the change in the velocity of the puck with respect to the observer is that the observer standing on the surface of the earth is accelerating. Almost all of this acceleration is due to the (daily) rotation of the earth about its axis. But there are other sources of acceleration, too. In order of decreasing magnitude, these arise from the (annual) revolution of the earth about the sun, the revolution of the sun about the center of our galaxy (one revolution takes approximately 2×10^8 years), and the motions, as yet not completely known, of our galaxy relative to the universe as a whole.

If measurements are made on a body experiencing no net force from a reference frame which does not partake of the rotation of the earth about its axis, the observed deviation of the motion from constant-velocity motion is much reduced. (Experimental evidence justifying this statement is presented in Sec. 5-4.) Following this line of thought, we are led to consider a reference frame that also does not revolve about the sun, then one that in addition does not revolve about the center of the galaxy, and finally one that does not even move with the galaxy. We believe that from such a frame of reference, completely motionless with respect to the universe as a whole, the velocity of a body would be observed to be exactly constant when it had no net force at all acting on it. The reference frame is called an **inertial reference frame.**

Newton modified Galileo's law of inertia by specifying that it describes the constant velocity of a body experiencing no net force as seen by an observer in an inertial reference frame, not as seen by an observer fixed on the earth's surface. (Actually, at the time of Newton the motions of stars through galaxies, and of galaxies through the universe, were not known. So Newton considered an inertial frame to be one that is motionless with respect to the "fixed stars." But there is no question that he would find the modern definition of an inertial frame completely acceptable.) The modified version of the **law of inertia** is: *If no net force is applied to a body, it maintains a constant velocity with respect to an observer fixed to an inertial frame.* (Included is the case in which the magnitude of the constant velocity is zero, so that the body remains at rest with respect to the observer.) This statement was incorporated by Newton into his theory of mechanics as its first axiom. Consequently, it is also called **Newton's first law of motion.**

A reference frame that is motionless with respect to the universe as a whole is an inertial frame. Newton's first law says that a body not acted on by a net force maintains a constant velocity with respect to an inertial frame. So the first law says that a body which has no net force acting on it maintains a constant velocity with respect to the universe as a whole. But the first law uses the concept of an inertial frame to make the statement indirectly. The reason is that the expression of Newton's *second* law of motion also involves the concept of an inertial frame. The first law serves to introduce the concept and define it in context. Thus the first law has two functions. One is to tell you that inertial frames exist. The other is to tell you how to determine whether you are stationed in such a reference frame without the need of knowing what the entire universe is doing in order to know your state of motion with respect to the universe. What you do to make the determination is to find a body, like an air table puck, that you have good reason to believe is not acted on by a net force. Then you observe it and determine whether its velocity relative to you remains constant. If this velocity is constant, you are in an inertial frame.

You will never be in a precisely inertial reference frame, even if you become an astronaut, because you will never be able to free yourself completely from the motions of the sun and of our galaxy. But when stationed on the earth's surface, you are in a good approximation to an inertial reference frame. It may or may not be a sufficiently good approximation for your purposes, depending on the system you are studying and the accuracy of your studies. If too much error would arise from ignoring the fact that a reference frame fixed to the earth's surface is not exactly an inertial frame, there is a procedure which can be used to correct accurately for this fact. (The procedure is developed in Sec. 5-4.) For most practical studies in physics, and for nearly every application of physics to engineering, you are completely justified in treating a reference frame fixed to the surface of the earth as an inertial frame.

Any reference frame which moves at constant velocity relative to an inertial frame is also an inertial frame. This is a consequence of Newton's first law and of a conclusion obtained in Sec. 3-8. There two reference frames were considered. One was the x, y, z frame and the other the x', y', z' frame which moved at an arbitrary constant velocity relative to the unprimed frame. We wrote the Galilean position transformation equation relating the position vector in the primed frame to the position vector in the unprimed frame. We then calculated the time derivative of each term in the equation and obtained the Galilean velocity transformation equation. From this equation we concluded that a body observed to move relative to the unprimed frame at constant velocity will be observed to move relative to the primed frame with a different, but still constant, velocity. Now let the unprimed frame be an inertial frame. Then Newton's first law requires that there be no net force acting on the body. To find out whether the primed frame is also an inertial frame, we note that a body on which no net force acts is observed from that frame to move with constant velocity. So the test of an inertial frame given by the first law is satisfied, and the primed frame therefore is also an inertial frame. This proves the italicized statement.

4-2 NEWTON'S SECOND AND THIRD LAWS

We begin the preliminary consideration of Newton's second and third laws of motion, with which this section is concerned, by drawing an implication from Newton's first law of motion. The first law implies that if a body does *not* maintain a constant velocity with respect to an inertial frame of reference, then there *is* a net force applied to it. Since a body with a changing velocity is accelerating, the implication is that a body with a net force applied to it is accelerating with respect to an inertial frame. *Newton's second law of motion* gives the *precise* connection. It asserts that the relation between the net force and the acceleration is a *direct proportionality*.

Everyone has at least an intuitive feeling for the fact that the net force applied to an object is directly proportional to its acceleration, as observed from the approximately inertial reference frame of the earth's surface. For instance, the drag racer knows that he doubles the acceleration of his vehicle by doubling the propulsive force produced by its engine, providing the tires maintain their grip.

Figure 4-4 indicates how a quantitative measurement could be made to verify the proportionality between the acceleration of an object, observed from the surface of the earth, and the strength of the net force applied to it. An experimenter pulls a block across the frictionless top of an air table by applying a force to it through a spring scale. As is discussed in detail near the end of this chapter, the extension of the spring is proportional to

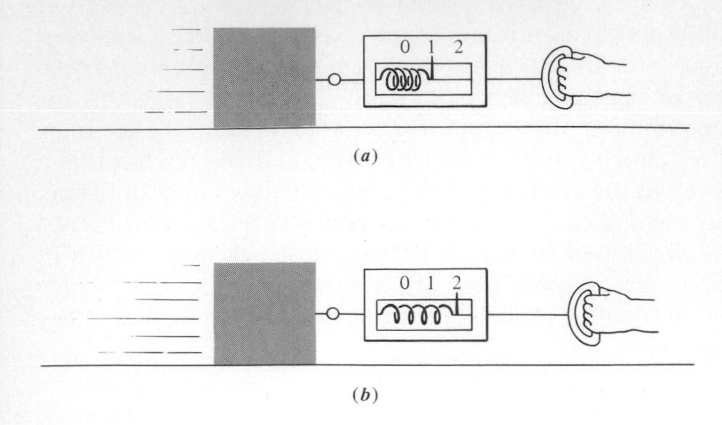

Fig. 4-4 An experiment which shows the proportionality between the acceleration of an object and the strength of the net force applied to it. In the upper part of the figure, an experimenter applies a constant horizontal force to the object consisting of the block plus the spring scale. The constancy of the force is verified by the constant reading on the scale. If friction is negligible because the block is on an air table, a strobe photo taken by a camera fixed to the earth will show that the object experiences a constant acceleration, within experimental accuracy. In the lower part of the figure, the experimenter doubles the force applied to the object. A strobe photo will show that its acceleration is doubled.

the force acting on it. Consequently, the scale reading provides a measure of the strength of the force transmitted through the scale to the block. The experimenter applies a constant force to the block. Meanwhile a strobe photo is made of its motion. Then the experiment is performed a second time, with the experimenter pulling harder so that the measured strength of the applied force is doubled. Analysis of the photographs will show that the acceleration is doubled in the second experiment.

Many measurements such as these show that the net force **F** applied to a body is proportional to its acceleration **a** with respect to an inertial frame of reference. We write the relation as

$$\mathbf{F} \propto \mathbf{a}$$

where the symbol $\propto$ means "proportional to." It is usually more convenient to treat equations than proportionalities. We therefore introduce a proportionality constant m and convert the proportionality into the equation

$$\mathbf{F} = m\mathbf{a} \tag{4-1}$$

This is **Newton's second law of motion.**

What is the *physical* meaning of m? From experience we know that not all bodies are equally easy to set into motion. It appears that the difficulty of accelerating them increases with the amount of matter they contain. This "amount of matter" is called **mass.** All material bodies in the universe have mass. Qualitatively, the mass of a body measures its inertia—in other words, its reluctance to accelerate. That is exactly what Eq. (4-1) says quantitatively.

The value of m in Eq. (4-1) specifies how difficult it is to make a body of that mass accelerate. The equation says that as viewed from an inertial reference frame the strength F of the force required to produce an acceleration of magnitude a will be proportional to the mass m of the body it acts on. An experiment verifying the proportionality of F to m, for a fixed value of a, can be performed along the lines suggested in Fig. 4-5. If the mass of the spring scale is negligible, the experimenter finds that for the double block it takes a doubling of the applied force (determined by the spring scale reading) to produce the same given acceleration (determined by a strobe photo). And since the total mass of two identical blocks must certainly be twice the mass of each, the proportionality of F to m is verified.

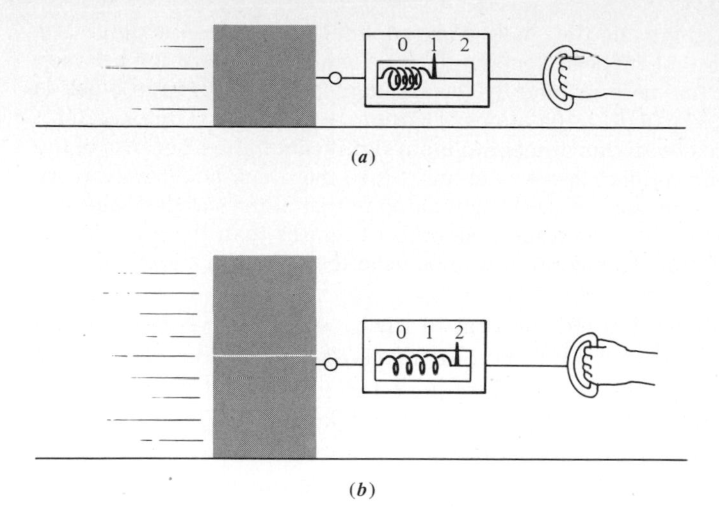

Fig. 4-5 An experiment which shows the proportionality between the mass of an object and the net force which must be applied to give the object a certain value of acceleration. In the upper part of the figure, the experimenter applied a force to a spring scale and a block supported by an air table. A strobe photo taken by a camera fixed to the earth will show that the block has a certain acceleration. In the lower part of the figure, a second *identical* block is attached to the first one. If the mass of the spring scale can be neglected, this doubles the mass of the object to which the force is applied. A strobe photo will now show that, within experimental accuracy, giving the object of doubled mass the same acceleration as before requires applying a force of doubled strength.

An important feature of Eq. (4-1) is that the symbol for force is written as a vector and the symbol for mass is written as a (positive) scalar. Thus the equation states that the force **F** is a vector having the same direction as the acceleration **a** which it produces. Intuition certainly indicates that this statement is correct. Can you give an example drawn from everyday experience?

Newton's second law of motion is the cornerstone of newtonian mechanics and is perhaps the most important relation in all physics. To help emphasize the law, we write again the equation representing it, Eq. (4-1),

$$\mathbf{F} = m\mathbf{a}$$

And remembering that the experiments used to establish Newton's second law are carried out in an approximation to an inertial frame, we express the law in words: *The net force* **F** *acting on a body equals the product of the mass m of the body and the acceleration* **a** *which it gives to that body, if the motion is observed from an inertial frame.* The net force acting on a body is the vector sum of *all* the forces applied *to* the body.

If we are to make Eq. (4-1) useful, we must define a standard unit of mass. This unit is the **kilogram** (abbreviated kg). It is the mass of a certain block of platinum-iridium alloy, kept for safety in a vault in Sèvres, a suburb of Paris. Its mass was originally chosen to be as close as possible to that of the amount of pure water, at a pressure of 1 atmosphere (atm) and a temperature of 4°C, which occupies a volume of 10^{-3} m^3 (that is, 1 L). In fact, the makers of the block came fairly close to this goal. But the standard is the block, and not the water. Once the standard had been established, many copies were made for use elsewhere. A subsidiary unit of mass is the **gram** (abbreviated g, or often gm in the older literature). It is defined to be exactly

$$1 \text{ g} = 0.001 \text{ kg}$$

The mass of this book is about 1.5 kg. A strong athlete can lift over his head a barbell of mass of about 200 kg.

The unit of mass for the system of units still frequently used in engineering practice in the United States is named the **slug.** To three decimal places, its value is

$$1 \text{ slug} = 14.594 \text{ kg}$$

The SI units of length and time have been redefined in terms of the properties of atoms, but this has not yet been done for the unit of mass. The relation between atomic masses and the mass defined by the standard platinum-iridium block is known. Specifically, 5.01848×10^{25} atoms of carbon-12 have a mass of 1 kg. But it is not yet possible to obtain this number to more significant figures because of the difficulty in determining the masses of atoms. Since the much larger masses we deal with in most circumstances can be compared to that of the standard block in measurements accurate to several more significant figures than there are in the number just quoted, the block continues to be used to define the kilogram.

Now that we have defined the unit of mass, we can use Newton's second law to define the unit of force. In terms of magnitudes, the law is $F = ma$. It tells us that the net force required to give a unit mass (1 kg) a unit acceleration (1 m/s^2) is $1\ kg \times 1\ m/s^2 = 1\ kg{\cdot}m/s^2$. This quantity is the unit of force. The force unit is important enough to warrant its own name. It is called the **newton** (N). That is, the force unit is

$$1\ N \equiv 1\ kg{\cdot}m/s^2$$

If you hold this book in the palm of your hand, it presses down with a force of about 15 N. The barbell supported by the athlete is exerting a downward force on his arms of about 2000 N.

Note that it is not necessary to construct a standard newton. The force unit is specified in terms of the three **fundamental units:** length, mass, and time. Consequently, the force unit is known as a **derived unit.** All the other units used in mechanics are also derived units. For instance, the unit of velocity, 1 m/s, is specified in terms of two of the three fundamental units. The special role of the units for length, mass, and time was made apparent by the name for the immediate ancestor of SI. It was called the mks system, the letters standing for meters, kilograms, and seconds.

Example 4-1 illustrates a simple application of Newton's second law of motion to an accelerating object.

EXAMPLE 4-1

The speed of a 1000-kg automobile on a straight road increases uniformly from 0 m/s to 30.0 m/s in 10.0 s. What net force is acting on the automobile?

▪ This is a problem in one dimension, so you can express directed quantities by using either the signed scalar symbols of Chap. 2 or the vector symbols of Chap. 3. If you choose vector symbolism to gain experience with it, you can evaluate the automobile's constant acceleration **a** in terms of the final velocity **v** by writing the relation for constant acceleration of Eq. (2-29) in vector form:

$$\mathbf{v} = \mathbf{v}_i + \mathbf{a}t$$

Setting the initial velocity $\mathbf{v}_i$ equal to zero and solving for **a**, you obtain

$$\mathbf{a} = \frac{\mathbf{v}}{t}$$

Since t is a positive scalar, it is evident that **a** has the same direction as **v**. Newton's second law states that the net force **F** acting on the automobile of mass m is

$$\mathbf{F} = m\mathbf{a}$$

Substituting for **a**, you have

$$\mathbf{F} = \frac{m\mathbf{v}}{t}$$

The direction of this net force is the same as that of the automobile's acceleration **a**, and thus of its final velocity **v**; it is the direction of the automobile's motion. The magnitude F of the net force is

$$F = \frac{mv}{t}$$

where v is the final speed of the automobile. Inserting the numerical values given for m, v, and t, you obtain

$$F = \frac{1000 \text{ kg} \times 30.0 \text{ m/s}}{10.0 \text{ s}} = 3000 \text{ kg·m/s}^2$$

or

$$F = 3000 \text{ N}$$

How have we gotten to where we are now? We began with the familiar intuitive concept of force. Combining this concept with the defined concept of acceleration, we arrived at the less familiar concept of mass. But although mass was not a familiar concept, there was no difficulty in defining a standard mass, the kilogram. Once this was done, we backtracked and put the intuitive concept of force on a sounder logical basis through the definition of the unit of force in terms of the units of mass and acceleration.

While mass may not be a concept in everyday use, it has an intuitive connection with the familiar concept of weight. This is demonstrated in Example 4-2.

EXAMPLE 4-2

A 0.70-kg billiard ball and a 7.0-kg bowling ball both fall toward the ground with the same downward acceleration **g** of magnitude 9.8 m/s². Find the force, symbolized as **W**, exerted by gravity on each ball.

■ Figure 4-6 shows the two falling balls. For each ball the only force acting on it is the gravitational force **W**. Newton's second law says that this net force must be in the direction of its acceleration **g**, namely downward. The second law also says that the magnitude W of the net force acting on the ball must equal its mass m times the magnitude g of its acceleration. So you have for the billiard ball

$$W = mg = 0.70 \text{ kg} \times 9.8 \text{ m/s}^2 = 6.9 \text{ N}$$

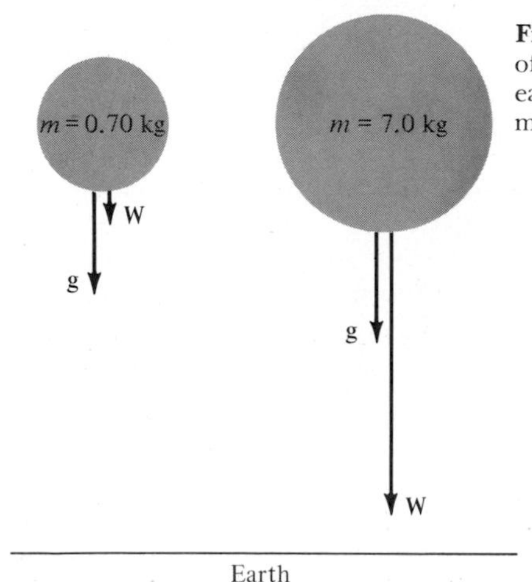

Fig. 4-6 The gravitational forces **W** exerted by the earth on two balls of different mass m and their accelerations **g** when falling toward the earth. Observation shows that both accelerations have the same magnitude, providing that air resistance is negligible.

For the bowling ball,

$$W = mg = 7.0 \text{ kg} \times 9.8 \text{ m/s}^2 = 69 \text{ N}$$

Although the masses of the two balls considered in Example 4-2 are different, experiment shows that near the surface of the earth they fall with accelerations of the same magnitude g (neglecting air resistance). According to Newton's second law as applied in the example, this can be true *only* if it is also true that the magnitude W of the gravitational force exerted on each ball is directly proportional to its mass m, with the proportionality constant being g. That is,

$$W = mg \tag{4-2}$$

This equation allows us to understand why the two balls fall with the same acceleration, despite the difference in their masses. The bowling ball, having 10 times as much mass as the billiard ball, feels a gravitational force which is 10 times as strong. But because its mass is 10 times that of the billiard ball, Newton's second law says it takes the 10 times stronger force to give the bowling ball the same acceleration as the billard ball. In other words, since force equals mass times acceleration, it follows that acceleration equals force divided by mass. And for each ball the gravitational force mg acting on it, divided by its mass m, yields the same acceleration g.

Frequent use will be made of Eq. (4-2) since it applies to all bodies near the surface of the earth, where the magnitude of their acceleration when falling with negligible air resistance is essentially equal to the standard value g. (In Chap. 11 you will see that in this context "near" means that the separation between a body and the earth's surface is small compared to the earth's radius.) We call the magnitude W of the gravitational force $\mathbf{W}$ acting on a body near the earth's surface its **weight.** With this terminology, Eq. (4-2) says that *the weight of a body is its mass multiplied by the magnitude of the gravitational acceleration.* The gravitational force exerted on the body is a manifestation of the gravitational attraction between the earth and the body. Its direction is always toward the center of the earth, that is, in the direction called downward.

Example 4-3 will show you how Newton's second law can be applied to stationary bodies and will demonstrate one method for finding the weight of a body.

EXAMPLE 4-3

The billiard ball and the bowling ball of Example 4-2 are suspended from spring scales and hang motionless. See Fig. 4-7. Find the magnitude S of the force exerted by the scale holding up each ball. In other words, find the force registered on each spring scale.

■ You see from the figure that acting on each ball is a force $\mathbf{W}$ exerted by gravity and a force $\mathbf{S}$ exerted by the scale. The net force $\mathbf{F}$ acting on the ball is the vector sum of the two:

$$\mathbf{F} = \mathbf{W} + \mathbf{S}$$

Since the ball is *not* accelerating, $\mathbf{a} = 0$ and so $\mathbf{F} = m\mathbf{a} = 0$. Thus you have in each case

$$\mathbf{W} + \mathbf{S} = 0$$

or

$$\mathbf{S} = -\mathbf{W} \tag{4-3}$$

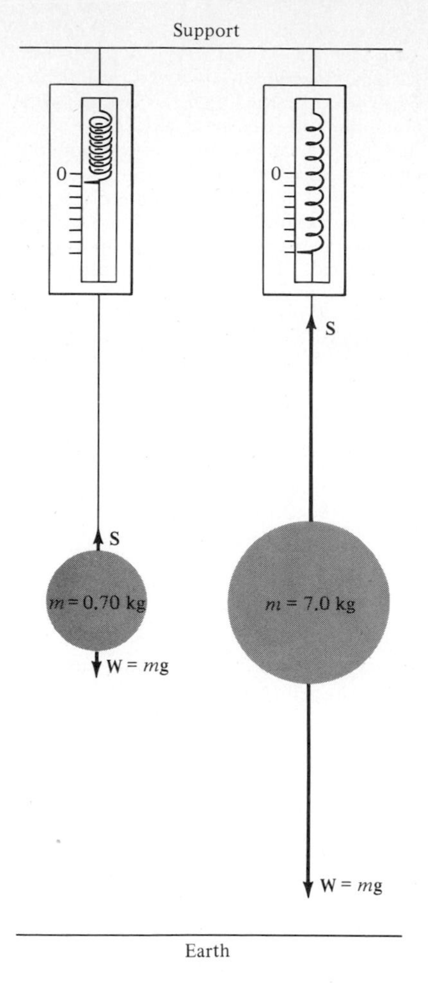

Fig. 4-7 The balls of Fig. 4-6 hanging motionless from cords connected to spring scales. Although the earth still exerts downward gravitational forces $\mathbf{W} = m\mathbf{g}$ on them, the balls are prevented from falling by the upward forces $\mathbf{S}$ exerted on them by the scales from which they are suspended.

Since the force exerted on the ball by gravity is downward, Eq. (4-3) shows that the force exerted on the ball by the scale is upward, in agreement with Fig. 4-7. As for magnitudes, the equation shows that

$$S = W$$

Using the values of W obtained from Example 4-2, you have for the billiard ball

$$S = 6.9 \text{ N}$$

For the bowling ball you have

$$S = 69 \text{ N}$$

A spring scale is calibrated so that the magnitude S of the force it exerts can be read from the markings on the scale. Because $S = W$, the scale reading gives the weight W of the ball it supports.

A spring scale measures the weight W of an object, not its mass m. That is, the scale measures how much gravitational force is exerted on the object by the earth, not how much matter the object contains. So although we do not measure mass directly in everyday life, we can infer it. For most purposes, the acceleration of gravity can be taken to have the same magnitude everywhere on the surface of the earth. Using the value $g = 9.8 \text{ m/s}^2$, we infer the mass to be $m = W/g$. It is, in fact, the masses of things, rather than their weights, which usually interests us. When we buy an expensive steak, for example, it is the quantity of matter—the mass—which we care about, not the gravitational force exerted on the steak by the earth.

The proportionality between mass and weight under everyday conditions has given rise to a universal but confusing practice. Since it is the mass that is really of interest, the result of the weighing process is almost always expressed directly in terms of the corresponding mass. Thus we speak of an object as "weighing" 5.0 kg. What we mean, precisely speaking, is that the object is supported against the gravitational force acting on it by a scale exerting a force of magnitude $S = W = mg = 5.0 \text{ kg} \times 9.8 \text{ m/s}^2 = 49 \text{ N}$. This is just the force that we expect to be exerted to support an object of 5.0-kg mass located near the earth's surface. So we can correctly infer that the object has a mass of 5.0 kg. Some spring scales are calibrated to read mass directly. This makes them convenient to use, in normal circumstances. How useful would they be if taken to the moon?

In Example 4-4 the mass of a body results in a gravitational force being exerted on it, and the mass also results in the body having inertia. These two different aspects of mass also are involved in treating a falling body, as in Example 4-2. But in the situation considered next it is easier to distinguish between the gravitational role of mass and its inertial role. Example 4-4 is the first to involve more than one dimension, and therefore the first in which the true vector nature of force and acceleration must be taken into account.

EXAMPLE 4-4

A block slides down a long, frictionless plane which is supported from the earth at an angle of inclination $\theta = 37°$ with respect to the horizontal. Find the acceleration of the block and the distance the block has traveled 3.0 s after it starts from rest.

■ Your first step is to make a sketch of the block and plane and to indicate on the sketch an appropriate set of coordinates. A good choice for the coordinates is shown in Fig. 4-8. The x axis lies in the inclined plane, with its positive direction in the direction of the block's motion. The y axis is constructed perpendicular to the

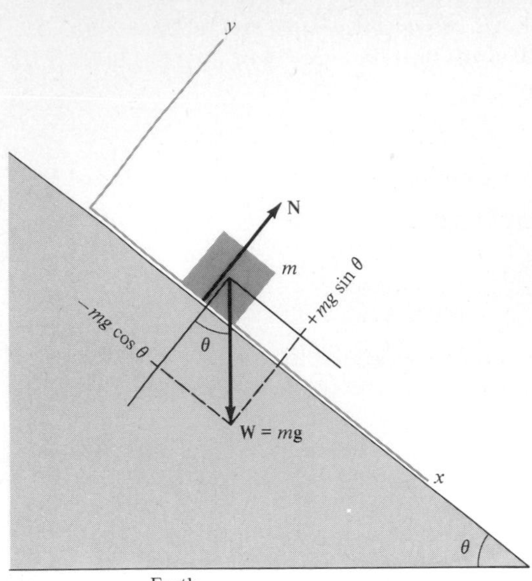

Fig. 4-8 The forces acting on a block of mass m sliding down a frictionless plane inclined to the horizontal at angle θ. Constructed parallel to the plane in the direction of motion, and normal to the plane in the generally upward direction, are x and y axes. The gravitational force **W** exerted by the earth has components $W_x = +mg \sin \theta$ and $W_y = -mg \cos \theta$ along the positive directions of these axes. The force **N** exerted by the plane has only a single nonzero component, $N_y = +N$. The ticks, labelled $+ mg \sin \theta$ and $-mg \cos \theta$, on the axes drawn parallel to the x and y axes illustrate the scheme used here and subsequently to depict the components of a vector. The tick on the axis which is parallel to the x axis marks a distance from the origin of the former whose magnitude is proportional to the magnitude of the x component of the vector. The distance is measured in the positive x direction if the x component has a positive value (as it does in this case), and in the negative x direction otherwise. The same scheme is used to represent pictorially the y component of the vector (which has a negative value in this case). This scheme is compatible with the scheme that is used to depict the vector itself by starting at the origin and measuring a distance, in the direction of the vector, whose magnitude is proportional to the magnitude of the vector.

x axis. Furthermore, the y axis is normal to the inclined plane. (The word **normal** means that *any* plane containing the y axis—not just the plane of the page—is perpendicular to the inclined plane.) The positive direction of the y axis is generally upward. These rectangular coordinates are better to choose than ones in which the x and y axes are horizontal and vertical because they make the block's acceleration vector have only a single nonzero component. With this choice, the vector describing the gravitational force acting on the block has two nonzero components. But only the one along the direction of the block's motion is important since it is the only one which leads to acceleration of the block.

If the mass of the block is m, the gravitational force acting on it has the magnitude $W = mg$. The figure shows that the x and y components of this force are $W_x = mg \sin \theta$ and $W_y = -mg \cos \theta$. There is also a supporting force exerted on the block by the inclined plane. The plane is assumed to be frictionless, and a frictionless plane cannot exert any force in a direction parallel to itself. So this force is directed normal to the inclined plane, just as the force exerted on a puck by an air table is normal to the tabletop. Representing the magnitude of the force by N, you have for its components $N_x = 0$ and $N_y = N$. As for the acceleration of the block, you know that whatever it may be, it has no component in the y direction. This is true since the block neither rises off the plane nor descends through it, so the block does not accelerate in the y direction. Also, the positive x direction is in the direction of the block's acceleration since that is the direction of its motion. Thus if the magnitude of the acceleration is a, then $a_x = a$ and, as just argued, $a_y = 0$.

The vector equation expressing Newton's second law, $\mathbf{F} = m\mathbf{a}$, is equivalent to two scalar equations: $F_x = ma_x$ and $F_y = ma_y$. (Can you explain why?) Consider the one involving x components. The x component F_x of the net force acting on the block is just equal to W_x since $N_x = 0$. Thus you have

$$W_x = ma_x$$

Evaluating W_x and a_x, you obtain

$$mg \sin \theta = ma \tag{4-4a}$$

Canceling the m appearing on both sides of this equality, and then solving for a, gives you

$$a = g \sin \theta \tag{4-4b}$$

For the values specified, you find

$$a = 9.8 \text{ m/s}^2 \times \sin 37° = 5.9 \text{ m/s}^2$$

Knowing a, you can find the distance x traveled in 3.0 s from Eq. (2-30), $x = x_i + v_i t + at^2/2$. With $x_i = v_i = 0$, that equation gives

$$x = \frac{at^2}{2}$$

or

$$x = \frac{5.9 \text{ m/s}^2 \times (3.0 \text{ s})^2}{2} = 27 \text{ m}$$

If you also wanted to determine the magnitude N of the force exerted on the block by the plane, you could find it by considering the equation $F_y = ma_y$. Since $a_y = 0$, it must be that F_y, the y component of the net force acting on the block, is also zero. Thus you have

$$F_y = W_y + N_y = 0$$

Evaluating W_y and N_y, you obtain

$$-mg \cos \theta + N = 0$$

or

$$N = mg \cos \theta \tag{4-5}$$

Inserting the particular values of m and θ, you will find N.

In more general terms, note that for a level plane ($\theta = 0$) this equation yields $N = mg = W$, while for a vertical plane ($\theta = 90°$) it yields $N = 0$. In the first case the frictionless plane fully supports the block, while in the second it gives the block no support at all. These results certainly conform to the results predicted by Eq. (4-4b) for the magnitude a of the acceleration down the plane. With $\theta = 0$ (a level plane) the block is fully supported, and so it will not accelerate, in agreement with the $a = 0$ predicted for this angle. With $\theta = 90°$ (a vertical plane) the plane might as well not be there at all since it gives no support at all to the block. The block will fall freely, in agreement with the predicted value $a = g$.

In the preceding example the inclined plane partially supports the block against gravity. The strength of the "effective gravitational force" acting on the block is the x component, $mg \sin \theta$. This is less than the weight mg of the block, so the inclined plane has "diluted" the effect of gravity. But the mass which the force $mg \sin \theta$ must accelerate is the full mass m of the block. This is why the magnitude a of the acceleration of the block is less than the value g that would be found if the block were falling freely.

Note that the value of the mass m in the left side of Eq. (4-4a), $mg \sin \theta = ma$, determines the strength of the effective gravitational force acting on the block, while the value of the mass m in the right side determines its inertia. In Example 4-5 the role of mass in determining the inertia of a body is even easier to distinguish from its role in determining the gravitational force acting on body. This is because two bodies are involved. For one of them all that matters is its inertia; for the other all that matters is the gravitational force acting on it.

EXAMPLE 4-5

Figure 4-9a is a reproduction of Fig. 3-30. It is a strobe photo showing a puck moving under the influence of a force exerted on it by one end of a string. The string extends from the puck to a swiveling pulley at the center of the table, over the pulley, and through a hole in the table down to a large washer hanging from the other end of the string, as in Fig. 4-9b. The gravitational force acting on the washer is transmitted by the string (whose mass is negligible) to the puck. The puck has

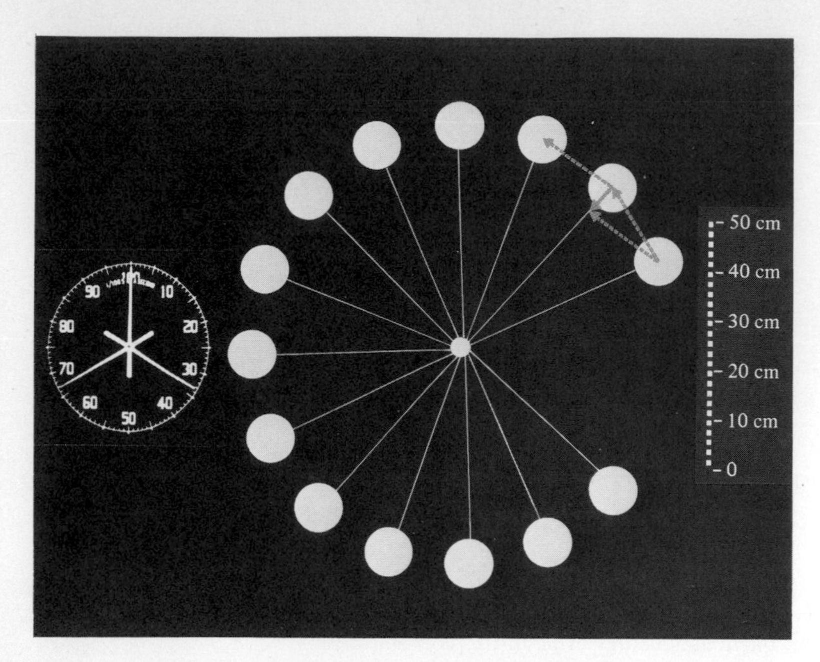

Fig. 4-9 (*a*) Strobe photo of a puck in a circular orbit on an air table. (*b*) A sketch of the apparatus. Shown acting on the stationary washer of mass m is a downward force of strength mg applied to it by the gravitational pull of the earth and an upward force of equal strength applied to the washer by the string. The other end of the string applies an inward force of the same strength mg to the puck of mass M. This force results in an inward acceleration of the puck. The acceleration has magnitude $a = v^2/r$, where v is the speed of the puck and r is its orbit radius. The force of gravity exerted by the earth on the puck, and the canceling normal force exerted on it by the air table top are not shown because they are not of interest. (*c*) The forces acting on the two ends of the string.

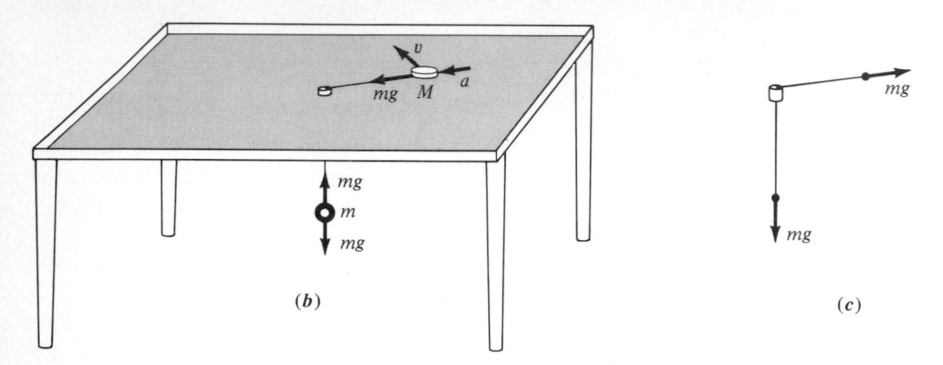

been set into motion in a circular orbit of radius r, and it moves around this orbit at speed v. Measurements made on Fig. 3-30 showed that $r = 0.44$ m and $v = 0.54$ m/s. Additional measurements showed the mass of the puck to be $M = 0.33$ kg. What is the mass m of the washer hanging from the string?

■ As you learned in Sec. 3-5, the acceleration **a** of the puck moving at constant speed around a circular orbit is always directed from the puck to the center of its orbit. Since **F** = M**a**, where M is the puck's mass, the net force **F** acting on it is also directed to the orbit's center. Any set of cartesian (x, y) coordinates would be difficult for you to use in this case, since **F** and **a** would have time-varying components. But polar (r, θ) coordinates with an origin at the center of the orbit are already being used, since the dimensions of the orbit are given in terms of the radial coordinate r. In these coordinates the acceleration of the puck has a component only along the inward radial direction, and the same is therefore true of the net force acting on it. According to Eq. (3-41*a*), the magnitude of this centripetal acceleration of the puck is

$$a = \frac{v^2}{r}$$

Newton's second law tells you that the magnitude F of the net force which produces the acceleration is

$$F = Ma$$

or

$$F = \frac{Mv^2}{r}$$

The net force acting on the puck is directed inward along the string and is exerted by the string connected to the puck. The magnitude of F is the weight mg of the washer hanging from the lower end of the string. Thus you have

$$F = mg$$

Combining the last two equations, you obtain

$$mg = \frac{Mv^2}{r} \tag{4-6a}$$

or

$$m = \frac{Mv^2}{gr} \tag{4-6b}$$

The numerical value of the mass of the washer is

$$m = \frac{0.33 \text{ kg} \times (0.54 \text{ m/s})^2}{9.8 \text{ m/s}^2 \times 0.44 \text{ m}} = 0.022 \text{ kg} = 22 \text{ g}$$

The mass m in the left side of Eq. (4-6a) determines the force that *gravity* exerts on the washer suspended from one end of the string. Since for the situation considered in the example the washer never accelerates anyway, its inertia is of no real consequence. In contrast, the mass M in the right side of the equation determines the *inertia* of the puck connected to the other end of the string. Its inertia is significant because it is accelerating. But since the puck is supported by the air table, whatever force is exerted downward on it by gravity is exactly canceled by the upward force exerted on it by the air film on which the puck rides. So the force of gravity on the puck is of no consequence. You can see now why it is particularly easy to distinguish the gravitational function of mass from its inertial function, for the system analyzed in Example 4-5. Can you invent another system which allows this to be done?

All the systems we have studied to illustrate Newton's second law also contain many examples of *Newton's third law of motion.* For instance, the washer at the lower end of the string in Example 4-5 applies a downward force to that end of the string. (It is the force, shown in Fig. 4-9c, pulling on the lower end of the string.) At the same time the lower end of the string applies a force to the washer of equal magnitude but directed upward. (It is the force, shown in Fig. 4-9b, supporting the washer.) The string and the washer interact because they are connected. The *interaction* between the two involves a *pair of forces:* the force which the string exerts on the washer and the force which the washer exerts on the string. These two forces have *equal magnitude* but *opposite direction.*

Whenever a force is applied to an object, something else must be interacting with this object in some way. Furthermore, if body 1 is interacting with body 2, then body 2 necessarily must be interacting with body 1. (That is what we mean by *inter*action.) So if a force is applied to body 1 as a result of its interaction with body 2, there must also be a force applied to body 2 as a result of its interaction with body 1. Forces come in pairs. The two forces in each pair are related by **Newton's third law of motion:** *If there is a force exerted on body 1 by body 2, there is also a force exerted on body 2 by body 1. The two forces have equal magnitudes but opposite directions.* Expressed in symbols, the third law is

$$\mathbf{F}_{\text{on 1 by 2}} = -\mathbf{F}_{\text{on 2 by 1}} \tag{4-7}$$

The pair of forces is called an **action-reaction pair.** (Which force is the action and which the reaction is arbitrary; it depends on which member of the pair you think about first.)

Note that the action force is exerted on *one* of the interacting bodies and the reaction force on the *other.* The two forces are *never* exerted on the same body. So you should *never add them* to reach the false conclusion that they produce a zero net force on the same body because they cancel.

Note also that Newton's third law is not restricted to inertial reference frames. Motions are not well defined unless there is a specification of the reference frame from which they are observed. But this is not true of forces. Despite the fact that it is called the third law of motion, the law refers to forces, not to motions. In fact, the third law applies no matter what reference frame is used to observe the interacting bodies that exert forces on one another.

In illustrating the third law, Newton wrote: Whatever draws or presses another is as much drawn or pressed by that other. If you press a stone with your finger, the finger is also pressed by the stone. If a horse draws a stone tied to a rope, the horse, if I may say so, will be equally drawn back towards the stone; for the distended rope, by the same endeavour to relax or unbend itself, will draw the horse as much towards the stone as it does the stone towards the horse, and will obstruct the progress of the one as much as it advances that of the other.

Example 4-5 provides yet another illustration of an action-reaction pair. The puck's end of the string applies a radially inward force to the puck. (This **centripetal force** produces the puck's centripetal acceleration; it is shown in Fig. 4-9*b*.) The puck applies a radially outward force of the same magnitude to its end of the string. (This **centrifugal force** is the force pulling on the puck's end of the string, shown in Fig. 4-9*c*, which prevents the string from running over the pulley in response to the force pulling on the washer's end.) What other instances of action-reaction pairs can you find in the system considered in Examples 4-1 through 4-5?

You should not let the cases of action-reaction pairs that have been cited give you the impression that actual contact between the two interacting bodies is necessary. Example 4-6 concerns a pair of forces exerted between two bodies which interact while separated.

EXAMPLE 4-6

The 7.0-kg bowling ball of Example 4-2 is falling toward the earth because of the gravitational force exerted on it by the earth. The two interacting bodies are separated. But the separation is assumed to be small compared to the radius of the earth. Hence, the gravitational acceleration of the bowling ball can be assumed to have the magnitude measured near the surface of the earth; that is, $a = g = 9.8\ \text{m/s}^2$. What is the force exerted *on* the earth *by* the bowling ball? If this force were the only force acting on the earth, what would be the earth's acceleration? Measurements and calculations described in Chap. 11 show that the mass of the earth is 6.0×10^{24} kg.

■ According to Newton's third law, you know that

$$\mathbf{F}_{\text{on earth by ball}} = -\mathbf{F}_{\text{on ball by earth}}$$

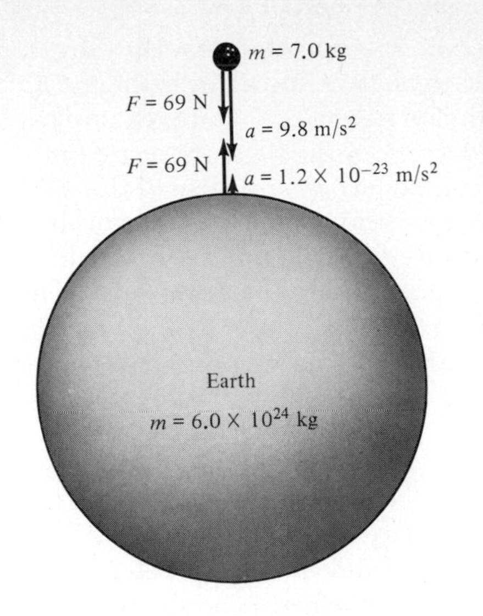

Fig. 4-10 A schematic illustration of a bowling ball and the earth, ignoring the presence of all other bodies. The size of the ball has been much exaggerated for the sake of clarity. The same is true of its separation from the earth's surface; the actual separation is small compared to the earth's radius. The magnitude of the acceleration of the ball toward the earth has the standard value of the gravitational acceleration, and the magnitude of the force exerted on the ball by the earth is the ball's mass times that value. The force exerted on the earth by the ball is also shown. According to Newton's third law, the two forces have equal magnitude but opposite direction. The force exerted on the earth causes it to have an acceleration toward the ball. But since this force has the same magnitude as the force exerted on the ball, Newton's second law requires that the mass of the earth times the magnitude of the earth's acceleration be equal to the mass of the ball times the magnitude of the ball's acceleration. This means the acceleration of the earth has an extremely small value compared to the acceleration of the ball, because the mass of the earth is extremely large compared to the mass of the ball. If the separation between the ball and the earth is comparable to, or larger than, the earth's radius, then the magnitude of the gravitational force exerted on the ball by the earth is reduced. But Newton's third law still applies, so there is still a force of equal magnitude but opposite direction exerted on the earth by the ball. Although no bowling balls are known to be out there, an example of such a situation is provided by the moon and the earth. Each exerts a gravitational force on the other, and the forces have equal magnitudes but opposite directions. The action of the earth on the moon provides the force required to keep it in its approximately circular orbit about the earth. The reaction force of the moon on the earth is best known through the ocean tides, of which it is the principal underlying cause.

The magnitude of the force on the right side of this equation is just the weight W of the bowling ball. Its value is

$$W = mg = 7.0 \text{ kg} \times 9.8 \text{ m/s}^2 = 69 \text{ N}$$

Thus you have for the magnitude of the force on the left side

$$F_{\text{on earth by ball}} = 69 \text{ N}$$

The direction of the force is opposite to the direction of the force exerted on the bowling ball by the earth. That is, the force exerted on the earth by the bowling ball is in the direction from the earth's center to the bowling ball's center. See Fig. 4-10.

If the only force acting on the earth were the one exerted by the bowling ball, then it would be the net force acting on the earth. You can determine the magnitude of the acceleration it would give to the earth by using Newton's second law:

$$a_{\text{of earth}} = \frac{F_{\text{on earth}}}{m_{\text{of earth}}}$$

Setting $F_{\text{on earth}} = 69$ N and using the quoted value $m_{\text{of earth}} = 6.0 \times 10^{24}$ kg, you obtain

$$a_{\text{of earth}} = \frac{69 \text{ N}}{6.0 \times 10^{24} \text{ kg}} = 1.2 \times 10^{-23} \text{ m/s}^2$$

The direction of this acceleration is also from the center of the earth to the center of the bowling ball.

A force of magnitude 69 N is a very significant one in affecting the motion of a body having the inertia of the 7.0-kg ball. It gives the body an acceleration of 9.8 m/s². But such a force is completely negligible, as judged by the effect it would produce on the motion of a body having the inertia of the earth. The acceleration would be only 1.2×10^{-23} m/s². Strictly speaking, a bowling ball does not accelerate toward an immovable earth when it falls. If these were the only bodies present, the earth would accelerate toward the bowling ball. But practically speaking, the acceleration of the earth could be neglected because it would be immeasurably small.

Newton's second law in the *special* form, $\mathbf{F} = m\mathbf{a}$, and Newton's third law, $\mathbf{F}_{\text{on 1 by 2}} = -\mathbf{F}_{\text{on 2 by 1}}$, make it possible to analyze successfully the great majority of systems considered in newtonian mechanics. Many examples involving their use have been presented in this section. Many more are presented in the next chapter. But there is a completely *general* form of the second law. This general form must be used to study some very important systems. It can be obtained by following a more fundamental, and sounder, logical approach than we have followed in this section. The same approach will clarify the origin of the third law. This approach, presented in the next three sections, is based on the experimental law of momentum conservation.

4-3 MASS AND MOMENTUM CONSERVATION

We obtained a clearer understanding of what happens when a bowling ball falls toward the earth in Example 4-6 than we did in Example 4-2. The reason is that we considered the behavior of both the earth and the bowling ball in Example 4-6, whereas the earth was ignored in Example 4-2. This situation is typical. We can go only so far by treating a single body with a force acting on it. After all, that force must be exerted *by* something—that is, by another body. To analyze matters thoroughly, we must at the very least consider the reaction force exerted on this other body by the first one.

In this section we study the simplest possible system of bodies in which Newton's laws can be applied in a complete fashion. The system consists of two bodies that interact with one another but in effect do not interact with anything else. The two bodies are effectively isolated from their environment because no *net* force is exerted on either of them by their environment. They form what is called an **isolated system.** We will view the system from a reference frame that we can consider to be an inertial frame. Specifically, we will investigate a system of two moving air table pucks which interact with each other if and when they collide. But in effect neither of them interacts with the nearby earth. This is because they are on the top of the air table, and the gravitational force exerted on each puck by the earth is canceled by the force exerted on it by the supporting air film. Since, moreover, the top of the air table is almost friction-free, there is no appreciable frictional force acting on the pucks. So essentially no net force is applied to either puck from outside the system of two pucks, and they can be considered to form an isolated system. We will observe the pucks with a strobe camera supported from the earth. Thus the system will be viewed from a reference frame that can be considered to be an inertial frame.

We could accept the largely intuitive justification given for Newton's second and third laws in the preceding section and then *apply* the laws to predict the behavior of this system. But we will not do so. Instead, we will start afresh and will *obtain* Newton's second and third laws from an analysis of experiments performed on the system. The analysis will be carried out in a series of steps. The first step is to use the experimental observations to give a rigorous definition of the basic concept of *mass.* At the same time we will be led to consider the important quantity *momentum,* which is the product of mass and velocity, and will find the fundamental *law of momentum conservation:* the total momentum of an isolated system viewed from an inertial frame remains constant. In subsequent steps, which we carry out in Secs. 4-4 and 4-5, the observations are used to formulate Newton's second and third laws in a logically consistent way. (We do not need to reconsider

Newton's first law. We already have a satisfactory understanding of this law, which has to do with the much simpler situation of an isolated system containing only one body.)

The analysis we will go through in these three sections involves only the *inertial properties of mass,* not the gravitational properties. The relation between mass and inertia will enter the analysis in a very significant way because the velocity of each puck will change when the pucks collide. But just as it was for the puck in orbit on the top of an air table in Example 4-5, the gravitational force is of no consequence here in analyzing the behavior of the pucks. Whatever the force of gravity acting on an air table puck may be, it is automatically canceled by the force provided by the air film supporting the puck. Therefore the relation between mass and gravitational force will not enter the analysis. The experiments considered in Sec. 4-2 have taught us already most of what we need to know about the gravitational role of mass. The remainder of what we need to know we will learn about from experiments discussed in Sec. 11-2.

The system in Fig. 4-11 consists of two *identical* pucks, supported on an air table. The pucks are made out of a relatively hard plastic. Puck 1 is set into motion by a launcher at the upper right. It strikes puck 2, which is initially at rest near the center of the table. After puck 1 leaves the launcher, the system of two pucks is effectively isolated from its environment.

The two pucks interact with each other when puck 1 strikes puck 2. The actual interaction takes place over a very short time. But while the collision time is short, the collision has a great effect on the motion of both pucks. After the collision occurs, puck 2 moves toward the bottom of the strobe photo, and puck 1 moves in a changed direction. You can see that the speed of puck 1 has also been changed because it travels a shorter distance in the constant strobe flash interval. Can you see any relation between the initial velocity of puck 1 and the final velocities of pucks 1 and 2?

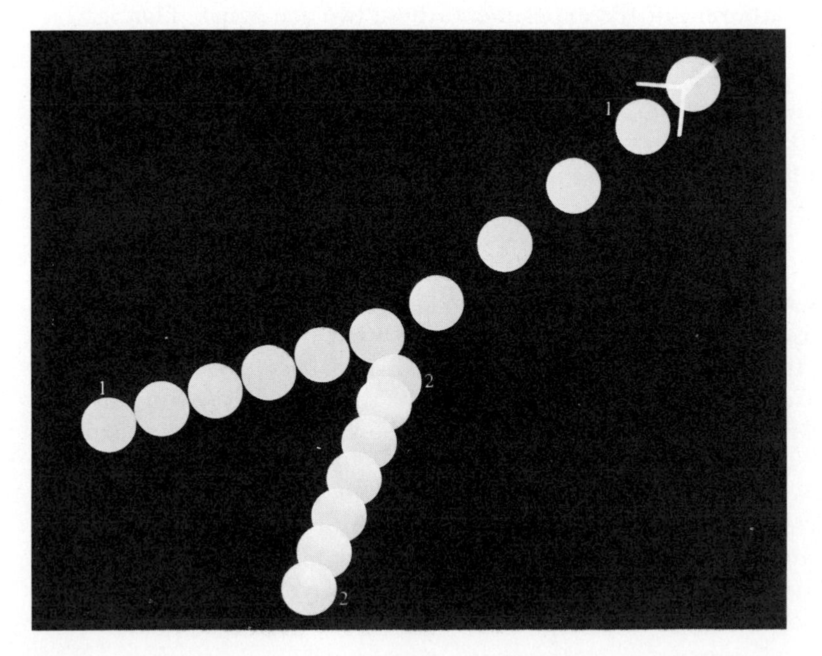

Fig. 4-11 Strobe photo of a collision between two identical plastic pucks on an air table. Puck 1 is incident from the upper right on puck 2, which is initially at rest near the center of the table.

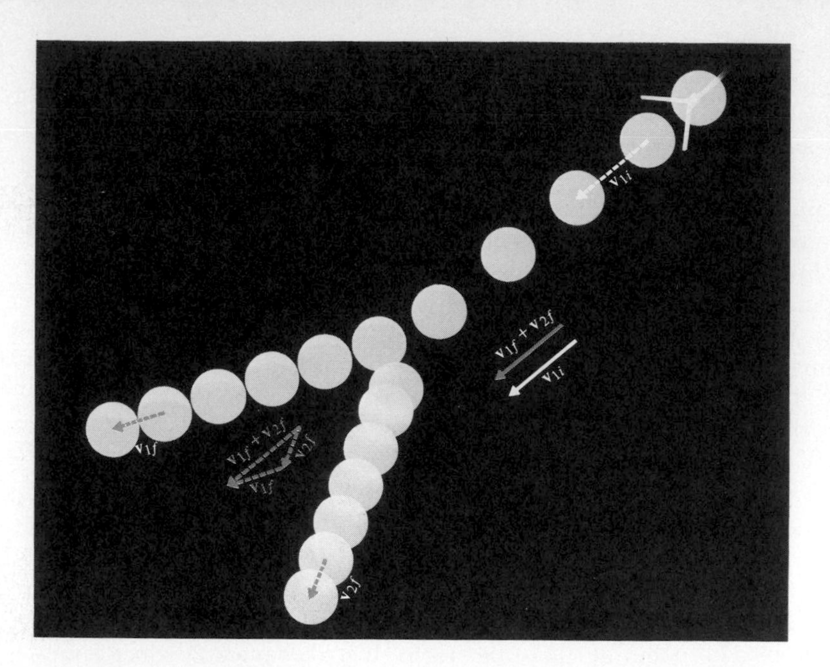

Fig. 4-12 Analysis of the initial and final velocities of the collision between identical plastic pucks shown in Fig. 4-11. Defining the time unit as the duration of the interval between successive strobe light flashes, as in Fig. 4-3, allows the velocity vectors to be constructed simply by connecting adjacent positions of the puck centers.

Figure 4-12 demonstrates that there is such a relation. Initial and final velocity vectors are constructed in exactly the same way as in Fig. 4-3. The strobe flash interval is taken to be the unit of time. As a consequence of this choice, vectors connecting adjacent puck positions are velocity vectors because they give the displacements per unit time. The initial velocity of puck 1 is given by the dashed white vector labeled $\mathbf{v}_{1i}$. Its final velocity is the dashed gray vector $\mathbf{v}_{1f}$. Puck 2 has zero initial velocity. Its final velocity is the dashed gray vector $\mathbf{v}_{2f}$. The relation between the initial and final velocities is seen by moving $\mathbf{v}_{1f}$ and $\mathbf{v}_{2f}$ together to construct their sum $\mathbf{v}_{1f} + \mathbf{v}_{2f}$. Then this vector and the vector $\mathbf{v}_{1i}$ are moved together so that they can be compared. The comparison is shown by the solid gray and white vectors. Within the accuracy of the experiment, these two are equal. That is,

$$\mathbf{v}_{1f} + \mathbf{v}_{2f} = \mathbf{v}_{1i}$$

We use the symbol $\mathbf{v}_{2i}$ for the initial velocity of puck 2 (which has zero value), and write this experimental relation between the initial and final velocities of the identical pucks in the form

$$\mathbf{v}_{1f} + \mathbf{v}_{2f} = \mathbf{v}_{1i} + \mathbf{v}_{2i} \qquad \text{where } \mathbf{v}_{2i} = 0$$

We would like to see whether this relation might be valid for the more general case where $\mathbf{v}_{2i} \neq 0$, that is, where both pucks are moving before the collision. The experiment recorded in Fig. 4-13 is a collision, between the same two identical plastic pucks, which tests the hypothesis. But this time puck 1 is launched from the upper right of the photograph with velocity $\mathbf{v}_{1i}$ to collide with puck 2 as it is moving from the left with velocity $\mathbf{v}_{2i}$. After the collision, puck 1 recoils to the bottom of the picture with velocity $\mathbf{v}_{1f}$, and puck 2 recoils to the lower left with velocity $\mathbf{v}_{2f}$.

The analysis of the velocity vectors is like that in Fig. 4-12. The two initial velocities are added. The same is done for the two final velocities. Then the sums of the initial and final velocities are compared. Again, we find

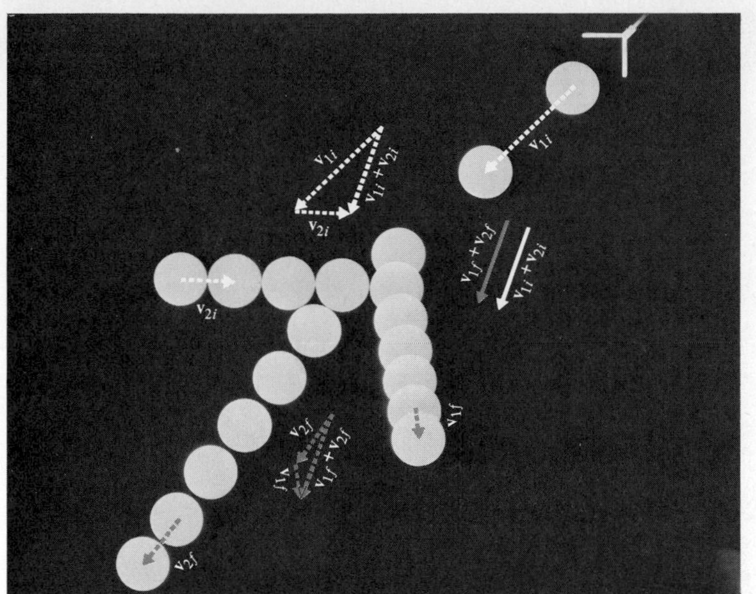

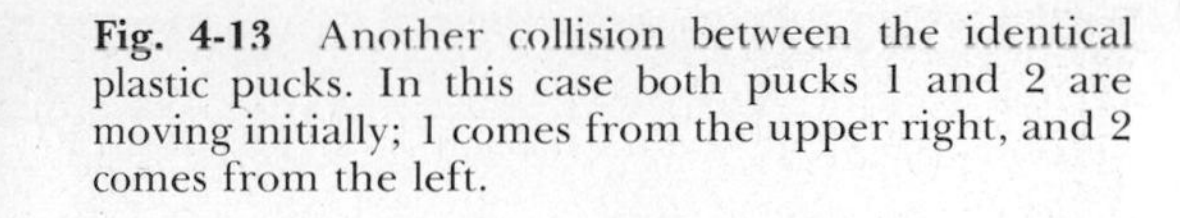

Fig. 4-13 Another collision between the identical plastic pucks. In this case both pucks 1 and 2 are moving initially; 1 comes from the upper right, and 2 comes from the left.

these sums are equal within experimental accuracy. So the rule for this more general collision between the identical plastic pucks is

$$\mathbf{v}_{1f} + \mathbf{v}_{2f} = \mathbf{v}_{1i} + \mathbf{v}_{2i} \tag{4-8}$$

The collision does not change the *sum* of the velocities of the identical pucks in this isolated system viewed from an inertial frame.

If we investigate any other collision between identical plastic pucks, we find that the same relation holds. It is also valid for collisions between identical pucks which interact in ways that are quite different from the way plastic pucks interact when they collide. We show this by studying a collision between two pucks with magnets mounted in them so that they repel each other. Thus they "collide" without actually touching. They have a *magnetic interaction* instead of the *contact interaction* that takes place between plastic pucks when touching.

The magnetic puck collision is of the kind called elastic. An elastic collision is defined as a collision in which the *speed* of one body *relative* to the other as they move apart after the collision has the same value as the *speed* of one *relative* to the other as the two bodies move together before the collision. That is, the condition for an **elastic** collision is

$$|\mathbf{v}_{1f} - \mathbf{v}_{2f}| = |\mathbf{v}_{1i} - \mathbf{v}_{2i}| \tag{4-9a}$$

The quantity $\mathbf{v}_{1f} - \mathbf{v}_{2f}$ is the final velocity of puck 1 relative to puck 2, since it is the velocity of puck 1 as seen by an (imaginary) observer moving with puck 2. You can verify this by letting $\mathbf{v}_{2f} = \mathbf{V}$ in the Galilean velocity transformation, Eq. (3-57). The magnitude of this quantity, $|\mathbf{v}_{1f} - \mathbf{v}_{2f}|$, is therefore the final relative speed of the two pucks. Similarly, $|\mathbf{v}_{1i} - \mathbf{v}_{2i}|$ is their initial relative speed. The plastic puck collisions are said to be **inelastic** since

$$|\mathbf{v}_{1f} - \mathbf{v}_{2f}| < |\mathbf{v}_{1i} - \mathbf{v}_{2i}| \tag{4-9b}$$

(You can easily show the inelastic character of a plastic puck collision by

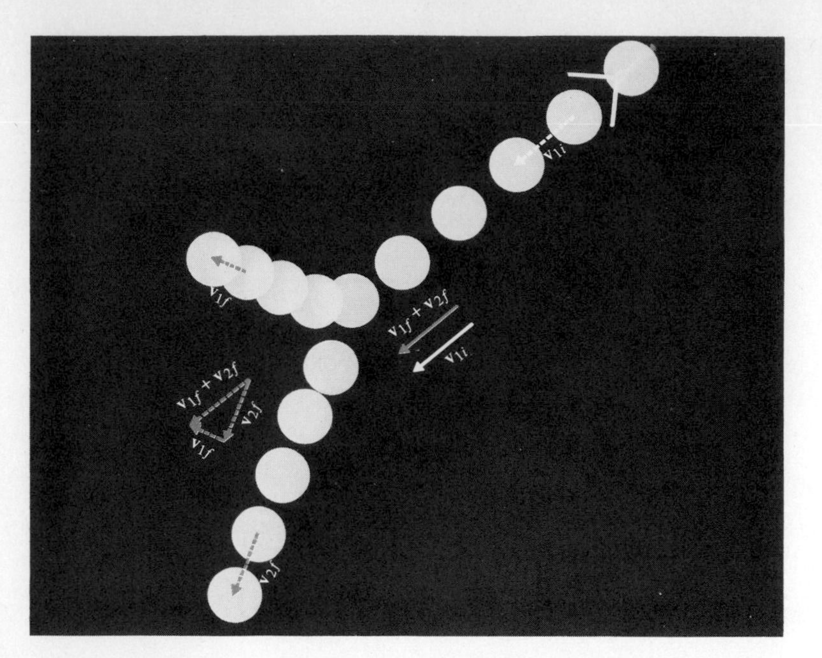

Fig. 4-14 A "collision" between two identical magnetic pucks. Puck 1 comes from the upper right, and puck 2 is initially at rest near the center of the air table. Note that there is never any actual contact between the pucks.

going back to Fig. 4-12. Construct graphically the velocity at which puck 1 recedes from puck 2 after the collision by taking the vector difference $\mathbf{v}_{1f} - \mathbf{v}_{2f}$ between their two velocities. Then compare its magnitude to the magnitude of $\mathbf{v}_{1i}$, the velocity at which puck 1 approaches the initially stationary puck 2 before the collision.)

In Fig. 4-14 puck 1 is a magnetic puck launched with velocity $\mathbf{v}_{1i}$ from the upper right of the photograph. It collides (without actually touching) with puck 2, an identical initially stationary magnetic puck. After the collision, puck 2 moves with velocity $\mathbf{v}_{2f}$ toward the bottom of the picture, while puck 1 moves with velocity $\mathbf{v}_{1f}$ toward the left. Note that the trajectories of the pucks after the collision make a 90° angle. This is a hallmark of elastic collisions at nonrelativistic speeds between identical bodies, one of which is initially stationary. If you look back at Fig. 4-12, you will see that there the angle between the trajectories is less than 90°. The angle is smaller because the pucks do not move apart as rapidly after an inelastic collision. [You can show that the collision in Fig. 4-14 is elastic by following the instructions given in parentheses below Eq. (4-9*b*).]

The manipulations of the velocity vectors in the elastic collision of Fig. 4-14 show you that Eq. (4-8), $\mathbf{v}_{1f} + \mathbf{v}_{2f} = \mathbf{v}_{1i} + \mathbf{v}_{2i}$, is valid. Thus the equation applies to this elastic collision between identical pucks that interact magnetically, just as it applies to inelastic collisions between identical pucks that interact on contact.

So far we have considered only collisions between identical pucks. Now we will investigate what happens when the pucks are not identical. We will find that the velocity relation for identical pucks does not hold. But we will also find a complete general relation, only slightly more complicated, which is satisfied, and which includes the velocity relation as a special case.

On the basis of common experience and the discussion in Sec. 4-2, it is easy to guess that the relation for the collision of nonidentical pucks will involve their masses. But in the discussion which follows we will abandon the definition of mass given in Sec. 4-2, and begin anew. While that definition is perfectly satisfactory for most practical purposes, it has a serious difficulty. We gave the name "mass" to the proportionality constant m which appears in Newton's second law, $\mathbf{F} = m\mathbf{a}$. But the "force" $\mathbf{F}$ had not been very adequately defined. Indeed, we later defined it in terms of mass—the procedure smacks of being a logically circular argument. Out of the following discussion will come a resolution of the difficulty.

Before the next experiment was performed, three identical magnetic pucks were procured. Two of them were glued together, one on top of the other, to form a double puck, called puck 2. The collision investigated took place between the double puck and the remaining single puck, called puck 1.

Figure 4-15 shows the collision process. Puck 1 is launched with velocity $\mathbf{v}_{1i}$ at the initially stationary puck 2. After they collide, puck 1 has velocity $\mathbf{v}_{1f}$, and puck 2 has velocity $\mathbf{v}_{2f}$. You can tell immediately that things are different from the experiment of Fig. 4-14, which used two identical magnetic pucks. Because the pucks are magnetic in the present experiment, the collision is elastic. This can be verified by constructing the vector differences of Eq. (4-9). But the angle between the two final trajectories is no longer 90°. The comparison in Fig. 4-15 of the sums of the initial and final velocities shows that things are indeed different in this case of colliding nonidentical pucks. Contrary to all the identical-puck cases, the sum of the velocities after the collision is *not* the same as the sum of the velocities before the collision. Can we find a quantity that *is* unchanged by the collision?

We may guess that what is missing from the "after = before" equation is the factor *mass*. Whatever mass may be, puck 2, the double puck, has twice as much of it as puck 1, the single puck. That is, $m_2/m_1 = 2$. Therefore it seems reasonable to try multiplying the vector $\mathbf{v}_{2f}$ of the double puck by 2 before adding it to the vector $\mathbf{v}_{1f}$ of the single puck. This is done in Fig. 4-16, and it works quite well!

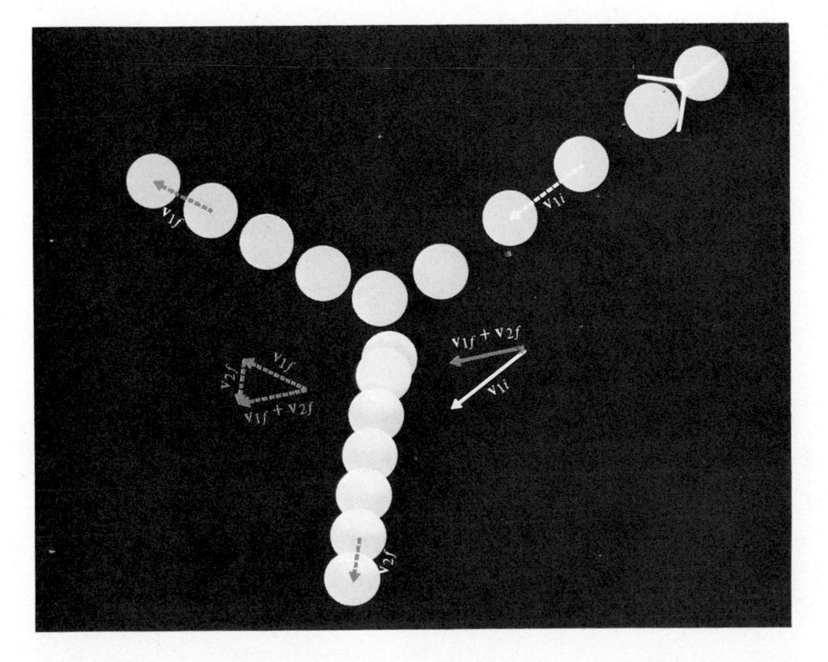

Fig. 4-15 A collision between a single and double magnetic puck. The single puck, called puck 1, comes from the upper right, and the double puck, called puck 2, is initially at rest near the center of the air table.

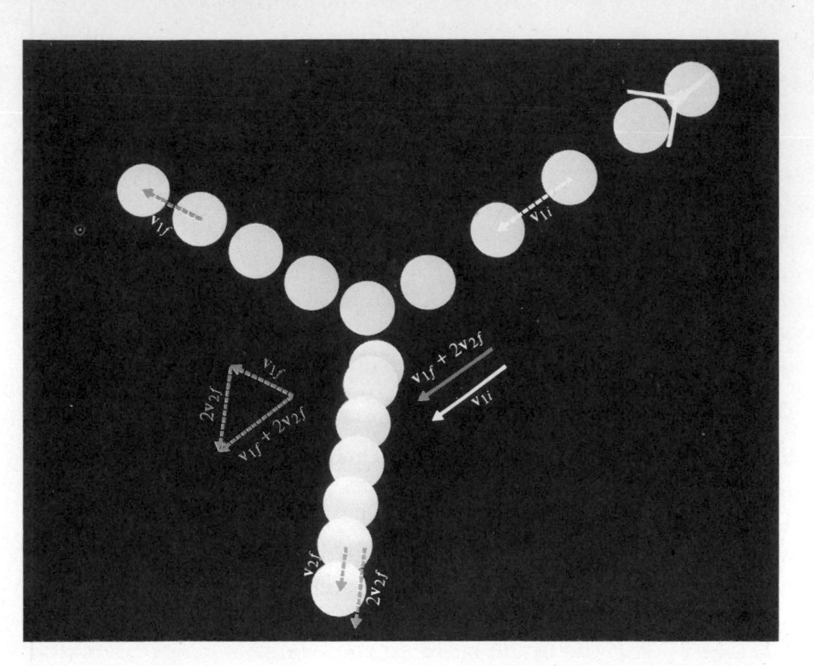

Fig. 4-16 An analysis of the collision between the single and double magnetic pucks, shown in Fig. 4-15, in which the velocity vector of the double puck is doubled in magnitude.

The guess is tried again in the experiment shown in Fig. 4-17. Puck 1, the single magnetic puck, comes in initially from the upper right with velocity $\mathbf{v}_{1i}$. It recoils after the collision to the upper left with velocity $\mathbf{v}_{1f}$. Puck 2, the double magnetic puck, comes in from the left with velocity $\mathbf{v}_{2i}$ and recoils to the lower right with velocity $\mathbf{v}_{2f}$. The initial and final velocity vectors for the double puck are doubled to account for the double mass, and then each is added to the corresponding velocity vector for the single puck. Here again the sum of the initial vectors is equal to the sum of the final vectors, within the accuracy of the measurement.

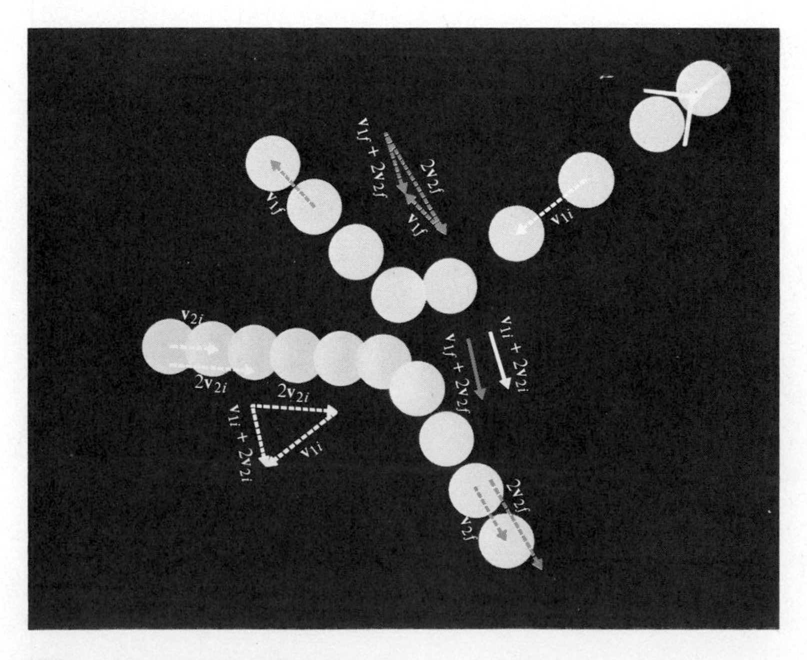

Fig. 4-17 A collision between a single and a double magnetic puck. In this case both puck 1, the single puck, and puck 2, the double puck, are moving initially; 1 comes from the upper right, and 2 comes from the left.

Thus the results of both of the nonidentical puck experiments that we have studied are in agreement with the equation

$$\mathbf{v}_{1f} + \frac{m_2}{m_1}\mathbf{v}_{2f} = \mathbf{v}_{1i} + \frac{m_2}{m_1}\mathbf{v}_{2i} \tag{4-10}$$

We therefore take a bold step and *define* the mass ratio by Eq. (4-10). In other words, we define **mass ratio** as follows: *Make bodies 1 and 2 collide while effectively isolated from their external environment, and use an inertial frame to measure their initial and final velocities. Then find the value of* m_2/m_1 *for which* $\mathbf{v}_{1f} + (m_2/m_1)\mathbf{v}_{2f}$ *equals* $\mathbf{v}_{1i} + (m_2/m_1)\mathbf{v}_{2i}$. *This value is the ratio of the mass of body 2 to the mass of body 1.* The definition of mass ratio that we have given is founded on just two things: (1) the definition of velocity (which itself depends only on the definitions of distance and time) and (2) a vast amount of experimental evidence exemplified by the two experiments we have discussed.

Once a way of defining mass ratio is available, the procedure for defining mass itself is just a matter of agreeing on the unit of mass. The unit could be the mass of a standard air table puck. Instead it usually is the mass of a standard kilogram. Thus we have the following definition for **mass**: *Perform a collision experiment between bodies 1 and 2 in which body 1 has a mass* m_1 *equal to* 1 kg, *and determine the mass ratio* m_2/m_1 *from the preceding definition. The mass* m_2 *of body 2 is* m_2/m_1 kg. With this procedure for defining mass we avoid the circular reasoning involved in the more intuitive definition of Sec. 4-2.

Multiplying Eq. (4-10) through by the mass m_1, we have

$$m_1\mathbf{v}_{1f} + m_2\mathbf{v}_{2f} = m_1\mathbf{v}_{1i} + m_2\mathbf{v}_{2i} \tag{4-11}$$

It is convenient to have a name for the quantity mass times velocity. So we give it the name **momentum** and the symbol **p**. That is, by definition *the momentum* **p** *of a body equals its mass m times its velocity* **v**:

$$\mathbf{p} \equiv m\mathbf{v} \tag{4-12}$$

The momentum is a vector since it is the product of a scalar (the mass) and a vector (the velocity).

The fact that the sum of the momenta of a system is not affected by a collision of its constituent parts does not depend on the collision being elastic. In order to demonstrate this, we do an experiment similar to the last one, but using plastic pucks which collide inelastically. Three identical plastic pucks are obtained. One of them is used as puck 1, and the other two are glued together to form puck 2. Figure 4-18 shows puck 1 coming from the upper right with momentum $\mathbf{p}_{1i}$ and colliding with puck 2 coming from the left with momentum $\mathbf{p}_{2i}$. For convenience, these momentum vectors are drawn to such a scale that the length of the vector showing the initial momentum of puck 1 is the same as the length of the vector that would be drawn in the manner of the preceding figures to show its initial velocity. With this scale, the initial momentum vector of puck 2 is twice as long as its initial velocity vector would be. This is because puck 2 has twice as much mass as puck 1. After the collision pucks 1 and 2 move away from each other with the momenta labeled $\mathbf{p}_{1f}$ and $\mathbf{p}_{2f}$, respectively. The momentum vector analysis is shown in the figure. As in all experimental situations, the results are not perfect. But they are certainly good enough to give

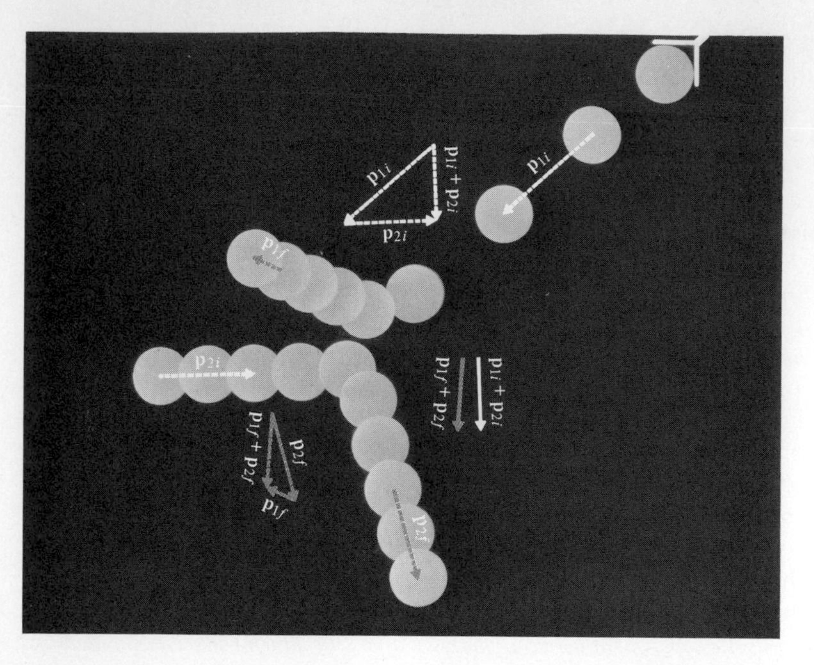

Fig. 4-18 The momenta in a collision between a single and a double plastic puck. Puck 1, the single puck, comes from the upper right, and puck 2, the double puck, comes from the left. The pucks appear to collide without quite touching. But this is only because the strobe light did not happen to flash at the instant of collision. A click heard by the experimenter at the moment of collision made it clear that there was actual contact.

one more verification of the relation obtained by using Eq. (4-12) in Eq. (4-11):

$$\mathbf{p}_{1f} + \mathbf{p}_{2f} = \mathbf{p}_{1i} + \mathbf{p}_{2i}$$

or

$$(\mathbf{p}_{\text{total}})_{\text{final}} = (\mathbf{p}_{\text{total}})_{\text{initial}} \tag{4-13}$$

We have found the **law of momentum conservation.** This *fundamental* law of physics can be stated as follows: *An observer in an inertial frame sees a system experiencing no net interaction with its environment maintain a constant total momentum.* An equivalent statement is: *The total momentum is conserved in an isolated system observed from an inertial frame.*

The momentum conservation law certainly is in agreement with the results of all the identical-puck experiments. To see this, write Eq. (4-11) explicitly for the case $m_1 = m_2$. Then cancel the masses, and you will obtain Eq. (4-8). In fact, there is a tremendous amount of experimental evidence taken from many different fields of physics which shows that the law of momentum conservation holds in *any* system of two or *more* particles of *any* type that interact with one another in *any* way, provided the system is effectively isolated from all external influences and is viewed from an inertial reference frame. Our search for a rigorous definition of mass has been very successful. Not only have we found the desired definition, but also we have found the most broadly applicable conservation law in mechanics.

The next example illustrates the use of momentum conservation in a simple collision problem.

EXAMPLE 4-7

A puck of mass $m_1 = 2$ kg moves with a speed $v_{1i} = 3$ m/s in the positive x direction. It hits "head-on" a second puck of mass $m_2 = 4$ kg, which is initially at rest. There is a drop of instant-setting glue on the second puck, so that the two stick together in a **totally inelastic** collision. Find the final velocities of the two pucks.

■ Puck 2 is initially stationary. Thus you have $\mathbf{v}_{2i} = 0$, and Eq. (4-11) reads

$$m_1\mathbf{v}_{1f} + m_2\mathbf{v}_{2f} = m_1\mathbf{v}_{1i}$$

Since the two pucks are joined after they collide, they have the same final velocity, so that

$$\mathbf{v}_{1f} = \mathbf{v}_{2f}$$

Using this equation in the equation above and then factoring, you obtain

$$(m_1 + m_2)\mathbf{v}_{1f} = m_1\mathbf{v}_{1i}$$

You can see that the direction of $\mathbf{v}_{1f}$ must be the same as that of $\mathbf{v}_{1i}$. As for magnitudes, you have

$$(m_1 + m_2)v_{1f} = m_1 v_{1i}$$

or

$$v_{1f} = \frac{m_1}{m_1 + m_2} v_{1i}$$

Inserting the numerical values, you obtain

$$v_{1f} = \frac{2 \text{ kg}}{2 \text{ kg} + 4 \text{ kg}} 3 \text{ m/s} = 1 \text{ m/s}$$

Because physicists are convinced that the law of momentum conservation is universally applicable, they can use it in nuclear and elementary-particle physics to determine experimentally the mass of a newly discovered

Fig. 4-19 Bubble-chamber photographs of a collision between a pion and a proton. The pion enters the chamber, which contains liquid hydrogen, from the left. As it passes near the hydrogen atoms, its electric charge results in ionization of some of them—that is, removal of their electrons. Under proper conditions this leads to boiling of the liquid in the vicinity of the ions, and the track consists of the tiny bubbles visible in the photo. Because the plane containing the track of the entering pion and the tracks of the scattered pion and proton is arbitrary, a pair of photos taken simultaneously from two different angles is required to make quantitative measurements. In this pair, the lower photo is a view looking down on the collision through the flat top of the glass chamber. The upper photo was taken through the flat left side of the chamber, as seen by an observer looking along the path of the incident pion. (The upper photo may be regarded as having been "folded upward." That is, the bottom edge of the left side of the chamber is seen at the top of the photo.) Liquid hydrogen consists entirely of protons and electrons, the latter having negligible mass compared to the pion. After traveling through about four-fifths of the 10-cm length of the bubble chamber without having come very close to any of the relatively massive but tiny protons, the pion by chance collides with a proton whose initial velocity is essentially zero and whose mass is an order of magnitude greater than that of the pion. The pion is deflected through a large angle; it moves downward (as seen in the upper photo), to the right (as seen in the lower photo) and back the way it came (as seen in both photos). It leaves the bubble chamber through the bottom surface. The proton recoils through a relatively short distance before coming to rest as a result of repeated collisions with hydrogen atoms. A few large bubbles of hydrogen are visible in the photos, as well as the tracks of energetic electrically charged particles not involved in the pion-proton collision. The faint grid lines are 1 cm apart. These historic 1956 photographs were made in the course of the first experiment ever performed using a liquid hydrogen bubble chamber. (*Courtesy Roger H. Hildebrand, Enrico Fermi Institute, University of Chicago.*)

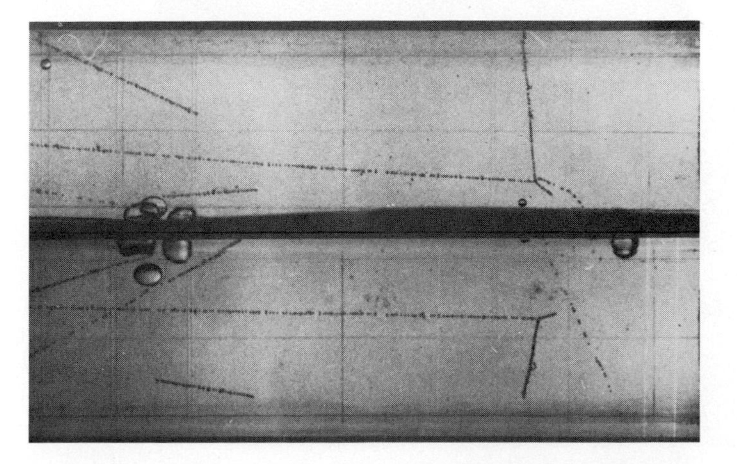

particle. This is done by first measuring the initial and final velocities in a collision between it and a particle of known mass and then determining the mass ratio which will satisfy momentum conservation. Such a collision between microscopic particles is shown in the bubble chamber picture of Fig. 4-19. Example 4-8 carries through the analysis for a collision between macroscopic particles.

EXAMPLE 4-8

Figure 4-20 shows a collision between an incident puck of unknown mass approaching from the upper left and an initially stationary puck which it strikes. Using the mass of the struck puck to define a unit mass and employing momentum conservation, determine the mass of the incident puck in terms of this unit.

▪ First, take advantage of your intuition to guess whether the incident puck is more or less massive than the struck puck. Imagine a collision between a moving billiard ball (a body of small mass) and a stationary bowling ball (a body of large mass). Then imagine a collision between a moving bowling ball and a stationary billiard ball. In which case would you expect that both balls move off after the collision in the same general direction as the direction of motion of the incident ball?

Now that you have guessed that the incident puck is more massive than the struck puck, try different values for the unknown mass of the incident puck until you find a value which satisfies the condition that the total momentum of the system be conserved.

In Fig. 4-21 the momentum conservation analysis is carried out using $m_{str} = 1$ mass unit for the mass of the struck puck and the three trial values $m_{inc} = 2$ mass units, 3 mass units, 4 mass units for the mass of the incident puck. For each value a construction compares the final momenta of the two pucks (with the vectors arranged so that their sum is apparent) and the initial momentum of the incident puck (the only one that has an initial momentum). This is done, as in Fig. 4-18, by drawing the momentum vectors to a scale in which the length of the momentum vector of the struck puck is the same as the length of its velocity vector. The constructions show that the total final momentum of the system equals its total initial momentum, within experimental accuracy, only for the choice $m_{inc} = 3$ mass units. That is, (3 mass units)$\mathbf{v}_{inc,f}$ + (1 mass unit)$\mathbf{v}_{str,f} \simeq$ (3 mass units)$\mathbf{v}_{inc,i}$. Thus the

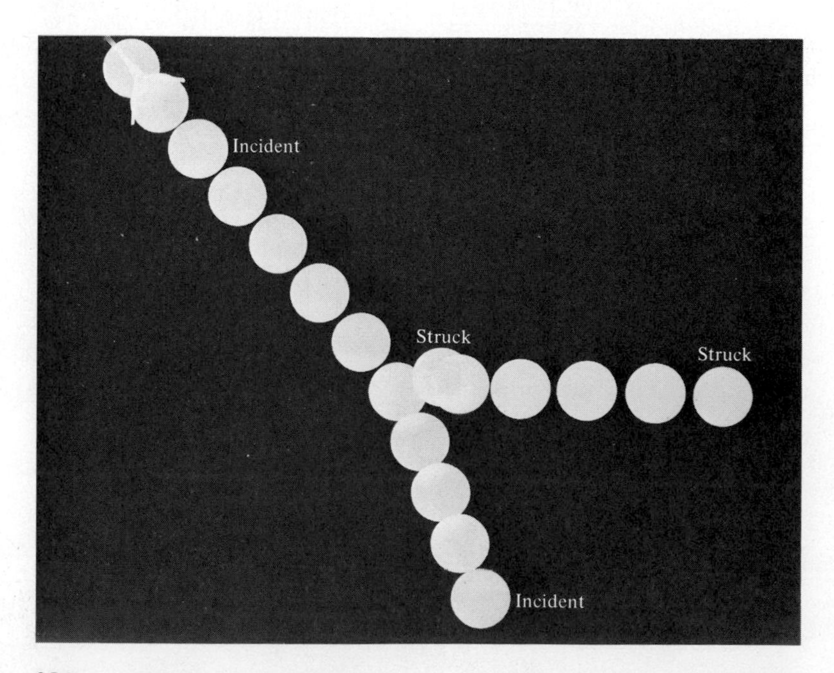

Fig. 4-20 A collision between a puck of unknown mass incident from the upper left on a puck of unit mass that is initially at rest near the center of the air table. The incident puck is labeled "Incident" and the struck puck is labeled "Struck."

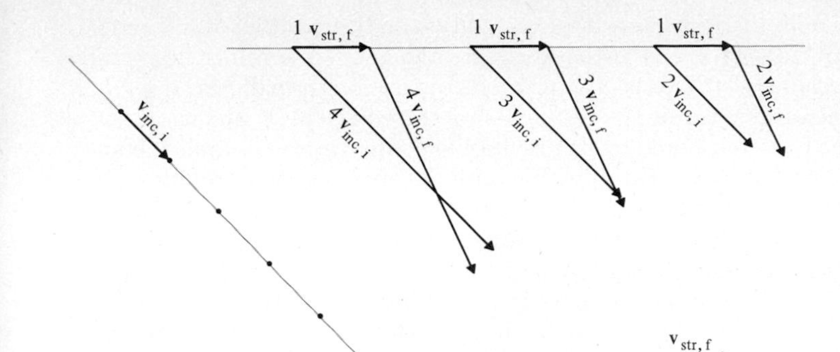

Fig. 4-21 Analysis of the initial and final momenta in the puck collision shown in Fig. 4-20. The dots used to construct the velocity vectors are at the centers of the puck images in that figure.

mass of the incident puck is determined to be approximately 3 mass units, where the mass unit is the mass of the struck puck. Of course, more accurate velocity measuring techniques would lead to a more accurate determination of the mass.

You may wish to analyze the initial and final relative speeds to determine whether the collision was elastic or inelastic. But it makes no difference as far as the mass determination is concerned because momentum conservation always holds, independent of the type of interaction involved in the collision.

The procedure used for a particular case in Example 4-8, and stated for the general case below Eq. (4-10), constitutes an **operational definition** of mass. It defines the quantity mass, in terms of an agreed-upon unit of mass, by specifying a set of operations for measuring the quantity. As you continue your study of physics, you will find that operational definitions are used quite frequently.

Operational definitions deliberately avoid answering such questions as: What is mass, really? They may therefore strike you at first as unsatisfactory skirtings of the really interesting questions in physics. But the answer to such a question as What is mass, really? is something like: "Mass is the amount of matter." This answer only convinces you that you understand something, when you have merely swept the issue under the rug by defining the quantity in question in terms of still another undefined quantity. It does not help to say that "Mass is the product of density and volume." Since density is mass divided by volume, such a definition commits the logical error of defining a quantity in terms of itself. In contrast, the operational definition is logically consistent, and it tells you what you really need to know to do physics: How to measure the quantity defined.

A question of a quite different nature is: What is the physical effect of mass? This question has two satisfactory answers. The first one is: "The mass of a body is a measure of its inertia, that is, how much it resists changes in its velocity." In an interaction between two bodies of the same mass, any change in the velocity of one is exactly compensated for by an equal but oppositely directed change in the velocity of the other. Equation (4-11) shows that when bodies of unequal mass interact, the ratio of the changes in their velocities is the negative of the reciprocal of their mass ratio. The more massive body experiences the smaller velocity change because it has more inertia.

The second answer to the question about the physical effect of mass is: "The mass of a body is a measure of the strength of the gravitational force that some other particular body will exert on it when the two bodies have a certain separation." For instance, Eq. (4-2) shows that the mass of a body determines how much gravitational force the earth exerts on it when its separation from the earth is small compared to the earth's radius. Experiments of extremely high precision show that the mass of a body measured by its gravitational effect always equals the mass measured by its inertial effect. Thus the gravitational effect of mass provides a *second way* of defining mass operationally. As an example, the mass of the earth is measured by (in other words, defined in terms of) the strength of the gravitational force that the earth exerts on some other body of known mass when the two bodies have a certain separation. We have more to say about these matters in Chap. 11.

To summarize, the procedure stated below Eq. (4-10) gives us a rigorous way of defining the basic quantity mass in terms of its inertial effect. This definition is of great theoretical significance because it puts newtonian mechanics on a firm logical foundation and, as you will see in Chap. 15, because it plays the central role in the development of relativistic mechanics. In the macroscopic world the definition is not of great practical significance since it is rarely used to measure the mass of an object. Instead, the mass of an object of macroscopic size (say a billiard ball) is almost always measured by some technique that involves its gravitational effect. It is usually much easier to measure accurately the mass of such a body by weighing it than by studying how it behaves in a collision with some other body. But in the microscopic world the way we have found of defining mass in terms of its inertial effect (or some related way) *must* be used to measure the mass of a body. The reason is that for a microscopic body (say an electron) the gravitational effect of its mass is so minute that it is impractical to measure.

4-4 FORCE AND NEWTON'S SECOND LAW

Let us consider again two interacting, but otherwise isolated, bodies viewed from an inertial frame—two pucks on an air table. The law of momentum conservation applies to the *total* momentum of the two bodies. But it does not apply to the *individual* momenta of each of these bodies since momentum is transferred between them when they interact. We investigate this fact by fixing our particular attention on *one* of the bodies. That is, we respecify the system of interest so that it now contains only *one* puck. If we follow that puck through a collision, we note that its momentum does not remain constant, even though we observe it from an inertial frame. The momentum of the puck that we are concerned with changes because it is not isolated from its environment. The system containing one puck is not an isolated system because something external to the system acts on the body it contains and causes its momentum to change. An agency acting on a body which leads to a change in its momentum is called a **force.** Thus we say that *the momentum of a body changes when it is acted on by a force.*

Figure 4-22 is a special strobe photo depicting a collision between two magnetic pucks. In order to show what happens when they interact in the collision, the camera shutter was opened for only a short period spanning the collision, and a very short strobe flash interval was used. To prevent confusion from multiple overlapping puck images, the pucks were painted black except for a white dot at the center of each. Taking the puck launched from the top of the photo to be the one of interest, you can see that it initially moved toward the bottom of the photograph with approximately constant momentum. Then it slowed and also changed direction. Next it increased its speed while continuing to change direction. Finally it moved off to the right with an approximately constant momentum, different from its original momentum in both direction and magnitude. The

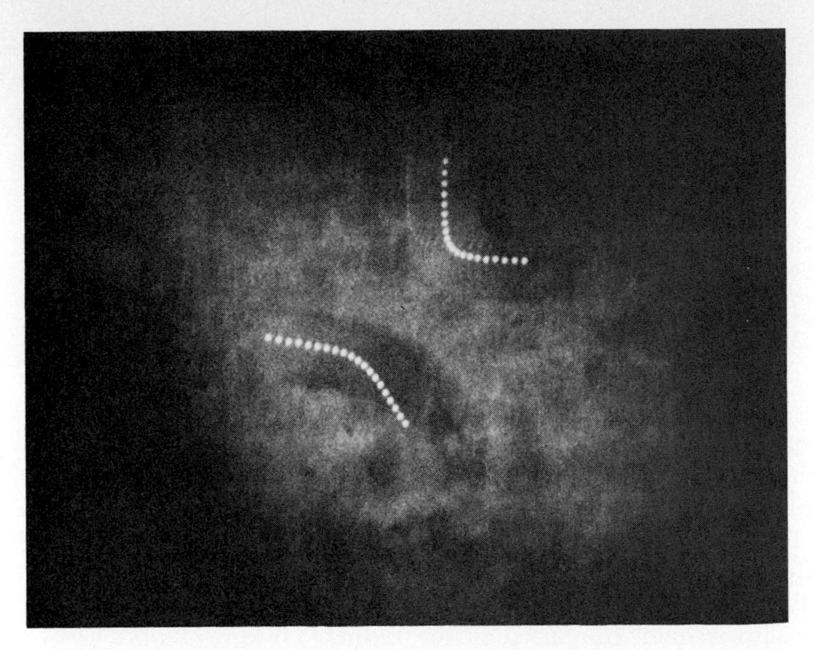

Fig. 4-22 A short-flash-interval strobe photo of a collision between two magnetic pucks. Multiple overlapping images were avoided by using pucks which were black except for a central white dot. The technique used caused more light to be reflected from the air table into the camera than in the other strobe photos.

change in momentum developed gradually while the puck was in proximity to the other puck. In fact, careful inspection will show you that the greatest momentum change in one strobe flash interval occurred when the puck was closest to the other puck. The reason is that the force exerted on the puck of interest by the other puck was strongest when the two pucks were closest.

A quantitative measure of the strength and direction of the force acting on the puck of interest is obtained by measuring the *rate* at which it caused the momentum of that puck to change. That is, the **net force F** acting on a body at some instant is specified in magnitude and direction by the time derivative $d\mathbf{p}/dt$ of the momentum $\mathbf{p}$ of the body:

$$\mathbf{F} = \frac{d\mathbf{p}}{dt} \tag{4-14}$$

This is **Newton's second law of motion,** in its most basic form: *Net force equals rate of change of momentum.* Remember that it assumes the observer is in an inertial frame of reference.

In the approach we are taking, this fundamental "law" is actually a definition. *Newton's second law,* $\mathbf{F} = d\mathbf{p}/dt$, *is the definition of force.* The definition follows from, but is distinct from, the definitions of mass and momentum. Note that the development leads necessarily to the conclusion that force is a vector quantity and therefore must satisfy the rules for the addition of vectors. This is the case since force is defined as the limiting value of a vector (the momentum change) divided by a scalar (the time change) and so must be a vector. Figure 4-23, and the analysis explained in the caption, lends conviction to the consistency of the definition by presenting experimental evidence for the vectorial nature of force. It shows that *the net force acting on a body is the vector sum of all the forces acting on it.*

(*a*)

Fig. 4-23 A demonstration of the vectorial nature of force. (*a*) Bodies whose weights are in the ratio 2 to 3 to 4 exert forces with magnitudes equal to their weights on the lower ends of three strings. The strings pass over pulleys at the rim of the circular table, and transmit these forces to the ring to which they are tied. The locations of the pulleys are adjusted until the ring remains stationary at the center of the table with the three forces acting on it. (*b*) A top view with vectors depicting the forces. Their magnitudes are in the ratio 2 to 3 to 4. (*c*) A graphic construction which sums the three force vectors to give the net force exerted on the ring. To within the accuracy of the technique, the net force is seen to have the value $\mathbf{F} = 0$. This value agrees with the prediction of Newton's second law, $\mathbf{F} = d\mathbf{p}/dt$, since the momentum of the ring maintains the constant value $\mathbf{p} = 0$ and so $d\mathbf{p}/dt = 0$. The agreement confirms the vectorial nature of force because the graphic construction is based on treating forces as vectors when summing them. It also shows that the net force acting on a body is the vector sum of all the forces acting on it.

(*b*)

(*c*)

For a definition to be worthwhile, it must be not only consistent but also convenient. Equation (4-14) passes the second test too. It is convenient because the forces that occur in nature have relatively simple physical and mathematical descriptions when that equation is taken to be the definition of force. For example, consider a body located anywhere in a region extending upward from the earth's surface through a distance small compared to the earth's radius, and extending along the surface through a distance that also is small compared to the radius. The gravitational force acting on the body is constant in both magnitude and direction. You could not ask a force to have a simpler description than that. Another example is the force exerted on a body by a stretched spring to which the body is connected. The force is constant in direction, the direction lying along the spring axis. The magnitude of the force varies, but the variation is simply in direct proportion to the stretch of the spring. As we proceed in our study of various fields of physics, we will come across numerous other examples of the simplicity of natural forces, when force is defined by the equation $\mathbf{F} = d\mathbf{p}/dt$.

The relation between that equation and $\mathbf{F} = m\mathbf{a}$ is easy to obtain. Since we have defined $\mathbf{p}$ as

$$\mathbf{p} = m\mathbf{v}$$

it is apparent that

$$\mathbf{F} = \frac{d\mathbf{p}}{dt} = \frac{d(m\mathbf{v})}{dt}$$

If, and only if, the mass m is a constant, this immediately yields

$$\mathbf{F} = m\frac{d\mathbf{v}}{dt} = m\mathbf{a}$$

The case of constant mass is the one usually encountered in newtonian mechanics. The great majority of mechanical systems occurring in practical studies certainly comprise bodies with constant mass. But there are some important exceptions. For instance, an engineer studying the motion of a rocket burning its fuel finds it expedient to treat the rocket as a system which is losing mass. In relativistic mechanics mass cannot be treated as a constant in *any* situation, and $\mathbf{F} = m\mathbf{a}$ is *not* correct. But $\mathbf{F} = d\mathbf{p}/dt$ remains valid in relativity, as we will see in Chap. 15.

In Sec. 4-2 we considered a number of examples of the application of $\mathbf{F} = d\mathbf{p}/dt$ for constant m, where the relation can be expressed as $\mathbf{F} = m\mathbf{a}$. Example 4-9 considers a case where m is not constant so that $\mathbf{F}$ is not equal to $m\mathbf{a}$.

EXAMPLE 4-9

Figure 4-24 shows crushed ore from a mine dropping onto a very long conveyor belt at a rate of 300 kg/s. The belt moves at a speed of 2.00 m/s. Find the net force that must be acting on the belt to keep it moving at this constant speed, neglecting friction in the rollers supporting the belt.

▪ If you take the belt plus the ore lying on it to be the system, then you have a simple example of a system with varying mass to which you can apply $\mathbf{F} = d\mathbf{p}/dt$. The mass at time t is m, and the momentum at that instant is $m\mathbf{v}$, where $\mathbf{v}$ is the velocity of the belt. At time $t + dt$ the mass is $m + dm$, and momentum is thus $(m + dm)\mathbf{v}$. Therefore the change in momentum in time dt is

$$d\mathbf{p} = (m + dm)\mathbf{v} - m\mathbf{v}$$

or

$$d\mathbf{p} = \mathbf{v}\, dm$$

Dividing by dt gives you

$$\frac{d\mathbf{p}}{dt} = \mathbf{v}\frac{dm}{dt}$$

Equation (4-14) tells you that $\mathbf{F} = d\mathbf{p}/dt$. That is, the rate of change of the system's momentum (due to its increasing mass) equals the force applied to it (to keep its velocity constant). So, equating $d\mathbf{p}/dt$ to the necessary force $\mathbf{F}$, you obtain

$$\mathbf{F} = \mathbf{v}\frac{dm}{dt} \tag{4-15}$$

This result shows that the force is in the direction of the belt's velocity, and its magnitude is the product of the velocity of the belt and the rate at which mass is added to it.

Since the speed v of the belt is 2.00 m/s and the rate dm/dt at which mass is added to it by the falling ore is 300 kg/s, the magnitude of the force, $F = v\, dm/dt$, is

$$F = 2.00 \text{ m/s} \times 300 \text{ kg/s} = 600 \text{ kg·m/s}^2 = 600 \text{ N}$$

In Example 4-9, $\mathbf{v}$ is constant. But this need not always be the case. To obtain an expression for $\mathbf{F}$ that applies more generally, we consider Newton's second law

$$\mathbf{F} = \frac{d\mathbf{p}}{dt} = \frac{d(m\mathbf{v})}{dt}$$

when neither m nor $\mathbf{v}$ is constant. Then we apply Eq. (2-15), the rule for differentiating a product of two variables. The result obtained immediately is

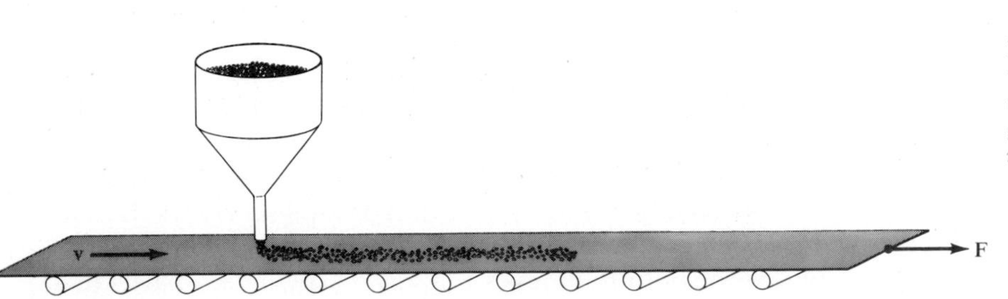

Fig. 4-24 Material dropping onto a very long conveyor belt. The belt plus the material on it form a system which is gaining mass. The equation $\mathbf{F} = m\mathbf{a}$ does *not* apply to such a system.

$$\mathbf{F} = m\frac{d\mathbf{v}}{dt} + \mathbf{v}\frac{dm}{dt} \tag{4-16}$$

Note that the second term on the right side of this equation is the only term present on the right side of Eq. (4-15), the expression for **F** developed in the example. The reason is that $d\mathbf{v}/dt = 0$ there. But in the great majority of systems treated in newtonian mechanics $dm/dt = 0$, and only the first term on the right side is present in Eq. (4-16). Then it becomes $\mathbf{F} = m\, d\mathbf{v}/dt = m\mathbf{a}$.

4-5 MOMENTUM CONSERVATION AND NEWTON'S THIRD LAW

When we previously studied the strobe photo of a collision between two magnetic pucks shown in Fig. 4-22, we singled out the puck in the upper part of the photograph to be the one of interest. Then we discussed the relation between its rate of change of momentum and the force acting on it. But we could just as well have focused our attention on the other puck. That puck also experiences a change in momentum, and therefore a force equal to the rate of change of its momentum is acting on the puck. It is apparent from the photograph that there is some symmetry between what happens to one puck and what happens to the other. The first puck exerts a repulsive force on the second, which changes its momentum; at the same time the second puck exerts a repulsive force on the first, thereby changing *its* momentum. We know from momentum conservation measurements that we have made on many such collisions that there is, in fact, a perfect symmetry between the two momentum changes. Since the two pucks form an effectively isolated system when considered together, each momentum change must equal the negative of the other because the sum of their two momenta must remain constant. An intimately related manifestation of this same symmetry is that each force is the negative of the other—the pair of forces obeys Newton's third law.

It is easy to derive Newton's third law mathematically by combining the experimental law of momentum conservation with the definition of force given by Newton's second law. Consider two bodies, 1 and 2, which interact with each other in any way but which have no net interactions with anything else. In an inertial reference frame, their momenta obey the momentum conservation law of Eq. (4-13):

$$\mathbf{p}_{1f} + \mathbf{p}_{2f} = \mathbf{p}_{1i} + \mathbf{p}_{2i}$$

or

$$\mathbf{p}_1 + \mathbf{p}_2 = \text{constant}$$

Take the time derivative of all terms in this equation, to obtain

$$\frac{d\mathbf{p}_1}{dt} + \frac{d\mathbf{p}_2}{dt} = 0 \tag{4-17a}$$

or

$$\frac{d\mathbf{p}_1}{dt} = -\frac{d\mathbf{p}_2}{dt} \tag{4-17b}$$

The rate at which the momentum of body 1 is changing must be the exact opposite of the rate at which the momentum of body 2 is changing because the momentum of the entire isolated system remains constant.

Since there is a rate of change $d\mathbf{p}_1/dt$ in the momentum of body 1, as viewed from an inertial frame, the definition of Eq. (4-14) says there is a net force acting on the body. This force is $\mathbf{F}_{\text{on 1 by 2}}$, the force exerted on it by body 2. Applying the definition gives $\mathbf{F}_{\text{on 1 by 2}} = d\mathbf{p}_1/dt$. Similarly, the force exerted on body 2 by body 1 is determined by the definition of force to be $\mathbf{F}_{\text{on 2 by 1}} = d\mathbf{p}_2/dt$. Using these two applications of the definition in Eq. (4-17*b*) produces immediately the result

$$\mathbf{F}_{\text{on 1 by 2}} = -\mathbf{F}_{\text{on 2 by 1}} \tag{4-18}$$

The forces which the bodies exert on each other have *equal magnitude but opposite direction*. Equation (4-18) is **Newton's third law of motion.** Thus Newton's third law does not describe a separate property of nature; it is contained in the experimental fact of momentum conservation and in the way force is defined.

Interesting questions arise concerning the applicability of the third law to situations in which objects interact at an appreciable distance, if one object or the other has a characteristic important to the interaction that is changing abruptly in time. An example is the interaction between electrons in two separated radio antennas. When the electrons in the first antenna are briefly put into oscillation, the electrons in the second will experience a corresponding pulse of oscillation, but only after a certain time delay. In turn, the pulsed oscillation of the electrons in the second antenna will induce a pulsed oscillation of the electrons in the first, after an additional time delay. Thus it appears that the electrons in each antenna exert forces on those in the other antenna, but only after a certain time delay. Such forces cannot satisfy Eq. (4-18) because they are not of equal magnitude but opposite direction at any instant.

When we study the electromagnetic force we will see that electrons in the two antennas do not interact directly with each other. Instead they interact *indirectly*. As a result of their oscillatory motion, the electrons in one antenna emit what is called electromagnetic radiation. The radiation moves away from the antenna at the speed of light. Part of it reaches the other antenna, after the time it takes for the radiation to travel between the two. There it is absorbed by the electrons in that antenna and causes them to oscillate. The radiation does convey forces from one set of electrons to the other because it carries momentum which goes from the emitting electrons to the absorbing electrons. But each of the steps of the process, emission and absorption, must be considered separately. Then Newton's third law is satisfied. The electrons emitting the radiation exert forces on it, and the radiation exerts reaction forces back on them. This interaction takes place during a particular interval of time and in a particular region of space, just like a collision between two pucks. And in the pair of forces involved in the interaction the two forces have equal magnitude but opposite direction. At a later time and in a different region, there is another interaction, the one between the radiation being absorbed and the electrons absorbing it. Newton's third law is also satisfied in this step of the process.

Thus there is actually no difficulty with the third law, even in this apparently difficult case. For the cases generally treated in newtonian mechanics, these questions never even arise.

4-6 FORCES IN MECHANICAL SYSTEMS

Now we will make a preliminary inquiry into the properties of some of the forces commonly involved in the behavior of mechanical systems. Our purpose is to develop enough information about these forces so that in the next chapter we can use them in Newton's laws to study the motion of mechanical systems. The approach we use in this section is mostly **empirical.** That is, we concentrate on the experimentally observed properties of certain types of forces, putting aside until later in the book any serious attempt to understand the origins of these properties in a fundamental way. Nevertheless, we will learn enough about the forces to be able to solve many practical problems involving the behavior of systems governed by them.

As an example, consider the force whose magnitude is called weight. At this juncture we have not really addressed the question of *why* a body has weight. We have merely indicated it is a manifestation of the gravitational interaction between the earth and the body. We inquire more deeply into gravitational force in Chap. 11. In the meantime, however, we can use to advantage the experimental fact considered in Example 4-2 and the associated discussion: All bodies near the surface of the earth fall with acceleration of the same magnitude g (when air resistance is negligible). This fact, taken together with the relation among force, mass, and acceleration imposed by Newton's second law, led us to a now familiar conclusion: The **gravitational force** exerted by the earth on a body of mass m has a magnitude (the **weight** of the body) given by

$$W = mg \tag{4-19}$$

if the separation between the body and the earth's surface is small compared to the radius of the earth. The direction of the force is downward, that is, toward the center of the earth. We have already used this empirical description of the gravitational force to explain the behavior of mechanical systems. We continue to use it in subsequent chapters.

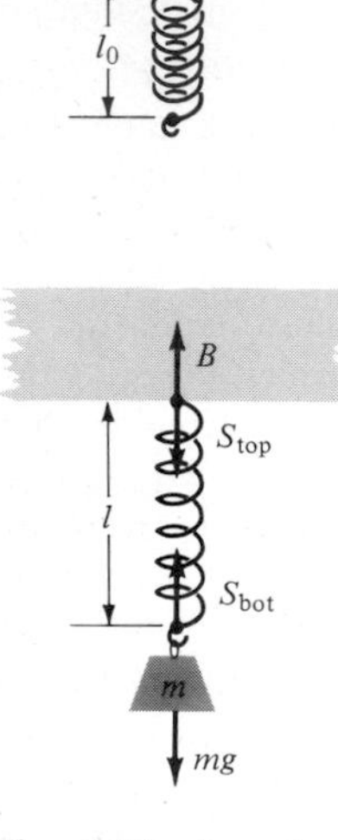

Fig. 4-25 A spring hanging from a beam has a length l_0. It is then stretched to a length l by suspending a body of mass m from its lower end. The forces shown are the important ones acting when the spring is stretched.

Another force which has a simple, and very useful, empirical description is the force exerted at either end of a compressed or stretched spring on whatever is connected to the end. Consider a coil spring whose coils are not in contact with each other when the spring is relaxed. The spring can be compressed by pushing on both ends, or it can be stretched by pulling on both ends. A convenient way to stretch the spring is to attach one end to a rigid beam and hang a body of mass m from the other end, as in Fig. 4-25. When the body hangs at rest from the extended spring, the magnitude S_{bot} of the force that the bottom end of the spring applies to the body must just equal the magnitude mg of the gravitational force acting on the body. This equality, $S_{bot} = mg$, must hold because the body is motionless, and therefore Newton's second law says there is no net force acting on it.

If we assume the mass of the spring itself to be negligible, the total gravitational force exerted on the system body-plus-spring also has the magnitude mg. Since this system is motionless, the second law requires the magnitude B of the force exerted on it by the beam to have the value $B = mg$. But the force of magnitude S_{top} applied to the beam by the top end of the spring and the force of magnitude B applied to the top end of the spring by the beam form an action-reaction pair. Therefore Newton's third law requires that $S_{top} = B$. Thus $S_{top} = mg$. Since in the preceding paragraph we showed that $S_{bot} = mg$, we conclude that the two ends of the spring exert forces of the same magnitude on the objects to which they are connected. Note that these forces exerted by the two ends of the *extended* spring are both directed *inward* along the axis of the spring.

The common magnitude S of the spring forces is related to the amount of extension of the spring. The relation can be studied experimentally by hanging bodies of different mass m from the spring, thereby varying the value of S. For each value of S, the length l of the spring is measured. Then the amount of extension is found by subtracting from l the length l_0 of the spring when it is relaxed. Results of measurements on a typical spring look like those in Table 4-1.

From the data we can conclude that the magnitude S of the forces pro-

Table 4-1

The Magnitude S of the Forces Exerted by a Spring at Each End When the Extension of the Spring Is $l - l_0$

Measurement	Magnitude S of spring force (in N)	Extension $l - l_0$ of spring (in m)
1	10	0.008
2	20	0.016
3	30	0.024
4	40	0.032
5	100	0.080
6	200	0.146
7	20	0.016

duced by the extended spring is directly proportional to its extension $l - l_0$, through measurement 5. For the large extension obtained in measurement 6 the rule seems to fail because there is a twofold increase in S from measurement 5 to measurement 6 but there is not a corresponding twofold increase in $l - l_0$. Measurement 7 serves the purpose of checking the reproducibility of the measurements and assures us that stretching the spring in measurement 6 did not permanently alter its characteristics.

The data obtained through measurement 5 can be summarized by an equation relating the magnitude S of the forces exerted by the ends of the extended spring to the extension $l - l_0$ of the spring. The equation is

$$S = k(l - l_0) \tag{4-20a}$$

where k is a constant. Its value can be determined by requiring the equation to conform to the results obtained in measurement 5. Solving for k and using these results, we obtain $k = S/(l - l_0) = 100\ \text{N}/0.080\ \text{m} = 1.2 \times 10^3\ \text{N/m}$ for the particular spring used in the measurements.

What makes these results interesting is that practically all springs behave this way, regardless of the details of their construction, provided they are not stretched too far. Furthermore, when a spring is compressed to a length l shorter than its relaxed length l_0 but not compressed enough to make its coils touch one another, then the magnitude S of the forces exerted by the spring at its end is related to the compression $l_0 - l$ by an equation having almost the same form as Eq. (4-20a). The equation is

$$S = k(l_0 - l) \tag{4-20b}$$

The value of k in Eq. (4-20b) for a certain spring is the same as the value of k in Eq. (4-20a) for that spring. The forces that a *compressed* spring exerts at its ends are directed *outward* along the spring axis.

A single equation can be used to describe the magnitude S of the **spring force.** This is the magnitude of the force exerted on whatever is connected to an end of a spring of relaxed length l_0 when it is compressed or extended to length l. The equation has the form

$$S = k|l - l_0| \tag{4-21}$$

The quantity $|l - l_0|$ is the **distortion** of the spring, the magnitude of the change in its length from the length when it is relaxed. The constant k is called the **force constant** (or sometimes the **spring constant**). It is a property of the spring which specifies how stiff the spring is. The relation is known as **Hooke's law,** after Robert Hooke, a contemporary—and longtime adversary—of Newton. Since the forces produced by a spring are directed inward at each end if the spring is extended, and outward if it is compressed, in all circumstances the forces act on the objects attached to the ends of the spring in directions which tend to make them move so as to restore the spring to its relaxed length. Because of their directional properties, the forces produced by springs are often called **restoring forces.** Can you write a vector form of Hooke's law, giving both the magnitude and direction of the restoring force exerted by the bottom end of the spring in Fig. 4-25?

The proportionality between restoring force and distortion expressed in Eq. (4-21) is not restricted to springs. It is a common property of many kinds of mechanical objects, particularly those composed of crystalline solids. We will gain insight into the fundamental cause of this behavior in Chap. 8. But until then Hooke's law can be considered to be an empirical relation.

Example 4-10 gives you some experience applying Hooke's law.

EXAMPLE 4-10

Two identical coil springs, each having force constant k, are connected end to end. Find the effective force constant k' of the spring pair. That is, find the value of k' in the equation $S = k'|l - l_0|$ relating the magnitude S of the restoring forces exerted by the spring pair at its two ends and the distortion $|l - l_0|$, the magnitude of the change in length of the spring pair from its relaxed length. Then let $k = 1.2 \times 10^3$ N/m, the value found for the spring in Table 4-1, and evaluate k'.

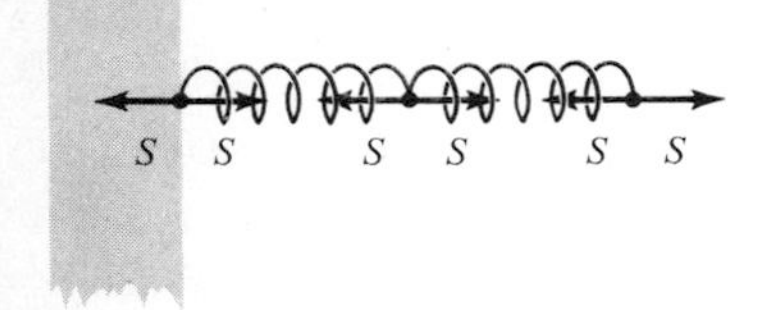

Fig. 4-26 Forces acting when two identical springs, connected end to end, are stretched.

▪ You should make a sketch, as in Fig. 4-26, showing the spring pair when it is distorted from its relaxed length so that forces are being exerted. The distortion can be either an extension or a compression. Suppose it is an extension.

The figure shows four restoring forces, one exerted at each end of each individual spring. Two of these are the restoring forces exerted by the spring pair. They are the force exerted to the right at its left end and the force exerted to the left at its right end. Also shown are a force directed to the right applied to the right end of the spring pair and a force directed to the left applied to its left end. These two are the forces producing the distortion. All the forces in the figure have the same magnitude S. You can prove this by an argument similar to the one used to prove that $S_{top} = S_{bot}$ in Fig. 4-25.

The restoring forces exerted by the spring pair have the same magnitude S as those exerted by each individual spring of the pair. But the change in length of each individual spring is only one-half the change in length of the spring pair. So when $|l - l_0|$ is the distortion of the spring pair, the distortion of each individual spring is $|l - l_0|/2$.

Applying Hooke's law to an individual spring, you have

$$\frac{|l - l_0|}{2} = \frac{S}{k}$$

or

$$|l - l_0| = \frac{2S}{k}$$

Applying the law to the entire spring pair gives you

$$k' = \frac{S}{|l - l_0|}$$

Using the evaluation of $|l - l_0|$ just obtained, you find

$$k' = \frac{S}{2S/k}$$

or

$$k' = \frac{1}{2/k} = \frac{k}{2}$$

This calculation shows that two springs connected in series comprise a weaker spring than either spring alone. The reason is that each contributes its own extension, or compression, to the total extension, or compression, of the pair, yet each exerts forces of the same magnitude as the forces exerted by the pair. In particular, if $k = 1.2 \times 10^3$ N/m, you will have $k' = 0.60 \times 10^3$ N/m.

The results of the example suggest that any spring made by using part of a longer spring will be stiffer than the longer spring. You can easily verify this by stretching a coil spring first from both ends and then from one end and the middle.

Now let us consider *frictional forces.* Seen from the point of view of their effects, frictional forces acting between two objects always do the same thing—they resist any attempt to put one object into motion relative to the other, and tend to slow the motion once the objects are moving relative to each other. Thus they are always directed "backward." The causes of friction are varied, and almost all of them are extremely complicated when studied in detail. Nevertheless, it is possible to give a quantitative account of the *effects* of friction in several cases of great practical importance. It turns out that simple empirical equations often give an accurate enough description of these effects for practical purposes.

There are two broad categories of friction. One is called contact friction, and the other is called fluid friction. **Contact friction** is present when an attempt is made to set one solid object into motion across the surface of another, and also when such motion takes place. **Fluid friction** is present when a solid object moves through a fluid. Frictional forces are the first ones we have encountered which are **velocity-dependent.** For both categories, the direction of the frictional force acting on an object depends on the direction of the velocity it would have if it moved, or actually has if it is moving, since friction always opposes relative motion. And for fluid friction there is also a dependence on the magnitude of the object's velocity, as you will soon see.

We begin by considering *contact friction.* If you apply a weak force in a horizontal direction to a heavy box resting on the floor, the box does not move. According to Newton's second law, the net force on the box must be zero while it remains stationary. Apparently the floor is able to apply a frictional force to the box, equal in strength and opposite in direction to *whatever* horizontal force you apply to it, providing the force you apply is weak. Now gradually increase the strength of the force you apply to the box. Nothing happens until the force you apply to the box exceeds the limiting strength of the frictional force that the floor can apply to the box. When it does so, the box "breaks free" and begins to move.

If the surfaces of two objects in contact are not so rough that there is a gross interlocking of projections on one with depressions on the other, the limiting strength of the contact friction force that can be developed between the surfaces depends mainly on two factors. The first factor is the

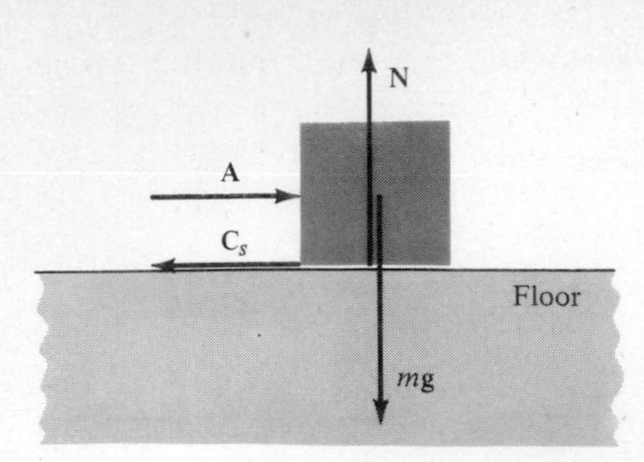

Fig. 4-27 Showing all the forces acting on a box of mass m resting on the floor, when the strength A of a force applied to it in some direction parallel to the plane of the floor just equals the limiting strength C_s of the force that the floor can apply to the box in the opposite direction. In these circumstances the box is on the verge of moving. In all circumstances the floor also applies to the box a force of strength N in the direction normal to its plane, and opposite to that of the gravitational force of equal strength mg.

magnitude of the forces which the objects exert on each other because they are pressed together. These forces act in opposite directions and are of the same magnitude, in agreement with Newton's third law. Each of the directions is normal to the plane containing the surfaces in contact. So each force is called a **normal force,** and its magnitude is represented by the symbol N. In the case of the box resting on the floor, illustrated in Fig. 4-27, the magnitude N of the normal force applied to it by the floor equals the weight mg of the box. The box will be just at the point of moving over the floor when the magnitude A of the force you apply to it in some direction parallel to the plane of contact equals the limiting magnitude of the oppositely directed force which contact friction can apply to the box to prevent its motion over the plane. This limiting frictional force is called the **static contact friction force,** and we designate its magnitude as C_s. Experiment shows that over a wide range of conditions

$$C_s \propto N \tag{4-22a}$$

The static contact friction force is proportional to the normal force. In other words, the greatest force that can be applied to one object without causing it to slide over another is proportional to the force pressing the objects together. Perhaps surprisingly, C_s does *not* depend at all strongly on the area of the surfaces in contact. You might guess that the more frictional surface there is, the more friction there should be. But this is not so, for reasons soon to be discussed.

The other major factor determining how much force can be applied to one object before it starts sliding over another is the specific nature and condition of the surfaces in contact. An approximate representation of the degree of "stickiness" of a pair of surfaces is given by their **coefficient of static friction** μ_s. (The symbol μ is the Greek letter mu.) This quantity is the proportionality constant which will convert the relation between C_s and N into an equality. That is, the relation among the magnitude C_s of the static contact friction force, the magnitude N of the normal force, and the coefficient of static friction μ_s is

$$C_s = \mu_s N \tag{4-22b}$$

The direction of the static contact friction force is always opposite to the direction of the motion it opposes. Since $\mu_s = C_s/N$ is the ratio of two forces, it is a dimensionless number. Some representative values are shown in Table 4-2.

When you set an object into motion, it is often possible to feel it "break loose," as we remarked earlier. In any case, it is usually easier to keep an object in motion against contact friction than it is to start the motion. Experi-

Table 4-2

Coefficients of Static and Kinetic Friction

Surfaces in contact	Approximate value of μ_s	Approximate value of μ_k
Copper and cast iron	1.1	0.3
Steel and steel	0.7	0.5
Steel and wood	0.4	0.2
Steel and Teflon	0.04	0.04

ment shows that the magnitude C_k of the **kinetic contact friction force** acting on an object that is moving does not depend on its speed, if the speed is not too large or too small. But C_k does depend on the magnitude N of the normal force applied to the object by the object in contact with it, and on the **coefficient of kinetic friction** μ_k for the surfaces in contact, according to the relation

$$C_k = \mu_k N \tag{4-23}$$

As is always true for a frictional force, the direction of the kinetic contact friction force exerted by one object on another is opposite to the direction of the relative motion it opposes. The value of μ_k for a given pair of surfaces is usually less than the value of μ_s for that pair. You can see this in the values quoted in Table 4-2.

EXAMPLE 4-11

A block of mass m slides down a plane supported from the earth at an incline. The angle θ_k between the plane and the horizontal is adjusted until the block slides with constant speed. Find the coefficient of kinetic friction μ_k. If θ_k is measured to be 35°, what is the value of μ_k?

■ Since the block is moving, the applicable equation is $C_k = \mu_k N$. In order to use this equation, you must first determine N. To do so, draw a diagram like Fig. 4-28 and define x and y coordinate axes. Just as in Fig. 4-8, it is convenient for you to make these axes parallel to and normal to the plane. The forces acting on the block are the gravitational force of magnitude $W = mg$ applied to it by the earth, the normal force of magnitude N applied to it by the plane, and the frictional force of magnitude C_k that the plane also applies to the block. The figure shows that the gravitational force has two components, $W_x = mg \sin \theta_k$ and $W_y = -mg \cos \theta_k$. The normal force has only a y component, $N_y = N$. The only component of the frictional force is an x component, $C_{k_x} = -C_k$.

Since the block does not accelerate in the y direction, the sum of the y components of the forces acting on it must be zero. Thus you have

$$W_y + N_y = 0$$

or

$$-mg \cos \theta_k + N = 0$$

So you have

$$N = mg \cos \theta_k$$

Using this in Eq. (4-23), $C_k = \mu_k N$, you obtain

$$C_k = \mu_k mg \cos \theta_k$$

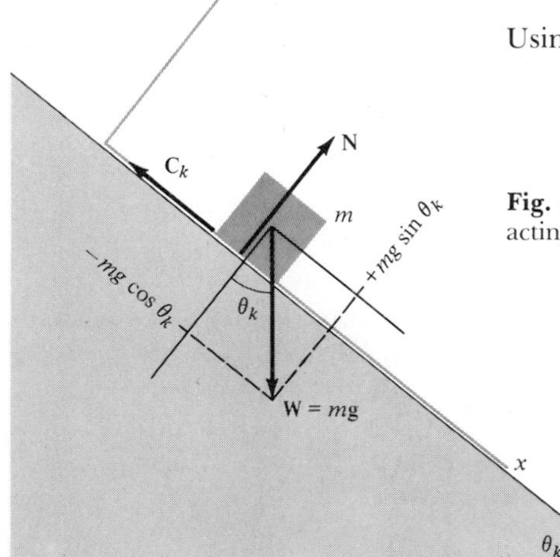

Fig. 4-28 A block sliding at constant speed down a rough plane. The forces acting on the block are shown.

The block does not accelerate in the x direction either, since the inclination of the plane is adjusted so that its speed sliding along the plane is constant. Thus the sum of the x components of the forces acting on it must be zero, too. Therefore you have

$$W_x + C_{k_x} = 0$$

or, since $C_{k_x} = -C_k$,

$$mg \sin \theta_k - \mu_k mg \cos \theta_k = 0$$

Solving for μ_k, you obtain

$$\mu_k = \frac{\sin \theta_k}{\cos \theta_k}$$

or

$$\mu_k = \tan \theta_k \tag{4-24}$$

For the particular case $\theta_k = 35°$, you can immediately calculate

$$\mu_k = \tan 35° = 0.70$$

Example 4-11 suggests a convenient way of measuring approximately the coefficient of kinetic friction for a pair of materials. You use one for the surface of a plane and the other for the surface of a block, and then you measure the angle θ_k for which the block slides down the plane at constant speed. This angle is called the **critical angle** for kinetic friction. Then Eq. (4-24) is used to determine the coefficient of kinetic friction μ_k. Try it, using a coin and the cover of this book for an inclined plane.

A similar method can be used to determine the coefficient of static friction. The block is placed on the plane, and the inclination of the plane is increased until the angle reaches the critical value θ_s just before motion commences. At this point the component of the gravitational force pulling the block down the inclined plane just equals the magnitude of the static contact friction force opposing the component, and you have

$$\mu_s = \tan \theta_s \tag{4-25}$$

It is easy to understand the origin of contact friction, at least in a general way. Imagine that two blocks of metal (say steel with surfaces of ordinary cleanliness and smoothness) are put together. Most surfaces, even those we call smooth in the usual sense, are extremely rough on a molecular-size scale; see Fig. 4-29. Hence the "peaks" of the two surfaces which first come into mutual contact comprise a very small part of the total surface area. If one block of steel is gently lowered onto the other, this very small initial contact area cannot possibly support the weight of the upper block; the pressure is well past the yield strength of steel. The peaks thus deform, allowing a larger and larger (but still very small) portion of the total area of one block to contact the adjacent peaks of the other.

Under the still high pressure present at the areas of contact between the deformed peaks, metals (especially similar metals) tend to "cold-weld" together unless the surfaces are quite dirty. That is, the two surfaces are joined by a multitude of tiny welds, in which attractive electric forces have bonded molecules from one surface to molecules in the other. These welds must be broken if sliding is to take place. This is one of the sources of static friction. In addition, there is always some interlocking of the peaks and valleys of the surfaces. If sliding is to begin, these protuberances must be deformed.

(a)

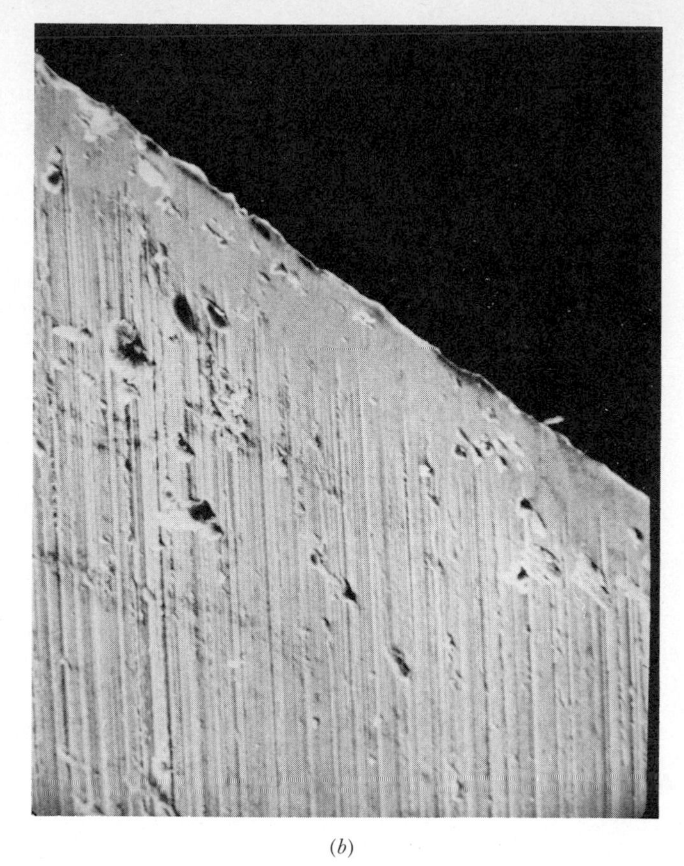

(b)

Fig. 4-29 (a) Photograph of a single-edged razor blade (×10). (b) Photograph of one of the surfaces forming the edge (×200). (*Courtesy General Electric Research Center, Schenectady, N.Y.*)

With this picture in mind, we can understand why the measured frictional force is independent of the total surface area in contact friction. The reason is that the frictional force depends on the area that is *actually* in contact. If the block in Fig. 4-30*a* is turned so that it rests on a side of greater area, as in Fig. 4-30*b*, the total area of the surface peaks actually in contact with those of the plate it rests on remains constant. The inserts in the figures indicate how the deformations in the two cases automatically arrange for this to be true. Pressing down on the block increases the actual contact area with the plate, and thus the frictional force. So the frictional force is proportional to the normal force. When the block has been set into motion across the plate, cold welds are constantly being broken and reformed. But the amount of cold welding present at any time is reduced below the static value, so the coefficient of kinetic friction is smaller than the coefficient of static friction. Finally, the presence of oil or grease at the surfaces in contact prevents welding by coating them with an inert material. This reduces both coefficients of friction. (Note in Table 4-2 that μ_s and μ_k are equal, and very small, for steel in contact with Teflon. Teflon is extremely inert chemically and bonds very weakly to other substances. Thus there is very little cold welding.)

When two objects press together, the force which one exerts on the other has a component in a direction parallel to the plane of contact—the frictional force—if the objects are sliding or attempting to slide past each other. There is also a component of the force in a direction normal to the plane of contact—the *normal force*—which is *always* present. The picture that provides some understanding of the properties of the frictional force also helps in understanding those of the normal force.

When you apply a force to an object by pushing on it in a direction normal to its surface, the reaction force you feel is the sum of the repulsive electric forces exerted on very many molecules at the surface of your hand by the very many adjacent molecules at the surface of the object. Newton's third law requires that the sum of these forces always be *just* enough to have the same magnitude as the magnitude of the force you apply. How does this come about? It is the same mechanism of yielding of the protuberances that does the trick. The two surfaces in contact deform until the actual contact area is sufficient to produce the required force. The forces which the two objects exert on each other when they are in contact, and which act in directions normal to the plane of contact, are often called **contact forces** instead of normal forces.

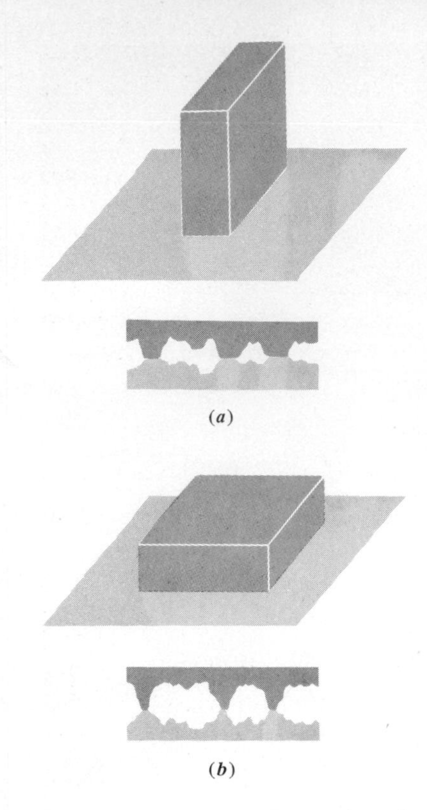

Fig. 4-30 (*a*) The smaller surface of a block rests on a plate. Below is a microscopic view showing large deformations of the peaks of both surfaces which are in contact. For each square centimeter of the block's surface, the area in actual contact is relatively large (though it is usually a small fraction of 1 cm^2). (*b*) The larger surface of the block rests on the plate. The microscopic view below shows that the deformations of the peaks in contact are smaller. So a relatively smaller area is in actual contact for each square centimeter of the block's surface. But the block's surface has more square centimeters. In both cases illustrated in parts (*a*) and (*b*) the *total* area in actual contact is essentially the same.

Finally, we give a brief empirical description of the properties of *fluid friction,* which exists when a solid object moves through a fluid. A treatment of its origin is presented in Chap. 16. The **fluid friction force,** also called **drag,** exerted by a liquid or a gas on a body moving through it always has a direction opposite to the direction of the body's velocity relative to the fluid, and a magnitude D which depends on the magnitude of the velocity. Experiment shows that for small objects moving slowly (like a marble dropping through oil) the value of D is proportional to the value of v, the speed of the body through the fluid. For spherical objects, the measured values of D are also proportional to their radii r. In fact, the measured magnitude D of the fluid friction force can be described by an equation known as **Stokes' law:**

$$D = 6\pi\eta r v \tag{4-26}$$

The proportionality constant η (the Greek letter eta) is called the **coefficient of viscosity.** The SI unit for η is newton seconds per meter squared ($N{\cdot}s/m^2$). Since 1 N/m^2 is the unit of pressure, which is called the *pascal* (Pa), the unit for η is often called the pascal-second and written as Pa·s. Some typical values of η are given in Table 4-3. Note the enormous range of viscosities encountered in practice. An application of Stokes' law is given in Example 4-12.

Table 4-3

Viscosity of Liquids and Gases

Substance	Temperature, (in °C)	η (in $N{\cdot}s/m^2$, or Pa·s)
Liquids		
Acetone	0	4.0×10^{-4}
Glycerine	0	1.2×10^{1}
Helium	−272	0
Pitch	0	6×10^{10}
Water	20	1.0×10^{-3}
Gases		
Air	20	1.8×10^{-5}
Hydrogen	0	8.3×10^{-6}
Steam	100	1.2×10^{-5}

EXAMPLE 4-12

The water droplets in a certain cloud in the atmosphere have radii $r = 5.0 \times 10^{-5}$ m. How fast can such droplets fall through the atmosphere? Assume the atmospheric temperature to be 20°C.

■ According to Stokes' law, the magnitude D of the fluid friction force acting on a falling droplet is proportional to its speed v. The direction of this force is upward since the velocity of the droplet is downward. As the droplet falls, its speed builds up until it is large enough that the value of D becomes equal to the magnitude mg of the downward-directed gravitational force acting on the droplet, whose mass is m. The net force acting on the droplet is then zero, and it will not fall any faster because Newton's second law then requires its acceleration to be zero. When this is the case, you have $D = mg$ and, using Stokes' law to evaluate D,

$$6\pi\eta r v = mg \tag{4-27}$$

For a spherical water droplet of radius r and density ρ (the Greek letter rho) the mass is ρ times its volume $\frac{4}{3}\pi r^3$. So

$$mg = \rho\frac{4\pi r^3 g}{3}$$

Combining this with Eq. (4-27), you obtain

$$6\pi\eta r v = \frac{4\pi\rho r^3 g}{3}$$

or

$$v = \frac{2\rho r^2 g}{9\eta} \tag{4-28}$$

Table 4-3 gives the viscosity η of air at 20°C to be 1.8×10^{-5} N·s/m². The density ρ of water at that temperature is 1.0×10^3 kg/m³. Using these values, the value 5.0×10^{-5} m given for r, and the standard value of g, you obtain

$$v = \frac{2 \times 1.0 \times 10^3 \text{ kg/m}^3 \times (5.0 \times 10^{-5} \text{ m})^2 \times 9.8 \text{ m/s}^2}{9 \times 1.8 \times 10^{-5} \text{ N·s/m}^2} = 3.0 \times 10^{-1} \text{ m/s}$$

The water droplet will acquire this low **terminal speed** quickly.

Since the terminal speed v is small, a cloud made of droplets this size will not descend if there is much of an updraft because the droplets will be moving more slowly downward with respect to the air than the air is moving upward with respect to the ground. This is why there must be coalescence of cloud droplets before an appreciable amount of rain can fall.

Stokes' law is valid only when a small enough body is moving slowly enough that the fluid through which it moves flows past the body in a smooth, orderly way. As the size and/or speed of the body increases, in due course the flow of fluid past the body becomes disorderly and turbulent. For example, the flow of air past an automobile moving at 100 km/h is quite turbulent. Turbulence leads to a much larger fluid friction force, in other words, a much larger drag. In fact, experiment shows that when turbulence is present, the magnitude D of the force is approximately proportional to v^2, the square of the body's speed. The experimental results are described to a good approximation by the empirical **law for turbulent flow:**

$$D = \frac{\rho_f A \delta v^2}{2} \tag{4-29}$$

Here ρ_f is the density of the fluid, A is the cross-sectional area which the body presents to the fluid, and δ is its **coefficient of drag.** The quantity δ

Table 4-4

Coefficient of Drag for Various Objects

Shape of object	Approximate value of δ
Circular disk (broadside to stream)	1.2
Sphere	0.4
Streamlined airplane body	0.06

(lowercase Greek delta) is a dimensionless, empirical constant which depends on the shape of the body, but is reasonably independent of A, ρ_f, and v over a fairly large range of these parameters. Some values of δ are given in Table 4-4. At the end of Chap. 5 we go through a detailed analysis of the motion of a skydiver falling through the air and experiencing a fluid friction force given by Eq. (4-29).

EXERCISES

Group A

4-1. *Mass versus weight; slugs versus pounds.* In the British engineering system of units, 1 *slug* is defined as the mass of a body which experiences an acceleration of 1 ft/s² when a force of 1 lb acts on it. Neglecting air resistance, a body falls under gravity with an acceleration of magnitude $g = 32$ ft/s².

a. A standard loaf of bread weighs 1 lb. What is its mass in slugs?

b. If a person weighs 130 lb, what is her mass in slugs?

c. Describe how to calculate the mass in slugs of an object of known weight.

d. What is the weight of an object whose mass is 1 slug?

e. If a 1-lb loaf of bread were transferred to the moon, its weight would be about $\frac{1}{6}$ lb. What would be its mass in slugs? Explain your answer.

4-2. *Poetry in motion.* Figure 4E-2 shows a piece of equipment commonly used in lecture demonstrations. The car has a funnel containing a compressed spring. A ball is set on top of the spring in the funnel. The car moves along a straight track with constant velocity. When the cart passes over an upward projection on the track, a trigger releases the spring and the ball is shot upward. Meanwhile the cart continues along the track at the same speed. Where will the ball land? Explain your answer

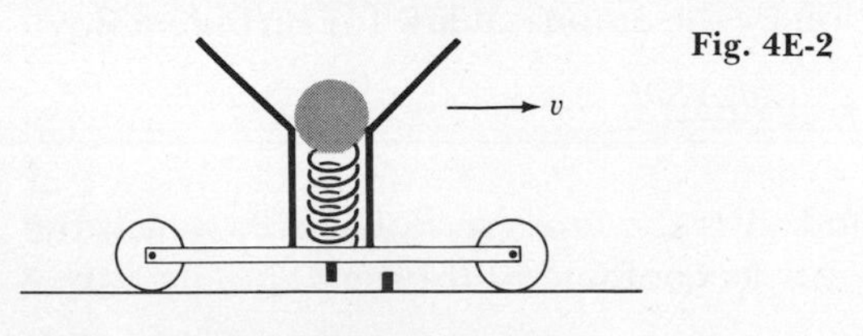

Fig. 4E-2

4-3. *Force, mass, and acceleration.* A constant force of magnitude 1 N acts on a body of mass 1 kg for 1 s. If the body is initially at rest, how far has it moved at the end of the second?

4-4. *Cart on a string.* A toy cart of mass 100 g lies on a variable-speed turntable. It is tied to the central shaft of the turntable by a string whose breaking strength is 5.0 N. The distance from the center of the shaft to the center of the cart is 10 cm.

a. What is the greatest speed with which the center of the cart can move as the turntable turns, if the string is not to break? (Rolling friction is negligible.)

b. What rate of turning (in revolutions per minute) would produce the speed obtained in part *a*?

4-5. *On the head.* A carpenter swings a 3-kg hammer so that its speed is 5 m/s just before it strikes a nail. The nail is driven 6 mm into a block of wood.

a. Assuming that a constant force resisted the motion of the nail, what must have been the magnitude of that force?

b. Compare the force to the weight of the hammer.

4-6. *Using signed scalars.* Repeat the calculation of Example 4-3 to find the magnitude of the force exerted by a spring scale from which a ball of a given weight is suspended, using signed scalar symbolism instead of vectors. Be sure to define an appropriate coordinate axis, take proper account of signs, and make clear distinctions between directed quantities and their magnitudes.

4-7. *Mass versus weight.* Near the surface of the moon an object falls under the influence of the moon's gravitational attraction with an acceleration of magnitude 1.62 m/s². The mass of an astronaut standing on the moon is 115 kg, including his life-support equipment. How much force does the moon exert on the astronaut and the equipment? How much would the astronaut and equipment weigh on earth?

4-8. *Evaluating normal force.* Determine the magnitude of the force exerted on the block by the smooth inclined plane of Example 4-4, if the mass of the block is 2.0 kg.

4-9. *All aboard!* A train, including its locomotive, has a mass of 2000 metric tons = 2000×10^3 kg. Starting from rest, the train acquires a speed of 2.0 m/s in 5.0 s.

a. Assuming that a constant net force acts on the train, what is the magnitude of that force?

b. Describe the forces that act on the locomotive itself. Describe the forces that act on one of the other cars of the train.

c. Compare the net force in part *a* with the weight of the train.

4-10. *Investigating a collision, I.* Show that the puck collision in Fig. 4-12 is inelastic by making the appropriate graphical subtraction of velocity vectors.

4-11. *Investigating a collision, II.* Show that the puck collision in Fig. 4-14 is elastic by making the appropriate graphical subtraction of velocity vectors.

4-12. *Investigating a collision, III.* Use graphical methods to determine whether the puck collision in Fig. 4-20 is elastic or inelastic.

4-13. *Snowslide.* A gondola (open freight car) of mass 20,000 kg is coasting on a siding at a speed of 10 m/s. There is negligible rolling friction. All of a sudden 10,000 kg of snow falls from a snowbank into the car. What is the speed of the gondola after this snowslide?

4-14. *Muzzle "velocity."* A rifle is mounted vertically with a large wooden ball balanced on its muzzle (the end of the barrel). When the gun is fired, the bullet embeds itself in the ball, which rises 1.00 m. If the mass of the wooden ball is 1.00 kg and that of the bullet is 10.0 g, with what speed did the bullet leave the muzzle?

4-15. *Momentum conservation, I.*

a. A 3.0-kg rifle is suspended to hang freely with its barrel supported in a horizontal position. With what speed does it recoil if it shoots an 11-g bullet with a speed of 300 m/s?

b. Why should a rifle be held firmly against the shoulder when it is fired?

4-16. *Momentum conservation, II.* A 1.0-kg car and a car of unknown mass M are pushed together to compress a spring between them (Fig. 4E-16). The cars are then released. A movie is made of the experiment. It is determined from the movie that the speed of the second car immediately after release is one-fourth that of the 1.0-kg car. What is the mass M?

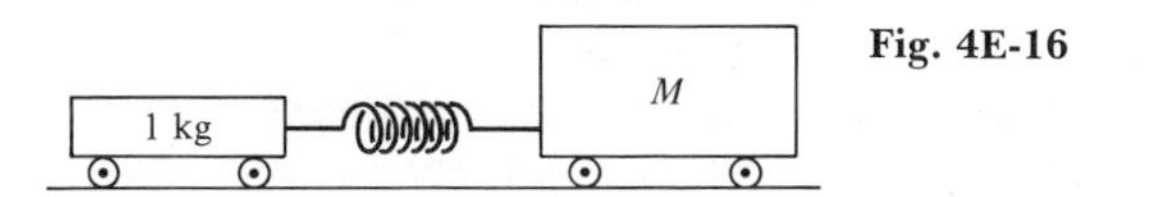

Fig. 4E-16

4-17. *Momentum nonconservation?* A tennis ball is dropped. Before it strikes the ground the value of its momentum mv is negative, if the upward direction is taken as positive. After it rebounds, it is moving upward so that its momentum has a positive value. Its momentum is not constant. Explain why this does not violate the law of conservation of momentum.

4-18. *When bat meets ball.* A baseball thrown with a speed of 40 m/s is hit by the batter. It leaves the bat with a speed of 70 m/s. If it remains in contact with the bat for 0.025 s, what is the average force exerted by the bat on the ball? The mass of the baseball is 145 g.

4-19. *Blasting off.* A rocket of mass 10,000 kg is set for takeoff. The speed of its exhaust gas is 1500 m/s.

a. At what rate must burned gas be ejected to give the rocket an initial upward acceleration of magnitude g?

b. Although the exhaust speed and ejection rate remain constant, the acceleration of the rocket during a later part of the burning is greater than its initial value. Why?

4-20. *Force and momentum change*

a. What are the magnitude and direction of a constant force that changes the momentum of a body from 10 kg·m/s east to 10 kg·m/s north in 2.0 s?

b. If the same force continues to act, what will be the momentum of the body after an additional time interval of 2.0 s?

4-21. *Ice follies.* Two young boys on ice skates are on a frozen pond. One boy has a rope around his waist. The second boy, initially at distance L from the first boy, holds the rope taut and pulls it continually hand over hand.

a. If the two boys have the same mass, what happens?

b. When the two boys meet, how much rope will have passed through the second boy's hands?

4-22. *Stopping distance.* A car is moving with speed v.

a. Prove that the minimum distance for stopping the car without skidding is given by the expression

$$s_{\text{min}} = \frac{v^2}{2\,\mu_s g}$$

where μ_s is the coefficient of static friction between the tires and the road.

b. Find s_{min} if $v = 25$ m/s and $\mu_s = 0.75$.

c. Calculate the time required to bring the car to rest.

4-23. *Those sudden stops.* Explain why an unsecured block of stone will shift forward on a flatbed trailer if the tractor-trailer undergoes gradual starts and sudden stops.

4-24. *Terminal speed.* A stainless-steel ball bearing of diameter 1.0 mm is dropped into a deep tank of glycerine whose temperature is 0°C. The density of stainless steel is

7.8×10^3 kg/m^3. Assuming that viscous friction obeys Stokes' law, use Eq. (4-28) to predict the speed of the ball bearing when it hits the bottom of the tank.

Group B

4-25. *Collision: before and after.* A freely moving ball strikes a stationary ball of equal mass. The initial velocities of the balls are $\mathbf{v}_{1i}$ and $\mathbf{v}_{2i} = 0$, respectively. After the collision, the velocities are $\mathbf{v}_{1f}$ and $\mathbf{v}_{2f}$, respectively.

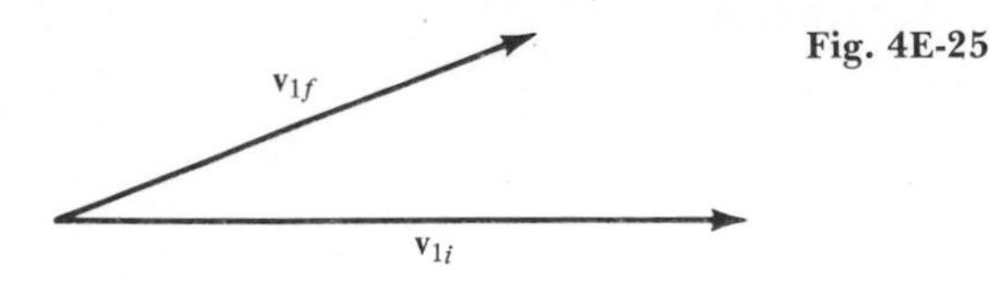

Fig. 4E-25

a. Use the law of momentum conservation to obtain an equation relating $\mathbf{v}_{1i}$, $\mathbf{v}_{1f}$, and $\mathbf{v}_{2f}$.

b. Show that $\mathbf{v}_{1i}$, $\mathbf{v}_{1f}$, and $\mathbf{v}_{2f}$ are confined to a single plane.

c. Figure 4E-25 shows the initial and final velocities of ball 1, $\mathbf{v}_{1i}$ and $\mathbf{v}_{1f}$. Copy this figure and use it to construct graphically the vector $\mathbf{v}_{2f}$. Also construct the parallelogram determined by $\mathbf{v}_{1f}$ and $\mathbf{v}_{2f}$.

d. What is the geometrical significance of $\mathbf{v}_{1i}$ in the parallelogram which you have constructed?

e. What is the geometrical significance of the vector $\mathbf{v}_{1f} - \mathbf{v}_{2f}$ in the parallelogram?

f. If the collision is elastic, then $|\mathbf{v}_{1f} - \mathbf{v}_{2f}| = |\mathbf{v}_{1i} - \mathbf{v}_{2i}|$. What restriction does this place on the shape of the parallelogram? What does the restriction imply about $\mathbf{v}_{2f}$?

4-26. *Cruising.* A jet plane is cruising at a steady speed of 200 m/s. It is burning 3.0 kg/s of fuel. This requires the intake of 80 kg/s of air. The products of combustion are exhausted with a speed of 500 m/s relative to the plane.

a. What is the thrust (propulsive force) provided by the jet engines?

b. How large is the drag force acting on the plane?

4-27. *Pitching and catching.* A standard baseball has a mass of 145 g.

a. A typical major league pitcher can throw a baseball at a speed of about 40 m/s. During the final part of the pitching motion (the delivery stage), the ball is accelerated from very low speed to its final speed in about 0.25 s. What is the average force exerted on the ball during the delivery stage?

b. Assume a catcher is going to stop the ball by decelerating it with the application of a constant retarding force of strength 200 N. How much time would be required to bring the ball to rest? How far would the ball travel before coming to rest? In your opinion, is the assumed value of 200 N reasonable?

4-28. *Great leap upward.* A physicist of mass m finds that by crouching and then springing upward he can (temporarily) elevate himself a distance h above his tiptoe standing height. The launch time (from the beginning of the spring until his feet leave the floor) is Δt.

a. What must be the velocity of the "springer" at the instant his toes leave the floor?

b. What is the average *net* force on the springer during launch? What average force does the *floor* exert on the springer during launch? Express these forces in SI units and also as multiples of the physicist's weight.

c. A not-very-springy adult (whose mass m is 70 kg) finds that his spring height is 0.40 m and his launch time is 0.10 s. Obtain numerical values for the quantities obtained in parts *a* and *b*.

4-29. *Dueling spheres.* Two steel spheres are suspended on very long cords from a common point. The ratio of their masses is 1:3.

a. The smaller sphere is pulled to the left and released. When it strikes the larger sphere, the small sphere has velocity $v_0\hat{\mathbf{x}}$, and the line joining the centers of the spheres is horizontal. Collisions between steel spheres are almost perfectly elastic. With what velocities do the two spheres rebound from the collision?

b. Suppose the larger sphere were pulled aside and allowed to strike the (stationary) small sphere with velocity $-v_0\hat{\mathbf{x}}$. What would the velocities of the spheres be immediately after the collision?

4-30. *Head-on and elastic.* An object of mass m_1 strikes a stationary object of mass m_2. The initial velocity of object 1 is $\mathbf{v}_{1i}$. The collision is perfectly elastic and is also a head-on collision, so that the initial and final paths of the objects all lie along a single line.

a. Prove that the final velocities $\mathbf{v}_{1f}$ and $\mathbf{v}_{2f}$ are given by

$$\mathbf{v}_{1f} = \frac{m_1 - m_2}{m_1 + m_2}\mathbf{v}_{1i} \quad \text{and} \quad \mathbf{v}_{2f} = \frac{2m_1}{m_1 + m_2}\mathbf{v}_{1i}$$

b. Evaluate $\mathbf{v}_{1f}$ and $\mathbf{v}_{2f}$ for the case $m_1 = m_2$.

c. If $m_1 < m_2$, how does the direction of $\mathbf{v}_{1f}$ compare with that of $\mathbf{v}_{1i}$?

d. If $m_1 \ll m_2$, what are the final velocities?

e. If $m_1 > m_2$, how do the directions of $\mathbf{v}_{1f}$ and $\mathbf{v}_{2f}$ compare with that of $\mathbf{v}_{1i}$?

4-31. *Head-on collision.* A massive object M moving with speed V_i encounters a small object of mass $m \ll M$ moving in the opposite direction with speed v_i. The collision is perfectly elastic, and after the collision both objects move in the original direction of M. Prove that the final speed of the small object exceeds its initial speed by $2V_i$.

4-32. *Puck collision, I.* Dots 1 and 2 in Fig. 4E-32 represent the positions of the centers of air table pucks 1 and 2 in a strobe photo of a collision between the pucks. Puck 1 was incident from the left, striking puck 2, which was initially stationary.

a. Use your intuition to guess whether the mass of puck 1 is larger than or smaller than the mass of puck 2.

b. Determine, as accurately as you can from a graphical analysis, the ratio of the puck masses.

c. Is the collision elastic or inelastic?

Fig. 4E-32

Fig. 4E-33

4-33. *Puck collision, II.* Using Fig. 4E-33, carry out the analyses called for in Exercise 4-32.

4-34. *Measuring mass without a balance.* A ball C is allowed to roll freely down a curved incline, as in Fig. 4E-34. It strikes the floor at B. The point A is vertically below the point where the ball leaves the incline. The line AB therefore gives the direction of the velocity of the ball when it leaves the incline. Another ball, D, is placed at rest at the end of the incline directly above A, and ball C is again released from the top of the incline. It strikes ball D slightly off center. Ball C now lands on the floor at C', and D lands at D'. The perpendicular distances $C'C''$ and $D'D''$ from the straight line $AD''C''B$ are measured. The ratio $C'C'':D'D''$ is found to be 2:1. What is the ratio of the mass of C to D? Justify your result. (This experiment exemplifies a method for determining mass from its operational definition.)

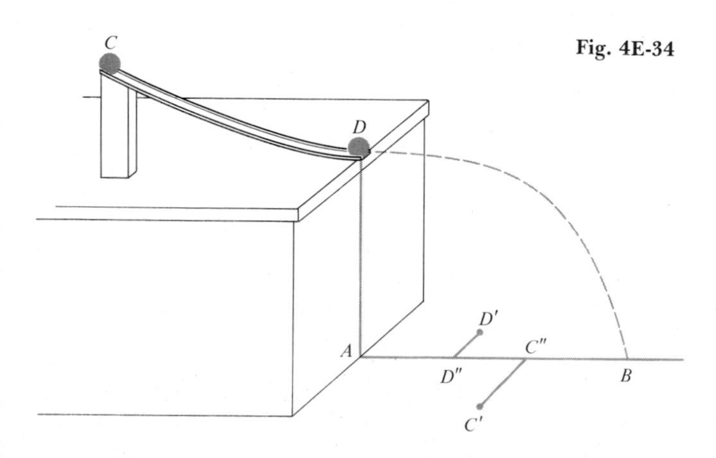

Fig. 4E-34

4-35. *The bouncing ball.* A 0.10-kg ball falls from a height of 1.0 m onto a disk equipped with a microswitch which activates an electronic timer. The ball bounces back up from the disk, reaching a height of 0.50 m. The timer indicates that the ball was in contact with the disk for 2.0 milliseconds $= 2 \times 10^{-3}$ s.

a. With what speed did the ball strike the disk?

b. With what speed did the ball leave the disk?

c. What was the average acceleration of the ball during its contact with the disk?

d. What average force did the disk exert on the ball?

e. What is the ratio of the force found in part *d* to the weight of the ball?

4-36. *Springs end to end.* A spring of force constant k_1 and one of different force constant k_2 are connected end to end. Find an expression for the effective force constant k' of the spring pair. Let S be the magnitude of the total restoring force exerted by the pair. Then you have $S = k'|l - l_0|$, where $|l - l_0|$ is the magnitude of its total change in length.

4-37. *Springs side by side.* The ends of two identical springs, each having force constant k, are joined by two rigid bars, as illustrated in Fig. 4E-37. Find an expression for the effective force constant k'' of the spring pair, as suggested in Exercise 4-36.

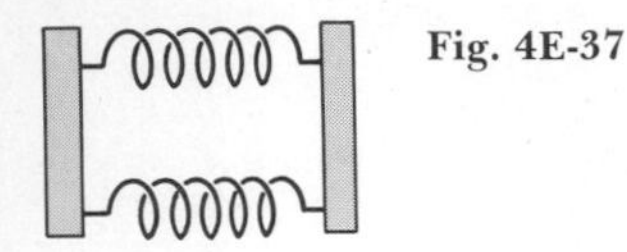

Fig. 4E-37

4-38. *An unusual sundeck.* An eccentric astronomer plans to use the dome of his observatory as a sun deck. Let μ_s be the coefficient of static friction between the dome surface and the blanket he plans to use.

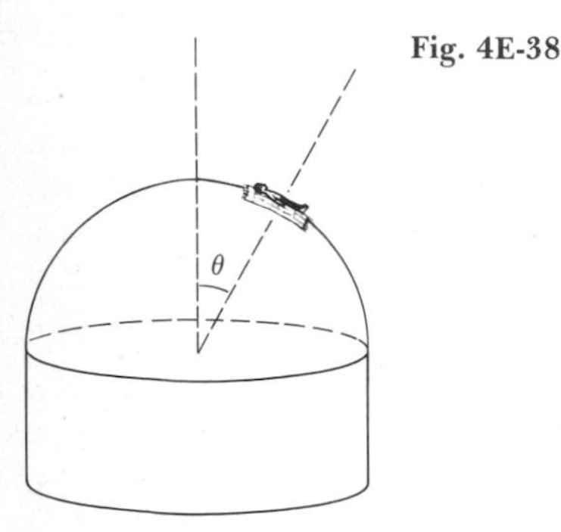

Fig. 4E-38

a. How far from the top of the dome can he place his blanket without sliding off the dome? Give your answer as the maximum angular distance θ from the "pole" of the dome. See Fig. 4E-38.

b. Evaluate your result for $\mu_s = 0.30$.

4-39. *On the ramp.* A ramp is constructed with a parabolic shape such that the height y of any point on its surface is given in terms of the point's horizontal distance x from the bottom of the ramp by $y = x^2/2L$. A block of granite is to be set on the ramp; the coefficient of static friction is μ_s.

a. What is the maximum x coordinate x_M at which the block can be placed on the ramp and remain at rest? What is the corresponding height y_M?

b. Evaluate your answers for the case $L = 10$ m and $\mu_s = 0.80$.

4-40. *Terminal speed when friction involves turbulence.* Derive an equation for the terminal speed of a body moving through a fluid under circumstances in which viscous friction obeys the law for turbulent flow. Compare your results with Eq. (4-28).

4-41. *Geronimo!* Use the equation derived in Exercise 4-40 to estimate the speed at which a parachutist hits the earth at a location where the altitude is approximately that of sea level. Take the mass of the parachutist plus the parachute to be 100 kg, the diameter of the parachute to be 10 m, and the coefficient of drag of the parachute to have the approximate value 1.2 quoted in Table 4-4 for a circular disk. The density of air at sea level is about 1.2 kg/m^3.

Group C

4-42. *Trapped!* In Fig. 4E-42 the vertical block rests on a wedge and is held in equilibrium between two guides G and G', with the help of a horizontal push **P** on the wedge. All surfaces are highly lubricated, so frictional forces can be neglected. The magnitude of the force **W**, is the weight of the vertical block and the wedge has opening angle θ, as shown.

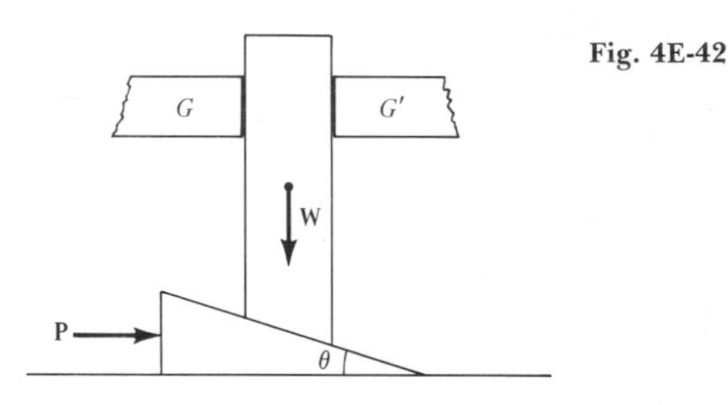

Fig. 4E-42

a. Find the required force **P** in terms of **W** and θ.

b. What prevents the wedge from accelerating to the right?

c. What prevents the vertical block from accelerating to the right?

4-43. *Cart-walking, I.* A 50-kg girl stands at the left end of the platform of a 100-kg cart with small, well-lubricated wheels. The cart is initially motionless, but rolling friction between the wheels and the ground is negligible.

a. The girl begins walking steadily toward the right end of the cart. Her velocity relative to the cart is 2.0 m/s. What is the velocity of the cart with respect to the ground?

b. What is the velocity of the girl with respect to the ground?

c. How long will it take her to reach the right end of the cart, which is 4.0 m long?

d. How far will she have moved with respect to the ground?

e. How far will the cart have moved with respect to the ground?

f. If the girl stops walking when she reaches the right end of the cart, what will be the final velocity of the cart (and girl) with respect to the ground?

4-44. *Cart-walking, II.* Consider again the girl and cart described in Exercise 4-43. The cart is initially stationary, and the girl approaches the cart from the left, walking steadily along the ground at a speed of 2.0 m/s. When she reaches the cart, she steps up without hesitating. She stands at the left end of the platform very briefly and then begins walking to the right at a speed of 2.0

m/s with respect to the cart surface. When she reaches the right end, she steps off without hesitating and continues walking away.

Carefully analyze the motion of the girl and the cart in order to answer the questions below. As you work, state your assumptions clearly. Comment on any differences between your results here and the corresponding results for Exercise 4-43.

a. Just after the girl steps up, but before she starts walking across the cart, what is the velocity of the cart relative to the ground?

b. While the girl is walking across the cart, what is the cart's velocity relative to the ground?

c. While she is walking across the cart, what is the girl's velocity relative to the ground?

d. How long does it take for the girl to reach the right end of the cart?

e. Relative to the ground, how far does the girl move during her walk across the cart?

f. Relative to the ground, how far does the cart move while the girl is on it?

g. After the girl has stepped down, what is the velocity of the cart relative to the ground?

4-45. *Collision course.* Bodies A and B of mass m_A and m_B, respectively, approach each other and collide. The initial velocities are given by $\mathbf{v}_{Ai}$ and $\mathbf{v}_{Bi}$, respectively. The final velocities are $\mathbf{v}_{Af}$ and $\mathbf{v}_{Bf}$. All velocities are confined to the xy plane, and velocity measurements indicate the following speeds and directions. The angles ϕ are the angles between the positive x axis and the vectors, with the positive sense being toward the positive y axis.

$$\mathbf{v}_{Ai}: v_{Ai} = 10.0 \text{ m/s}; \ \phi_{Ai} = 110.0°$$

$$\mathbf{v}_{Bi}: v_{Bi} = 20.0 \text{ m/s}; \ \phi_{Bi} = 50.0°$$

$$\mathbf{v}_{Af}: v_{Af} = 16.0 \text{ m/s}; \ \phi_{Af} = 70.0°$$

$$\mathbf{v}_{Bf}: v_{Bf} = 16.0 \text{ m/s}; \ \phi_{Bf} = 54.6°$$

a. Evaluate $|\mathbf{v}_{Bf} - \mathbf{v}_{Af}|$ and compare it with $|\mathbf{v}_{Bi} - \mathbf{v}_{Ai}|$. Was the collision elastic?

b. What is the mass ratio m_B/m_A?

c. Based on your result for part *b*, what would have been the common final velocity if bodies A and B had experienced a completely inelastic collison, starting from the initial velocities given?

4-46. *Recoil in a cloud chamber.* Particle A in Fig. 4E-46 is a fast alpha particle (helium nucleus of 4 atomic mass units) that is traveling through a "cloud chamber," which records the trajectories of electrically charged particles. At point P, particle A collides with B, a particle of unknown mass. As shown in the figure, the recoiling particles A and B produce tracks PA' and PB' at angles with the original direction of motion of A. The speeds v_{Bf} and v_{Af} of the particles after collision, as inferred from the appearance of their tracks, have the ratio $v_{Bf}/v_{Af} \simeq 0.63$. Determine the mass of particle B in atomic mass units. Can you guess the nature of the particle from your result?

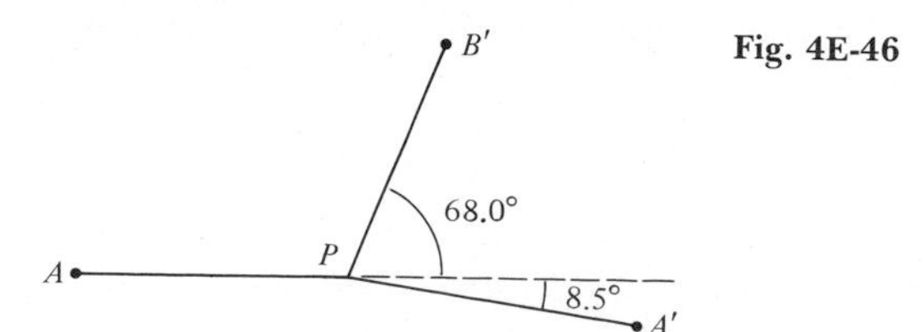

Fig. 4E-46

4-47. *Hanging by a thread.* A sphere of mass M is suspended by a thread. A second thread is attached to the bottom of the sphere. Two experiments are performed. First, the lower thread is pulled downward with a very gradually increasing force, so that the sphere has negligible acceleration. It is observed that the upper thread breaks first. Then the sphere is suspended just as before. Now the lower thread is jerked downward, so that if the lower thread did not break, the sphere would have a large downward acceleration. However, it is found that the lower thread breaks.

a. Write an equation relating the acceleration of the sphere to the forces acting on it. Let T_u and T_l represent the tensions in the upper and lower threads, respectively.

b. Use this equation to explain the differing outcomes of the two experiments.

4-48. *The sand is running out.* Figure 4E-48 shows an equal-arm balance suspended from a knife edge. The upper part of the vessel on the right contains sand which is pouring out of the opening at A at a steady rate c. (The dimensions of the constant c are mass per unit time.) Before the stopcock at A was opened, the balance was leveled by placing the appropriate weights in the left pan. The balance was then clamped, and the stopcock was opened. Once the sand was flowing at a steady rate, the balance was unclamped. It was observed that the balance remained level, as shown.

Fig. 4E-48

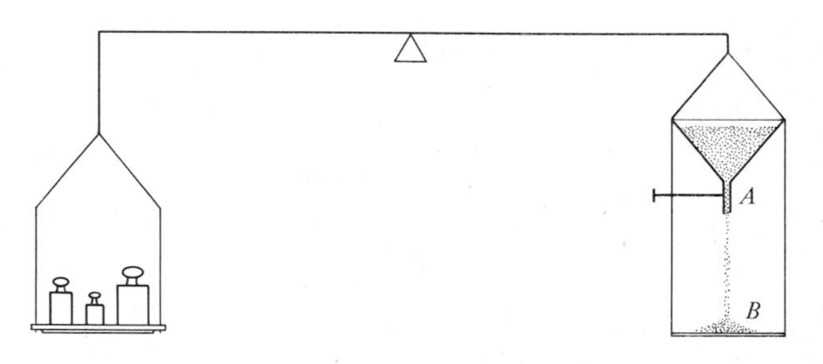

a. How long does it take each grain of the sand to fall from A to B, a distance h?

b. What is the mass of the falling stream of sand between A and B?

c. What is the weight of the stream?

d. What is the velocity of the falling sand as it reaches point B?

e. What momentum per unit time is given by the sand stream to the bottom of the vessel at B?

f. What is the force exerted on the bottom of the vessel by the sand stream as the sand grains come to rest?

g. Why does the balance remain level?

h. Describe qualitatively the motion of the balance as the sand runs out.

4-49. *Where will the slippage be?* Three numbered blocks (1 is a rectangular slab, and 2 and 3 are triangular wedges) are stacked as shown in Fig. 4E-49. The blocks have masses m_1, m_2, and m_3, respectively. Interface *A* (between block 1 and block 2) is characterized by coefficient of static friction μ_A, and interface *B* (between block 2 and block 3) is characterized by μ_B. The surfaces comprising interface *B* are inclined at an angle θ_B with the horizontal.

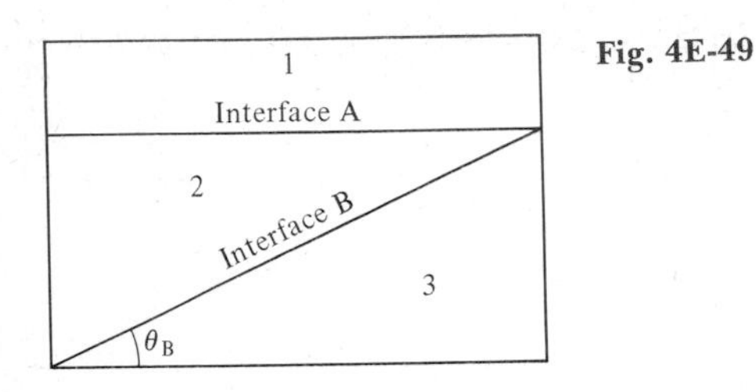

Fig. 4E-49

a. Block 3 is pulled to the right with a small acceleration of magnitude *a*. No slippage occurs along either surface. What is the magnitude of the frictional force along interface *A*? What is the magnitude of the frictional force along interface *B*? For each interface, find the ratio of the actual frictional force to the maximum available force of static friction.

b. Block 3 is now pulled to the right with a gradually increasing acceleration, until slippage occurs along one of the two interfaces. Where does the slippage occur? That is, under what conditions does the slippage first occur along interface *A*, and under what conditions does it occur along interface *B*?

c. Starting once more from rest, block 3 is pushed to the left with gradually increasing acceleration until slippage occurs. Describe slippage conditions in this case.

d. Suppose $\mu_A = 0.50$ and $\mu_B = 0.80$. For what values of θ_B does slippage first occur along interface *B* when block 3 is pulled to the right and along interface *A* when block 3 is pushed to the left?

4-50. *Down the incline.* A line of *n* identical rectangular wooden blocks (each of mass *m*) is sitting on a smooth incline at an angle α with the horizontal. (See Fig. 4E-50.) The evenly spaced blocks are of length *d* and are separated by gaps of length *l*, which is also the distance from the *n*th block to the bottom of the ramp. The coefficient of static friction between each block and the incline is $\mu_s > \tan\alpha$, but the coefficient of kinetic friction is $\mu_k < \tan\alpha$.

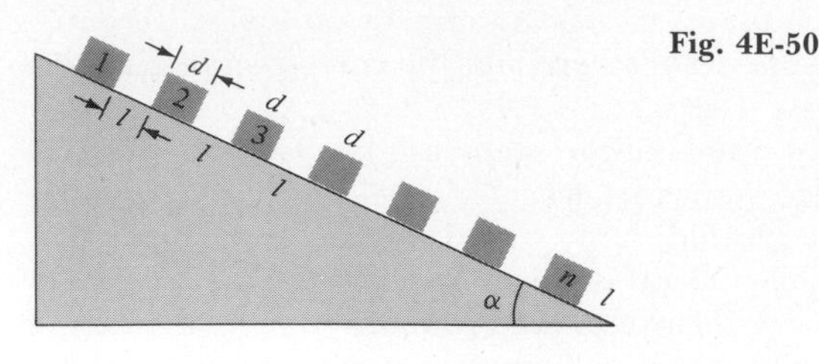

Fig. 4E-50

a. Suppose the top block (block 1) is given a small initial velocity to start it down the incline. What is its acceleration during the time interval before it strikes block 2? How much time elapses before it strikes block 2? With what speed does it strike block 2?

b. If the collisions between blocks are completely elastic, what is the speed of the *j*th block just after it has been struck, and again just before it strikes block $j + 1$? How much time elapses between the start of block 1 and the instant when block *n* reaches the end of the ramp?

c. Compare and contrast your results for part *b* with the motion of a solitary block started down an empty ramp. In which case does a block reach the bottom end of the ramp in a shorter time? In which case is the speed of the block greater as its front end reaches the bottom of the ramp?

d. Suppose the collisions between blocks are totally inelastic, so that after block *j* is struck, blocks 1 through *j* move as a unit. What is the velocity of block *j* just after it is struck, and again just before it strikes block $j + 1$? How much time elapses before the front end of the *n*th block reaches the bottom of the ramp?

e. Compare and contrast the results of part *d* with the results for parts *b* and *c*.

4-51. *Unscheduled stop.* A flatbed truck driver is hauling a large block of granite. He is driving along on level ground at speed v_0 when he rounds a curve and sees a disabled car a distance S_0 ahead blocking the road.

a. Can the trucker stop without causing the load to shift forward? The coefficient of static friction between the granite and the flatbed is μ_s.

b. Evaluate your results for $v_0 = 30$ m/s, $S_0 = 100$ m, and $\mu_s = 0.50$.

c. Suppose the trucker requires reaction time Δt before the brakes are applied. Modify the result of part *a* to allow for this.

d. Evaluate your results for the values of v_0, S_0, and μ_s given in part *b* if $\Delta t = 0.50$ s.

5

Applications of Newton's Laws

5-1 THE FREE-BODY DIAGRAM

In this chapter we apply Newton's laws, developed in Chap. 4, to the analysis of the motion of systems of practical interest. The techniques to be introduced here are widely useful in almost every field of physics and engineering. As in Chaps. 2 and 3, we restrict our attention to systems consisting of bodies whose motion can be studied in terms of changes in location only. The study of motion which must be described in terms of changes in orientation is deferred until Chaps. 9 and 10.

Specifically, in this chapter we apply Newton's second law, usually in the form

$$\mathbf{F} = m\mathbf{a} \tag{5-1}$$

to a study of the motion of bodies in a variety of systems. In doing so, it is always necessary to have the answers to two questions clearly in mind: (1) A system may consist of several parts, each of which has a certain mass and acceleration. Equation (5-1) must be applied separately to each part. Precisely what are these parts? (2) The quantity $\mathbf{F}$ in Eq. (5-1) is the *net* force acting on each of the parts of the system which have been precisely defined in answering question 1. Thus for each part $\mathbf{F}$ is the vector sum of all the forces acting *on* the part. What are the magnitudes and directions of all these forces?

In this section, we develop a systematic way of answering these questions. To begin with, consider the stationary *system* consisting of body 1 and body 2 in Fig. 5-1. Body 1 rests on body 2, which in turn rests on a tabletop. The mass of body 1 is m_1, and the mass of body 2 is m_2. Even in this relatively simple case there is the possibility of confusion if we are not careful. All the forces having to do with the system are shown, and there are eight of them. First, there is the force $\mathbf{W}_1$ exerted on body 1 by the gravitational

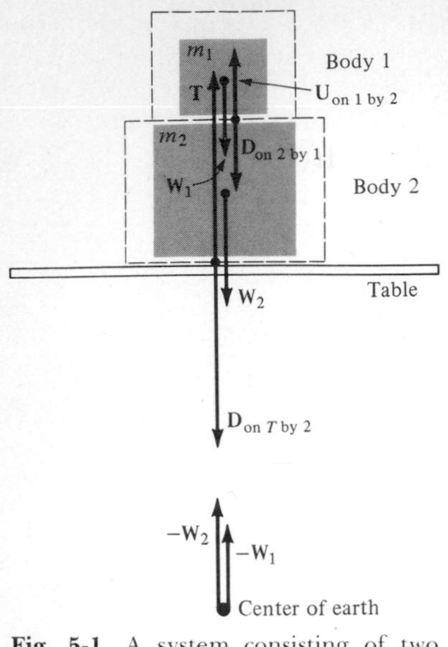

Fig. 5-1 A system consisting of two bodies rests on a table. The gravitational force exerted on body 1 by the earth is $\mathbf{W}_1$; its reaction force is $-\mathbf{W}_1$, which is exerted on the earth by body 1. The corresponding forces $\mathbf{W}_2$ and $-\mathbf{W}_2$ for body 2 are also shown. Exerted on body 2 by body 1, due to its weight, is a downward force $\mathbf{D}_{on\ 2\ by\ 1}$; the reaction force is the upward support force $\mathbf{U}_{on\ 1\ by\ 2}$. Exerted on the table by body 2 is a downward force $\mathbf{D}_{on\ T\ by\ 2}$; the reaction force is the upward support force $\mathbf{T}$ exerted by the table. The dashed boxes indicate the division of the system into two parts for separate treatment, as shown in Fig. 5-2.

attraction of the earth. The reaction force to $\mathbf{W}_1$ is the force $-\mathbf{W}_1$ acting on the earth. Body 1 is supported by body 2, which exerts on it an upward force $\mathbf{U}_{on\ 1\ by\ 2}$, and there is a downward reaction force $\mathbf{D}_{on\ 2\ by\ 1}$ exerted on body 2 by body 1. The other forces shown in the diagram, and defined in the figure caption, can be accounted for similarly.

In order to determine precisely which forces are acting on what body, we must answer question 1 posed above: What are the parts of the system? One possible division into parts is indicated by the dashed boxes in Fig. 5-1, where we choose to consider each of the two bodies separately.

Once this choice has been made, it is useful to emphasize it by separating the bodies which make up the entire system and drawing an individual force diagram for each. This is done in Fig. 5-2, which thus provides an answer for question 2 above. The two diagrams are called **free-body diagrams.** Each of the bodies can be idealized as a point, since only its mass, not its size or shape, is significant in applying Newton's second law to treat the change in location of the body. Figure 5-2*a* shows all the forces acting *on* body 1. It does *not* show the reaction forces associated with these forces, since they are not relevant to the motion (or, in this special case, the nonmotion) of body 1. There are just two forces of interest for body 1. These are the gravitational force $\mathbf{W}_1 = m_1\mathbf{g}$ which acts downward and the upward support force $\mathbf{U}_{on\ 1\ by\ 2}$ exerted by body 2.

Figure 5-2*b* is a similar picture of all the forces acting *on* body 2. There are three. They are the gravitational force $\mathbf{W}_2 = m_2\mathbf{g}$, the upward support force $\mathbf{T}$ exerted by the table, and the downward force $\mathbf{D}_{on\ 2\ by\ 1}$ exerted by body 1. (Note that while $\mathbf{U}_{on\ 1\ by\ 2}$ and $\mathbf{D}_{on\ 2\ by\ 1}$ are an action-reaction pair, only one of the pair is significant in each free-body diagram.)

Once the free-body diagram or diagrams have been constructed for a system, it is possible to use Newton's second law to analyze the motion of the system. In the present case this is relatively simple, since the entire system is at rest. We can therefore write $\mathbf{a} = 0$ for either body. The only re-

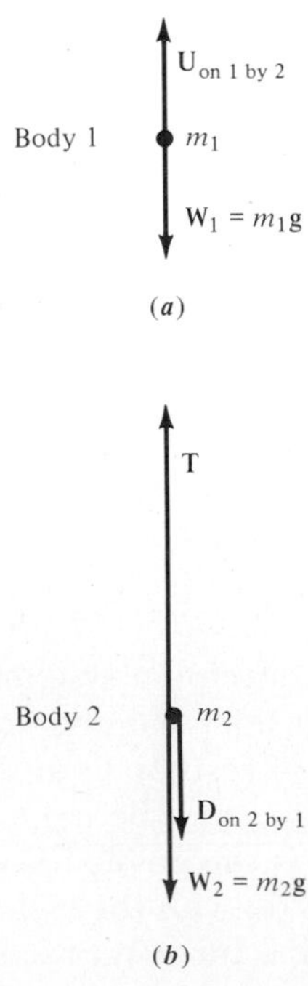

Fig. 5-2 Separate free-body diagrams for the two bodies comprising the system of Fig. 5-1. The construction of the diagrams is explained in the text.

maining task, then, is to find the magnitudes of the unknown forces $\mathbf{U}_{\text{on 1 by 2}}$, $\mathbf{D}_{\text{on 2 by 1}}$, and $\mathbf{T}$. Applying Newton's second law to body 1 gives

$$\mathbf{F}_1 = m_1\mathbf{a} = 0$$

for the net force $\mathbf{F}_1$ acting on body 1. Evaluating $\mathbf{F}_1$ from the free-body diagram of Fig. 5-2a leads to the equation

$$\mathbf{U}_{\text{on 1 by 2}} + \mathbf{W}_1 = 0$$

or

$$\mathbf{U}_{\text{on 1 by 2}} = -\mathbf{W}_1 = -m_1\mathbf{g}$$

Likewise, the net force $\mathbf{F}_2$ acting on body 2 is $\mathbf{F}_2 = m_2\mathbf{a} = 0$. Thus the free-body diagram of Fig. 5-2b yields $\mathbf{T} + \mathbf{W}_2 + \mathbf{D}_{\text{on 2 by 1}} = 0$, or

$$\mathbf{T} = -\mathbf{W}_2 - \mathbf{D}_{\text{on 2 by 1}}$$

The two bodies, considered in terms of their free-body diagrams, are independent. However, Newton's third law provides a link between them through the fact that $\mathbf{U}_{\text{on 1 by 2}}$ and $\mathbf{D}_{\text{on 2 by 1}}$ comprise an action-reaction pair. We thus have

$$\mathbf{D}_{\text{on 2 by 1}} = -\mathbf{U}_{\text{on 1 by 2}}$$

or

$$\mathbf{D}_{\text{on 2 by 1}} = +\mathbf{W}_1$$

The force $\mathbf{T}$ exerted by the table can thus be written

$$\mathbf{T} = -\mathbf{W}_2 - \mathbf{W}_1 = -m_2\mathbf{g} - m_1\mathbf{g}$$

or

$$\mathbf{T} = -(m_2 + m_1)\mathbf{g}$$

The results thus obtained in this simple case are consistent with intuition. The magnitude of the upward force $\mathbf{U}_{\text{on 1 by 2}}$ is equal to the weight m_1g of body 1, which is supported by this force. Body 2 is subject to a downward force $\mathbf{D}_{\text{on 2 by 1}}$ whose magnitude is also equal to the weight of body 1. And the table exerts an upward force $\mathbf{T}$ on body 2 whose magnitude is equal to the combined weights $(m_2 + m_1)g$ of the two bodies. Beginning with Fig. 5-1, how can you draw a free-body diagram so as to obtain this result in a single step?

EXAMPLE 5-1

The two bodies discussed above are laid in contact with each other on an air table, as shown in Fig. 5-3a. The experimenter pushes them to the right by exerting on body 2 a contact force $\mathbf{C}$ of magnitude 7.0 N. If the mass of body 1 is $m_1 = 3.0$ kg and the mass of body 2 is $m_2 = 5.0$ kg, find the force $\mathbf{R}_{\text{on 1 by 2}}$ exerted to the right on body 1 by body 2 and the acceleration $\mathbf{a}$ of the two bodies.

■ You begin by separating the system in imagination into the two separate bodies, and you construct a free-body diagram for each, as in Fig. 5-3b and c. In these diagrams, the gravitational forces exerted on each body and the support forces exerted by the air table on each are shown as dashed vectors. The two bodies remain always on the air table surface, and thus do not accelerate in the vertical direction. Both intuition and the discussion immediately preceding this example tell you that the vertical forces on each body add to zero. Since the forces of interest in this example are all horizontal, you need not consider the vertical forces further.

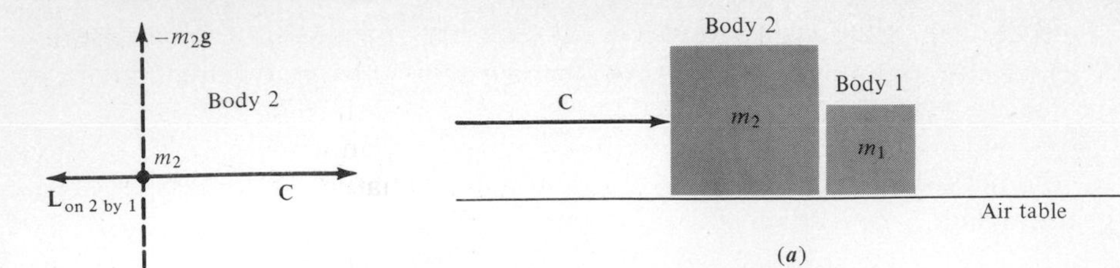

Fig. 5-3 (*a*) Illustration of the system discussed in Example 5-1. (*b*) Free-body diagram for body 2. (*c*) Free-body diagram for body 1.

Just as in the discussion of the stationary system immediately preceding this example, you next apply Newton's second law to body 1. In this case, however, the acceleration is *not* zero, and you have

$$\mathbf{F}_1 = \mathbf{R}_{\text{on 1 by 2}} = m_1\mathbf{a}$$

for the net force $\mathbf{F}_1$ acting on body 1. Similarly, the free-body diagram for body 2 gives you

$$\mathbf{F}_2 = \mathbf{C} + \mathbf{L}_{\text{on 2 by 1}} = m_2\mathbf{a}$$

for the net force $\mathbf{F}_2$ acting on body 2, where $\mathbf{L}_{\text{on 2 by 1}}$ is the force exerted to the left on body 2 by body 1. You next link the equations obtained from the two free-body diagrams by noting that $\mathbf{L}_{\text{on 2 by 1}}$ and $\mathbf{R}_{\text{on 1 by 2}}$ comprise an action-reaction pair, so that

$$\mathbf{L}_{\text{on 2 by 1}} = -\mathbf{R}_{\text{on 1 by 2}} = -m_1\mathbf{a}$$

Thus you have for $\mathbf{F}_2$ the equation

$$\mathbf{F}_2 = \mathbf{C} - m_1\mathbf{a} = m_2\mathbf{a}$$

This can be solved for the acceleration $\mathbf{a}$, giving

$$\mathbf{a} = \frac{\mathbf{C}}{m_2 + m_1} \tag{5-2}$$

Inserting the given numerical values into Eq. (5-2), you obtain

$$\mathbf{a} = \frac{7.0\text{ N}}{5.0\text{ kg} + 3.0\text{ kg}}\,\hat{\mathbf{C}} = 0.88\text{ m/s}^2 \text{ in the direction of } \mathbf{C}.$$

You can now work backward and evaluate $\mathbf{R}_{\text{on 1 by 2}}$, the force which body 2, the directly pushed body, exerts on body 1 to accelerate it. You have

$$\begin{aligned}\mathbf{R}_{\text{on 1 by 2}} &= m_1\mathbf{a} = 3.0\text{ kg} \times 0.88\text{ m/s}^2 \times \hat{\mathbf{C}}\\ &= 2.6\text{ N in the direction of } \mathbf{C}\end{aligned}$$

In the following two sections, the method of the free-body diagram is applied to more complicated systems, where its usefulness becomes still more apparent.

5-2 ATWOOD'S MACHINE AND SIMILAR SYSTEMS

Many important physical systems consist of several parts, each of which is acted on by one or more external forces. However, the parts of such a system are linked so that they cannot move independently of one another. In this section, we begin the study of such systems by considering the relatively simple device illustrated in Fig. 5-4. This device is called **Atwood's machine** in honor of its inventor. (In the nomenclature of the time, the word "machine" was applied to any mechanical device, whether or not it performed a useful task.)

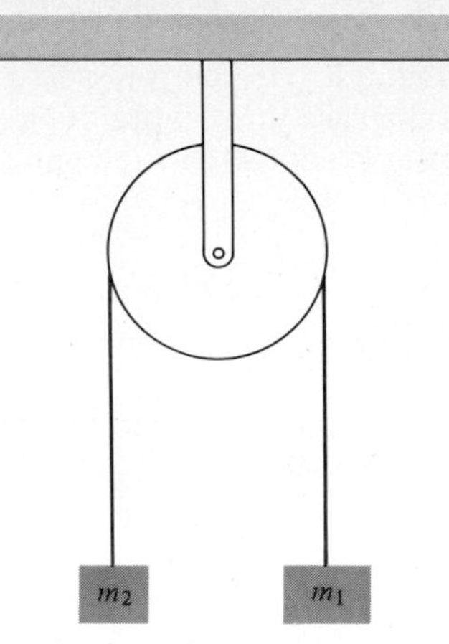

Fig. 5-4 Atwood's machine. Ideally, the system is frictionless, the string is inextensible, and the string and pulley are both massless.

The machine consists of two bodies of mass m_1 and m_2, attached to the ends of a limp but inextensible string whose mass is negligible compared to theirs. The string runs over a pulley. It is assumed that the pulley also has negligible mass and that the friction of the bearing on which it turns is negligible. (These ideal conditions can be approximated reasonably well in actual systems.)

If m_1 and m_2 are not equal, the system will begin to move as soon as it is released. The essential problem is to find the accelerations of the two bodies. Once the accelerations are known, the rules of kinematics developed in Chaps. 2 and 3 can be applied to find the velocity and position of the bodies at any time.

We begin the analysis of the motion of the Atwood machine, as viewed in an inertial reference frame, by drawing the necessary free-body diagrams, as in Fig. 5-5. Since the two bodies will move in different directions, it is best to treat each as a separate free body. In the ideal Atwood machine, the string and the pulley are massless, and we need not draw free-body diagrams for them. In Fig. 5-5, $\mathbf{S}_1$ and $\mathbf{S}_2$ are the forces exerted on m_1 and m_2, respectively, by the string. The gravitational forces exerted on the two bodies are also shown; they are, respectively, $m_1\mathbf{g}$ and $m_2\mathbf{g}$. (At this point we do not know the magnitudes of the various forces, so only the directions of the vectors should be taken literally.)

We can now write Newton's second law, Eq. (5-1), for each of the two bodies separately. Assigning the symbol $\mathbf{a}_1$ to the acceleration of m_1 and the symbol $\mathbf{a}_2$ to the acceleration of m_2, we have

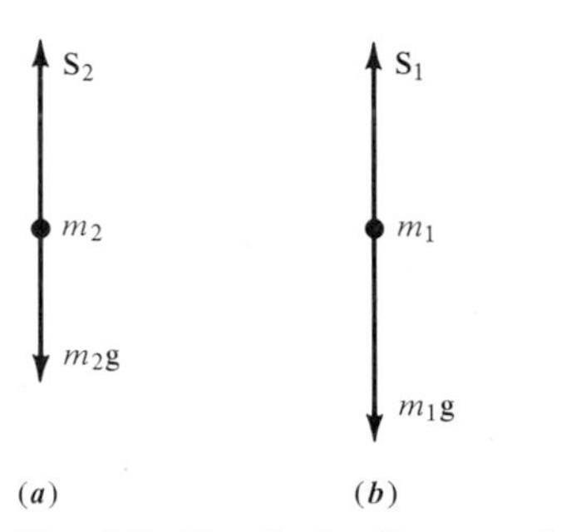

Fig. 5-5 Free-body diagrams for the bodies of mass m_1 and m_2 which are part of the Atwood machine of Fig. 5-4.

$$\mathbf{S}_1 + m_1\mathbf{g} = m_1\mathbf{a}_1 \tag{5-3a}$$

and

$$\mathbf{S}_2 + m_2\mathbf{g} = m_2\mathbf{a}_2 \tag{5-3b}$$

These two equations contain four unknowns, $\mathbf{S}_1$, $\mathbf{S}_2$, $\mathbf{a}_1$, and $\mathbf{a}_2$. We thus need two more equations to solve the problem of finding the accelerations. These are provided by the physical situation. First, the length of the string remains constant as the system moves. The string and the pulley taken together ensure that if one body has an upward acceleration, the other body must have a downward acceleration of equal magnitude. Thus we have

$$\mathbf{a}_2 = -\mathbf{a}_1 \tag{5-4}$$

Second, the forces $\mathbf{S}_1$ and $\mathbf{S}_2$ exerted by the ends of the massless string passing over the massless, frictionless pulley are equal in magnitude and have the same direction, so that

$$\mathbf{S}_1 = \mathbf{S}_2 \tag{5-5}$$

This equation depends directly on the assumption that the string and the pulley have zero (or negligible) mass. The string is not a source of gravitational "driving" force for the Atwood machine even when one body is lower than the other, so that different lengths of string hang from the two sides of the pulley. And since the string and the pulley have no inertia, none of the driving force is "used up" in accelerating them. (To see exactly what is meant by the vivid but imprecise term "used up," consider what would happen if a massless string were replaced by a chain, each of whose links possesses mass. Suppose a force exerted on the chain at its right end makes it accelerate toward the right. Each link in the chain exerts a force on the link to its left. But since that link is accelerating, the net force acting on it is not zero. Thus the force it exerts on the link to its left is less than the force exerted on it by the link to its right. This is the case throughout the chain. Hence the chain exerts a smaller force on some object attached to its left end than that exerted on it at its right end.)

Equations (5-3*a*), (5-3*b*), (5-4) and (5-5) comprise a set of four simultaneous equations in four unknowns. We first use Eqs. (5-4) and (5-5) to eliminate $\mathbf{a}_2$ and $\mathbf{S}_2$ from Eq. (5-3*b*), which thus becomes

$$\mathbf{S}_1 + m_2\mathbf{g} = -m_2\mathbf{a}_1$$

We next eliminate the unknown quantity $\mathbf{S}_1$ by subtracting this equation from Eq. (5-3a),

$$\mathbf{S}_1 + m_1\mathbf{g} = m_1\mathbf{a}_1$$

Doing so gives

$$(m_1 - m_2)\mathbf{g} = (m_1 + m_2)\mathbf{a}_1 \tag{5-6}$$

The acceleration $\mathbf{a}_1$ of m_1 is thus given by the equation

$$\mathbf{a}_1 = \frac{m_1 - m_2}{m_1 + m_2}\,\mathbf{g} \tag{5-7a}$$

Again using Eq. (5-4), $\mathbf{a}_2 = -\mathbf{a}_1$, we find the acceleration $\mathbf{a}_2$ of m_2 to be

$$\mathbf{a}_2 = -\frac{m_1 - m_2}{m_1 + m_2}\,\mathbf{g} \tag{5-7b}$$

The accelerations $\mathbf{a}_1$ and $\mathbf{a}_2$ have opposite directions. But the magnitudes of the accelerations of both bodies in Atwood's machine have a common value equal to the magnitude of the acceleration of gravity, multiplied by a fraction whose absolute value is always less than or equal to 1. Whether m_1 accelerates downward or upward depends on whether m_1 is greater than or less than m_2; this determines the sign of the numerator of the fraction in Eqs. (5-7*a*) and (5-7*b*).

It is particularly easy to compare the predictions of these equations with what we expect intuitively in the two extreme cases $m_2 = 0$ and $m_2 = m_1$. If $m_2 = 0$, Eq. (5-7*a*) yields $\mathbf{a}_1 = \mathbf{g}$. That is, in the absence of the retarding force exerted on m_1 because m_2 must be accelerated upward, the body of mass m_1 descends in free fall. If $m_2 = m_1$, the system is balanced, as evidenced by the fact that Eqs. (5-7*a*) and (5-7*b*) yield $a_1 = a_2 = 0$.

Let us return to Eq. (5-6) and write the magnitudes so as to show clearly that it is a special case of Newton's second law:

$$(m_1 - m_2)g = (m_1 + m_2)a_1 \tag{5-8a}$$

$$F \quad = \quad M \quad a \tag{5-8b}$$

We call Eq. (5-8*a*) the **equation of motion** of the entire system, consisting of the two bodies and the connecting string. We obtained it by using the **constraints** of the system, Eqs. (5-4) and (5-5), to conjoin the two separate equations of motion for the parts of the system, Eqs. (5-3*a*) and (5-3*b*). As comparison of Eqs. (5-8*a*) and (5-8*b*) makes clear, M is the total mass of the system, while F is the magnitude of the net force on the system. The two connected bodies move in unison with acceleration of magnitude a as a result of the application of F to M. As far as the system is concerned, the forces $\mathbf{S}_1$ and $\mathbf{S}_2$ exerted on the individual bodies by the ends of the string are *internal* forces. They therefore do not appear in the equation of motion of the system. But they do appear in the equations of motion for the individual bodies, for which they are *external* forces. The magnitude of the net force acting on the entire system is thus the difference in weight of the two bodies. Because of the configuration of the system, with the string bent around the pulley, the gravitational forces acting on the two bodies pull against each other.

Equations (5-8*a*) and (5-8*b*) also make apparent the distinction between the gravitational and inertial aspects of mass. On the left, or "*F*," side of these equations, it is the gravitational aspect which is significant, since it is from the masses of the bodies in their gravitational role, taken together with the presence of the earth, that there arises the force which propels the system. On the right, or "*Ma*," side of the equations, it is the inertial masses which provide the resistance of the system to acceleration. Note especially, in this connection, that it is the *sum* of the masses (that is, the total mass of the system) which appears on the right side of the equation, and the *difference* of the masses appears on the left side.

Atwood taught physics at Cambridge (Newton's university). He invented his machine as a classroom demonstration, and it has been so used ever since. Atwood's machine seems to have been the very first laboratory demonstration of the validity of Newton's laws. Yet it was not invented until about 1730, or almost half a century after the publication of the *Principia*, and was not even then regarded as an important experimental confirmation of newtonian mechanics. The reason lies in the absolutely central role which Newton's laws assume, directly or indirectly, in the description of every natural phenomenon involving motion of bodies whose speeds are small compared to the speed of light and whose sizes are large compared to those of atoms. By Atwood's time, Newton and others had piled up such a vast fund of evidence and examples (such as planetary motion) that Atwood's machine became an interesting demonstration rather than a significant experimental proof.

EXAMPLE 5-2

Find the force exerted by the string on either of the bodies in Atwood's machine.

■ According to Eq. (5-5), the forces $\mathbf{S}_1$ and $\mathbf{S}_2$ exerted by the two ends of the string are equal. So you may as well evaluate $\mathbf{S}_1$. One of several ways you can do this is to begin with Eq. (5-3*a*), which you can rewrite in the form

$$\mathbf{S}_1 = -m_1\mathbf{g} + m_1\mathbf{a}_1 = m_1(\mathbf{a}_1 - \mathbf{g})$$

If you substitute into this equation the value of $\mathbf{a}_1$ given by Eq. (5-7*a*), you get

$$\mathbf{S}_1 = m_1\left(\frac{m_1 - m_2}{m_1 + m_2}\mathbf{g} - \mathbf{g}\right)$$

Factoring the quantity $\mathbf{g}$ out of the term in parentheses in this equation gives you

$$\mathbf{S}_1 = m_1\mathbf{g}\left(\frac{m_1 - m_2}{m_1 + m_2} - 1\right)$$

Note that the fraction $(m_1 + m_2)/(m_1 + m_2)$ can be substituted for 1 in this equation, which may thus be rewritten

$$\mathbf{S}_1 = m_1\mathbf{g}\left(\frac{m_1 - m_2}{m_1 + m_2} - \frac{m_1 + m_2}{m_1 + m_2}\right)$$

or

$$\mathbf{S}_1 = -m_1\mathbf{g}\left(\frac{2m_2}{m_1 + m_2}\right) \tag{5-9a}$$

According to Eq. (5-5), this is also the value of $\mathbf{S}_2$.

What is the physical meaning of Eq. (5-9*a*), which gives the force $\mathbf{S}_1$ exerted by the string on m_1? Suppose that $m_1 \geqslant m_2$. The magnitude S_1 is the strength of the force exerted by the string on either body. The equation shows that it is equal to the weight m_1g of the body of greater mass multiplied by the quantity in parentheses. This quantity is a fraction smaller than 1. Thus the string does not exert a force strong enough to hold up the body of greater mass m_1, and it accelerates downward.

On the other hand, since $\mathbf{S}_1 = \mathbf{S}_2$ the force exerted by the other end of the string on the body of smaller mass m_2 also has the value given by the right side of Eq. (5-9*a*). Rearranging the terms of that equation slightly, and setting $\mathbf{S}_1 = \mathbf{S}_2$, thus gives

$$\mathbf{S}_2 = -\frac{m_1\mathbf{g}\,2m_2}{m_1 + m_2}$$

or

$$\mathbf{S}_2 = -m_2\mathbf{g}\left(\frac{2m_1}{m_1 + m_2}\right) \tag{5-9b}$$

Equation (5-9*b*) tells you that the force exerted by the string on the body of smaller mass m_2 has a magnitude equal to the weight m_2g of that body multiplied by the quantity in parentheses, which is *greater* than 1. Thus the force exerted by the string on the body of smaller mass is more than sufficient to support it, and the body accelerates upward.

Equations (5-9*a*) and (5-9*b*) both tell you that the force exerted by the string on either body is strongest when the masses are equal. In this case, each body can exert a force on the string sufficient to "hold the other one up." On the other hand, the force exerted by the string on one of the bodies will be zero if the mass of the other body is zero.

How would the discussion following Example 5-2 proceed if you assumed that $m_2 \geqslant m_1$?

In Example 5-3, the accelerations of the bodies in an Atwood machine are calculated. This information is then used, together with the rules of kinematics, to develop information concerning position, time, and velocity.

EXAMPLE 5-3

In a certain Atwood machine the masses of the two bodies have the values $m_1 = 2.10$ kg and $m_2 = 2.00$ kg. The string and the pulley have negligible mass, and the friction of the system is negligible as well. If the two bodies are initially at rest at the same level, how long will it be before the vertical separation between them is 1.5 m? How fast will they then be going?

■ It is most convenient here to work in terms of signed scalar quantities, rather than the vector quantities we used to derive the equations of motion for the Atwood machine. In these terms, the constant accelerations of the bodies bear the relation

$$a_2 = -a_1$$

which is the signed scalar form of Eq. (5-4). So you may as well focus your attention

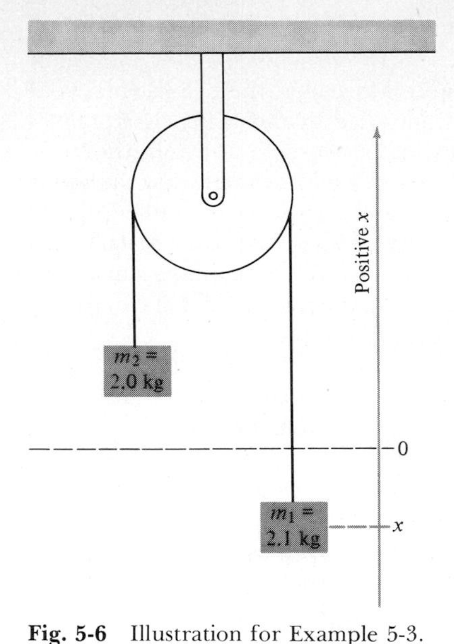

Fig. 5-6 Illustration for Example 5-3.

on m_1. Choose the upward direction as that of the positive x axis, with the origin at the starting level of the two bodies, as shown in Fig. 5-6. The initial position of m_1 is then given by $x_1 = x_{1i} = 0$. Since the system starts at rest, you also have the initial velocity $v_1 = v_{1i} = 0$ for m_1. Equation (2-30), written in the notation of the present discussion, then reduces to the simple form

$$x_1 = \frac{a_1 t^2}{2} \tag{5-10}$$

In order to use this equation, you must first use Eq. (5-7a) to find the acceleration a_1 of m_1. In signed scalar terms, this gives you

$$a_1 = \frac{m_1 - m_2}{m_1 + m_2} g$$

Since g is directed downward, it has the value $g = -9.8\ \text{m/s}^2$. You thus have

$$a_1 = \frac{2.10\ \text{kg} - 2.00\ \text{kg}}{2.10\ \text{kg} + 2.00\ \text{kg}} \times (-9.8\ \text{m/s}^2)$$

or

$$a_1 = -0.24\ \text{m/s}^2$$

The minus sign tells you that m_1 accelerates downward, as you would expect for the more massive of the two bodies. (You immediately have $a_2 = -a_1 = +0.24\ \text{m/s}^2$ as well.) Note that by making the difference between the masses small you can reduce the acceleration to a value which is not difficult to measure.

When the position of m_1 is given by x_1, the vertical separation of the two bodies is $2|x_1|$. Thus when the two bodies are 1.5 m apart, you have $x_1 = -0.75$ m. (The sign of x_1 is evident from Fig. 5-6.) You solve Eq. (5-10) for the elapsed time t to obtain

$$t = \sqrt{\frac{2x_1}{a_1}}$$

Then you find its numerical value by setting $a_1 = -0.24\ \text{m/s}^2$ and $x_1 = -0.75$ m;

$$t = \sqrt{\frac{2 \times (-0.75\ \text{m})}{-0.24\ \text{m/s}^2}} = 2.5\ \text{s}$$

Since the acceleration a_1 is constant, you can use Eq. (2-29), $v_1 = v_{1i} + at$, to find the velocity v_1 of m_1 at the instant when the two bodies are 1.5 m apart. Since $v_{1i} = 0$, you have

$$v_1 = a_1 t = -0.24\ \text{m/s}^2 \times 2.5\ \text{s} = -0.60\ \text{m/s}$$

Again the negative sign signifies the downward direction of v_1. The velocity v_2 of m_2 is immediately given by

$$v_2 = a_2 t = -a_1 t = +0.60\ \text{m/s}$$

(But you already knew that $v_2 = -v_1$.)

Example 5-4 concerns a system which amounts to an Atwood machine modified by the introduction of an inclined plane.

EXAMPLE 5-4

In Fig. 5-7a, the body of mass m_2 can slide without friction on the inclined plane, which makes an angle θ with the horizontal. A light, inextensible string is attached to m_2 and passes over a light, frictionless pulley. A body of mass m_1 hangs from the other end of the string. Find an algebraic expression for the acceleration of the system. Solve in particular for the case $m_1 = 7.0$ kg, $m_2 = 14.0$ kg, $\theta = 28°$.

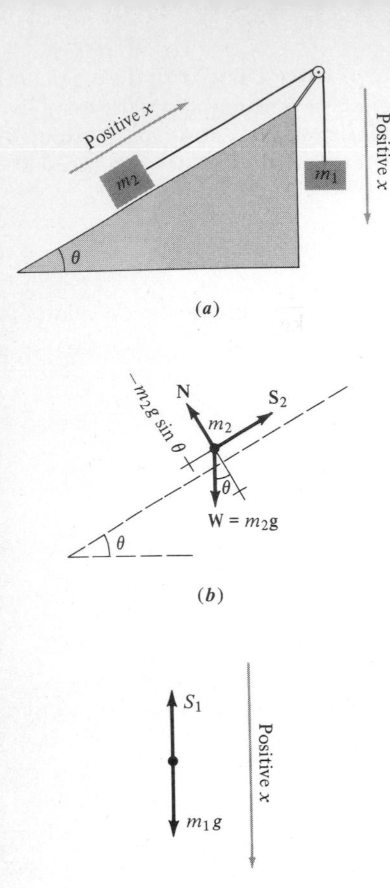

Fig. 5-7 (*a*) Sketch of the system discussed in Example 5-4, showing the directions chosen as positive for the motion of the two bodies. (*b*) Free-body diagram for the body m_2 on the inclined plane. (*c*) Free-body diagram for the hanging body. (*d*) Free-body diagram for m_2 simplified to one dimension by the method developed in Example 4-4.

Here again it is more convenient to work in terms of signed scalar quantities rather than vectors. While the system as a whole is two-dimensional, the motion of each of the bodies is restricted to one dimension. Moreover, the displacements of the two bodies are connected by the constraint imposed by the string. If m_1 moves through a certain distance vertically downward, m_2 must move through the same distance upward along the plane. Assume for the sake of argument that the **sense** of the motion will be clockwise; that is, m_1 will descend and m_2 will be pulled up the plane. If you then represent all quantities directed *clockwise* as *positive* scalars, and quantities directed *counterclockwise* as *negative* scalars, your results will be consistent. (If the actual motion turns out to be counterclockwise, the results will tell you this by yielding a negative acceleration.)

The significant difference between the system in Fig. 5-7*a* and the Atwood machine is that the weight of m_2 is partially supported by the inclined plane. You have already dealt with a simpler but similar situation in Example 4-4, in which a body slid freely down the plane.

You begin by drawing free-body diagrams for the two bodies, as shown in Fig. 5-7*b* and *c*. The latter diagram, which represents the hanging body of mass m_1, is one-dimensional. But the former, representing the body of mass m_2 on the inclined plane, is not. The vectors $\mathbf{S}_2$, $\mathbf{W}$, and $\mathbf{N}$ are, respectively, the force exerted on m_2 by the string, the gravitational force on m_2, and the normal force exerted on m_2 by the plane. They require two dimensions to represent them. However, as explained in Example 4-4, the normal force cancels the normal component of the gravitational force. Hence the free-body diagram of m_2 reduces to the simpler one-dimensional form of Fig. 5-7*d*. The quantity $W_{\|}$ is the component of the gravitational force along the x axis drawn parallel to the plane. If you take the sign convention into consideration, you see that Fig. 5-7*d* shows that $W_{\|}$ has the value

$$W_{\|} = -m_2 g \sin\theta \tag{5-11}$$

In terms of the same sign convention, the accelerations of the two bodies are related by the expression

$$a_2 = a_1 = a \tag{5-12a}$$

And the forces exerted on the two bodies by the string bear the relation

$$S_2 = -S_1 \tag{5-12b}$$

Newton's second law gives you the equations of motion for the two bodies. For m_1 you have

$$m_1 g + S_1 = m_1 a_1 \tag{5-13a}$$

and for m_2 you have

$$S_2 + W_{\|} = m_2 a_2 \tag{5-13b}$$

Substituting the values of $S_{\|}$, a_2, S_2, and a_1 from Eqs. (5-11), (5-12*a*), and (5-12*b*), you obtain the equations of motion in the form

$$m_1 g + S_1 = m_1 a$$

and

$$-S_1 - m_2 g \sin\theta = m_2 a$$

Adding these two equations gives you

$$(m_1 - m_2 \sin\theta) g = (m_1 + m_2) a$$

Solving for the acceleration a of either of the bodies, you have

$$a = g\,\frac{m_1 - m_2 \sin\theta}{m_1 + m_2} \tag{5-14}$$

This result is very much like the solution for the Atwood machine given by Eq. (5-7*a*). The main difference is that the term in the numerator containing m_2 is diminished by the factor sin θ, which is always less than 1. Note that the system will move clockwise if $m_1 > m_2 \sin\theta$ and counterclockwise if the reverse is true. Why does the factor sin θ not appear in the denominator of Eq. (5-14)?

Now you can use Eq. (5-14) to find the specific numerical solution. In terms of the sign convention adopted, $g = +9.8\ \text{m/s}^2$. Thus you have

$$a = 9.8\ \text{m/s}^2 \times \frac{7.0\ \text{kg} - 14.0\ \text{kg} \times \sin 28°}{7.0\ \text{kg} + 14.0\ \text{kg}}$$

or

$$a = 0.020\ \text{m/s}^2$$

The body of mass m_1 descends even though m_1 is smaller than m_2. This is the reason why inclined planes are useful devices; they make possible the raising of large weights with relatively small forces.

Example 5-5 applies the Atwood machine analysis to a system which has been discussed twice before, in Sec. 3-5 and in Example 4-5. In our experimental approach to centripetal acceleration in Sec. 3-5, we began by placing an air puck in a circular orbit on the horizontal top of an air table. The necessary centripetal force was supplied by the weight of a washer hanging from a string which passed through a small hole in the center of the air table, as shown in Fig. 5-8. A crucial part of the quantitative argument involved the fact that the centripetal acceleration of the puck (which was measured using strobe photos) was equal to the radially inward acceleration of the same puck when it was released from rest and was pulled toward the hole by the string. The tension in the string was the same in both cases. In Example 5-5, you will see that keeping the tension the same involves using washers of slightly different weights in the two cases. You will determine the difference and thus achieve a complete analysis of the system.

EXAMPLE 5-5

In Example 4-5, a puck of mass $M = 0.33$ kg was made to move at constant speed in a circular orbit. The necessary centripetal force was supplied by a string, from the other end of which was hung a washer of mass $m = 0.022$ kg. The washer did not move. Next, the same puck was released from rest, so that the washer as well as the puck experienced acceleration. If the force exerted by the string on the puck is to have the *same* magnitude S in the two experiments, what must be the mass m' of the washer in the second experiment?

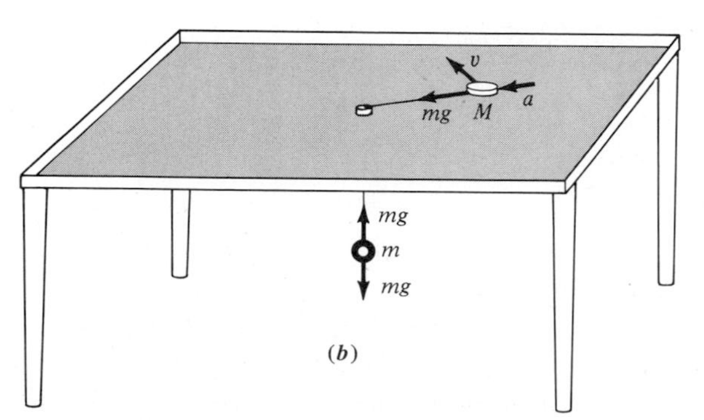

Fig. 5-8 Sketch of the air table experiment discussed in detail in Sec. 3-5. When the puck moves in a circular orbit, the weight of the washer supplies the necessary centripetal force. When the puck is released from rest, the weight of the washer accelerates the system consisting of washer, string, and puck. This system is very similar to the system of Fig. 5-7*a*.

■ When you apply Newton's second law to the puck in the two experiments, you obtain $S = Ma_c$ and $S = Ma_r$, respectively, where a_c and a_r are the magnitudes of the centripetal and radial accelerations of the puck. Hence you have $a_c = a_r$; this was verified by actual measurement of the accelerations.

The net force exerted on the washer is the sum of the force **S** exerted by the string and the gravitational force **W**. When the puck is whirling in a circle, the washer does not move. Hence the net force on the washer must be $\mathbf{F} = \mathbf{S} + \mathbf{W} = 0$. You thus have $S = W = mg$. This gives you $Ma_c = mg$, or

$$a_c = \frac{S}{M} = \frac{m}{M} g \tag{5-15}$$

When the puck is released from rest and moves radially inward, the system is the same as that just considered in Example 5-4, with the simplification that the inclination angle of the plane is now $\theta = 0°$. You can therefore apply Eq. (5-14). Since here $\sin\theta = 0$, $m_1 = m'$, and $m_2 = M$, and $a = a_r$, that equation assumes the particularly simple form

$$a_r = \frac{m'}{m' + M} g \tag{5-16}$$

Compare this with Eq. (5-15). You can see that the condition of equal string tensions, which led to the equality of a_c and a_r, requires that $m' \neq m$. (Since $m \ll M$, however, the difference between m and m' will turn out not to be very large.) Equating the right sides of Eqs. (5-15) and (5-16), you have

$$\frac{m}{M} = \frac{m'}{m' + M}$$

Multiplying both sides of this equation by $M(m' + M)$, in order to clear fractions, gives you

$$Mm' = mm' + mM$$

Collecting all terms containing m' on one side of the equation and factoring, you obtain

$$m'(M - m) = mM$$

You thus find for m' the result

$$m' = \frac{mM}{M - m} \tag{5-17}$$

Using the known values of the puck mass M and the mass m of the washer used in the circular-orbit experiment, you can now find the mass of the washer required for the radial-acceleration experiment. You have

$$m' = \frac{0.022 \text{ kg} \times 0.33 \text{ kg}}{0.33 \text{ kg} - 0.022 \text{ kg}} = 0.024 \text{ kg}$$

The difference between m' and m is thus 0.002 kg = 2 g. The fractional difference is 2 parts in 24, or about 8 percent.

5-3 MOTION WITH CONTACT FRICTION

In the last section, we deliberately ignored the effects of friction. In a situation such as Example 5-4, where a block is pulled up an inclined plane, this is not usually realistic. We are now ready to incorporate the effects of frictional forces into the analysis of such systems. All we need do in principle is to include the frictional force in the free-body diagram of the block on the plane. The method of finding the magnitude of that force is the one used in Example 4-11, where a block slid down an inclined plane under the influence of gravitation and friction alone. Example 5-6 combines the essential features of Example 4-11 with those of Example 5-4.

EXAMPLE 5-6

In the system of Fig. 5-7a the body of mass m_2 now slides on a well-greased plane; the coefficient of kinetic friction is $\mu_k = 0.030$. As in Example 5-4, let $m_1 = 7.0$ kg, $m_2 = 14.0$ kg, and $\theta = 28°$. Find the acceleration of the system. Compare the result with that found in Example 5-4, where friction was neglected.

■ To begin with, assume that the frictional force is not so large as to keep the system from moving altogether. A frictional force can never *change* the direction of motion (why not?), so you can assume that the acceleration is clockwise, as it was in Example 5-4, if it is not zero. The frictional force C_k must therefore act downward to the left, in the same counterclockwise sense as $W_{\parallel}$. See Fig. 5-9, which is simply the free-body diagram of Fig. 5-7d with the frictional force added.

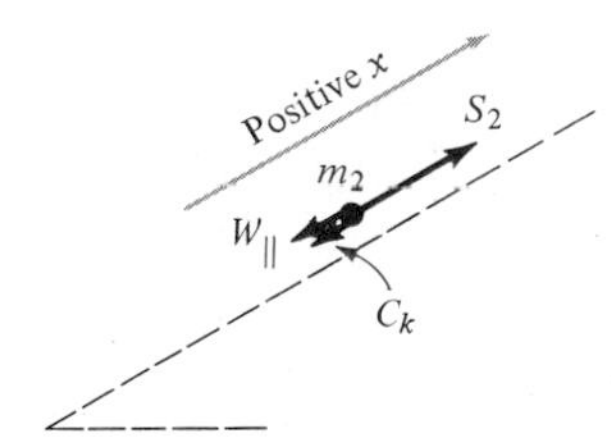

Fig. 5-9 Illustration for Example 5-6.

From Example 4-11 you have

$$C_k = -\mu_k m_2 g \cos\theta$$

The negative sign is needed to conform to the convention of taking the clockwise sense to be positive, which is the same here as in Example 5-4. Just as in that example, you can write the equations of motion for the two bodies:

$$m_1 g + S_1 = m_1 a_1 \qquad (5\text{-}18a)$$

$$S_2 + W_{\parallel} + C_k = m_2 a_2 \qquad (5\text{-}18b)$$

These are the same as Eqs. (5-13a) and (5-13b) except that the contact friction force C_k has been included in the net force acting on m_2. You proceed to solve these equations just as in Example 5-4, and you obtain

$$g[m_1 - m_2(\sin\theta + \mu_k \cos\theta)] = a(m_1 + m_2)$$

or

$$a = g\frac{m_1 - m_2(\sin\theta + \mu_k\cos\theta)}{m_1 + m_2} \qquad (5\text{-}19)$$

which you should compare with Eq. (5-14).

When you insert the numerical values, you find the result

$$a = 9.8\ \text{m/s}^2\,\frac{7.0\ \text{kg} - 14.0\ \text{kg} \times (\sin 28° + 0.030\cos 28°)}{7.0\ \text{kg} + 14.0\ \text{kg}}$$

or

$$a = 0.0027\ \text{m/s}^2 \qquad (5\text{-}20)$$

This is only about one-tenth as great as the acceleration in the frictionless case. The large effect of a quite small frictional force in this example is due to the fact that the system was nearly balanced without friction. If C_k had been a little bigger, the system would not have moved at all. However, Eq. (5-19) does not give you this information automatically, and it is important to check your initial assumption that a is not zero. The general method for doing this is discussed immediately following this example. However, it is easy to check a particular numerical result such as Eq. (5-20), provided you are careful to keep your signs straight. You assumed at the outset that a was positive. If you had obtained a negative numerical result, you would have to see whether a would be negative even in the absence of friction. To do this, repeat the calculation with $\mu_k = 0$. (In fact, this was done in Example 5-4.) If a turns out to be positive without friction, although it was negative with friction, the true result is that the system does not move, or will come to rest if it is started with a push. Once at rest, the system is under the influence of the even larger *static* coefficient of friction μ_s, and Eq. (5-19) does not correctly describe the motion. If the kinetic frictional force was large enough to bring the system to rest (assuming it was moving in the first place), the static friction will certainly provide a force sufficient to balance the other forces on the body and keep it at rest.

A system of the sort treated in Example 5-6 may involve arbitrary masses m_1 and m_2, an arbitrary inclination angle θ, and arbitrary coefficients of friction μ_k and/or μ_s. Given the values of these quantities, the system may fall into one of two general classes, each of which has three subclasses. First you must determine the general class: whether the system tends to move clockwise or counterclockwise. This is determined by what the system would do in the absence of friction, as given by Eq. (5-14). Once the direction of friction-free motion has been found, taking friction into consideration leads to one of the three following subclasses: (*a*) the system may move under all circumstances; (*b*) it may move only if it is initially in motion or if it is given a push to start it; or (*c*) it may come to a stop even if it is initially in motion. The various possibilities, enumerated in Table 5-1, are the subject of an exercise at the end of this chapter.

Table 5-1

Summary of Cases for the System of Example 5-6

Case	Condition
I. Clockwise motion	$m_2 g \sin\theta < m_1 g$
a. System always moves	$m_2 g(\sin\theta + \mu_s \cos\theta) < m_1 g$
b. System moves if started	$m_2 g(\sin\theta + \mu_k \cos\theta) < m_1 g < m_2 g(\sin\theta + \mu_s \cos\theta)$
c. System comes to rest if moving	$m_2 g \sin\theta < m_1 g < m_2 g(\sin\theta + \mu_k \cos\theta)$
System balanced in absence of friction	$m_1 g = m_2 g \sin\theta$
II. Counterclockwise motion	$m_1 g < m_2 g \sin\theta$
c′. System comes to rest if moving	$m_2 g(\sin\theta - \mu_k \cos\theta) < m_1 g < m_2 g \sin\theta$
b′. System moves if started	$m_2 g(\sin\theta - \mu_s \cos\theta) < m_1 g < m_2 g(\sin\theta - \mu_k \cos\theta)$
a′. System always moves	$m_1 g < m_2 g(\sin\theta - \mu_s \cos\theta)$

5-4 FICTITIOUS FORCES

There is a kind of force familiar to everyone, to which we have so far given only the briefest mention. Everyone knows that on rounding a curve the passengers in a car feel an *outward* force, called the **centrifugal** (that is, center-fleeing) **force.** No such force appeared in our analysis of circular motion in Examples 4-5 and 5-5; the only radial force dealt with was the inward centripetal force.

This apparently paradoxical situation arises from one of the most fundamental and subtle points in newtonian mechanics. We could settle the matter by simply saying that the passengers in the car are not in an inertial reference frame, so that they cannot apply Newton's laws. (Remember that Newton's laws of motion were developed in Chap. 4 to account for the observations of *observers stationed in inertial reference frames.* They cannot validly be used by observers in **noninertial frames.**) What we wish to do now, however, is broaden our point of view so that *any* observer can predict what *any other* observer will see, *regardless of the reference frames in which the observers are located.* To put it another way, we want to see how Newton's laws need to be supplemented or modified so that any observer can apply them to what she or he sees.

Consider a car traveling around a circular curve of radius r. If you stand by the roadside, it is clear that the car is undergoing a centripetal acceleration. At an instant when the velocity of the car is **V** and the radius of the curve is r, the instantaneous acceleration is given by Eq. (3-41b),

$$\mathbf{a}_c = -\frac{V^2}{r}\hat{\mathbf{r}} \tag{5-21}$$

where $\hat{\mathbf{r}}$ is the unit vector directed from the center of the curve to the car. The minus sign in the equation specifies that the direction of $\mathbf{a}_c$ is inward.

A passenger O' in the car will agree with you that the radius of the curve is r. He will even agree with you that the velocity of the car is **V**, although that is not the raw product of his observation. What he actually sees is the ground moving past him with velocity $-\mathbf{V}$, but he has learned from experience to make the transformation of velocity to your reference frame, which makes it possible for him to speak to you in your own language.

As far as casual observation is concerned, this tends to be the limit of the passenger's broadmindedness. His position relative to the car in which he is sitting does not change, and he therefore finds it difficult to think of himself as being accelerated. Instead, he takes a more natural but more narrow-minded point of view. He insists that he is experiencing no acceleration. But if he tries to verify Newton's laws by means of an experiment, he will find that they fail. For example, suppose that he has a small air table and is holding a puck at a certain position on it. At a particular moment, as the car is rounding the curve, he releases the puck. As soon as he does so, there can be no further net force acting on the puck. First consider what you, observer O, see from your observation point at the roadside. As shown in Fig. 5-10a, b, and c, you see the puck continue moving at constant velocity after it is released, tangent to the curve at the point of release. And you see that the car, passenger, and air table are accelerated out from under the puck with a centripetal acceleration

$$\mathbf{a}_c = -\frac{V^2}{r}\hat{\mathbf{r}}$$

(This acceleration is produced by the sidewise frictional force of the tires on the road.)

But the passenger, observer O', observes things from his own point of view. As suggested by Fig. 5-10d, he must explain why it is that *he* sees the puck accelerating *outward* with instantaneous acceleration

$$\mathbf{a}' = +\frac{V^2}{r}\hat{\mathbf{r}} \tag{5-22a}$$

He decides to apply Newton's laws, ignoring the fact that he is not justified in doing so because he is not in an inertial frame. Knowing that an acceleration implies the existence of a force, and knowing the mass m of the puck, he postulates the existence of a centrifugal force $\mathbf{F}'$ given by

$$\mathbf{F}' = m\mathbf{a}' = +\frac{mV^2}{r}\hat{\mathbf{r}} \tag{5-22b}$$

It is nothing new to invent a force in order to account for an acceleration; we have done precisely this in accounting for the gravitational acceleration of falling bodies by means of a gravitational force. But we believe that we can assign a tangible source or cause to the gravitational force—namely the presence of the earth—whereas there is no such explanation at hand

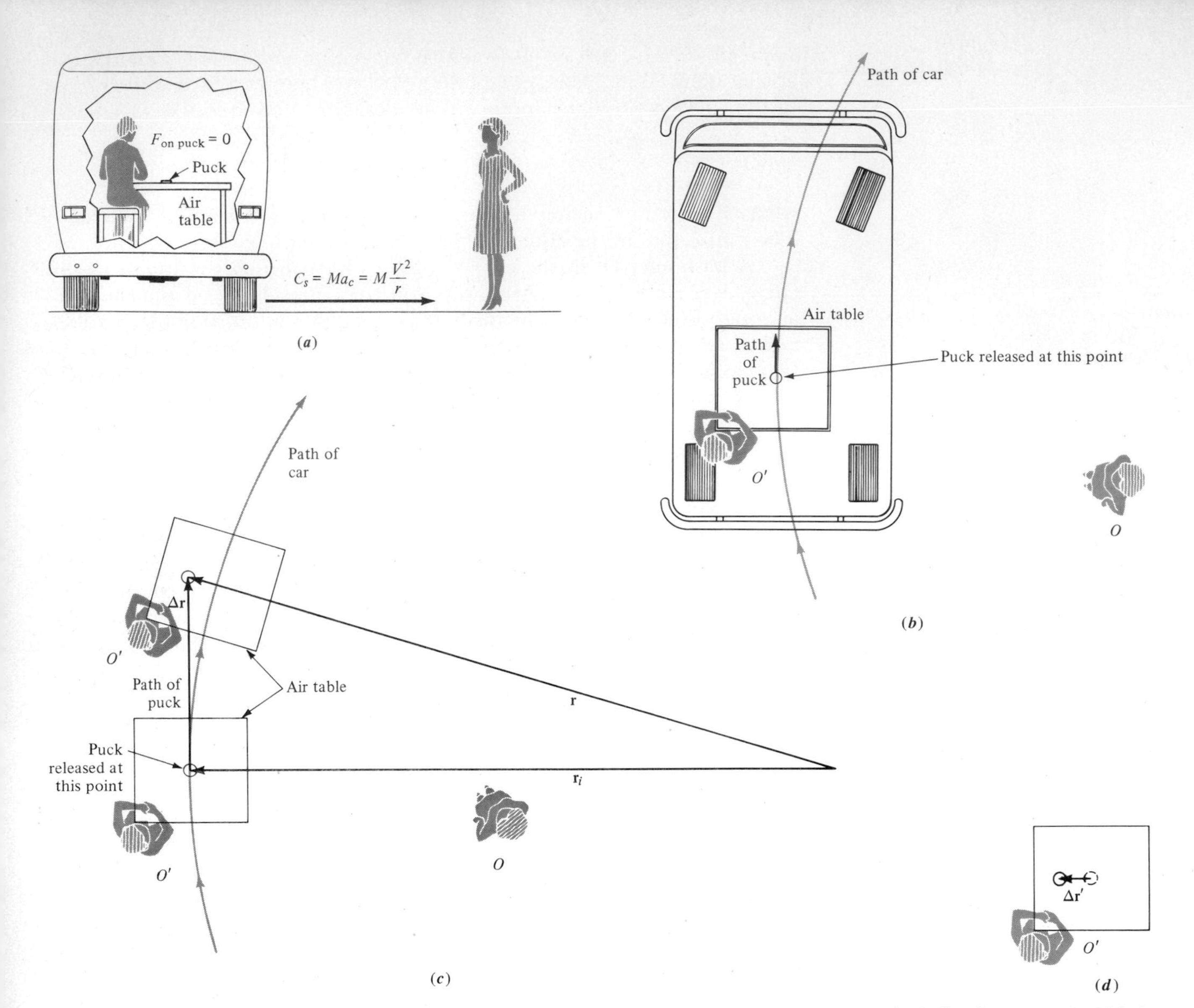

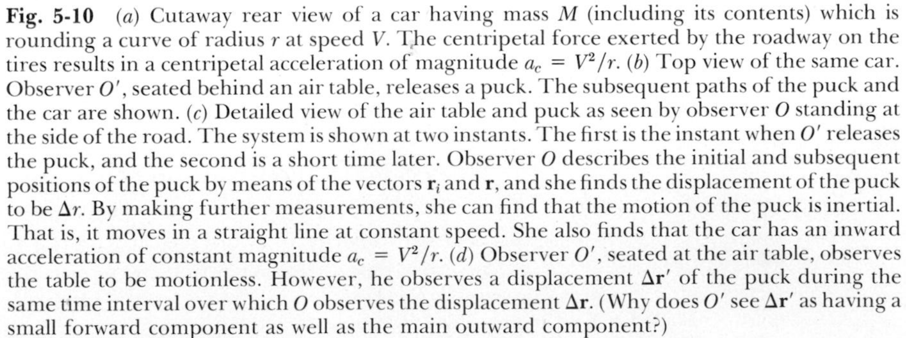

Fig. 5-10 (*a*) Cutaway rear view of a car having mass M (including its contents) which is rounding a curve of radius r at speed V. The centripetal force exerted by the roadway on the tires results in a centripetal acceleration of magnitude $a_c = V^2/r$. (*b*) Top view of the same car. Observer O', seated behind an air table, releases a puck. The subsequent paths of the puck and the car are shown. (*c*) Detailed view of the air table and puck as seen by observer O standing at the side of the road. The system is shown at two instants. The first is the instant when O' releases the puck, and the second is a short time later. Observer O describes the initial and subsequent positions of the puck by means of the vectors $\mathbf{r}_i$ and $\mathbf{r}$, and she finds the displacement of the puck to be Δr. By making further measurements, she can find that the motion of the puck is inertial. That is, it moves in a straight line at constant speed. She also finds that the car has an inward acceleration of constant magnitude $a_c = V^2/r$. (*d*) Observer O', seated at the air table, observes the table to be motionless. However, he observes a displacement $\Delta\mathbf{r}'$ of the puck during the same time interval over which O observes the displacement $\Delta\mathbf{r}$. (Why does O' see $\Delta\mathbf{r}'$ as having a small forward component as well as the main outward component?)

for the centrifugal force. It is purely an artifice, invented by the passenger in the car who wishes to use Newton's laws in a noninertial frame. Such a force is called a **fictitious force.** [It is also sometimes called a **d'Alembert force,** after the French physicist Jean le Rond d'Alembert (1717–1783) who first systematically clarified the rules for application of the laws of physics by an accelerated observer.] Note that only the observer in the noninertial frame must use the fictitious force to account for what he observes.

With these ideas in mind, let us derive the rules for comparison of observations made by an observer in a noninertial frame with simultaneous observations made in an inertial frame where we know Newton's laws to be correct. Observer O in the inertial frame is doing some of the air table-puck experiments we described in Chap. 4, in order to verify that she is indeed in an inertial frame. She finds that pucks accelerate only when they interact with other bodies. Otherwise, they maintain constant velocity. She thus knows that the principle of momentum conservation and Newton's laws are applicable to the observations she will make subsequently.

As O works, she is observed by observer O', who rides on a magic carpet. In contrast to the situation described in Sec. 3-8, where O' made his observations from a car moving at constant velocity, *the magic carpet is accelerated with a constant acceleration* $\mathbf{A}$ relative to observer O. That is, $\mathbf{A}$ is the acceleration of the frame of reference O', as measured by observer O. Using suitable telescopes and so forth, observer O' makes for himself every measurement made by O. For every velocity $\mathbf{v}$ determined by O at a certain time, O' simultaneously determines a velocity $\mathbf{v}'$.

In order to compare his results with those of O, observer O' must find the proper velocity transformation equation, which relates to $\mathbf{v}$ the value of $\mathbf{v}'$ that he observes from his nonrotating but noninertial reference frame. He begins by making the sketch of Fig. 5-11*a*. The relative positions of the two reference frames are shown at a certain arbitrary time t. In frame O, the position of the origin of frame O' is described by the vector $\mathbf{R}$. At this instant, observers O and O' simultaneously measure the position of an arbitrary body B. The results of their respective measurements are described by the vectors $\mathbf{r}$ and $\mathbf{r}'$.

It is evident from inspection of Fig. 5-11*a* that the vectors $\mathbf{r}'$ and $\mathbf{r}$ are related by the equation

$$\mathbf{r}' = \mathbf{r} - \mathbf{R} \tag{5-23a}$$

The desired velocity transformation equation can be obtained by differentiating both sides of this equation with respect to time. This gives

$$\frac{d\mathbf{r}'}{dt} = \frac{d\mathbf{r}}{dt} - \frac{d\mathbf{R}}{dt}$$

Employing the definitions $\mathbf{v}' = d\mathbf{r}'/dt$, $\mathbf{v} = d\mathbf{r}/dt$, and $\mathbf{V} = d\mathbf{R}/dt$, we have

$$\mathbf{v}' = \mathbf{v} - \mathbf{V} \tag{5-23b}$$

This equation appears identical to Eq. (3-57), the Galilean velocity transformation which applies when $\mathbf{V}$ is constant. Now, however, the velocity $\mathbf{V}$ of the reference frame O' as observed by O is *not* constant. The value of $\mathbf{V}$ which must be used in Eq. (5-23*b*) is its value *at the same instant* at which O measures the velocity $\mathbf{v}$ and O' measures the velocity $\mathbf{v}'$ for the same body.

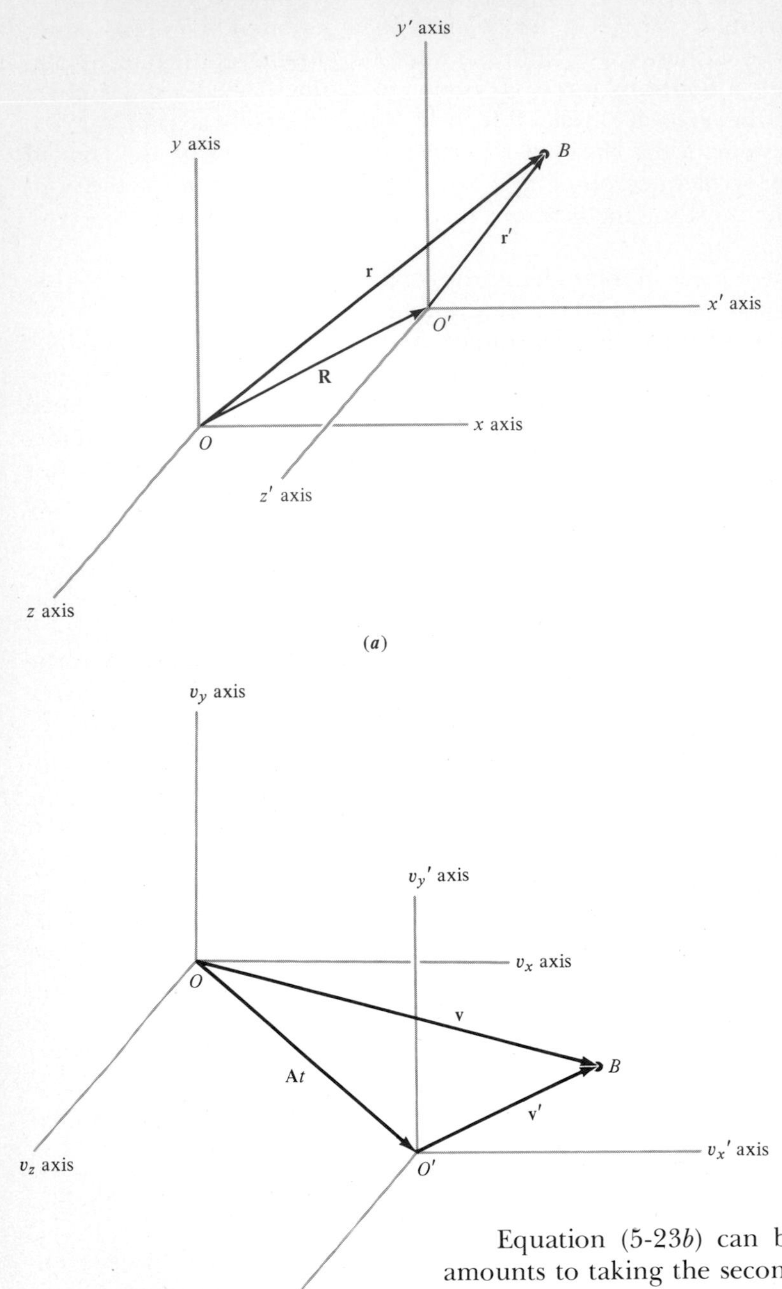

Fig. 5-11 (*a*) At a certain time t, the position of the origin O' of a noninertial frame of reference is described with respect to the inertial frame O by means of the vector $\mathbf{R}$. At the same instant, the position of body B is described by the vector $\mathbf{r}$ in the inertial frame O and by the vector $\mathbf{r}'$ in the noninertial frame O'. (*b*) A representation of the same two frames of reference in velocity space. Frame O' has constant acceleration $\mathbf{A}$ relative to inertial frame O. The two frames coincide at $t = 0$; that is, the instantaneous velocity of O' with respect to O is $\mathbf{V} = \mathbf{A}t$. At some arbitrary instant t, the velocity of body B is measured simultaneously by observers in both frames of reference. Its value is $\mathbf{v}$ as measured from O and $\mathbf{v}'$ as measured from O'. The vector diagram shows that $\mathbf{v}' = \mathbf{v} - \mathbf{A}t$. You should compare this velocity-space representation with the position-space representation of two inertial frames in Fig. 3-42.

Equation (5-23*b*) can be differentiated with respect to time. [This amounts to taking the second derivative of Eq. (5-23*a*).] We obtain

$$\frac{d\mathbf{v}'}{dt} = \frac{d\mathbf{v}}{dt} - \frac{d\mathbf{V}}{dt}$$

Using the definitions $\mathbf{a}' = d\mathbf{v}'/dt$, $\mathbf{a} = d\mathbf{v}/dt$, and $\mathbf{A} = d\mathbf{V}/dt$, we have

$$\mathbf{a}' = \mathbf{a} - \mathbf{A} \tag{5-23c}$$

This result is *not* the same as that for the Galilean acceleration transformation, where $\mathbf{a}' = \mathbf{a}$.

We assume for convenience that the two reference frames are instantaneously at rest with respect to each other at a certain instant which we call $t = 0$. That is, at the time $t = 0$ (but *only* then) any vector $\mathbf{v}$ which describes the velocity of a body B with respect to frame O is identical to the vector $\mathbf{v}'$ which describes its velocity with respect to frame O'.

At any other time t, the velocity $\mathbf{V}$ of frame O' relative to frame O is given by the vector $\mathbf{A}t$, as shown in Fig. 5-11b. Inspection of the figure shows that the relation between $\mathbf{v}'$ and $\mathbf{v}$ is

$$\mathbf{v}' = \mathbf{v} - \mathbf{A}t \tag{5-24}$$

This can be seen in algebraic terms as well, by substituting $\mathbf{V} = \mathbf{A}t$ into Eq. (5-23b).

Let us suppose that the two observers have determined the mass m of body B. (This might have been done at an earlier time when they were at rest with respect to each other and the body and could easily measure its weight mg.) Once this has been done, both observers can use their observations of the velocity of body B at a certain instant to determine its momentum at that instant. Observer O finds the momentum to be

$$\mathbf{p} = m\mathbf{v}$$

Observer O' finds the momentum to have the different value

$$\mathbf{p}' = m\mathbf{v}'$$

Using Eq. (5-24), observer O' can reexpress the momentum which he determines by measurement on body B in terms of the quantities $\mathbf{v}$ and $\mathbf{A}$ measured by observer O. He has

$$\mathbf{p}' = m\mathbf{v}' = m(\mathbf{v} - \mathbf{A}t) \tag{5-25}$$

In her inertial frame, observer O defines the net force acting on body B by means of Eq. (4-14), according to which this force is the rate of change of the momentum $\mathbf{p}$ of the body. With a slight change in notation, this equation can be written

$$\mathbf{f} \equiv \frac{d\mathbf{p}}{dt} \tag{5-26a}$$

Given the assumption that body B has a constant mass m, the force can be expressed in terms of the acceleration $\mathbf{a}$ of body B as seen by O:

$$\mathbf{f} = \frac{d}{dt}(m\mathbf{v}) = m\frac{d\mathbf{v}}{dt} = m\mathbf{a} \tag{5-26b}$$

If O observes body B to be accelerating with acceleration $\mathbf{a}$, she attributes this to a force $\mathbf{f}$ which is applied to body B by some external agent (such as a spring or a contact with another body or the gravitational attraction of the earth).

In like manner, observer O' defines the force acting on body B in terms of the rate of change of momentum which *he* observes:

$$\mathbf{f}' \equiv \frac{d\mathbf{p}'}{dt} \tag{5-27}$$

In order to express this force in terms of the force $\mathbf{f}$ which he knows O has calculated, O' first uses Eq. (5-25) to express $\mathbf{p}'$ in terms of the quantities O measures. He inserts this value of $\mathbf{p}'$ into Eq. (5-27) to obtain

$$\mathbf{f}' = \frac{d}{dt}[m(\mathbf{v} - \mathbf{A}t)]$$

Carrying out the differentiation, he obtains

$$\mathbf{f}' = m \frac{d\mathbf{v}}{dt} - m \frac{d}{dt} (\mathbf{A}t)$$

And since $\mathbf{A}$ is constant, this yields

$$\mathbf{f}' = m\mathbf{a} - m\mathbf{A} \tag{5-28a}$$

which can also be written

$$\mathbf{f}' = m(\mathbf{a} - \mathbf{A}) \tag{5-28b}$$

Let us consider the latter form of this equation first. According to Eq. (5-23*c*), the vector difference $\mathbf{a} - \mathbf{A}$ is equal to $\mathbf{a}'$, the acceleration of body B as seen by O'. We make this substitution, and Eq. (5-28*b*) becomes

$$\mathbf{f}' = m\mathbf{a}' \tag{5-28c}$$

This equation has the form of Newton's second law. That is, Newton's second law is valid even though observer O' is *not* in an inertial frame. How can this be, when Newton's laws of motion are valid only in inertial frames? The answer is that *the force* $\mathbf{f}'$ *acting on body* B*, as determined by* O'*, is not the same as that observed by* O *in her inertial frame.*

To see this, consider again the same equation written in the form of Eq. (5-28*a*), $\mathbf{f}' = m\mathbf{a} - m\mathbf{A}$. The term $m\mathbf{a}$ is simply the force $\mathbf{f}$ which O finds to be acting on body B, as indicated by Eq. (5-26*b*). It is the product of the mass of B and its observed acceleration. The extra term $-m\mathbf{A}$ also has the form of a force, since it is the product of a mass and an acceleration. We use it to define the **fictitious force F** by means of the equation

$$\mathbf{F} \equiv m(-\mathbf{A}) \tag{5-29}$$

The fictitious force acting on body B is a peculiar force indeed. It is the mass of the body, multiplied not by its own acceleration as observed either by O or by O', but by the negative of the acceleration of O' as observed by O! From a purely formal point of view $\mathbf{F}$ is the algebraic term needed to make the connection between the force $\mathbf{f}'$ observed by O' and the force $\mathbf{f}$ observed by O. Using $\mathbf{F}$, we find this connection to be

$$\mathbf{f}' = \mathbf{f} + \mathbf{F} \tag{5-30}$$

The fictitious force $\mathbf{F}$ *makes it possible to adjust Newton's laws of motion so that they can be used in a noninertial frame.* As we have seen in the example of the air table in the car (Fig. 5-10), O' invents the fictitious force to account for his being "pulled out from under" the system he observes. However, as mathematical formalisms and fictitious entities go, the force $\mathbf{F}$ has a very convincing perceptual existence, as anyone who has ever ridden in a car around a sharp curve can attest!

In developing the concept of fictitious force, we restricted our consideration to a noninertial frame of reference O' which does not rotate with respect to the inertial frame O. This was done for simplicity. Every point fixed to the noninertial frame has the same acceleration $\mathbf{A}$ as the origin O', as seen by an observer in O. This is not the case for a rotating frame. In such a frame, the acceleration of a point depends on its position (x', y', z') as well as on the acceleration $\mathbf{A}$ of the origin O' with respect to O. (We will develop vector methods for dealing with such rotation in Chaps. 9 and 10.)

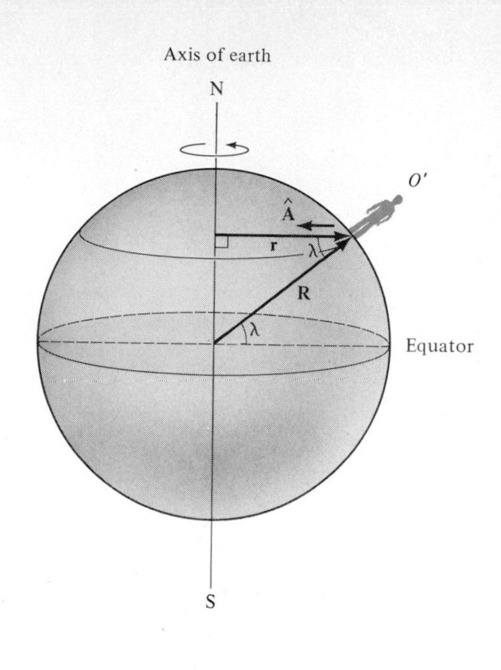

Fig. 5-12 Observer O' stands on the surface of the rotating earth at a location having latitude λ. The radius vector from the center of the earth to O' is $\mathbf{R}$. The vector along the perpendicular from the axis of the earth to O' is $\mathbf{r}$. From the point of view of an observer in an inertial frame, O' experiences a centripetal acceleration whose direction is $\hat{\mathbf{A}} = -\hat{\mathbf{r}}$.

Nevertheless, the concept of fictitious force applies to rotating reference frames as well as to nonrotating ones. Although it may be more complicated to determine the acceleration $\mathbf{a}$ of a body as seen by O and the corresponding acceleration $\mathbf{a}'$ as seen by O', there is still associated with each of these accelerations a force. If the mass of the body is m, those forces are, respectively, $\mathbf{f} = m\mathbf{a}$ and $\mathbf{f}' = m\mathbf{a}'$. And Eq. (5-23c), $\mathbf{a}' = \mathbf{a} - \mathbf{A}$, still yields a value of $-\mathbf{A}$ which is related to the fictitious force $\mathbf{F}$ by the equation $\mathbf{F} = -m\mathbf{A}$, as in Eq. (5-29).

The surface of the earth is a noninertial frame. Rotating as it does on its axis, the earth continually accelerates us centripetally with acceleration $\mathbf{A} = -(v^2/r)\hat{\mathbf{r}}$, where v is the surface speed of the earth at the latitude λ (the Greek letter lambda) of the observer and $\mathbf{r}$ is the vector from the axis of the earth to the observer, along the perpendicular to the axis, shown in Fig. 5-12.

EXAMPLE 5-7

Find the centrifugal acceleration experienced by an observer O' at the equator ($\lambda = 0°$), the observed magnitude g' of the gravitational acceleration, and the observed weight of a standard 2-kg mass. Take the value of g, which would be observed if the earth were not rotating, to be 9.832 m/s² and the radius of the earth as 6380 km.

▪ An observer in an inertial frame (say, in space) sees an observer O' standing at the equator to have a centripetal acceleration $\mathbf{A} = -(v^2/R)\hat{\mathbf{r}}$, where $\hat{\mathbf{r}}$ is the outward unit vector along the perpendicular from the axis of the earth to O' (see Fig. 5-12), R is the radius of the earth, and v is the speed of O' as he is carried around by the rotation of the earth. For an observer at the equator, the vectors $\mathbf{R}$ and $\mathbf{r}$ coincide, so $\hat{\mathbf{R}} = \hat{\mathbf{r}}$. The centrifugal acceleration $\mathbf{A}'$ is the negative of $\mathbf{A}$, so you have

$$\mathbf{A}' = -\mathbf{A} = +\frac{v^2}{R}\hat{\mathbf{R}}$$

A point on the equator goes around the circumference of the earth with a period T = 23 h 56 min (see Example 3-10) and a speed

$$v = \frac{2\pi R}{T}$$

Its centrifugal acceleration is

$$\mathbf{A}' = \frac{4\pi^2 R}{T^2}\,\hat{\mathbf{R}}$$

directed outward. Thus you have

$$A' = \frac{4\pi^2 \times 6.380 \times 10^6\ \text{m}}{(23\ \text{h} \times 3600\ \text{s/h} + 56\ \text{min} \times 60\ \text{s/min})^2}$$
$$= 3.393 \times 10^{-2}\ \text{m/s}^2$$

or about 0.35 percent of the magnitude of the gravitational acceleration. Since the direction of $\mathbf{A}'$ is opposite to that of $\mathbf{g}$, you have

$$g' = g - A'$$
$$= 9.832\ \text{m/s}^2 - 0.034\ \text{m/s}^2 = 9.798\ \text{m/s}^2$$

A 2.000-kg mass, which would weigh mg = 19.66 N in the absence of rotation, will weigh 19.60 N at the equator. An observer standing on the earth and noting this slight reduction in weight can attribute it to the fictitious centrifugal force $\mathbf{F} = m\mathbf{A}' = 0.06\ \text{N}\ \hat{\mathbf{R}}$ acting on it. But you can see why, for many practical purposes, we can consider ourselves to be situated in an inertial reference frame.

Even though weight reduction due to the earth's rotation is quite small, there are important consequences of the fact that the surface of the earth is a noninertial frame. It is well known that ocean currents and moving air masses tend to curve in a clockwise direction in the Northern Hemisphere and in a counterclockwise direction in the Southern Hemisphere. We will here give a qualitative account of this phenomenon in terms of fictitious forces.

Consider a quantity of water in the ocean at the equator. It participates in the daily rotation of the earth as a whole, moving eastward with a speed equal to that of the surface of the earth at the equator, as shown in Fig. 5-13*a*. Now suppose that for some reason (say because of the upwelling of water beneath it) this water moves northward. As it does so, it remains on the surface of the earth, and consequently finds itself closer to the axis of the earth than before. From the point of view of an observer in space, in an inertial frame, the law of inertia must apply. That is, the water tends to continue to move eastward with the same speed as when it was at the equator. But the surface speed of the earth in the new latitude is less than that at the equator, so that the water tends to outpace the earth. That is, it moves eastward with respect to the surface of the earth.

Now consider the reverse situation. A quantity of water somewhere north of the equator is moving at the surface speed of the earth at this latitude—that is, it is at rest with respect to the earth's surface. Suppose that for some reason it moves southward, toward the equator. Again, an observer in an inertial frame will use the law of inertia to predict that the water will tend to move eastward at its original speed. But as it moves southward, it moves farther from the earth's axis, into a region where the surface speed is higher. Therefore, the water tends to lag, or to move westward with respect to the surface of the earth. Taken together, these ten-

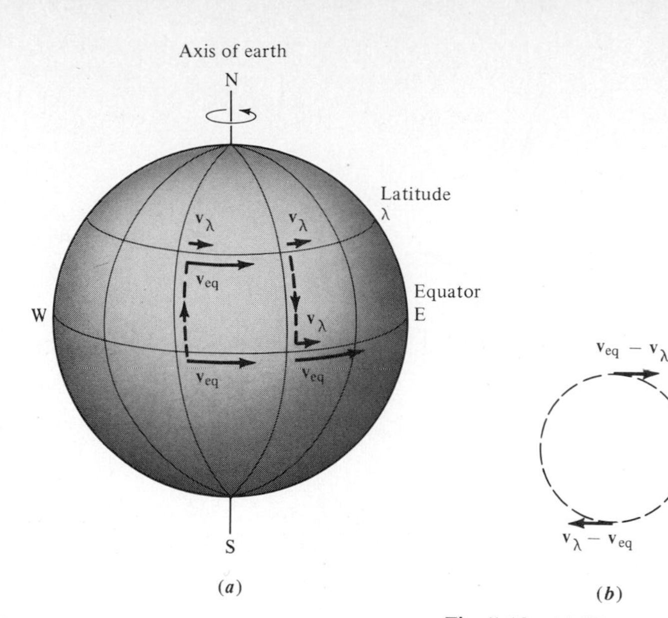

Fig. 5-13 (*a*) Water moving northward from the equator tends to maintain its surface speed v_{eq}, which is greater than v_λ, the characteristic surface speed at latitude λ. An observer on the surface of the earth who does not perceive the rotation of the earth attributes the tendency of the water to move relatively eastward to the fictitious Coriolis force. The same argument in reverse applies to water moving southward toward the equator. (*b*) The eastward velocity $\mathbf{v}_{eq} - \mathbf{v}_\lambda$ relative to the earth of water which has moved northward and the westward velocity $\mathbf{v}_\lambda - \mathbf{v}_{eq}$ relative to the earth of water which has moved southward in the Northern Hemisphere. The overall effect is a tendency for moving water to circulate clockwise in the Northern Hemisphere.

dencies result in a clockwise rotation of water masses, as shown in Fig. 5-13*b*.

But now let us look at the situation from the point of view of an observer O' on the surface of the earth. This observer is not (or chooses not to be) conscious of the rotation of the earth. Nevertheless, he must account for the tendency of the ocean currents to rotate clockwise. He therefore "invents" a fictitious force to explain the motion of the water. This force is peculiar in that it acts only on *moving* matter (here water), and in that it acts at right angles to the direction of motion (eastward on northward-moving water and westward on southward-moving water). This fictitious force is called the **Coriolis force,** after the French physicist who first studied it in detail.

Using an argument similar to that above, you can show that the Coriolis force in the Southern Hemisphere acts so as to produce counterclockwise motion of ocean currents. (Contrary to popular belief, the Coriolis force is far too small to account for the vortex sometimes seen in draining bathtubs and in similar situations. It is not true that the direction of rotation of this vortex is characteristic of the hemisphere in which the tub is located.)

It is a little harder to see that there is also a Coriolis force acting on a body which moves relative to the surface of the earth in an east-west direction. In order to understand its origin, consider first the simple case of a plumb bob (a body hanging at rest from a string) on a nonrotating earth. Figure 5-14*a* is a free-body diagram of the bob, whose mass is m. It is at rest because the upward force **S** ex-

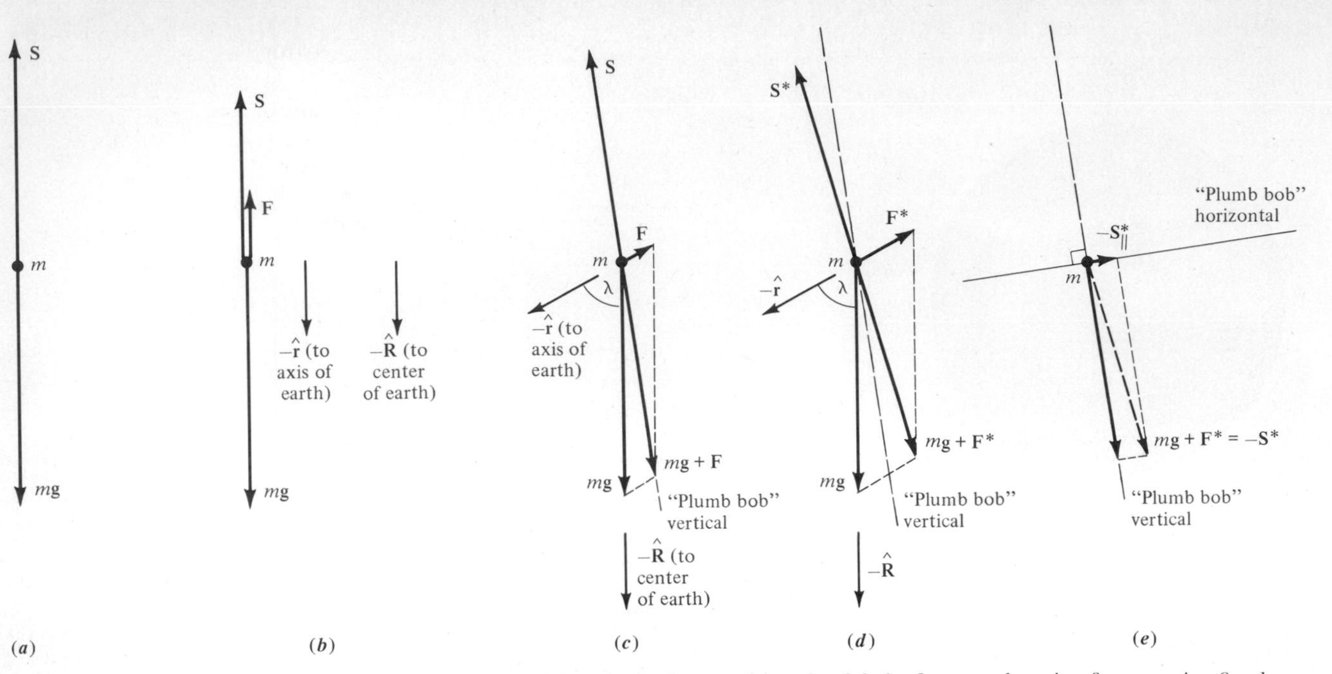

Fig. 5-14 (*a*) Free-body diagram for a plumb bob of mass m, hanging from a string fixed near the surface of a nonrotating earth. (*b*) The same bob located on the equator of a rotating earth. An observer standing next to the bob attributes the decrease in the magnitude of the force $\mathbf{S}$ exerted by the string to the fictitious centrifugal force $\mathbf{F}$ (much exaggerated). The vector $-\hat{\mathbf{r}}$ points toward the axis of the earth, and the vector $-\hat{\mathbf{R}}$ points toward the center of the earth (see Fig. 5-12). (*c*) The same plumb bob located at latitude λ in the Northern Hemisphere. (Why is the magnitude F of the centrifugal force less than that shown in part *b*?) (*d*) An identical plumb bob moving smoothly and steadily eastward in a submarine does not hang along the vertical established by the stationary plumb bob in part *c*, but makes a larger angle with $-\hat{\mathbf{R}}$. (*e*) The downward force $-\mathbf{S}^*$ on the moving plumb bob is replaced by its constituent vectors along the horizontal and vertical established by the stationary plumb bob. The southward-directed horizontal force $-\mathbf{S}^*_\parallel$ is the Coriolis force.

erted by the string is exactly equal in magnitude to the downward gravitational force $m\mathbf{g}$. Next, let the earth rotate and consider the free-body diagram, Fig. 5-14*b*, of a plumb bob hanging at the equator. The gravitational force $m\mathbf{g}$ is unchanged in magnitude and still has the direction $-\hat{\mathbf{R}}$ toward the center of the earth (as it always does). But as you saw in Example 5-7, the magnitude of the force exerted by the string is slightly reduced from that in the case of the nonrotating earth. An observer O in space sees this as a consequence of the centripetal acceleration of the bob as it rotates with the earth. But an observer O' standing on the surface of the earth is in a noninertial reference frame and attributes the observed reduction in $\mathbf{S}$ to the presence of the fictitious centrifugal force $\mathbf{F}$ shown in the figure. The direction of $\mathbf{F}$ is that of $\hat{\mathbf{r}}$, the unit vector along the perpendicular to the axis of the earth to the plumb bob (see Fig. 5-12). For a point on the equator, $\hat{\mathbf{r}}$ coincides with $\hat{\mathbf{R}}$, the unit vector along the earth's radius. The force exerted by the string is given by the condition

$$\mathbf{S} + m\mathbf{g} + \mathbf{F} = 0$$

or

$$\mathbf{S} = -(m\mathbf{g} + \mathbf{F})$$

Now let the plumb bob hang just above the surface of the earth at a location having latitude λ. (We will suppose that this location is in the Northern Hemisphere.) The free-body diagram of the bob is shown in Fig. 5-14*c*. The gravitational force $m\mathbf{g}$ again has direction $-\hat{\mathbf{R}}$, and the centrifugal force again has direction $\hat{\mathbf{r}}$. But as you can see from Fig. 5-12, $\hat{\mathbf{r}}$ and $\hat{\mathbf{R}}$ are no longer coincident. Instead they make an angle λ with each other. This angle is shown in Fig. 5-14*c*. In order to keep the plumb bob motionless, the string must again exert a force $\mathbf{S} = -(m\mathbf{g} + \mathbf{F})$, as the figure shows. Since plumb bobs are normally used to establish the direction called "vertical," we call the line along which $\mathbf{S}$ lies the "plumb bob" vertical. (Note that $\hat{\mathbf{S}}$ differs from $\hat{\mathbf{R}}$, which can be determined by means of astronomical measurements.)

We are now ready to consider the Coriolis force which is exerted (from the point of view of observer O') on bodies moving *with respect to the surface of the earth*. Suppose that a second, identical plumb bob is mounted in a submerged submarine, which can move very smoothly. The submarine heads eastward at constant speed at a constant depth just great enough to avoid wave action. The observer O, in an inertial frame in space, notes that the plumb bob is now circling the earth's axis faster than O', who remains stationary with respect to the surface of the earth. She therefore expects that the centripetal acceleration of the bob is increased and that the new magnitude S^* of the force exerted by the string is slightly less than S. This is indeed the case.

Let us consider the situation in more detail from the point of view of O'. (We must assume that he has some means of observing the plumb bob in the moving submarine from his fixed position on the surface of the earth.) His new observations are shown in Fig. 5-14*d*. His own plumb bob still hangs along the plumb bob vertical, which is shown as a dashed line. But the bob in the submarine hangs at a different angle. Observer O' argues that the centrifugal force exerted on the bob in the submarine is greater in magnitude than that exerted on his own, because of the greater speed at which the bob in the submarine circles the earth's axis. He calls the new force $\mathbf{F}^*$. Its direction is still $\hat{\mathbf{r}}$; that is, it is parallel to $\mathbf{F}$. The force exerted by the string on the moving bob, which holds it motionless with respect to the submarine, is given by

$$\mathbf{S}^* = -(m\mathbf{g} + \mathbf{F}^*)$$

as shown in Fig. 5-14*d*.

We now show that $\mathbf{F}^*$ tends to produce southward motion. In Fig. 5-14*e*, the "plumb bob" horizontal is constructed by drawing the perpendicular to the plumb bob vertical determined in Fig. 5-14*c*. If the ocean water through which the submarine is traveling is stationary with respect to the surface of the earth, the ocean surface will conform to the plumb bob horizontal. (Not only does the ocean adjust to this horizontal, but so does the "solid" earth. This is the source of the equatorial bulge of the earth, whose radius is some 21 km greater at the equator than at the poles.) The downward vector $-\mathbf{S}^*$ specifies the direction in which the moving bob hangs. It is not parallel to $-\mathbf{S}$. We can replace $-\mathbf{S}^*$ by its constituent vectors along $-\mathbf{S}$, the plumb bob vertical, and perpendicular to $-\mathbf{S}$, along the plumb bob horizontal. The horizontal constituent vector $-\mathbf{S}_{\parallel}^*$ lies along the surface in the southward direction. The plumb bob is held in place by the string and merely inclines toward this direction. But an unconstrained body, such as a moving mass of water in an ocean current or a long-distance artillery shell, will be accelerated southward by this force, which is also given the name Coriolis force. Can you explain the direction of this acceleration from the point of view of inertial observer O, who does not "invent" a fictitious force?

A similar argument shows that a westward-moving body in the Northern Hemisphere is likewise deflected to the right, that is, northward. A quantitative discussion is beyond the scope of this book, but it turns out that the magnitude of the horizontal component of the Coriolis force is independent of the direction of motion of the body on which it is exerted, as long as the body moves parallel to the surface of the earth.

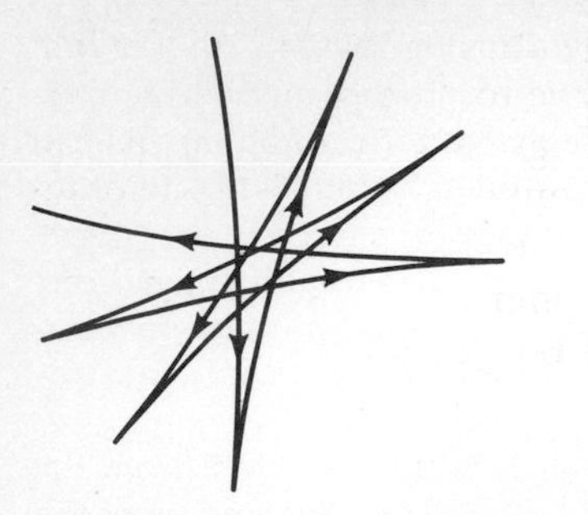

Fig. 5-15 The rotation of the path described by a Foucault pendulum in successive swings, resulting from the Coriolis force. The path of the bob, located in the Northern Hemisphere, is seen as it would be by a viewer looking down on it. The path always has clockwise curvature. (The curvature is grossly exaggerated. In actuality, the path during a single swing is not distinguishable to the unaided eye from a straight line.)

A spectacular demonstration of the Coriolis effect is the **Foucault pendulum,** named after the distinguished French physicist who was the first to set one up in 1851. A very long pendulum with a heavy bob will swing for a long time once it is started and will not be much disturbed by drafts and similar effects. The pendulum describes a vertical plane as it oscillates. As time passes, an observer standing near the pendulum (and thus rotating with the earth) observes that the plane of oscillation rotates slowly about a vertical axis. That is, if the pendulum is at first swinging in an east-west direction, it will later swing between southeast and northwest (in the Northern Hemisphere) and later south-north, and so on. See Fig. 5-15, where the effect is grossly exaggerated. This phenomenon is often explained by saying that "the earth rotates under the pendulum," but this is not a complete explanation except at the earth's poles. What does happen is that the Coriolis force exerts a small but consistent force toward the right (in the Northern Hemisphere) no matter in what direction the pendulum is swinging. This leads to the described effect. It turns out that the rotation rate of the plane of the pendulum is

$$\frac{2\pi}{24} \sin \lambda \text{ rad/h} = 15^\circ \sin \lambda \text{ h}^{-1}$$

where λ is the latitude. You can see that this expression gives the expected result at the earth's poles ($\lambda = 90^\circ$) where the earth does indeed "turn beneath" the pendulum and the plane of the pendulum takes 24 h to rotate with respect to the earth. At the equator ($\lambda = 0^\circ$) the centrifugal force is vertically upward. Thus there is no horizontal component, and the pendulum plane does not rotate at all, as the expression predicts.

5-5 ROCKETS

Now that people have begun the exploration of space, rockets have become a subject of general public interest. The central feature of the rocket, which underlies its unique capabilities, is its **reaction engine.** The rocket carries its own supply of fuel and oxidizer, which the engine ejects backward as hot gas. (Fuel and oxidizer together are frequently referred to as "fuel.") Unlike a jet plane, a rocket has no dependence on an external supply of air.

As the rocket engine operates, the mass of the rocket changes rapidly. Consequently, Newton's second law in the general form $F = dp/dt$ does not lead to $F = ma$. We must therefore begin the analysis of the motion of a rocket from the basic principle of momentum conservation.

We consider a rocket in deep space, far from any gravitating body. It moves in a straight line, subject only to the influence of its main engine. Figure 5-16 shows two consecutive views of the rocket, as seen from an inertial frame of reference. In Fig. 5-16*a*, the rocket is shown at some arbitrary time t, when its mass is m. It is most convenient to describe the motion of the system in terms of signed scalars. An inertial observer sees the rocket moving with instantaneous velocity V, the forward direction of its motion being used to define the positive direction. The engine ejects a continuous stream of gas backward from the tail of the rocket. As a result, the mass of the rocket changes at a constant rate dm/dt. This quantity has a *negative* value, because the mass is decreasing.

The velocity at which gas is ejected from the rocket is v_g'. But this is the velocity of the gas *relative to the rocket,* and *not* relative to the inertial frame. In order to find the velocity v_g of the gas relative to the inertial frame, we use the velocity transformation, Eq. (5-23*b*). In signed scalar notation this is written $v' = v - V$, or $v = V + v'$. For the present case, v' is the velocity of the gas (the body B of Sec. 5-4) as seen from the rocket, so that $v' \equiv v'_g$. And v is the velocity of the gas as seen from the inertial frame of reference, so that $v \equiv v_g$. The velocity transformation of Eq. (5-23*b*) thus gives

$$v_g = V + v'_g$$

This result is shown in Fig. 5-16*b*. (Unlike v'_g, which always has a negative value since gas is always ejected backward from the point of view of an observer on the rocket, v_g can be either positive or negative depending on how fast the rocket is going. All observers will see the rocket leaving its exhaust gas behind, but not all observers will see the rocket and the exhaust gas moving in opposite directions!)

We now consider the entire rocket-plus-gas system from the inertial frame. As an isolated system seen from an inertial frame, its momentum will remain constant as time passes. We can thus write

$$\begin{pmatrix}\text{momentum of rocket}\\ \text{at time } t\end{pmatrix} = \begin{pmatrix}\text{momentum of rocket}\\ \text{at time } t + dt\end{pmatrix} + \begin{pmatrix}\text{momentum of gas}\\ \text{ejected during time}\\ \text{interval } dt\end{pmatrix}$$

In order to make use of this equation, we next calculate the various momenta.

At time t, shown in Fig. 5-16*a*, the total mass of the system is the mass of the rocket, m, and its velocity is V. Its momentum is thus mV. At the infinitesimally later time $t + dt$, shown in Fig. 5-16*b*, the mass of the rocket has changed by an amount given by the product of the rate of change dm/dt and the time dt which has elapsed. Thus the rocket has mass

$$m + \frac{dm}{dt}\,dt = m + dm$$

(The change in mass dm has a negative value, since dm/dt has a negative value.)

During the same time interval dt, the velocity of the rocket has changed (increased) by an infinitesimal amount dV, and is thus $V + dV$. The new momentum of the rocket is

$$(m + dm)(V + dV)$$

The ejected gas has mass $-dm$ (since any change in the mass of the rocket must correspond to a change of the same magnitude but opposite sign in the mass outside the rocket). As seen from the inertial frame, the gas has emerged from the engine at velocities changing uniformly from $V + v'_g$ to $V + dV + v'_g$. The average velocity of the gas ejected during the time dt is thus $V + v'_g + dV/2$. The momentum of the ejected gas is consequently

$$-dm\left(V + v'_g + \frac{dV}{2}\right)$$

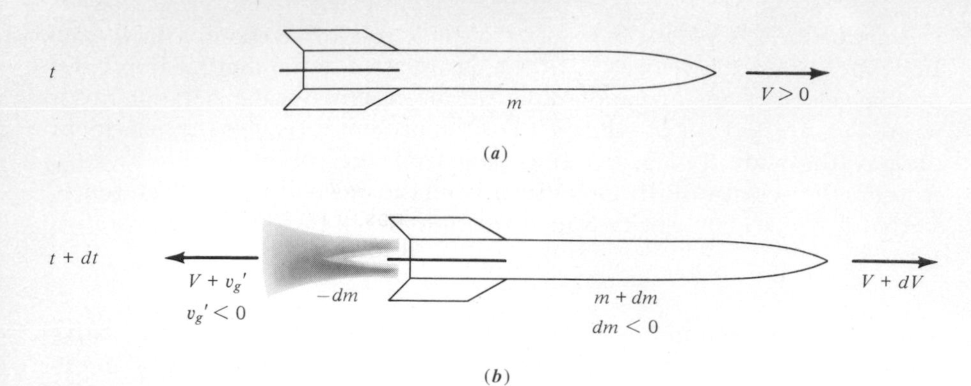

Fig. 5-16 A rocket shown at two instants separated by an infinitesimal time interval dt. The engine is operating continuously, but the gas expelled before the time t is not shown in either view. (a) The rocket at time t. Its velocity as seen from an inertial frame is V, and its mass is m. (b) The rocket at time $t + dt$. Its velocity is $V + dV$. The engine has expelled gas having a mass $-dm$ during the interval dt, and the mass of the rocket is now $m + dm$. As seen from the inertial frame, the gas expelled at the beginning of the interval dt has velocity $V + v_g'$. In the case illustrated, the rocket has not yet attained a very large velocity V. Consequently, the quantity $v_g = V + v_g'$ has a negative value. That is, an observer at rest sees the ejected gas to be moving backward. The analysis in the text shows that the direction of v_g is immaterial.

Equating the total momentum at the beginning of the time interval with the total momentum at its end, we have

$$mV = (m + dm)(V + dV) + (-dm)\left(V + v_g' + \frac{dV}{2}\right)$$

Expanding the terms on the right side of this equation gives

$$mV = mV + m\,dV + dm\,V + dm\,dV - dm\,V - dm\,v_g' - dm\,\frac{dV}{2}$$

which reduces to

$$0 = m\,dV + dm\,dV - dm\,v_g' - dm\,\frac{dV}{2}$$

All the terms on the right side of this equation are infinitesimal. The first and the third are products of an infinitesimal quantity with a finite quantity. But the second and fourth terms are each the product of two infinitesimals. They are therefore negligibly small compared to the other two. We thus neglect them and have

$$0 = m\,dV - dm\,v_g'$$

Dividing through by dt, we obtain

$$0 = m\,\frac{dV}{dt} - \frac{dm}{dt}\,v_g'$$

or

$$m\,\frac{dV}{dt} = v_g'\,\frac{dm}{dt} \tag{5-31a}$$

Since V is the velocity of the rocket, dV/dt is its acceleration A. So we obtain the result

$$mA = v_g'\,\frac{dm}{dt} \tag{5-31b}$$

(Note that A has a positive value, as expected, since the right side of this equation is the product of two terms with negative value.)

In Eq. (5-31a) we have expressed the product of the mass m of the rocket and its acceleration A as the quantity $v_g' \, dm/dt$. Since the equation is in the form of Newton's second law

$$mA = F$$

we can identify $v_g' \, dm/dt$ as a force. That is, *we now take as a system the rocket alone,* instead of the rocket plus the ejected gas. This new system is not an isolated system. Rather, there is an external force

$$F = v_g' \frac{dm}{dt} \tag{5-32}$$

acting on it, which causes it to accelerate. This force is called the **thrust** of the rocket engine. It comprises one member of a Newton's third law action-reaction pair, and is the force exerted on the rocket engine by the ejected gas. Hence the name "reaction engine." (The other member of the pair is the force exerted on the ejected gas by the engine. It has no direct effect on the system of interest, which is the rocket.) It is reasonable that the magnitude of the thrust be proportional to the speed $|v_g'|$ at which the engine ejects gas, and also to $|dm/dt|$, the mass of gas ejected per unit time, called the **fuel consumption rate.**

Equation (5-32) can be written in the form

$$v_g' = \frac{F}{dm/dt} \tag{5-33}$$

The term on the right is the thrust of the rocket, divided by the fuel consumption rate required to obtain it. This is an important figure of merit in the design of rocket systems; the rocket engineer would like to have the maximum possible thrust for the minimum possible fuel consumption rate.

The quantity $F/(dm/dt)$ on the right side of Eq. (5-33) is called the **specific thrust** or, more commonly, the **specific impulse.** Its SI units are usually quoted in the form newton-seconds per kilogram (N·s/kg). Equation (5-33) shows that specific impulse is equal to the velocity, relative to the rocket, at which the engine can eject the gas. It turns out that for well-designed engines this is determined mainly by the chemical energy per unit mass stored in the fuel. The largest value of specific impulse theoretically possible for chemical fuels is about 3300 N·s/kg, using a hydrogen-fluorine mixture. This mixture is rarely (if ever) used because of the great difficulty and danger of handling fluorine and the extreme noxiousness of the principal exhaust product, hydrogen fluoride. Fortunately the hydrogen-oxygen mixture has a theoretical specific impulse only a few percent smaller than hydrogen-fluorine, and the exhaust is mainly steam. The rockets used as the main engines in launching space vehicles nearly all use hydrogen and oxygen, which are stored in their liquid form.

Even if v_g' and dm/dt are constant while the rocket engine is burning, so that the thrust F is constant, the acceleration $A = F/m$ of the rocket is *not* constant. The acceleration increases with time because the mass m of the rocket decreases. This decrease is due to the fact that the engine burns fuel to produce the ejected gas, so that the value of dm/dt is negative. The mass decrease is by no means a small effect; more than 80 percent of the launch mass of a rocket intended for space flight is fuel.

To determine quantitatively the motion of the rocket—that is, to find its velocity and position from its acceleration—involves complications we have not had to deal with in previous cases, where the acceleration was always constant. One procedure for handling systems with nonconstant acceleration is developed in Sec. 5-6.

5-6 THE SKYDIVER

The rocket discussed in Sec. 5-5 is an important, though not typical, example of a physical system in which a body moves with nonconstant acceleration. While its engine is burning, the acceleration of a rocket increases even though the force acting on it remains constant, because its mass is decreasing. Much more commonly, varying acceleration is the result of a varying force acting on a body whose mass remains constant. But regardless of the cause of variation, *if the acceleration of a body varies, its velocity and position cannot be found by making direct use of the familiar kinematic equations of Sec. 2-7, since they were derived for the case of constant acceleration.*

In this section we develop a method for determining the velocity and position of a body which moves with *nonconstant* acceleration. The method involves a sequence of numerical calculations. We will use the method to study an object falling vertically downward through the air near the surface of the earth, making the physically realistic assumption that the strength of the retarding drag force due to air resistance is proportional to the square of the velocity. That is, we will assume that the air flow around the relatively large, rapidly moving object is the turbulent flow discussed in Sec. 4-6. In particular, the method will be applied to study the motion of a **skydiver**—a person who jumps from an airplane to enjoy the sensation of fall for some time before opening a parachute. See Fig. 5-17.

In terms of the signed scalar representation, with the downward direction taken as positive, the net force acting on the skydiver is positive. But as the velocity of this daring athlete builds up during the fall, the negative drag force also builds up and the net force decreases. The acceleration a =

Fig. 5-17 A skydiver.

F/m therefore decreases, so the constant-acceleration kinematic equations we have frequently used up to this point cannot be applied in a direct manner. However, you will soon see that they can be applied in a restricted way which nevertheless yields useful results.

As a preliminary, let us reconsider a one-dimensional situation in which the acceleration a of a body *is* constant. Suppose that you know the value of a and also know the velocity v_0 and the position x_0 of the body at some initial time $t_0 = 0$. One way that you can determine its velocity and position at subsequent times is to select a time increment Δt and then carry out the following set of calculations:

Use the *constant* value of a to calculate

$$v_{1/2} = v_0 + a\,\frac{\Delta t}{2} \tag{5-34a}$$

from the given value of v_0. Use the result to calculate

$$x_1 = x_0 + v_{1/2}\,\Delta t \tag{5-34b}$$

from the given value of x_0. Then set $t_1 = \Delta t$.

Next use a, and the value of $v_{1/2}$ just obtained, to calculate

$$v_{3/2} = v_{1/2} + a\,\Delta t \tag{5-34c}$$

Use the result, and the value of x_1 just obtained, to calculate

$$x_2 = x_1 + v_{3/2}\,\Delta t \tag{5-34d}$$

Then set $t_2 = 2\Delta t$.

Next use a, and the value of $v_{3/2}$ just obtained, to calculate

$$v_{5/2} = v_{3/2} + a\,\Delta t \tag{5-34e}$$

Use the result, and the value of x_2 just obtained, to calculate

$$x_3 = x_2 + v_{5/2}\,\Delta t \tag{5-34f}$$

Then set $t_3 = 3\Delta t$.

Continue these calculations until t reaches whatever value is required.

Equations (5-34) are exact for the case of a constant acceleration a, regardless of the size of the chosen time increment Δt. The value of Δt you actually choose is determined by how far apart in time you want to know the successive values of velocity and position.

Let us consider in detail how the calculations produce the desired results. First you find $v_{1/2}$, which is the velocity of the body at a time $\frac{1}{2}\Delta t$ later than the initial time. To do this, you add to the initial velocity v_0 the product of the acceleration a and the time $\frac{1}{2}\Delta t$ during which that acceleration has acted. The equation used, Eq. (5-34a), is just a reexpression of the definition of acceleration for the case of constant acceleration.

Now that you have found $v_{1/2}$, you use it to find x_1, which is the position at a time $1\Delta t$ later than the initial time. The quantity x_1 is obtained from Eq. (5-34*b*) by adding to the initial value x_0 the product of the velocity $v_{1/2}$ and the time increment Δt. The validity of doing so hinges on two facts: (1) Since the acceleration is constant, the velocity changes uniformly during the time interval starting at $t = 0$ and ending at $t = 1\Delta t$. Therefore its *average* value $\langle v \rangle$ over that interval is equal to its *instantaneous* value halfway through the interval, which is $v_{1/2}$. (See Fig. 2-22*a*.) (2) Since total change in position is the average velocity multiplied by the elapsed time, it follows that the change in position during the time interval is $\langle v \rangle \, \Delta t = v_{1/2} \, \Delta t$.

To complete the first cycle of calculations, you set $t_1 = \Delta t$ in order to update the time.

In the next cycle of calculations, you first evaluate the velocity $v_{3/2}$ at time $\frac{3}{2}\Delta t$, using Eq. (5-34*c*) to do so. This equation is very much the same as Eq. (5-34*a*). The only differences are in the subscripts (which specify the particular values of v involved) and in the fact that the time increment used is now Δt. Then you use Eq. (5-34*d*) to obtain x_2 from x_1 by employing the just-obtained value of $v_{3/2}$. The equation is identical to Eq. (5-34*b*) except for the subscripts. Then you again update time by setting $t_2 = 2\Delta t$.

The next cycle of calculations yields first $v_{5/2}$ and then x_3 in exactly the same way that the two preceding values $v_{3/2}$ and x_2 were obtained, and then t_3 is evaluated.

You continue these sequential calculation cycles as long as is required. They will yield values of v and x over any desired range of time following the initial time. The method works for a time increment Δt of any size. If you choose Δt sufficiently small, you will, for all practical purposes, obtain v and x for *any* desired value of t. That is, you will determine the *functions* $v(t)$ and $x(t)$ for the given values of a, v_0, and x_0. (These functions can be displayed in either tabular or graphical form.)

For most cases of interest, the method involves a large number of repetitive numerical calculations. Consequently, it is not very practical if the calculations must be carried out by hand, even though each calculation is itself quite simple. However, programmable pocket calculators are perfectly suited to performing repetitive numerical calculations, as are electronic computers in general. With such devices the numerical method becomes quite feasible, as you will see in Example 5-8. Even so, the method is not competitive with the analytical method for treating the special case of constant acceleration, because the analytical method is so easy to use in this case.

Nevertheless, the numerical method has an *extremely* valuable attribute. Namely, it can easily be modified into a method for treating situations in which the acceleration a is *not* constant. Acceleration can vary because it depends explicitly on any of or all the quantities v, x, and t.

The method used to treat varying acceleration is almost the same as that used to treat constant acceleration. To be specific, this is what you do. Always employing the equation that expresses a in terms of the quantities on which it depends, you select a *small* time increment Δt and then carry out the following set of calculations:

Determine the value of a from the values at $t_0 = 0$ of the quantities on which it depends, and use it to calculate

$$v_{1/2} \simeq v_0 + a \frac{\Delta t}{2} \tag{5-35a}$$

from the given value of v_0. Use the result to calculate

$$x_1 \simeq x_0 + v_{1/2}\,\Delta t \tag{5-35b}$$

from the given value of x_0. Then set $t_1 = \Delta t$.

Next determine the value of a from the new values of the quantities on which it depends. Use it, and the value of $v_{1/2}$ just obtained, to calculate

$$v_{3/2} \simeq v_{1/2} + a\,\Delta t \tag{5-35c}$$

Use the result, and the value of x_1 just obtained, to calculate

$$x_2 \simeq x_1 + v_{3/2}\,\Delta t \tag{5-35d}$$

Then set $t_2 = 2\Delta t$.

Next determine the value of a from the new values of the quantities on which it depends. Use it, and the value of $v_{3/2}$ just obtained, to calculate

$$v_{5/2} \simeq v_{3/2} + a\,\Delta t \tag{5-35e}$$

Use the result, and the value of x_2 just obtained, to calculate

$$x_3 \simeq x_2 + v_{5/2}\,\Delta t \tag{5-35f}$$

Then set $t_3 = 3\Delta t$.

Continue these calculations until t reaches whatever value is required.

The idea is as follows: If the time increment Δt is small, the change in the acceleration a over each time interval will be small. Thus, although a varies, it is possible *to approximate* the effect of the acceleration on the motion by using in each time interval a fixed value of a. The value is changed from one time interval to the next. For a particular interval, the value of a to be used is calculated from the latest available values of v, x, and t. That value of a is then used to calculate the change in v over the time interval. The new value of v is then used to calculate the new value of x. Updating t completes this cycle of calculations. Then a new value of a, appropriate for the next cycle of calculations, is calculated from these updated values of v, x, and t, and that cycle is then carried out in the same manner as the preceding one. Thus, by following sequentially the instructions given by Eqs. (5.35), you obtain a sequence of values of v and x which describes the actual physical situation to good approximation. That is, they approximate the actual velocity and position corresponding to the actual varying acceleration, and to the given initial values v_0 and x_0. *The approximation becomes more accurate the smaller you choose the time increment* Δt. This is because the shorter the time increment, the smaller the range of actual values of a in a particular interval of time and the smaller the error introduced by using a single constant value of a to approximate these values.

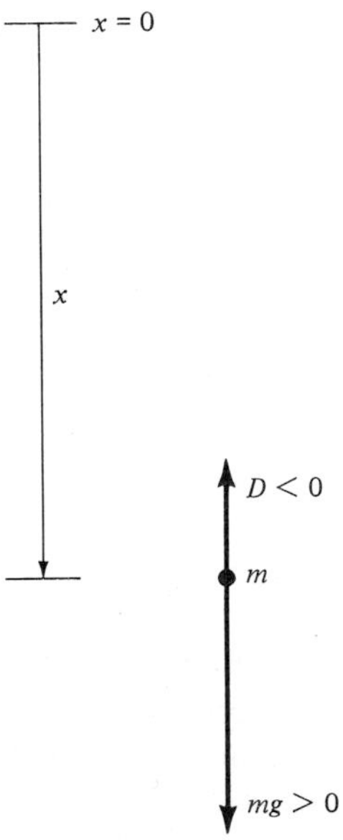

Fig. 5-18 Free-body diagram for a body of mass m when it has fallen through air a distance x, having started at rest. The forces acting on the body are the gravitational force mg and the drag force due to air resistance $D = -rv^2$.

Let us try this method on the fall of a skydiver of mass m. The free-body diagram of Fig. 5-18 shows the forces acting on her at any moment. Again we will use signed scalars and take the downward direction as positive. Then the constant gravitational force will have the positive value mg. There is also a varying drag force with a negative value. As noted at the beginning of this section, the size and speed of a skydiver are such that the flow of air is turbulent. According to Eq. (4-29), the value of the drag force D is therefore given by

$$D = -\frac{\rho A \delta}{2} v^2 \tag{5-36a}$$

In this equation, ρ is the density of the air, A is the frontal cross-sectional area presented to the air by the skydiver, δ is the coefficient of drag, which depends on her shape, and v is her velocity. It will be convenient to write Eq. (5-36*a*) in the compact form

$$D = -rv^2 \tag{5-36b}$$

Here the constant r is the product of all the constants on the right side of Eq. (5-36*a*). That is,

$$r \equiv \frac{\rho A \delta}{2} \tag{5-36c}$$

At a moment when the skydiver's velocity is v, the net force acting on her is

$$F = mg + D = mg - rv^2 \tag{5-37}$$

Her acceleration is given by Newton's second law,

$$a = \frac{F}{m} = \frac{mg - rv^2}{m}$$

This is more conveniently written in the form

$$a = g - \gamma v^2 \tag{5-38a}$$

where the constant γ (lowercase Greek gamma) is defined to be

$$\gamma \equiv \frac{r}{m} = \frac{\rho A \delta}{2m} \tag{5-38b}$$

In order to calculate the numerical values of a needed to use in Eqs. (5-35), we must have a numerical value of the constant γ on which all the values of a depend. A value of γ for a skydiver can be measured in a "wind tunnel," or estimated from Eq. (5-38*b*). We will do the latter. The equation shows that γ depends on both the frontal area A and the coefficient of drag δ, as well as on the mass m of a particular skydiver. But a skydiver does not usually maintain a constant orientation as she falls. Consequently, both A and δ will vary. Let us take rough average values and assume that the skydiver does not engage in acrobatics while falling. Suppose that $m = 60$ kg and $A = 0.4\ \text{m}^2$. Since a skydiver is less streamlined than a sphere but more streamlined than a circular disk facing broadside, we take for her coefficient of drag the average of the values given for those two shapes in Table

4-4. This gives $\delta = 0.8$. Although the density of air varies with altitude, we will use in this rough calculation the sea-level value $\rho = 1.2$ kg/m³. Inserting these numerical values into Eq. (5-38*b*), we obtain a one-significant-figure estimate for γ:

$$\gamma = \frac{1.2 \text{ kg/m}^3 \times 0.4 \text{ m}^2 \times 0.8}{2 \times 60 \text{ kg}} = 0.003 \text{ m}^{-1}$$

Thus Eq. (5-38*a*) becomes

$$a = 9.8 \text{ m/s}^2 - (0.003 \text{ m}^{-1})v^2 \tag{5-39a}$$

The equivalent expression, suitable for use in numerical calculations, is

$$a = 9.8 - 0.003v^2 \qquad (a \text{ in m/s}^2,\ v \text{ in m/s}) \tag{5-39b}$$

Having an equation which expresses the acceleration of the skydiver in terms of her velocity, you are now ready to use the numerical method of Eqs. (5-35) to find her position x and velocity v at subsequent values of the time t, for given values of her position x_0 and velocity v_0 at the initial time $t_0 = 0$.

Use of the numerical method is explored in Examples 5-8 and 5-9. The skydiver program, which will direct a programmable calculator or computer through the required sequence of computations, is found in the Numerical Calculation Supplement. In Example 5-8 the program is run with the value of γ set to zero. This means that there is no air resistance and the skydiver is falling freely. The purpose is to check the numerical method and its implementation on the calculating device. With $\gamma = 0$, you have $a = g =$ constant. Thus you are dealing with a case in which the approximate equations, Eqs. (5-35), reduce to the exact equations, Eqs. (5-34). Consequently, the results of the numerical method should be in complete agreement with those obtained by the analytical method for the familiar case of free fall from rest:

$$x = \frac{gt^2}{2} \tag{5-40}$$

EXAMPLE 5-8

Run the skydiver program with the following set of initial conditions on x_0, v_0, and t_0 and values for Δt, γ, and g:

$$x_0 = 0;\ v_0 = 0;\ t_0 = 0;\ \Delta t = 0.5 \text{ (in s)};\ \gamma = 0;\ g = 9.8 \text{ (in m/s}^2)$$

In this run the skydiver falls from rest at $t = 0$ and falls without air resistance.

■ You enter the listed values in the programmed calculating device and then start the run. The device interprets each value you have entered as being precise. For instance, it interprets the numerical value of g to be $9.80000 \cdots$, to as many decimal places as it carries.

The results produced for the skydiver's position x are plotted versus the elapsed time t as dots in Fig. 5-19. The crosses are the values of x obtained from the analytical method by evaluating Eq. (5-40) for a few values of t. The numerical and analytical results agree beautifully.

Fig. 5-19 Numerical and analytical results for the position x of a body falling freely from rest versus the time t. Here $\gamma = 0$ and $g = 9.8$ m/s^2.

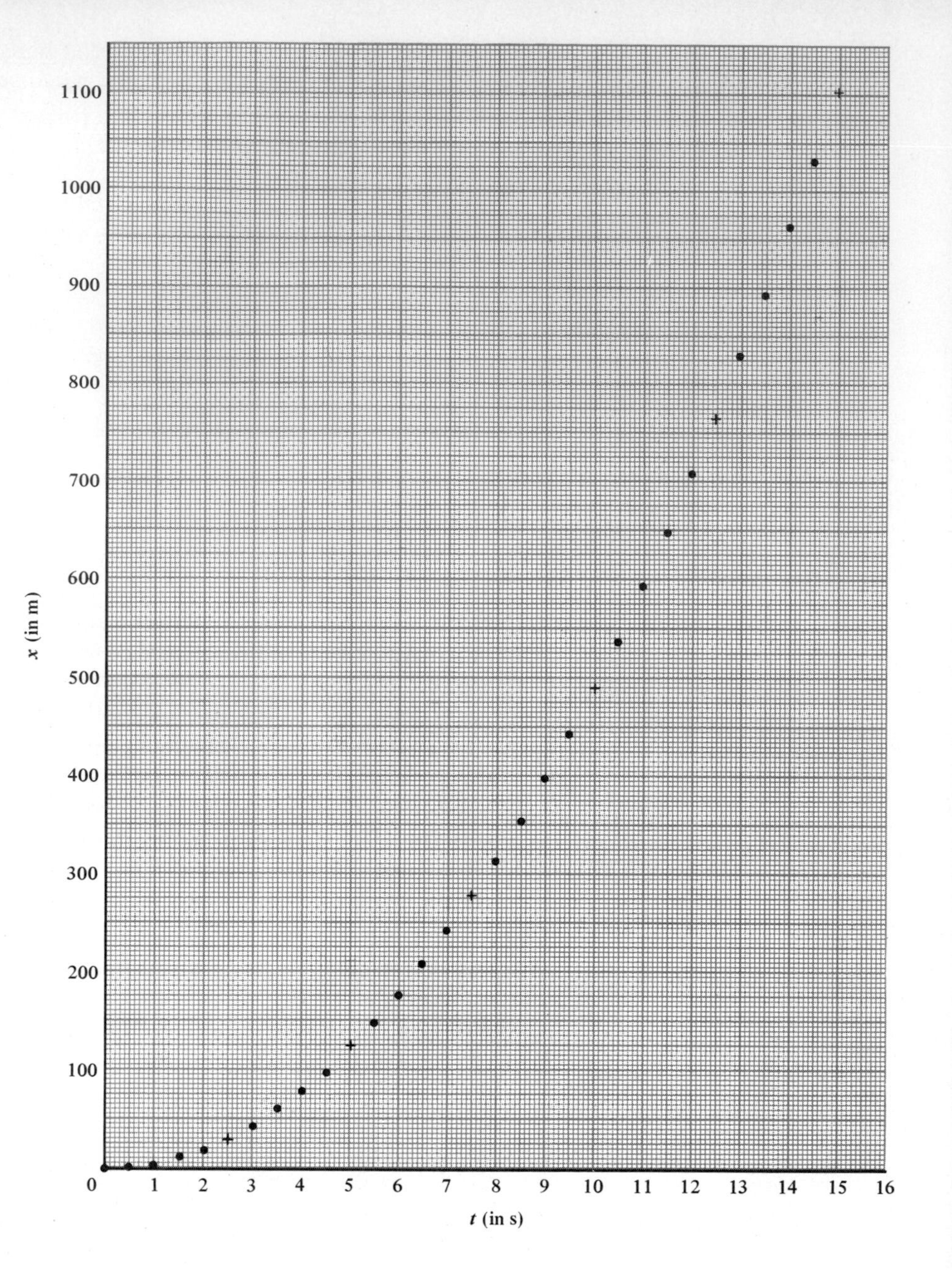

In Example 5-9, the constant γ specifying the amount of air resistance is set equal to 0.003 m^{-1}, the value estimated for a skydiver.

EXAMPLE 5-9

Run the skydiver program with the following set of initial conditions on x_0, v_0, and t_0 values for Δt, γ, and g:

$$x_0 = 0;\ v_0 = 0;\ t_0 = 0;\ \Delta t = 0.5 \text{ (in s)};\ \gamma = 0.003 \text{ (in m}^{-1}\text{)};\ g = 9.8 \text{ (in m/s}^2\text{)}$$

Here the skydiver again starts from rest. But now the effect of air resistance on her fall is taken into account.

Table 5-2

Numbers Appearing in Skydiver Program Calculations with $x_0 = 0, v_0 = 0, t_0 = 0, \Delta t = 0.5$ (in s), $\gamma = 0.003$ (in m^{-1}), $g = 9.8$ (in m/s^2)

$$v_{1/2} \simeq 2.45 = 0 + [9.8 - 0.003(0)^2]0.5/2$$
$$x_1 \simeq 1.23 = 0 + (2.45)0.5$$
$$v_{3/2} \simeq 7.34 = 2.45 + [9.8 - 0.003(2.45)^2]0.5$$
$$x_2 \simeq 4.90 = 1.23 + (7.34)0.5$$
$$v_{5/2} \simeq 12.16 = 7.34 + [9.8 - 0.003(7.34)^2]0.5$$
$$x_3 \simeq 10.98 = 4.90 + (12.16)0.5$$

Note: The calculating device treats the input numbers as exact, to as many decimal places as it carries, and also produces values of x (in m) and v (in m/s) to that many decimal places. The values of x and v listed here have been rounded off to two decimal places.

■ Entering the listed values, and running the program, you will obtain the results plotted versus time in Fig. 5-20. In this figure the values of the position are shown as dots.

Table 5-2 shows the numbers which the calculator or computer used in the first three cycles of the calculation. The values of all the terms in the first six of Eqs. (5-35) are given in the table. In ordinary operation, the device displays only the consecutive values of v and x. Table 5-2 should make clear that there is nothing magical in what the calculator or computer does. In principle, you could do the same thing yourself with pencil and paper—if you had enough patience! In practice, you would soon tire. And you would surely soon make a mistake. In such a sequential procedure, a mistake made *anywhere* invalidates everything that follows. Nevertheless, before the advent of programmable calculators and computers, professional scientists and engineers often devoted large amounts of time to doing such tedious calculations *by hand,* because for some important physical problems they provide the only way to obtain solutions. Now that programmable calculators and computers are commonplace, numerical methods are widely used in everyday professional work. They are used for solving many types of problems, including those involving variable acceleration.

The crosses in Fig. 5-20 were obtained by evaluating

$$x = \frac{1}{\gamma} \ln \left[\frac{e^{(\gamma g)^{1/2} t} + e^{-(\gamma g)^{1/2} t}}{2} \right] \qquad (5\text{-}41)$$

This exact formula is found by means of a sophisticated and quite complicated mathematical analysis of the problem of the skydiver. It is quoted (without proof since the proof involves mathematical techniques above the level assumed in this book) because it is useful in testing the accuracy of the numerical method. [In case you are unfamiliar with the symbols ln and e in Eq. (5-41), $e = 2.718 \cdots$ is a number whose properties are developed in Chap. 7, where it is put to actual use; the symbol ln means the logarithm to the base e.]

The numerical results are slightly larger than the corresponding analytical ones. For the conditions of this example, the error in the numerical result is 1.7 percent at $t = 5$ s and 1.3 percent at $t = 15$ s. But the error can be reduced by using a smaller value of Δt. For instance, with $\Delta t = 0.25$ s, the error drops to 0.87 percent at $t = 5$ s.

The squares in Fig. 5-20 represent the skydiver's velocity v.

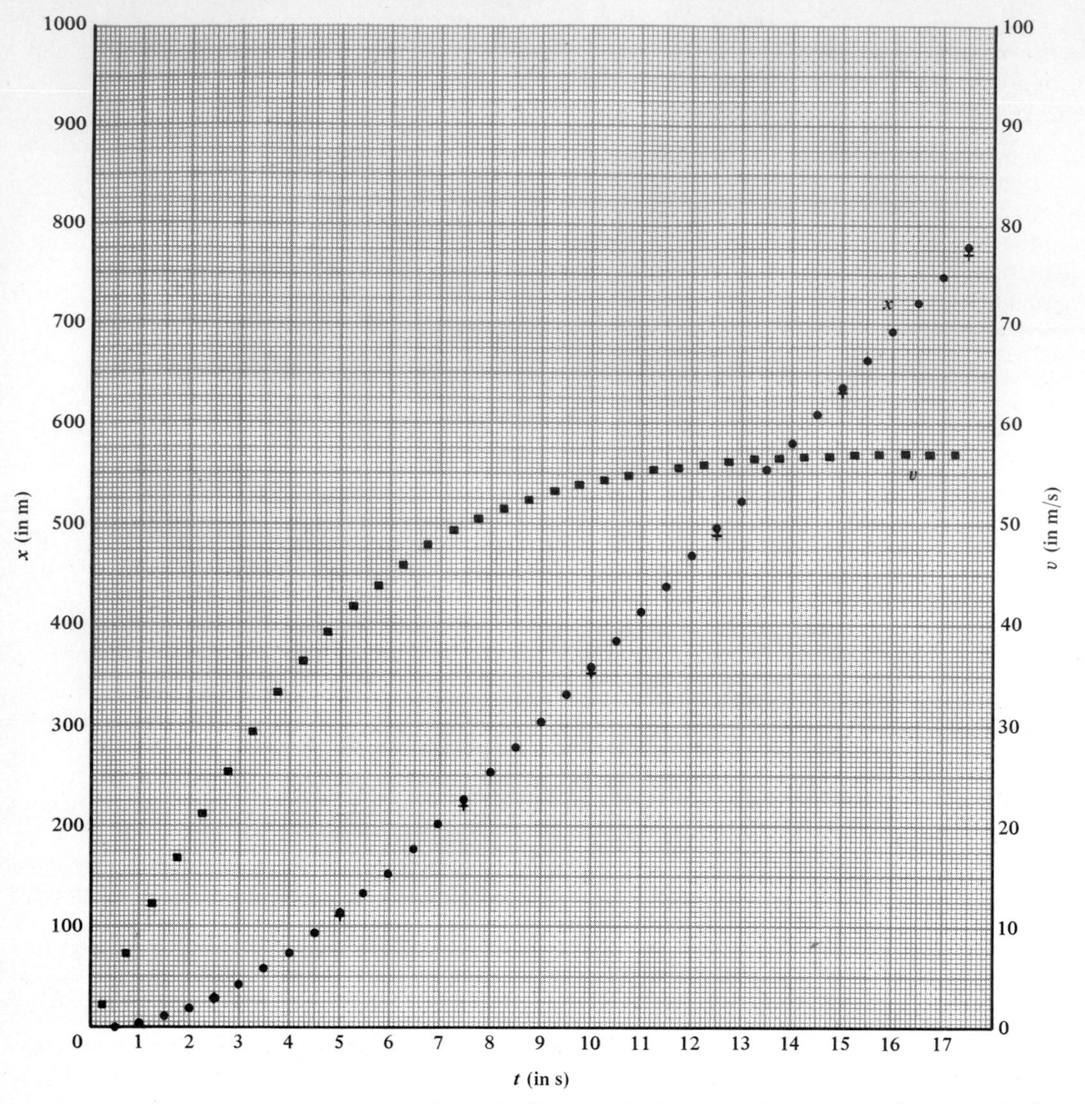

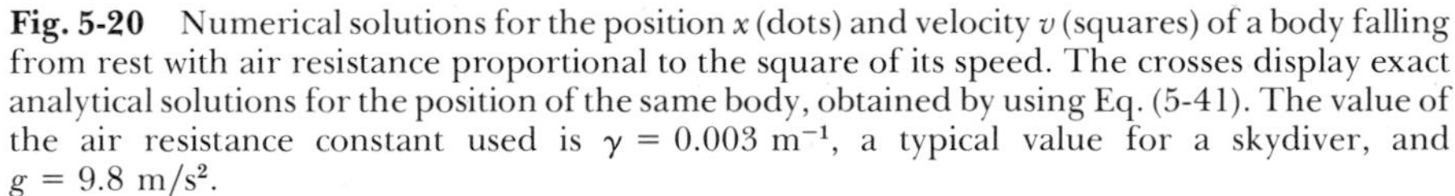

Fig. 5-20 Numerical solutions for the position x (dots) and velocity v (squares) of a body falling from rest with air resistance proportional to the square of its speed. The crosses display exact analytical solutions for the position of the same body, obtained by using Eq. (5-41). The value of the air resistance constant used is $\gamma = 0.003\ \text{m}^{-1}$, a typical value for a skydiver, and $g = 9.8\ \text{m/s}^2$.

Note how the skydiver's velocity builds up rapidly for the first few seconds of the jump, but soon approaches a constant value a little less than 60 m/s. This velocity is called the **terminal velocity.** (Its magnitude is the terminal speed, used in Sec. 4-6.) The skydiver can adjust the terminal velocity to some extent by extending or retracting her arms and legs, and by turning with respect to the direction of fall—and to a very considerable extent (fortunately!) by opening the parachute. All these actions change γ by changing both the frontal cross-sectional area A and the coefficient of drag δ. Skilled skydivers deliberately use such changes to carry out spectacular maneuvers.

Again referring to Fig. 5-20, note that the x versus t curve looks approximately parabolic (as in Fig. 5-19) while the velocity is relatively small, but becomes linear with the approach to terminal velocity. The physical reasons for this behavior are as follows. In the first second or two of falling from rest, the velocity is so small that the resulting air resistance is small compared to mg. Thus the only significant force in action is the gravitational force. The acceleration at the beginning of the jump is almost equal to g, and the distance traveled increases in approximate proportion to the square of the elapsed time, as is exactly the case in free fall. But with continued acceleration, the velocity increases rapidly. The drag force of air resistance builds up even more rapidly, because it is proportional to the square of the velocity. Since this force opposes the gravitational force, the net force acting on the skydiver becomes smaller and smaller. The effect is to reduce the acceleration, which is the rate of change of velocity. The skydiver's velocity stops changing when it is essentially equal to just that value for which the frictional force is equal in magnitude to the gravitational force. According to Eq. (5-37), this happens when

$$mg = rv^2$$

Solving for the terminal velocity v, we have

$$v = \sqrt{\frac{mg}{r}} = \sqrt{\frac{g}{\gamma}} \tag{5-42}$$

For the numerical values used in Example 5-9, this yields

$$v = \sqrt{\frac{9.8\ \text{m/s}^2}{0.003\ \text{m}^{-1}}} = 57\ \text{m/s}$$

While v is still increasing slowly at the last time recorded in Fig. 5-20, its value at that time, $v = 56.8$ m/s, is in good agreement with the value just calculated. This value of terminal velocity lies within the range of values actually observed for skydivers of various masses falling in various attitudes. The assumptions made in estimating γ for a skydiver are thus justified.

Terminal velocity is attained quite rapidly when a body such as a stone falls through a relatively resistive medium such as water. Its magnitude depends on the mass of the body, among other things. It is on the basis of this observation that physicists before Galileo (and many people to this day) came to believe that an object in free fall moves with a constant velocity proportional to its weight.

The numerical method which we have just developed has extremely broad applicability. It almost always allows you to determine the motion of any body whose acceleration varies. It does not matter whether its acceleration varies because the forces acting on the body depend on its velocity (as for a skydiver), or on its position (as for a pendulum bob), or on the time (as for a child on a swing being pushed by another child), or on all these things. And by means of a generalization which amounts to nothing more than a change of symbols, the same method can be applied to electrical systems, or systems of waves, or even quantum mechanical systems. We will use this method for these purposes throughout the remainder of this book.

EXERCISES

Group A

5-1. *Force, mass, and acceleration.* What is the magnitude of the acceleration of a body if the magnitude of the net force acting on the body is equal to its weight?

5-2. *Determining inertial mass.* A spring is compressed by squeezing it between a standard kilogram and a body of unknown mass. The system is placed on an air table and released. At a certain instant the acceleration of the standard mass is 2 m/s^2, whereas the acceleration of the body of unknown mass is -1 m/s^2. What is the magnitude of the unknown mass? Assume that the mass of the spring is negligible.

5-3. *Going up?* An elevator car whose mass is 1 metric ton $\equiv 10^3$ kg is pulled upward with a force of 11,000 N. What is its acceleration?

5-4. *Going down?* The elevator car described in Exercise 5-3 descends with an acceleration of 2.00 m/s^2. What is the tension in its supporting cable?

5-5 *Pulleys, bodies, and string.* Two systems are shown in Fig. 5E-5. Assume the systems to be frictionless and the pulleys to be massless.

a. What is the acceleration of the 2-kg body in Fig. 5E-5*a*?

b. What is the acceleration of the 2-kg body in Fig. 5E-5*b*?

c. What is the tension in the string in part *a*?

d. What is the tension in the string in part *b*?

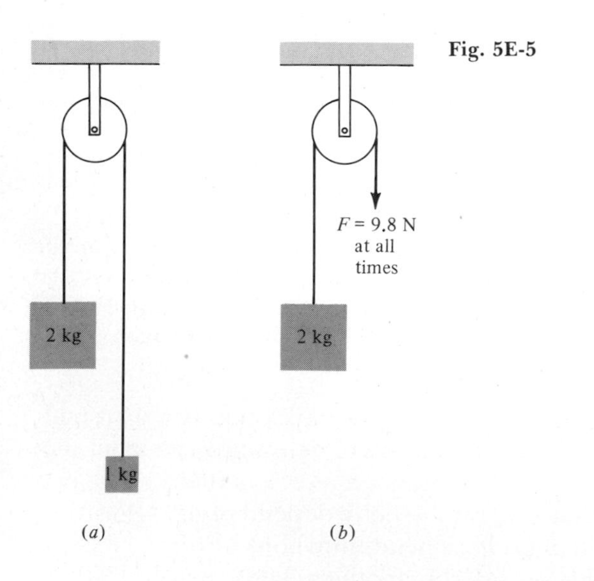

Fig. 5E-5

5-6. *Down the incline.* An object of mass m is placed on a smooth incline. Starting from rest, the object moves 2.00 m in 2.00 s. What is the angle of inclination of the plane?

5-7. *Sliding to a halt.* A block of mass 1.00 kg slides over the rough surface of a table. From the point where its speed is 1.00 m/s, it slides 0.20 m before coming to rest.

a. What is the frictional force (assumed constant)?

b. What is the coefficient of kinetic friction between the block and the table?

5-8. *Looping the loop.* A stunt flier in an open-cockpit biplane loops the loop in a vertical circle of radius 400 m. What must her minimum speed be at the top of the loop if a loose pencil in the cockpit does not fall out?

5-9. *Merry-go-round.* A child on a merry-go-round holds a ball on the floor. When the merry-go-round is moving counterclockwise at constant speed, he releases the ball. Neglect any friction between the ball and the floor.

a. Describe qualitatively the motion of the ball as seen by an observer standing on the ground near the merry-go-round.

b. Describe qualitatively the motion of the ball as seen by the child.

5-10. *Rounding a corner.* An automobile rounds a corner in a circle of radius R at constant speed v. The roadway is horizontal.

a. What must be the minimum value of the coefficient of friction, if the car is not to skid?

b. Which coefficient of friction is relevant here, the static coefficient or the kinetic coefficient? Explain your choice.

5-11. *On the railroad, I.* A train is moving along a straight and level section of track. The conductor stands in the aisle of the train, dangling his ticket punch from a chain in his hand. The chain is inclined backward, making an angle of 5° with the vertical. What are the magnitude and direction of the train's acceleration?

5-12. *On the railroad, II.*

a. Suppose that you are in a windowless train and are using a watch and watch chain as a makeshift plumb bob. You observe that the chain is inclined toward the rear of the train. This inclination could be due to an acceleration of the train, an upward slope of the railroad tracks, or a combination of the two. Can you describe a way to distinguish among those possibilities without using any evidence other than the inclination of the watch chain?

b. Suppose that you are in the situation described in part *a*, but that you also have access to an accurate pan balance and a set of standard masses. Can you distinguish among the various possible reasons for the bob's inclination?

c. Suppose that you are in the situation described in part *a*, but that you have an accurate spring scale and a set of standard masses in addition to the plumb bob. What could you determine about the cause of the bob's inclination?

5-13. *Oh, how I love to go up in a swing!* Estimate the magnitude of the fictitious force experienced by a child on a playground swing as she passes through the bottom of the swing's arc. This fictitious force is perceived by the child as an increase in her weight. Compare this increase to her actual weight.

5-14. *Ski jumping.*

a. As a ski jumper passes the low point on the approach ramp, his speed is v_0. The radius of curvature of the ramp is R. What is the upward acceleration of the skier?

b. What upward force must the ramp exert on the skier, whose mass is m? What multiple of his normal weight is this?

c. Evaluate your results for $v_0 = 22$ m/s (approximately 50 mi/h), $R = 15$ m, and $m = 70$ kg.

5-15. *A rocket in action.* A rocket has a takeoff mass of 13,000 kg. It burns fuel at the rate of 125 kg/s, with an exhaust velocity of 1800 m/s.

a. What is the thrust of the rocket?

b. What is its initial acceleration in a vertical takeoff?

Group B

5-16. *Tension, tension, tension.* A locomotive pulls three freight cars with an acceleration of 2.0 m/s^2. Each car has mass 1.0×10^4 kg.

a. What is the tension in the drawbar between the locomotive and the first car?

b. Between the first car and the second?

c. Between the second car and the third?

5-17. *Hoisting a chain.* A chain of mass 10 kg hangs vertically. It is pulled upward with an acceleration of 3.0 m/s^2.

a. What is the tension at the top of the chain?

b. What is the tension in the middle of the chain?

5-18. *Atwood's heirs, I.* An Atwood machine is suspended from a spring scale. It is motionless, with two 1-kg standard masses on each side.

a. Find the reading of the spring scale.

b. One of the 1-kg masses is transferred from the right side to the left side of the system. The Atwood machine proceeds to move. What is the reading of the spring scale now?

5-19. *Atwood's heirs, II.* Starting from rest, the system illustrated in Fig. 5E-5*a* is accelerated clockwise by a force F applied downward to the body on the right. The mass of the pulley is negligible. Find the acceleration of each of the two bodies if

a. $F = 29.4$ N

b. $F = 49.0$ N

5-20. *Downhill run.* A man stands on a board of mass m. The board rests on frictionless, massless rollers, which in turn lie on an inclined plane making an angle θ with the horizontal. If the man's mass is M, with what acceleration must he move if he wishes the board to remain stationary?

5-21. *Bead on a wire.* This problem was devised by Galileo Galilei. From a point A, which is the highest point on a vertical circle, three wires are stretched to B, B', and B''. AB is a diameter, while AB' and AB'' are arbitrary chords of the circle. On each wire is a bead which can slide with negligible friction. The beads are held at A and then released simultaneously. Prove that the three beads arrive at the circumference of the circle simultaneously in a time $t = \sqrt{2d/g}$, where d is the diameter of the circle.

5-22. *Block on a table top.* As shown in Fig. 5E-22, a 2.0-kg body is attached by a massless string to a 400-g body. The pulley is frictionless and massless. The system is initially motionless.

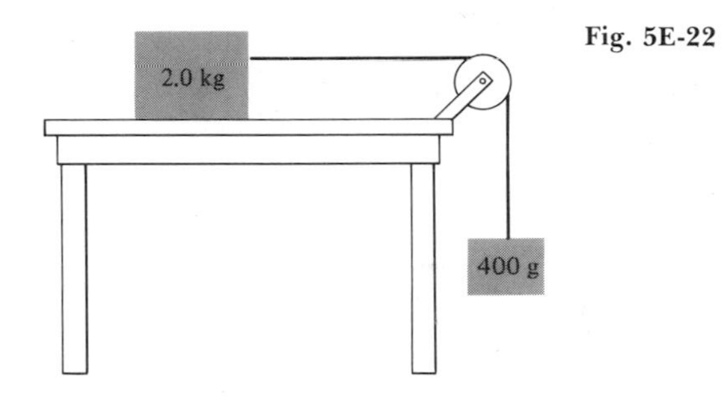

Fig. 5E-22

a. If the table surface is frictionless, find the acceleration of the 2.0-kg body.

b. Suppose the table is reasonably smooth but unlubricated, so that the coefficient of static friction is $\mu_s = 0.25$ and the coefficient of kinetic friction is $\mu_k = 0.17$. Will the system begin to move? If so, find the acceleration of the 2.0-kg body after it has started moving.

c. Suppose that the table top is lubricated, so that $\mu_s = 0.15$ and $\mu_k = 0.10$. Will the system begin to move? If so, find the acceleration of the system once it has started moving.

5-23. *Up and back.* A block is given an initial speed v_i upward along a very long incline. The incline makes an angle θ with the horizontal. The coefficient of static friction between the block and the surface of the incline is μ_s, and the coefficient of kinetic friction is μ_k.

a. What is the condition that determines whether the block will simply come to rest on the plane or begin to slide back down again?

b. Assuming that the block does slide back down again, what is its speed v_f when it returns to the point from which it started? Find an expression for the ratio v_f/v_i in terms of $\tan\theta$ and μ_k.

c. Find v_f if $\mu_k = 0.20$, $\theta = 35°$, and $v_i = 25$ m/s.

5-24. *Down and out.* Starting from rest at a height h above the base of a frictionless incline which curves gently toward the bottom until it is horizontal, a block slides down the incline and onto a level surface where the coefficient of kinetic friction is μ_k. Prove that $\mu_k = h/s$, where s is the distance traveled on the level surface by the block before it comes to rest.

5-25. *How will it go?* Figure 5-7a depicts two blocks connected by a string passing over a pulley. One block hangs from the string while the other lies on an inclined plane. It is pointed out in the discussion following Example 5-6 that the behavior of the system depends on the masses of the blocks, the angle of the inclined plane, and the coefficient of friction between the sliding block and the plane. Table 5-1 summarizes the conditions which lead to all possible cases of behavior of the system. Verify these conditions.

5-26. *Blocked shot.* A bullet with mass $m = 10$ g is fired with speed $v = 500$ m/s into a wooden block of mass $M = 1000$ g that is at rest on the top of a very long table. The bullet buries itself in the block, which slides a distance $s = 5.0$ m before coming to rest. What is the coefficient of kinetic friction between the block and the table?

5-27. *Spinning like a top.* What would be the length of the day if the earth were rotating fast enough for objects at the equator to appear weightless? Take the radius of the earth to be 6400 km and the acceleration due to gravity to be 9.8 m/s².

5-28. *Trick cycling.* A trick motorcyclist rides in a horizontal circle around the vertical walls of a cylindrical pit of radius R.

a. What is the minimum speed with which she must ride if the coefficient of static friction between the tires and the wall is μ_s?

b. Evaluate this speed if $R = 5.0$ m and $\mu_s = 0.90$.

5-29. *Round and round.* A string is threaded through a smooth glass tube. Bodies of mass M and m are tied to its ends, with $M > m$. As shown in Fig. 5E-29, the body m is made to revolve around the tube in a horizontal circle, so that body M neither rises nor descends. The period of the circular motion is T.

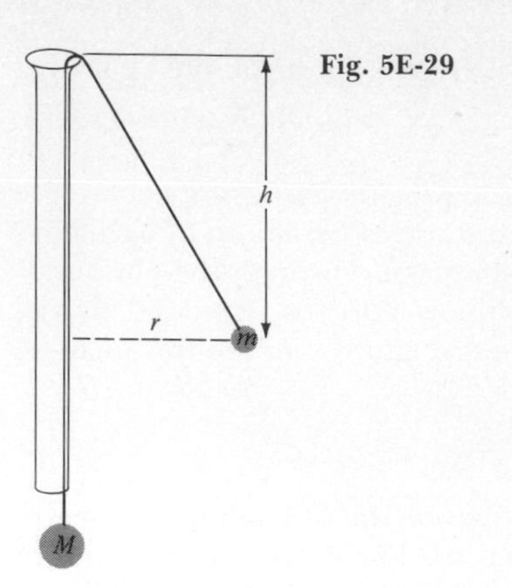

Fig. 5E-29

a. What is the angle between the string and the tube?

b. Express the free string length L in terms of T, m, M, and g.

c. Express T in terms of g and h, where h is the vertical distance from the top of the tube to the body m.

5-30. *Sudden stop.* A motorist's brakes fail and his automobile strikes a brick wall at speed v_0. He is wearing a seat belt which keeps his body fixed with respect to the car. The front end of the car is compressed by an amount d as the car comes to rest.

a. Assuming a constant deceleration during the impact, find an expression for the acceleration a of the driver. Express a in terms of v_0 and d.

b. In his (highly noninertial) personal frame of reference, the driver perceives that he is acted on by two balanced forces: the "real" restraining force supplied by the seat belt and the "fictitious" force, equal in magnitude and oppositely directed. Express the magnitude of these forces as a multiple of the driver's weight W. This ratio is the "number of g's" of horizontal acceleration to which the driver is subjected.

c. Evaluate the result of part b for $v_0 = 9.0$ m/s ($\simeq$20 mi/h) and $d = 0.30$ m.

d. During the deceleration, the driver will perceive that his head has an "added weight" that is directed forward. Estimate this apparent extra head weight.

5-31. *The bigger they come, the harder they fall.* Two spheres are made of the same material, but one has twice the diameter of the other. They are dropped from an airplane, and each sphere ultimately reaches its terminal velocity. What is the ratio of the two terminal velocities, assuming that the resistance of the air is proportional to the square of the velocity, as in Eq. (5-36a)?

5-32. *Togetherness.* Two blocks of mass m_1 and m_2 tied together with a string are sliding down the incline shown in Fig. 5E-32. The coefficient of kinetic friction between m_2 and the plane is $\mu_2 = 0.10$. The coefficient of kinetic friction between m_1 and the plane is $\mu_1 = 0.20$.

a. Calculate the acceleration of the blocks.

b. Calculate the tension in the string connecting the blocks.

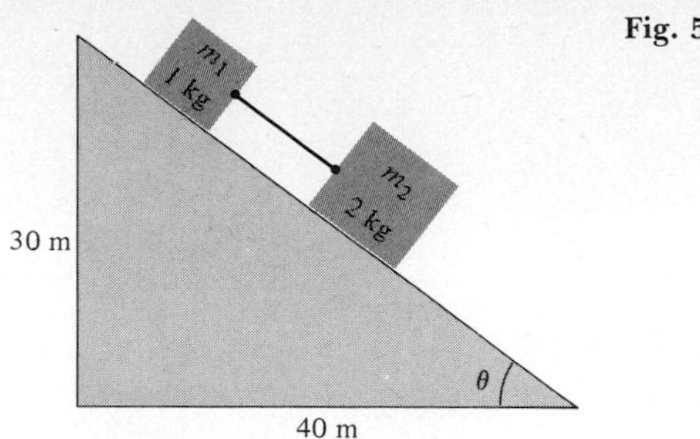

Fig. 5E-32

5-33. *Table for two.* As shown in Fig. 5E-33, a body of mass m_1 is supported by the level, frictionless surface of a table. The pulleys have negligible mass, and the system is initially motionless.

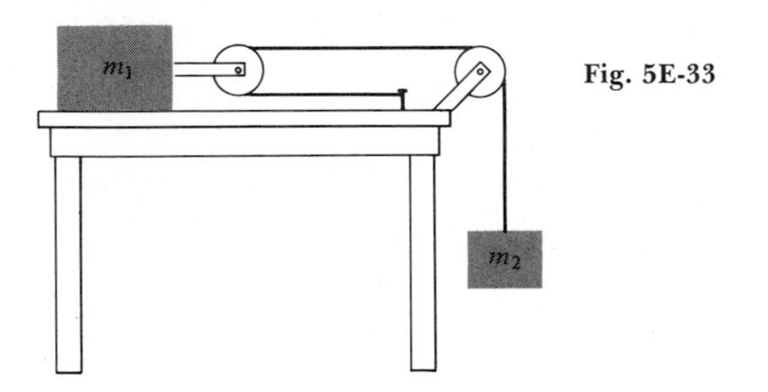

Fig. 5E-33

a. As the system begins to move, find the relationship that must exist between the distances d_1 and d_2 traveled by m_1 and m_2.

b. If $m_1 = 500$ g and $m_2 = 100$ g, find the magnitudes a_1 and a_2 of their accelerations.

c. For the masses given in part *b*, find the tension in the string.

Group C

5-34. *Slipping and sliding.* A block of mass m_1 is placed on an incline whose mass is m_2, as shown in Fig. 5E-34. The incline rests on a level surface, and the system is initially motionless. All surfaces are frictionless, so both block and incline are free to move.

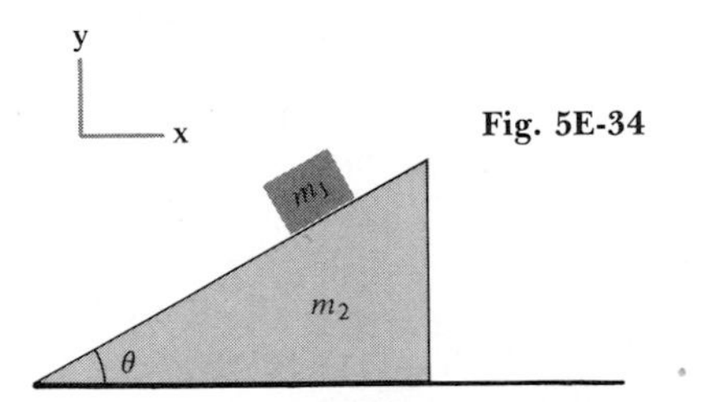

Fig. 5E-34

a. Show that the *x* component of the acceleration of the block is

$$a_{1x} = -\frac{m_2 g \tan\theta}{m_2 \sec^2\theta + m_1 \tan^2\theta}$$

b. Show that the *x* component of the acceleration of the incline (its only component) is

$$a_{2x} = \frac{m_1 g \tan\theta}{m_2 \sec^2\theta + m_1 \tan^2\theta}$$

c. Show that the *y* component of the acceleration of the block is

$$a_{1y} = -\frac{(m_1 + m_2) g \tan^2\theta}{m_2 \sec^2\theta + m_1 \tan^2\theta}$$

d. Find the magnitude N_B of the normal force between the block and the incline.

e. Find the magnitude N_T of the normal force between the incline and the supporting surface.

5-35. *Monkey business.* A rope is hanging over a frictionless pulley. One end of the rope is attached to a bunch of bananas, and the other end is held by a monkey at a lower level. The weight of the bananas is equal to the weight of the monkey. Describe what happens when the monkey climbs the rope to reach the bananas.

5-36. *Atwood's heirs, III.* As shown in Fig. 5E-36, body 1 of mass m_1 is suspended from a pulley, while body 2 of mass m_2 is attached to the end of the string. The pulleys have negligible mass, and the system is initially motionless.

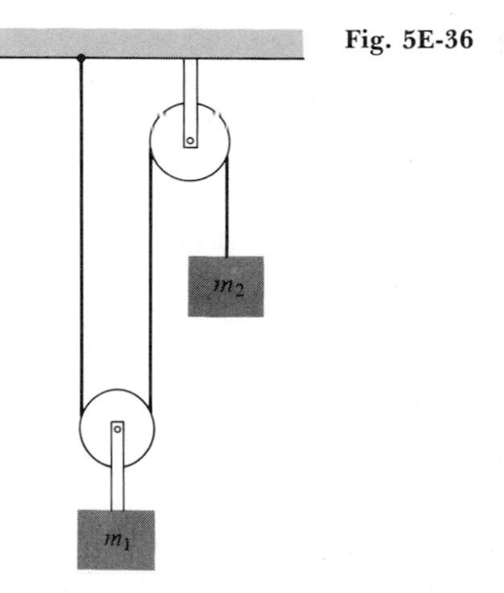

Fig. 5E-36

a. Find the relationship that must exist between the distances d_1 and d_2 traveled by body 1 and body 2 as the system begins to move.

b. If $m_1 = 0.30$ kg and $m_2 = 0.50$ kg, find the accelerations a_1 and a_2 of the bodies.

c. For the masses given in part *b*, find the tension in the string.

5-37. *Atwood's heirs, IV.* The system shown in Fig. 5E-37 is initially motionless, and the pulleys are of negligible mass. Bodies 1 and 2 are of mass 1.0 and 2.0 kg, respectively. The mass m_3 of body 3 is unspecified. When the system is released, it is found that body 1 remains stationary.

a. What is the tension in the string supporting body 1?

b. What are the magnitude and direction of the acceleration of body 2? Of pulley *A*? Of body 3?

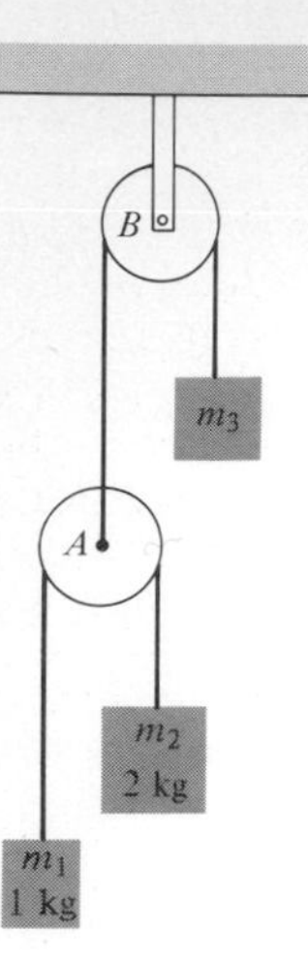

Fig. 5E-37

c. What is the tension in the string supporting pulley A?

d. What is the mass m_3 of body 3?

5-38. *Turntable.* A body of mass m rests on a turntable. The coefficient of static friction between the body and the table is μ_s. Another body of mass M is attached to a string that passes through a hole in the center of the turntable and over a massless, frictionless pulley and is then attached to m.

a. What are the shortest and longest periods of revolution of the turntable for which m can remain fixed on the turntable at a distance r from its center?

b. If $M = 2m$ and $\mu_s = 0.50$, find the ratio of the longest period to the shortest.

5-39. *Which way will they go?* The coefficient of static friction between the blocks and the planes shown in Fig. 5E-39 is 0.30. The coefficient of kinetic friction is 0.25. The pulley is frictionless and massless. The triangular base is fixed, and the system is initially motionless.

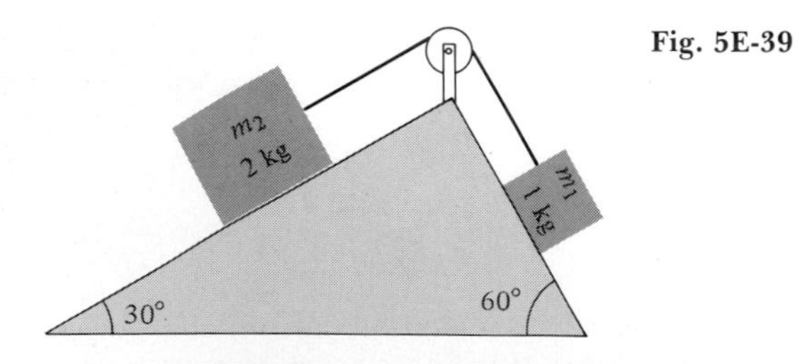

Fig. 5E-39

a. For the masses and angles shown, will the system begin to move?

b. If the system does move, in which direction will it move and with what acceleration?

5-40. *Bob was framed.* As shown in Fig. 5E-40, a frame is sliding down a frictionless inclined plane. A plumb bob suspended from the frame has settled to a steady direction.

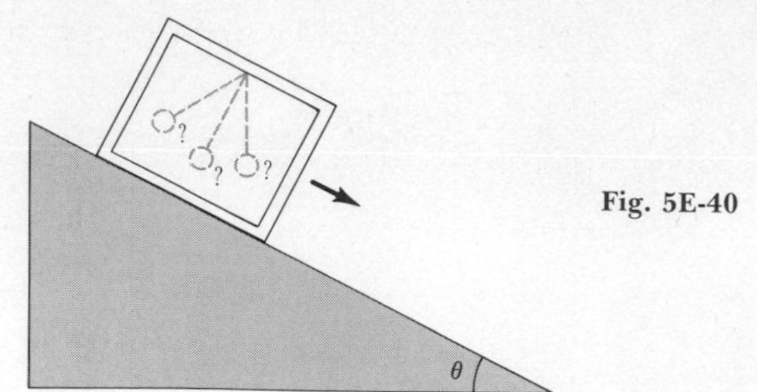

Fig. 5E-40

a. Which way does the plumb bob incline with respect to the vertical?

b. Let α be the angle between the steady direction of the plumb bob and the vertical. Prove that $\alpha = \theta$.

5-41. *Caught by the camera.* A long inclined plane is fixed on a level tabletop. A partially filled aquarium tank is sliding down the incline. The water in the tank has stopped sloshing and settled down to a plane surface. A snapshot is taken of the tabletop, incline, and the sliding tank. Describe how measurements on the single snapshot can be used to determine the coefficient of kinetic friction between the tank and the incline.

5-42. *Effective weight on a rotating earth.* As discussed in Example 5-7, a person standing at the earth's equator has a slightly smaller effective weight than he or she would have if the earth did not rotate. This is due to the fictitious centrifugal force; at the equator, this force acts in a direction opposite to that of the true gravitational force.

a. Show that the fictitious centrifugal force $\mathbf{F}_c$ observed by a person of mass M standing at latitude λ has magnitude $(4\pi^2RM/T^2)\cos\lambda M$, and that it points outward in a direction perpendicular to the earth's axis of rotation. Here R is the earth's radius, and T is the earth's period of rotation.

b. Construct and label a diagram that shows the true gravitational force $\mathbf{W} = M\mathbf{g}$ and the centrifugal force $\mathbf{F}_c$ that act on a person at latitude λ.

c. Find an expression for effective weight $\mathbf{W}'$ of a person standing at latitude λ. Remember that $\mathbf{W}' = \mathbf{W} + \mathbf{F}_c$.

Numerical

5-43. *Skydiver, I.* Run the skydiver program with the same values of x_0, v_0, t_0, γ, and g as in Example 5-9, but with $\Delta t = 0.125$ s, and obtain values for x at $t = 5$ s and $t = 15$ s. Determine the error in these values by using Eq. (5-41). Compare these errors with the ones in the values obtained in Example 5-9, and with the ones quoted below Eq. (5-41). What relation do you find between the error in the numerical method for treating skydiver motion and the size of Δt?

5-44. *Skydiver, II.* Run the skydiver program with the same values of x_0, v_0, t_0, Δt, and g as in Example 5-9, but with $\gamma = 0.001\ \text{m}^{-1}$. Plot x and v. Compare with the plots obtained in Example 5-9, and comment on the similarities and differences. Also use Eq. (5-42) to predict the terminal velocity, and use it to check the value you obtained from running the program.

5-45. *Skydiver, III.* A skydiver falls approximately 500 m with arms and legs bent so that $\gamma = 0.0025$. He then extends his arms and legs, increasing the frictional coefficient to $\gamma = 0.0035$, and continues falling. When the total distance fallen is 750 m, he opens his parachute. Determine the time dependence of his velocity v and distance fallen x through the period ending when the parachute opens. Also determine the elapsed time when the parachute is opened. Do this by running the program with the initial value of γ until x has a value near 500 m, stopping the calculating device, changing the stored value of γ, and then continuing. Briefly discuss your results.

5-46. *Skydiver, IV.* A skydiver falls for 25 s with her body adjusted so as to make $\gamma = 0.0015\ \text{m}^{-1}$. She then adjusts it to make $\gamma = 0.003\ \text{m}^{-1}$. Run the skydiver program, using $\Delta t = 0.5$ s, to obtain a plot of x and v for the first 40 s of her fall.

5-47. *Skydiver, V.* Change the skydiver program to make the magnitude of the frictional drag force proportional to the first power of the velocity by deleting the step in the program in which the velocity is squared. Then find a value of γ that will lead to the same terminal velocity found in Example 5-9, from an equation obtained by an analytical argument analogous to the one leading to Eq. (5-42). Run your modified program with this γ, plot x and v, and compare with the plot in Example 5-9. Discuss the similarities and differences.

5-48. *Skydiver, VI.* Change the skydiver program as in Exercise 5-47, but this time modify the program so that the frictional drag force is proportional to the cube of the velocity. This is of practical importance in certain cases, but there is no way to handle the calculation other than numerically since the equation determining the motion of the falling body has no analytical solution.

5-49. *Model rocket, I.* Modify the skydiver program so that it can predict the vertical upward flight of a model rocket launched from the ground. The engine of a model rocket produces an upward thrust force on the rocket which, to a good approximation, is constant during the burn period and zero afterward. Furthermore, the decrease in the mass of the model during engine burn is small compared to its total mass, so it is a good approximation to take the total mass of the model during engine burn to be a constant equal to the average value of the mass during this time. After engine burn, the model continues to coast vertically upward, with constant mass, until it comes to the top of its trajectory. At all times gravity exerts a downward force on the model, of magnitude equal to its mass multiplied by g (use average mass during engine burn), and air resistance exerts a downward force of magnitude equal to the drag constant r multiplied by the square of its velocity.

5-50. *Model rocket, II.* Run the model rocket program obtained in Exercise 5-49 with the following set of values: Engine thrust = 4.17 N. Engine burn time = 1.2 s. Average mass of model during engine burn = 0.0792 kg. Final mass of model = 0.0751 kg. Drag constant = 0.000646 kg/m. Time increment = 0.05 s. Watch the displayed values of the rocket's velocity, without plotting, until the first negative value obtained signals that the rocket has passed the top of its trajectory. Then obtain its altitude above ground level and the time required for it to reach that altitude.

5-51. *Optimizing peak altitude.* A certain model rocket has specified total mass, fuel mass, and exhaust speed, but the burn rate can be set to any desired value prior to launch.

a. If the rocket is to be fired in an airless environment (such as from the lunar surface), how does its peak altitude depend on the burn rate that is used? What burn rate provides maximum altitude?

b. If there *is* a drag force that acts on the rocket, how is your argument in part *a* affected?

c. Run the model rocket program of Exercise 5-49 with the following sets of values for the thrust, burn time, and burn rate.

	Engine thrust (in N)	Burn time (in s)	Burn rate (in 10^{-3} kg/s)
1.	4.17	1.2	6.83
2.	8.34	0.6	13.66
3.	2.085	2.4	3.415

Use the values given in Exercise 5-50 for the other quantities. (Note that the set 1 exactly reproduces the conditions of Exercise 5-50.)

d. Based on your results in part *c*, is the optimum burn rate faster or slower than 6.83×10^{-3} kg/s?

5-52. *Rockets away, I.* At $t = 0$ s, a rocket with initial mass $m_0 = 1 \times 10^4$ kg is fired vertically. The speed of the exhaust gases (relative to the rocket) is $v_g = 2500$ m/s. The fuel is burned at a steady rate such that $0.05m_0$ is ejected each second until the fuel is exhausted, when $m = m_{\text{final}} = 0.1m_0$. For the purposes of this exercise, ignore the pull of gravity.

a. For how long does the fuel burn?

b. What is the value of $m_{1/2}$, the mass of the rocket at $t = \frac{1}{2}\,\Delta t$?

c. Write an expression for the velocity increment Δv during a small time increment Δt, in terms of dm/dt, v_g, Δt, and the mass average m during the time increment.

d. In the present exercise, find Δv_1, the increase during the first time increment. Use the mass value $m_{1/2}$ in your calculation.

e. Find v_1, the velocity at the end of the first time increment. Find the ratio v_1/v_g.

f. What is the value of $m_{3/2}$, the rocket mass at $t = \frac{3}{2}\,\Delta t$?

g. Find Δv_2, the velocity change between $t = 1\,\Delta t$ and $t = 2\,\Delta t$.

h. Find v_2, the velocity at $t = 2\ \Delta t$. Find v_2/v_g.
i. Using the general expressions

$$\frac{\Delta v_{j+1}}{v_g} = \frac{|dm/dt|}{m_{j+1/2}} \Delta t$$

and

$$\frac{v_{j+1}}{v_g} = \frac{v_j}{v_g} + \frac{\Delta v_j}{v_g}$$

write a program which enables a calculating device to calculate the velocity at the end of each time increment from $t = 0$ s up to burnout. Use an increment $\Delta t = 1$ s in your calculations.

j. Construct a graph of v_j/v_g versus t_j.
k. Use the graph of velocity versus time to determine the distance traveled by the rocket before burnout.

5-53. *Rockets away, II.* Repeat the analysis of Exercise 5-52, taking gravity into account. Assume a constant value of 9.8 m/s^2 for g.

6

Oscillatory Motion

6-1 STABLE EQUILIBRIUM AND OSCILLATORY MOTION

You are surrounded by things whose motion is oscillatory. Some obvious examples are the pendulum in a grandfather clock and a guitar string when it is plucked. Less obvious, but at least as important, are such microscopic examples as the oscillating air molecules carrying the sound wave produced by the guitar. Furthermore, many very important nonmechanical phenomena involve oscillations. A set of related examples is found in the alternating-current electric power that runs a television receiver, the electromagnetic signal that it receives, the currents in the electric circuits which decode the signal to produce the pictures, and the light waves by which you see these pictures. In this chapter you will study the oscillatory motion of several simple mechanical systems. Almost all that you learn will be applied later in the book to more complex mechanical, and nonmechanical, forms of oscillatory motion.

Any object which has a position of stable equilibrium is capable of performing oscillations about that position. Figure 6-1 will be used first to explain the meaning of the expression "position of stable equilibrium" and then used again to discuss the oscillation of an object about a position of stable equilibrium. The spring in the figure is of negligible mass, and its top end is attached to a rigid support. From its bottom end hangs a body of mass m. We employ signed scalar symbols in discussing the directed quantities, such as forces, which enter into the analysis of this one-dimensional system. The downward direction is taken to be positive.

In Fig. 6-1a, the body is shown at the position where the spring is stretched from its completely relaxed length by such an amount that the negative (upward) force S exerted on the body by the spring just cancels the positive (downward) force mg exerted on the body by the earth's gravity. At this position the net force $F = mg + S$ acting on the body is zero. When a body is at such a position, where zero net force acts on it, it is said to be at a **position of equilibrium.** A body at rest at an equilibrium position will remain at rest because there is no net force acting to change its velocity from zero and thereby set it into motion.

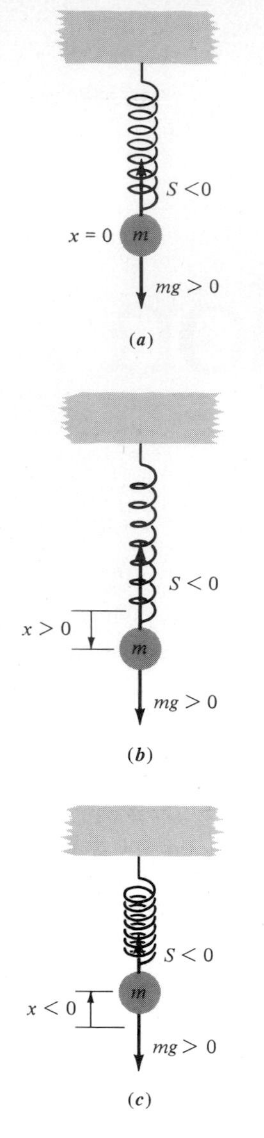

Fig. 6-1 A body of mass m is connected to the bottom end of a spring whose top end is attached to a rigid support. The two forces acting on the body are shown. One is the force, labeled mg, which is produced by the gravitational attraction of the earth. This force is always directed downward. The other is the force, labeled S, which is produced by the spring. In each part of the figure it is assumed that the length of the spring is greater than its length when completely relaxed. So in each the spring force is directed upward. The coordinate x of the body is measured from its position in part a, in which the body hangs motionless with the gravitational force just canceling the spring force. The downward direction is taken as positive.

The body shown in Fig. 6-1a is at a position of equilibrium of a special kind called a **position of stable equilibrium.** That is, if the body is displaced a small distance from the position in Fig. 6-1a in either direction, it will experience a net force in a direction which tends to return the body to that position. We now show that this is so. Let us introduce the coordinate x to specify the position of the body, with the origin, $x = 0$, at the body's equilibrium position. In Fig. 6-1b the body is at a position where x is positive, so that the spring is stretched more than in Fig. 6-1a. The additional stretch increases the magnitude of the negative force S that the spring exerts on the body. But there is no change in the positive force mg which the earth exerts on the body. Thus the net force $F = mg + S$ acting on the body has a negative value, and so acts upward. In Fig. 6-1c the body is at a position where x is negative, and the spring is stretched less than in Fig. 6-1a. Consequently the negative force S exerted on the body by the spring is of reduced magnitude and the net force $F = mg + S$ acting on the body has a positive value, and so acts downward. Thus if the body is displaced in either direction from the equilibrium position, it experiences a net force which acts in the direction that tends to move the body back to the equilibrium position. This makes the position of equilibrium a position of *stable* equilibrium.

Figure 6-2 is a qualitative plot of the net force F, which acts on the body at the end of the spring, versus the coordinate x specifying the position of the body. Since $x = 0$ has been chosen to be at a position of equilibrium, the value of F at that point must be zero, as shown. In other words, *the $F(x)$ curve passes through the x axis at a position of equilibrium.* In the immediate vicinity of that position, F is negative where x is positive and F is positive where x is negative, as must be the case in the vicinity of a position of stable equilibrium. In other words, *the slope of the $F(x)$ curve is negative in the vicinity of a position of stable equilibrium.*

In this particular case, the $F(x)$ curve is a straight line in the region near $x = 0$. This is because the spring obeys Hooke's law, Eq. (4-21), providing its length is not too different from its relaxed length. That is, the magnitude of the spring force S depends linearly on x since x specifies the distortion of the spring, and so the magnitude of the net force $F = mg + S$ is proportional to x.

Now let us discuss oscillations about a position of stable equilibrium. If you connect a body which you are holding to the lower end of an unstretched spring whose upper end is fixed, and then gradually remove the supporting force you apply to the body, the body will settle into the position of stable equilibrium shown in Fig. 6-1a. If you again take hold of the body, pull it to the position shown in Fig. 6-1b, and then release it, it will begin oscillating about its stable equilibrium position. Immediately after you release the body in the position illustrated in Fig. 6-1b, it feels a net force tending to make it move back to the stable equilibrium position shown in Fig. 6-1a. But the body does not simply move back to that position and then stop. All the while it is moving toward the stable equilibrium position it is acted on by a net force directed toward that position. Thus the

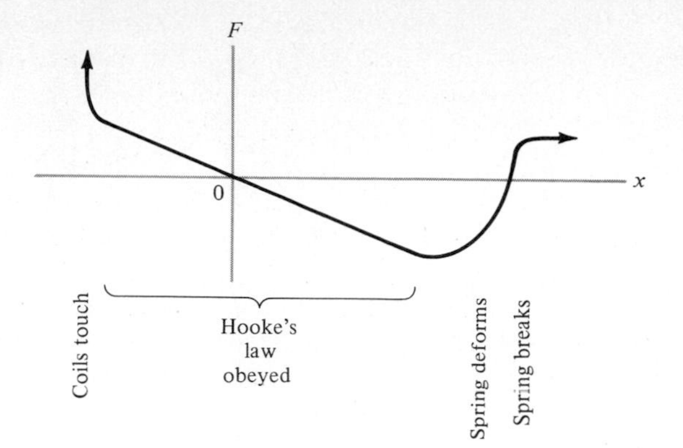

Fig. 6-2 The net force $F = mg + S$ acting on the body in Fig. 6-1 is plotted versus the coordinate x specifying its position. Since $x = 0$ is defined to be the position at which $mg + S = 0$, the net force has the value $F = 0$ for $x = 0$. A positive value of F means that the force acts in the direction of positive x; that is, in the downward direction in which it tends to move the body so as to extend the spring. A negative value of F means that the force acts in the opposite direction. Over the range of x where the spring obeys Hooke's law, the plot is linear. This range includes values of x for which the length of the spring is less than its length when completely relaxed, although such a situation is not illustrated in Fig. 6-1. Since the net force acting on the body is always in the direction in which it tends to move the body to the position $x = 0$, the sign of F is opposite to the sign of x. This requires that the slope be negative when x is in the linear range. But if x becomes too negative, the coils of the spring touch and F very rapidly becomes more positive to prevent them from penetrating each other. If x becomes too positive, Hooke's law is no longer satisfied. At even more positive values of x, the coils are stretched so much that they begin to yield to permanent deformation. Ultimately, the spring breaks, and the net force acting on the body becomes the constant gravitational force, which is positive because it is directed downward. For values of x in the region beyond the point where permanent deformation begins, the curve is not really a plot of a function, $F(x)$. The reason is that it describes the dependence of F on x in this region only as x becomes more positive and only for the first time x enters the region. So in this region the value of F depends not only on the value of x but also on what happened before x reached the value.

body picks up speed while moving toward the stable equilibrium position. As it approaches, the net force gradually diminishes, until at the stable equilibrium position the net force is zero. The body keeps moving, however, because it has mass and cannot change its speed when no net force acts on it. So it moves past the position of stable equilibrium. As the body continues moving, a net force directed back toward the equilibrium position gradually develops. This slows the body until it comes momentarily to rest at the position shown in Fig. 6-1*c*. The first half-cycle of an oscillation has now been completed. The body immediately starts the next half-cycle of oscillation. This half-cycle is just the reverse of the first half-cycle. In the absence of friction the oscillation would continue indefinitely.

All other examples of mechanical oscillation involve the same interplay between the same two factors: (1) a force, acting on a body, that always is directed toward a certain position—the position of stable equilibrium—and (2) the inertia of the body on which the force acts. The force always pushes the body toward the stable equilibrium position, and its inertia always makes it "overshoot."

A system which is particularly interesting in regard to stable equilibrium and oscillatory motion is shown in Fig. 6-3. There are two equilibrium positions for the body of the system which has appreciable mass. One is a position of stable equilibrium, and the other is not. The body will execute oscillations about the position of stable equilibrium, but not about the other equilibrium position.

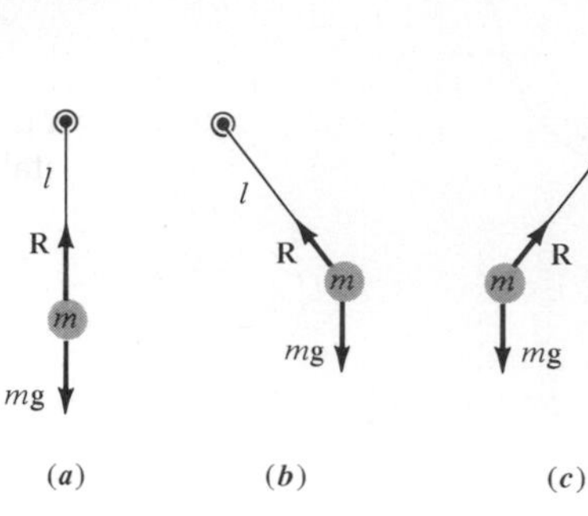

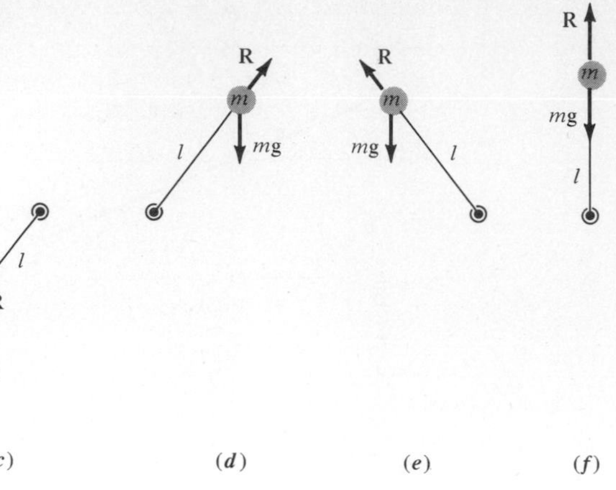

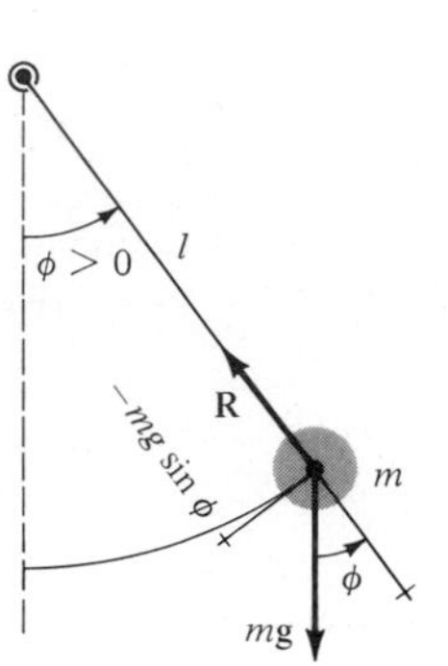

Fig. 6-3 A body of mass m is connected to one end of a light rod whose other end is attached to an axle in horizontal bearings. The two forces acting on the body are shown. One is the gravitational force $m\mathbf{g}$ exerted in the downward direction by the earth. The other is the force $\mathbf{R}$ exerted by the rod in a direction along its length. Parts g and h show the angular coordinate ϕ of the body. It is measured from the downward vertical, with the counterclockwise direction positive. Also shown in g and h is the tangential component of the net force acting on the body. This quantity has the value $F_t = -mg \sin \phi$. The sign correctly expresses the fact that, in part g, $\sin \phi$ is positive and F_t is negative, while, in part h, $\sin \phi$ is negative and F_t is positive.

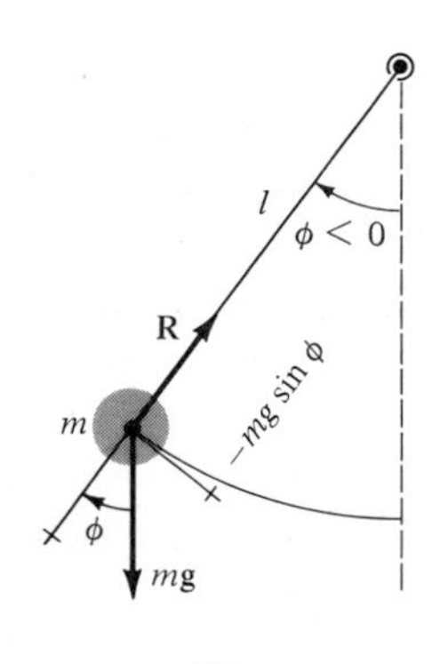

The system consists of a compact body of mass m at one end of a rod having length l and negligible mass. The other end of the rod is connected to a horizontal axle which is supported by bearings of negligible friction. Thus the body can move through a circle of radius l in a vertical plane.

Your intuition will probably tell you that the body has two equilibrium positions, one when the rod is oriented vertically downward and the other with the rod vertically upward. In both cases, the rod can exert a force on the body which exactly cancels the downward gravitational force acting on it. But only the lower position is one of *stable* equilibrium. If the body is displaced slightly from the upper position and released, it will not experience a net force which makes it tend to return to that position, as would be the case for the lower position.

We now discuss the forces acting on the body in detail. Whatever the position of the body, there are two forces acting on it. One is the downward-directed gravitational force $m\mathbf{g}$ exerted on it by the earth. The other is the force $\mathbf{R}$ exerted on the body by the rod. The direction of $\mathbf{R}$ is either outward or inward along the rod, depending on the body's position and speed. The magnitude of $\mathbf{R}$ also depends on the position and speed of the body.

The speed dependence of **R** results from the fact that when the body is moving, its acceleration has a component along the direction of the rod. This is the inward-directed centripetal acceleration whose magnitude depends on the speed of the body. The centripetal acceleration is the result of a centripetal force exerted on the body. This force is supplied in part by the rod and in part by the gravitational force acting on the body, which has a component along the direction of the rod. In each part of the figure the force **R** is drawn for the simple case in which there is no centripetal acceleration because the body is motionless at the position shown. It is the force that would be exerted on the body by the rod immediately after you moved the body to that position and then released it.

In considering the stability of the equilibrium positions of the body (and also its oscillations about a position of stable equilibrium) the speed dependence of **R** does not matter. If the body is placed at rest near a position of stable equilibrium and then released, it will begin to move toward that position. In other words, the magnitude of its velocity will be changed. But we saw in Sec. 3-4 that the only component of the net force acting on a body which is effective in changing the magnitude of its velocity is the component along the direction of its path. Since the force **R** never has a component along this direction, we are not concerned with the value of **R** in our present considerations.

The most convenient way to specify the position of the body at the end of the rotatable rod is to use the value of the angular coordinate ϕ shown in Fig. 6-3*g* and *h*. It is the angle, measured in radians, from the downward vertical to the rod. The value of ϕ is chosen to be positive when the rod is rotated counterclockwise from the downward vertical. You can see from the figure that the component F_t of the net force acting on the body along a direction tangent to its path is given by the equation $F_t = -mg \sin \phi$. The minus sign takes into account the fact that when the value of $\sin \phi$ is positive (the rod is rotated not more than half a revolution in the counterclockwise direction), the value of F_t is negative (the tangential force acts on the body in the clockwise direction). And when the value of $\sin \phi$ is negative, the value of F_t is positive.

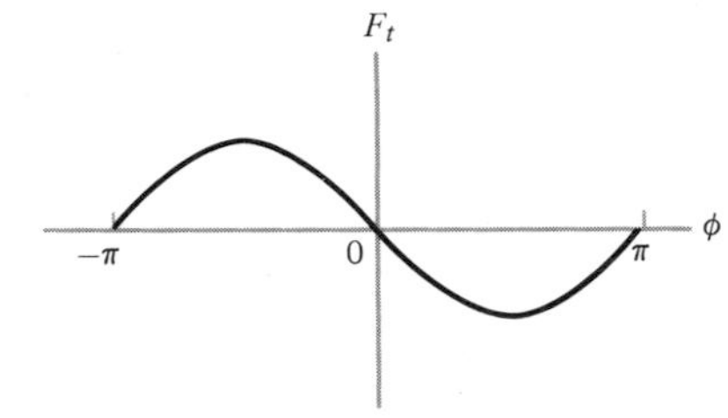

Fig. 6-4 The tangential component F_t of the net force acting on the body in Fig. 6-3 is plotted versus the coordinate ϕ specifying its position. This component of the net force is the one which tends to make the body move along its circular path, and so it is the one that determines the body's positions of equilibrium. There are two such positions, at which $F_t = 0$. The first is where $\phi = 0$, and the second is where $\phi = \pm\pi$. In the first position the body is stable, and in the second it is unstable. Does this statement agree with the physical intuition you have for the system shown in Fig. 6-3?

The relation $F_t = -mg \sin \phi$ is plotted in Fig. 6-4. The $F_t(\phi)$ curve passes through the ϕ axis at $\phi = 0$, consistent with the fact that $\phi = 0$ specifies an equilibrium position. Furthermore, the slope of the curve is negative as it passes through the axis since $\phi = 0$ is a position of stable equilibrium. This conclusion certainly makes sense. If the body is directly below the axle, as in Fig. 6-3*a*, there is no net force acting to make it move along its path, and it is at an equilibrium position. If the body is not far from this position in either direction, as in Fig. 6-3*b* or *c*, then there is a net force exerted on the body with a component in a direction tangent to its path. In either case this force is directed so as to tend to move the body toward the equilibrium position. Thus the position $\phi = 0$ is indeed a position of stable equilibrium.

The body also has another equilibrium position directly above the axle. This position corresponds to the coordinate value $\phi = \pi$ or, equally well, to the value $\phi = -\pi$ (since the two values describe the same position). It is an equilibrium position because the $F_t(\phi)$ curve passes through the ϕ axis at these values. So there is no tangential force component tending to move the body away from its position when that position is directly above the axle. You can see this is true by inspecting Fig. 6-3*f*. But this equilibrium position is not one of stable equilibrium. If the body is at the positions shown in Fig. 6-3*d* or *e*, there is a net force component tangent to the path of the body acting on the body in the direction which tends to move it away from—not toward—the equilibrium position directly above the axle. This is true not only of the positions shown, but of *any* position close to $\phi = \pm\pi$. That equilibrium position is thus a **position of unstable equilibrium.** With great care (and if there is a little friction in the axle bearings), the body can be "balanced" directly above the axle. But the least disturbance that moves it slightly away will put it at a position where a force acts to move it farther away. There it will feel a stronger tangential force that acts to move it even farther, and so forth. The behavior of the $F_t(\phi)$ curve at the position of unstable equilibrium is shown in Fig. 6-4 at $\phi = \pi$ or $\phi = -\pi$. The curve passes through the ϕ axis with a *positive* slope—in contrast to the negative slope characteristic of stable equilibrium.

The body at the end of the spring shown in Fig. 6-1 also has a position of unstable equilibrium. But this is not as evident as it is for the body at the end of a rotatable rod. Use Fig. 6-2 to identify the unstable equilibrium position. Can you explain on physical grounds why the equilibrium is unstable?

If you put the body at the end of the rotatable rod near its position of stable equilibrium and then release it, it will oscillate about that position. The explanation involves the same interplay between force and inertia as in the body and spring case. Here also there is a force acting on the body always in the direction toward the position of stable equilibrium and a body, having inertia, on which the force acts. You should explain to yourself in detail what happens through one cycle of an oscillation, patterning the explanation on the one given earlier for the body at the end of the spring.

In contrast, if you put the body at the end of the rotatable rod near its position of unstable equilibrium and then release it, it will *not* oscillate about that position. Why not? What will it do?

In an oscillating system the force involved is neither constant in direction nor constant in magnitude. So the acceleration of the body on which it acts is not constant. However, at the end of Chap. 5 we developed a quite general numerical method for treating systems with nonconstant accelerations. In the next two sections we will apply this method to analyze in detail oscillatory motion in the two systems we have discussed qualitatively here: a body at the end of a spring and a body at the end of a rotatable rod. There is also an analytical method for treating oscillatory motion which we will use later in this chapter to reanalyze the oscillations of a body at the end of a spring. In some cases the analytical method does not work. (It cannot be used for the body at the end of the rod if the rod rotates through an appreciable angle when the body oscillates.) But when the analytical method works, it has important advantages over the numerical method. The two methods complement each other in that the strong points of each generally correspond to the weak points of the other.

6-2 THE BODY AT THE END OF A SPRING

If a body of mass m hangs from a spring whose top is fixed to a rigid support, as in Fig. 6-1, then two forces act on the body. One is the force exerted by the spring, and the other is the force exerted by gravity. But the constant gravitational force does not play a role in the oscillatory motion of the body. It does serve to orient the system along the vertical direction. It also determines the location of the equilibrium position of the body because the weight of the body stretches the spring. But when the body is pulled from the position of stable equilibrium and then released, it is the *variable* net force acting on it that produces its oscillatory motion. This variable net force arises from the variation in the spring force from its value at the stable equilibrium position. Thus the oscillation of the body is due to the variable force exerted on it by the spring.

To focus attention on what is important, we will analyze the oscillatory motion of the body in the system shown in Fig. 6-5. It is connected to one end of a horizontally oriented spring of negligible mass whose other end is fixed. We can support the body against gravity by placing it on a frictionless track. (This ideal condition can be well approximated by using a one-dimensional air table called an **air track.** It is made from a rail pierced with many tiny holes through which air is blown to form a film of air that supports the body.) Here we are concerned only with the horizontally directed force, shown in the figure, which is exerted on the body by the spring when the spring is longer or shorter than its relaxed length. This is the force which makes the body oscillate.

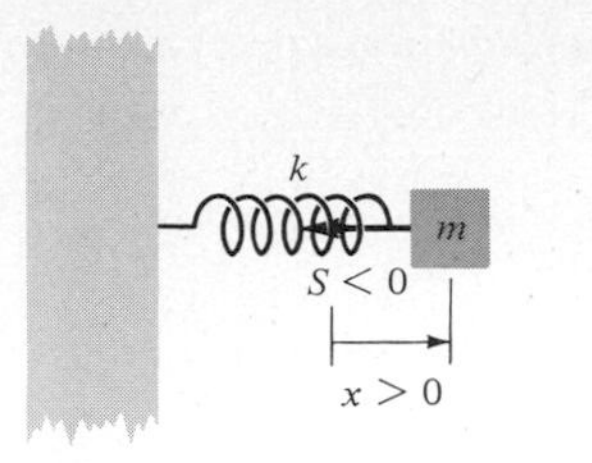

Fig. 6-5 A body of mass m is connected to the movable end of a spring of force constant k. The body is supported by an upward-acting force which cancels the downward-acting force of gravity. The supporting force is supplied by an air track, not shown in the figure, on which the body slides. When the body is moved from its position of stable equilibrium in which the spring has its relaxed length and then released, it oscillates along a horizontal line. The figure shows the body at an instant in its oscillation cycle when it happens to be to the right of its stable equilibrium position. In these circumstances, its position coordinate x, measured from the stable equilibrium position, is defined to have a positive value. The spring is extended and so exerts a horizontal force S on the body. This force acts to the left, and therefore its value is negative. The relation between S and x is given by Hooke's law. $S = -kx$, where k is a positive constant.

We specify the position of the body by the coordinate x, shown in the figure. This coordinate is measured along a horizontal axis, with the positive direction to the right. The origin of the axis is taken to be the body's position of stable equilibrium. In other words, when $x = 0$, the spring is relaxed. The force exerted on the body by the spring is indicated in the figure, at an instant when the spring is stretched. We represent this spring force by the signed scalar S. Provided that the maximum stretch, or compression, of the spring occurring during the oscillation is not *too* large, both the magnitude and the direction of S at any position x of the body can be expressed by the equation

$$S = -kx \qquad \text{for } |x| \text{ not too large} \tag{6-1}$$

This is just a form of **Hooke's law** pertaining to the present situation. Since the magnitude of x gives the magnitude of the change in the length of the spring from its relaxed length (the distortion of the spring), Eq. (6-1) says that the magnitude of the spring force is proportional to the distortion of the spring. Hooke's law in the form of Eq. (4-21) says the same thing. The positive proportionality constant in Eq. (6-1) is the **force constant** k. It is identical to the k in Eq. (4-21), and it specifies the stiffness of the spring. In contrast to Eq. (4-21), the direction of the spring force is also given by Eq. (6-1). For a positive value of x the spring is extended because the body is to the right of its equilibrium position. Thanks to the minus sign, Eq. (6-1) predicts correctly that in such a case the value of S is negative, so that the force exerted on the body by the spring acts to the left. And when x is negative because the body is to the left of its equilibrium position, the equation correctly predicts that the force which the compressed spring exerts on the body acts to the right since its sign is positive.

When Hooke's law was presented in Sec. 4-6, it was emphasized that the law applies not only to springs but to a wide variety of mechanical objects. When distorted moderately, almost anything composed of crystalline solids produces a force that is proportional to the distortion, in agreement with Eq. (6-1). Thus the body-and-spring system is really a prototype of a very large class of mechanical systems. They all act in essentially the same way as far as their oscillatory motion is concerned, because in all of them the force producing the motion is essentially that of Eq. (6-1).

We study this oscillatory motion by combining Newton's second law with Hooke's law and then finding solutions to the resulting equation. According to Newton's second law, the net force F acting on the body at the end of the spring is related to the mass m of the body and its acceleration a by the equation

$$F = ma \tag{6-2}$$

But the net force F is just the spring force S. So $F = S = -kx$, and we have

$$-kx = ma$$

Solving for the acceleration, we obtain

$$a = -\frac{k}{m}x$$

We write this as

$$a = -\alpha x \tag{6-3}$$

where we define

$$\alpha \equiv \frac{k}{m} \tag{6-4}$$

The quantity α is a **parameter** specifying the mechanical properties of the system. In other words, α is a quantity whose value is constant for a given system, and whose value determines the behavior of the system. In particular, α is the constant k specifying the stiffness of the spring divided by the constant m specifying the mass of the body connected to its movable end. Its SI units are newtons per meter-kilogram, that is N/(m·kg).

We will employ the numerical method of Sec. 5-6 to find solutions to Eq. (6-3) for several different sets of initial conditions, and for various values of α. That is, we will adapt the method to handle this case where the nonconstant acceleration a of a body has the mathematical form $a = -\alpha x$. Then we will make numerical calculations which determine how the body moves when it is started in several different ways, and how the motion depends on the stiffness of the spring and the mass of the body. The relations of the numerical method, Eqs. (5-35), apply immediately to this case, providing that Eq. (6-3) is used at each step to evaluate a. So all that is required is to make the appropriate changes in the part of the calculator or computer program of Sec. 5-6 where a is evaluated. The resulting body-and-spring program is listed in the Numerical Calculation Supplement. Some solutions to Eq. (6-3) obtained by using it are given in the following examples.

EXAMPLES 6-1 AND 6-2

Run the body-and-spring program with the following two sets of initial conditions and parameters:

$x_0 = 0.25$ (in m); $v_0 = 0$; $t_0 = 0$; $\Delta t = 0.2$ (in s); $\alpha = 1$ [in N/(m·kg)]

$x_0 = 1.5$ (in m); $v_0 = 0$; $t_0 = 0$; $\Delta t = 0.2$ (in s); $\alpha = 1$ [in N/(m·kg)]

The results obtained for both sets are plotted in Fig. 6-6. In Example 6-1 the body is initially displaced to precisely $x = 0.25$ m and then let go from rest. The plot shows that it starts to move slowly toward $x = 0$, picks up speed, and before long passes through $x = 0$. Then it slows down as it approaches $x = -0.25$ m. There it turns around and repeats this motion in reverse, until it returns to $x = 0.25$ m. It has completed the first **cycle** of its oscillation. But it immediately starts the next identical cycle.

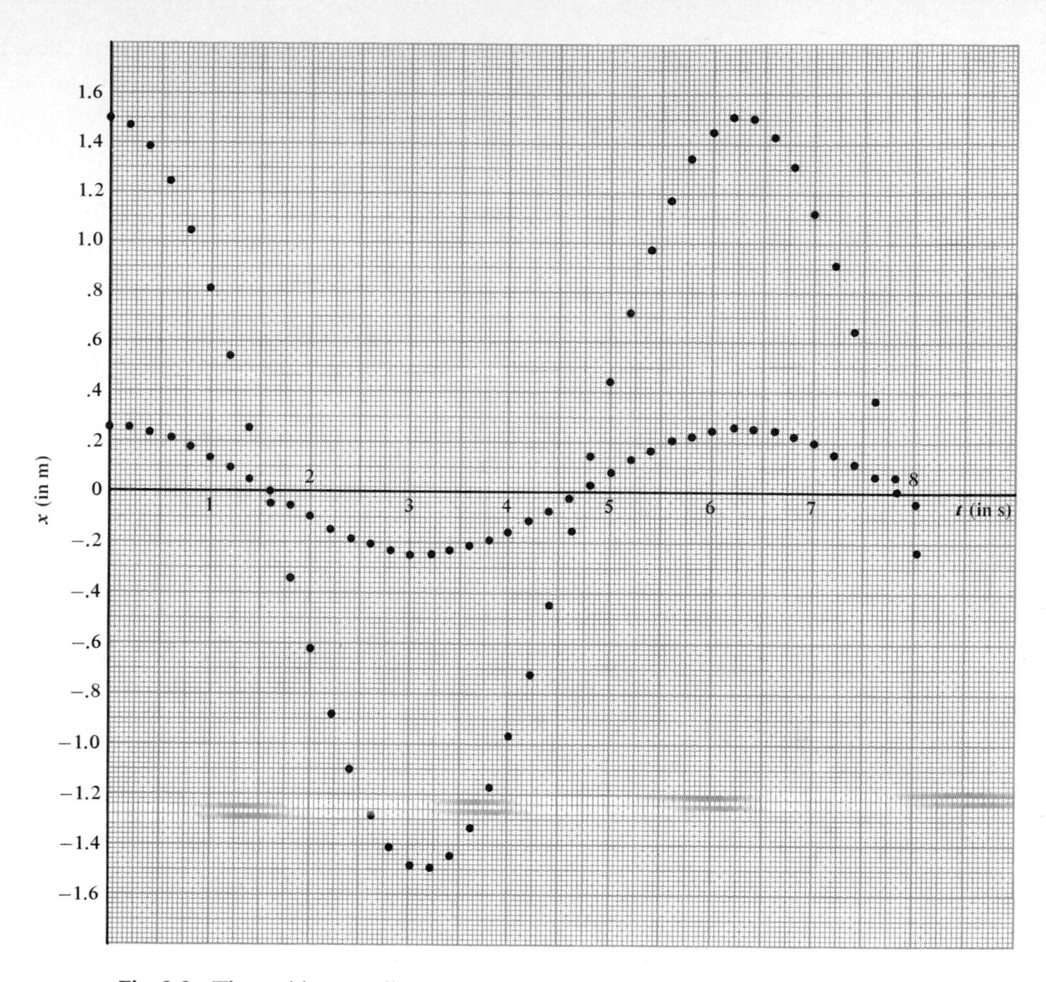

Fig. 6-6 The position coordinate x versus the time t for small- and large-amplitude oscillations of a body of mass m at the end of a spring of force constant k. The parameter $\alpha = k/m$ has the value 1 N/(m·kg) for both oscillations.

In Example 6-2 the body was initially displaced to precisely $x = 1.5$ m and released with zero initial velocity. Its subsequent oscillations look very much like those in Example 6-1, except that all the values of x are *scaled up* in magnitude by a factor of 1.5/0.25.

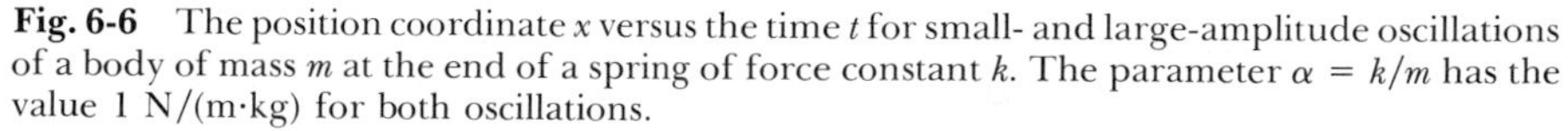

The comparison of the results obtained in Examples 6-1 and 6-2, displayed in Fig. 6-6, shows that the oscillations of a body acted on by a Hooke's law force have a **scaling property.** That is, the general "shape" of a plot of the coordinate x versus the time t does not depend on the maximum magnitude assumed by x during a cycle of oscillation. This maximum magnitude of x is called the **amplitude** of the oscillation. It is 0.25 m for the oscillation treated in Example 6-1 and 1.5 m for the one treated in Example 6-2.

As a result of the scaling property, two important characteristics of the oscillation are independent of its amplitude. These are its period and its frequency. The **period** T of an oscillation is the time required for the oscillation to go through one full cycle. For instance, it is the time required for the body in Example 6-2 to go from $x = +1.5$ m in one cycle to $x =$

+ 1.5 m in the next. In more general terms, T is the time elapsed from the passage of the body in a certain direction through any position to its next passage in the same direction through the same position. The **frequency** ν of an oscillation is the reciprocal of its period, that is, the number of passages per unit time through a given position in a given direction. Thus T is the number of seconds per cycle, ν is the number of cycles per second, and

$$\nu = \frac{1}{T} \tag{6-5}$$

The unit for T is the second. The unit for ν is called the **hertz** (Hz). A frequency of, say, $\nu = 5$ Hz means that a body completes 5 full cycles of oscillation each second.

You can measure the period T for the two oscillations plotted in Fig. 6-6. Careful inspection will show that for both $T \simeq 6.28$ s, despite their considerable difference in amplitude. The corresponding value of the frequency is $\nu \simeq (1/6.28)$ Hz $= 0.159$ Hz. Note that the parameter specifying the mechanical properties of the system has the same value $\alpha = 1$ N/(m·kg) in both examples.

You may be surprised to find that a body of a particular mass at the end of a particular spring oscillates with the same period and frequency whether the amplitude of the oscillation is large or small. The body certainly travels a longer path in one cycle of a larger amplitude oscillation, and this tends to increase the time required to complete a cycle. On the other hand, when the body has completed a certain fraction of its cycle of oscillation, the force exerted on it will be larger, the larger the amplitude, since it will be at a greater distance from its equilibrium position and hence the deformation of the spring will be greater. So with increasing amplitude the accelerations experienced by the body at various points in an oscillation cycle increase. Thus the speed it has at a certain fraction through the cycle increases also, and this tends to decrease the time required to complete a cycle. It appears from Examples 6-1 and 6-2 that these opposing effects cancel, so that the period and frequency of the oscillation are independent of its amplitude, if the force exerted on the body obeys Hooke's law. When we study the analytical treatment of the oscillatory motion in Sec. 6-5, we will see exactly how this comes about.

Figure 6-7 shows values of the period T for oscillations obtained by running the program with several different values of the parameter α. The data were obtained by using a quick procedure explained in the Numerical Calculation Supplement. The period T decreases as the ratio α of the force constant to the mass increases. This is what you might expect. A stiffer spring exerts a larger force. A body with smaller mass has less inertia. Both effects tend to speed up the oscillation.

When you are trying to understand unfamiliar data showing the dependence of one quantity on another, such as the dependence of T on α, it may be profitable to try to describe the dependence with some simple mathematical function. One good way of doing this is to *guess* at a functional dependence, and then use it to replot the data in such a way that, if the guess is correct, the points lie on a straight line.

Often it is possible to guess the functional dependence by a dimensional analysis argument. It certainly is possible here. The parameter α is defined by the equation $\alpha = k/m$, where k is a force divided by a length (the force constant of the spring) and m is a mass (the mass of the body at the free end of the spring). Ac-

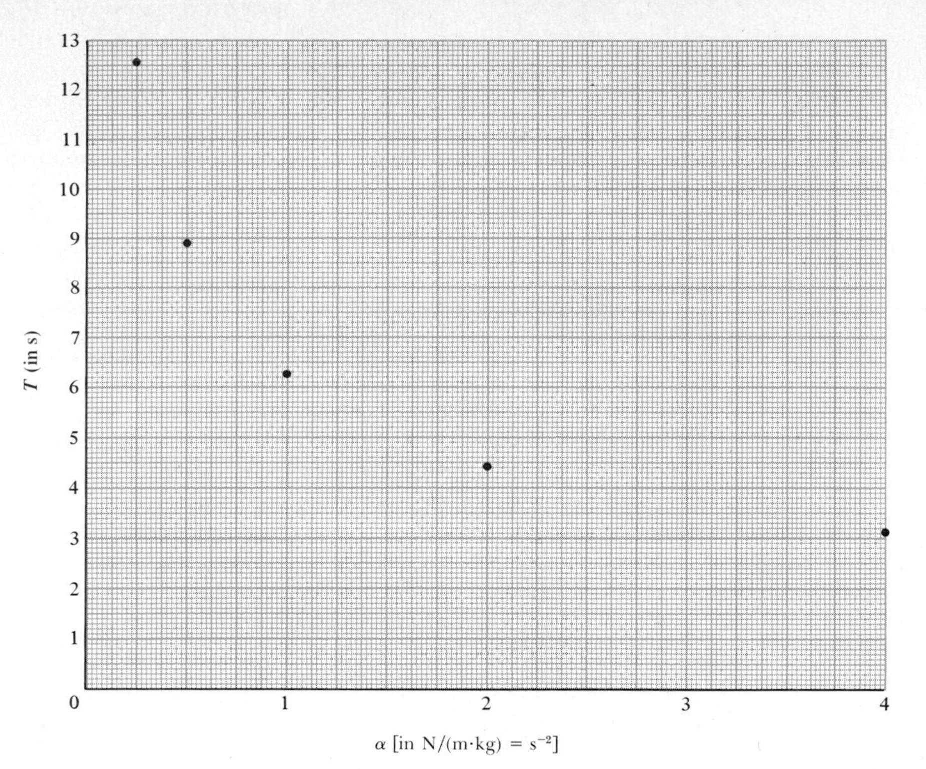

Fig. 6-7 The period T of the oscillations of a body at the end of a spring, for several values of its stiffness-to-mass ratio α.

cording to Newton's second law, the dimensions of force are the same as those of mass multiplied by acceleration; in other words, the dimensions of force are mass multiplied by length divided by time squared. The dimensions of k are therefore (mass)(length)(time)$^{-2}$/(length), that is, (mass)(time)$^{-2}$. The dimensions of α are consequently (mass)(time)$^{-2}$/(mass). Thus the dimensions of the parameter α are (time)$^{-2}$. Of course the dimensions of the period T are (time). Now, any functional relation correctly describing the dependence of T on α must be dimensionally

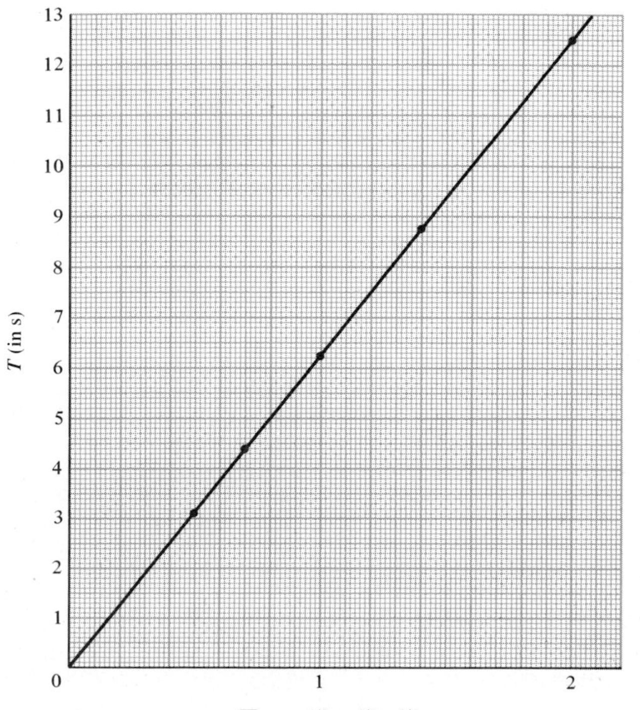

Fig. 6-8 A plot of T versus $1/\sqrt{\alpha}$ for the oscillations of a body at the end of a spring, using the data presented in Fig. 6-7. The way the points fall on a straight line passing through the origin shows that T is proportional to $1/\sqrt{\alpha}$. The slope is 6.28, which suggests the relation $T = 2\pi/\sqrt{\alpha}$

consistent. A simple relation that has this property is one in which T is proportional to $1/\sqrt{\alpha}$, with the proportionality constant being a pure number.

Figure 6-8 shows the same data points as those shown in Fig. 6-7, but with T plotted versus $1/\sqrt{\alpha}$. Since the points do appear to lie on a straight line in Fig. 6-8, the data confirm the guess, based on dimensional analysis, that T is proportional to $1/\sqrt{\alpha}$.

If you go one step further and measure the slope of the straight line to determine the proportionality constant, you will obtain the value 6.28. Within the accuracy that can be expected from the numerical work, this value is 2π. Thus it seems there is an unexplained but intriguing relation $T = 2\pi/\sqrt{\alpha}$. A complete explanation is given in Sec. 6-5.

Example 6-3 investigates the motion of the body at the end of the spring when it is started in a way quite different from the way it was started before.

EXAMPLE 6-3

Run the body-and-spring program with the following set of initial conditions and parameters:

$$x_0 = 0;\ v_0 = 1.5 \text{ (in m/s)};\ t_0 = 0;\ \Delta t = 0.2 \text{ (in s)};\ \alpha = 1 \text{ [in N/(m·kg)]}$$

■ Here the body at the end of the spring is given an initial velocity of precisely 1.5 m/s in the positive x direction when it is at its equilibrium position. (These initial

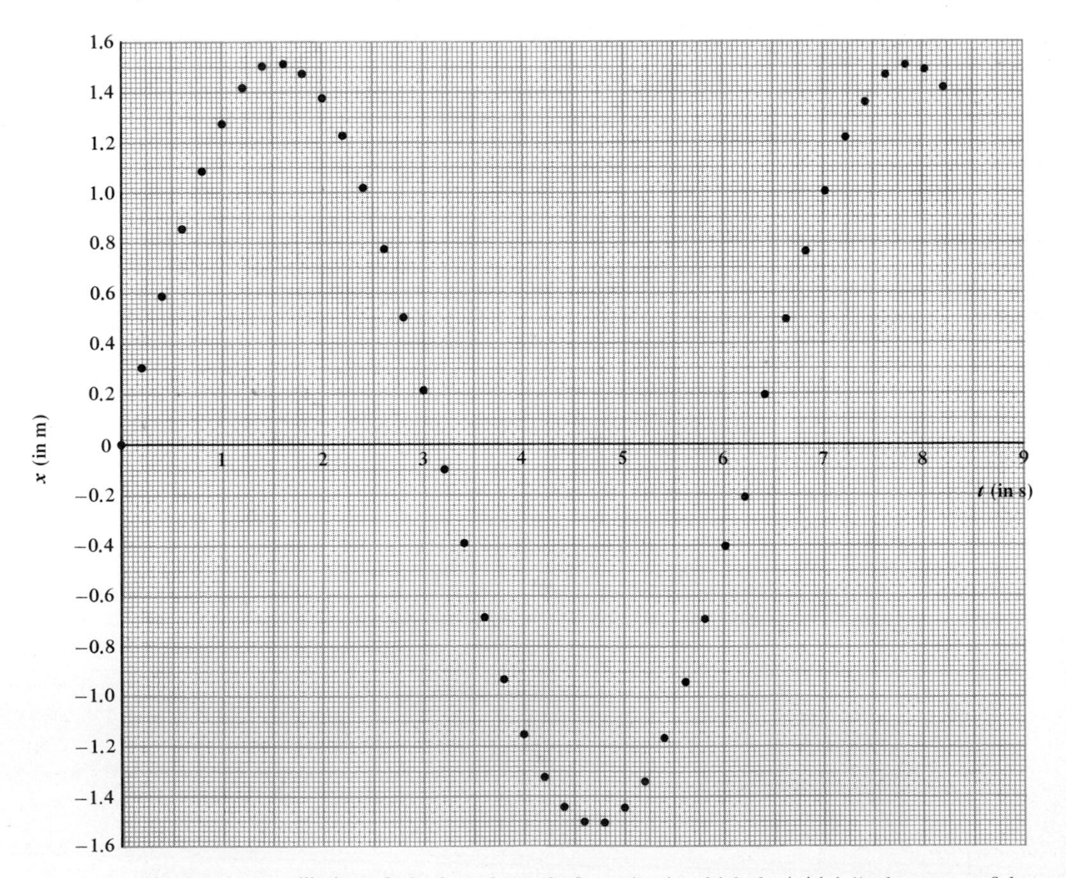

Fig. 6-9 An oscillation of a body at the end of a spring in which the initial displacement of the body is zero and its initial velocity is nonzero. The parameter $\alpha = k/m$ has the same value 1 N/(m·kg) as in the oscillations plotted in Fig. 6-6.

conditions could be achieved approximately by starting the body with a sharp blow.) Its subsequent motion is plotted in Fig. 6-9.

Note that with the parameter α having the same value as in Examples 6-1 and 6-2, the same value $T \simeq 6.28$ s is found for the period of the oscillation. So again you find that the period is governed by the value of α, and not by other considerations.

What do you think would happen if you ran the program, using initial conditions in which neither x_0 nor v_0 were zero? Try it.

6-3 THE SIMPLE PENDULUM

Now we will analyze quantitatively the oscillatory motion of the system shown in Fig. 6-10, which we discussed qualitatively in Sec. 6-1. One end of a rod of length l is attached to an axle supported by bearings in such a way that the rod can rotate with no appreciable friction in a vertical plane. The other end of the rod is attached to a body of mass m. The mass of the rod is negligible compared to the mass of the body, and the size of the body is negligible compared to the length of the rod. The body's position is specified by its angular coordinate ϕ, measured in radians from the downward vertical, with counterclockwise rotations of the body corresponding to positive values of ϕ. Two forces are acting on the body. They are the force $m\mathbf{g}$ exerted by gravity and the force $\mathbf{R}$ exerted by the rod. The direction of $m\mathbf{g}$ is downward. As was discussed in Sec. 6-1, the direction of $\mathbf{R}$ is either inward or outward along the rod. Which it is depends in general on both the speed of the body and its position. But if the magnitude of the angular coordinate ϕ never exceeds $\pi/2$ in an oscillation, so that the rod is never inclined above the horizontal, then $\mathbf{R}$ will never be directed outward along the rod. (Use your mechanical intuition to justify this statement.) In such a case the rod can just as well be a cord of the same length. Both a cord and a rod can supply whatever force is required to make the body move in a circular path, if that force is always directed *inward* to the center of the path. A cord can pull on something just as well as a rod can. But if the force is directed *outward* from the center, it can be supplied only by a rod. A cord is not capable of pushing on something. In summary, we will analyze oscillatory motion in a vertical plane of a friction-free system consisting of a compact, massive body at one end of a rod whose other end rotates about a fixed point, or at one end of a cord if the body is never higher than the fixed point about which the other end of the cord rotates. In either case, the system is called a **simple pendulum,** and the body is called a **bob.**

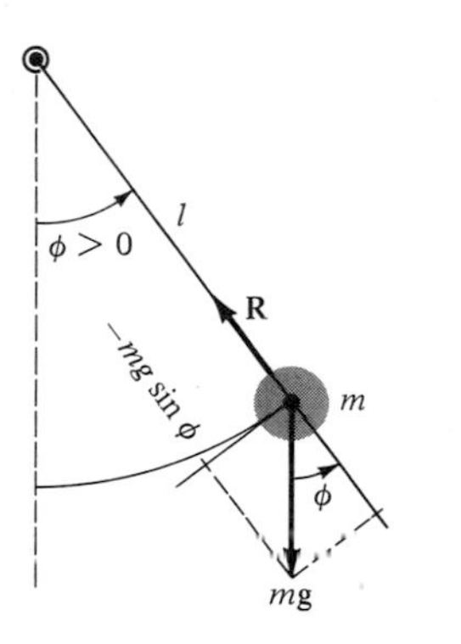

Fig. 6-10 A simple pendulum, comprising a compact body of mass m oscillating at the end of a light rod or cord of length l. The pendulum is shown at an instant when the angular coordinate ϕ, measured from the downward vertical, is positive. This choice of sign agrees with the usual convention of making counterclockwise rotations positive. At the instant depicted, the tangential component of the net force acting on the body has the value $F_t = -mg \sin \phi$. But Fig. 6-3 shows that this relation holds at any instant, that is, no matter what the sign or value of ϕ. The oscillatory motion results from the action of the tangential force component.

The oscillatory motion of the pendulum bob back and forth through its curved path is driven by the component tangent to its path of the net force acting on the bob. We know from the considerations of Sec. 6-1 that this tangential net force component F_t is what always tends to move the bob toward its equilibrium position, where $\phi = 0$. The net force acting on the bob also has a component which is perpendicular to its path (except at the extremes of the oscillation where the body is motionless). This component produces the centripetal acceleration arising from the curvature of the path. But here we are not interested in the centripetal acceleration, so we are not interested in the perpendicular component of the net force.

By referring to Sec. 6-1, or by inspecting Fig. 6-10, we see that the tangential component of the net force acting on the pendulum bob is given by

$$F_t = -mg \sin \phi \tag{6-6}$$

The minus sign makes this expression correctly describe the fact that the tangential force is in the clockwise direction (F_t is negative) when the bob lies in the counterclockwise direction from its stable equilibrium position ($\sin \phi$ is positive), and vice versa.

To analyze pendulum oscillations, we need to obtain an expression for a_t, the tangential component of the acceleration of the pendulum bob. This acceleration component *along* its path is associated with the change in the *magnitude* of its velocity, that is, with the change in its speed. (See Sec. 3-4, where this fact is developed and contrasted with the fact that the acceleration component *perpendicular* to the path is associated with the change in the *direction* of the velocity.) In fact, the bob's rate of change of speed gives the value of the acceleration along its path, just as the rate of change of speed of a body moving on a straight line gives the value of the acceleration of the body along its path. If we use the infinitesimal displacement vector $\mathbf{ds}$ to describe the change in position of the bob during an infinitesimal time interval dt, we can write its velocity as $\mathbf{v} = d\mathbf{s}/dt$. (Again see Sec. 3-4.) The speed v of the bob can then be written as $v = ds/dt$, where ds is the magnitude of the vector $\mathbf{ds}$. Figure 6-11 shows that ds can be expressed in terms of the length l of the pendulum rod or cord and the infinitesimal change $d\phi$ in the coordinate of the pendulum bob during the time dt. Specifically, it shows that

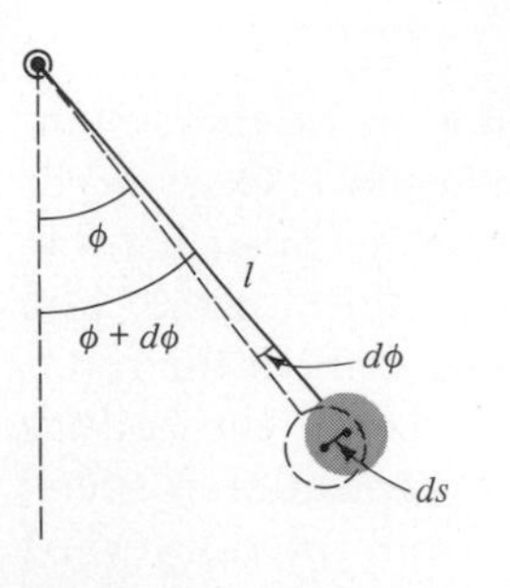

Fig. 6-11 The infinitesimal displacement of the pendulum bob when there is an infinitesimal change in its angular coordinate. No matter what units are used to measure angles, the magnitude ds of the displacement will be proportional to the magnitude $d\phi$ of the angular change and also proportional to the length l of the pendulum rod or cord. That is $ds \propto l\, d\phi$. But when angles are measured in radians, the proportionality constant has the convenient value 1, so that $ds = l\, d\phi$. This is why we measure angles in radians.

$$ds = l\, d\phi$$

so

$$v = \frac{ds}{dt} = \frac{l\, d\phi}{dt} = l \frac{d\phi}{dt}$$

The rate of change of speed is

$$\frac{dv}{dt} = \frac{d}{dt}\left(l \frac{d\phi}{dt}\right)$$

Since l is a constant, this is

$$\frac{dv}{dt} = l \frac{d}{dt} \frac{d\phi}{dt}$$

Expressed in terms of a second derivative, the rate of change of speed is

$$\frac{dv}{dt} = l \frac{d^2\phi}{dt^2}$$

Since the rate of change of its speed gives the value of the pendulum bob's acceleration along its path, and since that quantity is its tangential acceleration component a_t, we have

$$a_t = l \frac{d^2\phi}{dt^2} \tag{6-7}$$

This expression gives the sign of a_t as well as its magnitude. For instance, if the bob is moving counterclockwise with increasing speed, then $d\phi/dt$ is becoming more positive. So $d^2\phi/dt^2$ is positive and a_t is positive, in agreement with the fact that the tangential acceleration is directed counterclockwise. If the bob is moving counterclockwise but the speed is decreasing, then $d\phi/dt$ is becoming less positive, so $d^2\phi/dt^2$ and a_t are negative. This agrees with the fact that the tangential acceleration is directed clockwise. (You should go through similar arguments for the two situations in which the bob is moving clockwise.)

Newton's second law requires that the tangential component of the net force acting on the pendulum bob equal its mass times the tangential component of its acceleration. That is,

$$F_t = ma_t$$

Using Eqs. (6-6) and (6-7), we have

$$-mg \sin \phi = ml \frac{d^2\phi}{dt^2} \tag{6-8a}$$

or, canceling and transposing,

$$\frac{d^2\phi}{dt^2} = -\frac{g}{l} \sin \phi \tag{6-8b}$$

If we define the parameter α to be

$$\alpha \equiv \frac{g}{l} \tag{6-9}$$

we can write the **pendulum equation** as

$$\frac{d^2\phi}{dt^2} = -\alpha \sin \phi \tag{6-10}$$

The parameter α contained in the pendulum equation has a value determined by the gravitational acceleration g and the length l of the pendulum rod or cord. The equation itself determines how the pendulum bob will move, once it is started in a particular way. Note that the equation does not involve the mass of the bob. The m on the left side of Eq. (6-8*a*), which is a manifestation of the gravitational role of the body's mass, cancels the m on the right side, which is a manifestation of its inertial role. The pendulum is yet another example of a system where motion does not depend on mass.

In order to compare the pendulum equation with Eq. (6-3), $a = -\alpha x$, which determines the motion of a body at the end of a spring, we will rewrite the latter in calculus notation. Using the definition of acceleration, $a = d^2x/dt^2$, the body-and-spring equation becomes

$$\frac{d^2x}{dt^2} = -\alpha x \tag{6-11}$$

where

$$\alpha = \frac{k}{m}$$

Two differences between Eqs. (6-10) and (6-11) are that they involve the different dependent variables ϕ and x, and that the meaning of the constant α is not the same. But these differences are mathematically (though not physically) unimportant. The substantial difference is the presence of the sine function in Eq. (6-10). If $\sin \phi$ were equal to ϕ (with ϕ expressed in radians), then Eq. (6-10) would be mathematically identical to Eq. (6-11). Of course, it is not true that $\sin \phi = \phi$. But it is *approximately* true if ϕ is small compared to 1 rad. (See Example 2-7.)

Thus when a pendulum is performing oscillations of sufficiently small amplitude that $\sin \phi = \phi$ can be used as a good approximation, the pendulum equation can be written

$$\frac{d^2\phi}{dt^2} = -\alpha\phi \qquad \text{for } \phi \ll 1 \tag{6-12}$$

where

$$\alpha = \frac{g}{l}$$

Providing the restriction on ϕ is satisfied through the swing of the pendulum, this is the mathematical equivalent of the body-and-spring equation. But the dependent variable is now the angular coordinate ϕ, and the constant α is now the gravitational acceleration g divided by the length l of the pendulum rod or cord. In both equations α has the same units. For Eq. (6-11) the units are N/(m·kg). If we use the definition 1 N = 1 kg·m·s^{-2}, this reduces to s^{-2}. For Eq. (6-12) the units are m·s^{-2}/m, which reduces immediately to s^{-2}.

Because the equations determining their motion are mathematically identical, everything we learned in Sec. 6-2 about the motion of the body at the end of the spring carries over directly to the motion of a pendulum bob undergoing *small oscillations.* So we can conclude immediately that the period T (or frequency ν) of a pendulum does not depend on the amplitude of its oscillation, *providing* the amplitude remains small. But T does depend on g and l, since these quantities determine the value of α. An equation relating T to α for the body and spring, and therefore for the *small oscillations* of a pendulum, is the one found from numerical calculations in Sec. 6-2: $T = 2\pi/\sqrt{\alpha}$. It is obtained in Sec. 6-5 from analytical calculations.

What about *large oscillations* of a pendulum? It is very easy to modify the calculator or computer program so that it will solve Eq. (6-10), which applies to pendulum oscillations of *any* amplitude, instead of Eqs. (6-11) or (6-12), which apply to the body and spring or to a pendulum executing small oscillations. As you will see by looking at the pendulum program listed in the Numerical Calculation Supplement, doing so is just a matter of inserting an instruction to take the sine of the dependent variable at the appropriate place in the program, and then to use it instead of the dependent variable itself. The calculating device is set to run in the radian mode, and the results displayed are interpreted as the angular coordinate ϕ.

EXAMPLES 6-4 AND 6-5

Run the pendulum program with the following two sets of initial conditions and parameters:

$$\phi_0 = 0.25 \text{ (in rad)};\ (d\phi/dt)_0 = 0;\ t_0 = 0;\ \Delta t = 0.2 \text{ (in s)};\ \alpha = 1 \text{ (in s}^{-2})$$

$$\phi_0 = 1.5 \text{ (in rad)};\ (d\phi/dt)_0 = 0;\ t_0 = 0;\ \Delta t = 0.2 \text{ (in s)};\ \alpha = 1 \text{ (in s}^{-2})$$

■ In both examples the pendulum is given a certain initial displacement ϕ_0 from the position of stable equilibrium and then released with no motion, that is, with $(d\phi/dt)_0 = 0$. The subsequent oscillations are plotted in Fig. 6-12.

Compare the pendulum oscillation in ϕ of amplitude 0.25 rad, shown in Fig. 6-12, with the body-and-spring oscillation in x of amplitude 0.25 m shown in Fig. 6-6. Both plots are for the same value $\alpha = 1\ s^{-2}$. The comparison will make it clear that for the reasonably small amplitude 0.25 rad (about 14°) the plot describing the oscillation of a pendulum bob is indistinguishable from the one describing the oscillation of a body at the end of a spring. This confirms our earlier conclusion that the behavior of the two systems should be essentially identical, providing the amplitude of the pendulum oscillations is small.

But this is not so for a pendulum executing large oscillations. Compare the plot shown in Fig. 6-12 for a pendulum oscillation of amplitude 1.5 rad (approximately 86°) with the plot in that figure for the pendulum oscillation of amplitude 0.25 rad. Both are for the value $\alpha = 1\ s^{-2}$. But they do *not* have the same period T. The period is increased (and so the frequency is reduced) for the large-amplitude pendulum oscillation. Thus the period of a pendulum *does* depend on the amplitude of its oscillation if the amplitude becomes large. This contrasts sharply with the way the period of oscil-

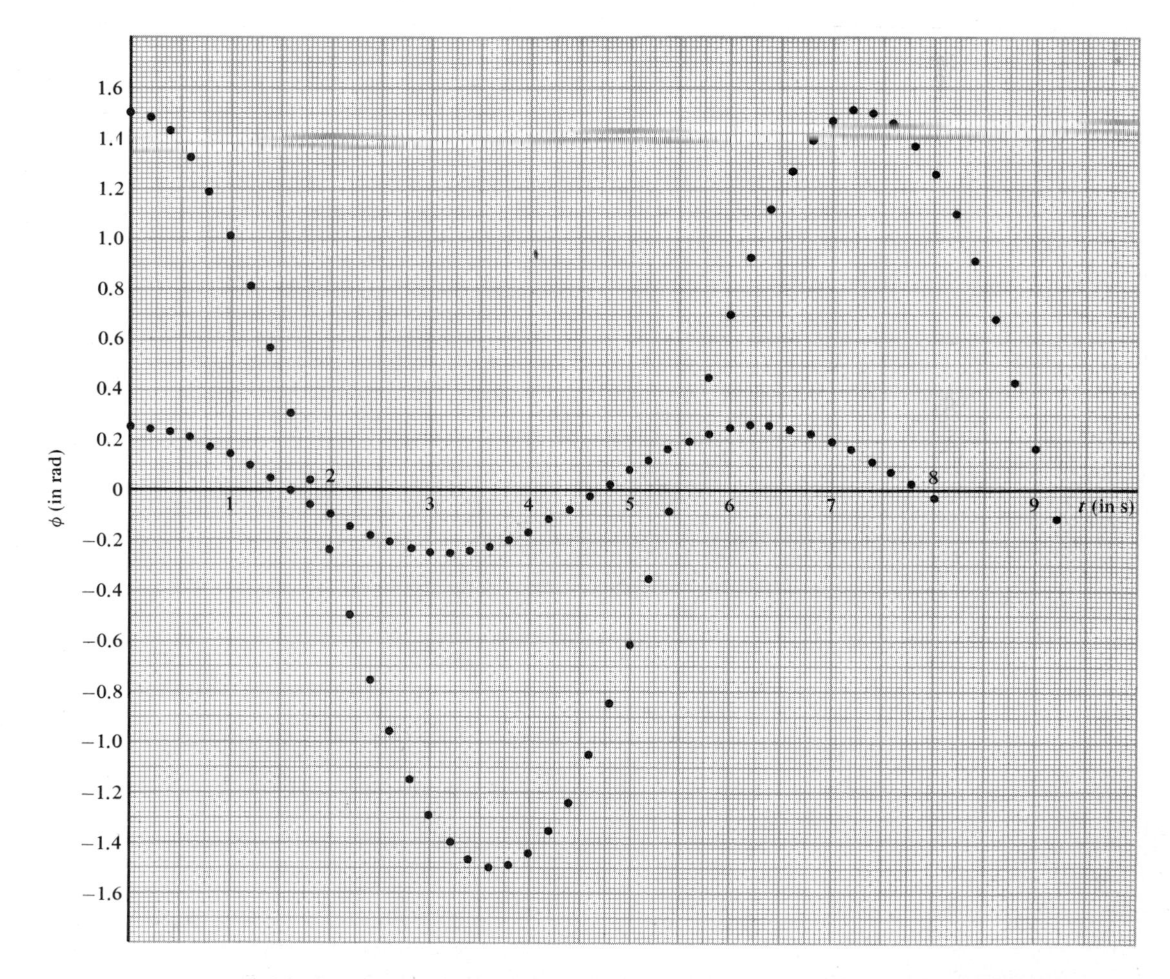

Fig. 6-12 The angular coordinate ϕ versus the time t for small- and large-amplitude oscillations of a pendulum. The parameter $\alpha = g/l$, the ratio of the magnitude of the gravitational acceleration to the length of the pendulum rod or cord, has the value $1\ s^{-2}$.

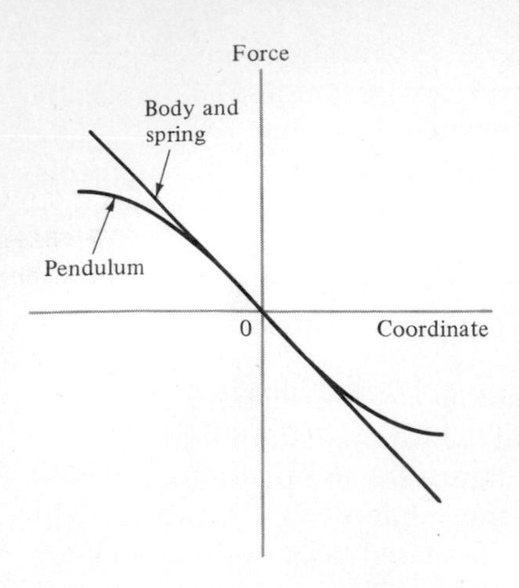

Fig. 6-13 The force acting in the direction of motion on a pendulum bob and the Hooke's-law force acting on a body at the end of a spring. For the sake of comparison, the parameter α is assumed to have the same numerical value in both systems. Also, the pendulum oscillations are assumed to be confined to the angular range $-\pi/2 \leq \phi \leq +\pi/2$, so that the pendulum can be constructed with a cord as well as with a rod.

lation of a body at the end of a spring remains independent of the amplitude as long as the spring continues to obey Hooke's law. If you compare the force acting in the direction of motion on a pendulum bob with the Hooke's-law force acting on a body at the end of a spring, you can understand qualitatively the reason for the difference in behavior of the two systems. The comparison is shown in Fig. 6-13 for a case in which α has the same value in both systems. This means the slope of both force versus coordinate curves will be equal at the position of stable equilibrium where the value of the coordinate is zero. For small oscillations of the pendulum bob about the stable equilibrium position, its behavior is indistinguishable from the behavior of the body at the end of a spring. The reason is that for small ϕ, where $\sin \phi = \phi$ to a very good approximation, the dependence of the force on the coordinate is essentially the same in both systems, and so the motion produced by the force is essentially the same. But for larger oscillations the force acting on the pendulum bob becomes weaker than the force acting on the body at the end of a spring. This makes the pendulum bob take more time to complete a cycle of its oscillation. In other words, the period increases. It is a simple matter to run the pendulum program for a variety of amplitudes, and thereby determine how the period depends on the amplitude. Results are shown in Fig. 6-14.

The curve in Fig. 6-12 for the larger-amplitude oscillation of a pendulum is not simply a scaled-up version of the smaller-amplitude curve in

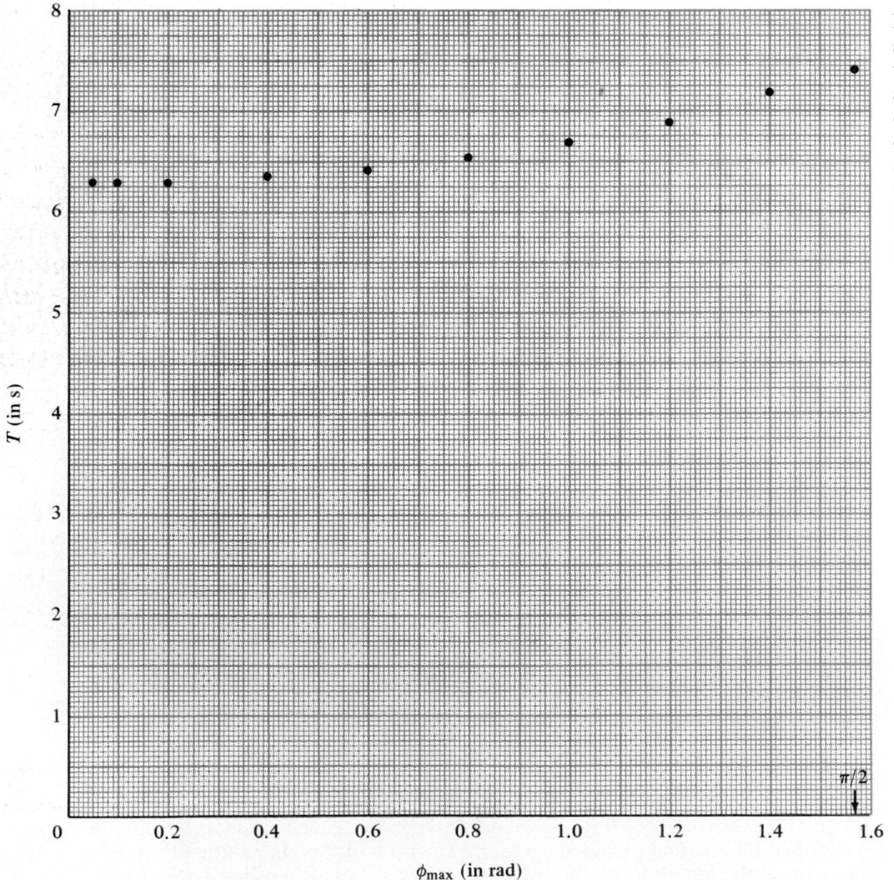

Fig. 6-14 The period T of a pendulum for several values of its amplitude ϕ_{max}. The value of the parameter $\alpha = g/l$ is 1 s^{-2}. The amplitudes do not exceed $\pi/2$ rad, so the data apply to a pendulum constructed with either a cord or a rod. For larger amplitudes a rod must be used. But there the dependence of T on ϕ_{max} becomes much more pronounced. You will see this if you use the pendulum program to obtain one or two data points for ϕ_{max} near π rad.

that figure. It cannot be a scaled-up version since the periods for the two amplitudes are not the same. Thus the scaling property, seen in Fig. 6-6 for the body-and-spring oscillations, does not apply to pendulum oscillations. The preceding paragraph gives a qualitative physical explanation of this fact; a quantitative mathematical explanation is given in Sec. 6-5.

6-4 NUMERICAL SOLUTION OF DIFFERENTIAL EQUATIONS

The numerical method for determining the motion of a body whose acceleration varies according to a known rule, when its position and velocity at some initial instant are given, was introduced in Sec. 5-6. There the method was used on the skydiver equation,

$$a = g - \gamma v^2$$

It was treated as an algebraic equation, and the numerical method for handling it was explained as a sequence of algebraic calculations. The same point of view was used in Sec. 6-2 for the numerical treatment of the equation for a body and spring,

$$a = -\alpha x$$

But in Sec. 6-3 the development of the pendulum equation led naturally to its expression in terms of a derivative in Eq. (6-10):

$$\frac{d^2\phi}{dt^2} = -\alpha \sin \phi$$

We then found it convenient to express the body-and-spring equation in terms of a derivative in Eq. (6-11):

$$\frac{d^2x}{dt^2} = -\alpha x$$

The skydiver equation can also be written in terms of derivatives. In this form it is

$$\frac{d^2x}{dt^2} = g - \gamma \left(\frac{dx}{dt}\right)^2 \qquad (6\text{-}13)$$

Equations (6-10), (6-11), and (6-13) are called **differential equations** because they are equations in which the "unknown," the dependent variable ϕ or x, occurs in derivatives. Specifically, they are called **second-order, ordinary** differential equations, with "second order" meaning that no third- or higher-order derivatives are found in the equations and "ordinary" meaning that each equation involves only a single independent variable, such as t. Equation (6-11) is said to be a **linear** differential equation since each of its terms is proportional to the first power of the dependent variable x, while Eq. (6-10), or Eq. (6-13), is called a **nonlinear** equation since some of the terms are not proportional to the first power of the dependent variable ϕ, or x. Thus you have actually been using the numerical method to obtain **numerical solutions** of linear and nonlinear second-order, ordinary differential equations!

The fundamental relation of newtonian mechanics is the second-order, ordinary differential equation

$$\frac{d^2x}{dt^2} = \frac{F}{m}$$

It may be linear or nonlinear, depending on the mathematical form of F. Furthermore, a great variety of phenomena of the physical universe lead through their analysis to second-order, ordinary differential equations. You will see many examples in this book—particularly in connection with planetary and satellite motion, mechanical and electromagnetic wave motion, electric circuits, and quantum mechanics.

Even though their analytical solutions may be difficult or impossible, all these equations, and many more, can be handled by applying a single numerical method. This method is a generalization of the numerical method used in Secs. 5-6, 6-2, and 6-3. The generalized method is applied in the *same* way to *any* second-order, ordinary differential equation.

The numerical method for solving second-order, ordinary differential equations goes as follows. You manipulate the differential equation so as to isolate the second derivative on the left side of the equality. If x and t are the dependent and independent variables, respectively, you then have

$$\frac{d^2x}{dt^2} = Q \tag{6-14}$$

The quantity Q represents whatever remains on the right side of the equality. Thus Q may involve any functions of any of or all the quantities x, t, and dx/dt, as well as any constants. For instance, in the body-and-spring equation Q is $-\alpha x$. In the skydiver equation Q is $g - \gamma(dx/dt)^2$. The symbol Q is used, instead of the symbol a, since in general the second derivative need not be an acceleration. Next, you convert Eqs. (5-35) to more general forms by writing Q for a and dx/dt for v.

The new equations are used in the same way that you used Eqs. (5-35). To be specific, this is what you do. Always employing the differential equation to express Q in terms of the quantities on which it depends, you select a *small* time increment Δt and then carry out the following set of calculations:

Determine the value of Q from the values at $t_0 = 0$ of the quantities on which it depends, and use it to calculate

$$\left(\frac{dx}{dt}\right)_{1/2} \simeq \left(\frac{dx}{dt}\right)_0 + Q\,\frac{\Delta t}{2} \tag{6-15a}$$

from the given value of $(dx/dt)_0$. Use the result to calculate

$$x_1 \simeq x_0 + \left(\frac{dx}{dt}\right)_{1/2} \Delta t \tag{6-15b}$$

from the given value of x_0. Then set $t_1 = \Delta t$.

Next determine the value of Q from the new values of the quantities on which it depends. Use it, and the value of $(dx/dt)_{1/2}$ just obtained, to calculate

$$\left(\frac{dx}{dt}\right)_{3/2} \simeq \left(\frac{dx}{dt}\right)_{1/2} + Q\,\Delta t \tag{6-15c}$$

Use the result, and the value of x_1 just obtained, to calculate

$$x_2 \simeq x_1 + \left(\frac{dx}{dt}\right)_{3/2} \Delta t \tag{6-15d}$$

Then set $t_2 = 2\Delta t$.

Next determine the value of Q from the new values of the quantities on which it depends. Use it, and the value of $(dx/dt)_{3/2}$ just obtained, to calculate

$$\left(\frac{dx}{dt}\right)_{5/2} \simeq \left(\frac{dx}{dt}\right)_{3/2} + Q\,\Delta t \tag{6-15e}$$

Use the result, and the value of x_2 just obtained, to calculate

$$x_3 \simeq x_2 + \left(\frac{dx}{dt}\right)_{5/2} \Delta t \tag{6-15f}$$

Then set $t_3 = 3\Delta t$.

Continue these calculations until t reaches whatever value is required.

The justification of these equations for a small value of Δt is apparent if you remember that the first derivative is the rate of change of the dependent variable, that the second derivative is the rate of change of the first derivative, and that Q equals the second derivative. Just read what the equations say, in words. Equations (6-15*a*), (6-15*c*), (6-15*e*), and so forth say the new value of the first derivative is approximated by the old value plus the rate of change of the first derivative, Q, multiplied by the increment in the independent variable. Equations (6-15*b*), (6-15*d*), (6-15*f*), and so forth say that the new value of the dependent variable is approximated by the old value plus the rate of change of the dependent variable, dx/dt, multiplied by the increment in the independent variable.

If the dependent and/or independent variables are other than x and t, all you have to do is rewrite these equations with the appropriate symbols replacing x and/or t. For example, in the pendulum equation the dependent variable is ϕ, instead of x, and Q is $-\alpha \sin \phi$. The numerical method is applicable to almost any second-order, ordinary differential equation. Its only general limitation is that Q must be finite everywhere within the range over which it must be evaluated.

It is difficult to predict the accuracy of the method in any particular application. However, it is almost always true that (1) the accuracy increases as you decrease the size of the increment in the independent variable; (2) the larger the value of Q, the smaller the size of the increment you must use to achieve a certain accuracy; (3) the accuracy decreases when Q involves a first derivative. You can understand the third comment by noting that the dependent and independent variables are both evaluated in Q at the middle of the interval where Q is used. But if the first derivative is present in Q, as is the case in the skydiver equation, then the first derivative is evaluated at the beginning of the interval. This makes the fixed value used to represent Q in the interval a poorer representation of the range of values actually assumed by Q through the interval. Fortunately, it is often true that where the first derivative is large, its rate of change is small, so that the value of Q is not too sensitive to where the first derivative is evaluated. The skydiver equation is an example.

All the numerical calculations given in Secs. 5-6, 6-2, and 6-3 are really examples of the numerical method for solving second-order, ordinary differential equations. We will consider several more examples in Sec. 6-6 and many others later in the book. But now we turn our attention to an analytical solution of the equation that we solved numerically in Sec. 6-2.

6-5 ANALYTICAL SOLUTION OF THE HARMONIC OSCILLATOR EQUATION

For many of the differential equations that arise in physics, solutions can be found by carrying out mathematical analyses that do not involve numerical procedures. In some cases it is quite easy to find such **analytical solutions,** while in others the analytical solutions are very difficult to obtain. But where there are analytical solutions, they can be expressed in a very concise form (often in terms of the simple and familiar functions that can be evaluated by pressing one or two keys on a scientific calculator). So for many purposes analytical solutions are much more convenient to use than the graphs or tables which must be employed to express numerical solutions of differential equations.

A major obstacle to obtaining analytical solutions to differential equations is that there is no single method that works on all those equations that do have such solutions. Instead, there are many different analytical procedures, each suited only to certain differential equations. Almost all the methods do, however, have a common approach: (1) You use whatever intuition you have available to *guess* at the mathematical form for the solutions to the equation. If you are wise, you will build flexibility into your guess by including in the assumed mathematical form constants whose values can be subsequently adjusted to make the form fit the specific requirements of the differential equation. (2) Next you evaluate the appropriate derivatives of the solutions whose form you have assumed, and substitute them into the equation. (3) Then you explore the consequences of this substitution. That is, you see if it is possible to adjust the values of one or more of the constants so as to obtain results which are mathematically consistent. (You will see immediately below an example of what is meant by "mathematically consistent.") (4) If you succeed, you have found solutions to the differential equation. (5) If you fail, because substituting your guess into the equation leads to an irreconcilable inconsistency, you must make another guess aimed at removing the inconsistency and then try again.

Let us see if we can use this approach to find analytical solutions of the equation determining the motion of a body at the end of a spring:

$$\frac{d^2x}{dt^2} = -\alpha x \tag{6-16}$$

First we will sharpen our intuition by looking back at the results we obtained in solving the equation numerically. Figure 6-6 shows that the equation has oscillatory solutions which look like cosine functions of t. Figure 6-9 shows it also has solutions that look like sine functions of t. Thus its solutions appear to be **sinusoids** (a word that includes both sine and cosine) which depend on t. Figure 6-7 shows that the sinusoid oscillates more or less rapidly depending on the value of α. These properties lead us to guess that solutions to the differential equation might be of the form

$$x \propto \cos(\omega t + \delta)$$

The value of the *constant* δ can be adjusted to make this single form represent any sinusoid that is required. For example, if $\delta = 0$, then the form describes a cosine, as in Fig. 6-6. If $\delta = -\pi/2$, then it describes a sine, as in Fig. 6-9, since $\cos(\omega t - \pi/2) = \cos(\pi/2 - \omega t) = \sin(\omega t)$. The value of the *constant* ω can be adjusted to make the rapidity of the oscillation fit any requirement. For instance, a large ω will make a small change in t take the sinusoid through a large part of a cycle, and thereby lead to a rapid oscillation.

The proportionality can be written as an equality,

$$x = A\cos(\omega t + \delta) \tag{6-17}$$

by introducing the adjustable *constant* A. Note that the form of Eq. (6-17) is consistent with what we learned from Fig. 6-6 about the scaling property of the solutions to the equation. That is, using different values of A will produce solutions which differ only in their amplitude. The form in Eq. (6-17) is the one we will assume for the solutions of the differential equation, Eq. (6-16). If we had no prior mathematical intuition about that equation, because we had not already studied some of its numerical solutions, we would have to base our assumption on physical intuition obtained from experimental study of the oscillations of a body at the end of a spring.

Now we will prepare to test the validity of the form in Eq. (6-17) by computing its second derivative. We begin by computing the first derivative:

$$\frac{dx}{dt} = \frac{d[A\cos(\omega t + \delta)]}{dt}$$

Since A is a constant, this is

$$\frac{dx}{dt} = A\,\frac{d[\cos(\omega t + \delta)]}{dt} \tag{6-18}$$

To proceed, we must employ the "chain rule" of differential calculus. Given any function f of any variable u, the rule states that

$$\frac{d[f(u)]}{dt} = \frac{d[f(u)]}{du}\,\frac{du}{dt} \tag{6-19}$$

We set $f(u) = \cos u$. Then the rule says

$$\frac{d(\cos u)}{dt} = \frac{d(\cos u)}{du}\,\frac{du}{dt}$$

Now $d(\cos u)/du = -\sin u$. [If you are not familiar with this relation, you can obtain it by taking $\omega = 1$ and $t = u$ in Eq. (2-18).] So we have

$$\frac{d(\cos u)}{dt} = -\sin u\,\frac{du}{dt}$$

Next we let $u = \omega t + \delta$ and obtain

$$\frac{d[\cos(\omega t + \delta)]}{dt} = -\sin(\omega t + \delta)\,\frac{d(\omega t + \delta)}{dt}$$

Since ω and δ are constants, the derivative on the right side yields ω. Thus

$$\frac{d[\cos(\omega t + \delta)]}{dt} = -\omega\sin(\omega t + \delta) \tag{6-20}$$

Using this relation in Eq. (6-18) produces immediately the result

$$\frac{dx}{dt} = -A\omega\sin(\omega t + \delta) \tag{6-21a}$$

In a very similar computation we use this result to find the second derivative. We have

$$\frac{d^2x}{dt^2} = \frac{d}{dt}\left(\frac{dx}{dt}\right) = \frac{d[-A\omega \sin(\omega t + \delta)]}{dt}$$

$$= -A\omega \frac{d[\sin(\omega t + \delta)]}{dt}$$

$$= -A\omega \cos(\omega t + \delta) \frac{d(\omega t + \delta)}{dt}$$

This gives

$$\frac{d^2x}{dt^2} = -A\omega^2 \cos(\omega t + \delta) \tag{6-21b}$$

Next we substitute Eqs. (6-17) and (6-21*b*) into the differential equation we hope that they satisfy. That is, we substitute the expressions for x and for d^2x/dt^2 into Eq. (6-16), $d^2x/dt^2 = -\alpha x$. We obtain

$$-A\omega^2 \cos(\omega t + \delta) = -\alpha A \cos(\omega t + \delta) \tag{6-22}$$

Can this equation be mathematically consistent? Indeed it can! If we adjust the value of ω so that

$$\omega^2 = \alpha$$

or

$$\omega = \sqrt{\alpha} \tag{6-23}$$

then Eq. (6-22) is satisfied *identically*. That is, both sides of Eq. (6-22) have the same value under all circumstances. This means that the form in Eq. (6-17) will then be a solution to Eq. (6-16). [The positive root is taken in Eq. (6-23) since, by convention, the quantity ω is always positive. This is because ω specifies the rapidity of the oscillation; a negative ω would have no physical meaning.]

Thus we have found solutions to the differential equation, Eq. (6-16),

$$\frac{d^2x}{dt^2} = -\alpha x$$

The solutions are of the form given by Eq. (6-17) with the constant ω adjusted so that $\omega = \sqrt{\alpha}$. That is, they are

$$x = A \cos(\sqrt{\alpha}\, t + \delta) \tag{6-24}$$

The solutions are sinusoidal functions, which are also called **harmonic functions.** For this reason the physical system whose motion is described by the solutions—the system we have exemplified by a body at the movable end of a spring whose other end is fixed—is called a **harmonic oscillator.** And the differential equation determining the motion, Eq. (6-16), is called the **harmonic oscillator equation.**

The plural is used in referring to Eq. (6-24) since this actually represents a whole *family* of different solutions, each corresponding to a particular set of values of the constants A and δ. To determine the significance of these constants, we will evaluate the position and velocity of the oscillating body at the instant $t = 0$. From Eq. (6-24) we have for the initial position

$$x_0 = A \cos \delta$$

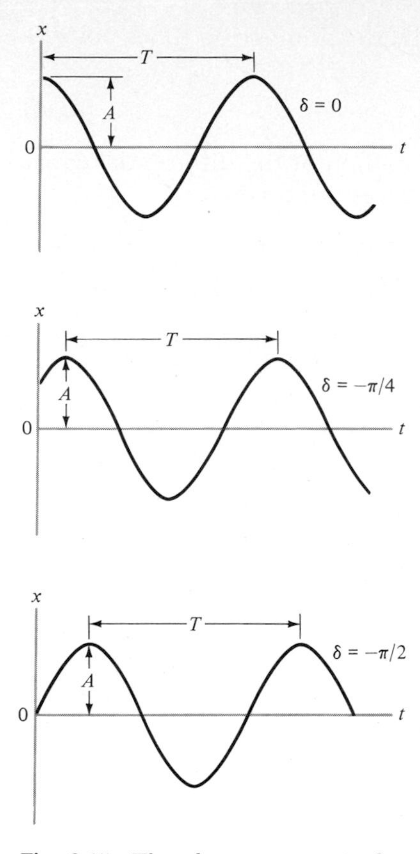

Fig. 6-15 The phase constant δ of a sinusoidal function $x(t)$. Changing δ has no effect on its amplitude A or period T, but slides the function along the t axis. An alternative point of view is to say that changing δ has no effect at all on the function, but it does change the way that $t = 0$ is defined. Can you draw a sketch which illustrates this point of view?

The initial velocity is given by Eq. (6-21a), with $\omega = \sqrt{\alpha}$ and $t = 0$:

$$\left(\frac{dx}{dt}\right)_0 = -A\sqrt{\alpha}\,\sin\,\delta$$

Now note that when $\delta = 0$, then $x_0 = A$ and $(dx/dt)_0 = 0$. When $\delta = -\pi/2$, then $x_0 = 0$ and $(dx/dt)_0 = A\sqrt{\alpha}$. For general values of δ, neither x_0 nor $(dx/dt)_0$ is zero. As an example, if $\delta = -\pi/4$, then $x_0 = 0.707A$ and $(dx/dt)_0 = 0.707A\sqrt{\alpha}$. Figure 6-15 illustrates the oscillations described by Eq. (6-24) for these three values of the constant δ, which is called the **phase constant.** You can see that changing the phase constant corresponds to taking the same sinusoidal solution and moving it along the t axis. If δ is made more negative, the sinusoid shifts in the direction of increasing t; if δ is made more positive, the sinusoid shifts in the opposite direction.

Figure 6-15 also shows that the constant A is the **amplitude** of the oscillation because A is the maximum magnitude assumed by x. You can see this directly from Eq. (6-24) since the value of the cosine oscillates between -1 and $+1$ as t increases. Thus x, which represents the position of the body, will oscillate between $-A$ and $+A$.

The two constants A and δ allow you to adjust the right side of Eq. (6-24) so that it is the particular solution to the harmonic oscillator equation appropriate to any values of x_0 and $(dx/dt)_0$. These values are called the **initial conditions** on the equation. In fact, the general expression for the solutions to any second-order, ordinary differential equation must contain two constants so that the expression can be tailored to fit the values of the dependent variable and its first derivative at some initial value of the independent variable.

The **period** T of the harmonic oscillator, that is, the time for it to complete one full oscillation, is indicated in Fig. 6-15. Its value is determined by the coefficient of t in Eq. (6-24):

$$x = A\cos(\sqrt{\alpha}\,t + \delta)$$

Since $(\sqrt{\alpha}\,t + \delta)$ is the "argument" of a cosine (in other words, a quantity whose cosine is to be calculated), that quantity is an angle. While the angle in radians increases by 2π, the values of the cosine oscillate through one cycle and the time increase is one period. Thus T is such that

$$[\sqrt{\alpha}\,t + \delta] + 2\pi = [\sqrt{\alpha}(t + T) + \delta]$$

or

$$\sqrt{\alpha}\,T = 2\pi$$

So

$$T = \frac{2\pi}{\sqrt{\alpha}} \qquad (6\text{-}25a)$$

The **frequency** ν has the value $\nu = 1/T$, or

$$\nu = \frac{\sqrt{\alpha}}{2\pi} \qquad (6\text{-}25b)$$

The quantity ω that we used earlier is the **angular frequency** of the oscillation. It measures the rate of increase, in radians per second, of the angle specified by the argument of the cosine. Since the angle increases by

2π in each cycle, and since the frequency is the number of cycles per second, we have the relation

$$\omega = 2\pi\nu \tag{6-26}$$

In terms of the frequency ν, the general solution of the differential equation for a harmonic oscillator, Eq. (6-17), takes the form

$$x = A\cos(2\pi\nu t + \delta) \tag{6-27}$$

For the harmonic oscillator the value of ω is, from Eqs. (6-25*b*) and (6-26),

$$\omega = 2\pi\frac{\sqrt{\alpha}}{2\pi} = \sqrt{\alpha}$$

as we already know from Eq. (6-23). These results from T, ν, and ω apply to both the body and spring and the small oscillations of a pendulum, providing the proper expression for α is used. For the body and spring it is

$$\alpha = \frac{k}{m}$$

so that

$$\nu = \frac{1}{T} = \frac{1}{2\pi}\sqrt{\frac{k}{m}} \tag{6-28a}$$

For the pendulum it is

$$\alpha = \frac{g}{l}$$

so that

$$\nu = \frac{1}{T} = \frac{1}{2\pi}\sqrt{\frac{g}{l}} \qquad \text{where } \phi \ll 1 \tag{6-28b}$$

Compare the results obtained from the analytical solution of the harmonic oscillator differential equation with those obtained earlier from solving the equation numerically. You will find the following: (1) The sinusoidal forms of the analytical solutions of Eq. (6-24) agree with the forms of the numerical solutions plotted in Figs. 6-6 and 6-9. When appropriate values of the amplitude and phase constants A and δ are used, the solutions agree in every detail. (2) The presence of the multiplicative constant A in Eq. (6-24) agrees with the scaling property seen by reading the numerical values of the two solutions plotted in Fig. 6-6. (3) The relation of Eq. (6-25*a*) between T, the period of the oscillator, and α, the quantity describing its mechanical properties, agrees quantitatively with the values plotted in Figs. 6-7 and 6-8, as well as with the equation inferred from those values. Of course, it is often much more convenient to have the solutions to the harmonic oscillator differential equation in the one equation provided by the analytical work than it is to have them only in the many graphs provided by the numerical work.

Equation (6-28*b*) predicts that the period for the small oscillations of a simple pendulum with a certain length l is the same as the period for small rotations of a conical pendulum with the same length l. The latter is given by Eq. (3-53) with $\theta \ll 1$ so that $\cos\theta = 1$. Make a pendulum and test this prediction.

EXAMPLE 6-6

a. An object is connected to one end of a horizontal spring whose other end is fixed, as in Fig. 6-16. The object is pulled to the right (in the positive x direction) by an externally applied force of magnitude 20.00 N, causing the spring to stretch 1.000 cm. Determine the value of the force constant.

■ The force constant is the k in Hooke's law, Eq. (6-1):

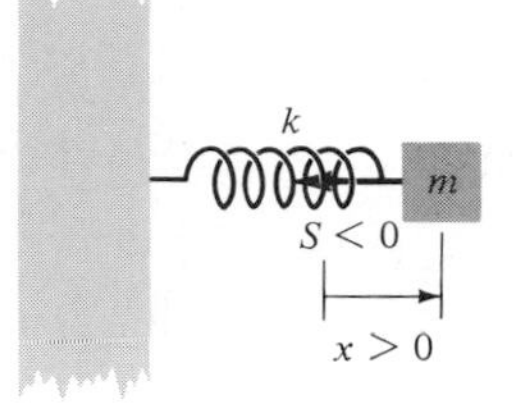

Fig. 6-16 A harmonic oscillator.

$$F = -kx$$

The force F produced by the spring is $F = -20.00$ N, where the minus sign means that the force acts to the left (in the negative x direction). Since $x = 1.000 \times 10^{-2}$ m, you have

$$k = -\frac{F}{x} = -\frac{-20.00 \text{ N}}{1.000 \times 10^{-2} \text{ m}} = 2.000 \times 10^{3} \text{ N/m}$$ ■

b. The mass of the object is 4.000 kg. Determine the period with which it oscillates if the applied force is suddenly removed.

■ The ratio α of the force constant to the mass determines the period T, since Eq. (6-25a) shows that $T = 2\pi/\sqrt{\alpha}$. According to Eq. (6-4),

$$\alpha = \frac{k}{m} = \frac{2.000 \times 10^{3} \text{ N/m}}{4.000 \text{ kg}} = 5.000 \times 10^{2} \text{ N/(m·kg)}$$

Thus the period is

$$T = \frac{2\pi}{\sqrt{\alpha}} = \frac{2\pi}{\sqrt{5.000 \times 10^{2} \text{ N/(m·kg)}}} = 0.2810 \text{ s}$$ ■

c. Determine the frequency of the oscillation.

■ Since $\nu = 1/T$, you have

$$\nu = \frac{1}{0.2810 \text{ s}} = 3.559 \text{ cycles/s} = 3.559 \text{ Hz}$$

(Since a cycle is simply a count, it is a dimensionless number and can be inserted at will in the units of the answer.) ■

d. Determine the angular frequency of the oscillation.

■ Applying Eq. (6-26), you obtain

$$\omega = 2\pi\nu = 2\pi \text{ rad/cycle} \times 3.559 \text{ cycles/s} = 22.36 \text{ rad/s}$$ ■

e. Determine the position of the object 0.7500 s after it begins its oscillation.

■ The oscillation is described by Eq. (6-17),

$$x = A\cos(\omega t + \delta)$$

if A and δ are adjusted to fit the initial conditions and if the proper value of ω is used. As you have seen before from Eqs. (6-21a) and (6-24), the initial conditions $(dx/dt)_0 = 0$ and $x_0 = 1.000 \times 10^{-2}$ m require that

$$\delta = 0 \qquad \text{and} \qquad A = 1.000 \times 10^{-2} \text{ m}$$

So

$$x = 1.000 \times 10^{-2} \text{ m} \times \cos(22.36 \text{ rad/s} \times t)$$

where the value of ω is that found in part d. The position at $t = 0.7500$ s is

$$\begin{aligned} x &= 1.000 \times 10^{-2} \text{ m} \times \cos(22.36 \text{ rad/s} \times 0.7500 \text{ s}) \\ &= 1.000 \times 10^{-2} \text{ m} \times \cos(16.77 \text{ rad}) \\ &= 1.000 \times 10^{-2} \text{ m} \times (-0.4871) = -4.871 \times 10^{-3} \text{ m} = -0.4871 \text{ cm} \end{aligned}$$

The argument of the cosine is larger than 2π rad because the oscillation has passed beyond its first cycle. The minus sign means that the body is to the left of its equilibrium position, and the spring is compressed. Note that for the purpose of evaluating x it is most convenient to express the cosine in terms of the angular frequency ω rather than the frequency ν. ■

f. Determine the velocity of the object at $t = 0.7500$ s.

■ Evaluating the terms in Eq. (6-21a), you have

$$\begin{aligned}\frac{dx}{dt} &= -A\omega \sin(\omega t)\\ &= -1.000 \times 10^{-2}\text{ m} \times 22.36\text{ rad/s} \times \sin(16.77\text{ rad})\\ &= -1.000 \times 10^{-2}\text{ m} \times 22.36\text{ rad/s} \times (-0.8733)\\ &= 0.1953\text{ m/s} = 19.53\text{ cm/s}\end{aligned}$$

(A radian is a dimensionless number, being defined as the ratio of two lengths, so it can be deleted at will from the units.) The positive velocity means the body is moving to the right. Thus although the spring is compressed, the compression is decreasing. ■

g. Determine the acceleration of the object at $t = 0.7500$ s.

■ From Eq. (6-21*b*) the acceleration is

$$\begin{aligned}\frac{d^2x}{dt^2} &= -A\omega^2 \cos(\omega t)\\ &= -1.000 \times 10^{-2}\text{ m} \times (22.36\text{ rad/s})^2 \times \cos(16.77\text{ rad})\\ &= -1.000 \times 10^{-2}\text{ m} \times (22.36\text{ rad/s})^2 \times (-0.4871)\\ &= 2.435\text{ m/s}^2 = 243\text{ cm/s}^2\end{aligned}$$

The body is accelerating to the right since the acceleration is positive. ■

h. Determine the force exerted on the body by the spring at $t = 0.7500$ s.

■ One way you can do this is to apply Newton's second law:

$$\begin{aligned}F &= m\frac{d^2x}{dt^2}\\ &= 4.000\text{ kg} \times 2.435\text{ m/s}^2\\ &= 9.740\text{ N}\end{aligned}$$

where the value of d^2x/dt^2 is that found in part *g*. The force has a positive value, and so acts to the right since the acceleration is to the right.

Another way of finding F is to apply Hooke's law·

$$\begin{aligned}F &= -kx\\ &= -2.000 \times 10^3\text{ N/m} \times (-4.871 \times 10^{-3}\text{ m})\\ &= 9.740\text{ N}\end{aligned}$$

where the values of k and x are those found in parts *a* and *e*. The force acts to the right since the spring is compressed when the body is to the left of its equilibrium position. The agreement between these two evaluations of F provides a good check on the numerical work of this example.

Figure 6-17 shows the quantitative time dependence of the position, velocity, and acceleration of the harmonic oscillator treated in Example 6-6. The principal features of these three interrelated curves were discussed before in Sec. 2-6, using the similar curves of Fig. 2-21. But it is worthwhile to reemphasize their most important feature: the acceleration d^2x/dt^2 is always proportional in magnitude but opposite in sign to the position coordinate x. The physical reason for this is that *the acceleration is proportional to the force* (Newton's second law) and *the force is proportional to the*

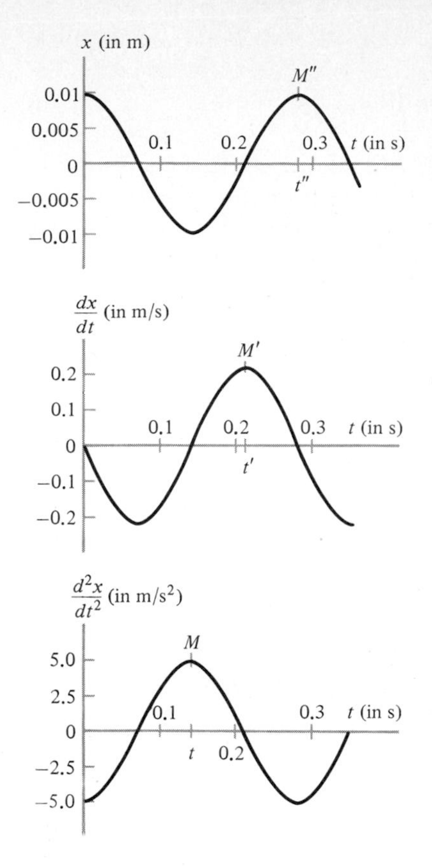

Fig. 6-17 The position x, velocity dx/dt, and acceleration d^2x/dt^2 of the harmonic oscillator treated in Example 6-6 are plotted as a function of the time t.

negative of the position coordinate (Hooke's law). This relation can be seen directly from the harmonic oscillator differential, $d^2x/dt^2 = -\alpha x$.

A related feature of the quantities x, dx/dt, and d^2x/dt^2 can be seen in Fig. 6-17. Consider any particular relative value on the acceleration curve, say the maximum M. If this value is attained at a time t, the velocity curve will not achieve its maximum value M' until a later time t'. From the figure you can see that $t' - t = T/4$, where T is the period. We say that the velocity *lags* the acceleration by one-quarter of a cycle. In like manner, the position lags the velocity by one-quarter cycle, as you can see by noting the location of the corresponding maximum M'' on the position curve. Since the acceleration is directly proportional to the force, we can say that there is a **phase lag** of one-half cycle, between the force acting on the body and the response of the body as expressed by its position. This is a result of the body's inertia.

We close the analytical treatment of the harmonic oscillator differential equation by discussing briefly the bases of the two most important properties of its solutions. These are the oscillatory nature of the solutions and their scaling property. It can be discerned from direct inspection of the equation, $d^2x/dt^2 = -\alpha x$, that its solutions must be *oscillatory functions* of t. As explained in Sec. 2-7, d^2x/dt^2 is a measure of the curvature of a plot of x versus t. If d^2x/dt^2 is positive, the curvature is concave upward, and if it is negative, the curvature is concave downward. Since the quantity α is positive, the differential equation says that the sign of the second derivative of x is always opposite to the sign of x itself. Thus if in some region of t the position curve traced out by plotting x versus t lies above the t axis, then x is positive, d^2x/dt^2 is negative, and the curve is concave downward. If the position curve lies below the t axis, the curve is concave upward. Simply put, the position curve is under all circumstances concave toward the t axis. This means that the position must oscillate about the axis, and so x is an oscillatory function of t.

The *scaling property* of the solutions to the harmonic oscillator equation is a direct consequence of the *linearity* of the differential equation. The equation, $d^2x/dt^2 = -\alpha x$, is said to be linear since each of its terms is proportional to the first power of the dependent variable x. As a consequence, if a certain dependence of x on t satisfies the harmonic oscillator equation, then that dependence, with x scaled up or down by any constant factor C, will also be a solution. In other words, if $x(t)$ satisfies the equation, then $Cx(t)$ also satisfies the equation, as we will now prove. Given that $x(t)$ is a solution to the harmonic oscillator equation, we test whether $Cx(t)$ is also a solution by substituting it into the equation, obtaining

$$\frac{d^2[Cx(t)]}{dt^2} = -\alpha[Cx(t)]$$

Since C is a constant, the derivative can be simplified to yield

$$C\,\frac{d^2[x(t)]}{dt^2} = -\alpha Cx(t)$$

Now divide through by C (for any solution of interest, $C \neq 0$). The result is

$$\frac{d^2[x(t)]}{dt^2} = -\alpha x(t)$$

Is this result mathematically consistent? Yes it is. Since $x(t)$ is a solution to the harmonic oscillator equation, the left side always equals the right side. So the equation from which we obtained the result is valid, and we have confirmed that $Cx(t)$ is also a solution to the harmonic oscillator equation.

The basic properties of the solutions to the pendulum differential equation,

$$\frac{d^2\phi}{dt^2} = -\alpha \sin \phi$$

can be understood from considerations similar to those used in discussing the harmonic oscillator differential equation. The sign of $\sin \phi$ is the same as the sign of ϕ, within the range $-\pi \leq \phi \leq \pi$ that ϕ assumes for a pendulum. So the solutions to the pendulum equation must also be oscillatory functions of t. But the scaling property does not hold for these solutions because one of the terms in the equation is not proportional to the first power of ϕ. Rather it is proportional to $\sin \phi$, and so the equation is nonlinear. The solutions are not sinusoidal, or harmonic, functions. A pendulum is consequently said to be an **anharmonic oscillator.**

The pendulum equation does not have analytical solutions which can be expressed by a finite number of terms (in contrast to an infinite series), each of which involves only elementary functions (the ones you find on the keys of a scientific calculator). The differential equation is best dealt with by numerical methods. As a rule of thumb, you can expect that the same is true of most nonlinear differential equations. And there are many nonlinear differential equations which can be solved only by numerical methods. There are also some lucky exceptions, like the nonlinear differential equation for the skydiver, Eq. (6-13). It has an analytical solution, quoted in Eq. (5-41), involving a small number of elementary functions. It should also be said that there are many linear differential equations which can be solved only by numerical methods.

6-6 THE DAMPED OSCILLATOR

Although we have ignored the fact until now, real macroscopic oscillators always experience at least a small frictional force that tends to dissipate or, as it is said, **damp** their motion. In certain situations the frictional force is large, and introduced on purpose. Think of the springing system of an automobile. If there were only a small frictional **damping force,** a single bump in the road would set the automobile into a vertical oscillation that would persist long enough to make its occupants quite uncomfortable. This is prevented by the shock absorbers, which are designed to produce a frictional force that damps out the oscillation as rapidly as possible. In this section we will study the motion of a **damped oscillator** by using Newton's second law to set up the differential equation governing its motion and then solving the equation. Both numerical and analytical methods will be employed.

Consider an oscillator in which the force producing the oscillation obeys Hooke's law and in which the oscillating body experiences fluid friction. Since oscillating bodies very often are not large enough, or fail to move rapidly enough, to produce much turbulence, we will assume that the damping force of fluid friction has a magnitude proportional to the first power of the speed of the body, as in Stokes' law, Eq. (4-26). The case in which the magnitude of the damping force is proportional to the first power of the speed is of special interest for another reason. The differential equation governing the motion of the body has exactly the same mathematical form as an important one that will arise when we study alternating current electric circuits in Chap. 26. So everything that we learn here about the properties of its solutions will carry over directly to the study of electric circuits.

To obtain the differential equation governing the motion of the damped oscillator, we write Newton's second law of motion

$$\frac{d^2x}{dt^2} = \frac{F}{m}$$

and then evaluate the net force F acting on the body. Here the net force is the algebraic sum of the Hooke's-law force, $-kx$, and the damping force. The magnitude of the damping force is proportional to the magnitude of the velocity dx/dt and is always directed opposite to the direction of the velocity. Thus the damping force can be written as $-r\ dx/dt$. The minus sign gives the damping force the proper direction. The constant r governs the magnitude of the damping force. The value of r depends on the size and shape of the oscillating body and on the coefficient of viscosity of the substance through which the body moves. Its numerical value is the strength of the damping force for unit speed. Thus the net force acting on the body is

$$F = -kx - r\frac{dx}{dt}$$

and Newton's second law yields

$$\frac{d^2x}{dt^2} = \frac{-kx - r\ dx/dt}{m} \tag{6-29}$$

We define the constants

$$\alpha \equiv \frac{k}{m} \qquad \text{and} \qquad \beta \equiv \frac{r}{m} \tag{6-30}$$

This allows us to express Eq. (6-29) as

$$\frac{d^2x}{dt^2} = -\alpha x - \beta\frac{dx}{dt} \tag{6-31}$$

This is the **damped oscillator equation.** We will write it in the form

$$\frac{d^2x}{dt^2} = Q \qquad \text{where } Q = -\alpha x - \beta\frac{dx}{dt} \tag{6-32}$$

Now that the differential equation governing the motion has been written in the form of Eq. (6-14), the numerical method for solving it according to Eqs. (6-15) can be applied immediately. Doing this involves nothing more than adding to the body-and-spring program several steps which will generate the value of Q specified in Eq. (6-32) and then running the calculator or computer with this new program. The damped oscillator program is listed in the Numerical Calculation Supplement. It is used in Examples 6-7 and 6-8.

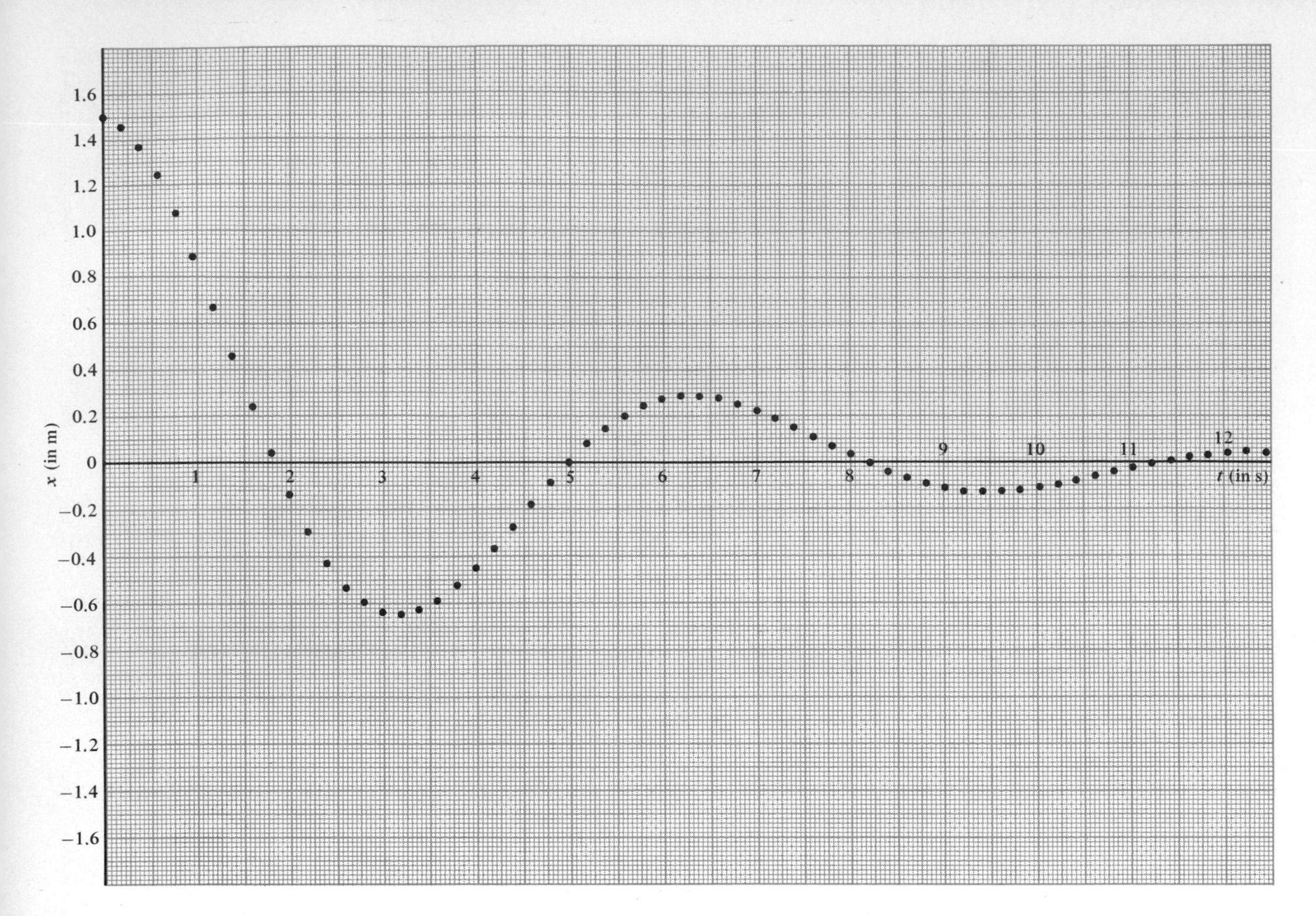

Fig. 6-18 A lightly damped oscillation. The parameters specifying the mechanical properties of the oscillator have the values $\alpha = 1\ \text{N/(m·kg)} = 1\ \text{s}^{-2}$ and $\beta = 0.5\ \text{N/(m·s}^{-1}\text{·kg)} = 0.5\ \text{s}^{-1}$.

EXAMPLE 6-7

Run the damped oscillator program with the following set of initial conditions and parameters:

$x_0 = 1.5$ (in m); $(dx/dt)_0 = 0$; $t_0 = 0$; $\Delta t = 0.2$ (in s); $\alpha = 1$ [in N/(m·kg)]; $\beta = 0.5$ [in $\text{N/(m·s}^{-1}\text{·kg)}$]

■ The oscillator has the same stiffness-to-mass ratio α and the same initial conditions x_0 and $(dx/dt)_0$ as the undamped oscillator treated in Example 6-2 and plotted in Fig. 6-6. The motion of the damped oscillator is plotted in Fig. 6-18. You can see that the motion is oscillatory, but with an amplitude which decreases, that is to say, "dies down," with each succeeding cycle. If you compare Figs. 6-18 and 6-6 carefully, you will see that the period of the damped oscillator is slightly longer than the period of the undamped oscillator with the same value of α. The damping inhibits the motion of the oscillator and makes it oscillate a bit less rapidly.

EXAMPLE 6-8

Run the damped oscillator program with the following set of initial conditions and parameters:

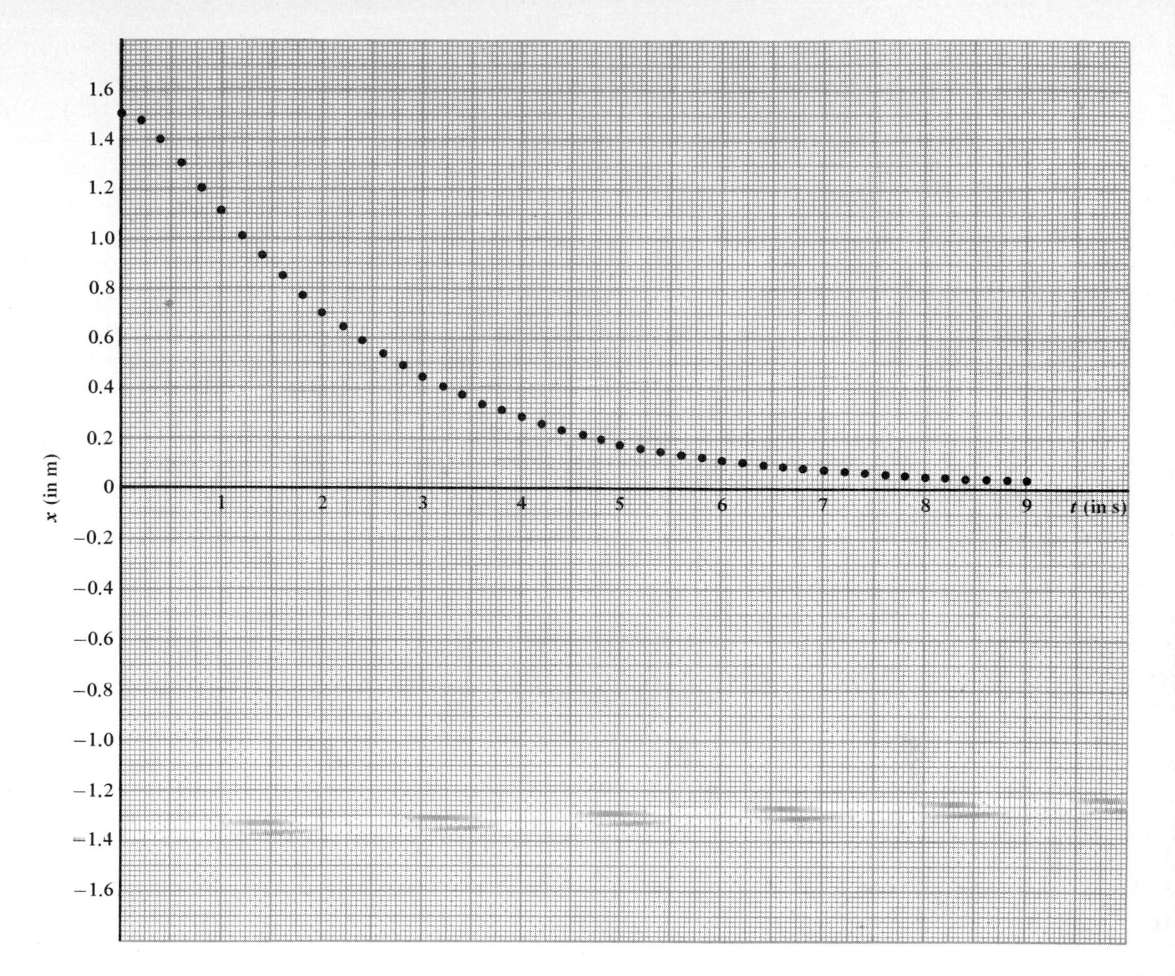

Fig. 6-19 A heavily damped oscillation. Here $\alpha = 1$ N/(m·kg), as in the lightly damped oscillation of Fig. 6-18. But the damping parameter has the larger value $\beta = 2.5$ N/(m·s^{-1}·kg) $=$ 2.5 s^{-1}.

$x_0 = 1.5$ (in m); $(dx/dt)_0 = 0$; $t_0 = 0$; $\Delta t = 0.2$ (in s); $\alpha = 1$ [in N/(m·kg)]; $\beta = 2.5$ [in N/(m·s^{-1}·kg)]

■ In this example the damping-constant-to-mass ratio β is 5 times larger than it was in Example 6-7. Nothing else is changed. The motion of the oscillator is plotted in Fig. 6-19. The motion is not oscillatory, and so a period cannot be defined. Because of the large value of β there is so much damping that the body never passes through the coordinate origin. Indeed, it appears as if its position x will reach zero only asymptotically.

The oscillatory motion in Fig. 6-18 is said to be **lightly damped,** or **underdamped,** because the oscillation persists through several cycles despite the damping. The motion shown in Fig. 6-19 is said to be **heavily damped,** or **overdamped,** since the damping has completely suppressed the oscillation. Does the behavior of a lightly or heavily damped oscillator, predicted from obtaining numerical solutions to the differential equation, agree with the behavior your physical intuition would lead you to expect?

Now we will try to obtain analytical solutions to Eq. (6-31), the differential equation for a damped oscillator. As usual in solving a differential equation, the first step is to guess a mathematical form for the analytical solutions. Since we have already obtained numerical solutions, we can use them to supplement our physical intuition and thereby enhance the chance that our guess will prove to be valid. One thing that Figs. 6-18 and 6-19 suggest immediately is that we treat light damping and heavy damping separately, as far as analytical solutions are concerned. Quite different mathematical expressions for the assumed form of the analytical solutions are suggested by the numerical solutions plotted in the figures for those two cases. We will focus our attention on getting solutions for *light damping* and then indicate briefly what is done to treat heavy damping.

Figure 6-18 also suggests that when the value of β is small enough so that the damping is light, the solutions to the differential equation can be written as the product of two functions. The first is an oscillatory function of time which describes the oscillation. The second is a function which decreases smoothly with increasing values of time and thus describes the diminution in amplitude resulting from damping. A reasonable guess at a form which has these properties is

$$x = Ae^{-\mu t}\cos(\omega t + \delta) \tag{6-33}$$

The factor $e^{-\mu t}$ is a decreasing *exponential function* of the quantity μt. The symbol μ (the Greek letter mu) is a positive constant called the **damping coefficient,** and t is the variable time. If you are not familiar with exponential functions or do not know how to evaluate their derivatives, you should read the following material set in small type.

The **exponential function** e^u of the variable u is a certain number e with the exponent u. For integral values of u, the function is just the uth power of e. But e^u is defined for all values of u, not just integral values, by means of an equation involving an infinite series. Specifically, the definition is

$$e^u \equiv 1 + u + \frac{u^2}{2} + \frac{u^3}{3 \times 2} + \frac{u^4}{4 \times 3 \times 2} + \frac{u^5}{5 \times 4 \times 3 \times 2} + \cdots \tag{6-34}$$

The motivation behind this definition can be seen by evaluating the derivative of e^u with respect to u. Taking the derivative with respect to u of the term on the left side of Eq. (6-34), and of every term on the right side, produces

$$\frac{d(e^u)}{du} = 0 + 1 + \frac{2u}{2} + \frac{3u^2}{3 \times 2} + \frac{4u^3}{4 \times 3 \times 2} + \frac{5u^4}{5 \times 4 \times 3 \times 2} + \cdots$$

Cancellation gives

$$\frac{d(e^u)}{du} = 1 + u + \frac{u^2}{2} + \frac{u^3}{3 \times 2} + \frac{u^4}{4 \times 3 \times 2} + \cdots$$

Comparison of the right side of this equation with the right side of Eq. (6-34) shows that

$$\frac{d(e^u)}{du} = e^u \tag{6-35}$$

Thus the function e^u defined by Eq. (6-34) has the property that *it equals its own first derivative* with respect to u. No other function of u has this very useful property.

The value of the number e is obtained from the definition of Eq. (6-34) by setting u = 1. This yields

$$e = 1 + 1 + \frac{1}{2} + \frac{1}{3 \times 2} + \frac{1}{4 \times 3 \times 2} + \frac{1}{5 \times 4 \times 3 \times 2} + \cdots$$

The series converges quite rapidly. With a calculator it will be easy for you to evaluate and sum the first few of its terms. Doing this so as to obtain results to five-decimal-place accuracy, you will find that

$$e = 2.71828 \tag{6-36}$$

Thus e^u has the value 2.71828 for $u = 1$. The value of e^u for any other value of u is obtained by summing the series in Eq. (6-34) for that value of u. Results for a limited range of positive and negative values of u are plotted in Fig. 6-20a and b. Accurate tables of values of e^u are available for much wider ranges of u. Using almost any calculator intended for scientific work, you can, in effect, sum the series that evaluates the exponential function of any number entered in the calculator by pressing only one or two keys.

With values obtained from Fig. 6-20 or a calculator, you can verify that

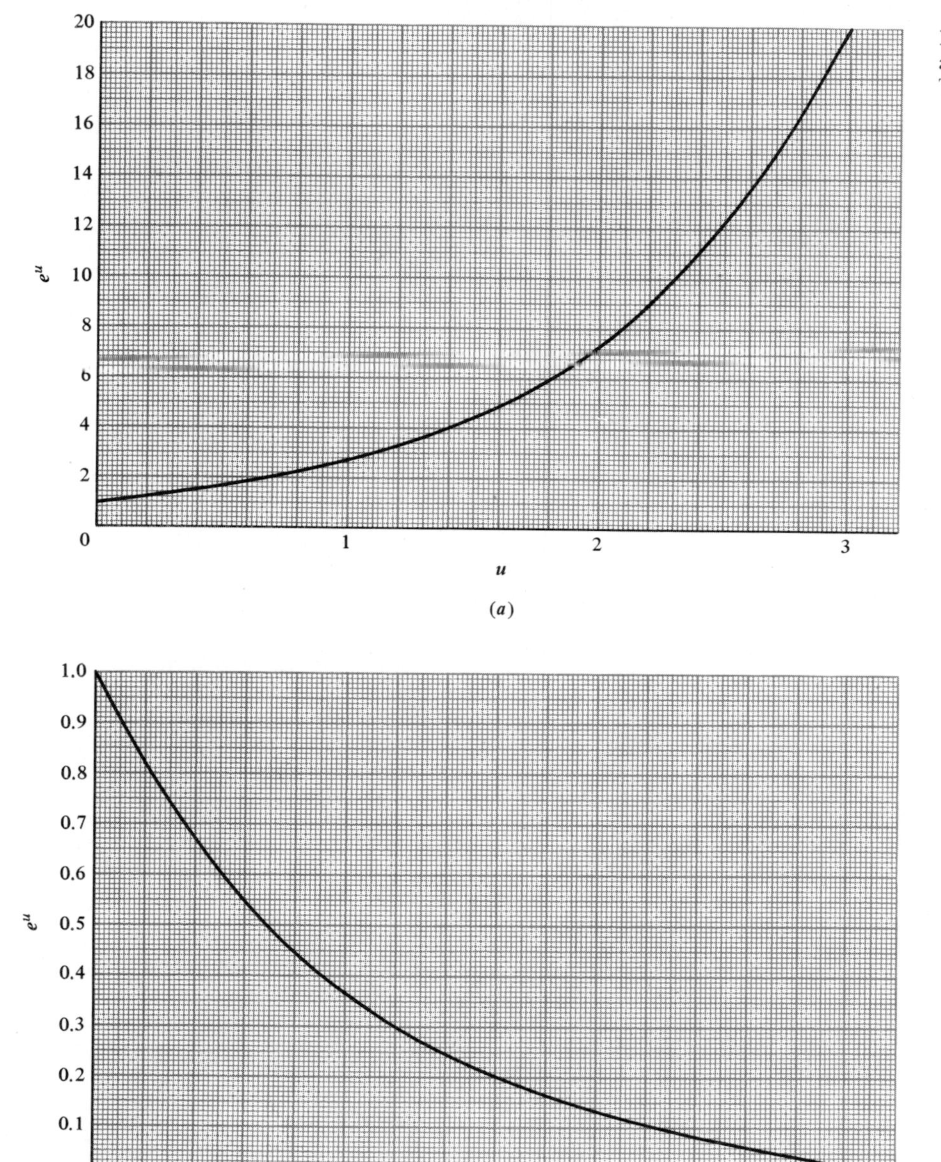

Fig. 6-20 (a) The function e^u plotted for a limited range of positive values of u. (b) The same for negative values of u.

$e^{u_1}e^{u_2} = e^{u_1+u_2}$ and that $e^{-u} = 1/e^u$. Thus the definition of e^u given by Eq. (6-34) conforms to the familiar properties of exponents.

For the exponential function occurring in Eq. (6-33), the exponent u is negative, and its magnitude is written as the product of a positive constant μ and the variable t, which is always positive since its initial value is $t_0 = 0$. That is $u = -\mu t$. To verify that the form given by Eq. (6-33) is a solution to the differential equation for a lightly damped oscillator, it will be necessary to differentiate the exponential function with respect to t alone. This is done by using the chain rule, Eq. (6-19). If we set $f(u) = e^u$, the rule says that

$$\frac{d(e^u)}{dt} = \frac{d(e^u)}{du}\frac{du}{dt}$$

According to Eq. (6-35), the first term on the right side has the value

$$\frac{d(e^u)}{du} = e^u$$

For $u = -\mu t$, with μ a constant, the second term gives

$$\frac{du}{dt} = -\mu$$

Therefore

$$\frac{d(e^{-\mu t})}{dt} = -\mu e^{-\mu t} \tag{6-37}$$

Now we return to considering Eq. (6-33).

In Eq. (6-33), the value of the exponential factor $e^{-\mu t}$ decreases smoothly from 1 with increasing values of t, the decrease being more or less rapid depending on the size of the damping-to-mass ratio μ. The behavior of this factor is indicated in the top part of Fig. 6-21 for a typical small value μ. The middle part of that figure illustrates the factor $\cos(\omega t + \delta)$ for typi-

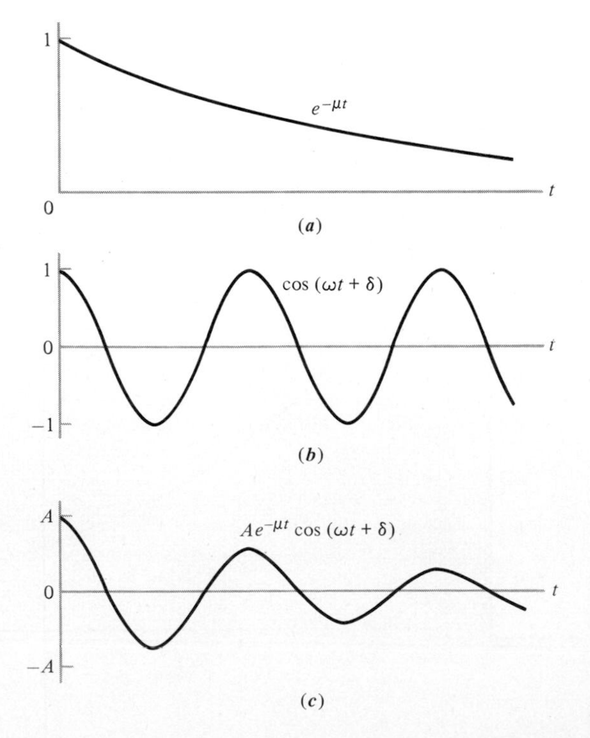

Fig. 6-21 Illustrating an argument justifying a form assumed for the analytical solution to the differential equation for a lightly damped oscillator.

cal values of the constants ω and δ, and the lower part shows the product $Ae^{-\mu t}\cos(\omega t + \delta)$ for some value of the amplitude A. Comparison of the lower part of the figure with Fig. 6-18 gives justification to the guess that Eq. (6-33),

$$x = Ae^{-\mu t}\cos(\omega t + \delta)$$

might represent a family of analytical solutions to Eq. (6-31),

$$\frac{d^2x}{dt^2} = -\alpha x - \beta\frac{dx}{dt}$$

This is the differential equation which produced the numerical solution shown in Fig. 6-18. To find out whether Eq. (6-33) really does describe the solutions, we will use it to evaluate dx/dt and d^2x/dt^2 from the form it specifies for x. Then we will substitute these derivatives and x into Eq. (6-31) and see whether they satisfy the differential equation. You will see that while the process is somewhat messy, it is quite straightforward. It is worthwhile to carry out the process not only to verify the correctness of the solutions, but also because it leads to values of the constants ω (which gives the angular frequency of the oscillator) and μ (which gives its rate of damping) in terms of the mechanical constants α and β. Compare this with the procedure used to solve the harmonic oscillator equation analytically, which led to the relation $\omega = \sqrt{\alpha}$.

First we must evaluate

$$\frac{dx}{dt} = \frac{d}{dt}[Ae^{-\mu t}\cos(\omega t + \delta)]$$

Since A is a constant, this is

$$\frac{dx}{dt} = A\frac{d}{dt}[e^{-\mu t}\cos(\omega t + \delta)]$$

According to Eq. (2-15), the derivative of the product of two functions equals the first function times the derivative of the second function plus the second function times the derivative of the first function. Using this property and employing Eq. (6-20) to differentiate the cosine and Eq. (6-37) to differentiate the exponential, we obtain directly

$$\frac{dx}{dt} = -A\omega e^{-\mu t}\sin(\omega t + \delta) - A\mu e^{-\mu t}\cos(\omega t + \delta) \qquad (6\text{-}38a)$$

Next we use this result to evaluate

$$\frac{d^2x}{dt^2} = \frac{d}{dt}\left(\frac{dx}{dt}\right) = \frac{d}{dt}[-A\omega e^{-\mu t}\sin(\omega t + \delta) - A\mu e^{-\mu t}\cos(\omega t + \delta)]$$

The computation is very similar to the one which evaluated dx/dt from x. But now there are two terms to differentiate, so there are four terms in the result. The result that we obtain is

$$\frac{d^2x}{dt^2} = -A\omega^2 e^{-\mu t}\cos(\omega t + \delta) + A\mu\omega e^{-\mu t}\sin(\omega t + \delta)$$
$$+ A\mu\omega e^{-\mu t}\sin(\omega t + \delta) + A\mu^2 e^{-\mu t}\cos(\omega t + \delta)$$

Since the second and third terms on the right side are identical, this simplifies to

$$\frac{d^2x}{dt^2} = A(\mu^2 - \omega^2)e^{-\mu t}\cos(\omega t + \delta) + 2A\mu\omega e^{-\mu t}\sin(\omega t + \delta) \quad \text{(6-38}b\text{)}$$

Now we test the validity of the assumed form for x by substituting it, and the expressions for dx/dt and d^2x/dt^2 obtained from it, into the differential equation that x is supposed to satisfy, Eq. (6-31). This produces the equation

$$A(\mu^2 - \omega^2)e^{-\mu t}\cos(\omega t + \delta) + 2A\mu\omega e^{-\mu t}\sin(\omega t + \delta) =$$
$$-\alpha A e^{-\mu t}\cos(\omega t + \delta) + \beta A\omega e^{-\mu t}\sin(\omega t + \delta) + \beta A\mu e^{-\mu t}\cos(\omega t + \delta)$$

While this equation is lengthy, each term is either $e^{-\mu t}\cos(\omega t + \delta)$ or $e^{-\mu t}\sin(\omega t + \delta)$ multiplied by a combination of constants. Gathering the coefficients of the cosine terms, and then of the sine terms, and transposing, we have

$$A(\mu^2 - \omega^2 + \alpha - \beta\mu)e^{-\mu t}\cos(\omega t + \delta)$$
$$+ A(2\mu\omega - \beta\omega)e^{-\mu t}\sin(\omega t + \delta) = 0 \quad \text{(6-39)}$$

If the form for x given by Eq. (6-33) is, in fact, a solution to the differential equation given by Eq. (6-31), then Eq. (6-39) must be satisfied. That is, the sum of the first and second terms of Eq. (6-39) must be zero at *all* times. But the first term in Eq. (6-39) has a dependence on t which is different from that of the second term. Therefore the *only* way that the sum of the two terms can equal zero for *all* values of t is for each of the two terms individually to be equal to zero. Thus for Eq. (6-39) to be mathematically consistent, two equations must be satisfied. The first is

$$A(\mu^2 - \omega^2 + \alpha - \beta\mu)e^{-\mu t}\cos(\omega t + \delta) = 0 \quad \text{(6-40}a\text{)}$$

and the second is

$$A(2\mu\omega - \beta\omega)e^{-\mu t}\sin(\omega t + \delta) = 0 \quad \text{(6-40}b\text{)}$$

We can achieve satisfaction of Eqs. (6-40*a*) and (6-40*b*) by properly adjusting the values of the constants A, μ, ω, and δ. One way is to set $A = 0$. But doing so makes Eq. (6-33) yield the solution $x = 0$ of the differential equation. That solution is correct but uninteresting, because it describes the body remaining fixed at its stable equilibrium position.

The other way of satisfying Eqs. (6-40*a*) and (6-40*b*) is to adjust the values of μ and ω so that two relations hold. These are

$$\mu^2 - \omega^2 + \alpha - \beta\mu = 0$$

and

$$2\mu\omega - \beta\omega = 0$$

The second of these relations immediately gives the value of μ, which determines the rate of damping. It is

$$\mu = \frac{\beta}{2} \quad \text{(6-41)}$$

Substituting this value of μ into the first relation, we have

$$\frac{\beta^2}{4} - \omega^2 + \alpha - \frac{\beta^2}{2} = 0$$

This can now be solved for ω^2 to give

$$\omega^2 = \alpha - \frac{\beta^2}{4}$$

and we find the angular frequency of the lightly damped oscillator to be

$$\omega = \sqrt{\alpha - \frac{\beta^2}{4}} \tag{6-42}$$

Substitution into the differential equation, Eq. (6-31), of the assumed form of its solutions, Eq. (6-33), has shown that this form is correct, *providing* that Eqs. (6-41) and (6-42) are obeyed. In other words, *if* the constants μ and ω which specify the motion of the damped oscillator are related by Eqs. (6-41) and (6-42) to the constants α and β that specify its mechanical properties, *then* the form we have assumed for the solutions is correct. Using these values of μ and ω in Eq. (6-33), we therefore have the solutions describing the motion. They are:

For lightly damped motion

$$x = Ae^{-\beta t/2} \cos\left(\sqrt{\alpha - \frac{\beta^2}{4}}\, t + \delta\right) \qquad \text{for} \quad \frac{\beta^2}{4} < \alpha \tag{6-43}$$

As in the case of the solutions to the undamped harmonic oscillator, the constants A and δ remain to be adjusted so that Eq. (6-43) can be fitted to any pair of initial conditions, x_0 and $(dx/dt)_0$. This adjustment of the amplitude and the phase constant produces a solution which describes the motion particular to the initial conditions.

When the damping-to-mass ratio β is zero, Eq. (6-43) reduces, as it should, to Eq. (6-24) describing the solutions for the undamped harmonic oscillator. As β increases from zero, the coefficient of t in the decreasing exponential becomes larger. This damps the oscillation. Also the coefficient of t in the sinusoid becomes smaller. This reduces the oscillation frequency and thereby increases the oscillation period. Because of the form of the dependence of the frequency on β, there is very little change from the frequency of an undamped oscillator until β^2 becomes large enough compared to α to approach the condition

$$\frac{\beta^2}{4} = \alpha \tag{6-44}$$

For instance, with the values used in the numerical solution of Example 6-7, $\alpha = 1.00$ N/(m·kg) and $\beta = 0.50$ N/(m·s^{-1}·kg), the angular frequency of the damped oscillator is $\omega = \sqrt{\alpha - \beta^2/4} = 0.97$ s^{-1}. This may be compared to the value $\omega = \sqrt{\alpha} = 1.00$ s^{-1} that would be obtained for the same α if there were no damping, so that $\beta = 0$. As mentioned earlier, the physical reason why increased damping leads to a decreased frequency is that damping inhibits the motion of the oscillator, thereby making it oscillate less rapidly.

When β is large enough in comparison to α that the condition of Eq. (6-44) is satisfied, the angular frequency ω calculated from Eq. (6-42) is zero. When it is even larger, the calculated value of ω is the square root of a negative number; that is, ω is an "imaginary" number. It is apparent that there is something special about the condition specified by Eq. (6-44). In fact, this is the condition for a **critically damped** "oscillation." At critical damping the motion is not oscillatory at all. Nor is it simply a decreasing exponential. In other words, the pure decreasing exponential that Eq. (6-43) becomes when $\beta^2/4 = \alpha$ does *not* correctly describe solutions to the differential equation for any possible pair of initial conditions x_0 and $(dx/dt)_0$. You can see this physically by visualizing what the motion is like when a critically damped oscillator is released with $x_0 \neq 0$ and $(dx/dt)_0 = 0$. Its motion at the very first looks like the beginning of a cosine curve, as in the beginning part of Figs. 6-18 and 6-19. This is true, independent of the degree of damping, since while the velocity is very low (as it will be at the beginning), the damping is not effective in any case. This is not the behavior exhibited by a decreasing exponential, such as the one plotted in the top part of Fig. 6-21.

A separate investigation must be made of the analytical solutions to the differential equation for critically damped motion, and it must be done again for heavily damped motion, since these solutions are of quite different forms from the oscillatory form initially assumed in Eq. (6-33). This can be done, and the following solutions are obtained:

For critically damped motion

$$x = (A + Bt)e^{-\beta t/2} \qquad \text{for } \frac{\beta^2}{4} = \alpha \tag{6-45}$$

For heavily damped motion

$$x = (Ae^{\sqrt{\beta^2/4-\alpha}\,t} + Be^{-\sqrt{\beta^2/4-\alpha}\,t})e^{-\beta t/2} \qquad \text{for } \frac{\beta^2}{4} > \alpha \tag{6-46}$$

In both expressions, A and B are constants whose values can be adjusted to describe any particular motion.

In this section you have studied both numerical and analytical solutions to a differential equation typical of those that arise in physics. If you consider the procedures used and results obtained from the two methods, you will get a good idea of their comparative advantages and disadvantages, when applied to an equation that can be solved by either method. As you proceed through this book, you will find both methods employed from time to time, but usually not on the same equation. We will solve a differential equation analytically, if it has an analytical solution, *and* if it is not too difficult to obtain the analytical solution. In practice, this means that we will solve most differential equations numerically.

EXAMPLE 6-9

You are an engineer working on the design of an automobile springing system. With springs but no shock absorbers installed on the preliminary model, you give the automobile body an initial downward push to set it into vertical oscillation. You observe that the oscillation persists for a large number of cycles and has a period of $T = 0.72$ s. Next you install trial shock absorbers, give the body another push, and find that discernible oscillation still persists for several cycles—enough to allow you to determine that the period is now $T = 0.81$ s. Thus the trial shocks are not adequate to produce the desirable condition of critical damping, at which the automobile recovers from the effect of a bump in the road as rapidly as possible without oscillating. You cannot afford to find the proper shocks by continuing to experiment by trial and error. How can you use your data to determine specifications for shocks which will render the springing system critically damped?

▪ Shock absorbers utilize fluid friction arising from a fluid passing through a number of narrow orifices connecting two chambers, as shown schematically in Fig.

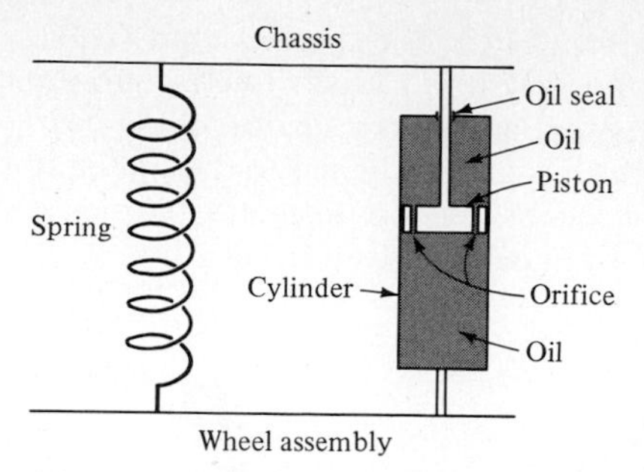

Fig. 6-22 A schematic drawing of a spring and shock absorber system.

6-22. The differential equation treated in this section applies quite accurately to an automobile springing system, since the damping is proportional to the first power of the speed of vertical motion. Thus you may use the relation $\omega = 2\pi/T$ to write Eq. (6-42) as

$$\omega = \frac{2\pi}{T} = \sqrt{\alpha - \frac{\beta^2}{4}} \tag{6-47}$$

The almost undamped oscillatory motion for the case without shocks implied that to a good approximation $\beta = 0$ for that case. Thus the value of α characteristic of the springing system and the mass of the automobile can be found from the value $T = 0.72$ s observed for the oscillation without shocks. Setting $\beta = 0$ in Eq. (6-47), solving for $\sqrt{\alpha}$ in terms of T, and then squaring, you find

$$\alpha = \left(\frac{2\pi}{T}\right)^2$$

Inserting the value of T gives you

$$\alpha = \left(\frac{2\pi}{0.72\ \text{s}}\right)^2 = 76\ \text{s}^{-2}$$

The effectiveness of the trial shocks is characterized by the value of the quantity β. It can be obtained by solving Eq. (6-47) for β:

$$\left(\frac{2\pi}{T}\right)^2 = \alpha - \frac{\beta^2}{4}$$

$$\frac{\beta^2}{4} = \alpha - \left(\frac{2\pi}{T}\right)^2$$

$$\beta = 2\sqrt{\alpha - \left(\frac{2\pi}{T}\right)^2}$$

Setting $\alpha = 76\ \text{s}^{-2}$ and $T = 0.81$ s, you find

$$\beta = 2\sqrt{76\ \text{s}^{-2} - \left(\frac{2\pi}{0.81\ \text{s}}\right)^2} = 8.0\ \text{s}^{-1}$$

This is the value of β for the trial shocks.

For critical damping, Eq. (6-42) or (6-44) shows you need

$$\frac{\beta^2}{4} = \alpha$$

or

$$\beta_{\text{crit}} = 2\sqrt{\alpha} = 2\sqrt{76\ \text{s}^{-2}} = 17.4\ \text{s}^{-1}$$

Thus to obtain critical damping, you must specify that the car be equipped with shocks which produce a damping force stronger by a factor of

$$\frac{\beta_{\text{crit}}}{\beta} = \frac{17.4\ \text{s}^{-1}}{8.0\ \text{s}^{-1}} = 2.2$$

than is produced by the trial shocks. This can be achieved by reducing the number of orifices by that factor.

EXERCISES

Group A

6-1. *An inertial method for determining mass.* A body of mass 1.00 kg is suspended from a spring. When the body is pulled down from its equilibrium position and then released, the resulting oscillations have a period of 2.00 s. When the 1.00-kg body is replaced by a body of unknown mass, the oscillations have a period of 1.00 s. Assuming that the mass of the spring is negligible, determine the mass of the unknown body.

6-2. *Body on a vertical spring.* A vertically suspended spring of negligible mass and force constant k is stretched by an amount l when a body of mass m is hung on it. The body is pulled by hand an additional distance y (positive direction downward) and then released. Show that the motion of the body is governed by the equation $a = -ky/m$, so that the body executes harmonic motion about its equilibrium position.

6-3. *Bird on a spring.* A bird cage hung on a spring extends the spring by 10.0 cm. What is the frequency of oscillation of the cage about its equilibrium position?

6-4. *Spring versus pendulum.* A body hung from a spring stretches the spring a distance l.

a. What is the period of the oscillations that this system can exhibit?

b. Show that the period found in part a is the same as the period of small oscillations of a pendulum of length l.

6-5. *How much stretching?* A body hanging from a spring is set into motion, and the period of oscillation is found to be 0.50 s. After the body has come to rest, it is removed. When the spring comes to rest, how much shorter will it be?

6-6. *Springs on both sides.* Two anchored springs, each of force constant k, are attached to opposite sides of a block of mass m. The springs are initially unstretched. The block is displaced to the right and then released. There is negligible friction between the block and the supporting surface.

a. What is the period of oscillation of the block?

b. What would the period be if each spring were stretched by an amount s when the block was in equilibrium?

6-7. *Harmonic motion, I.* An object is executing harmonic motion with a frequency of 5.00 Hz. At $t = 0$ s its displacement is $x(0) = 10.0$ cm and its velocity is $v(0) = -314$ cm/s.

a. Use the given information to obtain an analytical expression for the object's displacement $x(t)$, velocity $v(t)$, acceleration $a(t)$.

b. If you have not already done so, express the displacement in the form $x(t) = A\cos(\omega t + \delta)$. That is, determine the values of A and δ which are appropriate for the given information.

c. Find the maximum value of the object's displacement $x(t)$, velocity $v(t)$, acceleration $a(t)$.

6-8. *Harmonic motion, II.* A body executes harmonic motion with an amplitude of 2.00 cm and a frequency of 3.00 Hz. At $t = 0$ s, the displacement is $x(0) = 0$ cm and the velocity $v(0)$ is positive.

a. Obtain an analytical expression for the displacement $x(t)$, the velocity $v(t)$, the acceleration $a(t)$.

b. Evaluate the expressions found in part a for $t = 5.00 \times 10^{-2}$ s.

6-9. *A harmonic shadow.* A circus bear rides a unicycle at constant speed y around a circle of radius r (see Fig. 6E-9). A distant ground-level spotlight casts a shadow of the bear onto a vertical wall which is perpendicular to the spotlight beam. Show that the bear's shadow executes harmonic motion with angular frequency ω given by $\omega = v/r$.

Fig. 6E-9

Group B

6-10. *A search for equilibrium.* The pulleys shown in Fig. 6E-10 are frictionless, the string is massless and inextensible, and $M = 10m$.

a. Do you expect that the body of mass m has an equilibrium position in which the string from A to B is not pulled into a straight line? Explain your answer.

b. If you believe that there is a position of equilibrium, is the equilibrium stable? Explain your answer.

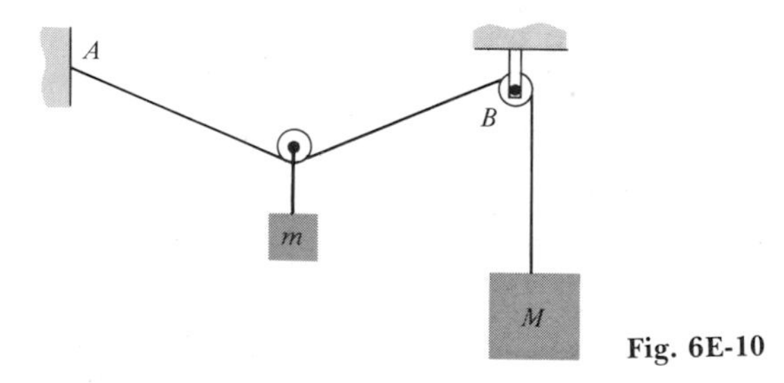

Fig. 6E-10

6-11. *Object on a board.* A massive object is placed on the middle of a thin board supported at both ends, causing the board to sag by a distance d at its midpoint. Assume that the board has negligible mass and that it exerts an upward force proportional to its sag. Find the period of oscillation of the object.

6-12. *Cut in two.* An object suspended from a spring exhibits oscillations of period T. Now the spring is cut in half, and the two halves are used to support the same object, as shown in Fig. 6E-12. Show that the new period of oscillation is $T/2$.

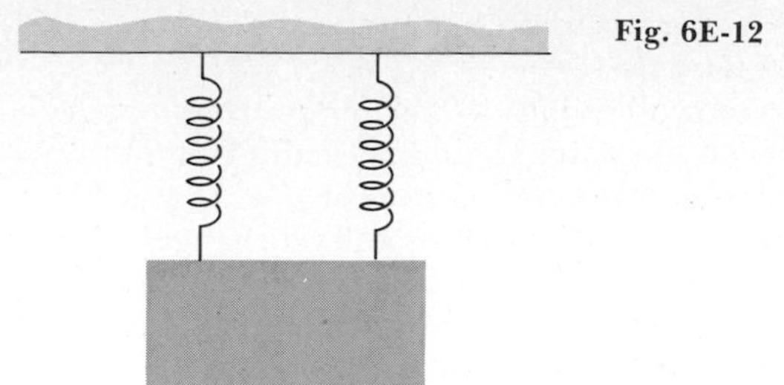
Fig. 6E-12

6-13. *A compound spring.* Springs A and B have spring constants of 2000 N/m and 1000 N/m, respectively. Spring A is hung from a rigid horizontal beam and its other end is attached to an end of spring B. The pair of springs is then used to suspend a body of mass 50 kg from the lower end of spring B. What is the period of harmonic oscillation of the system?

6-14. *Swinging around an obstacle.* A pendulum has a period T for small oscillations. An obstacle is placed directly beneath the pivot, so that only the lowest one-quarter of the string can follow the pendulum bob when it swings to the left of its resting position. The pendulum is released from rest at a certain point. How long will it take to return to that point? In answering this question, you may assume that the angle between the moving string and the vertical is a small angle throughout the motion.

6-15. *Population growth.* When nutrients are not a limiting factor, the number of bacteria produced in each generation of growth is proportional to the number present in that generation.

a. Show that the number N present at any time t satisfies the differential equation $dN/dt = RN$, where R is a positive constant.

b. Find an analytical solution to this differential equation.

6-16. *Radioactive decay.* In a sample of radioactive nuclei, the total number decaying per second is proportional to the number present.

a. Show that the number N of undecayed nuclei remaining at any time t satisfies the differential equation $dN/dt = -RN$, where R is a positive constant.

b. Find an analytical solution to the differential equation.

6-17. *Suspended.* As shown in Fig. 6E-10, a body of mass m is attached to the center of a string which is kept taut by the weight of a body of mass $M \gg m$. There is a length l of string between the fixed end A and the pulley B; the mass of the string itself is negligible. The body of mass m is pushed downward a distance $d \ll l$ and then released. Show that it executes a harmonic up-and-down motion whose period is $\pi\sqrt{ml/Mg}$.

6-18. *Body on a string on a spring.* A body of mass m is attached by a string to a suspended spring of spring constant k. Both the string and the spring have negligible mass, and the string is inextensible (it has a fixed length). The body is pulled down a distance A and then released.

a. Assuming that the string remains taut throughout the motion, find the maximum (downward) acceleration of the oscillating body.

b. The string will remain taut only as long as it remains under tension. Determine the largest amplitude A_{max} for which the string will remain taut throughout the motion.

c. Evaluate A_{max} for $m = 0.10$ kg and $k = 10$ N/m.

6-19. *Hold on!* A massive block resting on a table top is attached to an anchored horizontal spring. There is negligible friction between the massive block and the tabletop. The oscillation frequency is ν. A much less massive block is placed on top of the large block. The coefficient of static friction between the two blocks is μ_s.

a. What is the largest amplitude of oscillation of the large block which permits the small block to ride without slipping?

b. Evaluate your result for $\nu = 3.0$ Hz and $\mu_s = 0.60$.

6-20. *The effects of damping.* The parameter α for a damped oscillator has the value 1.00 s^{-2}. At $t = 0$ s, the oscillator is set into motion with $x_0 = 0$ m and $v_0 = 5.00$ m/s. Use the appropriate equation to write an analytical expression for $x(t)$ in each of the cases listed.

a. No damping, $\beta = 0$ s^{-1}

b. Light damping, $\beta = 0.10$ s^{-1}, 1.00 s^{-1}

c. Critical damping, $\beta = 2.00$ s^{-1}

d. Heavy damping, $\beta = 3.00$ s^{-1}, 10.0 s^{-1}

e. Use the expressions obtained in parts *a* to *d* to find the maximum value of $x(t)$ attained in each case. Express each result as a fraction of the undamped amplitude $v_0/\sqrt{\alpha} = 5.00$ m.

Group C

6-21. *Two bodies, joined by a spring.* Body 1 and body 2 of masses m_1 and m_2 respectively are resting on a frictionless horizontal surface. They are joined by a spring of spring constant k. Initially the spring is relaxed. The two bodies are pushed closer together, compressing the spring by an amount A, and then they are simultaneously released from rest.

a. Determine the subsequent motion of the system, including the period of the oscillations, the relative velocities of bodies 1 and 2, and the amplitudes of motion of bodies 1 and 2.

b. Show that your results take the correct form for $m_1 \gg m_2$ and for $m_1 \ll m_2$.

c. Evaluate your results for $m_1 = m_2 = m$. Why is the period shorter than $2\pi\sqrt{m/k}$?

6-22. *Piggyback.* A block of mass M_1 resting on a frictionless horizontal surface is connected to a spring of spring constant k that is anchored in a nearby wall. A block of mass $M_2 = \alpha M_1$ is placed on top of the first block. The coefficient of static friction between the two bodies is μ_s.

a. Assuming that the two bodies move as a unit, find the period of oscillation of the system.

b. What is the maximum oscillation amplitude A_{max} that permits the two bodies to move as a unit?

c. Evaluate your results of parts *a* and *b* for $k = 6.0$ N/m, $M_1 = 1.0$ kg, $\alpha = 0.50$, and $\mu_s = 0.40$.

d. Describe qualitatively what happens if the two bodies are released from rest at a coordinate value somewhat greater than A_{max}.

6-23. *Time flies.* A recording pendulum clock is installed in a rocket, which is then fired vertically with a constant acceleration $a = 3.0g$ for 15 s, when burnout occurs. The rocket continues upward with diminishing speed until its velocity is zero. It then reverses its path, falling freely to the ground. It is possible to neglect the variation of the earth's gravity with altitude, and air resistance is negligible throughout the flight. Although the clock is smashed upon impact, the record is recovered.

a. What is the actual duration of the flight?

b. What duration is indicated by the record?

6-24. *A springy spherical pendulum.* As shown in Fig. 6E-24, a body of mass m is suspended from a frictionless pivot by a massless spring of relaxed length l_0 and spring constant k. Determine the possible horizontal circular motions of the body.

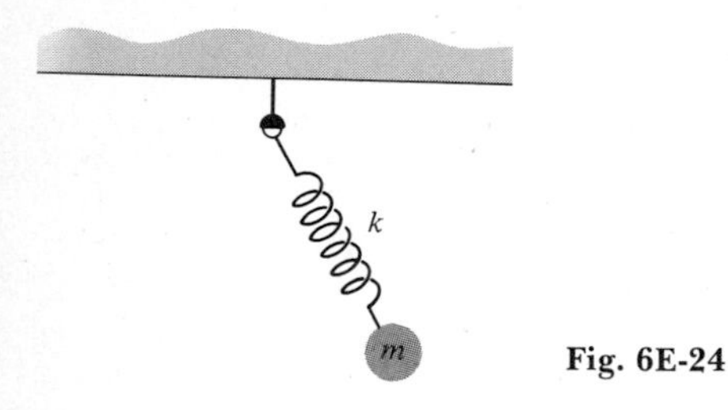

Fig. 6E-24

6-25. *Oscillations of a water column.* A U-tube of uniform cross section A contains a length l of water. Initially the water is in equilibrium, as shown in Fig. 6E-25*a*. Then the tube is tilted to the left until the water attains a new equilibrium. Finally it is turned upright very quickly. As a result, the water level on the right is a distance y below its equilibrium level, as shown in Fig. 6E-25*b*. At that instant the fluid is motionless, but evidently it will not remain so.

a. What is the difference D in water level between the left and right sides of the tube, for the situation of Fig. 6E-25*b*?

Fig. 6E-25

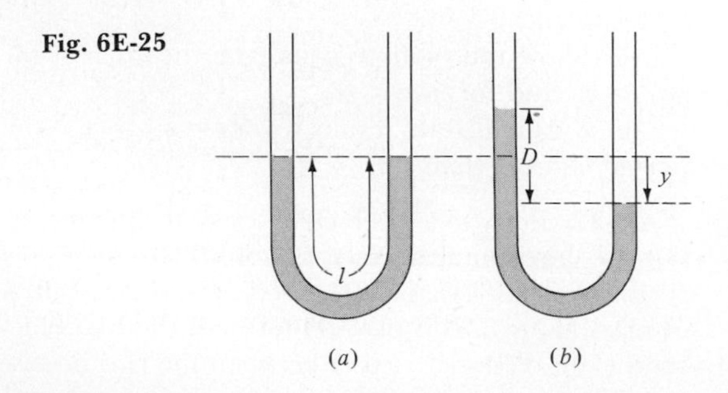

b. The water has uniform density ρ. What is the total mass M_w in the tube?

c. It is shown in Chap. 16 that a restoring force $F = -\rho g A D$ acts on the water column. Use this fact and your result for part *a* to write F in terms of ρ, g, A, and y.

d. It we assume that when the level changes, *all* the water moves with speed $|dy/dt|$, then Newton's law takes the form $M_w\, d^2y/dt^2 = F$. Use this to determine the period of oscillation of the water column.

6-26. *Buoy, oh buoy!* A rectangular block of wood floating in a large pool of water. It is shown in Chap. 16 that the water exerts an upward force on the bottom face of the block whose strength is $Ad\rho g$, where A is the area of that face, d is its depth beneath the surface of the water, ρ is the density of water, and g is gravitational acceleration. (This is *Archimedes' principle:* The loss of weight of a body immersed partly or completely in a liquid equals the weight of the liquid displaced by the body.)

a. The mass of the block is m. Find the value of d for which the block is in equilibrium.

b. Find an expression that describes the net force acting on the block for values of d that differ from the equilibrium value. Use it to show that the equilibrium is stable.

c. Show that if the block is depressed below its equilibrium depth (but not beneath the surface of the water) and then released, it will execute harmonic oscillations.

d. Determine the frequency of the oscillations

6-27. *Critical Damping, I.* Prove by substitution that the analytical form given in Eq. (6-45) is a solution to the damped oscillator equation, Eq. (6-31), for the case of critically damped motion.

6-28. *Heavy Damping.* Prove by substitution that the analytical form given in Eq. (6-46) is a solution to the damped oscillator equation, Eq. (6-31), for the case of heavily damped motion.

Numerical

6-29. *Body-and-spring program, I.*

a. Use the analytical result in Eq. (6-17), with the initial conditions and parameters of Example 6-1, and evaluate to five decimal places the position x of the body at the end of the spring when $t = 4$ s.

b. Run the body-and-spring program as in Example 6-1 and obtain x to five decimal places.

c. Compare the results of parts *a* and *b* to determine the error in the numerical method.

6-30. *Body-and-spring program, II.*

a. Follow the procedure of Exercise 6-29, using a reduced time increment $\Delta t = 0.1$ s in the body-and-spring program. Determine the error in the numerical method for this value of Δt.

b. Compare with the error obtained in Exercise 6-29. What relation do you find between the error in the numerical method for treating the body and spring and the size of Δt? In Exercise 5-43 the error in the numerical method for treating the skydiver was found to be inversely proportional to Δt. If this is not the case here, can you explain why?

6-31. *Body-and-spring program, III.* Run the body-and-spring program with $x_0 = 0.5$ m, $v_0 = 0.5$ m/s, $t_0 = 0$ s, $\Delta t = 0.2$ s, and $\alpha = 1$ N/(m·kg). Discuss your results.

6-32. *Body-and-spring program, IV.* Run the body-and-spring program with the same initial conditions as in Example 6-1, but with $\alpha = 0.125$ N/(m·kg), and use the results to add another point to Figs. 6-7 and 6-8. Should you use the same value of Δt as in Example 6-1?

6-33. *Body-and-spring program, V.* Run the body-and-spring program with the same initial conditions as in Example 6-1, but with $\alpha = 8$ N/(m·kg), and use the results to add another point to Figs. 6-7 and 6-8. Should you use the same value of Δt as in Example 6-1?

6-34. *Pendulum program, I.* Run the pendulum program with the same initial conditions and parameters as in Example 6-4, except that the initial angle is $\phi_0 = 2$ rad. Use the results to add another point to Fig. 6-14.

6-35. *Pendulum program, II.* Run the pendulum program with the same initial conditions and parameters as in Example 6-4, except that the initial angle is $\phi_0 = 3$ rad. Use the results to add another point to Fig. 6-14.

6-36. *Pendulum versus harmonic oscillator.*

a. Run the pendulum program with the same initial conditions and parameters as in Example 6-4, except that the initial angle is $\phi_0 = 3$ rad. Plot ϕ for the first quarter-cycle, and determine the period of the pendulum.

b. Use Eq. (6-25*a*) to find the value of the parameter α for a harmonic oscillator that will make it have the same period as the pendulum motion found in part *a*. Plot x for the first quarter-cycle of a harmonic oscillator with that value of α and with $x_0 = 3$ m and $(dx/dt)_0 = 0$. (The quickest way to do this is to run the body-and-spring program with the appropriate initial conditions and parameters.)

c. Compare the sinusoidal function describing the harmonic oscillator motion with the oscillatory, but non-sinusoidal, function describing the pendulum motion. Explain their similarities and their differences.

6-37. *Pendulum program, III.* A pendulum is released from rest with the cord horizontal. The length of the cord is 0.5 m. Run the pendulum program to determine the speed of the pendulum bob at an instant when the cord is vertical.

6-38. *Reaching the top.* As shown in Fig. 6E-38, a pendulum bob is fixed to one end of a rod of length 2 m. The other end of the rod is mounted on a frictionless axle. The mass of the rod is negligible.

a. The bob is struck sharply, giving it an initial speed v. Use the pendulum program to determine, by trial and error, the value of v for which the bob will almost reach the point X (directly above the axle) before it reverses its motion.

b. What is the motion of the bob for an initial speed greater than that found in part *a*?

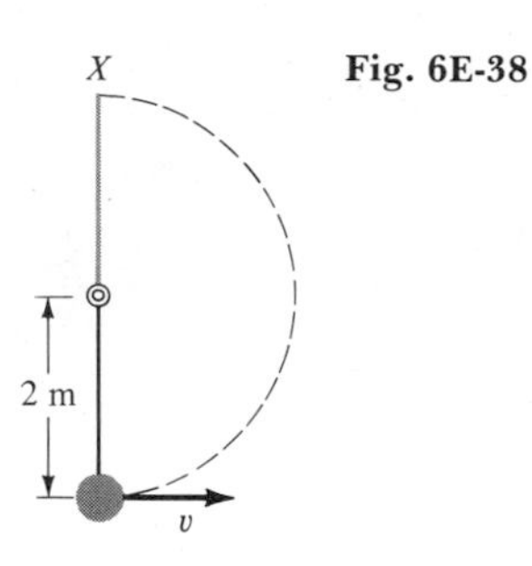

Fig. 6E-38

6-39. *Critical damping, II.* Run the damped oscillator program with $\alpha = 1$ N/(m·kg) and various values of β. Using $x_0 = 1$ m and $(dx/dt)_0 = 0$, find the largest β for which $x(t)$ makes just one, barely perceptible, negative swing. Plot x versus t for this so-called critically damped case, and compare it with the heavily damped case plotted in Fig. 6-19.

6-40. *Near-critical damping.* Use a value of β that is 10 percent smaller than the value found in Exercise 6-39, and measure the period of the damped oscillations. How does it compare with the undamped oscillator having the same value of β?

6-41. *Anharmonic oscillations, I.* Modify the body-and-spring program to obtain a numerical solution to the equation of motion of an anharmonic oscillator in which the force acting on the body is $-kx - px^3$. This describes a situation where the spring to which the body is connected becomes more stiff (if $p > 0$) or less stiff (if $p < 0$) with increasing extension or compression. Run an example. There is no analytical solution to the equation.

6-42. *Anharmonic oscillations, II.*

a. Modify the body-and-spring program to solve numerically the equation of motion of a body acted on by the anharmonic force $-kx + px^2$, where $p > 0$. This force law describes a spring that is stiffer when compressed ($x < 0$) than when extended ($x > 0$) by the same amount. There is no analytical solution to this equation of motion, which is used to model the oscillations in the center-to-center separation of the atoms of a diatomic molecule. Insert into the program steps which evaluate the time average of x over an entire cycle of the motion.

b. Run the program for several different maximum values of x. Show that the average value of x increases with increasing amplitude of the motion. This feature plus the fact that the amplitude of oscillation of a molecule increases with temperature makes the model useful in describing the thermal expansion of solids made up of such molecules.

7

Energy Relations

7-1 A PREVIEW OF ENERGY RELATIONS

Two of the basic sets of tools of mechanics are now at our disposal. The first set consists of Newton's laws of motion. We can use these laws to write an equation for the acceleration of any body acted on by any system of forces. The second set of tools consists of the techniques for finding the solutions to that equation. The solutions determine how the body moves when it is started in a particular way. We can almost always obtain these solutions by using numerical methods, and in many cases we can apply analytical methods. Thus we can now, in principle, study the motion of almost any mechanical system by direct application of Newton's laws.

In practice, however, it is often much easier to analyze mechanical behavior by applying other relations. A very important set of such relations involves a quantity called *energy*. As you will see, using energy relations is really a matter of using Newton's laws indirectly. The energy relations are not independent of Newton's laws. Rather they reexpress these basic laws in a way which makes it possible to answer certain questions about mechanical systems very easily.

Consider a pendulum. As straightforward as it might seem to be from a physical point of view, the motion of a pendulum can be quite complicated mathematically. If you apply Newton's laws directly to obtain the equation determining the motion, you find a very difficult equation—a nonlinear differential equation which can be solved to obtain the position of the pendulum as a function of time best by numerical methods on a programmable pocket calculator or a computer. When it is solved, you can predict *all* aspects of the behavior of the pendulum. But that takes some doing.

There are *certain* aspects of the behavior of the pendulum which can be treated in a simple way by applying energy relations. Let's say you give the pendulum bob some large initial displacement from its stable equilibrium position

Fig. 7-1 Strobe photo of a pendulum.

and then release it when it is at rest. How fast will the bob be moving when it goes through the bottom of its swing? The situation is shown in the strobe photo of Fig. 7-1.

One way you can answer the question is to obtain the appropriate numerical solution to the pendulum differential equation. But there is a much easier way. You will soon see that energy relations can be used to predict quite directly how fast the pendulum bob moves through the bottom of its swing if you know its initial displacement and speed. On the other hand, these relations have limitations. You cannot use them, for instance, to answer the question: How much time does it take for the bob to go from its initial position to the bottom of its swing? That can be answered only by solving the differential equation.

A question which can be answered even more easily by applying the energy relations is the following: Given the same initial conditions as above, what is the displacement of the pendulum bob when it is instantaneously at rest at the top of its swing on the side opposite the initial side? By using energy relations you can say that its final height will be the same as its initial height, and hence its final displacement will be of the same magnitude as its initial displacement, although of opposite sign. Of course, you can also make that prediction, without knowing anything about energy, by invoking the symmetry of the system.

But consider an *interrupted* pendulum, as shown in Fig. 7-2. When the bob is at the bottom of its swing, the cord supporting it is intercepted by a horizontal rod. The bob continues its swing, with the effective length of the cord reduced. The question remains the same: Given the initial displacement of the bob, what is its extreme displacement on the other side? Symmetry will not help you here. However, the energy relations remain useful. You will see that here again they require

Fig. 7-2 Strobe photo of an interrupted pendulum.

the final displacement to be related to the initial displacement in such a way that the final height of the bob equals its initial height.

You could also answer the question by joining numerical solutions to the differential equations for a long pendulum and a short pendulum. But the complications of such a direct application of Newton's laws would be great, and are unnecessary to answer the question posed.

Now that we have given a hint of the utility of the energy relations, we will begin to develop them. In this section we take a special case in which an object moves while a force acts on it that is *constant* in both direction and magnitude. Taking the force to be constant makes the physics and mathematics as simple as possible. This will allow us to go through the developments of energy concepts and their relations quickly. The primary purpose of this section is to acquaint you with the energy relations and to give you an idea of how they fit into an overall pattern. This will help you follow the detailed treatments of general cases presented in subsequent sections.

The concept of *work* is basic to the energy relations. When you apply a constant, horizontally directed force to push a heavy box across the floor, the box moves in the same direction as the direction of the force. See Fig. 7-3. The force you apply is said to do an amount of work on the box equal to the product of the applied force and the displacement of the object to which the force is applied. In this one-dimensional situation we can use signed scalars to specify directed quantities, like force and displacement. Let the positive x axis extend along the direction of motion of the box, with its origin at the initial location of the box. Then the displacement of the box when it has arrived at some subsequent location is given by its coordinate x, which is positive. The force applied to the box is specified by the value of F.

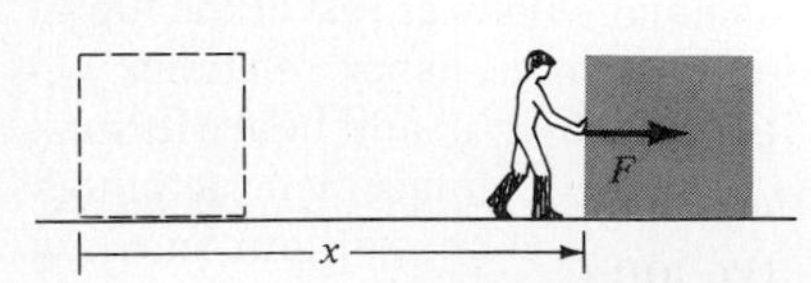

Fig. 7-3 A box being pushed across the floor. The force applied to produce the motion is labeled F, but none of the other forces acting on the box are shown. The displacement of the box resulting from the application of the force is labeled x.

This quantity is also positive because the force is applied in the positive x direction. If we use these symbols, and the symbol W for work, the **work** that the force does on the box is defined to be

$$W = Fx \qquad \text{for constant force acting along straight path} \tag{7-1}$$

Say the force you apply is $F = 100$ N and the displacement of the box is $x = 10$ m. Then the work that this force does on the box is $W = Fx = 100 \text{ N} \times 10 \text{ m} = 1000 \text{ N·m}$. The work unit, a newton-meter, is used so frequently that it is given its own name. It is called a **joule** (abbreviated J), in honor of the English physicist James Prescott Joule (1818–1889). So

$$1\text{J} \equiv 1\text{N·m} \tag{7-2}$$

and the work done in moving the box is written $W = 1000$ J.

Most people will agree that this is a reasonable definition of work because it conforms to the everyday use of the word. If you continue pushing on the box, you will soon begin to feel the physiological sensation of having performed work. Furthermore, your degree of fatigue will be proportional, roughly speaking, to the force applied and also to the distance moved.

But consider a counterexample. If you push vigorously on a rigid wall, and keep pushing, you will also soon feel fatigue. To use common language, you will have been working. But according to the physical definition you will have done no work. Although a force is applied, no displacement occurs. So the x in the equation $W = Fx$ has the value zero, and therefore W is zero. This example emphasizes that in physics "work" means the quantity W defined (in the circumstances considered here) by the equation $W = Fx$. It should not be confused with the common use of the word.

The work $W = Fx$ done by the force F you apply to the box to give it a displacement x across the floor is positive because both F and x have the same sign. But there are many situations in which a force is applied to an object in the direction opposite to the direction of the object's displacement. An example is found in the force of kinetic contact friction C_k which the floor applies to the box. This force acts in the direction opposite to the direction in which the box is displaced. Thus C_k has a negative value because the displacement x has a positive value. As a consequence, the work $W = C_k x$ done by the frictional force will be negative. For a specific example, say that you push the box across the floor at a low and constant speed. Then the force you apply to the box is of the same strength as the frictional force that the floor applies to it, except at the very beginning of the displacement where you apply a bit of excess force to start the box moving. This must be so since the box does not accelerate, except at the beginning. The work done by the frictional force thus has the value $W = C_k x = -100 \text{ N} \times 10 \text{ m} = -1000$ J.

When the box is moving across the floor with you pushing on it with a force in the direction of its displacement, and the floor applying a frictional force to it of the same magnitude in the direction opposite to its displacement, no net work is being done on the box. The negative work done on the box by the frictional force just cancels the positive work done by the force you apply. Furthermore, since there is no net force acting on the box in these circumstances, the box has no acceleration and so continues to move slowly across the floor.

On the other hand, if you push on the box with a force of magnitude greater than that of the frictional force, then the force you apply will do more positive work than the frictional force does negative work. In such a case, net positive work is done on the box because there is a net force acting on it in the direction of its displacement. In fact, you can show easily that the net work done is just the constant net force in the direction of the displacement multiplied by the displacement. Furthermore, with a net force applied to the box Newton's second law says it has an acceleration. The acceleration causes the speed of the box to increase throughout the displacement, and at the end of the displacement the box will be moving rapidly. Thus the effect of the positive work done on the box is to increase its speed.

We will find the relation between the work done on an object and the change in its speed by considering not a box on a floor, but a puck on the top of an air table. The advantage of the air table is the absence of friction. When you apply a horizontally directed force to an air table puck, this force *is* the net force acting on the puck in that direction. As is illustrated in Fig. 7-4, we again take the positive x axis along the direction of the force you apply, and of the displacement it produces. So the acceleration of the puck is the positive quantity a, because the acceleration is in the direction of the applied force. We also take the origin of the axis to be at the position of the puck at the instant $t = 0$ when you begin to apply the force to the puck, and we assume that before this instant the puck was at rest. Then its position coordinate and velocity have the values $x = 0$ and $v = 0$ at $t = 0$. When you have pushed the puck through a displacement x—that is, when its coordinate has the positive value x—the positive acceleration a has resulted in an increase in its velocity to the positive value v.

The work you do in this process can be evaluated by using the relations that describe the motion resulting from a constant net force. According to Newton's second law, the acceleration a resulting from this force is constant. So we can employ the kinematical equations for motion with constant acceleration a to find the puck's velocity v when the displacement of the puck from its initial position is x. Setting $x_i = 0$ and $v_i = 0$ in Eq. (2-32), we obtain

$$x = \frac{v^2}{2a} \tag{7-3}$$

The work you do on the puck by applying the constant net force F to give it the displacement x in the direction of the force is, by definition,

$$W = Fx$$

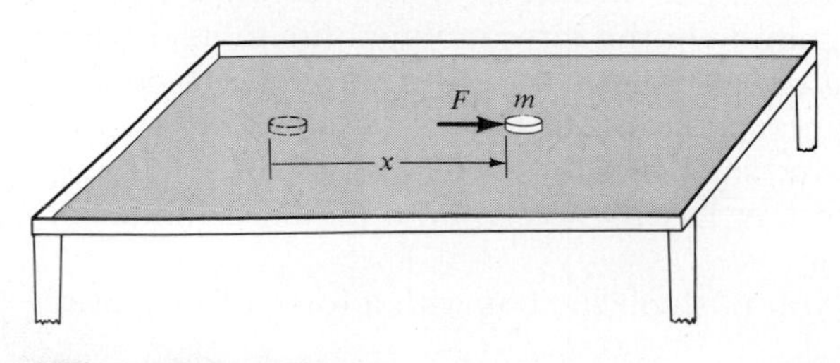

Fig. 7-4 A puck being pushed across the top of an air table by a force labeled **F**. This is the only force acting in a horizontal direction, so the puck accelerates in the direction of the force. The quantity x gives the position of the puck relative to its initial position, $x = 0$. In other words, x is the displacement of the puck from the position it had before the force was applied. Forces acting in vertical directions are not shown.

Employing Eq. (7-3), we can write this as

$$W = \frac{Fv^2}{2a}$$

Now we make explicit use of Newton's second law to express a in terms of the net force F applied to the puck and the mass m of the puck. That is, we write

$$F = ma$$

Thus the work done on the puck can be expressed as

$$W = \frac{mav^2}{2a}$$

or

$$W = \frac{mv^2}{2} \tag{7-4}$$

We have expressed the work done by the net force acting on the initially stationary puck during its displacement in terms of the mass of the puck and its speed at the end of the displacement. The word "speed" is used, instead of "velocity," because the value of v^2 is independent of the sign of v, and so the value of W depends on only the magnitude of v. The work done depends on the magnitude of the velocity, but not on its direction.

When the puck is speeding across the top of the air table after you have done work on it, the puck has an attribute that it did not have when it was at rest. The new attribute is its ability to do work on some other object, if it interacts with that object. For instance, if the puck hits the head of a nail sticking out of a block of wood fastened to the edge of the air table, the puck can drive the nail into the block. By assuming, for simplicity, that the force exerted by the puck on the nail is constant while the latter is being driven into the block, and that a drop of instant-setting glue on the nail head prevents the puck from rebounding, you should be able to calculate how much work the puck does on the nail. You use Newton's third law to equate the negative of the force exerted by the puck on the nail to the force exerted by the nail on the puck. Then you make a calculation involving the mass and acceleration of the puck which is just like the one leading to Eq. (7-4), except for the sign and magnitude of the acceleration. The result is identical to that quoted in Eq. (7-4) because the acceleration cancels in the equation immediately preceding it. Thus you will find that the work done *by* the puck as it comes to rest is just equal to the work done *on* the puck in bringing it up to speed.

Whenever something has the ability to do work, we say that it has **energy**. There is more than one reason why something can have this ability. For the puck we are considering, the reason is that it is moving. The energy that a body has by virtue of its motion is called its **kinetic energy**. The word "kinetic" comes from the Greek *kinetikos,* which means "moving." We will use the symbol K for kinetic energy. Its value equals the work done by the net force acting on the body to set it into motion. That is, K equals the quantity W evaluated in Eq. (7-4). So the expression

$$K = \frac{mv^2}{2} \tag{7-5}$$

gives the value of the *kinetic energy K of a body of mass m that is moving with velocity v*. Note again that the sign of v is of no consequence in determining the value of K. The kinetic energy of a body depends on its speed, but not on the direction of its motion.

Now we consider a second experiment in which you apply a constant force to a puck and it is displaced in the direction of that force. But the work that the force does on the puck will give the puck a different kind of energy because the circumstances are different from what they were in the first puck experiment.

What you do is to raise the puck vertically from the floor to a certain height and then hold it there. At the beginning of the displacement you momentarily apply an upward force to the puck which is slightly stronger than the downward force that the earth applies to it through gravity. This starts the puck moving upward very slowly. Then you keep the strength of the force you apply just equal to that of the gravitational force, so that the puck maintains its very low-speed, upward motion. At the end of the displacement you stop the puck by momentarily making the force you apply to it slightly weaker than the gravitational force. The experiment is depicted in Fig. 7-5.

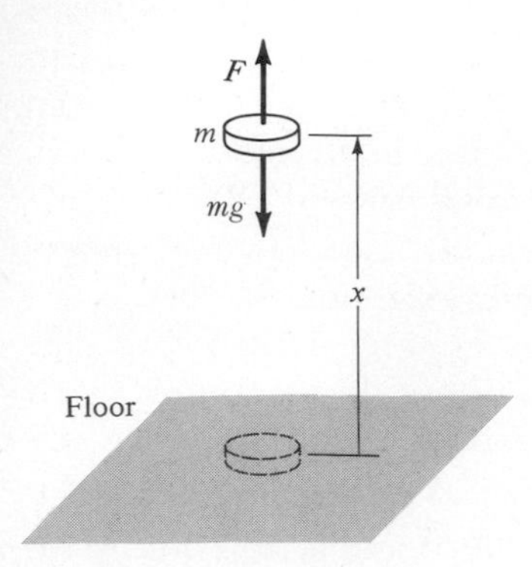

Fig. 7-5. A puck being raised very slowly from the floor by the application of an upward force F of the same magnitude as the downward gravitational force mg exerted on the puck by the earth. The quantity x gives the position of the puck relative to its initial position on the floor, where $x = 0$, and also its displacement from that position.

As is indicated in the figure, we define an x axis whose positive direction is upward and whose origin is at the level of the floor. Then the displacement of the puck is just the positive quantity x, the value of its coordinate at the end of the displacement. The magnitude of x is the height of the puck above floor level at the end. Except at the very beginning and end of the displacement, the upward force that you apply to the puck in order to cancel the downward force of gravity has the positive value $F = mg$. Here m is the mass of the puck, and g is the magnitude of the gravitational acceleration. So the positive work you do on the puck in elevating it is given by the expression

$$W = mgx \tag{7-6}$$

The elevated puck has a property that it did not have when resting on the floor. The new property is the ability of the puck to do work on some other object in the process of returning to the floor. In other words, the puck has energy. Because of its energy, the puck can now do as much work on you as you did on it in elevating it. Reverse the experiment by very slowly letting the puck push your hand down to the floor. The downward force that the puck applies to your hand has the value $-mg$, and the downward displacement of your hand has the value $-x$. So the positive work done by the puck on you is $W = (-mg)(-x) = mgx$. The work the puck does on you when it descends just equals the work you do on it when it ascends. Thus when you lift the puck against the gravitational attraction of the earth, you give it energy, and it then has the ability to do as much work as you did on it to lift it. The work can be done on you, as in the experiment just described, or on some other object.

The energy that a body has by virtue of its position is called its **potential energy**. It is said that the puck has **gravitational potential energy** when it is at a position above the floor, instead of at its **reference position** on the floor. The word "potential" is used because the puck is potentially able to do work in returning from its position above the floor to its reference position on the floor.

A more complete statement is that the *system* of the puck plus the earth has gravitational potential energy when the puck is at a position higher than its reference position. *The potential energy is really stored in the puck-plus-earth system,* not in the puck alone. The reason is that the force which the stationary (or very slowly moving) puck can apply to some other object is a result of the gravitational force which the earth exerts on the puck. So the work that can be done on an object by the force which the puck can exert on it is a consequence of two things: (1) The puck is a member of a system whose other member, the earth, exerts a force on it; hence the puck can, in turn, exert a force on an object *external* to the system. (2) The position of the puck is such that the force which it exerts on the external object does positive work as the puck returns to its reference position.

The symbol that we will use for potential energy is U. Thus we say that when the puck is at vertical position x above the floor, the puck-plus-earth system has gravitational potential energy U, with reference to the situation when $x = 0$. The value of U just equals the work W done in elevating the puck from position $x = 0$ to position x. Using Eq. (7-6) to evaluate W, we have

$$U = mgx \tag{7-7}$$

In this expression the floor plays the role of a reference position. That is, the vertical position x is measured relative to the floor-level position $x = 0$, and the potential energy U is measured relative to the value $U = 0$ that it has when $x = 0$.

No matter how the puck got to the position x—whether you or someone else put it there, whether you lifted it directly or overshot and then lowered it back to that position—the system has the same amount of gravitational potential energy. This must be so because in all cases the gravitational force that the system produces will do the same amount of work if the puck returns from position x to the reference position $x = 0$. It can be so because the work done by the force that is applied to the system to displace the puck from position $x = 0$ to position x is independent of the path the puck follows between these two positions. Show that this is true for a case in which you raise the puck vertically upward to a position x'' that is higher than position x, and then lower it vertically downward to position x. You should have no difficulty if you take into account the fact that you do negative work on the puck in the displacement from x'' to x.

Before continuing this preview of energy relations, we will summarize the most important features of kinetic and potential energy saying that *kinetic energy is energy of motion and potential energy is energy of position.*

Now we consider a final experiment which will bring out a very important relation between potential energy and kinetic energy. Begin with the puck resting on the floor. You do work on the puck-plus-earth system by very slowly raising the puck to position x above the floor, with positive x measured upward. The positive work you do is mgx, and this equals the gravitational potential energy you have given the system. Then you release the puck. The system immediately becomes an *isolated* one. This isolated system has an initial positive gravitational potential energy $U = mgx$, with reference to the floor, and an initial kinetic energy $K = 0$. As the puck falls, the system loses potential energy since the height of the puck decreases. But it gains kinetic energy since the speed of the puck increases.

Let's evaluate U and K at an instant when the puck falls past a position

x' that is lower than x but higher than the floor. At x' the potential energy of the system relative to the floor has the smaller, but still positive, value

$$U = mgx' \tag{7-8}$$

The reason is that you would have given the system that much energy if you had slowly raised the puck directly from the floor to x'. In falling from x to x', the puck has acquired a negative velocity v'. According to Eq. (7-5), it has a positive kinetic energy

$$K = \frac{mv'^2}{2}$$

In order to compare these values of U and K, we must express them in comparable terms.

To do this, we relate the displacement of the puck traveling at constant acceleration to the square of the velocity it acquires during the displacement. This can be done, if the effect of air resistance is neglected, by again making use of Eq. (2-32). If we set the initial position coordinate equal to x, the final position coordinate equal to x', the initial velocity equal to zero, the final velocity equal to v', and the acceleration equal to $-g$, Eq. (2-32) gives us the relation

$$x' = x + \frac{v'^2}{-2g}$$

We transpose, to obtain

$$\frac{v'^2}{2g} = x - x'$$

Solving for v'^2 produces

$$v'^2 = 2g(x - x')$$

We can therefore express the value of K, in these particular circumstances, as

$$K = \frac{mv'^2}{2} = \frac{m[2g(x - x')]}{2}$$

or

$$K = mg(x - x') \tag{7-9}$$

Figure 7-6 is a plot of the kinetic and potential energies of the system, given by Eqs. (7-8) and (7-9), as a function of the instantaneous position x'. The fall begins with $x' = x$ and ends with $x' \simeq 0$—just *before* the puck strikes the floor. At first (when $x' = x$) the kinetic energy K is zero because the puck is just beginning to move, and the potential energy U has the initial value mgx. When the puck has fallen to some intermediate height x', the kinetic energy has increased, since the puck has gained speed. But this increase has been at the expense of a loss of potential energy, since the puck has lost height. Just before the puck hits the floor (when $x' \simeq 0$), the potential energy is essentially zero and the kinetic energy has increased to a value essentially equal to the initial value of the potential energy. In fact, Fig. 7-6 makes it clear that the decrease in potential energy is exactly compensated for by an increase in kinetic energy at *all* points on the fall.

We are therefore led to introduce the **total mechanical energy** E. This

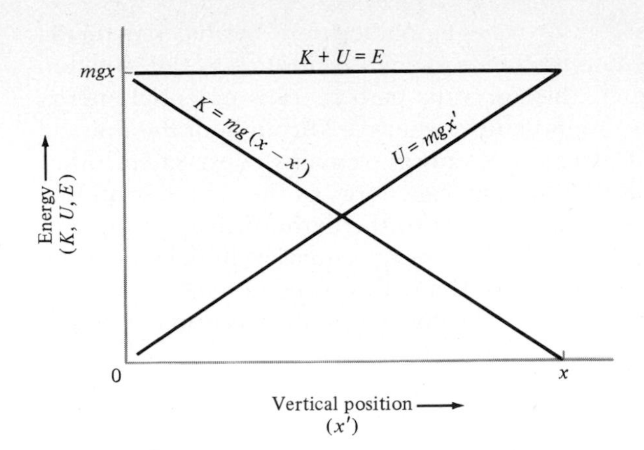

Fig. 7-6 The kinetic energy K, potential energy U, and total mechanical energy E of a puck falling from rest at height x to the floor at height 0. These quantities are plotted versus its height x' at intermediate points. During the fall the value of x' goes to the *left* from $x' = x$ to $x' \simeq 0$.

quantity is defined as *the sum of the kinetic and potential energies.* That is,

$$E = K + U \tag{7-10a}$$

As the puck falls, the energy of the system gradually changes from one form, potential, to another, kinetic. But their sum, the total mechanical energy, remains constant. In particular, we have from Eqs. (7-8) and (7-9)

$$E = K + U = mg(x - x') + mgx' = mgx$$

The quantity mgx is a constant because x is the fixed value specifying the initial height of the puck. This simple experiment provides an example of the **law of conservation of total mechanical energy** of an isolated system in which there is no friction. The law can be expressed symbolically by the equation

$$E_{\text{final}} = E_{\text{initial}} \tag{7-10b}$$

It is one of the most important conservation laws of physics.

Let us recapitulate. First the puck was stationary on the floor. Then you did positive work on the puck-plus-earth system by applying an upward-directed external force to the puck in such a way that it was very slowly displaced upward. In doing this, you gave the system potential energy only. Throughout almost all its upward displacement, the external force you applied to the puck just canceled the internal force applied to it by the earth through gravity. So the puck experienced no net force, did not accelerate, and did not gain kinetic energy from the work you did on it. After the puck was released at its elevated position, the system was isolated. Then the only force acting on the puck was the internal force exerted on it by the earth. The puck accelerated downward under the influence of this net force. As it fell with increasing speed, the puck gained kinetic energy. Therefore the system gained kinetic energy. But at the same time the system lost potential energy because the elevation of the puck was decreasing. The gain in kinetic energy was exactly compensated for by the loss of potential energy, so that the total mechanical energy of the friction-free system remained constant as long as the system remained isolated. When the puck hit the floor, the system was no longer isolated.

As another example, suppose you throw a ball vertically upward from floor level. Immediately after the ball leaves your hand, the ball plus the earth forms an isolated system with negligible friction that has kinetic energy, but no gravitational potential energy with reference to the situation with the ball on the floor. At the top of its path there is potential energy with reference to the floor, but no kinetic energy. Throughout the upward part of its flight the total mechanical energy remains constant; in other words, it is conserved. The total mechanical energy of the system continues to remain constant for the downward part of the flight of the ball. Just before the ball hits the floor, its final kinetic energy equals its initial kinetic energy. (Why?) Thus the final velocity of the ball is equal in magnitude to its initial velocity. The sign reversal of the velocity has no effect on the kinetic energy, which depends on the square of the velocity.

Consider the ball-plus-earth system *after* the ball hits the floor, assuming it does not rebound. The ball is at rest at floor level. Therefore the system has neither kinetic energy nor potential energy. The constant total mechanical energy which the system had during the flight of the ball has vanished. What has happened to the mechanical energy?

This is what happens. When the ball hits the floor, the molecules in the surface of the ball collide with the molecules in the surface of the floor, setting the surface molecules in both bodies into vibrational motion. Their vibration sets adjacent molecules into vibration, and so on. So the vibrational motion propagates into the two bodies. The molecules are not vibrating in unison, however. There is still energy present, even though the ball as a whole is at rest on the floor. But it is an energy of random vibrational motion.

Just before the ball hits the floor the system has mechanical energy, which is all in the form of kinetic energy. Kinetic energy is an *organized* energy of motion. It is organized in the sense that all the molecules of the ball are moving in unison—in the same direction at the same speed. After the ball hits the floor there is still motion of molecules, and there is energy associated with this motion. But it is the *disorganized* energy of randomly vibrating molecules. This energy is called **thermal energy**. So when the ball hits the floor, the mechanical energy of the system is transformed into thermal energy.

The same sort of process happens in a more gradual fashion whenever any form of *friction* is present in a system. Contact friction transfers part of the mechanical energy of a system of two objects sliding over each other into thermal energy associated with random motion of the molecules near the surfaces in contact. In the case of fluid friction, some of the mechanical energy is lost to heating the moving body and the material through which it moves.

Thus the law of mechanical energy conservation does *not* hold when frictional effects are significant. But even when they are, experiment shows that the *sum* of the mechanical and thermal energies of an isolated system *is* constant. The lost mechanical energy appears as thermal energy. It is also possible for some (but not all) of the thermal energy of a system to be converted into mechanical energy; devices which do this are called heat engines. These matters are discussed at length in Chaps. 17 through 19.

We close this preview of the energy relations by using them in Example 7-1.

EXAMPLE 7-1

a. An elevator car is moving upward at a constant speed of 2.00 m/s. When it is at an elevation 10.00 m above ground, the lifting cable breaks. Calculate the maximum elevation attained by the car, ignoring friction.

■ After the cable is broken, the car-plus-earth system is isolated. Therefore, since the system is assumed to be frictionless, you can apply the law of conservation of mechanical energy, $E_f = E_i$. In particular, you can equate the initial total mechanical energy $E_i = K_i + U_i$ of the system at the instant after the cable breaks to the final total mechanical energy $E_f = K_f + U_f$ at the instant when the car is stationary at its maximum elevation. Use an x axis whose origin is at ground level and whose positive direction is vertically upward, and measure the gravitational potential energy of the system with reference to the car at ground level. Then the initial kinetic and potential energies of the system are $K_i = mv_i^2/2$ and $U_i = mgx_i$, where v_i and x_i are the initial velocity and position of the car and m is its mass. The final kinetic and potential energies are $K_f = 0$ and $U_f = mgx_f$. Equating the sum of the final energies to the sum of the initial energies, you have

$$E_f = E_i$$

or

$$K_f + U_f = K_i + U_i$$

or

$$0 + mgx_f = \frac{mv_i^2}{2} + mgx_i$$

Solving for x_f gives you

$$\begin{aligned} x_f &= x_i + \frac{v_i^2}{2g} \\ &= 10.00 \text{ m} + \frac{(2.00 \text{ m/s})^2}{2 \times 9.80 \text{ m/s}^2} \\ &= 10.20 \text{ m} \end{aligned}$$

■

b. The safety brakes automatically engage when the downward speed reaches 4.00 m/s. Determine the elevation at which this happens.

■ Now you take the initial conditions to be those when the car is instantaneously at rest at its maximum elevation, and the final conditions to be those at the instant before the safety brakes engage. Energy conservation gives

$$K_f + U_f = K_i + U_i$$

or

$$\frac{mv_f^2}{2} + mgx_f = 0 + mgx_i$$

or

$$\begin{aligned} x_f &= x_i - \frac{v_f^2}{2g} \\ &= 10.20 \text{ m} - \frac{(4.00 \text{ m/s})^2}{2 \times 9.80 \text{ m/s}^2} \\ &= 9.39 \text{ m} \end{aligned}$$

This is 0.81 m below the highest point reached, or 0.61 m below the point where the cable broke. ■

c. The brakes apply a constant upward frictional force to the car, with a magnitude $\frac{3}{2}$ times the weight of the car. Determine the elevation of the car when it stops.

■ Here the mechanical energy is not constant since energy is lost to friction in the brakes. To say it another way, the car-plus-earth system is no longer isolated, since the brakes do work by applying an external force to the car. If you take the initial conditions to be those when the brakes engage, the work done by the brakes on the car is the product on the force they apply and the displacement of the car as it comes to a stop under the influence of this force. The upward force has a positive value $\frac{3}{2} mg$, whereas the downward displacement $(x_f - x_i)$ has a negative value. So the work done by the brakes on the system is the negative quantity $\frac{3}{2} mg(x_f - x_i)$. It is negative because the force acts in the direction opposite to the displacement, and thus mechanical energy is removed from the system. You can equate the total initial mechanical energy of the system, plus the negative work W done on it by the brakes, to its total final mechanical energy. This gives

$$K_f + U_f = K_i + U_i + W$$

or

$$0 + mgx_f = \frac{mv_i^2}{2} + mgx_i + \frac{3mg}{2}(x_f - x_i)$$

or

$$-\frac{mgx_f}{2} = -\frac{mgx_i}{2} + \frac{mv_i^2}{2}$$

So

$$\begin{aligned} x_f &= x_i - \frac{v_i^2}{g} \\ &= 9.39 \text{ m} - \frac{(4.00 \text{ m/s})^2}{9.80 \text{ m/s}^2} \\ &= 7.76 \text{ m} \end{aligned}$$

The braking distance is 1.63 m. Can you explain why this is twice the distance the elevator fell from its maximum height before the brakes engaged?

7-2 WORK DONE BY A VARIABLE FORCE

Now we will begin generalizing the energy relations by extending the definition of work to a one-dimensional case in which the force doing work is not of constant magnitude. The force and the displacement of the body to which it is applied both lie along the x axis, as before. But here the strength of the force depends on the position of the body, and so it must be written as $F(x)$. Nevertheless, an equation very much like Eq. (7-1) can still be used to evaluate approximately the amount of work done by the force in a *small* displacement of the body. If the body moves from x_i to $x_i + \Delta x$, then the small amount of work ΔW done by the force applied to it will be

$$\Delta W \simeq F(x_i + \Delta x/2)\,\Delta x$$

The idea is that in the small displacement from x_i to $x_i + \Delta x$ none of the values actually assumed by the force $F(x)$ are very different from the particular value $F(x_i + \Delta x/2)$ of the force at the *midpoint* of the displacement. Hence we can well approximate $F(x)$ over the displacement by using the constant value $F(x_i + \Delta x/2)$. Doing so, we can then apply the basic definition of Eq. (7-1)—work equals (constant) force times displacement—to obtain a good approximation to the work actually done in the displacement. Figure 7-7 indicates that the same procedure can be used to estimate the work done by the force when the body moves through the next displace-

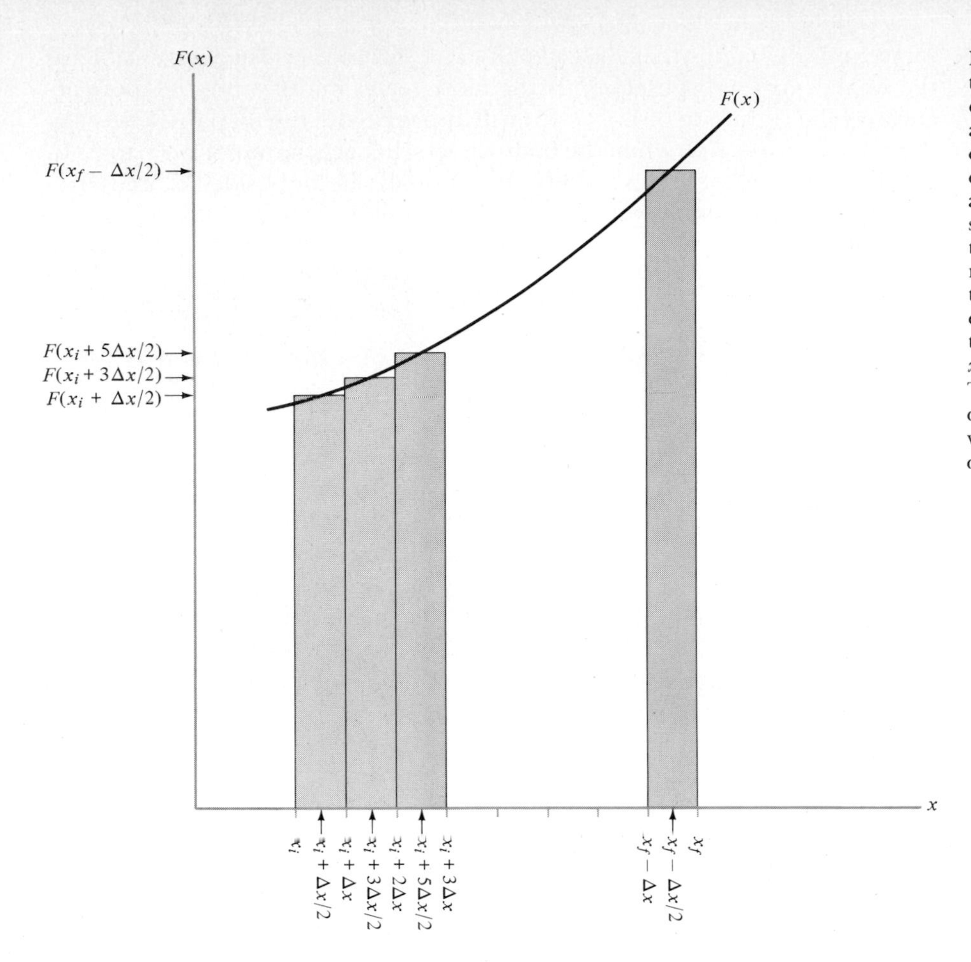

Fig. 7-7 Illustration of a procedure that can be used to evaluate the work done by a variable force $F(x)$ acting on a body that is displaced in the direction of the force from x_i to x_f. The total area enclosed by the rectangles gives an approximate value of the work. The smaller the increment Δx, and therefore the larger the number of rectangles, the more accurate the approximation. In the limit as Δx goes to zero, the area enclosed by the rectangles approaches the area enclosed by the curve $F(x)$, the x axis, and the vertical lines at x_i and x_f. This "area under the curve" is called the definite integral of $F(x)$ from x_i to x_f. Its value is precisely equal to the work done.

ment. This one is from $x_i + \Delta x$ to $x_i + 2\Delta x$. Its midpoint is at $x_i + 3\Delta x/2$. The force at the midpoint is $F(x_i + 3\Delta x/2)$. And the work done by the force in the displacement is

$$\Delta W \simeq F(x_i + 3\Delta x/2)\ \Delta x$$

Continuing in this manner, we can obtain an approximation to the total work W done when the body moves from x_i to x_f by summing the contributions ΔW obtained for each range of width Δx. Thus

$$W \simeq F(x_i + \Delta x/2)\ \Delta x + F(x_i + 3\Delta x/2)\ \Delta x + F(x_i + 5\Delta x/2)\ \Delta x + \cdots + F(x_f - \Delta x/2)\ \Delta x \quad (7\text{-}14)$$

This can be written more concisely as

$$W \simeq \sum_{x_i+\Delta x/2}^{x_f-\Delta x/2} F(x_j)\ \Delta x \quad (7\text{-}15)$$

The large Greek letter Σ (sigma) means the operation of taking a sum of terms of the quantity to its right, $F(x_j)\ \Delta x$. The quantity x_j represents successively the points at which F is evaluated, with x_j starting at $x_i + \Delta x/2$ and increasing each time by Δx until the value $x_f - \Delta x/2$ is reached. Thus Eq. (7-15) has exactly the same meaning as Eq. (7-14).

The smaller you take the width of the displacements Δx, the smaller the difference between the value of $F(x)$ at the midpoint of each displacement and the values that $F(x)$ actually has within the displacement. Thus the smaller Δx (and consequently the more terms in the sum), the more accurately the right side of Eq. (7-15) will approximate the total work actually done by the force $F(x)$ when the body on which it acts moves from x_i to x_f. In the limit when Δx becomes infinitesimally small, the right side becomes precisely equal to the total work done. Therefore

$$W = \operatorname*{limit}_{\Delta x \to 0} \sum_{x_i + \Delta x/2}^{x_f - \Delta x/2} F(x_j)\, \Delta x \tag{7-16}$$

In the notation of integral calculus, this limit of a sum is written as

$$\operatorname*{limit}_{\Delta x \to 0} \sum_{x_i + \Delta x/2}^{x_f - \Delta x/2} F(x_j)\, \Delta x \equiv \int_{x_i}^{x_f} F(x)\, dx \tag{7-17}$$

The distorted letter S is called an **integral sign,** and the right side of Eq. (7-17) is called the **definite integral over** x **of the function** $F(x)$ **from the lower limit** x_i **to the upper limit** x_f. Thus the **work** W done by the force acting on the body over the entire displacement from x_i to x_f is, by definition, the definite integral

$$W \equiv \int_{x_i}^{x_f} F(x)\, dx \qquad \text{for force acting along straight path} \tag{7-18}$$

7-3 INTEGRATION

This section develops the integral calculus that we will need in continuing our investigation of work and energy and elsewhere in the book. If you are already familiar with integral calculus, you may wish to go directly to Sec. 7-4.

First we will obtain a geometrical interpretation of a definite integral. By definition, a definite integral has the value specified in Eq. (7-17),

$$\int_{x_i}^{x_f} F(x)\, dx \equiv \operatorname*{limit}_{\Delta x \to 0} \sum_{x_i + \Delta x/2}^{x_f - \Delta x/2} F(x_j)\, \Delta x$$

The value of the definite integral is the limiting value of a sum of terms. Now the value of each of these terms equals the area under the corresponding rectangle in Fig. 7-7, since the height of each is the midpoint value of $F(x)$ and the width is Δx. Each of these rectangles is a good approximation to the area under the part of the $F(x)$ curve passing through the rectangle. This is because the overestimate that the first half of the rectangle makes to the actual area under the $F(x)$ curve is largely compensated for by the underestimate that the second half of the rectangle makes. In the limit as the width Δx of each rectangle approaches zero, the area under the entire set of rectangles approaches the actual area under the $F(x)$ curve between x_i and x_f. Therefore we conclude that *the value of the definite integral of* $F(x)$ *from* x_i *to* x_f *equals the area under the curve described by* $F(x)$ *between these limits.* This suggests that an approximate evaluation of a particular definite integral can be obtained by carefully plotting $F(x)$ between x_i and x_f on graph paper with closely spaced grid lines, and then measuring the area under the curve by counting the number of grid squares in this area.

A much more accurate and convenient way to evaluate a particular definite integral directly from the definition is to employ a programmable pocket calculator, or a small computer. The Numerical Calculation Supplement contains a numerical integration program which makes such a device calculate the sum for successively smaller values of Δx. By inspecting the sequence of sums it is possible to determine when a limiting value has been reached, to within the required accuracy. This limiting value is the value of the definite integral, to that accuracy. Example 7-2 uses the numerical integration program.

EXAMPLE 7-2

Use the numerical integration program to evaluate the definite integral

$$\int_{x_i}^{x_f} F(x)\,dx \qquad \text{where } F(x) = x^2, \text{ with } x_i = 1 \text{ and } x_f = 2$$

That is, evaluate the definite integral

$$\int_1^2 x^2\,dx$$

Obtain results for each value of Δx to three-decimal-place accuracy.

■ To do this, you calculate a sequence of values of the summation in Eq. (7-17), with each value in the sequence obtained by using a successively smaller value of Δx. Then you determine by inspection the limiting value to the sequence. The program starts by taking $\Delta x = x_f - x_i$, so that there is only one term in the summation. Then it takes $\Delta x = (x_f - x_i)/2$, and evaluates the corresponding two-term summation. Continuing to reduce Δx successively by halving, it generates a sequence of increasingly accurate approximations to the integral, which converges to the actual value of the integral.

The sequence you obtain on running the program is

$$\sum_{1+\Delta x/2}^{2-\Delta x/2} x_j^2\,\Delta x = 2.250,\ 2.313,\ 2.328,\ 2.332,\ 2.333,\ 2.333$$

These values evidently converge to the limit 2.333. Thus you can conclude that to three decimal places the value of the definite integral is

$$\int_1^2 x^2\,dx = 2.333$$

There is an analytical method for evaluating integrals, which has the usual advantages over numerical methods. It is not always applicable—the integrals of some not very complicated, but very important, functions can be evaluated only numerically. But nearly all the integrals that are of interest in elementary physics can be found by the analytical method. The method amounts to applying the **fundamental theorem of calculus:**

$$\int_{u_i}^{u_f} du = u_f - u_i \tag{7-19}$$

A proof of this theorem is given in the small-print material that follows.

Consider a quantity u and a uniformly increasing set of its values in the interval ranging from u_i to u_f. These values are displayed as ticks plotted along a line in Fig. 7-8, but the line itself has no particular significance. The set of values can be written

$$u_i,\ u_{i+1},\ u_{i+2},\ u_{i+3},\ \ldots,\ u_{f-3},\ u_{f-2},\ u_{f-1},\ u_f$$

u_i u_{i+1} u_{i+2} u_{i+3} u_{f-3} u_{f-2} u_{f-1} u_f

Fig. 7-8 A uniformly increasing set of values of a quantity u, ranging from u_i to u_f, displayed by plotting them along a line.

The meaning of the subscript notation is made apparent in the figure. The differences between adjacent values of the set form another set

$$u_{i+1} - u_i, u_{i+2} - u_{i+1}, u_{i+3} - u_{i+2}, \ldots, u_{f-2} - u_{f-3}, u_{f-1} - u_{f-2}, u_f - u_{f-1}$$

But each element of this difference set has the same numerical value, Δu, because of the uniform separation of the values in the first set. Take the sum of all the elements in the difference set:

$$\sum_{u_i+\Delta u/2}^{u_f-\Delta u/2} \Delta u = (u_{i+1} - u_i) + (u_{i+2} - u_{i+1}) + (u_{i+3} - u_{i+2}) + \cdots + (u_{f-2} - u_{f-3}) + (u_{f-1} - u_{f-2}) + (u_f - u_{f-1})$$

The limits of the summation are indicated by the values of u at the midpoints of the first and last terms. Note that each intermediate value of u (such as u_{i+1} or u_{f-1}) occurs twice in the sum, once with a positive sign and once with a negative sign. Thus the intermediate values cancel out of the sum. Because of these cancellations the right side of the equality simplifies drastically, and we have

$$\sum_{u_i+\Delta u/2}^{u_f-\Delta u/2} \Delta u = u_f - u_i$$

Now take the limit as $\Delta u \to 0$ on both sides. This produces

$$\lim_{\Delta u \to 0} \sum_{u_i+\Delta u/2}^{u_f-\Delta u/2} \Delta u = u_f - u_i$$

By the definition of Eq. (7-17), the left side of this equality is the definite integral of the rudimentary function $F(u) = 1$ from u_i to u_f. Therefore we have

$$\int_{u_i}^{u_f} 1\, du = u_f - u_i$$

or

$$\int_{u_i}^{u_f} du = u_f - u_i$$

This is the fundamental theorem of calculus, Eq. (7-19).

As is sometimes the case with theorems, the formal notation used to prove the fundamental theorem of calculus may make the result seem more complicated than it really is. To appreciate its simplicity, say in words what Eq. (7-19) means, while looking at Fig. 7-8: "The (limit of the) sum of the changes in u from one value to the next equals the total change in u over the interval from u_i to u_f." Example 7-3 shows how to use the theorem to evaluate definite integrals.

EXAMPLE 7-3

Use the fundamental theorem of calculus to evaluate the definite integral

$$\int_{x_i}^{x_f} F(x)\, dx \qquad \text{where } F(x) = x^2, \text{ with } x_i = 1 \text{ and } x_f = 2$$

That is, evaluate the definite integral

$$\int_1^2 x^2\, dx$$

■ To apply the theorem, you must find an expression for u which has the property that

$$du = x^2\, dx$$

By trial and error, based on your knowledge of differentiation, you will eventually find

$$u = \frac{x^3}{3}$$

This is correct since

$$\frac{du}{dx} = \frac{d(x^3/3)}{dx} = \frac{1}{3}\frac{d(x^3)}{dx} = \frac{3}{3}x^2 = x^2$$

Multiplying through by dx verifies that

$$du = x^2\, dx$$

Using this relation, you can write

$$\int_{x_i}^{x_f} x^2\, dx = \int_{u_i}^{u_f} du$$

Here u_i and u_f are the limiting values of u that correspond to the limiting values x_i and x_f of x. Since $u = x^3/3$, $x_i = 1$, and $x_f = 2$, they are

$$u_i = \frac{x_i^3}{3} = \frac{1}{3} \qquad \text{and} \qquad u_f = \frac{x_f^3}{3} = \frac{8}{3}$$

The fundamental theorem immediately gives

$$\int_{u_i}^{u_f} du = u_f - u_i = \frac{8}{3} - \frac{1}{3} = \frac{7}{3}$$

So you have

$$\int_1^2 x^2\, dx = \frac{7}{3}$$

in agreement with the results obtained numerically in Example 7-2.

Example 7-3 shows that integration by use of the fundamental theorem is a matter of finding a function which, when differentiated, yields the function to be integrated. Thus *the process of integration is the inverse of the process of differentiation.*

A generalization of the calculation carried out in Example 7-3 will lead to the evaluation of an integral that we will find very useful. The generalization is

$$\int_{x_i}^{x_f} x^n\, dx = \frac{x_f^{n+1}}{n+1} - \frac{x_i^{n+1}}{n+1} \qquad \text{for } n \neq -1$$

Using an abbreviated notation, we can write this as

$$\int x^n\, dx = \frac{x^{n+1}}{n+1} \qquad \text{for } n \neq -1 \tag{7-20}$$

In this notation it is understood that the integral on the left side of the equation is taken between the lower limit x_i and the upper limit x_f, and that the quantity on the right side is evaluated at the upper limit and then from this is subtracted the result obtained by evaluating it at the lower limit. In the same notation, some other important integrals are:

$$\int \frac{dx}{x} = \ln |x| \qquad \text{for } x \neq 0 \tag{7-21}$$

$$\int e^x \, dx = e^x \tag{7-22}$$

$$\int \sin x \, dx = -\cos x \tag{7-23}$$

$$\int \cos x \, dx = \sin x \tag{7-24}$$

And for *any* functions of x, $f(x)$ and $g(x)$,

$$\int f(x) \frac{dg(x)}{dx} \, dx + \int g(x) \frac{df(x)}{dx} \, dx = f(x)g(x) \tag{7-25}$$

These integrals were evaluated by applying the fundamental theorem. Direct application of the definition of an integral proves the following basic properties of integrals:

$$\int cf(x) \, dx = c \int f(x) \, dx \qquad \text{for constant } c \tag{7-26}$$

$$\int [f(x) + g(x)] \, dx = \int f(x) \, dx + \int g(x) \, dx \tag{7-27}$$

$$\left[\text{Caution: Note that } \int f(x)g(x) \, dx \neq \int f(x) \, dx \int g(x) \, dx.\right]$$

7-4 WORK AND KINETIC ENERGY

We now continue to generalize the concepts introduced in Sec. 7-1. Assuming that a certain variable force is the only force acting on a body, we will establish the relation between the work done by that force and the change in kinetic energy of the body. We will do this for one-dimensional motion, using signed scalars, and then extend the argument to two or three dimensions.

Consider a body moving on the x axis as a result of the net force $F(x)$ acting on it, the force being directed along the x axis. According to the definition of Eq. (7-18), when the body moves from x_i to x_f, the work done on the body by the force is

$$W = \int_{x_i}^{x_f} F(x) \, dx$$

The work-kinetic energy relation that we will obtain is founded on Newton's second law because the first step in obtaining the relation is to apply the law

$$F(x) = m \frac{d^2x}{dt^2}$$

in order to evaluate $F(x)$ in terms of the mass m of the body and its acceleration d^2x/dt^2. Using it in the integral, we have

$$W = \int_{x_i}^{x_f} m \frac{d^2x}{dt^2} \, dx \tag{7-28}$$

Next, the quantity which must be integrated is manipulated into a form that will allow the integral to be evaluated immediately from the fundamental theorem of calculus. This is done as follows:

$$m \frac{d^2x}{dt^2} dx = m \frac{d}{dt}\left(\frac{dx}{dt}\right) dx = m \frac{dv}{dt} dx$$

As usual, v represents the velocity of the body. Now $dx = v\ dt$, so we have

$$m \frac{d^2x}{dt^2} dx = m \frac{dv}{dt} v\ dt = mv\ dv$$

But $v\ dv = d(v^2)/2$, as can be seen by noting that $d(v^2)/dt = 2v\ dv/dt$ and then multiplying through by $dt/2$. Thus

$$m \frac{d^2x}{dt^2} dx = \frac{m}{2} d(v^2) \tag{7-29}$$

Using Eq. (7-29) in Eq. (7-28), we obtain

$$W = \int_{v_i^2}^{v_f^2} \frac{m}{2} d(v^2)$$

Here v_i^2 is the value of the square of the velocity of the body when it is at the initial position x_i, and v_f^2 is the corresponding value at the final position x_f. Since $m/2$ is a constant, the integral becomes

$$W = \frac{m}{2} \int_{v_i^2}^{v_f^2} d(v^2)$$

In this form, the fundamental theorem of Eq. (7-19) can be applied to evaluate the integral. We set $u = v^2$ so that $du = d(v^2)$, $u_i = v_i^2$, and $u_f = v_f^2$. We then obtain the result

$$W = \frac{m}{2} (v_f^2 - v_i^2) = \frac{mv_f^2}{2} - \frac{mv_i^2}{2}$$

The right side of this equation is the difference between the final and initial values of a quantity we recognize from Sec. 7-1. It is the kinetic energy

$$K = \frac{mv^2}{2} \tag{7-30}$$

Expressed in these terms, our results can be written

$$W = K_f - K_i \tag{7-31}$$

We have shown that for one-dimensional motion the work done on a body by a net force of varying strength acting on it equals the change in its kinetic energy. If the force acts in the same direction as the displacement, then positive work is done and the kinetic energy increases in proportion to the increase in v^2, the square of the velocity of the body. (Remember that here we represent velocity by a signed scalar. So v^2 is just the square of the scalar v. Soon, however, we will treat velocity as a vector. Then we will have to define what we mean by the square of a velocity vector.) Since v^2 is always positive, its value is always the same as the square of the speed of the body. Thus the kinetic energy of a body increases in proportion to the increase in the square of its speed when the net force acting on it does positive work. The situation is just the same as that discussed in Sec. 7-1, where the strength of the net force was constant. When the variable net force acts in the direction opposite to that of the displacement, it does negative work and also reduces the speed of the body—just as a constant net force does. The reduction in kinetic energy is in proportion to the decrease in the square of the speed.

We would like to generalize Eq. (7-31) to two or three dimensions. A way to do this can be found by noting that the work done by the force acting on the body is related to the change in its speed, that is, to the change in the magnitude of its velocity. In Sec. 3-3 we saw that a velocity vector **v** changes its magnitude during the infinitesimal time interval dt only when the infinitesimal change in velocity $d\mathbf{v}$ has a component in the direction of **v**. The point is seen again in Fig. 7-9, which shows that the change in the magnitude of **v** just equals the parallel component of $d\mathbf{v}$. Now, $d\mathbf{v}$ is parallel to the force **F** acting on the body, since $\mathbf{F} = m\, d\mathbf{v}/dt$. So it is only $F_\parallel$, the component of **F** parallel to the velocity **v** of the body, which is effective in changing the speed of the body. But **v** is always in the direction of the infinitesimal displacement $d\mathbf{s}$ of the body during dt, since $\mathbf{v} = d\mathbf{s}/dt$. Therefore $F_\parallel$ is also the component of **F** parallel to $d\mathbf{s}$. And therefore the change in the speed of a body arises from the component $F_\parallel$ of the force acting on it which is parallel to its displacement—and from that component only. Since the change in speed is related to the change in kinetic energy, we are led to conclude that $F_\parallel$ is what does work on the body in the case of two- or-three-dimensional motion.

In one dimension, the work W done by any force F acting along the x axis is

$$W = \int_{x_i}^{x_f} F\, dx$$

when the body it acts on moves in a succession of displacements dx from a point on that axis with coordinate x_i to another point on the axis with coordinate x_f. The preceding discussion leads us to generalize this definition to several dimensions, as follows. Suppose that a body acted on by a force **F** moves in a succession of displacements $d\mathbf{s}$, from a point on the path of the body with coordinate s_i to a point elsewhere on the path with coordinate s_f. We take the work W done by the force to be, by definition,

$$W = \int_{s_i}^{s_f} F_\parallel\, ds$$

The symbol $F_\parallel$ represents the component of the force **F** that is parallel to each small displacement $d\mathbf{s}$ along the path of the body. The symbol ds represents the magnitude of the displacement. And the coordinate s is *measured along the path.* This definition is illustrated in Fig. 7-10. If θ is the angle between the vectors **F** and $d\mathbf{s}$, the same figure shows that

$$F_\parallel = F \cos\theta \tag{7-32}$$

Thus the definition of the work W done by the force in the complete displacement from s_i to s_f can be written as

$$W = \int_{s_i}^{s_f} F \cos\theta\, ds \tag{7-33a}$$

A definition equivalent to Eq. (7-33*a*) is given by the equation

$$dW = F\cos\theta\, ds \tag{7-33b}$$

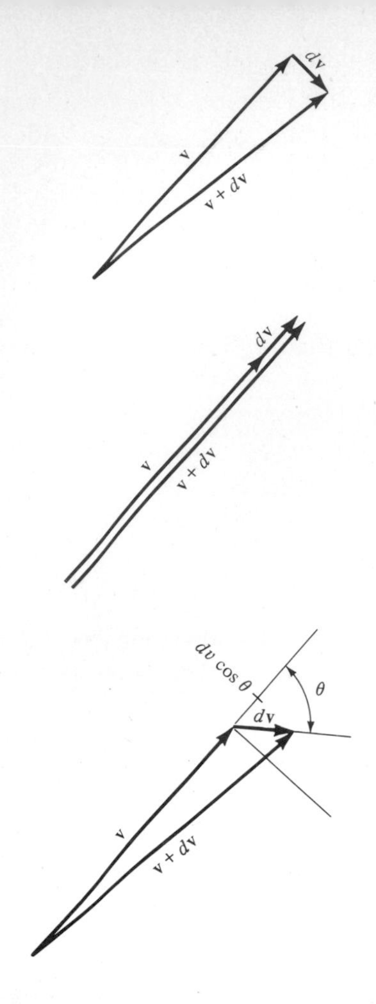

Fig. 7-9 The magnitude of a vector **v** will be changed by the addition of a vector $d\mathbf{v}$ only to the extent that $d\mathbf{v}$ has a component along the direction of **v**. On the top, a perpendicular velocity increment $d\mathbf{v}$ is added to the velocity **v**. If $d\mathbf{v}$ is infinitesimal, the sum $\mathbf{v} + d\mathbf{v}$ has the same magnitude as **v**. So adding $d\mathbf{v}$ to **v** does not change the magnitude of the velocity when $d\mathbf{v}$ is perpendicular to **v**. In the center $d\mathbf{v}$ is parallel to **v**. Here adding $d\mathbf{v}$ to **v** increases the magnitude of the velocity by an amount equal to dv, the magnitude of $d\mathbf{v}$. On the bottom $d\mathbf{v}$ is at an angle θ to **v**. In this case adding $d\mathbf{v}$ to **v** increases the magnitude of the velocity by an amount equal to $dv \cos\theta$, the component of $d\mathbf{v}$ along the direction of **v**.

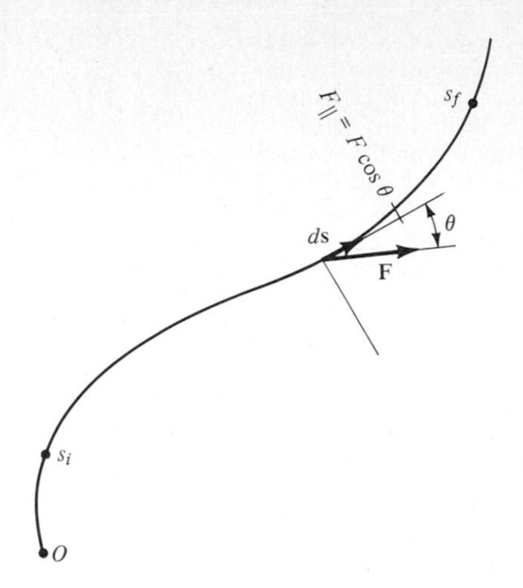

Fig. 7-10 The general definition of work. A body follows a path from an initial location specified by the coordinate s_i to a final location specified by the coordinate s_f, while a force **F** is applied to it. The coordinate s is measured *along* the path from some origin O. In an infinitesimal part of the path **ds**, for which **F** and **ds** are at the angle θ, the infinitesimal amount of work dW done by the force on the body has the value $dW = F_{\|}\, ds$. Here $F_{\|} = F \cos\theta$ is the component of **F** parallel to **ds**, and ds is the magnitude of **ds**. The total work W done by the force on the body is the sum of these infinitesimal contributions. That is, the total work is the integral $W = \int_{s_i}^{s_f} F \cos\theta\, ds$.

This says that in an infinitesimal displacement of magnitude ds, the infinitesimal element of work done dW is obtained by taking the product of ds and $F \cos\theta$. The quantity $F \cos\theta$ is used, instead of just F, since only the parallel component of the force vector is wanted. To calculate the total work W done when the body moves along its path between the points s_i and s_f, the right side of Eq. (7-33*b*) is integrated between these limits, and the left side is integrated between the corresponding limits W_i and W_f. Doing so, we have

$$\int_{W_i}^{W_f} dW = \int_{s_i}^{s_f} F \cos\theta\, ds$$

According to the fundamental theorem of calculus, the integral on the left side of this equality has the value $W_f - W_i$. We write the value as W, the total work done. Hence

$$\int_{W_i}^{W_f} dW = W_f - W_i = W$$

We therefore have

$$W = \int_{s_i}^{s_f} F \cos\theta\, ds$$

in agreement with Eq. (7-33*a*).

The real justification of Eq. (7-33*a*) is that it leads to the same relation between work and change in kinetic energy as is obtained in one dimension. We will show this shortly. But first we must make a small digression for the purpose of introducing a vector notation which is very convenient to use in manipulating the expression $F \cos\theta\, ds$.

The **scalar product** of two vector quantities **A** and **B** is a scalar quantity equal to $A (\cos\theta) B$, which we will write as $A \cos\theta\, B$, where A and B are their magnitudes and θ is the angle between their directions. The angle θ is always counted as positive, and is always the smaller of the two angles formed by the directions of **A** and **B.** Thus θ always lies in the range $0 \leq \theta \leq \pi$. Since A and B are positive, the sign of $A \cos\theta\, B$ is determined by the sign of $\cos\theta$. It is positive in the range $0 \leq \theta < \pi/2$ and negative in the range $\pi/2 < \theta \leq \pi$.

The scalar product is written symbolically as $\mathbf{A} \cdot \mathbf{B}$, with the bold dot being used to indicate the specific mathematical operation involved in the scalar product. In spoken language the scalar product is expressed as "A dot B," and consequently the scalar product is often called the **dot product.** Thus, by definition

$$\mathbf{A} \cdot \mathbf{B} = A \cos\theta\, B \tag{7-34}$$

It should be emphasized that the complete symbol $\mathbf{A} \cdot \mathbf{B}$ represents a scalar, even though it contains the two vector symbols **A** and **B.** This is so because its third part is the symbol $\cdot$ which stands for the operation of using these two vectors to evaluate the scalar on the right side of Eq. (7-34). Figure 7-11 illustrates $\mathbf{A} \cdot \mathbf{B}$ for a variety of cases. Note that $\mathbf{A} \cdot \mathbf{B} = AB$ if **A** is parallel to **B,** that $\mathbf{A} \cdot \mathbf{B} = -AB$ if **A** is antiparallel (in other words, is oppositely directed) to **B,** and that $\mathbf{A} \cdot \mathbf{B} = 0$ if **A** is perpendicular to **B.** The definition makes it clear that $\mathbf{A} \cdot \mathbf{B} = \mathbf{B} \cdot \mathbf{A}$. That is, the scalar product is *commutative.*

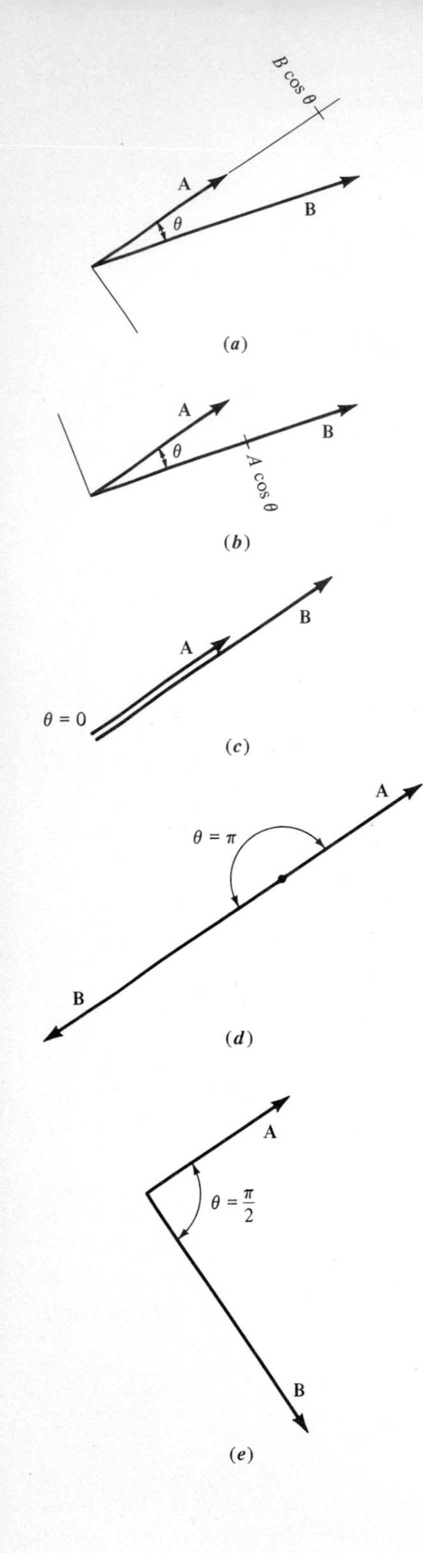

Fig. 7-11 (*a*) The dot product $\mathbf{A} \cdot \mathbf{B}$ of two vectors **A** and **B**, whose directions are separated by the angle θ, can be obtained by multiplying the magnitude of **A** and the component of **B** along the direction of **A**. In other words, the value is $A(\cos \theta B)$. (*b*) Alternatively, $\mathbf{A} \cdot \mathbf{B}$ can be obtained by multiplying the component of **A** along the direction of **B** and the magnitude of **B**. In other words, the value is $(A \cos \theta)B$. (*c*) If **A** is parallel to **B**, then $\theta = 0$ and $\cos \theta = 1$, so $\mathbf{A} \cdot \mathbf{B} = A \cos \theta \, B = AB$. (*d*) If **A** is antiparallel to **B**, then $\theta = 180°$ and $\cos \theta = -1$, so $\mathbf{A} \cdot \mathbf{B} = A \cos \theta \, B = -AB$. (*e*) If **A** is perpendicular to **B**, then $\theta = 90°$ and $\cos \theta = 0$, so $\mathbf{A} \cdot \mathbf{B} = A \cos \theta \, B = 0$.

The scalar, or dot, product has an obvious application in the definition of work, as well as in many other places in physics where there is a need to generate the component of one vector along the direction of another. It also has useful applications in geometry and trigonometry, as Example 7-4 shows.

EXAMPLE 7-4

The lengths of sides A and B of a triangle and the angle ϕ between these sides are known. Find the length C of the third side in terms of A, B, and ϕ.

■ The triangle is drawn in Fig. 7-12*a*. You must find a rule for this arbitrary triangle analogous to the pythagorean theorem for right triangles. For this purpose, it makes sense to generate an expression for the square of the length of side C. This can be done by considering Fig. 7-12*b*, which converts the triangle into a diagram for the vector addition

$$\mathbf{C} = \mathbf{A} + \mathbf{B}$$

You can generate an expression for C^2 by writing this equation again

$$\mathbf{C} = \mathbf{A} + \mathbf{B}$$

and then setting the dot product of the left sides of these two equations equal to the dot product of their right sides. You obtain

$$\mathbf{C} \cdot \mathbf{C} = (\mathbf{A} + \mathbf{B}) \cdot (\mathbf{A} + \mathbf{B})$$

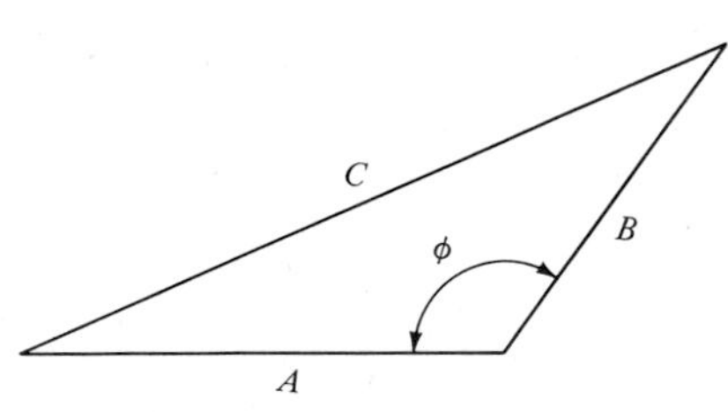

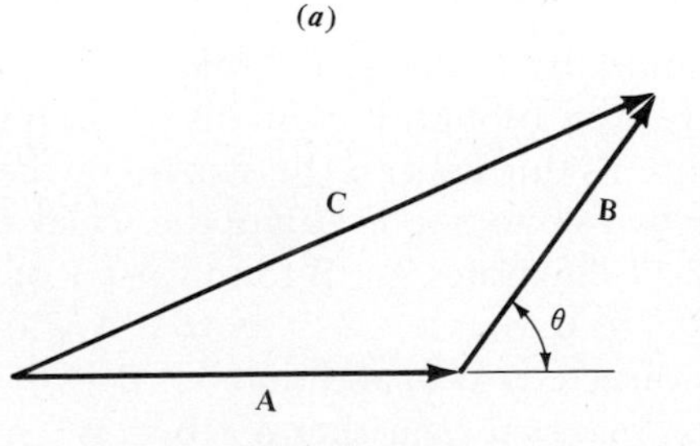

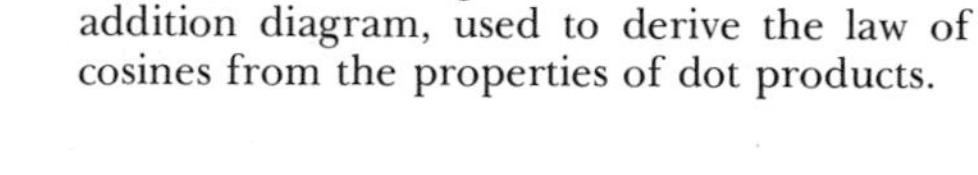

Fig. 7-12 A triangle, and associated vector addition diagram, used to derive the law of cosines from the properties of dot products.

Since $\mathbf{C} \cdot \mathbf{C} = C^2$, you have just what you want on the left side of the equality. Expanding the expression on the right side you obtain

$$C^2 = \mathbf{A} \cdot \mathbf{A} + \mathbf{B} \cdot \mathbf{B} + \mathbf{A} \cdot \mathbf{B} + \mathbf{B} \cdot \mathbf{A}$$

or, since $\mathbf{A} \cdot \mathbf{B} = \mathbf{B} \cdot \mathbf{A}$,

$$C^2 = A^2 + B^2 + 2\mathbf{A} \cdot \mathbf{B}$$

Evaluating the dot product, you get

$$\begin{aligned} C^2 &= A^2 + B^2 + 2A \cos \theta \, B \\ &= A^2 + B^2 + 2AB \cos (\pi - \phi) \\ &= A^2 + B^2 - 2AB \cos \phi \end{aligned}$$

So

$$C = \sqrt{A^2 + B^2 - 2AB \cos \phi}$$

You may remember that this result is called the law of cosines, and that its proof by the standard techniques of trigonometry is *considerably* more involved than its proof by using the dot product.

The most general definition of **work** is Eq. (7-33*a*). Written in terms of a scalar product, the definition is

$$W = \int_{s_i}^{s_f} \mathbf{F} \cdot d\mathbf{s} \tag{7-35}$$

The work done on a body by a force acting on it is the integral from its initial to final position of the dot product of the force and the infinitesimal displacements of the body along its path.

Now we will use this definition to show that in two or three dimensions the work done by the net force acting on a body equals the change in kinetic energy of the body, just as Eq. (7-31) says is the case for one dimension. The procedure is analogous to that used to obtain Eq. (7-31). In following it through you will see how to handle the differential calculus of scalar products.

Again we start with Newton's second law, thereby making it the basis for the relation we will obtain. The second law

$$\mathbf{F} = m \frac{d^2\mathbf{s}}{dt^2}$$

is used to evaluate $\mathbf{F}$ in the work integral. This gives

$$W = \int_{s_i}^{s_f} m \frac{d^2\mathbf{s}}{dt^2} \cdot d\mathbf{s}$$

in analogy to Eq. (7-28). Next we manipulate the quantity being integrated in very much the same way as in the calculation leading to Eq. (7-29):

$$m \frac{d^2\mathbf{s}}{dt^2} \cdot d\mathbf{s} = m \frac{d}{dt}\left(\frac{d\mathbf{s}}{dt}\right) \cdot d\mathbf{s} = m \frac{d\mathbf{v}}{dt} \cdot d\mathbf{s}$$

Now $d\mathbf{s} = \mathbf{v}\, dt$, so we have

$$m \frac{d^2\mathbf{s}}{dt^2} \cdot d\mathbf{s} = m \frac{d\mathbf{v}}{dt} \cdot \mathbf{v}\, dt = m\, d\mathbf{v} \cdot \mathbf{v}$$

This can be put into a more useful form by using the rule for differentiating the product of two functions to write

$$\frac{d(\mathbf{v} \cdot \mathbf{v})}{dt} = \mathbf{v} \cdot \frac{d\mathbf{v}}{dt} + \frac{d\mathbf{v}}{dt} \cdot \mathbf{v}$$

But the dot product is commutative, so

$$\frac{d(\mathbf{v} \cdot \mathbf{v})}{dt} = \frac{d\mathbf{v}}{dt} \cdot \mathbf{v} + \frac{d\mathbf{v}}{dt} \cdot \mathbf{v} = 2\frac{d\mathbf{v}}{dt} \cdot \mathbf{v}$$

Multiplying through by $dt/2$ yields

$$\frac{d(\mathbf{v} \cdot \mathbf{v})}{2} = d\mathbf{v} \cdot \mathbf{v}$$

Then multiplying through by m, and interchanging the sides, we obtain

$$m\, d\mathbf{v} \cdot \mathbf{v} = \frac{m}{2} d(\mathbf{v} \cdot \mathbf{v})$$

Substituting this into the expression for $m\, d^2\mathbf{s}/dt^2 \cdot d\mathbf{s}$ gives

$$m \frac{d^2\mathbf{s}}{dt^2} \cdot d\mathbf{s} = \frac{m}{2} d(\mathbf{v} \cdot \mathbf{v}) = \frac{m}{2} d(v^2)$$

From here on the calculation is identical to that leading to Eq. (7-30). We write the work integral as

$$W = \int_{v_i^2}^{v_f^2} \frac{m}{2} d(v^2) = \frac{m}{2} \int_{v_i^2}^{v_f^2} d(v^2)$$

and then use the fundamental theorem to evaluate the integral. We obtain

$$W = \frac{mv_f^2}{2} - \frac{mv_i^2}{2} \tag{7-36}$$

Just as it did in one dimension, this result prompts us to define the **kinetic energy** K of a body of mass m moving with speed v to be

$$K = \frac{mv^2}{2} \tag{7-37a}$$

The kinetic energy of a body is one-half its mass times the square of its speed. The value of K can be expressed in terms of the velocity $\mathbf{v}$ of a body by using the relation $\mathbf{v} \cdot \mathbf{v} = v^2$ to write

$$K = \frac{m\mathbf{v} \cdot \mathbf{v}}{2} \tag{7-37b}$$

Introducing the definition of kinetic energy into Eq. (7-36), we obtain an important relation between the work W done by the net force acting on a body and the initial and final values of the kinetic energy of the body:

$$W = K_f - K_i \tag{7-38a}$$

If we write the change in kinetic energy as $\Delta K = K_f - K_i$, the relation becomes

$$W = \Delta K \tag{7-38b}$$

That is, *the change in the kinetic energy of a body equals the work done by the net force acting on it during its motion.* This is the **work-kinetic energy relation.** It applies in one, two, or three dimensions. But since the relation was obtained by using Newton's second law, it is valid only in an inertial frame of reference.

Example 7-5 applies the work-kinetic energy relation in a two-dimensional situation.

EXAMPLE 7-5

A projectile of mass 1.25 kg is fired horizontally with an initial speed of 30.0 m/s at an initial elevation above the ground of 10.0 m. Moving with negligible air resistance under the influence of gravity, it follows the parabolic trajectory shown in Fig. 7-13 until it strikes the ground.

a. Calculate the work done on the projectile by the gravitational force while it is in flight.

b. Then use the work-kinetic energy relation to determine the speed of the projectile just before it strikes the ground.

■ **a.** The gravitational force acting on the projectile is $\mathbf{F} = m\mathbf{g}$, where $\mathbf{g}$ is the gravitational acceleration. The work done by this force on the projectile while it is in flight is, according to Eq. (7-35),

$$W = \int_{s_i}^{s_f} \mathbf{F} \cdot d\mathbf{s}$$

At any point in the trajectory

$$\mathbf{F} \cdot d\mathbf{s} = m\mathbf{g} \cdot d\mathbf{s} = mg \cos \theta \, ds$$

Here θ is the angle, illustrated in Fig. 7-13, between the downward gravitational force $m\mathbf{g}$ and the infinitesimal displacement $d\mathbf{s}$ along the trajectory at that point. Choosing x and y axes, with positive directions to the right and upward as in the figure, allows you to relate $d\mathbf{s}$ to the infinitesimal change dy in the projectile's y coordinate. The figure shows that the relation is

$$dy = -\cos \theta \, ds$$

Using this relation in the preceding equation, you have

$$\mathbf{F} \cdot d\mathbf{s} = -mg \, dy$$

Thus you can write the work integral as

$$W = \int_{s_i}^{s_f} \mathbf{F} \cdot d\mathbf{s} = \int_{y_i}^{y_f} -mg \, dy$$

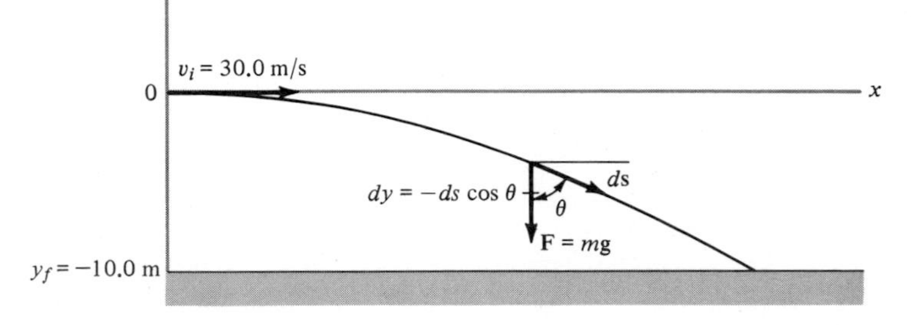

Fig. 7-13 A projectile moving without air resistance along a parabolic trajectory.

where y_i and y_f are the initial and final values of the coordinate y that correspond to the initial and final values of the coordinate s measured along the trajectory. Since mg is a constant, you have

$$W = -mg \int_{y_i}^{y_f} dy$$

The fundamental theorem of calculus then immediately gives you the result

$$W = -mg(y_f - y_i)$$

Setting $y_i = 0$, $y_f = -10.0$ m, $m = 1.25$ kg, and $g = 9.80$ m/s², you obtain the numerical value

$$W = -1.25 \text{ kg} \times 9.80 \text{ m/s}^2 \times (-10.0 \text{ m})$$
$$= 123 \text{ kg·m}^2/\text{s}^2$$

or

$$W = 123 \text{ J}$$

This is the work done by the gravitational force acting on the projectile while it is in flight. How does it compare with the work done if the projectile is simply dropped from the same height?

b. Since the only force acting on the projectile while it is in flight is the gravitational force, this is the net force acting on it. Therefore you can use the work-kinetic energy relation of Eq. (7-38*a*) to write

$$K_f - K_i = W$$

According to Eq. (7-36), the initial and final values of the projectile's kinetic energy are

$$K_i = \frac{mv_i^2}{2} \quad \text{and} \quad K_f = \frac{mv_f^2}{2}$$

where m is its mass and v_i and v_f are the initial and final values of its speed. So you have

$$\frac{mv_f^2}{2} - \frac{mv_i^2}{2} = W$$

Using the expression for W obtained in part *a*, with $y_i = 0$, gives you

$$\frac{mv_f^2}{2} - \frac{mv_i^2}{2} = -mgy_f$$

or

$$v_f^2 = v_i^2 - 2gy_f$$

Taking square roots of both sides of this equality, you obtain the result

$$v_f = \sqrt{v_i^2 - 2gy_f}$$

Positive roots are used since a speed is necessarily positive.

Setting $v_i = 30.0$ m/s, $g = 9.80$ m/s², and $y_f = -10.0$ m, you find that the final speed of the projectile is

$$v_f = \sqrt{(30.0 \text{ m/s})^2 - 2 \times 9.80 \text{ m/s}^2 \times (-10.0 \text{ m})}$$

or

$$v_f = 33.1 \text{ m/s}$$

You can verify that this value is correct by obtaining it from the equations developed in Sec. 3-1.

7-5 CONSERVATIVE FORCES

We saw in Sec. 7-1 that energy relations are particularly useful in analyzing systems whose total mechanical energy remains constant or, in other words, whose total mechanical energy is conserved. When a system is observed from an inertial reference frame, its total mechanical energy will be conserved if two conditions are satisfied:

1. The system must be isolated. For present purposes, this means that either no external forces act on its constituent bodies *or* every external force that acts on these bodies does zero work on them in the course of any possible motions of the bodies.

2. Each force internal to the system—that is, a force exerted on one body of the system by another body of the system—has the following property: The force does zero total work when the bodies contained in the system move from any given arrangement to any other arrangement and then back to the original arrangement. If this is the case, the bodies comprising the system complete their "round trips" with zero total work having been done on them by the forces that act within the system.

It is easy to understand why condition 1 must be satisfied by the forces acting on a system from the outside if there is to be conservation of a system's total mechanical energy. If the condition is not satisfied, then work will be done on the system by the external forces, and so the total mechanical energy of the system cannot remain constant. But it will take some effort to understand why condition 2 must be satisfied by the forces acting from within the system if its total mechanical energy is to be conserved. In this section we will study the properties of forces which meet condition 2, as well as the properties of forces which do not. Forces which do meet condition 2 are called **conservative forces** because of the role they play in the conservation of total mechanical energy.

A simple example of a system which conserves its mechanical energy is depicted in Fig. 7-14. A body of mass m is connected to one end of a spring of negligible mass. The other end is fixed. The body is supported against gravity by the frictionless surface of an air track. The forces acting on the body are a downward gravitational force of magnitude mg produced by the attraction of the earth, an upward supporting force of the same magnitude exerted by the air track, and a Hooke's-law force exerted by the spring. Choosing an x axis along the length of the air track, we write the spring force as $F(x) = -kx$. The coordinate x measures the extension ($x > 0$) or compression ($x < 0$) of the spring. If the body is pulled from its stable equilibrium position ($x = 0$) and then released, it will oscillate along the x axis about the equilibrium position in harmonic motion.

We can take the body-plus-spring system as an isolated system containing a single moving body acted on by the single internal force produced by the spring. In other words, we consider the essentially massless spring

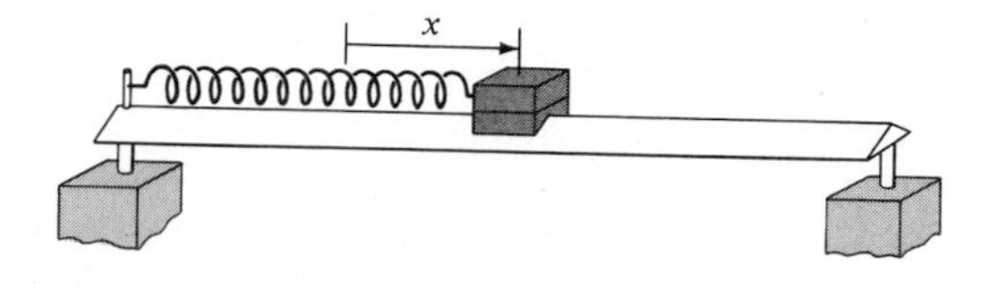

Fig. 7-14 A block supported by an air track and connected to one end of a spring. The other end of the spring is attached to a fixed pin.

not as a body in its own right, but as the source of the internal force acting on the body connected to its movable end. What about the forces exerted on the moving body by the earth and the air track? The downward force exerted on the body by the earth and the upward force exerted on it by the air track have the function of constraining the body to move in the horizontal plane. They are external to the body-plus-spring system. But since each is always acting in a direction perpendicular to the displacement of the body, *neither* does work on the body. Such forces are called **workless constraints.** Another workless constraint is the force exerted on the fixed end of the spring by the pin in the air track. No work is done by this force since its point of application is not displaced. So the body-plus-spring system is isolated, in the sense of condition 1, because the only external forces acting on it are workless constraints. Another significant advantage of using energy relations to analyze mechanical systems, instead of using Newton's laws of motion directly, is that *workless constraint forces can be ignored.*

The internal force acting on the body in the system, mentioned in condition 2, is the force exerted on it by the spring. This force does work on the body because it acts along the same line as the body's displacement. However, the total work it does is zero when the body moves from some location to some other location and then back to the original location. We will show this by following the body through such a journey.

Consider the oscillating body at an instant when it is moving to the right and passes the point $x = 0$ so that the spring is relaxed. The round trip will consist of its journey from that point to the point where the spring has its maximum extension, and then back to the point $x = 0$. In the first half of the trip, the spring force does a certain amount of negative work on the body. The work is negative since the force exerted by the spring acts to the left and the displacement of the body it acts on is to the right. In the second half of the trip, the spring force does positive work on the body because the force acts to the left and the displacement is also to the left.

Focus attention on the infinitesimal segment of the path traversed by the body which is located at x and whose length is $|dx|$. The body passes through this segment twice, once in each half of the round trip. Since the body travels in the positive direction during the first half of the trip, its displacement $dx_{\text{first half}}$ as it passes through the segment the first time has a positive value. And since the body travels in the negative direction during the second half of the trip, its displacement $dx_{\text{second half}}$ as it passes through the segment the second time has a negative value. That is,

$$dx_{\text{second half}} = -dx_{\text{first half}}$$

In each passage of the body through the path segment, the spring exerts the same force $F(x)$ on the body, since x has the same value both times. The contributions to the work integral from the two passages are

$$dW_{\text{first half}} = F(x)\ dx_{\text{first half}}$$

and

$$dW_{\text{second half}} = F(x)\ dx_{\text{second half}} = -F(x)\ dx_{\text{first half}} = -dW_{\text{first half}}$$

Thus the two contributions cancel:

$$dW_{\text{first half}} + dW_{\text{second half}} = 0$$

This argument does not depend on the particular value of x which specifies the location of the path segment. Consequently, it holds for *all* values of x within the region traversed by the body. The total work done by the spring force in the round trip is therefore zero.

By applying Eq. (7-38*b*),

$$W = \Delta K$$

to the round trip, we can conclude immediately that the kinetic energy K of the body when it returns to $x = 0$ must equal its kinetic energy when it left that point. This is true because the equation says that if the work done on the body is $W = 0$, then its change in kinetic energy is $\Delta K = 0$. The same conclusion will be reached for the next round trip made by the oscillating body, and so on. Therefore the body always has the *same* value of K when it passes through the point $x = 0$. When the body is at $x = 0$, the spring is at its relaxed length. At that length the spring cannot exert a force on anything. So it has no energy content because it has no ability to do work. Hence, the total mechanical energy of the system equals the kinetic energy of the body at $x = 0$. We have thus shown that the total mechanical energy of the system has the same value each time the body passes $x = 0$, because the spring force does zero total work on the body in each of its round trips from $x = 0$. This is consistent with condition 2 stated at the beginning of this section.

Soon we will learn how to evaluate the energy content of the spring when $x \neq 0$. We will find that the total mechanical energy of the body plus-spring system has the same value *whatever* the location of the body, so that the system conserves its total mechanical energy. This will prove to be a consequence of the fact that the spring force does zero total work in a round trip from *any* point $x \neq 0$ back to the same point.

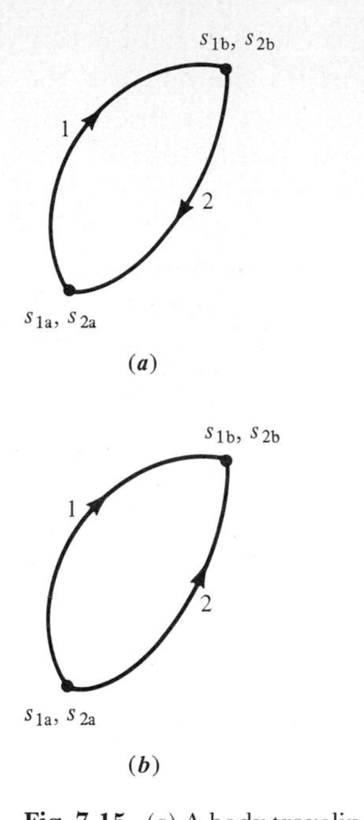

Fig. 7-15 (*a*) A body traveling from one point to another along path 1 and then traveling back to the starting point along path 2. The coordinate measured along path 1 of the starting point is s_{1a}, and that of the intermediate point is s_{1b}. The coordinate measured along path 2 of the intermediate point is s_{2b}, and that of the starting point is s_{2a}. (*b*) For a conservative force, the work it does on a body when the body moves from a point specified by coordinate s_a to a point specified by coordinate s_b does not depend on the path followed by the body in moving between the points. Thus the work done is the same whether the body follows path 1 or path 2.

We now consider conservative forces in systems where bodies move in more than one dimension. *A general test for a conservative force is that it must do zero total work on a body when the body moves through any closed path.* Expressed mathematically, a **conservative force F** is defined as one that satisfies the relation

$$\int_{s_{1a}}^{s_{1b}} \mathbf{F} \cdot d\mathbf{s}_1 + \int_{s_{2b}}^{s_{2a}} \mathbf{F} \cdot d\mathbf{s}_2 = 0 \qquad (7\text{-}39a)$$

The body on which the force acts makes the round trip depicted in Fig. 7-15*a*. The first integral is the work done along some path labeled 1 as the body moves from some position whose coordinate, measured along that path, is s_{1a} to some other position whose coordinate is s_{1b}. The second integral is the work done in the body's return along some other path 2 from the position on that path whose coordinate, measured along it, is s_{2b} to the position whose coordinate is s_{2a}. Note that s_{1b} and s_{2b} are corresponding values of the coordinates s_1 and s_2; that is, they specify the same position. The same is true for s_{1a} and s_{2a}. We can write the relation defining a conservative force in the compact form

$$W_{a \text{ to } b \text{ on } 1} + W_{b \text{ to } a \text{ on } 2} = 0 \qquad (7\text{-}39b)$$

by introducing the notation

$$\int_{s_{1a}}^{s_{1b}} \mathbf{F} \cdot d\mathbf{s}_1 \equiv W_{a \text{ to } b \text{ on } 1} \qquad (7\text{-}39c)$$

and similarly for the integral along path 2. That there is a connection between conservative forces and energy conservation is evident from the discussion, immediately above, of the body and spring, since the discussion led to a special case of Eq. (7-39*b*). We develop this connection fully in Sec. 7-6, but first we must learn more about conservative forces.

A completely equivalent form of Eq. (7-39*b*), which is more conveniently used to test for a conservative force, can be obtained by considering a body acted on by a conservative force which makes two separate round trips from the point whose label is *a* to the point whose label is *b*, and then back. In the first trip it goes out along path 1 and then back along path 1. By the definition of a conservative force, zero total work is done in this round trip. So

$$W_{a \text{ to } b \text{ on } 1} + W_{b \text{ to } a \text{ on } 1} = 0$$

or

$$W_{a \text{ to } b \text{ on } 1} = -W_{b \text{ to } a \text{ on } 1} \tag{7-40}$$

Thus we see that *reversing the direction in which a body traverses any path between any two points reverses the sign of the work done on the body by a conservative force.* In the second trip the body goes out along path 2 and then back along path 1. Again, the conservative force does zero total work in the round trip, and we have

$$W_{a \text{ to } b \text{ on } 2} + W_{b \text{ to } a \text{ on } 1} = 0$$

Applying Eq. (7-40) to the second term on the left side of this equation allows us to rewrite it as

$$W_{a \text{ to } b \text{ on } 2} - W_{a \text{ to } b \text{ on } 1} = 0$$

or

$$W_{a \text{ to } b \text{ on } 1} = W_{a \text{ to } b \text{ on } 2} \tag{7-41}$$

This relation must be satisfied if the force acting on the body is conservative. The relation tells us that *for a force to be conservative, the work it does when the body it acts on moves from any position to any other position must depend only on the two positions.* See Fig. 7-15*b*. In particular, *the work that a conservative force does cannot depend on any other specific characteristics, such as the path followed by the body, the speed of the body when making the trip, or the time when it does so.* Examples 7-6 through 7-8 demonstrate how Eq. (7-41) is used to find out whether a force is conservative. (The results of the first two examples are also used later to evaluate potential energies.)

EXAMPLE 7-6

A bead moves along the path shown in Fig. 7-16 between the position whose coordinate is s_a and the position whose coordinate is s_b. It is guided in the path because it is threaded through a wire bent to the shape shown, and it moves along the path because of its inertia. The bead is connected to one end of a very extensible spring, whose other end can rotate freely around a fixed pin. Assume that the spring is so extensible that its relaxed length is negligible compared to its length when the bead is anywhere along the part of the wire being considered. This will simplify the mathematics to be used, without affecting the point of physics to be learned. Evaluate the work $W_{a \text{ to } b}$ done by the force that the spring exerts on the bead. Then determine if the spring force is conservative. (Other forces may be exerted on the

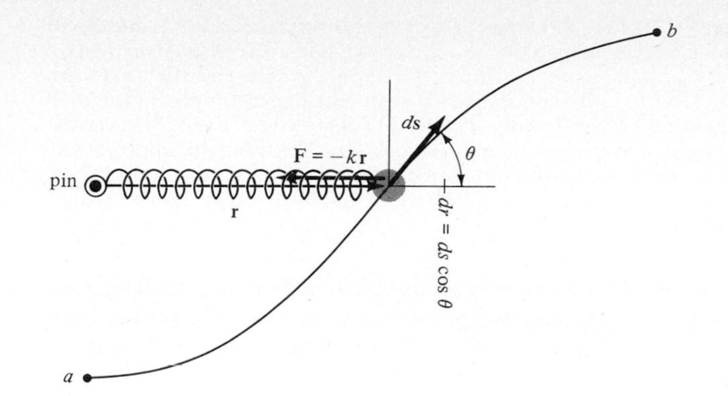

Fig. 7-16 A bead connected to one end of a spring and guided by a wire from point a to point b. The other end of the spring is connected to a fixed pin.

bead. In particular, the wire exerts a force on it. But only the spring force is being investigated here.)

■ If you let **r** be a vector extending from the pin to the bead, then its magnitude r is essentially equal to the extension of the spring. The Hooke's-law force which the spring exerts on the bead will be of magnitude kr, where k is the force constant. Since the force always acts in the direction toward the pin, the expression

$$\mathbf{F} = -k\mathbf{r}$$

gives both its direction and its magnitude. Thus

$$W_{a\text{ to }b} = \int_{s_a}^{s_b} \mathbf{F} \cdot d\mathbf{s} = -k \int_{s_a}^{s_b} \mathbf{r} \cdot d\mathbf{s}$$

where $d\mathbf{s}$ is an infinitesimal element of the path followed by the bead. Application of the definition of the dot product to Fig. 7-16 shows you that

$$\mathbf{r} \cdot d\mathbf{s} = r \cos \theta \, ds$$

But $\cos \theta \, ds$ is equal to dr, the change in the distance from the pin to the bead during the displacement. Thus

$$\mathbf{r} \cdot d\mathbf{s} = r \, dr$$

and so

$$W_{a\text{ to }b} = -k \int_{r_a}^{r_b} r \, dr$$

Using Eq. (7-20) with $n = 1$ to evaluate the integral, you obtain

$$\int_{r_a}^{r_b} r \, dr = \frac{r_b^2}{2} - \frac{r_a^2}{2}$$

Therefore you have

$$W_{a\text{ to }b} = -k \left(\frac{r_b^2}{2} - \frac{r_a^2}{2} \right) \tag{7-42}$$

The value of $W_{a\text{ to }b}$ in Eq. (7-42) was obtained without specifying the shape of the path followed by the bead in going from a to b. The same amount of work would be done by the force, no matter what path the wire made the bead follow, providing the bead went from position a to position b. So you have shown that the spring force is conservative. In this particular case, $W_{a\text{ to }b}$ actually depends not on the complete specification of positions a and b, but only on their distances r_a and r_b from the pin.

The physical reason for these results is as follows. In any displacement $d\mathbf{s}$, one component of the displacement lies in a direction parallel to the spring axis, and the other component lies in a direction perpendicular to that axis. The parallel component stretches the spring along its axis, and the perpendicular component rotates it

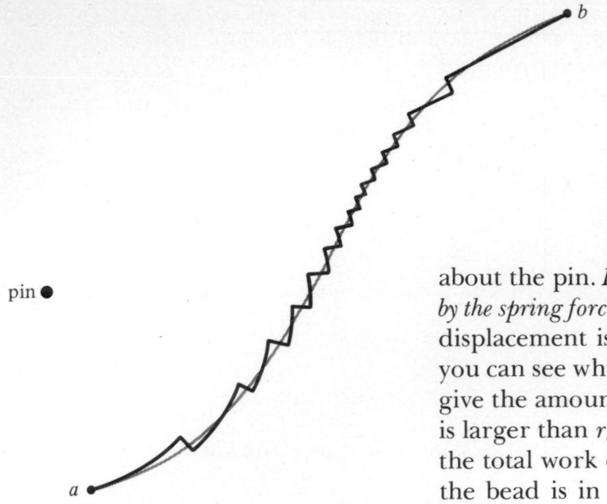

Fig. 7-17 The path from a to b is approximated by a sequence of radial line segments along the spring axis and arcs centered on the pin. No work is done by the spring force on the arcs, and the total work done by that force on the radial segments depends only on the total extension of the spring. The result obtained in Eq. (7-42) can be thought of as the limit of the result obtained by using this approximate path, since the approximation to the actual path is improved by making the arcs and radial line segments shorter and increasing their number.

about the pin. *But only the parallel component of the displacement leads to work being done by the spring force, because the force acts along the spring axis.* Since the work done in each displacement is related to how much the spring is stretched by the displacement, you can see why the total work done depends on only the quantities r_a and r_b, which give the amount of stretch at the beginning and end of the total displacement. If r_b is larger than r_a, as in the figure, the motion increases the stretch of the spring and the total work done is negative. This is because the force exerted by the spring on the bead is in the general direction opposing that motion. Figure 7-17, and the explanation in its caption, presents a somewhat different version of this argument.

EXAMPLE 7-7

The wire of Example 7-6 lies in a vertical plane near the surface of the earth. Thus the bead experiences a downward force $m\mathbf{g}$, where m is its mass and $\mathbf{g}$ is the gravitational acceleration. Calculate the work $W_{a \text{ to } b}$ done by the uniform gravitational force when the bead moves from a to b. Then decide whether the gravitational force is conservative.

■ The calculation that must be performed is similar to the one in Example 7-5. First you express the gravitational force as

$$\mathbf{F} = m\mathbf{g}$$

Then you note from Fig. 7-18 that

$$\begin{aligned} \mathbf{F} \cdot d\mathbf{s} &= F \cos \theta \, ds \\ &= F \cos (\pi - \phi) \, ds \end{aligned}$$

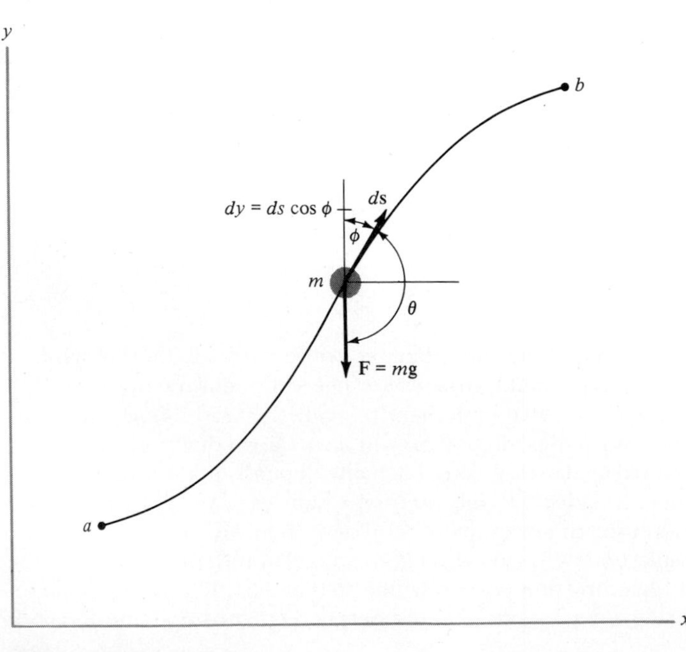

Fig. 7-18 The evaluation of the dot product $\mathbf{F} \cdot d\mathbf{s}$ for a particle moving under the influence of a gravitational force $\mathbf{F} = m\mathbf{g}$.

Now $\cos(\pi - \phi) = -\cos\phi$, so

$$\mathbf{F} \cdot d\mathbf{s} = -F \cos\phi \, ds$$

The figure also shows that $\cos\phi \, ds = dy$, where y is the vertical coordinate of the bead whose upward direction is defined to be positive. Thus

$$\mathbf{F} \cdot d\mathbf{s} = -F \, dy = -mg \, dy$$

The work done by the force is

$$W_{a \text{ to } b} = \int_{s_a}^{s_b} \mathbf{F} \cdot d\mathbf{s} = \int_{y_a}^{y_b} -mg \, dy = -mg \int_{y_a}^{y_b} dy$$

The integral immediately yields

$$W_{a \text{ to } b} = -mg(y_b - y_a) = mgy_a - mgy_b \tag{7-43}$$

Since $W_{a \text{ to } b}$ depends only on the positions of the beginning and end of the path—in fact, only on the heights of these positions—you can conclude that the uniform gravitational force is conservative.

For a case such as is illustrated in the figure, where y_b represents a higher elevation than y_a, the work $W_{a \text{ to } b}$ will be negative because the force acts downward and the general motion is upward. You should be able to give a physical explanation comparable to those at the end of Example 7-6 of why the work done depends only on the (heights of the) beginning and end points of the motion of the bead, and not on the particular path it follows in going between those points.

If the wire is perfectly smooth, then the force it exerts on the bead always is normal to the surface of the wire, and therefore always perpendicular to the bead's displacement along the wire. Hence this force constraining the bead to move along the path determined by the shape of the wire can do no work at all on the bead—it is a workless constraint and can be ignored. So in such a case only two forces are acting on the bead which do work—the spring force and the gravitational force. Both are conservative. Is the net force acting on the bead conservative?

EXAMPLE 7-8

The wire in Examples 7-6 and 7-7 is not perfectly smooth. It exerts a contact friction force of constant magnitude F on the bead. Evaluate the work $W_{a \text{ to } b}$ done by this force as the bead moves from position a to position b. Is the frictional force conservative?

▪ Since the contact friction force will always act on the bead in such a direction as to oppose its motion, when the bead makes the displacement $d\mathbf{s}$ along the wire, the direction of the contact friction force $\mathbf{F}$ will be opposite to the direction of $d\mathbf{s}$. So you can write the force in the form

$$\mathbf{F} = -F \, \widehat{d\mathbf{s}}$$

Here $\widehat{d\mathbf{s}}$ is a unit vector in the direction of $d\mathbf{s}$. (Note that the "hat" covers the entire symbol $d\mathbf{s}$ to remind you that the vector $\widehat{d\mathbf{s}}$ has a unit magnitude, and not an infinitesimal magnitude.) Using this form, you have

$$W_{a \text{ to } b} = \int_{s_a}^{s_b} \mathbf{F} \cdot d\mathbf{s} = \int_{s_a}^{s_b} -F \, \widehat{d\mathbf{s}} \cdot d\mathbf{s} = -F \int_{s_a}^{s_b} \widehat{d\mathbf{s}} \cdot d\mathbf{s}$$

since F is constant. But $\widehat{d\mathbf{s}} \cdot d\mathbf{s} = 1(\cos 0) \, ds = ds$. So

$$W_{a \text{ to } b} = -F \int_{s_a}^{s_b} ds$$

The integral can be evaluated immediately, and it yields the result

$$W_{a \text{ to } b} = -F(s_b - s_a) \tag{7-44}$$

At first glance, this may seem much like the results obtained in Examples 7-6 and 7-7. Certainly the mathematical expression in Eq. (7-44) looks quite similar to the one in Eq. (7-43). But its physical significance is *very* different. Since the infinitesimal element of displacement $\mathbf{ds}$ always lies *along* the path from a to b, the integral of its magnitude ds represents the total length of the path taken between those positions. That is, in the relation

$$\int_{s_a}^{s_b} ds = s_b - s_a$$

the quantity s_b is the distance measured along the path from some reference position on the path to position b, s_a is the distance along the path from the reference position to position a, and $s_b - s_a$ is therefore the distance along the path from a to b. Thus the work done depends on the *length* of the path followed by the bead on which the frictional force acts. The quantity $W_{a\text{ to }b}$ is *not* the same for all paths connecting a and b, and so the frictional force is *not* conservative. The point is illustrated in Fig. 7-19.

Equation (7-44) shows that the work done by the frictional force is always negative because the path length is always positive. The physical reason is that the frictional force always acts on the bead in the direction opposite to its displacement, and consequently the force does negative work in any motion of the bead. Thus when the bead makes a round trip, from a to b and then from b back to a, this force does negative work on the bead throughout. Friction continually removes mechanical energy from the system. So mechanical energy is *not* conserved in a system involving frictional forces, even if it is an isolated system. To put it another way, since one of the forces contributing to the net force acting on the bead is the nonconservative frictional force, the net force itself is not conservative.

It can be said that the contact friction force is not a conservative force because it depends on the direction of the velocity $\mathbf{v}$ of the object on which it acts. This is so since the direction of $\mathbf{v} = d\mathbf{s}/dt$ is the direction of $d\mathbf{s}$. The fluid friction force depends on both the direction and the magnitude of the velocity, and it also is not a conservative force. A force which does work cannot be conservative if the force varies with the velocity of the object on which it acts. The work done by such a force on the object does not depend only on the positions between which the object travels. Rather the work depends both on *where* the object it acts on goes and on *how* it gets there (on its direction and/or speed when making the trip), and so it is not conservative. Furthermore, any force that depends explicitly on time—that is, on *when* the trip is taken—is not conservative. Can you see why?

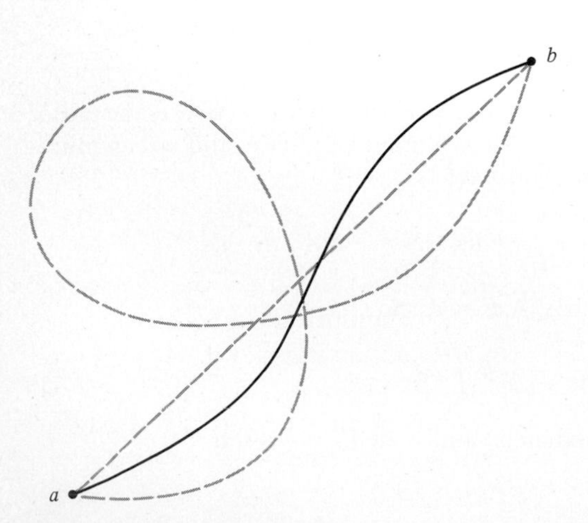

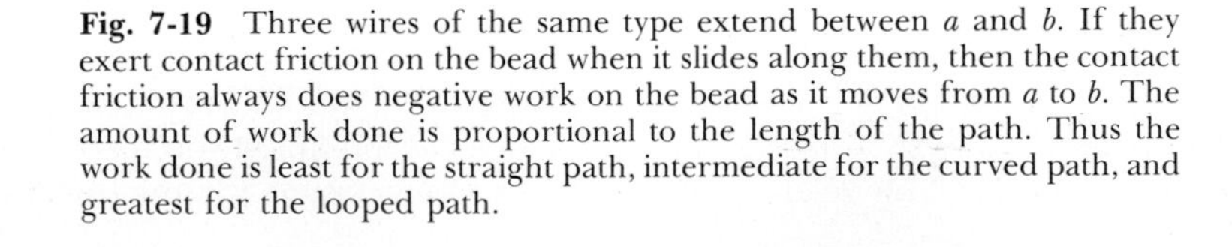

Fig. 7-19 Three wires of the same type extend between a and b. If they exert contact friction on the bead when it slides along them, then the contact friction always does negative work on the bead as it moves from a to b. The amount of work done is proportional to the length of the path. Thus the work done is least for the straight path, intermediate for the curved path, and greatest for the looped path.

7-6 POTENTIAL ENERGY AND ENERGY CONSERVATION

In this section we will continue to generalize the concepts introduced in Sec. 7-1 by developing the connection between a conservative force and the potential energy associated with that force. Then we will use this connection to obtain a general statement of one of the most important laws of physics, *the law of conservation of total mechanical energy.*

Suppose that a system contains only a single body whose position we must consider. Suppose also that the only force acting on the body is a *conservative* force exerted because of the presence of something else in the system. This force is an *internal* force—it is not an external force applied to the body from outside the system. An example is a system consisting of the earth and a brick. The brick can move, and so its position must be considered. But seen from a reference frame fixed to the ground, the earth cannot move. The conservative force acting on the movable body is the force of gravity. It is a force internal to the system, and it is exerted on the brick because of the presence in the system of the earth. Another example is a system comprising the brick supported by a horizontal air track and connected to one end of a spring, whose other end is attached to a pin in the track. As seen by an observer stationed at the air track, the position of the brick completely describes the appearance of the system at any instant. The conservative internal force acting on the brick is the force exerted on it by the spring. (The forces exerted by gravity and the air track are workless constraints.)

Consider the system when the movable body is at some position a. The conservative internal force acting on it will do work if the body moves to some other position o. Thus when the body is at position a, there is the potentiality that work can be done by this internal force, and so it is said that the system has potential energy. Specifically, the **potential energy** of the system is defined as *the work which will be done on the body by the internal force acting on it if the body moves from position a to some agreed-upon position o.* The symbol U is used to represent potential energy. By the definition just stated, its value is

$$U \equiv W_{a \text{ to } o} \tag{7-45}$$

The position o is called the **reference position.** It is chosen on the basis of convenience. In due course you will see many examples of how this is done. But note here that *the potential energy is zero if the body happens to be at the reference position.* That is, $U = 0$ if a is o, since $W_{o \text{ to } o} = 0$. Note also that U is *well defined,* even though the path from a to o is not specified, because the value of $W_{a \text{ to } o}$ is independent of the path for a conservative force. This comment should make it apparent to you that *there is no such thing as a potential energy for a nonconservative force.* It is impossible to associate a unique energy U with each position a in a situation where the work done if the body moves from a to o does not have a unique value.

We now embark on an argument which will lead to the law of mechanical energy conservation. Consider a system containing a single movable body. The system is *isolated* from all external forces that can do work on it. (This allows there to be external forces from workless constraints—such as a force exerted by a completely frictionless track that guides the movable body—since such forces do no work on the system.) Let the *net internal force* acting on the movable body be *conservative,* so that the system has a poten-

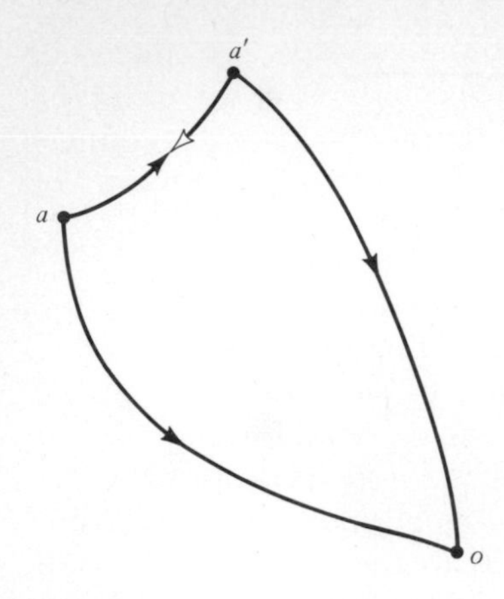

Fig. 7-20 Paths followed by a body which are used to establish the work-potential energy relation.

tial energy corresponding to this force. Suppose the body moves from position a, where the potential energy has the value U, to some other position a', where the potential energy is U', as shown in Fig. 7-20. The potential energy changes by the amount

$$\Delta U = U' - U = W_{a' \text{ to } o} - W_{a \text{ to } o}$$

But

$$W_{a' \text{ to } o} = W_{a' \text{ to } a} + W_{a \text{ to } o}$$

The reason is that the work done by the conservative net force is path-independent. So it will be the same for either of the paths indicated in the figure connecting a' with the reference position o. Substituting the value of $W_{a' \text{ to } o}$ in the equation for ΔU, we find the change in the potential energy of the system is

$$\Delta U = W_{a' \text{ to } a} + W_{a \text{ to } o} - W_{a \text{ to } o} = W_{a' \text{ to } a}$$

Now we can write

$$W_{a' \text{ to } a} = -W_{a \text{ to } a'}$$

because Eq. (7-40) tells us that reversing the direction in which a body traverses any path between any two points reverses the sign of the work done on the body by a conservative force. Using this in the equation for ΔU, we obtain a relation between the potential energy change when the body moves from a to a' and the work done on the body by the conservative force in this motion. It is

$$\Delta U = -W_{a \text{ to } a'} \tag{7-46}$$

This is the **work-potential energy relation:** *The change in the potential energy of an isolated system equals the negative of the work done by the conservative net internal force acting on the body during its motion.*

According to Eq. (7-38*b*), the work done by the net force acting on the body as it moves from a to a' is also equal to the change ΔK in its kinetic energy if, as we assume, the system is viewed from an inertial reference frame. Thus, using our present notation,

$$W_{a \text{ to } a'} = \Delta K \tag{7-47}$$

Combining this with Eq. (7-46), we have

$$\Delta U = -\Delta K$$

or

$$\Delta K + \Delta U = 0 \tag{7-48}$$

Since the sum of the changes in two quantities equals the change in their sum, we can write

$$\Delta K + \Delta U = \Delta(K + U)$$

So Eq. (7-48) can be written as

$$\Delta(K + U) = 0 \tag{7-49}$$

This result suggests the utility of defining the sum of the kinetic and potential energies of the system to be its **total mechanical energy** E. In symbols,

$$E \equiv K + U \tag{7-50}$$

Then Eq. (7-49) says

$$\Delta E = 0 \tag{7-51}$$

The total mechanical energy of the isolated system does not change when the body moves under the influence of a conservative net force. The reason is that any change in the kinetic energy is exactly compensated for by a change in the potential energy so that there is no change in their sum, the total mechanical energy. This observation leads us to conclude that Eq. (7-51) is equivalent to the statement

$$E = K + U = \text{constant} \tag{7-52}$$

This is the very important **law of conservation of total mechanical energy:** *If a system is isolated except for workless constraints, and all its internal forces are conservative so that the net internal force acting on a body of the system is conservative, then its total mechanical energy remains constant when it is viewed from an inertial reference frame.* We developed the law by considering a system with only one moving body, but an analogous development shows it to be true no matter how many moving bodies the system contains.

The law of conservation of total mechanical energy applies to a wide variety of systems. In each of them this conservation law plays as vital a role, in governing the behavior of the system, as does the law of total momentum conservation. The law of total mechanical energy conservation is less important in physics than the total momentum conservation law only in that the latter applies to *every* isolated system viewed from an inertial frame, while the former does not apply if there are any forces which are not conservative acting within such a system. In particular, any frictional force will prevent the law of total mechanical energy conservation from applying to a system, since all frictional forces are nonconservative.

The total mechanical energy of an isolated system involving only conservative forces is *not* constant when the system is viewed from a reference frame that is not inertial. Imagine a block stationary on the floor (considered to be an inertial frame). Think what the kinetic and gravitational potential energies will be when these quantities are evaluated from a frame of reference which is accelerating in some horizontal direction with respect to the floor (and which therefore is not an inertial frame). Since the speed of the block changes continually, as seen from this frame, its kinetic energy changes. But the horizontal motion of the noninertial frame does not affect the potential energy, and so the total mechanical energy does not appear to be constant.

An example of the application of mechanical energy conservation in analyzing the behavior of a simple system was given in Sec. 7-1. Another simple example is given immediately below. Chapter 8 is devoted to working out examples using the energy relations, particularly the law of conservation of total mechanical energy.

EXAMPLE 7-9

A pendulum bob of mass m at the end of a cord of length l is displaced from its stable equilibrium position so that the cord is at right angles to the vertical (see Fig. 7-21). The bob is then released from rest. How fast will it be moving when it goes through the bottom of its swing? This is the first question posed at the beginning of Sec. 7-1.

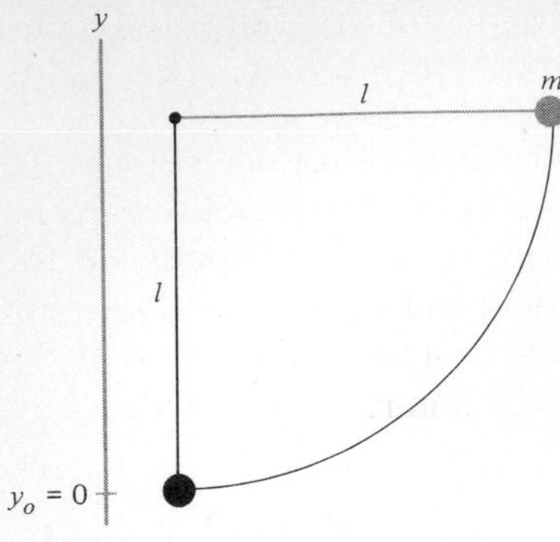

Fig. 7-21 A pendulum with the cord initially horizontal and the bob released from rest.

■ You can answer the question easily by applying the law of conservation of total mechanical energy. Take the frictionless, isolated system to be the bob plus the earth, which you view from the essentially inertial frame of the ground. The cord acts as a workless constraint, since the force it exerts on the bob is always perpendicular to the bob's instantaneous motion. The only internal force to be considered is the gravitational attraction exerted by the earth on the bob. To obtain an expression for the potential energy U of the system, combine its definition in Eq. (7-45),

$$U = W_{a \text{ to } o}$$

with Eq. (7-43)

$$W_{a \text{ to } b} = mgy_a - mgy_b$$

to obtain

$$U = mgy_a - mgy_o$$

Then, as indicated in Fig. 7-21, take the reference height y_o to be the height of the bob at its stable equilibrium position; also use that height to fix the origin of the upward-directed y axis. With these choices you have $y_o = 0$, so that

$$U = mgy_a$$

Dropping the now-unneeded subscript, you get

$$U = mgy \tag{7-53}$$

This is the potential energy of the system when the bob is at any height y above the reference height y_o. The kinetic energy is that of the motion of the bob (the earth being motionless in your reference frame).

The initial values of the potential, kinetic, and total mechanical energies are, since $y = l$ initially,

$$U = mgl$$

$$K = 0$$

and

$$E = K + U = mgl$$

The final values of these energies are

$$U = 0$$

$$K = \frac{mv^2}{2}$$

and

$$E = K + U = \frac{mv^2}{2}$$

where v is the speed of the bob when $y = 0$. Equating the initial and final values of the total mechanical energy, you have

$$mgl = \frac{mv^2}{2}$$

Solving for v produces the equation you set out to find:

$$v = \sqrt{2gl} \tag{7-54}$$

You should use similar, but even simpler, energy considerations to answer the questions posed in Sec. 7-1 about the maximum displacement of the bob on the opposite side of its swing, and about the swing of the interrupted pendulum.

7-7 EVALUATION OF FORCE FROM POTENTIAL ENERGY

If both the conservative net force F and the motion of the body it acts on lie along the x axis, the definition of Eq. (7-45) for the potential energy U at position x_a reduces to the one-dimensional form

$$U = W_{x_a \text{ to } x_o} = \int_{x_a}^{x_o} F\, dx \tag{7-55}$$

where x_o is the reference position. This relation can be used to evaluate the potential energy from the force. But sometimes it is necessary to go the other way, that is, to evaluate the force from the potential energy. This can be done by using the relation

$$F = -\frac{dU}{dx} \tag{7-56}$$

To show that Eq. (7-56) is consistent with the definition of Eq. (7-55), we substitute it into the definition and then show the validity of the result. We have

$$\begin{aligned} U &= \int_{x_a}^{x_o} F\, dx = -\int_{x_a}^{x_o} \frac{dU}{dx}\, dx = -\int_{U_a}^{U_o} dU = -(U_o - U_a) \\ &= U_a - U_o \end{aligned}$$

Here U_a is the value of the potential energy at position x_a, and U_o is its value at the reference position x_o. As explained immediately below Eq. (7-45), the potential energy is zero at the reference position. So $U_o = 0$, and we have

$$U = U_a$$

This is certainly valid since the symbol on the left is just an abbreviation for the symbol on the right. Therefore Eq. (7-56) is consistent with Eq. (7-55). Example 7-10 makes use of Eq. (7-56).

EXAMPLE 7-10

The results of Example 7-6, and particularly Eq. (7-42), show that the potential energy associated with the force produced by the spring in Fig. 7-16 can be expressed as

$$U = \frac{kr_a^2}{2} \tag{7-57}$$

For this expression the fixed pin at one end of the spring is chosen to be the reference position of the body at the other end, and r_a is the length of the spring when extended. In obtaining Eq. (7-42) it was assumed that the length of the spring is negligible when relaxed, so that r_a is also equal to the extension of the spring. Figure 7-22 shows a spring with the same force constant k whose relaxed length is *not* negligible. One end is fixed, and the other is attached to a body which can move along the x axis. In light of Eq. (7-57), it is reasonable to guess that the potential energy in this case can be expressed as

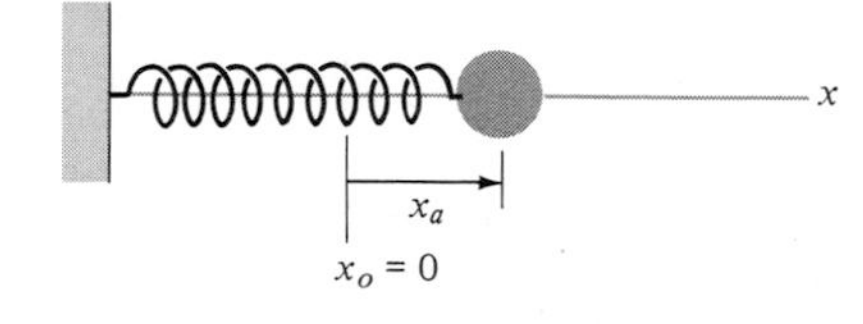

Fig. 7-22 A spring with one end fixed and the other end at x_a, a coordinate measured from the position of the free end when the spring has its relaxed length.

$$U = \frac{kx_a^2}{2}$$

or, without the subscript,

$$U = \frac{kx^2}{2} \tag{7-58}$$

Here x is the location of the free end of the spring, measured from an origin located at the free end of the spring when it is relaxed. Thus x is also the extension of the spring if x is positive or its compression if x is negative. Show that the force corresponding to this potential agrees with Hooke's law, and thereby verify Eq. (7-58). Also find the reference position chosen for the potential energy in Eq. (7-58).

■ You employ Eq. (7-56)

$$F = -\frac{dU}{dx}$$

and obtain

$$F = -\frac{d}{dx}\left(\frac{kx^2}{2}\right) = -\frac{k}{2}\frac{d(x^2)}{dx} = -\frac{k}{2}2x = -kx$$

Since this expression for F is precisely Hooke's law, the correctness of the expression for U in Eq. (7-58) is proved.

The reference position used in specifying the values of a potential energy is that position where the potential energy is defined to be zero. Since Eq. (7-58) yields $U = 0$ when $x = 0$, it is apparent that the reference position chosen in the definition of this potential energy is the origin of the x axis. This is the location of the free end of the spring when it is at its relaxed length.

An important point can be made in connection with Example 7-10. If the reference position used to define the potential energy associated with the spring force is changed (without changing the way the coordinate x is defined), the effect will be to add a constant to the right side of Eq. (7-58). This constant is, physically, the work that the spring force does on a body connected to the end of the spring if the body moves to the new reference position from the original reference position. But adding a constant to the potential energy U will have no effect at all on the force, $F = -dU/dx$, calculated from U, because the derivative of a constant is zero. The same situation occurs for any type of force and its associated potential energy. The actual value of the potential energy of a system is *arbitrary,* in the sense that a change in the choice of the reference position has the effect of adding a constant to the potential energy. But this makes no change in the force associated with the potential energy. And since motion is produced by force, it makes no change in the motion of the body on which the force acts.

The choice actually made for a reference position is dictated by convenience. Usually, it is chosen so as to make the form of the expression for U as simple as possible. For instance, in Example 7-10 the reference position was chosen in such a way that Eq. (7-58) had the form $U = kx^2/2$. Any other choice would lead to an expression of the form $U = kx^2/2 + C$. For the spring force the simplest expression results from choosing $U = 0$ at the origin $x = 0$. But in other cases a different choice is required to produce the simplest expression.

In more complicated situations the potential energy may depend on all three space coordinates x, y, x, so that it is written as $U(x, y, z)$. The components F_x, F_y, F_z of the force $\mathbf{F}$ along the corresponding axes can be evaluated from the relations

$$F_x = -\frac{\partial U(x, y, z)}{\partial x} \tag{7-59a}$$

$$F_y = -\frac{\partial U(x, y, z)}{\partial y} \tag{7-59b}$$

$$F_z = -\frac{\partial U(x, y, z)}{\partial z} \tag{7-59c}$$

The quantities in these relations are **partial derivatives.** A partial derivative of a function of several independent variables is a derivative evaluated by allowing one of the variables to vary while treating the others as if they were constants. For instance, the meaning of the quantity on the right side of the first relation is simply

$$\frac{\partial U(x, y, z)}{\partial x} \equiv \left[\frac{dU(x, y, z)}{dx}\right]_{\text{evaluated by treating } y \text{ and } z \text{ as constants}} \tag{7-60}$$

The validity of Eqs. (7-59) is established in much the same way as we verified Eq. (7-56). Example 7-11 uses them.

EXAMPLE 7-11

Figure 7-23 shows a two-dimensional system consisting of a spring extending from the point (0, 0) to a body at the point (x, y) and a second spring extending from the point $(c, 0)$ to the same body. The force constant of the first is k_1; the force constant of the second is k_2. As in Example 7-6, both springs are assumed to be so extensible that their lengths when relaxed are negligible. Thus vectors $\mathbf{r}_1$ and $\mathbf{r}_2$ specify the magnitudes and directions of the extensions of the two springs. (This idealization simplifies the mathematics without affecting the significant physical concepts.) Determine the x and y components of the total force exerted on the body.

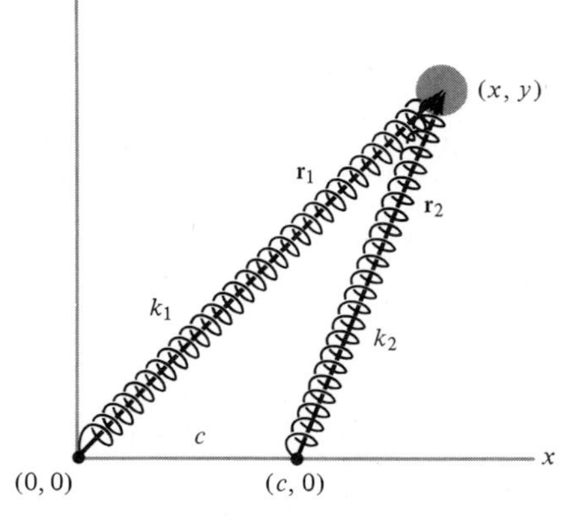

Fig. 7-23 Springs of force constant k_1 and k_2 connected to a body located somewhere in the xy plane. The other end of the first spring is at $x = 0$, $y = 0$, and the other end of the second spring is at $x = c$, $y = 0$.

■ Using Eq. (7-57), you can write the potential energy arising from the first spring as

$$U_1 = \frac{k_1 r_1^2}{2} = \frac{k_1}{2}(x^2 + y^2)$$

and the potential energy arising from the second spring as

$$U_2 = \frac{k_2 r_2^2}{2} = \frac{k_2}{2}[(x - c)^2 + y^2]$$

The total potential energy of the system is

$$U = U_1 + U_2 = \frac{k_1}{2}(x^2 + y^2) + \frac{k_2}{2}[(x - c)^2 + y^2]$$

Knowing U, you can immediately obtain the components F_x and F_y of the total force acting on the body by using Eqs. (7-59a) and (7-59b) for $U = U(x, y)$. That is,

$$F_x = -\frac{\partial U}{\partial x} = -\frac{k_1}{2}\frac{\partial}{\partial x}(x^2 + y^2) - \frac{k_2}{2}\frac{\partial}{\partial x}[(x - c)^2 + y^2]$$
$$= -\frac{k_1 2x}{2} - \frac{k_2 2(x - c)}{2} = -(k_1 + k_2)x + k_2 c$$

and

$$F_y = -\frac{\partial U}{\partial y} = -\frac{k_1}{2}\frac{\partial}{\partial y}(x^2 + y^2) - \frac{k_2}{2}\frac{\partial}{\partial y}[(x - c)^2 + y^2]$$
$$= -\frac{k_1 2y}{2} - \frac{k_2 2y}{2} = -(k_1 + k_2)y$$

It is easy to evaluate the magnitude and direction of the total force from its components F_x and F_y. But it is likely that the reason you want to know the total force is that you want to set up the equations governing the motion of the body. If so, then it is the components that you need, since the equations are

$$F_x = m\frac{d^2x}{dt^2}$$

and

$$F_y = m\frac{d^2y}{dt^2}$$

You can also determine the components of the total force by adding the forces produced by the two springs and then taking components, or more easily by taking the components of the two forces and then adding them. Do this, and show that the results obtained are the same as those obtained here.

EXERCISES

Group A

7-1. *Kinetic energy and stopping distance.*

a. Find the kinetic energy of each of the following objects:

(1) A pitched baseball: mass = 0.15 kg, speed = 40 m/s

(2) Rifle bullet: 0.002 kg, 500 m/s

(3) Jogger: 70 kg, 3.0 m/s

(4) Automobile on the highway: 2000 kg, 25 m/s

(5) Medium-sized cargo ship approaching dock: 3×10^7 kg, 1.0 m/s

b. Suppose each of the objects listed in part *a* were acted on by a steady retarding force equal in magnitude to the weight of a typical adult human, 700 N. What distance would be required to bring each to rest?

7-2. *Sprinting.*

a. Estimate your top sprinting speed.

b. Starting from rest, what is the minimum amount of work you must do to reach your top speed?

7-3. *Gunsmoke.* A gun of mass M fires a bullet of mass m. Therefore $MV = mv$, where V and v are the speeds of the gun and bullet, respectively, immediately after firing. Calculate the ratio of the kinetic energy of the gun to the kinetic energy of the bullet at that moment.

7-4. *Energy and the pendulum, I.* As shown in Fig. 7E-4, a pendulum bob is hanging at one end of a rod of length 2 m. The other end of the rod is mounted on a frictionless axle. The mass of the rod is negligible.

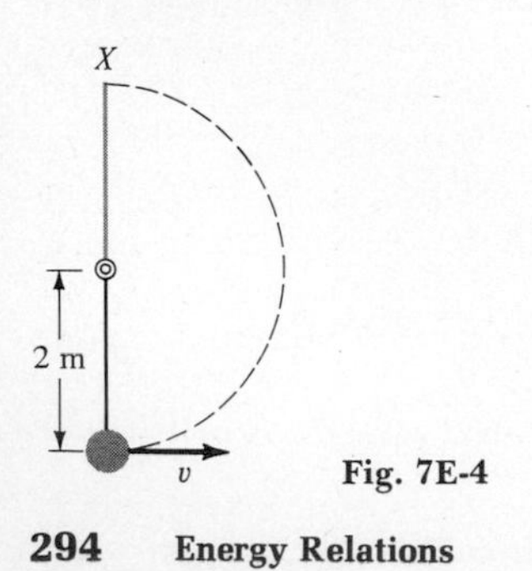

Fig. 7E-4

a. The bob is struck sharply, which gives it an initial speed v. Use energy relations to find the value of v for which the bob will almost reach the point X (directly above the axle) before it reverses its motion.

b. Compare your result with the result obtained by direct application of Newton's laws in Exercise 6-38.

7-5. *What's the angle?* A 1.00-m long pendulum is tied to the top of a cupboard (Fig. 7E-5). The bob is raised so that the string makes an angle of 30° with the vertical. The bob is released. If the side of the cupboard is 0.50 m long, what angle will the string make with the vertical when the bob is at its highest point under the cupboard? Assume all frictional effects to be negligible.

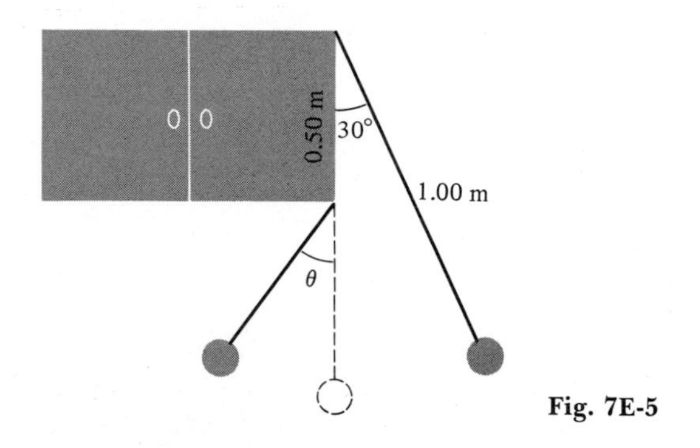

Fig. 7E-5

7-6. *Pushing a block.* Starting from rest, a 10-kg block is pushed along a horizontal surface for a distance of 5.0 m. The horizontal force used to push the block is 30 N, and there is a resisting frictional force of 25 N.

a. What is the total work done by the applied force?

b. How much work is done in overcoming friction?

c. What is the speed of the block when it has covered 5.0 m?

7-7. *Speed check.* Verify the final projectile speed obtained in Example 7-5 by using the equations developed in Sec. 3-1.

7-8. *The effect of work on speed.* The net force acting on an automobile of mass 2000 kg does 100,000 J of work. Find the final speed of the automobile, if its initial speed is

a. Zero
b. 10.0 m/s (about 22 mi/h)
c. 25.0 m/s (about 56 mi/h)
d. 40.0 m/s (about 89 mi/h)

7-9. *Workforce.* Under the action of a net force whose direction is along the direction of motion, a particle of mass m increases its speed from v_i to v_f in covering a straight-line path of length s.

a. How much work is done by the net force?
b. If the net force is constant, what is its magnitude?
c. Evaluate your results for parts a and b for $m = 10$ kg, $v_i = 2.0$ m/s, $v_f = 7.0$ m/s, and $s = 20$ m.

7-10. *Working out.*

a. How much work is done against gravity by a weightlifter who lifts 100 kg a vertical distance of 2.0 m?
b. Compare the result of part a with the work done against gravity by a 70-kg person climbing four flights of stairs (total vertical distance 12 m).

7-11. *Filled to the brim.* A cylindrical water tank 10 m high has a capacity of 1000 m^3. It is to be filled with water from a lake whose surface is at the same elevation as the bottom of the tank. (In working this exercise, ignore frictional losses in the pipes and the kinetic energy of the water as it leaves the hose.)

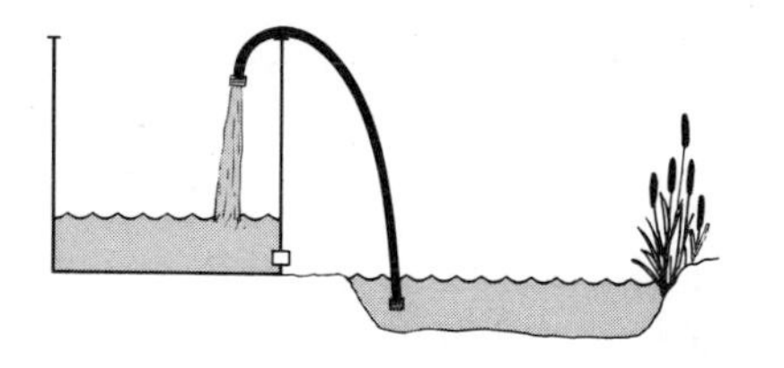

Fig. 7E-11

a. Suppose that the tank is filled by raising a hose to the top and pumping in water, as shown in Fig. 7E-11. How much work is done in filling the tank in this manner?
b. Suppose that the tank is filled by connecting the hose to an inlet at the bottom of the tank. How much work is done in filling the tank in this manner?
c. What happens to the extra energy in part a?

7-12. *Chair lift.* The sailor in the bosun's chair shown in Fig. 7E-12 has a mass m. He plans to raise himself a distance h to the upper pulley by pulling on the rope on the right.

a. By how much will his gravitational potential energy increase?
b. What force must he exert to lift himself?

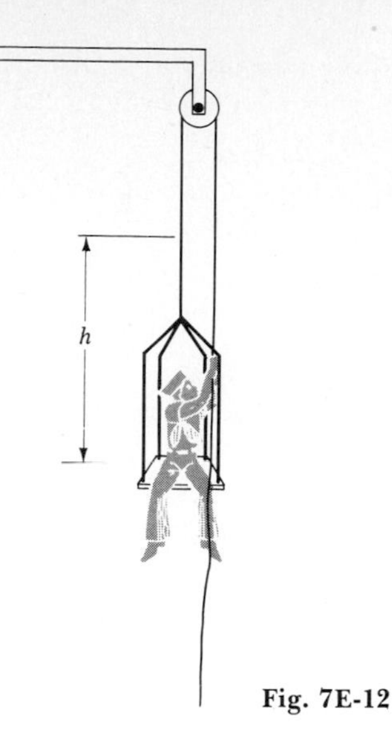

Fig. 7E-12

c. What length of rope must he pull to get to the upper position?
d. How much work will he do? Neglect friction.

7-13. *Hauling a sled.* As shown in Fig. 7E-13, a force of 100 N at 30° to the horizontal is required to draw a sled at uniform speed along a horizontal sidewalk.

a. How much work is done by the applied force in pulling the sled a distance of 10 m?
b. What is the magnitude of the frictional force exerted on the sled by the sidewalk?
c. How much work does the frictional force do when the sled is pulled a distance of 10 m?
d. What is the net work done on the sled?

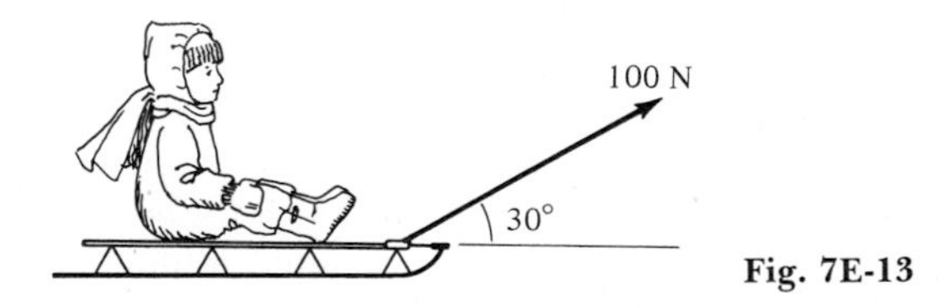

Fig. 7E-13

7-14. *Heavy work.* You are pushing a heavy box across the floor by applying a horizontally directed force of magnitude 150 N. The force of kinetic contact friction applied by the floor to the box has a magnitude of 140 N. The box moves 8.00 m.

a. How much work is done on the box by the force you apply? By the force the floor applies? By the net force acting on the box?
b. Describe qualitatively the motion of the box.

7-15. *Down and around.* A small sphere of mass 1.00 kg is attached to one end of a 1.00-m rod of negligible mass. The other end is mounted on an axle having negligible friction. Initially, the sphere is directly above the axle, as shown in Fig. 7E-15.

a. When the sphere falls, how fast will it be moving as it passes its lowest point?

b. What will be the tension in the rod at that instant?

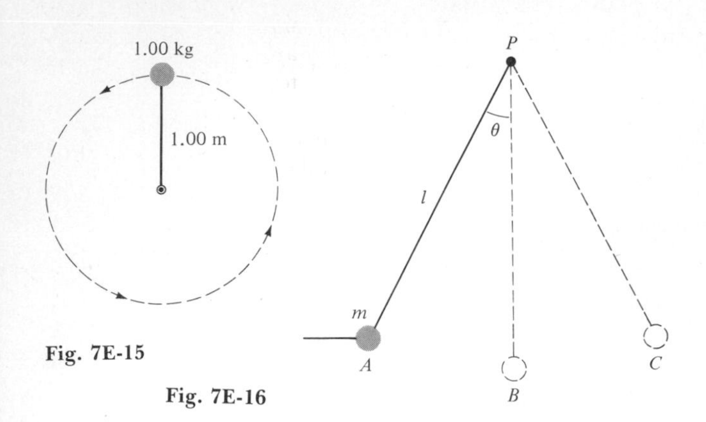

Fig. 7E-15

Fig. 7E-16

7-16. *A captive pendulum.* Figure 7E-16 shows a pendulum consisting of a bob of mass m attached to a cord of negligible mass having length l. The pendulum bob is held by a horizontal string at A, so that the pendulum cord is inclined at an angle $\theta_i = 30°$ with the vertical.

a. What is the tension in the pendulum cord?

b. What is the tension in the horizontal string?

c. The horizontal string is cut, releasing the pendulum. What is the speed of the bob as it passes through the lowest point, B?

d. What is the tension in the pendulum cord at B?

e. What is the tension in the cord when the bob reaches the highest point, C?

7-17. *Net force exerted by two springs.* Verify the expressions obtained in Example 7-11 for the components of the net force exerted by the two springs. Do this by finding the components of each spring force and then adding them.

Group B

7-18. *Driving a nail.* Refer to the discussion in the text following Eq. (7-4). A procedure is suggested in the second paragraph after Eq. (7-4) for evaluating the work done by a puck when it drives a nail into a block of wood fastened to the edge of an air table. Follow the procedure and obtain an expression for the work done, in terms of the mass of the puck and its initial speed.

7-19. *Lifting and lowering.* By following the procedure suggested in the small-print section after Eq. (7-7), calculate the net work done by the force you apply in slowly raising a puck of mass m vertically upward from the floor at $x = 0$ to a position x'' and then lowering it vertically downward to a position x lower than x'' but higher than $x = 0$.

7-20. *Energy and the pendulum, II.* A pendulum is released from rest with the cord horizontal. The length of the cord is 0.50 m. Use energy relations to determine the speed of the pendulum bob at an instant when the cord is vertical. Compare your results with the results obtained by direct application of Newton's laws in Exercise 6-37.

7-21. *Across the table, I.* In the system shown in Fig. 7E-21, the pulley and string have negligible mass, and the pulley and tabletop are frictionless. The system is released from rest.

a. Find the speed of bodies A and B when B has descended a distance D.

b. Evaluate your result for $m_A = 20$ kg, $m_B = 30$ kg, and $D = 5.0$ m.

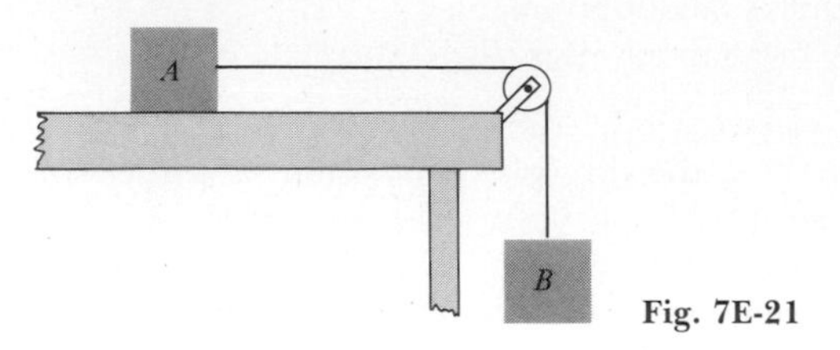

Fig. 7E-21

7-22. *An unbalanced rod.* A light stick of length l is pivoted at its center (Fig. 7E-22). Bodies of mass $2m$ and m are attached at its ends. The stick is held horizontally and released. What will be the speed of either end when the stick is vertical?

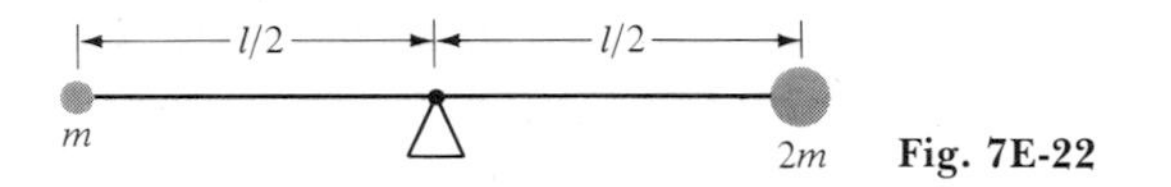

Fig. 7E-22

7-23. *Kinetic energy in athletics.* Outstanding performances for a number of athletic events are listed below. Neglecting air resistance and assuming that each projectile is launched at the optimum 45° elevation angle, calculate the initial kinetic energy for each case.

a. Shot put: mass = 7.26 kg, distance thrown = 22.0 m

b. Discus throw: 2.00 kg, 70.9 m

c. Hammer throw: 7.26 kg, 79.3 m

d. Javelin throw: 0.800 kg, 94.6 m

e. Long jump: 60.0 kg, 8.90 m

f. Baseball throw: 0.145 kg, 130 m

7-24. *To the top.* A child at play wishes to launch a 2.0-kg block up an inclined plane with sufficient speed to reach the top of the incline. The plane is 3.0 m long and is inclined at 20°. The coefficient of kinetic friction between the block and the plane is 0.40. What minimum initial kinetic energy must the child supply to the block?

7-25. *Rocket mass and velocity.* The motion of a rocket moving in free space is governed by Eq. (5-31a), $m(dV/dt) = v_g'(dm/dt)$, where m is the mass of the rocket, V is its velocity, and v_g' is the constant velocity of the gas ejected from the rocket engine, as seen from the rocket, so that $v_g' < 0$.

a. The infinitesimal change dV in the velocity of the rocket can be written $dV = v_g'\, dm/m$. Integrate both sides of this equation between initial and final values of V and m to find a relation between the change in the rocket's velocity and the change in its mass necessary to produce it.

b. A rocket starts from rest with mass m_i. What is the mass of the rocket when it reaches a speed $|V| = 1.5|v_g'|$?

7-26. *Stopping distance.*

a. The speed of a car is v, and the coefficient of static friction between the tires and the road is μ_s. Use energy considerations to derive an expression for the shortest distance in which the car can come to a full stop. Neglect the reaction time of the driver and the effect of the idling engine.

b. Using the expression derived in part *a*, calculate the stopping distance for a car traveling at 30 m/s (about 70 mi/h) with $\mu_s = 0.5$.

7-27. *Calculating the work done, I.* One of the forces acting on a certain particle depends on the particle's position in the *xy* plane. This force $\mathbf{F}_1$, expressed in newtons, is given by the expression $\mathbf{F}_1 = (2x^2\hat{\mathbf{x}} + 3y^2\hat{\mathbf{y}})(1\ \text{N/m}^2)$, where x and y are expressed in meters. Calculate the work $\int_A^D \mathbf{F}_1 \cdot d\mathbf{s}$ done by this force when the particle moves from point A to point D in Fig. 7E-27.

a. along the straight line AD

b. along the path ABD, which consists of two straight lines

c. along the path ACD, which consists of the straight line AC followed by the circular quadrant CD

d. Is $\mathbf{F}_1$ a conservative force?

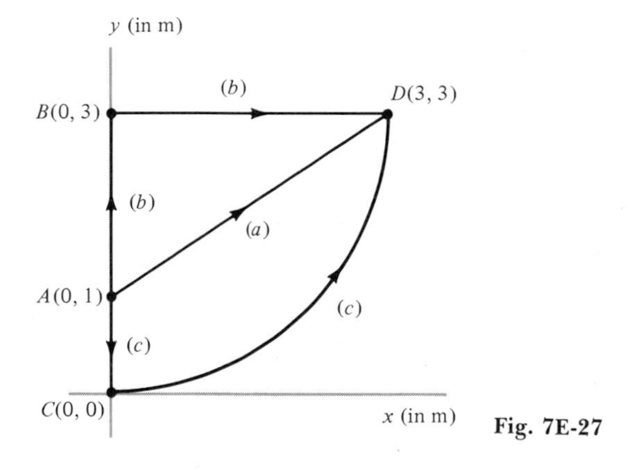

Fig. 7E-27

7-28. *Down the incline.* A crate of mass 50 kg slides down a 30° incline. The crate's acceleration is 2.0 m/s², and the incline is 10 m long.

a. What is the kinetic energy of the crate as it reaches the bottom of the incline?

b. How much work is spent in overcoming friction?

c. What is the magnitude of the frictional force that acts on the crate as it slides down the incline?

d. What is the coefficient of kinetic friction between the crate and the incline?

e. At the base of the incline there is a horizontal surface with the same coefficient of kinetic friction. How far will the crate slide before coming to rest?

7-29. *Up the slope.* As shown in Fig. 7E-29, a body of mass 1.00 kg is pulled slowly up a 30° slope 1.00 m long by a force directed parallel to the plane. The coefficient of kinetic friction between the body and the plane is 0.30.

a. How much work is done to increase the gravitational potential energy?

b. How much work is done against friction?

c. If the body is released and slides down the incline, what is its kinetic energy at the bottom?

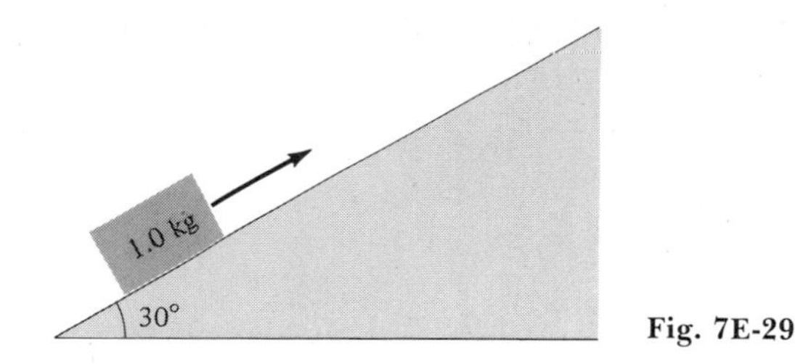

Fig. 7E-29

7-30. *Table pounding.* A body of mass 100 g is attached to a hanging spring whose force constant is 10 N/m. The body is lifted until the spring is in its unstretched state. The body is then released. Using the law of conservation of total mechanical energy, calculate the speed of the body when it strikes a table 15 cm below the release point.

7-31. *Down the track.* In the track shown in Fig. 7E-31, section AB is a quadrant of a circle of 1.0-m radius. A block is released at A and slides without friction until it reaches point B.

a. How fast is it moving at B, the bottom of the quadrant?

b. The horizontal part is not smooth. If the block comes to rest 3.0 m from B, what is the coefficient of kinetic friction?

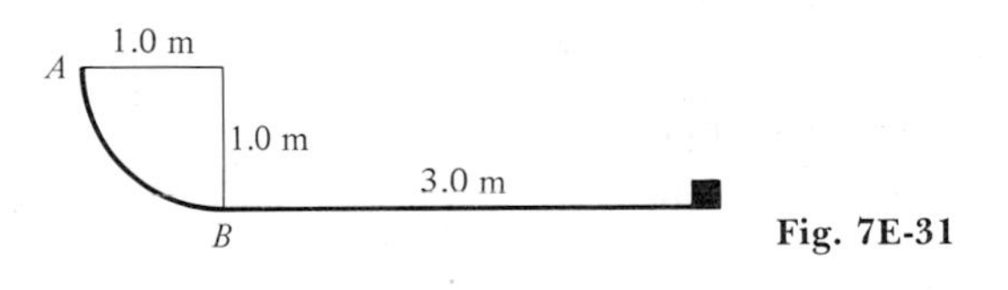

Fig. 7E-31

7-32. *Spring power.* A spring with negligible mass and a force constant of 600 N/m is kept straight by confining it within a smooth-walled guiding tube. The tube is anchored in a horizontal position on a tabletop. The spring is compressed by 10.0 cm and held there by a latch pin inserted through the wall of the tube. A 200-g ball of the same diameter as the spring is placed in contact with the spring, as shown in Fig. 7E-32. Then the latch pin is removed releasing the spring.

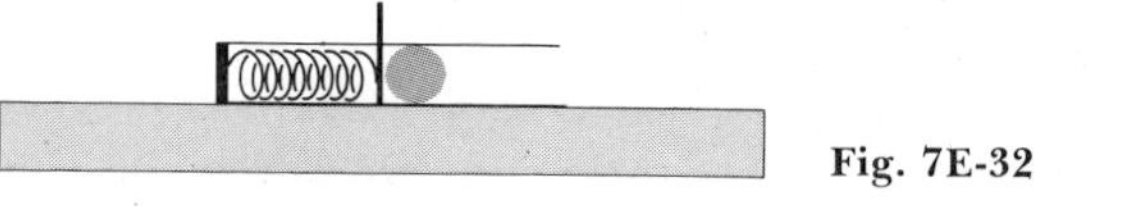

Fig. 7E-32

a. What speed does the ball acquire?

b. If the same procedure is followed with the tube pointing vertically upward, what will be the speed of the ball as it leaves contact with the spring?

7-33. *Roller coaster.* Figure 7E-33 shows the plan for a proposed roller coaster track. Each car will start from rest at point A and will roll with negligible friction. For safety, it is important that there be at least some small positive normal force (that is, a push) exerted by the track on the car at all points. (Why?) What is the minimum safe value for the radius of curvature at point B?

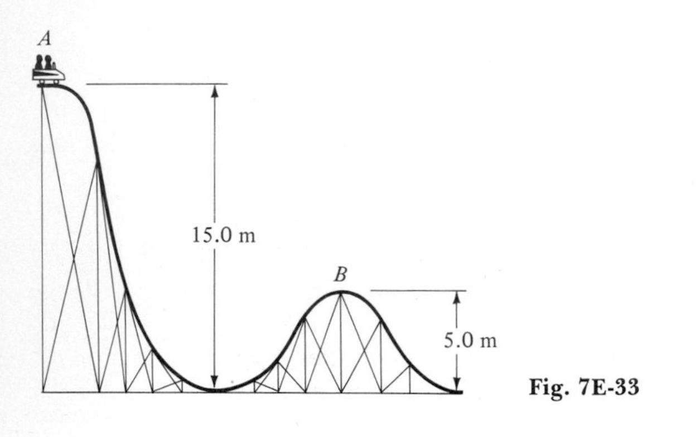

Fig. 7E-33

7-34. *Bob in a vertical circle.* A bob of mass m is revolving in a vertical circle at the end of a freely pivoted rod of length R and negligible mass. The bob's speed at the lowest point of the circle is v_0, but the speed varies with position as a result of the pull of gravity.

a. What is the tension in the rod as the bob passes through the lowest point?

b. How fast is the bob moving as it passes through the highest point of the circle?

c. What is the tension in the rod when the bob is at the highest point?

d. What is the difference between the tensions found in parts a and c? Express the difference as a multiple of the bob's weight.

e. Interpret your result for part c for the case of v_0^2 less than $5gR$.

f. What minimum value is implied for v_0 by the fact that the bob is traversing complete circles?

7-35. *Looping the loop.* In the toy illustrated in Fig. 7E-35 a small car loops the loop.

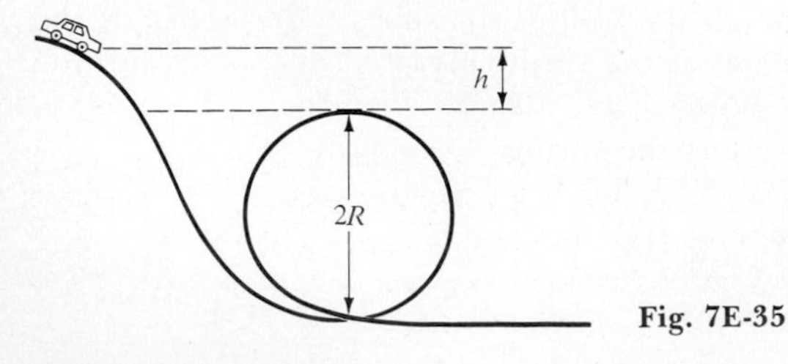

Fig. 7E-35

a. What is the minimum speed required at the top of the loop?

b. Assuming energy losses to be negligible, what is the minimum height h above the top of the loop from which the car must start?

c. Experience with a particular toy suggests that the actual minimum height that allows the car to loop the loop is 1.3 times the value found in part b. Compare the actual kinetic energy of the car as it passes the top of the loop with the kinetic energy the car would have when released from the same point if there were no frictional losses.

7-36. *Across the table, II.* In the system shown in Fig. 7E-36, friction is negligible and the string and pulleys are of negligible mass. The system is released from rest.

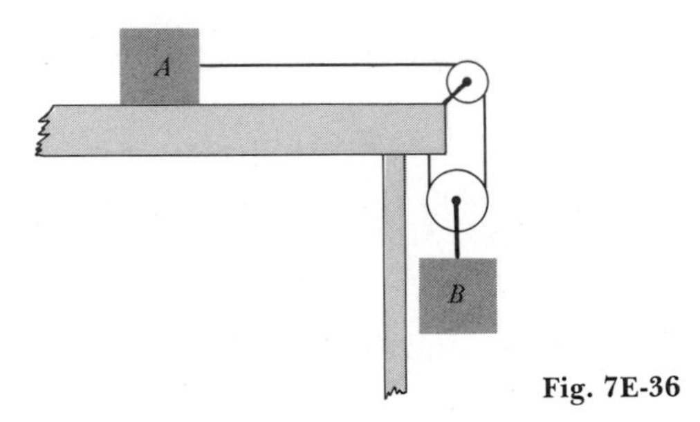

Fig. 7E-36

a. Find a relationship between the vertical drop of body B and the horizontal displacement of body A.

b. Find a relationship between the speeds v_A and v_B of bodies A and B.

c. After body B has descended a distance D, what is the speed v_A of body A?

d. Evaluate the result of part c for $m_A = m_B$ and $D = 2.0$ m.

7-37. *Block, spring, and kinetic friction.* As shown in Fig. 7E-37, a block of mass m is resting on a horizontal surface. The coefficients of static and kinetic friction between the block and the surface are μ_s and μ_k, respectively. The block is attached to a spring of negligible mass having spring constant k. Initially the block is at rest, and the spring is relaxed. Then the block is struck sharply, so that it begins moving to the right with speed v_0.

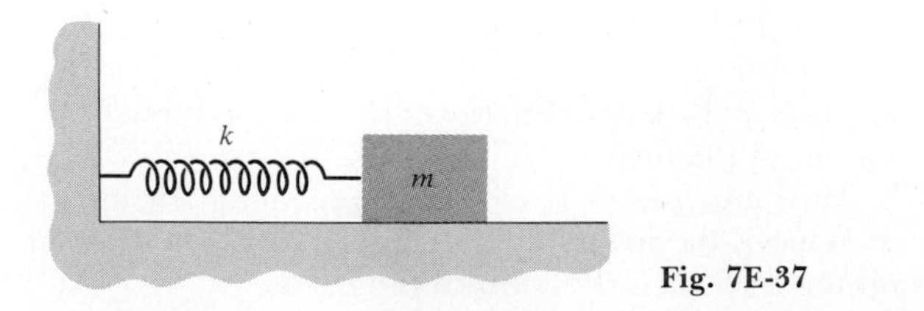

Fig. 7E-37

a. How far does the spring extend before the rightward motion is arrested?

b. Find a criterion that can be used to determine whether the block begins to move back to the left or simply remains at the point of maximum extension found in part *a*.

c. Evaluate the expressions you obtained in parts *a* and *b* for the case $m = 10$ kg, $k = 100$ N/m, $\mu_s = 0.30$, $\mu_k = 0.15$, and $v_0 = 1.0$ m/s.

7-38. *Launch speed versus elevation angle.* A spring gun uses a spring of negligible mass having spring constant k to launch a ball of mass m. The spring is initially compressed through a distance s. Use energy considerations to show that the launch speed v depends on the launch angle θ. Specifically, show that $v^2 = ks^2/m - 2gs \sin \theta$.

Group C

7-39. *A loss of support.* A body of mass m is attached to the hook of a stationary spring scale, as shown in Fig. 7E-39. The body is supported so that the reading of the scale is zero. The support is then removed. What is the maximum momentary reading of the spring balance? (Assume the damping is very slight.)

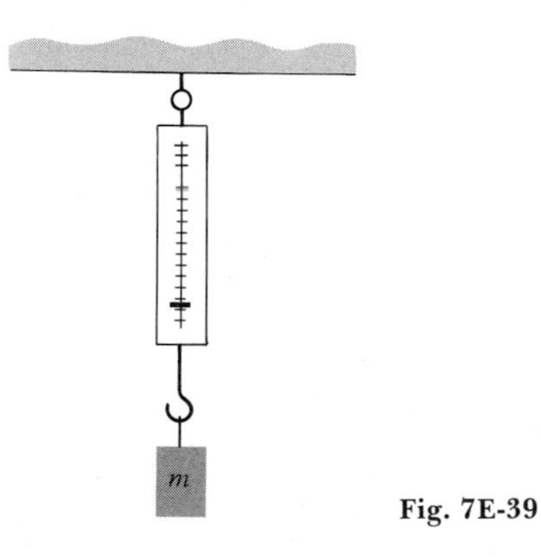

Fig. 7E-39

7-40. *A sliding launch.* Starting from rest at the top, a body slides down a frictionless hemispherical dome, as shown in Fig. 7E-40. Show that the body leaves the dome surface when $\theta = \cos^{-1} \frac{2}{3}$.

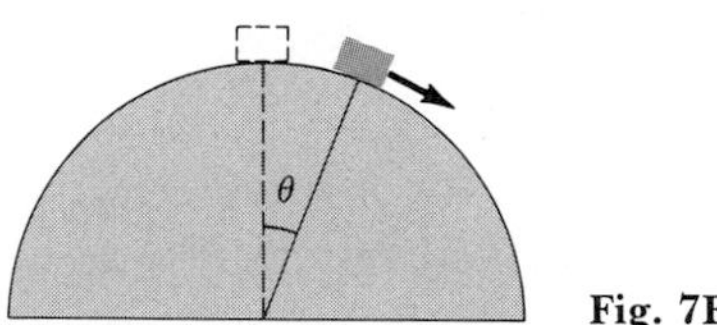

Fig. 7E-40

7-41. *Calculating the work done, II.* One of the forces acting on a certain particle depends on the particle's position in the xy plane. This force $\mathbf{F}_2$, expressed in newtons, is given by the expression $\mathbf{F}_2 = (xy\hat{\mathbf{x}} + xy\hat{\mathbf{y}})(1 \text{ N/m}^2)$, where x and y are expressed in meters. Calculate the work $\int_O^C \mathbf{F}_2 \cdot d\mathbf{s}$ done by this force when the particle moves

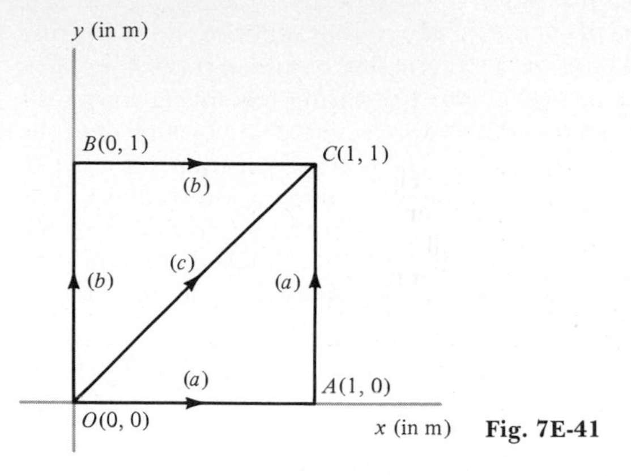

Fig. 7E-41

from point O to point C in Fig. 7E-41 along

a. The path OAC, which consists of two straight lines

b. The path OBC, which consists of two straight lines

c. The straight line OC

d. Is $\mathbf{F}_2$ a conservative force? Explain your answer. Compare it with the results obtained in Exercise 7-27.

7-42. *Oscillations of a leaky cart.* A small lab cart is able to roll without friction on a linear horizontal track. The cart is connected to an anchored spring of negligible mass having spring constant k, as shown in Fig. 7E-42. The cart is designed like a railroad hopper car. It is initially full of sand and has total mass M_i. The spring is extended by an amount A_i and is then released at $t = 0$. At the same time, the outlet on the bottom of the cart is opened, so that sand leaks out of the cart at a constant rate dM/dt whose value is negative. The leak is very slow and the cart is heavy, so that the fractional decrease of the cart mass in a time $2\pi\sqrt{M/k}$ is very small. Under these conditions, the motion of the cart is approximately harmonic, but with gradually changing frequency and amplitude.

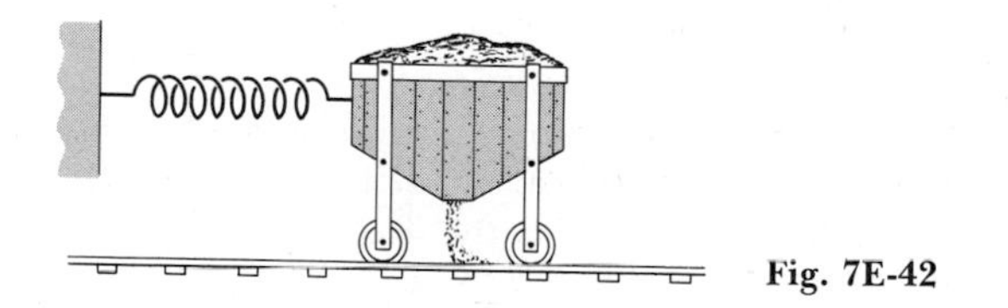

Fig. 7E-42

a. Find the angular frequency of oscillation $\omega(t)$ at time t, when $M(t) = M_i + t\, dM/dt$.

b. Find the kinetic energy carried out of the system by the sand leaking out in any one cycle of oscillation. Express your result in terms of dM/dt, $\omega(t)$, and the time-dependent amplitude $A(t)$.

c. Use energy considerations to determine how the amplitude of the oscillation varies with time. That is, find an expression for $A(t)$.

d. The leak stops when the cart is empty. Its final (unloaded) mass is M_f. With what frequency, amplitude, and total energy does the empty cart oscillate?

e. Compare inital and final values of ω, A, and total energy E for the case $M_f = 0.10M_i$.

7-43. *Body, track, spring, and pivots.* One end of a spring is attached to a pivot O at the end of a fixed vertical support, as shown in Fig. 7E-43. The spring has spring constant k and relaxed length l_0. A body is attached by a pivot to the free end of the spring. The body is constrained by a yoke to a horizontal, frictionless circular track of radius r. The track is fixed, with its center a distance h from O. In completing parts a through d, take the relaxed configuration $l = l_0$ to define the reference position for potential energy.

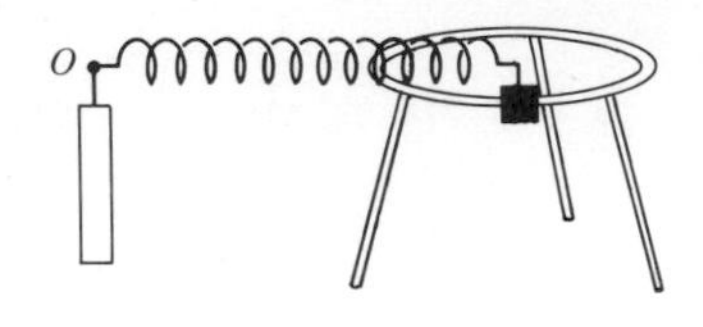

(*a*) Perspective view

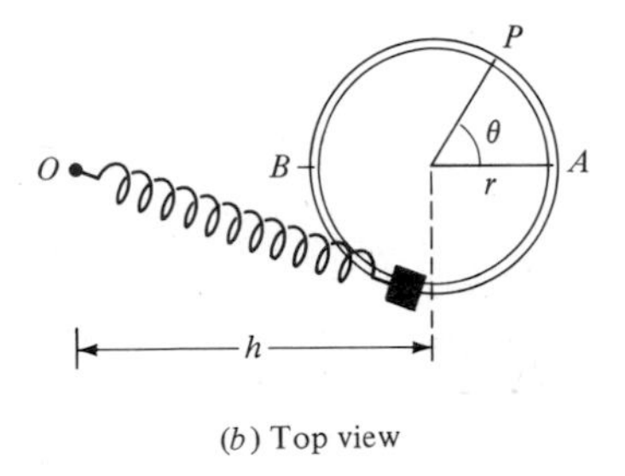

(*b*) Top view

Fig. 7E-43

a. What is the potential energy of the system when the body is at point A? Express your result in terms of k, h, r, and l_0.

b. What is the potential energy when the body is at point B?

c. How much work is done by the spring if the body moves from point A to point B?

d. What is the potential energy of the system when the body is located at the point P?

e. Suppose the body is started from point A in the counterclockwise sense with initial speed v_A. Describe the possible subsequent motion on the following cases: (1) $h + r < l_0$; (2) $|h - r| < l_0 < h + r$; (3) $|h - r| > l_0$; (4) $h = 0$.

f. Which of your answers in parts a through e would have to be changed if a different reference position were chosen for potential energy? Explain your answer.

7-44. *Two bodies, two tracks, and a spring.* Two bodies of equal mass m are constrained by yokes to move on identical horizontal frictionless circular tracks of radius r. As shown in Fig. 7E-44, the centers of the tracks are separated by a distance h. The bodies are linked by pivots to a spring of spring constant k and relaxed length l_0. The two bodies are initially at maximum separation as shown; each body is given a small initial speed. The body on the left is started in the clockwise sense, and that on the right in the counterclockwise sense.

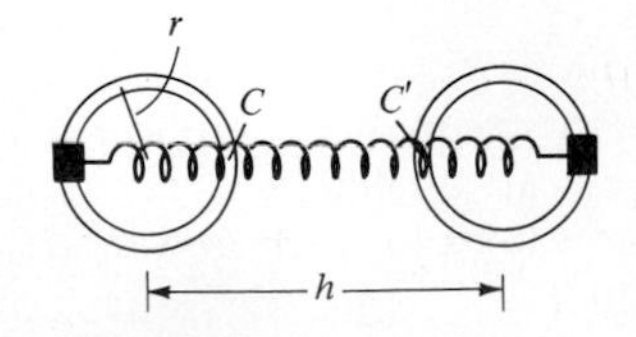

Fig. 7E-44

a. Under what condition will the bodies reach the points C and C'?

b. Assuming the condition in part a is satisfied, find the magnitude and direction of the force exerted by each track on its body as the bodies pass through C and C'.

c. For given values of r and l_0, find the value of h for which the force found in part b is zero.

7-45. *Frictionless but restrained.* As shown in Fig. 7E-45, a smooth rod is mounted horizontally just above a table top. A 10-kg collar, which is able to slide on the rod with negligible friction, is fastened to a spring whose other end is attached to a pivot at O. The spring has negligible mass, a relaxed length of 10 cm, and a spring constant of 500 N/m. The collar is released from rest at point S.

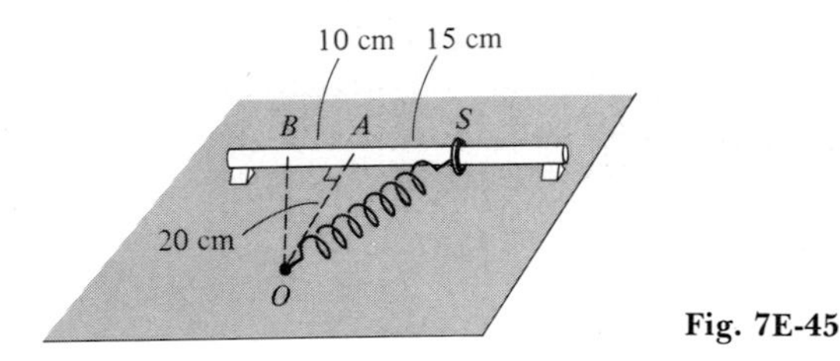

Fig. 7E-45

a. What is the speed of the collar as it passes A the closest point to O?

b. What is the speed of the collar as it passes point B?

Numerical

7-46. *Integration: A comparison of methods.* For each of the integrals below, use the numerical method of Example 7-2 to obtain a result to two-place accuracy. Then evaluate the integral using the appropriate equation(s) from Eqs. (7-20) to (7-27). Compare the two results for each integral.

a. $\int_1^3 x^2\,dx$ **d.** $\int_0^1 x(1 - x)\,dx$

b. $\int_1^2 x^3\,dx$ **e.** $\int_0^2 e^{-2x}\,dx$

c. $\int_0^\pi \sin\theta\,d\theta$ **f.** $\int_2^3 \frac{dt}{t}$

7-47. *An important integral.*

a. Use the numerical method of Example 7-2 to evaluate to three significant figures the integral

$$\sqrt{\frac{2}{\pi}}\int_0^{x_f} e^{-x^2/2}\,dx$$

for $x_f = 0.5, 1, 1.5$, and 2. This integral is called the **gaussian integral** or **normal probability integral,** and it plays a very important role in the error analysis of experimental data. It can be evaluated only by numerical methods; there is no analytical expression for the value of the integral.

b. Compare the values you obtained in part *a* with values that can be found in almost any table of mathematical data.

7-48. *Integrating over a spectrum.* The surface of the sun emits radiation over a wide range of wavelengths. The power radiated by 1 m^2 of the surface is different at different wavelengths λ. It is specified by the emitted power per unit wavelength, $R(\lambda)$, given by the so-called **Planck function**

$$R(\lambda) = \frac{3.74 \times 10^{-16}}{\lambda^5(e^{2.52\times10^{-6}/\lambda} - 1)} \text{ W/m}$$

where the wavelength λ is expressed in meters. The wavelength $\lambda = 3.50 \times 10^{-7}$ m represents approximately the extreme blue end of the visible spectrum, and the wavelength $\lambda = 7.00 \times 10^{-7}$ m represents approximately the extreme red end. The integral

$$I = \int_{3.50\times10^{-7}}^{7.00\times10^{-7}} R(\lambda)\, d\lambda$$

gives the power per unit area radiated by the surface of the sun in the visible range of wavelengths. Use the numerical method of Example 7-2 to evaluate the integral to an accuracy of three significant figures. The integral can be evaluated only by numerical methods. (*Note:* You may find it convenient to factor out a numerical constant before integrating.)

7-49. *Trapezoidal integration procedure.* There is another numerical integration procedure that converges rapidly enough to be useful and is simple enough to run on any calculating device capable of running the program used in Example 7-2. It amounts to calculating the total area under the set of trapezoids in Fig. 7E-49. The sides of the trapezoids are perpendicular to the x axis and intersect it at x_i, $x_i + \Delta x$, $x_i + 2\Delta x$, . . . , $x_f - \Delta x$, x_f. The tops of the trapezoids are straight lines joining the intersections of the sides and the curve $F(x)$.

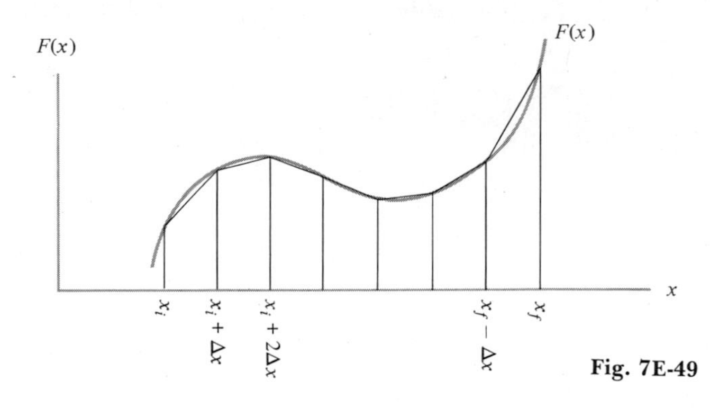

Fig. 7E-49

a. Show that the total area will be $\frac{1}{2}F(x_i)\,\Delta x + F(x_i + \Delta x)\,\Delta x + F(x_i + 2\Delta x)\,\Delta x + \cdots + F(x_f - \Delta x)\,\Delta x + \frac{1}{2}F(x_f)\,\Delta x$.

b. Write a program to carry out this trapezoidal integration procedure. Use it to evaluate $\int_1^2 x^2\, dx$ to three-decimal-place accuracy, and compare the result with that obtained in Example 7-2.

8

Applications of Energy Relations

8-1 POWER

The principal purpose of this chapter is to present a number of examples demonstrating the application of the theory of energy to interesting and important physical systems. A secondary purpose is to introduce several points of the theory that were not raised in Chap. 7 because they were not essential to the development of the main theme. One of these is the concept of *power*.

In many circumstances it is important to distinguish between work done rapidly and work done slowly. For an automobile engine to bring a car from rest up to a certain speed in a short time, the engine must be able in that time to do the work required to give the automobile the corresponding amount of kinetic energy. Such an engine is one of high power. To be more specific, *power measures the rate of doing work.* If work is being done at a constant rate, the **power** P is, by definition,

$$P \equiv \frac{W}{t} \qquad \text{for constant } P \tag{8-1a}$$

where W is the total amount of work done in the total time t. If work is being done at a varying rate, the **power** P expended at any instant is defined as

$$P \equiv \frac{dW}{dt} \tag{8-1b}$$

The unit of power in the SI system is the **watt** (W), named after James Watt (1736–1819), the inventor of the most important form of the steam engine. If work is being done at the rate of 1 joule per second (J/s), the power is 1 W:

$$1 \text{ W} \equiv 1 \text{ J/s} \tag{8-2}$$

A commonly used, non-SI unit of power is the **horsepower** (hp). The conversion factor is

$$1 \text{ hp} = 746 \text{ W} \tag{8-3}$$

The unit was introduced by Watt so that he could specify the power of his engines in terms that would be understood by his contemporaries. It was supposed to represent the rate at which a horse can do work (for prolonged periods), but in fact it overestimates the ability of an average horse by about 50 percent. A human can do work at an appreciable fraction of this rate (but not for prolonged periods), as you will see in Example 8-1.

EXAMPLE 8-1

A bicyclist is pedaling up a hill at a speed of 3.0 m/s. The slope of the road is 4.0°, the mass of the bicycle is 15 kg, and the mass of the bicyclist is 65 kg. Estimate the power he is expending over and above the rather low power he would use to maintain this speed against friction if the road were level.

■ Work can result in an increase in potential energy, in kinetic energy, or in thermal energy as a result of frictional losses. The various frictional losses (in the bearings, drive chain, flexing of the tire against the road, and air resistance) are the same as those encountered in riding at the same speed on a level road. Since the bicycle's speed is constant during the trip up the hill, there is no increase in kinetic energy. Thus all the additional work done by the bicyclist goes to increase the potential energy of the system comprising the bicycle plus the bicyclist plus the earth. The additional work per unit time dW/dt equals the increase in potential energy per unit time dU/dt. If you measure the elevation of the bicyclist by the vertical coordinate y, as in Fig. 8-1, U can be written

$$U = mgy$$

Here m is the combined mass of the bicycle and bicyclist, and the reference position in the definition of the potential energy U is whatever elevation is used to define $y = 0$. So you have

$$\frac{dW}{dt} = \frac{dU}{dt} = mg\frac{dy}{dt} = mgv_y$$

where v_y is the vertical component of the velocity of the bicyclist. Thus the additional power required to climb the hill is

$$P = \frac{dW}{dt} = mgv_y$$

The figure shows that

$$v_y = v \sin \theta$$

where v is the speed of the bicycle and θ is the slope of the road. So

$$P = mgv \sin \theta$$

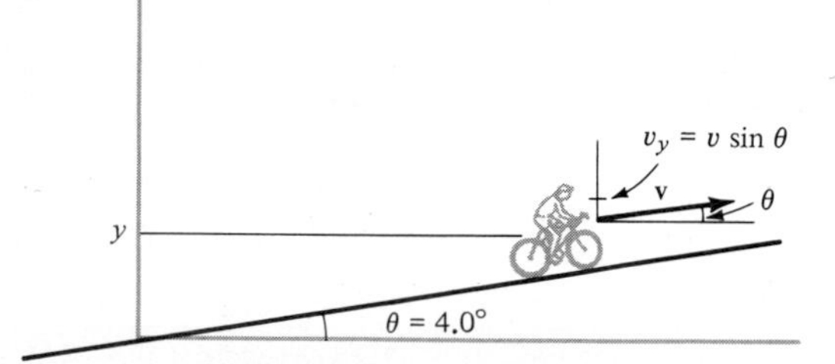

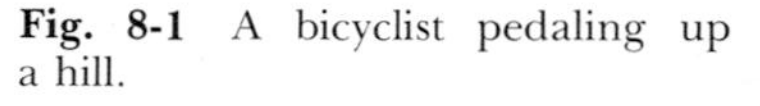
Fig. 8-1 A bicyclist pedaling up a hill.

The numerical values specified lead to the result

$$P = (15 \text{ kg} + 65 \text{ kg}) \times 9.8 \text{ m/s}^2 \times 3.0 \text{ m/s} \times \sin 4.0^\circ$$
$$= 1.7 \times 10^2 \text{ W} = 0.22 \text{ hp}$$

Although the power is constant in this example, so that it could be computed from Eq. (8-1*a*), it is really more convenient to use Eq. (8-1*b*) in the computation. Then you do not have to bother specifying the conditions at which U, W, and t are zero. If the power varies, you must use Eq. (8-1*b*).

EXAMPLE 8-2

In Example 4-9 you saw how to evaluate the force that must be applied to pull a very long conveyor belt at a constant speed of 2.00 m/s while crushed ore from a mine drops on the belt, thereby adding mass to the belt at a rate $dm/dt = 300$ kg/s. The situation is depicted again in Fig. 8-2. Evaluate the power P expended in moving the conveyor belt.

▪ Define an x axis whose positive direction is in the direction of motion of the belt, and use signed scalars. You can then write the work dW done by the force F applied to move the belt during a displacement dx of the belt as

$$dW = F\,dx$$

The power P is

$$P = \frac{dW}{dt} = F\frac{dx}{dt}$$

or, in terms of the velocity v of the belt,

$$P = Fv \tag{8-4}$$

That is, the power expended equals the force applied times the velocity of the object to which it is applied.

According to Eq. (4-15), the force applied is

$$F = v\frac{dm}{dt}$$

So you have

$$P = v\frac{dm}{dt}v = v^2\frac{dm}{dt}$$

Since v and dm/dt are constants, F and P are also constants.

The numerical value of P is

$$P = (2.00 \text{ m/s})^2 \times 300 \text{ kg/s} = 1.20 \times 10^3 \text{ W} = 1.20 \text{ kW}$$

Since v is a constant, the expression for the power P can be written as

$$P = \frac{d(mv^2)}{dt}$$

Fig. 8-2 Material dropping onto a very long conveyor belt.

(Evaluate the derivative for constant v, and you will see immediately that it gives $P = v^2\, dm/dt$, as before.) Introducing a factor of $\frac{1}{2}$ in the quantity being differentiated, and a compensating factor of 2 in front of the derivative, you can also write the new expression for P as

$$P = 2\,\frac{d(mv^2/2)}{dt}$$

But $mv^2/2$ is just the kinetic energy K of the moving belt and the moving ore lying on it. So

$$P = 2\,\frac{dK}{dt}$$

where dK/dt is the rate at which this energy increases as the amount of ore on the belt increases. It may surprise you that P is twice as large as dK/dt, because this means that only half the expended power goes into increasing the kinetic energy of the system comprising the belt plus the ore lying on it. Where does the other half go?

Inspection of Fig. 8-2 will verify that the "missing" power certainly does not go into increasing the potential energy of the system. In fact, the other half of the expended power is lost to frictional effects. These occur in the inelastic collisions of the rocks striking the moving belt. When the rocks first strike the belt, they skid and bump until they have come up to speed. Are you surprised to find that this power loss is calculable?

8-2 MACHINES

In this section we will apply the energy relations in a set of examples which analyze a number of simple mechanical devices or, as it is said, **machines.** The most elementary of these analyses will serve mainly to exemplify the energy approach. In others, the use of energy relations will make the analysis easier than it would be if Newton's second law were applied directly. Perhaps more important is that these relations will provide a quite different point of view which can add substantially to your depth of understanding of mechanical systems.

EXAMPLE 8-3

Use the mechanical-energy conservation law to find the force required to raise the weight with the **lever** shown in Fig. 8-3.

■ You begin by imagining that the free end of the lever is slowly raised through a small angle $\Delta\phi$, which you define to be positive. As a result, the weight is slowly given a vertical displacement that you can express by the signed scalar $\Delta y_1 = r_1\,\Delta\phi$. The quantity r_1 is the distance along the lever from the fixed axis, called the **fulcrum** (from the Latin word meaning a support), to the point from which the weight is suspended.

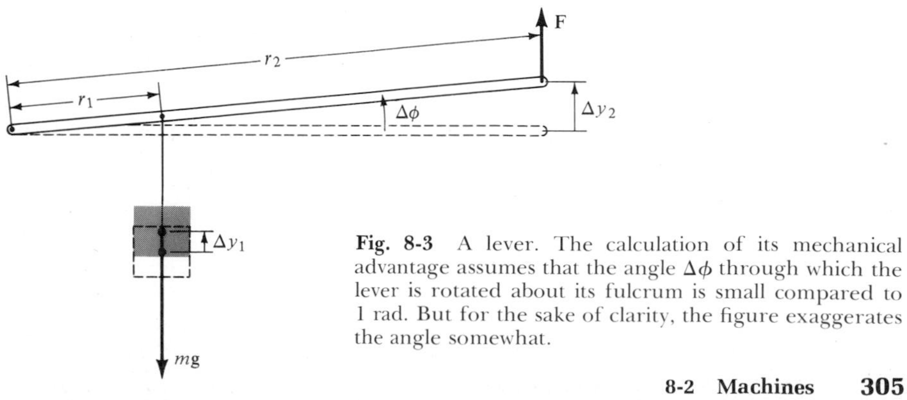

Fig. 8-3 A lever. The calculation of its mechanical advantage assumes that the angle $\Delta\phi$ through which the lever is rotated about its fulcrum is small compared to 1 rad. But for the sake of clarity, the figure exaggerates the angle somewhat.

In performing this task, you must give the end of the lever a vertical displacement $\Delta y_2 = r_2 \, \Delta\phi$, where r_2 is the distance from the fulcrum to the point of application of the vertical force F you exert on the lever.

Since everything is moving slowly, no appreciable kinetic energy is involved. If you assume friction in the fulcrum is negligible, it follows that the work you do by applying an external force to the weight goes into increasing the potential energy of the weight-plus-earth system. The work done is

$$W = F \, \Delta y_2 = F r_2 \, \Delta\phi$$

Write the gravitational potential energy of the system in terms of the vertical coordinate y_1 of the weight, measured from the reference position, as

$$U = mgy_1$$

Then the increase in potential energy is

$$\Delta U = mg \, \Delta y_1 = mgr_1 \, \Delta\phi$$

Equating W to ΔU, you have

$$Fr_2 \, \Delta\phi = mgr_1 \, \Delta\phi$$

Thus you find

$$F = \frac{mgr_1}{r_2}$$

The quantity mg/F is the ratio of the weight lifted to the force applied. It is called the **mechanical advantage** of the machine. For the lever, the mechanical advantage is

$$\frac{mg}{F} = \frac{r_2}{r_1} \tag{8-5}$$

Since r_2 is greater than r_1, the weight mg lifted by the lever is greater than the force F applied to the lever. To put it the other way, a lever allows you to lift a given weight by applying a smaller force than you would apply if lifting it unassisted. Of course, you do not get something for nothing. The price paid for reducing the necessary force is an increase in the distance that the far end of the lever must be moved by this force.

EXAMPLE 8-4

Use energy conservation to find the mechanical advantage of the **block and tackle,** shown in Fig. 8-4*a* and *b*.

■ Suppose that you do an amount of work W on the weight-plus-earth system by pulling on the free end of the rope so that the weight is raised, through a vertical displacement Δy, slowly enough that its kinetic energy can be neglected. If friction in the pulley wheels is negligible, W will be equal to the increase $\Delta U = mg \, \Delta y$ in the system's potential energy. Thus

$$W = mg \, \Delta y$$

In raising the weight, each of the n segments of the rope which support it must be shortened by an amount equal in magnitude to Δy. The total length of rope you must pull through the pulleys at the free end is therefore $n \, \Delta y$. This is the magnitude of the displacement of the end of the rope to which you apply a force. Since the force, of magnitude F, acts in the direction of this displacement, the work it does is

$$W = Fn \, \Delta y$$

Equating the two expressions for W yields

$$mg \, \Delta y = Fn \, \Delta y$$

(*a*)

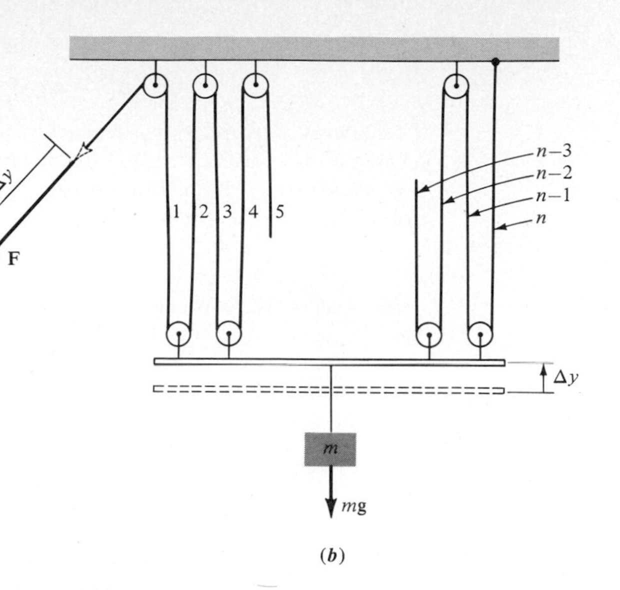

(*b*)

Fig. 8-4 (*a*) A block and tackle with six pulleys. (*b*) A schematic drawing of a block and tackle. Since the number of pulleys is arbitrary, only the first few and the last few are shown. (In practice, friction in the pulley bearings limits their total number to a maximum of about 10.)

or

$$\frac{mg}{F} = n \tag{8-6}$$

Thus the mechanical advantage of the block and tackle, neglecting friction, equals the number n of rope segments supporting the weight. In a real block and tackle the friction in the pulley bearings can be significant, and this reduces the mechanical advantage to an appreciably lower value.

EXAMPLE 8-5

A system often used in place of the block and tackle is the **differential pulley.** It is shown in Fig. 8-5. The two pulleys of radii r_1 and r_2 are joined and rotate together. Their grooves have teeth which engage the links of a continuous loop of chain so that it passes over them without slipping. As you pull on the chain at point P, the segment labeled 2 is taken up more than the segment labeled 1 is let out because r_2 is larger than r_1. Thus the weight supported by the chain is raised. The convenience of this machine lies in the fact that there are only two pulleys and one axle, plus the auxiliary pulley beneath. Yet by properly choosing the radii r_1 and r_2, any desired mechanical advantage can be obtained. Find the strength F of the force you must apply to raise slowly an object of mass m = 1000 kg if the radii have the values r_1 = 9.50 cm and r_2 = 10.00 cm. Neglect friction.

■ Consider what happens when you slowly pull the chain at the point P and the differential pulley goes around once. The segment of chain labeled 2 in the figure is taken up by an amount $2\pi r_2$. However, the segment labeled 1 is let out by an amount $2\pi r_1$. So the loop supporting the object is shortened by the amount $2\pi(r_2 - r_1)$. The object is raised by half this amount. (Why?) Thus its upward displacement is

$$\Delta y = \pi(r_2 - r_1)$$

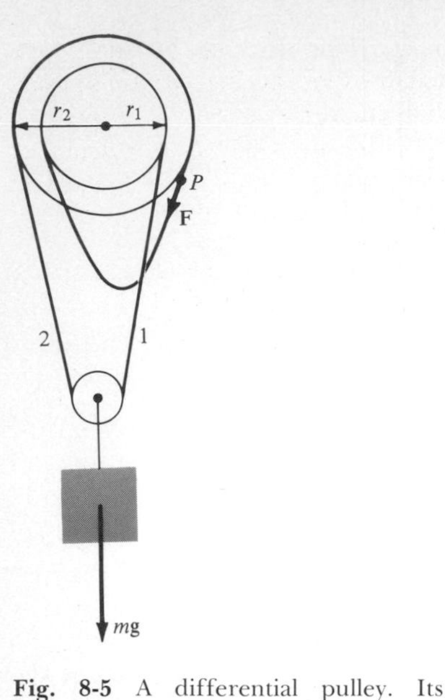

Fig. 8-5 A differential pulley. Its operation is explained in Example 8-5.

The increase in potential energy of the weight-plus-earth system is

$$\Delta U = mg\ \Delta y = mg\pi(r_2 - r_1)$$

This energy is supplied by the work you do on the system in pulling at point P with a force of magnitude F. The force is applied through a displacement of magnitude $2\pi r_2$ because the chain comes off the large pulley. Since the force acts in the direction of this displacement, the work done is

$$W = F\ 2\pi r_2$$

Setting $\Delta U = W$, you get

$$mg\pi(r_2 - r_1) = F\ 2\pi r_2$$

or

$$\frac{mg}{F} = \frac{2r_2}{r_2 - r_1}$$

So you have

$$\frac{mg}{F} = \frac{2}{1 - r_1/r_2} \tag{8-7}$$

This is the mechanical advantage of a differential pulley. If r_1 is nearly equal to r_2, the denominator of the fraction on the right side of Eq. (8-7) will be very small and the mechanical advantage will be very large.

To find the force required to lift the object, you write Eq. (8-7) as

$$F = \frac{mg}{2}\left(1 - \frac{r_1}{r_2}\right)$$

and then substitute the values given to obtain

$$F = \frac{1000\ \text{kg} \times 9.80\ \text{m/s}^2}{2}\left(1 - \frac{0.0950\ \text{m}}{0.1000\ \text{m}}\right)$$
$$= 245\ \text{N}$$

It takes a force of only 245 N to lift an object whose weight is 9800 N because this differential pulley has a mechanical advantage of 40.

In Examples 8-3 through 8-5 we have used energy relations to solve what are essentially static problems. That is, we were not concerned with acceleration because the systems we studied are generally used to move objects slowly and steadily, with an applied force just large enough to do the necessary useful work. Consequently, the kinetic energy was negligible in every case. In Example 8-6 we will again consider Atwood's machine, a system in which acceleration is important and the kinetic energy is significant.

EXAMPLE 8-6

In the Atwood machine shown in Fig. 8-6, the bodies at the ends of the cord of negligible mass have masses $m_1 = 2.10$ kg and $m_2 = 2.00$ kg, and the mass of the friction-free pulley wheel is assumed to be negligible. The bodies are released with zero velocity at the same height. At a certain later instant, when they are separated by a vertical distance $2|\Delta y| = 1.5$ m, their accelerations have given them certain velocities. Find these velocities.

■ You take as an isolated, friction-free system the two bodies and the earth. The cord and pulley act as workless constraints that always make one body move up at the same speed as the other body moves down. Since the bodies accelerate, they gain speed; therefore the system gains kinetic energy. This gain is at the expense of the system's gravitational potential energy. Choose the initial height of the bodies as the origin of an upward-directed y axis, and also as the reference level for gravitational potential energy. Then the potential energy of the system at the initial instant has the convenient value $U = 0$. Since both bodies start from rest, the system has kinetic energy $K = 0$ initially. Thus the total mechanical energy E of the system has the initial value $E = K + U = 0$. It will maintain this value.

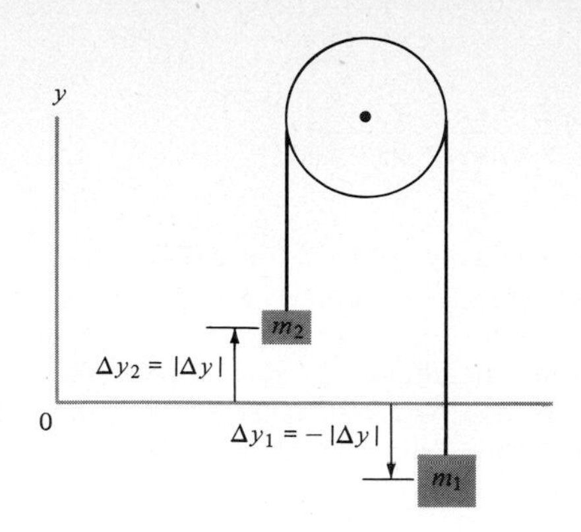

Fig. 8-6 Atwood's machine.

At the instant depicted in the figure, the change in the potential energy of the system from its initial value of zero is

$$\Delta U = m_2 g\ \Delta y_2 + m_1 g\ \Delta y_1 = m_2 g|\Delta y| - m_1 g|\Delta y|$$

or

$$\Delta U = -(m_1 - m_2)\ g|\Delta y|$$

Note that the displacement of the body with the smaller mass m_2 has caused the system to gain potential energy, but the displacement of the body with the larger mass m_1 has caused the system to lose a larger amount of potential energy. So there is a net loss of potential energy. This loss of potential energy must show up as a gain in kinetic energy, in order that the total mechanical energy remain constant. That is, you must have

$$\Delta K + \Delta U = 0$$

where ΔK is the change in the potential energy of the system from its initial value of zero.

Now the change in the kinetic energy of the initially motionless system is

$$\Delta K = \frac{m_1 v_1^2}{2} + \frac{m_2 v_2^2}{2}$$

where v_1 and v_2 are the velocities of the two bodies at the instant being considered. Writing the common value of their speeds as $|v|$, you have

$$\Delta K = \frac{m_1|v|^2}{2} + \frac{m_2|v|^2}{2}$$

Therefore the mechanical-energy conservation relation demands that

$$\frac{m_1|v|^2}{2} + \frac{m_2|v|^2}{2} - (m_1 - m_2)\ g|\Delta y| = 0$$

This simplifies to

$$|v|^2\frac{(m_1 + m_2)}{2} - (m_1 - m_2)\ g|\Delta y| = 0$$

or

$$|v|^2 = 2g|\Delta y|\left(\frac{m_1 - m_2}{m_1 + m_2}\right)$$

Thus you obtain

$$|v| = \left[2g|\Delta y|\left(\frac{m_1 - m_2}{m_1 + m_2}\right)\right]^{1/2} \qquad (8\text{-}8)$$

The body with the larger mass m_1 is moving downward and so has velocity $v_1 = -|v|$. The body of smaller mass m_2 is moving upward with the velocity $v_2 = |v|$. The speed of the downward-moving body is the speed it would acquire in free fall through the same distance, $(2g|\Delta y|)^{1/2}$, multiplied by the quantity $[(m_1 - m_2)/(m_1 + m_2)]^{1/2}$, which is always less than 1.

Inserting the numerical values given, you find

$$|v| = \left[2 \times 9.8\ \text{m/s}^2 \times 0.75\ \text{m} \times \left(\frac{2.10\ \text{kg} - 2.00\ \text{kg}}{2.10\ \text{kg} + 2.00\ \text{kg}}\right)\right]^{1/2}$$

$$= 0.60\ \text{m/s}$$

The velocities of the bodies are therefore $v_1 = -0.60$ m/s and $v_2 = 0.60$ m/s. They are the same as the velocities evaluated by direct application of Newton's second law in Example 5-3.

Suppose you had chosen some other reference height for the zero of potential energy. Would there be any difference in the result? As a test, take $U = 0$ at the final position of the body of mass m_2, and rework the calculation. You will find that this choice of reference height, or any other choice, will lead to the same result. It is only the *change* ΔU in potential energy which counts.

Note how the approach to the Atwood machine through the energy relations completely avoids a detailed analysis of the internal and external forces acting on the system. You need to know only the initial and final conditions in order to find the final velocities of the bodies.

8-3 IMPULSE AND COLLISIONS

In Chap. 4 we studied the behavior of bodies experiencing *collisions* by analyzing the changes in their momenta. In this section we will learn more about the behavior of colliding bodies by analyzing also the changes in the kinetic energies of the bodies. But first we will introduce a quantity called *impulse*. It is related to change in momentum, and the impulse-momentum relation will give us some additional insight into the momentum changes taking place in a collision.

The motivation and procedure leading to the impulse-momentum relation have a strong analogy to those leading to the work-kinetic energy relation in Chap. 7. Let us review very briefly how and why the work-kinetic energy relation is obtained. This relation is derived by calculating a work integral, that is, the integral of the component of the net force acting on a body, along the directions of its infinitesimal position changes, multiplied by these position changes. Newton's second law is used in the calculation to express the net force in terms of the mass of the body and the derivative of its velocity. Since integration is the inverse of differentiation, the calculation produces results which involve the velocity itself rather than the derivative of the velocity. This is why the work-kinetic energy relation involves only the velocity, although Newton's second law involves its derivative, the acceleration. Application of the relation often leads much more rapidly to a description of the behavior of the body than does application of Newton's second law. The reason is that in working directly with the velocity, you are dealing with a quantity that is closer to the goal—finding the position of the body as a function of time—than is the acceleration. Another advantage of the work-kinetic energy relation over Newton's second law is that the former involves scalars such as W, while the latter involves vectors such as **F**, and scalars are easier to handle than vectors.

Since consideration of the integral of the net force over the change in position proves to be so fruitful, it is reasonable to ask: Will an integration of the net force over the change in time also yield a useful quantity? Let's try it and see. Consider the integral

$$\mathbf{I} = \int_{t_i}^{t_f} \mathbf{F}\, dt \qquad (8\text{-}9)$$

Here the integral, whose value is represented by the symbol **I**, is a sum of terms each of which is the product of a vector, the net force **F** acting on a body, and a scalar dt, the infinitesimal increment of time. Thus **I** is a vector. We use Newton's second law in its most basic form, $\mathbf{F} = d\mathbf{p}/dt$, to evaluate **F** in terms of the time derivative of the momentum **p** of the body. This gives us

$$\mathbf{I} = \int_{t_i}^{t_f} \frac{d\mathbf{p}}{dt}\, dt = \int_{\mathbf{p}_i}^{\mathbf{p}_f} d\mathbf{p}$$

The fundamental theorem of calculus immediately yields

$$\mathbf{I} = \mathbf{p}_f - \mathbf{p}_i = \Delta\mathbf{p} \tag{8-10}$$

The quantity **I** is called the **impulse,** and Eq. (8-10) is called the **impulse-momentum relation.** It says that *the time integral of the net force acting on a body, called the impulse* **I**, *equals the change in its momentum* $\Delta\mathbf{p}$. Since Newton's second law was employed in deriving the impulse-momentum relation, it applies only when the body is viewed from an inertial frame of reference.

Is this relation useful? Not very. For the special case in which $\mathbf{F} = 0$, we have $\mathbf{I} = 0$ and $\Delta\mathbf{p} = 0$, or $\mathbf{p}_f = \mathbf{p}_i$. But this is just the familiar law of momentum conservation for the uninteresting situation in which a system contains only a single body. For the general case where $\mathbf{F} \neq 0$, we have $\mathbf{I} \neq 0$ so $\Delta\mathbf{p} \neq 0$, and we would like to know just what the value of $\Delta\mathbf{p}$ is. But we can obtain the value of **I** only if we can evaluate the integral in Eq. (8-9). To do this, we must know how **F** depends on t. That is, we must be able to express the force as $\mathbf{F}(t)$. However, we usually do *not* know $\mathbf{F}(t)$. (That is, we do not know it until the behavior of the body has been completely determined so that we know the position of the body at all times. But when this has been done, there is no motivation for additional calculations.) The point is that the forces in nature usually do not depend explicitly on time. Instead, they depend explicitly on position. If one of these typical forces is acting on the body of interest, we can immediately evaluate its integral over change in position—the work. But we cannot immediately evaluate its integral over change in time—the impulse.

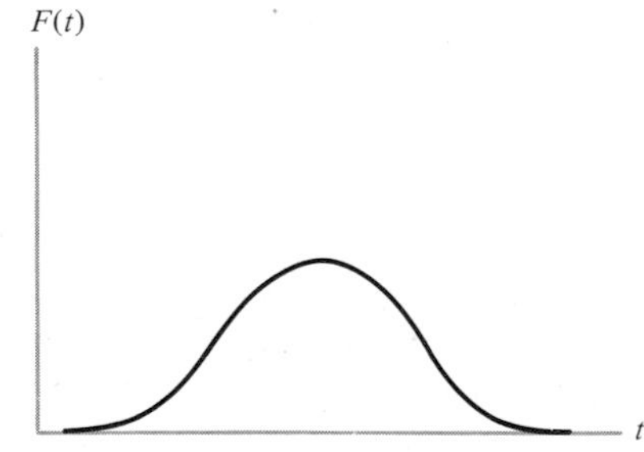

Fig. 8-7 Qualitative representation of the time dependence of the strength of the force acting on a magnetic puck when it collides head-on (without actually touching) with another magnetic puck.

Although it has limited practical application, impulse is useful conceptually. In particular, it helps us think about what happens during collisions. A **collision** is an event in which two objects approach each other, interact strongly during the short time that they are in proximity, and then (generally) move apart. A number of examples of collisions between pucks on an air table were studied in Chap. 4.

Figure 8-7 illustrates qualitatively the time dependence of the repulsive force acting on one of two colliding magnetic pucks, assuming for simplicity a "head-on" collision so that the force can be represented by a signed scalar. The direction of the force is defined to be positive. The force builds up rapidly in strength as the other puck approaches, reaches a maximum when that puck is at the point of closest approach, and then drops rapidly as it recedes. The effect of the force is to change the momentum of the puck on which it acts. According to the impulse-momentum relation, the net momentum change is just equal to the time integral of the force, which is measured by the area under the $F(t)$ curve. Of course, the *total* momentum of the system of two pucks is unchanged by the collision. The

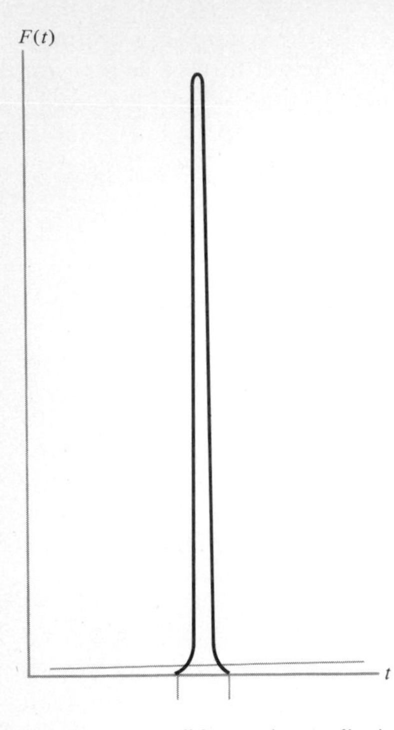

Fig. 8-8 The solid curve is a qualitative representation of the time dependence of the strength of the force acting on a plastic puck when it collides head-on with another plastic puck. Compared to the magnetic force illustrated in Fig. 8-7, the time during which the contact force acts on the plastic puck is very short, and its maximum strength is very large. The gray curve represents a constant, or slowly varying, force of moderate strength which could also be acting on the puck. In studying the collision, the impulse produced by this force can be neglected. The impulse produced during the collision is the area under the gray curve between the two marks on the time axis. This is very small compared to the area under the solid curve between the two marks, which is the impulse produced by the contact force.

other puck always experiences a force of the same magnitude but opposite direction. So it receives an equal but opposite impulse and experiences an equal but opposite net momentum change. As in all collisions, the impulses acting on the two colliding magnetic pucks cause momentum to be transferred from one to the other, without changing the total momentum of the system.

Figure 8-8 indicates the much more abrupt $F(t)$ curve for a "head-on" collision between two plastic pucks on an air table. The collision is abrupt because the pucks interact only when they are actually touching. When this happens, very strong repulsive contact forces develop for a very short time. But if the numerical value of the impulse happens to be the same as in the magnetic puck collision—that is, if the areas under the $F(t)$ curves in the two figures happen to be equal—the momentum transfer will be the same in the two collisions. What counts in determining the momentum transfer in a collision is the impulse, not the detailed behavior of $F(t)$.

Another thing that is unimportant in a collision is the action of any force of moderate strength. As is illustrated in Fig. 8-8, if the collision force is strong and the collision time is short, any much weaker force can be ignored because its contribution to the impulse will be negligible. For instance, the momentum transferred between two plastic pucks that collide while flying through the air, instead of while moving across the surface of an air table, is virtually unaffected by the uncompensated gravitational forces acting on them during the collision because these forces are very much weaker than the contact forces.

Examples 8-7 through 8-10 analyze some of the air table puck collisions that were studied in Chap. 4. Both momentum and energy are taken into account in the analyses. As a consequence, the examples provide a more thorough understanding of the collisions than was obtained in Chap. 4, where only momentum was considered.

EXAMPLE 8-7

The strobe photo in Fig. 4-14, reproduced here as Fig. 8-9, shows a collision between two equal-mass magnetic pucks on an air table. Puck 2 was initially stationary near the center of the air table, and puck 1 was initially moving toward the center from the upper right. The strobe photo shows that the final trajectories of the pucks make a 90° angle. Prove that such behavior is consistent with the assumption that mechanical energy is conserved in the collision. Do this by assuming that the mechanical-energy conservation law holds and then predicting the value of the angle between the final trajectories of the pucks.

▪ The magnetic pucks do not exert forces on each other of appreciable strength until they are quite close. You can see this from the photo. It shows that puck 1 maintains an essentially constant velocity during its approach to puck 2. (The word "approach" is used to mean that part of the motion of puck 1 toward puck 2 which ends when the separation between the pucks is less than something like one puck diameter.) Thus there can be no appreciable force exerted between pucks during the approach. As a consequence, the kinetic energy of puck 1 is essentially all the energy stored in the system when puck 1 is at any position in the approach. Additional energy could result only from work done by a force acting on puck 1 while it moved up to that position. But there is no such force. In other words, you can say that the system has essentially no potential energy relative to the potential energy reference value zero for puck 1 at a reference position chosen to be very far from puck 2, when puck 1 is at any position in the approach. (When puck 1 is very close

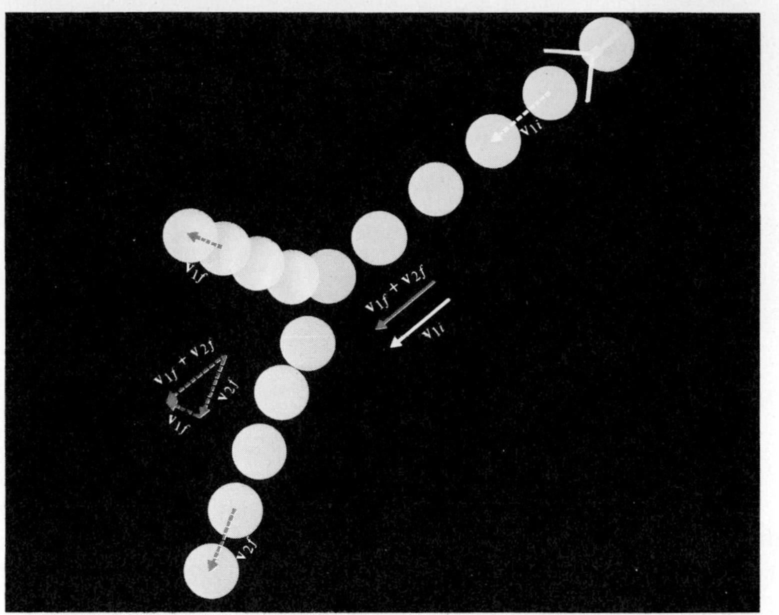

Fig. 8-9 Strobe photo of a collision between two identical magnetic pucks on an air table. Puck 1 comes from the right, and puck 2 is initially at rest near the center of the table.

to puck 2, the system does have potential energy. However, you will see that there is no need to consider it.) A similar argument allows you to say that the system has essentially no potential energy during the part of the motion when the two pucks are receding from each other and their separation exceeds something like one puck diameter.

Taking advantage of the conclusions just drawn, you need consider the energy content of the system only in the initial time interval when the velocity vector $\mathbf{v}_{1i}$ in the photo was measured and in the final time interval when the velocity vectors $\mathbf{v}_{1f}$ and $\mathbf{v}_{2f}$ were measured. When the pucks have these initial or final velocities, the system has initial or final kinetic energies K_i or K_f. But the corresponding values of both the initial and the final potential energies of the system are zero. Therefore, what you do in following the instruction to assume that mechanical energy is conserved in the collision is to equate the initial and final values of the kinetic energy:

$$K_i = K_f$$

To obtain information about the angle between the final trajectories of the pucks, you need to deal with directions. For this reason, you write the expressions for the kinetic energies in their vector forms. Using Eq. (7-37*b*) to express K_i and K_f in terms of $\mathbf{v}_{1i}$, $\mathbf{v}_{1f}$, $\mathbf{v}_{2f}$, and the mass m of either identical puck, you have

$$\frac{m}{2}\,\mathbf{v}_{1i}\cdot\mathbf{v}_{1i} = \frac{m}{2}\,\mathbf{v}_{1f}\cdot\mathbf{v}_{1f} + \frac{m}{2}\,\mathbf{v}_{2f}\cdot\mathbf{v}_{2f} \tag{8-11a}$$

The left side of this equality is the initial total kinetic energy of the system, since puck 2 is initially stationary. It is also the initial total mechanical energy, since initially there is zero potential energy in the system. The right side is the final total kinetic energy of the system, and also its final total mechanical energy. For the particular case at hand where the masses are equal, the factor $m/2$ can be divided out of each term to yield the simplified form

$$\mathbf{v}_{1i}\cdot\mathbf{v}_{1i} = \mathbf{v}_{1f}\cdot\mathbf{v}_{1f} + \mathbf{v}_{2f}\cdot\mathbf{v}_{2f} \tag{8-11b}$$

You have assumed that the total mechanical energy of the system is unchanged in the collision. But it is not necessary to *assume* that the total momentum of the system is unchanged. You *know* that to be true because the system is an isolated one viewed from an (approximately) inertial frame. Thus you can also write

$$m\mathbf{v}_{1i} = m\mathbf{v}_{1f} + m\mathbf{v}_{2f} \tag{8-12a}$$

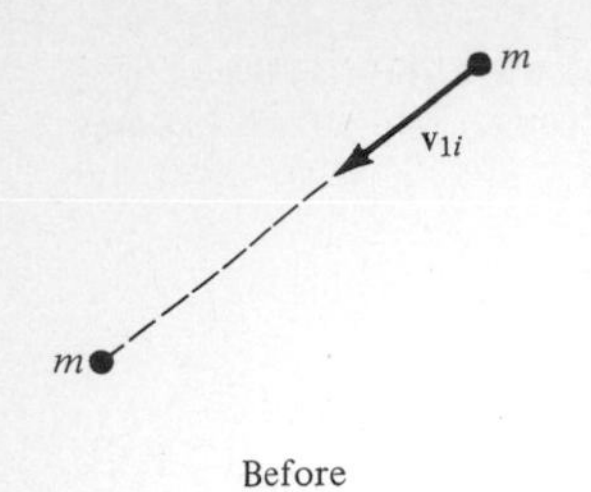

Fig. 8-10 An analysis of the collision of Fig. 8-9 between identical magnetic pucks. Before the collision puck 1 has velocity $\mathbf{v}_{1i}$ and approaches the stationary puck 2. After the collision, puck 1 moves away from the collision point with velocity $\mathbf{v}_{1f}$ and puck 2 moves away from that point with velocity $\mathbf{v}_{2f}$. The angle between the two final velocity vectors $\mathbf{v}_{1f}$ and $\mathbf{v}_{2f}$ is θ.

Here the left side is the initial total momentum of the system, and the right side is the final total momentum. This equation is quite independent of Eq. (8-11*b*), because it has nothing to do with energy considerations. It can be simplified immediately to the form

$$\mathbf{v}_{1i} = \mathbf{v}_{1f} + \mathbf{v}_{2f} \tag{8-12b}$$

You now have two simultaneous equations, Eqs. (8-11*b*) and (8-12*b*), involving the three quantities $\mathbf{v}_{1i}$, $\mathbf{v}_{1f}$, and $\mathbf{v}_{2f}$. You want to prove that the angle between the two final trajectories is 90°. In other words, you want to prove that $\mathbf{v}_{1f}$ and $\mathbf{v}_{2f}$ form a 90° angle. To obtain a relation between these two quantities, you eliminate $\mathbf{v}_{1i}$ from the two equations. To do this, you take the dot product of the left side of Eq. (8-12*b*) with itself to obtain $\mathbf{v}_{1i} \cdot \mathbf{v}_{1i}$. You maintain the equality by also taking the dot product of the right side with itself, producing

$$\mathbf{v}_{1i} \cdot \mathbf{v}_{1i} = (\mathbf{v}_{1f} + \mathbf{v}_{2f}) \cdot (\mathbf{v}_{1f} + \mathbf{v}_{2f})$$

or

$$\mathbf{v}_{1i} \cdot \mathbf{v}_{1i} = \mathbf{v}_{1f} \cdot \mathbf{v}_{1f} + \mathbf{v}_{2f} \cdot \mathbf{v}_{2f} + 2\mathbf{v}_{1f} \cdot \mathbf{v}_{2f} \tag{8-13}$$

This is the vector equivalent of squaring both sides of an equation. Subtracting Eq. (8-11*b*) from Eq. (8-13), you obtain

$$0 = 2\mathbf{v}_{1f} \cdot \mathbf{v}_{2f} = 2v_{1f} \cos\theta \, v_{2f} \tag{8-14}$$

Here θ is the angle between the final trajectories of the two pucks, as shown in Fig. 8-10. In general, neither v_{1f} nor v_{2f} is zero, and thus it must be true that

$$\cos\theta = 0$$

So you have predicted that if the collision is energy-conserving, the angle between the final puck trajectories must be $\theta = 90°$. Your prediction is borne out by Fig. 8-9. An interesting special case, familiar to billiard players, is a head-on collision where $v_{1f} = 0$. How does this case fit into the analysis you have just gone through?

EXAMPLE 8-8

The collision between identical magnetic pucks of Fig. 8-9 was called an **elastic collision** in Chap. 4. There an elastic collision was defined as one in which the pucks move apart after the collision with a relative speed equal to the relative speed with which they came together before the collision. That is, in an elastic collision the velocities satisfy the condition of Eq. (4-9a), which is

$$|\mathbf{v}_{1f} - \mathbf{v}_{2f}| = |\mathbf{v}_{1i} - \mathbf{v}_{2i}| \tag{8-15}$$

Show that this condition is consistent with the assumption that mechanical energy is conserved in the collision.

■ You want to evaluate the final relative speed $|\mathbf{v}_{1f} - \mathbf{v}_{2f}|$. You can do this by considering the final relative velocity, $\mathbf{v}_{1f} - \mathbf{v}_{2f}$, and then generating the square of the speed from the velocity by taking the dot product of the velocity into itself. You obtain, on expanding,

$$(\mathbf{v}_{1f} - \mathbf{v}_{2f}) \cdot (\mathbf{v}_{1f} - \mathbf{v}_{2f}) = \mathbf{v}_{1f} \cdot \mathbf{v}_{1f} + \mathbf{v}_{2f} \cdot \mathbf{v}_{2f} - 2\mathbf{v}_{1f} \cdot \mathbf{v}_{2f}$$

By using Eq. (8-14), you simplify this to

$$(\mathbf{v}_{1f} - \mathbf{v}_{2f}) \cdot (\mathbf{v}_{1f} - \mathbf{v}_{2f}) = \mathbf{v}_{1f} \cdot \mathbf{v}_{1f} + \mathbf{v}_{2f} \cdot \mathbf{v}_{2f}$$

And by using Eq. (8-11b), you simplify it further to

$$(\mathbf{v}_{1f} - \mathbf{v}_{2f}) \cdot (\mathbf{v}_{1f} - \mathbf{v}_{2f}) = \mathbf{v}_{1i} \cdot \mathbf{v}_{1i}$$

This result can be written in the form

$$|\mathbf{v}_{1f} - \mathbf{v}_{2f}|^2 = |\mathbf{v}_{1i}|^2$$

The terms on both sides of the equality are scalars. Taking their square roots, you have

$$|\mathbf{v}_{1f} - \mathbf{v}_{2f}| = |\mathbf{v}_{1i}|$$

For the case at hand, $\mathbf{v}_{2i} = 0$. So the equation you have just obtained by using the conservation of mechanical energy is consistent with Eq. (8-15), the kinematical condition defining elastic collisions.

The conclusion reached in Example 8-8 concerning an **elastic collision** is valid in general. Even when both colliding bodies are moving before the collision, and even when their masses are unequal, the kinematical definition for elastic collisions (the bodies move apart with a final relative speed equal to the initial relative speed at which they came together) is equivalent to the statement that total mechanical energy is conserved in elastic collisions (the bodies move apart with a final total kinetic energy equal to the initial total kinetic energy they have as they come together). Briefly put, *an elastic collision is one in which mechanical energy is conserved.*

An inelastic collision is one in which there is a loss of mechanical energy. There is no potential energy in the system of colliding bodies in either the initial or the final state where they are well separated and so do not exert forces on each other, as was explained in Example 8-7. Thus the loss of mechanical energy in an inelastic collision must be a result of a loss of kinetic energy. The speeds of both bodies are generally lower after an **inelastic collision** because their kinetic energies are generally smaller. Thus the colliding bodies rebound from an inelastic collision with a relative speed that is lower than the relative speed with which they approach before the collision, in agreement with the kinematical definition of Eq. (4-9b). An example is found in Fig. 4-12, reproduced here as Fig. 8-11. The strobe photo shows a collision between an incident plastic puck and an initially stationary plastic puck of the same mass. When the pucks come into contact, they are deformed by the very strong contact force. Most of, but not all, the work done in producing the deformation is recovered when the pucks move apart. Some of the associated mechanical energy remains in the pucks in the form

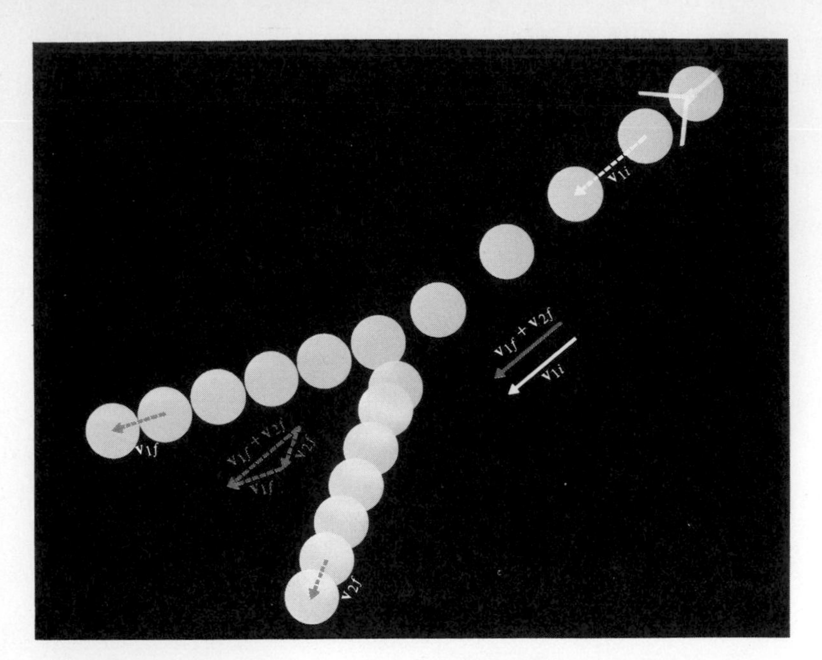

Fig. 8-11 Strobe photo of a collision between two identical plastic pucks. Puck 1 is incident from the upper right on puck 2, which is initially at rest near the center of the table.

of thermal energy of vibration of their constituent molecules (and a bit is lost to the acoustical energy in the "click" that can be heard when they collide). But it would be very difficult to determine how much mechanical energy is lost by making thermal (and acoustical) measurements. Example 8-9 shows that there is a much easier way.

EXAMPLE 8-9

Develop an equation which can be used as a basis for a *convenient* mechanical measurement of the amount of mechanical energy lost in an inelastic collision. Do this for a case in which one of the colliding bodies is initially stationary but the masses of the two bodies are unequal.

■ For an inelastic collision, an energy conservation equation like Eq. (8-11*a*) is invalid since the left side is actually larger than the right side. The difference between their values is ΔK, the change in kinetic energy between the initial and final states. It equals ΔE, the change in the total mechanical energy of the system between the initial and final states, because there is no potential energy in the system in either state. Using labels defined in Fig. 8-12 to specify the masses of the colliding bodies, you can express these energy changes as follows:

$$\Delta E = \Delta K = \left(\frac{m_1}{2}\mathbf{v}_{1f}\cdot\mathbf{v}_{1f} + \frac{m_2}{2}\mathbf{v}_{2f}\cdot\mathbf{v}_{2f}\right) - \frac{m_1}{2}\mathbf{v}_{1i}\cdot\mathbf{v}_{1i} \tag{8-16}$$

In an inelastic collision the energy changes are negative since kinetic and total mechanical energies are lost. If you know both masses, a measurement of all three velocities in Eq. (8-16) would allow a measurement of the energy loss.

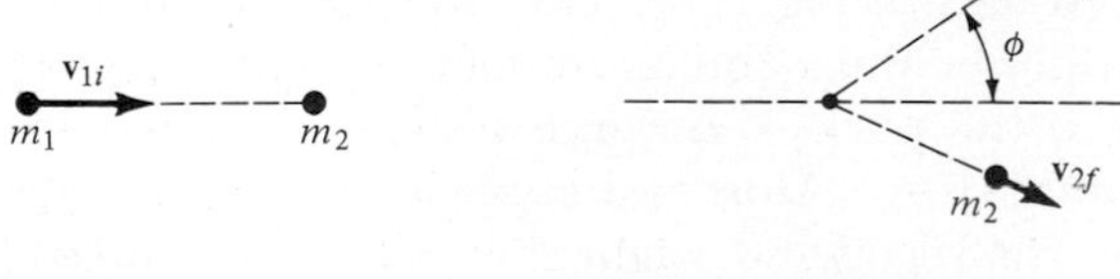

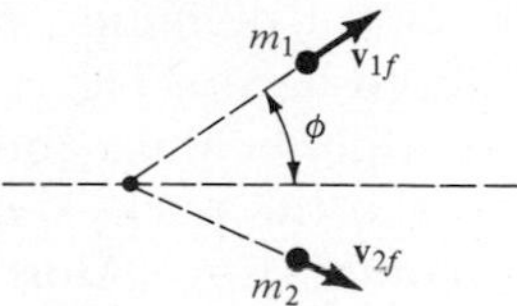

Fig. 8-12 Analysis of a collision. Before the collision, the body of mass m_1 has velocity $\mathbf{v}_{1i}$ and approaches the stationary body of mass m_2. After the collision, the body of mass m_1 moves from the collision point with velocity $\mathbf{v}_{1f}$, and the body of mass m_2 moves away from that point with velocity $\mathbf{v}_{2f}$. The angle between $\mathbf{v}_{1i}$ and $\mathbf{v}_{1f}$, the initial and final velocity vectors of the body of mass m_1, is ϕ.

However, by combining Eq. (8-16) with the momentum conservation equation, the need for measuring one of these velocities can be eliminated. This is a very great convenience because it considerably simplifies the experimental technique that must be used to measure the energy loss. The momentum conservation equation is

$$m_1\mathbf{v}_{1i} = m_1\mathbf{v}_{1f} + m_2\mathbf{v}_{2f} \tag{8-17}$$

It applies whether the collision is elastic or inelastic.

What you do to use Eq. (8-17) is to solve it for $\mathbf{v}_{2f}$, obtaining

$$\mathbf{v}_{2f} = \frac{m_1}{m_2}(\mathbf{v}_{1i} - \mathbf{v}_{1f})$$

Then you evaluate

$$\mathbf{v}_{2f} \cdot \mathbf{v}_{2f} = \frac{m_1}{m_2}(\mathbf{v}_{1i} - \mathbf{v}_{1f}) \cdot \frac{m_1}{m_2}(\mathbf{v}_{1i} - \mathbf{v}_{1f})$$

Expanding, you obtain

$$\mathbf{v}_{2f} \cdot \mathbf{v}_{2f} = \left(\frac{m_1}{m_2}\right)^2 (\mathbf{v}_{1i} \cdot \mathbf{v}_{1i} + \mathbf{v}_{1f} \cdot \mathbf{v}_{1f} - 2\mathbf{v}_{1i} \cdot \mathbf{v}_{1f})$$

Now you substitute this expression for $\mathbf{v}_{2f} \cdot \mathbf{v}_{2f}$ into Eq. (8-16) and find

$$\Delta K = \frac{m_1}{2}\mathbf{v}_{1f} \cdot \mathbf{v}_{1f} + \frac{m_1^2}{2m_2}(\mathbf{v}_{1i} \cdot \mathbf{v}_i + \mathbf{v}_{1f} \cdot \mathbf{v}_{1f} - 2\mathbf{v}_{1i} \cdot \mathbf{v}_{1f}) - \frac{m_1}{2}\mathbf{v}_{1i} \cdot \mathbf{v}_{1i}$$

All the dot products can be evaluated immediately, except for the one involving $\mathbf{v}_{1i}$ and $\mathbf{v}_{1f}$. But it can be expressed in terms of the magnitudes of these velocities and the angle ϕ between them. This is also the angle between the initial and final trajectories of the incident body (see Fig. 8-12). Expressing the value of $\mathbf{v}_{1i} \cdot \mathbf{v}_{1f}$ in terms of v_{1i}, v_{1f}, and $\cos\phi$, and placing the $\cos\phi$ factor at the end of the expression for convenience, you have

$$\Delta K = \frac{m_1}{2}v_{1f}^2 + \frac{m_1^2}{2m_2}v_{1i}^2 + \frac{m_1^2}{2m_2}v_{1f}^2 - \frac{m_1^2}{m_2}v_{1i}v_{1f}\cos\phi - \frac{m_1}{2}v_{1i}^2$$

Gathering the terms in v_{1i}^2 and v_{1f}^2, you get the result

$$\Delta K = \frac{m_1 v_{1f}^2}{2}\left(1 + \frac{m_1}{m_2}\right) - \frac{m_1 v_{1i}^2}{2}\left(1 - \frac{m_1}{m_2}\right) - \frac{m_1^2}{m_2}v_{1i}v_{1f}\cos\phi \tag{8-18a}$$

An equivalent form, expressed in terms of the initial and final kinetic energies K_{1i} and K_{1f}, is

$$\Delta K = K_{1f}\left(1 + \frac{m_1}{m_2}\right) - K_{1i}\left(1 - \frac{m_1}{m_2}\right) - 2\frac{m_1}{m_2}\sqrt{K_{1i}K_{1f}}\cos\phi \tag{8-18b}$$

Thus a measurement of the initial and final kinetic energies K_{1i} and K_{1f} of the so-called **scattered** body, and of its **scattering angle** ϕ, gives you the kinetic energy change ΔK of the system. From this you immediately find the change in total mechanical energy ΔE, since $\Delta E = \Delta K$.

Either form of Eqs. (8-18) provides a very convenient way of analyzing inelastic collisions. These equations are also very useful for elastic collisions. To apply them to such a collision, you just set $\Delta K = 0$. This is done for a specific case in Example 8-10, and the predictions obtained are tested against experiment.

EXAMPLE 8-10

A body of mass 1.00 kg, moving at a speed of 1.00 m/s, collides elastically with a stationary body of mass 2.00 kg. The angle between the final trajectory of the 1.00-kg body and its initial trajectory is $\phi = 65°$. Calculate its final speed.

■ Since the collision is elastic, you set $\Delta K = 0$ in Eq. (8-18*a*). Dividing through by $m_1/2$ and reordering terms, you have

$$v_{1f}^2\left(1 + \frac{m_1}{m_2}\right) - 2\,\frac{m_1}{m_2}\,v_{1i}v_{1f}\cos\phi - v_{1i}^2\left(1 - \frac{m_1}{m_2}\right) = 0$$

Using $m_1/m_2 = 1.00\text{ kg}/2.00\text{ kg} = 0.50$, $v_{1i} = 1.00$ m/s, and $\cos\phi = \cos 65° = 0.42$, you obtain

$$1.50\,v_{1f}^2 - (0.42\text{ m/s})v_{1f} - 0.50\text{ m}^2/\text{s}^2 = 0$$

This is a quadratic equation in the unknown v_{1f}. The standard expression for the solutions to such an equation tells you that

$$v_{1f} = \frac{0.42\text{ m/s} \pm \sqrt{(0.42\text{ m/s})^2 + 4 \times 1.50 \times 0.50\text{ m}^2/\text{s}^2}}{2 \times 1.50}$$

$$= \frac{0.42\text{ m/s} \pm 1.78\text{ m/s}}{2 \times 1.50}$$

Since v_{1f} is a speed, it cannot be negative. So the physically significant solution is found by choosing the positive sign that precedes the square root. With this choice of sign, you obtain the result

$$v_{1f} = 0.74\text{ m/s}$$

Figure 8-13 is a reproduction of the strobe photo of Fig. 4-16. It shows an elastic collision between a single (that is, $m_1 = 1$) magnetic puck coming from the launcher and a double (that is, $m_2 = 2$) magnetic puck which is initially stationary at the center of the air table. The collision scatters the incident puck through an angle ϕ, and it moves off to the upper left of the photo. Measurement with a protractor will show you that $\phi \simeq 65°$. You thus have all the information you need to make a comparison of the puck collision with the collision calculated in this example. Mea-

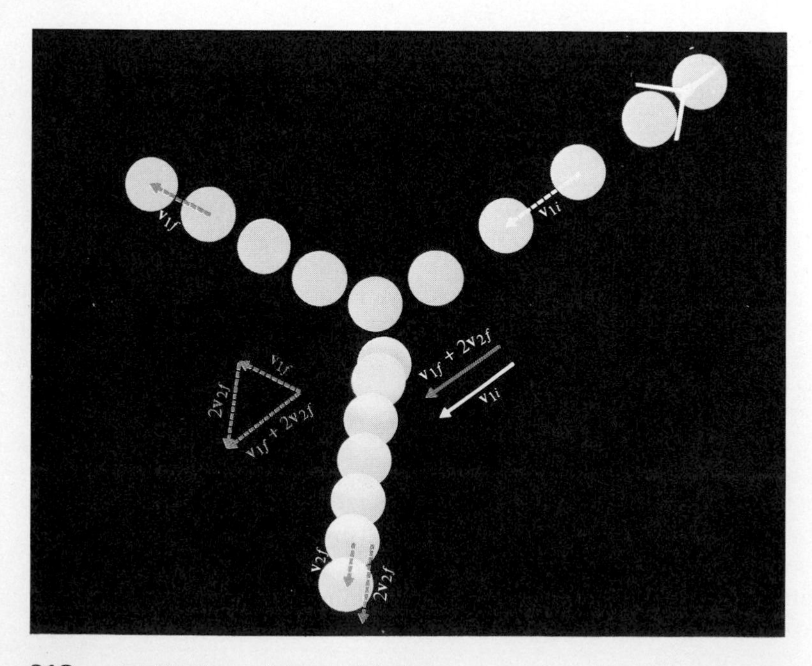

Fig. 8-13 Strobe photo of a collision between a single and a double magnetic puck. The single puck, labeled 1, comes from the upper right; the double puck, labeled 2, is initially at rest near the center of the air table. If the mass of the single puck is taken as the unit of mass, that is, $m_1 = 1$, then the mass of the double puck is $m_2 = 2$.

sure the center-to-center separations of adjacent images of the incident puck, after the collision and before the collision, to obtain the ratio of its final and initial speeds. Then compare this ratio obtained from measurement with the calculated ratio $v_{1f}/v_{1i} = (0.74\ \text{m/s})/(1.00\ \text{m/s}) = 0.74$. You will find the comparison to be quite satisfactory.

It should be emphasized that Eq. (8-18*a*) or (8-18*b*), having been obtained by using only conservation laws, can relate only possible final conditions of a system of colliding bodies to possible initial conditions of the system. Since the conservation laws do not contain complete information about the forces acting between the bodies, the equations obtained from them cannot answer all the questions that can be asked about a particular collision. For instance, the question posed in Example 8-10 was: Given that the incident body is scattered through an angle of 65°, what will its final speed be? This question can be answered by using Eqs. (8-18*a*) and (8-18*b*). Another interesting question is: In what circumstances will the incident body be scattered through an angle of 65°? The equations will not provide an answer to this question. It can be answered only by a calculation which fully takes into account the properties of the forces acting between the colliding bodies. An example of such a calculation is considered in Chap. 20, when we study the scattering of an alpha particle by an atomic nucleus.

An equation completely equivalent to Eq. (8-18*b*) is very frequently used in atomic and nuclear physics to analyze measurements of the scattering of microscopic particles. And a slightly modified equation is used in the analysis of reactions between such particles. We will obtain the modified equation in Chap. 15, which treats relativistic mechanics. The measurement of scattering and reactions is one of the most widely used experimental techniques of contemporary physics. It is used to study the forces which microscopic particles exert on each other when they interact.

Example 8-11 deals with a completely inelastic collision.

EXAMPLE 8-11

Figure 8-14 depicts the **ballistic pendulum** technique that can be used to measure the speed of a bullet. The bullet, of mass m, is fired into a wood block of mass M which is sufficiently thick to stop the bullet. The wood block is suspended by a wire of length l and mass negligible compared to m and M. After the bullet enters the block, the pendulum swings to a maximum angle ϕ, which is measured. Derive an expression giving the speed v of the bullet in terms of the measured value of ϕ.

Fig. 8-14 A ballistic pendulum.

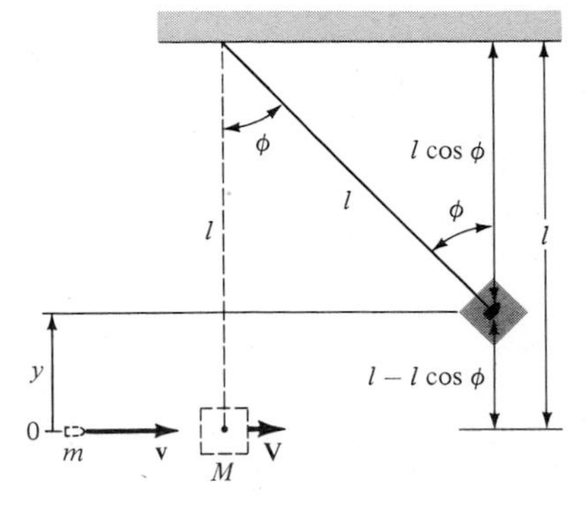

■ You must consider, in sequence, two processes. In the first, the bullet collides with the wood block, penetrating it until the bullet comes to rest with respect to the block. The total momentum of the bullet-plus-block system is conserved in the collision. The reason is that no forces act on this system during the collision which have components along the direction of motion of the bullet. (And even if there were such a force, its effect on the momentum of the system could be ignored, to a good approximation, since it could produce no significant impulse during the very short duration of the collision.) Writing the velocity of the bullet immediately before colliding with the block as $\mathbf{v}$, you get by momentum conservation the equation

$$m\mathbf{v} = (m + M)\mathbf{V} \tag{8-19}$$

Here $m\mathbf{v}$ is the initial total momentum of the system, since the block initially was stationary. And $(m + M)\mathbf{V}$ is the final total momentum of the system with the block and embedded bullet moving at velocity $\mathbf{V}$ immediately after the collision.

In the second process, the pendulum swings to ϕ, where it is instantaneously at rest. In this process the mechanical energy of the bullet-plus-block-plus-earth system is conserved. The reason is that the force exerted on this system by the wire connected to the block does no work on the system; the wire is a workless constraint. Take the potential energy to be zero when the bullet and block are at the height $y = 0$ at the start of the swing, by defining that to be the reference height. Then equate the initial kinetic energy of the system

$$K = \frac{(m + M)V^2}{2} \tag{8-20}$$

to its final potential energy

$$U = (m + M)gy = (m + M)g(l - l \cos \phi)$$

Doing so, you obtain

$$\frac{m + M}{2} V^2 = (m + M)g(l - l \cos \phi)$$

or

$$V^2 = 2gl(1 - \cos \phi)$$

Using Eq. (8-19) to evaluate V^2 in terms of v^2, you then have

$$\left(\frac{m}{m + M}\right)^2 v^2 = 2gl\,(1 - \cos \phi)$$

So

$$v = \frac{m + M}{m} \sqrt{2gl(1 - \cos \phi)} \tag{8-21}$$

The result obtained in Eq. (8-21) allows v to be determined from the known values of m, M, g, l, and ϕ.

The collision between the bullet and the block is completely inelastic, in that the colliding objects remain together after the collision. Thus you would expect mechanical energy to be lost in the collision. But it is apparent that the mechanical energy is not completely lost in the completely inelastic collision—the block and bullet are moving immediately afterward. A portion of the mechanical energy must remain because momentum must *not* be lost in the collision. Use Eq. (8-19) to express the K of Eq. (8-20) in terms of the initial kinetic energy of the bullet, and thereby determine just how much mechanical energy has disappeared in the collision. Where did it go?

8-4 HARMONIC OSCILLATIONS

In Examples 7-9 and 8-11 we have applied energy relations to predict certain aspects of the motion of a pendulum. In this section we will apply energy relations to a harmonic oscillator, such as the body at the end of a spring illustrated in Fig. 8-15. The purpose is not to predict the motion—we already know about the motion of a harmonic oscillator from the detailed study we made in Chap. 6 with the aid of Newton's second law. Rather, the purpose is to gain the additional insight that the energy relations have to offer about properties of this important system.

EXAMPLE 8-12

Using the analytical solutions of the harmonic oscillator differential equation obtained in Sec. 6-4, evaluate the kinetic, potential, and total mechanical energies of the oscillator.

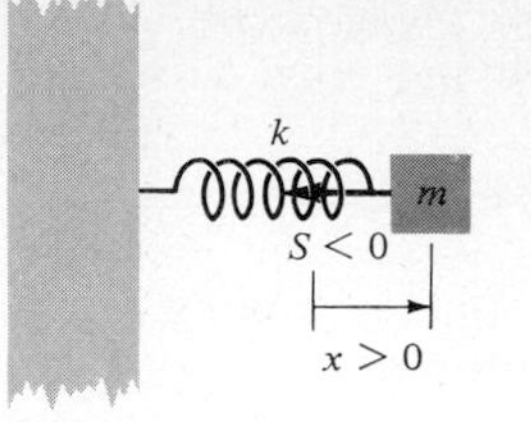

Fig. 8-15 A harmonic oscillator.

■ According to Eq. (6-17), the solutions are

$$x = A \cos(\omega t + \delta) \tag{8-22}$$

Here x is the displacement of the oscillating body at time t, measured from its equilibrium position at $x = 0$. See Fig. 8-15. The angular frequency ω is determined by the mechanical properties of the oscillator through the relation

$$\omega = \sqrt{\frac{k}{m}} \tag{8-23}$$

In this expression k specifies the stiffness of the spring connected to the body, and m specifies the mass of the body. But the amplitude constant A and the phase constant δ can have whatever values are required to describe a particular oscillation. In other words, the values of A and δ are determined by the initial conditions.

To evaluate the kinetic energy K, you need to compute the velocity dx/dt of the body since

$$K = \frac{m}{2}\left(\frac{dx}{dt}\right)^2$$

Differentiating Eq. (8-22), or copying Eq. (6-21a), you immediately obtain

$$\frac{dx}{dt} = -A\omega \sin(\omega t + \delta) \tag{8-24}$$

Substituting this into the expression for K, you have

$$K = \frac{mA^2\omega^2}{2} \sin^2(\omega t + \delta) \tag{8-25}$$

According to Eq. (7-58), the potential energy stored in the spring is

$$U = \frac{kx^2}{2} \tag{8-26}$$

Using Eq. (8-22), you get

$$U = \frac{kA^2}{2} \cos^2(\omega t + \delta)$$

To facilitate comparison with K, you can use Eq. (8-23) to write $k = m\omega^2$. Then you have

$$U = \frac{mA^2\omega^2}{2} \cos^2(\omega t + \delta) \tag{8-27}$$

The total mechanical energy E of the oscillator is

$$E = K + U$$

Evaluating K and U from Eqs. (8-25) and (8-27), you have

$$E = \frac{mA^2\omega^2}{2} [\sin^2(\omega t + \delta) + \cos^2(\omega t + \delta)] \tag{8-28a}$$

The trigonometric identity $\sin^2 \theta + \cos^2 \theta = 1$ holds for any angle θ and for the angle $\omega t + \delta$ in particular. Thus the identity allows you to simplify Eq. (8-28a) to

$$E = \frac{mA^2\omega^2}{2} \tag{8-28b}$$

Although the kinetic and potential energies of the harmonic oscillator vary in time, these results show that its total mechanical energy has *no* time dependence.

Thus you can conclude that built into the solutions of the harmonic oscillator differential equation is the information that the total energy of the oscillator is a constant. The solutions "know" that the oscillator conserves mechanical energy, although they came directly from Newton's laws of motion, and were found before we developed the energy relations and proved that the force produced by a spring is conservative! Equation (8-28b) provides a convincing demonstration of the consistency of all the theory which has led to it.

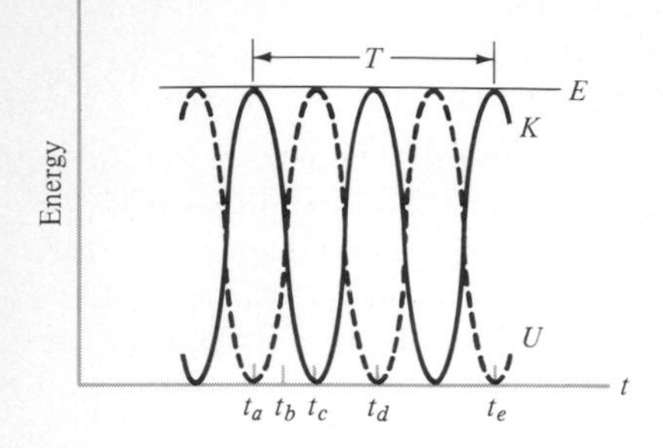

Fig. 8-16 The kinetic energy K, potential energy U, and total mechanical energy E of a harmonic oscillator plotted as a function of the time t.

The way the mechanical energy of the oscillating system is transformed back and forth between its kinetic manifestation and its potential manifestation is made very evident by Eq. (8-28a). Figure 8-16 plots the total, kinetic, and potential energies of a harmonic oscillator, E, K, and U, from the term on the left side of the equation and the two terms on the right side. At the time labeled t_a in the figure, U is zero and K is a maximum. At this instant the spring has its relaxed length, and the body at its end is moving with maximum speed through its position of stable equilibrium, say in the direction which will subsequently result in compression of the spring. As time passes, the spring is compressed and the body slows down. The increase in U is exactly as great as the decrease in K, so that at t_b both of these energies have half their maximum value. At t_c the spring has maximum compression and the body is instantaneously at rest, so that U is a maximum and K is zero. The time interval from t_a to t_c is one-quarter of the oscillation period $T = 2\pi/\omega$. At t_d the body is again moving through the equilibrium position, but now in the direction which will result in extension of the spring. Thus one-half of a period has elapsed. The time t_e corresponds to the end of a full period of the oscillation.

The total mechanical energy E of the harmonic oscillator is conserved because of the way any change in its kinetic energy K is always exactly compensated for by an opposite change in its potential energy U.

The figure also makes clear that $\langle K \rangle$, the average of the kinetic energy K over any full oscillation period, is equal to $\langle U \rangle$, the average of the potential energy U over the same period. That is

$$\langle K \rangle = \langle U \rangle$$

since both K and U vary in the same way between a minimum value of 0 and a maximum value of E. An easy way to determine the actual value of $\langle K \rangle$ or $\langle U \rangle$ is to note that the definition

$$K + U = E$$

tells us that

$$\langle K + U \rangle = \langle E \rangle$$

This, in turn, tells us that

$$\langle K \rangle + \langle U \rangle = \langle E \rangle \tag{8-29}$$

since the average value of the sum of two quantities equals the sum of their average values. Equating $\langle U \rangle$ to $\langle K \rangle$, we have

$$2\langle K \rangle = \langle E \rangle \tag{8-30}$$

We can write this as

$$\langle K \rangle = \frac{E}{2}$$

since $\langle E \rangle = E$ because the total energy E is constant. Evaluating E from Eq. (8-28*b*), we obtain

$$\langle K \rangle = \langle U \rangle = \frac{mA^2\omega^2}{4} \tag{8-31}$$

(A more formal proof of this result can be obtained by direct calculation of $\langle K \rangle$ and $\langle U \rangle$. The calculation requires evaluating the integral over one period of the square of a sinusoid.)

A number of different energies can be associated with a harmonic oscillator: K, U, E, $\langle K \rangle$, and $\langle U \rangle$. But Eqs. (8-25) through (8-31) show that all these energies have a common feature: *the energy is proportional to both the square of the amplitude of the harmonic oscillator and the square of its frequency.*

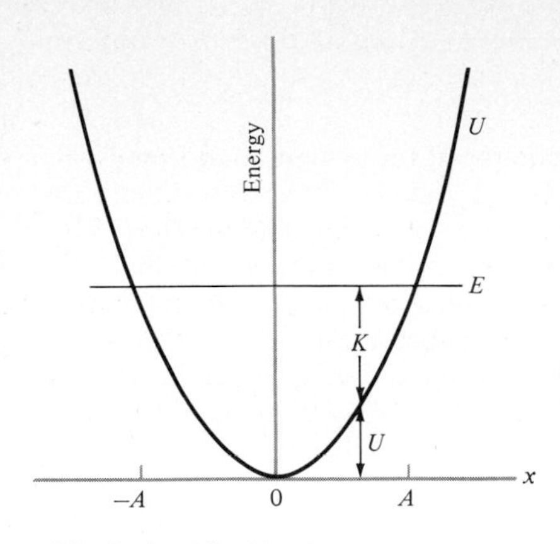

Fig. 8-17 The kinetic energy K, potential energy U, and total mechanical energy E of a harmonic oscillator plotted as a function of the displacement x of the oscillating body from its equilibrium position.

A different, and even more useful, representation of the relation among K, U, and E for a harmonic oscillator is shown in Fig. 8-17, which was obtained by plotting U and E versus x, the position of the oscillating body. You can see from inspection of Eq. (8-26) why the curve of U versus x is a vertically oriented parabola, with "vertex" at the origin and "width" that is determined by the force constant k of the spring. The curve of E versus x is a horizontal line, since E has no x dependence. The possible values of x lie in the range $-A \leq x \leq A$, where A is the amplitude of the oscillator. Also note that the limits of this range are the values of x where $U = E$. The oscillating body cannot be found outside these limits because it is necessary to have $U \leq E$ in order to have $K = E - U \geq 0$. (Since $K = mv^2/2$, it is not possible to have $K < 0$.) When the oscillating body is anywhere within the allowed range of x, the value of U is measured by the vertical distance from the point on the x axis characterizing the body's location to the U curve, and the value of K is measured by the remainder of the vertical distance to the E line.

If you begin watching the oscillating body when it is moving in the positive direction through $x = 0$, you will see it continue to move to $x = A$, slowing down all the while because its kinetic energy is decreasing. At the **turning point** $x = A$, it reverses its direction of motion to turn around and start back toward $x = 0$. It picks up speed as its kinetic energy increases, until it passes $x = 0$. At that point it has completed one-half of a cycle of oscillation. The second half-cycle is just the reverse of the first half, taking it to the other turning point at $x = -A$ and then back to $x = 0$.

Giving the oscillating body an appropriately timed blow will increase its total energy. If this happens, the E line in Fig. 8-17 will move up and the turning points, where $U = E$, will move farther from $x = 0$. This means the amplitude A of the oscillation will increase. Because of the parabolic shape of the curve described by U, the relation between E and A is $E \propto A^2$, in agreement with Eq. (8-28*b*).

Figure 8-17 can also be used to give you information about the force F acting on the body when it is at location x. To obtain this information, use Eq. (7-56), the relation between force and potential energy:

$$F = -\frac{dU}{dx} \tag{8-32}$$

Since dU/dx is the slope of the U curve, this tells you that F is the negative of the slope. Thus $F = 0$ at $x = 0$ because the U curve has zero slope there. As the body moves toward $x = A$, the slope of the U curve becomes ever more positive. The body feels an ever-increasing negative force, that is, a force acting back toward $x = 0$. The signs are reversed when the body moves from $x = 0$ toward $x = -A$.

Another example of a plot of the x dependence of the potential and total energies of a system is shown in Fig. 8-18. The system consists of two atoms that can bind together and form a molecule, say, sodium and chlorine. The potential energy U is plotted versus the separation x between the centers of the two atoms. The minimum in U occurs at the **equilibrium separation** x_e of the two atoms in the molecule. At that separation, the force F acting on either atom is $F = -dU/dx = 0$. If the separation is smaller than the equilibrium separation, a strong force $F = -dU/dx > 0$ develops. The force acts in the direction tending to increase the separation; in other words, the force is repulsive. If x is larger than x_e, the atoms feel an attractive force $F = -dU/dx < 0$. For values of x close to x_e, the magnitude of the force tending to restore the separation to its equilibrium value is a function of x which is symmetrical about x_e. In fact, near x_e the potential energy curve approximates a parabola—as any smooth curve with a minimum at x_e must. The corresponding force obeys Hooke's law for x near x_e. The potential energy of a crystal, such as in a metal, also is a parabolic function of the center-to-center separation of the atoms comprising the crystal, providing the separation is near its equilibrium value. *This microscopic property is what causes metals, and many other materials, to obey Hooke's law on the macroscopic level.* Its macroscopic aspect is studied in Chap. 16.

But for a sufficiently large increase in x, a point is reached where the attractive force begins to increase less rapidly than Hooke's law would predict. This is the microscopic equivalent of the point at which a crystalline material ceases to obey Hooke's law. At an even larger value of x, the attractive force between the atoms begins to become weaker with increasing x. At this point the attractive force has its ultimate strength. It is the microscopic equivalent of the yield strength of a material. If x exceeds the **dissociation separation** x_d, the force drops to zero. This happens because the two atoms have become so widely separated that they no longer interact. A plot of F versus x is shown in Fig. 8-19 for the U versus x plot in Fig. 8-18. Inspecting it will help clarify the points made in this paragraph.

If the value of the total energy E is less than the **dissociation energy** E_d, as in the case E_1 illustrated in Fig. 8-18, then the two atoms are **bound** in the molecule. That is, their separation distance x will oscillate within the range $x_1 \leq x \leq x_1'$. At a somewhat higher value E_2, the allowed range increases to $x_2 \leq x \leq x_2'$. Note that because U is not symmetrical about x_e, the outer limit of the range has moved out more than the inner limit has moved in. Thus the oscillations in x are not symmetrical about x_e. Averaging x over a cycle of oscillation, the molecule has expanded as its total energy has become higher. This is the essential mechanism operating in the thermal expansion of most materials, since the total energy of a molecule tends to increase as the temperature of its surroundings increases.

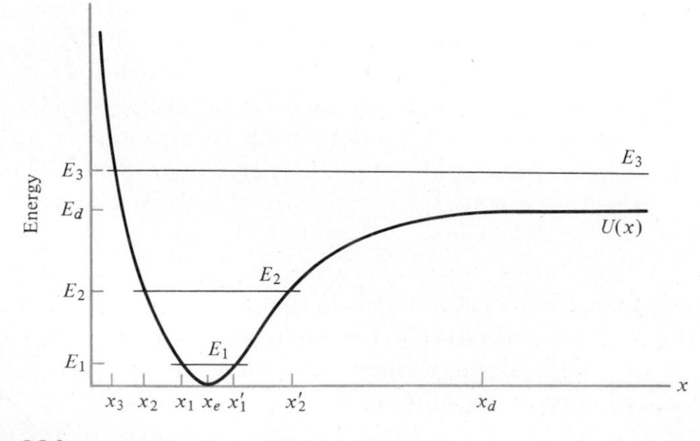

Fig. 8-18 This curve shows qualitatively the dependence of the potential energy U of a diatomic molecule on the separation x between the centers of its two atoms. The horizontal lines represent three possible values of the constant total energy E of the molecule. The equilibrium separation of the atoms is x_e, and the dissociation separation is x_d. The dissociation energy is E_d.

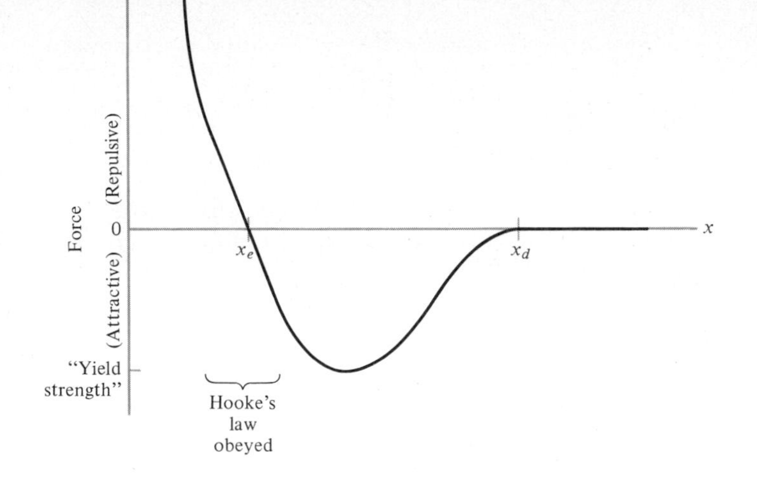

Fig. 8-19 Qualitative representation of the force F acting on an atom of a diatomic molecule along the line between the centers of its atoms, as a function of the center-to-center separation x. This curve is drawn from Fig. 8-18 by using the relation $F = -dU/dx$.

For a total mechanical energy higher than the dissociation energy, such as E_3, the separation x can be any value in an infinite range beginning at x_3, and the two atoms are **unbound.**

How do the atoms become bound? If they approach each other from a large initial separation, with total energy E_3, they move together with constant relative speed until $x < x_d$. Their relative speed then increases until $x < x_e$ and subsequently decreases until the atoms are instantaneously motionless with respect to each other at $x = x_3$. They then retrace their relative motion in the opposite direction. But suppose that while x is near x_e, the system gets rid of enough energy that the total energy drops to a value less than E_d. Then the two atoms will form a bound molecule because they will no longer have enough energy to separate. This can happen if the system emits energy in the form of electromagnetic radiation. Such radiation emission characterizes many inelastic collisions on the atomic level.

8-5 LIGHTLY DAMPED OSCILLATIONS

In this section we investigate the decrease in the mechanical energy of a *lightly damped* oscillator, whose motion was treated in Sec. 6-5. One way we could approach this task is to follow the method of Example 8-12. Beginning with the analytical solutions to the differential equation for a lightly damped oscillator given in Eq. (6-33),

$$x = Ae^{-\mu t}\cos(\omega t + \delta)$$

we could differentiate to find dx/dt and then find $K = m(dx/dt)^2/2$. Adding the result to $U = kx^2/2$, we would obtain $E = K + U$. But these expressions would have rather complicated forms. It is easier and more instructive to work directly with the differential equation.

According to Eq. (6-29), the differential equation for a damped oscillator is

$$m\frac{d^2x}{dt^2} = -kx - r\frac{dx}{dt} \tag{8-33}$$

where $-r(dx/dt)$ is the frictional damping force. From the point of view of energy, the main difference between the damped oscillator and the undamped oscillator is that in the former the total mechanical energy E of the system diminishes as a result of the frictional term $-r(dx/dt)$. To stress this point, we will recast the differential equation in the form

$$\frac{d}{dt}\,(\text{total mechanical energy}) = \text{frictional power drain}$$

First, we rewrite Eq. (8-33) with the frictional term isolated on the right side:

$$m\,\frac{d^2x}{dt^2} + kx = -r\,\frac{dx}{dt} \tag{8-34}$$

Each term in this equation has the dimensions of a force. Thus if the term on the right side were multiplied by dx/dt, the product would then have the desired dimensions of power because Eq. (8-4) shows us that

$$(\text{force})(\text{velocity}) = \text{power}$$

This suggests that we multiply Eq. (8-34) through by dx/dt to obtain

$$m\,\frac{dx}{dt}\,\frac{d^2x}{dt^2} + kx\,\frac{dx}{dt} = -r\left(\frac{dx}{dt}\right)^2 \tag{8-35}$$

Then we note that the left side of this equation can be rewritten so that it becomes

$$\frac{d}{dt}\left[\frac{m}{2}\left(\frac{dx}{dt}\right)^2 + \frac{k}{2}\,x^2\right] = -r\left(\frac{dx}{dt}\right)^2 \tag{8-36}$$

This can be verified immediately by working out the derivative of the two terms within the brackets.

Now, the first of the terms in the brackets is the kinetic energy K of the oscillating body, and Eq. (8-26) says that the second term is the potential energy U associated with the spring force. So we have

$$\frac{d}{dt}\,[K + U] = -r\left(\frac{dx}{dt}\right)^2$$

or

$$\frac{dE}{dt} = -r\left(\frac{dx}{dt}\right)^2 \tag{8-37}$$

Here E is the total mechanical energy in the body-and-spring system, and dE/dt is its rate of change. Thus the differential equation has been recast into the desired form, and we can identify its right side $-r(dx/dt)^2$ as the frictional power drain. For an undamped oscillator $r = 0$, and Eq. (8-37) shows immediately that in such a system $dE/dt = 0$ and so $E = \text{constant}$.

In a lightly damped oscillator the friction term removes mechanical energy from the system at a rate which varies throughout each oscillation cycle because $(dx/dt)^2$ varies. Consequently, it is useful to calculate the average of this rate over one full cycle. To do this, we first solve the equation $K = m(dx/dt)^2/2$ for $(dx/dt)^2$, obtaining

$$\left(\frac{dx}{dt}\right)^2 = \frac{2K}{m}$$

We substitute this expression into the right side of Eq. (8-37), which yields

$$\frac{dE}{dt} = -\frac{2r}{m}\,K$$

Taking averages over one cycle and remembering that $-2r/m$ is a constant, we have

$$\left\langle \frac{dE}{dt} \right\rangle = -\frac{2r}{m} \langle K \rangle$$

Let us assume that the oscillator is quite lightly damped. Then its amplitude will decrease only slightly from one cycle to the next, and its motion over any one cycle will not be significantly different from the motion of an undamped oscillator that happens to be oscillating with the same amplitude. Thus when the damping is small, it is a good approximation to use Eq. (8-30):

$$2\langle K \rangle = \langle E \rangle$$

with $\langle E \rangle$ representing the slowly decreasing average total mechanical energy of the damped oscillator. Then we have

$$\left\langle \frac{dE}{dt} \right\rangle = -\frac{r}{m} \langle E \rangle \qquad (8\text{-}38)$$

For the purpose of determining how the mechanical energy decreases over many cycles, we can replace the average value of its derivative by the derivative of its average value. That is, we can go from Eq. (8-38) to

$$\frac{d\langle E \rangle}{dt} = -\frac{r}{m} \langle E \rangle \qquad (8\text{-}39)$$

In actuality, the rate of decrease of E fluctuates through each cycle of oscillation, since energy is lost most rapidly during the parts of the cycle when the motion is most rapid. This detailed behavior is indicated schematically by the solid curve in Fig. 8-20. The shaded curve represents the averaged behavior of E. The justification of replacing the average of the derivative with the derivative of the average is that the slope of the solid curve, averaged over a particular cycle, is accurately represented by the slope of the shaded curve at that cycle. In using Eq. (8-39) we are trying to describe only the overall time dependence of the mechanical energy of the oscillator, not the fine structure of this time dependence.

Equation (8-39) says that the average mechanical energy of a lightly damped oscillator is a function of time having the property that *its first derivative is proportional to its value.* What function has this property? According to Eq. (6-37), an exponential function does. In fact, a form for $\langle E \rangle$ which satisfies Eq. (8-39) is

$$\langle E \rangle = \langle E \rangle_0 \, e^{-(r/m)t} \qquad (8\text{-}40)$$

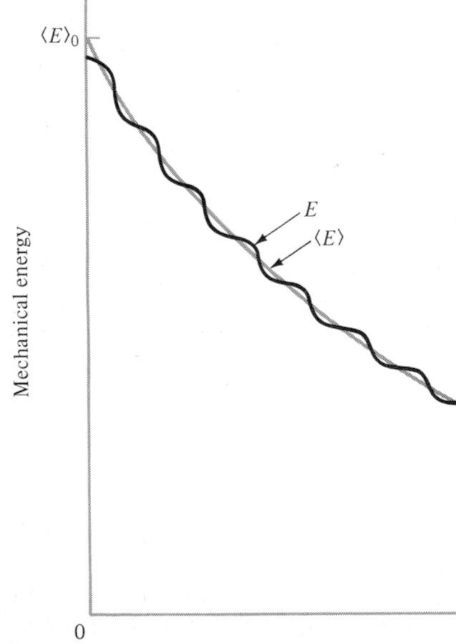

Fig. 8-20 The time dependence of the total mechanical energy of a lightly damped oscillator.

The quantity $\langle E \rangle_0$ is a constant whose value we can adjust to fit any particular case by setting it equal to the value of $\langle E \rangle$ at $t = 0$. To verify that Eq. (8-40) satisfies Eq. (8-39), we differentiate Eq. (8-40) by applying Eq. (6-37), the rule for differentiating an exponential. We have

$$\frac{d\langle E \rangle}{dt} = \frac{d}{dt}\left[\langle E \rangle_0 \, e^{-(r/m)t}\right] = \langle E \rangle_0 \frac{d}{dt}\left[e^{-(r/m)t}\right] = -\frac{r}{m}\langle E \rangle_0 \, e^{-(r/m)t}$$

Substituting this and Eq. (8-40) itself into Eq. (8-39), we obtain

$$-\frac{r}{m}\langle E \rangle_0 \, e^{-(r/m)t} = -\frac{r}{m}\langle E \rangle_0 \, e^{-(r/m)t}$$

Since this equation is certainly valid, and since it was obtained from Eq. (8-40), we have proved that Eq. (8-40) is valid.

Our conclusion is that $\langle E \rangle$, the average mechanical energy stored in a lightly damped oscillator, decreases exponentially from one cycle to the next as frictional effects convert the mechanical energy into thermal energy. The rapidity of the decrease is governed by the value of r/m, the coefficient of t in the negative exponent. The shaded curve showing the behavior of $\langle E \rangle$ in Fig. 8-20 was obtained by plotting Eq. (8-40).

In a damped oscillator (and in any equivalent system such as an oscillating electrical system), the ratio of the energy stored to the energy loss in one cycle of oscillation, multiplied by 2π, is called the **quality factor,** or **Q factor.** That is, we define

$$Q \equiv 2\pi \frac{\text{energy stored}}{\text{energy loss per cycle}} \qquad (8\text{-}41a)$$

The energy stored in the oscillator is given by the quantity $\langle E \rangle$, and the energy loss per cycle is the magnitude of its average change per unit time $|\langle dE/dt \rangle|$ multiplied by the time per cycle, the period T. Thus we can write

$$Q = 2\pi \frac{\langle E \rangle}{|\langle dE/dt \rangle| \, T} \qquad (8\text{-}41b)$$

The Q factor is a figure of merit that indicates how thorough a job has been done in eliminating frictional sources of energy loss from an oscillating mechanical system (or resistive losses in an electrical system). A high Q factor means a low loss.

To evaluate Q for a lightly damped oscillator, we use Eq. (8-38) for $\langle dE/dT \rangle$. This yields

$$Q = \frac{2\pi \langle E \rangle}{(r/m)\langle E \rangle T}$$

or

$$Q = \frac{2\pi \, m}{rT} \qquad (8\text{-}42a)$$

Expressed in terms of the angular frequency $\omega = 2\pi/T$, it is

$$Q = \frac{m\omega}{r} \qquad (8\text{-}42b)$$

Since r specifies the strength of the frictional force acting on the oscillating body, it is not surprising to find that Q is proportional to $1/r$. Can you explain on physical grounds why Q is proportional to m and also to ω?

It is difficult to make the Q factor in practical mechanical systems much larger than 10^2. (In high-quality electrical systems Q factors in excess of 10^6 can be attained, while in superconducting circuits $Q \to \infty$.) We will find in Chap. 26 that the Q factor has a very important bearing on the sharpness of the response of oscillatory systems (such as those in radio receivers) when they are driven by an applied oscillation.

Example 8-13 evaluates the Q factor of a lightly damped oscillator.

EXAMPLE 8-13

Again you are the engineer working on the design of the automobile springing system considered in Example 6-9. With springs but no shock absorbers installed on the preliminary model, you give the automobile body an initial downward push to set it into vertical oscillation. Your measurements show that from each cycle to the next the amplitude of the oscillation decreases by 16 percent. What is the Q factor of the system?

■ You can find the Q factor from the results of your measurements by using Eqs. (6-30), (6-33), and (6-38) to write an expression for the position x of a lightly damped oscillator that involves the quantities m and r. It is

$$x = Ae^{-(r/2m)t}\cos(\omega t + \delta) \tag{8-43}$$

Let t_1 be the time at which the body reaches the highest point x_1 in its first oscillation cycle. For this value of t, you know that $\cos(\omega t_1 + \delta) = 1$. So

$$x_1 = Ae^{-(r/2m)t_1}$$

Similarly, if t_2 is the time for the highest point x_2 in the second cycle, you have $\cos(\omega t_2 + \delta) = 1$, and

$$x_2 = Ae^{-(r/2m)t_2}$$

Divide the second equation by the first:

$$\frac{x_2}{x_1} = \frac{e^{-(r/2m)t_2}}{e^{-(r/2m)t_1}} = e^{-(r/2m)(t_2-t_1)}$$

Your measurements show that

$$x_2 = (1 - 0.16)x_1 = 0.84\ x_1$$

or

$$\frac{x_2}{x_1} = 0.84$$

Also, $t_2 - t_1 = T$, the period of the oscillation. So you have

$$0.84 = e^{-(r/2m)T}$$

Now you solve for m/r in terms of T. First you transpose and take reciprocals, to obtain

$$e^{(r/2m)T} = \frac{1}{0.84} = 1.19$$

Taking logarithms to the base e, you then have

$$\ln e^{(r/2m)T} = \ln 1.19$$

By the definition of a logarithm, the left side of this equation gives you simply $(r/2m)T = rT/2m$. So you have

$$\frac{rT}{2m} = \ln 1.19$$

Evaluating the logarithm, you find

$$\frac{rT}{2m} = 0.174$$

or

$$\frac{2m}{rT} = 5.74$$

To two significant figures, you obtain the result

$$\frac{m}{r} = 2.9T$$

Would you obtain the same result if you used the amplitudes for the fifth and sixth cycles to evaluate m/r, instead of the amplitudes for the first and second cycles? Why? What is the advantage of using the first and second cycles for measurement? Also, can you explain on physical grounds why the coefficient of t in the negative exponent of Eq. (8-43) is $r/2m$, whereas it is r/m in Eq. (8-40)?

From Eq. (8-28), you know that

$$Q = \frac{2\pi\, m}{rT}$$

Using the result you obtained for m/r, you have

$$Q = \frac{2\pi}{T} \times 2.9T$$

or

$$Q = 18$$

The system does not have a high Q factor. But it is not intended to be a "high-Q" system. Indeed, you must make the Q factor close to 2π in order to make the automobile have a comfortable ride. As Q approaches 2π, the energy loss in one cycle of oscillation becomes the same as the energy content of the oscillation, and critical damping is approached. How do you reduce the Q factor of the springing system?

EXERCISES

Group A

8-1. *Upstairs.* You increase your elevation by 15 m in 50 s by running up the stairs of a multistory building. Assuming that all the work you do goes into increasing the gravitational potential energy of the system you-plus-earth and that your mass is 70 kg, calculate your power output in watts and in horsepower.

8-2. *Niagara Falls.* About 180,000 metric tons (1.8×10^8 kg) of water per minute plunges over Niagara Falls, which is 48 m high. Find the rate of release of gravitational potential energy. Express your result in kilowatts.

8-3. *Rain power.* A heavy rainstorm can drop 10 mm of water per hour on a locality, and good-sized raindrops have a terminal speed of 8 m/s.

a. If the rain cloud is at an average altitude of 600 m, find the rate at which the rainstorm releases gravitational potential energy per unit of land area. Express your result in watts per square meter.

b. What fraction of the power per unit area found in part *a* is in the form of the kinetic energy of the impacting raindrops? What has happened to the rest of the initial gravitational potential energy?

8-4. *Rain on the roof.* A gently sloped roof of area 100 m^2 is exposed to the rainstorm described in Exercise 8-3.

a. What is the time-averaged force the roof must exert to stop the impacting drops?

b. What force per unit area is this?

c. Express the result of part *b* as a fraction of the standard atmospheric pressure, 1.013×10^5 N/m^2.

8-5. *Measuring an engine's brake horsepower.* The device shown in Fig. 8E-5 can be used to measure the **brake horsepower** of an engine. The engine is used to drive the drum at a steady rate against the frictional resistance provided by a heat-resistant band that is held against the drum by the two spring balances.

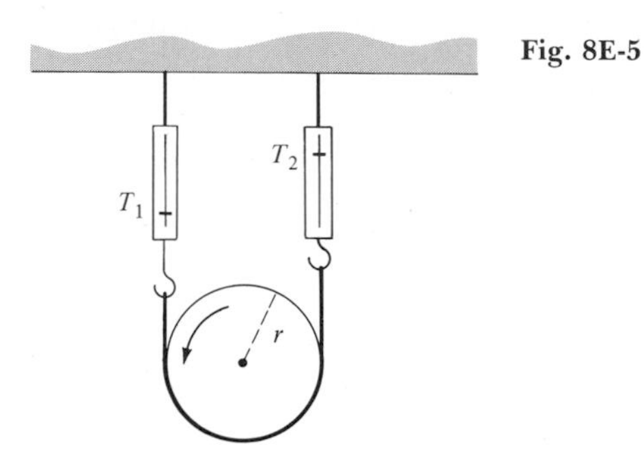

Fig. 8E-5

a. If the drum is turning at a steady rate, how much work is done against friction each time the drum rotates? Express your result in terms of the drum radius r and the tensions T_1 and T_2.

b. If the drum is turning steadily at n rotations per second, find the power P being delivered by the engine.

c. With a drum of radius 15 cm, an engine is tested at a speed of 3000 rotations per minute (rot/min). With the engine at full throttle, tensions T_1 = 1100 N and T_2 = 200 N are required to keep the drum from turning faster. What maximum power can the engine provide at 3000 rot/min? Express your result in horsepower. This is the engine's brake horsepower at 3000 rot/min.

8-6. *A long, hard climb.* According to the seventeenth edition of the *Guinness Book of World Records,* a New York City resident once climbed the 1575 steps of the Empire State Building in 12 min 32 s. Assume that the climber's mass was 70 kg. If the stairstep height is 24 cm, at what average rate did the climber do work against gravity? Express your result in watts and in horsepower.

8-7. *The advantage of a lever.* Sketch a lever with a fulcrum located at a point *between* the two ends. The distance from the fulcrum to the end of the lever where the load being lifted is connected is r_1, and the distance from the fulcrum to the end where the lifting force is applied is r_2. Find the mechanical advantage of this lever. Compare your results with Eq. (8-5).

8-8. *The advantage of a crank.* What is the mechanical advantage of the crank mechanism shown in Fig. 8E-8?

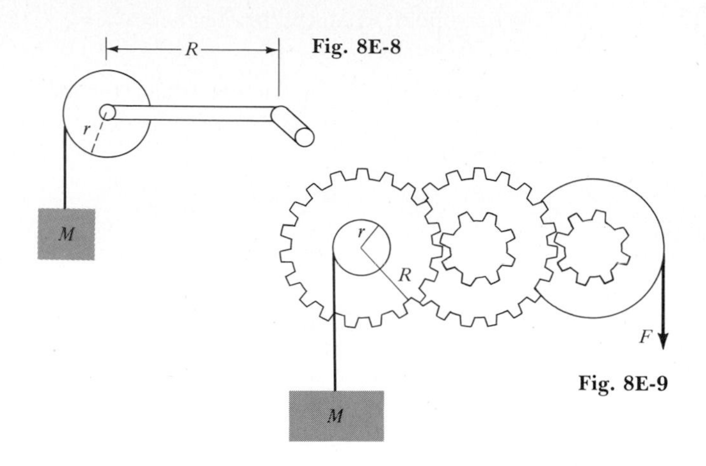

Fig. 8E-8

Fig. 8E-9

8-9. *The advantage of a gear train.* A train of gears is used to lift a block of mass M, as shown in Fig. 8E-9. In each gear, the radius of the central section is r, and the overall radius is R.

a. Neglecting friction, what is the mechanical advantage of the gear train?

b. Suppose $R = 5r$. Evaluate the mechanical advantage. What applied force F would be required to lift a 100-kg block?

8-10. *A painter pulls a pulley.* A painter raises himself by pulling on rope A in the pulley system shown in Fig. 8E-10.

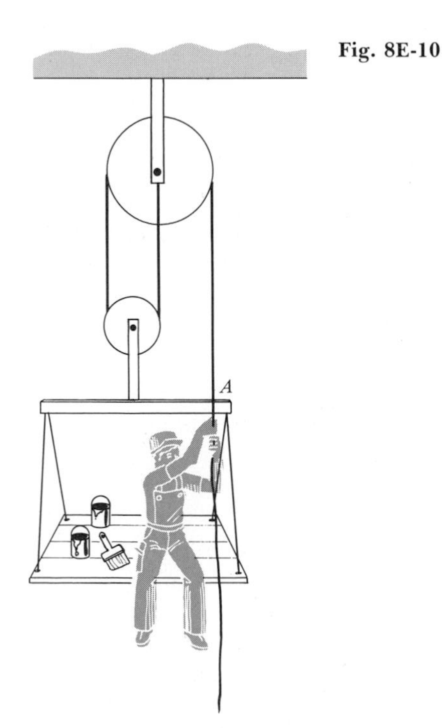

Fig. 8E-10

a. Neglecting friction, what is the mechanical advantage of the system?

b. If the painter lets someone on the ground hoist him by pulling on rope A, what will the mechanical advantage be?

8-11. *More pulleys.* Neglecting friction, what is the mechanical advantage of each of the pulley systems shown in Fig. 8E-11.

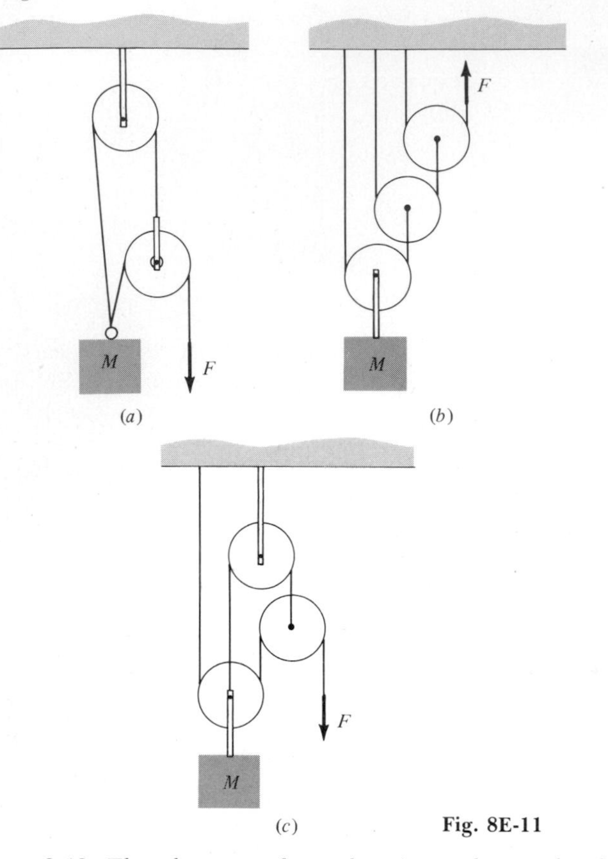

Fig. 8E-11

8-12. *The advantage of a wedge.* A wooden wedge is pushed horizontally to raise a heavy object. All surfaces are frictionless. Apply the law of conservation of mechanical energy to determine the mechanical advantage in terms of the wedge angle θ.

8-13. *Winding a windlass.* A windlass mechanism is shown in Fig. 8E-13.

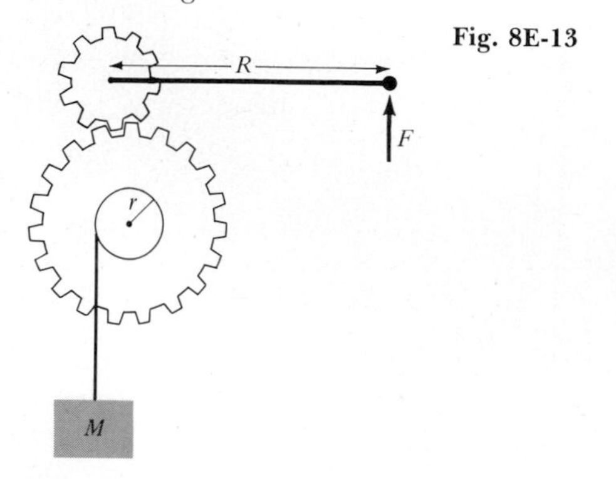

Fig. 8E-13

a. Neglecting friction, what is its mechanical advantage if the radius of the lower gear is N times the radius of the upper gear?

b. Evaluate your results for $N = 3.0$, $R = 40$ cm, and $r = 5.0$ cm.

8-14. *The advantage of a jackscrew.* In a jackscrew, a horizontal arm of length R extends the screw and raises the jack a distance p for each complete turn of the arm. The quantity p is called the **pitch** of the screw.

a. Neglecting friction, what is the mechanical advantage of the jackscrew in terms of p and R?

b. Evaluate the mechanical advantage for $R = 1.0$ m and $p = 1.0$ cm.

8-15. *A safety valve.* A safety valve for a steam boiler is shown in Fig. 8E-15. The total mass of the arm and plug is small compared to M.

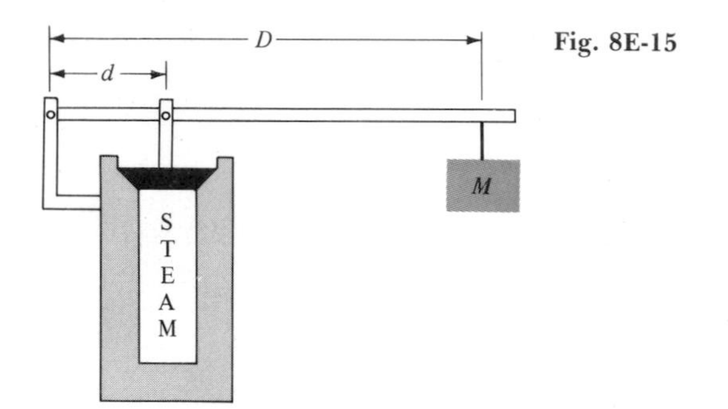

Fig. 8E-15

a. What force must the steam exert on the plug to open the valve?

b. Evaluate your result for $d = 3.0$ cm, $D = 30$ cm, and $M = 5.0$ kg.

c. The plug bottom has an area of 3.0 cm^2 exposed to the steam. What is the required force per unit area?

d. The force per unit area exerted by a fluid is called the *fluid pressure.* A common unit for pressure is the standard atmosphere (atm), whose SI equivalent is 1.013×10^5 N/m^2. Express the result of part c in atmospheres.

8-16. *Head-on collision with a stationary object.* A billiard ball moving with velocity $\mathbf{v}_{1i}$ makes a head-on collision with another billiard ball. The second ball is stationary before the collision; that is, $\mathbf{v}_{2i} = 0$. Both balls have the same mass. Assuming the collision to be elastic, predict the final velocities of both balls, $\mathbf{v}_{1f}$ and $\mathbf{v}_{2f}$. Explain how your results fit in with the analysis of Eq. (8-14) given in the text.

8-17. *Where did it go?* What fraction of the initial kinetic energy was lost when the bullet in Example 8-11 stopped in the wood block of the ballistic pendulum? Where did the lost kinetic energy go?

8-18. *Splat!* Imagine that you are hit by a rotten tomato thrown by a prankster. The tomato has the same density as water and is roughly spherical, with a radius of 3.0 cm. It strikes you at a speed of 10 m/s and flattens, so that it comes to rest in approximately the time taken to travel its own diameter at its initial speed.

a. What impulse must be supplied to the tomato?

b. During what time interval is this impulse delivered?

c. What is the average contact force between you and the tomato during this interval?

8-19. *A hailstone strikes.* A hailstone of radius 5.0 mm and density 0.90 g/cm³ strikes the roof of a parked car. The hailstone hits with a speed of 10 m/s and rebounds to a height of 0.20 m.

a. What fraction of its initial kinetic energy is lost in the impact?

b. Assume that during the collision the hailstone is decelerated to rest in the time it would take to travel its own diameter. What average force must the roof exert?

c. Assuming that this force must be provided by a circular area equal in radius to the hailstone, what is the pressure? Express your result in newtons per square meter and in atmospheres. (See Exercise 8-15.)

8-20. *Experimental determination of a pendulum's Q factor.* Construct a pendulum by suspending any conveniently available compact object of mass about 0.25 kg from the lower end of a string approximately 0.5 m long. When the amplitude of the oscillation is about 0.10 m, make measurements that will allow you to determine approximately the Q factor of the pendulum, in the manner employed in Example 8-13. Compare your result with the Q factor of the system studied in the example, and explain the difference in the values of the two Q factors.

Group B

8-21. *Flea power.* According to the *Guinness Book of World Records,* the common flea can perform vertical jumps as high as 20 cm (or about 130 times its own height!). It is reported that in the process the flea subjects itself to an upward acceleration of about $200g$.

a. What launch speed is required to reach the quoted height neglecting air resistance? Neglect the small vertical displacement during launch.

b. If the launch involves a uniform $200g$ acceleration, how much time does the flea use to launch itself? How far does it move during launch? Is your answer consistent with the size of a flea, as given implicitly above?

c. Make a reasoned estimate for the mass of a flea.

d. Use the mass estimate to calculate the flea's initial kinetic energy and the average power expenditure during launch.

e. Compare the power output per kilogram of a jumping flea with the power output per kilogram of the champion stair climber of Exercise 8-6.

8-22. *Rolling freight.* Each car of a 50-car freight train has a mass of 4.0×10^4 kg. The coefficient of static friction for iron wheels on iron rails is 0.0040. The train is being pulled up a one percent grade, so that the elevation increases by 10 m per kilometer of horizontal displacement. Find the power required to pull the train at a steady speed of 15 m/s. Express your result in kilowatts.

8-23. *Atwood's machine and energy conservation.* Use energy considerations to analyze the motion of the Atwood machine shown in Fig. 8E-23. The string and pulley have negligible mass, the pulley is frictionless, and the mass m_1 is greater than m_2. The system is initially held at rest.

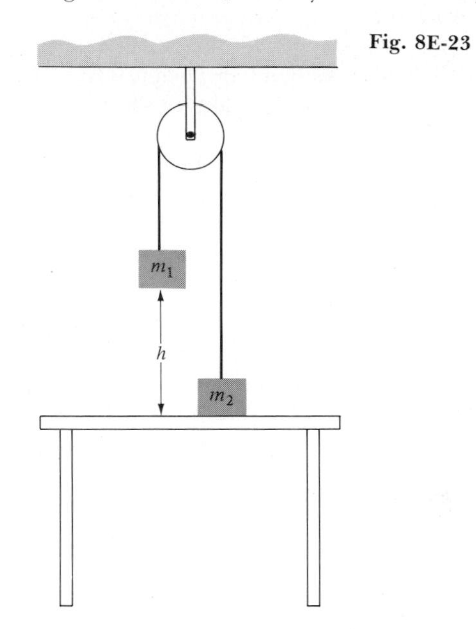

Fig. 8E-23

a. If you take the tabletop on which m_2 rests as the reference level, what is the initial total energy of the system?

b. The system is released, and m_1 descends to the table. Write an expression for the total energy of the system just before m_1 strikes the table.

c. Use the results of parts *a* and *b* to determine the speed of the bodies just before m_1 strikes the table.

d. When m_1 strikes the table, the string goes slack. Use energy considerations to determine how far m_2 rises after the string goes slack.

8-24. *A self-locking machine.* An inclined plane can be regarded as a machine for doing work against gravity. Because of the kinetic friction between the object being elevated and the surface of the incline, the applied force must do an amount of work W that exceeds the increase ΔU of gravitational potential energy. Consider the motion of a load up an inclined plane. Prove that if $W > 2\Delta U$, the incline is self-locking. That is, if the applied force is removed, the object will not slip back down the incline.

8-25. *An elastic head-on collision.* A body of mass m_1 traveling with velocity $\mathbf{v}_{1i}$ makes a head-on elastic collision with a stationary body of mass m_2. The velocities after collision are $\mathbf{v}_{1f}$ and $\mathbf{v}_{2f}$.

a. Calculate $\mathbf{v}_{1f}$ and $\mathbf{v}_{2f}$ in terms of $\mathbf{v}_{1i}$.

b. Calculate the ratio of the kinetic energy transferred to m_2 to the original kinetic energy.

c. For what value of m_2 is all the energy transferred to the stationary body?

d. If m_1 were a neutron and m_2 were a carbon atom whose mass is about 12 times the neutron's mass, what fraction of the neutron's energy would be transferred to the carbon atom in the head-on collision?

8-26. *Elastic collisions within an isolated system.* Three bodies form an isolated system. Their masses are m_1, $m_2 = 2m_1$, and $m_3 = 3m_1$. They have different directions of motion, but they all have the same initial speed v_0. One or more elastic collisions occur between pairs of the bodies which otherwise do not interact. Use energy considerations to find the maximum possible final speed of each of the three bodies. (It should be quite evident that not all the bodies could have their maximum speeds in the same situation.)

8-27. *Colliding pucks.*

a. Make measurements on the strobe photo of the collision between identical plastic pucks in Fig. 8-11, to obtain the initial and final kinetic energies of the two pucks. Assume that the time interval between strobe light flashes is 0.200 s, that the distance between adjacent positions of the center of puck 1 before the collision is 0.100 m, and that each puck has a mass of 0.250 kg. Evaluate the kinetic energy loss ΔK from your measurements of K_{1i}, K_{2i}, K_{1f}, and K_{2f}.

b. Apply Eq. (8-18*b*) to calculate ΔK from K_{1i}, K_{1f}, and the value you measure for the scattering angle ϕ. Compare your results with those you obtained in part *a*.

8-28. *Coefficient of restitution.* For head-on collisions, the **coefficient of restitution** ϵ for a pair of bodies is defined as the ratio of the magnitude of the relative velocity $\mathbf{v}_f$ after the collision to the magnitude of $\mathbf{v}_i$, that before the collision. Thus, $\epsilon = v_f/v_i$. The usefulness of the coefficient ϵ lies in the fact that, for given materials and/or objects, it is approximately constant over a reasonable range of impact speeds. The value of the coefficient can be determined by clamping one body, dropping the other object onto it from a known height h_i, and measuring the height of rebound h_f.

a. Find an expression for ϵ in terms of h_i and h_f.

b. It is observed that a Ping-Pong ball rebounds 24 cm when dropped onto a hardwood table from a height of 30 cm. What is the coefficient of restitution?

c. A ball is dropped onto a fixed horizontal surface from height h_1. The coefficient of restitution is ϵ. Find the total distance traveled by the ball as it bounces before it comes to rest on the surface. Express your result in terms of h_1 and ϵ.

d. Evaluate the total distance traveled by the Ping-Pong ball described in part *b*.

8-29. *Bouncing down the stairs.* A ball is bouncing down a flight of stairs. The coefficient of restitution, as defined in Exercise 8-28, is ϵ. The height of each step is d, and the ball descends one step at each bounce. After each bounce it rebounds to a height h above the next lower step. The height h is large enough compared with the width of a step that the impacts are effectively head on. Show that $h = d/(1 - \epsilon^2)$.

8-30. *On the rebound.* A body of mass m collides with a frictionless surface. Its initial speed is v_i, and it strikes the surface at an angle θ_i. It bounces from the surface, but the collision is not elastic, so that after the impact the magnitude of the normal component of the velocity is only a fraction ϵ of the original value $v_i \sin \theta_i$.

a. Find the impulse delivered by the surface to the body.

b. Find the angle θ_f at which the body leaves the surface.

c. Find the speed at which the body leaves the surface.

d. Express the ratio of final to initial kinetic energy in terms of ϵ and θ_i.

8-31. *Tennis power.* A powerful tennis player can serve the ball at 50 m/s (more than 110 mi/h!). The mass of a tennis ball is 57 g.

a. What is the impulse delivered to a tennis ball as it is served?

b. Estimate the duration of contact between the tennis racket and the ball. To do this, estimate the diameter of a tennis ball and assume that it is (temporarily) squashed to half this value by the racket.

c. Use your results for parts *a* and *b* to obtain estimates for the contact force between ball and racket and for the rate of increase of kinetic energy of the ball during the serve. Express your answer in kilowatts.

8-32. *The right impulse.* An astronaut is doing maintenance work outside a space station. She is coasting along the station at a speed of 1.00 m/s. She wishes to change her direction of motion by 90° and to increase her speed to 2.00 m/s. Her total mass is 100 kg, including her spacesuit and rocket belt, which provides a thrust of 50 N.

a. Find the magnitude and direction of the impulse needed to accomplish the desired change in motion.

b. What is the shortest time in which the astronaut can complete the change in motion? How must the rocket be pointed?

c. Suppose the astronaut makes the change by decelerating to rest, turning the rocket exhaust by 90°, and then accelerating up to the desired final speed. How long would this take? How much rocket fuel is used, compared to the minimum?

Group C

8-33. *Energy of a harmonic oscillator.*

a. Find an expression for the average kinetic energy $\langle K \rangle$, over one period T, of the total mechanical energy of a harmonic oscillator by evaluating the integral in the expression

$$\langle K \rangle = \frac{\int_0^T K(t)\, dt}{T}$$

where $K(t)$ is given by Eq. (8-25). The trigonometric identity $\sin^2 \theta = (1 - \cos 2\theta)/2$ will be useful in evaluating the integral. Compare your result with the expression for $\langle K \rangle$ given in Eq. (8-31).

b. Carry out a similar evaluation of $\langle U \rangle$ by using the expression for $U(t)$ given by Eq. (8-27). The identity $\cos^2 \theta = (1 + \cos 2\theta)/2$ will be useful. Compare this result with the expression for $\langle U \rangle$ given in Eq. (8-31).

8-34. *Energy loss in a lightly damped oscillator.* Apply the method of Example 8-12 to investigate the decrease in mechanical energy of a lightly damped oscillator whose coordinate x is given by the expression

$$x = Ae^{-(r/2m)t} \cos(\omega t + \delta)$$

That is, evaluate $E = K + U = m(dx/dt)^2/2 + kx^2/2$, and then compute $\langle E \rangle$, the average of E over one period of the oscillation. Compare your result with Eq. (8-40). What assumptions must you make in order to bring your result into complete agreement with Eq. (8-40)? How do these assumptions relate to those used in deriving Eq. (8-40)?

8-35. *Collisions, collisions.* A body of mass m_1 and velocity $\mathbf{v}_{1i}$ approaches a stationary body of mass $m_2 = Cm_1$. After the collision, the two bodies have a relative speed $|\mathbf{v}_{2f} - \mathbf{v}_{1f}| = \epsilon\, |\mathbf{v}_{1i}|$, where ϵ is some constant.

a. For perfectly elastic collisions, in which the total kinetic energy K does not change, what is the value of ϵ? What values of ϵ describe inelastic collisions, in which $K_f < K_i$? Justify your answers.

b. Show that

$$\mathbf{v}_{1f} \cdot \mathbf{v}_{2f} \equiv v_{1f} v_{2f} \cos \theta = \frac{(1 - \epsilon^2)v_{1i}^2 + (1 - C^2)v_{2f}^2}{(1 + C)}$$

Here θ is the angle between the final velocities, $\mathbf{v}_{1f}$ and $\mathbf{v}_{2f}$.

c. Use the result of part *b* to prove the following statements:

(1) When a body collides elastically with a more massive body that is initially at rest, the final velocities form an obtuse angle.

(2) When a body collides elastically with a less massive body that is initially at rest, the final velocities form an acute angle.

(3) When a body collides inelastically with an initially stationary body of equal or smaller mass, the final velocities form an acute angle.

8-36. *Collision with a "massive" body.* An alpha particle (helium nucleus) enters a chamber containing mercury vapor. The alpha particle has an initial kinetic energy of 3.0×10^6 electron volts (eV). (The electron volt is a unit of energy commonly used in describing motion in the microscopic world: 1 eV $\equiv 1.602 \times 10^{-19}$ J.) The alpha particle is scattered through 90° by an elastic collision with the nucleus of a mercury atom. The mercury atom was essentially at rest before being struck. Its mass is 50.0 times that of the alpha particle.

a. In what direction does the mercury atom recoil after the collision?

b. Find the final kinetic energy of the mercury atom (1) as a fraction of the initial kinetic energy of the alpha particle and (2) in electron volts.

8-37. *Collision with a "light" body.* An alpha particle enters a chamber containing atomic hydrogen. The alpha particle has an initial kinetic energy of 3.0×10^6 eV ($= 4.8 \times 10^{-13}$ J). It collides elastically with the nucleus of a hydrogen atom (a proton). The hydrogen atom is essentially at rest prior to the collision. The mass of a proton is 0.25 times the mass of the alpha particle.

a. What is the maximum possible angle through which the alpha particle can be scattered? If this maximum deflection occurs, in what direction does the proton recoil?

b. For maximum deflection of the alpha particle, find the recoil energy of the proton (1) as a fraction of the initial kinetic energy of the alpha particle and (2) in electron volts.

8-38. *Hockey puck.* A hockey puck strikes the base of a vertical wall. It impacts at speed v_i and angle θ_i. In addition to the normal force $N(t)$ that the wall exerts on the puck, there is a kinetic frictional force $\mu_k N(t)$ acting parallel to the wall surface. Neglect any effects related to the spinning of the puck, and assume that the final magnitude of the normal velocity component $v_f \sin \theta_f$ is equal to $\gamma\, v_i \sin \theta_i$, where γ is a constant whose value is less than 1.

a. Show that if $\theta_i > \tan^{-1}\left[\dfrac{1}{\mu_k(1 + \gamma)}\right]$, then $\theta_f = 90°$.

b. Suppose that $\theta_i < \tan^{-1}\left[\dfrac{1}{\mu_k(1 + \gamma)}\right]$. Show that θ_f satisfies the equation

$$\gamma \cot \theta_f = \cot \theta_i - (1 + \gamma)\mu_k$$

c. The largest reasonable value of γ is $\gamma = 1$. Assuming that $\gamma = 1$, independent of v_i and θ_i, find the ratio of the final kinetic energy to the initial kinetic energy as a function of θ_i and μ_k.

8-39. *Pushed apart.* A spring of negligible mass having spring constant k is held by a latch between two bodies, 1 and 2, having masses m_1 and m_2, as shown in Fig. 8E-39*a*. The spring is compressed through a distance d. Initially the entire assembly is moving with velocity $\mathbf{v}_i = v_i\hat{\mathbf{x}}$, and the line joining body 1 to body 2 makes an angle α with $\mathbf{v}_i$. At $t = 0$ the latch breaks, allowing the spring to push apart the bodies.

Fig. 8E-39

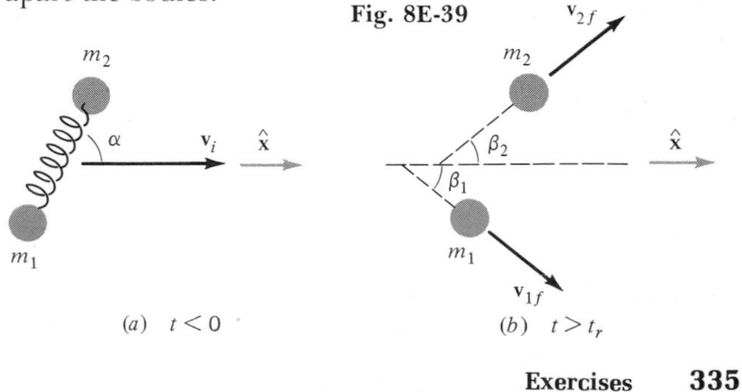

a. Show that the spring reaches its relaxed length at $t_r = \frac{1}{4}(2\pi/\omega)$, where $\omega^2 = k(m_1 + m_2)/m_1 m_2$.

b. The spring is not bonded to the two bodies, so they move freely after $t = t_r$. Find the total kinetic energy K_f of the bodies for $t \geq t_r$.

c. Find the magnitude and direction of the final relative velocity $\mathbf{u}_f = \mathbf{v}_{2f} - \mathbf{v}_{1f}$ in terms of ω, d, and α. Express $\mathbf{u}_f$ in magnitude-and-direction form.

d. The final motion of the two bodies is illustrated in Fig. 8E-39*b*. Solve for the speeds v_{1f} and v_{2f} and the angles β_1 and β_2 in terms of m_1, m_2, ω, d, α, and v_i.

e. Carefully evaluate t_r, K_f, $\mathbf{u}_f$, $\mathbf{v}_{1f}$, $\mathbf{v}_{2f}$, β_1, and β_2 for the following case: $k = 480$ N/m, $m_1 = 3.00$ kg, $m_2 = 2.00$ kg, $d = 0.200$ m, $\alpha = 110°$, and $v_i = 2.00$ m/s. When you have obtained the numerical values, check these by confirming that the final kinetic energy is numerically equal to the initial total energy.

8-40. *Oscillating system with constant frictional force.* Consider an oscillating system which is subject to a frictional retarding force $\mathbf{F}_r$ whose magnitude is independent of the speed, that is $\mathbf{F}_r = -f\,\hat{\mathbf{v}}$, with f a constant. Assume that the damping is small, so that the motion is oscillatory with gradually decreasing amplitude $A(t)$. That is the position of the body is given by $x(t) = A(t)\cos(\omega t + \delta)$.

a. Obtain an expression for $dE(t)/dt$ for the oscillator. In obtaining an expression for $v(t)$, neglect the term containing the small factor $dA(t)/dt$.

b. Integrate your expression over one cycle to show that $\langle dE/dt\rangle = -\gamma_0 \langle E\rangle^{1/2}$ where $\gamma_0 = (2\sqrt{2}/\pi m^{1/2})f$.

c. Show that the equation given in part *b* is satisfied by $\langle E\rangle = \langle E\rangle_0\,(1 - \gamma_0\, t/2\langle E\rangle_0^{1/2})^2$.

d. Using the definition

$$Q \equiv \frac{\omega\langle E\rangle}{|\langle dE/dt\rangle|}$$

show that the Q factor of this oscillator depends on the amplitude of the oscillation.

e. Does the Q factor of this oscillator increase or decrease as the oscillation is gradually damped down?

f. The results obtained above are dependent on the assumption of light damping. This assumption is correct only if $Q \gg 1$. What restriction does this inequality place on the amplitude of oscillation? Interpret this restriction in terms of a comparison between the damping force and the oscillator's restoring force.

9

Rotational Motion, I

9-1 ROTATIONAL KINEMATICS FOR A FIXED AXIS

As was discussed briefly in Sec. 2-1, there are two basically different ways an object can move: one is to change its location, the other is to change its orientation. In some cases that we have considered, both the location and the orientation of an object change. A puck tied to a string and orbiting on an air table is an example. The orientation of the puck changes, as well as its location. This is because as it moves around the air table, the string to which one side of the puck is attached is continually changing orientation. But we concentrated on the changes in location of the puck, paying no attention to its changes in orientation. Many of the objects that we have studied move by changing *only* their location. We now give the name **translation** (from the Latin word meaning "carried across") to motion involving change in location with no change in orientation. When an object experiences translational motion, *every point in the object has the same displacement in any small time interval as every other point.*

A brick flying through the air without changing its orientation provides an example of translational motion. See Fig. 9-1. Since every point in the brick moves in the same way as every other point, at each instant the location of a single point in the brick (such as its center) can be used to give the location of the brick. Up to now we have always described the motion of a body in terms of the motion of a single point in the body. This means that we have assumed either that the body moves in translation, or that if it also changes its orientation these changes can be ignored (as for the puck orbiting on the air table). Thus we have been using **translational kinematics** to describe motion and **translational mechanics** to explain it.

In this chapter and the next we will study the kind of motion in which the orientation of an object changes. An example of such motion is depicted in Fig. 9-2. The child's top shown there moves by changing its orien-

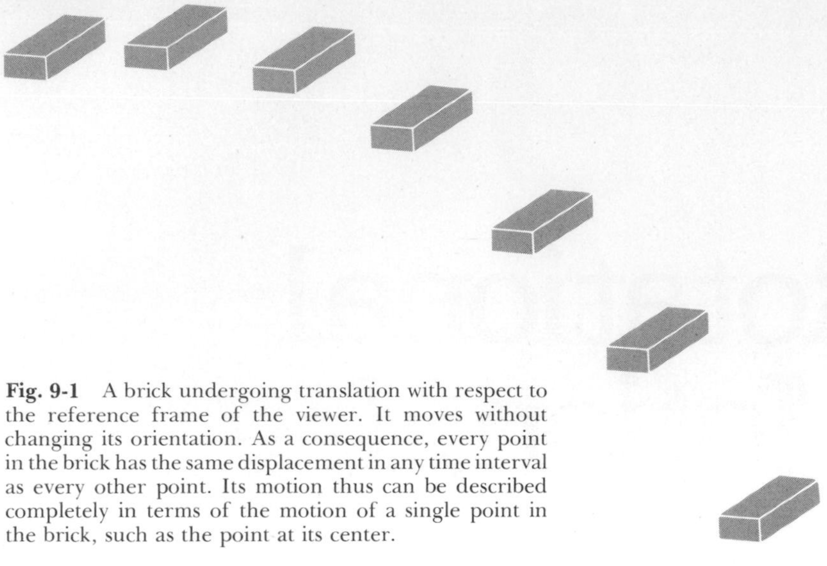

Fig. 9-1 A brick undergoing translation with respect to the reference frame of the viewer. It moves without changing its orientation. As a consequence, every point in the brick has the same displacement in any time interval as every other point. Its motion thus can be described completely in terms of the motion of a single point in the brick, such as the point at its center.

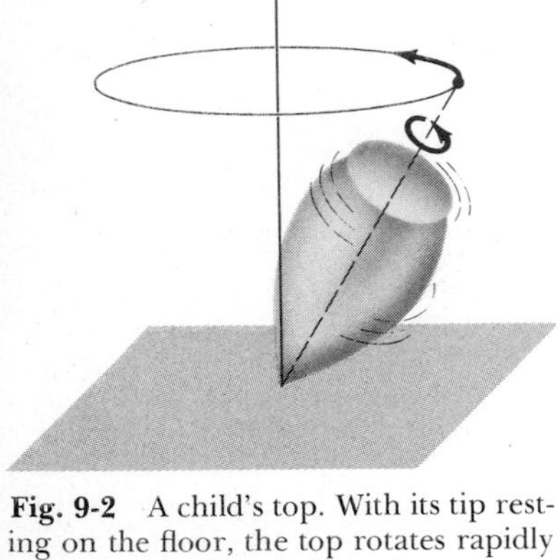

Fig. 9-2 A child's top. With its tip resting on the floor, the top rotates rapidly about an axis along its line of symmetry. The arrow circling the axis indicates the direction of rotation about the axis. A point on the rotation axis at the tip remains at a fixed location in the reference frame of the viewer. But the alignment of the axis in the reference frame changes slowly in such a way that the axis traces out a cone about the vertical, as indicated by the other arrow.

tation, while the location of one point—the tip that is supported by the floor—remains fixed. The object is said to be experiencing **rotation.** In rotational motion *every point in an object moves through an arc of a circle in any small time interval, and the centers of all these circles lie along a straight line.* The straight line is called the **rotation axis.** For a case like that in Fig. 9-2, one point on the rotation axis remains at a fixed location because the rotation axis must pass through the one fixed point in the object. But the rotation axis is free to change its alignment in space because no other point in the object is fixed. It is also possible for an object to move by rotating about an axis which changes its alignment in space, with no point on the rotation axis remaining at a fixed location. This is what the earth does. The earth rotates daily about its polar axis. The polar axis changes alignment in space cyclically, in much the same way as a child's top, taking 25,920 yr to complete each cycle. No point on this rotation axis remains fixed because the earth also moves in its annual orbit about the sun.

In Chap. 10 we consider the motion of the earth and of other objects rotating about an axis that passes through no fixed points. We will see that such motion can be treated by considering it to be a combination of rotation about an axis going through one fixed point and translation of that point. In Sec. 9-2 we take up the simpler case of rotation of an object about an axis that actually does pass through one fixed point. In this section we start with the simplest case, in which an object rotates about an axis going through two fixed points.

The flywheel in Fig. 9-3 exemplifies a body rotating about an axis that is completely fixed in the reference frame from which the body is observed. The axis is fixed because the bearings at the two ends of the flywheel axle are attached to a supporting structure. We consider the motion of the flywheel for the purpose of developing **rotational kinematics** for a fixed axis. That is, we develop the ideas, language, and relations needed to *describe* the motion, leaving until later the task of developing the **rotational mechanics** needed to explain it.

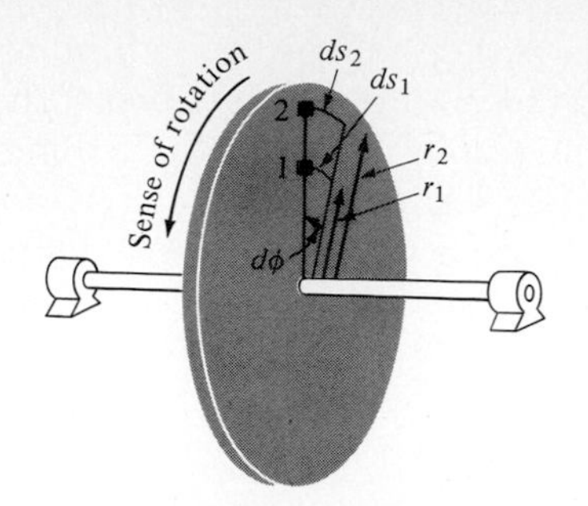

Fig. 9-3 A flywheel rotating about an axis which has a fixed alignment and is in a fixed location, in the reference frame of the viewer. The axis is fixed because the axle of the flywheel passes through rigidly supported bearings at its two ends. The figure shows the motion of two small pieces of the flywheel while it rotates through the angle $d\phi$ in a time dt. The direction of rotation, or, as it usually is said, the **sense** of rotation, is counterclockwise from the viewpoint illustrated. According to the standard sign convention for angles, the angle $d\phi$ specifying the counterclockwise rotation is counted as positive. (But note that as seen from the *other* side of the wheel it appears to be rotating *clockwise,* and the angle $d\phi$ would be counted as *negative.* Hence, if you are describing the sense of a rotation to someone else by giving the sign of the rotation angle, you must be sure that both of you agree on the side from which the rotating body is to be observed.)

A flywheel is also a fine example of a **rigid body**—a body whose constituent parts do not change their locations relative to one another while the body moves as a whole. Most of the objects we will be concerned with in studying rotational motion are rigid bodies.

When the flywheel rotates through a small angle $d\phi$ about its axis, the small piece of the flywheel labeled 1, located a distance r_1 from the axis, is displaced an amount ds_1 along an arc lying in the plane of the wheel and centered on its axis. Measuring the rotation angle in radians makes it possible to write

$$ds_1 = r_1\, d\phi$$

By continuing to follow the motion of piece 1, we could determine its velocity and acceleration. Then we could multiply the acceleration by the mass of piece 1 to determine the force which must be exerted on the piece. In principle, the same thing could be done for all the other small pieces of the body.

The difficulty with this approach is made clear if you consider the simultaneous motion of piece 2, which is located at a different distance r_2 from the axis of the flywheel. Although the rotation angle $d\phi$ is the same for all parts of the wheel, the displacement ds_2 is different from the displacement ds_1. So, too, are all the related kinematical quantities, such as the acceleration. As a consequence, if piece 1 and piece 2 have the same mass, they do not experience the same force. This means that it would take an unreasonable amount of labor to make direct application of the rules of translational mechanics to treat the motion of the many small pieces of the body, even in this simple case where the body is a highly symmetrical disk.

But while the displacement ds is different for parts of the wheel at different distances r from its axis, the **rotation angle** $d\phi$, given by

$$d\phi \equiv \frac{ds}{r} \tag{9-1}$$

is the same for every part of the wheel. Indeed, rotation is perceived intuitively in these very terms. It is the fact that all parts of a *rigid body* rotate simultaneously through the *same angle* that leads us to say that the flywheel is rotating as a whole. So what we will try to do is to develop a rotational kinematics—and then later a rotational mechanics—in which rotation angle, rather than displacement, is the basic quantity measured. In treating rotation about a fixed axis, we will represent rotation angles, and related quantities, as signed scalars.

As the flywheel rotates, it takes a certain time dt for any point on the wheel to pass through the angle $d\phi$. We define the **angular velocity** ω of the wheel to be

$$\omega \equiv \frac{d\phi}{dt} \tag{9-2}$$

in direct analogy to the definition of the familiar translational velocity $v = dx/dt$.

By convention, the rotation angle $d\phi$ has a positive value if the rotation described appears to be counterclockwise from an agreed-upon point of view—that is, as seen from a particular side of the wheel—and $d\phi$ is negative if the rotation appears to be clockwise from this point of view. Since the quantity dt has a positive value, the sign of ω is the same as the sign of $d\phi$. And since the angle $d\phi$ is measured in the dimensionless units radians, the units for ω must be reciprocal seconds (s^{-1}). What amounts to the same thing, the units for ω are often quoted as radians per second (rad/s).

If the wheel is speeding up or slowing down, ω will change. We define the wheel's **angular acceleration** α to be

$$\alpha \equiv \frac{d\omega}{dt} = \frac{d^2\phi}{dt^2} \tag{9-3}$$

The units for α are reciprocal seconds squared (s^{-2}), or radians per second squared (rad/s^2). Here again there is a complete analogy with the translational quantity $a = dv/dt = d^2x/dt^2$. Just as is true of the relation between v and a, when ω is positive and its magnitude is increasing, then α is positive. When ω is positive and its magnitude is decreasing, then α is negative. What is the sign of α when ω is negative and its magnitude is increasing? What is it if ω is negative and its magnitude is decreasing?

In Example 9-1 you will be led through a derivation of all the kinematical equations relating the angular coordinate ϕ of a body rotating about a fixed axis, its angular velocity ω, its angular acceleration α, and the time t for the cases of constant ω and of constant α. The process will be much easier and quicker than the derivation of the analogous translational equations in Chap. 2. First, you will be closely guided by analogy. Second, you will carry out the process by integration, in the order $\alpha \rightarrow \omega \rightarrow \phi$, rather than in the reverse order of Chap. 2.

EXAMPLE 9-1

a. Derive the relations among ϕ, ω, and t for *constant angular velocity*.

b. Derive the relations among ϕ, ω, α, and t for *constant angular acceleration*.

c. A wheel accelerates from rest with $\alpha = 1.53\ rad/s^2$. Use the relations developed in part b to find ω and ϕ after 3.00 s has elapsed.

■ **a.** From Eq. (9-2) you have

$$d\phi = \omega\, dt \tag{9-4}$$

Integrating this expression, you obtain

$$\int_{\phi_i}^{\phi_f} d\phi = \int_{t_i}^{t_f} \omega\, dt \tag{9-5}$$

where the subscripts i and f denote the initial and final values of the quantities. In general, ω will be a function of t which must be determined in order to evaluate the integral. Here, however, you have ω constant, so that

$$\int_{\phi_i}^{\phi_f} d\phi = \omega \int_{t_i}^{t_f} dt$$

The fundamental theorem of calculus then can be applied to produce the result

$$\phi_f - \phi_i = \omega\,(t_f - t_i) \tag{9-6}$$

If you set $t_i = 0$ and drop the subscript f since the relation holds for any later time t_f, the value of ϕ_f at that time is given by

$$\phi = \phi_i + \omega t \qquad \text{for constant } \omega \tag{9-7}$$

The angular coordinate ϕ is positive if the wheel appears in the agreed-upon view to have rotated counterclockwise from the orientation used to define $\phi = 0$. Equation (9-7) is analogous to the translational equation $x = x_i + vt$, which is valid for constant v.

b. From the first part of Eq. (9-3), you can write

$$d\omega = \alpha\, dt$$

Thus you have

$$\int_{\omega_i}^{\omega_f} d\omega = \int_{t_i}^{t_f} \alpha\, dt$$

Taking α constant and applying the fundamental theorem, you obtain

$$\omega_f - \omega_i = \alpha \int_{t_i}^{t_f} dt = \alpha(t_f - t_i)$$

By setting $t_i = 0$ and dropping the unneeded subscript f, this simplifies to

$$\omega = \omega_i + \alpha t \qquad \text{for constant } \alpha \tag{9-8}$$

Equation (9-8) is analogous to the equation of translational mechanics: $v = v_i + at$ for constant a.

Using Eq. (9-5) gives you

$$\int_{\phi_i}^{\phi_f} d\phi = \int_0^{t_f} \omega\, dt = \int_0^{t_f} (\omega_i + \alpha t)\, dt = \omega_i \int_0^{t_f} dt + \alpha \int_0^{t_f} t\, dt$$

Applying the fundamental theorem and Eq. (7-21), you get

$$\phi_f - \phi_i = \omega_i t_f + \frac{\alpha t_f^2}{2}$$

This is written as

$$\phi = \phi_i + \omega_i t + \frac{\alpha t^2}{2} \qquad \text{for constant } \alpha \tag{9-9}$$

where again you drop the subscript f. Equation (9-9) is analogous to the equation of translational kinematics: $x = x_i + v_i t + at^2/2$ for constant a.

The analogy that you have found between Eqs. (9-8) and (9-9), on the one hand, and the corresponding translational equations, on the other hand, allows you to write immediately the relations

$$\phi = \phi_i + \frac{\omega^2 - \omega_i^2}{2\alpha} \qquad \text{for constant } \alpha \tag{9-10}$$

and

$$\phi = \phi_i + \frac{(\omega + \omega_i)t}{2} \qquad \text{for constant } \alpha \tag{9-11}$$

The first of these is the rotational analogue to Eq. (2-32), and the second is the rotational analogue to Eq. (2-33).

c. Since the wheel starts from rest, you have $\omega_i = 0$. Equation (9-8) thus gives you

$$\omega = 0 + \alpha t = 1.53 \text{ rad/s}^2 \times 3.00 \text{ s} = 4.59 \text{ rad/s}$$

for the angular velocity after 3.00 s. The positive value means the wheel is rotating counterclockwise, as viewed from the agreed-upon side of the wheel.

To find the angle through which the wheel has turned, you can apply Eq. (9-9), setting $\phi_i = 0$ since no particular initial angular orientation is specified. You have

$$\phi = 0 + 0 + \frac{\alpha t^2}{2} = \frac{1.53 \text{ rad/s}^2 \times (3.00 \text{ s})^2}{2} = 6.89 \text{ rad}$$

You can obtain the same result by applying Eq. (9-10), using the final value of ω just obtained. What does the positive value of ϕ mean? The magnitude of ϕ is greater than 2π. What does this mean?

The relation between the velocity of a point in the flywheel and the angular velocity of the flywheel can be established by writing Eq. (9-1) in the form

$$ds = r\, d\phi \tag{9-12}$$

Here r is the distance of the point from the rotation axis, and ds is its infinitesimal displacement along its circular path when the wheel rotates through an infinitesimal angle $d\phi$. Dividing through by the time dt it takes for this to happen, we have

$$\frac{ds}{dt} = r\frac{d\phi}{dt}$$

or

$$\frac{ds}{dt} = r\omega$$

The magnitude of the quantity ds/dt gives the magnitude of the point's velocity. The direction of the velocity is always tangent to its path and is counterclockwise if the sign of ds/dt is positive. Thus we can specify the velocity of the point by saying that it has only a tangential component and that the **tangential velocity component** v_t has the value

$$v_t = r\omega \tag{9-13}$$

Since r is intrinsically positive, this means that v_t has the same sign as the angular velocity ω. That is, v_t is positive if the wheel is rotating in the counterclockwise sense.

If the value of v_t is changing, then the speed of the point along its circular path is not constant. It was shown in Sec. 3-3 that in these circumstances the point will have an acceleration along its path. The value of this tangential acceleration is given by the rate of change of the speed. Thus we can find an expression for the point's **tangential acceleration component** a_t by evaluating

$$a_t = \frac{dv_t}{dt}$$

Using Eq. (9-13), we have, since r is constant,

$$a_t = \frac{d(r\omega)}{dt} = r\frac{d\omega}{dt}$$

Then writing $d\omega/dt$ as α, we obtain the expression

$$a_t = r\alpha \tag{9-14}$$

Again, the sign of a_t is the same as the sign of the angular acceleration α. You should go through an analysis like the one immediately below Eq. (9-3) and relate the sign of a_t to the four cases enumerated there.

The calculation leading to Eq. (9-14) is very similar to the one leading to Eq. (6-7) for the tangential component of the acceleration of a bob at the end of a pendulum. But the argument and the symbolism used in the earlier calculation are somewhat different. It would be worthwhile for you to review the development of Eq. (6-7). If you do so, you will be reminded that whereas the v_t in Eq. (9-13) is the only component of the rotating point's velocity, the a_t in Eq. (9-14) is *not* the only component of its acceleration. The point also has an acceleration component in the centripetal direction, that is, in the direction toward the center of its circular path. We can evaluate the **centripetal acceleration component** a_c by combining Eq. (3-41), written as $a_c = v_t^2/r$, with Eq. (9-13), to obtain

$$a_c = \frac{(r\omega)^2}{r}$$

or

$$a_c = r\omega^2 \tag{9-15}$$

In Sec. 9-2 we rederive Eqs. (9-13), (9-14), and (9-15) from more general and more rigorous arguments. Example 9-2 gives an application of these equations.

EXAMPLE 9-2

The wheel of Example 9-1 accelerates from rest with a constant angular acceleration $\alpha = 1.53 \text{ rad/s}^2$. The distance from the axis of rotation to a certain point in the wheel is $r = 0.250$ m.

a. Find the velocity components of the point 3.00 s after the wheel begins to move.

b. Find the acceleration components of the point at that time.

■ **a.** The velocity of the point has only a tangential component. That is, the velocity vector is perpendicular to the direction from the axis to the point. According to Eq. (9-13), the value of the tangential velocity component is

$$v_t = r\omega$$

Setting $r = 0.250$ m and using the value of the angular velocity calculated in Example 9-1, $\omega = 4.59$ rad/s, you have

$$v_t = 0.250 \text{ m} \times 4.59 \text{ rad/s} = 1.15 \text{ m/s}$$

The positive sign tells you that the point appears to be moving counterclockwise, as seen from the agreed-upon side of the wheel.

b. The acceleration of the point has a tangential component given by Eq. (9-14),

$$a_t = r\alpha$$

Using the values quoted for r and α, you get

$$a_t = 0.250 \text{ m} \times 1.53 \text{ rad/s}^2 = 0.383 \text{ m/s}^2$$

The direction of the tangential acceleration is also counterclockwise.

From Eq. (9-15), you know that the centripetal component of the point's acceleration is

$$a_c = r\omega^2$$

With the given value of r and the previously calculated value of ω, you have

$$a_c = 0.250 \text{ m} \times (4.59 \text{ rad/s})^2 = 5.27 \text{ m/s}^2$$

The direction of the centripetal acceleration is that from the location of the point directly into the axis of rotation.

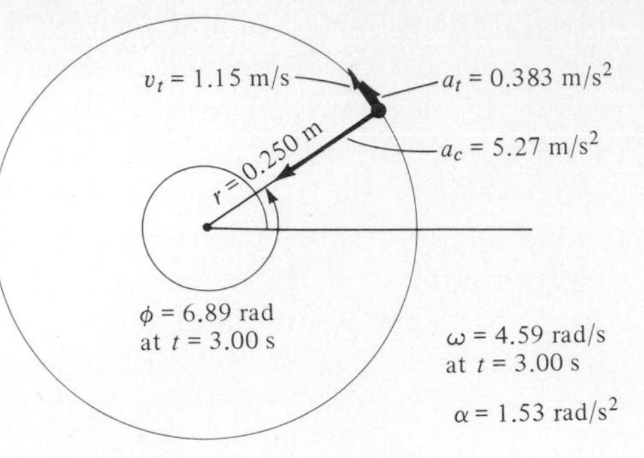

Fig. 9-4 The velocity v_t, tangential acceleration a_t, and centripetal acceleration a_c of a point on a wheel whose distance from the axis is r. The angular coordinate of the point is ϕ, and ω is the angular velocity of the wheel. The wheel started from rest at $t = 0$, rotated with constant angular acceleration α, and is depicted at a later time t.

The velocity of the point and its tangential and centripetal acceleration are indicated in Fig. 9-4. Also indicated are the angular coordinate of the point and the angular velocity and angular acceleration of the wheel. What is the total acceleration of the point? What would it be if the angular velocity were the same but the angular acceleration were -1.53 rad/s^2?

9-2 ROTATIONAL KINEMATICS FOR A FREE AXIS

In some of the most interesting cases of rotational motion, the axis about which a body rotates is free to change its alignment in the reference frame used to observe the body. An example is the child's top that was illustrated in Fig. 9-2. The top rotates rapidly about an axis that lies along the top's line of symmetry while that axis slowly changes its alignment, tracing out a cone whose apex is where the pointed end of the top rests on the floor. The top performs this motion in apparent defiance of the tendency for gravity to make it fall over. If we want to understand the mechanics that explains this motion, first we must extend our treatment of rotational kinematics to situations more general than that of rotation about a fixed axis.

In this section you will see that rotational kinematics for a free axis requires introducing *vector* quantities to describe rotations. You will also see that these vector quantities are so convenient that they are used not only in the kinematics of systems where the rotation axis is free to change its alignment, but also in the kinematics of systems, like flywheels, where the alignment of the rotation axis is fixed. In fact, the vector description of rotation is used mostly in this chapter for simple systems that have fixed rotation axes. But in Chap. 10 we treat more complicated systems that have free rotation axes, like the spinning top and the spinning earth. There the great power of the vector description of rotation becomes completely apparent.

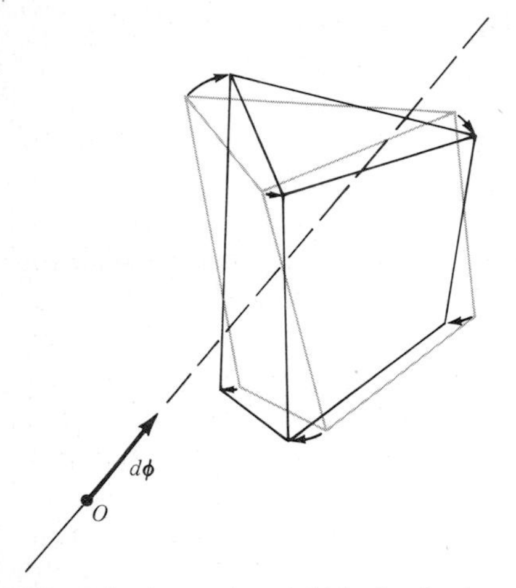

Fig. 9-5 A rotating rigid body. In the short time during which its orientation changes from the one given by the gray lines to the one given by the black lines, the body is rotating about the instantaneous axis shown as a dashed line. The instantaneous axis is defined by the property that all moving points in the body travel on short arcs centered on the axis. If there are points in the body which do not move, they lie on the instantaneous axis. The vector $d\boldsymbol{\phi}$ describes the rotation in a way explained in the text.

Figure 9-5 shows a rotating *rigid* body of arbitrary shape. It is depicted at some initial time and at a slightly later time. The dashed line is the *instantaneous* axis of rotation—the axis about which the body is rotating during this small time interval. All points in the body move in short arcs centered on that axis, except for those points in the body (if there are any) which lie on the axis and so do not move. The rotation axis may change alignment in space in subsequent time intervals. But at present we restrict our attention to situations in which at all times the instantaneous axis passes through a point O which is *fixed in the reference frame* used to observe the motion. In Sec. 10-4 we relax this restriction.

A very useful description of the *small* rotation of the body can be given in terms of the **rotation vector $d\boldsymbol{\phi}$.** By definition, this vector is aligned parallel to the instantaneous rotation axis, in the *direction* given by the **right-hand rule for rotation vectors:** *Place your right hand so that the fingers curl* *in the sense of rotation. The thumb then points in the direction of the rotation vector* $d\boldsymbol{\phi}$.

The choice of a right-hand rule, instead of a left-hand rule, is a matter of convention. It is important to remember the rule and to use it consistently. The rule is presented pictorially in Fig. 9-6. Note that the relation between the sense of rotation and the direction of the rotation vector, given by the right-hand rule, is the same as the relation between the sense of rotation of a wood screw and its direction of motion as it is screwed into a block of wood.

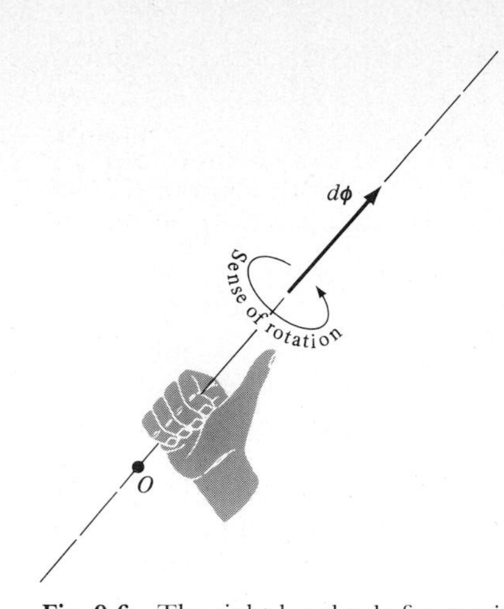

Fig. 9-6 The right-hand rule for specifying the direction of a rotation vector. The alignment of a rotation axis does not describe uniquely the rotation of a body because the body can be rotating about an axis either one way or the other. To specify the particular sense of rotation, the thumb of the right hand is pointed along the rotation axis with the fingers of the hand curling in the sense of rotation. Then, by definition, the thumb points along the rotation vector $d\boldsymbol{\phi}$.

The *magnitude* of the rotation vector is equal to the angle $d\phi$ through which the body has rotated about the axis, measured in radians. The vector describing the rotation illustrated in Fig. 9-5 is shown on the axis of rotation. Rotation vectors are sometimes called **axial vectors.** The many vectors that we will encounter, which have mathematical properties like those of rotation vectors, are also given this generic name.

It is only for a *rigid* body that there is a single rotation vector $d\boldsymbol{\phi}$ which completely describes a small rotation. If a body is not rigid, the rotation in one part may be different from the rotation in another part. We will consider only rigid bodies.

When a rigid body undergoes a small rotation, each point in the body experiences a small displacement (except for points on the rotation axis). The direction and magnitude of the displacement depend on the location of the point relative to the rotation axis, as well as on the magnitude of the rotation angle. We need to know the relation between these quantities in order to calculate the velocity and acceleration of each small piece of the rotating body. The relation can be obtained by considering Fig. 9-7.

In this figure the vector $d\boldsymbol{\phi}$ characterizing the rotation is drawn along the axis of rotation. The *fixed* point O on that axis is chosen as the origin of coordinates. Thus the position of a point P in the body is specified by a vector $\mathbf{r}$ extending from O to P. You can see from the figure that the shortest distance from P to the axis is the length of the perpendicular from that point to the axis. This is $r \sin \theta$, where θ is the angle between the vectors $\mathbf{r}$ and $d\boldsymbol{\phi}$. When the body rotates about the axis through an angle $d\phi$, expressed in radians, the distance moved by the tip of the line of length r is $d\phi$

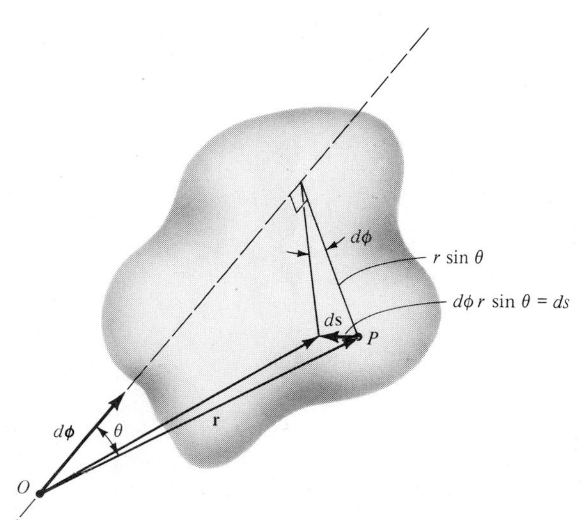

Fig. 9-7 A rigid body undergoing a small rotation about a dashed rotation axis. The L-shaped symbol at the intersection of the line from a typical point P in the body to the axis means that the angle formed at the intersection is a right angle. The displacement $d\mathbf{s}$ of the point is in a direction normal to the plane containing $\mathbf{r}$ and $d\boldsymbol{\phi}$. Thus $d\mathbf{s}$ is perpendicular to both $\mathbf{r}$ and $d\boldsymbol{\phi}$.

times $r \sin \theta$. This distance is ds, the magnitude of the displacement $d\mathbf{s}$ of the point P. That is,

$$ds = d\phi \, r \sin \theta \tag{9-16}$$

The direction of the vector $d\mathbf{s}$ is perpendicular to both $\mathbf{r}$ and $d\boldsymbol{\phi}$. This direction can be described trigonometrically, but the process and the result are cumbersome. However, a vector operation, called taking the vector product, directly specifies both the magnitude and the direction of $d\mathbf{s}$. The vector product will also be of great use in connection with a number of other vectors that also arise in the treatment of rotations. So we now digress for the purpose of introducing it.

The **vector product** of two vector quantities $\mathbf{A}$ and $\mathbf{B}$ is itself a vector quantity. In symbols, the vector product is written as $\mathbf{A} \times \mathbf{B}$. The bold cross is used both to indicate the specific mathematical operation of taking the vector product and to emphasize that this operation is completely different from the operation of taking the scalar, or dot, product of two vectors, $\mathbf{A} \cdot \mathbf{B}$. The expression $\mathbf{A} \times \mathbf{B}$ is read "**A** cross **B**." Consequently, the vector product is frequently called the **cross product.** By definition, the magnitude of $\mathbf{A} \times \mathbf{B}$ is A ($\sin \theta$) B. We write this as

$$|\mathbf{A} \times \mathbf{B}| \equiv A \sin \theta \, B \tag{9-17}$$

Here A and B are the magnitudes of $\mathbf{A}$ and $\mathbf{B}$, and θ is the angle between their directions. The angle θ is always counted as positive and is always the smaller of the two angles formed by the directions of $\mathbf{A}$ and $\mathbf{B}$. Thus θ always lies in the range $0 \leq \theta \leq \pi$. Absolute value signs are used in Eq. (9-17) to indicate the magnitude of the vector $\mathbf{A} \times \mathbf{B}$. So the left side of the equation is necessarily a positive quantity. The right side is too, since A and B are positive and $\sin \theta$ is everywhere positive in the range $0 \leq \theta \leq \pi$.

The direction of $\mathbf{A} \times \mathbf{B}$ is, by definition, perpendicular to both $\mathbf{A}$ and $\mathbf{B}$ and in the direction given by the **right-hand rule for cross products:** *Place your right hand so that the fingers curl in the sense that would sweep* **A** *into alignment with* **B** *through the smaller angle between them. The thumb then points in the direction of* $\mathbf{A} \times \mathbf{B}$. You will find illustrations of the use of this rule to determine the direction of $\mathbf{A} \times \mathbf{B}$ in Fig. 9-8. Can you state a "wood screw rule" for cross products?

In Fig. 9-8 you also find examples of the use of Eq. (9-17) to determine the magnitude of $\mathbf{A} \times \mathbf{B}$. The figure demonstrates some important properties of the cross product. First, $\mathbf{A} \times \mathbf{B} = 0$ if $\mathbf{A}$ is parallel or antiparallel (oppositely directed) to $\mathbf{B}$. Second, $|\mathbf{A} \times \mathbf{B}| = AB$ if $\mathbf{A}$ is perpendicular to $\mathbf{B}$. The directions of $\mathbf{A}$, $\mathbf{B}$, and $\mathbf{A} \times \mathbf{B}$ are related in this case just like those of the x, y, and z axes of a "right-handed" coordinate system. Third, $\mathbf{A} \times \mathbf{B} = -\mathbf{B} \times \mathbf{A}$. This means that the cross product does *not* obey the commutativity rule obeyed by either the product of two scalars, that is, $AB = BA$, or the dot product of two vectors, that is, $\mathbf{A} \cdot \mathbf{B} = \mathbf{B} \cdot \mathbf{A}$. Instead, the cross product obeys what may be called an *anticommutativity* rule.

The cross, or vector, product is in almost constant use in any general treatment of rotations because it enters in almost all the kinematical and mechanical quantities pertaining to rotational motion. It is also very useful in other branches of physics, such as electromagnetism, where rotationlike quantities are encountered. In geometry and trigonometry it can be employed to generate the component of one vector along a direction perpendicular to another, for the purpose of determining the shortest distance to a line or a plane.

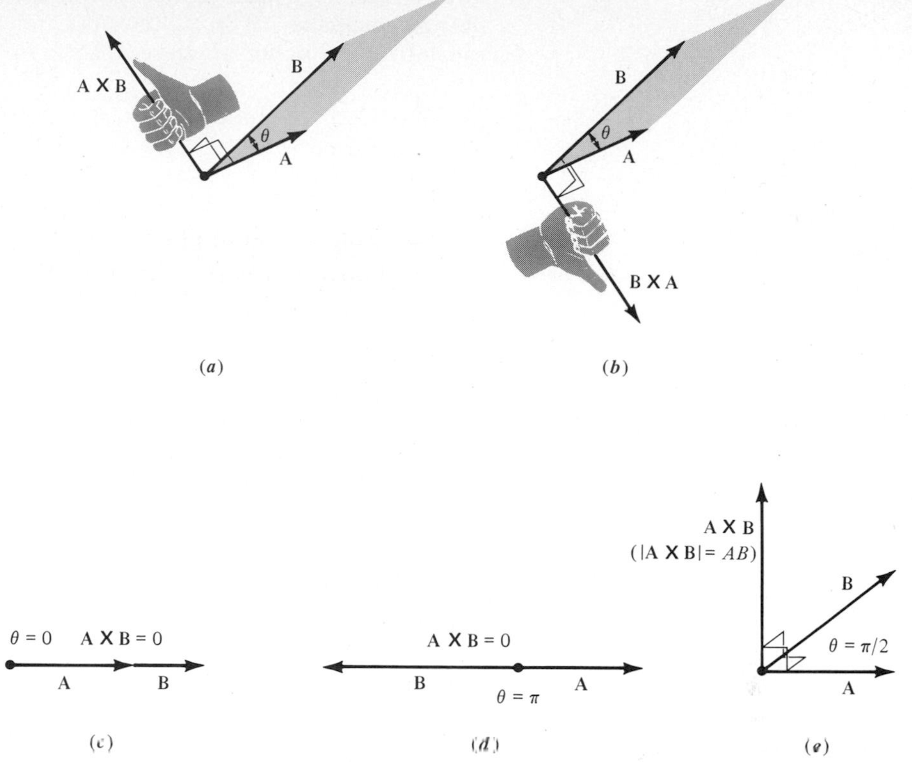

Fig. 9-8 (*a*) The cross product $\mathbf{A} \times \mathbf{B}$ of two vectors $\mathbf{A}$ and $\mathbf{B}$ is a third vector oriented normal to the plane containing $\mathbf{A}$ and $\mathbf{B}$, in the direction specified by the right-hand rule illustrated in the figure. In applying this rule, the fingers of the right hand must be curling in the sense that would cause the *first* vector in the cross product to sweep into alignment with the *second* vector, going through the *smaller* of the two angles between them. Here the curling fingers show the sense which would cause $\mathbf{A}$ to sweep into alignment with $\mathbf{B}$, going through an angle somewhat less than $\pi/2$. The magnitude of the cross product, $|\mathbf{A} \times \mathbf{B}|$, is $A \sin \theta \, B$. The angle θ is always positive and is the smaller of the two angles between $\mathbf{A}$ and $\mathbf{B}$. The quantity $A \sin \theta$ is the length of the perpendicular dropped from the tip of vector $\mathbf{A}$ to the line along vector $\mathbf{B}$. That is, $A \sin \theta$ is the "height" of the shaded parallelogram whose sides are defined by $\mathbf{A}$ and $\mathbf{B}$. The length B of the vector $\mathbf{B}$ is the "base" of this parallelogram. Since the area of the parallelogram is the product of its base and its height, the magnitude $|\mathbf{A} \times \mathbf{B}| = A \sin \theta \, B$ of the cross product is numerically equal to the area of the parallelogram. (*b*) The cross product $\mathbf{B} \times \mathbf{A}$ has the magnitude $|\mathbf{B} \times \mathbf{A}| = B \sin \theta \, A$. This is the same as $A \sin \theta \, B = |\mathbf{A} \times \mathbf{B}|$, the magnitude of $\mathbf{A} \times \mathbf{B}$. But the direction of $\mathbf{B} \times \mathbf{A}$ is opposite to the direction of $\mathbf{A} \times \mathbf{B}$. The direction of $\mathbf{B} \times \mathbf{A}$ is shown by the thumb of the right hand in the figure, the fingers of which curl in the sense that would cause $\mathbf{B}$ to sweep into alignment with $\mathbf{A}$, going through the smaller angle. The relation between $\mathbf{A} \times \mathbf{B}$ and $\mathbf{B} \times \mathbf{A}$ is specified completely by the equation $\mathbf{A} \times \mathbf{B} = -\mathbf{B} \times \mathbf{A}$. (*c*) If $\mathbf{A}$ is parallel to $\mathbf{B}$ then $\theta = 0$ and $\sin \theta = 0$, so $|\mathbf{A} \times \mathbf{B}| = A \sin \theta \, B = 0$. Since the magnitude of the vector $\mathbf{A} \times \mathbf{B}$ is zero in this case, we can write $\mathbf{A} \times \mathbf{B} = 0$. (*d*) If $\mathbf{A}$ is antiparallel to $\mathbf{B}$, then $\theta = \pi$ and $\sin \theta = 0$, so $|\mathbf{A} \times \mathbf{B}| = A \sin \theta \, B = 0$ and $\mathbf{A} \times \mathbf{B} = 0$. (*e*) If $\mathbf{A}$ is perpendicular to $\mathbf{B}$ then $\theta = \pi/2$ and $\sin \theta = 1$, so $|\mathbf{A} \times \mathbf{B}| = A \sin \theta \, B = AB$.

At first the cross product may strike you as artificial and unduly complicated. This feeling can arise out of the question: Why define an operation on two vectors in such a way that the product is a vector perpendicular to both? The answer is that the mathematical operation involved is *intrinsically three-dimensional*—in fact, the cross product cannot be defined in a space with any other number of dimensions. Now, the physical world is also intrinsically three-dimensional, and there are many physical phenomena which simply cannot exist in one or two dimensions. Rotation is the first of these phenomena which we will discuss thoroughly. The simplest case of rotation—a wheel spinning on a fixed axis—can be handled in two dimensions (just barely!). But even in this case, it is necessary to define an axis which lies in a third dimension. We need an intrinsically three-dimensional mathematical operation to describe three-dimensional physical phenomena. Believe it or not, the cross product is the simplest operation with this capability. (If you don't believe it, try to invent a simpler one.)

We can make immediate use of the cross product by applying it in Eq. (9-16) to write

$$d\mathbf{s} = d\boldsymbol{\phi} \times \mathbf{r} \tag{9-18}$$

for the displacement in Fig. 9-7. This equation describes at once the magnitude and direction of the vector $d\mathbf{s}$, which represents the displacement of a point at the position $\mathbf{r}$ when the body undergoes a rotation $d\boldsymbol{\phi}$. You can check this by inspecting Eq. (9-17) and Fig. 9-8.

If we divide Eq. (9-18) by the infinitesimal time increment dt, during which the infinitesimal rotation and displacement occur, we have

$$\frac{d\mathbf{s}}{dt} = \frac{d\boldsymbol{\phi}}{dt} \times \mathbf{r}$$

Because $d\mathbf{s}/dt$ gives the velocity $\mathbf{v}$ of the point, this can be written

$$\mathbf{v} = \boldsymbol{\omega} \times \mathbf{r} \tag{9-19}$$

where we define

$$\boldsymbol{\omega} \equiv \frac{d\boldsymbol{\phi}}{dt} \tag{9-20}$$

The quantity $\boldsymbol{\omega}$ is the **angular velocity** of the rotating body. Since $d\phi$ is an angle expressed in radians, the magnitude ω of the angular velocity, the **angular speed,** gives the rate of rotation in radians per second. The direction of $\boldsymbol{\omega}$ specifies the alignment of the rotation axis and the sense of rotation about that axis, just as $d\boldsymbol{\phi}$ does, because $\boldsymbol{\omega}$ has the same direction as $d\boldsymbol{\phi}$. Equation (9-19) allows you to calculate the vector describing the velocity $\mathbf{v}$ of the point P in the body at any instant, given its position vector $\mathbf{r}$ and the angular velocity vector $\boldsymbol{\omega}$ of the body at that instant.

Figure 9-9 illustrates the application of the general relation of Eq. (9-19) to calculate the velocity $\mathbf{v}$ of a point P on the rim of a flywheel of radius r and angular velocity $\boldsymbol{\omega}$, rotating in the sense shown in the figure. It seems natural to choose the origin O on the axis of the wheel at its intersection with the plane of the wheel. This choice makes the angular velocity $\boldsymbol{\omega}$ perpendicular to the position vector $\mathbf{r}$ that gives the location of the point. Thus, since the angle between $\boldsymbol{\omega}$ and $\mathbf{r}$ has the value $\theta = \pi/2$, we find for the magnitude of $\mathbf{v}$ the expression

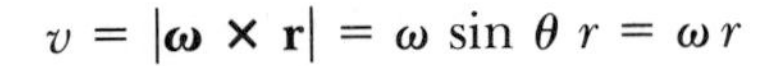

$$v = |\boldsymbol{\omega} \times \mathbf{r}| = \omega \sin\theta\, r = \omega r$$

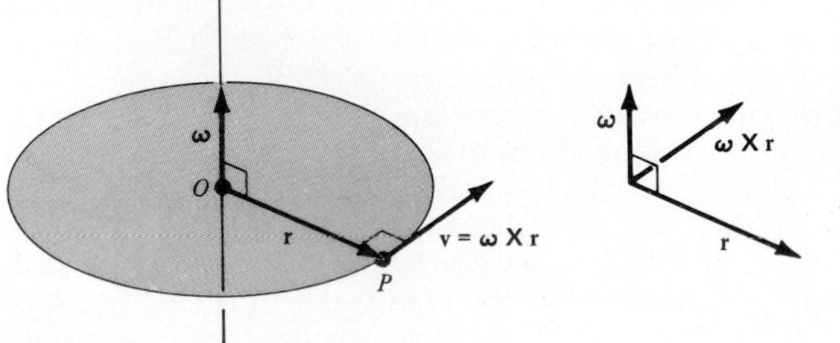

Fig. 9-9 Use of the equation $\mathbf{v} = \boldsymbol{\omega} \times \mathbf{r}$ to find the velocity $\mathbf{v}$ of a point P in a flywheel whose location relative to the coordinate origin O on the body's rotation axis is given by the vector $\mathbf{r}$. The angular velocity of rotation is $\boldsymbol{\omega}$. To the right is an auxiliary construction which determines the direction of $\boldsymbol{\omega} \times \mathbf{r}$.

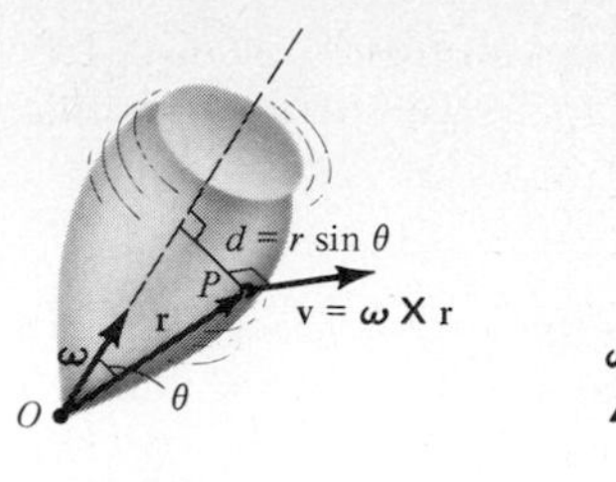

Fig. 9-10 The equation $\mathbf{v} = \boldsymbol{\omega} \times \mathbf{r}$ can be used to find the velocity $\mathbf{v}$ of a point P in a top, just as it was used in Fig. 9-9 for a point in a flywheel. The only difference is that in this figure the origin O is not in the plane of rotation of P. Consequently, the angular velocity $\boldsymbol{\omega}$ of the rotating body is not perpendicular to the position vector $\mathbf{r}$ of the point in the body.

This agrees with Eq. (9-13), the particular relation obtained earlier for such a case. The figure also shows that the direction of $\mathbf{v} = \boldsymbol{\omega} \times \mathbf{r}$ correctly describes the tangential motion of P at the instant depicted.

Let us use Eq. (9-19) to determine the velocity of a point in a spinning top, say a point on its surface at the widest part. See Fig. 9-10. We fix the coordinate origin O at the motionless tip of the top. The location of the point P at the instant shown is then given by the position vector $\mathbf{r}$ from O to P. The angular velocity vector $\boldsymbol{\omega}$ is parallel to the top's symmetry axis and is in the direction shown if the sense of rotation is as shown. Also, we choose to draw $\boldsymbol{\omega}$ along the symmetry axis of the top. The instantaneous direction of the velocity $\mathbf{v}$ of point P is tangential, and perpendicular to both $\boldsymbol{\omega}$ and $\mathbf{r}$, since this is the direction of $\boldsymbol{\omega} \times \mathbf{r}$. Its magnitude is

$$v = |\boldsymbol{\omega} \times \mathbf{r}| = \omega \sin \theta \, r = \omega d$$

Here $d = r \sin \theta$ is the perpendicular distance from P to the axis, as shown in the figure.

What is the rationale for our choice of the origin O of the vector $\mathbf{r}$ describing the position of a particular point P in the body? At any instant the angular velocity $\boldsymbol{\omega}$ has a certain direction, and therefore the axis of rotation has a certain alignment. At that instant any point on the rotation axis could be chosen as an origin of coordinates. Each different choice of O would lead to a different value of the vector $\mathbf{r}$ extending to the same point P. But the velocity $\mathbf{v}$ of that point, found from Eq. (9-19), is independent of the choice of O. All choices lead to the same value of $\boldsymbol{\omega} \times \mathbf{r}$. The reason is that the cross product, with its implicit term sin θ, produces the component of $\mathbf{r}$ perpendicular to the direction of the rotation axis (the quantity d in the equation immediately above), and this does not depend on where O lies on that axis. However, O should almost always be chosen at the single *fixed* point on the axis. This is the only choice of O which will *continue* to be on the rotation axis as the axis changes orientation.

Next we will develop an expression for the acceleration $\mathbf{a}$ of a point P in a rotating body. Taking the time derivative of Eq. (9-19), we have

$$\frac{d\mathbf{v}}{dt} = \frac{d}{dt}(\boldsymbol{\omega} \times \mathbf{r})$$

or

$$\mathbf{a} = \frac{d\boldsymbol{\omega}}{dt} \times \mathbf{r} + \boldsymbol{\omega} \times \frac{d\mathbf{r}}{dt}$$

The last step is an example of the rule for differentiating cross products. The rule combines the scalar calculus rule for differentiating a product of two quantities and the rule that the ordering of terms in a cross product *must not be changed,* because the cross product is not commutative.

Now $d\mathbf{r}/dt$ gives the velocity $\mathbf{v}$ of the point as well as $d\mathbf{s}/dt$, since $d\mathbf{s} = d\mathbf{r}$. (See Fig. 9-7.) So the point's acceleration $\mathbf{a}$ is

$$\mathbf{a} = \frac{d\boldsymbol{\omega}}{dt} \times \mathbf{r} + \boldsymbol{\omega} \times \mathbf{v}$$

This result can be written

$$\mathbf{a} = \boldsymbol{\alpha} \times \mathbf{r} + \boldsymbol{\omega} \times \mathbf{v} \tag{9-21}$$

where

$$\boldsymbol{\alpha} \equiv \frac{d\boldsymbol{\omega}}{dt} \tag{9-22}$$

We have introduced $\boldsymbol{\alpha}$ to represent the **angular acceleration** vector, defined as the rate of change of the body's angular velocity vector $\boldsymbol{\omega}$. Example 9-3 explores the physical meaning of each of the terms in Eq. (9-21).

EXAMPLE 9-3

a. Apply Eq. (9-21) to calculate the acceleration **a** of a point P on the rim of a flywheel rotating with constant angular speed ω.

■ The angular velocity $\boldsymbol{\omega}$ for a flywheel is in a constant direction pointing along the fixed rotation axis in the way that is related to the sense of rotation of the wheel by the right-hand rule for rotation vectors. If its magnitude ω is also constant, then the angular acceleration $\boldsymbol{\alpha} = d\boldsymbol{\omega}/dt$ is zero. Thus the only term contributing to **a** in Eq. (9-21) is the term $\boldsymbol{\omega} \times \mathbf{v}$. As you saw in Fig. 9-9, $\mathbf{v} = \boldsymbol{\omega} \times \mathbf{r}$ is in a tangential direction and has magnitude $v = \omega r$. You can see from Fig. 9-11*a* that since $\boldsymbol{\omega}$ is in an axial direction, its cross product with **v** must be in a radial direction; specifically, Fig. 9-11*a* shows you that $\boldsymbol{\omega} \times \mathbf{v}$ is in the inward radial direction—that is, in the centripetal direction. And since there is a right angle between $\boldsymbol{\omega}$ and **v**, the magnitude of their vector product is

$$a = |\boldsymbol{\omega} \times \mathbf{v}| = \omega v$$

Writing $v = \omega r$, and using the subscript c to indicate that the acceleration is centripetal, you have

$$a_c = \omega^2 r$$

This result is identical to Eq. (9-15) for the centripetal acceleration of a point moving with constant speed around a circle of radius r. It constitutes an independent derivation of that important equation. ■

b. Now use Eq. (9-21) to calculate the acceleration of a point on the rim when the angular speed ω of the flywheel is not constant, but is increasing at the rate α.

■ In this case there is an angular acceleration $\boldsymbol{\alpha} = d\boldsymbol{\omega}/dt$. Its magnitude is $\alpha = d\omega/dt$, and its direction is the same as the direction of $\boldsymbol{\omega}$. So, in addition to the centripetal acceleration found in part *a*, there is also a contribution to the acceleration of the point on the rim of the flywheel due to the term $\boldsymbol{\alpha} \times \mathbf{r}$. This vector quantity is in the same tangential direction as the velocity **v** of the point, as you can see by con-

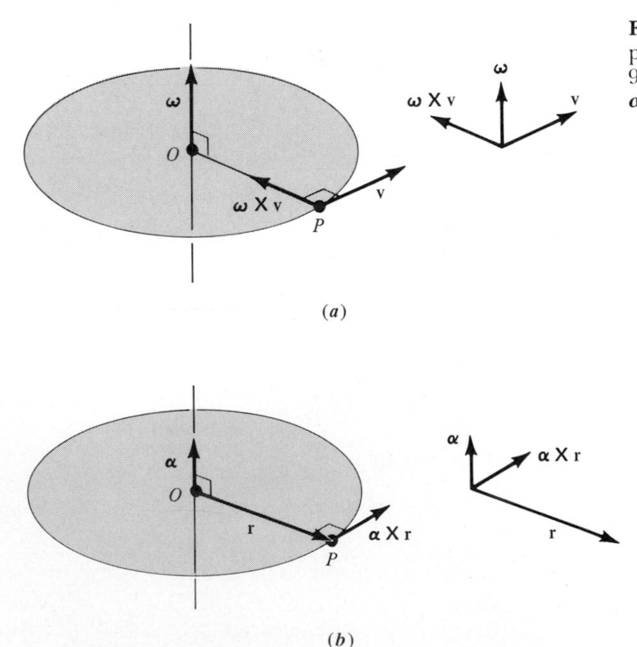

Fig. 9-11 (*a*) The cross product $\boldsymbol{\omega} \times \mathbf{v}$ in Example 9-3*a*. (*b*) The cross product $\boldsymbol{\alpha} \times \mathbf{r}$ in Example 9-3*b*.

structing Fig. 9-11*b*. So you find that the magnitude a_t of the tangential acceleration is

$$a_t = |\boldsymbol{\alpha} \times \mathbf{r}| = \alpha r$$

in agreement with Eq. (9-14).

To evaluate the total acceleration **a** of the point on the rim of the angularly accelerating flywheel, you take the vector sum of its tangential and radial constituents. This gives

$$\mathbf{a} = \alpha r \,\hat{\mathbf{v}} - \omega^2 r \,\hat{\mathbf{r}} \tag{9-23}$$

Here $\hat{\mathbf{v}}$ is a unit vector tangential to the path of the point and in its direction of motion, and $\hat{\mathbf{r}}$ is a unit vector in the outward radial direction. The first term on the right side of this equation is the acceleration resulting from the changing *magnitude* of the velocity of the point on the rim of the wheel when the angular speed of the wheel is increasing. The second term is the acceleration resulting from the changing *direction* of the velocity of the point on the rim as the point rotates around the axis of the wheel. How would Eq. (9-23) be modified if the angular speed of the flywheel were decreasing?

9-3 ANGULAR MOMENTUM

Angular momentum is a quantity that plays a role in rotational mechanics completely analogous to the role played by momentum in translational mechanics. The fundamental reason why momentum is important is that when a certain set of conditions is satisfied, the momentum of a body remains constant. Angular momentum is important for the same reason—when a certain other set of conditions is satisfied, a body maintains a constant angular momentum. The concept of angular momentum will be introduced by analyzing several strobe photos of a puck moving on an air table. In the first of these, the puck moves in such a way that both its momentum and its angular momentum are constant. Then photos will be considered which depict cases where the momentum of the puck changes as it moves, but its angular momentum does not change. In such cases it is more useful to consider the angular momentum of the moving body than to consider its momentum, because the angular momentum has a simpler behavior.

Figure 9-12 shows a puck moving freely across the air table, as viewed by a camera fixed to the earth's surface. The position of the puck at each instant when the strobe light flashes is specified by a position vector **r** drawn from an arbitrarily chosen point O to the center of the puck. This point is the origin of a reference frame which can be considered to be an inertial frame. As the puck moves, there is a continual change in its direction from O because the direction of the vector **r** changes. The situation bears a similarity to the continual change in the direction of a small piece of a flywheel from a point on the axis of the wheel as the small piece rotates about the point. Thus it can be said that the puck *rotates* about O, even though it never makes a complete circuit. Certainly there is an angular velocity $\boldsymbol{\omega}$ associated with the puck's motion about O since the angle ϕ from some fixed line through O to the line through **r** changes as time passes. The right-hand rule for rotation vectors shows that $\boldsymbol{\omega}$ is always directed outward (that is, normal to the page and toward you when you view the page). But its magnitude ω varies as the puck moves. Inspection of the photo shows that the largest change $\Delta\phi$ in the angle specifying the direction of **r**, during any of the equal time intervals Δt between consecutive strobe light flashes, occurs when the puck is closest to O. Thus the average value of ω for each time interval, which is given by $\Delta\phi/\Delta t$, increases as the puck approaches that point

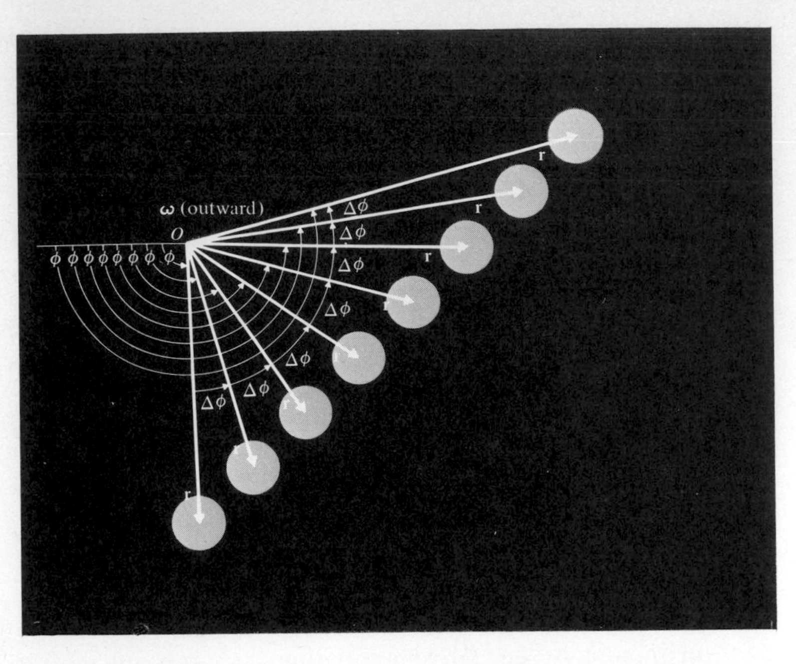

Fig. 9-12 A strobe photo of a puck moving freely across the top of an air table from lower left to upper right. The notation (outward), next to the vector **ω**, means that the vector is pointing directly outward from the page, toward the viewer.

and decreases as it recedes from it. Because the magnitude of the vector **ω** varies, that quantity is not particularly useful in describing the rotation of the puck about O.

A quantity that *is* particularly useful is the **angular momentum** vector, to which we assign the symbol **l** (the letter pronounced "el", not the number pronounced "one"). By definition,

$$\mathbf{l} \equiv \mathbf{r} \times \mathbf{p} \tag{9-24}$$

A particle's angular momentum vector about an origin O is the cross product of its position vector and its momentum vector. The magnitude of the angular momentum is $l = |\mathbf{r} \times \mathbf{p}| = r \sin \theta \, p$, where θ is the angle between **r** and **p.** Units used to measure l are $\text{m}\cdot\text{kg}\cdot\text{m/s} = \text{m}^2\cdot\text{kg/s}$. The direction of the angular momentum is given by the cross-product right-hand rule.

Angular momentum is defined in Eq. (9-24) for a *particle.* In everyday language, a particle is something that is very small. In the language of newtonian mechanics, a **particle** is a body having mass whose motion can be treated completely by considering only the motion of a single point. A particle is an idealization, not something actually found in nature. But approximations of actual bodies by particles are common. To be adequately approximated by a particle, the *size* of a body must be *sufficiently small,* and its *changes in orientation* must be *sufficiently slow.*

The puck in Fig. 9-12 is reasonably approximated by a particle. First, its radius is rather small compared to its distance from O, even at the closest point. Thus the position vector **r** from O to the center of the puck is much the same as the position vector from O to any piece of the puck. So the single vector **r** does an adequate job of representing the position of every piece, and hence the position of the puck as a whole. Second, the motion of any piece resulting from the changing orientation of the puck is rather slow compared to the motion resulting from the changing location of the puck. As a consequence, the velocity of the center of the puck is not very different from the velocity of any piece of the puck. This means that the momentum vector **p,** obtained by multiplying the puck's mass into the velocity of its center, is a fair representation of the momentum of the puck as a whole. Thus we can consider the puck to be a particle. And we can evaluate its angular momentum by using the equation $\mathbf{l} = \mathbf{r} \times \mathbf{p}$, with **r** the position vector of its center and **p** its mass times the velocity vector of its center. Let us do so.

Figure 9-13 reproduces the position vector **r** of the puck at the beginning of each of three time intervals. The momentum vector **p** of the puck for each of these time intervals is shown also. This is done, as in Fig. 4-18, by using a scale for the magnitude of the momentum such that the momentum vector in a time interval extends from the puck's position at the beginning of the interval (when the strobe light flashes) to its position at the end of the interval (when the strobe light flashes next). A vector construction in the figure applies the right-hand rule for cross products to determine the direction of the puck's angular momentum $\mathbf{l} = \mathbf{r} \times \mathbf{p}$ for a typical pair of **r** and **p** vectors. For that pair, and all the others, **l** has the same direction, namely, outward. So **l** has a constant direction.

The magnitude of **l** is also constant. To prove this, we evaluate

$$l = |\mathbf{r} \times \mathbf{p}| = r \sin \theta \, p$$

The figure shows that for a typical pair of vectors **r** and **p,** the quantity $r \sin \theta$ equals d_m, the *perpendicular distance* from O to an extension of the *line of motion* of the puck during the time interval used to determine **r** and **p.** Thus we have

$$l = d_m p \tag{9-25}$$

But since the puck is moving along a straight line, the value of d_m is the same for every pair of vectors. Furthermore, since the puck moves with momentum of constant magnitude, the value of p is the same for every pair. Therefore, the magnitude l of the puck's angular momentum about O

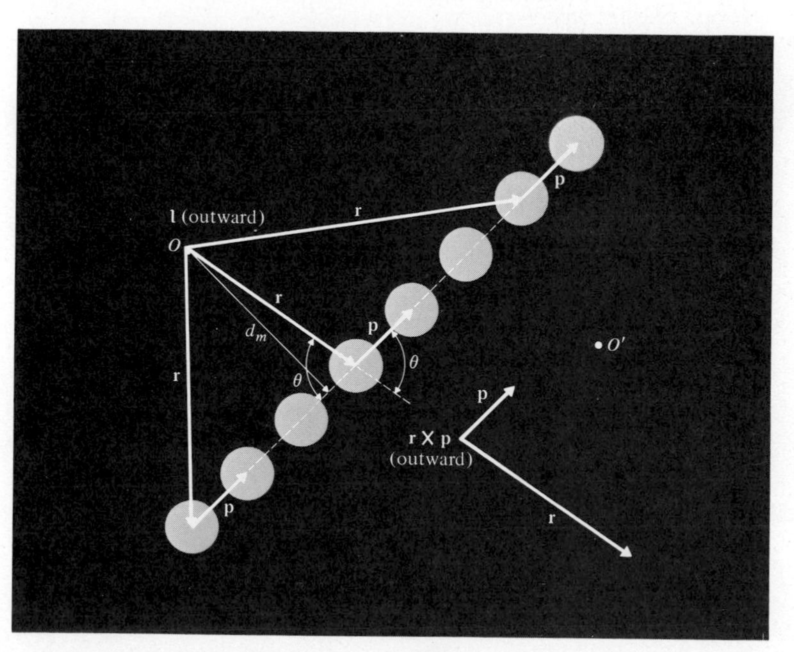

Fig. 9-13 The position vector **r** from origin O, and the momentum vector **p**, of a puck moving uniformly across an air table. Its angular momentum vector $\mathbf{l} = \mathbf{r} \times \mathbf{p}$ about O is constant. The perpendicular distance from O to the line of motion is d_m.

has the constant value $l = d_m p$. We conclude that the puck moving with constant momentum **p** maintains an angular momentum **l** about an arbitrary origin O which is constant in both direction and magnitude. The direction of **l** specifies the sense of rotation about O in much the same way that the direction of $\boldsymbol{\omega}$ does. Its magnitude l gives the value of the magnitude p of the puck's momentum times the perpendicular distance d_m from O to its line of motion.

The puck moving freely across the air table with constant momentum **p** and angular momentum **l** has no net force acting on it. What happens if there is a net force acting on the puck? Its momentum **p** will surely no longer be constant. But in certain circumstances, which apply in many very important cases, the puck can still have a constant angular momentum **l.** Figure 9-14 is a familiar strobe photo of a puck orbiting on an air table and viewed by a camera mounted on the ground. The net force acting on the puck is the force exerted on it by the string going from the puck, over a swiveling pulley in the middle of the table, to a weight hanging beneath the table. This force is *always directed toward a certain fixed point* (the pulley), no matter where the puck on which it acts is located. Such a force is called a **central force.** The puck's angular momentum **l** about the origin O of an essentially inertial reference frame is constant, *providing* the proper point is chosen for the origin. The origin must be chosen at the point, called the **force center,** toward which the central force is directed.

To see this, we again evaluate $\mathbf{l} = \mathbf{r} \times \mathbf{p}$. Application of the right-hand rule for cross products to each of the **r** and **p** pairs constructed in the figure demonstrates that, for each, **l** is in the direction outward. Its magnitude l can be determined by applying Eq. (9-25),

$$l = d_m p$$

[This relation is valid, no matter how the puck moves before or after a particular time interval during which its position and momentum are specified

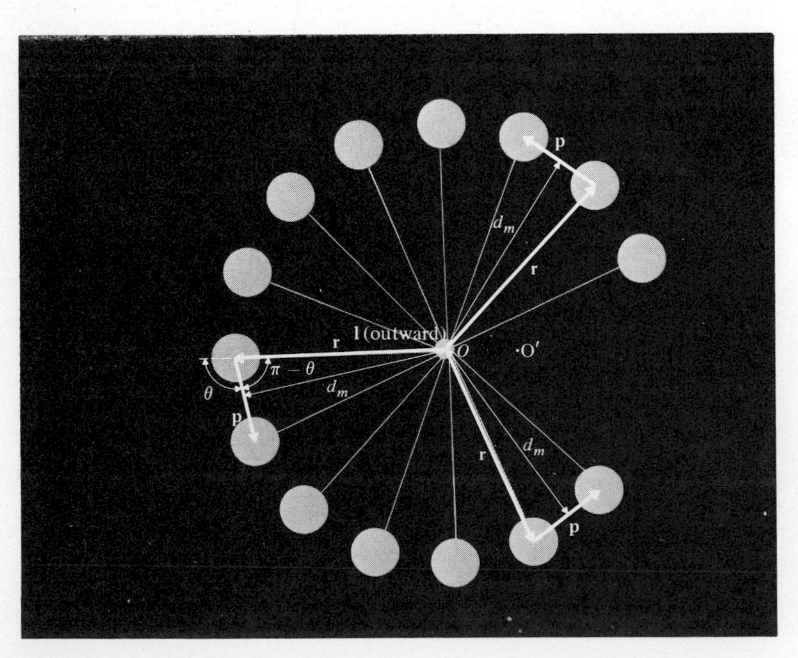

Fig. 9-14 A puck moving uniformly in a circular orbit on an air table, under the influence of a force always directed toward a fixed point called the force center. The puck maintains a constant angular momentum $\mathbf{l} = \mathbf{r} \times \mathbf{p}$ about an origin O at the force center.

by a certain pair of values **r** and **p**. This is because the relation involves only the magnitude p of the puck's momentum and the perpendicular distance d_m from the origin O to its line of motion, during that time interval. Thus the relation can be applied to the circular motion of interest here, as well as in the straight-line motion used to derive it. If this is not evident, use Fig. 9-14 to rederive the relation, using the fact that $\sin(\pi - \theta) = \sin\theta$.] Since the puck is moving in a circular path, and since the origin O is chosen to be at the center of the circle, the perpendicular distance d_m from O to the puck's line of motion has the same value for each pair of **r** and **p** vectors. Furthermore, because the puck moves uniformly through its path, the magnitude p of its momentum is also the same for every pair. Thus the magnitude $l = d_m p$ of its angular momentum about O is constant.

Another way to reach the same conclusion is to note that the angle θ between **r** and **p** is shown by the figure to be close to $\pi/2$. If the strobe flash time interval were reduced to an infinitesimal value, the angle would approach $\pi/2$. In this limit, l has the value $l = |\mathbf{r} \times \mathbf{p}| = r \sin\theta \, p = rp$. Since r is a constant equal to the radius of the circle, and since p is a constant equal to the magnitude of the puck's instantaneous momentum, it follows that $l = rp$ is constant. (It is better to determine the actual value of l in this way than to use the relation $l = d_m p$, with d_m and p constructed as in the figure. The construction in the figure underestimates the magnitude of l for two reasons. First, the d_m constructed in the figure is smaller than the actual value, the radius of the circular orbit. Second, the construction gives the magnitude of an average momentum p which is smaller than the magnitude of the actual instantaneous momentum, because the puck moves in a circular path and so has a higher speed than the distance measured on a straight line between consecutive positions indicates.)

We have found that a puck moving through an inertial frame in a circular orbit under the influence of a central force maintains an angular momentum **l** about an origin O at the force center which is constant in both direction and magnitude. The magnitude of its momentum **p** is also constant, but the direction of this vector is continually changing as the puck moves around the orbit. So in this case **l** is constant, but **p** is not. As a consequence, it may be easier to deal with **l** than with **p** in studying the motion of the puck. This advantage will become still greater for more complicated motions, as we will see soon.

For a given motion the value of the angular momentum $\mathbf{l} = \mathbf{r} \times \mathbf{p}$ depends on the choice of the origin O. This is because the value of the vector **r** from O to a given position changes if the choice of O is changed. For example, the angular momentum of the puck moving uniformly in a straight line in Fig. 9-13 is reversed in direction, and changed in magnitude, if the origin is chosen to be at the point in the figure labeled O'. However, the new angular momentum $\mathbf{l}'$ will also have a constant value, and so O' would probably be an equally appropriate choice for the origin. For uniform circular motion, the off-center origin O' shown in Fig. 9-14 would most likely be inappropriate, since the angular momentum $\mathbf{l}'$ for that origin does *not* have a constant value. The question of choice of origin arises repeatedly in the study of rotational motion. In some cases the location of the origin makes no practical difference. But in others considerable simplification in the analysis of the system can be achieved by a judicious choice. If the system has an obvious point of symmetry, or an obvious rotation axis, then the origin should usually be located at the symmetry point, or on the rotation axis (at a fixed point, if there is one).

A puck moving under the influence of a central force, as viewed from an inertial frame, maintains a constant angular momentum about an origin at the force center, even if its orbit is not circular. An example is shown in Fig. 9-15. The same apparatus was used to obtain this strobe photo as was used to obtain the photo of a puck in a circular orbit. But in the motion recorded in Fig. 9-15 the puck was given an initial velocity, perpendicular to the string, of smaller magnitude than that required to put it into a circular orbit.

The angular momentum of the puck in the noncircular orbit can be analyzed by applying the definition $\mathbf{l} = \mathbf{r} \times \mathbf{p}$. Doing so, you see once more that its angular momentum $\mathbf{l}$ about an origin O at the force center is in the fixed direction outward. In this case the magnitudes of both its position vector $\mathbf{r}$ and its momentum vector $\mathbf{p}$ vary, as does the angle θ between them. Nevertheless, the relation $l = d_m p$ applies. You can use it by measuring with a ruler the lengths of the lines labeled d_m and p in Fig. 9-15 and then calculating their product l. You will find the same value for each of the three $\mathbf{r}$ and $\mathbf{p}$ pairs in the figure, within the accuracy that can be expected of the technique. Thus you will show, within this accuracy, that the angular momentum of the puck about an origin is constant if the puck is acted on by a central force directed toward that origin. (The accuracy is only of the order of 10 percent. The source of error is the underestimate of l resulting from the use of the relation $l = d_m p$, with d_m and p constructed by employing consecutive puck positions for the finite strobe flash time interval in Fig. 9-15. See the parenthetical remark made about Fig. 9-14. The underestimate is quite small for the construction in Fig. 9-15 where the puck is farthest from O, but the underestimate is more than 10 percent for the other two constructions.)

Angular momentum is particularly important in systems containing particles at either of the two extreme ends of the scale of size. At the large end is a satellite or a planet. This "particle" is acted on by a gravitational

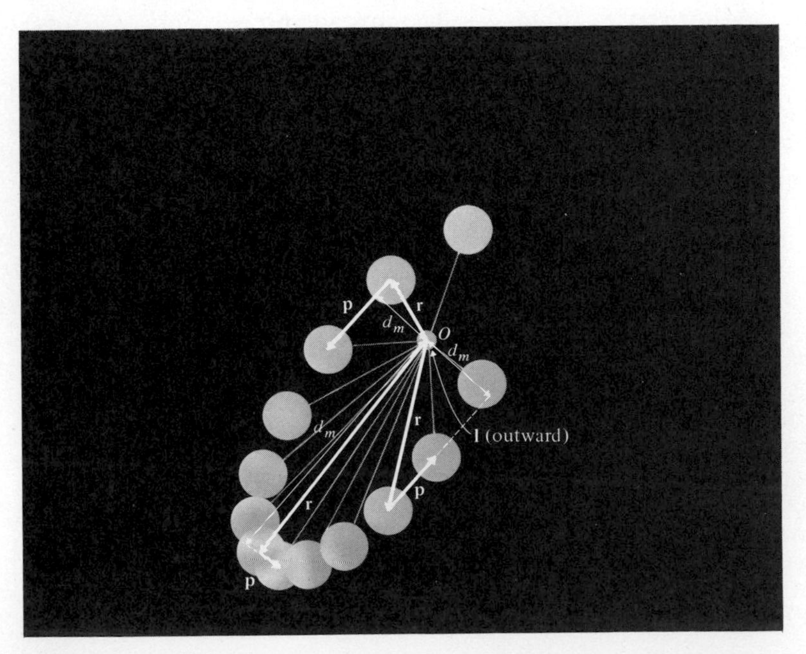

Fig. 9-15 A puck moving nonuniformly in a noncircular orbit on an air table under the influence of a central force. The puck maintains a constant angular momentum $\mathbf{l} = \mathbf{r} \times \mathbf{p}$ about an origin O at the force center.

force, which is a central force always directed toward a certain point in a reference frame that can be considered to be inertial. The angular momentum of the particle about an origin at that point is constant. For instance, Halley's comet is observed to move through its highly elliptical orbit with constant angular momentum about a certain origin. This origin is at the sun, the point toward which the gravitational force exerted on the comet is always directed.

At the small end of the size scale, the same situation is found for an electron in a hydrogen atom. The dominant force acting on the electron is an electric force that is always directed toward the nucleus of the atom, which can be taken as fixed in an inertial frame. Experiment shows that the electron maintains a constant angular momentum about an origin at the nucleus. In treating all these systems, it can be much more productive to deal with the constant angular momentum of the particle than to deal with its varying momentum—for the same reason that it can be very useful to make calculations involving the total mechanical energy in a system where it is constant.

This section has presented strobe photos and mentioned other experimental evidence which leads to the conclusion that a particle's angular momentum about the origin of some inertial frame is constant if no net force acts on it, or if the net force acting on the particle is a central force and the origin is taken at the force center. In Sec. 9-4 it is shown that this conclusion is not really a new law of physics. It can be obtained by combining the definition of angular momentum, $\mathbf{l} = \mathbf{r} \times \mathbf{p}$, with Newton's second law of motion, $\mathbf{F} = d\mathbf{p}/dt$. But before we get into these matters, the conclusion is used in Example 9-4 to solve a problem that would be very difficult to solve without it.

EXAMPLE 9-4

Figure 9-16 shows a student projecting a puck of mass 500 g into a circular orbit of radius 30.0 cm on an air table. The orbital period (the time required for one trip around the orbit) is 2.00 s. The necessary tension in the string is supplied by a second student crouched beneath the table, who applies a force to the lower end of the string.

a. Evaluate the magnitude of the angular momentum of the puck.

b. After a while, the second student increases the force exerted on the string,

Fig. 9-16 Experiment considered in Example 9-4.

pulling it until the puck goes into a circular orbit of radius 15.0 cm. What happens to the angular momentum of the puck? What happens to its orbital period?

■ **a.** In the initial circular orbit, the puck's position vector **r** relative to an origin at the center is always perpendicular to its momentum vector **p.** So its angular momentum has magnitude

$$l = rp = rmv$$

where r is the orbit radius, m is the mass of the puck, and v is its speed. Since

$$v = \frac{2\pi\, r}{t}$$

with t being the orbital period, you have

$$l = \frac{2\pi\, r^2 m}{t} \tag{9-26}$$

For the values given,

$$l = \frac{2\pi \times (0.300 \text{ m})^2 \times 0.500 \text{ kg}}{2.00 \text{ s}}$$
$$= 0.141 \text{ m}^2\text{·kg/s}$$

b. Because the string exerts a central force on the puck, the puck's angular momentum is constant, even when the radius of its orbit is reduced from 30.0 cm to 15.0 cm. This allows you to predict the period of the new orbit. Writing Eq. (9-26) as

$$t = \frac{2\pi\, m}{l} r^2$$

and noting that $2\pi\, m/l$ does not change, you can see that when r is one-half as large, t will be one-quarter as large. So the period of the new orbit will be 2.00 s/4 = 0.500 s.

Finding this answer by using angular momentum was a simple task. It is a *very* different matter if you use momentum and force. Try it!

9-4 TORQUE

Why is a particle's angular momentum **l** about the origin O of an inertial reference frame constant if no net force acts on the particle, or if the net force acting on it is always directed toward that origin? We can find out by starting from the definition of angular momentum, $\mathbf{l} = \mathbf{r} \times \mathbf{p}$, where **r** is the position of the particle relative to O and **p** is its momentum. To investigate the circumstances in which this quantity does not change in time, we evaluate its time derivative,

$$\frac{d\mathbf{l}}{dt} = \frac{d(\mathbf{r} \times \mathbf{p})}{dt}$$

Using the rule for finding the derivative of a cross product, we have

$$\frac{d\mathbf{l}}{dt} = \frac{d\mathbf{r}}{dt} \times \mathbf{p} + \mathbf{r} \times \frac{d\mathbf{p}}{dt} \tag{9-27}$$

For the moment, let us consider only the first term on the right side of this equation. Since $d\mathbf{r}/dt = \mathbf{v}$, where **v** is the velocity of the particle, and since $\mathbf{p} = m\mathbf{v}$, where m is its mass, the term can be written

$$\frac{d\mathbf{r}}{dt} \times \mathbf{p} = \mathbf{v} \times m\mathbf{v}$$

But $\mathbf{v} \times m\mathbf{v} = 0$ always, because the vector $\mathbf{v}$ is always parallel to the vector $m\mathbf{v}$. Therefore, we have

$$\frac{d\mathbf{r}}{dt} \times \mathbf{p} = 0$$

and so Eq. (9-27) simplifies to

$$\frac{d\mathbf{l}}{dt} = \mathbf{r} \times \frac{d\mathbf{p}}{dt}$$

Finally, we invoke Newton's second law to write $d\mathbf{p}/dt = \mathbf{F}$, where $\mathbf{F}$ is the net force acting on the particle. We can do this because we use the origin of an inertial frame for O. The result is

$$\frac{d\mathbf{l}}{dt} = \mathbf{r} \times \mathbf{F} \tag{9-28}$$

Equation (9-28) provides the answer to the question posed. Consider first the zero-force case. When no net force acts on a particle, $\mathbf{r} \times \mathbf{F} = 0$ because $\mathbf{F} = 0$. Then Eq. (9-28) says $d\mathbf{l}/dt = 0$, or $\mathbf{l} =$ constant. This result holds for any choice of the inertial frame origin O, since no choice has been specified.

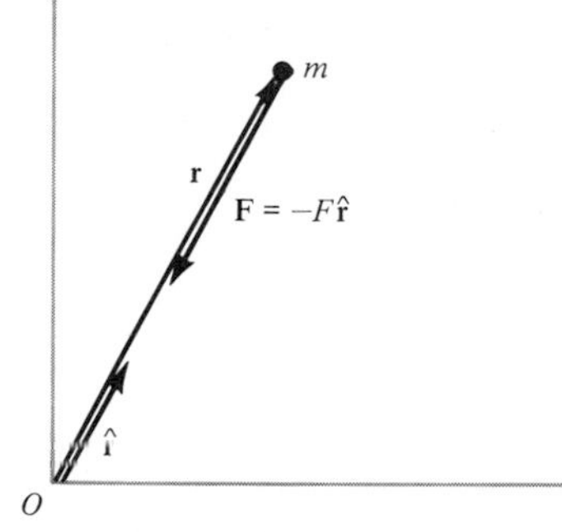

Fig. 9-17 An attractive central force, $\mathbf{F} = -F\hat{\mathbf{r}}$, acting on a particle of mass m that is viewed from an inertial frame whose origin O is at the force center.

Next consider the central-force case. Here there is a net force $\mathbf{F}$ acting on the particle, but it is always directed toward a certain point in the inertial frame, and that point has been chosen for the origin O. As is illustrated in Fig. 9-17, the force can be expressed in terms of a unit vector $\hat{\mathbf{r}}$ that is in the direction of the particle's position vector $\mathbf{r}$. The expression is

$$\mathbf{F} = -F\hat{\mathbf{r}} \tag{9-29}$$

Here F is the magnitude of the attractive central force, and it may or may not be a constant. Evaluating the right side of Eq. (9-28) with this $\mathbf{F}$, we have

$$\mathbf{r} \times \mathbf{F} = \mathbf{r} \times (-F\hat{\mathbf{r}}) = -F\mathbf{r} \times \hat{\mathbf{r}}$$

But $\mathbf{r} \times \hat{\mathbf{r}} = 0$ always, because the vector $\mathbf{r}$ is always parallel to its own unit vector $\hat{\mathbf{r}}$. Hence

$$\mathbf{r} \times \mathbf{F} = 0$$

and so Eq. (9-28) again says $d\mathbf{l}/dt = 0$, or $\mathbf{l} =$ constant.

The force expressed by Eq. (9-29) is called an **attractive** central force because its effect is to attract the particle on which it acts toward the force center. The force exerted by the string on the air table puck is of this form, as are the gravitational force exerted on a planet or a satellite and the dominant electric force exerted on an electron in an atom. In Chap. 20 we will be concerned with a **repulsive** central force, whose effect is to repel the particle on which it acts from the force center. Such a force is the electric force exerted on an alpha particle as it is scattered by an atomic nucleus. How would you express a repulsive central force? Can you use the expression to prove that the particle on which it acts must maintain a constant angular momentum about an origin at the force center?

If a particle is moving under the influence of a net force which is not a central force—or if the force is central but for some reason the origin O of the inertial frame used to observe the particle has not been chosen at the force center—then the right side of Eq. (9-28),

$$\frac{d\mathbf{l}}{dt} = \mathbf{r} \times \mathbf{F}$$

will not be zero. In such a case the particle's angular momentum $\mathbf{l}$ about O will have a nonzero rate of change $d\mathbf{l}/dt$, and $\mathbf{l}$ will change in time. It is said that the particle's angular momentum about O varies because the force acting on it produces a torque about O. To be specific, the **torque T** about an origin O produced by a force $\mathbf{F}$ applied at a point whose position with respect to O is $\mathbf{r}$ has, by definition, the value

$$\mathbf{T} \equiv \mathbf{r} \times \mathbf{F} \qquad (9\text{-}30)$$

The torque vector about an origin O is the cross product of the position vector of the point of application of the force and the force vector. The magnitude of the torque is $T = |\mathbf{r} \times \mathbf{F}| = r \sin\theta\, F$, where θ is the angle between $\mathbf{r}$ and $\mathbf{F}$. Units used to measure T are meter-newtons (m·N). The direction of the torque is given by the right-hand rule for cross products.

In common language the word "twist" is frequently used to convey the same idea as the technical word "torque." (In fact, torque is just the Latin word for twist.) Either implies the act of imparting rotation. Some examples of torque are illustrated in Fig. 9-18. Observe that the sense of the rotational effect produced by a torque about O is in agreement with the two right-hand rules. The magnitude of the torque is

$$T = |\mathbf{r} \times \mathbf{F}| = r \sin\theta\, F$$

The caption to Fig. 9-18*c* shows that this can be written in the form

$$T = d_a F \qquad (9\text{-}31)$$

where d_a is the *perpendicular distance* from O to an extension of the *line of action* of the force. For a force of given magnitude F, maximum torque is achieved by maximizing d_a.

In terms of the net torque acting on a particle when it experiences the net force occurring in Eq. (9-28), that equation can be written

$$\mathbf{T} = \frac{d\mathbf{l}}{dt} \qquad (9\text{-}32)$$

In words, *net torque equals rate of change of angular momentum.* This is the **rotational form of Newton's second law.** The relation is as basic to rotational mechanics as the relation "net force equals rate of change of momentum" is to translational mechanics. We have obtained the relation by considering the motion of a single particle viewed from an inertial frame. But soon we will see that a very similar relation applies to a system containing many particles.

There is a striking analogy between the rotational form of Newton's second law, Eq. (9-32), and its translational form,

$$\mathbf{F} = \frac{d\mathbf{p}}{dt}$$

The momentum $\mathbf{p}$ measures the inertial tendency of a particle to maintain its translational motion. If this measure of the translational motion changes, we say the reason is that a net force $\mathbf{F}$ acts on it. The value of $\mathbf{F}$ is equal to the rate of change of $\mathbf{p}$. Similarly, the angular momentum $\mathbf{l}$ mea-

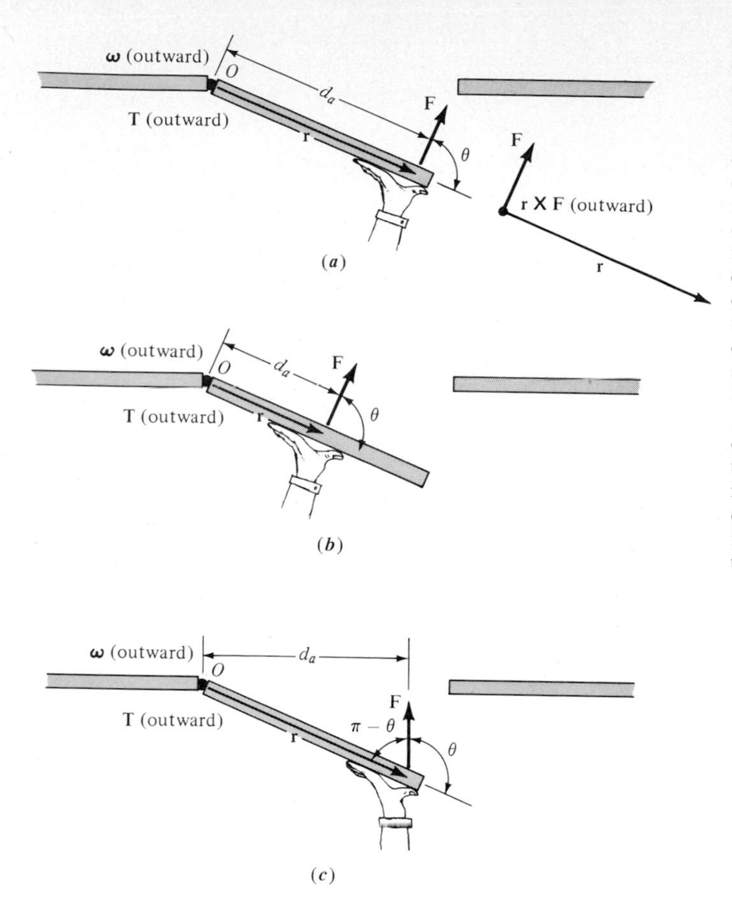

Fig. 9-18 (*a*) To set a heavy door into rotation, you must produce a torque $\mathbf{T} = \mathbf{r} \times \mathbf{F}$ about a point O on the hinge axis, directed parallel to that axis. Applying a force $\mathbf{F}$ of a given magnitude leads to a torque in the required direction that has the largest magnitude when the vector $\mathbf{r}$ from O to the point of application of the force has a maximum magnitude and a direction perpendicular to the direction of the force. (*b*) If the magnitude of the vector $\mathbf{r}$ from O to the point of application of the force $\mathbf{F}$ is reduced, the magnitude of the torque $\mathbf{T}$ is reduced. (*c*) Also, if the direction of the applied force $\mathbf{F}$ is changed, so that it is no longer perpendicular to the direction of the vector $\mathbf{r}$ from O to the point of application, the magnitude of the torque $\mathbf{T}$ is reduced. In all cases, the magnitude of the torque is given by $T = |\mathbf{r} \times \mathbf{F}| = r \sin \theta F$. Since $\sin(\pi - \theta) = \sin \theta$, this can be written $T = r \sin(\pi - \theta) F$. The figure shows that $r \sin(\pi - \theta) = d_a$. The quantity d_a is the perpendicular distance from O to a line passing through the point of application of the force and extending in its direction, called the line of action of the force. So the magnitude of the torque is $T = d_a F$. The three parts of this figure give three examples of the proportionality of T to d_a. Does this agree with your intuition? Do you agree, on an intuitive basis, that T is proportional to F?

sures the inertial tendency of a particle to maintain its rotational motion about an origin. If this measure of the rotational motion changes, we say the reason is that a net torque $\mathbf{T}$ about the origin acts on the particle. And the value of $\mathbf{T}$ is equal to the rate of change of $\mathbf{l}$. Of course, the analogy is not accidental; we defined torque and angular momentum so as to obtain it!

From the point of view of Eq. (9-32), the reason why the angular momentum about an inertial frame origin O does not change for a particle acted on by no net force, or by a force directed to or from O, is that in both cases no net torque is exerted about O. Example 9-5 applies Eq. (9-32) to investigate the motion of a particle in a situation in which the force acting on the particle does exert a torque about an origin O, and consequently changes its angular momentum about that origin.

EXAMPLE 9-5

a. The air table puck of mass 500 g in Example 9-4 is traveling in the circular orbit of radius 30.0 cm with period 2.00 s. Suddenly the power fails, the pump supplying air to the table stops, and the puck comes into direct contact with the tabletop. The coefficient of kinetic friction between the plastic puck and the aluminum table top is 0.13. How much time elapses from the moment when the puck contacts the top to the moment when it comes to rest?

■ First you should make a sketch, like Fig. 9-19, showing the puck at some point in its orbit while it is moving in contact with the tabletop. The forces acting on the

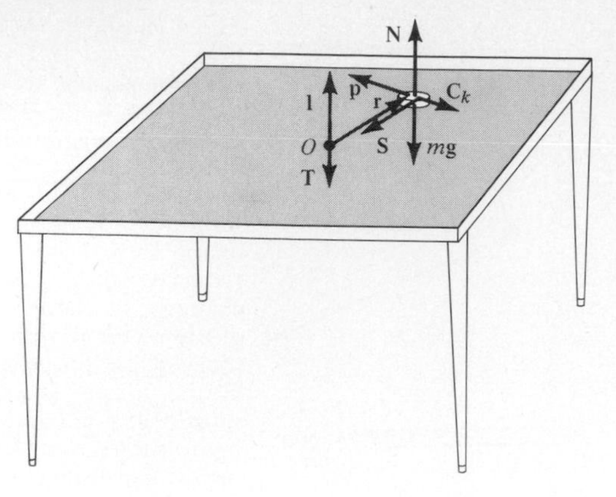

Fig. 9-19 Vectors associated with the motion of the air table puck considered in Example 9-5.

puck of mass m are the force $m\mathbf{g}$ exerted in the downward direction by gravity, the supporting force $\mathbf{N}$ exerted by the tabletop in the upward direction normal to the surface, the contact kinetic friction force $\mathbf{C}_k$ exerted by the tabletop in the direction parallel to the surface and opposite to its direction of motion (the direction of its momentum $\mathbf{p}$), and the force $\mathbf{S}$ exerted by the string in the inward radial direction to keep the puck in orbit.

The torque exerted about O by the force $\mathbf{N}$ is $\mathbf{r} \times \mathbf{N}$, where $\mathbf{r}$ is the position vector of the puck relative to the origin O. Similarly, the torque exerted about O by the force $m\mathbf{g}$ is $\mathbf{r} \times m\mathbf{g}$. The sum of these torques is $\mathbf{r} \times \mathbf{N} + \mathbf{r} \times m\mathbf{g} = \mathbf{r} \times (\mathbf{N} + m\mathbf{g}) = 0$. This is true because the puck does not accelerate in the vertical direction, and so $\mathbf{N} + m\mathbf{g}$, which is the net vertical force it feels, must be zero.

The torque about O produced by the force $\mathbf{S}$ is $\mathbf{r} \times \mathbf{S} = \mathbf{r} \times (-S\hat{\mathbf{r}}) = -S\mathbf{r} \times \hat{\mathbf{r}} = 0$, where $\hat{\mathbf{r}}$ is a unit vector in the direction of $\mathbf{r}$. In words, the force that the string exerts on the puck is a central force and so produces no torque about the origin located at the force center.

Consequently, the net torque about O acting on the puck is the torque produced by the force $\mathbf{C}_k$. This torque is

$$\mathbf{T} = \mathbf{r} \times \mathbf{C}_k$$

The right-hand rule for cross products shows you that the direction of $\mathbf{T}$ is downward, as indicated in the figure. Its magnitude is

$$T = |\mathbf{r} \times \mathbf{C}_k| = r \sin\theta\, C_k = rC_k$$

$d\mathbf{l} = \mathbf{T}\,dt$

l

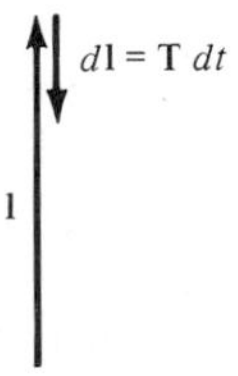

Fig. 9-20 The relation between the angular momentum $\mathbf{l}$ of the puck in Example 9-5 and the change $d\mathbf{l}$ in the angular momentum occurring during the time interval dt.

because the angle θ between $\mathbf{r}$ and $\mathbf{C}_k$ has the value $\pi/2$. According to Eq. (4-23), you have $C_k = \mu_k N$, with μ_k being the coefficient of kinetic friction. Since $N = mg$, this can be written $C_k = \mu_k mg$, and so you have

$$T = r\mu_k mg$$

The angular momentum of the puck about O is

$$\mathbf{l} = \mathbf{r} \times \mathbf{p}$$

where $\mathbf{p}$ is its momentum. The right-hand rule for cross products shows that $\mathbf{l}$ is in the upward direction, as the figure indicates.

Hence you find that the puck is acted on by a torque vector whose direction is opposite to that of its angular momentum vector. Since the torque is equal to the rate of change of the angular momentum, the fact that the torque is directed opposite to the angular momentum means that the angular momentum will have an ever-decreasing magnitude. You can see that this is the case by making a sketch, as in Fig. 9-20, showing the vector $\mathbf{l}$ at some instant and its change $d\mathbf{l}$ during the small time interval dt immediately following that instant. Since $d\mathbf{l}/dt = \mathbf{T}$, the vector $d\mathbf{l}$ is given by the expression $d\mathbf{l} = \mathbf{T}\,dt$. And since $\mathbf{T}$ is directed opposite to $\mathbf{l}$ and dt is positive, $d\mathbf{l}$ must be directed opposite to $\mathbf{l}$, as shown in the sketch. The sketch makes it clear that the magnitude of $\mathbf{l}$ is decreasing.

The magnitude of $\mathbf{T}$ gives the rate at which the magnitude of $\mathbf{l}$ decreases, because $\mathbf{T} = d\mathbf{l}/dt$. The rate has the constant value $r\mu_k mg$. If l decreases by this amount each second, in Δt seconds it will decrease by the amount Δl, where

$$\Delta l = r\mu_k mg \, \Delta t$$

Solving for Δt, you obtain

$$\Delta t = \frac{\Delta l}{r\mu_k mg}$$

When the puck comes to rest, the magnitude of its angular momentum has decreased from the value it had at the instant it first touched the tabletop to its final value, zero. The numerical result obtained in Example 9-4*c* shows you that the decrease is

$$\Delta l = 0.141 \text{ m}^2\cdot\text{kg/s}$$

Using this, the well-known value of g, and the values quoted for r, μ_k, and m, you find that the time required for the puck to come to rest is

$$\Delta t = \frac{0.141 \text{ m}^2\cdot\text{kg/s}}{0.300 \text{ m} \times 0.13 \times 0.500 \text{ kg} \times 9.80 \text{ m/s}^2}$$
$$= 0.74 \text{ s}$$

■

b. If the air pump stays on until after the student below the table pulls on the string to make the puck go into the orbit of radius 15.0 cm, and then the pump stops, how much time elapses from when the puck contacts the tabletop to when it comes to rest?

■ The puck has the same angular momentum about O in the orbit of 15.0 cm radius as it does in the orbit of 30.0 cm radius, because the string cannot exert a torque on the puck to change its angular momentum. Therefore the value of Δl is the same as in part *a*. The force of kinetic friction has the same value as in part *a* also. However, this force exerts only half as much torque about O on the puck because the radius r of its orbit is only half as large as in part *a*. This means the decrease per second in l is half as much as in part *a*, and consequently it takes twice as much time for the decrease to take place. Hence in this case you have

$$\Delta t = 2 \times 0.74 \text{ s} = 1.5 \text{ s}$$

9-5 ROTATION OF SYSTEMS AND ANGULAR MOMENTUM CONSERVATION

The next step in our study of rotation is to broaden the scope of the equation governing rotational motion so that it applies to a system comprising many particles. This will allow us to treat the rotation of a body which cannot be considered to be a single particle. It also will lead us to the very important law of the conservation of angular momentum of an isolated system.

Consider a system containing n ideal particles, viewed from an inertial reference frame whose origin is O. As shown in Fig. 9-21, the position of the first particle relative to O is specified by the position vector $\mathbf{r}_1$. Its momentum vector is $\mathbf{p}_1$. Subscripts ranging from 2 to n are used to identify the other particles of the system and to label their positions and momenta. The letter j is used to represent a typical subscript. Thus the position and momentum of a typical particle are $\mathbf{r}_j$ and $\mathbf{p}_j$. According to the definition of Eq. (9-24), the jth particle's angular momentum about O is

$$\mathbf{l}_j = \mathbf{r}_j \times \mathbf{p}_j$$

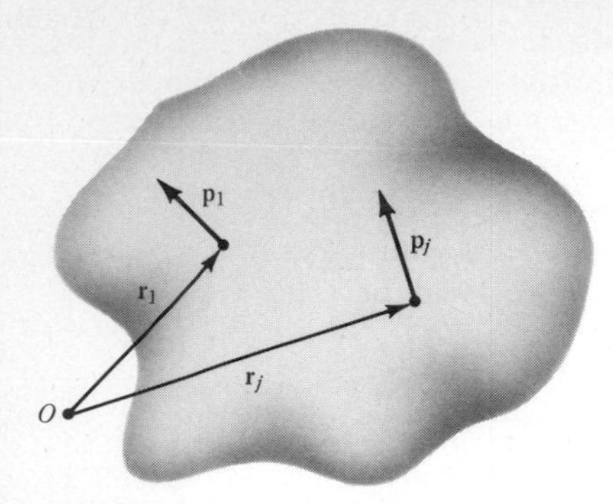

Fig. 9-21 A rotating body comprising a set of n particles labeled by the subscripts $1, 2, 3, \ldots, j, \ldots, n$. The momentum of a typical particle is given by the vector $\mathbf{p}_j$, and its position with respect to the origin O is given by the vector $\mathbf{r}_j$.

We define the system's total angular momentum as the sum of the angular momenta of its constituent particles. Thus the **total angular momentum L** about O is given by the vector sum

$$\mathbf{L} \equiv \sum_{j=1}^{n} \mathbf{l}_j \tag{9-33}$$

Evaluating $\mathbf{l}_j$, we have

$$\mathbf{L} = \sum_{j=1}^{n} \mathbf{r}_j \times \mathbf{p}_j$$

To develop the equation governing the rotation of the system as a whole, we proceed in exactly the same way as in developing Eq. (9-32), the equation governing the rotation of a single particle. That is, we calculate

$$\frac{d\mathbf{L}}{dt} = \frac{d}{dt}\left(\sum_{j=1}^{n} \mathbf{r}_j \times \mathbf{p}_j\right)$$

Since the derivative of a sum of terms equals the sum of their derivatives, this is

$$\frac{d\mathbf{L}}{dt} = \sum_{j=1}^{n} \frac{d}{dt}(\mathbf{r}_j \times \mathbf{p}_j)$$

Evaluating the derivative of the cross product, we have

$$\frac{d\mathbf{L}}{dt} = \sum_{j=1}^{n} \frac{d\mathbf{r}_j}{dt} \times \mathbf{p}_j + \sum_{j=1}^{n} \mathbf{r}_j \times \frac{d\mathbf{p}_j}{dt} \tag{9-34}$$

Now each of the terms in the first sum has the value

$$\frac{d\mathbf{r}_j}{dt} \times \mathbf{p}_j = \mathbf{v}_j \times m_j\mathbf{v}_j = 0$$

So the first sum on the right side of Eq. (9-34) is zero. Thus we have

$$\frac{d\mathbf{L}}{dt} = \sum_{j=1}^{n} \mathbf{r}_j \times \frac{d\mathbf{p}_j}{dt}$$

Since we are working in an inertial frame, we can employ Newton's second law to yield

$$\frac{d\mathbf{L}}{dt} = \sum_{j=1}^{n} \mathbf{r}_j \times \mathbf{F}_j \tag{9-35}$$

where $\mathbf{F}_j$ is the net force acting on the jth particle of the system.

We can achieve a tremendous simplification of Eq. (9-35) by realizing that the net force $\mathbf{F}_j$ felt by a typical particle arises from two distinct sources—external and internal. The first part of the net force comes from the force applied externally to the particle. This is the force acting on it from outside the system being considered, which we write as $\mathbf{F}_{\text{ext}\,j}$. The second part of the net force $\mathbf{F}_j$ experienced by a typical particle is the internally applied force $\mathbf{F}_{\text{int}\,j}$ acting on it from inside the system. The $\mathbf{F}_{\text{int}\,j}$ are the forces which the particles of the system exert on one another. As an example, consider a system containing only the particles of a child's top. Each particle of the system is acted on by an external force, the gravitational force which an object external to the system (the earth) exerts on the particle. The particle at the tip is also acted on by the external force exerted by the surface on which the top rests. The internal forces acting on the particles of the top are those which bind them rigidly to the other particles, so that the entire system remains a top. But a system containing only a single rigid body is not the only type of system for which a decomposition of total force into external and internal forces can be made. The decomposition can be done for any system. Thus the results we obtain apply as well to a system containing no rigid bodies, or several rigid bodies moving with respect to one another.

These considerations suggest that we write

$$\mathbf{F}_j = \mathbf{F}_{\text{ext}\,j} + \mathbf{F}_{\text{int}\,j}$$

Doing so makes the right side of Eq. (9-35) break into two parts, to yield

$$\frac{d\mathbf{L}}{dt} = \sum_{j=1}^{n} \mathbf{r}_j \times \mathbf{F}_{\text{ext}\,j} + \sum_{j=1}^{n} \mathbf{r}_j \times \mathbf{F}_{\text{int}\,j} \tag{9-36}$$

The simplification resulting from this decomposition is that the second summation on the right side of Eq. (9-36) has the value zero. Hence the rate of change of the total angular momentum of the system depends only on the forces applied externally to its particles.

To see this, we must realize that each of the terms

$$\mathbf{r}_j \times \mathbf{F}_{\text{int}\,j}$$

in the second summation of Eq. (9-36) is itself the sum of a series of terms. This is because $\mathbf{F}_{\text{int}\,j}$ is the vector sum of all the forces exerted on particle j by all the other particles. For instance, for particle 1

$$\mathbf{r}_1 \times \mathbf{F}_{\text{int}\,1} = \mathbf{r}_1 \times (\mathbf{F}_{\text{on 1 by 2}} + \mathbf{F}_{\text{on 1 by 3}} + \cdot \cdot \cdot + \mathbf{F}_{\text{on 1 by }n})$$

There is a similar series for each of the particles 2, 3, . . . , n which comprise the system.

Now note that we can arrange the terms in the sum

$$\sum_{j=1}^{n} \mathbf{r}_j \times \mathbf{F}_{\text{int}\,j}$$

by pairing each force exerted on particle j by particle k with the force exerted on particle k by particle j. For particles 1 and 2, as an example, the two terms associated with the pair of forces look like this:

$$\mathbf{r}_1 \times \mathbf{F}_{\text{on 1 by 2}} + \mathbf{r}_2 \times \mathbf{F}_{\text{on 2 by 1}}$$

The two forces in this pair are illustrated in Fig. 9-22. They constitute (as does every similar pair) an equal but oppositely directed action-reaction pair, according to Newton's third law:

$$\mathbf{F}_{\text{on 1 by 2}} = -\mathbf{F}_{\text{on 2 by 1}}$$

So we can write

$$\begin{aligned} \mathbf{r}_1 \times \mathbf{F}_{\text{on 1 by 2}} + \mathbf{r}_2 \times \mathbf{F}_{\text{on 2 by 1}} &= \mathbf{r}_1 \times \mathbf{F}_{\text{on 1 by 2}} - \mathbf{r}_2 \times \mathbf{F}_{\text{on 1 by 2}} \\ &= (\mathbf{r}_1 - \mathbf{r}_2) \times \mathbf{F}_{\text{on 1 by 2}} \\ &= 0 \end{aligned}$$

The reason for the zero also is illustrated in Fig. 9-22. The vector $\mathbf{r}_1 - \mathbf{r}_2$ is antiparallel to the vector $\mathbf{F}_{\text{on 1 by 2}}$ if the forces are attractive (or parallel if they are repulsive). So the sine factor determining the magnitude of the

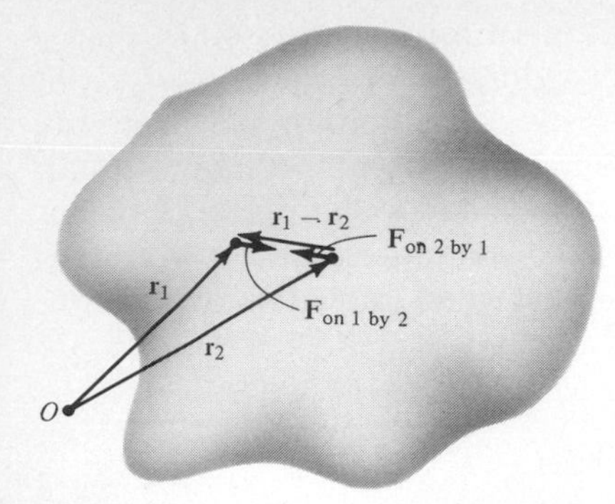

9-22 A pair of particles in a system and the action-reaction forces they exert on each other.

cross product is sin π (or sin 0), and the cross product is therefore zero. This is the case because in the newtonian domain the forces two particles exert on each other always act along the line passing through the two points specifying their positions, as the figure indicates. We will have more to say about this property of forces soon.

Since each of the pairs of the sum adds to zero and since there are no leftover terms, the entire sum adds to zero also. Hence we have

$$\sum_{j=1}^{n} \mathbf{r}_j \times \mathbf{F}_{\text{int}\,j} = 0 \tag{9-37}$$

This result can be explained by saying that while each internal force of an action-reaction pair produces a torque about the origin, the torque is exactly canceled by the torque produced by the equal but oppositely directed other force of that pair, because the two forces have a common line of action.

Returning to Eq. (9-36), we see that it simplifies to

$$\frac{d\mathbf{L}}{dt} = \sum_{j=1}^{n} \mathbf{r}_j \times \mathbf{F}_{\text{ext}\,j}$$

We now use the definition of torque in Eq. (9-30) to write

$$\mathbf{r}_j \times \mathbf{F}_{\text{ext}\,j} = \mathbf{T}_j$$

Here $\mathbf{T}_j$ is the torque applied about the origin O to the jth particle by the *external* force acting on it. Then we have

$$\frac{d\mathbf{L}}{dt} = \sum_{j=1}^{n} \mathbf{T}_j$$

Next, we define the **net torque T** about O exerted on the system as

$$\mathbf{T} \equiv \sum_{j=1}^{n} \mathbf{T}_j \tag{9-38}$$

[Of course, **T** is actually the net external torque, but we do not need to indicate this by using a subscript "ext" because Eq. (9-37) tells us that there is no net internal torque.] In terms of **T**, we have

$$\mathbf{T} = \frac{d\mathbf{L}}{dt} \tag{9-39}$$

Net torque equals the rate of change of angular momentum. The angular momentum in the mathematical and verbal statements of the equation is the total angular momentum of a system of particles. The system is viewed from an inertial reference frame, and the torque and angular momentum are both taken about its origin. This equation is the **rotational form of Newton's second law.** It is more general than Eq. (9-32) in that it applies to any system, and not only to a system containing a single particle.

We use Eq. (9-39) to determine how the angular momentum of a system changes when there is a net torque applied to it. But first let us consider the important special case of a system *isolated* from its environment in such a way that *no* net torque can be applied. We view the system from an inertial frame, as assumed in obtaining Eq. (9-39). For such a system the equation states that $d\mathbf{L}/dt = 0$, so that **L** must be constant. This is the **law of conservation of angular momentum:** *As seen by an observer in an inertial frame, a system to which no net torque is applied maintains a constant total angular momentum.*

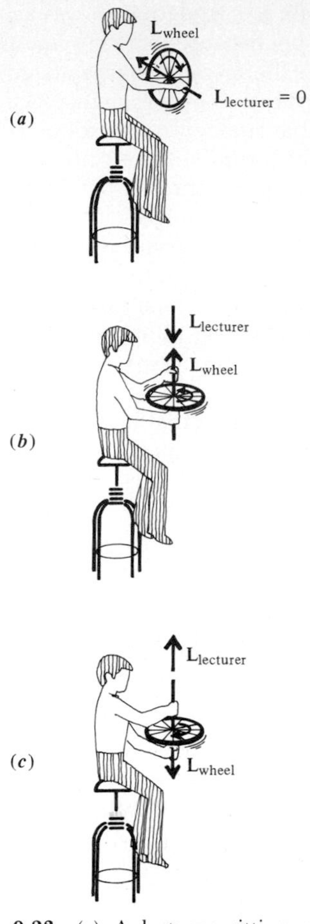

Fig. 9-23 (*a*) A lecturer sitting on a swiveling stool with low-friction bearings, while holding a spinning bicycle wheel by its axle with the axle horizontal. (*b*) As the lecturer turns the axle so that the angular momentum vector of the wheel points upward, his body begins to rotate so that its angular momentum vector points downward. (*c*) By turning the axle so as to reverse the direction of the wheel's angular momentum vector, the lecturer can reverse the direction of his body's angular momentum vector.

The law of conservation of angular momentum is a *fundamental* law of physics. It is on a completely equal footing with the law of conservation of momentum. Each of these basic conservation laws is founded firmly on experimental evidence. In Sec. 4-3 we obtained the momentum conservation law directly from experimental observation. Here also we obtained the angular momentum conservation law by using an experimental observation, namely that in the newtonian domain the forces which a pair of particles exert on each other always act along the line passing through the particles. Another way of describing the situation is to state that in newtonian mechanics the force which particle 1 exerts on particle 2 is a *central force* since it is always directed toward or from a point, lying somewhere along the line between the particles, that can be taken as fixed in a suitably chosen inertial frame. And similarly, the reaction force which particle 2 exerts on particle 1 is a central force acting toward or away from the same force center. There is experimental evidence for this statement. But there is even more direct experimental evidence for the *conclusion* reached by using the statement, that is, for the law of conservation of angular momentum.

An example of this evidence is pictured in Fig. 9-23*a*. A lecturer seats himself on a stool that can rotate freely about a vertical axis, while holding a bicycle wheel that is spinning about a horizontal axis so that the angular momentum vector of the wheel is horizontal. The tire is filled with sand, causing the wheel to have an appreciable amount of angular momentum when it spins. In Fig. 9-23*b* the lecturer turns the rotation axis of the wheel so that its angular momentum vector is upward. As he does this, he starts rotating on the stool, with his body's angular momentum vector downward. The lecturer is demonstrating conservation of the vertical component of the total angular momentum of the system consisting of the wheel plus his body and the seat of the stool. The system is isolated by the smooth bearings of the stool from the application of external torques with vertical components. Since the total vertical angular momentum of the system is initially zero, it must remain zero. In Fig. 9-23*c*, the lecturer turns the bicycle wheel so that its angular momentum vector points downward, which causes his body to reverse its sense of rotation in order to reverse the direction of its angular momentum. If you can carry out this demonstration yourself, you will literally get a seat-of-the-pants feeling for angular momentum conservation.

There is a tremendous variety of direct experimental evidence demonstrating that the total angular momentum of an isolated system is conserved. Observation shows this to be true of the solar system, as an example. In fact, *the most fundamental view is to take the angular momentum conservation law to be based on direct observation,* just as the momentum conservation law was justified in Sec. 4-3 on the basis of direct observation. From this point of view, the calculation leading to Eq. (9-39), and its special case $\mathbf{L}$ = constant for $\mathbf{T} = 0$, suggests that in the newtonian domain the forces acting between a pair of particles are central forces.

When you come to the study of electromagnetism, you will learn that there are situations outside the newtonian domain where noncentral forces arise. For instance, the forces acting between a pair of charged particles, such as a pair of electrons, depart significantly from being central forces if the particles are moving relative to each other at a speed comparable to the speed of light. The situation is illustrated in Fig. 9-24 from the point of view of an observer stationed at an origin of an inertial frame which remains midway between the two electrons. These forces cause a net torque about O to act on the electrons, and their angular momentum

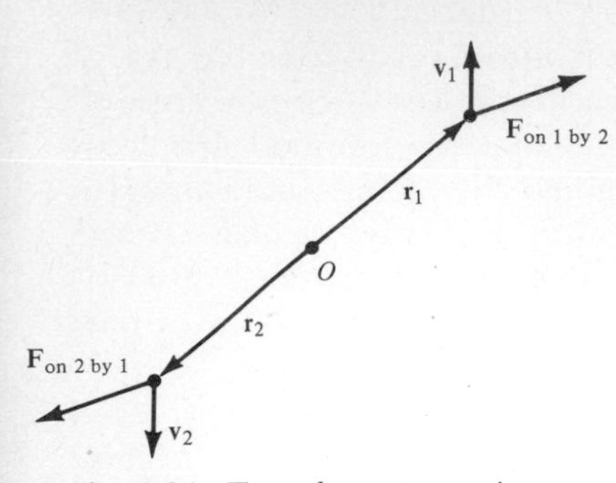

Fig. 9-24 Two electrons moving past each other at a speed comparable to the speed of light exert forces on each other which are not central forces. But this situation does not lead to a violation of the law of conservation of angular momentum, for reasons explained in the text.

does not remain constant. It would appear that this is a violation of the law of angular momentum conservation. But actually it is not because there is something in the isolated system not shown in Fig. 9-24. This is the so-called electromagnetic field, which also contains angular momentum. When the complete electron-pair-plus-field system is considered, its total angular momentum is found to be conserved because every change in the angular momentum of the electron pair is exactly compensated for by a change in the angular momentum of the field.

Thus even for microscopic systems containing particles moving at relativistic speeds, the total angular momentum is conserved, providing that the system is isolated, all its parts are taken into account, and it is viewed from an inertial frame. The uranium atom is an example. Its inner electrons move at speeds quite close to the speed of light, exchanging angular momentum between themselves and the electromagnetic field in the atom in a variety of complex processes. But measurement shows that the *total* angular momentum of the system remains constant.

The outer electrons of an atom are involved in the forces that bind atoms into molecules and molecules into solids. They move at speeds which are very small compared to the speed of light. As a consequence, the internal forces playing a significant role in keeping a rigid body rigid are central forces. The same is true of all the other forces of interest in newtonian mechanics.

Example 9-6 applies the law of angular momentum conservation.

EXAMPLE 9-6

Figure 9-25 shows the loaded bicycle wheel with its axle secured firmly in a vertical orientation and initially not rotating. A small-caliber rifle fires a bullet tangentially into the sand-filled tire, where it embeds itself. The mass of the bullet is $m = 8.1$ g, and its speed is $v = 370$ m/s. The mass of the sand, the tire, and the rim on which it is mounted is $M = 5.2$ kg. The radius to the center of the narrow tire is $r = 33$ cm. Determine the angular speed ω of the wheel after the impact.

■ Before using angular momentum conservation in the system comprising the bullet plus the wheel to solve this problem, you should note that it is the *only* conservation law applicable to the system. Mechanical energy is not conserved because the bullet undergoes a sequence of inelastic collisions when it embeds itself in the sand. Momentum is not conserved because the structure supporting the axle exerts forces on it—particularly during the impact—so the system is not isolated from external forces, and its total momentum changes. You can see that this is true by considering the total momentum of the system before impact and at several times after impact, using the fact that the symmetrical wheel itself has zero total momentum whether it is rotating or not.

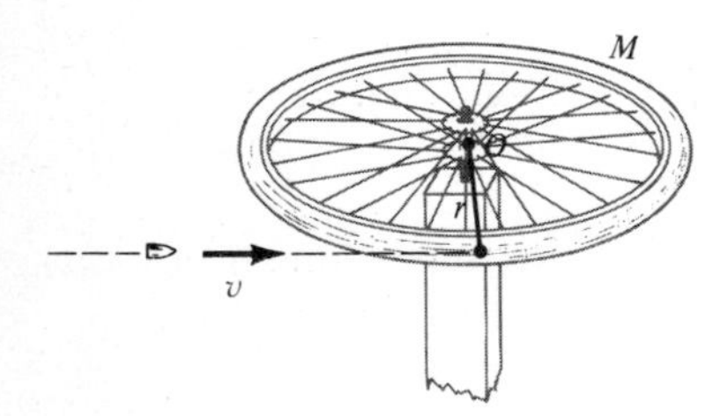

Fig. 9-25 Experiment analyzed in Example 9-6.

But the forces exerted on the axle by the support do not exert a torque about an origin O at the intersection of the axle and the plane of the wheel. Thus the total angular momentum about O of the system maintains a constant magnitude and a constant direction, the direction being vertically upward in Fig. 9-25.

You can obtain a simple but sufficiently accurate expression for the final angular momentum magnitude L in terms of the angular speed ω by assuming that all the mass M of the sand, tire, and rim of the wheel is rotating in a circle of the same radius r, with the same speed $v = \omega r$. Then each element m_j of this mass makes a contribution $\mathbf{L}_j$ to the total angular momentum vector $\mathbf{L}$. All the $\mathbf{L}_j$ are parallel to the axle, and each has magnitude

$$L_j = rm_jv = rm_j\omega r = r^2\omega m_j$$

Summing over the mass elements immediately gives

$$L = \sum_{j=1}^{n} L_j = \sum_{j=1}^{n} r^2\omega m_j = r^2\omega \sum_{j=1}^{n} m_j = r^2\omega M$$

The justification for ignoring the angular momentum of the spokes is that their mass is small compared to M and that much of this mass is rotating at a distance from the axis that is small compared to r. For the same two reasons, and particularly for the latter, you can ignore the angular momentum of the hub of the wheel. And the angular momentum of the bullet embedded in the rotating wheel can be ignored in evaluating the final total angular momentum magnitude L because its mass is very small compared to M.

The bullet contains all the angular momentum of the system before it hits the wheel. The bullet of mass m is moving at the high speed v along a line whose perpendicular distance from O is r. According to Eq. (9-25), the magnitude of its angular momentum is this perpendicular distance times the magnitude of its momentum mv. So the bullet has angular momentum of magnitude rmv. Equating the initial angular momentum to the final angular momentum $r^2\omega M$, you have

$$r^2\omega M = rmv$$

or

$$\omega = \frac{mv}{Mr}$$

The numerical value is

$$\omega = \frac{8.1 \times 10^{-3} \text{ kg} \times 3.7 \times 10^2 \text{m/s}}{5.2 \text{ kg} \times 3.3 \times 10^{-1} \text{ m}}$$
$$= 1.7 \text{ s}^{-1}$$

So, after the impact, the wheel rotates at 1.7 rad/s (or 0.44 rotations per second).

Example 9-7 treats a system whose angular momentum is not conserved because a net torque is acting on it.

EXAMPLE 9-7

The Atwood machine of Fig. 9-26 contains a pulley whose mass cannot be neglected. The pulley consists of a substantial rim, of mass M and radius r, supported by spokes of negligible mass. Compact bodies 1 and 2 hang from each end of a cord of negligible mass, which passes over the pulley. Their masses are m_1 and m_2, with $m_1 > m_2$. Find the magnitude of the downward acceleration of body 1.

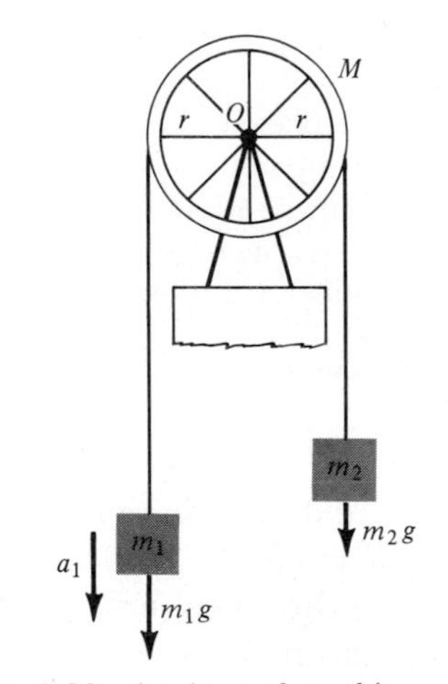

Fig. 9-26 An Atwood machine.

■ Choose the center of the pulley as the origin O, and then evaluate the magnitude of the torque exerted on the pulley-plus-bodies-plus-string system by the gravitational force acting on body 1. It is

$$T_1 = rm_1g$$

The direction of this torque is outward. The torque exerted by the gravitational force on body 2 is of magnitude

$$T_2 = rm_2g$$

and is directed inward (that is, normal to the page and away from you as you view the page). Since $m_1 > m_2$, the net torque is outward and has magnitude

$$T = T_1 - T_2 = r(m_1 - m_2)g$$

In this example the net torque applied to the system is *not* zero.

When the pulley is rotating with an angular velocity that is directed outward and is of magnitude ω, the speed of body 1 is ωr and its angular momentum has magnitude $rm_1\omega r = r^2\omega m_1$. The direction of this angular momentum is outward. The angular momentum of body 2 is in the same direction and has the magnitude $r^2\omega m_2$. The angular momentum of the pulley is also outward. As argued in Example

9-6, its magnitude is $r^2\omega M$. So the magnitude L of the total angular momentum of the system will be the sum of the angular momenta of bodies 1 and 2 and of the pulley:

$$L = r^2\omega m_1 + r^2\omega m_2 + r^2\omega M$$
$$= r^2\omega(m_1 + m_2 + M)$$

The total angular momentum is outward.

Now you write the rotational form of Newton's second law, Eq. (9-39), for a situation where both $\mathbf{T}$ and $\mathbf{L}$ have the same constant direction. It is

$$T = \frac{dL}{dt}$$

When this is applied to the case at hand, you have

$$r(m_1 - m_2)g = \frac{d}{dt}[r^2\omega(m_1 + m_2 + M)]$$
$$= r^2(m_1 + m_2 + M)\frac{d\omega}{dt}$$
$$= r^2(m_1 + m_2 + M)\alpha$$

where α is the magnitude of the angular acceleration. Solving for this quantity yields

$$\alpha = \frac{(m_1 - m_2)g}{(m_1 + m_2 + M)r}$$

The direction of the angular acceleration is also outward.

The acceleration of body 1 is the same as the tangential part of the acceleration of the point on the wheel to which the string to body 1 is tangent. Thus you can use the first term on the right side of Eq. (9-21) to evaluate its magnitude as

$$a_1 = \alpha r$$

Then you have the required result

$$a_1 = \frac{(m_1 - m_2)g}{m_1 + m_2 + M} \tag{9-40}$$

If M is so much smaller than $m_1 + m_2$ that it can be neglected, the value of a_1 predicted by this result agrees with Eq. (5-7a), the result obtained in Sec. 5-2 for an Atwood machine with a massless pulley. Comparison of the present calculation with the earlier one will show you that it easier to give a realistic treatment of an Atwood machine by considering torques than it is to give a simplified treatment, ignoring the mass of the pulley, by considering forces.

If you carry out the stool-and-bicycle-wheel demonstration of angular momentum conservation pictured in Fig. 9-23, you will experience the torque that your arms must apply to the wheel to change its angular momentum vector, as well as the **reaction torque** which the wheel must exert through your arms to change your angular momentum vector. Furthermore, you will appreciate that the behavior of the system can be explained either by the law of angular momentum conservation or by the **rotational form of Newton's third law.** The latter states that when two bodies exert torques on each other in their interaction, *the torques have equal magnitude but opposite direction.* That is,

$$\mathbf{T}_{\text{on 1 by 2}} = -\mathbf{T}_{\text{on 2 by 1}} \tag{9-41}$$

Table 9-1

Some Equations Used in Rotational and Translational Mechanics

Rotational equations	Connecting equations	Translational equations
$\omega = \frac{d\phi}{dt}$	$\mathbf{v} = \boldsymbol{\omega} \times \mathbf{r}$	$\mathbf{v} = \frac{d\mathbf{s}}{dt}$
$\boldsymbol{\alpha} = \frac{d\boldsymbol{\omega}}{dt}$	$\mathbf{a} = \boldsymbol{\alpha} \times \mathbf{r} + \boldsymbol{\omega} \times \mathbf{v}$	$\mathbf{a} = \frac{d\mathbf{v}}{dt}$
For a single particle:		
$\mathbf{T} = \frac{d\mathbf{l}}{dt}$	$\mathbf{T} = \mathbf{r} \times \mathbf{F}; \quad \mathbf{l} = \mathbf{r} \times \mathbf{p}$	$\mathbf{F} = \frac{d\mathbf{p}}{dt}$
For n particles:		
$\mathbf{T} = \frac{d\mathbf{L}}{dt}$	$\mathbf{T} = \sum_{j=1}^{n} \mathbf{r}_j \times \mathbf{F}_{\text{ext}\,j}; \quad \mathbf{L} = \sum_{j=1}^{n} \mathbf{r}_j \times \mathbf{p}_j$	$\mathbf{F} = \frac{d\mathbf{P}}{dt}$, where $\mathbf{F} = \sum_{j=1}^{n} \mathbf{F}_{\text{ext}\,j}$ and $\mathbf{P} = \sum_{j=1}^{n} \mathbf{p}_j$
$\mathbf{T}_{\text{on 1 by 2}} = -\mathbf{T}_{\text{on 2 by 1}}$		$\mathbf{F}_{\text{on 1 by 2}} = -\mathbf{F}_{\text{on 2 by 1}}$

In Sec. 4-5, the translational form of Newton's third law was derived by using the law of momentum conservation and the definition of force given in the translational form of Newton's second law. You should have no difficulty in modifying the derivation to obtain the rotational form of the third law by using the law of angular momentum conservation and the rotational form of the second law. What is the rotational form of Newton's first law?

Table 9-1 summarizes the most important equations obtained up to this point in our study of rotational motion, and it also shows their relationships to analogous equations that we employ when studying translational motion. In Chap. 10 we continue the development of equations that we will use in the investigation of how objects rotate. In particular, we will find the rotational analogue of Newton's second law in the form "force equals mass times acceleration" and the rotational analogues of the energy relations. We close this chapter by turning our attention in Secs. 9-6 and 9-7 to an important practical application of the theory already at hand.

9-6 STATIC EQUILIBRIUM OF RIGID BODIES AND CENTER OF MASS

In the design of structures such as buildings or bridges, it is essential to ensure that they will remain in place. In doing this work, structural engineers often begin their calculations by applying Newton's second law, in both the rotational form and the translational form, to a stationary rigid body. They are then working with the subject of this section: the *static equilibrium of rigid bodies.* It is relatively simple, at least in concept. At later stages of their calculations, structural engineers must frequently take into account the fact that no body is perfectly rigid. This leads to complications which we will not consider here.

The rotational and translational forms of Newton's second law show that for a body to be *stationary* with respect to the earth's surface (considered as an inertial frame), two conditions must be met: (1) *The net torque about any origin acting on the body must be zero* because the nonrotating body has the constant angular momentum zero about any origin. (2) *The net force acting on the body must be zero* because the nontranslating body has the constant momentum zero. Example 9-8 illustrates the application of these two conditions to a simple case.

EXAMPLE 9-8

Find the force exerted on the lever in Fig. 9-27 by its fulcrum, labeled O. The lever is stationary and is supporting a body of mass m because an upward force is applied to the free end of the lever. Assume that the mass of the lever is zero.

■ Consider first the condition that the net force acting *on* the lever must be zero. The force exerted on the lever by the body it supports certainly acts in the downward direction. And you are told that the force exerted on the lever at its free end acts in the upward direction. Since these forces have no horizontal components, neither can the force exerted on the lever by the fulcrum. This is because the net force acting on the lever must have a zero horizontal component. Thus the force applied by the fulcrum must act in either the upward or the downward direction. You can use signed scalars to specify the three forces. If you take the upward direction as positive, the force applied by the fulcrum is F_0 (whose value has unknown sign and magnitude), the force applied by the supported body is F_1 (whose value $-mg$ is negative), and the force applied at the free end is F_2 (whose value has a positive sign but an unknown magnitude). The condition that the net force acting on the lever is zero,

$$F_0 + F_1 + F_2 = 0$$

then reads

$$F_0 - mg + F_2 = 0 \tag{9-42a}$$

or

$$F_0 = mg - F_2 \tag{9-42b}$$

To find F_0, you need another equation so that you can determine the value of F_2. You can obtain the equation by considering the condition that the net torque about any origin acting on the lever must be zero. But before utilizing this condition, you must choose an origin. The best choice is to take the origin O at the fulcrum, as shown in the figure. The reason is that then the force F_0 can exert no torque about O, and hence F_0 will not appear in the equation that determines the value of F_2. With this choice, the torque about O exerted on the lever by the force F_2 is described by a vector which is directed outward. That is, this torque tends to rotate the lever in the counterclockwise sense, from the point of view illustrated in the figure. You can see that this is so by using the right-hand rules or, better, by using your intuition. Equation (9-31) shows that the magnitude of the torque equals the perpendicular distance r_2 from O to the line of action of the force, multiplied by its

Fig. 9-27 A lever.

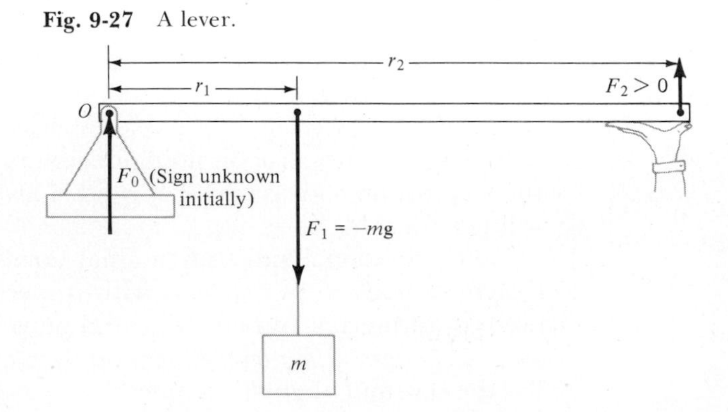

value F_2. Signed scalars can be used to specify torque vectors that all act either in one direction or in the opposite direction, just as signed scalars can be used for force vectors that are all directed in one way or in the opposite way. Choosing the direction outward as the direction of positive torque, the torque due to the force F_2 is written as $T_2 = r_2F_2$ (whose value is positive). Similarly, the torque due to the force F_1 is $T_1 = r_1F_1 = r_1(-mg) = -r_1mg$ (whose value is negative). Then the condition $T_1 + T_2 = 0$, that the net torque about O be zero, becomes

$$-r_1mg + r_2F_2 = 0 \tag{9-43a}$$

or

$$F_2 = \frac{r_1mg}{r_2} \tag{9-43b}$$

Using Eq. (9-43b) in Eq. (9-42b), you obtain

$$F_0 = mg - \frac{r_1}{r_2}mg$$

or

$$F_0 = mg\left(1 - \frac{r_1}{r_2}\right)$$

Since r_1 cannot be greater than r_2, the value of F_0 must always be positive. Thus the force exerted on the lever by the fulcrum must always be directed upward, as indicated in the figure. Explain to yourself the physical meaning of the value predicted for F_0 in the three cases $r_1 = 0$, $r_1 = r_2/2$, and $r_1 = r_2$.

How could you extend the calculation in Example 9-8 to take into account the fact that a lever is not actually massless? At first thought, this would seem to be a difficult thing to do because there is a gravitational force acting on every one of its particles. Each of these particles is at a different distance from the origin O, so each of the many forces produces a different torque about O. But the concept of *center of mass* makes such an extension easy, as we now show.

Consider the body of arbitrary shape and mass distribution in Fig. 9-28. A typical particle of the body has mass m_j and position $\mathbf{r}_j$, relative to some origin O. Let us evaluate the total torque $\mathbf{T}$ about O produced by all the gravitational forces which the earth exerts on the particles of the body. The gravitational force on a typical particle is

$$\mathbf{F}_j = m_j\,\mathbf{g}$$

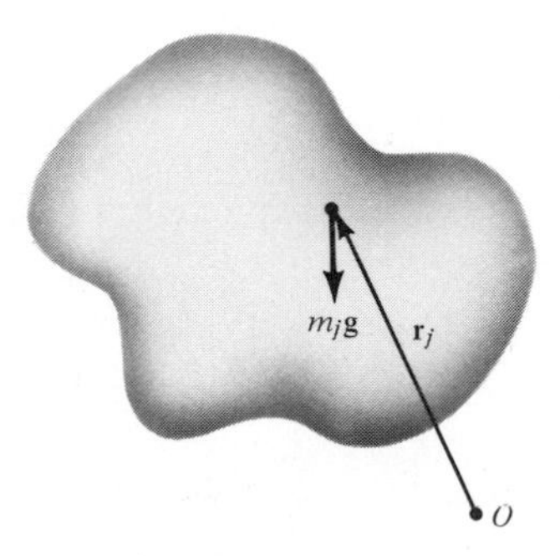

Fig. 9-28 The position $\mathbf{r}_j$, relative to an origin O, of a typical particle in an extended body. The gravitational force acting on that particle is $m_j\mathbf{g}$.

No label is needed on the gravitational acceleration vector **g.** This vector has the same magnitude and same direction over the entire body if, as we will assume to be the case, its size is small compared to the size of the earth. The torque about O produced by this gravitational force is

$$\mathbf{T}_j = \mathbf{r}_j \times \mathbf{F}_j = \mathbf{r}_j \times m_j\mathbf{g}$$

The total torque produced by all the gravitational forces acting on the n particles in the body is

$$\mathbf{T} = \sum_{j=1}^{n} \mathbf{T}_j = \sum_{j=1}^{n} r_j \times m_j\mathbf{g}$$

But since the m_j are scalars, their ordering in the vector product is of no consequence, and we can write

$$\mathbf{T} = \sum_{j=1}^{n} m_j\mathbf{r}_j \times \mathbf{g}$$

Using parentheses to indicate explicitly that the summation of terms $m_j\,\mathbf{r}_j$ can be performed first, and then the vector product of the result can be taken with **g,** we have

$$\mathbf{T} = \left(\sum_{j=1}^{n} m_j\mathbf{r}_j\right) \times \mathbf{g} \tag{9-44}$$

The purpose of this manipulation is to isolate the quantity in parentheses:

$$\sum_{j=1}^{n} m_j\mathbf{r}_j$$

It is a summation, over all the particles of the body, of the mass of each times its position vector. The value of this summation is not obvious, but it must have *some* value. Because the summation is taken over all the constituent masses of the body, the total mass M of the body should appear as a factor in the value. In other words, we should be able to write

$$\sum_{j=1}^{n} m_j\mathbf{r}_j = M\mathbf{r} \tag{9-45}$$

where **r** represents a vector whose value is such that when it is multiplied by M, the result has the same value as the summation. Deferring for the moment the question of just how we are going to determine **r,** we can use Eq. (9-45) to put Eq. (9-44) in the form

$$\mathbf{T} = M\mathbf{r} \times \mathbf{g}$$

or

$$\mathbf{T} = \mathbf{r} \times M\mathbf{g} \tag{9-46}$$

To interpret the meaning of this result, first we note that since the gravitational force acting on the jth particle of the body is

$$\mathbf{F}_j = m_j\mathbf{g}$$

the total gravitational force acting on the body is

$$\mathbf{F} = \sum_{j=1}^{n} \mathbf{F}_j = \sum_{j=1}^{n} m_j\mathbf{g} = \left(\sum_{j=1}^{n} m_j\right)\mathbf{g}$$

The summation in parentheses is M, the total mass of the body. So we have

$$\mathbf{F} = M\mathbf{g} \tag{9-47}$$

This equation makes the obviously correct statement that the net force produced by the gravitational forces acting on the particles of the body of total mass M is the same as the gravitational force acting on a particle of mass M. Next we use Eq. (9-47) to write Eq. (9-46) for the net torque about O acting on the body as $\mathbf{T} = \mathbf{r} \times \mathbf{F}$. We then conclude that the net torque $\mathbf{T}$ about O produced by the gravitational forces acting on the particles of the body of total mass M is the same as that which would be produced by the net gravitational force $\mathbf{F}$ if it were acting on a particle of mass M at position $\mathbf{r}$. In other words, Eqs. (9-46) and (9-47) show that *the effect of the gravitational forces acting on all the particles of the body is the same, with regard to both the net force and the net torque, as if the entire mass M of the body were concentrated at the position* $\mathbf{r}$. The vector $\mathbf{r}$ gives the location relative to O of a point in the body called its **center of mass.**

It remains, of course, to find $\mathbf{r}$. In general terms, this is a matter of solving Eq. (9-45) for $\mathbf{r}$. Doing so, we obtain the following expression for the vector $\mathbf{r}$ that gives the *location of the center of mass:*

$$\mathbf{r} = \frac{\sum_{j=1}^{n} m_j \mathbf{r}_j}{M} \tag{9-48a}$$

Here m_j is the mass of the jth particle of the body whose location is given by the vector $\mathbf{r}_j$. There are n particles in the body, and its total mass is M. An equivalent expression for the vector $\mathbf{r}$ is

$$\mathbf{r} = \frac{\sum_{j=1}^{n} m_j \mathbf{r}_j}{\sum_{j=1}^{n} m_j} \tag{9-48b}$$

The two expressions are equivalent because M is just the sum of the masses m_j. Before we discuss the use of Eqs. (9-48) to evaluate $\mathbf{r}$ in specific cases, it is worthwhile considering some general properties of the center of mass.

The location of a body's center of mass does not depend on the choice of the origin O, even though the vectors $\mathbf{r}_j$ and $\mathbf{r}$ change if the origin is changed. This statement is verified in Fig. 9-29 and the accompanying cap-

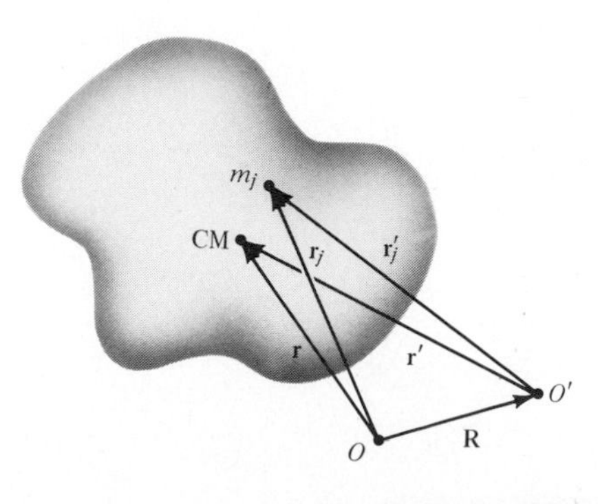

Fig. 9-29 If the origin O used in the definition of a body's center of mass (CM) is moved to O', the position vector $\mathbf{r}_j$ of a typical particle in the body becomes $\mathbf{r}_j'$. The relation between the two vectors is $\mathbf{r}_j = \mathbf{r}_j' + \mathbf{R}$, where $\mathbf{R}$ is the vector giving the position of O' relative to O. Using this relation in Eq. (9-48a) yields $\mathbf{r} = \sum_{j=1}^{n} m_j(\mathbf{r}_j' + \mathbf{R})/M = \sum_{j=1}^{n} m_j\mathbf{r}_j'/M + \mathbf{R}\sum_{j=1}^{n} m_j/M$. This immediately reduces to the result $\mathbf{r} = \mathbf{r}' + \mathbf{R}$, where $\mathbf{r}'$ gives the location of the body's CM relative to the origin O'. Inspection of the figure will show that this result is consistent with the fact that the CM is fixed in the body and does not depend on the location of the origin used in its definition.

tion. Rather, the center of mass of a body is at a position fixed relative to the body (though not necessarily within the body). The position does depend on the distribution of mass in the body. Indeed, the vector $\mathbf{r}$ specifies the *average* location of that mass. If each particle of the body has the same mass, that is, if all the m_j have a common value m, then Eq. (9-48b) simplifies to

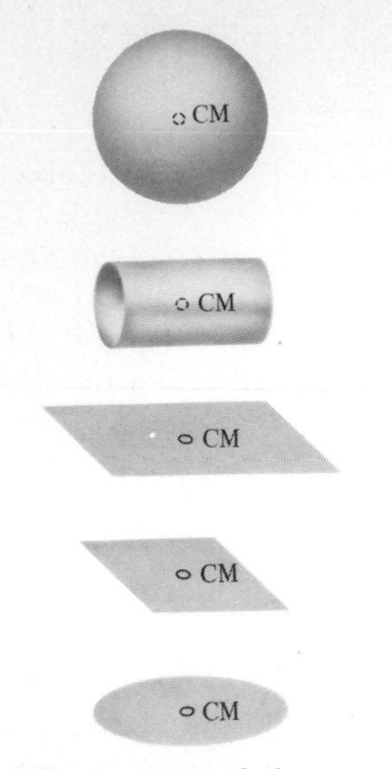

Fig. 9-30 Location of the center of mass of a sphere, a cylinder, a rectangular plate, a square plate, and a circular plate. In each case the body is made of material of uniform density.

$$\mathbf{r} = \frac{m \sum_{j=1}^{n} \mathbf{r}_j}{nm} = \frac{\sum_{j=1}^{n} \mathbf{r}_j}{n}$$

In this case, the computation of $\mathbf{r}$ is nothing more than a matter of computing the simple arithmetical average, over all the n particles of the body, of the position vectors $\mathbf{r}_j$ to each particle. You add all the $\mathbf{r}_j$ and then divide by the total number of terms in the sum, which is n. In a general case where the particle masses m_j are not all the same, in computing the average you "weight" the position $\mathbf{r}_j$ of each particle by multiplying it by the mass m_j of the particle. In fact, Eq. (9-48b) is a direct application of the general mathematical rule for calculating a weighted average, or mean. Figure 9-30 illustrates the location of the center of mass of some simple bodies. Note that if the mass distribution of a body is symmetrical about a point, then its center of mass will lie at that point. If the mass distribution is symmetrical about a line, then the center of mass will lie somewhere on the line. If it is symmetrical about a plane, the center of mass will lie somewhere in the plane. If there is more than one line or plane of symmetry, the center of mass must be where they intersect.

The center of mass of a body is often called its **center of gravity,** for reasons arising from the interpretation given to Eq. (9-46). The point at which it is located is sometimes called the **balance point.** The terminology is appropriate because when an object is balanced about a point O, its center of mass lies somewhere on a vertical line passing through O. The reason is that the body is balanced if there is no total gravitational torque about O. Equation (9-46) shows that there will be no such torque when the vector $\mathbf{r}$ from O to the center of mass is antiparallel, or parallel, to the downward-directed vector $\mathbf{g}$ because the center of mass is immediately

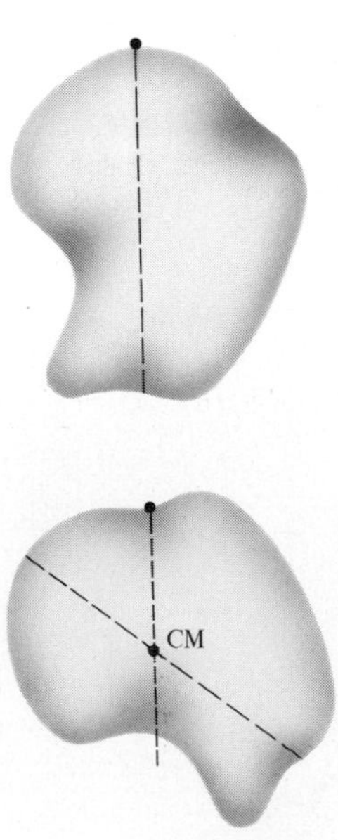

Fig. 9-31 The center of mass of an irregular body can be located experimentally by suspending it from a point on its rim, recording on the body the direction of a line extending vertically downward from the suspension point, and then repeating the procedure. The center of mass is at the intersection of the two lines. Why?

above, or below, O. This fact provides the basis of an experimental method, indicated in Fig. 9-31, for locating the center of mass of a body.

The theoretical method of finding the center of mass is to use Eq. (9-48a). If the body is not composed of a relatively small number of discrete particles, this is most easily done by replacing the sum by an integral. That is, an alternative expression for the position vector $\mathbf{r}$ of the center of mass is

$$\mathbf{r} = \frac{\int_{\text{body}} \mathbf{r}'\, dm}{M} \tag{9-49a}$$

Here dm is an infinitesimal element of mass of the body whose total mass is M, $\mathbf{r}'$ is the position vector of that mass element, and the integral is taken over the entire body. This definition can also be written as

$$\mathbf{r} = \frac{\int_{\text{body}} \mathbf{r}'(dm/dV)\, dV}{M} = \frac{\int_{\text{body}} \mathbf{r}'\rho\, dV}{M} \tag{9-49b}$$

where $\rho \equiv dm/dV$ is the density of the body and dV is an element of its volume. The integral over the volume of the body of the vector quantity $\mathbf{r}'\rho$ is evaluated by writing $\mathbf{r}'$ in terms of its components. For instance, in rectangular coordinates doing so gives $\mathbf{r}'\rho = x'\rho\,\hat{\mathbf{x}} + y'\rho\,\hat{\mathbf{y}} + z'\rho\,\hat{\mathbf{z}}$. Using this in Eq. (9-49b) leads to expressions for the components of $\mathbf{r}$, each of which involves an integral of a scalar quantity.

In practice, the center of mass of a body often can be obtained quite simply by invoking whatever symmetries it has. For instance, the symmetry of a uniform, straight rod makes it apparent that its center of mass lies at its midpoint. Therefore you can modify Example 9-8 to take account of the mass M of the lever by adding $-Mg$ to the left side of Eq. (9-42a) and by adding $-(r_2/2)Mg$ to the left side of Eq. (9-43a). Something very similar is done in Example 9-9.

To summarize, investigating the conditions in which bodies will remain stationary in an inertial reference frame involves applying the two conditions

$$\mathbf{F} = 0 \tag{9-50a}$$

and

$$\mathbf{T} = 0 \tag{9-50b}$$

The first states that there must be *zero net force* acting on a stationary body; the second states that that there must be *zero net torque* acting on the body, about any origin. The net contribution from a set of gravitational forces acting on an extended member of the body is evaluated by treating those forces as if they were all applied at the center of mass of the member. The choice of the origin about which the torques are computed is dictated by convenience, since the second condition must be satisfied for any origin. This is emphasized by Example 9-9.

EXAMPLE 9-9

A ladder of uniform construction, length 5.00 m, and mass 20.0 kg is supported by a rough floor. It is found that it remains leaning against a smooth wall if the distance from the wall to the bottom of the ladder is no greater than 3.00 m. If this distance is exceeded, the ladder slips. Determine the force acting between the wall and the top of the ladder when its bottom is at the critical location.

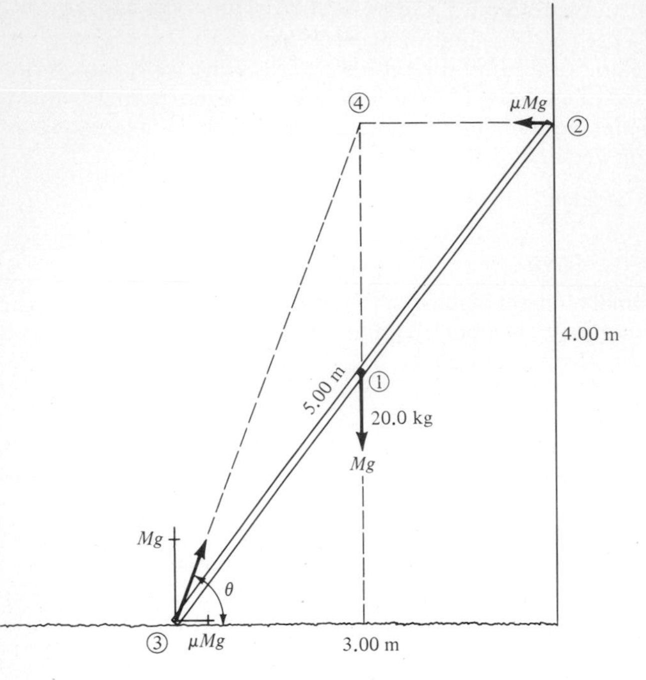

Fig. 9-32 The forces acting on a ladder resting on a rough floor and leaning against a perfectly smooth wall.

▪ The drawing in Fig. 9-32 shows the ladder when it is just at the point of slipping and the forces which then act on it. The gravitational forces exerted on it have been replaced by a single force of magnitude Mg, where M is its total mass. This force acts downward at the midpoint of the uniform ladder, which is the location of its center of mass. The force applied by the floor to the bottom of the ladder is not directed perpendicular to the floor, since the floor is not smooth. The figure shows the vertical and horizontal components of this force, taking the upward direction to be positive and the direction to the right to be positive. Since Eq. (9-50*a*) requires that the net force acting on the ladder have no vertical component, you find immediately that the vertical component is the positive quantity Mg, as shown in the figure. The figure also shows that the horizontal component of the force exerted by the floor is the positive quantity μMg, where μ is the coefficient of static friction between the floor and the bottom of the ladder. The sign is found to be positive by considering that this component is the frictional force which acts to the right in resisting the incipient leftward motion of the bottom of the ladder. The magnitude is found by using Eq. (4-22*b*), which relates the static friction force that the floor exerts on the ladder to the normal force exerted by the floor on the ladder. Finally, the force applied by the wall on the top of the ladder is drawn perpendicular to the wall. This is because the wall is frictionless and hence cannot apply a force parallel to its surface. Using Eq. (9-50*a*) again, you can see that this horizontal force applied by the vertical wall must act to the left and have magnitude μMg, as shown. The reason is that the net force exerted on the ladder can have no horizontal component, according to Eq. (9-50*a*).

Applying Eq. (9-50*b*) requires choosing a location for the origin O. Four possible choices are shown in the figure. Perhaps the most obvious choice is location 1, at the center of mass of the ladder. But with this choice there would be torques produced by two forces, the ones acting on the ends of the ladder. The same objection holds for location 2, at the top of the ladder. Location 3, at the bottom of the ladder, will simplify the calculation somewhat because the more complicated force acting at the bottom of the ladder exerts no torque about an origin at that location.

The least obvious, but most advantageous, choice of origin is location 4, at the intersection of the lines of action of the forces exerted on the top of the ladder and on its center of mass. Neither of these forces produces a torque about that origin.

Equation (9-50b) therefore requires that the same be true of the force acting at the bottom of the ladder, in order that there be zero net torque about the origin. Thus the line of action of the force on the bottom of the ladder must also pass through location 4.

The figure shows that this means the line of action is inclined to the horizontal at an angle θ such that $\cot\theta = 1.50 \text{ m}/4.00 \text{ m} = 0.375$. Furthermore, it shows that $\cot\theta = \mu Mg/Mg = \mu$. So you find that the coefficient of static friction is $\mu = 0.375$. Then you can immediately evaluate the force acting between the wall and the top of the ladder; it has the value $\mu Mg = 0.375 \times 20.0 \text{ kg} \times 9.80 \text{ m/s}^2 = 73.5 \text{ N}$.

Cases can arise in which it is impossible to treat the equilibrium of a *rigid* structure because there are more unknown force components than there are scalar equations involving these components. The structure is then said to be **underdetermined.** This would be so in Example 9-9 if both the floor and the wall were rough, with both coefficients of friction being unknown. For a less contrived case, consider the forces which the floor exerts on the four legs of a chair. One equation relating the four unknown forces, and the gravitational force acting at its center of mass, is obtained from the vertical component of $\mathbf{F} = 0$. Relative to any particular origin, two more equations relating these forces are obtained from the two horizontal components of $\mathbf{T} = 0$. The structure is underdetermined because there are four unknowns and only three independent equations. In contrast, a three-legged stool is not underdetermined.

The difficulty is removed by taking into account the fact that no real structure is completely rigid. When this is done, additional equations are found that involve the elastic properties of the members of the deformable structure. They make possible a complete analysis of the conditions for its equilibrium. But we leave these matters to specialized books on structural engineering.

We turn instead to a treatment of the stability of equilibrium. This treatment extends the one given in Chap. 6 by using energy relations developed in Chap. 7 and the concepts of center of mass and torque introduced in this chapter.

9-7 STABILITY OF EQUILIBRIUM

Every structure is continually subjected to influences which displace it slightly from its equilibrium position. Gusts of wind and vibrations in the ground have such an effect. Since the displaced position is not an equilibrium position, there will be a net force and/or net torque acting on the structure in that position. If for all positions near the equilibrium position they are directed so as to tend to return the structure to the equilibrium position, then it is a position of **stable equilibrium.**

In Sec. 6-1 we studied the equilibrium of a particle in a system in cases where the particle's position can be specified completely by one coordinate, and it is not necessary to consider torque. If we write the coordinate as x and the net force acting on the particle as $F(x)$, the conclusions reached there can be summarized as follows: (1) $F(x) = 0$ at $x = x_e$ if x_e is an equilibrium position. (2) In addition, $dF(x)/dx < 0$ at $x = x_e$ if x_e is a position of stable equilibrium. In Sec. 7-6 we learned that if $F(x)$ is a conservative force arising from an interaction between the particle and some other body in the system, then the potential energy $U(x)$ of the system is related to

$F(x)$ by the equation $F(x) = -dU(x)/dx$. Using this, we can write the condition for equilibrium at $x = x_e$ in the form

$$-\frac{dU(x)}{dx} = 0 \qquad \text{at } x = x_e$$

This is equivalent to

$$\frac{dU(x)}{dx} = 0 \qquad \text{at } x = x_e \quad \text{(equilibrium condition)} \tag{9-51}$$

And we can write the condition that the equilibrium at $x = x_e$ is stable in the form

$$\frac{d}{dx}\left[-\frac{dU(x)}{dx}\right] < 0 \qquad \text{at } x = x_e$$

or

$$-\frac{d^2U(x)}{dx^2} < 0 \qquad \text{at } x = x_e$$

If the negative of a quantity is less than zero, the quantity itself is greater than zero. Thus the condition for the equilibrium to be stable can be expressed as

$$\frac{d^2U(x)}{dx^2} > 0 \qquad \text{at } x = x_e \quad \text{(stability condition)} \tag{9-52}$$

Figure 9-33*a* illustrates qualitatively the behavior of the potential energy $U(x)$ near a position of stable equilibrium at x_e. The slope of the $U(x)$ curve [in other words, the first derivative of $U(x)$] is zero at x_e. This satisfies the mathematical requirement of Eq. (9-51). Physically, it means that the net force acting on the particle, $F(x) = -dU(x)/dx$, is zero at x_e, so that x_e is, indeed, an equilibrium position. The rate of change of slope of the $U(x)$ curve [in other words, the second derivative of $U(x)$] is positive at x_e. Thus the requirement of Eq. (9-52) is also satisfied. You can see the physical significance of this by using the relation $F(x) = -dU(x)/dx$ to determine from the negative of the slope of the $U(x)$ curve the direction of the net force that the particle feels when some influence has given the particle a small displacement from x_e. You will find that, whether the particle has been displaced to one side of x_e or to the other, the net force is always directed toward x_e. Hence x_e is, in fact, a position of stable equilibrium. The effect of Eq. (9-51) is to require the $U(x)$ curve to have zero slope at x_e. The effect of Eq. (9-52) is to require the curve to be concave upward at x_e. It is apparent from the figure that their combined effect is to require the curve to describe a potential energy $U(x)$ which has a *minimum value* at the position of *stable equilibrium* x_e.

Figure 9-33*b* sketches the potential energy $U(x)$ for a case where it has a maximum value at x_e. In this case x_e is still an equilibrium position since the relation $F(x) = -dU(x)/dx$ shows that the net force $F(x)$ acting on the particle is still zero at x_e. But if you apply this relation to determine the direction of the force that the particle feels when some influence has given it a small displacement from x_e, you will find that the force tends to push it even farther from that position. Thus x_e is a position of **unstable equilibrium** if $U(x)$ has a *maximum value* there.

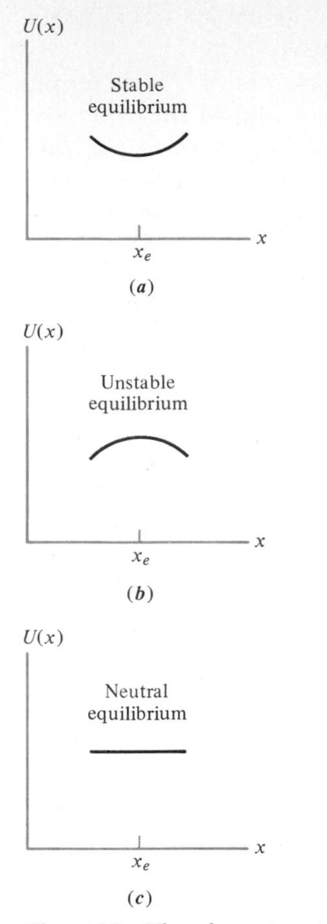

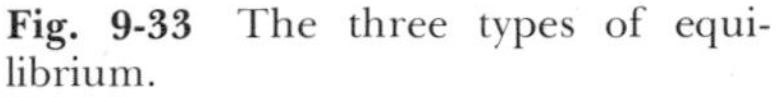

Fig. 9-33 The three types of equilibrium.

Figure 9-33c shows the *constant value* of the potential energy $U(x)$ near a position x_e of **neutral equilibrium.** The name is appropriate because if the particle is given a displacement from that position to any nearby position, there is no net force acting on it at the displaced position. This is because $F(x) = -dU(x)/dx = 0$ everywhere near x_e. As a consequence, there is no tendency for the displaced particle then to move either closer to or farther from x_e.

Although we have been considering a potential energy which is a function of the single coordinate x of a particle, our conclusions are perfectly general. Whether the movable object is a particle or an extended body, and no matter what coordinates are used to specify its position, *the potential energy U of a system has a minimum at a position of stable equilibrium, has a maximum at a position of unstable equilibrium, and is constant at a position of neutral equilibrium.* This statement applies to any type of potential energy. That is, it makes no difference what type of force gives rise to the potential energy—except that the force must be conservative if there is to be a potential energy.

A particularly important case is the one in which U represents the gravitational potential energy of a system containing the earth and a movable body near the earth's surface. To obtain an expression for U, we consider the body to be comprised of particles and calculate the gravitational potential energy associated with each of them. We then sum over the n particles in the body. With the subscript j employed to designate a typical particle, Eq. (7-53) states that the potential energy resulting from the gravitational force exerted on it by the earth is

$$U_j = m_j g y_j$$

Here m_j is the particle's mass, y_j is its coordinate measured vertically upward from the reference height chosen to be at $y = 0$, and g is the magnitude of the gravitational acceleration. The gravitational potential energy of the system is

$$U = \sum_{j=1}^{n} U_j = \sum_{j=1}^{n} m_j g y_j$$

Since g is a constant, this is

$$U = g \sum_{j=1}^{n} m_j y_j \tag{9-53}$$

The sum can be evaluated by writing Eq. (9-47*a*) as

$$\sum_{j=1}^{n} m_j \mathbf{r}_j = M\mathbf{r}$$

In this equation M represents the total mass of the body, and $\mathbf{r}$ the position vector of its center of mass. Considering the y component of both sides of the vector equation gives us the scalar equation

$$\sum_{j=1}^{n} m_j y_j = My$$

where y is the y coordinate of the body's center of mass. Using this relation in Eq. (9-53), we obtain

$$U = Mgy \tag{9-54}$$

Our result tells us that *the gravitational potential energy is the same as if the entire mass of the body were concentrated at its center of mass.*

Now look again at Fig. 9-33, considering the case where U is the gravitational potential energy given by Eq. (9-54). According to the equation, the vertical scale in each part of the figure is a measure of the *height of the body's center of mass.* Restate the conditions for stable equilibrium, unstable equilibrium, and neutral equilibrium in these terms. Do the restated conditions agree with your intuition?

Example 9-10 uses Eqs. (9-51), (9-52), and (9-54) to investigate the equilibrium of a rigid body acted on by gravity.

EXAMPLE 9-10

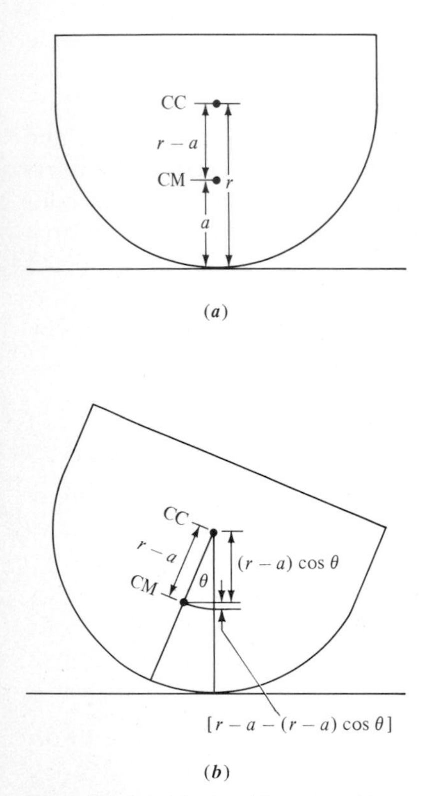

Fig. 9-34 (*a*) A body with a curved bottom supported in an upright position by a rough plane. (*b*) The body after it has rolled through a small angle θ.

The rigid body shown in Fig. 9-34*a* is supported by a rough horizontal plane. The density of the body is nonuniform, the mass being distributed in such a way that its center of mass (CM) is at a distance a above the point of contact between the curved surface and the plane, with a being less than the distance r from the point of contact to the center of curvature (CC) of the surface. Is the body in equilibrium? Is the equilibrium stable?

■ To answer these questions by using Eqs. (9-51) and (9-52), first you must obtain a mathematical expression for the potential energy of the body plus-earth system, when the body is in some position near the one shown in Fig. 9-33*a*. It gets to the new position, illustrated in Fig. 9-34*b*, by rolling over the rough plane from its original position. In the rolling process, the force which the plane exerts on the body is always applied to a point in the body that is at rest at the instant it is in contact with the plane. Therefore the force exerted by the plane does no work since there is no displacement of the point to which the force is applied. In other words, *rolling is a workless constraint.* This means the potential energy of the system arises entirely from the work done by the gravitational force exerted on the body by the earth. So you will be able to evaluate the potential energy by using Eq. (9-54) as soon as you have an expression for the height of the CM of the body.

To this end, you specify the position of the body in terms of the angle θ shown in Fig. 9-34*b*. It is the angle between the intersection at the CC of two lines. One is the radial line from the CC through the CM to the point that had been in contact with the plane. The other is the radial line from the CC to the point that now is in contact with the plane. Choose the reference height used to define the gravitational potential energy of the system as the height of the CM when the body is in the position shown in Fig. 9-34*a*, and use this height to locate the origin of a vertically directed y axis. Then inspection of the figure will show you that the y coordinate of the CM when the body has rolled to the position shown in Fig. 9-34*b* is

$$y = (r - a) - (r - a)\cos\theta$$

or

$$y = (r - a)(1 - \cos\theta)$$

Now use Eq. (9-54) to express the gravitational potential energy of the system as

$$U(\theta) = Mgy$$

or

$$U(\theta) = Mg(r - a)(1 - \cos\theta) \qquad (9\text{-}55)$$

where M is the mass of the body and g is the magnitude of the gravitational acceleration.

You can find an equilibrium position of the body by applying Eq. (9-51), if you write it in terms of the coordinate θ instead of the coordinate x. Doing so, you have

$$\frac{dU(\theta)}{d\theta} = 0 \qquad \text{for } \theta = \theta_e$$

where θ_e specifies an equilibrium position. Differentiating both sides of Eq. (9-55) with respect to θ, you obtain

$$\frac{dU(\theta)}{d\theta} = Mg(r - a) \sin \theta \tag{9-56}$$

Setting this equal to zero for $\theta = \theta_e$ gives you

$$Mg(r - a) \sin \theta_e = 0$$

Since $Mg(r - a) \neq 0$, it must be that $\sin \theta_e = 0$. Thus you find that

$$\theta_e = 0$$

describes an equilibrium position of the body. The body is, therefore, in *equilibrium* in the position shown in Fig. 9-34*a*.

To determine the stability of the equilibrium position at $\theta_e = 0$, you differentiate both sides of Eq. (9-56) with respect to θ, producing

$$\frac{d^2U(\theta)}{d\theta^2} = Mg(r - a) \cos \theta$$

Setting $\theta = \theta_e = 0$, you have

$$\frac{d^2U(\theta)}{d\theta^2} = Mg(r - a) \qquad \text{for } \theta = \theta_e = 0 \tag{9-57}$$

But $r > a$, so $Mg(r - a) > 0$, and therefore

$$\frac{d^2U(\theta)}{d\theta^2} > 0 \qquad \text{for } \theta = \theta_e = 0$$

Since Eq. (9-52) is satisfied, $\theta_e = 0$ describes a stable equilibrium position. The position shown in Fig. 9-34*a* is a position of *stable* equilibrium.

If the mass in the body is distributed so that in Fig. 9-34*a* the CM lies *above* the CC, you can show that Eqs. (9-55) and (9-57) still apply with $r < a$. In such a case, $d^2U(\theta)/d\theta^2 < 0$ for $\theta = \theta_e = 0$. This means that $\theta_e = 0$ is a position of *unstable* equilibrium. If the CM and the CC are *coincident*, then $r = a$, $d^2U(\theta)/d\theta^2 = 0$ for $\theta = \theta_e = 0$, and $\theta_e = 0$ is a position of *neutral* equilibrium. To verify these statements, and to see that they agree with your intuition, use Eq. (9-55) to sketch U versus θ for the three cases $r > a$, $r < a$, and $r = a$. Then compare your sketches with those for U versus x for the three cases in Fig. 9-33.

You can calculate the net torque about some origin acting on the body from the potential energy $U(\theta)$ of the system by writing the torque as the signed scalar $T(\theta)$ and employing the relation

$$T(\theta) = -\frac{dU(\theta)}{d\theta} \tag{9-58}$$

This relation is obtained by analogy to the one in Eq. (7-56), $F(x) = -dU(x)/dx$. Find $T(\theta)$, and then use it to explain to yourself the equilibrium properties of the body in the cases $r > a$, $r < a$, and $r = a$, in terms of the net torque acting on the body. To check the validity of Eq. (9-58), find an expression for $T(\theta)$ from a direct consideration of the two forces acting on the body and the geometry of Fig. 9-34*b*, taking account of the fact that both forces have the same magnitude Mg. Note that you get the same expression for $T(\theta)$, no matter where you choose the origin O. Can you explain why?

EXERCISES

Group A

9-1. *How much turning?* Evaluate the angle through which the wheel in Example 9-1 has turned by applying Eq. (9-10), and compare your results with those obtained in the example.

9-2. *A change of sign.* Compute the direction and magnitude of the total acceleration of the point on the wheel in Example 9-2, using $\alpha = 1.53\ \text{rad/s}^2$. Then repeat the calculation using $\alpha = -1.53\ \text{rad/s}^2$. Compare the two results.

9-3. *Plane facts.* If the rotation $d\phi$ in Fig. 9-3 is to be represented by a vector, you might be tempted to use a vector in the plane of the flywheel. Give two reasons why such a procedure would not give a determinate vector.

9-4. *Angular acceleration and tangential acceleration.* Starting from rest, a flywheel of 30-cm radius makes 6.0 rotations in 3.0 s. Assume that the angular acceleration is constant during this interval.

a. Evaluate the magnitude of the angular acceleration.

b. What is the magnitude of the tangential acceleration of a point on the rim of the flywheel?

9-5. *Grinding to a halt.* A motorized grindstone is turning at its full speed of 1800 rotations per minute when a worker switches off the motor. The grindstone comes to rest exactly 2 minutes later.

a. What is the average angular deceleration?

b. Assuming that the angular deceleration is constant, how many rotations does the grindstone make during its deceleration?

9-6. *Speed and acceleration in steady spinning.* A grindstone is turning at 1800 rotations per minute. Its diameter is 20 cm.

a. What is the speed of any point on its rim?

b. What is the acceleration of this point?

9-7. *Angular velocities.* A ball of radius r rolls along a loop-the-loop track of radius R. As the ball passes through the bottom of the loop, the magnitude of the velocity of the center of the ball is v_0.

a. What is the instantaneous angular velocity of the ball?

b. What is the instantaneous angular velocity of the line joining the center of the ball to the center of the loop?

c. Evaluate your results for $r = 1.0$ cm, $R = 5.0$ cm, and $v_0 = 20.0$ cm/s.

9-8. *Geometry of the cross product.* Two vectors **A** and **B** determine a parallelogram. Show that $|\mathbf{A} \times \mathbf{B}|$ is equal to the area of the parallelogram. Hence the area can be represented by a single vector $\mathbf{A} \times \mathbf{B}$. What is the spatial relationship between this vector and the plane of the parallelogram?

9-9. *What if the earth lost its caps?* The ice in the Greenland and Antarctic caps is almost entirely above the present sea level. Suppose the earth's temperature rose sufficiently to melt this ice. Neglecting any adjustments of the solid earth, how would the length of earth's day be affected? Explain your answer.

9-10. *Encounter with a merry-go-round.* A playground merry-go-round is shaped like a bicycle wheel, with a circular bench supported by rods arranged like spokes about the vertical central axle. Its radius is 1.0 m, and its mass of 50 kg is concentrated in the circumference. The merry-go-round is stationary. A 30-kg child runs along a line tangent to the circumference with a speed of 5.0 m/s and jumps onto the merry-go-round.

a. Neglecting friction, what angular velocity does the merry-go-round acquire?

b. The child jumps from the moving merry-go-round, being careful to push only along the radial direction. What is the final angular velocity of the merry-go-round?

9-11. *A student's turn.* A student volunteer is sitting stationary on a piano stool with her feet off the floor. The stool can turn freely on its axle.

a. The volunteer is handed a nonrotating bicycle wheel which has handles on the axle. Holding the axle vertically with one hand, she grasps the rim of the wheel with the other and spins the wheel clockwise (as seen from above). What happens to the volunteer as she does this?

b. She now grasps the ends of the vertical axle and turns the wheel until the axle is horizontal. What happens?

c. Next she gives the rotating wheel to the instructor, who turns the axle until it is vertical with the wheel rotating clockwise, as seen from above. The instructor now hands the wheel back to the volunteer. What happens?

d. The volunteer grasps the ends of the axle and turns the axle until it is horizontal. What happens now?

e. She continues turning the axle until it is vertical but with the wheel rotating counterclockwise as viewed from above. What is the result?

9-12. *A force couple.* A pair of equal, oppositely directed, but noncollinear forces is called a **couple.** Consider the couple shown in Fig. 9E-12.

a. What torque does the couple exert about point A?

b. What torque does the couple exert about point B?

c. If a force couple is applied to an initially stationary and nonrotating object, describe the resultant motion.

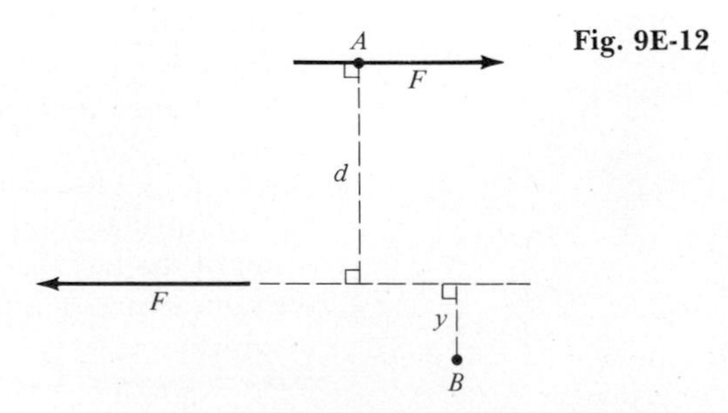

Fig. 9E-12

9-13. *Between earth and moon.* The mass of the earth is 81 times the mass of the moon. The distance between their centers is 384,000 km. Where is the center of mass of the earth-moon system in relation to the surface of the earth? Take the radius of the earth as 6400 km.

9-14. *Center of mass in a three-particle system, I.* Three particles of equal masses are placed so that they form a 3, 4, 5 right triangle as shown in Fig. 9E-14. Locate their center of mass using a cartesian system (x, y), with $\hat{\mathbf{x}}$ and $\hat{\mathbf{y}}$ as shown and with the origin located at the vertex of the right angle.

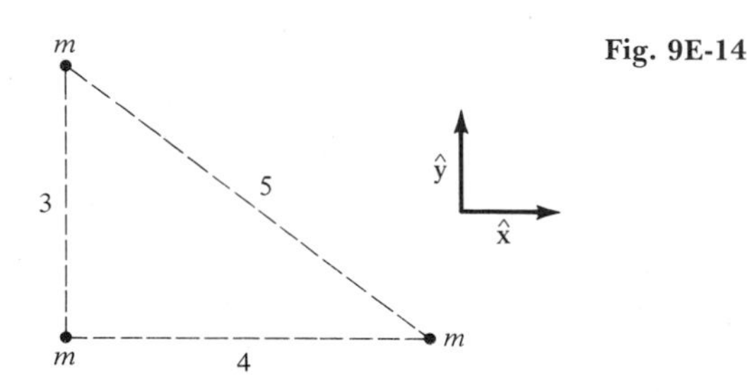

Fig. 9E-14

9-15. *Center of mass in a four-particle system.* Locate the center of mass for the system shown in Fig. 9E-15, in which the four particles form a square of side l. Use a cartesian system (x, y) whose origin coincides with the 40-g particle and whose orientation is as indicated.

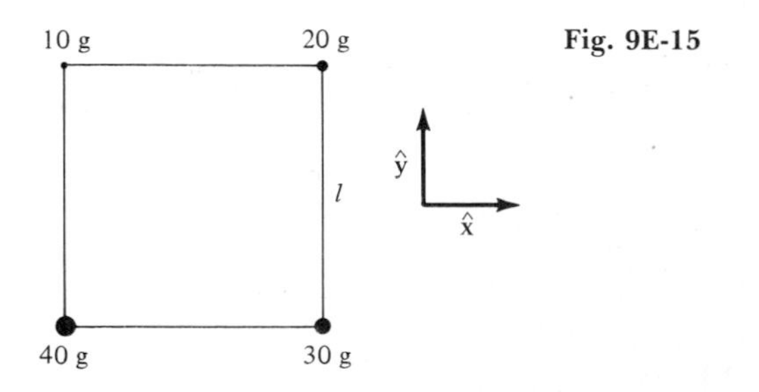

Fig. 9E-15

9-16. *How much work?* A plank of mass 25 kg and length 2.0 m lies flat on the floor with one end against a wall. A worker takes hold of the other end and raises the plank to the vertical position by using the end against the wall as a pivot. How much work does the worker do?

9-17. *When the mass of a lever cannot be neglected.* Modify the calculation in Example 9-8 to take into account the mass M of the lever.

9-18. *A delicate balance.* The beam of the balance in Fig. 9E-18 has mass of 25.0 g. The center of mass of the beam is at G, 4.00 mm below the point of support, O.

a. If a load of exactly 1 mg is placed in one pan, through what angle θ will the beam be deflected? Express your result in radians.

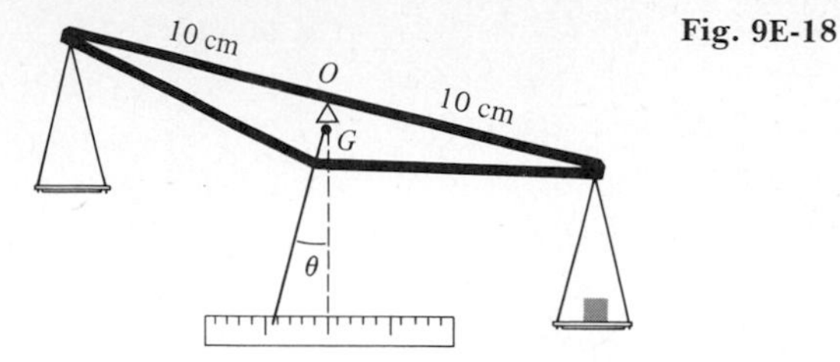

Fig. 9E-18

b. If the pointer (of negligible mass) attached to the beam is 20.0 cm long, through what distance in mm will 1 mg deflect the free end of the pointer? This is called the *sensitivity* of the balance (in millimeters per milligram).

9-19. *Surprisingly stable.* A half-dollar is partly embedded in a large cork into which two forks have been stuck, as shown in Fig. 9E-19. If the edge of the coin is placed on a needle, the system is stable. It can oscillate on the point of the needle without falling over. Account for the stability of this system.

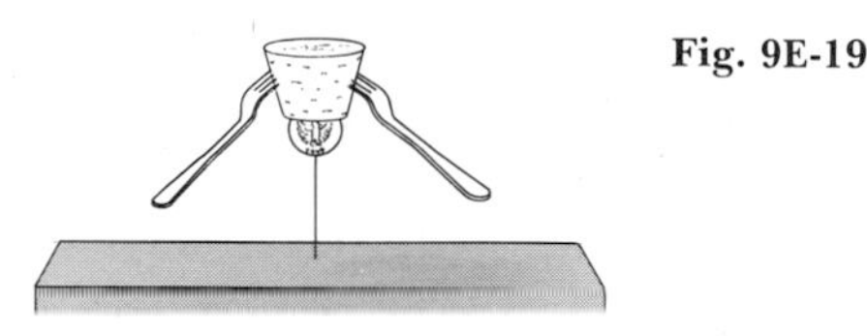

Fig. 9E-19

Group B

9-20. *Angular kinematics.*

a. Derive Eq. (9-10), $\phi = \phi_i + (\omega^2 - \omega_i^2)/2\alpha$ for constant α, directly from Eqs. (9-7) through (9-9).

b. Derive Eq. (9-11), $\phi = \phi_i + (\omega + \omega_i)t/2$ for constant α, directly from Eqs. (9-7) through (9-9).

9-21. *Running down.* Modify Eq. (9-23), $\mathbf{a} = \alpha r\hat{\mathbf{v}} - \omega^2 r\hat{\mathbf{r}}$, so that it describes the total acceleration of a point on the rim of a flywheel whose angular speed is decreasing. Remember that $\alpha \equiv |\boldsymbol{\alpha}| \equiv |d\boldsymbol{\omega}/dt|$.

9-22. *Central force and angular momentum conservation.* A particle is acted on by a repulsive central force. Write a vector expression for the force, and then use it to prove that the particle maintains a constant angular momentum in an inertial frame about an origin located at the force center.

9-23. *Alternative derivation of the rotational form of Newton's third law.* Modify the derivation of Sec. 4-5 to prove that the law of conservation of angular momentum and the rotational form of Newton's second law lead to the rotational form of Newton's third law.

9-24. *Which way will it go?* A spool of thread rests on a level tabletop, as shown in Fig. 9E-24. The thread is pulled gently, so that there is no slippage at P, the point of contact between the spool and the tabletop. For each of the

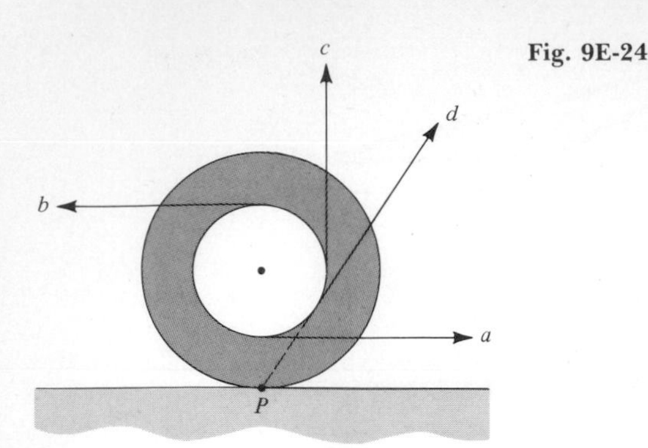

Fig. 9E-24

thread positions a through d, determine which way the spool will roll. Explain your answers. Notice that in position d the line determined by the thread passes through point P.

9-25. *Eight lives to go.* A uniform board 6.0 m long overhangs the edge of a table by 2.0 m. The mass of the board is 12.0 kg. An 8.0-kg cat walks out along the board. How far from the edge of the table will the cat be when the board begins to tip?

9-26. *Center of mass in a three-particle system, II.* As indicated in Fig. 9E-26, the position vectors of three particles of equal mass m are given by **a**, **b**, and **c**.

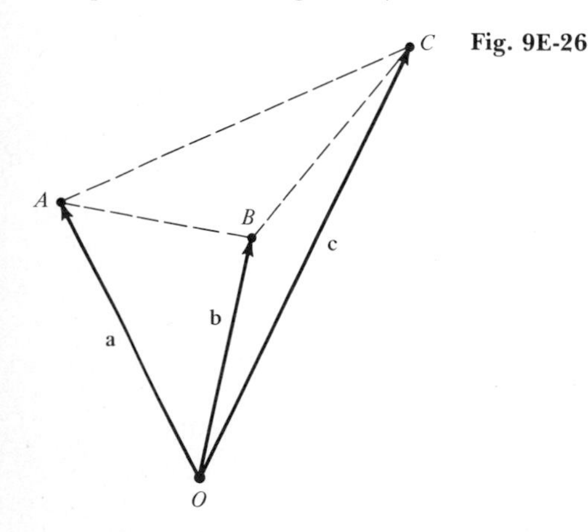

Fig. 9E-26

a. Find the position vector **r** which locates the center of mass of the system.

b. Show that the position specified by **r** lies precisely at the intersection of the medians of the triangle ABC of Fig. 9E-26.

9-27. *Center of mass of a homogeneous plane triangle.* Prove that the center of mass of a homogeneous sheet-metal triangle lies at the intersection of the medians. (This point is called the **barycenter.**)

9-28. *Center of mass of a homogeneous hemisphere.* A hemispherical object of uniform density has radius R. Prove by integration that its center of mass lies along its axis of symmetry at a distance $3R/8$ from the center of the plane surface.

9-29. *Center of mass of a homogeneous tetrahedron.* Prove that the center of mass of any tetrahedron of uniform density lies at the common intersection of four lines, each drawn from a vertex to the intersection of the medians of the opposite base. (It may be helpful to refer to Exercise 9-27.)

9-30. *Take the upper end!* A heavy trunk is uniformly filled so that its center of mass is at its center of volume. Its length is twice its height. As shown in Fig. 9E-30, it is being carried down a flight of stairs by two movers, A and B, with A in front. The trunk is at an angle of 45° with the horizontal, and A and B apply strictly vertical forces at opposite ends of the bottom. What fraction of the weight does A carry?

Fig. 9E-30

9-31. *No tipping, please.* A uniformly loaded crate of height h and width w rests on the floor. The coefficient of kinetic friction between crate and floor is μ_k.

a. Suppose the crate is pulled slowly along the floor by a horizontal force F applied at the level of the center of mass. How far from the center of the base does the resultant normal reaction force of the floor act?

b. What is the maximum height above the floor at which the horizontal force F can be applied without tipping the crate?

c. Evaluate your results of parts a and b for $h = 1.00$ m, $w = 0.50$ m, and $\mu_k = 0.33$.

9-32. *The tasks of a hinge.* A uniform door of height h, width b, and mass m is hung from two hinges, which are at a distance d from the upper and lower ends of the door, respectively.

a. What is the magnitude of the horizontal component of the force exerted on the door (1) by the upper hinge; (2) by the lower hinge?

b. What is the total magnitude of the vertical component of the force carried by both hinges? (*Note:* The separate vertical components of the force carried by each hinge cannot be determined from the information given.)

9-33. *Finding the center of mass of a rod.* A horizontal rod or beam of mass M (not necessarily uniform) is supported by a person using one finger from each hand. The points of support lie on opposite sides of the center of mass at distances d_1 and d_2 from it.

a. Show that the magnitudes of the normal forces N_1 and N_2 exerted by the fingers are given by $N_1 = Mgd_2/(d_1 + d_2)$ and $N_2 = Mgd_1/(d_1 + d_2)$.

b. Show that $N_1 > N_2$ if $d_1 < d_2$.

c. Suppose that the person supporting the rod moves his hands closer together by exerting inward-directed horizontal forces that are equal in magnitude and just strong enough to cause slippage between (at least) one of the support fingers and the rod. Assuming that the same coefficient of friction applies to the contact between each finger and the rod, show that if $d_1 < d_2$, then only finger 2 moves relative to the rod until $d_1 = d_2$, after which both fingers move equally relative to the rod. Show also that if $d_1 > d_2$, then only finger 1 moves relative to the rod, until $d_1 = d_2$, after which both fingers move equally.

d. Use the above results to explain why the following method can be used to locate the center of mass of any rod: "Use one finger from each hand to support the rod in a horizontal position. Using the smallest horizontal force that results in some motion, draw the support fingers inward. Continue until the fingers touch. When they do so, they are directly under the center of mass of the rod."

e. Suppose the coefficients of friction for the two supports differ slightly. Will the method still work? Explain your answer.

9-34. *Slippery job.* Consider once more the 20-kg ladder described in Example 9-9, but now suppose that both the ground and the wall are frictionless. The ladder is kept from slipping by a horizontal rope which is tied to the ladder's center and anchored to the wall. A 70-kg man is standing on the ladder at some point P above C, the midpont of the ladder, the distance from P to C is 1.50 m.

a. What force does the ground exert on the base of the ladder?

b. What force does the wall exert on the top end of the ladder?

c. What is the tension in the rope?

9-35. *The statics of a derrick.* As shown in Fig. 9E-35, a derrick consists of a uniform 6.00-m boom of mass 100 kg, hinged to a vertical mast. A load of mass 400 kg is fastened to the free end of the boom. A cable is attached to a hook 1.50 m from the end of the boom. The cable is fastened to the mast, and the angles between the various members are as shown.

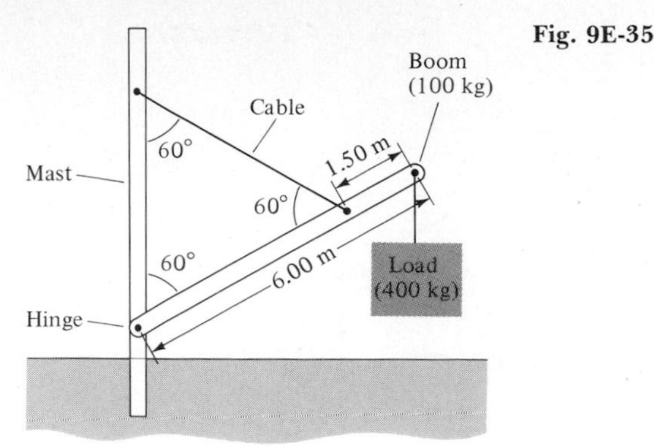

Fig. 9E-35

a. What is the tension in the cable?

b. What are the horizontal and vertical components of the force exerted by the hinge on the base of the boom?

c. Using an origin located at the hinge, evaluate the torque exerted on the mast by the cable. What prevents the mast from toppling?

9-36. *To maintain equilibrium.* As shown in Fig. 9E-36, a light, bent rod is pivoted at point B. The distances AB, BC, and CD are equal, and the angles at B and C are right angles. A force $\mathbf{F}_1$ of magnitude 100 N is applied at point A; the force is directed parallel to the line from B to C. A second force $\mathbf{F}_2$ is applied at point D in order to maintain the equilibrium of the rod. What is the minimum possible magnitude of $\mathbf{F}_2$?

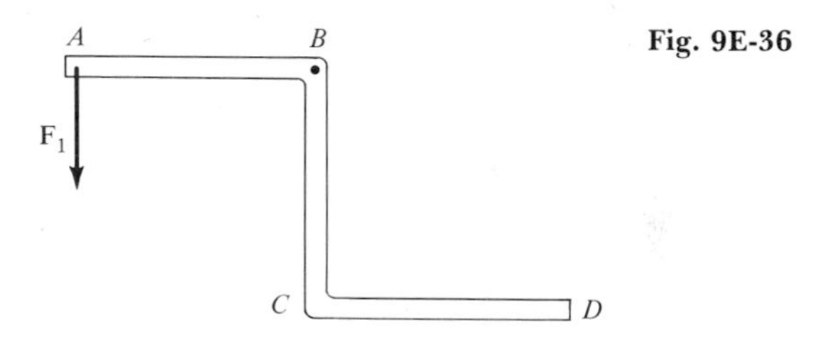

Fig. 9E-36

9-37. *Hanging by a thread, I.* A homogeneous sphere is suspended by a string from a wall. The sphere is set against the wall in such a way that the point of attachment A of the string to the sphere is vertically above the center of the sphere O. Prove that the coefficient of static friction between the sphere and wall must equal or exceed 1 if the sphere is to remain in the given position.

9-38. *Supporting a sign.* As shown in Fig. 9E-38 a 50-kg sign is supported by a light horizontal beam. The sign's mass is uniformly distributed. The beam is hinged to a wall bracket at one end. The other end of the beam is supported by a guy wire which makes an angle of 30° with the horizontal.

a. Show that the force $\mathbf{F}_B$ exerted by the wall bracket on the beam acts along a line that passes through the midpoint of the guy wire. What angle does $\mathbf{F}_B$ make with the horizontal?

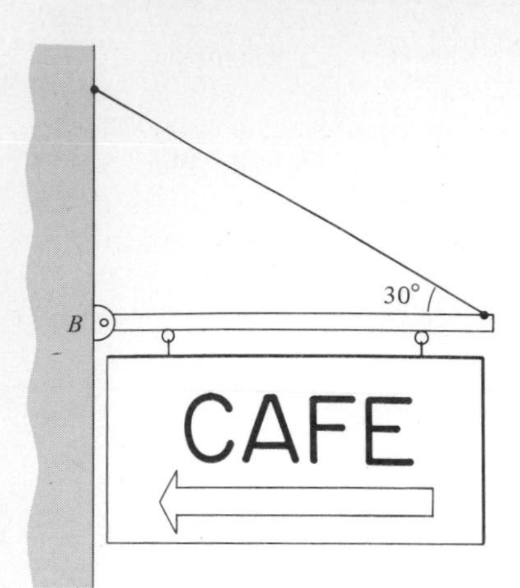

Fig. 9E-38

b. Determine the magnitude of the force $\mathbf{F}_B$ and its horizontal and vertical components.

c. What is the magnitude of the tension in the guy wire?

9-39. *A moving center of mass.* A body of mass m_1 is moving with velocity $\mathbf{v}_1$. A second body of mass m_2 is stationary.

a. What is the velocity of the center of mass of the system consisting of the two bodies?

b. What is the velocity $\mathbf{v}_1'$ of body 1 relative to the center of mass?

c. What is the velocity $\mathbf{v}_2'$ of body 2 relative to the center of mass?

d. What is the momentum of the system in a frame of reference attached to the moving center of mass?

9-40. *Frozen fast.* A hollow cylindrical tube is resting on its side on a horizontal surface, half-filled with water. The water freezes, so that the tube is half full of ice, which has a mass M_i. The tube has outer radius R, and inner radius r. The total mass of the tube and ice is $M_t + M_i$.

a. Determine the location of the center of mass of the system.

b. Suppose that the tube rolls to one side (without slipping) so that it turns through an angle θ. What is the increase in gravitational potential energy of the system? Assume that the ice is frozen fast to the inner wall of the tube.

9-41. *An important intersection.* Explain why the center of mass of the body in Fig. 9-31 is located at the intersection of the two dashed lines.

9-42. *Making the best of an unequal-arm balance.* The mass of an object is to be determined by using a balance. It is suspected that the arms of the balance are not exactly equal—that is, $l_1 \neq l_2$ in Fig. 9E-42. The body of unknown mass M is placed in the right pan, and a known mass m_1 is required in the left pan to level the beam. The body of unknown mass is now placed in the left pan. It is found that a known mass m_2 is required in the right pan for leveling.

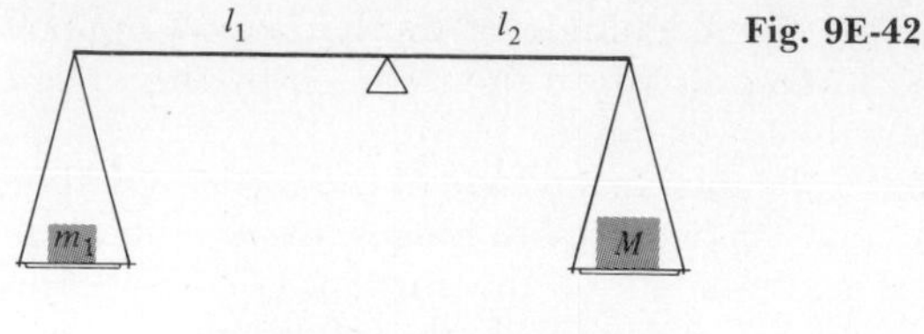

Fig. 9E-42

a. Show that $M = \sqrt{m_1 m_2}$.

b. Show that $l_2/l_1 = \sqrt{m_1/m_2}$.

9-43. *Torque and potential energy.* Verify Eq. (9-58), $T(\theta) = -dU(\theta)/d\theta$, by following the procedure suggested below that equation.

9-44. *Off her rocker.* A homogeneous, hemispherical solid of mass M and radius R is resting on a horizontal surface. An aluminum pole of negligible mass extends vertically upward from the center of the flat face of the hemisphere. As shown in Fig. 9E-44, an acrobat of mass m is climbing the pole. How far up the pole can she climb before the equilibrium becomes unstable?

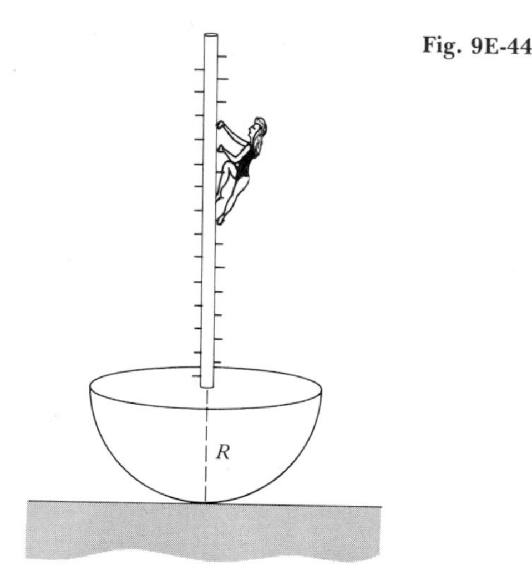

Fig. 9E-44

9-45. *Futuristic rocking chair?* A homogeneous half-cylinder of mass M and radius R is able to rock without slipping on a horizontal surface, as shown in Fig. 9E-45. Find the gravitational potential energy of the system as a function of the angle θ. (Assume that $\theta < 90°$.)

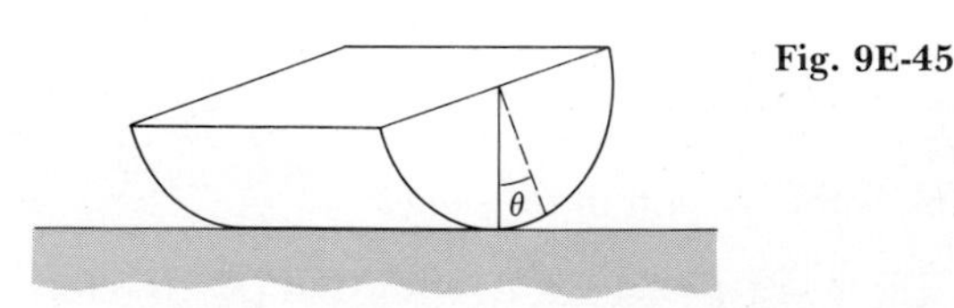

Fig. 9E-45

9-46. *Weighting the base.* A uniform rod of length 100 cm and mass 300 g protrudes from a cylindrical can, with its bottom end resting against the bottom edge of the can. The can has a diameter of 12.0 cm, a height of 16.0 cm, and negligible mass. What minimum depth of water must be poured into the can to prevent the can from tipping? The density of water is 1.00 g/cm^3.

9-47. *Point of no return.* Consider the uniform block shown in Fig. 9E-47.

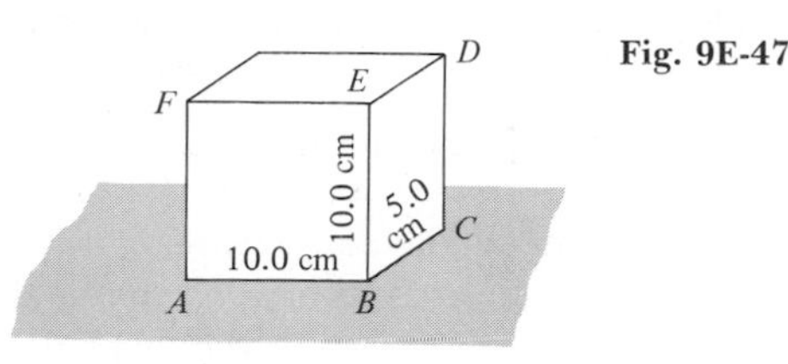

Fig. 9E-47

a. What is the maximum angle through which the block can be rotated about edge AB and still return to its original position when released?

b. What would the maximum angle be if the block were rotated about edge BC?

Group C

9-48. *Top spin.* Apply Eq. (9-21), $\mathbf{a} = \boldsymbol{\alpha} \times \mathbf{r} + \boldsymbol{\omega} \times \mathbf{v}$, to describe qualitatively the acceleration $\mathbf{a}$ of a point fixed on the surface of a child's top at the widest point of the top. The top spins rapidly about its axis of symmetry. The magnitude of its angular velocity $\boldsymbol{\omega}$ is constant, but the direction changes as the axis of symmetry moves slowly around a cone, as in Fig. 9-2. First make a sketch at an instant when $\boldsymbol{\omega}$ is just moving into the plane of the page, and determine the direction of the top's angular acceleration $\boldsymbol{\alpha}$. Next make two more sketches with $\boldsymbol{\omega}$ just moving into the plane of the page. In the first of these, the point fixed on the surface of the top lies outside the cone described by the rotation of $\boldsymbol{\omega}$. In the second, which illustrates the situation a short time later, the point has rotated around the axis of the top so that it lies inside the cone. Use Eq. (9-19), $\mathbf{v} = \boldsymbol{\omega} \times \mathbf{r}$, to find the direction of the velocity $\mathbf{v}$ of the point in both cases, measuring its position vector $\mathbf{r}$ from an origin at the pointed tip of the top. Then find the direction of $\mathbf{a}$ in both cases. Also compare the magnitude of $\mathbf{a}$ in the two cases. Can you relate the behavior of $\mathbf{a}$ to the path traced out in space by the point fixed on the surface of the top?

9-49. *Surmounting an obstacle.*

a. A roller is 1.0 m in diameter and has a mass of 50 kg. Find the magnitude, direction, and point of application of the smallest applied force that can get it on top of a flat block 0.30 m high. Assume that the roller does not slip at its line of contact with the upper edge of the block.

b. If the force found in part *a* is applied, what will be the magnitude and direction of the force $\mathbf{F}'$ that the edge of the block exerts on the roller as the roller loses contact with the floor?

c. For the force $\mathbf{F}'$ found in part *b*, find the component tangential to the roller surface as well as the normal component. Evaluate the ratio of the tangential component to the normal component.

9-50. *A common (and frustrating) occurrence.* A dresser drawer has depth l and width w. The handles are a distance d apart and are properly centered, as shown in Fig. 9E-50. The drawer binds when it is pushed with a force $\mathbf{F}$ at one handle, as shown in the figure. This happens because the corners A and B press against the drawer guides; the resulting frictional forces $\mathbf{f}_A$ and $\mathbf{f}_B$ immobilize the drawer. The coefficient of static friction between the drawer and the guides is μ_s.

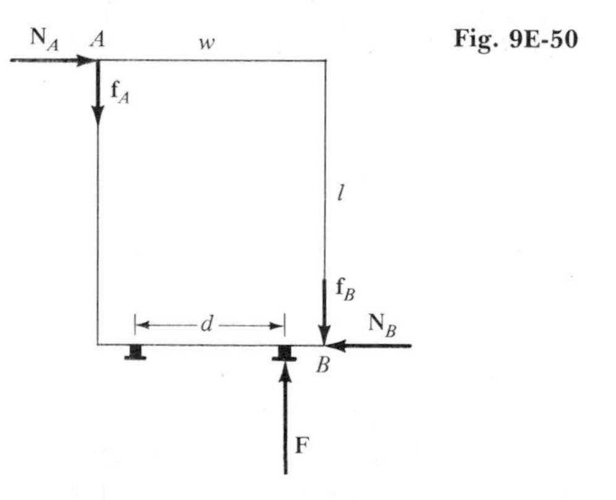

Fig. 9E-50

a. What is the relationship between the normal forces $\mathbf{N}_A$ and $\mathbf{N}_B$?

b. What is the relationship between the frictional forces $\mathbf{f}_A$ and $\mathbf{f}_B$ if the drawer is just barely stuck?

c. Find $\mathbf{f}_A$ and $\mathbf{f}_B$ in terms of $\mathbf{F}$ if the drawer is stuck.

d. Given that the drawer sticks when pushed at one handle, find a relationship among l, d, and μ_s.

e. Can the sticking be overcome simply by pushing harder—that is, by increasing the magnitude of $\mathbf{F}$? Explain your answer.

9-51. *The ground, the wall, and the ladder.* A uniform ladder is standing on a horizontal surface and leaning against a vertical wall. The coefficient of static friction between the ladder and the horizontal ground surface is μ_G, and the coefficient of static friction between the ladder and the wall is μ_W.

a. In terms of μ_G and μ_W, find the minimum elevation angle θ_0 at which the ladder can be leaned without beginning to slip.

b. Evaluate θ_0 for $\mu_G = \mu_W = 0.40$.

c. Evaluate θ_0 for $\mu_G = 0.40$ and $\mu_W = 0$.

d. Evaluate θ_0 for $\mu_G = 0$ and $\mu_W = 0.40$.

e. Based on your results for parts *b* to *d*, which coefficient of friction appears to be more important in determining the minimum elevation angle?

9-52. *Appropriate compensation.* An Atwood machine is set up on a lever, as shown in Fig. 9E-52. Frictional forces and pulley masses are negligible. Initially, the masses m_A and m_B are equal, and weights are placed in the pan until the beam is horizontal. Then an amount of mass m is transferred from m_A to m_B, so that $m_A' = m_A - m$ and $m_B' = m_A + m$.

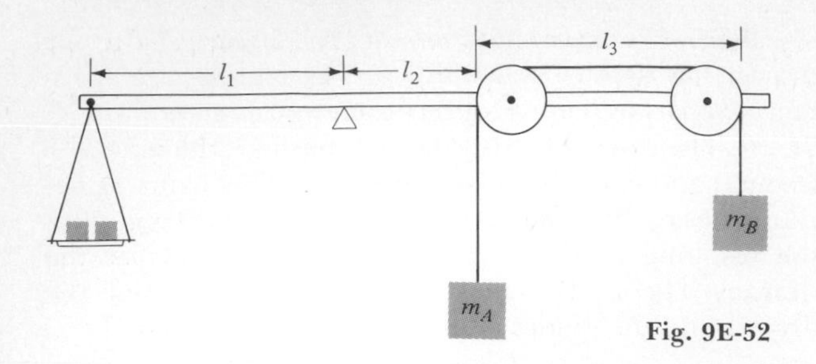

Fig. 9E-52

a. How much mass Δm must be simultaneously removed from the pan if the lever is to remain horizontal while the Atwood machine operates?

b. Evaluate the result of part *a* for the case $m_A = 0.100$ kg, $m = 0.010$ kg, $l_1 = l_3 = 0.100$ m, and $l_2 = 0.050$ m.

9-53. *Center of mass of a right-angle bracket.* Consider a uniform right-angle iron bracket, *ABC*, with equal arms of length *l*.

a. Where is the center of mass of the system relative to *B*, the angle vertex?

b. If end *A* is tied to a vertical string and the bracket is suspended, what will be the angle between the line *AB* and the downward vertical?

9-54. *A well-designed scale.* Figure 9E-54 is a schematic diagram of a platform scale similar to ones used by physicians. This type of scale has three advantages: (1) It can be used to measure a large mass with a small one; (2) the result is independent of the position of the body of mass *M* on the platform *EF*; (3) *EF* always remains horizontal. These advantages result from the construction of the scale so that the ratios BD/BC and GH/GF are equal. Call the ratio *n*.

a. If $EK = a$ and $KF = b$, what are the forces $\mathbf{F}_E$ pulling at *E* and $\mathbf{F}_F$ pushing at *F*?

b. What is the force $\mathbf{F}_H$ pulling at *H*?

c. Write the equation for the equilibrium of the beam *AD* and show that the value of the mass *m* required for equilibrium satisfies the condition $mg \times AB = Mg \times BC$,

d. If $AB/BC = 100$, what is the relation between *M* and *m*? Show that your result is independent of the position of the body of mass *M* along the platform *EF*.

e. Suppose *E* descends a distance *d*. How far will *H* descend? How far will *F* descend? Show that the platform *EF* which bears the load remains horizontal.

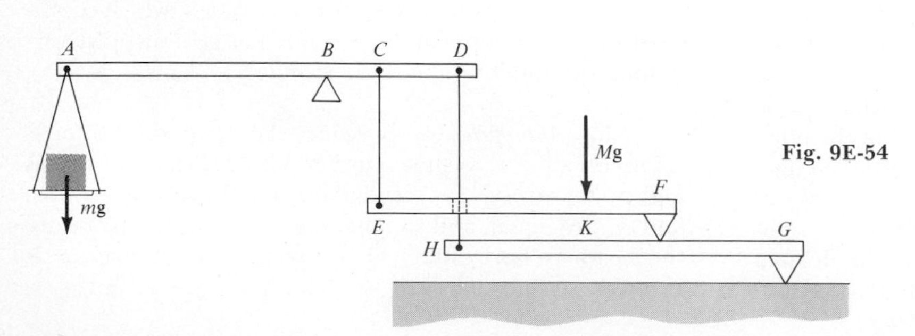

Fig. 9E-54

9-55. *The critical slope.* A cylinder lies on an incline of angle θ. A string is wrapped around it many times over, so as to cover its surface. The string continues above the cylinder parallel to the incline. The end of the string is tied to a nail. Let the coefficient of static friction between the string and the plane be μ_s. Prove that if $\tan\theta < 2\mu_s$, the cylinder will remain in place.

9-56. *Pappus's theorem.* A plane lamina (that is, a flat, thin sheet of material of uniform density and thickness) is shown in Fig. 9E-56. This lamina has surface area *A*, total mass *M*, and mass per unit area $\sigma = M/A$. The shape of the lamina is arbitrary. The lamina lies in the *xy* plane. The *x* axis does not pass through it, though it may be tangent to the lamina. The center of mass of the lamina lies at some point whose y coordinate is Y_c.

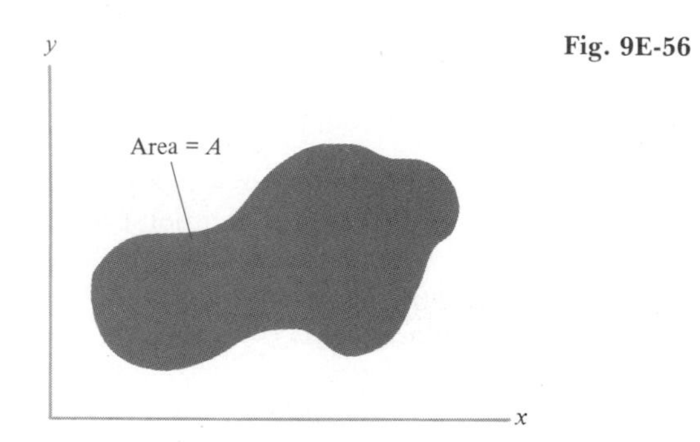

Fig. 9E-56

a. Suppose that the lamina is revolved about the *x* axis to sweep out a solid of revolution. Prove that the volume *V* of the solid (not including the axial hole that may pass through it) satisfies the equation $V = 2\pi A Y_c$. This result is known as *Pappus's theorem,* after the Alexandrian mathematician Pappus (fourth century A.D.).

b. To see one practical use of Pappus's theorem, apply it to determine the position of the center of mass of a semicircular piece of sheet metal whose radius is *R*.

9-57. *Precarious balance.* Consider a stack consisting of a large number of uniform and identical rectangular boards of length *l*. Initially the boards are neatly stacked one above another; starting from the top, they are labeled 1, 2, 3, and so on.

a. Board 1 is moved (along the length of the stack) as far as possible without causing it to tilt. What is this maximum overhang?

b. With board 1 overhanging board 2 by the distance found in part *a*, boards 2 and 1 are moved, in the same direction as in part *a*, as far as possible without causing board 2 to tilt. By how much does board 2 overhang board 3? What is the total overhang of boards 1 and 2? That is, by how much does board 1 overhang board 3?

c. The process described in parts *a* and *b* is continued until boards 1 through N are all displaced as far as possible. By how much does board N overhang board $N + 1$? What is the total overhang of boards 1 through N?

d. Show that the total overhang of boards 1 to 4 is greater than l, the length of each board.

e. Show that total overhang found in part *b* increases without limit as $N \to \infty$.

9-58. *Hanging by a thread, II.* A homogeneous sphere of mass M and radius R is suspended from a vertical wall by a string of length L. There is no friction between the wall and the sphere.

a. Find the tension S in the string.

b. Find the normal force N exerted on the sphere by the wall.

c. For what value of L/R does the tension equal twice the sphere's weight ($S = 2Mg$)? What is the corresponding value of N?

d. Find appropriate approximations for S and for N in the limiting cases $L \ll R$ and $L \gg R$.

9-59. *On top of the globe.* A homogeneous, solid hemisphere of radius r is sitting, flat side up, on top of a fixed sphere of radius R. The contact surface is sufficiently rough to prevent slippage, but the hemisphere can roll freely. What is the minimum value of R for which the equilibrium is a stable equilibrium?

10

Rotational Motion, II

10-1 MOMENT OF INERTIA

When you analyze the translational motion of a rigid body, it is not necessary to consider its individual particles because each has the same velocity $\mathbf{v}$. Furthermore, the total momentum of the body is related to $\mathbf{v}$ in a very simple way. If you write the total momentum as $\mathbf{P}$, then $\mathbf{P} = M\mathbf{v}$, where the scalar constant M is the total mass of the body. The translational form of Newton's second law gives you an equation for the rate of change of $\mathbf{v}$, in terms of M and the net force applied to the body. You have had considerable experience working with that equation, studying its solutions and their physical interpretations in a wide variety of cases.

It would be very desirable for you to be able to use an analogous procedure in analyzing the rotational motion of a rigid body. If this were possible, you could use analogy to carry over directly to rotational motion much of what you have learned from the study of translational motion. Is it possible? Each particle of a rigid body rotating about some axis has the same angular velocity $\boldsymbol{\omega}$, just as each particle of a translating rigid body has the same velocity. Furthermore, in certain circumstances the body's total angular momentum $\mathbf{L}$ about some origin can be written as

$$\mathbf{L} = I\boldsymbol{\omega} \tag{10-1}$$

The constant I is called the **moment of inertia** of the body for rotation about the axis. When Eq. (10-1) is valid, prediction of the rotational motion of the body from $\mathbf{T}$, the net torque about the origin that acts on the body, is completely analogous to the prediction of the translational motion of a body from $\mathbf{F}$, the net force acting on it. This is so because the rotational form of Newton's second law, $\mathbf{T} = d\mathbf{L}/dt$, is mathematically equivalent to its translational form $\mathbf{F} = d\mathbf{P}/dt$ and because the relation $\mathbf{L} = I\boldsymbol{\omega}$ is mathematically equivalent to the relation $\mathbf{P} = M\mathbf{v}$.

To see an example of the analogy, recall that for constant M the translational form of Newton's second law gives $\mathbf{F} = d\mathbf{P}/dt = d(M\mathbf{v})/dt = M\,d\mathbf{v}/dt$. Writing $d\mathbf{v}/dt = \mathbf{a}$, the body's acceleration, you have $\mathbf{F} = M\mathbf{a}$. This is the equation you have used most frequently in studying translational mechanics. If the relation $\mathbf{L} = I\boldsymbol{\omega}$ is valid and I is constant, then the rotational form of Newton's second law gives $\mathbf{T} = d\mathbf{L}/dt = d(I\boldsymbol{\omega})/dt = I\,d\boldsymbol{\omega}/dt$. Writing $d\boldsymbol{\omega}/dt = \mathbf{d}$, the angular acceleration of the body, you obtain

$$\mathbf{T} = I\boldsymbol{\alpha} \tag{10-2}$$

This equation allows you to find a body's angular acceleration from the net torque acting on it and its moment of inertia. Then you can use the angular acceleration to determine the angular velocity and the angular position of the body as functions of time. Since the procedure is identical, from a mathematical point of view, to the one starting with the equation $\mathbf{F} = M\mathbf{a}$, frequently you will not have to repeat the detailed mathematical arguments. Instead you can go directly to the physical insight you seek.

However, the relation $\mathbf{L} = I\boldsymbol{\omega}$, with I a scalar constant, is valid only in *certain cases.* While the angular velocity $\boldsymbol{\omega}$ is, by definition, always directed along the axis about which a rigid body rotates, you will soon see that the total angular momentum $\mathbf{L}$ of the body is not necessarily directed along the rotation axis. Thus $\mathbf{L}$ may be at an angle to $\boldsymbol{\omega}$. When this is the case, the equation $\mathbf{L} = I\boldsymbol{\omega}$ certainly cannot be correct. But you will see also that there are many cases in which $\mathbf{L} = I\boldsymbol{\omega}$ *is* a valid equation because $\mathbf{L}$ is in the direction of $\boldsymbol{\omega}$ and because the magnitude of $\mathbf{L}$ is proportional to the magnitude of $\boldsymbol{\omega}$. These cases, which fortunately include many of great practical interest, are ones in which the particles whose masses make up the total mass of the body are distributed in special ways relative to the rotation axis, or relative to the origin used to define $\mathbf{L}$.

Furthermore, you will see that even when the total angular momentum of a rigid body is not aligned parallel to the rotation axis, its component $L_{\|}$ in a particular direction along that axis has a value proportional to the value of the signed scalar ω specifying the angular velocity of the body. Thus in *all cases* the scalar relation $L_{\|} = I\omega$ is correct. The positive constant I appearing in this relation also is called the body's moment of inertia for rotation about the axis. The relation makes it possible to use the equation $T_{\|} = I\alpha$ to find the component $T_{\|}$ of the net torque applied to the body when it has an angular acceleration given by the signed scalar α. You cannot use a similar equation to determine the components of the net torque in directions perpendicular to the rotation axis. But in many practical circumstances $T_{\|}$ is the only component you need to know.

The actual value of the moment of inertia depends not just on the total mass of a rigid body but also on how that mass is distributed with respect to the axis about which the body rotates. Although in the newtonian domain a particular rigid body has a unique total mass M, it does *not* have a unique moment of inertia I. Changing the rotation axis changes the value of I. Thus the concept of moment of inertia is more complex and less general than the concept of mass. Nevertheless, it is a worthwhile concept because it leads to an efficient way of treating rotational motion.

We begin this treatment by investigating the relation between a body's angular momentum about an origin and its angular velocity. First we consider the simplest possible case, illustrated in Fig. 10-1*a*. A single particle of

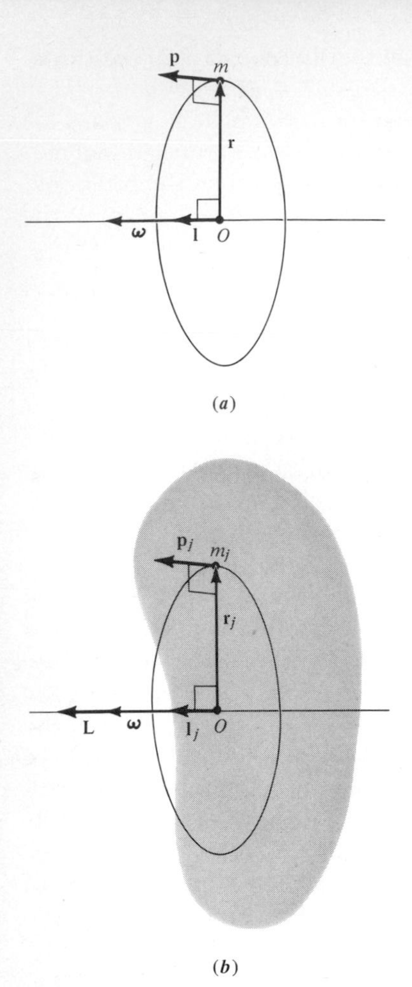

Fig. 10-1 (*a*) A single particle rotating in a circle centered on a point O, which is used for an origin. Its angular velocity is $\boldsymbol{\omega}$. (*b*) A thin, flat, rigid plate rotating in its own plane. Then particles of the body are all rotating in circles centered on the point O that is used for the origin, but only the jth particle is shown. They all have the same angular velocity $\boldsymbol{\omega}$.

mass m is rotating in a circle with angular velocity $\boldsymbol{\omega}$. We fix the origin O, used to define the particle's angular momentum $\mathbf{l}$, on the rotation axis at its intersection with the plane of the circle. Then the particle's position vector $\mathbf{r}$ is in the plane of rotation and is always perpendicular to its momentum vector $\mathbf{p}$. Since $\mathbf{p}$ also lies in that plane, the angular momentum $\mathbf{l} = \mathbf{r} \times \mathbf{p}$ is perpendicular to the plane. The right-hand rules show that $\mathbf{l}$ points in the same direction as $\boldsymbol{\omega}$. This parallelism of $\mathbf{l}$ and $\boldsymbol{\omega}$ is essential to the argument which follows.

The magnitude of $\mathbf{l}$ has the value

$$l = |\mathbf{r} \times \mathbf{p}| = rp$$

because $\mathbf{r}$ and $\mathbf{p}$ are perpendicular. Since $p = mv$, where v is the particle's speed, we can write this as

$$l = rmv$$

But the particle's velocity $\mathbf{v}$ is given by $\mathbf{v} = \boldsymbol{\omega} \times \mathbf{r}$. So $v = |\boldsymbol{\omega} \times \mathbf{r}| = \omega r$ because $\boldsymbol{\omega}$ and $\mathbf{r}$ are perpendicular. Thus we have

$$l = rm\omega r = mr^2\omega$$

The magnitude of $\mathbf{l}$ is proportional to the magnitude of $\boldsymbol{\omega}$, the proportionality constant being mr^2. Since the two vectors are in the same direction, we can write the vector equation

$$\mathbf{l} = mr^2\boldsymbol{\omega}$$

We express this result as

$$\mathbf{l} = i\boldsymbol{\omega} \tag{10-3}$$

where

$$i \equiv mr^2 \tag{10-4}$$

The quantity i is a rudimentary example of a **moment of inertia.** Specifically, it is the moment of inertia of the particle for rotation about an axis for which the perpendicular distance from the axis to the particle always has the same value r.

Now consider the case shown in Fig. 10-1*b*. There are n particles forming a thin, flat rigid plate of irregular shape and mass distribution. The plate is rotating in its own plane about a fixed origin O chosen in that plane. Since Eqs. (10-3) and (10-4) apply to each of its particles, a typical one has angular momentum

$$\mathbf{l}_j = m_j r_j^2\,\boldsymbol{\omega}$$

The total angular momentum about O of the body is

$$\mathbf{L} = \sum_{j=1}^{n} \mathbf{l}_j = \sum_{j=1}^{n} m_j r_j^2 \boldsymbol{\omega} = \left(\sum_{j=1}^{n} m_j r_j^2\right)\boldsymbol{\omega}$$

No subscript is needed on the $\boldsymbol{\omega}$ since the angular velocity is the same for all the particles in the rigid body. We can write this result as

$$\mathbf{L} = I\boldsymbol{\omega} \tag{10-5}$$

where

$$I \equiv \sum_{j=1}^{n} m_j r_j^2 \tag{10-6}$$

The quantity I is the **moment of inertia** of the rigid body for rotation about the axis.

Note that we can also get this result by evaluating

$$I = \sum_{j=1}^{n} \mathrm{i}_j$$

with

$$\mathrm{i}_j = m_j r_j^2$$

so that, again,

$$I = \sum_{j=1}^{n} m_j r_j^2$$

Thus the total moment of inertia of a rigid body for rotation about an axis equals the sum of the moments of inertia of its component particles for rotation about that axis, just as its total mass equals the sum of their masses.

We have shown that for a flat plate of arbitrary mass distribution the angular momentum **L** is proportional to the angular velocity **ω** if the plate is rotating about an axis perpendicular to its plane and if the origin O is chosen at the intersection of the axis and the plane of rotation. The proportionality constant I is the moment of inertia of the flat plate when rotating about the perpendicular axis.

There is no unique moment of inertia I for a particular flat plate. The value of I depends on where the perpendicular rotation axis passes through the plate. For different rotation axes, all the r_j^2 terms in Eq. (10-6) change, and so there is a different value of I.

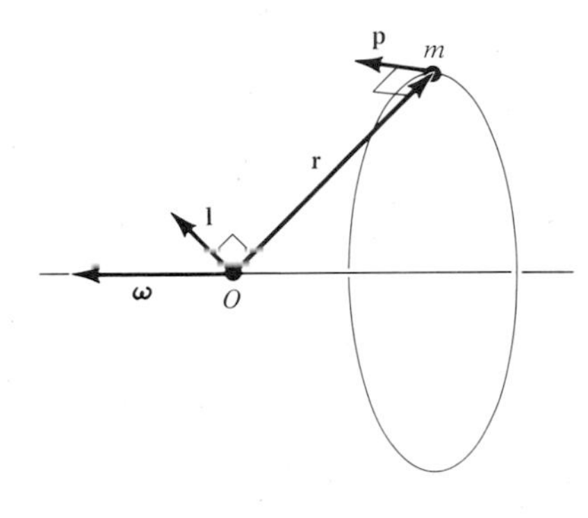

Fig. 10-2 The same rotating particle shown in Fig. 10-1*a*. If the origin O is not chosen to be in the plane of its rotation, its angular momentum **l** is not parallel to its angular velocity **ω**.

Furthermore, for rotation about a particular perpendicular axis, *it is essential to locate the origin on the axis at its intersection with the plane of rotation of the plate,* as in Fig. 10-1*b*. Otherwise, Eq. (10-5) may not be valid. The reason is that if the origin O is not chosen at the intersection of the rotation axis and the rotation plane, **L** may not be parallel to **ω.** In that case, **L** and **ω** cannot be connected by a scalar, as they are in Eq. (10-5). This can be seen from Fig. 10-2, which shows the same rotating particle depicted in Fig. 10-1*a* but with the origin O fixed at a point on the rotation axis that is not at the intersection with the rotation plane. The figure shows the position vector **r** from O to the particle, its momentum vector **p,** and its angular momentum vector **l** = **r** × **p.** The magnitude of **l** has the constant value $l = rp$, since **r** is perpendicular to **p** and both **r** and **p** have constant magnitudes. But the direction of **l,** being at all times perpendicular to the direction of **p,** is always changing. Thus the vector **l** rotates around the axis, in step with the rotation of the particle about the axis. In contrast, the direction of the angular velocity **ω** is constant since it always lies along the rotation axis. Therefore the angular momentum **l** is *not* parallel to the angular velocity **ω** if we use this choice of the origin O.

Since **l** is not parallel to **ω** for a single rotating particle when O is not in the rotation plane, typically it is true that the total angular momentum **L** of all the particles that form the rotating plate of Fig. 10-1*b* will not be parallel to its angular velocity **ω** if the origin O is not taken to be at the intersection of the rotation axis and the rotation plane. Of course, for a flat plate rotating in its own plane, we can always locate O in the plane and thereby obtain the very considerable simplification of having **L** parallel to **ω.** But the situation is more complicated for a rotating body whose extension in space

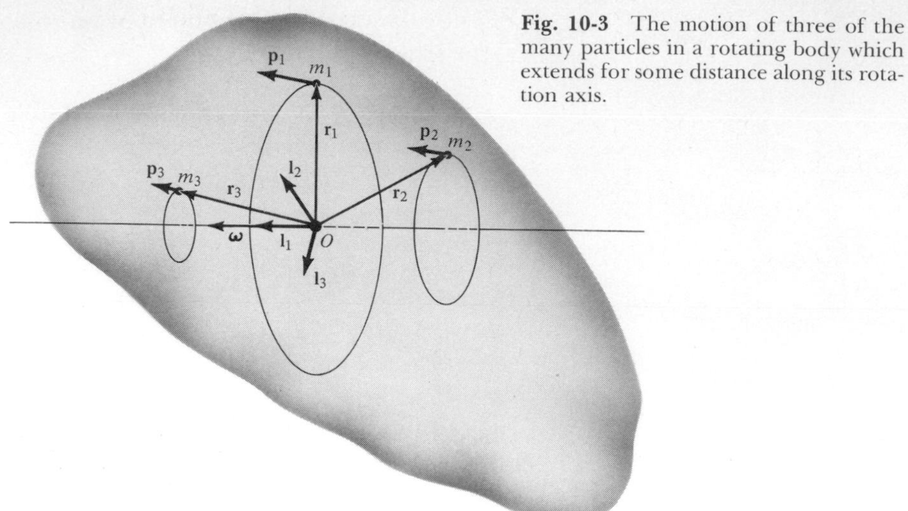

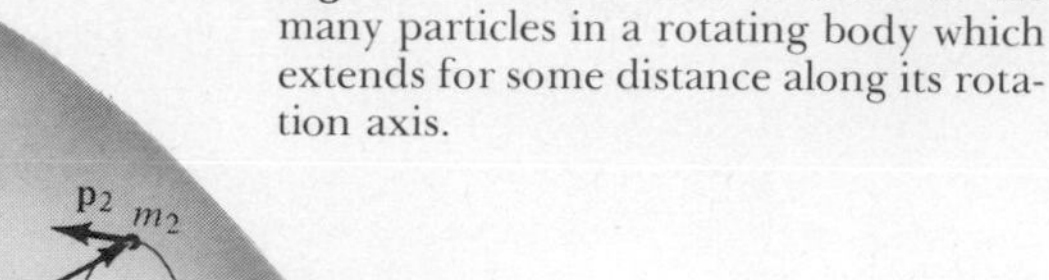
Fig. 10-3 The motion of three of the many particles in a rotating body which extends for some distance along its rotation axis.

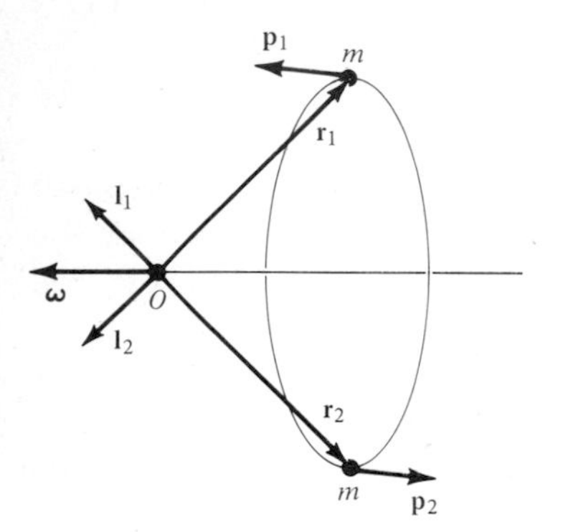

Fig. 10-4 A rotating body composed of two equal-mass particles. The body has enough symmetry to make its total angular momentum **L** parallel to its angular velocity $\boldsymbol{\omega}$, even though the origin O does not lie in the plane of rotation of the two particles.

is three-dimensional, instead of two-dimensional like a flat plate. If a body extends along the direction parallel to the rotation axis, as well as along the directions perpendicular to that axis, then no matter where on the axis O is located, most of the particles in the body do not rotate in a plane passing through O. This point is made in Fig. 10-3.

However, if an extended body has a sufficient degree of *symmetry* relative to the axis about which it is rotating, then its total angular momentum **L** *will* be parallel to its angular velocity $\boldsymbol{\omega}$, no matter where on the axis O is located. In addition, the magnitudes of these vectors will be proportional. Thus we will be able to write $\mathbf{L} = I\boldsymbol{\omega}$. To understand the nature of the required symmetry, we consider first the very simple example shown in Fig. 10-4. The figure shows a system consisting of two equal-mass particles rotating at opposite ends of a diameter of their common circle of rotation. Although neither of the individual angular momenta $\mathbf{l}_1$ or $\mathbf{l}_2$ is parallel to the angular velocity $\boldsymbol{\omega}$ (because the origin O does not lie in the plane of the circle), their components perpendicular to $\boldsymbol{\omega}$ cancel so that the total angular momentum $\mathbf{L} = \mathbf{l}_1 + \mathbf{l}_2$ is parallel to $\boldsymbol{\omega}$.

The same kind of construction shows that the same result will be obtained for three equal-mass particles distributed uniformly around a common circle of rotation, as in Fig. 10-5*a*. And Fig. 10-5*b* shows four equal-mass particles uniformly distributed around the circle that will produce a total **L** parallel to $\boldsymbol{\omega}$. Furthermore, for four particles distributed as shown in Fig. 10-5*c*, it will also be true that **L** is parallel to $\boldsymbol{\omega}$. As the number of equal-mass particles rotating around the common circle increases, there is an increasing number of ways that they can be distributed so that their total angular momentum will be parallel to their angular velocity. But in all cases of two or more particles, a simple way that this can be done is to distribute them uniformly around the circle. It is also the most commonly encountered distribution. In particular, a uniform ring of particles forming a rigid body (a hoop) has a total angular momentum **L** that is parallel to its angular velocity $\boldsymbol{\omega}$ if the body rotates about its symmetry axis, no matter where O is on that axis.

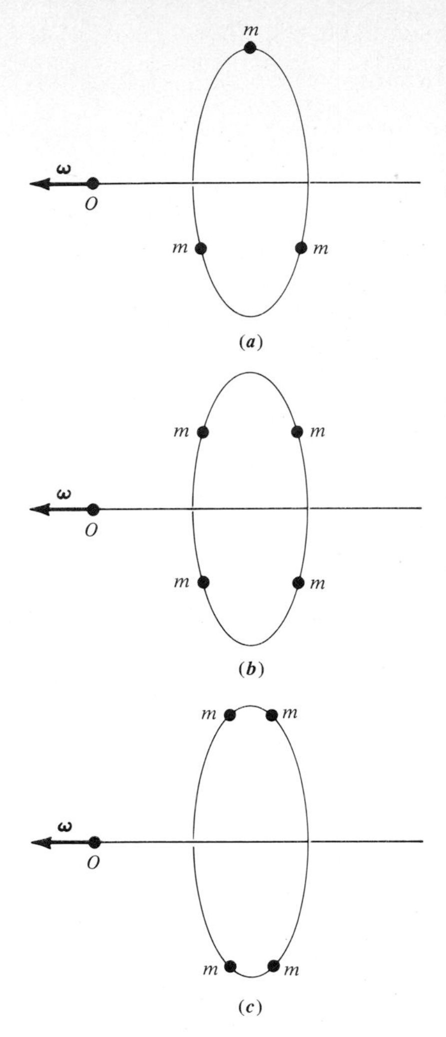

Fig. 10-5 Other simple examples of a body composed of a few equal-mass particles with symmetry such that the total angular momentum **L** is parallel to the angular velocity **ω**. This is true no matter where along the rotation axis the origin O is located.

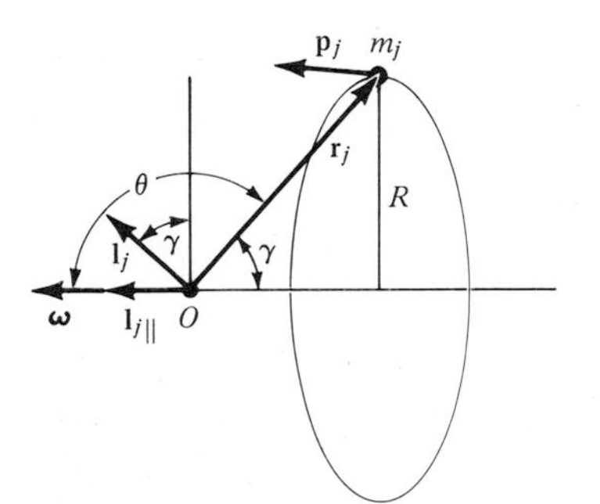

Fig. 10-6 A typical particle in a rotating body consisting of a set of n particles of equal mass distributed in some manner around a circle about the rotation axis.

Consider a set of particles rotating in a common circle about an axis, with angular velocity **ω**, and distributed in an arbitrary manner around the circle. Let us evaluate the component along the axis of rotation of their total angular momentum about an origin somewhere on the axis. In Fig. 10-6 the magnitude of the angular momentum $\mathbf{l}_j$ of the particle having position $\mathbf{r}_j$ and momentum $\mathbf{p}_j$ is

$$l_j = |\mathbf{r}_j \times \mathbf{p}_j| = r_j p_j$$

since $\mathbf{r}_j$ is perpendicular to $\mathbf{p}_j$. Now $p_j = m_j v_j$, where m_j and v_j are the particle's mass and speed. So we have

$$l_j = r_j m_j v_j$$

But

$$v_j = |\boldsymbol{\omega} \times \mathbf{r}_j| = \omega r_j \sin\theta$$

Here θ is the angle between **ω** and $\mathbf{r}_j$, and the $\sin\theta$ factor has been placed at the end of the expression for the sake of convenience. We will write this in terms of the angle $\gamma = \pi - \theta$. As the figure shows, γ is the angle between the rotation axis and $\mathbf{r}_j$, as well as the angle between the perpendicular to the axis and $\mathbf{l}_j$. Since $\sin(\pi - \theta) = \sin\theta$, we have $\sin\theta = \sin\gamma$ and so $v_j = \omega r_j \sin\gamma$. Using this in the equation for l_j, we obtain

$$l_j = m_j r_j^2 \omega \sin\gamma$$

The contribution of this particle to the total component of angular momentum along the rotation axis is the component of $\mathbf{l}_j$ that is parallel to that axis. Since $\pi/2 - \gamma$ is the angle between the rotation axis and $\mathbf{l}_j$, this is

$$l_{j\parallel} = l_j \cos(\pi/2 - \gamma) = l_j \sin\gamma$$

Summing the contributions of all n particles rotating in the circle, we have

$$L_\parallel = \sum_{j=1}^{n} l_{j\parallel} = \sum_{j=1}^{n} l_j \sin\gamma = \sum_{j=1}^{n} m_j r_j^2 \,\omega \sin^2\gamma$$

The figure shows that $r_j \sin\gamma = R$, where R is the distance from the jth particle to the axis of rotation—in other words, where R is the radius of the common circle of rotation. In terms of R, the expression for $L_\parallel$ can be written

$$L_\parallel = \sum_{j=1}^{n} m_j R^2 \omega$$

Since R and ω are the same for all terms in the sum, we have

$$L_\parallel = R^2 \omega \sum_{j=1}^{n} m_j$$

This is

$$L_\parallel = MR^2\omega \tag{10-7}$$

with M being the total mass of the rotating particles.

Now let the rotating particles be distributed around the circle of rotation with *any* symmetry that leads to a cancellation of the components of angular momenta perpendicular to the rotation axis. (See again the examples illustrated in Fig. 10-5.) Then the $L_\parallel$ just evaluated is the magnitude of the total angular momentum of the system. That is, we will have

$$L = L_{\parallel} = MR^2\omega$$

and we can write our results in the vector form

$$\mathbf{L} = I\boldsymbol{\omega}$$

In this expression the **moment of inertia** I for rotation about the axis has the value

$$I = MR^2 \tag{10-8}$$

These results are valid for any location of the origin O on the rotation axis. The moment of inertia of the rotating mass distribution depends only on its total mass M and the square of the common distance R from all its parts to the axis.

If the particles are distributed on their common circle of rotation in such a manner that $\mathbf{L}$ is not parallel to $\boldsymbol{\omega}$ for an origin O that does not lie in the plane of rotation, the basic results expressed in Eq. (10-7) are still valid. They can be used to calculate $L_{\parallel}$, the *component* of $\mathbf{L}$ along the direction of $\boldsymbol{\omega}$. This may be all that you are interested in calculating. For instance, say you have an old and rather asymmetrical grinding wheel that is manually operated. You want to operate it electrically, and you need to calculate how much torque an electric motor must supply to bring the wheel up to a certain angular speed in a certain amount of time. Then you do not care about the torque applied by the bearings supporting the axle of the wheel. This torque, which fixes the rotation axis, is always directed perpendicular to that axis. The torque you care about is the one applied by the motor in a direction parallel to the axis to change the angular speed of the wheel. If this is the case, you need only $L_{\parallel}$, the component of $\mathbf{L}$ along a direction parallel to $\boldsymbol{\omega}$, and you can use Eq. (10-7) to write

$$L_{\parallel} = I\omega \tag{10-9}$$

where

$$I = MR^2$$

just as in Eq. (10-8).

The basic point to understand is that Eq. (10-8) can be used in *all* circumstances to evaluate the proportionality constant I between the *component* $L_{\parallel}$ of the rotating body's total angular momentum parallel to the rotation axis and its angular *speed* ω. The only restriction is that all the mass elements be at the same distance R from the rotation axis and that M be the sum of their masses. If the mass distribution has sufficient symmetry, there will be no perpendicular components of the total angular momentum. Otherwise there will be. But whether there is or is not symmetry has no effect on the parallel component.

Another point to be made is that Eq. (10-8) can be applied immediately to particles which are distributed along a line parallel to the rotation axis at a constant distance R from it. The reason is that Eq. (10-7) is valid no matter where the origin O is located with respect to the plane of rotation of a constituent particle, that is, no matter where that plane is located with respect to the origin.

For a distribution of particles extending in the directions that are not along the rotation axis, the body's moment of inertia I about the axis can be obtained by using Eq. (10-8) to evaluate the contribution of each element of mass M_j, consisting of all the particles having essentially the same distance R_j from the axis, and then summing over all of its n mass elements. That is, the total moment of inertia of the body for rotation about the given axis is

$$I = \sum_{j=1}^{n} M_j R_j^2 \tag{10-10}$$

In actually evaluating I for an extended body, it is usually convenient to replace the sum over its elements of finite mass M_j by an integral over elements of infinitesimal mass dM. Each mass element has a moment of inertia $dI = R^2\, dM$, and the total moment of inertia of the body is, consequently,

$$I = \int_{\text{body}} dI = \int_{\text{body}} R^2\, dM \tag{10-11a}$$

The integrals are taken over the entire body, although the notation is abbreviated by not showing explicit limits on the integral signs. Equation (10-11a) can also be expressed as

$$I = \int_{\text{body}} R^2 \rho\, dV \tag{10-11b}$$

where $\rho \equiv dM/dV$ is the density of the body, with dV an element of its volume.

Before we work out some examples, it is worthwhile making a general comment about the presence in these equations of R^2, the square of the distance from a rotating mass element to the axis about which it rotates. Since I depends quadratically on the distance from the mass element to the axis, and only linearly on the mass of the element, the dominant factor in determining the moment of inertia of a rotating body tends to be not how much mass it has but how far the mass is from the axis. To put the matter another way, the part of the mass of the rotating body that is farthest from the axis of rotation makes the dominant contribution to its moment of inertia.

EXAMPLE 10-1

A uniform bar of length A and mass M is rotating about an axis passing through its center and oriented as shown in Fig. 10-7. The thickness t of the bar in the direction perpendicular to the axis is very small, but its width w along the axis is not. Determine whether the total angular momentum of the bar about the origin O indicated in the figure is parallel to its angular velocity. Then evaluate the bar's moment of inertia for rotation about the axis.

▪ You can think of the bar as consisting of a sum of pairs of elements of infinitesimal length dR, shown in the figure. Since the mass in each pair of elements has the same symmetry as that in Fig. 10-5c, their net contribution to the angular momentum will be parallel to the angular velocity. Thus the total angular momentum $\mathbf{L}$ will be parallel to the angular velocity $\boldsymbol{\omega}$, and you will have $\mathbf{L} = I\boldsymbol{\omega}$.

To evaluate I, you make use of the fact that the thickness of the bar is negligible, and thus all the mass in any pair of elements of length dR is at the same distance R from the rotation axis. The ratio of the mass dM of this pair of elements to

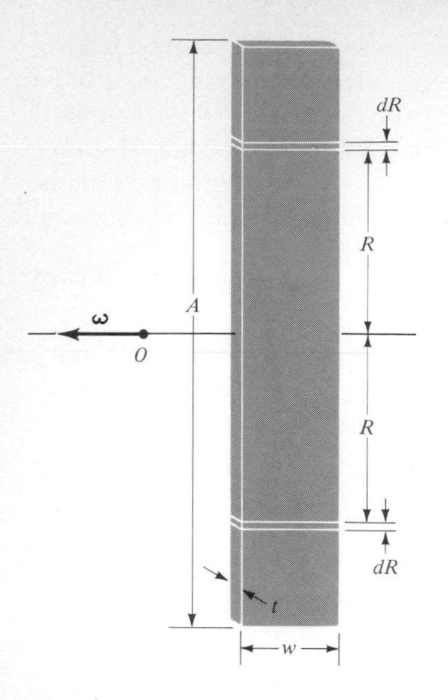

Fig. 10-7 A bar of uniform density rotating about an axis passing through its center.

the total mass M of the bar is the same as the ratio of their combined length $2\,dR$ to the total length A of the bar. That is,

$$\frac{dM}{M} = \frac{2\,dR}{A}$$

or

$$dM = \frac{2M\,dR}{A}$$

So Eq. (10-11a) becomes

$$I = \int_{\text{body}} R^2 \frac{2M\,dR}{A} = \frac{2M}{A}\int_{\text{body}} R^2\,dR$$

To evaluate the integral, its limits of integration must be made explicit. Since the smallest value of the variable R is 0 and the largest is $A/2$, you have

$$I = \frac{2M}{A}\int_0^{A/2} R^2\,dR$$

Using Eq. (7-20), you find

$$I = \frac{2M}{A}\left[\frac{(A/2)^3}{3} - \frac{(0)^3}{3}\right] = \frac{2M}{3A}\left(\frac{A^3}{8}\right)$$

or

$$I = \tfrac{1}{12}MA^2$$

The factor $\frac{1}{12}$ indicates how ineffectively the mass located near the rotation axis contributes to the moment of inertia.

EXAMPLE 10-2

A hollow cylinder of uniform density, shown in Fig. 10-8, is rotating about an axis along the center of the cylinder. The inner radius is R_1, the outer radius is R_2, and the length along the axis is A. Is $\mathbf{L}$ parallel to $\boldsymbol{\omega}$ for the origin O shown in the figure? What is the value of I for rotation about the axis?

■ The answer to the first question is yes. To see this, decompose the cylinder mentally into a set of thin, concentric tubes, as indicated. The mass in each is uniformly distributed around common circles of rotation, with the symmetry of Fig. 10-5b. Each tube contributes a net angular momentum parallel to the angular velocity, and the total angular momentum of the hollow cylinder is therefore also parallel to the angular velocity.

To evaluate I, you can first find the volume of an elemental tube of radius R and infinitesimal wall thickness dR. Since its periphery is $2\pi R$, its volume is

$$dV = A2\pi R\,dR$$

Using this expression in the integral of Eq. (10-11b) and writing explicit limits, you have

$$I = \int_{R_1}^{R_2} R^2\rho A2\pi R\,dR$$

or

$$I = 2\pi\rho A\int_{R_1}^{R_2} R^3\,dR$$

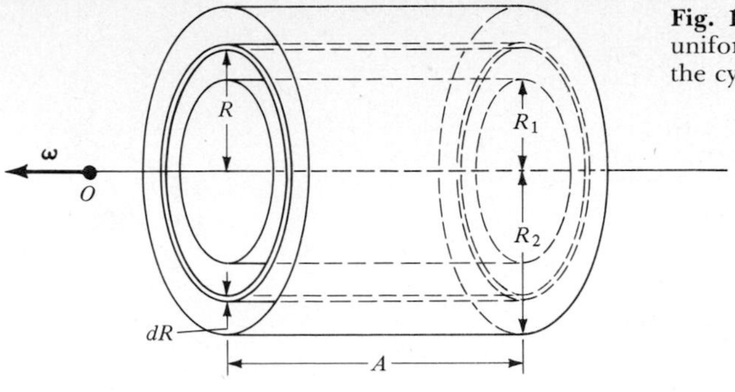

Fig. 10-8 A hollow cylinder of uniform density rotating about the cylinder axis.

since both ρ and A are constants. The limits of integration on the definite integral show that it is taken from the inner radius of the cylindrical body to its outer radius. Using Eq. (7-20) to evaluate the integral, you obtain

$$I = 2\pi\rho A\left(\frac{R_2^4}{4} - \frac{R_1^4}{4}\right)$$

It is convenient to factor this result into the expression

$$I = \frac{\pi\rho A}{2}(R_2^2 - R_1^2)(R_2^2 + R_1^2)$$

The reason is that the volume of the hollow cylinder is $\pi(R_2^2 - R_1^2)A$, and so its total mass M is

$$M = \rho\pi(R_2^2 - R_1^2)A$$

Therefore the moment of inertia of the hollow cylinder rotating about its axis of symmetry simplifies to

$$I = M\frac{R_2^2 + R_1^2}{2} \tag{10-12}$$

The moment of inertia I of the hollow cylinder evaluated in Eq. (10-12) is its mass times a quantity that can be interpreted as the average of the squares of its inner and outer radii. This result seems reasonable in consideration of the remarks made immediately before the examples about the significance of R^2. For a solid cylinder of radius R rotating about its axis, Eq. (10-12) can be used by setting $R_1 = 0$ and $R_2 = R$, producing

$$I = \tfrac{1}{2}MR^2 \tag{10-13}$$

Equation (10-12) contains also the case of a thin-walled tube. Just set $R_1 \simeq R_2 \simeq R$, and it yields

$$I \simeq M\frac{R^2 + R^2}{2} = MR^2$$

in agreement with Eq. (10-8).

The relation $\mathbf{L} = I\boldsymbol{\omega}$ is not restricted to the rotation of bodies that have some sort of symmetry in their mass distribution. In fact, it is possible to prove that through every point in any body, no matter how asymmetric, there pass three perpendicular axes, each having the property that $\mathbf{L}$ is proportional to $\boldsymbol{\omega}$ if the body

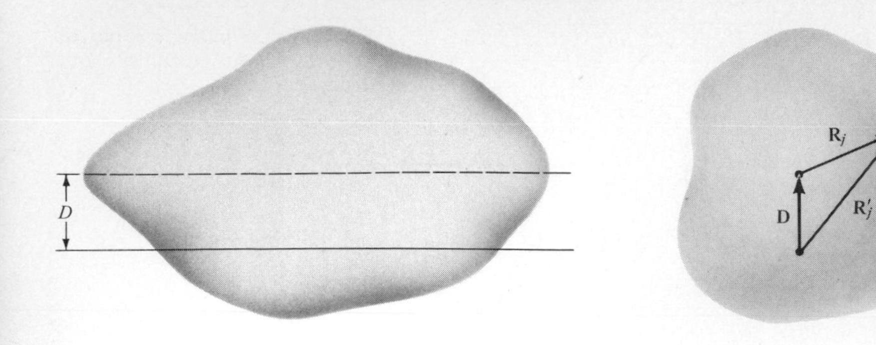

Fig. 10-9 (*a*) The parallel-axis theorem, Eq. (10-14). The dashed line is a rotation axis passing through the body's center of mass. The solid line is a parallel rotation axis at distance *D*. (*b*) A view of the body looking along the parallel axes.

rotates about it. The proportionality constant usually is not the same for rotation about each of these axes. In other words, it is possible to write $\mathbf{L} = I_x\boldsymbol{\omega}$ for rotation about the first so-called **principal axis,** $\mathbf{L} = I_y\boldsymbol{\omega}$ for rotation about the second principal axis, $\mathbf{L} = I_z\boldsymbol{\omega}$ for rotation about the third, and generally $I_x \neq I_y \neq I_z$. If the body rotates about an axis other than one of its principal axes, then **L** is not in the same direction as $\boldsymbol{\omega}$. A principal axis can be located experimentally by finding a rotation axis for which a constant $\boldsymbol{\omega}$ leads to a constant **L**, so that $d\mathbf{L}/dt = \mathbf{T} = 0$, and no torque **T** need be applied to maintain the constant $\boldsymbol{\omega}$. One way to do this is to throw the body in the air in such a way that it spins. If the rotation axis maintains a fixed alignment, you have found a principal axis. If it "wobbles," the body is not rotating about a principal axis.

When the moment of inertia of a body rotating about an axis passing through its *center of mass* is known to have the value I, it is easy to evaluate the moment of inertia I' it will have for rotation about a *parallel* axis passing through some other point, either inside or outside the body. The situation under consideration is shown in Fig. 10-9*a*. According to the **parallel-axis theorem,**

$$I' = I + MD^2 \tag{10-14}$$

where M is the mass of the body and D is the distance from the new axis to the parallel axis passing through its center of mass. We will prove this and then apply it in Example 10-3.

Figure 10-9*b* shows a view of Fig. 10-9*a* that would be seen by looking along the parallel axes; that is, it is Fig. 10-9*a* projected onto a plane perpendicular to the axes. The distance D is the magnitude of a vector **D** extending from the new axis to the original axis passing through the center of mass of the body. The distance from an element of the body with mass M_j to the original axis is the magnitude of a vector $\mathbf{R}_j$. The distance from the same element to the new axis is the magnitude of a vector $\mathbf{R}'_j$. The relation between these quantities is

$$\mathbf{R}'_j = \mathbf{R}_j + \mathbf{D} \tag{10-15}$$

The new moment of inertia is

$$I' = \sum_{j=1}^{n} M_j R'^2_j$$

To evaluate $R_j'^2$, we take the dot product of Eq. (10-15) into itself, producing

$$\mathbf{R}_j' \cdot \mathbf{R}_j' = (\mathbf{R}_j + \mathbf{D}) \cdot (\mathbf{R}_j + \mathbf{D})$$
$$= \mathbf{R}_j \cdot \mathbf{R}_j + \mathbf{D} \cdot \mathbf{D} + 2\mathbf{R}_j \cdot \mathbf{D}$$

or

$$R_j'^2 = R_j^2 + D^2 + 2\mathbf{R}_j \cdot \mathbf{D}$$

So we have

$$I' = \sum_{j=1}^{n} M_j(R_j^2 + D^2 + 2\mathbf{R}_j \cdot \mathbf{D})$$

or

$$I' = \sum_{j=1}^{n} M_j R_j^2 + \sum_{j=1}^{n} M_j D^2 + 2 \sum_{j=1}^{n} M_j \mathbf{R}_j \cdot \mathbf{D}$$

The first term on the right side of this equality is the moment of inertia I of the body for rotation about the original axis. The second term is a summation, giving the body's total mass M times the constant D^2. Thus we have

$$I' = I + MD^2 + 2\left(\sum_{j=1}^{n} M_j \mathbf{R}_j\right) \cdot \mathbf{D}$$

According to the definition of center of mass given by Eq. (9-48*a*), the summation in the third term (which has been put in parentheses for emphasis) has a value equal to M times a vector extending from the original axis to the center of mass of the body. But this vector has zero length since the center of mass is located on the original axis. Therefore the third term is zero, and the parallel-axis theorem, Eq. (10-14), is proved.

Determine I for a uniform, solid cylinder of mass $M = 10.0$ kg and radius $R = 5.00$ cm, about an axis parallel to its symmetry axis and tangent to its surface. See Fig. 10-10.

■ Using Eq. (10-13), obtained from Example 10-2, you know that the moment of inertia for rotation about the axis of symmetry of the cylinder is

$$I = \tfrac{1}{2} MR^2$$

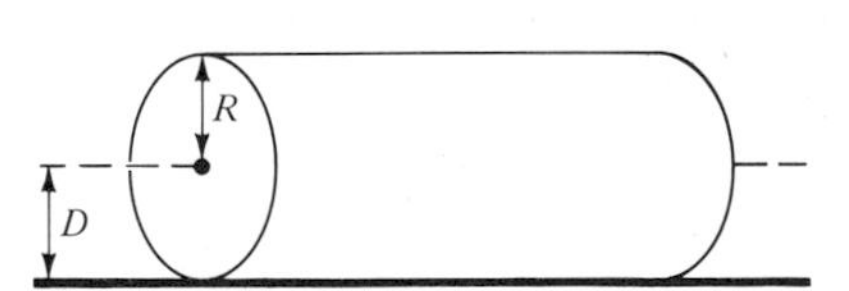

Fig. 10-10 A uniform, solid cylinder, to be rotated about the axis shown as a solid line.

The axis of symmetry passes through the center of mass. So you can apply Eq. (10-14), by setting the distance D between the parallel axes equal to the radius R. Then you have

$$I' = I + MD^2 = \tfrac{1}{2} MR^2 + MR^2$$

or

$$I' = \tfrac{3}{2} MR^2$$

It is very much easier to evaluate I' this way than to evaluate it by a direct application of Eq. (10-11).

The numerical value of I' is

$$I' = \frac{3 \times 10.0\ \text{kg} \times (0.0500\ \text{m})^2}{2}$$
$$= 3.75 \times 10^{-2}\ \text{kg}\cdot\text{m}^2$$

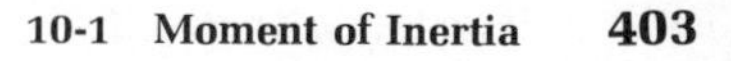

Table 10-1

Square of Gyration Radius for Uniform-Density Bodies

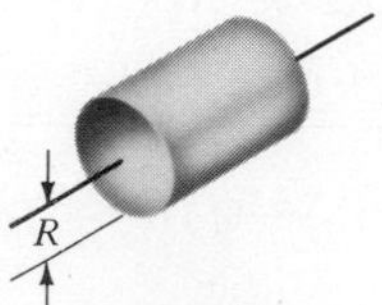

Body	Description	G^2
R	Thin-walled, cylindrical shell about axis of symmetry	$G^2 = R^2$
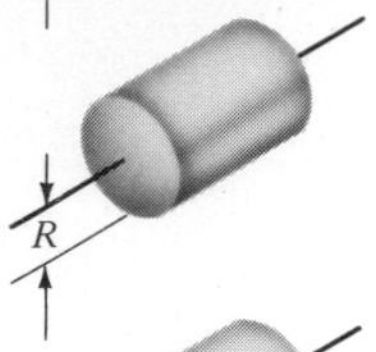 R	Solid cylinder about axis of symmetry	$G^2 = \frac{1}{2}R^2$
R_2, R_1	Hollow cylinder about axis of symmetry	$G^2 = \frac{1}{2}(R_1^2 + R_2^2)$
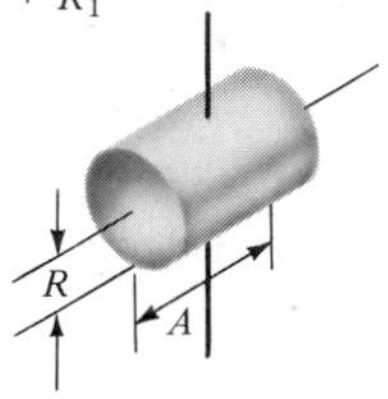 R, A	Thin-walled, cylindrical shell about axis through center of mass, perpendicular to axis of symmetry	$G^2 = \frac{1}{2}R^2 + \frac{1}{12}A^2$
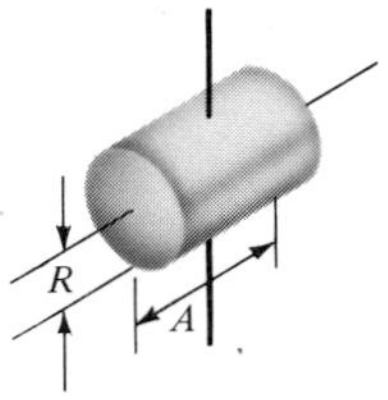 R, A	Solid cylinder about axis through center of mass, perpendicular to axis of symmetry (for thin rod, set $R = 0$)	$G^2 = \frac{1}{4}R^2 + \frac{1}{12}A^2$
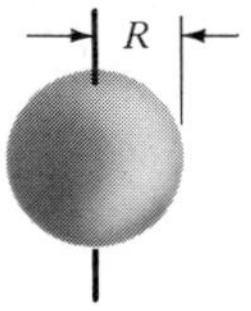 R	Thin-walled, spherical shell about diameter	$G^2 = \frac{2}{3}R^2$
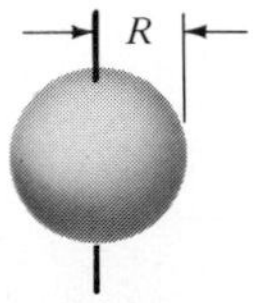 R	Solid sphere about diameter	$G^2 = \frac{2}{5}R^2$
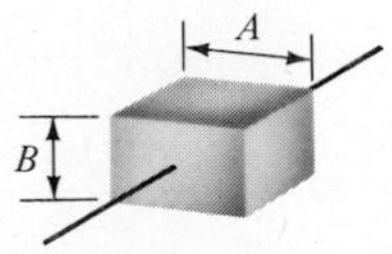 A, B	Rectangular block about axis through center of mass perpendicular to face	$G^2 = \frac{1}{12}(A^2 + B^2)$

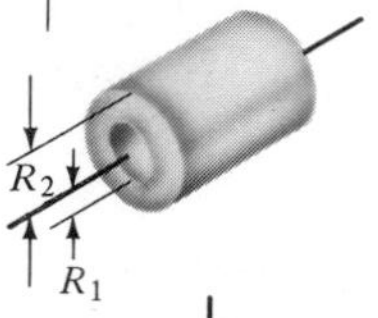

Table 10-1 shows a number of bodies of uniform density and commonly encountered shapes, rotating about the axes indicated. It lists for each body the quantity G^2, the square of the **gyration radius.** This quantity provides a convenient way of specifying the moment of inertia of a body for rotation about a certain axis, being defined so that

$$I = MG^2 \tag{10-16}$$

where M is the mass of the body. The gyration radius is just the radius of a hoop, having the same mass as the body, whose moment of inertia for rotation about its axis of symmetry is equal to the moment of inertia of the body. In terms of gyration radii, the parallel-axis theorem reads

$$G'^2 = G^2 + D^2 \tag{10-17}$$

Example 10-4 makes use of Eq. (10-2), $\mathbf{T} = I\boldsymbol{\alpha}$, and a gyration radius obtained from Table 10-1.

EXAMPLE 10-4

A grinding wheel of uniform density, radius 32 cm, and mass 100 kg is depicted in Fig. 10-11. When it is switched on, the electric motor, whose shaft forms the axle of the wheel, applies a torque which can be assumed to be of constant magnitude. (Certain types of electric motors do produce approximately constant torques.) What must be the magnitude of the torque if the motor is to be able to bring the wheel from rest to its operating speed of 200 rotations per minute in 3.0 s? Ignore friction.

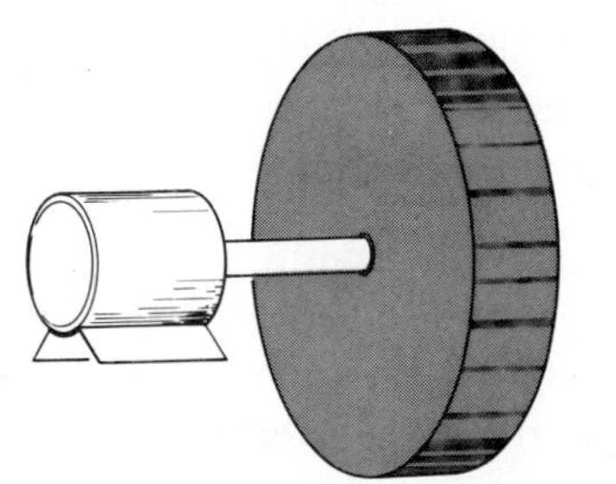

Fig. 10-11 A grinding wheel mounted on the shaft of an electric motor.

■ Because of the symmetry of the wheel about the rotation axis, its angular momentum about an origin on this axis is parallel to its angular velocity, no matter where along the axis you choose to locate the origin. Thus $\mathbf{L} = I\boldsymbol{\omega}$, and you can use Newton's second law for rotational motion in the form given by Eq. (10-2):

$$\mathbf{T} = I\boldsymbol{\alpha}$$

The direction of the torque vector **T** is the same as that of the angular acceleration vector $\boldsymbol{\alpha}$, namely along the shaft. Hence you can write the equation in terms of signed scalars:

$$T = I\alpha$$

To evaluate the wheel's moment of inertia I when rotating about the axle, you consult Table 10-1 and find from the second entry that the square of its gyration radius is

$$G^2 = \tfrac{1}{2}R^2$$

where R is the radius of the wheel. According to Eq. (10-16),

$$I = MG^2$$

or

$$I = \tfrac{1}{2}MR^2$$

where M is the mass of the wheel. Hence

$$T = \frac{MR^2\alpha}{2}$$

Since T is constant, α is also. So you can evaluate α in terms of the given values of the wheel's final angular velocity ω and the elapsed time t by using Eq. (9-8),

$$\omega = \omega_i + \alpha t$$

which pertains to constant angular acceleration. Setting $\omega_i = 0$ and solving for α, you have

$$\alpha = \frac{\omega}{t}$$

Thus

$$T = \frac{MR^2\omega}{2t}$$

To obtain a numerical value, first you must express $\omega = 200$ rotations per minute in terms of radians per second. This is done by writing

$$\omega = 200\,\frac{\text{rotations}}{\text{min}} \times \frac{2\pi\ \text{rad}}{1\ \text{rotation}} \times \frac{1\ \text{min}}{60\ \text{s}} = 21\ \text{rad/s}$$

Using this value and the other values quoted, you obtain

$$T = \frac{100\ \text{kg} \times (0.32\ \text{m})^2 \times 21\ \text{s}^{-1}}{2 \times 3.0\ \text{s}}$$

$$= 36\ \text{m}\cdot\text{kg}\cdot\text{m/s}^2$$

or

$$T = 36\ \text{m}\cdot\text{N}$$

10-2 THE PHYSICAL PENDULUM AND THE TORSION PENDULUM

A pendulum is a suspended body oscillating under the influence of the gravitational force acting on it. If the body is treated as a particle and if the mass of the suspending cord or rod is ignored, the system is called a simple pendulum. The system is called a **physical pendulum** if the body is not treated as a particle and/or if the mass of the suspending member is not ignored. To make *accurate* predictions concerning *any* pendulum, we must treat it as a physical pendulum. This can be done in an easy way that builds directly on the results of the detailed analysis of the simple pendulum carried out in Secs. 6-3 and 6-5. The procedure involves using a form of Newton's second law to relate the torque, moment of inertia, and angular acceleration of the physical pendulum in an equation completely analogous to the equation relating the force, mass, and acceleration in a simple pendulum. In fact, we will find the equations governing the oscillations of these two systems to be of identical mathematical form. We will thus be able to apply directly to the physical pendulum all the results obtained in Chap. 6 for the simple pendulum.

A system which must certainly be treated as a physical pendulum is shown in Fig. 10-12. A body of mass M and arbitrary shape is suspended from a fixed axle about which it can rotate, the axle being perpendicular to the vertical plane represented by the page. This plane passes through the body's center of mass. The origin O is located on the axle at its intersection with the plane. At the instant illustrated, a vector $\mathbf{D}$ extending from O to the center of mass is inclined at an angle ϕ. This angle is measured from the downward direction, and we take it to be positive when $\mathbf{D}$ is rotated counterclockwise from that direction, as in the figure.

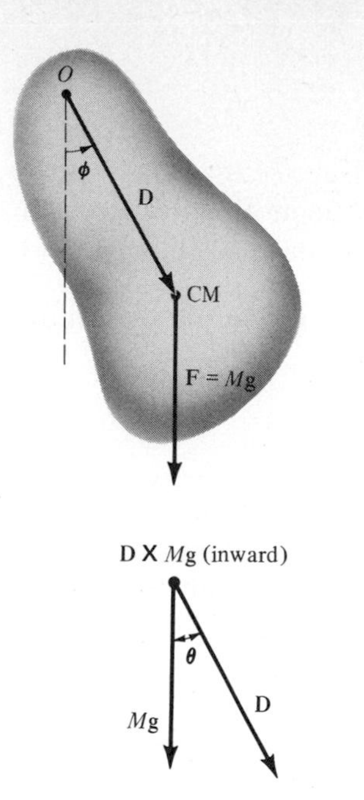

Fig. 10-12 A physical pendulum consisting of a body of mass M suspended from a horizontal axle. The origin O lies on the axle at its intersection with the vertical plane containing the body's center of mass. The position of the center of mass with respect to O is given by the vector **D**. In the auxiliary construction, the notation (inward) next to the vector $\mathbf{D} \times M\mathbf{g}$ means that the vector points directly into the page, away from the viewer.

The first thing we must do is to obtain a rotational form of Newton's second law that is applicable to the case at hand. Because of the arbitrary shape of the body, its angular momentum **L** about O may not be parallel to its angular velocity $\boldsymbol{\omega}$. Thus here there is a question about the validity of Eq. (10-1), $\mathbf{L} = I\boldsymbol{\omega}$. This puts into question the validity of Eq. (10-2), $\mathbf{T} = I\boldsymbol{\alpha}$, because the second equation depends on the first. However, we can always use Eq. (10-9) to write

$$L_{\parallel} = I\omega$$

where $L_{\parallel}$ is the component of **L** in a direction along the rotation axis. Then we take components along that direction of both sides of the always-valid equation $\mathbf{T} = d\mathbf{L}/dt$. We obtain $T_{\parallel} = dL_{\parallel}/dt = d(I\omega)/dt = I\, d\omega/dt = I\alpha$, or

$$T_{\parallel} = I\alpha \tag{10-18}$$

The quantity $T_{\parallel}$ in Eq. (10-18) is the component of **T**, the net torque about O acting on the body, in the direction along the rotation axis. We take that direction to be positive outward. Since the force applied to the body by the axle exerts no torque about O, **T** is the torque about O produced by the gravitational force $M\mathbf{g}$ applied to the body's center of mass. This torque is

$$\mathbf{T} = \mathbf{D} \times M\mathbf{g}$$

Its magnitude is

$$T = |\mathbf{D} \times M\mathbf{g}| = D \sin\theta\, Mg$$

The auxiliary construction in the figure shows that the angle θ between **D** and $M\mathbf{g}$ equals the magnitude of ϕ. Also, the construction shows that the cross-product right-hand rule gives the direction of **T** as inward at the instant illustrated. Thus

$$T_{\parallel} = -DMg \sin\phi \tag{10-19}$$

The minus sign reflects the fact that we have chosen the outward direction as positive.

The quantity α in Eq. (10-18) is a signed scalar describing the magnitude and sense of the body's angular acceleration about the axis. Its value is

$$\alpha = \frac{d\omega}{dt} = \frac{d}{dt}\left(\frac{d\phi}{dt}\right) = \frac{d^2\phi}{dt^2} \tag{10-20}$$

The correct sign is given by this expression because we have taken the positive sense of rotation (counterclockwise) and the positive direction along the rotation axis (outward) so as to conform with the rotation-vector right-hand rule.

The quantity I in Eq. (10-18) is the body's moment of inertia for rotation about the axis. It can be written as

$$I = MG^2 \tag{10-21}$$

where G is the body's gyration radius for that axis.

Substituting Eqs. (10-19) through (10-21) into Eq. (10-18), we obtain

$$-DMg \sin\phi = MG^2 \frac{d^2\phi}{dt^2}$$

After canceling and transposing, we have

$$\frac{d^2\phi}{dt^2} = -\frac{Dg}{G^2} \sin \phi \tag{10-22}$$

The solutions to this differential equation describe the motion of the physical pendulum by specifying how its angular coordinate ϕ changes with time t.

The differential equation governing the motion of a simple pendulum is Eq. (6-8*b*):

$$\frac{d^2\phi}{dt^2} = -\frac{g}{l} \sin \phi$$

Compare this with Eq. (10-22). The two seem to be identical, except that the constant factor on the right side of the physical-pendulum equation is Dg/G^2, whereas it is g/l in the simple-pendulum equation. This means that we can apply to the physical pendulum everything we learned in Chap. 6 about a simple pendulum from solving the simple-pendulum equation. All we have to do is replace g/l everywhere by Dg/G^2. For instance, by making the replacement in Eq. (6-28*b*), $\nu = (1/2\pi)\sqrt{g/l}$ where $\phi \ll 1$, we can conclude immediately that the frequency ν for *small-amplitude* oscillations of a physical pendulum is

$$\nu = \frac{1}{2\pi}\sqrt{\frac{Dg}{G^2}} \qquad \text{where } \phi \ll 1 \tag{10-23}$$

Note that this general result contains also the case of a simple pendulum of length l because in a simple pendulum $G = l$ and $D = l$, so $Dg/G^2 = g/l$.

Example 10-5 makes use of the relation between Dg/G^2 in a physical pendulum and g/l in a simple pendulum.

EXAMPLE 10-5

A physical pendulum is made by hanging a meter stick from a nail passing through a hole very near the end of the stick. Determine the length of the string of a simple pendulum that has approximately the same frequency for small oscillations as that of the physical pendulum.

■ Considering the meter stick to be a rod of length A and negligible width B, you find from the last entry in Table 10-1 that the square of the gyration radius for an axis perpendicular to the rod, and passing through its center of mass, has the value $\frac{1}{12}A^2$. Using the parallel-axis theorem in the form of Eq. (10-17), with

$$D = \frac{A}{2}$$

you find that for an axis at the end of the rod the square of the gyration radius is

$$G^2 = \frac{A^2}{12} + D^2 = \frac{A^2}{12} + \left(\frac{A}{2}\right)^2 = \frac{A^2}{3}$$

So

$$\frac{G^2}{Dg} = \frac{A^2 2}{3Ag} = \frac{2A}{3g}$$

The simple pendulum will have the same period if its length l is such that

$$\frac{l}{g} = \frac{2A}{3g}$$

so that

$$l = \frac{2A}{3}$$

For the particular case of a meter stick, you have

$$l = \frac{2 \times 1 \text{ m}}{3} = \tfrac{2}{3}\text{m}$$

You should test this result experimentally, using a meter stick (or a rod of any convenient length) and a string supporting a compact body.

For an extended body, the square of the gyration radius about a fixed axis which does not pass through its center of mass can be measured by using the body as a physical pendulum. First a measurement is made to determine the location of the center of mass by suspending the body from several points, preferably on its periphery, as in Fig. 9-31. The value of D can then be measured. Next the body is mounted on the axis and set into oscillation, and the value of the frequency ν is measured. Since the gravitational acceleration g is known, Eq. (10-23) can be used to evaluate G^2. The moment of inertia I can then be found by measuring the mass M of the body and setting $I = MG^2$.

Alternatively, if G^2 and D are known for a physical pendulum, a measurement of ν can be used to determine the value of g. This is the technique used in 1817 by Captain Henry Kater for making the first accurate measurements of the variation of the gravitational acceleration from one location on the surface of the earth to another. Such measurements of g are of great importance in the field of geology, where variations of the order of a few parts per million can yield information concerning the shape of the earth, the composition and structure of the subsurface, and the presence of valuable minerals.

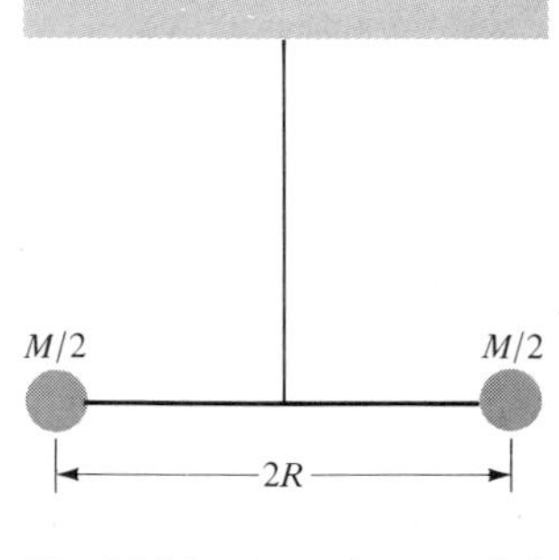

Fig. 10-13 A torsion pendulum whose total mass is M. It consists of two compact bodies, each of mass $M/2$, connected by a rigid rod of negligible mass having length $2R$. The rod is suspended from its central balance point by a torsion fiber.

Now we will apply rotational mechanics to analyze the behavior of an important system called a **torsion pendulum.** The simplest form of a torsion pendulum is illustrated in Fig. 10-13. It consists of a rod of length $2R$ having a negligible mass, with compact bodies of mass $M/2$ mounted on each end. The rod is suspended horizontally at its center from a wire called the **torsion fiber.** If the wire is twisted, it obeys a rotational version of Hooke's law quite analogous to the law $F = -kx$ obeyed by a wire when it is stretched by an amount x. The **rotational Hooke's law** is

$$T = -k\phi \tag{10-24}$$

In this expression the signed scalar ϕ is the angle through which the torsion fiber is twisted, the signed scalar T is the restoring torque the fiber exerts because it is twisted, and k is a positive constant known as its **torsion constant.**

If the rod is displaced from its equilibrium position through some angle in the horizontal plane and then released, the rod will start to oscillate about the equilibrium position. We can use the analogy between rotational and translational mechanics to show that the oscillation is harmonic. Consider the system at a moment when the angle of displacement from equilibrium is ϕ. The restoring torque $T = -k\phi$ produces an angular acceleration $d^2\phi/dt^2$ in accordance with the rotational form of Newton's second law, $T = I\alpha = I\, d^2\phi/dt^2$. Thus

$$\frac{d^2\phi}{dt^2} = \frac{T}{I}$$

or

$$\frac{d^2\phi}{dt^2} = -\frac{k}{I}\phi \tag{10-25}$$

Here I is the moment of inertia of the torsion pendulum for rotation about the perpendicular axis through its center of mass. This differential equation is to be compared to Eq. (6-11),

$$\frac{d^2x}{dt^2} = -\frac{k}{m}x$$

which pertains to a body of mass m at the free end of a spring of force constant k. The differences between these equations are that the dependent variable is a rotational coordinate in the first and a translational coordinate in the second, and that the constant on the right side is a rotational Hooke's-law constant divided by a moment of inertia in the first and a translational Hooke's-law constant divided by a mass in the second. But these differences are trivial from a mathematical point of view. Therefore we know that if the torsion pendulum is displaced from its equilibrium position and then released, the angular coordinate ϕ will henceforth be a sinusoidal function of time.

By analogy to Eq. (6-28*a*), the frequency of the harmonic oscillation is

$$\nu = \frac{1}{2\pi}\sqrt{\frac{k}{I}} \tag{10-26}$$

Note that this result is not limited to oscillations in which $\phi \ll 1$. The only restriction on the oscillation is that it must never be large enough that the restoring torque ceases to be proportional to ϕ, so that Eq. (10-24) ceases to apply. In the particular case of the ideal torsion pendulum in Fig. 10-13, whose entire mass M is concentrated at the ends of the rod of length $2R$, we have $I = MR^2$. The frequency in this case is

$$\nu = \frac{1}{2\pi}\sqrt{\frac{k}{MR^2}} \tag{10-27}$$

The timing of a mechanical watch is regulated by a torsion pendulum in the form of a wheel connected to a fine spiral spring which resists rotation of the wheel from its equilibrium position. The wheel oscillates about this position at a frequency governed by Eq. (10-26). Very good bearings are used to minimize friction. Also, the wheel is made from an alloy such as Invar, which expands and contracts very little as the temperature changes. The use of this metal reduces temperature variations of the wheel's moment of inertia and therefore stabilizes the oscillation frequency. Each time the wheel oscillates, it causes the motion of a ratchet arrangement that allows a gear driven by the mainspring of the watch to rotate by one tooth. When this happens, a small blow is transmitted from the gear through the ratchet to the oscillating wheel. The purpose is to compensate for friction in the wheel bearings, which tends to damp the oscillation. As the mainspring unwinds, the strength of the blows diminishes, and thus the amplitude of the oscillation diminishes also. But since the oscillation is harmonic, the oscillation frequency does not change. The watch therefore continues to run at a constant rate.

The torsion pendulum is often used to determine the torsion constant k of a torsion fiber. If the moment of inertia of the system is known from other measurements or can be calculated, a measurement of the oscillation frequency immediately yields the value of k through use of Eq. (10-26). Once this has been determined, the system can be used to measure very small torques (and thus very small forces as well). As we will see in Chap. 11, a torsion pendulum was used by Cavendish in the late 1790s to determine the so-called universal gravitational constant. This constant is a measure of the strength of the gravitational force exerted between any two bodies in the universe. Cavendish used its value and the value of the gravitational acceleration to obtain the first reliable evaluation of the mass of the earth.

10-3 THE TOP

In Chap. 9 the kinematics of rotation about a free axis was introduced by calling your attention to the gravity-defying motion of a child's top, and an explanation of the motion was promised. We give the explanation in this section. Then we give examples of systems, much more important than a child's top, whose motions have the same explanation.

A top performing what is called **uniform precessional motion** is illustrated in Fig. 10-14. The body spins rapidly about its axis of symmetry with a large **spin angular velocity $\boldsymbol{\omega}_s$**. We assume the sense of spin to be counterclockwise as seen from above, so that $\boldsymbol{\omega}_s$ is directed along the axis away from the pointed end. At the same time the axis of symmetry itself **precesses** slowly and uniformly about the vertical direction in the same sense as the spin, tracing out a cone whose apex lies where the tip (the pointed end) of the top rests in the small depression that the top's weight produces in the floor. Thus the top's center of mass rotates about a vertical axis with a small **precessional angular velocity $\boldsymbol{\omega}_p$**, directed vertically upward.

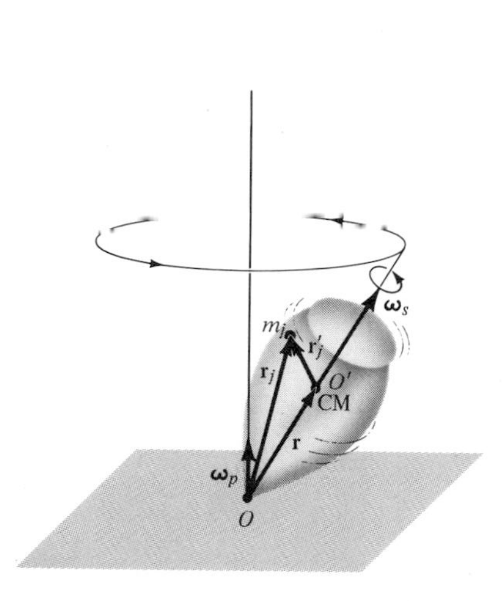

Fig. 10-14 The uniform precessional motion of a child's top. The symbols are explained in the text.

The total angular momentum of the top about an origin O at the fixed location of its tip is due principally to the rapid rotation of the individual particles of the top around the spin axis, that is, the $\boldsymbol{\omega}_s$ axis. But there is also a small contribution coming from the slow rotation of the center of mass of the top around the precession axis, that is, the $\boldsymbol{\omega}_p$ axis. We cannot begin our explanation by neglecting the latter, even though it is small, because the precessional motion is precisely what we want to explain.

There is a relation, that we will prove soon, which makes it easy to take into account both contributions to the top's total angular momentum. The relation shows that the total angular momentum $\mathbf{L}$ of a body about an origin O can be expressed as the sum of two parts. One part is $\mathbf{L}_{\text{about CM}}$, the angular momentum of the body about an origin moving with its center of mass. This angular momentum arises from the rotation of the body about its center of mass. The other part is $\mathbf{L}_{\text{of CM}}$, the angular momentum about O due to the rotation of the center of mass of the body about that origin. Thus the total angular momentum $\mathbf{L}$ has the value.

$$\mathbf{L} = \mathbf{L}_{\text{about CM}} + \mathbf{L}_{\text{of CM}} \tag{10-28}$$

A body's total angular momentum about an origin is the angular momentum due to its rotation about its center of mass plus the angular momentum due to the rotation of its center of mass about the origin. This statement sounds so plausible that you might think it to be true if "center of mass" were replaced by "any point in the body." The proof of Eq. (10-28) given below in small print shows, however, that it applies *only* to the center of mass.

Consider the top in Fig. 10-14 as an example of a body to which Eq. (10-28) applies. A typical particle of the body has mass m_j, as indicated in the figure. Its position relative to a fixed origin O is given by the vector $\mathbf{r}_j$. Its position relative to a moving origin O', located at the body's center of mass, is given by the vector $\mathbf{r}'_j$. If we write the position vector of O' relative to O as $\mathbf{r}$, the figure shows that

$$\mathbf{r}_j = \mathbf{r}'_j + \mathbf{r} \tag{10-29}$$

We calculate the velocity $\mathbf{v}_j = d\mathbf{r}_j/dt$ of the particle with respect to O by taking the time derivative of all terms in Eq. (10-29), obtaining

$$\mathbf{v}_j = \mathbf{v}'_j + \mathbf{v}$$

Here $\mathbf{v}'_j = d\mathbf{r}'_j/dt$ is the velocity of the particle with respect to O' and $\mathbf{v} = d\mathbf{r}/dt$ is the velocity of O' with respect to O. Next we multiply the velocity equation by m_j. This leads to a relation between $\mathbf{p}_j = m_j\mathbf{v}_j$, the momentum of the particle as seen from the origin O, and $\mathbf{p}'_j = m_j\mathbf{v}'_j$, its momentum as seen from O'. The relation is

$$\mathbf{p}_j = \mathbf{p}'_j + m_j\mathbf{v} \tag{10-30}$$

The angular momentum of the particle about O is

$$\mathbf{l}_j = \mathbf{r}_j \times \mathbf{p}_j$$

By using Eqs. (10-29) and (10-30), $\mathbf{l}_j$ can be expressed in terms of the primed quantities as

$$\mathbf{l}_j = (\mathbf{r}'_j + \mathbf{r}) \times (\mathbf{p}'_j + m_j\mathbf{v})$$

or

$$\mathbf{l}_j = \mathbf{r}'_j \times \mathbf{p}'_j + \mathbf{r}'_j \times m_j\mathbf{v} + \mathbf{r} \times \mathbf{p}'_j + \mathbf{r} \times m_j\mathbf{v}$$

To evaluate $\mathbf{L}$, the body's total angular momentum about O, we sum over the n particles it contains. Thus

$$\mathbf{L} = \sum_{j=1}^{n} \mathbf{l}_j$$

Using the evaluation of $\mathbf{l}_j$ just obtained, we have

$$\mathbf{L} = \sum_{j=1}^{n} \mathbf{r}'_j \times \mathbf{p}'_j + \sum_{j=1}^{n} \mathbf{r}'_j \times m_j\mathbf{v} + \sum_{j=1}^{n} \mathbf{r} \times \mathbf{p}'_j + \sum_{j=1}^{n} \mathbf{r} \times m_j\mathbf{v} \tag{10-31}$$

The first term on the right side of this equation has the value

$$\sum_{j=1}^{n} \mathbf{r}'_j \times \mathbf{p}'_j = \sum_{j=1}^{n} \mathbf{l}'_j = \mathbf{L}'$$

The quantity $\mathbf{L}'$ is the body's angular momentum about O' due to the motion of its particles about that point. Since O' is at the body's center of mass, $\mathbf{L}'$ can also be written as $\mathbf{L}_{\text{about CM}}$. Thus

$$\sum_{j=1}^{n} \mathbf{r}'_j \times \mathbf{p}'_j = \mathbf{L}' = \mathbf{L}_{\text{about CM}} \tag{10-32a}$$

The second term in Eq. (10-31) is

$$\sum_{j=1}^{n} \mathbf{r}'_j \times m_j\mathbf{v} = \left(\sum_{j=1}^{n} m_j\mathbf{r}'_j\right) \times \mathbf{v}$$

According to Eq. (9-48a), the summation over j of $m_j\,\mathbf{r}_j$ gives the total mass M of the body multiplied into a vector extending from O' to the body's center of mass. Since O' is *at* the center of mass, this vector has zero length. Thus the summation has the value zero, and so the second term in Eq. (10-31) is zero.

To evaluate the third term in Eq. (10-31), we write $\mathbf{p}_j' = m_j\mathbf{v}_j' = m_j d\mathbf{r}_j'/dt$ and use these equalities to obtain

$$\sum_{j=1}^{n} \mathbf{r} \times \mathbf{p}_j' = \sum_{j=1}^{n} \mathbf{r} \times m_j \frac{d\mathbf{r}_j'}{dt}$$

The summation on the right side of this equation can be rewritten by considering the following facts: (1) Since $\mathbf{r}$ is the same for all terms in the summation, it can be taken outside the summation. (2) Since all the m_j are constants, $m_j d\mathbf{r}_j'/dt = (d/dt)(m_j\mathbf{r}_j')$. (3) The sum of the derivatives equals the derivative of the sum; see Eq. (2-14). Hence the differentiation can be performed after the summation is carried out, instead of before. Thus the third term of Eq. (10-31) gives us

$$\sum_{j=1}^{n} \mathbf{r} \times \mathbf{p}_j' = \mathbf{r} \times \frac{d}{dt} \sum_{j=1}^{n} m_j \mathbf{r}_j'$$

Since we have just shown that the summation over j of $m_j \mathbf{r}_j'$ is zero, its time derivative is zero as well. Hence the third term in Eq. (10-31) is zero also.

The last term in Eq. (10-31) is

$$\sum_{j=1}^{n} \mathbf{r} \times m_j\mathbf{v} = \mathbf{r} \times \left(\sum_{j=1}^{n} m_j\right) \mathbf{v} = \mathbf{r} \times M\mathbf{v} = \mathbf{r} \times \mathbf{P} \equiv \mathbf{L}_{\text{of CM}} \qquad (10\text{-}32b)$$

Here M is the total mass of the body and $\mathbf{P}$ is the momentum which a particle of that mass would have as viewed from O if it moved with the velocity $\mathbf{v}$ of the center of mass relative to O. Since $\mathbf{r}$ gives the position of the particle relative to O, the quantity $\mathbf{r} \times \mathbf{P}$ is $\mathbf{L}_{\text{of CM}}$, the angular momentum about O due to the motion of the body's center of mass about that origin.

Using Eqs. (10-32a) and (10-32b) to evaluate the first and last terms on the right side of Eq. (10-31), together with the fact that the other two terms are zero, we obtain the proof of Eq. (10-28).

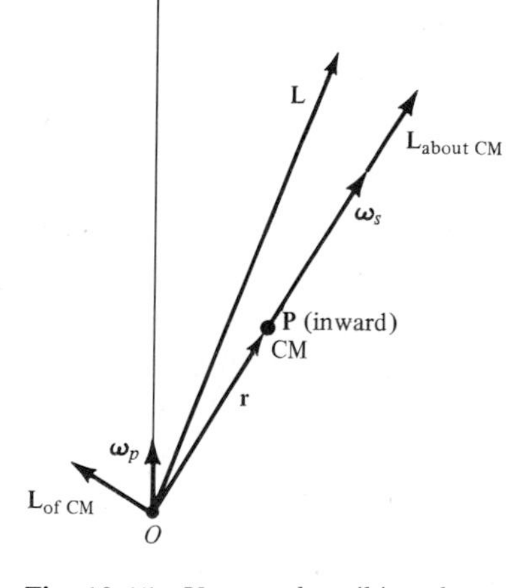

Fig. 10-15 Vectors describing the uniform precessional motion of a top. At the instant depicted, the top's center of mass is moving through the plane of the page, and all the vectors in the figure lie in that plane, except for the vector **P** which extends into the plane.

Figure 10-15 applies Eq. (10-28) to the uniform precessional motion of a top at an instant when its spin angular velocity vector $\boldsymbol{\omega}_s$ lies in the plane of the page. The angular momentum $\mathbf{L}_{\text{about CM}}$ arises from the top's rotation about its center of mass along the axis of $\boldsymbol{\omega}_s$. Since the top is symmetrical with respect to this axis, the direction of $\mathbf{L}_{\text{about CM}}$ is the same as that of $\boldsymbol{\omega}_s$. The magnitude of $\mathbf{L}_{\text{about CM}}$ is large because the magnitude of $\boldsymbol{\omega}_s$ is large. The vector $\mathbf{r}$ extends from the fixed origin O along the spin axis to the top's center of mass. At the instant depicted in the figure, the center of mass of the top of total mass M is moving into the page with velocity $\mathbf{v}$. So the associated momentum $\mathbf{P} = M\mathbf{v}$ is directed into the page. The angular momentum arising from the rotation about O of the center of mass at the precessional angular velocity $\boldsymbol{\omega}_p$ is $\mathbf{L}_{\text{of CM}} = \mathbf{r} \times \mathbf{P}$. The cross-product right-hand rule shows that $\mathbf{L}_{\text{of CM}}$ is in the plane of the page and inclined to the vertically directed axis of $\boldsymbol{\omega}_p$, as illustrated in the figure. The magnitude of $\mathbf{L}_{\text{of CM}}$ is small because the magnitude of $\boldsymbol{\omega}_p$ is small. The total angular momentum of the top about O is given by the relation $\mathbf{L} = \mathbf{L}_{\text{about CM}} + \mathbf{L}_{\text{of CM}}$. All three of these vectors lie in the plane of the page at the instant depicted. The construction shows that the direction of $\mathbf{L}$ is a little closer to the vertical than is the direction of $\mathbf{L}_{\text{about CM}}$ and that its magnitude is a little larger than is the magnitude of $\mathbf{L}_{\text{about CM}}$. As time passes, the plane containing all three angular momentum vectors rotates slowly about a vertical line with the precessional angular velocity $\boldsymbol{\omega}_p$. But in every other regard these vectors remain as in the construction of Fig. 10-15.

The only angular momentum vector shown in Fig. 10-16 is $\mathbf{L}$, the top's total angular momentum vector about the fixed origin O. Also shown is the

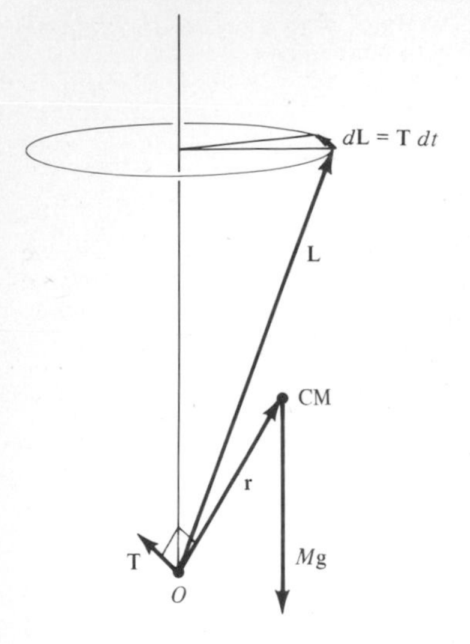

Fig. 10-16 A top is acted on by a net torque **T** about an origin at its pointed end, with **T** always in the horizontal direction. The reason is that **T** = **r** × *M***g**, so it is the cross product of two vectors which both always lie in a vertical plane. The first of these is the vector **r** from the origin to the top's center of mass, and the second is the gravitational force *M***g** acting on the center of mass. In any small time interval *dt*, the change *d***L** in the top's total angular momentum **L** is in the same horizontal direction as **T** because *d***L** = **T** *dt*. Since **L** itself is always in a vertical plane (the one containing **r** and *M***g**), this means that *d***L** is always perpendicular to **L**. Consequently, **L** cannot change its magnitude; it can change only its direction. At the end of the time interval *dt*, the tip of the vector **L** will have moved to the tip of the vector *d***L** shown in the figure. In the next time interval the process repeats itself. However, the plane containing **r**, *M***g**, and **L** has precessed through a small angle about the vertical line, causing **T** and *d***L** to do the same. As time continues to pass, the tip of the vector **L** slowly describes a circle about the vertical line, and the vector itself describes a cone about that line.

position vector **r**, from *O* to the center of mass of the top, and the gravitational force vector **F** = *M***g**. This downward-directed force applied at the center of mass is completely equivalent in its effects to the combined effects of all the gravitational forces applied to each of the particles of the top. In particular, it can be used to evaluate the net torque **T** about *O*, exerted on the top by gravity. This torque is **T** = **r** × **F**, or

$$\mathbf{T} = \mathbf{r} \times M\mathbf{g} \tag{10-33}$$

The cross-product right-hand rule shows that **T** is horizontally directed, as illustrated in the figure. (There is also an upward-directed force applied at *O* to the tip of the top by the table on which it rests. But this force exerts *no* torque about *O*. Why?)

The net torque **T** applied to the top causes its total angular momentum **L** to change in such a way as to satisfy the rotational form of Newton's second law,

$$\mathbf{T} = \frac{d\mathbf{L}}{dt} \tag{10-34}$$

We can apply the law because both **T** and **L** are specified about an origin *O* which can be considered fixed in an inertial frame.

The essential point in the explanation of the top's uniform precessional motion is illustrated in Fig. 10-16. Since the torque **T** applied to the top always acts in a horizontal direction, while its total angular momentum **L** is always in a vertical plane, **T** is always perpendicular to **L**. According to Eq. (10-34), the change in **L** in a small time interval *dt* is *d***L** = **T** *dt*. Thus *d***L** is always in the same direction as **T**. It follows that *d***L** is always perpendicular to **L** itself. (This is quite analogous to the fact that when a centripetal force **F** acts on a body whose momentum is **p**, the change in momentum *d***p** is perpendicular to **p** itself.) Inspection of the figure will clarify the directions of the vectors and show that their consequence is to make **L** change in time in such a way that its magnitude remains constant while its direction precesses in a cone about a vertical line. Thus **L** should behave just as it is observed to behave.

A crucial feature responsible for the behavior of a uniformly precessing top is its large spin angular momentum $\mathbf{L}_{\text{about CM}}$. If a top that is *not* spinning is placed with its pointed end on the floor, and its symmetry axis is inclined to the vertical, the top will certainly fall over when released. Since the top has no *initial* angular momentum, the horizontal increment in its angular momentum produced by the horizontally acting torque in the first time interval after release results in the top having a purely horizontal angular momentum at the end of the first time interval. This means that the top's center of mass is starting to rotate about the fixed origin *O* at its tip in a vertical plane containing the symmetry axis of the top. In other words, the top is beginning to fall, as illustrated in Fig. 10-17. In the next time interval the increment in angular momentum is in the same direction as in the first time interval because the direction of the torque is the same. So the magnitude of the angular momentum $\mathbf{L}_{\text{of CM}}$ increases as the fall accelerates.

But if the top has a large angular momentum along its symmetry axis when released, the situation is completely different. Since the increment in angular momentum is perpendicular to the initial angular momentum, the angular momentum at the end of the first time interval is *not* changed in magnitude, but *is* changed in direction. This causes the direction of the ap-

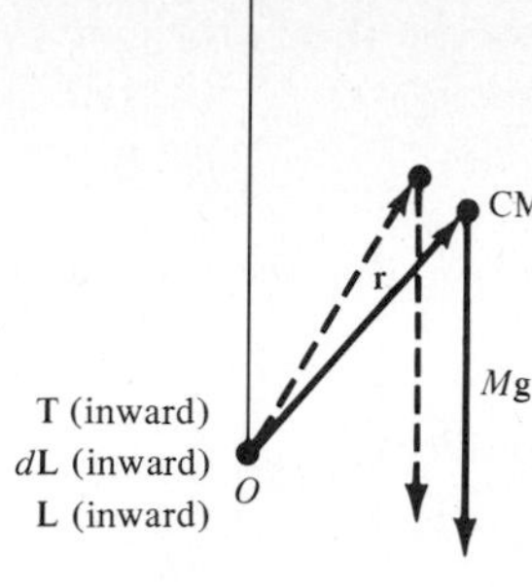

Fig. 10-17 If a top is released without spin, the gravitational torque about the origin O at its pointed end causes it to fall over. The dashed vector labeled $\mathbf{r}$ is the position vector of the top's center of mass relative to O at the instant of release. The dashed vector labeled $M\mathbf{g}$ is the gravitational force exerted at that point. The corresponding torque about O is $\mathbf{T} = \mathbf{r} \times M\mathbf{g}$, and its direction is into the page. The angular momentum change $d\mathbf{L}$ produced by the torque in a small time interval dt is $d\mathbf{L} = \mathbf{T}\, dt$, and so it also is directed into the page. Since the angular momentum itself was zero at the beginning of the time interval, $\mathbf{L}$ is directed into the page at the end of the time interval. For the beginning of the next time interval, $\mathbf{r}$ and $M\mathbf{g}$ are shown by solid vectors. The corresponding torque vector $\mathbf{T}$ is again directed into the page, and hence the vector $d\mathbf{L}$ is also. Thus at the end of that interval $\mathbf{L}$ is still directed into the page but has a larger magnitude. The process continues, with the magnitude of $\mathbf{L}$ continuing to increase, until the top strikes the floor. In this case, where $\mathbf{L}_{\text{about CM}} = 0$, we have $\mathbf{L} = \mathbf{L}_{\text{of CM}}$.

plied torque to change. So the angular momentum of the top precesses about the vertical, maintaining a *constant* magnitude. The top cannot fall over because such a motion corresponds to an ever-*increasing* magnitude of angular momentum, as we found in the preceding paragraph.

Example 10-6 develops a relation between magnitudes of the spin angular velocity and the precessional angular velocity and then applies it to obtain numerical values in a particular case.

EXAMPLE 10-6

A top is spinning about its axis of symmetry with angular speed $\omega_s = 200$ rad/s $\simeq$ (32 rotations per second). Its gyration radius about that axis is $G = 3.00$ cm, and the distance from its tip to its center of mass is $r = 4.00$ cm. Evaluate ω_p, the top's angular speed of precession, assuming it to be very small compared to ω_s. Then check the validity of the assumption.

■ First you should make a sketch like the one in Fig. 10-18, in which you assume the total angular momentum $\mathbf{L}$ of the top about an origin O at its tip to have the value $\mathbf{L} = \mathbf{L}_{\text{about CM}}$. That is, you assume $\mathbf{L}_{\text{of CM}}$ has such a small magnitude in comparison to the magnitude of $\mathbf{L}_{\text{about CM}}$ that you can ignore it in evaluating $\mathbf{L}$. This assumption is based on the fact that ω_p for a well-spun top is very small compared to ω_s. In the figure the angle between $\mathbf{L}$ and the axis of the precession cone is labeled γ, and $d\phi$ is the angle swept out in time dt by a line drawn from the tip of the vector $\mathbf{L}$

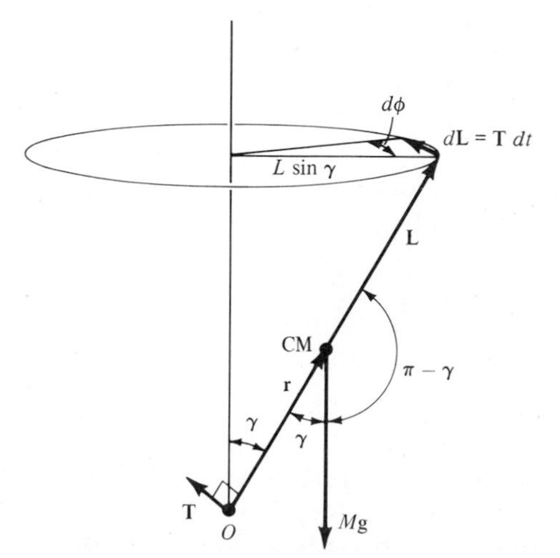

Fig. 10-18 A sketch used in Example 10-6 to evaluate the precessional angular velocity of a top. The vectors $\mathbf{T}$ and $d\mathbf{L}$ are parallel to each other, and both are perpendicular to the plane containing the vectors $\mathbf{r}$, $M\mathbf{g}$, and $\mathbf{L}$.

perpendicularly in to the cone's axis. The desired quantity ω_p is precisely $d\phi/dt$.
The figure shows you that

$$d\phi = \frac{dL}{L \sin \gamma}$$

And Eq. (10-34) tells you that the relation between dL and the torque T is

$$dL = T\,dt$$

So direct substitution gives

$$d\phi = \frac{T\,dt}{L \sin \gamma}$$

or

$$\omega_p = \frac{d\phi}{dt} = \frac{T}{L \sin \gamma} \tag{10-35}$$

You can use Eq. (10-33) to evaluate T. This gives

$$T = |\mathbf{r} \times M\mathbf{g}| = rMg \sin (\pi - \gamma)$$

where M is the mass of the top. Since $\sin (\pi - \gamma) = \sin \gamma$, this is

$$T = rMg \sin \gamma$$

Substitution into Eq. (10-35) gives you the following expression for the top's precessional angular speed:

$$\omega_p = \frac{rMg \sin \gamma}{L \sin \gamma} = \frac{rMg}{L}$$

Note that *the precessional angular speed is independent of the angle* γ *between the angular momentum of the top and the vertical direction.* Now $L = I\omega_s$ for the rotation of the symmetrical top about its axis of symmetry at angular speed ω_s. Also, its moment of inertia is $I = MG^2$, where G is its gyration radius. Hence you can write the result as

$$\omega_p = \frac{rMg}{MG^2\omega_s}$$

or

$$\omega_p = \frac{rg}{G^2}\frac{1}{\omega_s} \tag{10-36}$$

The precessional angular speed ω_p of the top is *inversely* proportional to its spin angular speed ω_s. (If you have ever played with a top, you probably have noticed the increase in ω_p as ω_s decreases as a result of friction.) The proportionality constant connecting ω_p and $1/\omega_s$ is the distance r from its tip to its center of mass, multiplied by the gravitational acceleration g, and divided by the square of its gyration radius G for rotation about its symmetry axis.

To find the numerical value of ω_p, you substitute the values given for r, G, and ω_s, and also the value of g, into Eq. (10-36). You obtain

$$\omega_p = \frac{4.00 \times 10^{-2}\ \text{m} \times 9.80\ \text{m/s}^2}{(3.00 \times 10^{-2}\ \text{m})^2 \times 2.00 \times 10^2\ \text{rad/s}}$$
$$= 2.18\ \text{rad/s} \quad (\simeq 0.35\ \text{rotations per second})$$

Since ω_p is about 1 percent of ω_s, this justifies the assumption made at the beginning of the example—that $L_{\text{of CM}}$ is very small compared to $L_{\text{about CM}}$. What would you do to obtain an accurate evaluation of ω_p if the result obtained by making the assumption $\omega_p \ll \omega_s$ led to a value of ω_p which was, say, half as large as the value of ω_s.

The behavior of a top can be more complicated than it is in uniform precessional motion. One thing that can happen is that the axis of symmetry can rise and fall periodically, while precessing at a nonuniform rate about the vertical direction. This nodding motion is called **nutation.** Another possible motion of a top is called **sleeping.** When a top sleeps, it rotates with its axis of symmetry remaining vertical. These motions are more difficult than uniform precession to observe experimentally and to analyze theoretically, so they will not be treated in this book.

Uniform precession is by far the most significant motion of a top because it is exactly analogous to a motion found in systems that are of much more interest to contemporary physics and related fields than the top. In most atoms, the nucleus has a spin angular momentum parallel to its axis of symmetry, much like the angular momentum of a top spinning about its axis of symmetry. Such a nucleus also has magnetic properties, as if it contained a microscopic bar magnet along the direction of the axis of symmetry. In the presence of a magnetic field, the nucleus experiences a torque that is always perpendicular to the direction of the axis of symmetry, and therefore always perpendicular to the direction of the angular momentum. This is just the same as the relation between the direction of the gravitational torque experienced by a top and the direction of its angular momentum. Therefore the nuclear angular momentum performs uniform precessional motion, like a top. In fact, the situation is somewhat simpler than that for a top. This is because the center of mass of the nucleus can be considered at rest in an inertial frame, so that the origin O can be taken at the center of mass. This means that $\mathbf{L} = \mathbf{L}_{\text{about CM}}$ exactly. Just as in Eq. (10-35), the angular speed of precession of the nucleus is proportional to the torque (now magnetic rather than gravitational) acting on it. Since the torque is proportional to the strength of the magnetic field applied to the nucleus, the precession speed is proportional to the strength of that magnetic field. Corresponding to the precession speed ω_p is a precession frequency $\nu_p = \omega_p/2\pi$. This precession frequency is also proportional to the strength of the magnetic field acting on the precessing nucleus.

The precessing magnetic nuclei in a sample of material can absorb electromagnetic waves, just as if they were tiny radio receivers, providing the frequency of the electromagnetic waves is the same as the precession frequency of the nuclei. By determining the frequency at which absorption occurs, and therefore determining the precession frequency, physicists and chemists can measure with great precision the strength of the magnetic field acting on the nuclei. This allows the scientists to study the magnetic field present at the nuclei of atoms, molecules, or solids as a result of the circulation of electrons about the nuclei. The magnetic field depends on the precise distribution of the electrons, and so its measurement provides detailed information about the structure of systems of prime importance in solid-state physics and chemistry. The experimental technique is called **nuclear magnetic resonance.**

Section 10-4 opens by considering one more example of uniform precessional motion.

10-4 ROTATION ABOUT AN ACCELERATING CENTER OF MASS

The earth is a majestic example of the topic of Sec. 10-3—uniform precessional motion. The precessional motion of the earth can be studied most simply by using a reference frame with an origin fixed at the earth's center of mass. We will do so, obtaining thereby an example of the topic of this section—the rotation of a body about an accelerating center of mass.

The earth spins daily about its polar axis. This axis is inclined at an angle of 23.45° to a perpendicular to the plane of the earth's annual orbit about the sun. The situation is depicted in Fig. 10-19 from the point of view

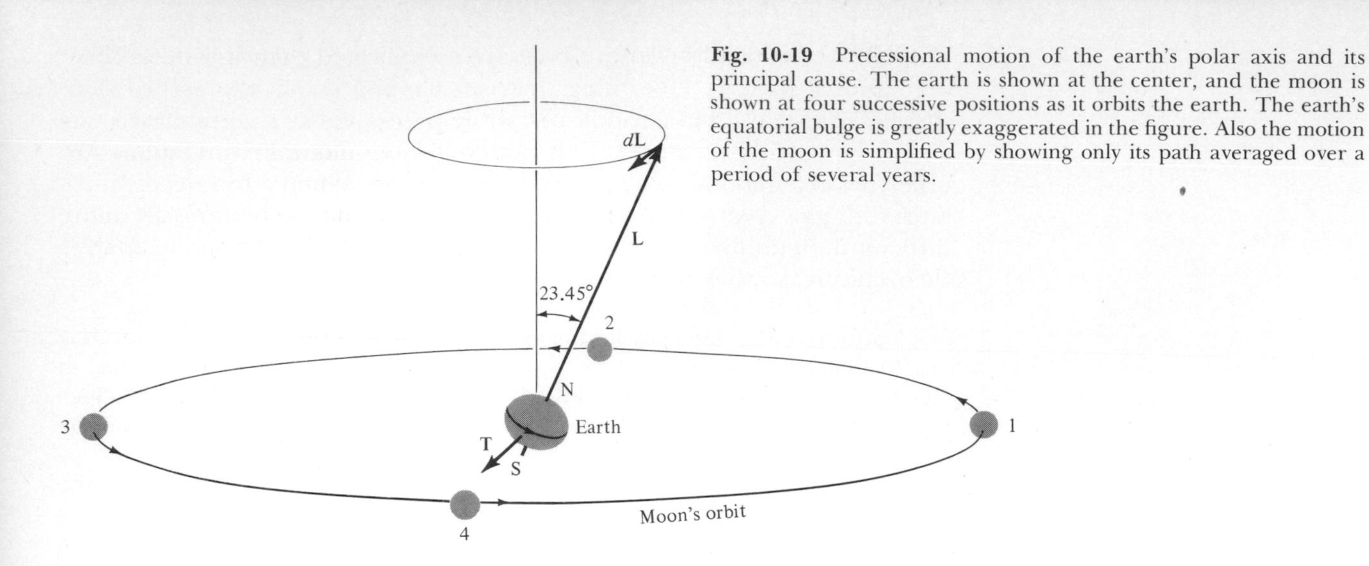

Fig. 10-19 Precessional motion of the earth's polar axis and its principal cause. The earth is shown at the center, and the moon is shown at four successive positions as it orbits the earth. The earth's equatorial bulge is greatly exaggerated in the figure. Also the motion of the moon is simplified by showing only its path averaged over a period of several years.

of a reference frame whose origin is fixed at the earth's center of mass. This reference frame does not participate in the daily rotation of the earth. That is, the earth is seen to spin about its polar axis when viewed from the reference frame.

Because of its rapid rotation about the polar axis, the earth's shape has deformed slightly from perfect sphericity. The earth bulges at the equator. A consequence of the equatorial bulge is that the gravitational forces exerted by the moon on various parts of the earth produce a torque on the earth. The torque varies in strength as the moon travels each month through its orbit about the earth. But the torque is always directed perpendicular to the earth's polar axis. You can see this by considering the forces exerted at the four moon positions shown in the figure. In position 1 the gravitational attraction of the moon for the nearer bulge of the earth, which is below the plane of the moon's orbit, is greater than its attraction for the farther bulge, which is above the plane. So the moon pulls upward on the bulge below the plane more than it pulls downward on the bulge above the plane. This produces a torque about an origin at the earth's center of mass, acting in the direction shown in Fig. 10-19. In position 3 the moon exerts a greater downward pull on the nearer bulge above the plane than on the farther bulge below the plane, and again the torque vector acts in the direction shown in the figure. At positions 1 and 4 there is no torque. If we average over each orbit of the moon about the earth, there is a net torque exerted on the earth by the moon. The torque is directed perpendicular to the earth's polar axis. It is quite weak on the scale of this system. A very similar, but somewhat weaker, torque is exerted on the earth by the sun. The sum of these two constitutes a net torque **T** acting on the earth about its center of mass. Averaged over any year, this torque has an essentially constant magnitude and a direction which is in the plane of the earth's annual orbit about the sun and perpendicular to the direction of the earth's polar axis during the particular year. That is, **T** is always perpendicular to $\boldsymbol{\omega}_s$, the spin angular velocity describing the daily rotation of the earth about its polar axis.

The earth has a large angular momentum about an origin at its center of mass because of the daily rotation. Since it is essentially symmetrical about the polar axis, this angular momentum $\mathbf{L} = \mathbf{L}_{\text{about CM}}$ is in the same direction as $\boldsymbol{\omega}_s$.

Thus there is a weak net torque $\mathbf{T}$ acting on the earth whose direction is always *perpendicular* to the earth's large angular momentum $\mathbf{L}$, just as in the case of a top. This results in a very slow precession of the angular momentum vector around a cone whose axis is normal to the plane of the earth's motion about the sun. The half-apex angle of the cone is 23.45°, and the period of the precession is 25,920 yr. The angular momentum vector is at present directed so that the North end of the polar axis of the earth points within 1° of a particular star that we call the North Star, or Polaris. People living 13,000 yr from now will find that the North pole misses pointing at that particular star by about 47°, and they will surely give it a different name. But after an additional 13,000 yr, Polaris will again be an appropriate name because the continued precessional motion will have brought the earth's polar axis back to its present alignment.

The above discussion of the precessional motion of the earth was simplified by taking the center of mass of the earth as the origin O used to define the torque and angular momentum under consideration. It was implied in the discussion that for this choice of origin the torque and angular momentum are related by Newton's second law for rotational motion—torque equals rate of change of angular momentum—just as they are in the case of a top. Is such an implication valid?

In the treatment of the precession of the top, the origin O was at the position where its tip rests on the floor. That choice of origin satisfies to a very good approximation the condition, used in deriving the rotational form of Newton's second law in Sec. 9-5, that O be the origin of an inertial reference frame. The reason is that the acceleration of O with respect to an exact inertial frame is completely negligible, compared to the acceleration associated with the motion of the top about O. But in the treatment of the precession of the earth, the origin O was at the earth's center of mass. In this case the acceleration of O with respect to an exact inertial frame, resulting from the annual motion of O about the sun, is not very much smaller than the acceleration associated with the motion of the earth about O. Thus it is not apparent that it is valid to use Newton's second law for rotational motion with an origin fixed at the earth's center of mass in discussing the precession of the earth. But, in fact, it is valid to do so. One of the surprising properties of the center of mass of any body is that when the net torque acting on a body, and its total angular momentum, are taken about an origin at the *center of mass* of the body, then the torque equals the rate of change of angular momentum, *no matter how the center of mass is moving*. This useful fact is proved in the material in small print which follows.

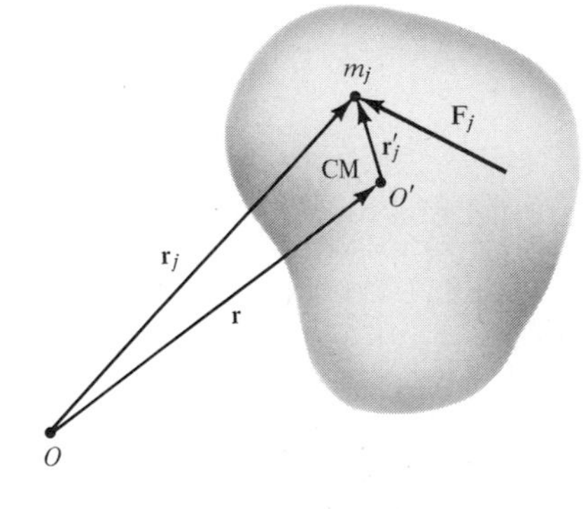

Fig. 10-20 A rotating body and one of its constituent particles. The center of mass lies at O'.

Consider the rotating body shown in Fig. 10-20 and two origins about which its angular momentum can be measured. One is the origin O of an an inertial reference frame. The other is the origin O′ fixed at the center of mass of the body. In general, the frame of which O′ is the origin is not inertial. We designate the angular momentum of the body about O by the symbol $\mathbf{L}$ and its angular momentum about O′ by the symbol $\mathbf{L}'$. A relation between these two quantities is given by Eq. (10-28). Using Eqs. (10-32) to write this relation in terms of our present notation, we find

$$\mathbf{L} = \mathbf{L}' + \mathbf{r} \times M\mathbf{v}$$

Here $\mathbf{r}$ is the vector from O to O', M is the total mass of the body, and $\mathbf{v}$ is the velocity of O' as seen from O.

Next we take the time derivative of all terms in the relation to obtain

$$\frac{d\mathbf{L}}{dt} = \frac{d\mathbf{L}'}{dt} + \frac{d\mathbf{r}}{dt} \times M\mathbf{v} + \mathbf{r} \times \frac{d}{dt}(M\mathbf{v}) \tag{10-37}$$

The second term on the right side of this equation has the value

$$\frac{d\mathbf{r}}{dt} \times M\mathbf{v} = \mathbf{v} \times M\mathbf{v} = 0 \tag{10-38a}$$

since $\mathbf{v}$ is parallel to $M\mathbf{v}$. Newton's second law for translational motion can be applied to evaluate the third term. This is because the momentum $M\mathbf{v}$ is measured relative to the inertial frame whose origin is O, so the law is applicable. It gives

$$\mathbf{r} \times \frac{d}{dt}(M\mathbf{v}) = \mathbf{r} \times \mathbf{F} \tag{10-38b}$$

where $\mathbf{F}$ is the net force applied to the body. Newton's second law for rotational motion can be applied to the term on the left side of Eq. (10-37) because the angular momentum $\mathbf{L}$ is measured about the origin O of an inertial frame. This form of the law gives

$$\frac{d\mathbf{L}}{dt} = \mathbf{T} \tag{10-38c}$$

where $\mathbf{T}$ is the net torque about O applied to the body. Using Eqs. (10-38) in Eq. (10-37), we have

$$\mathbf{T} = \frac{d\mathbf{L}'}{dt} + \mathbf{r} \times \mathbf{F} \tag{10-39}$$

Now the net torque about O applied to the body is the sum of the torques about that origin produced by the external forces applied to each of its n particles. The external force applied to the jth particle is $\mathbf{F}_j$, its position vector from O is $\mathbf{r}_j$, and the torque about O produced by the force is $\mathbf{T}_j = \mathbf{r}_j \times \mathbf{F}_j$. Thus

$$\mathbf{T} = \sum_{j=1}^{n} \mathbf{r}_j \times \mathbf{F}_j \tag{10-40a}$$

Furthermore, the net force applied to the body is the sum over the n particles of the individual forces $\mathbf{F}_j$. So

$$\mathbf{F} = \sum_{j=1}^{n} \mathbf{F}_j \tag{10-40b}$$

Substituting the values of $\mathbf{T}$ and $\mathbf{F}$ given by Eqs. (10-40) into Eq. (10-39), we have

$$\sum_{j=1}^{n} \mathbf{r}_j \times \mathbf{F}_j = \frac{d\mathbf{L}'}{dt} + \mathbf{r} \times \sum_{j=1}^{n} \mathbf{F}_j$$

Transposing the second summation on the right side of this equality, and using the fact that $\mathbf{r}$ is the same for all terms in this summation, we can rewrite the equality as

$$\sum_{j=1}^{n} \mathbf{r}_j \times \mathbf{F}_j - \sum_{j=1}^{n} \mathbf{r} \times \mathbf{F}_j = \frac{d\mathbf{L}'}{dt}$$

This can be expressed in terms of a single summation, as follows:

$$\sum_{j=1}^{n} (\mathbf{r}_j - \mathbf{r}) \times \mathbf{F}_j = \frac{d\mathbf{L}'}{dt}$$

Figure 10-20 shows that $\mathbf{r}_j - \mathbf{r} = \mathbf{r}'_j$, where $\mathbf{r}'_j$ is the position vector of the jth particle from O'. Thus we have

$$\sum_{j=1}^{n} \mathbf{r}'_j \times \mathbf{F}_j = \frac{d\mathbf{L}'}{dt}$$

This is

$$\sum_{j=1}^{n} \mathbf{T}'_j = \frac{d\mathbf{L}'}{dt}$$

where $\mathbf{T}'_j$ is the torque about O' applied to the jth particle. The sum of these torques is just $\mathbf{T}'$, the net torque about O' applied to the body. Hence we have proved that

$$\mathbf{T}' = \frac{d\mathbf{L}'}{dt} \tag{10-41}$$

This has the form of Newton's second law for rotational motion, even though O' may not be the origin of an inertial reference frame.

The physical reason for the property of the center of mass described by Eq. (10-41) can be explained in terms of the concept of fictitious forces, discussed in Sec. 5-4. Newton's second law involving force applies directly only in an inertial reference frame. But in an accelerating frame it can be made to apply if appropriate fictitious forces are added to the real forces. The sum of all the fictitious forces for a body is equivalent to a single fictitious force, and it turns out that this force acts at the body's center of mass. Therefore the fictitious force exerts no torque about an origin at the center of mass, and it can be neglected as far as rotation about the center of mass is concerned. The result is that Newton's second law for rotation applies *directly* to rotation about an origin at the center of mass, even if the center of mass is accelerating with respect to an inertial frame!

Equations (9-39), $\mathbf{T} = d\mathbf{L}/dt$, and (10-41), $\mathbf{T}' = d\mathbf{L}'/dt$, can be summarized by saying the following: *The net torque applied to a body is equal to the rate of change of its total angular momentum if either of the following conditions is met in choosing the origin about which the torque and angular momentum are taken:*

1. *The origin is located at any point, but is the origin of an inertial frame.*

2. *The origin is located at the body's center of mass, but is the origin of a reference frame which need not be inertial.*

Knowing this, we will henceforth not bother to make a distinction between Eqs. (9-39) and (10-41). That is, we will drop the primes in the latter and write both as

$$\mathbf{T} = \frac{d\mathbf{L}}{dt} \tag{10-42}$$

with the understanding that **T** and **L** are the net torque applied to the body and its total angular momentum, about an origin O of a frame which *does not need to be an inertial frame* if O lies at the body's center of mass. In many, but *not all,* circumstances, the total angular momentum of the body can be written in terms of its moment of inertia I about the rotation axis and its angular velocity $\boldsymbol{\omega}$ as

$$\mathbf{L} = I\boldsymbol{\omega} \tag{10-43}$$

The center of mass itself moves in a manner governed by the equation

$$\mathbf{F} = \frac{d\mathbf{P}}{dt} \tag{10-44}$$

where $\mathbf{F}$ and $\mathbf{P}$ are the net force applied to the body and its total momentum. In this equation the total momentum is measured from an origin O which *must be the origin of an inertial frame.* In *all* circumstances the total momentum of the body can be written in terms of its mass M and the velocity $\mathbf{v}$ of its center of mass, as

$$\mathbf{P} = M\mathbf{v} \tag{10-45}$$

This property of the center of mass can be summarized by saying the following: *The center of mass of a body moves in such a way that the net force acting on the body equals the rate of change of its total momentum relative to an inertial frame, the total momentum being the product of the body's mass and the velocity of its center of mass in the inertial frame.* The justification of this statement comes directly from experiment. We specified the position of each of the air table pucks in Chap. 4 by the position of its center of mass, and its velocity by the rate of change of that position. Then we took the momentum of the puck to be the product of its mass and the velocity of its center of mass, and we defined the net force acting on it as the rate of change of this total momentum. No matter what bodies are studied to deduce Newton's second law for translational motion, the same thing is done. So the direct results of the deduction are Eqs. (10-44) and (10-45).

If you wish, you can say that the most basic form of Newton's second law is $\mathbf{F} = d\mathbf{p}/dt$, where $\mathbf{p} = m\mathbf{v}$ and where all these quantities pertain to a particle. An exercise at the end of this chapter requires you to do a calculation in which these relations are applied to each of the particles in a body, and then a summation is taken over all its particles. The definition of center of mass is used to express the right side of the equation obtained in terms of the time derivative of the product of the body's mass and the velocity of its center of mass. Newton's law of action and reaction is used to show that the forces acting between the particles in the body to hold it together cancel in pairs, so that the left side of the equation reduces to the net force applied to the body. The equation is then seen to be precisely Eq. (10-44), in which the quantity being differentiated on the right side satisfies Eq. (10-45).

Since the translational form of Newton's second law, $\mathbf{F} = d\mathbf{P}/dt$, deals with different quantities than those dealt with in the rotational form, $\mathbf{T} = d\mathbf{L}/dt$, *the motion of the center of mass of a body can be considered separately from the rotation of the body about its center of mass.* We have done just this, quite successfully, in many of the treatments given prior to Chap. 9. These are the ones in which a body actually changes its orientation as well as its location as it moves, but the changes in orientation are ignored and only the changes in location are considered. We have done the same thing in Chap. 9, and up to this point in Chap. 10, by ignoring whatever changes in location a body might have as it moves and considering only its changes in orientation.

Now we investigate situations in which we must take into account *both* changes in orientation and changes in location of a body. We do this by describing the body's motion as a combination of rotation about its center of mass and motion of the center of mass. This is not the only way that a general motion of a body can be described. (If you give this book some gen-

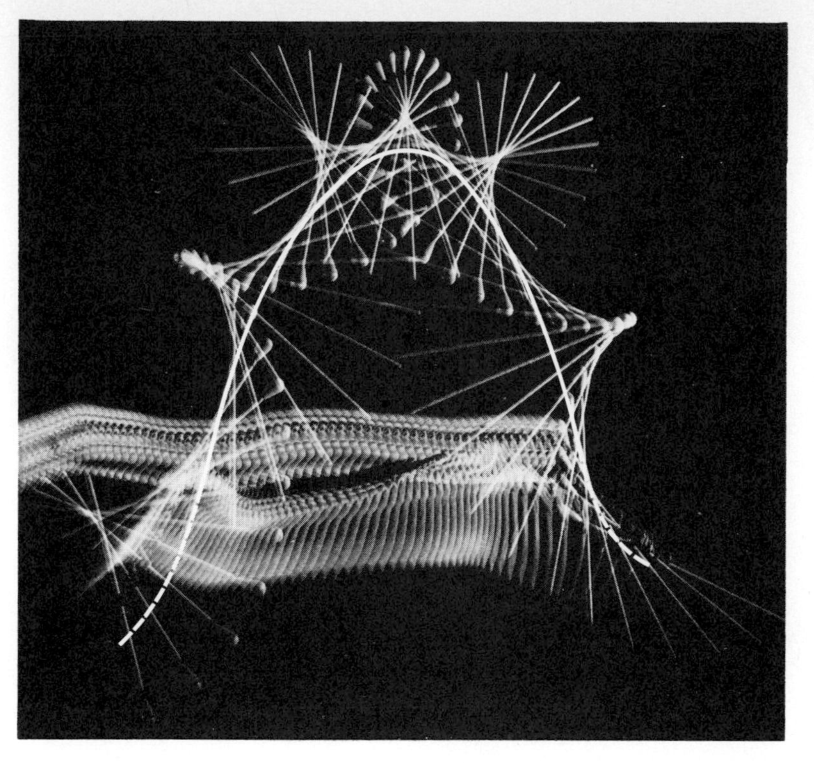

Fig. 10-21 Strobe photo of a drum majorette's baton. The path of its center of mass while in flight is shown by the solid part of the curve superimposed on the photo. (*Photo courtesy of Harold E. Edgerton.*)

eral motion and think about the motion for a moment, you will conclude that the motion can be described as a rotation about any point you choose, combined with a motion of that point.) We choose the point about which we consider the body to be rotating to be its center of mass because then we can be sure that the important equation $\mathbf{T} = d\mathbf{L}/dt$ will be valid, no matter how that point moves.

The strobe photo of a drum majorette's baton in Fig. 10-21 gives an example of a general motion of a body that can be understood by using the rotational form of Newton's second law, $\mathbf{T} = d\mathbf{L}/dt$, to analyze the motion of the body about its center of mass, and its translational form, $\mathbf{F} = d\mathbf{P}/dt$, to analyze the motion of the center of mass. The baton is a stick with one end weighted so that its center of mass is not at the center of the stick. After the baton is thrown, only the gravitational attraction of the earth acts on it. As a consequence, its center of mass moves along the parabolic trajectory shown by the solid curve in the photo. This trajectory is the path that any particle obeying the equations $d\mathbf{P}/dt = \mathbf{F} = M\mathbf{g}$ would follow. Because the baton is thrown with an initial angular momentum about the center of mass, the baton also rotates while in flight. In fact, it rotates with constant angular velocity $\boldsymbol{\omega}$ about the center of mass so as to maintain a constant angular momentum $\mathbf{L} = I\boldsymbol{\omega}$ about that point, its moment of inertia I about the rotation axis being constant. (The mass elements of the baton are distributed in what is essentially a single plane of rotation passing through the origin at the center of mass, so its angular momentum is always parallel to its angular velocity. Therefore it is valid to describe its rotation in terms of its moment of inertia.) The angular momentum cannot change because the net gravitational attraction of the earth for the body acts as if the attraction were applied at the center of mass. So the gravitational force exerts no torque about that point, and $d\mathbf{L}/dt = \mathbf{T} = 0$.

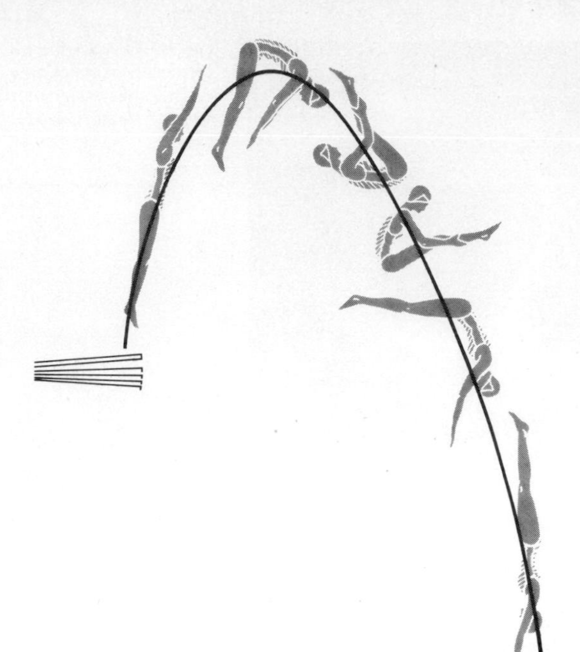

Fig. 10-22 The motion of a diver. Her center of mass moves in the parabolic trajectory of a particle, while her body rotates with constant angular momentum about the center of mass. But her angular velocity is not constant since she changes her moment of inertia about the rotation axis.

An artist's rendition of a strobe photo of a diver is shown in Fig. 10-22. The motion is similar to that of the baton because $d\mathbf{P}/dt = M\mathbf{g}$, so the diver's center of mass moves along a parabolic trajectory. Also, $d\mathbf{L}/dt = 0$, as for the baton. However, during the dive the diver temporarily decreases the moment of inertia I of her body about the rotation axis passing through her center of mass by doubling up. This causes her angular velocity $\boldsymbol{\omega}$ to increase, since the angular momentum $\mathbf{L} = I\boldsymbol{\omega}$ must be constant. The increased angular velocity helps her complete the forward turn before hitting the water. [It is valid to describe the rotation of the diver in terms of a moment of inertia because her angular momentum is always parallel to her angular velocity. The reason is that the diver is rotating about a *principal axis.* See the discussion in small print following Eq. (10-13).]

The motion of the earth exemplifies a case in which there are nonzero values both for the force **F** in the equation governing the motion of the center of mass of a body and for the torque **T** in the equation governing its rotation about the center of mass. As we will see in Chap. 11, the center of mass of the earth moves in an elliptical orbit about the sun under the influence of a gravitational force exerted on that point by the sun. The earth also spins about an axis passing through its center of mass. We saw in Sec. 10-3 that the angular momentum is not constant in direction because of a torque due primarily to the unequal strengths of the moon's gravitational attraction on the near and far parts of the earth's equatorial bulge. Example 10-7 treats a simpler case of motion with nonzero values of **F** and **T.**

EXAMPLE 10-7

A wheel in the form of a uniform, solid disk of radius R rolls without slipping down a plane inclined to the horizontal at an angle θ. If the wheel starts from rest at a height y above the bottom of the plane, determine the speed v of its center of mass when it reaches the bottom.

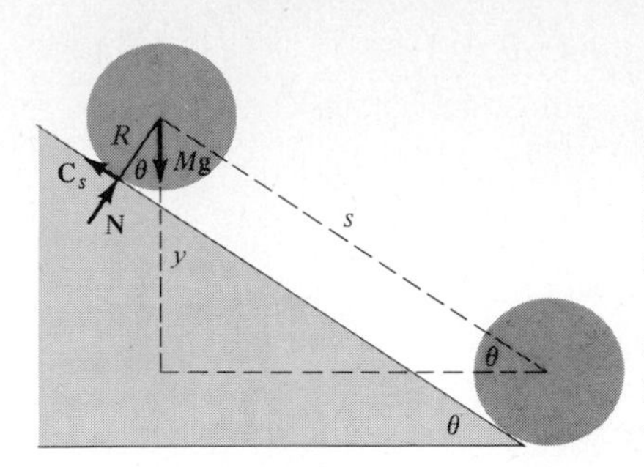

Fig. 10-23 A uniform solid disk rolling down an inclined plane.

■ Your first step should be to make a drawing of the system and the forces acting on the wheel, as in Fig. 10-23. These are the weight Mg acting vertically downward on the center of mass, the force **N** exerted by the plane on the wheel in the direction normal to its surface, and the static contact friction force $\mathbf{C}_s$ exerted by the plane on the wheel in the generally upward direction along the plane that prevents the wheel from slipping.

The motion of the wheel's center of mass along the inclined plane is found by taking the components in that direction of the equation stating Newton's second law in the translational form, $\mathbf{F} = d\mathbf{P}/dt$. Choosing the positive direction to be downward along the plane, you have

$$Mg \sin \theta - C_s = \frac{dP}{dt} = \frac{d(Mv)}{dt} = M\frac{dv}{dt}$$

Since dv/dt is the acceleration a of the center of mass down the plane, this can be written

$$Mg \sin \theta - C_s = Ma \tag{10-46}$$

When the wheel of radius R rotates through an angle $\phi = 2\pi$ to make one complete turn, its center of mass moves down the plane a distance $s = 2\pi R$. (This is the distance around its circumference.) In general, the relation between s and ϕ is $s = \phi R$, as long as the wheel does not slip. Differentiating with respect to time, you find the speed v of the center of mass of the rolling wheel to be

$$v = \frac{ds}{dt} = R\frac{d\phi}{dt}$$

Since $d\phi/dt$ is the wheel's angular speed of rotation ω, this can be written

$$v = R\omega \tag{10-47}$$

The rotational motion is found by taking the components along the rotation axis of the equation stating Newton's second law in the rotational form, $\mathbf{T} = d\mathbf{L}/dt$. As explained, this equation can be used for an origin O at the center of mass, even though the point is accelerating down the plane. The only force producing a torque about the center of mass is the frictional force, since the gravitational force is applied at that point and the normal force is directed toward it. Taking the inward direction as positive, you have

$$RC_s = \frac{dL}{dt} = \frac{d(I\omega)}{dt} = I\frac{d\omega}{dt} \tag{10-48}$$

Here I is the moment of inertia of the wheel for rotation about its symmetry axis.

Solving Eq. (10-48) for C_s gives

$$C_s = \frac{I}{R}\frac{d\omega}{dt}$$

Differentiating Eq. (10-47) with respect to t, you obtain

$$\frac{dv}{dt} = R\frac{d\omega}{dt}$$

or

$$a = R\frac{d\omega}{dt}$$

This allows you to write

$$C_s = \frac{I}{R}\frac{a}{R} = \frac{Ia}{R^2}$$

Then you substitute this into Eq. (10-46), producing

$$Mg \sin \theta - \frac{Ia}{R^2} = Ma$$

or

$$a = \frac{Mg \sin \theta}{M + I/R^2} \tag{10-49}$$

Table 10-1 shows that the square of the gyration radius of the uniform disk rotating about its symmetry axis is

$$G^2 = \tfrac{1}{2}R^2$$

So you have

$$\frac{I}{R^2} = \frac{MG^2}{R^2} = \tfrac{1}{2}M$$

Thus Eq. (10-49) becomes

$$a = \frac{Mg \sin \theta}{M + \frac{1}{2}M}$$

or

$$a = \tfrac{2}{3}g \sin \theta \tag{10-50}$$

This result for the acceleration a of the center of mass of a disk rolling down a plane inclined at an angle θ can be compared with the result $a = g \sin \theta$, obtained in Example 4-4, for a block sliding without friction down the plane. Can you give a qualitative physical explanation of why the acceleration is smaller here?

When the wheel rolls through a distance s down the plane, its center of mass drops through a height y. The figure shows you that $y = s \sin \theta$, or

$$s = \frac{y}{\sin \theta} \tag{10-51}$$

The speed v of its center of mass after moving from rest through a distance s with acceleration a is obtainable immediately from Eq. (2-32). Take the initial value of a coordinate extending down the plane in the direction of the acceleration a to have the value zero, and assume that the initial velocity in that direction has the value zero also. If the final value of the coordinate is s, the equation shows that $s = v^2/2a$. Thus

$$v^2 = 2as$$

Using Eqs. (10-50) and (10-51), you find

$$v^2 = 2\left(\frac{2g \sin \theta}{3}\right)\left(\frac{y}{\sin \theta}\right) = \frac{4gy}{3}$$

Hence

$$v = \sqrt{\frac{4gy}{3}} = \sqrt{\tfrac{2}{3}}\sqrt{2gy} \tag{10-52}$$

The final speed of the center of mass of the rolling wheel is less than the value $v = \sqrt{2gy}$ for the sliding block by the factor $\sqrt{\frac{2}{3}}$ because of its smaller acceleration. So if you have a physical explanation for why the acceleration is smaller, you should have one for why the final speed is smaller. Do you?

10-5 ENERGY IN ROTATIONAL MOTION

It can be quite difficult to give a qualitative physical explanation of the motion of a body which is both translating and rotating by considering the forces and torques acting on it. If you tried to do this for the wheel rolling down the inclined plane in Example 10-7, you will undoubtedly agree. The trouble is that your intuition must deal simultaneously with the two quite different types of *vector* quantities involved, the forces and the torques, and with the interplay between the effects that they produce.

In many cases it is much less difficult to think through the reasons for the behavior of a moving body by considering energy, instead of force and torque. You have seen this before for motion that involves only translation. It is even more true for the case of motion involving both translation and rotation. The simplicity of energy considerations results from the fact that they concern only *scalar* quantities, and scalars are much easier to keep track of than vectors.

The physical understanding that can frequently be found by considering energy deserves to be stressed. As important as it is to be able to make quantitative predictions concerning the behavior of a physical system, it is even more important to develop a qualitative feeling for the physical reasons underlying the behavior.

As you know, the energy relations also frequently have the additional advantage of leading much more easily than direct application of Newton's second law to quantitative predictions about physical systems. Thus we have ample motivation for extending the energy relations, developed in Chap. 7 for translational motion, to the case of a body whose motion is observed to be rotational or both translational and rotational.

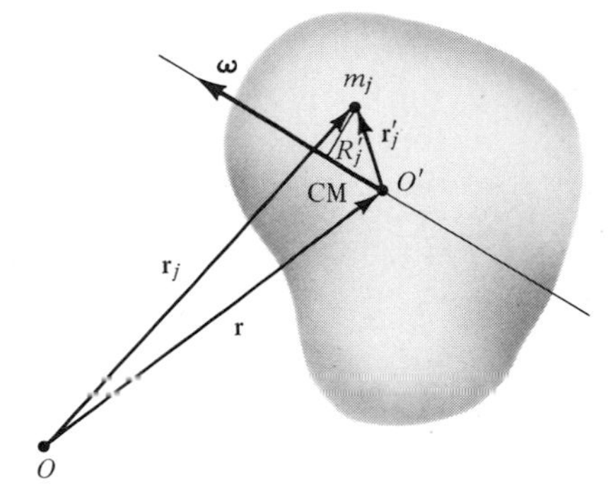

Fig. 10-24 A rotating body and one of its constituent particles.

A body and its jth constituent particle are shown in Fig. 10-24. The position of that particle relative to the center of mass at O' is given by the vector $\mathbf{r}'_j$, and its position relative to the origin O of the reference system is given by $\mathbf{r}_j$. The vector $\mathbf{r}$ specifies the position of O' relative to O. Writing

$$\mathbf{r}_j = \mathbf{r} + \mathbf{r}'_j$$

and then taking time derivatives, we obtain the relation

$$\mathbf{v}_j = \mathbf{v} + \mathbf{v}'_j \tag{10-53}$$

among the velocity $\mathbf{v}_j$ of the particle with respect to O, its velocity $\mathbf{v}'_j$ with respect to O', and the velocity $\mathbf{v}$ of O' with respect to O.

The kinetic energy K of the body, as seen from O, is the sum of the kinetic energies of its n particles. From the point of view of O, each particle of mass m_j has kinetic energy $m_j v_j^2/2$, since its speed is v_j. Therefore

$$K = \sum_{j=1}^{n} \frac{m_j v_j^2}{2}$$

It will be very useful to express K in terms of v and v'_j. To this end, we take the dot product of Eq. (10-53) into itself, producing

$$\mathbf{v}_j \cdot \mathbf{v}_j = \mathbf{v} \cdot \mathbf{v} + \mathbf{v}'_j \cdot \mathbf{v}'_j + 2\mathbf{v}'_j \cdot \mathbf{v}$$

or

$$v_j^2 = v^2 + v_j'^2 + 2\mathbf{v}'_j \cdot \mathbf{v}$$

Substitution then gives

$$K = \sum_{j=1}^{n} \frac{m_j v^2}{2} + \sum_{j=1}^{n} \frac{m_j v_j'^2}{2} + \sum_{j=1}^{n} m_j \mathbf{v}'_j \cdot \mathbf{v} \tag{10-54}$$

The first term on the right side of this equation is just

$$\sum_{j=1}^{n} \frac{m_j v^2}{2} = \frac{Mv^2}{2} \tag{10-55a}$$

where M is the total mass of the body. The third term can be evaluated by writing $\mathbf{v}_j' = d\mathbf{r}_j'/dt$, so as to obtain

$$\sum_{j=1}^{n} m_j \mathbf{v}_j' \cdot \mathbf{v} = \sum_{j=1}^{n} m_j \frac{d\mathbf{r}_j'}{dt} \cdot \mathbf{v}$$

But all the m_j are constants, so $m_j\, d\mathbf{r}_j'/dt = (d/dt)\,(m_j\mathbf{r}_j')$. Furthermore, the sum of the derivatives equals the derivative of the sum, as is shown by Eq. (2-14). So the differentiation can be performed after the summation is carried out, instead of before. Thus this term can be written as

$$\sum_{j=1}^{n} m_j \mathbf{v}_j' \cdot \mathbf{v} = \left(\frac{d}{dt} \sum_{j=1}^{n} m_j \mathbf{r}_j'\right) \cdot \mathbf{v}$$

Now according to the definition of center of mass, the summation over j of $m_j\mathbf{r}_j'$ equals M times a vector extending from O' to the center of mass of the body. But since O' is *at* the center of mass, the vector is of zero magnitude. Hence the value of the summation on the right side of the last equation is zero, and so we have

$$\sum_{j=1}^{n} m_j \mathbf{v}_j' \cdot \mathbf{v} = 0 \tag{10-55b}$$

Using Eqs. (10-55*a*) and (10-55*b*) in Eq. (10-54), we obtain

$$K = \frac{Mv^2}{2} + \sum_{j=1}^{n} \frac{m_j v_j'^2}{2} \tag{10-56}$$

Thus the kinetic energy K of the body, measured in the reference frame of origin O, is the sum of a term which is the kinetic energy of a single particle, whose mass M is the total mass of the body and whose speed v is the speed of its center of mass, and a term which is the summation of the kinetic energies of the particles of mass m_j of the body due to their motion with speeds v_j' relative to the center of mass. This motion relative to the center of mass is a result of the rotation of the body about its center of mass. So we can summarize the meaning of Eq. (10-56) by saying that *the kinetic energy of a body is the kinetic energy due to the motion of its center of mass plus the kinetic energy due to its rotation about the center of mass.*

The first term on the right side of Eq. (10-56) is the kinetic energy a body would have if its center of mass were moving as it actually does, but without any rotation of the body about its center of mass, so that the motion of the body were purely translational. The second term is the kinetic energy arising from the body's actual rotation about its center of mass. Thus we can also interpret the equation by saying that *the kinetic energy of a body is its kinetic energy of translation plus its kinetic energy of rotation.*

A much more concise expression for the kinetic energy of rotation can be found by writing the speed v_j' resulting from the rotation of the body about an axis passing through its center of mass as

$$v_j' = \omega R_j'$$

Here R_j' is the radius of the circle on which the jth particle is instantaneously moving about the rotation axis, and ω is the angular speed of rotation about that axis (see Fig. 10-24). Squaring both sides of this equation gives

$$v_j'^2 = \omega^2 R_j'^2$$

Therefore the second term in Eq. (10-56) can be written

$$\sum_{j=1}^{n} \frac{m_j v_j'^2}{2} = \sum_{j=1}^{n} \frac{m_j \omega^2 R_j'^2}{2} = \frac{\omega^2}{2} \sum_{j=1}^{n} m_j R_j'^2$$

According to Eq. (10-10), the summation on the far right is just the moment of inertia I of the body for rotation about the axis. So

$$\sum_{j=1}^{n} \frac{m_j v_j'^2}{2} = \frac{\omega^2 I}{2} \tag{10-57}$$

This allows us to write Eq. (10-56) in the simple and symmetrical form

$$K = \frac{Mv^2}{2} + \frac{I\omega^2}{2} \tag{10-58}$$

The second term on the right is the body's kinetic energy of rotation. *The kinetic energy of rotation of a body is one-half its moment of inertia times the square of its angular speed.* Equations (10-57) and (10-58) are valid even if the rotation axis passing through the body's center of mass is changing its orientation in the body, so that the value of I is changing. But the equations are most useful in circumstances (which, fortunately, are the most frequently encountered ones) where the rotation axis is fixed in the body so that I is constant.

The work-kinetic energy relation, derived in Chap. 7 for translational motion, applies just as well to motion that is rotational or a combination of translational and rotational. That relation says that *the work W done by the net force acting on a body during a motion viewed from an inertial reference frame equals the change K in the body's kinetic energy in that reference frame:*

$$W = \Delta K \tag{10-59}$$

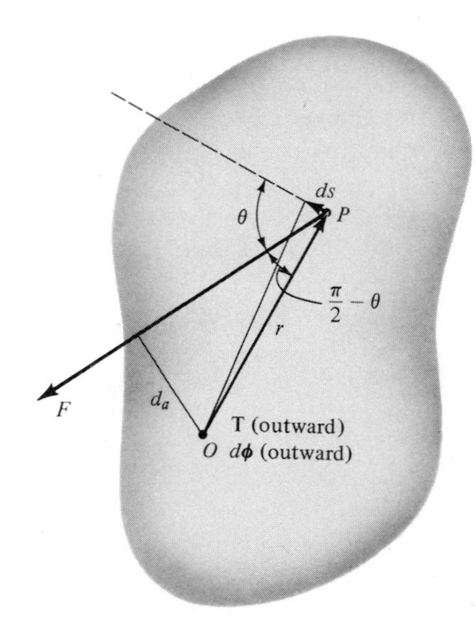

Fig. 10-25 A construction used to establish the relation $dW = T\,d\phi$.

Equation (10-58) can be used to calculate ΔK.

The work W done by a force applied to a point in a rotating body often can be calculated more easily by considering not the force and the displacement of the point of application but instead the torque produced by the force about some origin and the rotation of the point of application about that origin. Figure 10-25 shows a body rotating through an infinitesimal angle $d\phi$ about an axis extending outward through the origin O. During the infinitesimal rotation, the direction of the force $\mathbf{F}$ applied to the point P is assumed to be in the plane of the page, at an angle θ to the displacement $d\mathbf{s}$ of P. The work dW done by the force is

$$dW = \mathbf{F} \cdot d\mathbf{s} = F \cos \theta \, ds$$

Since $\cos \theta = \sin (\pi/2 - \theta)$, and since $ds = r\,d\phi$ with r being the distance from O to P, this expression can be written as

$$dW = F \sin (\pi/2 - \theta)\, r\, d\phi$$

The figure shows that $r \sin(\pi/2 - \theta) = d_a$, where d_a is the perpendicular distance from O to the line of action of the force. Thus

$$dW = Fd_a \, d\phi$$

But $Fd_a = T$, where T is the magnitude of the torque about O produced by the force. Therefore we have

$$dW = T \, d\phi \tag{10-60}$$

In the case leading to Eq. (10-60), **F** is in the plane of the page, and the vector **T** describing the torque about O applied to the body is directed outward, as is the vector $d\boldsymbol{\phi}$ describing the infinitesimal rotation of the body. If the direction of **F** forms an angle α with the plane, then only its component along the plane does work in the displacement $d\mathbf{s}$ along the plane. This component has the value $F \cos \alpha$. Thus in the equations leading to Eq. (10-60), F will be replaced by $F \cos \alpha$, and Eq. (10-60) will become $dW = T \cos \alpha \, d\phi$. But when **F** is at an angle α to the plane, **T** is at an angle α to $d\boldsymbol{\phi}$. Hence $T \cos \alpha \, d\phi = \mathbf{T} \cdot d\boldsymbol{\phi}$, and Eq. (10-60) becomes

$$dW = \mathbf{T} \cdot d\boldsymbol{\phi} \tag{10-61}$$

The total work done in a finite rotation is

$$W = \int_{\phi_i}^{\phi_f} \mathbf{T} \cdot d\boldsymbol{\phi} \tag{10-62}$$

Equations (10-60) through (10-62) are valid not only when **T** is a single torque applied to a body because of the application of a single force, but also when **T** is the net torque applied to a body which experiences a rotation $d\boldsymbol{\phi}$. Note the complete analogy between these equations and the corresponding ones involving the net force **F** and the displacement $d\mathbf{s}$.

Example 10-8 calculates the work done by a torque applied to a rotating body.

EXAMPLE 10-8

Calculate the work done by the constant-torque electric motor of Example 10-4 in bringing the grinding wheel from rest to its operating speed. The wheel has uniform density, radius 32 cm, mass 100 kg, and an operating speed of 200 rotations per minute = 21 rad/s. Also calculate the average power produced by the motor in bringing the wheel up to speed in 3.0 s.

▪ The easiest way to calculate the work done is to use the work-kinetic energy relation of Eq. (10-59),

$$W = \Delta K$$

Since the initial kinetic energy of the wheel is zero, its final kinetic energy K, when rotating at its operating speed, equals the change in kinetic energy ΔK. Hence you have

$$W = K$$

According to Eq. (10-58), K is given by

$$K = \frac{I\omega^2}{2}$$

where I is the wheel's moment of inertia for rotation about its fixed axle and ω is the operating value of its angular speed. In Example 10-4, the moment of inertia was found to have the value

$$I = \frac{MR^2}{2}$$

where M is the mass of the wheel and R is its radius. Putting it all together, you obtain

$$W = \frac{I\omega^2}{2} = \frac{MR^2\omega^2}{4}$$

The numerical value of the work done by the motor is

$$W = \frac{100 \text{ kg} \times (0.32 \text{ m})^2 \times (21 \text{ s}^{-1})^2}{4} = 1.1 \times 10^3 \text{ J}$$

Calculate this value by applying Eq. (10-62), and compare the result obtained with the one obtained here.

A constant-torque electric motor does not produce constant power. Why? The average power produced by this motor in the 3.0-s interval while it is bringing the wheel up to speed is

$$P = \frac{W}{t} = \frac{1.1 \times 10^3 \text{ J}}{3.0 \text{ s}} = 3.8 \times 10^2 \text{ W}$$

or

$$P = 3.8 \times 10^2 \text{ W} \times \frac{1 \text{ hp}}{746 \text{ W}} = 0.50 \text{ hp}$$

The law of conservation of total mechanical energy applies to a body whose motion is completely or partly rotational, just as it does to a body whose motion is completely translational. That is, if the body is in an isolated system with workless constraints and conservative internal forces, and if its motion is observed from an inertial frame, then

$$E = K + U$$

Here E is the *constant* total mechanical energy of the system, K is the kinetic energy of the body, and U is the potential energy associated with the forces internal to the system. Using Eq. (10-58) to evaluate K, we have

$$E = \frac{Mv^2}{2} + \frac{I\omega^2}{2} + U \tag{10-63}$$

If U represents the gravitational potential energy of a system containing a body of mass M near the surface of the earth, then Eq. (9-54) shows that

$$U = Mgy \tag{10-64}$$

In this expression y is used to represent the height of the body's center of mass above the reference height $y = 0$ where $U = 0$.

Example 10-9 applies the energy relations of Eqs. (10-63) and (10-64) to solve a problem previously solved by applying the translational and rotational forms of Newton's second law. It should make the advantages of using energy relations apparent to you.

EXAMPLE 10-9

Use energy relations to repeat the calculation of Example 10-7, in which a wheel in the form of a uniform, solid disk of radius R rolls without slipping down an inclined plane, starting from rest at a height y above the bottom. Determine the speed v of its center of mass when it reaches the bottom.

■ You take as an isolated system the wheel plus the earth. The inclined plane on which the wheel rolls acts as a workless constraint. Rolling is workless because even though a frictional force C_s must be exerted on the wheel by the plane to make it roll rather than slide, the force is applied to the instantaneous point of contact of the wheel to the plane, and that point is always instantaneously at rest. So the frictional force does no work because it does not act through a distance. The same is true for the normal force N exerted by the plane on the wheel. Since the gravitational force Mg acting internal to the system is conservative, you can apply Eqs. (10-63) and (10-64).

At the instant when the wheel is released, you have

$$E = \frac{Mv^2}{2} + \frac{I\omega^2}{2} + U$$

$$= 0 + 0 + Mgy$$

where y is the initial height of the center of mass.

At the bottom of the inclined plane, where $y = 0$, you have

$$E = \frac{Mv^2}{2} + \frac{I\omega^2}{2} + 0$$

Equating the two expressions for the constant total energy E gives

$$\frac{Mv^2}{2} + \frac{I\omega^2}{2} = Mgy$$

Now, you can write

$$I = MG^2$$

where G is the gyration radius of the rotating body. So you can write

$$\frac{Mv^2}{2} + \frac{MG^2\omega^2}{2} = Mgy$$

or

$$\frac{v^2}{2} + \frac{G^2\omega^2}{2} = gy \qquad (10\text{-}65)$$

Using the value of $G^2 = \frac{1}{2}R^2$ for a uniform, disk-shaped wheel of radius R from Table 10-1, you obtain

$$\frac{v^2}{2} + \frac{R^2\omega^2}{4} = gy$$

According to Eq. (10-47), the relation between the angular speed ω of the rolling wheel and the speed v of its center of mass is

$$\omega = \frac{v}{R}$$

Substitution into the equation immediately above gives

$$\frac{v^2}{2} + \frac{v^2}{4} = gy$$

or

$$\frac{3v^2}{4} = gy$$

and

$$v = \sqrt{\frac{4gy}{3}} \qquad (10\text{-}66)$$

This result is identical to the one expressed in Eq. (10-52) of Example 10-7.

Comparison with Example 10-7 will show that it is appreciably easier to evaluate the final speed by considering the energy content of the system than by considering the forces and torques acting in the system. However, there are always interesting questions to be asked about a system which cannot be answered by using only energy relations. For instance, you may want to determine the ratio C_s/N in order to verify that it is less than the coefficient of static friction between the wheel and the plane, so that it is actually possible for the wheel to roll without slipping. To do so, you *must* deal with forces and torques, not energies.

The energy relations certainly do provide a simple explanation of why the wheel rolls more slowly down an inclined plane than a block slides without friction down a plane inclined at the same angle. In both cases the lost gravitational potential energy appears as kinetic energy. If the body does not rotate about its center of mass, then all the lost potential energy appears in the form of kinetic energy of motion of the center of mass. But if the body rotates about the center of mass, then some of the lost potential energy appears as rotational kinetic energy, and so there is less available to appear as kinetic energy of motion of the center of mass. Hence the center of mass of the body moves more slowly down the plane.

You can see from Eq. (10-65) that the mass M of the rolling body does not affect the rapidity of its motion. But the gyration radius G does. In other words, what counts is not the mass, but how the mass is distributed about the rotation axis. If a hoop of the same mass and radius as the solid disk were started down the inclined plane at the same time as the disk, would it arrive at the bottom at the same time?

Table 10-2 lists some of the most important equations of rotational mechanics developed or extended since Table 9-1 was presented, and it also shows the analogous equations from translational mechanics.

Table 10-2

More Equations Used in Rotational and Translational Mechanics

Rotational motion	Translational motion
$\mathbf{T} = \frac{d\mathbf{L}}{dt}$	$\mathbf{F} = \frac{d\mathbf{P}}{dt}$
$\mathbf{L} = I\boldsymbol{\omega}$	$\mathbf{P} = M\mathbf{v}$
$\mathbf{T} = I\boldsymbol{\alpha}$	$\mathbf{F} = M\mathbf{a}$
$I = \sum_{j=1}^{n} m_j R_j^2$	$M = \sum_{j=1}^{n} m_j$
$W = \int_{\phi_i}^{\phi_f} \mathbf{T} \cdot d\boldsymbol{\phi}$	$W = \int_{s_i}^{s_f} \mathbf{F} \cdot d\mathbf{s}$
$T(\theta) = -\frac{dU(\theta)}{d\theta}$	$F(x) = -\frac{dU(x)}{dx}$
$K = \frac{I\omega^2}{2}$	$K = \frac{Mv^2}{2}$

Combined motion

$\mathbf{L} = \mathbf{L}_{\text{about CM}} + \mathbf{L}_{\text{of CM}}$

$E = \frac{Mv^2}{2} + \frac{I\omega^2}{2} + U$

EXERCISES

Group A

10-1. *Circle and square.* A hoop of radius R and a square frame of side $2R$ each have mass M.

a. Calculate the moment of inertia of each object about an axis through the center of mass normal to the plane in which it lies.

b. Why do the moments of inertia differ?

10-2. *Moments of inertia: rectangular solid.* The homogeneous rectangular block shown in Fig. 10E-2 has mass M. Find the moment of inertia of the block about each of the following axes:

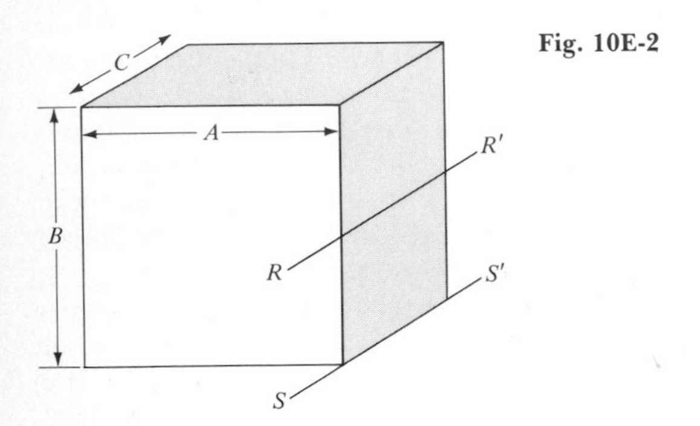

Fig. 10E-2

a. The axis RR', which passes through the center of two edges of length B.

b. The axis SS', which coincides with an edge of length C.

10-3. *Flywheel-powered cars, I.* A 200-kg flywheel 1.00 m in diameter is rotating at 6000 rotations per minute. If its kinetic energy were used to propel a car that requires an average propulsive force of 500 N, how far could the car travel before the flywheel stopped? For simplicity, assume that the entire mass is concentrated in the rim of the flywheel.

10-4. *Rolling along.* A uniform sphere of mass M and radius R rolls along a floor. Its center of mass travels with speed v_c.

a. What is the total energy of its motion?

b. Evaluate your result for $M = 10$ kg and $v_c = 5$ m/s.

c. Explain why it is not necessary to know the sphere's radius and angular speed, as long as its translational speed is given.

10-5. *A design problem.* How would you construct an object of specified mass M, having symmetry about its axis of rotation, so that it would roll down a given inclined plane as rapidly as possible? How would you construct it so that it would roll down the inclined plane as slowly as possible?

10-6. *Effective gravity and the precession of a top.* A space-age child is among the passengers aboard a space bus that is about to take off from a launch pad on the earth. He is playing with a toy top, which spins with angular velocity ω_s and is supported on a smooth, horizontal table. The top is inclined to the vertical, so that it precesses steadily at angular velocity $\omega_p \ll \omega_s$. Describe what happens to the precession velocity ω_p of the top during launch, when the space bus gradually develops an upward acceleration equal to $3g$. Assume that the time required for the vehicle to reach full acceleration is much longer than one precessional period, and very much longer than one spin period.

Group B

10-7. *Moment of inertia of a bar.* Find the moment of inertia for the bar in Fig. 10-7 when it rotates about an axis that passes through one end and is aligned parallel to the direction along which the width w is measured. Do this by direct integration, as in Example 10-1. Then do it again by applying the parallel-axis theorem to the moment of inertia obtained in the example for rotation about an axis passing through the center of the bar. Compare your two results.

10-8. *A broken wheel.* A chip of mass 4.0 kg is broken from the rim of the grinding wheel in Example 10-4. Where must the origin O be located so that the equation $\mathbf{T} = I\boldsymbol{\alpha}$ still applies to the asymmetrical wheel? Using that origin, recalculate the magnitude of the torque that the motor must apply to bring the wheel from rest to 200 rotations per minute in 3.0 s.

10-9. *The other way.* Construct a figure like Fig. 10-6, but with the direction of the angular velocity $\boldsymbol{\omega}$ reversed. Use it to make a calculation like the one in the text leading to Eq. (10-7), $L_{\|} = MR^2\omega$.

10-10. *Moments of inertia of a hoop.* A flat hoop of mass M and radius R is shown in Fig. 10E-10. The hoop lies in the xy plane and is centered at the origin O. In the figure, the z axis rises from O directly toward the viewer. Consider the mass element dm. It contributes an amount $dI_z = R^2\,dm$ to the moment of inertia about the z axis. The ele-

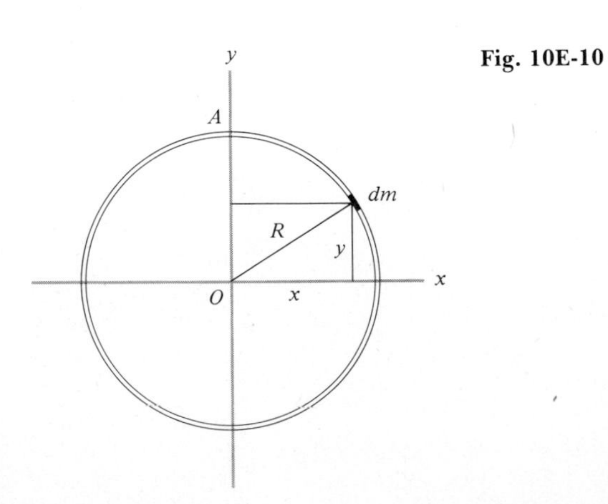

Fig. 10E-10

ment dm also contributes an amount $dI_x = y^2\, dm$ to the moment of inertia about the x axis; similarly, $dI_y = x^2\, dm$.

a. Find an equation relating dI_x, dI_y, and dI_z.

b. Find an equation relating I_x, I_y, and I_z.

c. Use symmetry considerations to obtain an equation relating I_x and I_y.

d. Determine I_x, I_y, and I_z for the hoop. Express your results in terms of M and R.

e. Determine the moment of inertia of the hoop about an axis that passes through point A and runs parallel to the x axis.

10-11. *A pivoted rod.* A slender, uniform rod of mass m and length l is pivoted at one end so that it can rotate in a vertical plane. There is negligible friction at the pivot. The free end is held almost vertically above the pivot and then released.

a. What is the rod's angular acceleration when it makes an angle θ with the vertical?

b. What is the magnitude of the translational acceleration of the free end of the rod for this angle?

c. What is the vertical component of this translational acceleration?

d. For what value of θ does the vertical component of the translational acceleration become equal to g?

10-12. *The perpendicular-axis theorem.* Generalize the arguments involved in Exercise 10-10*a* and *c* to prove the following theorem: Any *flat* object in the xy plane has a moment of inertia I_z about the z axis which is given by $I_z = I_x + I_y$, where I_x and I_y are the object's moments of inertia about the x axis and the y axis, respectively.

10-13. *A swinging ring.*

a. A ring of mass M and radius R is hung from a knife-edge, so that the ring can swing in its own plane as a physical pendulum. Find the period T_1 of small oscillations.

b. Suppose that an identical ring is pivoted from an axis PP' lying in the ring plane and tangent to the circumference. This ring can execute oscillations in and out of the plane. Find the period T_2 of those small oscillations.

c. Which oscillation has the longer period? How much longer?

10-14. *Moments of inertia of a thin disk.* Employ the perpendicular-axis theorem of Exercise 10-12 to relate the moment of inertia of a thin, circular disk (about any diameter) to its moment of inertia about its central axis. Check your results against Table 10-1 by using the tabulated results for solid cylinders, in the limiting case that the length of the cylinder approaches zero.

10-15. *A gaucho's life.* Two bodies, one of mass 1.0 kg and the other of 2.0 kg, are tied to the ends of a wire 60 cm long. The 1.0-kg body is held in the hand, and the 2.0-kg body is whirled in a horizontal circle at 2.0 rotations per second. Then the 1.0-kg body is released, sending the pair whirling horizontally through the air. What is the angular velocity of rotation of the system just after the release?

10-16. *No slipping, please.* Starting from rest, a sphere rolls down a 30° incline. What is the minimum value of the coefficient of static friction if there is to be no slipping?

10-17. *Descent of a Yo-Yo.* Consider a Yo-Yo with outside radius R equal to 10 times its spool radius r. The moment of inertia I_c of the Yo-Yo about its spool is given with good accuracy by $I_c = \frac{1}{2}MR^2$, where M is the total mass of the Yo-Yo. The upper end of the string is held motionless.

a. Compute the acceleration of the center of mass of the Yo-Yo. How does it compare with g?

b. Find the tension in the string as the Yo-Yo descends. How does it compare with Mg?

10-18. *Turning the wheel.* A heavy wheel of radius 20 cm is mounted on a horizonal axle. A rope wrapped around its rim is pulled straight downward with a constant force of 50 N. The rope moves a distance of 50 cm in 1.0 s.

a. What is the angular acceleration of the wheel?

b. What is the moment of inertia of the wheel?

c. Suppose that an object whose weight is 50 N is attached to the rope, and the system is released from rest. What would the angular acceleration of the wheel be in that case?

d. Account for the difference between the results of parts *a* and *c*.

e. The wheel is a homogeneous, solid disk. What is its mass?

10-19. *Cylinder versus pipe.* A solid cylinder and a thin-walled pipe are simultaneously released from rest at the upper end of a ramp of inclination θ. Each object rolls without slipping.

a. Find the acceleration of the center of mass of the solid cylinder.

b. Find the acceleration of the center of mass of the pipe.

c. When the cylinder has rolled a distance s_c, how far has the pipe rolled?

10-20. *From pure sliding to pure rolling.* An upright hoop is projected onto a pavement with an initial horizontal speed v_0 but without spin, so that it slides. The resulting frictional force causes the hoop to lose translational speed and to acquire an angular speed. Eventually the hoop rolls without slipping. Prove that when the hoop ceases to slip, it has speed $v_0/2$.

10-21. *Oscillations of an angle iron.* A uniform, right-angle iron is hung over a thin nail so that the iron pivots freely at the bend. Each arm of the iron has mass m and length l. Show that the period T of small oscillations (in the plane of the iron) is given by $T = 2\pi\sqrt{2\sqrt{2}\, l/3g}$.

10-22. *A family of physical pendulums, I.* Figure 10E-22 represents a three-dimensional object (not necessarily of uniform density) whose center of mass is at point C. The axis ZCZ' passing through point C has been chosen at random and is *not* necessarily one of the "principal" axes

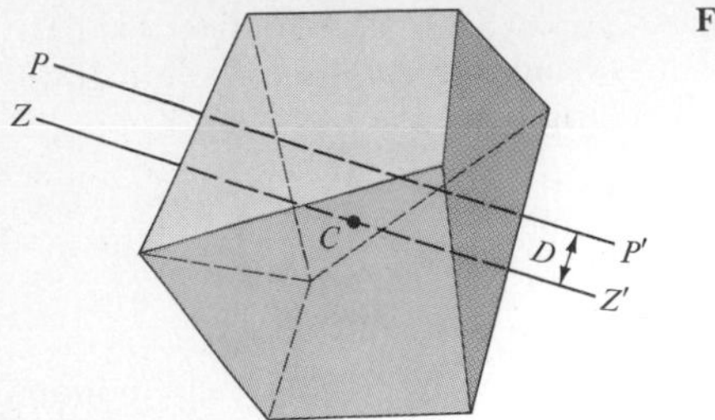

Fig. 10E-22

mentioned in the text. The body had gyration radius $G_C(ZZ')$ about the axis ZCZ': the notation $G_C(ZZ')$ makes explicit the fact that the gyration radius for an axis through C depends on the particular choice (ZZ') of axis.

Suppose that a physical pendulum is constructed by pivoting the body about the axis PP', which is parallel to ZCZ' at a distance D. Prove that the frequency ν of small, oscillations about equilibrium is given by

$$\nu = \frac{1}{2\pi}\sqrt{\frac{Dg}{G_C^2(ZZ') + D^2}}$$

This shows that the frequency of pendulum oscillations is the same for any choice of axis PP' on a cylinder of radius D centered on ZZ'.

10-23. *A family of physical pendulums, II.* Figure 10E-23 represents any thin and flat rigid object (not necessarily of uniform density). The center of mass is at C, and the indicated circle is centered at C. Imagine that the object is pivoted about an axis which (1) passes through some point P on the circle and (2) is perpendicular to the plane of the object. The object is allowed to move in a vertical plane about this axis, forming a physical pendulum. Prove that small oscillations about equilibrium have the same frequency for any choice of P on the indicated circle.

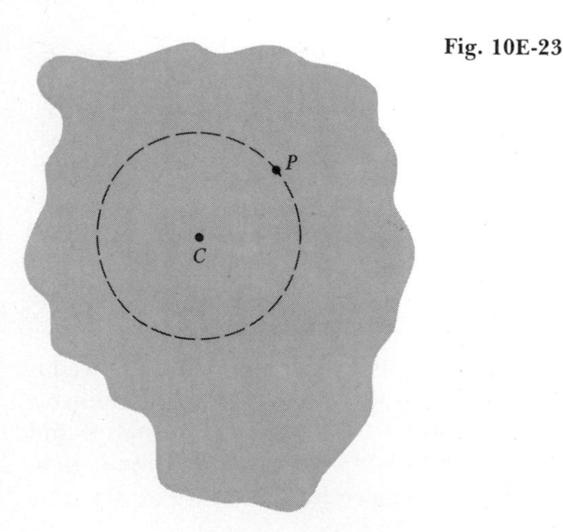

Fig. 10E-23

10-24. *The Kater pendulum.* Figure 10E-24 depicts a special kind of physical pendulum invented in 1817 by the British geodesist Captain Henry Kater. It is used as described in this exercise to measure the acceleration of gravity with high accuracy. (Until recently, it was the most accurate method available.) It consists of a rigid rod on which a bob is mounted. The mass of the bob is sufficiently large so that the center of mass of the pendulum is fairly far from the middle of the rod. The bob is mounted on a slide; by moving the bob and then clamping it in position, the location of the center of mass of the pendulum can be adjusted. There are two very precisely made knife-edges mounted on the rod. The pendulum can be swung from knife-edge 1, and then reversed and swung from knife-edge 2.

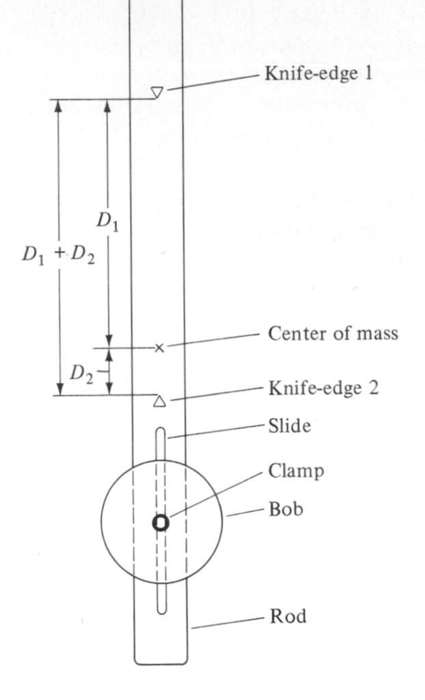

Fig. 10E-24

a. Let G_0 be the gyration radius of the pendulum about an axis through its center of mass. Show that if knife-edge 1 and knife-edge 2 are located respectively at distances D_1 and D_2 from the center of mass, the gyration radii G_1 and G_2 about the knife-edges satisfy the equations $G_1^2 = G_0^2 + D_1^2$ and $G_2^2 = G_0^2 + D_2^2$.

b. Beginning with Eq. (10-23), express T_1, the period of small oscillations of the pendulum about knife-edge 1, in terms of G_0, D_1, and the acceleration of gravity g. Likewise, express T_2, the period for knife-edge 2, in terms of G_0, D_2, and g.

c. Using the expressions obtained in part *b*, show that $(T_1^2 + T_2^2)/(D_1 + D_2) = (4\pi^2/g)(G_0^2/D_1D_2 + 1)$, and that $(T_1^2 - T_2^2)/(D_1 - D_2) = -(4\pi^2/g)(G_0^2/D_1D_2 - 1)$.

d. Using the results of part *c*, show that

$$\frac{8\pi^2}{g} = \frac{T_1^2 + T_2^2}{D_1 + D_2} + \frac{T_1^2 - T_2^2}{D_1 - D_2}$$

e. By counting pendulum swings over an accurately measured total time of many hours, the values of T_1 and T_2 can be determined with great accuracy. Similarly, the distance $D_1 + D_2$ between the knife-edges can be measured very accurately by means of a traveling microscope or a laser interferometer. But evaluation of the quantity $D_1 - D_2$ depends on separate measurements of the distances D_1 and D_2 between the knife-edges and the center of mass. It is not possible to determine the position of the center of mass with great accuracy. However, by moving the pendulum bob, the oscillation periods T_1 and T_2 can be adjusted. How would you adjust them to maximize the accuracy of the value of g obtained by using the result of part *d*?

10-25. *Taking the cue.* How far above its center should a billiard ball be struck in order to make it roll without any initial slippage? Denote the ball's radius by R, and assume that the impulse delivered by the cue is purely horizontal.

10-26. *A difficult way to hold up a cylinder.* A string is wound around an otherwise unsupported homogeneous, horizontal cylinder of mass M and radius R. As the string unwinds and the cylinder spins, the end of the string is continually pulled vertically upward with a force just sufficient to keep the cylinder from descending relative to the ground.

a. What is the tension in the vertical portion of the string?

b. What is the angular acceleration of the cylinder?

c. What is the upward acceleration of any given point along the vertical portion of the string?

10-27. *How much friction is needed for pure rolling?* A spool of mass M is resting on a horizontal surface. The spool has moment of inertia MG_C^2 about its axis of symmetry. The spool is subjected to a rightward horizontal force of magnitude F, applied at a distance r above the axis.

a. Show that if there is to be no slippage between the spool and the supporting surface, a leftward frictional force $f = F(G_C^2 - rR)/(G_C^2 + R^2)$ must act on the spool.

b. Show that the required frictional force f has the value zero for a particular value r_0 of the distance r.

c. Interpret the result of part *a* when r exceeds the value r_0 found in part *b*.

d. Show that the rightward translational acceleration of the center of the mass of the spool a_C exceeds F/M when $r > r_0$. Explain how this can happen.

10-28. *Karate chop.* A uniform, slender rod 1.00 m long is initially at rest on a smooth, horizontal surface. It is struck a sharp horizontal blow at one end, with the blow directed at right angles to the rod axis. As a result, the rod acquires an angular velocity of 3.00 rad/s.

a. What is the translational velocity of the center of mass of the rod after the blow?

b. Which point on the rod is stationary just after the blow?

10-29. *Unwinding.* Consider the system shown in Fig. 10E-29. The cylinder on the 30° incline has mass M and radius R. A string, which is wound around the cylinder, runs parallel to the incline and then passes over a pulley of negligible mass to a hanging body of mass m. The tension in the string and the kinetic frictional force exerted on the cylinder by the incline are just sufficient to keep the cylinder from descending the incline as it turns, while the string unwinds and the body of mass m descends with acceleration a. The coefficient of kinetic friction μ_k between the incline and the cylinder is given by $\mu_k = 0.25$. Determine the acceleration a of the hanging body and the mass ratio M/m.

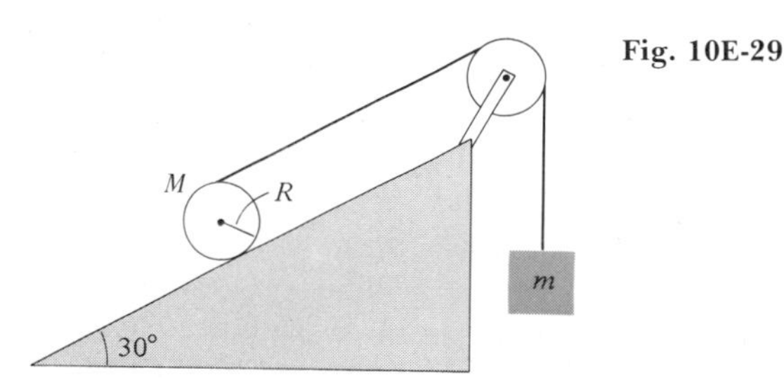

Fig. 10E-29

10-30. *The motion of a system's center of mass.* Consider the system composed of a 1-kg body and a 2-kg body initially at rest at a center to center distance of 1 m. All numerical values quoted in this exercise are to be considered exact, and the 2-kg body is to the right of the 1-kg body.

a. How far is the system's center of mass from the center of the 1-kg body?

b. Beginning at $t = 0$ s, a net rightward force of 2 N acts on the 2-kg body. What is its resultant acceleration?

c. How far does the 2-kg body move between $t = 0$ s and $t = 1$ s?

d. How far is the center of mass from the 1-kg body at $t = 1$ s?

e. How far did the center of mass move between $t = 0$ s and $t = 1$ s?

f. What is the acceleration of the center of mass, beginning at $t = 0$ s?

g. Suppose that all the mass in both objects was concentrated at the center of mass, and the net rightward force of 2 N acted on this concentrated mass. What would its acceleration be?

h. State the general theorem which is illustrated by the results of parts *f* and *g*.

10-31. *Work to be done.* Calculate the work done by the electric motor in Example 10-8 by applying Eq. (10-61), $W = \int_{\phi_i}^{\phi_f} \mathbf{T} \cdot d\boldsymbol{\phi}$. Compare your results with those obtained in the example.

10-32. *The three-dumbbell experiment?* In a common physics lecture demonstration, a lecturer sits on a stool that can rotate freely about a vertical axis on low-friction bearings. The lecturer holds with extended arms two dumbbells, each of mass m, and kicks the floor so as to achieve an initial angular speed ω_1. The lecturer then pulls in the dumbbells, so that their distances from the rotation axis decrease from the initial value R_1 to the final value R_2. Determine the final angular speed ω_2, assuming that the moment of inertia about the rotation axis of the lecturer's body plus the stool does not change in the process. Then evaluate K_1 and K_2, the initial and final kinetic energies in the system. What is the source of the additional kinetic energy?

10-33. *Toy top.* A top consists of a uniform disk of mass m_0 and radius r_0 rigidly attached to an axial rod of negligible mass. The top is placed on a smooth table and set spinning about its axis of symmetry with angular speed ω_s.

a. How much work must be done in setting the top spinning? Evaluate your result for $m_0 = 0.050$ kg, $r_0 = 2.0$ cm, and $\omega_s = 200\ \pi$ rad/s (or 6000 rotations per minute).

b. The center of the disk is a distance d from the top's point of contact with the table. The top is observed to precess steadily about the vertical axis with angular speed ω_p. Assuming that $\omega_p \ll \omega_s$, use Eq. (10-36) to write ω_p in terms of r_0, d, ω_s, and g. Evaluate ω_p for $d = 4.0$ cm and $g = 9.80\ \text{m/s}^2$, with the other quantities as given in part *a*. Is your result consistent with the assumption $\omega_p \ll \omega_s$?

10-34. *A loaded disk.* A disk of mass M and radius R is supported vertically by a pivot at its center. As shown in Fig. 10E-34, a small, dense object (also of mass M) is attached to the rim and raised to the highest point above the center. The system is then released. What is the angular speed of the system when the attached object passes directly beneath the pivot?

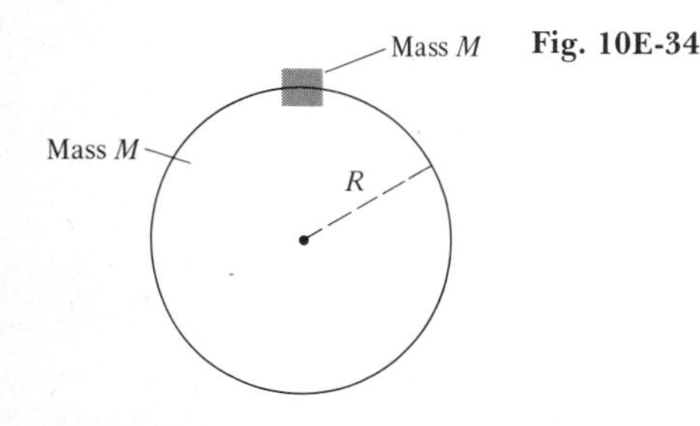

Fig. 10E-34

10-35. *Looping the loop.* A sphere of mass m and radius r rolls down an incline and acquires enough speed to loop a loop of radius R, as shown in Fig. 10E-35. The sphere rolls without slipping throughout its motion.

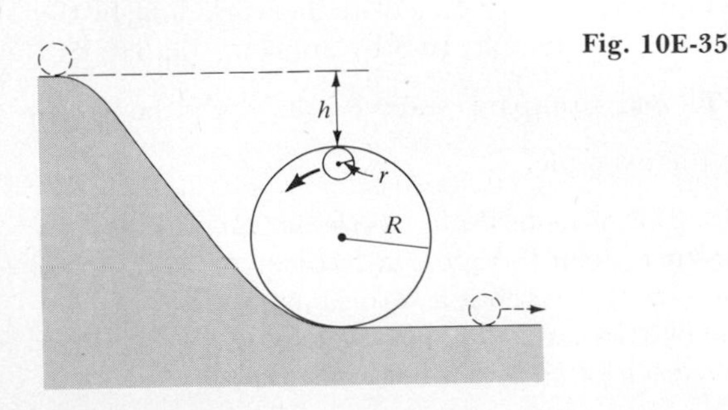

Fig. 10E-35

a. Find the minimum elevation h of the starting point, measured from the top of the loop. Be careful to take into account the rotational kinetic energy of the sphere.

b. Show that the minimum elevation is negative for $r > \frac{7}{27}R$. Explain how this can be correct.

10-36. *The end of the leash.* A horizontal, homogeneous cylinder of mass M and radius R is pivoted about its axis of symmetry. As shown in Fig. 10E-36*a*, a string is wrapped several times around the cylinder and tied to a body of mass m resting on a support positioned so that the string has no slack. The body m is carefully lifted vertically a distance h, and the support is then removed, as shown in Fig. 10E-36*b*.

a. Just *before* the string becomes taut, evaluate (1) the angular velocity ω_0 of the cylinder, (2) the speed v_0 of the falling body m, (3) the kinetic energy K_0 of the system.

b. Evaluate the corresponding quantities, ω_1, v_1, and K_1, for the instant just *after* the string becomes taut.

c. Why is K_1 less than K_0? Where does the energy go?

d. If $M = m$, what fraction of the kinetic energy is lost when the string becomes taut?

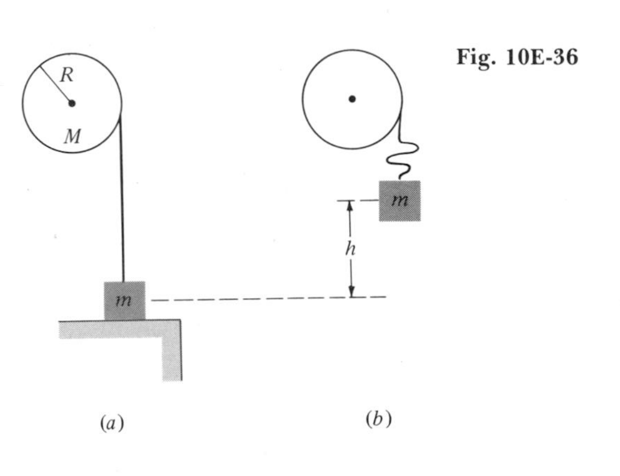

Fig. 10E-36

10-37. *A hollow sphere.* A thick-walled hollow sphere has outside radius R_0. It rolls down an incline without slipping, and its speed at the bottom is v_0. Now the incline is waxed, so that it is practically frictionless, and the sphere is observed to slide down (without rolling). Its speed at the bottom is observed to be $5v_0/4$.

a. Determine the radius of gyration of the hollow sphere about an axis through its center.

b. The central hollow has zero density and unknown radius R_i. For $R_i < r < R_0$, the density is uniform. Determine R_i/R_0. Compare the volume of the cavity to the total volume $4\pi R_0^3/3$.

10-38. *Flywheel-powered cars, II.* It has been proposed that the kinetic energy stored in a flywheel be used to propel an automobile. (A small electric motor located where the automobile is parked overnight could be used to spin up the flywheel each night and thus make up for the energy used during the day.) Calculate the energy content in a cylindrical flywheel of uniform density, radius 0.50 m, mass 200 kg, and angular speed 20,000 rotations per minute. (This angular speed is near the limit at which a steel flywheel would break apart.) With the flywheel installed, an automobile has a total mass of 1000 kg. When the car is traveling on a level road at a speed of 100 km/h, the total frictional force acting on it has a magnitude equal to 10 percent of the weight of the automobile. Calculate how far it could travel before using up all the energy stored in the flywheel.

10-39. *The earth in space.* The earth has a mass of 5.98×10^{24} kg, a radius of 6.37×10^6 m, and a moment of inertia of 8.04×10^{37} kg·m^2. It rotates around its axis once every 8.62×10^4 s. It travels at (nearly) constant speed in its (nearly) circular orbit about the sun, completing one orbit in 3.16×10^7 s. Its orbital radius is 1.50×10^{11} m. The sense of rotation of the earth about the sun is the same as the sense of its rotation about its own axis.

a. Compare the earth's moment of inertia about its own axis to that of a uniform sphere of the same mass and radius. Explain the difference.

b. Evaluate the earth's spin angular momentum $\mathbf{L}_s$ about its own axis.

c. Evaluate the earth's angular momentum of rotation about the sun $\mathbf{L}_r$ using the sun as the origin.

d. Of the total angular momentum $\mathbf{L}_s + \mathbf{L}_r$, what fraction is spin angular momentum?

e. Evaluate the earth's spin kinetic energy K_s.

f. Evaluate the earth's kinetic energy of rotation about the sun, K_r.

g. Of the total kinetic energy, $K_s + K_r$, what fraction is spin kinetic energy?

Group C

10-40. *Divide and conquer: the spherical shell.* Table 10-1 indicates that a homogeneous, thin-walled spherical shell of mass M and radius R has a moment of inertia I about a diameter given by $I = \frac{2}{3}MR^2$. In this exercise, you will confirm that expression by adding the moments of inertia of various small portions of the shell.

a. Show that the spherical shell has a mass per unit spherical surface area given by $M/4\pi R^2$.

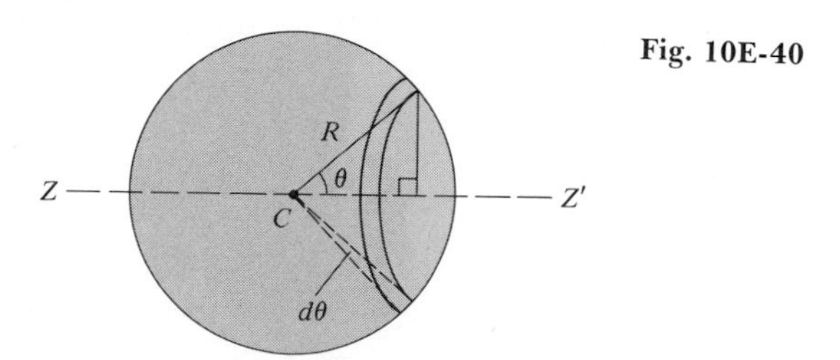

Fig. 10E-40

b. Imagine that the shell is divided into flat hoops of various sizes, as shown in Fig. 10E-40. Show that the hoop indicated there accounts for an area $dA = (2\pi R \sin \theta)(R\, d\theta)$ of the spherical surface. What is the mass dM of that hoop?

c. Obtain an expression for the hoop's moment of inertia dI (about the axis ZZ').

d. What range of values of θ is needed to account for the entire spherical shell?

e. Integrate the expression obtained in part c to obtain I. Your result should agree with that in Table 10-1.

10-41. *Divide and conquer: the sphere.* Table 10-1 indicates that a homogeneous sphere of mass M and radius R has a moment of inertia I (about a diameter) given by $I = \frac{2}{5}MR^2$. In this exercise, you can confirm that expression by adding the moments of inertia of spherical shells.

a. Show that the sphere has a mass density ρ given by $\rho = M/\frac{4}{3}\pi R^3$.

b. Imagine that the sphere is divided into concentric shells of various sizes. Show that a shell of radius r and thickness dr has volume $dV = 4\pi r^2\, dr$. What is the mass dM of that shell?

c. Use the expression for the moment of inertia of a spherical shell found from Table 10-1, to write the moment of inertia dI of the shell (about a diameter).

d. What range of values of r is needed to account for the entire sphere?

e. Integrate the expression obtained in part c to obtain I. Your result should agree with that found from Table 10-1.

10-42. *Tough on the bearings.* A rod of negligible mass and length d has a particle of mass m attached to each end. As shown in Fig. 10E-42, the midpoint O of the rod is rigidly attached to the shaft AB, also of negligible mass, making an angle γ with it. (Notice that the point O is *not* necessarily midway between A and B.) The shaft is supported by bearings at A and B, and the system is rotating about AB with angular velocity $\boldsymbol{\omega}$, as shown. At the instant shown, the particles of mass m are in the plane of the figure.

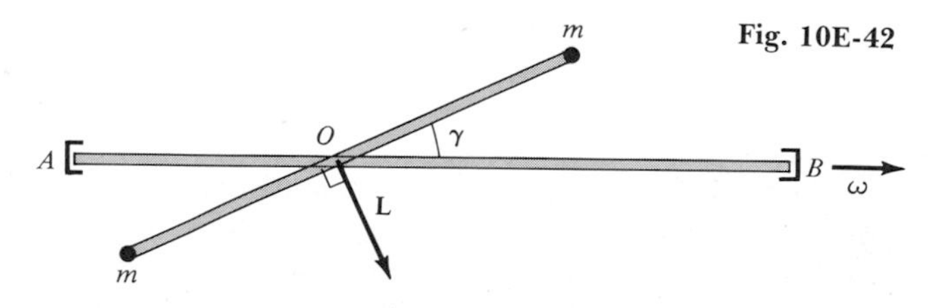

Fig. 10E-42

a. Show that the center of mass of the system is located at O and that it remains motionless.

b. Determine the angular momentum vector $\mathbf{L}$ of the rotating system, using O as the origin. Show that the orientation of $\mathbf{L}$ is correctly depicted, and show that its magnitude is $L = \frac{1}{2}m\omega\, d^2 \sin \gamma$.

c. Show that any other choice of origin would give the same result for $\mathbf{L}$.

d. What is the magnitude of the component $L_{\parallel}$ of **L** along AB? What is the magnitude of the component $L_{\perp}$ perpendicular to AB and into the page?

e. Show that $L_{\parallel}$ is constant but that $L_{\perp}$ changes with time. Find the magnitude and direction of $dL_{\perp}/dt$ for the instant shown in the figure.

f. If the shaft AB has length h, find the magnitudes and directions of the forces at A and at B for the instant shown in the figure. Does your result depend on exactly where the rod is attached along AB? Why or why not?

10-43. *What's inside?* Suppose someone hands you a sphere of radius R and tells you that it consists of an inner solid sphere of one material, covered by a concentric spherical outer covering of another material. You are required to determine the size and density of each part, without cutting or removing any material. Can you do it? If not, why not? If so, how?

10-44. *Pure rolling, or some slippage?* A thin cylindrical shell, a solid cylinder, a thin spherical shell, and a solid sphere are placed side by side at the top of an incline of angle θ. The four objects are all made of the same material, so that all have the same coefficients of friction with respect to the surface of the incline: static friction μ_s and kinetic friction μ_k, which is less than μ_s.

a. For *each* object, find the largest incline angle (call it θ_M) for which that object can roll down the incline without slipping. (Each object has its own value for θ_M.)

b. For *each* object, determine the translational motion in the case $\theta \leq \theta_M$.

c. For *each* object, determine the translational motion in the case $\theta > \theta_M$.

d. Set $\mu_s = 0.30$, $\mu_k = 0.10$, and $\theta = 40°$. In what order will the four objects reach the bottom of the incline?

10-45. *Trying to hit the spot.* A uniform bar of length l and mass m is suspended from a very thin axle that passes through a hole near the top end A of the bar.

a. How far from A should a blow be applied at right angles to the bar, in order to start the bar rotating about A without breaking the axle? (If the blow is properly placed, there will be no impulsive force on the axle as the blow is applied. In this case, the point of application of the blow is called the **center of percussion** relative to A.)

b. What is the period of oscillation of the rod when it is suspended from A?

c. What is the length of the simple pendulum having the same period? The length you obtain here should be the same as the distance you obtained in part *a*. The center of percussion relative to A is also called the *center of oscillation* relative to A.

10-46. *Not so fast!* Suppose the top described in Example 10-6 is spinning about its axis with an angular speed $\omega_s = 40$ rad/s. Evaluate its angular speed of precession ω_p. This will require modifying the procedure used in the example because ω_p is not very small compared to ω_s.

10-47. *Scaling: mass and moment of inertia.* Symmetrical tops 1 and 2 are built from the same material and are geometrically similar in shape, but top 2 is λ times as large as top 1 in *each* linear dimension.

a. Express the mass M_2 of top 2 as a multiple of the mass M_1 of top 1.

b. Designate the moments of inertia (about the axis of symmetry) of tops 1 and 2 by I_1 and I_2, respectively. Express I_2 as a multiple of I_1.

c. Evaluate the results of parts *a* and *b* for the following values of λ: 1.2, 2.0, 5.0.

10-48. *Scaling at constant angular speed.* Top A and top B are built from the same material and are geometrically similar in shape, but top B is λ times as large as top A in *each* linear dimension. Top A is set spinning with angular speed ω_{sA}; top B is set spinning with the same angular speed $\omega_{sB} = \omega_{sA}$.

a. Find the spin angular momentum L_{sB} of top B as a multiple of the spin angular momentum L_{sA} of top A. Your result should involve only L_{sA}, L_{sB}, and λ.

b. Find the spin kinetic energy K_{sB} of top B as a multiple of the rotational kinetic energy K_{sA} of top A.

c. Top A is inclined to the vertical as it spins on a smooth tabletop, and it is observed to precess with precessional angular speed ω_{pA}. This is much smaller than ω_{sA}, so Eq. (10-36) applies. Find the precessional angular speed ω_{pB} of top B. Assume that $\omega_{pB} \ll \omega_{sB}$. Express ω_{pB} as a multiple of ω_{pA}.

d. Express the ratio ω_{pB}/ω_{sB} as a multiple of ω_{pA}/ω_{sA}.

e. Suppose that $\omega_{pA}/\omega_{sA} = 1.0 \times 10^{-3}$ and $\lambda = 2.0$. Is the assumption that $\omega_{pB} \ll \omega_{sB}$ a valid one? Can you expect your result for ω_{pB} to be accurate?

10-49. *An adjustable top.* Consider the adjustable top shown in Fig. 10E-49. The main body of the top is a uniform hemisphere of mass M with radius R. A short, light

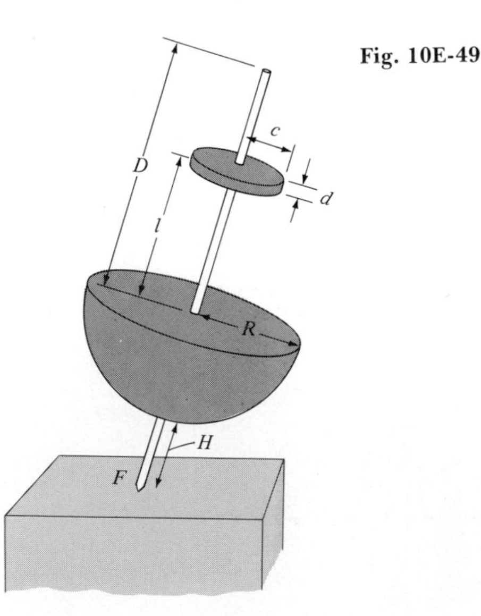

Fig. 10E-49

rod of length H serves as a "foot" for the top. A light rod of length D extends upward along the axis of symmetry of the top. A cylinder of mass m, radius c and thickness d can be clamped to the rod so that its center is at any desired distance l from the flat surface of the hemisphere. The extreme values for l are $l_{\min} = d/2$ and $l_{\max} = D - d/2$.

a. Show that the moment of inertia I of the entire top about the symmetry axis does not depend on l, and obtain an expression for I.

b. Obtain an expression for the distance r between the center of mass of the top and F, the point of contact with the support surface.

c. Using the definition $G^2 \equiv I/(M + m)$, giving the gyration radius of the top in terms of its moment of inertia I and its total mass M, apply Eq. (10-36) in order to obtain an expression for the angular speed of precession ω_p of the top, if it has spin angular speed $\omega_s \gg \omega_p$.

d. Where should the cylinder be placed in order to obtain the maximum angular speed of precession for a given spin angular speed? To obtain the minimum angular speed of precession rate for a given spin angular speed?

e. Evaluate the ratio of the maximum angular speed of precession to the minimum angular speed of precession, under the following conditions: $m/M = 0.20$, $H/R = 0.50$, $D/R = 3.0$, $c/R = 0.50$, and $d/R = 0.20$.

10-50. *Precession of a conical top.* As shown in Fig. 10E-50, a solid conical top of mass M, height h, and radius R is spinning about its symmetry axis OO' with spin angular speed ω_s. The axis OO' makes an angle α with the vertical.

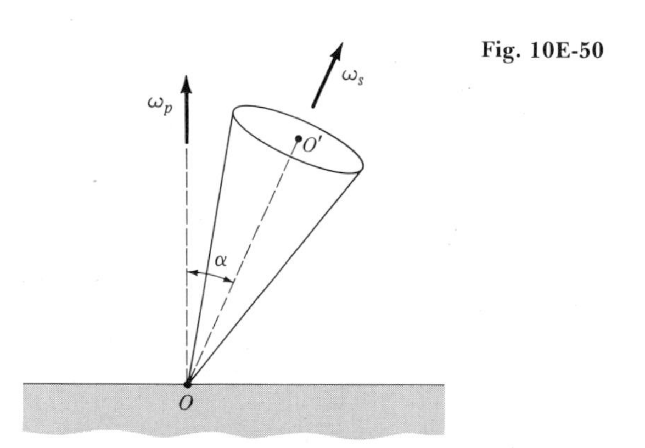

Fig. 10E-50

a. Show that the center of mass of the top is located along OO' at a distance $3h/4$ from the vertex O.

b. Show that the moment of inertia I about the axis OO' is given by $I = \frac{3}{10}MR^2$.

c. Find the angular speed ω_p at which the top precesses about the vertical.

d. Consider a top for which $h = 10.0$ cm and $R = 3.0$ cm. The top is spinning at 5800 rotations per minute. Using $g = 9.80$ m/s², evaluate the precession angular speed ω_p.

e. If the top were spinning at the same spin angular speed but in a more tilted position (that is, with a larger value of α than is shown), what would be the angular speed of precession?

10-51. *No slippage here, I.* Consider the system shown in Fig. 10E-51. Both the cylinder and the pulley are homogeneous, each having mass M and radius R. The hanging object also has mass M. The incline makes an angle θ with the horizontal. There is no slippage between the cylinder and the incline, nor between the pulley and the string. The pulley's axle is frictionless.

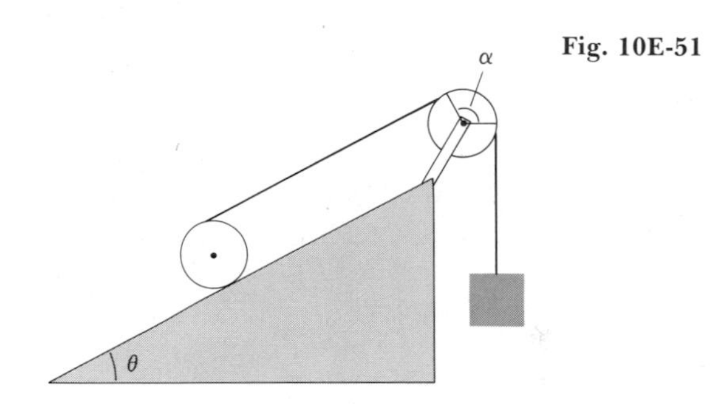

Fig. 10E-51

a. Determine the acceleration of the hanging weight, the angular acceleration of the pulley, and the translational and angular accelerations of the cylinder.

b. Determine the tensions in both straight sections of the string.

c. Find the minimum coefficient of static friction μ_s between the cylinder and the incline which is consistent with the lack of slippage.

10-52. *No slippage here, II.*

a. Consider the application of Newton's second law to a cord of negligible mass which lies along the surface of a pulley of radius R. Show that the maximum rate of change of tension with distance along the cord is given by $dT/ds \leq \mu_s T/R$, where μ_s is the coefficient of static friction between the cord and the pulley.

b. Use the result of part *a* to show that the ratio of the tension on either side of a pulley that is in contact with the cord over a central angle α is less than or equal to $e^{\mu_s \alpha}$.

c. Show that the pulley in Fig. 10E-51 has a central angle of contact given by $\alpha = \pi/2 + \theta$.

d. Find the minimum frictional coefficient μ_s that is consistent with the assumed lack of slippage in Exercise 10-51.

10-53. *The second law for a system of particles.* Newton's second law for a system of particles is stated in Eq. (10-44). Prove the validity of this law on the basis of Newton's laws for a single particle, in the manner suggested in small print below the equation.

11

Gravitation and Central Force Motion

11-1 UNIVERSAL GRAVITATION

This chapter is devoted largely to the application of the mechanics of rotational motion of particles, developed in Chap. 9, to a subject which has fascinated humankind from prehistoric times. This subject is the motion of heavenly bodies—and particularly the motion of the moon, the planets, and the sun—as they appear to sweep majestically across the sky.

The apparent motion of the stars is relatively simple, corresponding as it does to the daily rotation of the earth about its own axis. The moon, the planets, and the sun share in this daily apparent motion. In addition, however, they appear to move relative to the stars, each in its own way. The motions of the planets are particularly complex in appearance. In spite of these complexities, there are intriguing regularities about the motions and the interconnections among them.

An accurate quantitative description of the apparent motions of the moon, the planets, and the sun may be called the fundamental *kinematical* problem of astronomy. Thoughtful persons have risen to the challenge of describing heavenly motions since long before the invention of written arithmetic. One of the most dramatic evidences of this response is the complex at Stonehenge, in southern England. Figure 11-1*a* shows a view of Stonehenge, which functioned (perhaps among other purposes) as an astronomical observatory and calculator for measuring and predicting astronomical events. The carefully oriented stones appear to have been used in various combinations to establish sight lines for astronomical objects near the horizon. One such application is suggested in the figure caption.

Aside from such practical uses as predicting the onset of seasons for agricultural and other purposes, kinematical descriptions of astronomical motions made possible predictions of eclipses and other events which had

Fig. 11-1 (*a*) General view of Stonehenge. Besides the large stones, or megaliths, which are evident in the photograph, many smaller stones are hidden in the grass. The stones were transported to the site over a considerable distance by unknown means. Apparently, the whole was intended as an astronomical observation instrument and calculator which predicted eclipses among other purposes, not all of which are known. At the vernal equinox (the day in the spring when day and night are each 12 h long), the shadow of the rising sun was cast along a major row of stones. (*b*) An instrument used in the sixteenth century to determine the elevation and bearing angles of a heavenly body. The sights are at S and T. (*From Tycho Brahe Astronomiae Instrumentae Mechanica, 1598. Courtesy New York Public Library.*)

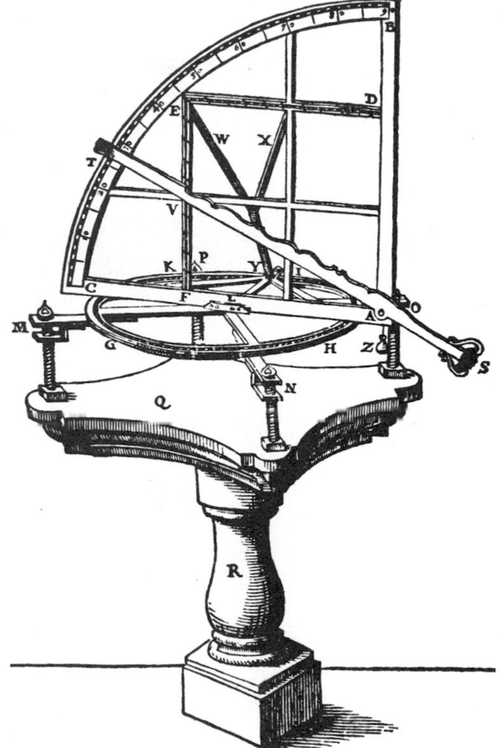

religious, astrological, or other mystical significance. But the intrinsic interest of the problem of making more accurate observations and predictions must always have been a strong motivation.

With the development of increasingly powerful mathematical and observational techniques, the arts of astronomical observation and prediction became more and more precise. Naked-eye observation reached its culmination in the work of the Danish astronomer Tycho Brahe (1546–1601). Figure 11-1*b* shows one of his sighting instruments, which were designed by him and were the best in the world at the time. While such instruments were far more precise and versatile than the sighting stones of Stonehenge, their underlying principles are not very different. To this day, indeed, the precise measurement of sighting angles is indispensable to observational astronomy.

Invention of the astronomical telescope by Galileo in 1609, and its subsequent improvement, made possible a precision of angular measurement better by orders of magnitude than the 30 arc seconds (0.008°) or so that represents the best the naked eye can do with the aid of sighting devices such as Tycho's. The improvement of the clock by Huygens and others in the mid-seventeenth century and afterward was also essential; it is much more useful to know precisely *where* a heavenly body is if you know precisely *when* it is there.

It is not surprising that these strides in the kinematical description of the heavens inspired attempts at *mechanical* explanations, that is, explanations of the observed motions in terms of the forces governing them. Modern physics may be said to have had its birth in these attempts.

As long as it was generally believed that the earth lay at the center of the universe, there was scant hope of even a semiquantitative mechanical explanation. But in the first half of the sixteenth century the Polish-Prussian physician, church official, and astronomer Nicolaus Copernicus (1473–1543) advocated an alternative model of the universe. In the Copernican system, the planets (*including* the earth) revolve around the sun, and the moon revolves around the earth. This arrangement—after considerable modification of its details had been made by others—ultimately made possible a relatively simple mechanical description in terms of the single, universal, fundamental force called *gravitation.*

But the Copernican system in its original form was not remarkably simple, and progress was not easy. Such outstanding men of genius as Johannes Kepler (1571–1630), Galileo, and René Descartes (1596–1650) speculated on the mechanical problem without real success.

It remained for Newton to find the key, using the groundwork which had been laid by others. More than a century after Copernicus, he established firmly the fact that the lunar and planetary motions, as seen from properly chosen reference frames, can be accounted for on the basis of the influence of gravitation, acting as a central force.

Why was the Copernican view of the solar system so important in Newton's development of the mechanics of that system on the basis of a single kind of force? You will see, as we follow Newton's development through this chapter, that the development "lifts itself by its own bootstraps." It begins by considering the behavior of an ordinary body near the surface of the earth, such as an apple, as the body moves under the influence of gravitational force. It then considers quantitatively the consequences of assuming that the motion of the moon around the earth is governed by the very same kind of force.

The next step depends heavily on the Copernican viewpoint. From that viewpoint, the motion of the moon about the earth is akin to the motion of the planets (including the earth) about the sun. On this basis, an understanding of the mechanics of lunar motion may well be expected to lead directly to insights into the mechanics of planetary motion.

A remark on nomenclature is appropriate here. In astronomy, it is conventional to call the rotation of one body about another **revolution.** The word **rotation** is restricted to the turning of a body about its own axis. We follow the astronomical convention in this chapter. Elsewhere in this book, the word "rotation" is used to denote any angular motion about any center, but "revolution" is used occasionally when the distinction must be made.

Newton began by applying Huygens' formula for centripetal acceleration to the revolution of the moon about the earth. In the form given by Eq. (3-41*b*), this is

$$\mathbf{a} = -\frac{v^2}{r}\,\hat{\mathbf{r}} \tag{11-1}$$

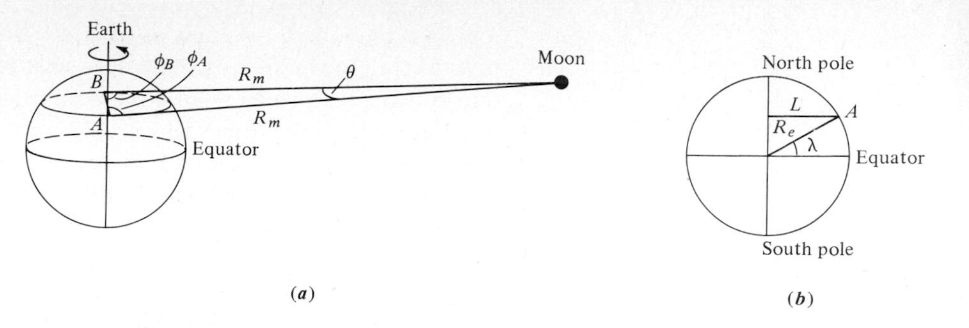

Fig. 11-2 Method for measuring R_m/R_e, the ratio of the distance from the earth to the moon to the radius of the earth. (*a*) Perspective view of the earth-moon system. Two observers A and B are located at the same latitude λ, on opposite sides of the earth. Their distance apart is $2L = 2R_e \cos \lambda$, where L is the axial distance shown in (*b*). If A and B simultaneously measure the angles ϕ_A and ϕ_B, they can determine the **parallax angle** θ, which is given by $\theta = \pi - (\phi_A + \phi_B)$. Since $R_m >> R_e$, the angle θ is small, and the approximation $\theta = \sin \theta$ is valid. Thus $\theta = 2L/R_m = 2R_e \cos \lambda/R_m$, which can be solved to give $R_m/R_e = (2 \cos \lambda)/\theta$. (In practice, the same observer makes two observations approximately 12 h 25 min apart, at the rising and the setting of the moon, and then corrects for the motion of the moon in the interim.)

On the basis of the approximation that the moon's orbit is circular, he then made a calculation of the centripetal acceleration of the moon about the earth. The details of this calculation are given in Example 11-1.

EXAMPLE 11-1

The average distance from the center of the earth to the center of the moon is $R_m = 3.84 \times 10^5$ km, or close to 60 times the radius of the earth, R_e. (The ratio of these two distances was first determined in classical antiquity. A more modern method is sketched in Fig. 11-2.) The **sidereal period** of the moon—that is, the time it takes to make one complete circuit of the earth, as seen by an observer fixed in space—is $T = 27.3$ days. Find the centripetal acceleration a_m of the moon.

▪ Note, to begin with, that this is simply the satellite problem of Example 3-10 with the numbers changed. Using the argument developed in Sec. 3-6, you express the tangential speed of the moon in terms of its orbit radius R_m and period T, and reexpress Eq. (11-1) in the form of Eq. (3-45) to produce

$$a_m = \frac{4\pi^2 R_m}{T^2} \tag{11-2}$$

You can then insert the numbers to obtain the magnitude of the centripetal acceleration of the moon,

$$a_m = 4\pi^2 \frac{3.84 \times 10^8 \text{ m}}{(27.3 \text{ days} \times 86{,}400 \text{ s/day})^2}$$
$$= 2.72 \times 10^{-3} \text{ m/s}^2$$

The value of a_m is very close to $1/3600 = 1/(60)^2$ that of the acceleration of gravity $g = 9.80$ m/s² at the surface of the earth. That is, it appears that $a_m/g = (R_e/R_m)^2$.

Is it a coincidence that $a_m/g = (R_e/R_m)^2$? Quite to the contrary, it is a pivotal clue in the line of reasoning that led Newton to his theory of gravitation. Like his predecessors, Newton asked himself: What is the source of the force which produces this acceleration? Kepler had speculated that the force was magnetic. Galileo had suggested tentatively that the inertial path of an isolated body is not a straight line but a large circle, so that no force is necessary.

Newton was thinking about these matters, and about mechanics in general, during the plague year of 1665. Cambridge University was closed, and Newton returned to his childhood home in the village of Woolsthorpe. As it turned out, it was about two years before he could go back to Cambridge, and those two years were the period in which Newton, who was in his middle twenties, accomplished the major portion of his truly prodigious creative life work. (He later wrote, ". . . for in those days I was in the prime of my age for invention, and minded Mathematicks and Philosophy more than at any time since.")

In later life Newton suggested to a friend that while sitting in an orchard he had seen an apple fall. He had been struck by the idea that *the very same force that made the apple fall was that which constrained the moon to its orbit—the gravitational force.* Taking this insight as far as it would go, Newton made the hypothesis that *every* body in the universe exerts a gravitational force on *every other* body.

What sort of mathematical form must such a *universal* gravitational force law take? The magnitude of the centripetal acceleration of the moon, calculated in Example 11-1, gives the first hint. If that centripetal acceleration is a *gravitational* acceleration, the apple would also experience a gravitational acceleration if it were placed a distance R_m from the center of the earth equal to that of the moon. What is more, we already know that the gravitational acceleration g experienced by bodies near the surface of the earth is independent of their masses. So it is fair to guess that the apple, if placed at a distance R_m from the center of the earth, would experience an acceleration of magnitude a_m equal to that calculated for the moon in Example 11-1. Thus, increasing the distance from the center of the earth to the apple 60-fold—from R_e, the radius of the earth, to R_m—would *decrease* the gravitational acceleration which it experiences $(60)^2$-fold—from g to a_m. Comparison of the gravitational acceleration of the apple falling from the tree with the gravitational acceleration of the moon "falling" in its orbit suggests that the acceleration of a given body, because of the gravitational force exerted on the body by the earth, depends *inversely on the square of the distance* from the center of the earth to the body.

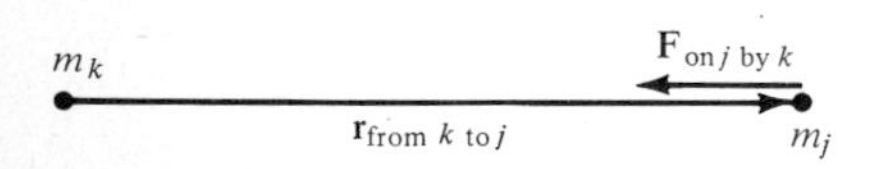

Fig. 11-3 Direction of the gravitational force in the case of two interacting bodies whose sizes are small enough, compared to the distance between them, that they can be considered as point masses m_j and m_k. The position of body j is given relative to body k by the vector $\mathbf{r}_{\text{from } k \text{ to } j}$. The gravitational force $\mathbf{F}_{\text{on } j \text{ by } k}$ exerted *on* body j due to the presence of body k is directed *toward* body k. That is, $\hat{\mathbf{F}}_{\text{on } j \text{ by } k} = -\hat{\mathbf{r}}_{\text{from } k \text{ to } j}$.

We are not yet in a position to verify this suggestion by considering further evidence. However, if it is to lead to a truly universal law of gravitation, it must apply to *any* pair of bodies j and k, and not just to the earth and the apple, or the earth and the moon. So let us make a tentative generalization, subject to later verification. In Fig. 11-3, the vector $\mathbf{r}_{\text{from } k \text{ to } j}$ describes the position of an arbitrary body j with respect to another body k. We express the distance dependence of the magnitude of the gravitational force $\mathbf{F}_{\text{on } j \text{ by } k}$ exerted on body j due to the presence of body k in the form

$$F_{\text{on } j \text{ by } k} \propto \frac{1}{(r_{\text{from } k \text{ to } j})^2} \tag{11-3}$$

This is the **inverse-square law.** Ultimately its validity will be justified by the consistency of predictions deduced from it with very large numbers of observations made over a very wide range of circumstances. For the moment, however, we continue with Newton's argument.

The magnitude $F_{\text{on } j \text{ by } k}$ of the gravitational force must be proportional to the mass m_j of the body *on which* it is acting, say, the apple. If this were not so, different bodies falling to earth from small heights above its surface would not experience the same gravitational acceleration. Stated mathematically, this is

$$F_{\text{on } j \text{ by } k} \propto m_j \tag{11-4a}$$

Since bodies j and k are chosen entirely arbitrarily, there must also be a force $\mathbf{F}_{\text{on } k \text{ by } j}$ exerted on body k due to the presence of body j. A repetition of the argument immediately above leads to the relation

$$F_{\text{on } k \text{ by } j} \propto m_k \tag{11-4b}$$

where m_k is the mass of body k.

According to Newton's third law of motion, the forces $\mathbf{F}_{\text{on } j \text{ by } k}$ and $\mathbf{F}_{\text{on } k \text{ by } j}$ must have equal magnitudes; that is,

$$F_{\text{on } j \text{ by } k} = F_{\text{on } k \text{ by } j}$$

The proportionality (11-4b) can therefore be written

$$F_{\text{on } j \text{ by } k} \propto m_k$$

This proportionality, taken together with that of (11-4a), shows that the force exerted on body j by body k is proportional to *both* masses. But if a quantity is proportional to two independent quantities, it is proportional to their product. Thus we can write

$$F_{\text{on } j \text{ by } k} \propto m_j m_k \tag{11-4c}$$

Combining this proportionality with the inverse-square law expressed in the proportionality of Eq. (11-3), we obtain

$$F_{\text{on } j \text{ by } k} \propto \frac{m_j m_k}{(r_{\text{from } k \text{ to } j})^2} \tag{11-5}$$

As usual, this relation is more usefully written as an equation. We define the **universal gravitational constant** G according to the equation

$$F_{\text{on } j \text{ by } k} = G \frac{m_j m_k}{(r_{\text{from } k \text{ to } j})^2} \tag{11-6a}$$

This equation can also be written in vector form, so as to express direction as well as magnitude. As shown in Fig. 11-3, the force exerted on body j by body k is always *toward* body k. Since the vector $\mathbf{r}_{\text{from } k \text{ to } j}$ is directed from body k toward body j, we have

$$\mathbf{F}_{\text{on } j \text{ by } k} = -G \frac{m_j m_k}{(r_{\text{from } k \text{ to } j})^2} \hat{\mathbf{r}}_{\text{from } k \text{ to } j} \tag{11-6b}$$

The bodies labeled j and k are quite arbitrary. Thus, if there is a force exerted on body j by body k, there must likewise be a force exerted on body k by body j. That force can be evaluated by repeating the entire argument leading to the equation immediately above, with the roles of bodies j and k interchanged. Doing so leads to the equation

$$\mathbf{F}_{\text{on } k \text{ by } j} = -G \frac{m_k m_j}{(r_{\text{from } j \text{ to } k})^2} \hat{\mathbf{r}}_{\text{from } j \text{ to } k} \tag{11-6c}$$

To compare this equation with the previous one, note that $(r_{\text{from } j \text{ to } k})^2 = (r_{\text{from } k \text{ to } j})^2$ and $\hat{\mathbf{r}}_{\text{from } j \text{ to } k} = -\hat{\mathbf{r}}_{\text{from } k \text{ to } j}$. Consequently we have

$$\mathbf{F}_{\text{on } k \text{ by } j} = -\mathbf{F}_{\text{on } j \text{ by } k}$$

These two gravitational forces comprise an action-reaction pair, in accordance with Newton's third law.

While the subscript notation used in the series of equations immediately above is explicit in its meaning, it has the disadvantage of being cumbersome. We may substitute a shorthand notation for the subscripts, in which the words are omitted and their meaning is implied by the order of the indices j and k. For example, we can define $\mathbf{F}_{jk} \equiv \mathbf{F}_{\text{on } j \text{ by } k}$ and $\mathbf{r}_{kj} \equiv \mathbf{r}_{\text{from } k \text{ to } j}$. In this more compact notation, the equation for the magnitude of the gravitational forces becomes

$$F_{jk} = F_{kj} = G \frac{m_j m_k}{r_{kj}^2} \tag{11-6d}$$

The vector equations assume the forms

$$\mathbf{F}_{jk} = -G \frac{m_j m_k}{r_{kj}^2} \hat{\mathbf{r}}_{kj} \tag{11-6e}$$

and

$$\mathbf{F}_{kj} = -G \frac{m_k m_j}{r_{jk}^2} \hat{\mathbf{r}}_{jk} \tag{11-6f}$$

Any of Eqs. (11-6*a*) through (11-6*f*) is called **Newton's law of universal gravitation.**

In some cases, it is necessary to consider the gravitational forces exerted on body j due to the presence of a number of other bodies. The net gravitational force on body j is found by taking the vector sum of the individual forces calculated by repeated application of Eq. (11-6*b*).

If the argument leading to Eq. (11-6*b*) stood alone, there would be little cause for surprise in the fact that the gravitational force obeys the inverse-square law. After all, the inverse-square law was "built into" the equation in the light of the known facts that the distance from the center of the earth to the moon is 60 times the distance from the center of the earth to its surface, and that the magnitude of the acceleration of the moon as it moves in its orbit is $1/(60)^2$ that of the falling apple. To see this "building in" explicitly, we write Eq. (11-6*a*) for the special cases of the apple of mass m_a located a distance R_e from the center of the earth, whose mass is m_e, and for the moon of mass m_m located a distance R_m from the center of the earth. The gravitational forces on the two bodies are, respectively, of magnitude

$$F_{\text{on apple by earth}} = G \frac{m_a m_e}{R_e^2} \quad \text{and} \quad F_{\text{on moon by earth}} = G \frac{m_m m_e}{R_m^2}$$

Using Newton's second law, we can immediately write

$$F_{\text{on apple by earth}} = m_a g$$

(since we know the magnitude of the acceleration of gravity at the surface of the earth to be g) and

$$F_{\text{on moon by earth}} = m_m a_m$$

where a_m is the centripetal acceleration of the moon, which we have determined in Example 11-1 on the basis of its orbit radius and period. When these forces are substituted into the two gravitational-force equations above, the masses m_a and m_m, respectively, cancel, and we have

$$g = G\frac{m_e}{R_e^2} \tag{11-7a}$$

and

$$a_m = G\frac{m_e}{R_m^2} \tag{11-7b}$$

While we know neither the value of the universal gravitational constant G nor the mass of the earth m_e, we can divide the second equation by the first to obtain the ratio

$$\frac{a_m}{g} = \left(\frac{R_e}{R_m}\right)^2 \tag{11-8}$$

We have already seen that the numerical values of the ratios on the two sides of the equation "answer pretty nearly," as Newton put it.

In deriving Eq. (11-8), we have used Newton's second law, which correctly describes the connection between force and motion for an observer fixed in an inertial frame. But the derivation was carried out in a reference frame fixed at the center of the earth. The earth experiences a centripetal acceleration as it moves in its orbit about the sun which is by no means negligible compared to a_m, the centripetal acceleration of the moon with respect to the earth. (Indeed, the acceleration of the earth is more than twice a_m in magnitude. Can you show this by direct calculation?) Moreover, if the earth exerts a gravitational force on the moon, the sun presumably does so as well. And the latter force has been ignored.

Why, then, do the numerical values inserted into Eq. (11-8) "answer pretty nearly"? To put it more generally, why is it valid to use Newton's second law in the noninertial frame fixed at the center of the earth in support of the hypothesis that the law of gravitation is an inverse-square law?

Let us begin to answer this question by looking at the earth-moon system from a frame of reference fixed with respect to the sun, which can surely be considered an inertial frame for the purpose at hand. Figure 11-4*a* is a free-body diagram of the moon as seen in this frame. The force $\mathbf{F}_{\text{on moon by earth}}$ is the gravitational force we have been discussing. In addition, there must be a force exerted on the moon $\mathbf{F}_{\text{on moon by sun}}$ which constrains it to follow the path around the sun that it shares, generally speaking, with the earth. (We have not established that this force arises from the same type of gravitational interaction which makes the moon orbit the earth. But for the sake of the present argument, that is not important.)

Next, consider the earth-moon system from the noninertial frame fixed at the center of mass of the earth-moon system itself. Figure 11-4*b* is a free-body diagram of the moon as seen in this frame. As was shown in Sec. 5-4, it is possible to use Newton's second law in a noninertial frame *if* a suitable fictitious force is added to the forces acting on the body whose motion is being studied—in the present case, the moon. If the acceleration of the noninertial frame of reference, as seen from an inertial frame, is $\mathbf{A}$ and the mass of the body being studied—here the moon—is m_m, Eq. (5-29) gives the fictitious force $\mathbf{F}_{\text{fict}}$ as

$$\mathbf{F}_{\text{fict}} = -m_m\mathbf{A}$$

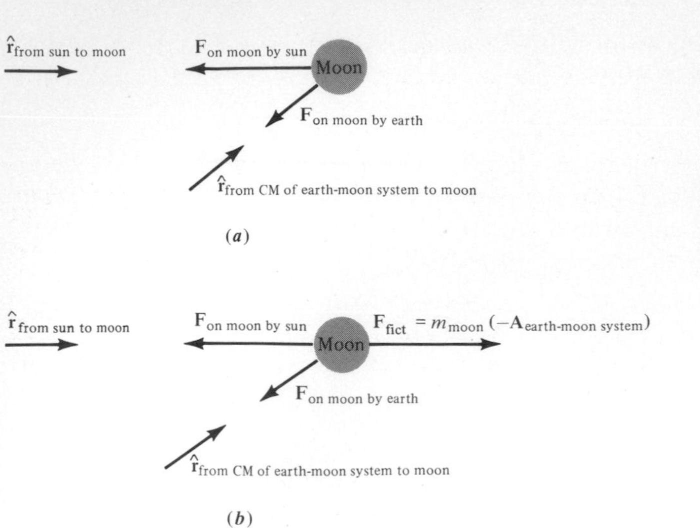

Fig. 11-4 Free-body diagram of the moon from the point of view of (*a*) an observer in the substantially inertial frame of reference fixed with respect to the sun and (*b*) an observer in the noninertial frame of reference fixed with respect to the center of mass of the earth-moon system. In part *a* the two forces shown acting on the moon are those leading to its acceleration about the sun and the earth respectively. In part *b* is shown the additional fictitious force $\mathbf{F}_{\text{fict}}$ which the accelerated observer requires in order to apply Newton's laws of motion.

The proper net force to be used in applying Newton's second law in the noninertial frame of Fig. 11-4*b* is thus

$$\mathbf{F} = \mathbf{F}_{\text{on moon by earth}} + \mathbf{F}_{\text{on moon by sun}} + \mathbf{F}_{\text{fict}}$$

If the moon were located exactly at the center of mass of the earth-moon system, the force exerted on the moon by the sun would produce an acceleration exactly equal to **A**, the acceleration of the earth-moon system about the sun. But, in fact, the distance from the sun to the moon varies as the moon orbits the earth. Its value oscillates about the distance from the sun to the center of mass of the earth-moon system, with an amplitude which is small compared to the latter distance.

But in spite of this slight variation, we know that the force exerted on the moon by the sun must be such that its *average* value would produce an acceleration of the moon of magnitude A. We know this because the period of the moon's orbit about the sun is 1 yr—the same as the period of the center of mass of the earth-moon system about the sun. We thus have

$$\mathbf{F}_{\text{on moon by sun}} = m_m\mathbf{A}$$

Consequently, the net force acting on the moon from the point of view of the observer in the noninertial frame fixed with respect to the center of mass of the earth-moon system is, on average,

$$\mathbf{F} = \mathbf{F}_{\text{on moon by earth}} + m_m\mathbf{A} - m_m\mathbf{A}$$

or

$$\mathbf{F} = \mathbf{F}_{\text{on moon by earth}}$$

Finally, we must note that the actual observer is located in a frame fixed not with respect to the center of mass of the earth-moon system, but with respect to the earth itself. But as you will see in Example 11-6, the center of mass is only 4700 km from the center of the earth—that is, inside the earth itself. The additional acceleration of the actual observer resulting from her or his monthly rotation around the center of mass of the earth-moon system is therefore negligible.

This is why it is possible to carry out Newton's comparison of the acceleration of the moon about the earth with the acceleration of the apple even though (1) the frame of reference is not inertial and (2) the force exerted on the moon by the sun is ignored.

For further evidence in support of the inverse-square law, Newton had to extend his vision—and the scope of validity of the law of universal gravitation—to more distant parts of the solar system. We will follow Newton in this shortly. But we have, as Newton did not, supporting "close-by" evidence provided by artificial earth satellites. Example 11-2 illustrates the use of such evidence.

EXAMPLE 11-2

If the inverse-square law is correct, Eq. (11-8), $a_m/g = (R_e/R_m)^2$, should be applicable to any earth satellite having a circular orbit of known radius and period. Such a satellite is the synchronous satellite, whose sidereal period is equal to the sidereal period of rotation of a point on the surface of the earth, T = 23 h 56 min. In Example 3-10 the altitude, or distance above the surface of the earth, was given without justification to be 3.58×10^7 m. Show that these data satisfy Eq. (11-8) and hence the inverse-square law as well.

Some comment is required concerning the value of g appropriate for use in Eq. (11-8). The synchronous satellite revolves about the earth as seen by an observer fixed with respect to the center of the earth. But the acceleration of gravity is actually measured by observers rotating with the earth. This rotation introduces a centrifugal acceleration which leads to a measured value of g smaller than that which would be measured by a nonrotating observer (assuming that such an observer could measure g while hovering just above the surface of the earth while the earth spun beneath). The rotation of the earth also introduces another complicating factor indirectly. Because of the rotation, the earth is not perfectly spherical; its radius is greater at the equator than at the poles. Thus the distance of the observer from the center of the earth depends on the latitude at which the observation is made, and this also influences the measured value of g. When these matters are taken into consideration, the proper value of g, for a nonrotating observer located at a distance from the center of the earth equal to the earth's average radius, $R_e = 6.367 \times 10^6$ m, is $g = 9.848\ \text{m/s}^2$ to four significant figures.

■ You modify the notation of Eq. (11-8) for application to an arbitrary satellite. Calling the magnitude of the satellite's centripetal acceleration under the influence of the earth's gravitation a_s and its orbit radius R_s, you have

$$\frac{a_s}{g} = \left(\frac{R_e}{R_s}\right)^2$$

The necessary value of a_s for a body moving in a circular orbit of radius R_s with period T is given by Eq. (11-2), and is $a_s = 4\pi^2 R_s/T^2$. Substituting this value into the equation immediately above, you obtain

$$\frac{4\pi^2 R_s}{gT^2} = \frac{R_e^2}{R_s^2}$$

or

$$R_s = \left(\frac{gR_e^2T^2}{4\pi^2}\right)^{1/3}$$

The numerical values give you

$$R_s = \left[\frac{9.848\ \text{m/s}^2 \times (6.367 \times 10^6\ \text{m})^2 \times (23\ \text{h} \times 3600\ \text{s/h} + 56\ \text{min} \times 60\ \text{s/min})^2}{4\pi^2}\right]^{1/3}$$

$$= 4.218 \times 10^7\ \text{m}$$

To find the satellite altitude h, you subtract the radius of the earth from R_s, which gives you

$$h = R_s - R_e = 4.218 \times 10^7\ \text{m} - 6.367 \times 10^6\ \text{m}$$

Rounded off to three significant figures, this yields

$$h = 3.58 \times 10^7\ \text{m}$$

It is very gratifying that Newton's law of gravitation, in the scalar form of Eq. (11-6*a*), agrees so well with observation. However, there is a weak link in the logic leading to this result, and we must now backtrack to discuss it. In applying Eq. (11-6*a*) to the moon, it is reasonably clear what is meant by R_m, the distance from the earth to the moon. Neither of them is a particle, but the sizes of the earth and of the moon are both small compared to the distance between them, and it cannot be far wrong to let R_m be the distance between their centers.

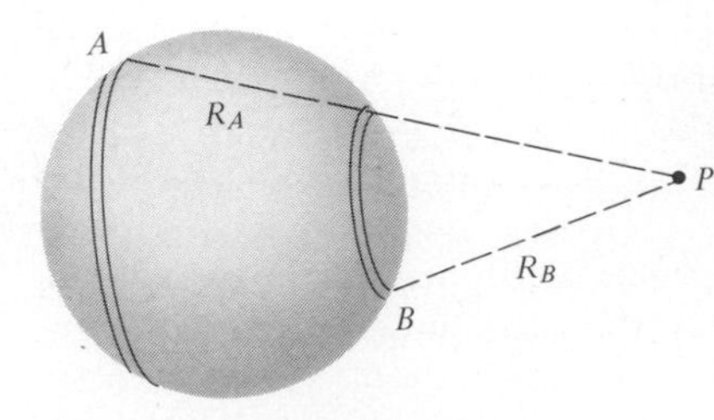

Fig. 11-5 Qualitative justification of Newton's conjecture that the gravitational attraction resulting from a uniform sphere of mass M, experienced by a body located at point P outside the sphere, is equal to that resulting from a point particle of mass M located at the center of the sphere. The uniform sphere is divided into a nest of infinitesimally thick shells, one of which is illustrated. Two infinitesimally wide zones A and B are shown on the shell. All points in zone A are essentially equidistant from P, and the same is true for the points in zone B. While zone B is closer to P than is zone A, it also contains less mass. Detailed analysis shows that the gravitational attractions of the various zones "average out" to the result stated above. If that result holds for a shell, it holds as well for a spherical solid made up of a nest of concentric shells. A rigorous proof is given in Chap. 20.

The apple is in a different situation. It is itself small enough to be regarded as a particle, but it is close to a very large earth. Some parts of the earth are just a few meters away, while others are 10^6 times more distant. To make a proper determination of the force exerted on the apple by the earth, it is necessary to divide the earth into small mass elements m_j and to use Eq. (11-6*c*) to find the sum of their individual effects. In the discussion above, we tacitly made the same guess as Newton initially did. According to this guess, illustrated in Fig. 11-5, the relatively strong attraction of nearby mass elements and the relatively weak attraction of distant ones "average out" in such a way that the earth as a whole attracts the apple *just as if all the earth's mass were concentrated at its own center.* It turns out that this plausible guess can be proved correct for an inverse-square law.

The need to verify this guess caused Newton great difficulty. The proof was one of the first major applications of the integral calculus which he invented. It is straightforward, though somewhat lengthy. We will not discuss it here, however. An argument based on symmetry, developed a century and a half after Newton, provides a very much simpler proof and leads to much deeper insights into the nature of the inverse-square forces. This very powerful approach, called Gauss' law, is discussed in detail in Chap. 20, in connection with the electric force. At that point, we also show explicitly how Gauss' law verifies the guess.

Artificial satellites provide experimental evidence, not available to Newton, that the guess is correct. The synchronous satellite discussed in Example 11-2 is only about six earth radii distant from the surface of the earth. This is close enough that the earth, when seen from the satellite, does not visually approximate a point mass. Thus any error in our guess would lead to an improper choice of the orbit radius for synchronous motion. A quite small deviation from the stationary appearance of the satellite over a fixed meridian of longitude would be readily detectable. But, in fact, the calculated synchronous orbit radius is quite accurate. That is, the earth *does* "look" like a point as "seen" gravitationally from the satellite. Even satellites much closer to the earth have orbit periods whose deviations from those calculated on the basis of our guess are due only to the very small departure of the earth's shape from perfect sphericity.

If the law of gravitation were not an inverse-square law, it would not be correct to assume that the distributed mass of the earth acts on external masses as if its own mass were concentrated at its center. In the long run, the real verification of the inverse-square nature of the law of gravitation is its consistency with observation. In Sec. 11-6, we study hypothetical planetary systems in which the gravitational force depends on other powers of the distance, and we compare them with what is actually observed.

The inverse-square law appears to be deeply embedded into the geometrical nature of the space in which we live. To see this in a simple way, we can make the following argument. The body j of mass m_j in Fig. 11-6*a* is attracted toward the

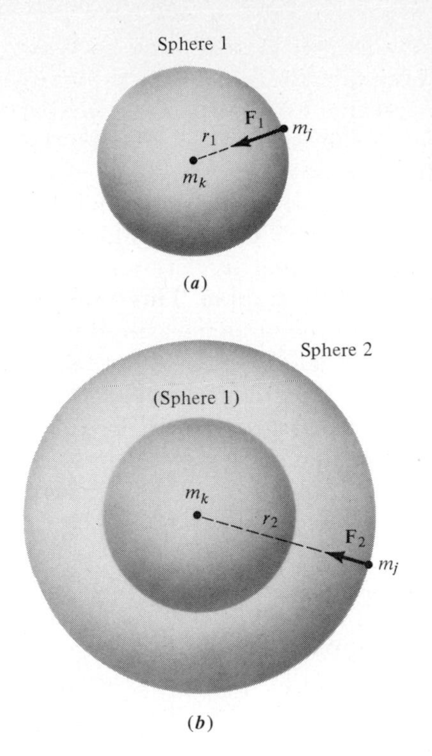

Fig. 11-6 Diagram for the geometrical plausibility argument, given in the text, for the inverse square form of the law of gravitation for two point masses.

body k of mass m_k "simply" as a result of the presence of body k in space. That is what we mean when we say that gravitational attraction is a fundamental observed property of bodies having mass. Now, j lies on the surface of an imaginary sphere 1 of radius r_1, whose center is at k. The magnitude F_1 of the force on j would be the same, no matter where j was located on sphere 1. That is, the gravitational influence of the mass m_k is the same everywhere on the surface of the sphere centered on k. So we can imagine some sort of "influence" emanating from m_k uniformly in all directions. The amount of the influence depends, as we have said above, on the magnitude of the mass m_k.

What happens if body j is relocated farther away from body k, that is, on the surface of the larger sphere 2 of radius r_2, as in Fig. 11-6*b*? The magnitude F_2 of the force on body j is the same no matter where j is located on the surface of the larger sphere. But F_2 is smaller than F_1. The influence emanating from body k still depends on the magnitude of m_k (which has not changed), but the influence has "spread out" over the outer sphere, which has a greater surface area than the inner one.

Although the gravitational force gets smaller and smaller as the distance between bodies increases, it never disappears entirely. Thus, while the influence of m_k "spreads out thinner and thinner," it does not vanish. Put another way, it is plausible to assume that all the influence which "passes through" sphere 1 passes through sphere 2 as well.

A direct measure of the spreading out of the influence of m_k is the ratio of the surface areas A_1 and A_2 of the two spheres. Making our plausibility argument quantitative, we write it in the form

$$\frac{F_2}{F_1} = \frac{m_k/A_2}{m_k/A_1}$$

Thus we have

$$\frac{F_2}{F_1} = \frac{m_k/4\pi r_2^2}{m_k/4\pi r_1^2}$$

or

$$\frac{F_2}{F_1} = \frac{r_1^2}{r_2^2} \tag{11-9}$$

which is the inverse-square law we set out to make plausible. When seen in this light, the inverse-square law is a fundamental *geometrical* property of the three-dimensional space with which we are familiar, *euclidean space*.

We will now consider the role of mass in more detail from a gravitational point of view. In Sec. 5-2, we dwelt briefly on the distinction between the *inertial* and *gravitational* aspects of mass. We defined mass in Chap. 4 as the tendency of a body to resist acceleration, that is, mass in its inertial aspect. To be specific, mass is defined inertially by either of the equations

$$\mathbf{p} = m\mathbf{v} \qquad \text{or} \qquad \mathbf{F} = m\mathbf{a}$$

In this chapter, mass appears in its gravitational aspect in Eq. (11-6*e*):

$$\mathbf{F}_{jk} = -G\,\frac{m_j m_k}{r_{kj}^2}\,\hat{\mathbf{r}}_{kj}$$

This equation may also be regarded as a definition of mass. But *mass so defined has an experimental (or observational) basis completely independent of its definition in terms of inertia.*

We are so used to sensing mass simultaneously in both these aspects, inertial and gravitational, that we do not even give them separate names in ordinary practice. This is so in spite of the fact that we must take care to distinguish the two in analyzing such systems as the Atwood machine or the block sliding down an inclined plane, with which we dealt in Chap. 5. But if you think about it for a while, you will agree that it is most striking that for any body, *the two masses appear to be equal!* It is entirely due to this fact that different objects dropped to the ground fall with equal accelerations, neglecting air resistance. Let us consider again the falling apple. This time, we make explicit the distinction between the mass of the apple measured inertially, m_{Ia}, and its mass measured gravitationally, m_{Ga}. (As far as the fall of the apple is concerned, we are interested in only the gravitational aspect of the mass of the earth, which we call m_{Ge}.) The magnitude of the force exerted on the apple by the earth depends on the gravitational aspect of the masses of the two bodies, and is found by applying Eq. (11-6*a*) to the present situation to obtain

$$F_{\text{on apple by earth}} = G\frac{m_{Ga}m_{Ge}}{R_e^2}$$

But it is the inertial aspect of the mass of the apple which must be used in Newton's second law, $F = ma$. Using the fact that the apple's acceleration is equal to g, the acceleration of gravity near the surface of the earth for any body, we can write Newton's second law in the form

$$F_{\text{on apple by earth}} = m_{Ia}g$$

Equating the two expressions for the force exerted on the apple by the earth, we have

$$m_{Ia}g = G\frac{m_{Ga}m_{Ge}}{R_e^2}$$

This equation can be solved for the acceleration of gravity g to yield

$$g = \frac{m_{Ga}}{m_{Ia}}\frac{Gm_{Ge}}{R_e^2}$$

Generalizing from the apple to any body j, we can write the acceleration of gravity for that particular body j in the form

$$g = \frac{m_{Gj}}{m_{Ij}}\frac{Gm_{Ge}}{R_e^2} \tag{11-10}$$

The observed fact that g is the same for all bodies falling without friction near the surface of the earth could not be true unless the ratio m_{Gj}/m_{Ij} were the same for all bodies.

Newton made a series of experiments to verify that this is indeed the case. He made a number of pendulums of the same length but having bobs of different materials. It is straightforward to show that the period T_j of a simple pendulum, whose bob is body j, is

$$T_j = 2\pi\sqrt{\frac{l}{g}}\sqrt{\frac{m_{Ij}}{m_{Gj}}}$$

if m_{Ij} and m_{Gj} are not assumed identical. Newton found that all his pendulums had equal periods within the experimental error of about 1 percent, which leads to the conclusion that m_{Gj}/m_{Ij} is substantially the same for all the materials tried.

In the late 1880s, the Hungarian physicist Roland von Eötvös (1848–1919) undertook a long series of experiments to determine, with the highest possible precision, just how close to the same value the ratio of the two masses is for different materials. Eötvös used a torsion pendulum in which the bodies at the two ends of the horizontal rod, one of platinum and the other of some other material, had the same gravitational mass m_G, as determined by precise weighing techniques. If the beam of the torsion pendulum is oriented east-west, the rotation of the earth results in a centrifugal force being exerted on each of the two bodies in the direction away from the earth's axis. Each of the two centrifugal forces is proportional to the inertial mass m_I of the body on which it is exerted. If m_I is the *same* for both bodies, the net torque is zero and there is no deflection of the torsion pendulum. However, if the two bodies, which have the same m_G, have *different* m_I, then the apparatus will be twisted by the resulting net torque, since centrifugal force depends on inertial mass. Eötvös could find no effect, although his apparatus could have detected a difference in m_I for the two bodies smaller than 1 part in 10^8.

The Eötvös experiment can be done in other ways. More recently, null results have been obtained with a margin of error almost a thousandfold smaller than that of Eötvös.

You may well suspect that when two quite different quantities coincide within 1 part in 10^{11} or better, it is no accident. The assertion that the two are really the *same* quantity, observed in different ways which obscure their fundamental identity, is the logical starting point of the general theory of relativity. For this reason we have usually spoken of the inertial and gravitational *aspects* of mass rather than separately of the inertial mass and the gravitational mass.

11-2 DETERMINATION OF THE UNIVERSAL GRAVITATIONAL CONSTANT *G*

In the calculations of Sec. 11-1, the gravitational constant G always appears as part of the product Gm_e, where m_e is the mass of the earth. We can determine the value of that product from the expression

$$g = \frac{Gm_e}{R_e^2} \qquad (11\text{-}11)$$

which is just Eq. (11-10) simplified by the assumption that the ratio of a gravitationally measured mass to an inertially measured mass has the value $m_G/m_I = 1$ for all bodies. If either G or m_e is known, the other can be obtained immediately from Eq. (11-11), since the gravitational acceleration g and the earth's radius R_e are known. Knowing G, we will be able to find the mass of any heavenly body which has a satellite, as you will see in Sec. 11-3.

A first rough guess at the value of G is made in Example 11-3 by using Eq. (11-11) together with an estimate of m_e.

EXAMPLE 11-3

Make the crude assumption that the density of the earth is uniform throughout, and estimate m_e. Then use the result to estimate G. Geological observations distributed all over the world suggest that the approximate average density, or mass per unit volume, of the solid material on or very near the surface of the earth is $\rho \simeq 4 \times 10^3$ kg/m^3.

▪ You have for the mass of the spherical earth, whose volume is $V = \frac{4}{3}\pi R_e^3$,

$$m_e = \rho V = \tfrac{4}{3}\pi\rho R_e^3$$

Insert this value into Eq. (11-1) to obtain

$$g = \tfrac{4}{3}\pi G\rho R_e$$

which you solve for G, obtaining

$$G = \frac{3g}{4\pi\rho R_e}$$

The numerical values give you

$$G \simeq \frac{3 \times 9.8 \text{ m/s}^2}{4\pi \times 4 \times 10^3 \text{ kg/m}^3 \times 6.4 \times 10^6 \text{ m}} = 9 \times 10^{-11} \text{ N}\cdot\text{m}^2/\text{kg}^2$$

This calculation must certainly overestimate the value of G. Even if there were no tendency for denser materials to lie deeper in the earth, compression resulting from the enormous pressure in the interior would guarantee a higher density there than on the surface. Thus the calculation underestimates the average density of the earth and so overestimates G. As you will see, the value of G obtained is about 50 percent too large. Still, that is not bad for a first guess.

The first laboratory measurement of G was made in 1797–1798 by Henry Cavendish (1731–1810), using a method still employed today with modifications only in detail. Something of an eccentric, Cavendish was one of the brilliant amateurs who made a very large share a significant British contributions to the physical sciences during the eighteenth and nineteenth centuries. Cavendish described his equipment as follows: "The apparatus is very simple; it consists of a wooden arm, 6 feet long, made so as to unite great strength with little weight. [See Fig. 11-7.] This arm is suspended in a horizontal position, by a slender wire 40 inches long, and to each extremity is hung a leaden ball, about 2 inches in diameter; and the whole is inclosed in a narrow wooden case, to defend it from the wind. As no more force is required to make this arm turn round on its centre, than what is necessary to twist the suspending wire, it is plain, that if the wire is sufficiently slender, the most minute force, such as the [gravitational] attraction of a leaden weight a few inches in diameter, will be sufficient to draw the arm sensibly aside."

Before the actual measurements leading to a determination of G can be begun, the torsion constant k of the so-called **torsion balance** must be determined. This is done by allowing it to oscillate and measuring the oscillation frequency ν. The moment of inertia I is then carefully calculated, with due allowance being made for the small contribution of the beam hh, shown in Fig. 11-7, which supports the bodies xx. Equation (10-25), $\nu = (2\pi)^{-1}(k/I)^{1/2}$, is then used to find k.

In the main part of the experiment, the equilibrium position of the torsion balance beam hh in Fig. 11-7b is measured with the large spheres W, which Cavendish called "weights," in the position shown by the solid lines. Then the displacement of the equilibrium position is measured when the large spheres are swung around into the position shown by the dotted lines. Needless to say, all kinds of precautions are necessary. A very small amount of electric charge on the spheres W, for example, will completely overwhelm the gravitational effect. (That is what we meant when we said in Sec. 1-2 that the gravitational interaction is very much weaker than the electric interaction.) Cavendish's biggest problem, however, was to minimize the effects of the air currents. Even minute differences in the temperature of different parts of the apparatus could produce sufficient motion of the air to lead to serious, disturbing effects. Besides enclosing the entire apparatus in a large case, he made all necessary manipulations remotely (to avoid the effect of his body heat) and made his readings with small telescopes.

The angular displacement of the torsion balance when the large spheres W are swung around, together with the known value of the torsion constant k, yields a value of the torque. This, together with the known

(a)

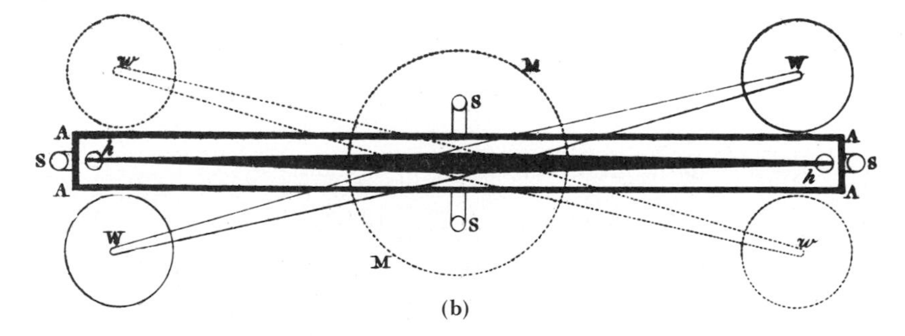

(b)

Fig. 11-7 Two views of Cavendish's torsion balance, taken from his 1798 paper. The torsion balance proper consists of the two small bodies x which are supported by the light but rigid structure *ghmh* suspended from the torsion fiber *lg*. The motion of the balance is observed by means of the light sources L, the small mirrors n on the ends of the balance beam, and the small telescopes T. Gravitational attraction is exerted on the bodies x by the large bodies W. The latter can be swung around, as shown in part *b*, by means of the large pulley M. (*Philosophical Transactions of the Royal Society, 1798. Courtesy of the New York Public Library.*)

length of the beam *hh*, can be used to calculate the gravitational force. Since the masses of *WW* and *xx*, and the distance between them, are known, Eq. (11-6*a*) can be solved to give a numerical value of the universal gravitational constant G.

Cavendish's result, incorporating all the many corrections required, was equivalent to $G = (6.65 \pm 0.48) \times 10^{-11}\ \text{N}\cdot\text{m}^2/\text{kg}^2$. More modern measurements are in general agreement on the value

$$G = (6.67 \pm 0.01) \times 10^{-11}\ \text{N}\cdot\text{m}^2/\text{kg}^2 \tag{11-12}$$

This is the least accurately known of the fundamental physical constants, because of the extreme weakness of the gravitational interaction.

As a by-product of the experiment, Cavendish also verified the inverse-square law at distances as small as a few millimeters. Up to that time, its validity had been supported only over much larger distances (such as those involved in the moon-apple calculation). While his measurements seem to have been reliable within about 1 percent, he did not consider them worthy of more than passing mention (possibly because Coulomb had obtained more definite results for the inverse-square nature of the electric force a decade earlier). Nevertheless, Cavendish's results lend further support to the use of center-to-center distances in applying Newton's law of gravitation to spherical bodies.

Cavendish and his contemporaries referred to his measuring of G as "weighing the earth," a dramatic if inaccurate phrase still sometimes used in speaking loosely of the experiment. Example 11-4 suggests that what is really meant by the phrase is measurement of the mass of the earth.

EXAMPLE 11-4

Using the modern value of G, find the mass and the mean density of the earth.

■ From Eq. (11-11) you have for the mass of the earth

$$m_e = \frac{gR_e^2}{G}$$

Using the values of g and R_e given in Example 11-2, you have

$$m_e = \frac{9.85\ \text{m/s}^2 \times (6.37 \times 10^6\ \text{m})^2}{6.67 \times 10^{-11}\ \text{N}\cdot\text{m}^2/\text{kg}^2}$$
$$= 5.99 \times 10^{24}\ \text{kg} = 5.99 \times 10^{21}\ \text{t}$$

[The ton t used here and occasionally elsewhere in the book is the metric ton, equal to 10^3 kg.] For the density you write $\rho = m_e/V$, or

$$\rho = \frac{3m_e}{4\pi R_e^3}$$
$$= \frac{3 \times 5.99 \times 10^{24}\ \text{kg}}{4\pi \times (6.37 \times 10^6\ \text{m})^3} = 5.53 \times 10^3\ \text{kg/m}^3$$

As expected, this is rather larger than the value $4 \times 10^3\ \text{kg/m}^3$ which comes from a survey of surface rocks, because of the general increase in density with depth. In fact, the core of the earth is probably made of metal, largely iron and nickel, whose density is well above $8 \times 10^3\ \text{kg/m}^3$ at the pressures involved.

11-3 THE MECHANICS OF CIRCULAR ORBITS: ANALYTICAL TREATMENT

In Sec. 11-1 you had a glimpse of the way in which Newton's consideration of the earth-moon system led him to combine his law of gravitation and his laws of mechanics, and thus to extend the mechanics of terrestrial objects so that it embraced the earth-moon system without essential change. The next step was to go beyond the moon and consider the entire solar system.

As the main basis for his attack on this more general problem, Newton used Kepler's analysis of the enormous volume of planetary observations made by Tycho Brahe. Although astronomical data had become very much more precise in the decades since Tycho's day, Kepler's empirical general rules of planetary motion remained valid.

A man of mystical bent who held a profound faith that the universe was ultimately based on number, Kepler had spent many years searching for every mathematical relationship he could find, in his efforts to demonstrate the regularity of the solar system. At least in part because of the religious significance he attached to the sun, he based his approach on the Copernican system with the sun at the center and the planets revolving around it. As you can imagine, he was able to find empirically many different numerical and mathematical relationships in a system as complex as the solar system. Most of these failed to hold up when more accurate data became available, or turned out to be trivial consequences of others, or came to be seen as mere accidental numerical coincidences of the sort in which numerologists delight.

But Kepler had used three such relationships in achieving "victory" in his "war on Mars," as he called his intense, decade-long effort to describe the orbit of the planet named after the Greek god of war. The relatively large deviation from circularity of that orbit made it a severe test of any proposed kinematical theory of planetary motion. What is more, Kepler had access to the unprecedentedly accurate observations of Tycho, for whom he had worked briefly.

Beginning with one of Tycho's observed angular positions of Mars, Kepler tried to predict an angular position at a later time. He knew that any fit between his predictions and Tycho's observed angular positions had to be good within 2 minutes of arc if his kinematical theory was to be a valid one. In this test a description based on his rules succeeded where all others had failed.

It was Newton who first called the crucial rules Kepler's three laws, leaving Kepler's many other empirical rules to fade into obscurity. Newton demonstrated that *the highly accurate but purely descriptive rules of Kepler were necessary logical consequences of his own law of gravitation and the laws of motion.* In doing so, Newton put both the science of terrestrial mechanics and the science of celestial mechanics on one and the same footing, and he set the stage for the transformation of astronomy into a branch of physics.

Kepler's laws are as follows:

1. The orbit of every planet in the solar system is an **ellipse** having the sun at one **focus.** See Fig. 11-8 and its caption.

2. As a planet moves in its orbit, its speed, angular speed, and orbit radius all vary. However, the vector from the sun to the planet sweeps out equal areas in equal times, as shown in Fig. 11-8.

3. If γ is half the length of the **major axis** (see Fig. 11-8) of the orbital ellipse—the **semimajor axis**—and T is the period of revolution of the planet around the sun as seen by an observer fixed in space (the sidereal period), then the ratio γ^3/T^2 is the same for all planets.

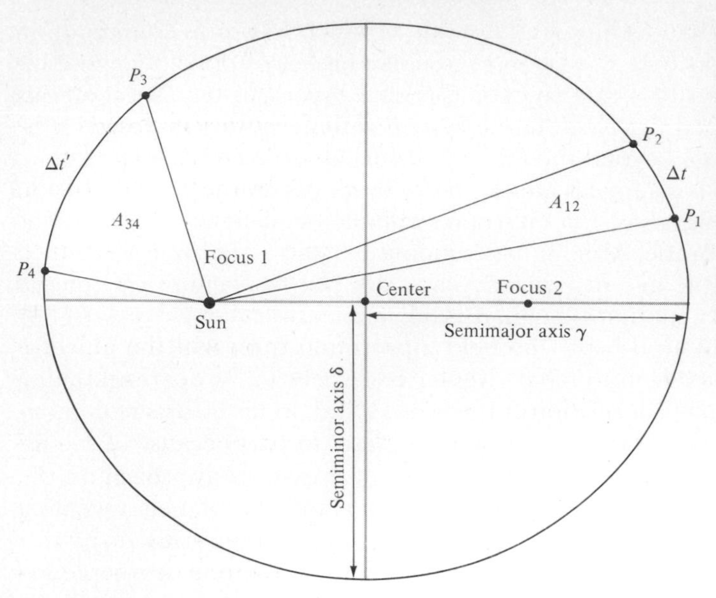

Fig. 11-8 Kepler's first and second laws. A planet P orbits the sun. The orbit is an ellipse. All the points on an ellipse have the property that their distances from two specified points, called foci, add to the same specified value. The sun lies at one of the two foci of the orbit. The line passing through the two foci is the longest that can be drawn through the ellipse; it is called the **major axis.** Its perpendicular bisector is the **minor axis.** The vector from the sun to the planet is the position vector of the planet. The planet passes from P_1 to P_2 in a time Δt. As it does so, the position vector sweeps out the area A_{12}. In a time $\Delta t'$ the planet passes from P_3 to P_4, the position vector sweeping out the area A_{34}. If Δt and $\Delta t'$ are equal times, then $A_{12} = A_{34}$.

With respect to the first law, Newton showed that the path of a particle under the influence of an inverse-square force emanating from a fixed point is always a **conic section,** that is, a circle, an ellipse, a parabola, or a hyperbola. Figure 11-9 makes clear the reason for calling these curves conic sections. The planets all have *orbits* which are ellipses differing only slightly from circles. But there are other objects in the solar system, and in other inverse-square-law systems, which demonstrate all the paths Newton showed to be possible. Kepler's first law is strictly true only if the force center—in the case of the solar system, the sun—is fixed in space. This is a good approximation for the solar system, because the sun is so much more massive than the rest of the system. For the time being, we will confine our attention to such systems.

The analytical derivation of Kepler's first law from Newton's laws is too complex mathematically to discuss in this book. In Sec. 11-4, however, we will accomplish the task numerically.

You can make an experimental check on the validity of Kepler's second law by direct measurement on Fig. 9-15, which is a strobe photo of a puck constrained by a string to move in a noncircular orbit on an air table.

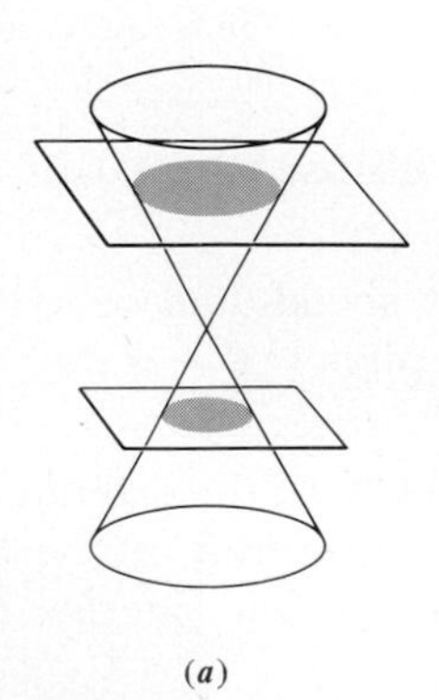

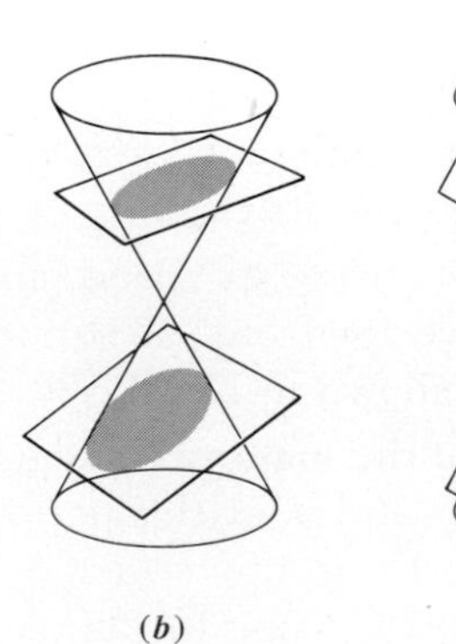

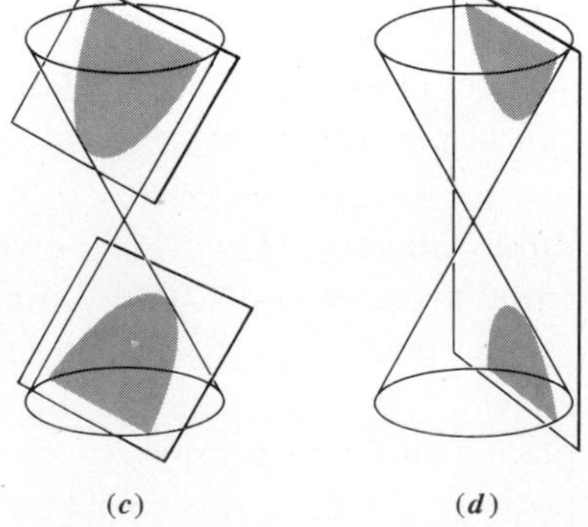

Fig. 11-9 Conic sections. The curved peripheries of the shaded regions are the intersections of the planes with the cones. These curves are: circles in (*a*), ellipses in (*b*), parabolas in (*c*), and hyperbolas in (*d*).

Like the gravitational force, the force exerted on the puck by the string is a central force, always directed toward the same point. The areas of the triangles formed by any two successive images of the puck and the hole at the center of the air table are equal, within the limits of accuracy imposed by the experimental conditions and by the error introduced in using the areas of triangles to approximate the areas actually swept out by the position vector.

We will now show that Kepler's second law can be derived from Newton's laws of motion for *any central force*. A central force is defined to be any force whose direction is always toward (or away from) a given fixed point in space, called the *force center*, and whose magnitude depends only on the distance from the center of force of the body on which it acts. For convenience, we will assume that the body is a planet, but that assumption will not affect the generality of the derivation.

In Eq. (9-24), we defined the angular momentum $\mathbf{l}$ of a body, about a specified origin, to be

$$\mathbf{l} \equiv \mathbf{r} \times \mathbf{p} \tag{11-13}$$

where $\mathbf{r}$ is the position vector of the body and $\mathbf{p}$ is its momentum. Since $\mathbf{p} = m\mathbf{v}$, the angular momentum of a planet (or similar body) can be written

$$\mathbf{l} = m\mathbf{r} \times \mathbf{v}$$

Here m is the mass of the planet, $\mathbf{v}$ is its velocity, and $\mathbf{r}$ is the position vector from the sun (fixed at the origin) to the planet. Multiplying both sides of this equation by dt, we have

$$\mathbf{l}\, dt = m\mathbf{r} \times \mathbf{v}\, dt$$

Since $d\mathbf{s} = \mathbf{v}\, dt$ is the displacement of the planet during the infinitesimal time interval dt, this can be written

$$\mathbf{l}\, dt = m\mathbf{r} \times d\mathbf{s} \tag{11-14}$$

In Fig. 9-8 you saw that the magnitude of the cross product of two vectors is geometrically equivalent to the area of the parallelogram whose adjacent sides are formed by the two vectors. In Fig. 11-10, the triangle of

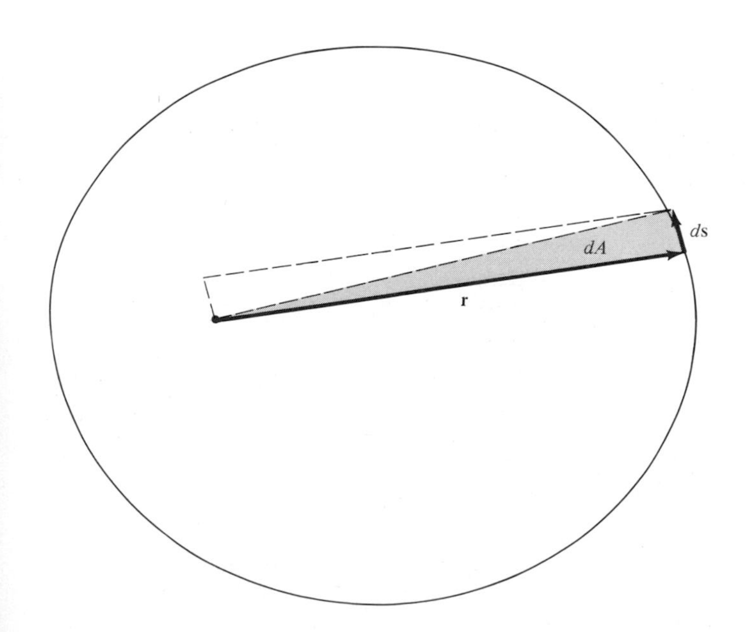

Fig. 11-10 Diagram for deriving Kepler's second law from the principle of conservation of angular momentum. When the planet experiences an infinitesimal displacement $d\mathbf{s}$ along its central-force orbit, the position vector $\mathbf{r}$ of the planet relative to the center of force sweeps out a triangular area dA. This is one-half the area $|\mathbf{r} \times d\mathbf{s}|$ of the parallelogram shown.

area dA, which is swept out by $\mathbf{r}$ in the time dt, is just one-half of such a parallelogram; the other half is shown by the dashed lines. Thus we have

$$dA = \tfrac{1}{2}|\mathbf{r} \times d\mathbf{s}|$$

Now the magnitude part of Eq. (11-14) is

$$l\,dt = m|\mathbf{r} \times d\mathbf{s}|$$

In terms of dA, this is

$$l\,dt = 2m\,dA$$

The equation may be integrated over an arbitrary time interval t_i to t_f. This gives

$$\int_{t_i}^{t_f} l\,dt = 2m \int_{\text{swept area}} dA = 2mA$$

where A is defined to be the entire area swept out by the position vector during the time interval. To evaluate the integral on the left side of the equation, we need to know how the angular momentum $\mathbf{l}$ of the planet, about an origin located at the sun, varies with time. The change in angular momentum is related to the force applied to the planet by means of Eq. (9-28),

$$\frac{d\mathbf{l}}{dt} = \mathbf{r} \times \mathbf{F}$$

Suppose (as is very closely true) that the only significant force exerted on the planet is the gravitational attraction $\mathbf{F}$ of the sun, which is a central force $\mathbf{F} = -F\hat{\mathbf{r}}$. We then have

$$\frac{d\mathbf{l}}{dt} = \mathbf{r} \times (-F\hat{\mathbf{r}}) = -F\mathbf{r} \times \hat{\mathbf{r}} = 0$$

or

$$\mathbf{l} = \text{constant}$$

so that the angular momentum of the planet is conserved.

The integration of $l\,dt$ can thus be carried out immediately, to yield

$$l(t_f - t_i) = 2mA$$

or

$$A = \frac{l}{2m}(t_f - t_i) \qquad (11\text{-}15)$$

Since $l/2m$ is constant, the area A swept out by $\mathbf{r}$ in equal time intervals $t_f - t_i$ will be equal, as Kepler had found through his empirical study of Tycho's astronomical data. But in this derivation of Kepler's second law from the rotational form of Newton's second law of motion, we used only the fact that $\mathbf{F}$ lies along $\mathbf{r}$, so that $\mathbf{r} \times \mathbf{F} = 0$. We never invoked the inverse-square dependence of the gravitational force on distance. *Kepler's second law therefore holds true for any central force, attractive or repulsive.*

We now derive Kepler's third law from Newton's laws for the special case of circular orbits. Consider a planet of mass m in a circular orbit about the sun, whose mass is M. In this case, the radius R of the orbit is equal to the semimajor axis γ, and can be substituted for it. Combining Newton's

Table 11-1

Planet	Semimajor axis γ (in 10^9 m)	Period T (in yr)	γ^3/T^2 (in 10^{18} m^3/s^2)
Mercury	57.90	0.2408	3.361
Venus	108.2	0.6152	3.357
Earth	149.6	1	3.362
Mars	228.0	1.881	3.363
Jupiter	778.4	11.86	3.366
Saturn	1427	29.46	3.362
Uranus	2869	84.01	3.361
Neptune	4497	164.8	3.363
Pluto	5900	248.4	3.343

second law of motion and his law of gravitation, we have for the magnitude a of the acceleration of the planet

$$a = \frac{F}{m} = G\frac{M}{R^2}$$

But according to Eq. (11-2), $a = 4\pi^2 R/T^2$, where T is the sidereal period of the planet. So we have

$$\frac{4\pi^2 R}{T^2} = G\frac{M}{R^2} \tag{11-16}$$

Rearranging terms yields the result

$$\frac{R^3}{T^2} = \frac{GM}{4\pi^2} \tag{11-17}$$

This satisfies Kepler's third law, since the terms on the right side are all constants for the solar system. Table 11-1 gives the length of γ of the semimajor axis, the period T, and the value of γ^3/T^2 for all the known planets. Except for Pluto, Mercury, and Mars, the planets all have orbits which are quite close to circular. Thus the derivation for circular orbits applies quite closely to them. However, Kepler's third law is true for all elliptical orbits, provided that γ^3/T^2 is used in lieu of R^3/T^2. We show this by numerical calculation in Sec. 11-5.

Since Kepler's laws depend only on the existence of an inverse-square force law, they are applicable to many more cases than that of the sun and the planets alone. Within the solar system, the validity of Kepler's laws can be demonstrated as well for any planet which has more than one satellite (artificial satellites included). As you will see in Chap. 20, the laws work equally well for pairs of charged particles, such as an electron revolving about a proton (so far as the newtonian visualization of the system in planetary terms is applicable).

In Example 11-5, Kepler's third law is used to determine the mass of the sun.

EXAMPLE 11-5

Find the mass of the sun from the data in Table 11-1.

■ You can rewrite Kepler's third law, Eq. (11-17), in the form

$$M = \frac{4\pi^2\gamma^3}{GT^2}$$

Using any of the values of γ^3/T^2 given in Table 11-1, you have

$$M = \frac{4\pi^2}{6.67 \times 10^{-11}\ \text{N}\cdot\text{m}^2/\text{kg}^2} \times 3.36 \times 10^{18}\ \text{m}^3/\text{s}^2$$

$$= 1.99 \times 10^{30}\ \text{kg} = 1.99 \times 10^{27}\ \text{t}$$

Thus the mass of the sun is more than 300,000 times greater than that of the earth.

The method of Example 11-5 can used to find the mass of any body having a satellite whose period and orbit radius can be determined. For this reason, the masses of such distant planets as Jupiter, which has several moons, have been known with considerable accuracy for centuries, while the mass of the nearby and highly visible but moonless planet Venus was known only poorly until very recently. In the past few years, a number of artificial satellites have been flown past Venus. Even in an open, fly-by orbit, a satellite yields information as to the planet's mass, by means of an extension of the above theory to noncircular orbits. However, the repetitive nature of a closed orbit, such as the orbits of the vehicles used to launch the Venus landers, allows it to yield much more precise information.

Jupiter is, by a considerable margin, the most massive planet in the solar system. Its mass is 1.900×10^{27} kg, or about 318 times that of the earth. More significantly, its mass is 9.55×10^{-4}, or about 0.1 percent, that of the sun. Since all the other eight planets are smaller than Jupiter, the sun contains practically all the mass in the solar system.

11-4 REDUCED MASS

In describing the motion of a body such as a planet about a force center, it is very useful to make the assumption that the force center is itself fixed in an inertial frame of reference. It then becomes possible to apply Newton's laws of motion directly to observations made with respect to the force center, without the need for dealing with fictitious forces. Assuming that the force center is fixed in an inertial frame is tantamount to making the approximation that the body whose motion is being studied has negligible mass compared to the central body. While this approximation can never be exactly true, it is not a bad one for the solar system. For instance, compare the mass m_e of the earth obtained in Example 11-4 with the mass M of the sun obtained in Example 11-5. The ratio of the two masses is

$$\frac{m_e}{M} = \frac{5.99 \times 10^{24}\ \text{kg}}{1.99 \times 10^{30}\ \text{kg}} = 3.01 \times 10^{-6}$$

However, observations made on the solar system are so accurate that the approximation cannot be made if full advantage is to be taken of their accuracy. Moreoever, there are many systems—the hydrogen atom at one end of the size scale and double star systems at the other—which resemble the solar system in many ways but where the approximation is not at all justified.

We now develop a method which makes the approximation unnecessary. Consider two bodies of arbitrary mass, each under the gravitational influence of the other but otherwise isolated. They are revolving about each other. Each follows an orbit which is most easily analyzed from the point of view of an observer located at the center of mass of the system, since that observer is in an inertial frame. This is because there is no external force applied to the isolated system; see Eq. (9-50a). The problem of analyzing this system is called the **two-body problem.** It would seem at first glance to be at least twice as complicated as the one-body problem, in which one of the bodies is stationary in an inertial frame. However, we will now derive a very direct method of reducing the two-body problem to a one-body problem.

Each of the two bodies 1 and 2, having masses m_1 and m_2, respectively, exerts some kind of force—it need not be gravitational or even inverse-square in nature—on the other. We must make the restriction, however, that both forces are central forces. If the force exerted on body 1 is **F,** we can write

$$m_1\mathbf{a}_1 = \mathbf{F}$$

Here $\mathbf{a}_1$ is the acceleration experienced by body 1, as seen by an observer in an inertial frame, say at the center of mass. According to Newton's third law, there must be an equal and opposite force $-\mathbf{F}$ exerted on body 2, so that

$$m_2\mathbf{a}_2 = -\mathbf{F}$$

where $\mathbf{a}_2$ is the acceleration of body 2, as seen by the observer in the inertial frame. If we now multiply the first of these equations by m_2 and the second one by m_1, we obtain the pair of equations

$$m_1m_2\mathbf{a}_1 = \mathbf{F}m_2$$

and

$$m_1m_2\mathbf{a}_2 = -\mathbf{F}m_1$$

Subtracting the second of these equations from the first yields

$$m_1m_2(\mathbf{a}_1 - \mathbf{a}_2) = \mathbf{F}(m_1 + m_2)$$

or

$$\mathbf{F} = \frac{m_1m_2}{m_1 + m_2}(\mathbf{a}_1 - \mathbf{a}_2) \tag{11-18}$$

The quantity $\mathbf{a}_1 - \mathbf{a}_2$ is the acceleration of body 1 as seen by an observer on body 2. We call this quantity the *relative acceleration* **a.** That is

$$\mathbf{a} = \mathbf{a}_1 - \mathbf{a}_2 \tag{11-19}$$

[This definition may be compared with Eq. (5-23c), $\mathbf{a}' = \mathbf{a} - \mathbf{A}$, derived in connection with the fictitious-force approach of Sec. 5-4.]

The fraction $m_1m_2/(m_1 + m_2)$ in Eq. (11-18) has the dimensions of a mass and is called the **reduced mass.** It is represented by the symbol μ which is defined by the expression

$$\mu \equiv \frac{m_1m_2}{m_1 + m_2} \tag{11-20a}$$

Written in terms of **a** and μ, Eq. (11-18) becomes

$$\mathbf{F} = \mu\mathbf{a} \tag{11-20b}$$

This has the *form* of Newton's second law of motion.

Equation (11-20b) is not really Newton's second law, strictly speaking. The force **F** is indeed the force exerted on body 1 by body 2. But the acceleration **a**, being that of body 1 as observed by an observer on body 2, is *not* an acceleration relative to an inertial frame. Thus the observer cannot directly apply Newton's second law without adding an appropriate fictitious force (see Sec. 5-4).

However, the use of the reduced mass eliminates the need for the fictitious force. In Eq. (11-20*b*), the reduced mass is substituted for the actual mass of body 1, while body 2 is assigned an *infinite* mass and thus a fixed position in an inertial frame in which an observer has no need for fictitious forces. Equations (11-20*a*) and (11-20*b*) show that this substitution is valid because the motion of body 1 relative to body 2 is exactly the same as the motion of the hypothetical body 1, with reduced mass μ, relative to the hypothetical body 2, which has infinite mass and is therefore fixed. That is, for the actual situation in which two bodies orbit about their common center of mass, we have substituted a single hypothetical body of mass μ in orbit about a fixed point in an inertial frame. We have thus reduced the two-body problem to the one-body problem, which we have already solved for circular orbits and will discuss further for other cases below.

The dependence of μ on the ratio m_1/m_2 is depicted in Fig. 11-11. Note that μ is in fact "reduced"; μ/m_2 is always less than m_1/m_2. That is, assigning an infinite mass to body 2 requires a reduction in the mass assigned to body 1.

Once the relative acceleration **a** has been found for the hypothetical particle of mass μ, it is possible to transform back to the original two-body system. For body 1, for example, we have $\mathbf{F} = m_1\mathbf{a}_1$. Substituting this value of **F** into Eq. (11-20*b*) gives

$$m_1\mathbf{a}_1 = \mu\mathbf{a} = \frac{m_1 m_2}{m_1 + m_2}\mathbf{a}$$

or

$$\mathbf{a}_1 = \frac{m_2}{m_1 + m_2}\mathbf{a} \tag{11-21}$$

The acceleration $\mathbf{a}_2$ can be found in terms of **a** in the same way. It is

$$\mathbf{a}_2 = -\frac{m_1}{m_1 + m_2}\mathbf{a} \tag{11-22}$$

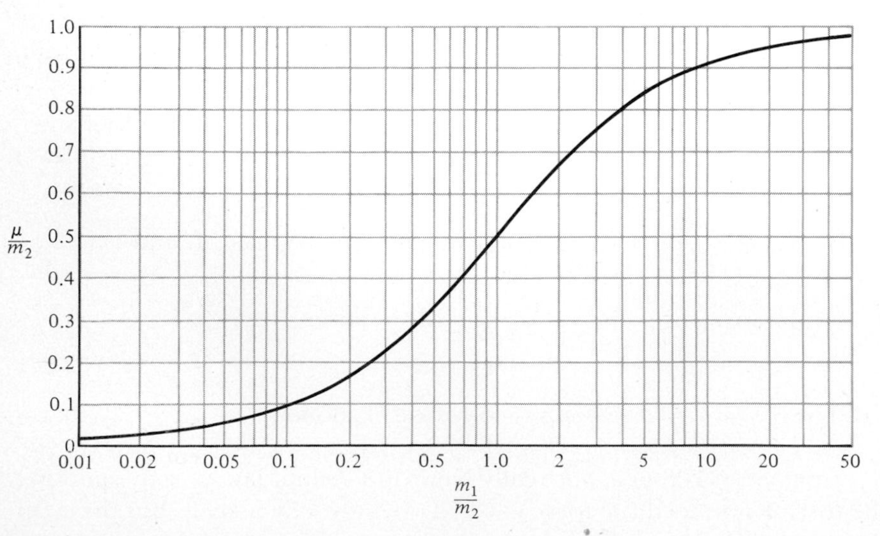

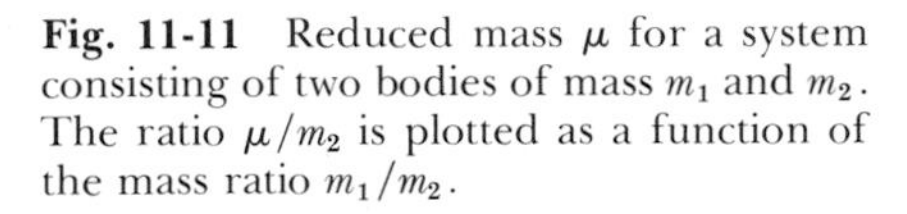
Fig. 11-11 Reduced mass μ for a system consisting of two bodies of mass m_1 and m_2. The ratio μ/m_2 is plotted as a function of the mass ratio m_1/m_2.

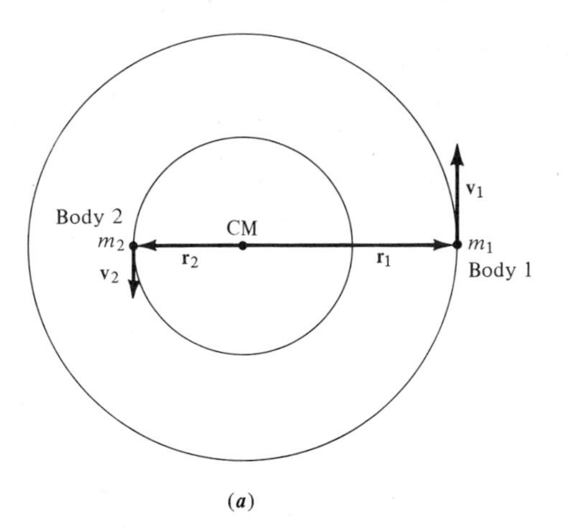

Fig. 11-12 The relation between (*a*) a two-body system and (*b*) the reduced-mass system equivalent to it. While the trajectories are shown as circles, they can be of any shape whatever. The only requirement is that the force exerted on each body be directed toward (or away from) the other. Note that the total distance between bodies 1 and 2 is the sum of their distances from the center of mass. It is equal to the distance between the hypothetical body of reduced mass μ and the fixed center about which it moves.

Equations (11-21) and (11-22) can be integrated to give the following transformation equations for the velocity and position vectors:

$$\mathbf{v}_1 = \frac{m_2}{m_1 + m_2}\mathbf{v} \quad \text{and} \quad \mathbf{v}_2 = -\frac{m_1}{m_1 + m_2}\mathbf{v} \tag{11-23}$$

$$\mathbf{r}_1 = \frac{m_2}{m_1 + m_2}\mathbf{r} \quad \text{and} \quad \mathbf{r}_2 = -\frac{m_1}{m_1 + m_2}\mathbf{r} \tag{11-24}$$

To verify the correctness of those integrations, differentiate Eqs. (11-24) once to obtain Eqs. (11-23), and again to obtain Eqs. (11-21) and (11-22). In the actual two-body system, the velocity of body 1 relative to an observer on body 2 is given by the vector difference

$$\mathbf{v} = \mathbf{v}_1 - \mathbf{v}_2$$

which is precisely the difference found by subtracting the second of Eqs. (11-23) from the first. Thus the velocity of the hypothetical body of mass μ about the force center, located in the fixed hypothetical body of infinite mass, is the same as the velocity of the actual body 1 relative to body 2.

Likewise, the vector difference

$$\mathbf{r} = \mathbf{r}_1 - \mathbf{r}_2$$

is the vector from the force center to the body of mass μ, and is also the vector from body 2 to body 1 in the actual system.

The relation between the actual and reduced-mass systems is illustrated in Fig. 11-12.

EXAMPLE 11-6

Find the location of the center of rotation of the earth-moon system, relative to the center of the earth. Then evaluate μ for the earth-moon system. The mass of the earth is 81 times that of the moon.

■ Using the second of Eqs. (11-24), you have

$$\mathbf{r}_2 = -\frac{m_m}{m_m + m_e}\mathbf{r} = -\frac{m_m}{82m_m}\mathbf{r} = -\frac{1}{82}\mathbf{r}$$

so that the distance r_2 from the center of the earth to the center of rotation is $\frac{1}{82}$ of the distance to the moon. Its numerical value is

$$r_2 = \frac{3.84 \times 10^5 \text{ km}}{82} = 4700 \text{ km}$$

Since the radius of the earth is 6370 km, the center of rotation is inside the earth. The value of μ is

$$\mu = \frac{m_m \times 81m_m}{m_m + 81m_m} = \frac{81}{82}m_m = \frac{1}{82}m_e$$

So μ is reduced from the value of the moon's mass $m_m = \frac{1}{81}m_e$.

11-5 THE MECHANICS OF ORBITS: NUMERICAL TREATMENT

Most of what has been done so far has been restricted to circular orbits, for which the mathematical analysis is relatively simple. We now tackle the problem of motion in orbits of any possible shape. The first task is to set up the general equations of motion for a central force of arbitrary form. Then we will solve these equations numerically for a variety of important special cases.

Figure 11-13a depicts a body of mass m moving under the influence of a central force arising from the presence of the central body of much larger mass M, which is taken to be fixed at the origin. (Even if M is not very much larger than m, the reduced-mass approach of Sec. 11-4 can be used to transform the problem into the one discussed here.) The central force $\mathbf{F}$ is always directed toward the origin, although its magnitude will vary as the magnitude of the position vector $\mathbf{R}$ varies.

The path of the moving body lies in a plane. To see this, consider an instant when the velocity of the body is $\mathbf{v}$, so that its momentum is $\mathbf{p} = m\mathbf{v}$. The vector $\mathbf{p}$ and the force center determine a plane. (It is the plane depicted in Fig. 11-13a.) The position vector $\mathbf{R}$ lies in this plane. And since a central force must always be parallel or antiparallel to the position vector, $\mathbf{F}$ also lies in the plane. But $\mathbf{F}$ is related to the rate of change of momentum by Newton's second law, $\mathbf{F} = d\mathbf{p}/dt$. After an infinitesimal time interval dt, the new momentum $\mathbf{p}'$ is given by the vector sum $\mathbf{p}' = \mathbf{p} + (d\mathbf{p}/dt)\, dt$. Since $\mathbf{p}$ and $d\mathbf{p}/dt$ lie in the same plane, so do the new momentum $\mathbf{p}'$, the new velocity $\mathbf{v}' = \mathbf{p}'/m$, and the new position $\mathbf{R}' = \mathbf{R} + \mathbf{v}\, dt$.

We begin the development of the equations of motion by writing Newton's second law. Equating the acceleration of the body to $\mathbf{F}/m$ gives

$$\frac{d^2\mathbf{R}}{dt^2} = \frac{\mathbf{F}}{m} \tag{11-25}$$

We take the plane of the orbit to be the xy plane. Writing Eq. (11-25) in component form, we have the pair of equations

$$\frac{d^2x}{dt^2} = \frac{F_x}{m} \tag{11-26a}$$

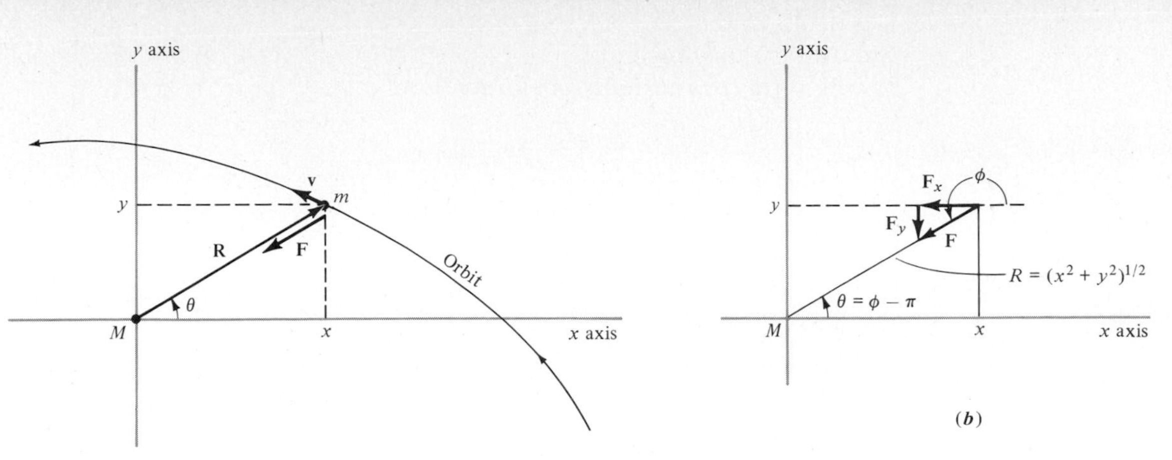

Fig. 11-13 A body of mass m moves in an unspecified orbit under the influence of a central force **F**. In the text, equations of motion are developed in a component form suitable for numerical treatment. Note that $\cos\phi = -\cos\theta = -x/(x^2 + y^2)^{1/2}$ and $\sin\phi = -\sin\theta = -y/(x^2 + y^2)^{1/2}$.

and

$$\frac{d^2y}{dt^2} = \frac{F_y}{m} \tag{11-26b}$$

The force components are shown in Fig. 11-13b, where the angle ϕ specifies the direction of **F** with respect to the positive x direction. The quantities $\cos\phi$ and $\sin\phi$ are evaluated in the figure and its caption. Using these values, we can write

$$F_x = F\cos\phi = -F\frac{x}{(x^2 + y^2)^{1/2}}$$

and

$$F_y = F\sin\phi = -F\frac{y}{(x^2 + y^2)^{1/2}}$$

Can you explain why the negative sign appears in these equations, regardless of the location of the moving body with respect to the origin?

Inserting the above expressions for the force components into Eqs. (11-26a) and (11-26b), we obtain

$$\frac{d^2x}{dt^2} = -\frac{F}{m}\frac{x}{(x^2 + y^2)^{1/2}} \tag{11-27a}$$

and

$$\frac{d^2y}{dt^2} = -\frac{F}{m}\frac{y}{(x^2 + y^2)^{1/2}} \tag{11-27b}$$

To be specific, let us assume that **F** is a gravitational force. According to Eq. (11-6a), the magnitude of the gravitational force is given by $F = GMm/R^2$. Since $R^2 = x^2 + y^2$, that equation can be written

$$F = GMm\frac{1}{x^2 + y^2}$$

Substituting this value of F into Eqs. (11-27a) and (11-27b) leads to the following form for the equations of motion:

$$\frac{d^2x}{dt^2} = -GM\frac{x}{(x^2 + y^2)^{3/2}} \tag{11-28a}$$

and

$$\frac{d^2y}{dt^2} = -GM\frac{y}{(x^2 + y^2)^{3/2}} \tag{11-28b}$$

This is the pair of differential equations which must be solved in order to describe the motion of the body of mass m. As noted in Sec. 11-3, analytical solutions do exist; however, a complete discussion involves mathematical techniques beyond the level of this book. But we can do just as well, as far as a physical picture of the motion is concerned, by means of *numerical solutions* which involve only a slight extension of the procedure developed in Sec. 6-4. What is equally significant, with almost no extra effort the procedure can be extended to deal with generalizations of Eqs. (11-28a) and (11-28b), for which no analytical solution exists. In particular, we will use numerical solutions to study force laws that deviate from the inverse-square form. We will also treat a variety of perturbations, such as that produced on the motion of the earth by the small but nonnegligible gravitational attraction of Jupiter.

To allow for subsequent generalization, it is most convenient to make a change in the notation of Eqs. (11-28a) and (11-28b). We define

$$\alpha \equiv GM$$

and give the name β to the exponent of the term $x^2 + y^2$. (In the case of an inverse-square force such as gravitation, we have $\beta = -\frac{3}{2}$.) Also, we use the notation introduced in Eq. (6-14), in connection with the general method for obtaining solutions for second-order differential equations. We write

$$\frac{d^2x}{dt^2} = Q_x \tag{11-29a}$$

and

$$\frac{d^2y}{dt^2} = Q_y \tag{11-29b}$$

where

$$Q_x \equiv -\alpha x(x^2 + y^2)^{\beta} \tag{11-30a}$$

and

$$Q_y \equiv -\alpha y(x^2 + y^2)^{\beta} \tag{11-30b}$$

Equations (11-27a) and (11-27b) or (11-28a) and (11-28b) are called **coupled differential equations.** They are "coupled" from a mathematical point of view by the fact that each differential equation of the pair contains *both* dependent variables. Physically, the acceleration d^2x/dt^2 in the x direction produces a change in $x^2 + y^2$ and thus a change in F_y. This changes the acceleration d^2y/dt^2 in the y direction, which influences the change in $x^2 + y^2$ and therefore F_x, and so forth. Thus the x and y motions are coupled in a physical sense.

Equations (11-29*a*), (11-29*b*), (11-30*a*), and (11-30*b*) are used in the same way that you used Eq. (6-14). The procedure is completely parallel to that described in Sec. 6-4. To be specific, this is what you do. Always employing Eqs. (11-30*a*) and (11-30*b*) to express Q_x and Q_y in terms of the quantities on which they depend, you select a *small* time increment Δt and then carry out the following double set of calculations:

Determine the values of Q_x and Q_y from the values at $t_0 = 0$ of the quantities on which they depend, and use them to calculate

$$\left(\frac{dx}{dt}\right)_{1/2} \simeq \left(\frac{dx}{dt}\right)_0 + Q_x \frac{\Delta t}{2} \quad \text{and} \quad \left(\frac{dy}{dt}\right)_{1/2} \simeq \left(\frac{dy}{dt}\right)_0 + Q_y \frac{\Delta t}{2} \tag{11-31a}$$

from the given values of $(dx/dt)_0$ and $(dy/dt)_0$. Use the results to calculate

$$x_1 \simeq x_0 + \left(\frac{dx}{dt}\right)_{1/2} \Delta t \quad \text{and} \quad y_1 \simeq y_0 + \left(\frac{dy}{dt}\right)_{1/2} \Delta t \tag{11-31b}$$

from the given values of x_0 and y_0. Then set $t_1 = \Delta t$.

Next determine the values of Q_x and Q_y from the new values of the quantities on which they depend. Use them, and the values of $(dx/dt)_{1/2}$ and $(dy/dt)_{1/2}$ just obtained, to calculate

$$\left(\frac{dx}{dt}\right)_{3/2} \simeq \left(\frac{dx}{dt}\right)_{1/2} + Q_x \,\Delta t \quad \text{and} \quad \left(\frac{dy}{dt}\right)_{3/2} \simeq \left(\frac{dy}{dt}\right)_{1/2} + Q_y \,\Delta t \tag{11-31c}$$

Use the results, and the values of x_1 and y_1 just obtained, to calculate

$$x_2 \simeq x_1 + \left(\frac{dx}{dt}\right)_{3/2} \Delta t \quad \text{and} \quad y_2 \simeq y_1 + \left(\frac{dy}{dt}\right)_{3/2} \Delta t \tag{11-31d}$$

Then set $t_2 = 2\Delta t$.

Next determine the values of Q_x and Q_y from the new values of the quantities on which they depend. Use them, and the values of $(dx/dt)_{3/2}$ and $(dy/dt)_{3/2}$ just obtained, to calculate

$$\left(\frac{dx}{dt}\right)_{5/2} \simeq \left(\frac{dx}{dt}\right)_{3/2} + Q_x \,\Delta t \quad \text{and} \quad \left(\frac{dy}{dt}\right)_{5/2} \simeq \left(\frac{dy}{dt}\right)_{3/2} + Q_y \,\Delta t \tag{11-31e}$$

Use the results, and the values of x_2 and y_2 just obtained, to calculate

$$x_3 \simeq x_2 + \left(\frac{dx}{dt}\right)_{5/2} \Delta t \quad \text{and} \quad y_3 \simeq y_2 + \left(\frac{dy}{dt}\right)_{5/2} \Delta t \tag{11-31f}$$

Then set $t_3 = 3\Delta t$.

Continue these calculations until t reaches whatever value is required.

The program required to carry out the numerical calculations described by Eqs. (11-31), called the central-force program, is listed in the Numerical Calculation Supplement. The remainder of this section is devoted to examples which carry out the calculations in a number of important cases. We begin with Example 11-7, which studies the motion of a planet very much like the earth whose orbit is, however, perfectly circular.

The motion of this earthlike planet is governed by Eqs. (11-28*a*) and (11-28*b*). Hence the value of Q_x given by Eq. (11-30*a*) and that of Q_y given by Eq. (11-30*b*) must be determined by setting $\alpha = GM$, where M is the mass of the sun, and by setting $\beta = -\frac{3}{2}$.

EXAMPLE 11-7

Run the central-force program with the following set of initial conditions and parameters. (The units used are discussed immediately below.)

$x_0 = 1$ (in AU); $(dx/dt)_0 = 0$; $y_0 = 0$; $(dy/dt)_0 = 2\pi$ (in AU/yr); $t_0 = 0$; $\Delta t = \frac{1}{52}$ (in yr); $\alpha = 39.5$ [in $(\text{AU})^3/(\text{yr})^2$]; $\beta = -1.5$.

Since $y_0 = 0$, the initial conditions represent the earth on the x axis, and you must set x_0 equal to the radius of the earth's orbit. The results are most transparent if you do as astronomers usually do, measuring the time in years and the distance from the sun in **astronomical units** (AU). The astronomical unit is defined to be the length of the semimajor axis of the earth's orbit. That is, you set $x_0 = 1\text{AU}$, where

$$1 \text{ AU} \equiv 149.6 \times 10^9 \text{ m}$$

Since the orbit is to be circular, the earth must cross the x axis at right angles. Therefore the initial velocity must have no x component, and so $(dx/dt)_0 = 0$. This is the reason for the above choice of the initial condition for $(dx/dt)_0$. Furthermore, since the speed will be constant in a circular orbit, you must have $(dy/dt)_0$ equal to the speed of an object which completes an orbit of radius of 1 AU, going a distance 2π AU, in 1 yr. This leads to the initial condition $(dy/dt)_0 = 2\pi$ AU/yr.

In such a numerical calculation, the time increment Δt must be chosen small enough that the orbit segments are reasonably well approximated by straight lines. (Can you explain why?) Equally important, the gravitational force over any one segment must be reasonably well approximated by a constant value. The choice $\Delta t = \frac{1}{52}$ yr makes each segment represent an elapsed time of a trifle more than 1 week (1 yr = 365.26 days = 52.180 weeks).

The value of $\alpha = GM$ is that which is consistent with a set of units in which length is measured in astronomical units and time in years. In converting from SI units, you have

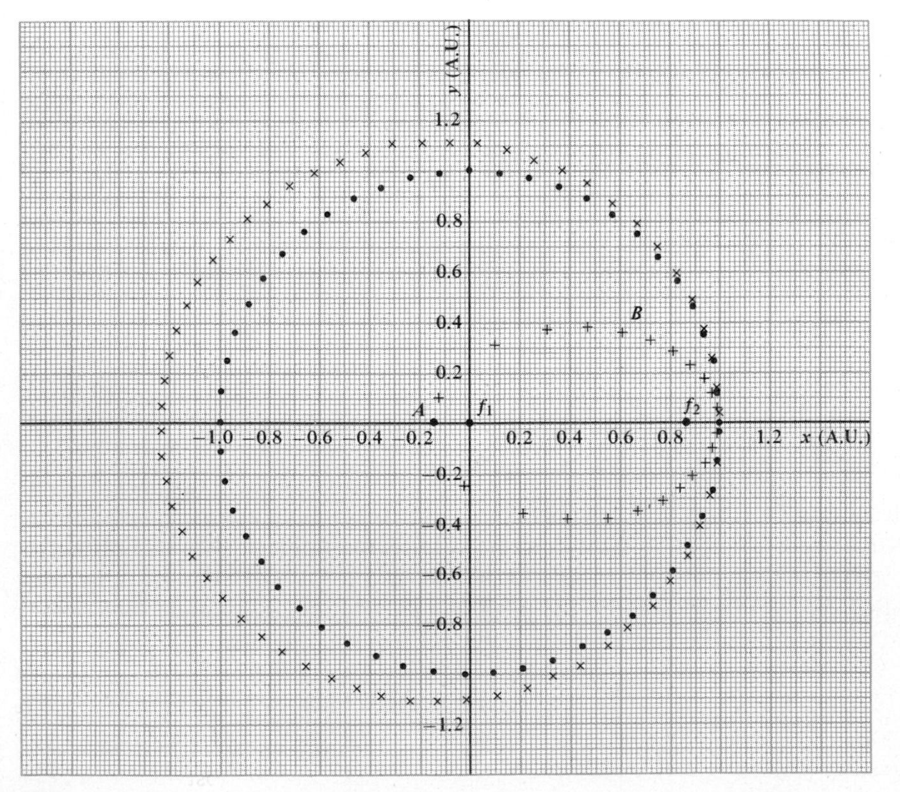

Fig. 11-14 Results of the numerical orbit calculations carried out in Examples 11-7, 11-8, and 11-9. The dots represent successive positions separated by equal time intervals, of an "idealized earth," which moves in a circular orbit at a distance of 1 AU from the sun. The x's represent successive positions of a planet which is initially at the same position ($x_0 = 1$ AU, $y_0 = 0$) and is moving in the same direction, but with a speed 5 percent greater than that which leads to a circular orbit. The resulting orbit is an ellipse with its aphelion at $x = -1.23$ AU, $y = 0$. The variation of the speed of the planet in this orbit is evidenced by the variation in the distances between adjacent points. The crosses represent successive positions of a planet which is initially at the same position and moving in the same direction, but with a speed 50 percent less than that which leads to a circular orbit. The resulting orbit is an ellipse with its perihelion at $x = -0.15$ AU, $y = 0$. In this case the noncircular character of the orbit is quite apparent, and the perihelion and aphelion distances markedly different from each other. The variation in orbital speed is likewise very marked. As discussed in the text, the sum of the focal distances f_1A and Af_2 is equal to the sum of the focal distances f_1B and Bf_2. Thus the highly noncircular orbit on which the two points A and B lie is, in fact, an ellipse, and Kepler's first law is satisfied.

$$G = 6.67 \times 10^{-11}\ \text{m}^3/(\text{s}^2\cdot\text{kg}) \left(\frac{1\ \text{AU}}{149.6 \times 10^9\ \text{m}}\right)^3 \left(\frac{8.640 \times 10^4\ \text{s}}{1\ \text{day}}\right)^2 \left(\frac{365.3\ \text{days}}{1\ \text{yr}}\right)^2$$

$$= 1.98 \times 10^{-29}\ (\text{AU})^3/[(\text{yr})^2\cdot\text{kg}]$$

Thus you have

$$\alpha = GM = 1.98 \times 10^{-29}\ (\text{AU})^3/[(\text{yr})^2\cdot\text{kg}] \times 1.99 \times 10^{30}\ \text{kg}$$

$$= 39.5\ (\text{AU})^3/(\text{yr})^2$$

■ The result of the calculation is the orbit represented by the dots in Fig. 11-14. The sequence of dots represents successive positions of the planet, beginning at $x_0 = 1$ AU, $y_0 = 0$ and proceeding counterclockwise. Since the time interval between adjacent positions is always the same, the graph is a "strobe photo" of the motion of the planet. You can see by inspection that the orbit is indeed circular, with the sun at the center. The uniform space between dots tells you that the speed of the planet is constant, as expected. Also as expected, the 52d dot almost coincides with the zeroth at $x = 1$ AU, $y = 0$, so the period is 1 yr.

On the scale of Fig. 11-14, the deviation of the earth's actual orbit from perfect circularity would be just visible, with a difference of about $1\frac{1}{2}$ grid divisions between the orbit radius at **aphelion,** where the earth is farthest from the sun, and the radius at **perihelion,** where it is closest.

A noncircular orbit is considered in Example 11-8. In it, the initial conditions of Example 11-7 are modified by increasing the initial tangential speed by 5 percent.

EXAMPLE 11-8

Run the central-force program with the following set of initial conditions and parameters.

$x_0 = 1$ (in AU); $(dx/dt)_0 = 0$; $y_0 = 0$; $(dy/dt)_0 = 2.1\pi$ (in AU/yr); $t_0 = 0$; $\Delta t = \frac{1}{52}$ (in yr); $\alpha = 39.5$ [in $(\text{AU})^3/(\text{yr})^2$]; $\beta = -1.5$.

■ The result of the calculation is the orbit represented by x's in Fig. 11-14. Now the gravitational force at $x_0 = 1$ AU is not strong enough to constrain the planet to an orbit of constant radius, because of its larger initial momentum. So it moves along a curve of smaller curvature, to a greater distance from the sun. To be more specific, the force $\mathbf{F} = d\mathbf{p}/dt$ acting on the planet at its initial position changes its momentum by the same amount per unit time as in Example 11-7. However, the initial momentum is greater in magnitude, so that the change per unit time is not great enough to bend the trajectory into a circular orbit. A somewhat different explanation of why the planet starts moving outside the circular orbit is given in Example 11-9.

You can tell by measuring the distance between successive points on the orbit that the planet slows down as it moves to greater distances from the sun. Since the orbit, unlike the one in Example 11-7, is not everywhere perpendicular to the position vector **R** along which the central force **F** acts, it follows that **F** has a component parallel to the path of the planet. This component acts to change the magnitude of the velocity, that is, the speed. As the planet comes closer to the sun in the second half of its orbit, it speeds up again. It reaches the starting point in about 62.4 "weeks," or $62.4/52 = 1.20$ terrestrial years.

Although the orbit of Example 11-8 is an ellipse, you cannot tell this by simple visual inspection of the graph. On the basis of such inspection, you might easily conclude that the orbit is a circle whose center lies at $x \simeq -0.115$ AU, $y = 0$. The shape of this orbit is very close to that of the orbit of Mars (though the size is only about two-thirds that of the orbit of Mars). Mars has the most noncircular orbit of the planets readily observable with the naked eye, and it was largely for that reason that Kepler chose Mars for his painstaking empirical study. You can see why highly precise observations are necessary to determine the shape of a planetary orbit in a unique manner, and why it would be easy to guess at some shape other than an ellipse. Indeed, Kepler initially believed that the orbit was an oval, that is, a curve composed of segments of four circles. But computations based on the oval turned out to be prohibitively difficult for a long series of trial-and-error calculations, and Kepler turned to the ellipse as a convenient approximation. Kepler intended, once he had determined the best-fitting approximate ellipse, to make final calculations to determine the "exact" oval. Imagine his surprise when he found that the ellipse fitted perfectly, within the limits of observational error!

In Example 11-9 we investigate the result of reducing the initial speed of the planet substantially—to one-half that required for a circular orbit at 1 AU.

EXAMPLE 11-9

Run the central-force program with the following set of initial conditions and parameters.

$x_0 = 1$ (in AU); $(dx/dt)_0 = 0$; $y_0 = 0$; $(dy/dt)_0 = \pi$ (in AU/yr); $t_0 = 0$; $\Delta t = \frac{1}{520}$ (in yr); $\alpha = 39.5$ [in $(\mathrm{AU})^3/(\mathrm{yr})^2$]; $\beta = -1.5$.

▪ As you can see from the resulting orbit, plotted with crosses in Fig. 11-14, the planet now "falls in" toward the sun, moving in an orbit inside the circular orbit of radius 1 AU. This is because the planet, initially moving more slowly than in Example 11-7, must travel along a path with a greater curvature than that of the circle through the initial position, in order for its acceleration to be equal to the gravitational force divided by the planet's mass. (How would you modify this explanation to discuss the case treated in Example 11-8? How would you modify the explanation given in Example 11-8 to discuss the case treated here?)

With the initial conditions given, the planet moves so close to the sun, and acquires such a great speed in doing so, that the interval of one "week" used in the previous examples is much too long to satisfy the condition of nearly constant gravitational force through each time interval. It is therefore appropriate to choose an interval one-tenth as long. But only every tenth point is plotted in the figure, in order to make comparison with the other two orbits easier.

In this example, the noncircularity of the orbit is evident on inspection. So is the variation in orbital speed, which is reflected in the dramatic change in the distance between successive points. This speed variation is in qualitative accord with Kepler's second law. Since the angular momentum about the origin at the force center $\mathbf{l} = m\mathbf{R} \times \mathbf{v}$ must remain constant, the magnitude of $\mathbf{v}$ must increase as the magnitude of $\mathbf{R}$ decreases, and vice versa.

The orbit period in this case is only about 22.5 "weeks," or 0.43 yr. There are two reasons for this. Not only is the length of the orbit smaller than that of the circular orbit discussed in Example 11-7, but also the average speed is greater even though the initial speed, at aphelion, is less.

The orbits calculated in Examples 11-7, 11-8, and 11-9 can be used to test the validity of all three of Kepler's laws. Table 11-2 concerns the third law. The first two rows of the table give the semimajor axis γ and the period T for each of the three orbits, measured directly from Fig. 11-14. The values vary quite widely. Nevertheless, the corresponding values of γ^3/T^2, given in the third row of the table, are equal to 1 $(\mathrm{AU})^3/(\mathrm{yr})^2$ within the limits of accuracy of the procedure.

The noncircular orbit of Example 11-9 makes it a useful test of Kepler's first and second laws. Let us verify the first law by showing that the orbit is

Table 11-2

Test of Kepler's Third Law Using the Results of Examples 11-7, 11-8, and 11-9

	Orbit of Example		
	11-7	**11-8**	**11-9**
Semimajor axis γ (in AU)	1.000	1.115	0.575
Orbital period T (in yr)	1.000	62.4/52	22.5/52
γ^3/T^2 [in $(\text{AU})^3/(\text{yr})^2$]	1.00	1.00	1.02

indeed an ellipse. According to the definition of an ellipse, the sum of the distances from the two foci to any point on the ellipse is the same. The sun lies at one focus f_1, whose coordinates are $x = 0$, $y = 0$. Compare with Fig. 11-8. The planet crosses the negative x axis at perihelion (the point labeled A in Fig. 11-14) with $x = -0.14$ AU. By symmetry, the other focus must be 0.14 AU from aphelion at $x = 0.86$ AU, $y = 0$, as indicated by the point marked f_2 in the figure.

Let us take as test points the perihelion point A and the point labeled B in the figure, whose coordinates are $x = 0.61$ AU, $y = 0.36$ AU. The sum of the distances from point A to the two foci is found by reading directly off the x axis, and it is

$$d_1 = 0.14 \text{ AU} + 1.00 \text{ AU} = 1.14 \text{ AU}$$

We can find the distances from point B to the two foci by using the pythagorean theorem twice. The distance to the focus f_1 at the origin is $[(0.61 \text{ AU})^2 + (0.36 \text{ AU})^2]^{1/2}$, where the two numbers are the x and y coordinates of the point B, read directly off Fig. 11-13. The distance to the other focus f_2 is found from both the difference between the x coordinates for f_2 and B and the difference between their y coordinates. Using these numbers in the pythagorean theorem gives $[(0.86 \text{ AU} - 0.61 \text{ AU})^2 + (0.00 - 0.36 \text{ AU})^2]^{1/2}$. Thus the sum of the distances to the foci for point B is

$$\begin{aligned} d_2 &= [(0.61 \text{ AU})^2 + (0.36 \text{ AU})^2]^{1/2} + [(0.25 \text{ AU})^2 + (-0.36 \text{ AU})^2]^{1/2} \\ &= 1.14 \text{ AU} \end{aligned}$$

Since $d_1 = d_2$ within the accuracy of the procedure, Kepler's first law is satisfied.

Figure 11-15 is a copy from Fig. 11-14 of the same orbit. In order to verify Kepler's second law, two sectors have been marked out, each one swept out in a time interval of one "week." To measure the areas swept out, chords uv and wz of the two sectors are drawn, dividing each into a triangle containing most of the area and a smaller region bounded by the orbit and the chord. The area of the triangles can be found by measuring their bases and altitudes in terms of the unit of the graph grid. This can be done with either a ruler or a pair of dividers. The smaller areas are best measured by counting grid squares, with estimates made of the contributions of partial squares. The error in estimation is not too serious, since the total contribution of the area involved is relatively small.

When this is done, the area labeled A in Fig. 11-15 is found to contain 78 grid squares, while the area labeled B contains 77. Thus in spite of their

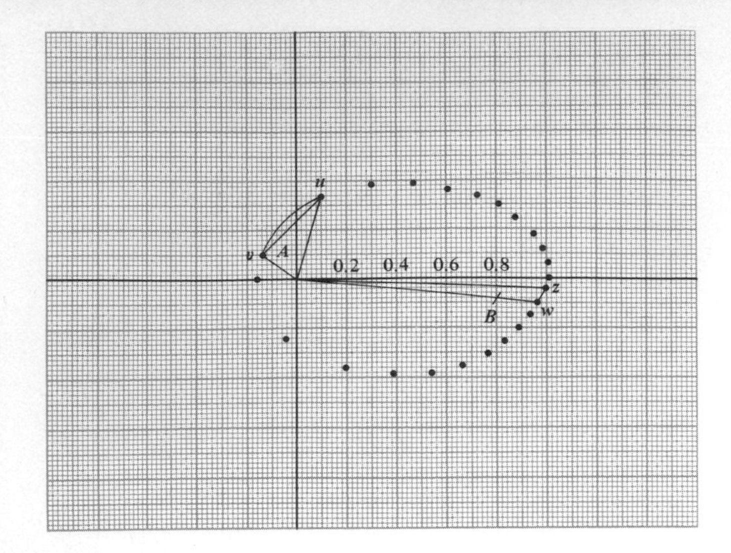

Fig. 11-15 Test of Kepler's second law. The orbit is the same one denoted by crosses in Fig. 11-14. The areas A and B are each swept out in $\frac{1}{52}$ yr. In spite of their quite different shapes and the quite different speeds of the planet in the two regions of the orbit, the two areas contain 78 and 77 grid squares, respectively, and are thus equal within the limits of accuracy of this method of area determination.

very different shapes, the areas are substantially equal, in accord with Kepler's second law.

You can make a more accurate verification of Kepler's second law by using the central-force program. In the course of each cycle of calculation the velocity components $(dx/dt)_{j+1/2} = v_x$ and $(dy/dt)_{j+1/2} = v_y$ are computed [see Eqs. (11-31)]. You can obtain these values by stopping the calculating device during any cycle and recalling the current values from memory. You can then use them, together with corresponding values of x_j and y_j, the most recent position coordinates of the planet, to calculate the magnitude of the vector product $\mathbf{R} \times \mathbf{v}$.

As seen in Sec. 11-3, both the law of equal areas and the constancy of $\mathbf{R} \times \mathbf{v}$ are direct consequences of the law of conservation of angular momentum. Hence if $\mathbf{R} \times \mathbf{v}$ is found to have a constant magnitude, the law of equal areas is satisfied. If you compare values of $|\mathbf{R} \times \mathbf{v}|$ obtained at different points on the orbit, you will find them to be quite close to equal.

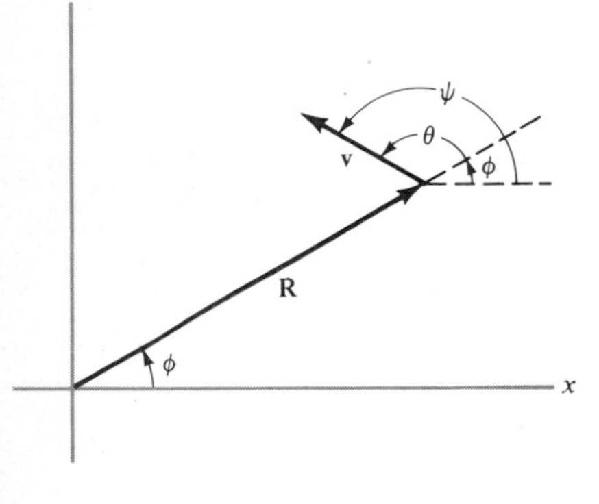

Fig. 11-16 Diagram for the proof, given in the text, that $|\mathbf{R} \times \mathbf{v}| = xv_y - yv_x$.

The easiest way to calculate $|\mathbf{R} \times \mathbf{v}|$ is directly from the components of $\mathbf{R}$ and $\mathbf{v}$. This avoids the necessity of calculating the angle between $\mathbf{R}$ and $\mathbf{v}$ explicitly. For two vectors $\mathbf{R}$ and $\mathbf{v}$ which lie in the xy plane, the magnitude of the vector product is given by

$$|\mathbf{R} \times \mathbf{v}| = xv_y - yv_x \tag{11-32}$$

The proof of Eq. (11-32) is as follows. In Fig. 11-16, the angle between the positive x axis and $\mathbf{R}$ is called ϕ, and the angle between the positive x axis and $\mathbf{v}$ is called ψ. The angle between $\mathbf{R}$ and $\mathbf{v}$ is called θ. It is evident from the diagram that

$$\psi = \phi + \theta$$

so that

$$\theta = \psi - \phi$$

According to the definition of the vector product, we have

$$|\mathbf{R} \times \mathbf{v}| = Rv \sin \theta$$

and thus

$$|\mathbf{R} \times \mathbf{v}| = Rv \sin(\psi - \phi)$$

Using the trigonometric identity for the sine of the difference of two angles yields

$$\begin{aligned}|\mathbf{R} \times \mathbf{v}| &= Rv(\sin \psi \cos \phi - \cos \psi \sin \phi)\\ &= (R \cos \phi)(v \sin \psi) - (R \sin \phi)(v \cos \psi)\end{aligned}$$

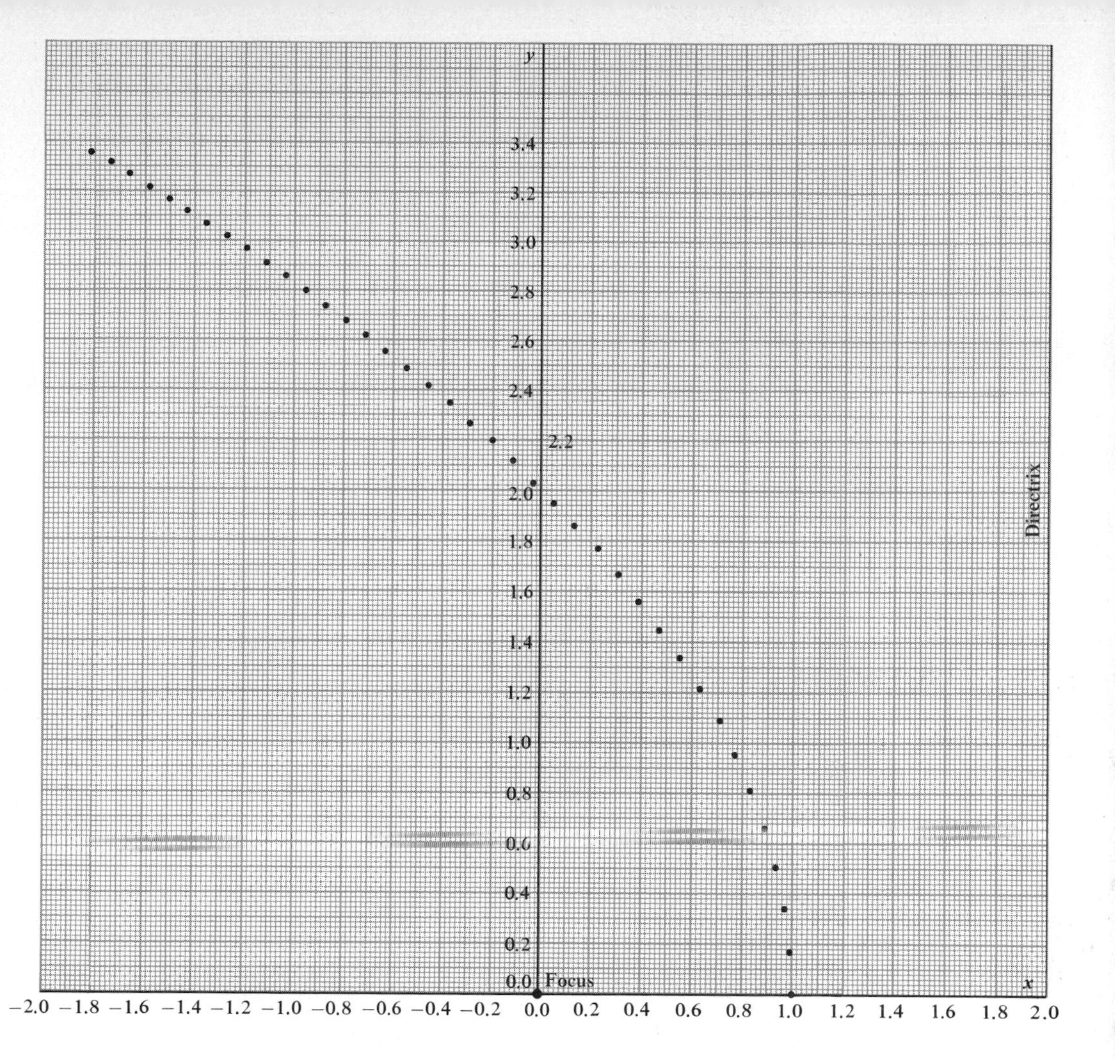

Fig. 11-17 Results of the numerical orbit calculation carried out in Example 11-10. The conditions are the same as those which led to the circular orbit of Example 11-7, except that the initial velocity is greater by the factor $\sqrt{2}$. The resulting orbit is parabolic. The fundamental geometric definition of the parabola can be used to verify this statement. According to this definition, a parabola is the locus of all points equidistant from the focus and a straight line called the **directrix.** Kepler's first law states that the force center is located at one of the two foci of an elliptical orbit. Since the orbit in this figure is a limiting case of the elliptical orbit, the focus of interest must be at the origin, where the force center is located. (The other focus lies at $x = -\infty$.) If the points in the figure do lie on a parabola, the definition of a parabola requires that the directrix cross the y axis at $x = 2.0$, so that the point on the parabola at $x = 1.0$, $y = 0$ can be equidistant from the focus and the directrix. If you imagine the other half of the curve described by the points below the x axis, symmetry dictates that the directrix lie parallel to the y axis, since the curve—if it is a parabola—is symmetrical with respect to the x axis. You can see from the figure that the curve crosses the y axis at $y = 2.0$, that is, 2.0 units from the focus. It is also 2.0 units from the directrix and thus satisfies the criterion for a parabola. You can verify that other points on the curve satisfy this criterion as well, by using a ruler.

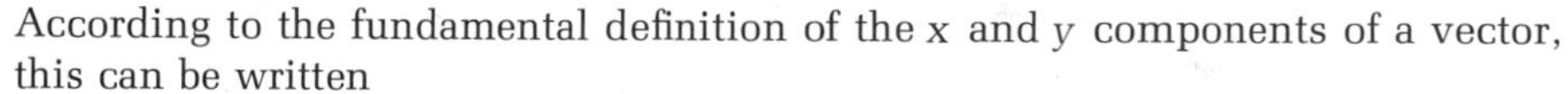

According to the fundamental definition of the x and y components of a vector, this can be written

$$|\mathbf{R} \times \mathbf{v}| = xv_y - yv_x$$

which is what we set out to prove.

Example 11-8 investigated the consequences of increasing the initial speed of a planet by 5 percent over that required to produce a circular orbit. The result was an elliptical orbit, albeit one whose deviation from circularity was not dramatic. In Example 11-10 we retain the initial conditions and parameters leading to the circular orbit of Example 11-7, except that we increase the initial speed over that required for a circular orbit by a factor of $\sqrt{2}$.

EXAMPLE 11-10

Run the central-force program with the following set of initial conditions and parameters.

$x_0 = 1$ (in AU); $(dx/dt)_0 = 0$; $y_0 = 0$; $(dy/dt)_0 = \sqrt{2}(2\pi)$ (in AU/yr); $t_0 = 0$; $\Delta t = \frac{1}{52}$ (in yr); $\alpha = 39.5$ [in $(\mathrm{AU})^3/(\mathrm{yr})^2$]; $\beta = -1.5$.

■ The resulting motion of the "planet" is plotted in Fig. 11-17. The initial momentum is now so great that the gravitational force is inadequate to bend the path into an orbit which closes on itself as an ellipse does. Thus the path is open. As the planet moves away from the sun, the path's curvature continually decreases, approaching a straight line. Once past the point of closest approach to the force center (which is the initial point in this example), the planet forever increases its distance from the force center. As the force **F** decreases with increasing distance, $d\mathbf{p}/dt$ decreases and the path approaches the straight-line path for which **p** is constant.

As the planet moves away from the sun, the force acting on it continually slows it down. You can see this from the spacing of the points on the path.

The path of Example 11-10, in fact, is a *parabola.* This statement is verified in Fig. 11-17 and its caption. The parabola forms the boundary in the hierarchy of conic sections between the family of closed ellipses and that of open hyperbolas, as can be seen in Fig. 11-9.

The considerable variations in the paths of the last four examples were produced by varying the initial speed—and hence the initial kinetic energy—of the moving body. But the initial position of the moving body with respect to the body fixed at the force center was the same in all cases. You may well guess on this basis that the system always had the same initial potential energy. In Sec. 11-6 we consider in detail the kinetic and potential energies of a system containing a moving body acted on by a central force, and we develop criteria for determining the form of paths in terms of energy.

11-6 ENERGY IN GRAVITATIONAL ORBITS

The very nearly constant properties of the orbits in which the planets move, over time spans of hundreds of millions of years, suggest that mechanical energy is very nearly conserved as a planet orbits its sun. However, you saw in Sec. 11-5 that the speed of a planet in a noncircular orbit varies significantly through each orbital circuit. Thus in each circuit there must be an interchange of mechanical energy between its kinetic and potential forms, much as there is such an interchange for oscillating systems in each cycle of oscillation. The kinetic energy of the sun-planet system (which resides almost entirely in the planet, since the much more massive sun is nearly motionless) decreases because the speed of the planet decreases as the planet "climbs outward" toward aphelion. As this happens, the system must gain potential energy in order for its total mechanical energy to remain constant. The inverse process takes place as the planet "falls inward" toward perihelion, with the system losing potential energy and gaining kinetic energy.

The conservation of energy which is observed to hold in the solar system is attributable to the isolation of the system and the fact that the only significant forces operating within the system are conservative gravitational forces. As far as any particular planet is concerned, the solar gravitational force is by far the largest of these. We therefore consider the principle of conservation of mechanical energy as it applies to an idealized system consisting of a single planet of mass m and a very much more massive sun of mass M. The results will apply directly, however, to any system in which a body is acted on by a central gravitational force (including systems treated by means of the reduced-mass method). In Chap. 21 you will see that only a trivial modification is required to apply the results to a system in which the gravitational force is replaced by any other central inverse-square force.

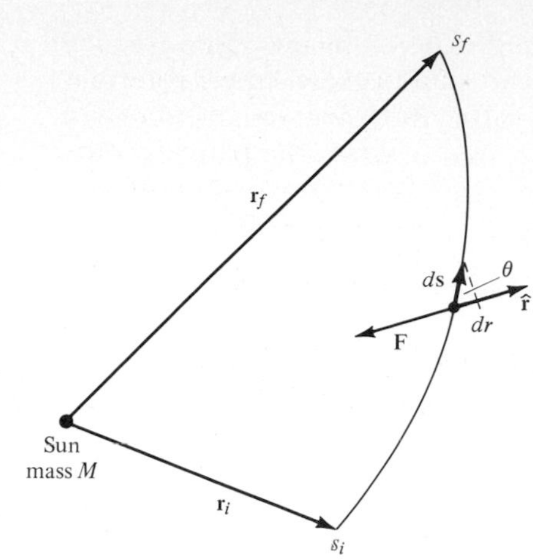

Fig. 11-18 Diagram for evaluating the change in potential energy of a sun-planet system as the planet moves along a segment of its orbit from initial position s_i to final position s_f. The coordinates s_i and s_f are measured *along* the orbit from some fixed origin lying on the orbit. At any point along the orbit, the direction of the gravitational force is $\hat{\mathbf{F}} = -\hat{\mathbf{r}}$, where $\mathbf{r}$ is the position vector from the sun to the planet. As the planet moves through the infinitesimal displacement $d\mathbf{s}$, the gravitational force does work $dW = \mathbf{F} \cdot d\mathbf{s}$.

We want to apply the conservation principle of mechanical energy in the form of Eq. (7-52), which is

$$E = K + U = \text{constant} \tag{11-33}$$

To do this, we must develop an explicit expression for the gravitational potential energy U. Figure 11-18 depicts a planet as it passes along a segment of its orbit. Specifically, let the planet move from the initial position, specified by the coordinate s_i measured along its path from an arbitrary point on the path, to the final position specified by the coordinate s_f. According to Eq. (7-35), the work W done by the gravitational force $\mathbf{F}$ is

$$W = \int_{s_i}^{s_f} \mathbf{F} \cdot d\mathbf{s} \tag{11-34}$$

Here $d\mathbf{s}$ represents an infinitesimal path element, as shown in the figure.

The work integral defined by Eq. (11-34) can be evaluated by substituting into that equation the value of $\mathbf{F}$ given by Eq. (11-6*b*). This yields

$$W = \int_{s_i}^{s_f} -\frac{GMm}{r^2}\,\hat{\mathbf{r}} \cdot d\mathbf{s}$$

As can be seen from Fig. 11-18, the scalar product is $\hat{\mathbf{r}} \cdot d\mathbf{s} = 1 \cos\theta\, ds = dr$. That is, $\hat{\mathbf{r}} \cdot d\mathbf{s}$ is the value dr of the radial component of $d\mathbf{s}$. Thus by using the rule of Eq. (7-20) to perform the actual integration, we obtain

$$W = -GMm \int_{r_i}^{r_f} \frac{dr}{r^2} = GMm \left(\frac{1}{r_f} - \frac{1}{r_i}\right) \tag{11-35}$$

In this expression, r_i and r_f are the magnitudes of the vectors $\mathbf{r}_i$ and $\mathbf{r}_f$, which describe the initial and final positions of the planet with respect to the sun. The vectors correspond to the coordinates s_i and s_f measured along the path. The fact that the work W depends only on the distances r_i and r_f, and not on the specific positions s_i and s_f, arises directly from the fact that the gravitational force under consideration is a central force.

Equation (11-35) can be used to confirm that the gravitational force is a conservative force. If the planet follows an arbitrary closed path (for example, once around its orbit) so that $r_f = r_i$, then the work done by the force is zero. As was discussed in Sec. 7-4, this is a necessary and sufficient condition for a conservative force.

We now use the work-potential energy relation of Eq. (7-46) to express the change ΔU in the potential energy of the sun-planet system as the planet passes from its initial to its final position. In the present notation, that relation is

$$\Delta U = -W \tag{11-36}$$

Hence the potential energy change is

$$\Delta U = GMm \left(\frac{1}{r_i} - \frac{1}{r_f}\right) \tag{11-37}$$

This expression tells you that the potential energy increases as the planet moves outward. For outward motion, $r_f > r_i$ and hence $1/r_i > 1/r_f$, leading to a positive value for ΔU. This is in accord with the intuitive notion that such outward motion is "uphill"—that is, in a direction opposed by the attractive gravitational force. Conversely, the potential energy decreases as the planet moves inward.

As is true of any potential energy, the potential energy U of the sun-planet system is defined with respect to some agreed-upon reference position at which U has the value 0. It is most often convenient to choose this reference position as one where $r = \infty$, that is, when the planet is separated from the sun by an infinite distance. Suppose that the planet is initially at an infinite distance from the sun (at $r_i = \infty$) and moves to a final position whose distance from the sun has the arbitrary value r (at $r_f = r$). Then Eq. (11-37) yields

$$\Delta U = -\frac{GMm}{r}$$

But $U = 0$ at the initial position because the planet is then at the reference position. Thus the value of U at the final position is equal to the change ΔU. So we have $U = \Delta U$, or

$$U = -\frac{GMm}{r} \tag{11-38}$$

It may seem awkward that the potential energy of a sun-planet system always has a negative value. But you will soon see that this is outweighed by the advantages when the total mechanical energy E of the system is considered. In any case, note that the potential energy of the system increases (becomes less negative) as r increases. Note also that while the (hypothetical) path through which the planet was moved in deriving Eq. (11-38) was of infinite length, the result is *not* an infinite change in potential energy. Can you explain why?

The value of U would be $-\infty$ if the planet were to move precisely to the force center, where r is 0. However, this is not a physical possibility. If there is to be a gravitational force, the center of force must be occupied by a body possessing mass. Such a body always occupies space, and that space is not available to the body on which the gravitational force is exerted.

In Example 11-11 the concept of gravitational potential energy is applied to an artificial earth satellite.

EXAMPLE 11-11

How much work is necessary to raise a satellite of mass $m = 100.0$ kg slowly from the surface of the earth to the altitude of a synchronous orbit? How much work would be required to remove the satellite entirely from the earth's gravitational influence?

■ Referring to Example 11-2, you have $r_i = R_e = 6367$ km (the radius of the earth) and $r_f = R_s = 42{,}180$ km (the orbit radius). Since the center of the earth is the center of force for the satellite, you can set the mass of the earth equal to M in Eq. (11-37) and use that equation to determine the potential energy increase ΔU of the earth-satellite system as the satellite is raised. This energy increase must be accomplished by doing work W' *against* the gravitational force, using a rocket. Thus $W' = -W = \Delta U$, and you have

$$W' = GMm\left(\frac{1}{R_e} - \frac{1}{R_s}\right)$$

This equation can be solved directly, by using the numerical values given. However, you can also proceed as follows. In the present notation, Eq. (11-11) becomes $g = GM/R_e^2$. Substituting into the equation for W' gives

$$W' = mgR_e - mg\frac{R_e^2}{R_s} = mgR_e\left(1 - \frac{R_e}{R_s}\right)$$

Inserting the numerical values gives you

$$W' = 100.0 \text{ kg} \times 9.80 \text{ m/s}^2 \times 6.367 \times 10^6 \text{ m} \times \left(1 - \frac{6.367 \times 10^6 \text{ m}}{4.218 \times 10^7 \text{ m}}\right)$$

$$= 6.24 \times 10^9 \times (1 - 0.1509) \text{ J}$$

$$= 5.30 \times 10^9 \text{ J}$$

The term 1 − 0.1509 in the next-to-last line of the calculation suggests that most of the work required to remove the satellite entirely from the earth has already been done by the time it has been raised to the altitude of the synchronous orbit. Indeed, you need only do approximately an additional 15 percent of this work to complete the removal. The total work W'' required to raise the satellite to an infinite altitude, where $r_f = \infty$ and the satellite is removed entirely from the earth's gravitational influence, is given by

$$W'' = mgR_e = 6.24 \times 10^9 \text{ J}$$

Example 11-12 considers the kinetic as well as the potential energy of a system bound by the gravitational force.

EXAMPLE 11-12

If a small body of mass m is dropped to the earth from an initial state of rest at a very great altitude, how fast will it be going when it strikes the earth? Ignore air resistance.

▪ The initial kinetic energy of the body is $K = 0$. If you take its initial distance from the center of the earth to be infinite, its initial potential energy is $U = 0$. Thus its initial total mechanical energy is $E = K + U = 0 + 0 = 0$. Since the system is a conservative one, E is constant and so must maintain the value $E = 0$. Thus the kinetic energy $K = E - U$ at the moment of collision with the earth is

$$K = 0 + \frac{GMm}{R_e}$$

where R_e is the radius of the earth and M is its mass. Since the mass m of the small body is very much less than that of the earth, M, it possesses essentially all the kinetic energy of the system at the moment of collision. In terms of the speed v of the body at that moment, you have

$$K = \frac{mv^2}{2} = \frac{GMm}{R_e} \tag{11-39}$$

Solving for v gives you

$$v = \left(\frac{2GM}{R_e}\right)^{1/2} \tag{11-40}$$

Substituting the numerical values of these quantities, you obtain

$$v = \left(\frac{2 \times 6.67 \times 10^{-11} \text{ N·m}^2/\text{kg}^2 \times 5.97 \times 10^{24} \text{ kg}}{6.37 \times 10^6 \text{ m}}\right)^{1/2}$$

$$= 1.12 \times 10^4 \text{ m/s} = 11.2 \text{ km/s}$$

The use of energy conservation in Example 11-12 implies that the process works both ways. If you were able to shoot a projectile upward from the surface of the earth with a speed of 11.2 km/s, and if there were no air resistance, the projectile would never come completely to rest, but would continue outward indefinitely as its kinetic energy and speed decreased asymptotically to zero. That is, it would never return to earth. But any projectile with lower initial speed would ultimately do so. The speed $v = 11.2$ km/s is therefore called the **escape speed** from the surface of the earth. You can see from Eq. (11-40) that for a planet other than the earth the escape speed is proportional to the square root of the planet's mass M and inversely proportional to the square root of its radius R.

The escape speed is often loosely called the escape velocity. This is misleading, however, since the direction in which the projectile is launched with escape speed is immaterial (provided, of course, it is not aimed into the earth). This is implicit in the scalar nature of energy. Only the scalar v^2 (and not the vector $\mathbf{v}$) appears in the energy relation, Eq. (11-39), used to evaluate the escape speed.

Let us again drop a body from $r = \infty$. But this time let us prevent collision with the earth by giving the body a very small initial velocity perpendicular to the radial (earth-body) direction at the moment of release. Its total energy will now be only negligibly greater than zero. But as a result of this initial tangential velocity, the body will travel along the path displayed in Fig. 11-19. The body will pass the center of the earth at some minimum distance R, called the distance of closest approach, or **perigee.** Continuing further along this path, the body will ultimately arrive at a very large distance from the earth with negligible velocity. According to Eqs. (11-39) and (11-40), the kinetic energy and the speed of the body at perigee have the values

$$K = \frac{GMm}{R} \qquad \text{for zero total energy path}$$

and

$$v = \left(\frac{2GM}{R}\right)^{1/2} \qquad \text{for zero total energy path}$$

The kinetic energy may be compared with that of the same body when it is in a *circular* orbit about the earth at distance R. For a circular orbit, the acceleration is purely centripetal and has magnitude $a = v^2/R$. The magni-

Fig. 11-19 A small body is allowed to fall toward the earth from a large initial distance. As it is released, it is given a very small initial velocity in the direction perpendicular to that toward the earth, whose mass is M. As a result, it "falls" in a parabolic orbit, missing the center of the earth by a distance R, called the distance of closest approach, or perigee.

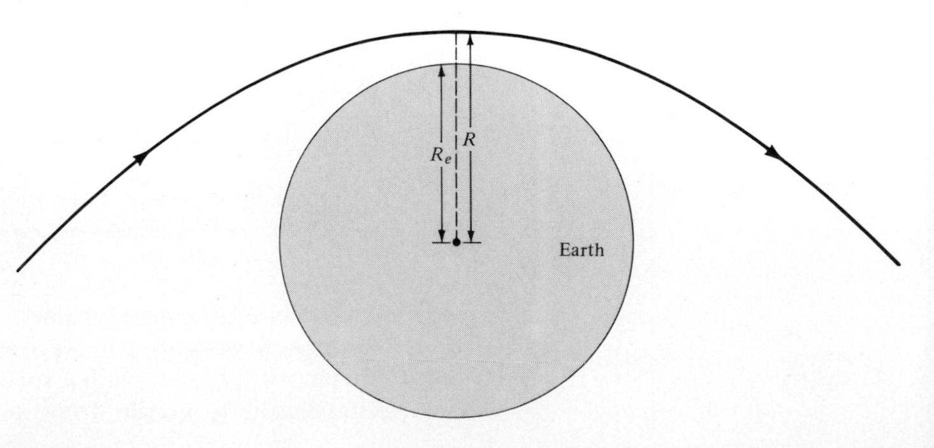

tude of the necessary centripetal force is $ma = mv^2/R$, and this is provided by the gravitational attraction $F = GMm/R^2$. Thus we have $GMm/R^2 = mv^2/R$, or

$$\frac{GMm}{R} = mv^2$$

Since $mv^2 = 2K$, this leads to the relation

$$K = \frac{GMm}{2R} \qquad \text{for circular orbit}$$

For a circular orbit of radius R, the kinetic energy is just one-half that of the same body in the path having a perigee distance R and for which the total energy of the system is essentially zero. The circular-orbit speed is

$$v = \left(\frac{GM}{R}\right)^{1/2} \qquad \text{for circular orbit}$$

Comparing this circular orbit speed with the speed at perigee of the zero total energy path, we see that the latter is greater by a factor of $\sqrt{2}$. Then looking again at Example 11-10, we find that the initial conditions used in the numerical calculation done there are ones in which a body has a speed at the perigee distance R which is greater by a factor of $\sqrt{2}$ than the speed it has in a circular orbit of radius R. Thus the path produced in the numerical calculation, and displayed in Fig. 11-17, is a zero total energy path. An analysis carried out in the figure caption shows that the path is a parabola. Hence we can conclude that *a path of zero total energy is a parabola.*

Another conclusion that can be drawn from these considerations is that the speed of a minimum-orbit earth satellite, 7.9 km/s (see Sec. 3-6), is $1/\sqrt{2}$ times the escape speed from the surface of the earth, for which $R = R_e$. The numerical result for the escape speed, obtained in Example 11-12, is 11.2 km/s, in agreement with this conclusion.

For a circular orbit, the total energy E is found by adding the value of the kinetic energy K just derived to the value of the potential energy U given by Eq. (11-38). This gives

$$E = K + U = \frac{GMm}{2R} - \frac{GMm}{R} \tag{11-41a}$$

or

$$E = -\frac{GMm}{2R} \qquad \text{for circular orbit} \tag{11-41b}$$

Equation (11-41a) shows that the kinetic energy in the case of a circular orbit has the value

$$K = -\tfrac{1}{2}U \qquad \text{for circular orbit} \tag{11-42}$$

All closed orbits—orbits in which the distance to the moving body from the center of force increases and decreases periodically—have associated with them total energies E less than zero. These orbits are the circle and the ellipses of Fig. 11-9. The open parabolic path is characterized by zero (or negligible) total energy. *Paths with nonnegligible positive total energy are hyperbolic.* Figure 11-9 illustrates the geometric connection among these paths, all of which are conic sections. A body in such a path has significant speed when it is a very large distance from the central body. A

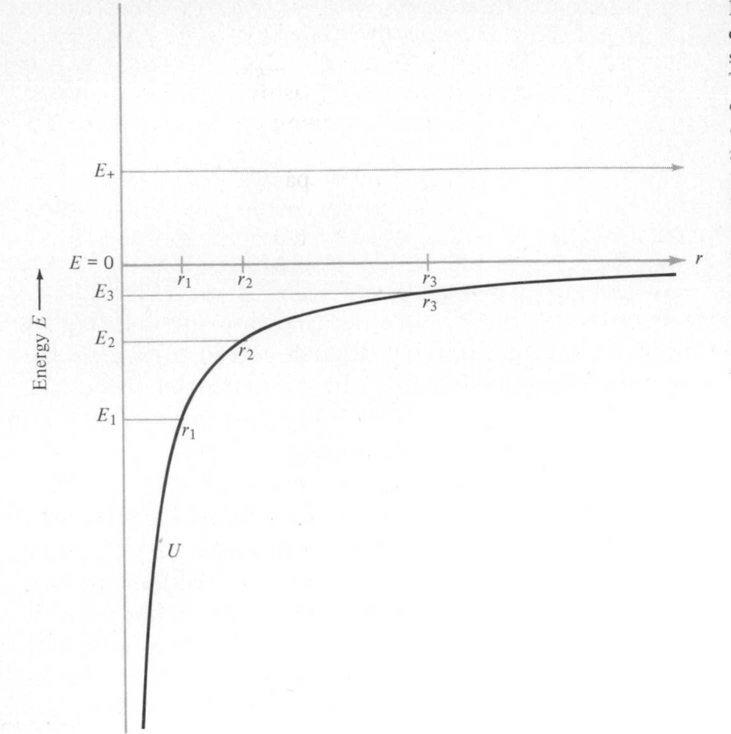

Fig. 11-20 Representation of the dependence of the potential energy U of a sun-planet system on the distance r between the sun and the planet. The curve represents the relation $U \propto -r^{-1}$. The negative quantities E_1, E_2, and E_3 represent possible total energies of the system for which the planet is bound, that is, confined to a closed (elliptical or circular) orbit. The distances r_1, r_2, and r_3 are the *maximum possible* aphelia (extreme sun-planet distances) for the three cases, respectively. In each case, they correspond to extremely elongated ellipses with essentially zero perihelion distances (semiminor axes). In each case, orbits are possible which pass through any lesser distance. But a greater distance (represented in each case by the shaded extension of the horizontal line denoting the total energy) would impose the requirement that the potential energy of the system exceed its total energy. Since $K = E - U$, this would require that $K = mv^2/2 < 0$, which is impossible. If the total energy of the system is $E = 0$, the sun-"planet" distance r can have any value at all, since the corresponding potential energy can never exceed the reference value $U = 0$. The possible paths are parabolas. If the total energy of the system has any positive value E_+, the system is unbound and the possible paths are hyperbolic. Compare with Fig. 8-18, which is the analogous plot for a diatomic molecule.

two-body system with negative total energy is called a **bound system;** a system with zero or positive total energy is called an **unbound system.** Figure 11-20 illustrates this distinction. This figure should be compared with Figs. 8-17 and 8-18, which describe the relation between potential energy and total energy for systems with different force laws.

In the solar system, practically all (if not all) bodies have elliptical orbits, either around the sun or (in the case of satellites) around their planets. With a few exceptions, the planetary and satellite orbits are ellipses of small "eccentricity"—that is, they are not very different from circles. Some of the asteroids, and nearly all the comets, have highly eccentric elliptical orbits. In the case of some comets, the presumably elliptical orbits are indistinguishable from parabolas on the basis of observation. If the aphelion of a comet is very far from the sun—beyond the orbit of Pluto—the acceleration and velocity of the comet near aphelion will be so small that the period may be millions of years.

It is unlikely that any celestial bodies are observed in hyperbolic paths. Such a path would imply that the body had entered the solar system with nonzero velocity—that is, it is not bound to, and is not a member of, the solar system. But the space outside the solar system is so empty (compared to the solar system itself) that such encounters with bodies of observable size cannot occur with any frequency.

However, particles moving in unbound paths are of great importance in atomic and nuclear phenomena. The attractive force in this case is electric rather than gravitational, but the general principles of orbit mechanics are the same except for scale, because the electric force obeys an inverse-square law just as the gravitational force does. It is quite easy to alter the total energy of an atom so as to ionize it—that is, to supply energy to the atom and transfer an atomic electron from a bound orbit to an unbound path taking it away from the atom. It is equally simple to observe the reverse process, called electron capture, in which the system emits energy in the form of light or otherwise as an electron moving along an unbound path approaching the atom jumps onto a bound orbit.

A final case, which we discuss in detail in Chap. 20, is the one in which the inverse-square force is repulsive rather than attractive. This is the case for two bodies with electric charges of the same sign, such as an alpha particle and a uranium nucleus. In this case, the total energy is *always* positive, since the potential energy *increases* from a minimum value of zero as one particle approaches the other from a large distance.

Experiments in which a particle is shot at a target particle with positive total energy are called scattering experiments. Since the path of the projected particle is necessarily not a closed one, the particle can be collected by a detector after undergoing some deflection (as in Fig. 11-19) because of the force exerted on it by the target particle. A study of the details of the deflection as a function of the energy of the particle and its distance of closest approach to the target particle can yield a tremendous amount of information about the details of the force, and hence about the structure of the target particle.

11-7 PERTURBATIONS AND ORBIT STABILITY

So far our study of orbits has been restricted to the case of central forces, in which the gravitational interaction takes place between only two bodies. We made the further simplification in Sec. 11-6 that one of the bodies has a much greater mass than the other, so that only the motion of the less massive body need be considered. No great complication is introduced by relaxing this latter restriction, since the reduced-mass procedure can be used to transform any two-body system into a system in which one of the bodies has infinite mass. However, the solar system is not a two-body system, and the planets exert gravitational forces on one another. These forces are small compared to the forces due to the sun, but they result in measurable disturbances of the planetary motions described by the two-body treatment. These disturbances are called **perturbations.** Perturbations are especially significant in certain cases where the forces, which vary periodically as the planets move around the sun, are in step with the periodic motion of the body on which they are acting. Such perturbing forces produce large cumulative effects over long periods. A complication is introduced by the fact that the orbits of the planets do not lie quite in the same plane. Thus the analysis, which has been two-dimensional to this point, must be extended to three dimensions if accurate calculations are required. We will not consider this extension because it is quite complicated. Nevertheless, a good deal of insight can be gained into the effects of perturbing forces by carrying out a quite simple numerical calculation in two dimensions.

The largest perturbations in the solar system are produced by Jupiter, whose mass is about 0.1 percent that of the sun. To give an example of the effect of Jupiter on other planets, consider its effect on the earth. It is about 4 times more distant than the sun when it is at its distance of closest approach to the earth. Thus the magnitude of the force it exerts on the earth at that point is about $10^{-3}/4^2 \simeq 6 \times 10^{-5}$ times that produced by the sun.

This closest approach of Jupiter to the earth takes place when the sun, the earth, and Jupiter lie in a straight line in that order. Astronomers call this alignment *opposition.* It takes place about every 399 days, which is the time required for the earth to circle the sun and catch up with Jupiter again. Hence the distance between the earth and Jupiter varies with this 399-day period, called the *synodic period* of Jupiter. The perturbing force exerted on the earth by Jupiter must also vary with the synodic period. But the perturbing force is relatively small, and it falls off rather rapidly ($\propto r^{-2}$)

with increasing distance. Hence it is possible to make a respectable first approximation to the effect of Jupiter on the earth by substituting for the periodically varying force an instantaneous impulse of appropriate magnitude, applied to the earth at the moment of opposition.

According to Eqs. (8-9) and (8-10), the impulse **I** can be written

$$\mathbf{I} = \int_{t_i}^{t_f} \mathbf{F}\, dt = \Delta\mathbf{p}$$

Thus the impulse **I** can be substituted for $\Delta\mathbf{p}$, the entire change in the momentum of the earth produced by the attraction of Jupiter over an entire synodic period $t_f - t_i = 399$ days. During most of that period, the force **F** is negligible, at least in first approximation. Thus **F** has the character of an impulsive force, something like the one displayed in Fig. 8-8. To a first approximation, therefore, the effect of Jupiter on the motion of the earth can be found by assuming that the impulse **I**—and hence the change in momentum $\Delta\mathbf{p}$—takes place entirely at the instant of closest approach between Jupiter and the earth. This method was first used by Newton. It is typical of his penetrating insight into physical phenomena.

Example 11-13 gives a good general idea of what happens. It employs an extension of the numerical orbit calculation technique developed in Sec. 11-5. We first allow the earth to make a complete, unperturbed circular orbit. At week 52, when the earth returns to its starting point, we assume that it passes Jupiter and experiences an instantaneous impulse directed away from the sun. (This direction would be precisely correct if the orbits of the earth and Jupiter lay in exactly the same plane.) The effect of this instantaneous impulse is represented by adding to the earth's velocity a new velocity $\Delta\mathbf{v}$ directed radially outward. In the interest of clarity in demonstrating the effect, we exaggerate greatly and let $\Delta\mathbf{v}$ have a magnitude equal to 5 percent of the earth's unperturbed speed. The radially outward impulse **I** is thus represented by giving the earth a radially outward momentum change $\Delta\mathbf{p} = m\,\Delta\mathbf{v}$. At week 109, when the earth catches up with Jupiter again 399 days later, we add a second velocity increment, of magnitude equal to the first and again directed radially outward (a direction which is now approximately 35° counterclockwise from the x axis).

EXAMPLE 11-13

Run the central-force program through 1 yr (52 "weeks" or calculation cycles) with the following set of initial conditions and parameters, which are identical with those of Example 11-7:

$x_0 = 1$ (in AU); $(dx/dt)_0 = 0$; $y_0 = 0$; $(dy/dt)_0 = 2\pi$ (in AU/yr); $t_0 = 0$; $\Delta t = \frac{1}{52}$ (in yr); $\alpha = 39.5$ [in $(\text{AU})^3/(\text{yr})^2$]; $\beta = -1.5$.

Stop the calculating device when the 52d cycle of calculation is completed. Add 0.1π to the dx/dt storage register. (This gives the earth a radial velocity toward Jupiter whose magnitude is 5 percent of the magnitude of the tangential velocity $dy/dt = 2\pi$.) Then continue running for 57 more "weeks" or cycles, for a total of 109 "weeks." At this time the sun, the earth, and Jupiter are again in line.

Stop the calculating device when the 109th cycle is completed. Add $0.1\pi \cos 35°$ to the dx/dt storage register and $0.1\pi \sin 35°$ to the dy/dt storage register. [This again gives the earth a radial velocity toward Jupiter whose magnitude is 5 percent of 2π. Now, however, the line from the sun through the earth to Jupiter lies 35° counterclockwise from the x axis. To see this, note that 109 weeks is 5 weeks more than 2 yr. And (5 weeks/52 weeks) $\times$ 360° = 35°.] Finally, run the calculation for 52 more "weeks" to complete another orbit.

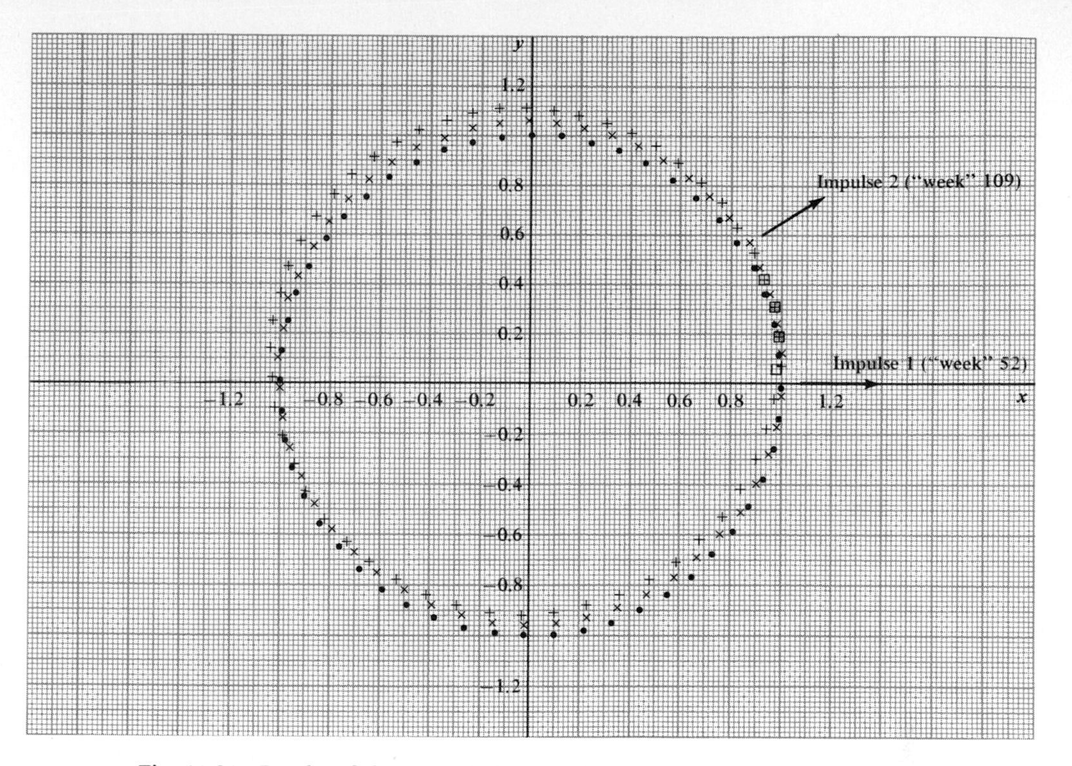

Fig. 11-21 Results of the numerical orbit calculations carried out in Example 11-13. The first orbit, shown by dots, is identical to the circular orbit of Example 11-7 and represents an idealized earthlike planet with constant orbit radius 1 AU. As the first orbit is completed at $x = 1$ AU, $y = 0$, the planet is given an outward radial velocity equal to 5 percent of its tangential velocity, in order to simulate (in exaggerated form) the gravitational attraction of Jupiter. The subsequent orbit, represented by x's, is both elongated into an ellipse whose major axis lies along the y axis (perpendicular to the direction of the perturbation) and shifted in the positive y direction (so that the sun remains at the focus of the ellipse). Fifty-seven "weeks" later, a second outward radial velocity component is added to simulate the effect of Jupiter. The third orbit, represented by crosses, is again elongated from that point onward and shifted in the direction perpendicular to the perturbational effect. The beginning of a fourth orbit is represented by squares.

■ The results are shown in Fig. 11-21. The initial circular orbit is denoted by dots, the second orbit by x's, the third by crosses, and the fourth (partial) orbit by squares.

It is clear that the first impulsive force, or *blow,* distorts the circular orbit into an ellipse. It may seem paradoxical that a blow in the positive x direction, delivered at $x = 1$ AU, $y = 0$ both shifts the orbit in the positive y direction and elongates it along the y axis. Likewise, the second blow, delivered radially at about a 35° angle counterclockwise from the x axis, shifts and elongates the ellipse along the perpendicular direction. Can you explain this effect qualitatively?

Perhaps the most remarkable thing about Example 11-13 is the way it illustrates the **stability** of the orbit. A substantial blow does indeed change the orbit. But the new orbit is not very different from the old one and is itself a stable orbit which would retrace itself indefinitely if no further perturbations were applied. In order to emphasize the significance of this point, and the special quality which it gives to inverse-square forces, we will again investigate the result of applying an impulse of the same magnitude and direction as the first one given to the planet in Example 11-13. However, this time we will apply it to a planet in a circular orbit under the influence of an attractive inverse-*cube* force, whose magnitude is given by $F = G'Mm/R^3$. The calculation will show that the orbit is *not stable.* That is, the slightest disturbance disrupts it into an open spiral.

It is quite easy to show that a circular orbit is possible for an inverse-cube force, in the total absence of perturbations. We have $ma_c = F$, or

$$\frac{mv^2}{R} = G'Mm\frac{1}{R^3}$$

where G' is the proportionality constant in the hypothetical "inverse-cube law of gravitation." Solving for v, we find that if the speed of the planet is

$$v = \left(\frac{G'M}{R^2}\right)^{1/2}$$

and its direction of motion is perpendicular to the position vector from the center of force, the centripetal acceleration a_c will have the proper relation to the force and the mass, and the motion will be circular.

In Example 11-14 we follow the unperturbed planet through one-quarter of a revolution to verify the circularity of the orbit, and then we impose the blow. The centripetal force is

$$\mathbf{F} = -G'Mm\frac{1}{R^3}\,\hat{\mathbf{R}} = -G'Mm\frac{1}{(x^2 + y^2)^{3/2}}\,\hat{\mathbf{R}}$$

Thus Eq. (11-26*a*) shows that the x component of the acceleration is

$$\frac{d^2x}{dt^2} = \frac{F_x}{m} = -\frac{F}{m}\frac{x}{(x^2 + y^2)^{1/2}} = -G'M\frac{x}{(x^2 + y^2)^2}$$

In the notation of Eq. (11-29*a*), this can be written

$$Q_x = -G'M\frac{x}{(x^2 + y^2)^2}$$

Likewise, the y component of the acceleration is

$$\frac{d^2y}{dt^2} = \frac{F_y}{m} = -\frac{F}{m}\frac{y}{(x^2 + y^2)^{1/2}} = -G'M\frac{y}{(x^2 + y^2)^2}$$

In the notation of Eq. (11-29*b*), this becomes

$$Q_y = -G'M\frac{y}{(x^2 + y^2)^2}$$

The general definitions of Q_x and Q_y appropriate to all central-force calculations are given by Eqs. (11-30*a*) and (11-30*b*), respectively. These are

$$Q_x \equiv -\alpha x(x^2 + y^2)^{\beta} \qquad \text{and} \qquad Q_y \equiv -\alpha y(x^2 + y^2)^{\beta}$$

The values of α and β applicable to the case of an inverse-cube force can be found by comparing the specific expressions for Q_x and Q_y immediately above with the definitions. For the inverse-cube case we have

$$\alpha = G'M \qquad \text{and} \qquad \beta = -2$$

The choice of the numerical value of the constant G' in the above inverse-cube equations is arbitrary. For convenience, we let its *numerical* value equal that of the universal gravitational constant G. If this is done, the numerical value of the constant α required for the calculation will again be 39.5. Since the calculating device manipulates numbers only, and not units, the substitution of G' for G will not alter the calculation. However, the units of G' must be different from those of G. Can you show that for the system of units used in the central-force calculations, the proper units for G' are $(\text{AU})^4/(\text{yr})^2$?

EXAMPLE 11-14

Run the central-force program for 13 "weeks," or calculation cycles, with the following set of initial conditions and parameters:

$x_0 = 1$ (in AU); $(dx/dt)_0 = 0$; $y_0 = 0$; $(dy/dt)_0 = 2\pi$ (in AU/yr); $t_0 = 0$; $\Delta t = \frac{1}{52}$ (in yr); $\alpha = 39.5$ [in $(\text{AU})^4/(\text{yr})^2$]; $\beta = -2$.

Stop the calculating device when the 13th cycle is completed. Add 0.1π to the dy/dt storage register. (This gives the planet an outward radial velocity increment whose magnitude is 5 percent of the magnitude of the tangential velocity $dx/dt = 2\pi$.) Run the calculating device through a sufficient number of further cycles to show clearly what kind of an effect the perturbation has on the original circular orbit.

■ The results are plotted in Fig. 11-22. The orbit is clearly not stable. The single outward blow causes the planet to leave its circular orbit and begin to spiral ever outward. The speed of the planet decreases as the distance from the center of force increases, and the curvature of the orbit continues to decrease indefinitely.

A stable planetary system would not be possible if the law of gravitation were an inverse-cube law. Even given perfectly circular orbits to begin with, the very slightest perturbation destroys the system. In other words, circular orbits in an inverse-cube force "universe" are not stable in the sense discussed in Sec. 9-7. If the perturbation is very small, the spiral will be very tight and it will take some time for the planet to leave the system; but it will inevitably happen. An inward impulse will produce an inward spiral which culminates in a collision of the planet with the star.

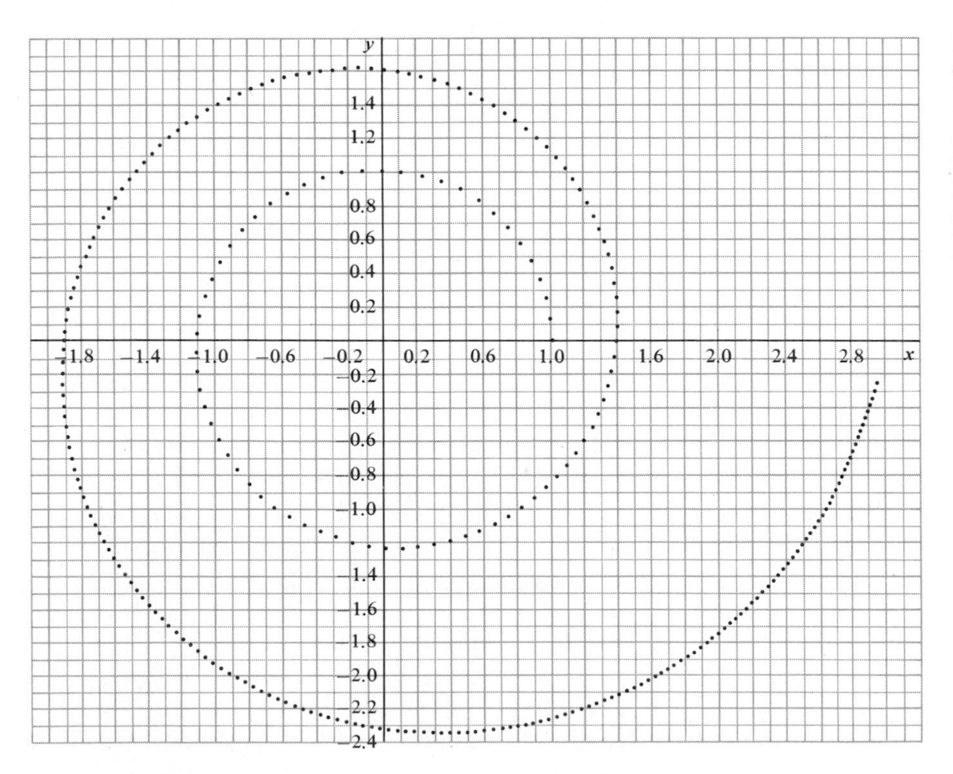

Fig. 11-22 Results of the numerical calculations of Example 11-14, showing the instability of the orbit of a planet under the influence of a hypothetical inverse-cube force. The first quarter-revolution is unperturbed and is circular. An outward radial velocity increment equal to 5 percent of the tangential velocity imposed at that point then leads to the open spiral trajectory.

By using the techniques of advanced mechanics, it can be shown quite generally that the orbit of a body under the influence of a force law of the form $\mathbf{F} \propto -(1/r^n)\hat{\mathbf{r}}$ will be *stable* if $n < 3$. The orbit will be *unstable* if $n > 3$. If $n = 3$, the stability is *neutral*. It is not difficult to see how this result comes about in the special case of a planet in a circular orbit. To do this, we place ourselves in the frame of reference of an observer located on the planet. As was the case for the observer in the discussion accompanying Fig. 11-4*b*, it is necessary to allow for the fact that this frame of reference is not inertial, because the planet is subjected to a centripetal acceleration. Regardless of the nature of the central force producing this acceleration, it can be written as the signed scalar $A = -v^2/R$, where the positive direction is taken to be the direction from the force center to the planet. The quantity v represents the speed of the planet, and R is its distance from the force center.

In order to apply Newton's laws in his noninertial frame, the observer must claim that there is a fictitious force acting on the planet, whose mass is m. He calls this force the centrifugal force F_{centrif}. According to Eq. (5-29), written in signed-scalar notation, this force is $F_{\text{centrif}} = m(-A)$. And since $A = -v^2/R$, we have

$$F_{\text{centrif}} = \frac{mv^2}{R}$$

In a circular orbit, both v and R are constant. From the point of view of the observer on the planet, the centrifugal force is equal in magnitude and opposite in direction to the central force F_{central} when the distance of the planet from the force center has the equilibrium value R_e. That is, the net force F_{net} exerted on the planet is

$$F_{\text{net}} = F_{\text{centrif}} + F_{\text{central}} = 0 \qquad \text{for } R = R_e$$

What happens if the planet is disturbed slightly from its equilibrium orbit radius $R = R_e$? Will the direction of F_{net} be such as to tend to return the planet to its original orbit or move it still farther from that orbit? In order to answer this question, we must express the centrifugal force in the form $F_{\text{centrif}} \propto 1/R^\nu$ and determine the value of the constant ν. Once we have done this, we can compare the way in which F_{centrif} varies with R to the way in which the central force $F_{\text{central}} \propto -1/R^n$ varies with R, and obtain the desired result.

However, the equation $F_{\text{centrif}} = mv^2/R$ is not in the desired form. This is because the speed v itself depends on the distance R. So we must eliminate v from the equation. To do this, we multiply the right side of the equation by the quantity mR^2/mR^2, which is equal to 1. This yields

$$F_{\text{centrif}} = \frac{m^2v^2R^2}{mR^3}$$

The quantity in the numerator of the fraction on the right side of this equation is equal to l^2, the square of the magnitude of the angular momentum of the planet about the force center. The reason is that the orbit is circular, so that the planet's momentum $\mathbf{p}$ is perpendicular to the vector $\mathbf{R}$ from the force center to the planet. As in Example 9-4, we therefore have

$$l = Rp = Rmv$$

or

$$l^2 = m^2v^2R^2$$

Using this result, we can write the centrifugal force in the form

$$F_{\text{centrif}} = \frac{l^2}{m}\frac{1}{R^3}$$

Since the angular momentum of the planet about the force center is constant, the quantity l^2/m is constant, and we have the desired proportionality

$$F_{\text{centrif}} \propto \frac{1}{R^3}$$

That is, the exponent ν is equal to 3.

In Fig. 11-23*a*,*b*, and *c*, this centrifugal force is compared with central forces proportional to $-1/R^2$, $-1/R^3$, and $-1/R^4$, respectively. In each case, the stability of the orbit is determined by the sign of $F_{\text{net}} = F_{\text{centrif}} + F_{\text{central}}$ when R deviates slightly from the value $R = R_e$. The details are given in the figure caption.

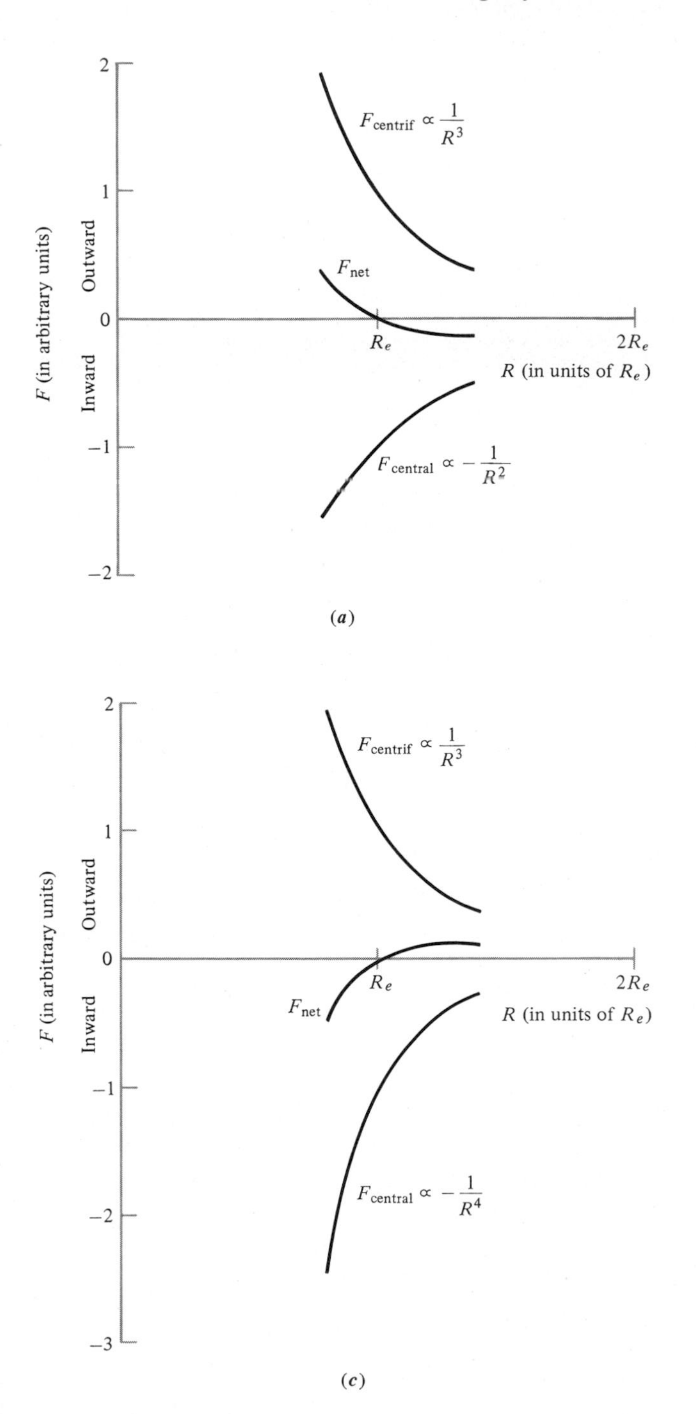

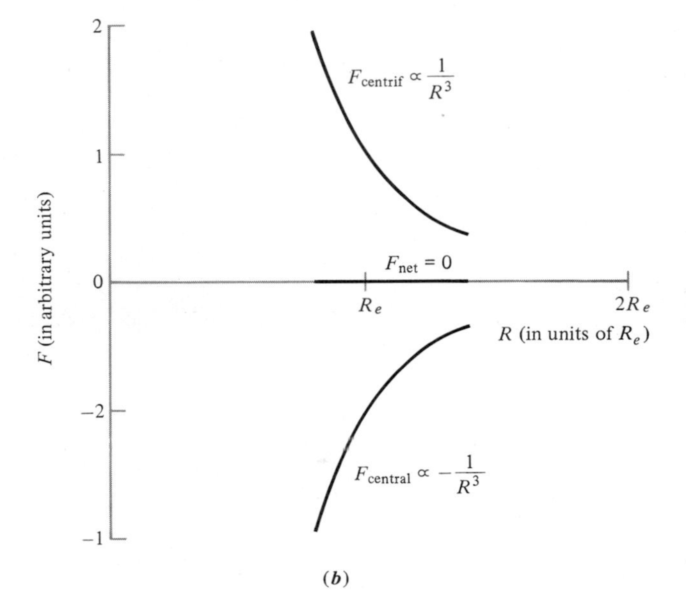

Fig. 11-23 Plots of the dependence of the centrifugal force $F_{\text{centrif}} \propto 1/R^3$ as a function of the distance R from the force center to the planet. The force is plotted in arbitrary units. The scale of force units has been adjusted so that $F_{\text{centrif}} = 1$ unit when $R = R_e$. (*a*) In this case, the central force conforms to the rule $F_{\text{central}} \propto -1/R^2$, as does Newton's law of gravitation. Its magnitude has been adjusted so that $F_{\text{central}} = -1$ unit when $R = R_e$. Thus $F_{\text{net}} = 0$ when $R = R_e$, in conformity with the condition that the planet be in a circular orbit of radius R_e. Note that $F_{\text{net}} = F_{\text{centrif}} + F_{\text{central}}$ is positive (outward) when $R < R_e$; and that it is negative (inward) when $R > R_e$. That is, the net force tends always to restore the planet to the orbit radius R_e if the planet is disturbed slightly, and the orbit is stable. (*b*) Here the central force conforms to the rule $F_{\text{central}} \propto -1/R^3$. Again, its magnitude has been adjusted so that $F_{\text{central}} = -1$ unit when $R = R_e$. The net force is zero regardless of the value of R, and there is no tendency to restore the planet to the orbit radius R_e if it is disturbed. The orbit is neutral—neither stable nor unstable. (*c*) Here the central-force law is $F_{\text{central}} \propto -1/R^4$. We have again set $F_{\text{central}} = -1$ unit when $R = R_e$. In this case, F_{net} is negative (inward) when $R < R_e$ and positive (outward) when $R > R_e$. Thus any slight disturbance in the orbit radius leads to a force which tends to remove the planet still farther from its initial orbit radius R_e, and the orbit is unstable.

There is special interest in the behavior of an orbit under the action of an attractive force whose magnitude is given by

$$F = GMm \frac{1}{R^{2+\delta}}$$

where δ is a number small compared to 1. The behavior induced by the slight deviation from the inverse-square law of newtonian gravitation is very important in the general theory of relativity, as is explained briefly in discussing Example 11-15. In Example 11-15 we exaggerate the deviation greatly by letting $\delta = 0.1$, so that $F \propto R^{-2.1}$. The corresponding value of β is seen from Eqs. (11-30*a*) and (11-30*b*) to be $\beta = -3.1/2 = -1.55$. We also choose the initial tangential velocity large enough to produce a rather noncircular orbit, to make the effect of the deviation of δ from zero more evident.

Fig. 11-24 Results of the numerical calculations of Example 11-15, showing in exaggerated form the consequences of a slight deviation from conformity of the law of gravitation to an inverse-square law. Such a deviation is predicted by the general theory of relativity. The resulting path does not quite close on itself, but can be thought of as an elliptical orbit which rotates or precesses slowly as the planet moves along it. The first four orbits are shown as dots, x's, crosses, and circles, respectively. Their perihelia are, respectively, P_1, P_2, P_3, and P_4. The precession of the perihelion about the sun is a convenient measure of the precession of the entire orbit.

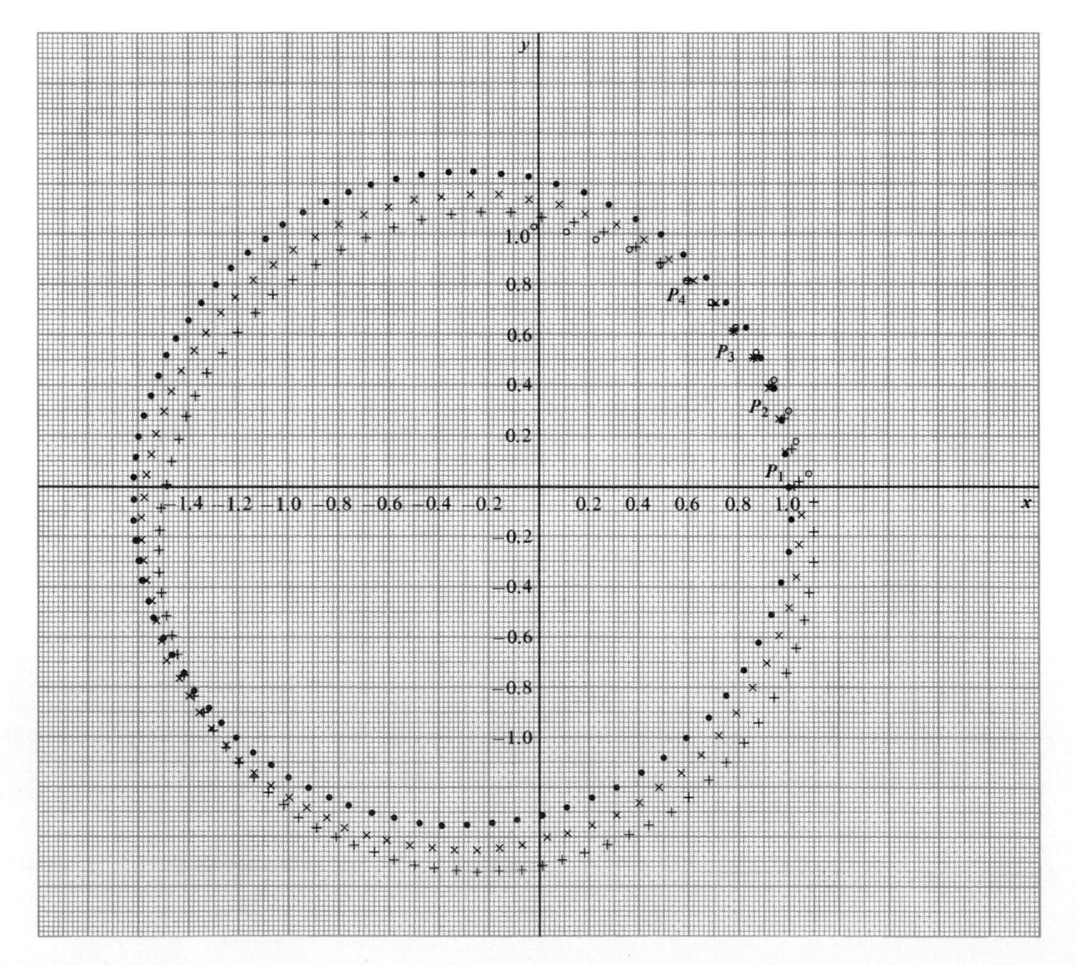

EXAMPLE 11-15

Run the central-force program with the following set of initial conditions and parameters:

$x_0 = 1$ (in AU); $(dx/dt)_0 = 0$; $y_0 = 0$; $(dy/dt)_0 = 2.2\pi$ (in AU/yr); $t_0 = 0$; $\Delta t = \frac{1}{52}$ (in yr); $\alpha = 39.5$ [in $(\text{AU})^3/(\text{yr})^2$]; $\beta = -1.55$.

▪ The results are plotted in Fig. 11-24. Here again, as in Example 11-14, this orbit does not retrace itself. However, it does not fail to close in the sense that the planet spirals away from (or into) the sun. Rather, the entire ellipse of the orbit appears to *advance* slowly, that is, to rotate in the same (counterclockwise) sense that the planet moves. This **precession** of the orbit is conventionally measured by noting the position of perihelion during each revolution. In Fig. 11-24, successive perihelia were determined by measurement on the graph, and are denoted by P_1, P_2, P_3, and P_4. For the orbit parameters chosen, the rate of precession is about $\frac{1}{18}$ revolution, or about 20°, per revolution of the planet.

Every planetary orbit displays precession. Nearly all the effect is due to the perturbations of the other planets, as seen in exaggerated form in Example 11-13. In the case of Mercury, the precession rate is about 9.6 minutes of arc (0.16°) per century. When the effects of all the other planets are subtracted, however, there remains a tiny effect—a precession rate of 43 seconds of arc (0.012°) per century. In his general theory of relativity, Einstein argued that this arises from "warping of space" in the vicinity of the sun, owing to its very large mass. The warping makes Mercury move past the sun along a path which bends more than it would according to the laws of newtonian physics, with an inverse-square force law and space remaining unwarped near the sun. The increased bending of the orbit near the sun makes the orbit "whip around" the sun, leading to a precession that makes the perihelion advance. The calculation carried out in Example 11-15 uses an inverse-2.1-power law in an unwarped space to simulate (in exaggerated fashion) the effect of an inverse-square law in a warped space, and thus to produce a perihelion advance. The simulation works because the warping of space in Einstein's theory affects the motion in much the same way as does making the gravitational attraction of the sun relatively stronger near the sun than it is for the inverse-square law, and this is just what is done by modifying the force law to an inverse-2.1-power law.

The orbit precession effect has been observed since 1974 in a system where its magnitude is very much greater than it is for Mercury. This system consists of two stars which orbit about their common center of mass. One of the stars is a so-called radio pulsar, designated PSR 1913 + 16. The radio telescope "sees" this star emit brief bursts of radio waves at time intervals measurable to 1 part in 10^{11}. This makes possible an accurate description of the pulsar orbit in spite of the very great distance of the double-star system from the earth.

The observed orbit period of the pulsar is approximately 7.75 h, compared to the 88 days of the period of Mercury. The distance between the pulsar and its companion star is considerably smaller than the average radius of Mercury's orbit about the sun. In addition, the orbit of the pulsar is considerably more eccentric (that is, noncircular) than the orbit of Mercury. For all these reasons, the precession of the orbit of the pulsar is very much more rapid than that of the orbit of Mercury. The value is 4.226 ± 0.002 degrees/yr compared to 43 arc seconds per century or 1.2×10^{-4} degrees/yr.

EXERCISES

Group A

11-1. *Finding the altitude from the period.* If a satellite is in a circular orbit above the earth with a period of 2.00 h, how far is it from the center of the earth? How high is it above the earth's surface?

11-2. *Orbital period of an asteroid.* An asteroid (a small planet) is in a circular orbit about the sun whose radius is 4.0 times the radius of the nearly circular orbit of the earth. What is the asteroid's period of revolution? Express your answer in years.

11-3. *Hot g.* The radius of the earth's orbit around the sun is 1.50×10^8 km, and its orbital period is 3.16×10^7 s. Taking the radius of the sun to be 6.96×10^5 km, find the acceleration of gravity at the surface of the sun.

11-4. *When the altitude equals the radius.*

a. What is the value of g at an altitude equal to the radius of the earth?

b. What is the period of an earth satellite in a circular orbit at this height?

c. What is the weight of an astronaut inside the satellite if his weight on earth is 70×9.8 N?

d. What is the astronaut's mass inside the satellite?

11-5. *Acceleration due to lunar gravity.* Let g_e be the acceleration of falling bodies near the earth's surface and g_m be that near the moon's surface. If the mass of the earth is 81 times the mass of the moon and the radii are $r_e = 6370$ km and $r_m = 1740$ km, respectively, compute g_e/g_m.

11-6. *How high would it go?* Kathy finds that if she throws a baseball directly upward, it falls back to earth after 6.00 s. (Neglect air resistance throughout this exercise.)

a. What is the initial speed of the baseball? What is its peak altitude? Take the acceleration of gravity at the earth's surface to be $g = 9.80$ m/s^2.

b. Suppose that Kathy could throw the ball upward with the same initial speed from the surface of some other astronomical object, where the local acceleration due to gravity is g'. Find the peak altitude and the total flight time.

c. Evaluate the results of part *b* for the following objects:

(i) Mars (its mass is 0.107 times that of the earth; its radius is 0.533 times that of the earth)

(ii) the moon (its mass is 0.0123 times that of the earth; its radius is 0.272 times that of the earth)

11-7. *A transplanted pendulum.* The mass of Mars is 0.107 times the mass of the earth. Its radius is 0.533 times the earth's radius. What would be the period of a pendulum on Mars if its period on the earth were 2.00 s?

11-8. *Mass ratios in the solar system, I.* Io is a satellite of Jupiter whose distance from the planet is 422,000 km and whose period is 1.77 days. Rhea is a satellite of Saturn. Its distance from Saturn is 527,000 km, and its period is 4.52 days. What is the ratio of the mass of Jupiter to the mass of Saturn?

11-9. *Finding the neutral gravity point and the equal gravity point.* The mass of the earth is 81 times the mass of the moon. At what two points along the line joining the earth and the moon does the magnitude of the pull of the earth on a body equal that of the moon? What is the distance between the two points?

11-10. *Planetary values for the acceleration due to gravity.*

a. Prove that the value of g', the gravitational acceleration of a falling body at the surface of any planet, is given by $g' = 4\pi G\rho r/3$ where ρ is the average density of the planet and r is its radius. Neglect the effect of rotation.

b. The ratio of the average density of Jupiter to that of the earth is 0.243. The ratio of their radii is 11.2. What is the ratio of g' at the surface of Jupiter to g at the earth's surface? Neglect rotational effects.

11-11. *Earth satellites obey Kepler!* The following data were published in newspapers for an artificial satellite launched on March 18, 1958:

Minimum altitude: 655 km
Maximum altitude: 4044 km
Period of revolution: 135 min = 2.25 h

Show that these data satisfy Kepler's third law. (*Hint:* The period of this orbit is the same as the period of a circular orbit whose radius equals the average of the minimum and maximum distances from the earth's center.)

Auxiliary data:

Average radius of the moon s orbit	3.84×10^5 km
Period of the moon	27.3 days
Radius of the earth	6.37×10^3 km

11-12. *Dirty snowball.* The nucleus of a typical comet (that is, the small, relatively dense part of its "head") is believed to be a roughly spherical solid body of frozen water and "gases" (sometimes called a "dirty snowball") with a density of about 0.8 g/cm^3 and a radius of about 5 km.

a. Find the mass of the comet's nucleus. Express your result in kilograms and as a fraction of the mass of the earth.

b. Find the gravitational acceleration g' at the surface of the comet's nucleus. Express your result in meters per second squared and as a fraction of the terrestrial gravitational acceleration g.

c. Find the escape speed at the surface of the comet's nucleus. Express your result in meters per second and as a fraction of the terrestrial escape speed.

11-13. *Thrown away.* Estimate the size of a rocky sphere with a density of 3.0 g/cm^3 from the surface of which you could just barely throw away a golf ball (and have it never return).

11-14. *Up and away?* A certain rifle has a muzzle velocity of magnitude 600 m/s.

a. If it were fired vertically upward from the moon's surface, would the bullet escape from the moon's gravitational field? (Take the moon's mass to be 7.35×10^{22} kg and the moon's radius to be 1738 km.)

b. If you find that the bullet would escape, what would its final speed be? (Use the principle of the conservation of mechanical energy, and ignore the other bodies in the solar system.)

c. If you find that the bullet would *not* escape, what would be its maximum distance from the center of the moon? From the surface of the moon? Find the total flight time of the bullet.

Group B

11-15. *Earth, moon, and sun.*

a. The sun's mass is about 320,000 times the earth's mass. The sun is about 400 times as far from the earth as the moon is. What is the ratio of the magnitude of the pull of the sun on the moon to that of the pull of the earth on the moon? (For the purpose of this exercise, it may be assumed that the sun-moon distance is constant and equal to the sun-earth distance.)

b. From the result of part *a,* it is seen that the pull on the moon is always directed toward the sun. What is the direction of the curvature of the moon's orbit as seen from the sun? A qualitative answer will suffice.

11-16. *Restoring the balance.* A standard 1-kg mass is suspended from each side of a sensitive beam balance, as in Fig. 11E-16. The wire supporting the right mass goes through an opening in the floor so that it is 10.00 m below the left mass.

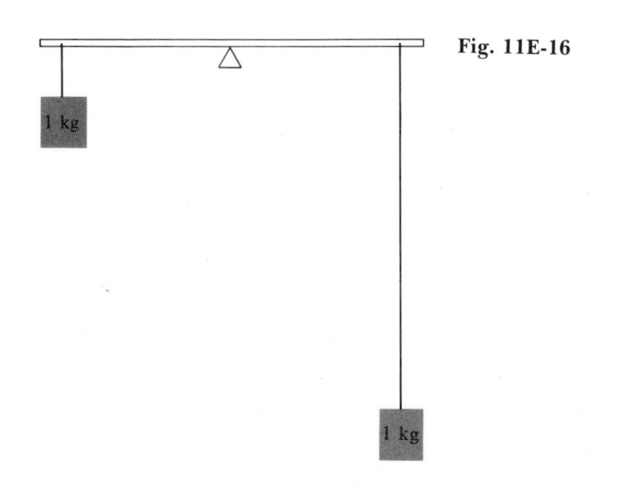

Fig. 11E-16

a. What is the fractional excess in the weight of the right mass over that of the left mass? (This is done most easily by using differentials.)

b. How many milligrams must be placed on the left mass to restore the balance?

11-17. *No equivalence.* Assume that the gravitationally measured mass m_G of a small body is not necessarily equal to its inertially measured mass m_I. Show that if a simple pendulum of length l is made, with the body used for a bob, the period T for small oscillations of the pendulum is given by the equation displayed immediately below Eq. (11-10), which can be written in the form

$$T = 2\pi \sqrt{\frac{l}{g}} \sqrt{\frac{m_I}{m_G}}$$

11-18. *Saturn's rings.* The rings of Saturn consist of myriad small particles, with each particle following its own circular orbit in Saturn's equatorial plane. The inner edge of the innermost ring is about 70,000 km from Saturn's center; the outer edge of the outermost ring is about 135,000 km from the center.

a. Find the orbital period of the outermost particles as a multiple of the orbital period of the innermost particles.

b. Spectroscopic studies indicate that the outermost particles have a speed of 17 km/s. Find the mass of Saturn. Express your result in kilograms and as a multiple of the earth's mass.

11-19. *Mass ratios in the solar system, II.* Newton, without knowledge of the numerical value of the gravitational constant G, was nevertheless able to calculate the ratio of the mass of the sun to the mass of any planet, provided the planet has a moon.

a. Show that for circular orbits

$$\frac{M_s}{M_p} = \left(\frac{R_p}{R_m}\right)^3 \left(\frac{T_m}{T_p}\right)^2$$

where M_s is the mass of the sun, M_p the mass of the planet, R_p the distance of the planet from the sun, R_m the distance of the moon from the planet, T_m the period of the moon around the planet, and T_p the period of the planet around the sun

b. If the planet is the earth, $R_p = 1.50 \times 10^8$ km, $R_m = 3.85 \times 10^5$ km, $T_m = 27.3$ days, and $T_p = 365.2$ days. Calculate M_s/M_p.

11-20. *Period of a low-altitude satellite.*

a. Find the orbital period of a satellite in a circular orbit just above the surface of a spherical planet of mass M and radius r.

b. Rewrite the result of part a in terms of the average density $\langle\rho\rangle \equiv 3M/4\pi r^3$ to show that for a given average planetary density the satellite's orbital period is independent of the size of the planet.

11-21. *Reduced mass via fictitious force.* In the inertial reference frame O of Fig. 11E-21, two bodies of mass m_1 and m_2 exert gravitational forces on each other along the line of $\mathbf{r}$. These forces are $-F\hat{\mathbf{r}} = m_1\mathbf{a}_1$ and $F\hat{\mathbf{r}} = m_2\mathbf{a}_2$. Studying the system from a noninertial reference frame

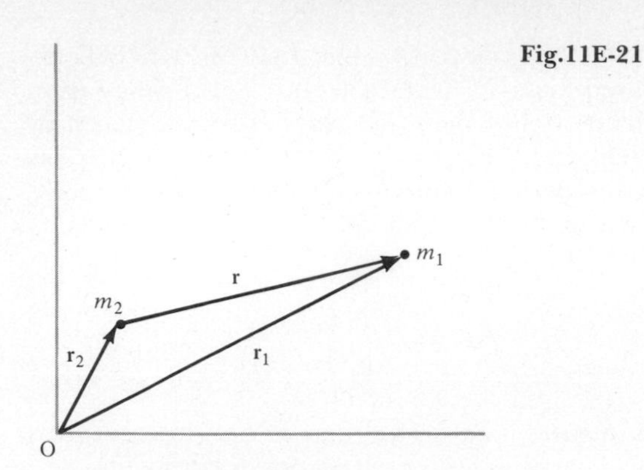

Fig.11E-21

O attached to body 2 requires the introduction of a fictitious force $-m_1\mathbf{a}_2$ acting on body 1. Show that this leads to Eq. (11-18).

11-22. *Harmonic motion of a two-body system.* Two bodies of unequal masses m_1 and m_2 are attached to the ends of an unstretched spring of negligible mass, length l and force constant k. See Fig. 11E-22. The system is placed on a frictionless table, and the bodies are brought closer together by compressing the spring. The bodies are released simultaneously. Since the system is isolated, its center of mass remains fixed.

Fig. 11E-22

a. What is the force constant k_1 of the part of the spring between the center of mass and body 1?

b. What is the period of oscillation T_1 of body 1?

c. Repeat parts *a* and *b* for body 2.

d. Show that the period of oscillation of the system can be obtained by using the reduced mass of the system.

11-23. *Reduced mass and the internal kinetic energy of a two-body system.* Two particles of mass m_1 and m_2 constitute an isolated system in rotation with common angular velocity ω about a fixed center of mass C. The distance from particle 1 to C is r_1; that from particle 2 to C is r_2.

a. What is the kinetic energy of particle 1? Of particle 2? What is the total kinetic energy of the system?

b. Show that the total kinetic energy equals $\frac{1}{2}\mu\omega^2r^2$, where μ is the reduced mass of the system and r is the distance between the two particles. This confirms that the kinetic energy of the system can be obtained by regarding one particle as fixed and the other as rotating about the fixed one, provided the mass assigned to the moving particle is the reduced mass.

11-24. *A central force proportional to $1/r$.* Consider an attractive force which is central but inversely proportional to the first power of the distance. Prove that if a particle is in a circular orbit with such a force, its speed is independent of the orbital radius, but its period is proportional to the radius.

11-25. *Energy of motion vs. energy of elevation.*

a. If a satellite is raised to a height h above the earth's surface and then placed in a circular orbit, what is the ratio $K/\Delta U$ of its kinetic energy to the increase in potential energy when it is raised from the surface?

b. For an orbit whose height is $h = R_e$, the radius of the earth, what is the numerical value of $K/\Delta U$?

11-26. *The Bohr atom.* A hydrogen atom consists of an electron of a very small mass and a massive proton. In Bohr's model of 1913, the electron revolves about the proton in an orbit of radius R. To first approximation, the proton can be considered as fixed. There is an electric force of attraction between the two of magnitude given by k/R^2, where k is a constant.

a. What is the expression for the potential energy of the system in terms of R?

b. For a circular orbit, what is the kinetic energy K of the electron? What is the total mechanical energy E of the system?

c. For a circular orbit, the angular momentum of the electron about the proton is given by the expression $l = mvR$, where m is the mass of the electron and v is its speed. What is the expression for R in terms of l, m, and k? What is the value of E in terms of these quantities?

d. Consider two circular orbits for which $l_2 = 2l_1$. What is the ratio R_2/R_1? E_2/E_1?

Group C

11-27. *Tied together.* Two earth satellites, with masses M_1 and M_2, are linked by a tether of length L, as shown in Fig. 11E-27. The satellites are traveling in circular orbits of radius R_1 and $R_2 = R_1 + L$.

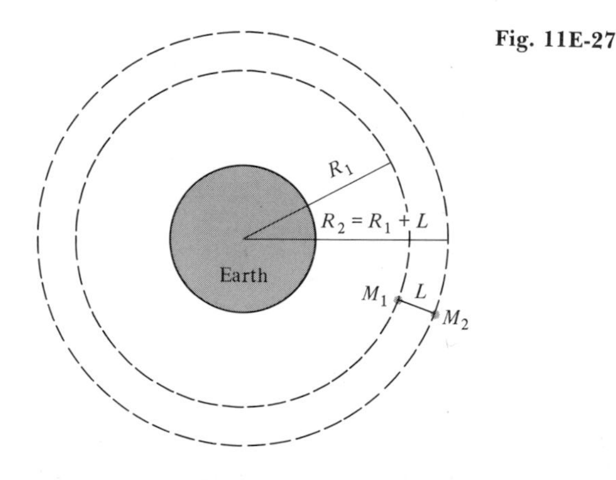

Fig. 11E-27

a. Find the (common) orbital period of the linked pair.

b. Determine the tension in the tether.

c. Evaluate your results for the case of an astronaut at the end of a 10-m tether attached to a craft comparable to Skylab: $M_1 = 50{,}000$ kg, $M_2 = 70$ kg, $L = 10$ m, and $R_1 = 6800$ km. Find the orbital period in minutes, and express the tension in newtons and also as a multiple of M_2g.

d. Evaluate the required tension for two large craft linked by a 1.0-km tether: $M_1 = M_2 = 50{,}000$ kg, $R_1 = 6800$ km, $L = 1.0$ km.

11-28. *Roche's limit.* A planet of mass M has a satellite in an orbit of radius R. The density of the satellite is ρ_s.

a. Show that the satellite will be able to hold itself together by internal gravitational forces only if the orbit radius R is greater than a certain minimum distance R_r, which is given approximately by $R_r = (3/2\pi)^{1/3}(M/\rho_s)^{1/3}$. This minimum orbit radius is called **Roche's limit.** If the orbit radius is smaller than Roche's limit, the satellite must hold itself together by nongravitational forces.

b. The orbit radius of the moon is increasing very slowly, owing to the effects of friction resulting from tidal forces. Find the value of Roche's limit for the moon. Take the average density of the moon to be $\rho_s = 3.4 \times 10^3$ kg/m^3.

c. Where in the solar system can you find observational confirmation for the theory you have developed in part *a*?

11-29. *A binary star system.* A pair of stars close to each other revolve about their common center of mass.

a. Using their reduced mass, show that

$$M + m = \frac{4\pi^2 R^3}{GT^2}$$

where M is the mass of one star, m the mass of the other, R the distance between the stars, and T their common period of revolution.

b. If M_s is the mass of the sun, R_s the radius of the earth's orbit, and T_s the earth's period of revolution, show that

$$\frac{M + m}{M_s} = \left(\frac{R}{R_s}\right)^3 \left(\frac{T_s}{T}\right)^2$$

c. For the double star 70 Ophiuchi, $R = 23R_s$ and $T = 88$ yr. Calculate $(M + m)/M_s$, the ratio of the mass of 70 Ophiuchi to that of the sun.

11-30. *Peak altitude of a satellite.* A satellite is launched from the surface of the earth by firing it horizontally at a speed of 10 km/s. Assume that the effect of air resistance is negligible.

a. Apply the laws of conservation of angular momentum and mechanical energy to find the satellite's maximum distance above the earth's surface.

b. If the satellite were fired vertically upward with the same speed, what would be its maximum height?

c. Account for the difference between the results of parts *a* and *b*. (Note that $GM_e = gR_e^2$, where M_e is the mass of the earth and R_e is its radius.)

11-31. *Elliptical orbits, I.*

a. An earth satellite in an elliptical orbit is 250 km above the earth's surface at its closest point, and its speed there is 9 km/s. What is its height above the earth's surface at its farthest point? (Use conservation of mechanical energy and angular momentum, and see the note in Exercise 11-30*c*.)

b. Let γ and δ be the semimajor and semiminor axes of the ellipse. Let c be the distance of the focus from the center of the ellipse. Calculate the values of γ, δ, and c for the ellipitical orbit in part *a*.

11-32. *Elliptical orbits, II.* Figure 11E-32 represents the elliptical orbit of a planet about the sun. The major axis AA' has length 2γ, the minor axis BB' has length 2δ, and the sun lies at one of the foci, F, a distance c from the center C of the ellipse.

a. When the planet is located at aphelion A, its velocity $\mathbf{v}$ is perpendicular to the major axis. The velocity $\mathbf{v}'$ at perihelion A' is likewise perpendicular to the major axis. Call the mass of the planet m, and use the principle of conservation of angular momentum l to write an equation relating the quantities of m, v, v', c, and γ.

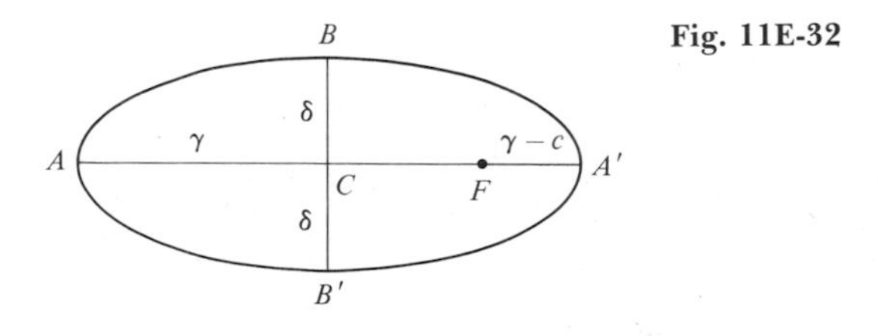

Fig. 11E-32

b. In terms of these quantities and the gravitational constant G, what is the total mechanical energy of the sun-planet system when the planet is at perihelion? At aphelion? What is the relation between these two energies?

c. From the results of parts *a* and *b*, show that the total energy is $E = -GMm/2\gamma$, where M is the mass of the sun, and that therefore E depends on only the length of the major axis of the ellipse, and not on its shape.

d. The area of an ellipse is given by $\pi\gamma\delta$. Use Eq. (11-15) to express Kepler's second law in the form $\pi\gamma\delta/T = l/2m$, where T is the period of the planet.

e. Another general property of ellipses is the relation $\delta^2 = \gamma^2 - c^2$. Show that $T^2 = 4\pi^2\gamma^3/GM$, so that the period of a planet depends on only the major axis of its orbit.

11-33. *Transfer orbit.* An astronaut who is in a circular orbit about the earth at 7000 km from its center, wishes to dock with a space station which is in a circular orbit about the earth at 10,000 km from its center. See Fig. 11E-33. He fires booster rockets briefly, to increase his speed so as to arrive at the point S_2 on the space station orbit where it is on the opposite side of the earth from the astronaut's position when he fires the rockets. He also wishes to approach S_2 along the tangent to the space station orbit.

a. What must be his speed after the completion of the burning?

b. What was the increase in his speed over the short rocket burning period?

c. With what speed does he arrive at S_2?

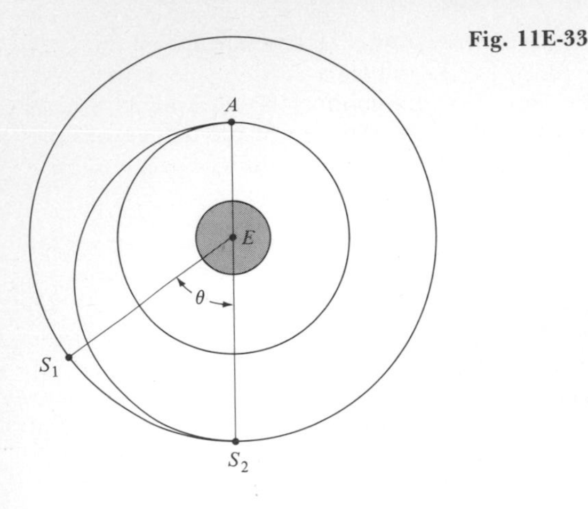

Fig. 11E-33

d. What must he do to dock with the space station?

e. Find the value of θ giving S_1, the position of the space station relative to S_2, when the astronaut fires his booster rockets.

(Hint: See Exercise 11-32, parts *c*, *d*, and *e*; also note that $GM_e = gr_e^2$, where M_e is the mass of the earth and r_e its radius.)

11-34. *Satellite watch.* A satellite is placed in a circular orbit traveling from west to east in the plane of the equator of the earth at an altitude of 800 km. How long will it remain above the horizon at any one place?

Numerical

11-35. *A numerical test of Kepler's first law.* Run the central force program with initial conditions and parameters as in Example 11-9, but use $(dy/dt)_0 = 1.5\ \pi$ (in AU/yr). Sufficient accuracy will be obtained if you take $\Delta t = 1/(4 \times 52) = 1/208$ (in yr), and plot every fourth point. (While plotting, pick some calculational cycle arbitrarily and record the numerical values of x, y, dx/dt, and dy/dt stored in the registers in that cycle. These data will be employed in another exercise.) Use your plot to test Kepler's first law.

11-36. *A numerical test of Kepler's second law.* Use the plot obtained in Exercise 11-35 to test Kepler's second law. Employ the geometrical procedure of Fig. 11-15.

11-37. *A numerical test of angular momentum conservation.* Use the values of x, y, dx/dt, and dy/dt from Exercise 11-35 for the initial calculational cycle, and for the arbitrary calculational cycle for which you recorded these values, to test the conservation of angular momentum in central force motion. Employ Eq. (11-32).

11-38. *A numerical test of Kepler's third law.* Combine the results obtained in Exercise 11-35 for the semimajor axis of the orbit and the orbital period, with those listed in a row of Table 11-2, to test Kepler's third law.

11-39. *A numerical test of energy conservation.* Use the values of x, y, dx/dt, and dy/dt from Exercise 11-35 for the initial calculational cycle, and for the arbitrary calculational cycle for which you recorded these values, to evaluate the potential and kinetic energy of the planet at the two points in its orbit. Then use these data to test the conservation of total mechanical energy. Explain the significance of the sign of the total energy.

11-40. *Orbit of a comet.* A comet is located at $x = 4$ AU, $y = 3$ AU with velocity components $dx/dt = 4.5$ AU/yr and $dy/dt = 0$. The sun is at the origin. Take $\alpha = GM = 39.5(\text{AU})^3/(\text{yr})^2$

a. Is the total mechanical energy of the sun-comet system positive, zero, or negative?

b. Is this a bound or unbound system?

c. Taking $\Delta t = 0.08$ yr, run the central force program given in the Numerical Calculation Supplement. At several points along the orbit record the values of x, y, dx/dt, and dy/dt stored in the registers. Use these values and Eqs. (11-13) and (11-32) to verify Kepler's second law.

d. Plot the orbit.

e. Repeat parts *a* through *d* assuming gravitational repulsion. While there is no evidence that gravitational repulsion exists, inverse-square *electric* repulsion is just as common as attraction, and the gravitational calculation is essentially the same as the electrical one. To use the central force program for repulsion, the value of α is taken to be negative; specifically $\alpha = -39.5$. Plot the orbit you obtain on the same graph as part *d*.

f. Compare the position of the center of force in relation to the orbit in the two cases.

11-41. *Numerical construction of a hyperbolic orbit.* Run the central force program as in Example 11-10, but use $(dy/dt)_0 = 2(2\pi)$ (in AU/yr). (While plotting, pick some calculational cycle arbitrarily and record the numerical values of x, y, dx/dt, and dy/dt stored in the registers in that cycle. These data will be employed in another exercise.) Compare the hyperbolic trajectory you obtain with the parabolic one obtained in Example 11-10 for $(dy/dt)_0 = \sqrt{2}(2\pi)$ (in AU/yr). Can you make a geometrical test to show that your trajectory actually is a hyperbola?

11-42. *Angular momentum conservation in a hyperbolic orbit.* Use the values of x, y, dx/dt, and dy/dt from Exercise 11-41 for the initial calculational cycle, and for the arbitrary calculational cycle for which you recorded these values, to test the conservation of angular momentum in central force motion. Employ Eq. (11-32).

11-43. *Energy conservation in a hyperbolic orbit.* Use the values of x, y, dx/dt, and dy/dt from Exercise 11-41 for the initial calculational cycle, and for the arbitrary calculational cycle for which you recorded these values, to evaluate the potential and kinetic energy of the celestial body at the two points in its trajectory. Then use these data to test the conservation of total mechanical energy. Explain the significance of the sign of the total energy.

11-44. *Stability test for a circular orbit with a central force of constant magnitude.* Test the stability of circular orbits for an attractive central force obeying the law $F \propto r^0$ (that is, F is a constant). Do this by running the central force program with a perturbation given to the orbiting body as in Example 11-14, but use the value of the parameter β corresponding to this force law. (An example of a system involving the force law is a puck on an air table connected by a string over a swiveling pulley at the center of the table to a suspended weight, as in Fig. 3-30. The force acting on the puck is an attractive central force of essentially constant magnitude.)

11-45. *Stability test for a circular orbit with a cental force proportional to distance.* Test the stability of circular orbits for an attractive central force obeying the law $F \propto r^1$. Do this by running the central force program with a perturbation given to the orbiting body as in Example 11-14, but use the value of the parameter β corresponding to this force law. (An example of a system involving the force law is a puck on an air table connected to one end of a spring whose other end is attached to a fixed swivel at the center of the table. If the spring is very extensible, and of negligible length when unextended, the magnitude of the attractive central force acting on the puck will obey Hooke's law; that is it will be proportional to the distance r from the center.)

11-46. *Stability test for a circular orbit with an inverse fourth power central force.* Test the stability of circular orbits for an attractive central force obeying the law $F \propto r^{-4}$. Do this by running the central force program with a perturbation given to the orbiting body as in Example 11-14, but use the value of the parameter β corresponding to this force law. Compare the spiral trajectory obtained with the one shown in Fig. 11-22, and explain the difference between them.

11-47. *Noncircular orbits with a central force of constant magnitude.* Run the central force program as in Example 11-15, but use the parameter β corresponding to an attractive central force obeying the law $F \propto r^0$. (See Exercise 11-44.) Compare the precession of the orbit with that found for the $F \propto r^{-2.1}$ orbit plotted in Fig. 11-24.

11-48. *Noncircular orbits for a central force proportional to distance.* Run the central force program as in Example 11-15, but use the parameter β corresponding to an attractive central force obeying the law $F \propto r^1$. (See Exercise 11-45). Can you explain why there is no precession of the orbit?

11-49. *Effect of solar wind on a satellite-orbit.* Use the central force program to study the effect of the "solar wind." This is a stream of ionized gas flowing at high speed outward from the sun. One of its effects is to sweep comet tails away from the sun. Another is to disturb the motion of earth satellites. To study the effect on satellites, modify the program in such a way that the number 0.001 is added to the register holding dx/dt in each calculational cycle—above and beyond whatever normally happens to dx/dt in the cycle. This continual addition of a small velocity in the x direction represents the effect of a weak force continually exerted on the satellite in that direction, as seen from a reference frame in which the earth is always at the origin and the sun is always in the negative x direction. Run the program with the following set of initial conditions and parameters: $x_0 = 1$; $(dx/dt)_0 = 0$; $y_0 = 0$; $(dy/dt)_0 = 1$; $t_0 = 0$; $\Delta t = 0.1$; $\alpha = 1$; $\beta = -1.5$. (These simple dimensionless values are used since the AU and the year are not appropriate units of distance and time here.) Can you explain why the orbit shifts in a direction perpendicular to the solar wind?

12 Mechanical Traveling Waves

12-1 TRAVELING WAVES

Imagine a cork floating on the surface of a calm pond. You throw a pebble into the pond at some point remote from the cork and then watch the cork. After a time, you see the cork move up and down. Some of the mechanical energy you originally gave to the pebble has appeared on the cork. This energy is carried from the pebble to the cork by the ripple that spreads over the surface of the pond from where the pebble struck the surface. If you put other corks elsewhere on the surface of the pond, they also will be set into motion by the ripple. A small fraction of the energy carried by the ripple is deposited on any cork that the ripple happens to meet. And even if there are no corks, still *the ripple itself carries energy.* This transport of *energy* by the ripple does not involve the transport of *matter.* That is, particular water molecules do not continually move along with the ripple. Instead, water molecules make a small up-and-down motion as the ripple passes, returning to their original positions after it has passed. What is moving across the pond, and carrying energy, is not water molecules but a *disturbance in the position* of water molecules.

A ripple traveling over the surface of a pond is a particular case of a general phenomenon called a **traveling wave.** A traveling wave in a mechanical system is a disturbance in the positions of the particles of the system from their normal positions, which moves through the system. Energy is transported over long distances by the moving disturbance, even though the particles of the system move only very short distances as the wave passes them, and end up where they started.

The topic of this chapter is traveling waves in mechanical systems. Sound waves traveling through the air are the most important example found in nature. But there are many other important examples, such as seismic waves traveling through the earth.

In many cases our interest in the *energy carried* by a mechanical traveling wave seems secondary to our interest in the *information carried* by the wave. Consider sound waves. Their importance in carrying information from the mouth of a speaker to the ear of a listener is obvious. Nevertheless, if the sound wave did not carry enough energy to set the listener's eardrum into vibration, it could carry no information. Geologists produce seismic waves by detonating explosives buried in the earth. From measurements of the speeds at which the waves travel through the earth along different paths, geologists obtain information which is invaluable in finding oil. However, the method would not work if seismic waves did not carry enough energy to be able to actuate the geologists' detectors. In some cases the amount of energy transported by a mechanical wave is so large that the energy itself is of primary interest. A very loud sound wave is painful because it imparts so much energy to the eardrum. A seismic wave produced by an earthquake is dangerous because of the tremendous amount of energy it carries.

Later in this chapter we apply newtonian mechanics to a simple system for the purpose of *explaining* the observed properties of mechanical traveling waves. But first we must build up the linguistic and mathematical tools that are needed in *describing* these properties. That is, first we must develop **traveling-wave kinematics.** We do so in this section for a wave consisting of a single, isolated disturbance traveling through a system. Such a traveling wave is known as a **wave pulse.** In Sec. 12-2 we extend our considerations to a traveling wave comprising a series of adjacent pulses.

The familiar separation between kinematics and mechanics is particularly useful in connection with wave motion. The reason is that the universe is, literally, filled with waves that travel through nonmechanical systems. These are light waves, radio waves, and the many other members of the family of electromagnetic waves. In most regards the "mechanics" of electromagnetic waves is completely different from the mechanics of waves traveling through mechanical systems. This is because the properties of electromagnetic waves must be explained in terms of laws of nature that are quite distinct from Newton's laws of motion. But the kinematics of electromagnetic waves is exactly the same as the kinematics of mechanical traveling waves, in most of its features. So we will be able to make use of much of what we develop here on a number of occasions later in the book when we study light and other electromagnetic waves.

We begin developing traveling-wave kinematics by considering a wave pulse traveling along a stretched rope, or heavy string. One end of the string is tied to a rigid support. The other end is pulled by the hand of an experimenter, with tension being applied by the hand stretching the string. The experimenter gives the end being held a single, sharp up-and-down jerk, forming a bulge in the string as illustrated in the first part of Fig. 12-1. The bulge does not stay in place. Instead it travels uniformly along the string as a wave pulse, in the manner illustrated in subsequent parts of the figure. A description of this situation is simpler than a description of a ripple traveling over the surface of a pond. As the ripple spreads over the surface of the pond in two dimensions from the point where it is produced, its height diminishes. This is because the energy carried by the ripple is spread over a larger and larger region. But a wave pulse traveling in one dimension along a string does not spread. In fact, experiment shows that the pulse itself maintains the *same shape* as it travels along the string,

Fig. 12-1 A wave pulse being produced in a stretched string. The top diagram shows the string stretched between the fixed support at its right end and the hand pulling on it at its left end. The diagrams below show the position of the hand and the shape of the string at subsequent times. The time interval between successive diagrams is the same.

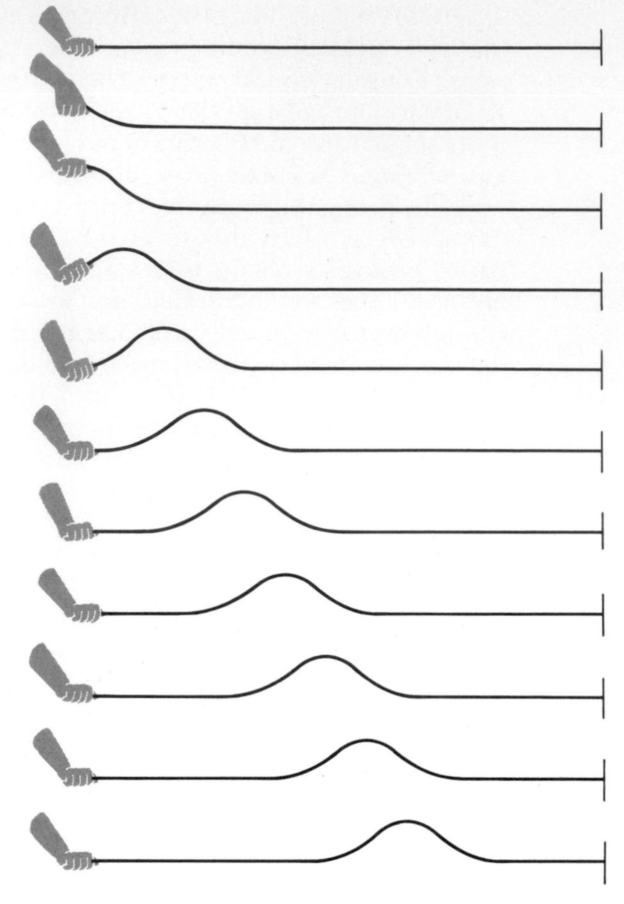

provided that frictional effects are not significant and that the height of the pulse is not so large that the tension in the string is affected significantly by the presence of the pulse. (Experimental evidence is presented in Fig. 12-2.) Another reason why the pulse traveling along the stretched string is simpler to describe is just that there is one less dimension that must be dealt with, and so one less coordinate that must be used.

You should note that what is traveling horizontally along the string is *not* the particles constituting the string. Each of these particles only moves up a little and then back down. This behavior is typical of wave motion. Although they transport energy through the system, the individual particles of the system move very little. In fact, the total displacement of each particle is zero. Furthermore, what motion the particles do go through is not even along the string (the way the pulse moves) but perpendicular to it. When the motion of the particles of the system is everywhere perpendicular to the direction in which the wave travels, it is said to be a **transverse wave.** So the wave pulse moving along the stretched string is a transverse traveling wave.

How can we describe the wave pulse traveling along the stretched string? At any instant the complete shape of the string has a unique and distinct character. Just after the wave pulse is generated, for instance, the string has a bulge at its left end and is otherwise straight. At some *particu-*

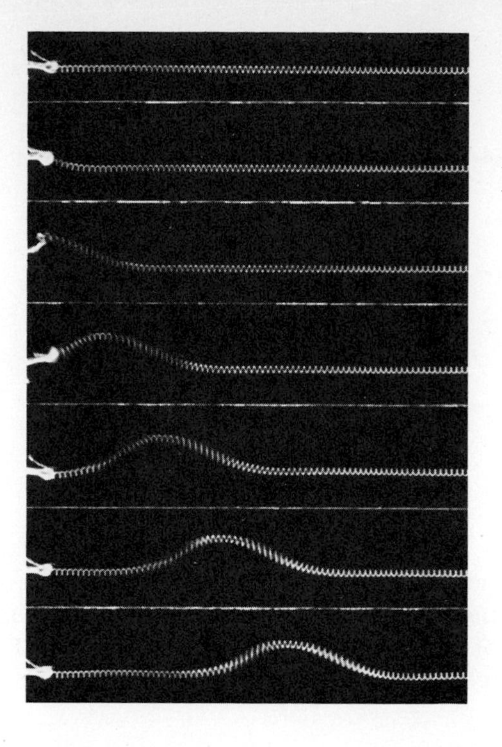

Fig. 12-2 Frames from a motion picture of the generation of a wave pulse in a stretched spring. A spring was used, instead of a string, because it was easier to photograph. (*From PSSC Physics, 2d ed., D. C. Heath, Boston, 1965. Courtesy Educational Development Center.*)

lar time, the complete shape of the string can be specified by giving the position of all the elements of the string in terms of the mathematical function

$$y = f(x) \tag{12-1}$$

In this expression y is the vertical coordinate of an element of string, measured from its undisturbed position at $y = 0$. The location of the element along the string is denoted by its horizontal coordinate x. The specific mathematical form of the function is whatever is required to describe the specific shape of the string at the time being considered. Figure 12-3 depicts such a function. Its graphical representation is, indeed, nothing but a "snapshot" of the string, at the time to which the function applies, with the proper coordinate axes superimposed.

The function in Eq. (12-1) describes the shape of the string at only one instant of time. But the vertical coordinate y of an element of the string whose horizontal coordinate is x also depends on the time t, since the wave pulse moves along the string. Thus, in order to describe the positions of all the elements of the string at *all times,* we need a function which depends on *both x and t.* Symbolically, such a function is written as

$$y = f(x, t) \tag{12-2}$$

and is called a **wave function.**

Fig. 12-3 A plot of the function $y = f(x)$ giving, at some instant of time, the dependence on the coordinate x measured along a string of the transverse coordinate y of elements of the string.

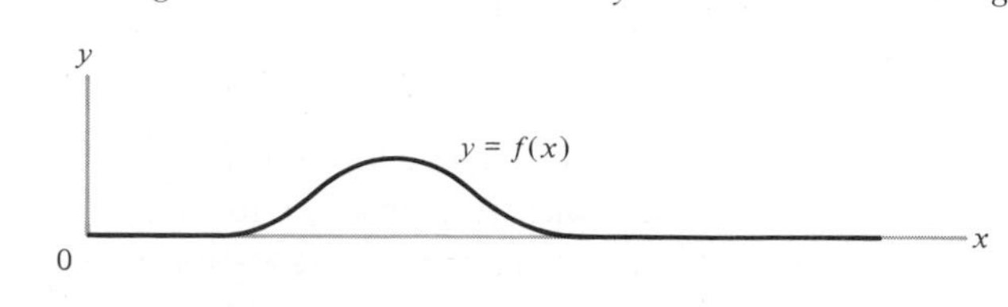

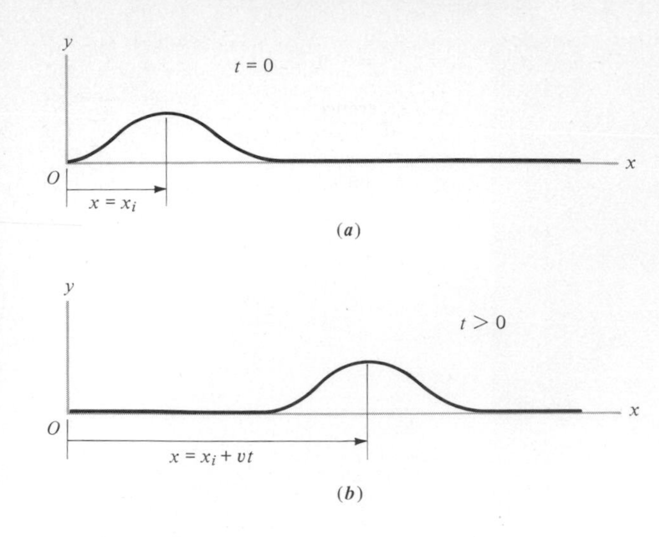

Fig. 12-4 (*a*) A plot of the wave function $y = f(x, t)$ versus x for $t = 0$. It shows a profile at that instant of a wave pulse traveling along a string. (*b*) A plot for the later instant $t > 0$ shows the profile of the wave pulse at that instant. The profile is the same as at the earlier instant, but it has traveled along the string. It moves an amount vt, where v is the velocity of the wave pulse.

We can find a more explicit form for this abstract expression by considering the physical situation in more detail. As time passes, the wave pulse moves uniformly along the string in some direction, say the direction of positive x. It makes perfect sense to describe the motion of the pulse in terms of a **wave velocity** v, whose value is given by

$$v = \frac{dx}{dt} \tag{12-3}$$

Here x is the horizontal coordinate at time t of any characteristic point on the pulse, say its maximum. Be sure to note that v is not the velocity of any material object. *The wave velocity is the velocity of the moving disturbance in the elements of the string, and not the velocity of any element of the string.* The significance of the wave velocity is illustrated in Fig. 12-4. The wave velocity is a signed-scalar quantity v, whose value is a positive constant for the case depicted in the figure.

The next step in finding a more explicit form for a wave function describing a wave pulse traveling along the string is to consider what the string looks like from the points of view of two different observers, one fixed with respect to the string and the other moving with the pulse. This consideration will show us that the variables x and t can enter into the function $f(x, t)$ only in a certain combination. A wave pulse is moving along the string in the positive x direction. An observer O remains fixed at the end of the string where the pulse was produced. She measures the vertical coordinate y of each element of the string with horizontal coordinate x at various times t. The summary of her observations is precisely what we mean by the wave function of Eq. (12-2):

$$y = f(x, t)$$

Another observer O' is moving with respect to O at a constant velocity which he adjusts to be exactly equal to the wave velocity v. That is, O sees O' moving at a velocity just equal to the wave velocity she observes for the pulse. The situation at time $t = 0$ is depicted from the viewpoint of O in Fig. 12-5*a*. At that time O' passes O, and the production of the pulse has just been completed.

Observer O' also watches the wave pulse in the string. His observations are made from the primed coordinate system, which moves along with the wave pulse. This moving coordinate system and the pulse are shown from

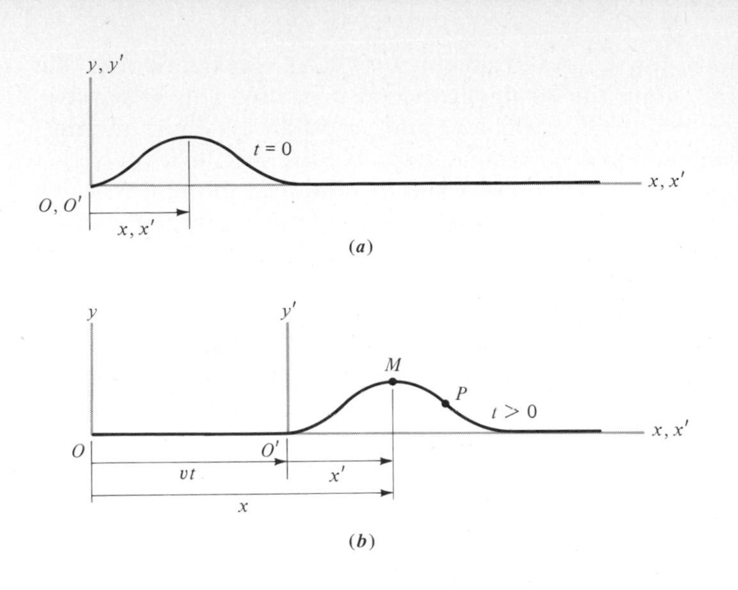

Fig. 12-5 (*a*) A wave pulse propagating along a string with velocity v to the right, as seen at time $t = 0$ in the reference frame of an observer O who is stationary with respect to the string. Also shown is the reference frame of an observer O' who is moving with velocity v. (*b*) The situation at a later time $t > 0$, as seen in the reference frame of O. Both the wave pulse and the reference frame of O' have moved by the same amount vt with respect to O.

the viewpoint of O at the time $t > 0$ in Fig. 12-5*b*. Since O' moves along with the wave pulse, its position does not change from his viewpoint. So the function which summarizes his observations of the vertical coordinates y' of elements of the string depends only on their horizontal coordinates x' measured from the origin of his coordinate system. There is no time dependence because the pulse does not move, as he sees it, and also because the shape of the pulse itself does not change. Thus O' writes the function as

$$y' = f(x') \tag{12-4}$$

In order to compare the observations of O' with her own, O must convert data expressed in terms of x' and y' to data expressed in terms of her own coordinates x and y. Figure 12-5*b* shows that the conversion is given by the two equations

$$x' = x - vt$$

and

$$y' = y$$

Applying the first of these to Eq. (12-4), she has

$$y' = f(x') = f(x - vt)$$

Applying the second, she finds

$$y = y' = f(x - vt)$$

or

$$y = f(x - vt) \tag{12-5}$$

The function on the right side of Eq. (12-2), $y = f(x, t)$, describes the observations of O. The function on the right side of Eq. (12-5), $y = f(x - vt)$, describes the observations of O' in terms of the coordinates used by O. Since O and O' have observed exactly the same phenomenon, these two functions are describing exactly the same thing, and so can be equated to obtain

$$f(x, t) = f(x - vt) \tag{12-6}$$

The *only possible* form for the wave function $y = f(x, t)$, which describes the wave pulse moving along the string in the positive direction with wave velocity v, is $y = f(x - vt)$. The variables x and t must always enter into this function in the combination $x - vt$, no matter what the specific form of the function is. Any other combination of x and t would lead to the physically impossible situation where different observers could not reconcile their observations. For instance, this would be the case for the dimensionally consistent, but physically impossible, combination $x - 2vt$.

The *specific* form of the function of $x - vt$ which must be used depends on the specific shape of the wave pulse that the function describes. An example of such a function (to be used early in Sec. 12-2 to describe a series of adjacent pulses extending above and below the undisturbed string) is $y = A \cos[B(x - vt)]$, where A and B are constants.

Because of the importance of Eq. (12-6), and because its interpretation may cause you some difficulty at first, it is worthwhile to go through an argument which verifies that the equation actually does describe a wave pulse that travels in the positive x direction, at the wave velocity v, without changing its shape. A particular point on the pulse—say the point marked P in Fig. 12-5b—is a point where the function $f(x - vt)$ always maintains a particular value y. If the function is always to maintain a particular value, it is necessary that the argument $x - vt$ of the function always maintain a particular value. That is, for any particular point on the wave pulse we must have

$$x - vt = \text{constant}$$

Different points are associated with different values of the constant. For instance, the point marked M in the figure is associated with the constant whose value leads to the function $f(x - vt)$ having a maximum value. As the time t increases, the coordinate x of a particular point on the pulse must become more positive in order to compensate for the fact that the quantity $-vt$ becomes more negative, since v has a positive value. This must be so if $x - vt$ is to maintain a constant value. Thus the point on the pulse moves in the positive x direction. You can calculate its velocity dx/dt by taking the time derivative of each term in the equation $x - vt =$ constant, remembering that v is constant, to obtain

$$\frac{dx}{dt} - v = 0$$

or

$$\frac{dx}{dt} = v$$

Thus the point moves with a velocity v that is, according to the definition of Eq. (12-3), the wave velocity. Since you will obtain the same result for dx/dt no matter what the value of the constant, *every* point on the pulse moves with that same velocity. As a consequence, the pulse maintains its shape as it moves.

If a wave pulse is moving along a stretched string in the *negative* direction, then its wave velocity v will have a negative value. Let us write it as $v = -|v|$, where $|v|$ represents the *speed* of the wave, and then substitute into Eq. (12-6). We obtain the wave function $f(x, t) = f(x + |v|t)$. From the viewpoint of an observer O fixed with respect to the string, this describes a wave traveling along the string in the negative x direction at speed $|v|$. You can check this wave function in either of two ways. First, you can derive it directly from an argument like the one leading to Eq. (12-6), but with the wave pulse and the observer O' moving in the negative x direction. Second, you can go through an argument like the one in small print above.

It is worthwhile to write an expression which is completely equivalent to Eq. (12-6), but which makes more apparent the fact that the sign of the numerical value of the second term in the argument of the wave function depends on the direction in which the wave travels. The expression is

$$f(x, t) = f(x \mp |v|t) \tag{12-7}$$

The minus sign is used for a wave traveling in the positive x direction ($v > 0$), and the plus sign is used for a wave traveling in the opposite direction ($v < 0$).

12-2 WAVE TRAINS

We have spoken so far of a single traveling wave pulse, produced by a single cycle of agitation of the end of the string. There are two excellent reasons for considering the case of repeated agitations, where an oscillating source moves the end of the string up and down continually.

One reason is that there are very many important physical situations in which waves are excited by an oscillating system. For example, most sources of sound and of radio waves are of this type.

Another reason is that we will see in Chap. 13 that *any* wave, no matter how complicated its shape, can be built up of separate waves, each of which is produced by a source (that is, a device) executing harmonic motion of a particular frequency. Thus an understanding of the relatively simple situation in which one end of a stretched string is set into motion by such an oscillator is a long step toward the analysis of waves of arbitrary shape.

For these reasons, we consider the traveling waves produced in a stretched string whose movable end is shaken repeatedly up and down by a source which itself moves vertically in harmonic motion. Using Eq. (6-27), we can write the vertical position y of the end of the string connected to the source as

$$y = A \cos(2\pi\nu t) \tag{12-8}$$

Here $y = 0$ corresponds to the harmonic oscillator being at its equilibrium position, which is also the undisturbed position of the end of the string. The amplitude A is the maximum displacement of the oscillator from that position. The frequency ν is the number of times per second that the oscillator goes through its cycle of oscillation. And the zero of the time scale is chosen so as to make $y = A$ at $t = 0$.

How does the stretched string behave as it is shaken by the source? Let us assume that the string is very long, so that we need not worry about what happens at the other end. Each half-cycle of the harmonic oscillator is very much like the single shake we have already discussed, which put a wave pulse into the string. Here, however, each upward shake is immediately followed by a downward shake, and so on. Each pulse travels down the string just as before, followed immediately by the next one. The result is that there is a series of pulses traveling down the string, called a **wave train.** If we take a snapshot of the string at some particular instant, it will look like Fig. 12-6*a*. Since the excitation of the end of the string is a repetitive process, a point on the string at some characteristic location on a wave—say, a maximum—will have exactly the same coordinate y as any other

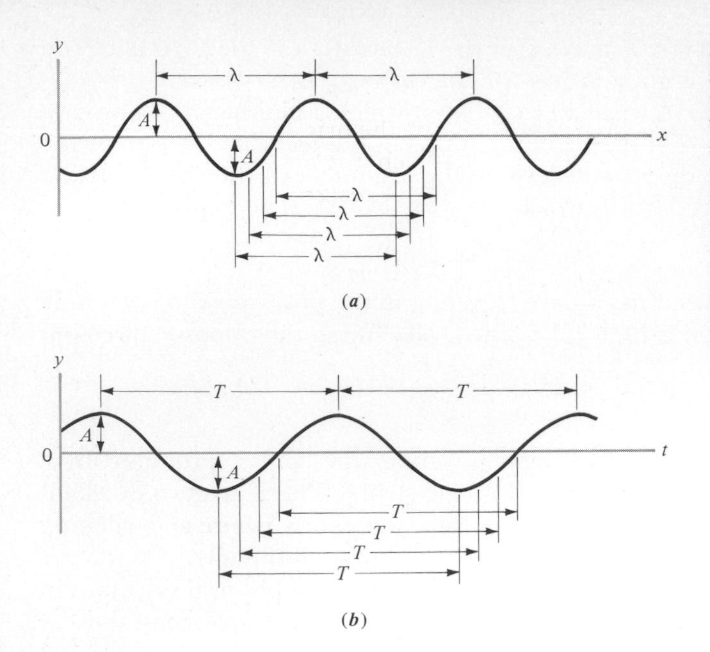

Fig. 12-6 (*a*) A continuous, sinusoidal wave propagating along a string. The transverse coordinate y of elements of the string is plotted versus the coordinate x measured along the string at a fixed time t. This plot represents the result of taking a "snapshot" of the string at one time t, over a range of values of x. The plot defines the wavelength λ of the wave and its amplitude A. (*b*) The same wave with the transverse coordinate y plotted versus the time t for the element of string at a fixed coordinate x. This plot represents the result of taking a "movie" of the string at one location x, over a range of values of t. The plot defines the period T of the wave and its amplitude A. The reciprocal of T is the frequency ν of the wave; that is, $\nu = 1/T$.

point at a corresponding location on the wave train. Furthermore, the corresponding locations are spaced at equal distances along the string. This distance is called the **wavelength** λ (Greek lambda). The figure shows that it does not matter at what point on the wave we begin the measurement of the wavelength. All that is necessary is that we end at the next point having a corresponding location on the wave train.

Figure 12-6*a* illustrates what the entire string, extending over space, looks like at a fixed instant of time t. How does a single particle of the string at a fixed horizontal coordinate x appear to move when it is observed over an extended time? We can imagine making a motion picture of a little piece of the string, using a vertical slot in a mask in front of the string to select a fixed value of x. Then we use the film to measure the y coordinate of the piece as a function of the time t as the piece oscillates up and down. A plot of the results would look like Fig. 12-6*b*. Here the horizontal axis is a t axis, rather than an x axis. But otherwise there is a very strong resemblance between Fig. 12-6*b* and 12-6*a*. The reason is that there is a strong connection between the snapshot of a wave train and the motion picture of a particle of the system through which the wave train travels. It takes a time T (the period of the particle's oscillation) for a particle of the string to go through one complete cycle of displacements y in the fixed-x movie of Fig. 12-6*b*. Analogously, it takes a length λ (the wavelength of the wave train) for the wave train to go through one complete cycle of displacements y in the fixed-t snapshot of Fig. 12-6*a*.

Since the source oscillator is moving in a sinusoidal fashion, and since the wave it generates in the string moves off in the positive x direction at a constant velocity, *the string itself has at any instant the shape of a sinusoidal curve.* The proof of this statement is as follows. At a time t, the y coordinate—given by $y = f(x - vt)$—of a particle of the string with a certain x coordinate is the same as the y coordinate—given by $y = f(0 - vt')$—of a particle at the origin $x = 0$ at an earlier time $t' = t - x/v$. This is true because

$$f(x - vt) = f(0 - vt') \tag{12-9a}$$

if $$t' = t - \frac{x}{v} \qquad (12\text{-}9b)$$

But the y coordinate of a particle of string at the origin is determined by the y coordinate of the source oscillator, to which it is firmly attached. At time t', the latter is given by Eq. (12-8) to be $y = A\cos(2\pi\nu t')$. Thus we have the relation

$$f(0 - vt') = A\cos(2\pi\nu t')$$

Using Eq. (12-9a) and then Eq. (12-9b), we have from this relation

$$f(x - vt) = A\cos(2\pi\nu t') = A\cos\left[2\pi\nu\left(t - \frac{x}{v}\right)\right]$$

Since multiplying the argument of a cosine by -1 does not affect the value of the cosine, this equation for $y = f(x - vt)$ can be rewritten in the form

$$y = A\cos\left[\frac{2\pi\nu}{v}(x - vt)\right] \qquad (12\text{-}10)$$

At any instant—that is, for any fixed value of t—the vertical position y of a particle of the string is a sinusoidal function of its horizontal position x, which is what we set out to prove. Incidentally, notice that Eq. (12-10) provides a specific example of a function which possesses the general form required for a traveling wave by Eq. (12-6): $f(x, t) = f(x - vt)$.

Equation (12-10) can be used to obtain a very important connection between the frequency ν of a wave train, its wave velocity v, and its wavelength λ. As defined by Fig. 12-6a, λ is the distance along the x axis in which the y coordinate passes through one complete cycle of oscillation while t remains fixed. Consider Eq. (12-10) for any fixed value of t. For y to go through one complete cycle, the quantity $2\pi\nu x/v$ must increase or decrease by the amount 2π as x increases by the amount defined to be λ. That is,

$$\frac{2\pi\nu x}{v} \pm 2\pi = \frac{2\pi\nu(x + \lambda)}{v}$$

Multiplying through by $v/2\pi$ and transposing, we have

$$\nu(x + \lambda) = \nu x \pm v$$

Canceling and solving for v produce

$$v = \pm\nu\lambda \qquad (12\text{-}11a)$$

The dual signs allow for the velocity v to have either a positive value or a negative value. In terms of the speed $|v|$, the equation can be written as

$$|v| = \nu\lambda \qquad (12\text{-}11b)$$

The speed $|v|$ of a wave equals the product of its frequency ν and its wavelength λ.

While we have derived Eq. (12-11b) for the particular case of a transverse sinusoidal wave train, and have considered in particular a wave train traveling along a string, the equation actually is valid for *any* type of repetitive wave train traveling through *any* medium. This can be proved by an argument that you may consider to be simpler than the one just given. Imagine you are at a fixed position, watching any repetitive wave train moving by you to the right. At a certain instant one maximum of the wave train is at your position, and an adjacent maximum is a distance to the left of your position equal to λ, the wavelength of the wave. In a time T the points on the wave at your fixed position go through one complete cycle of oscillation, where T is the period of the oscillation. And as this happens, the maximum of the wave that had been to your left moves up to your position. The maximum moved a distance λ in a time T, so its speed is $|v| = \lambda/T$. This is the speed of the wave itself. But the period T of the oscillation equals the reciprocal of ν, the frequency of the oscillation. Setting $T = 1/\nu$, you obtain $|v| = \nu\lambda$, in agreement with Eq. (12-11*b*).

Examples 12-1 and 12-2 apply Eq. (12-11*b*) to a mechanical traveling wave and to a nonmechanical traveling wave.

EXAMPLE 12-1

The frequency of the musical note called middle C is 261.6 Hz. Find the wavelength of the sound wave traveling through air when a flutist plays middle C (the lowest note on the flute). The speed of sound in air at room temperature (20°C) has the value $|v| = 344$ m/s.

■ From Eq. (12-11*b*) you have

$$\lambda = \frac{|v|}{\nu}$$

The numerical value is

$$\lambda = \frac{344 \text{ m/s}}{261.6 \text{ s}^{-1}}$$

or

$$\lambda = 1.31 \text{ m}$$

EXAMPLE 12-2

The frequency and wavelength of the radio wave emitted by a broadcasting station are measured to be $\nu = 1.200 \times 10^6$ Hz and $\lambda = 2.498 \times 10^2$ m. Evaluate the speed of radio waves.

■ Equation (12-11*b*) shows the speed to be

$$|v| = \nu\lambda = 1.200 \times 10^6 \text{ s}^{-1} \times 2.498 \times 10^2 \text{ m}$$

or

$$|v| = 2.998 \times 10^8 \text{ m/s}$$

Now we will express the sinusoidal wave function of Eq. (12-10), used to describe a sinusoidal wave train, in several simpler and very useful forms. Let us consider such a wave train traveling in the direction of positive x so that Eq. (12-11*a*) gives for its velocity v the positive value $v = \nu\lambda$. Substituting this value into Eq. (12-10), we obtain

$$y = A \cos\left(2\pi \frac{x}{\lambda} - 2\pi\nu t\right) \tag{12-12}$$

If we investigate the wave as it looks frozen in time—that is, with t arbitrary but fixed—then the term $2\pi\nu t$ will have a fixed value. Writing it as $2\pi\nu t = -\delta$, we have

$$y = A \cos\left(2\pi \frac{x}{\lambda} + \delta\right) \tag{12-13}$$

This equation has a directly evident physical meaning. When x varies in such a way as to change the value of the fraction x/λ by 1, the argument of the cosine function changes by 2π, and the cosine function itself passes through one cycle. This will be true regardless of the initial value of the argument of the cosine. The arbitrariness is expressed mathematically by the presence of the phase constant δ. It can have any value at all, positive or negative. Its value determines the "starting point" of the wave function, that is, the value of x for which the value of y is A.

Equation (12-13) completely describes the y coordinates of all points on the wave, provided it is frozen in time. That is, it is the algebraic equivalent of the snapshot of Fig. 12-6*a*. It is therefore called a **time-independent wave function.**

We can begin again with Eq. (12-12), this time fixing x and allowing t to vary. An argument completely analogous to the one which led to Eq. (12-13) leads to the expression

$$y = A\cos(2\pi\nu t + \delta') \tag{12-14}$$

in which δ' is a phase constant. Now the frequency ν is related to the period T by the definition in Eq. (6-5):

$$\nu = \frac{1}{T}$$

Substituting this value of ν into Eq. (12-14) yields the **space-independent wave function**

$$y = A\cos\left(2\pi\frac{t}{T} + \delta'\right) \tag{12-15}$$

Equation (12-15) completely describes the y coordinate of a particular point on the string for all times. It is thus the algebraic equivalent of the "movie through a slot" of Fig. 12-6*b*. The equation tells you that the cosine function goes through one complete cycle when t varies so as to change the fraction t/T by 1. Just as in the discussion leading to Eq. (12-13) for the time-independent case, this statement does not depend on the initial value of t.

Let us now return to the more general wave function, Eq. (12-12),

$$y = A\cos\left(\frac{2\pi}{\lambda}x - 2\pi\nu t\right)$$

from which the time- and space-independent wave functions [Eqs. (12-13) and (12-15), respectively] were derived. We can cast it into a more symmetrical form by using again the relation $\nu = 1/T$ to obtain

$$y = A\cos\left[2\pi\left(\frac{x}{\lambda} - \frac{t}{T}\right)\right] \tag{12-16}$$

This expression still does not possess, however, the most general form possible for a function representing sinusoidal waves traveling in the positive x direction. That is, it does not represent mathematically all such possible waves. In particular, it represents only those waves for which the y coordinate of the point on the string at the origin happened to have the maximum possible value A at time $t = 0$. As we have done before, we can remove this rather artificial constraint on the choice of the zeros for x and t by inserting a phase constant δ, the value of which may be adjusted to specify any desired coordinate y of the string, in the range $-A \leq y \leq A$, at any one location and time. We thus arrive at the wave function

$$y = A \cos\left[2\pi\left(\frac{x}{\lambda} - \frac{t}{T}\right) + \delta\right] \qquad \text{for } v > 0 \qquad (12\text{-}17)$$

With proper choices of A, λ, T, and δ, this will describe *any* sinusoidal wave traveling in the positive x direction. Note the symmetry of Eq. (12-17) with respect to x and t. Both of them are numerators of dimensionless fractions whose denominators express constant physical properties of the wave.

For a wave traveling in the negative x direction, the velocity v has the negative value $v = -\nu\lambda$. If you go again through the argument leading to Eq. (12-17), you will see that the effect of this sign change is to change the minus sign in that equation to a plus sign, so that it becomes

$$y = A \cos\left[2\pi\left(\frac{x}{\lambda} + \frac{t}{T}\right) + \delta\right] \qquad \text{for } v < 0 \qquad (12\text{-}18)$$

There is a more compact way of writing Eqs. (12-17) and (12-18), which will come in handy later. Note first that in both these equations the variable t is multiplied by the factor $2\pi/T$. Since $1/T = \nu$, we can rewrite this factor in the form $2\pi/T = 2\pi\nu$. Then we can introduce the angular frequency ω, defined in Eq. (6-26) to be $\omega \equiv 2\pi\nu$, so that the factor becomes

$$\frac{2\pi}{T} = \omega \qquad (12\text{-}19)$$

Physically, the angular frequency is the number of times that the source (or any point on the string) goes through its oscillation cycle in 2π s, just as the frequency ν is the number of times that the source goes through its oscillation cycle in 1 s. Then we define the **wave number** k by the equation

$$\frac{2\pi}{\lambda} \equiv k \qquad (12\text{-}20)$$

The wave number is the number of waves contained in 2π m, just as the angular frequency is the number of cycles contained in 2π s. In terms of the wave number and the angular frequency, Eqs. (12-17) and (12-18) can be rewritten as

$$y = A \cos(kx \mp \omega t + \delta) \qquad (12\text{-}21)$$

The minus sign is used for a wave traveling in the positive x direction ($v > 0$), and the plus sign is used for a wave traveling in the opposite direction ($v < 0$).

12-3 THE WAVE EQUATION

Up to this point we have concentrated on *describing* some of the important properties of traveling waves. Now we undertake the task of *explaining* the origin of these properties. Since the waves we are considering are waves traveling through mechanical systems, we can expect that the explanation will be based on the laws governing the behavior of mechanical systems—Newton's laws of motion. These laws predict the motion of a particle, not a wave. But a wave in a mechanical system—which will be exemplified by a stretched string—is just an organized motion traveling through the particles comprising the string. So we can use Newton's laws to treat the motion of these particles or, more practically, of sets of adjacent particles that form very short segments of the string. Each segment moves under the influence of the forces exerted on it by the segments on either side. By studying their motion we should be able to obtain predictions concerning the wave.

First let us look at what happens in a qualitative way. An experimenter holds the end of a long, uniform string extending to the right along the x axis, maintaining tension in the string by pulling on its end. The experimenter then begins to generate a wave pulse in the end of the string by rapidly moving his or her hand upward in the y direction. This applies a force in that direction to the left end of the segment of string next to the hand. But since the string segment has mass, and therefore inertia, its acceleration is finite, and so it can begin to move upward only gradually, in response to the force applied to it. So some time must pass before the segment is displaced upward an appreciable amount. When it has been, it in turn exerts a force in the upward direction on the left end of the next segment of string. After more time passes, this segment is displaced upward and then exerts an upward force on the left end of the next segment. The process continues. Its net effect is that the "leading edge" of a wave travels at a certain speed along the string. If the experimenter's hand starts to move down to the x axis immediately after it finishes moving up, the hand starts to exert a force on the segment of string next to the hand in the downward direction. After enough time has passed for the inertia of that segment to be overcome, the segment is displaced downward and then exerts a downward force on the next segment. After more time passes, this segment is displaced downward, and then it exerts a downward force on the next segment. The downward motion propagates from segment to segment at the same speed as the speed at which the leading edge of the wave travels. It constitutes the "trailing edge" of the single wave pulse which the experimenter generates in the string.

It is said that the wave **propagates** through its **propagation medium,** the string. The process involves two key factors: the *force* that each particle of the medium exerts on its neighbor, which tends to make the neighbor follow its own motion, and the *mass* of each particle, which tends to prevent the neighbor from following the motion instantly.

Our qualitative analysis suggests that the speed of the pulse traveling along the string will depend on the tension in the string, since the strength of the force which each segment of the string exerts on the neighboring segment increases as the tension increases. Also the speed of the pulse will depend on the mass per unit length of the string, since the mass of each segment will be proportional to this quantity. Would you guess that increasing the tension will increase or decrease the speed of the pulse? What about the mass per unit length? The quantitative analysis we go through next will develop the exact relation among these quantities.

Figure 12-7a shows the uniform string extending along the x axis. Its right end is fixed to a rigid support, and the experimenter's hand is pulling on its left end. The force applied by the hand stretches the string a certain amount because it puts the string under tension. That is, each segment of the string applies a force to the neighboring segment directed in such a way as to stretch the neighbor. These forces are all of the same magnitude, and the common magnitude equals that of the force exerted by the hand. (The

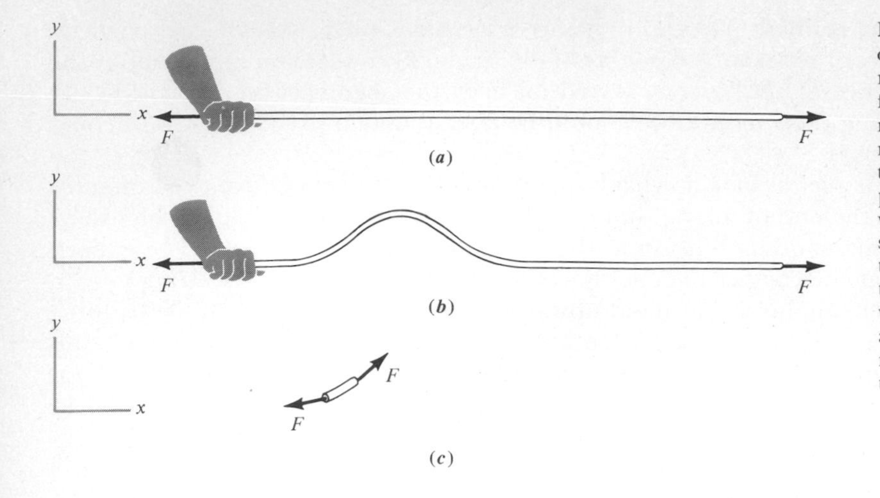

Fig. 12-7 (*a*) By pulling horizontally on the end of a long string, whose other end is attached to a rigid support, an experimenter produces a tension force of magnitude F in the string. (*b*) The experimenter's hand executes a quick up-and-down motion, thereby inducing a single wave pulse in the end of the string being held. The wave pulse propagates along the string and is shown at a time when it has moved away from the hand. (*c*) A very short segment of the string, in the region where the segments have transverse displacements, at the same instant depicted in part *b*. If all the transverse displacements are small, the tension force acting on both ends of the segment will have a magnitude F which is the same as the tension in the undisturbed string.

situation is the same as the one for the system of connected springs, illustrated in Fig. 4-26.) The forces are called the **tension forces** in the string. We will use the symbol F to represent their *magnitude.*

If the experimenter gives the end of the string a quick up-and-down motion in the y direction, while maintaining the tension, a single wave pulse will be generated in the string. Figure 12-7*b* shows the pulse after it has traveled some distance along the string. Figure 12-7*c* depicts the situation at the same instant as in Fig. 12-7*b*, but it shows only a very short segment of the string at a location where the pulse is passing. An enlarged view of this segment is given in Fig. 12-8.

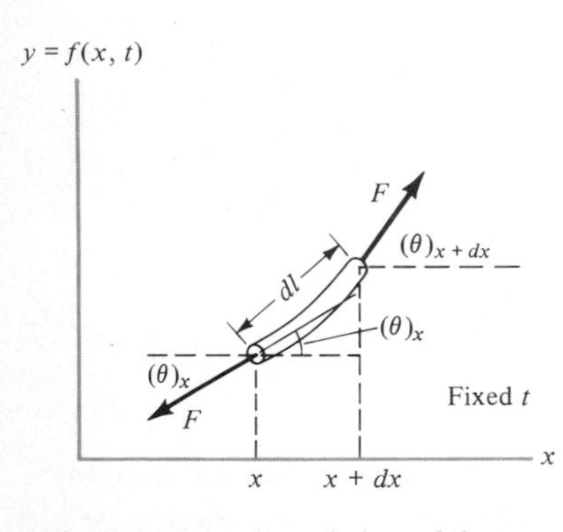

Fig. 12-8 An enlarged view of the segment of string shown in Fig. 12-7*c*. The sizes of the displacements, and therefore the sizes of the angles, have been exaggerated for the sake of clarity.

The profile of the string shown in Fig. 12-7*b* is just a plot of the x dependence of the wave function $y = f(x, t)$ describing the transverse displacements along the string, at the particular fixed value of time t illustrated in the figure. And the segment of that profile shown in Figs. 12-7*c* and 12-8 is a plot of a segment of that function versus x, for that fixed t, extending from the coordinate x of one end of the segment of string to the coordinate $x + dx$ of the other end. These plots are like the one considered in Fig. 12-5*b*. But there we concentrated our attention on describing the motion of the wave pulse. Now we must explain the motion by applying Newton's second law. Thus we must now stipulate that the reference frame containing the string, and used to measure x and y, be an *inertial frame* so that the second law can be applied.

The first step in applying Newton's second law is to find the forces acting on a typical segment of the string, such as the one shown in Fig. 12-8. If we ignore gravity, these are the forces exerted on each of its ends by the adjacent parts of the string. The part of the string to the left of the segment pulls on it with a force of magnitude F. The part of the string to the right of the segment pulls on it with a force of the same magnitude, acting in a direction which is not quite opposite to the direction of the other force because of the curvature of the string. We will treat the segment of string as a particle that moves under the influence of these two applied forces.

Also, we will treat only waves of *small transverse displacement* in a string which, when undisturbed, is stretched by a *large tension.* The effect of these two restrictions is to ensure that the further stretch occasioned by the transverse wave in the string does not increase the tension significantly. Thus we deal with a case in which we can take the magnitude F of the forces acting at the ends of the segment of string to be constant.

When this is not the case, the wave itself affects the mechanical properties of the system through which it travels by increasing the tension. This leads to considerable mathematical complications, as well as to much more complicated physical behavior. In many practical circumstances (such as most waves in stringed musical instruments) the tension in the undisturbed string is large enough, and the transverse displacements in the wave small enough, to satisfy very well the condition of constant tension.

To apply Newton's second law to the segment of string, we must determine the *net* force acting on the segment. The net force is the vector sum of the forces acting on the left and right ends. The x component of the net force has the value

$$F\cos(\theta)_{x+dx} - F\cos(\theta)_x$$

Here $(\theta)_{x+dx}$ is the angle between a line tangent to the segment at its right end and a line parallel to the x axis, and $(\theta)_x$ is the angle at its left end. Since we have assumed small transverse displacements, both angles are small (although they are exaggerated in the figure for clarity), and so the values of both cosines are very nearly equal to 1. Therefore

$$F\cos(\theta)_{x+dx} - F\cos(\theta)_x \simeq F - F = 0 \tag{12-22}$$

The negligibly small magnitude of the component of the net force acting on the segment in the direction parallel to the string is consistent with the fact that it does not move appreciably in that direction when the wave propagates through it. The segment, and all other segments of the string, move only in the perpendicular direction—the wave is transverse.

To treat the transverse motion of the segment, we determine the y component of the net force acting on it. This perpendicular component of the net force is

$$F\sin(\theta)_{x+dx} - F\sin(\theta)_x$$

Since each angle θ is small, it is an excellent approximation to replace $\sin\theta$ by $\tan\theta$. But $\tan\theta$ is just the slope of the segment. The slope is the derivative of $y = f(x, t)$ with respect to x, evaluated for the fixed value of t used in the figure. Such a derivative is, by a definition analogous to that of Eq. (7-60), the *partial derivative* of $f(x, t)$ with respect to x. That is,

$$\sin\theta \simeq \tan\theta = \left[\frac{df(x,t)}{dx}\right]_{\text{evaluated by treating } t \text{ as a constant}} \equiv \frac{\partial f(x,t)}{\partial x}$$

Thus the net perpendicular force on the string segment is approximately

$$F\sin(\theta)_{x+dx} - F\sin(\theta)_x = F\left\{\left[\frac{\partial f(x,t)}{\partial x}\right]_{x+dx} - \left[\frac{\partial f(x,t)}{\partial x}\right]_x\right\}$$

The quantity in braces is the difference between $\partial f(x, t)/\partial x$ at the right end of the segment, where the x coordinate has the value $x + dx$, and $\partial f(x, t)/\partial x$ at the left end, where that coordinate has the value x. It can be expressed as the *rate of change* of $\partial f(x, t)/\partial x$ with respect to the coordinate x, evaluated for fixed t, times the change dx in the coordinate. Thus

$$\left[\frac{\partial f(x,t)}{\partial x}\right]_{x+dx} - \left[\frac{\partial f(x,t)}{\partial x}\right]_x = \frac{\partial}{\partial x}\left[\frac{\partial f(x,t)}{\partial x}\right]dx$$

The term multiplying dx is the partial derivative with respect to x of the quantity $\partial f(x, t)/\partial x$. In the concise notation of calculus, it is written as the *second partial derivative* with respect to x of $f(x, t)$. That is, we define

$$\frac{\partial}{\partial x}\left[\frac{\partial f(x, t)}{\partial x}\right] \equiv \frac{\partial^2 f(x, t)}{\partial x^2} \tag{12-23}$$

Hence we have

$$F \sin(\theta)_{x+dx} - F \sin(\theta)_x = F \frac{\partial^2 f(x, t)}{\partial x^2} dx \tag{12-24}$$

This result gives a good approximation to the net perpendicular force acting on the segment of string.

Newton's second law requires that this net force equal the mass of the segment times its acceleration in the perpendicular direction. The mass is $dm = \mu\, dl$. The quantity μ is the **linear density** of the uniform string, that is, its mass per unit length. Since the segment of string is not parallel to the x axis, we have $dx = dl \cos \theta$. But θ is small. So we can take $\cos \theta = 1$ and write the mass of the segment as

$$dm = \mu\, dx \tag{12-25}$$

to a good approximation.

The acceleration of the segment of string is the second time derivative of the coordinate y, giving its transverse displacement from its undisturbed position on the x axis. The value of x must be held fixed when the second time derivative of $y = f(x, t)$ is computed since the value of that derivative varies significantly with the location along the string of the segment being considered. Therefore we compute the perpendicular acceleration of the segment at a fixed location x by taking the second partial derivative with respect to t of $f(x, t)$:

$$\frac{\partial^2 f(x, t)}{\partial t^2}$$

Using this with Eqs. (12-24) and (12-25) in Newton's second law, we obtain

$$F \frac{\partial^2 f(x, t)}{\partial x^2} dx = (\mu\, dx) \frac{\partial^2 f(x, t)}{\partial t^2}$$

The term on the left side of this equation is the net force applied to the segment of the string, which we have seen to be in the direction perpendicular to the undisturbed string. The term in parentheses on the right side is the mass of the segment. The remaining term on the right side is its acceleration, which is also in the perpendicular direction. (In the particular situation illustrated in Fig. 12-8 the acceleration of the segment is in the positive y direction since the net force acting on it is in that direction.) Canceling the common factor dx, and then dividing through by F, yields the final result

$$\frac{\partial^2 f(x, t)}{\partial x^2} = \frac{\mu}{F} \frac{\partial^2 f(x, t)}{\partial t^2} \tag{12-26}$$

This is the **wave equation.** The equation applies in an inertial reference frame to small transverse waves in the string. It says that the coordinate in the transverse direction of an element of the string, $y = f(x, t)$, has a second partial derivative with respect to its coordinate x along the axis of the undisturbed string that is proportional to its second partial derivative with respect to the time t. The proportionality constant depends on the mechanical properties of the system, being the mass per unit length μ of the string divided by the tension F in the string. The wave equation is the fundamental equation governing the behavior of transverse waves in the string. All such waves can be studied by analyzing the various solutions to this equation.

An equation of *exactly* the same mathematical form as Eq. (12-26) governs the propagation of *all* types of mechanical waves in essentially one-dimensional systems which, like the string, have uniform mass distributions before being disturbed and whose resistance to being disturbed is proportional to the disturbance. For pressure waves in air enclosed in a tubing (the common feature of most non-stringed musical instruments) the second partial derivative of the pressure with respect to position along the tube equals a constant, depending on the mechanical properties of air, times the second partial derivative of the pressure with respect to time. And the behavior of the electric or magnetic field in a one-dimensional wave of electromagnetic radiation (such as a light wave) obeys a mathematically equivalent equation.

For mechanical waves, the wave equation is obtained from Newton's second law. But for electromagnetic waves, essentially the same wave equation is obtained from laws specifying the properties of electric and magnetic fields. Since these laws are completely unrelated to Newton's laws, it is not universally true that the wave equation is only a reexpression of Newton's equation. From the perspective of physics as a whole, it is reasonable to say that the fundamental laws of particle motion and wave motion—Newton's equation and the wave equation—stand on an equal footing. In its own domain, each is the basic governing relation.

Example 12-3 employs the wave equation to help you develop a physical understanding of the mechanics underlying the motion of a traveling wave in a stretched string.

EXAMPLE 12-3

Make direct use of the wave equation to discuss qualitatively both the motion of a segment of a string along which a wave pulse is propagating to the right and the relation of this motion to the motion of the wave pulse. First relate the acceleration of the segment to its curvature. Then use the acceleration to describe the motion of the segment and the motion of the wave pulse.

▪ The wave equation tells you that the acceleration [measured by $\partial^2 f(x, t)/\partial t^2$] of a segment at any location along the string at any time is proportional to the curvature [measured by $\partial^2 f(x, t)/\partial x^2$] of that segment. The physical reason is that the curvature is proportional to the net force acting on the segment. For instance, if the segment is perfectly straight, so that there is no curvature, a construction analogous to that in Fig. 12-8 shows that there is a perfect cancellation of the perpendicular force components acting on its two ends.

In Fig. 12-9*a* the pulse is just beginning to move over the segment under consideration (the shaded part of the string). The segment is concave upward [$\partial^2 f(x, t)/\partial x^2 > 0$], and so its acceleration is upward [$\partial^2 f(x, t)/\partial t^2 > 0$]. The direction of this acceleration is indicated by the small arrow. As a result of its upward acceleration, the segment under consideration moves away from its initial location on the x axis, with an upward-directed velocity.

But as the wave continues to advance to the right, the segment is soon in the situation illustrated in Fig. 12-9*b*. It becomes concave downward and so experiences a

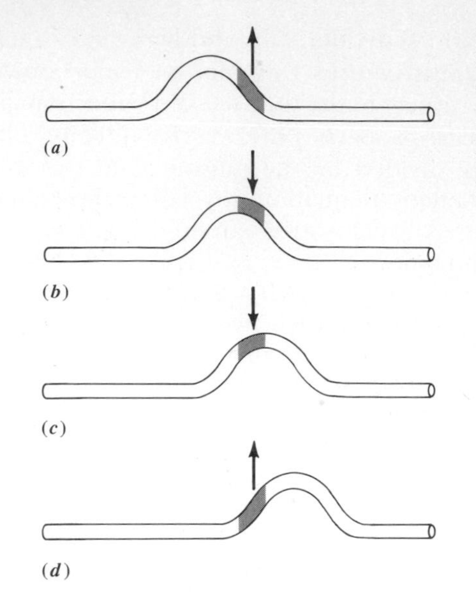

Fig. 12-9 A wave pulse traveling past a particular segment of a string. The segment is indicated by shading. Parts *a, b, c,* and *d* show its position at four successively later times. The arrows give the direction of the acceleration of the segment. The displacements have been exaggerated.

downward acceleration. This reduces its upward velocity to zero at the instant when the segment has its maximum transverse displacement.

As the wave continues its motion to the right, the segment being considered is next in the situation depicted in Fig. 12-9*c*. It continues to have a concave-downward curvature and therefore a downward acceleration. This acceleration develops a downward velocity, and the segment of string begins to return to its undisturbed location on the x axis.

In Fig. 12-9*d* the wave has advanced to such a point that the segment is concave upward. So it has an upward acceleration, which reduces its downward velocity to zero just as it comes back to the x axis.

Of course the wave pulse could have been moving to the left, instead of to the right. You should repeat the discussion for such a case.

12-4 TRAVELING-WAVE SOLUTIONS TO THE WAVE EQUATION

In this section we use analytical methods to prove that the wave equation has solutions describing traveling waves. In other words, we show that it has solutions of the form

$$f(x, t) = f(x - vt) \tag{12-27}$$

We obtained this form in Sec. 12-1 by analyzing observations of waves traveling along a stretched string. So we know already that waves of this form *actually can travel* along the string. In what follows, we use the wave equation to show that waves of this form *should be able to travel* along the string. There are two reasons why this is very much worth doing. First, since the wave equation applies to a wide variety of mechanical systems, showing that it has solutions of the form given by Eq. (12-23) amounts to showing that waves of this form can propagate through each of these systems. Second, in the course of the calculation we will obtain a very important relation which tells us how to evaluate the speed of waves traveling along a stretched string in terms of the tension in the string and its linear density.

The wave equation

$$\frac{\partial^2 f(x, t)}{\partial x^2} = \frac{\mu}{F} \frac{\partial^2 f(x, t)}{\partial t^2} \tag{12-28}$$

is a **partial differential equation,** that is, a differential equation containing partial derivatives. Although it is likely that you have not yet studied such equations in a mathematics course, this will cause no difficulty. All that you will need to know about the analytical methods that we will use for solving partial differential equations will be developed fully here. (It is also possible to solve partial differential equations numerically. An example is given in Chap. 21.) Actually, this book has already introduced you to the basic idea of the analytical method. It is the same for differential equations containing partial derivatives as it is for those containing ordinary derivatives.

As was explained in Sec. 6-5, the idea is that you use whatever prior knowledge you have to *guess* at the form of the solution to the differential equation. Then you substitute the form into the equation and see whether it is possible to obtain a consistent result. The assumed form of the solution can be based on observation and qualitative consideration of the behavior of the system whose physical properties are represented by the partial differential equation. This is just what we have done in Secs. 12-1 and 12-2. The discussion there leads us to assume that Eq. (12-27) is a solution to the partial differential equation. We will verify the assumption by substituting the second partial derivatives of Eq. (12-27) into Eq. (12-28).

To facilitate computing the required derivatives of the function in Eq. (12-27),

$$f(x, t) = f(x - vt)$$

we define a quantity h to be

$$h \equiv x - vt \tag{12-29}$$

Then we can write, for example,

$$\frac{\partial f(x - vt)}{\partial t} = \frac{df(h)}{dh}\frac{\partial h}{\partial t} \tag{12-30}$$

This is a form of the "chain rule" of differential calculus. Its validity is almost self-evident if you express it in words: The rate of change of f with respect to t equals the rate of change of f with respect to h times the rate of change of h with respect to t. Note that when f is considered to be a function of $x - vt$, its derivative with respect to t must be written as a partial derivative, with x held constant. But when it is considered to be a function of the single quantity h, its derivative with respect to that quantity is an ordinary derivative. The derivative of h with respect to t is a partial derivative.

If we assume that the wave is traveling along a *uniform* string, its velocity v will be a constant. Thus Eq. (12-29) gives

$$\frac{\partial h}{\partial t} = -v$$

and Eq. (12-30) produces

$$\frac{\partial f(x - vt)}{\partial t} = -v\frac{df(h)}{dh} \tag{12-31}$$

The same procedure, applied to calculating the partial derivative with respect to t of the quantity $\partial f(x - vt)/\partial t$, yields

$$\frac{\partial}{\partial t}\left[\frac{\partial f(x-vt)}{\partial t}\right]=\frac{d}{dh}\left[-v\frac{df(h)}{dh}\right]\frac{\partial h}{\partial t}=-v\frac{d}{dh}\left[\frac{df(h)}{dh}\right](-v)$$
$$=v^2\frac{d}{dh}\left[\frac{df(h)}{dh}\right]$$

Using second derivative notation, we can write this as

$$\frac{\partial^2 f(x-vt)}{\partial t^2}=v^2\frac{d^2 f(h)}{dh^2} \tag{12-32}$$

Now that we have found the required partial derivative with respect to t, we use the chain rule in the same way to find the partial derivative with respect to x. It reads

$$\frac{\partial f(x-vt)}{\partial x}=\frac{df(h)}{dh}\frac{\partial h}{\partial x}$$

Since Eq. (12-29) shows that

$$\frac{\partial h}{\partial x}=1$$

we obtain

$$\frac{\partial f(x-vt)}{\partial x}=\frac{df(h)}{dh} \tag{12-33}$$

Differentiating with respect to x again produces

$$\frac{\partial}{\partial x}\left[\frac{\partial f(x-vt)}{\partial x}\right]=\frac{d}{dh}\left[\frac{df(h)}{dh}\right]\frac{\partial h}{\partial x}=\frac{d}{dh}\left[\frac{df(h)}{dh}\right]$$

or

$$\frac{\partial^2 f(x-vt)}{\partial x^2}=\frac{d^2 f(h)}{dh^2} \tag{12-34}$$

Having evaluated the second partial derivatives of the function

$$f(x,t)=f(x-vt)$$

which we *guessed* to be a solution to the wave equation, we next substitute them into the wave equation,

$$\frac{\partial^2 f(x-vt)}{\partial x^2}=\frac{\mu}{F}\frac{\partial^2 f(x-vt)}{\partial t^2}$$

The purpose is to see whether the function actually *is* a solution to the equation. By substituting Eqs. (12-32) and (12-34) into the wave equation, we obtain

$$\frac{d^2 f(h)}{dh^2}=\frac{\mu}{F}v^2\frac{d^2 f(h)}{dh^2}$$

If this can be satisfied, then we have proved that the wave equation has the traveling-wave solution. Can it be? Certainly, *providing* the velocity v of the traveling wave is such that

$$\frac{\mu}{F}v^2=1$$

or

$$v=\pm\sqrt{\frac{F}{\mu}} \tag{12-35a}$$

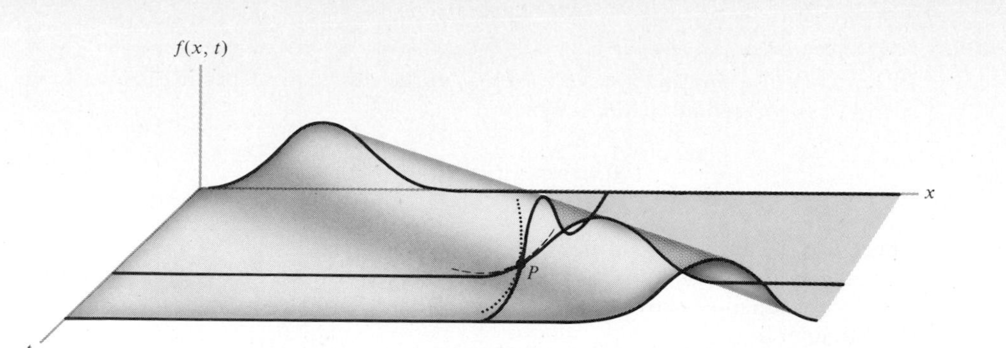

Fig. 12-10 A geometrical interpretation of the equation

$$\frac{\partial^2 f(x, t)}{\partial x^2} = \frac{1}{v^2} \frac{\partial^2 f(x, t)}{\partial t^2}$$

obtained by using Eq. (12-35*a*) to write the factor μ/F in the wave equation for a stretched string as $1/v^2$. The surface plots the function $f(x, t)$ versus x and t for the simple case of a single pulse traveling in the direction of increasing x. It represents the same thing shown by the last seven diagrams in Fig. 12-1. That is, the "wrinkle"-shaped surface intersects a plane perpendicular to the t axis in a curve that gives the shape and location of the pulse at the value of t for that plane, and as t increases, the pulse moves in the positive x direction. Its velocity v specifies the angle between the wrinkle and the t axis—the higher the velocity, the greater the angle. For any point P with a particular set of x and t values, the quantity $\partial^2 f(x, t)/\partial x^2$ measures the curvature at that point of the intersection of the surface with the plane perpendicular to the t axis passing through the point. This "curvature in the x direction" is indicated by the dashed line. The quantity $\partial^2 f(x, t)/\partial t^2$ measures the curvature at P of the surface's intersection with a plane perpendicular to the x axis passing through the point. The dotted line indicates this "curvature in the t direction." The wave equation requires that at each point the curvature in the x direction be the product of $1/v^2$ and the curvature in the t direction. If you consider several points other than P, you will see qualitatively that the two curvatures are everywhere proportional. And you can verify qualitatively the role played by $1/v^2$ if you visualize the surface formed by keeping the shape of the pulse the same—thus keeping the curvatures in the x direction the same—while increasing v, so as to increase the angle between the wrinkle to the t axis and decrease $1/v^2$. The greater the angle, the more abrupt the changes in slope of the intersection of the surface with any plane perpendicular to the x axis. Hence the curvatures in the t direction increase as $1/v^2$ decreases, in agreement with the fact that their product—the curvatures in the x direction—is unchanged.

Either sign is allowed by the wave equation. The positive sign corresponds to a wave traveling in the direction of positive x, and the negative sign corresponds to a wave traveling in the opposite direction. Thus what the wave equation actually determines is the speed $|v|$ of the wave. In these terms, Eq. (12-35*a*) can be written

$$|v| = \sqrt{\frac{F}{\mu}} \tag{12-35b}$$

In addition to verifying that the wave equation has traveling-wave solutions of the form $f(x, t) = f(x - vt)$, we have found something new and important. This is Eq. (12-35*b*), which shows how the speed $|v|$ of the traveling wave depends on the tension F and the mass per unit length μ of the string along which the wave is traveling. Be sure to remember that the speed $|v|$ we have calculated is the speed of the traveling wave measured with respect to an inertial reference frame containing the string. Thus $|v|$ *is the speed of the wave with respect to the inertial reference frame of the medium through which it propagates.*

Figure 12-10 provides a geometrical interpretation of the result obtained in the calculation just carried out. And Example 12-4 is intended to help you understand the calculation by going through a similar one in which the function $f(x, t) = f(x - vt)$ is given an explicit form.

EXAMPLE 12-4

In Eq. (12-16) a sinusoidal wave traveling in the direction of positive x on a long string was represented by the function

$$f(x, t) = A \cos\left[2\pi\left(\frac{x}{\lambda} - \frac{t}{T}\right)\right]$$

where A is the amplitude of the wave, λ is its wavelength, and T is its period. (For convenience the phase constant has been chosen to be $\delta = 0$.) By direct substitution in the wave equation, show that this particular traveling wave is, in fact, a solution to the equation if

$$\frac{\lambda}{T} = \sqrt{\frac{F}{\mu}} \qquad (12\text{-}36)$$

where F and μ are the tension and linear density of the string.

■ In principle, the calculation asked for is unnecessary because the form to be verified can be written

$$f(x, t) = A \cos\left[\frac{2\pi}{\lambda}\left(x - \frac{\lambda t}{T}\right)\right]$$

Using Eq. (12-11a) with a positive sign, we have $\lambda\nu = \lambda/T = v$, where v is the positive velocity of the wave. Thus we can write

$$f(x, t) = A \cos\left[\frac{2\pi}{\lambda}(x - vt)\right]$$

or

$$f(x, t) = f(x - vt)$$

Since we have given a general proof that $f(x - vt)$, with $v = \sqrt{F/\mu}$, is a solution for any form of the function f, it surely is for the case where the general symbol f represents the particular operation of taking the cosine of the constant $2\pi/\lambda$ times the quantity $x - vt$. But it is worthwhile to make an independent verification of this important particular case since it is easy to do and may clarify the general proof.

First you evaluate

$$\frac{\partial f}{\partial t} = -\left(-\frac{2\pi}{T}\right) A \sin\left[2\pi\left(\frac{x}{\lambda} - \frac{t}{T}\right)\right] \qquad (12\text{-}37a)$$

and

$$\frac{\partial^2 f}{\partial t^2} = -\left(-\frac{2\pi}{T}\right)^2 A \cos\left[2\pi\left(\frac{x}{\lambda} - \frac{t}{T}\right)\right]$$

Then you evaluate

$$\frac{\partial f}{\partial x} = -\left(\frac{2\pi}{\lambda}\right) A \sin\left[2\pi\left(\frac{x}{\lambda} - \frac{t}{T}\right)\right] \qquad (12\text{-}37b)$$

and

$$\frac{\partial^2 f}{\partial x^2} = -\left(\frac{2\pi}{\lambda}\right)^2 A \cos\left[2\pi\left(\frac{x}{\lambda} - \frac{t}{T}\right)\right]$$

You can now substitute into the wave equation,

$$\frac{\partial^2 f}{\partial x^2} = \frac{\mu}{F}\frac{\partial^2 f}{\partial t^2}$$

to get

$$-\frac{4\pi^2}{\lambda^2} A \cos\left[2\pi\left(\frac{x}{\lambda} - \frac{t}{T}\right)\right] = -\frac{\mu}{F}\frac{4\pi^2}{T^2} A \cos\left[2\pi\left(\frac{x}{\lambda} - \frac{t}{T}\right)\right]$$

Cancelling, you obtain

$$\frac{T^2}{\lambda^2} = \frac{\mu}{F}$$

or

$$\frac{\lambda}{T} = \sqrt{\frac{F}{\mu}}$$

The positive sign is taken in the square root because all four quantities involved are intrinsically positive. If this equation is satisfied, then the sinusoidal traveling wave certainly satisfies the wave equation. Comparison with Eq. (12-36) shows you have proved what you were required to prove.

Example 12-5 uses Eq. (12-35*b*), $|v| = \sqrt{F/\mu}$.

EXAMPLE 12-5

An apparatus for lecture demonstration of transverse wave pulses is shown in Fig. 12-11. One end of a long piece of rubber tubing is fastened to a wall, and the other end passes over a pulley and then down to a suspended weight. Plucking the tubing near one end produces a transverse pulse which travels along the tubing to the other end. If the distance from the wall to the pulley is l = 8.0 m, the mass of the tubing in that length is m = 0.65 kg, and the mass of the suspended weight is M = 5.0 kg, what is the time t required for the pulse to travel from the wall to the pulley?

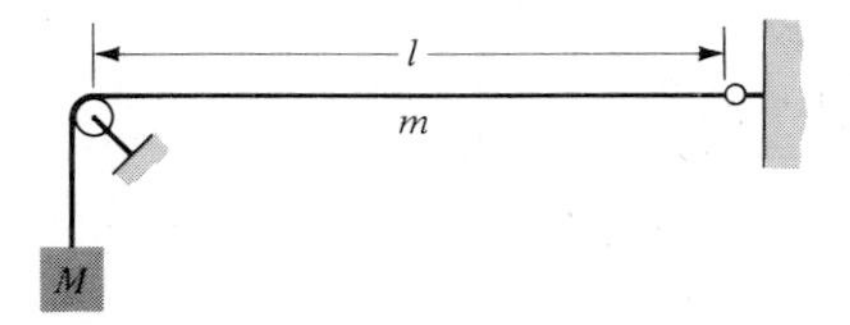

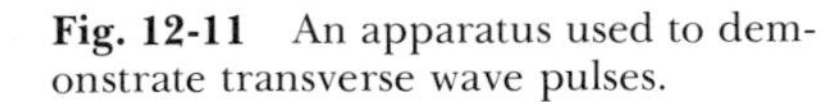
Fig. 12-11 An apparatus used to demonstrate transverse wave pulses.

■ The time t for the pulse to travel a distance l has the value

$$t = \frac{l}{|v|}$$

where the speed of the pulse is $|v|$. According to Eq. (12-35*b*),

$$|v| = \sqrt{\frac{F}{\mu}}$$

The tension F in the tubing is the force applied by the weight, or

$$F = Mg$$

The linear density μ of the tubing is

$$\mu = \frac{m}{l}$$

since a length l contains mass m. Combining these equations, you have

$$t = \frac{l}{\sqrt{F/\mu}} = \frac{l\sqrt{m/l}}{\sqrt{Mg}} = \sqrt{\frac{ml}{Mg}}$$

Inserting the numerical values gives you

$$t = \sqrt{\frac{0.65 \text{ kg} \times 8.0 \text{ m}}{5.0 \text{ kg} \times 9.8 \text{ m/s}^2}} = 0.33 \text{ s}$$

12-5 ENERGY IN WAVES

One of the fundamental properties of a wave is that it contains energy. If it is a traveling wave, the energy is carried along by the wave as it moves. Thus energy can be transported through the system in which the wave travels, from the location at which the energy is put into the system in the process of producing the wave to a location at which energy is absorbed from the system by an interaction between the wave and some object. An example discussed qualitatively in Sec. 12-1 is the transport of energy from a pebble dropped into a pond to a distant cork floating on the pond, by means of the water wave produced by the pebble. A more important example is found in the transport of energy by seismic waves in an earthquake. Even more important is the transport of energy from the sun to the earth by the electromagnetic waves called sunlight. This process is the original source of almost all the energy available to the earth.

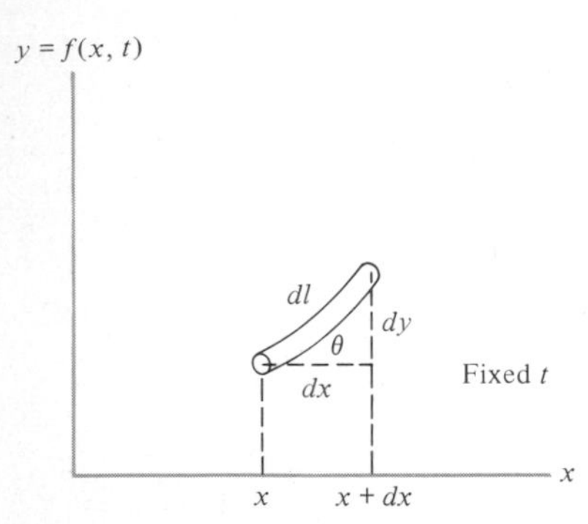

Fig. 12-12 A short segment of string at a time when a wave pulse is passing the segment.

Let us evaluate quantitatively the energy content, and then the energy transport, in the simplest type of mechanical wave—a transverse wave traveling along a stretched string. Figure 12-12 shows a short segment of the string at a time and place for which the wave has displaced it from its normal location on the x axis. The length of the segment is dl, and its mass per unit length is μ. Thus its mass dm is $\mu\ dl$. But since we assume, as before, that transverse displacements are small, so that the angle θ between the segment and the x axis is also small, for the purpose of calculating dm we again use the approximation $\cos\theta = 1$. Then

$$dl = dx \tag{12-38}$$

and

$$dm = \mu\ dx \tag{12-39}$$

The kinetic energy dK of the segment can be calculated from its mass dm and its velocity. Since the motion of the segment is entirely in the transverse direction, the velocity is the time rate of change of its transverse coordinate $y = f(x, t)$, for fixed x. Thus the velocity of the segment at the particular location x is

$$\frac{\partial f(x, t)}{\partial t}$$

Its kinetic energy is one-half its mass times the square of its velocity, or

$$dK = \frac{dm}{2}\left[\frac{\partial f(x, t)}{\partial t}\right]^2$$

Evaluating the mass dm from Eq. (12-39), we obtain

$$dK = \frac{\mu\ dx}{2}\left[\frac{\partial f(x, t)}{\partial t}\right]^2$$

Dividing through by dx yields

$$\frac{dK}{dx} = \frac{\mu}{2}\left[\frac{\partial f(x, t)}{\partial t}\right]^2$$

The quantity dK/dx is the kinetic energy per unit length along the x axis. It is called the **kinetic energy density** and is written ρ_K. Using this symbol, we have

$$\rho_K = \frac{\mu}{2}\left[\frac{\partial f(x, t)}{\partial t}\right]^2 \tag{12-40}$$

The velocity imparted to the segment of string as the wave passes it is proportional to its maximum transverse displacement. [You can see this for a sinusoidal traveling wave from Eq. (12-37*a*). Note that the velocity $\partial f/\partial t$ is directly proportional to the amplitude A, which is the maximum transverse displacement.] Since the displacements are assumed to be small, the proportionality means that the velocity of the segment is also a small quantity. Thus the factor $[\partial f(x, t)/\partial t]^2$ in Eq. (12-40) is the square of a small quantity, and therefore ρ_K is very small. We must take this fact into account in the next step of the calculation.

The next step is to evaluate the potential energy dU of the segment of the string, so that we can find the potential energy density $\rho_U = dU/dx$ associated with the passage of the wave pulse through the segment. Because the kinetic energy density involves the square of a small quantity, it seems likely that the same will be true of this potential energy density. So in the calculation we must be careful to keep track of terms involving even the squares of small quantities. The very small potential energy produced in the segment by the passage of the wave results from the very small stretch occurring in the segment when this happens. Since the segment is under the tension F even before the wave passes, it is already longer than its relaxed length. The situation is the same as when a spring which is already longer than its relaxed length, because a force F is applied to it, stretches a very small amount with that force applied. During the process the magnitude of the force remains essentially constant because the length of the spring changes only a very small amount. The force does work equal to its magnitude times the stretch, since the force acts through a distance equal to the stretch. This work is stored as potential energy.

Before the wave arrives, the segment of string illustrated in Fig. 12-12 lies along the x axis from x to $x + dx$. So its length is then dx. As its length is stretched to dl by the passage of the wave, the amount of stretch is $dl - dx$. To evaluate this, we use the pythagorean theorem

$$[(dx)^2 + (dy)^2]^{1/2}$$

where dy is the change in the y coordinate from one end of the segment to the other, at the particular instant t illustrated in the figure. Now,

$$dy = \frac{\partial f(x, t)}{\partial x} dx$$

That is, dy can be calculated by taking the rate of change of $y = f(x, t)$ with respect to x for fixed t times dx, the change in x along the segment. Therefore we have

$$dl = \left\{(dx)^2 + \left[\frac{\partial f(x, t)}{\partial x}\right]^2 (dx)^2\right\}^{1/2}$$

or

$$dl = dx \left\{1 + \left[\frac{\partial f(x, t)}{\partial x}\right]^2\right\}^{1/2} \tag{12-41}$$

Because we assume small transverse displacements, the slope $\partial f(x, t)/\partial x$ of the string is always small. This suggests we simplify the expression for dl by using an approximation obtained from the binomial expansion. The approximation has been used before; it is Eq. (3-36),

$$(1 + z)^{1/2} \simeq 1 + \tfrac{1}{2}z \qquad \text{where } z << 1$$

with z representing any quantity. We apply the approximation to Eq. (12-41) by setting $z = [\partial f(x, t)/\partial x]^2$. This gives us

$$dl = dx \left\{1 + \frac{1}{2}\left[\frac{\partial f(x, t)}{\partial x}\right]^2\right\} \tag{12-42}$$

Using this result to compute the amount of stretch, $dl - dx$, we find that to a good approximation it is equal to

$$dl - dx = \frac{1}{2}\left[\frac{\partial f(x, t)}{\partial x}\right]^2 dx$$

Thus the work done during the stretch is

$$F(dl - dx) = \frac{F}{2}\left[\frac{\partial f(x, t)}{\partial x}\right]^2 dx$$

This is also the potential energy dU produced by the work. Setting the left side of this equation equal to dU, dividing through by dx, and then writing dU/dx as ρ_U, we have

$$\rho_U = \frac{F}{2}\left[\frac{\partial f(x, t)}{\partial x}\right]^2 \tag{12-43}$$

The quantity ρ_U is the **potential energy density** in the wave.

Can you explain why we were justified in ignoring the squared term in Eq. (12-42) when we evaluated the dl in Eq. (12-38), although we could not ignore it in obtaining Eq. (12-43)? Why is it that in obtaining Eq. (12-43) we could ignore the fact that the tension F in the segment of string increases as the segment stretches?

It is easy to show that for any traveling wave the value of the kinetic energy density equals the value of the potential energy density. To do so, we consider the fact that for such a wave $f(x, t) = f(x - vt)$, with v the velocity of the wave. Equations (12-31) and (12-33) state that for the traveling wave

$$\frac{\partial f(x - vt)}{\partial t} = -v\frac{df(h)}{dh}$$

and

$$\frac{\partial f(x - vt)}{\partial x} = \frac{df(h)}{dh}$$

where $h = x - vt$. Therefore Eqs. (12-40) and (12-43) give

$$\rho_K = \frac{\mu}{2}\left[\frac{\partial f(x - vt)}{\partial t}\right]^2 = \frac{\mu v^2}{2}\left[\frac{df(h)}{dh}\right]^2 \tag{12-44a}$$

and

$$\rho_U = \frac{F}{2}\left[\frac{\partial f(x - vt)}{\partial x}\right]^2 = \frac{F}{2}\left[\frac{df(h)}{dh}\right]^2 \tag{12-44b}$$

Dividing Eq. (12-44a) by Eq. (12-44b) produces

$$\frac{\rho_K}{\rho_U} = \frac{\mu v^2}{F}$$

But Eq. (12-35) says that $v^2 = F/\mu$. So

$$\rho_K = \rho_U \tag{12-45}$$

The total mechanical energy per unit length of the wave is the sum of its kinetic energy per unit length and its potential energy per unit length. Thus the **total energy density** ρ_E of the wave is

$$\rho_E = \rho_K + \rho_U \tag{12-46}$$

Using Eq. (12-45), we have

$$\rho_E = 2\rho_K = 2\rho_U \tag{12-47}$$

We therefore can determine the total energy density of a traveling wave by taking either twice its kinetic energy density or twice its potential energy density.

Example 12-6 applies the theory just developed to a sinusoidal traveling wave.

EXAMPLE 12-6

A sinusoidal wave

$$f(x, t) = A \cos\left[2\pi\left(\frac{x}{\lambda} - \frac{t}{T}\right)\right]$$

of amplitude A, wavelength λ, and period T is traveling along a stretched string. Calculate the kinetic, potential, and total energies in a length of the wave equal to one wavelength. Interpret the results physically.

■ Since calculations involving this wave were carried out in Example 12-4, you already have expressions for $\partial f/\partial t$ and $\partial f/\partial x$. According to Eqs. (12-37*a*) and (12-37*b*),

$$\frac{\partial f}{\partial t} = \frac{2\pi}{T} A \sin\left[2\pi\left(\frac{x}{\lambda} - \frac{t}{T}\right)\right]$$

and

$$\frac{\partial f}{\partial x} = -\frac{2\pi}{\lambda} A \sin\left[2\pi\left(\frac{x}{\lambda} - \frac{t}{T}\right)\right]$$

So the kinetic and potential energy densities are

$$\rho_K = \frac{\mu}{2}\left(\frac{\partial f}{\partial t}\right)^2 = \frac{2\pi^2\mu A^2}{T^2} \sin^2\left[2\pi\left(\frac{x}{\lambda} - \frac{t}{T}\right)\right] \tag{12-48a}$$

and

$$\rho_U = \frac{F}{2}\left(\frac{\partial f}{\partial x}\right)^2 = \frac{2\pi^2 F A^2}{\lambda^2} \sin^2\left[2\pi\left(\frac{x}{\lambda} - \frac{t}{T}\right)\right] \tag{12-48b}$$

Equation (12-36) shows that

$$\frac{\mu}{T^2} = \frac{F}{\lambda^2}$$

If you substitute this expression for μ/T^2 into the right side of Eq. (12-48*a*) and compare the result with the right side of Eq. (12-48*b*), it becomes apparent immediately that

$$\rho_K = \rho_U \tag{12-49}$$

For this traveling wave the kinetic and potential energy densities are, indeed, equal.

You can evaluate the kinetic energy K for one wavelength of the wave by integrating $dK/dx = \rho_K$ over one wavelength λ. That is,

$$K = \int_0^\lambda \frac{dK}{dx}\,dx = \int_0^\lambda \rho_K\,dx$$

For simplicity, make the integration at $t = 0$. Then Eq. (12-48a) gives

$$K = \frac{2\pi^2\mu A^2}{T^2}\int_0^\lambda \sin^2\left(\frac{2\pi x}{\lambda}\right)dx$$

The value of the integral is

$$\int_0^\lambda \sin^2\left(\frac{2\pi x}{\lambda}\right)dx = \frac{\lambda}{2} \tag{12-50}$$

[You can verify this by performing the integration. You can also do it by noting that the value of the integral is the average value over one wavelength of $\sin^2(2\pi x/\lambda)$ times the integral from 0 to λ of dx. The latter factor equals λ. The former is easy to evaluate: Since $\langle\sin^2(2\pi x/\lambda)\rangle = \langle\cos^2(2\pi x/\lambda)\rangle$ and since $\sin^2(2\pi x/\lambda) + \cos^2(2\pi x/\lambda) = 1$, it follows that $\langle\sin^2(2\pi x/\lambda)\rangle = \frac{1}{2}$.] With the integral evaluated, the kinetic energy for one wavelength of the wave is seen to be

$$K = \frac{\pi^2\mu\lambda A^2}{T^2} \tag{12-51}$$

The potential energy U for one wavelength is

$$U = \int_0^\lambda \frac{dU}{dx}\,dx = \int_0^\lambda \rho_U\,dx$$

But Eq. (12-49) shows that $\rho_U = \rho_K$. So you have

$$U = \int_0^\lambda \rho_K\,dx = K$$

and Eq. (12-51) gives

$$U = \frac{\pi^2\mu\lambda A^2}{T^2} \tag{12-52}$$

The total energy content $E = K + U$ in one wavelength of the wave is

$$E = \frac{2\pi^2\mu\lambda A^2}{T^2} \tag{12-53a}$$

Expressed in terms of the frequency $\nu = 1/T$, this is

$$E = 2\pi^2\mu\lambda\nu^2A^2 \tag{12-53b}$$

You can give a physical interpretation of Eqs. (12-53a) and (12-53b) by noting that μ is the mass per unit length of the string. The quantity $\mu\lambda$ in, say, Eq. (12-53a), is therefore the mass involved in one wavelength. The amplitude A is proportional to the transverse distance traveled by that mass in the period of time T required for one oscillation. So A/T is a measure of the speed of transverse motion of the mass. Thus Eq. (12-53a) is seen to be of the form of any kinetic energy expression: kinetic energy $\propto$ mass $\times$ (speed)^2. The left side of the equation contains a total energy E, not a kinetic energy K. But the interpretation remains valid since Eq. (12-47) shows these energies are proportional.

Now we will evaluate the *energy transport* in a wave traveling along a stretched string. This is the energy carried per second by the wave past a fixed location. When undisturbed, the string lies along the x axis. Consider some fixed location on that axis at an initial time t when the wave is passing the location. In the short time interval from t to $t + dt$ the wave moves an amount

$$dx = v\, dt$$

where v is its velocity. As shown in Fig. 12-13*a*, the part of the wave initially contained within an adjacent region of length dx travels past the fixed location during the time dt. As this happens, the total energy dE of that part of the wave is carried past the location. The energy dE has the value

$$dE = \frac{dE}{dx}\, dx = \frac{dE}{dx}\frac{dx}{dt}\, dt$$

This is

$$dE = \rho_E v\, dt \tag{12-54}$$

where $\rho_E = dE/dx$ is the total energy density of the wave. The rate at which energy is transported by the wave past the fixed location is dE/dt. This rate at which energy flows by the fixed location is called the **energy flux** S. (*Flux* is the Latin word for *flow*.) Thus, by definition, the energy flux is

$$S \equiv \frac{dE}{dt} \tag{12-55}$$

Dividing both sides of Eq. (12-54) by dt and then applying this definition, we obtain

$$S = \rho_E v \tag{12-56}$$

The energy flux equals the energy density times the velocity at which energy is being transported.

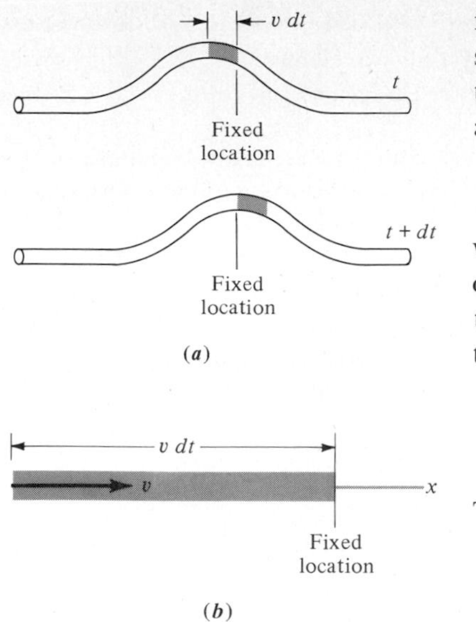

Fig. 12-13 (*a*) A wave pulse on a string at two successive instants separated by a time interval dt. During this interval the wave moves an amount $dx = v\, dt$, where v is its velocity. All the energy contained in a region of the wave of length equal to this amount is carried past a fixed location. Thus in the time interval dt the energy in the shaded region flows past the fixed location. (*b*) A schematic representation of energy flowing at velocity v past some fixed location. In time dt the energy contained in a region of length $v\, dt$ flows past this location. This energy is the energy per unit length ρ_E multiplied by the length $v\, dt$. So it is $\rho_E\, v\, dt$. The energy flow per unit time is $\rho_E v$, which is defined to be the energy flux S. Therefore $S = \rho_E v$. Since in this argument no specification has been made of *how* the energy is transported, the result $S = \rho_E v$ is of general applicability. It applies not only to the energy carried by all types of waves, but also to situations where energy is being carried by a completely different mechanism. For instance, the schematic could represent a conveyor belt carrying a uniformly distributed set of charged automobile batteries at velocity v past a fixed location, with ρ_E being the energy content of the batteries per unit length of the conveyor belt and S the associated energy flux.

Equation (12-56) is a general relation applying to the flow of energy in all waves—and in all other situations in which energy is flowing. Figure 12-13*b* is a schematic representation of the energy flow. In the caption of the figure the representation is used to show that Eq. (12-56) has general applicability. While energy is not a material substance, it is still quite useful to think of it as flowing past a fixed location, just as we think of water flowing past a fixed point on the bank of a river.

If the traveling wave consists of a single pulse, then there is a nonzero energy flux S at a given location only when there is a nonzero energy density ρ_E at that location. Before or after the wave pulse passes the location, there is zero energy density there and hence zero energy flux. But if the wave is a sinusoidal traveling wave, there is almost always a flux of energy passing a given location. The energy flux S is not constant in such a case, however, because the energy density ρ_E is not the same at all points in a sinusoidal traveling wave. On the other hand, the energy flux will be constant if it is averaged over the time required for one wavelength of the wave to pass the location. That is, the average flux $\langle S \rangle$ will be the same from one oscillation to the next.

Examples 12-7 and 12-8 involve evaluating $\langle S \rangle$.

EXAMPLE 12-7

Calculate the average, over the passage of one wavelength, of the energy flux carried by a sinusoidal wave of amplitude A, wavelength λ, and frequency ν which is traveling along a stretched string.

▪ This is easy to do because the total energy content E in one wavelength of the wave was obtained in Example 12-6. All you have to do is to take that energy from Eq. (12-53*b*), which is

$$E = 2\pi^2 \mu \lambda \nu^2 A^2$$

where μ is the linear density of the string. Dividing the total energy E for one wavelength by the wavelength λ, you obtain the average energy density $\langle \rho_E \rangle$ in a wavelength. It is

$$\langle \rho_E \rangle = \frac{E}{\lambda} = 2\pi^2 \mu \nu^2 A^2 \qquad (12\text{-}57)$$

The average flux is this quantity times the constant velocity v. Thus

$$\langle S \rangle = \langle \rho_E \rangle v$$

And so $\langle S \rangle$ has the value

$$\langle S \rangle = 2\pi^2 \mu \nu^2 A^2 v \qquad (12\text{-}58)$$

You can make an interesting comparison between this result and the expression for the total energy of a single particle of mass m that is executing harmonic oscillations of amplitude A and frequency ν. Writing Eq. (8-28*b*) in terms of ν (instead of $\omega = 2\pi\nu$), you will find that the total energy E of the oscillator is

$$E = 2\pi^2 m \nu^2 A^2 \qquad (12\text{-}59)$$

Then use the fact that each segment of the string acts like a transverse harmonic oscillator, when a sinusoidal wave travels along it, to explain the relation between Eqs. (12-58) and (12-59).

EXAMPLE 12-8

Using the demonstration apparatus considered in Example 12-5, a lecturer sends a sinusoidal wave of amplitude $A = 4.0$ cm and frequency $\nu = 5.0$ Hz traveling down the 8.0-m-long rubber tubing by touching a vibrator to it near one end. Evaluate the average energy flux $\langle S \rangle$ passing the center of the tubing during the time before waves reflected from its other end return to the center and complicate the issue.

▪ Using the equations developed in Example 12-5 and the numerical values given there, you have

$$\mu = \frac{m}{l} = \frac{0.65 \text{ kg}}{8.0 \text{ m}}$$
$$= 8.1 \times 10^{-2} \text{ kg/m}$$

and

$$v = \sqrt{\frac{F}{\mu}} = \sqrt{\frac{Mg}{\mu}} = \sqrt{\frac{5.0 \text{ kg} \times 9.8 \text{ m/s}^2}{8.1 \times 10^{-2} \text{ kg/m}}}$$
$$= 2.5 \times 10^1 \text{ m/s}$$

So

$$\langle S \rangle = 2\pi^2 \mu \nu^2 A^2 v$$
$$= 2\pi^2 \times 8.1 \times 10^{-2} \text{ kg/m} \times (5.0 \text{ s}^{-1})^2 \times (4.0 \times 10^{-2} \text{ m})^2 \times 2.5 \times 10^1 \text{ m/s}$$
$$= 1.6 \text{ J/s} = 1.6 \text{ W}$$

The results obtained in Example 12-7 for the average energy density $\langle \rho_E \rangle$ and average energy flux $\langle S \rangle$ of a sinusoidal wave traveling along a stretched string, and applied in Example 12-8 to a particular case, are actually of much wider validity. For all one-, two-, or three-dimensional waves continually traveling through any type of mechanical medium, the average energy density $\langle \rho_E \rangle$ and average energy flux $\langle S \rangle$ are both proportional to the square of the amplitude of the wave at any location, and also proportional to the square of the frequency of the wave. Furthermore, in Chap. 27 we will see that closely analogous relations apply to electromagnetic traveling waves.

12-6 LONGITUDINAL WAVES AND MULTIDIMENSIONAL WAVES

In Sec. 12-3 we derived for the case of a stretched string an equation governing the propagation of transverse waves (waves in which the particles of the propagation medium move perpendicular to the line of motion of the wave). This wave equation is Eq. (12-26):

$$\frac{\partial^2 f(x, t)}{\partial x^2} = \frac{\mu}{F} \frac{\partial^2 f(x, t)}{\partial t^2}$$

The mass per unit length μ of the string and the tension F in the string are quantities which have physical meaning only in connection with a stretched string. However, Eq. (12-35*a*) tells us that $\mu/F = 1/v^2$, where v is the velocity of the waves with respect to their propagation medium, the string. Thus the wave equation can be written in the form

$$\frac{\partial^2 f(x, t)}{\partial x^2} = \frac{1}{v^2} \frac{\partial^2 f(x, t)}{\partial t^2} \qquad (12\text{-}60)$$

where the wave velocity v does not depend on x or t. Even though v does depend on the specific properties of the stretched string, it has a meaning independent of them. An observer can measure the value of v directly and reach substantial conclusions about the behavior of the wave without a knowledge of either F or μ.

These considerations suggest that Eq. (12-60) is more general than the system from which we derived it, and this is indeed the case. The equation governs the propagation in any one-dimensional mechanical medium not only of transverse waves but also of longitudinal waves. **Longitudinal waves** are waves in which the particles of the propagation medium oscillate parallel to the line of motion of the wave. For mechanical waves in one dimension, Eq. (12-60) is applicable in all circumstances, providing two simple criteria are met:

1. All regions of the medium resist disturbance in the same linear (that is, "Hooke's law") fashion. In the case of the uniform stretched string, the force required to give a small transverse displacement to a segment of the string is proportional to the displacement, and the proportionality constant is everywhere the same. But there are many mechanical systems which conform to this criterion in quite different ways.

2. The mass in the medium is uniformly distributed, before the medium is disturbed. This means the inertial tendency of each region of the medium to avoid changing its state of motion is the same as that of every other region of equal size. In the string this criterion is met when the mass per unit length is constant.

The propagation of *sound* through gases or liquids provides a very important example of the applicability of Eq. (12-60) to propagation of longitudinal mechanical waves in one dimension. It is possible to derive a wave equation for sound traveling in one dimension directly from the elastic and thermodynamic properties of fluids. The equation proves to be mathematically identical to Eq. (12-60). But the derivation is quite complicated, so here we consider the matter qualitatively.

Figure 12-14a shows a long, air-filled tube, with a metal diaphragm (a thin plate) at one end that can be made to move back and forth, into and out of the tube. A disturbance can be generated in the air by pushing the diaphragm in once rapidly and letting it return equally rapidly to its original position. The air in the tube satisfies the two criteria just enumerated: (1) The rapid inward motion of the diaphragm causes a decrease in the volume available to the air in the region near the diaphragm and thus an increase in the air pressure in this region. The resistance of the air to this disturbance is directly proportional to the disturbance. (2) The air is uniformly distributed before it is disturbed.

The motion of a wave of slightly higher air pressure along the tube is quite analogous to the motion of a wave of small transverse displacement along a stretched string. When the diaphragm has completed its inward travel, the higher air pressure in the first region beyond the diaphragm begins to compress the air in the next region by forcing air molecules to move. The molecules in this second region cannot respond by changing their positions instantaneously, however, because molecules have inertia.

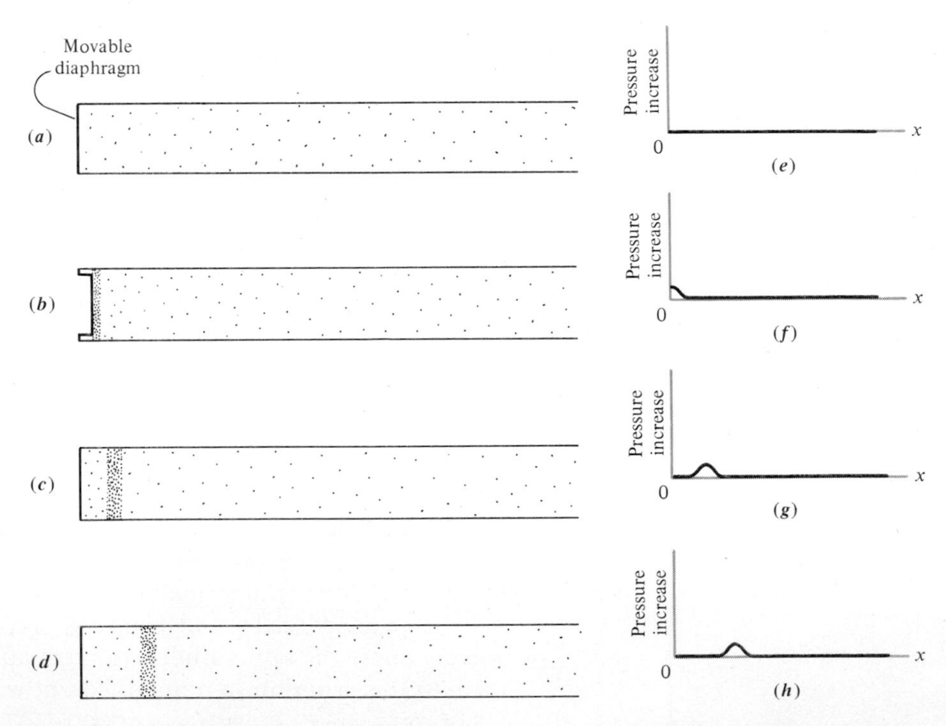

Fig. 12-14 Production of a sound wave pulse in a gas. On the left is the apparatus producing the pulse: a tube filled with gas and closed on one end by a movable diaphragm. The motion of the diaphragm and its effect on the density of the gas in its vicinity are shown in parts a, b, c, and d at four successively later times. The plots on the right sides of these figures represent the increase in gas pressure above the undisturbed pressure versus position at each of the times illustrated. That is, part e corresponds to part a, and so forth.

But soon they do change their positions, and the air pressure in the second region increases. Meanwhile, the diaphragm has returned to its original position, and the air pressure in the first region has returned to its undisturbed value. The increase in air pressure next propagates to the third region, then to the fourth, and so on along the tube. The analogy between this pulse of increased air pressure traveling along the air-filled tube and the pulse of transverse displacement traveling along a stretched string is made evident in Fig. 12-14*e* through 12-14*g*. The traveling air-pressure pulse is a rudimentary **sound wave.**

The generalization to the propagation of periodic sound waves is straightforward. Although it is not directly visible, the air in the tube behaves very much like the spring in Fig. 12-15*a*. The corresponding wave in air is visualized in Fig. 12-15*b*.

There are several different but closely related ways to describe the air through which a sound wave passes. The wave function may describe the change p in the pressure of the air as a function of position and time; that is,

$$f_1(x, t) = p(x, t) \tag{12-61a}$$

Or it may describe the *longitudinal* displacement s of small volume elements of the air from their undisturbed positions as they oscillate:

$$f_2(x, t) = s(x, t) \tag{12-61b}$$

A sound wave is a longitudinal wave because the volume elements of air move in directions parallel to the line of motion of the wave.

If the two wave functions of Eqs. (12-61*a*) and (12-61*b*) describe the same wave, there must be a relation between them. From a physical point of view, it is not surprising that the change p in the air pressure is related to the change s in the longitudinal position of the volume elements of air. It is the latter which leads to the "squeezing" of the air and thus to the change in pressure. However, the two wave functions $p(x, t)$ and $s(x, t)$ are not in step with each other. This can be seen from considering Fig. 12-16. Shown in

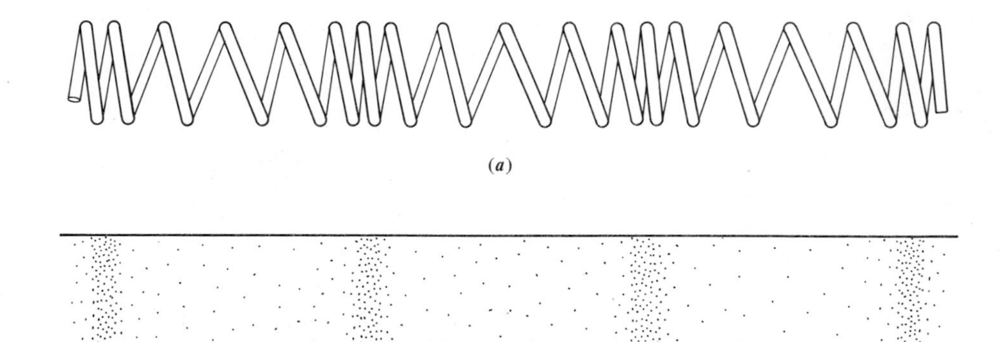

Fig. 12-15 (*a*) A view at a particular instant of a longitudinal periodic wave traveling along a spring. (Note that a spring can act as a propagation medium for longitudinal waves as well as for transverse waves, as in Fig. 12-2. Which type of wave occurs depends on how the spring is disturbed by the source inducing the wave.) (*b*) A longitudinal periodic wave traveling along a gas-filled tube at a particular instant. (Only longitudinal waves can propagate through a gas-filled tube, just as only transverse waves can propagate through a stretched string. What are the physical reasons for these observations?)

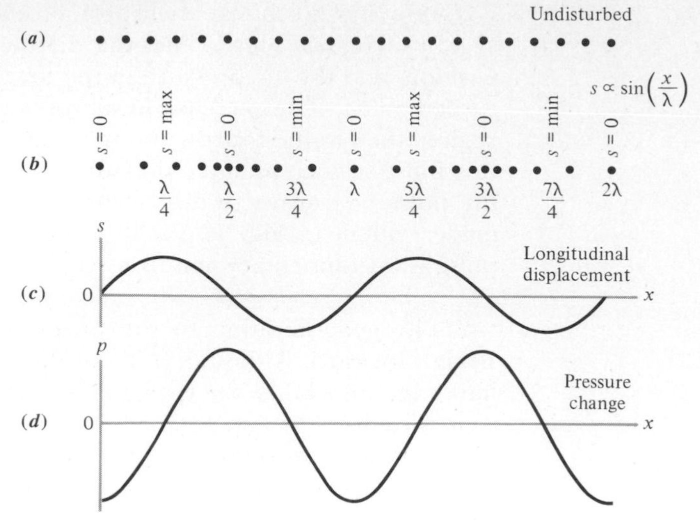

Fig. 12-16 The relation between longitudinal displacement of volume elements of air in a tube through which a sound wave propagates and the pressure changes in the tube.

Fig. 12-16*a* are evenly spaced points representing small volume elements in the undisturbed air. In Fig. 12-16*b* a sinusoidal longitudinal displacement wave is passing through the air, and the position of each element is shown at a given instant of time. Each element is displaced by an amount s from its own equilibrium position x. You can see that the density change—and thus the pressure change p—is largest at the locations where the displacement is zero. Just at these locations the neighboring elements have crowded in from both left and right, resulting in a maximum pressure, or else have moved away to both left and right, resulting in a minimum pressure. The position change s is plotted versus x in Fig. 12-16*c*, and the pressure change p is plotted versus x in Fig. 12-16*d*.

Longitudinal waves carry energy just as transverse waves do. In Sec. 12-5 we considered in detail the energy content and energy transport in waves of transverse displacement traveling along a stretched string. Nearly everything we learned there applies directly to waves of longitudinal displacement traveling along an air-filled tube. We can define an **average energy density** $\langle \rho_E \rangle$ for these one-dimensional sound waves to be their total energy content per unit length, averaged over one wavelength. Also, we can define the **average energy flux** $\langle S \rangle$ as the energy the waves carry past a fixed point per unit time, averaged over one wavelength. These two quantities are related to the sound wave velocity v just as in Eq. (12-56). That is,

$$\langle S \rangle = \langle \rho_E \rangle v \tag{12-62}$$

As we have explained, the relation *energy flux equals energy density times the velocity at which energy is transported* applies no matter how the energy is being transported. And since it applies to the unaveraged quantities S and ρ_E, it certainly applies to their averages $\langle S \rangle$ and $\langle \rho_E \rangle$. You will soon see that Eq. (12-62) holds even when it is necessary to change the precise definitions of $\langle S \rangle$ and $\langle \rho_E \rangle$ in dealing with waves traveling in two or three dimensions.

Sound waves, and certain other types of longitudinal waves, typically travel in three dimensions. The same is true of some types of transverse waves. But we will introduce a discussion of *multidimensional waves* by con-

Fig. 12-17 Photograph of circular water waves propagating away from a point where a vertically oscillating rod touches the surface of a shallow tank of water. The apparatus is called a **ripple tank.** (*From PSSC Physics, 2d ed., D. C. Heath, Boston, 1965. Courtesy Education Development Center.*)

sidering first a wave traveling in two dimensions. An example is shown in the ripple-tank photo of Fig. 12-17. A vertically oscillating rod of circular cross section touches the surface of a shallow layer of water in a transparent tank and produces transverse waves in the form of ripples in the surface. These waves are illuminated from below by a light source and photographed from above by a camera. The waves propagate outward from the source in all directions.

The most obvious feature of these waves traveling in two dimensions can be specified by the shapes of curves which, at any instant, connect all comparable points of the waves. Such curves are called **wave fronts.** For example, a curve that passes through the points where the crest of a particular ripple is located at any instant is a wave front. At the same instant there is also a wave front, lying outside the one just mentioned, which connects all the points where a trough is located. There is another wave front, lying inside, where another trough is located, and so forth. Since all directions are equivalent, the waves travel away from their source at a speed which is the same in every direction. As a consequence, all the wave fronts are circles centered on the source. For this reason the waves are called **circular waves.**

Careful inspection of the photo will show you that the amplitude of the waves (the "height" of the ripples) decreases slowly with increasing distance from the source. A small part of this decrease is due to frictional energy loss. But for the most part the decrease in amplitude is explained by applying energy conservation to the geometry of circular waves in an argument, starting in the next paragraph, that pertains to water waves, sound waves, or waves of any other type.

Consider the circular wave system sketched in Fig. 12-18. The source applies a force to the water at the surface of the ripple tank, and this water

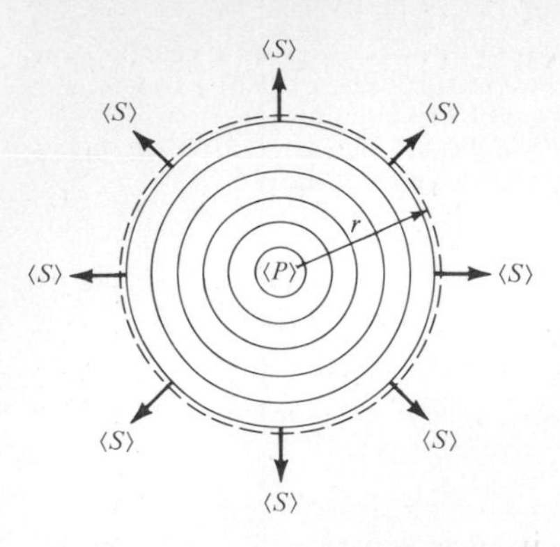

Fig. 12-18 The set of circular wave crests pictured in Fig. 12-17. The dashed circle of radius r is used in an argument concerning the energy flowing away from the source of circular waves.

is displaced by the action of the force. Thus the source does work on the water, supplying energy to it. The average rate at which the source continually supplies energy is the **source power** $\langle P \rangle$. The energy is not *accumulated* in the region near the source. Instead, the energy *flows* outward through this region because it is transported by the waves that the source produces. This means that the average energy per unit time crossing *any* closed curve surrounding the source must be equal to the average source power $\langle P \rangle$. Consider such a curve in the form of a circle of radius r centered on the source. For the two-dimensional situation we are dealing with here, the average energy flowing per unit time across a *unit length* of the periphery of the circle is defined as the **average energy flux** $\langle S \rangle$. Since by symmetry the value of the average energy flux is the same everywhere on the circle, the value of $\langle S \rangle$ is the source power $\langle P \rangle$ divided by the circumference $2\pi r$ of the circle. Thus

$$\langle S \rangle = \frac{\langle P \rangle}{2\pi} \frac{1}{r} \qquad \text{for uniform circular waves} \qquad (12\text{-}63)$$

In this case the flux is an energy flow per unit time per unit length. Its value is the constant $\langle P \rangle / 2\pi$ multiplied by the reciprocal of the distance r from the source. *The flux is inversely proportional to the distance from the source* because the same energy per unit time flows across a greater periphery the greater the distance.

Next we employ Eq. (12-62), $\langle S \rangle = \langle \rho_E \rangle v$. Here the **average energy density** $\langle \rho_E \rangle$ of the waves is their energy content per *unit area,* and v is their *speed.* A justification of Eq. (12-62) for a two-dimensional case is given in Fig. 12-19 and its caption. Solving for $\langle \rho_E \rangle$ and using Eq. (12-63), we obtain

$$\langle \rho_E \rangle = \frac{\langle P \rangle}{2\pi v} \frac{1}{r} \qquad \text{for uniform circular waves} \qquad (12\text{-}64)$$

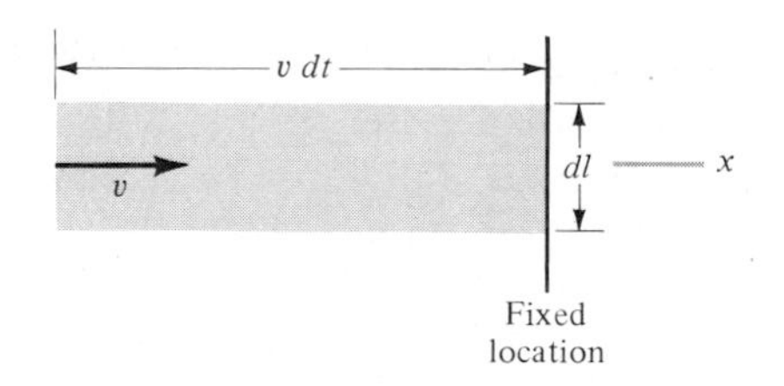

Fig. 12-19 A schematic representation of the energy, flowing at speed v, which in the infinitesimal time interval dt passes across an infinitesimal length dl of a fixed marker line that is perpendicular to the direction of flow. In that time all the energy in the shaded region flows past the indicated length of the marker line. The shaded region extends a distance along the direction of flow equal to $v\ dt$. Its area is $v\ dt\ dl$. The average total energy content in the region is the product of its average total energy per unit area $\langle \rho_E \rangle$ and its area $v\ dt\ dl$. Since all this energy flows past the length dl of the perpendicular marker line, the average energy flow per unit time per unit length of the marker line is $\langle \rho_E \rangle\ v$. This quantity is the average energy flux $\langle S \rangle$. Therefore $\langle S \rangle = \langle \rho_E \rangle\ v$. Why is the result valid when applied to waves that spread, like circular waves, even though no indication of spreading is seen in the figure? When you come to the discussion of waves in three dimensions, read the remainder of this caption. For a three-dimensional situation, the shaded region could represent the projection on the page of a rectangular volume of infinitesimal cross-sectional area da, extending a distance $v\ dt$ along the direction of flow. Its volume is $v\ dt\ da$. The average total energy content in the region is the product of its average total energy per unit volume $\langle \rho_E \rangle$ and its volume $v\ dt\ da$. Since all this energy flows past the area da of the perpendicular marker surface, the average energy flow per unit time per unit area of the marker surface is $\langle \rho_E \rangle\ v$. This quantity is the average energy flux $\langle S \rangle$. Therefore $\langle S \rangle = \langle \rho_E \rangle\ v$.

Since everywhere on the circular wave v has the same magnitude, the quantity $\langle P \rangle / 2\pi v$ is constant. Thus this equation leads to the proportionality $\langle \rho_E \rangle \propto 1/r$. The energy density is inversely proportional to the distance from the source because the greater the distance, the more thinly spread is the energy. Why is it inversely proportional to v?

Finally, we use the proportionality exhibited in Eq. (12-57) between $\langle \rho_E \rangle$ and A^2, the square of the amplitude of the waves. As discussed briefly after Example 12-8, this proportionality applies to waves of any nature. Since $A^2 \propto \langle \rho_E \rangle$, while Eq. (12-64) shows that $\langle \rho_E \rangle \propto 1/r$, it follows that $A^2 \propto 1/r$, or

$$A \propto \frac{1}{\sqrt{r}} \qquad \text{for uniform circular waves} \qquad (12\text{-}65)$$

The gradual decrease in the amplitude of the circular waves in Fig. 12-17 is thus seen to be a consequence of energy conservation and geometry.

If a source of power $\langle P \rangle$ is emitting waves which propagate away from it in all directions in a uniform, three-dimensional medium, the wave fronts form spheres centered on the source. Consequently, the traveling waves are called **spherical waves.** The energy in any given spherical wave will be distributed uniformly over the wave if the source is injecting energy into the system uniformly in all directions. (This is what the symmetrical ripple-tank source is doing in the two-dimensional case shown in Fig. 12-17.) In these circumstances the arguments just gone through for circular waves can be modified immediately to apply to spherical waves. In three dimensions, the **average energy flux** $\langle S \rangle$ is the energy flowing per unit time across a *unit area,* and the **average energy density** $\langle \rho_E \rangle$ is the energy content per *unit volume.* The latter part of the caption of Fig. 12-19 shows that these quantities are still related by Eq. (12-62), $\langle S \rangle = \langle \rho_E \rangle v$, where v is the speed at which the waves transport energy.

If you repeat the arguments, using a sphere of radius r and surface area $4\pi r^2$ centered on the source instead of a circle of radius r and circumference $2\pi r$, you will have no difficulty in showing that

$$\langle S \rangle = \frac{\langle P \rangle}{4\pi} \frac{1}{r^2} \qquad \text{for uniform spherical waves} \qquad (12\text{-}66)$$

and

$$\langle \rho_E \rangle = \frac{\langle P \rangle}{4\pi v} \frac{1}{r^2} \qquad \text{for uniform spherical waves} \qquad (12\text{-}67)$$

and that the amplitude A obeys the relation

$$A \propto \frac{1}{r} \qquad \text{for uniform spherical waves} \qquad (12\text{-}68)$$

What is the physical interpretation of these results?

The relations quoted in Eqs. (12-65) and (12-67) show that waves traveling in two and three dimensions through a medium in which frictional losses are negligible (so that mechanical energy is conserved) do not maintain constant amplitudes. This is in contrast to the constant amplitude of the one-dimensional traveling waves found by solving the one-dimensional wave equation, Eq. (12-60). The wave equations in two and three dimensions are similar to Eq. (12-60), but not identical to it or to each other. The differences in the geometry in the three cases lead to differences in the exact forms of the wave equations. In fact, they make the multidimensional wave equations considerably more complicated than the one-dimensional wave equation. So we will do nothing with them other than to state that direct solutions of these wave equations lead directly to the dependence of A on r seen in Eqs. (12-65) and (12-68).

The relations given by Eqs. (12-63) through (12-68) are valid for waves of any nature, because they are based on considerations of geometry and energy conservation that pertain to all waves. Example 12-9 applies Eq. (12-66) to light waves.

EXAMPLE 12-9

A 100-W light bulb uses about that much electric power. But the power in the visible light produced by the bulb is no more than about 10 W.

a. Estimate how much power in the form of visible light flows onto the pupil of an eye, of radius 1.0 mm, located a distance of 5.0 m from a bulb emitting 10 W of visible light, assuming light is emitted from the bulb uniformly in all directions.

b. What would be the effect of increasing the distance from the bulb to the eye to 10.0 m, without changing the orientation of the bulb?

■ **a.** Using Eq. (12-66), you first evaluate the energy flux

$$\langle S \rangle = \frac{\langle P \rangle}{4\pi r^2}$$

The numerical value is

$$\langle S \rangle = \frac{10 \text{ W}}{4\pi \times (5.0 \text{ m})^2} = 0.032 \text{ W/m}^2$$

Multiplying this energy flow per unit time per unit area by the area πR^2 of the pupil of radius R, you have

$$\begin{aligned}\text{Energy flow per second} &= \langle S \rangle \pi R^2 \\ &= 0.032 \text{ W/m}^2 \times \pi \times (1.0 \times 10^{-3} \text{ m})^2 \\ &= 1.0 \times 10^{-7} \text{ W}\end{aligned}$$

This can be only an estimate because light is not emitted uniformly in all directions by a light bulb. (For instance, no light comes past the base of the bulb.)

b. If the orientation of the bulb is held constant while its distance from the pupil is made larger by a factor of 2, the inverse-square law of Eq. (12-67) predicts, with good accuracy, that the energy flowing onto the pupil of fixed radius will become smaller by a factor of $\frac{1}{4}$. Why are nonuniformities in the emission of light into various directions not of importance here?

12-7 THE DOPPLER EFFECT

The Doppler effect concerns properties of mechanical traveling waves which result from the fact that when the waves are propagating through a particular medium *fixed in an inertial frame,* they maintain a constant speed *with respect to the propagation medium.* We proved this in Eq. (12-35*b*) for waves in a stretched string. But it is true of all mechanical traveling waves.

You have met one example of the Doppler effect if you have noticed how the pitch, or frequency, of the horn of an approaching automobile appears to drop as the automobile passes. (The "pitch" of a musical tone—the "highness" or "lowness" you perceive—is determined by the frequency of the sound wave striking your ear. The higher the frequency, the higher the pitch.) The downward shift in the frequency that you detect when stationary with respect to the air through which the sound waves travel arises because when their source is approaching you, the detected frequency is higher than the frequency of the source, and when the source is receding from you, the detected frequency is lower than that of the source. This property of mechanical traveling waves, and related properties, are called the **Doppler effect** after the physicist Christian Doppler (1803–1853), who first worked out the theory in detail for the case of sound and pointed out its implications for the case of light.

Let us begin our study of the Doppler effect by explaining qualitatively what happens when the automobile mentioned above approaches a stationary observer through still air. At a certain moment the oscillating diaphragm of the horn produces a region of maximum air pressure. This region immediately begins to propagate as a wave outward in all directions. Since the medium through which it propagates is uniform and stationary with respect to the observer fixed in an essentially inertial frame, the speed of the wave with respect to the observer is the same in all directions. From the observer's viewpoint, the wave must therefore propagate as a spherical wave. That is, the wave front passing through all points where the air pressure maximizes is an expanding sphere. This sphere is centered on where the horn *was* when it emitted the wave front. The position of this wave front at some subsequent time is indicated by the curve labeled 1 in Fig. 12-20*a*. It is a circle centered on the position labeled 1 and is supposed to represent a sphere centered on that position.

Later the horn emits the next region of maximum air pressure, which forms wave front 2. But from the viewpoint of the observer, the horn has *moved* from position 1 to position 2 by the time this happens. Thus from the observer's viewpoint wave front 2 propagates as a spherical wave whose center is at position 2. The figure shows wave front 2 at the same instant as it shows wave front 1. The process continues, and the result is the pattern of spherical waves indicated in Fig. 12-20*a*. The motion of the car has led to a spacing of the wave fronts which is different in different directions. This is shown for circular water waves emitted by a moving source in the ripple tank photo of Fig. 12-20*b*.

If the observer is at the location labeled A in Fig. 12-20*a*, to which the car is approaching, then consecutive wave fronts sweep by the observer's ear at a rate greater than that at which they were emitted because they are "bunched up." The observer consequently hears sound of frequency higher than the frequency of oscillation of the diaphragm of the horn. But if the observer is at location R, from which the car is receding, then consecutive wave fronts sweep by at a rate less than that at which they were emitted because they are "spread out," and the observer hears sound of frequency lower than the frequency of oscillation of the diaphragm.

In general, the observed frequency of the sound depends on not only the velocity of the source relative to a reference frame fixed to the ground, but also on the velocities with respect to that reference frame of the propagation medium and of the observer. Let us restrict ourselves to one-dimensional motion—that is, to the case where the source, the medium, and the observer all move with respect to a reference frame fixed to the ground along a common line, though at different speeds and/or directions. We will derive a general expression for the Doppler effect in this case. The deriva-

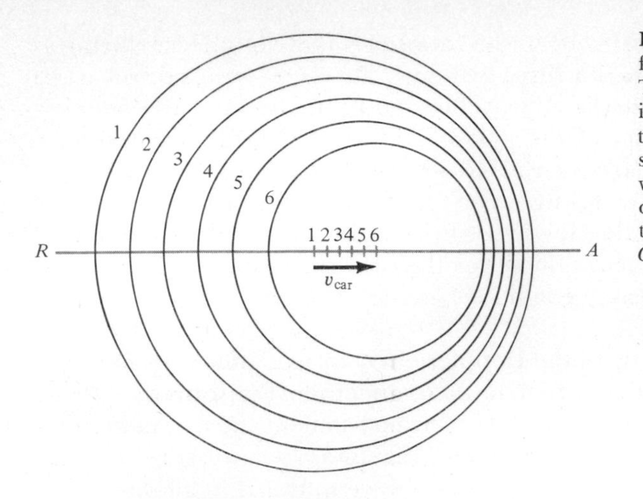

Fig. 12-20 (*a*) A constant-frequency sound source is moving with uniform velocity with respect to the air through which the sound propagates. The figure represents the locations of the wave fronts at a particular instant of time. Each wave front is a sphere centered on the location of the sound source at the instant when the front wave emitted. Since the source is moving to the right, so are the centers of the successive spherical wave fronts. (*b*) A ripple-tank photo showing circular wave fronts produced by a constant-frequency source moving uniformly to the right over the surface of the water in the tank. (*Courtesy Educational Development Center.*)

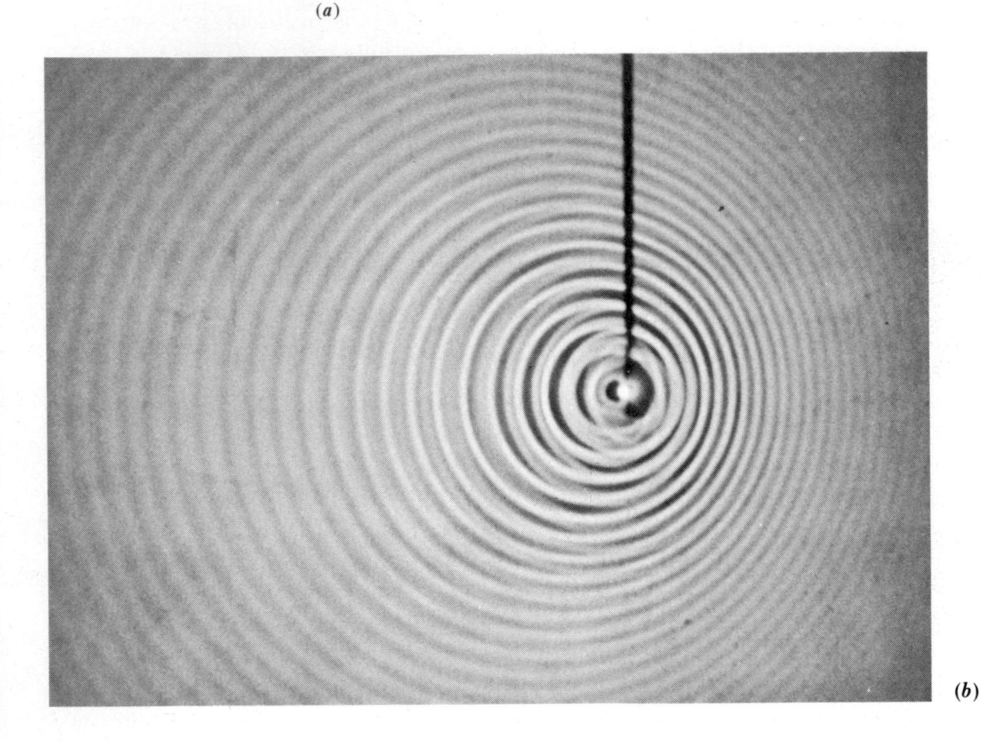

(*b*)

tion makes repeated use of the Galilean transformations developed in Sec. 3-8.

Figure 12-21 depicts the situation as seen from the essentially inertial reference frame of the ground. The wind blows—that is, the air which is the propagation medium moves—at a constant velocity with respect to the ground given by the signed scalar v_m. Because v_m is constant, the propagation medium itself is at rest in some other essentially inertial reference frame. Thus the speed of sound *with respect to the propagation medium* has the fixed value $|v|$. The source of the sound waves moves at a constant velocity v_s, and the observer who detects the waves moves at a constant velocity v_o. Both v_s and v_o are signed scalars, and both are measured with respect to the ground. We consider only the parts of the wave fronts that are near the common line of motion. And we use their direction of motion to define the positive direction of the x axis extending along the line of motion and fixed

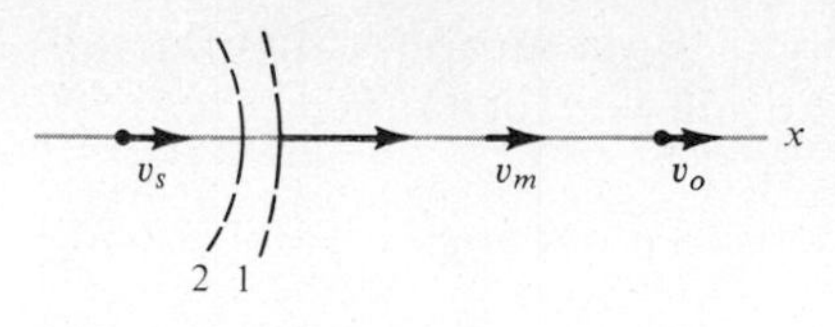

Fig. 12-21 A view, from a reference frame fixed to the ground, of a moving medium, moving source, and moving observer considered in deriving the general one-dimensional Doppler-effect formula. Only the solid parts of wave fronts 1 and 2 are taken into account. Their direction of motion through this reference frame defines the positive direction of its x axis.

to the ground. (Note carefully how the positive direction is defined, and be sure to conform to it when assigning numerical values to v_m, v_s, and v_o in applying the formula we will obtain to a specific calculation.) For the sake of convenience, the figure shows the velocities of the medium, the source, and the observer to be all in the positive x direction.

Let the frequency of the source be ν. It emits wave front 1, and at a time $T = 1/\nu$ later it emits wave front 2. In that time the displacement of wave front 1 with respect to the propagation medium is $|v|T$. But the propagation medium is itself moving with respect to the ground at velocity v_m and so has been displaced with respect to the ground by an amount v_mT in the same time. Thus wave front 1 has been displaced with respect to the ground by the amount $(|v| + v_m)T$. In the same time, the source has been displaced with respect to the ground, by the amount v_sT, to the location where it emits wave front 2. Thus the separation between wave front 1 and wave front 2 is

$$(|v| + v_m)T - v_sT = [|v| - (v_s - v_m)]T$$

But this separation is precisely what is meant by the wavelength λ of the waves. So

$$\lambda = [|v| - (v_s - v_m)]T$$

Since $T = 1/\nu$, this is

$$\lambda = \frac{|v| - (v_s - v_m)}{\nu} \tag{12-69}$$

The parts of the wave fronts traveling in the positive x direction along the ground are measured from the ground to move with a velocity given by $|v| + v_m$. But the observer is moving with respect to the ground as well, at velocity v_o. So the velocity of the wave fronts with respect to the observer is $|v| + v_m - v_o = |v| - (v_o - v_m)$. We must assume that the speed $|v|$ of the waves with respect to the medium is greater than the speed $|v_o - v_m|$ of the observer with respect to the medium. Then the velocity $|v| - (v_o - v_m)$ of the wave fronts with respect to the observer will have a positive value. This means that the wave fronts will catch up with the observer, moving at a speed with respect to him whose numerical value is given by $|v| - (v_o - v_m)$. After wave front 1 has caught up with the observer, it takes a time T' for wave front 2 to catch up. This time is just the separation λ between consecutive wave fronts, whose value is given by Eq. (12-69), divided by the speed whose value is given by $|v| - (v_o - v_m)$. Thus

$$T' = \frac{1}{\nu}\frac{|v| - (v_s - v_m)}{|v| - (v_o - v_m)}$$

The frequency $\nu' = 1/T'$ measured by the observer is therefore obtained by inverting numerator and denominator of this equation, to yield

$$\nu' = \nu\frac{|v| - (v_o - v_m)}{|v| - (v_s - v_m)} \tag{12-70}$$

For this general **Doppler-effect formula** to be valid, we must assume also that the speed $|v|$ of the wave with respect to the medium is greater than the speed $|v_s - v_m|$ of the source with respect to the medium. Then we can be sure that the denominator in Eq. (12-70) will never be zero, so that ν' is never infinite, and that the denominator will never be negative, so that ν' is never negative.

If only the source is moving with respect to the ground, then v_o and v_m are zero, and the formula reduces to the simpler form

$$\nu' = \nu \frac{|v|}{|v| - v_s} \qquad \text{for only source moving} \qquad (12\text{-}71)$$

If only the observer is moving, the corresponding expression is

$$\nu' = \nu \frac{|v| - v_o}{|v|} \qquad \text{for only observer moving} \qquad (12\text{-}72)$$

Even though the propagation medium is not moving with respect to the ground in the cases to which Eqs. (12-71) and (12-72) pertain, its presence is still essential. Not only is the medium responsible for the very existence of the waves, but also it specifies the frame of reference with respect to which $|v|$ is measured. It was thinking about this "privileged" frame of reference in connection with the propagation of light that led Einstein to the theory of relativity, as you will see in Chap. 14. There is a Doppler effect for light, and other electromagnetic waves, which is qualitatively similar to the Doppler effect for mechanical waves. But the Doppler-effect formula for electromagnetic waves is not the same as Eq. (12-70), and the physical explanation of the effect is completely different, because there is no privileged reference frame for electromagnetic waves.

A problem in which none of the terms in Eq. (12-70) is zero is worked out in Example 12-10.

EXAMPLE 12-10

Excursion boats A and B are moving toward each other, both traveling at speeds of 7 m/s with respect to the reference frame fixed to the earth. The wind is blowing at a speed of 5 m/s parallel to the direction of motion of boat A, as seen from the reference frame fixed to the earth. There is a band on each boat. The oboist on boat A plays the musical note A, which corresponds to the frequency 440 Hz. The musicians on boat B tune to that note and then play it back. What frequency does the bandleader on boat A hear? Take the speed of sound in air to be 344 m/s.

■ You must be very careful in applying Eq. (12-70) to assign the proper signs to the various velocities. To help in doing this, you first draw a sketch like that in Fig. 12-22*a*. It represents the first part of the process, in which the source is on boat A and the observers on boat B, and takes the positive x direction in the direction of motion of the parts of the wave fronts that carry the sound received on boat B. You can then calculate ν'_B, the frequency heard on boat B. It is

$$\nu'_B = \nu \frac{|v| - (v_o - v_m)}{|v| - (v_s - v_m)}$$

$$= 440 \text{ Hz} \times \frac{344 \text{ m/s} - [(-7 \text{ m/s}) - (+5 \text{ m/s})]}{344 \text{ m/s} - [(+7 \text{ m/s}) - (+5 \text{ m/s})]} = 458 \text{ Hz}$$

The **Doppler shift** $\Delta\nu = \nu'_B - \nu$ has the value $\Delta\nu = 458 \text{ Hz} - 440 \text{ Hz} = +18 \text{ Hz}$.

Next you sketch the situation for the second part of the process, in which the musicians on boat B act as the source and the bandleader on boat A is the observer. In so doing, you must remember that in the analysis leading to Eq. (12-70), the positive x direction was defined to be in the direction of motion of the parts of the wave fronts which carry the sound from source to observer. Hence the positive x direction must now be reversed from that in the first part of the analysis, and you make

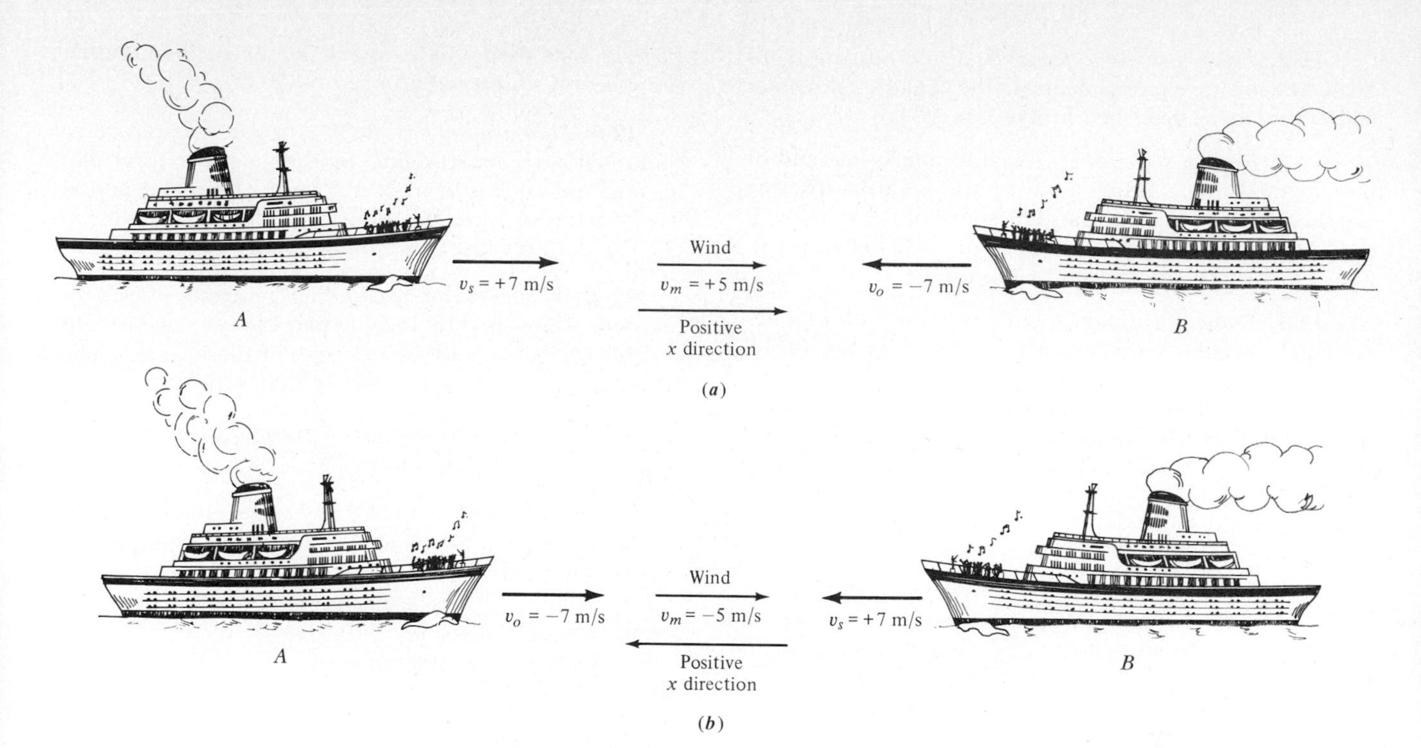

Fig. 12-22 Illustration for Example 12-10. (*a*) The musicians on boat *B* hear the oboist on boat *A*. The positive *x* direction is defined as shown. (*b*) The bandleader on boat *A* hears the musicians on boat *B*. The positive *x* direction is redefined as shown.

your sketch as in Fig. 12-22*b*. Then you use Eq. (12-70), again, this time setting $\nu = \nu'_B = 458$ Hz and calculating ν'_A, the frequency heard on boat *A*. You find

$$\nu'_A = \nu \frac{|v| - (v_o - v_m)}{|v| - (v_s - v_m)}$$

$$= 458 \text{ Hz} \times \frac{344 \text{ m/s} - [(-7 \text{ m/s}) - (-5 \text{ m/s})]}{344 \text{ m/s} - [(+7 \text{ m/s}) - (-5 \text{ m/s})]} = 477 \text{ Hz}$$

The second Doppler shift is $\Delta\nu = 477 \text{ Hz} - 458 \text{ Hz} = +19$ Hz. The bandleader thus hears a frequency differing from that of his oboist by $477 \text{ Hz} - 440 \text{ Hz} = +37$ Hz. In musical terminology, the bandleader would say that the note he hears from the other boat is more than half a tone sharp. The difference is very easily detectable, even though the speeds are all quite modest.

EXERCISES

Group A

12-1. *Triangular wave pulse, I.* A transverse wave pulse is traveling in the positive x direction along a long stretched string. The origin is taken at a point on the string which is far from the ends. The speed of the wave is $|v|$. At time $t = 0$, the displacement of the string is described by the function $y = f(x, 0) = A(1 - |x|/l)$ for $|x| \leq l$, and $y = f(x, 0) = 0$ for $|x| > l$.

a. Construct a graph that portrays the actual shape of the string at $t = 0$ for the case $A = l/2$.

b. Sketch a graph of the wave form at the following times:

1. $t = l/|v|$
2. $t = 2l/|v|$
3. $t = -3l/2|v|$

12-2. *Triangular wave pulse, II.* Carry out the construction of a wave pulse moving in the negative x direction but otherwise as described in Exercise 12-1.

12-3. *Bird on a clothesline.* A child plucks one end of a taut clothesline, sending a wave pulse toward a sparrow perched 6.0 m down the line. The speed of the wave is 8.0 m/s. How much time does the sparrow have to fly away in order to avoid being shaken by the pulse?

12-4. *Tsunami warning!* A burst of seismic waves is detected at a seismic station located at point S in Fig. 12E-4. Seismologists analyze the record and are able to determine that a submarine earthquake has occurred at point Q, 900 km from the station and 200 km offshore from a seaport P.

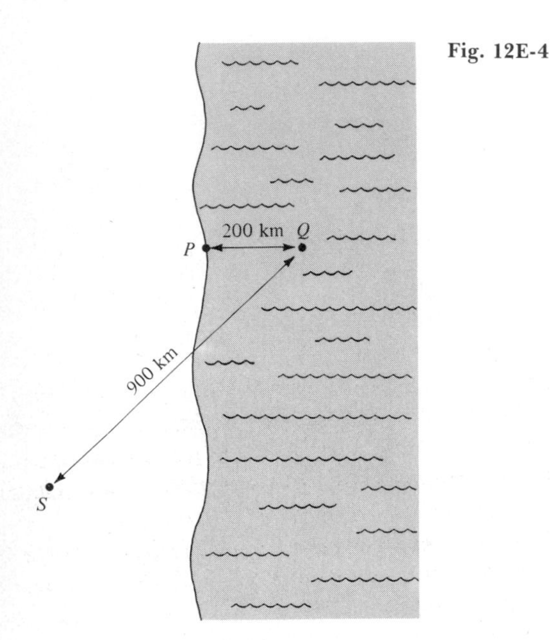

Fig. 12E-4

a. How much time was required for the seismic waves from the quake to reach the seismic station? Assume that the average speed of seismic waves is 10 km/s, and that the actual path followed by the waves in going from Q to S is 900 km in length.

b. The seismologists realize that the quake must have produced a potentially destructive tsunami (often inaccurately called a "tidal wave"). This is a water wave propagating along the ocean surface with a speed of 150 m/s. Once the seismic waves have been detected at point S, how much time is available for evacuation of low-lying areas of the port P? (Assume that the seismic record is interpreted with a negligible loss of time.)

12-5. *A prodigious explosion!* On June 30, 1908, a huge explosion occurred over Siberia. The sound from this explosion (which was apparently due to a very large meteorite) was heard 500 km away! How much time elapsed between the actual explosion and the arrival of the sound waves at such distant locations? Make the assumption that the speed of sound was a uniform 340 m/s.

12-6. *Mathematical description of a wave.* A transverse sinusoidal wave is traveling in the negative direction, having amplitude 0.10 m, wavelength 2.0 m, and period 0.50 s, write its wave function in the forms of both Eq. (12-18) and Eq. (12-21) assuming that $\delta = 0$.

12-7. *What's the wavelength?* The frequency of electromagnetic waves used to transmit television signals is about 1×10^8 Hz. What is the wavelength of these waves? The speed of electromagnetic waves is 3.00×10^8 m/s.

12-8. *Reading a wave function.* The equation for a sinusoidal traveling wave is given by

$$y = 0.00500 \text{ m} \times \cos[(10.0 \text{ m}^{-1})\pi x - (40.0 \text{ s}^{-1})\pi t]$$

a. What are the direction of propagation, amplitude, wavelength, and velocity of the wave?

b. What is the displacement y at $x = 0.0500$ m at time $t = 0$ s; $t = 0.0125$ s; and $t = 0.0375$ s?

c. What is the displacement y at $t = 0.0125$ s at position $x = 0$ m; $x = 0.0500$ m; and $x = 0.150$ m?

12-9. *A traveling wave, I.* A sinusoidal wave is traveling along a stretched string. The point on the string at the origin ($x = 0$) oscillates in a manner described by the equation $y = A\cos(\omega t)$, where $A = 0.10$ m and $\omega = 20\pi \text{ s}^{-1}$. The point at $x = 0.050$ m oscillates in a manner described by the equation $y = A\cos(\omega t - \pi/4)$.

a. What is the frequency of the wave?

b. Assume that the wave is traveling in the positive x direction and that its wavelength is greater than 0.050 m. Find the wavelength and the wave velocity. Write an equation $y = f(x, t)$ that describes this traveling wave.

12-10. *To preserve the wave speed.* A stretched string is replaced by another of the same material but with twice the diameter. What should be the ratio of the new tension to the old if the wave speed is to remain the same?

12-11. *Transverse motion of the string itself.* The equation of a certain one-dimensional transverse traveling wave along a string is given by $y = A\cos(kx - \omega t)$, where $A = 0.0050$ m, $k = 10\pi \text{ m}^{-1}$, and $\omega = 40\pi \text{ s}^{-1}$.

a. What is the expression for the *tranverse* velocity dy/dt of any part of the string?

b. At $t = 0$, what is the value of this velocity at $x = 0.050$ m? What is the value of the displacement at the same point at $t = 0$?

12-12. *Waves on a nylon cord.* A nylon cord whose diameter is 0.50 cm is held under a tension of 500 N. The density of nylon is 1.1 g/cm³. Find the speed at which a transverse wave will propagate along the cord.

12-13. *Energy flow in a sinusoidal traveling wave.* A transverse sinusoidal wave is traveling at a speed of 20 m/s along a cable whose linear density is $\mu = 0.20$ kg/m. The wave amplitude is 0.10 m and the frequency is 5.0 Hz.

a. Determine the wavelength.
b. Determine the tension in the cable.
c. Find the average energy density in the wave.
d. Find the average energy flux carried by the wave.

12-14. *Across the bay.* A wave source of power 100 W, located in an otherwise calm bay, creates spreading circular water waves of frequency 1.0 Hz.

a. What average energy flux (in W/m) is carried past an observation point P_1 located 30 m from the source?

b. What is the average energy flux passing point P_2, which is located 60 m from the source?

c. Express the wave amplitude at point P_2 as a fraction of the amplitude at point P_1.

12-15. *Albatross.* A shipwrecked sailor adrift in a lifeboat can just barely hear the cries of an albatross. The bird is flying high above the ocean, and the straight-line distance between the bird and the sailor is 3 km. As you will learn in Chapter 13, the threshold (minimum energy flux) for human hearing is about 1×10^{-12} W/m^2.

a. Assuming that the albatross emits sound isotropically (equally in all directions), estimate the power it emits during its cries.

b. Suppose that the albatross cries 1000 times daily and that each cry lasts 1 s. How much energy does it emit in sound waves each day?

12-16. *Beep beep.* Two cars are traveling in the same direction with the same speed, 30 m/s. The driver of the rear car blows her horn, which has a frequency of 200 Hz. What frequency does the driver of the front car hear?

12-17. *Doppler demonstration.* In a lecture demonstration a whistle is mounted at the end of a 0.500 m arm of a rotator which is turning at 2.00 revolutions per second. The whistle emits a sound whose frequency is 300 Hz. What are the highest and lowest frequencies that the audience hears? Take the velocity of sound to be 340 m/s.

Group B

12-18. *A leftward-moving wave pulse, I.* Derive directly the wave function $f(x, t) = f(x + |v|t)$ for a wave pulse traveling at speed $|v|$ in the negative x direction. Use an argument similar to the one leading to Eq. (12-6), but with observer O' moving in the negative x direction.

12-19. *A leftward-moving wave pulse, II.* Verify directly that the wave function $f(x, t) = f(x + |v|t)$ describes a wave pulse traveling at speed $|v|$ in the negative x direction. Use an argument like the one in small print following Eq. (12-6).

12-20. *Wave function and wave shape.* The wave function

$$f(x, t) = Ae^{-B(x-vt)^2}$$

describes a single wave pulse, traveling at velocity v, which has a shape quite like the one shown in Fig. 12-4. To verify this, choose whatever dimensionally consistent values you wish for the arbitrary constants A and B, and for the wave velocity v, and then plotting the wave function versus x for two different values of t. How do you determine the wave velocity from your plots?

12-21. *A leftward-moving sinusoidal wave.* Derive the wave function

$$f(x, t) = A \cos\left[2\pi\left(\frac{x}{\lambda} + \frac{t}{T}\right) + \delta\right]$$

for a sinusoidal wave traveling with velocity $v = -\nu\lambda = -\lambda/T$. Use an argument like the one leading to Eq. (12-17).

12-22. *Slope and velocities.* Suppose that a transverse sinusoidal wave is traveling along a string. Prove that at any time t the slope $\partial y/\partial x$ of the string at any point x is equal to the negative of the instantaneous transverse velocity $\partial y/\partial t$ of the string at x divided by the wave velocity v. That is, show that $\dfrac{\partial y}{\partial x} = -\dfrac{\partial y/\partial t}{v}$.

12-23. *Find the wave function.* A sinusoidal wave traveling in the positive direction on a stretched string has amplitude 2.0 cm, wavelength 1.0 m, and wave velocity 5.0 m/s. At $x = 0$ and $t = 0$, you have $y = 0$ and $\partial y/\partial t < 0$. Find the wave function $y = f(x, t)$ which describes the wave.

12-24. *Confirming a solution.* By direct substitution into the wave equation for a stretched string, Eq. (12-28), show that the wave function

$$f(x, t) = Ae^{-B(x-vt)^2}$$

is a solution to the equation if $v = \pm\sqrt{F/\mu}$, where F is the tension in the string and μ is its linear density.

12-25. *Waves on a steel cable.* A gondola that is used to carry sightseers out over a deep gorge is suspended from a number of steel cables. Each cable is 3.0 cm in diameter and is kept under a tension of 1.0×10^4 N. The density of steel is 7.8 g/cm^3. With what speed would transverse waves propagate along the cables?

12-26. *Railroad crossing.* A train sounds its whistle as it approaches and leaves a railroad crossing. An observer at the crossing measures a frequency of 219 Hz as the train approaches and a frequency of 184 Hz as the train leaves. The speed of sound is 340 m/s. Find the speed of the train and the frequency of its whistle.

12-27. *On wings of song.* A hawk is flying directly away from a birdwatcher and directly toward a distant cliff at a speed of 15 m/s. The hawk produces a shrill cry whose frequency is 800 Hz.

a. What is the frequency in the sound that the birdwatcher hears directly from the bird?

b. What is the frequency that the birdwatcher hears in the echo that is reflected from the cliff?

12-28. *On the track.* As shown in Fig. 12E-28, an observer P is standing between two parallel train tracks when two trains approach from opposite directions. Locomotive A has a speed $|v_A| = 15$ m/s. It toots its whistle, which has a frequency $\nu_0 = 200$ Hz. Locomotive B has a speed $|v_B| = 30$ m/s. The speed of sound in the air is 340 m/s, and no breeze is blowing.

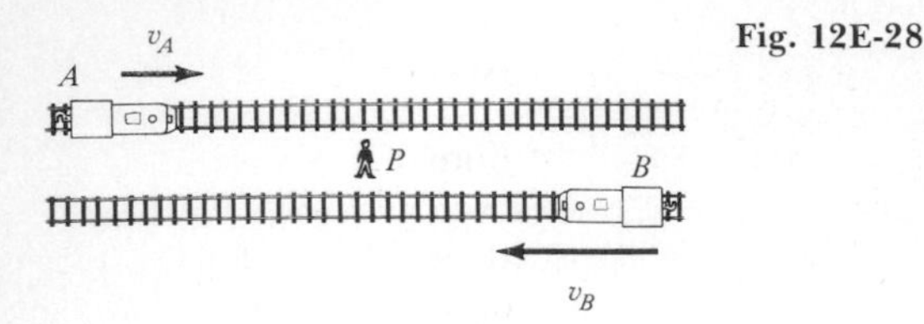

Fig. 12E-28

a. Find the wavelength λ_1 and frequency ν_1 of the sound waves observer P receives from locomotive A.

b. What frequency ν_2 is heard by the engineer on locomotive B?

Some of the sound waves reaching locomotive B are reflected back toward observer P and locomotive A.

c. Find the wavelength λ_3 and the frequency ν_3 of the reflected sound waves that observer P hears.

d. What frequency ν_4 does the engineer on locomotive A hear in the reflected waves?

12-29. *Songs of the whales.* Several whale species are apparently able to communicate over distances of up to 50 km or more. The speed of sound in water is 1400 m/s.

a. If two whales located 50 km apart begin to communicate, what is the minimum time lag experienced by each whale between the emission of its call and its reception of a response?

b. The maximum swimming speed for a whale is about 8 m/s. Suppose that one whale emits a sound of frequency precisely 100 Hz. Find the maximum and minimum possible frequencies heard by another whale.

12-30. *Moving medium.* A wave source of frequency ν and an observer are located a fixed distance apart. Both the source and the observer are stationary. However, the propagation medium (through which the waves travel at speed v) is moving at a uniform velocity $\mathbf{v}_m$ in an arbitrary direction. Find the frequency ν' received by the observer. Explain your result physically.

12-31. *A traveling wave, II.* Consider once again the sinusoidal traveling wave described in Exercise 12-9.

a. Find the frequency of the wave.

b. Suppose that no information is given about the direction of propagation or the wavelength. Show that for each direction of propagation, there are an infinite number of possible wavelengths. Determine those wavelengths and the corresponding wave speed. [*Hint*: $\cos\varphi = \cos(\varphi + 2n\pi)$, where n is an integer.]

12-32. *Rope trick.* A loop of rope is whirled at a high angular velocity, ω, so that it becomes a taut circle of radius R. A kink develops in the whirling rope; see Fig. 12E-32.

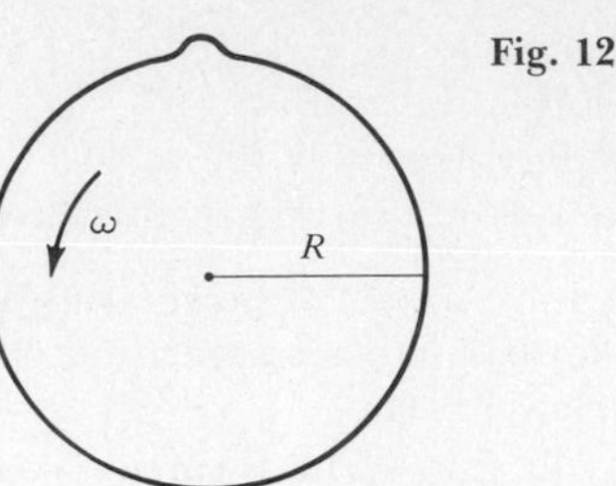

Fig. 12E-32

a. Show that the tension in the rope is $F = \mu\omega^2R^2$, where μ is the linear density of the rope.

b. Under what conditions does the kink remain stationary relative to an observer on the ground?

Group C

12-33. *The wave equation for a moving observer, I.* The wave equation for a stretched string can be written

$$\frac{\partial^2 f(x, t)}{\partial x^2} = \frac{1}{v^2}\frac{\partial^2 f(x, t)}{\partial t^2}$$

where $v = \pm\sqrt{F/\mu}$ is the velocity of waves measured by an observer O in an inertial reference frame motionless with respect to the string. An observer O' is stationed in an inertial reference frame moving at a constant velocity V with respect to the frame of observer O. The position and time variables used by O' are x' and t'. Their relations to the variables x and t used by O are given by the Galilean equations for transforming position and time, $x' = x - Vt$ and $t' = t$. Use these relations to show that when transformed into the variables x' and t' the wave equation becomes

$$\left(1 - \frac{V^2}{v^2}\right)\frac{\partial^2 f(x', t')}{\partial x'^2} = \frac{1}{v^2}\frac{\partial^2 f(x', t')}{\partial t'^2} - \frac{2V}{v^2}\frac{\partial^2 f(x', t')}{\partial x'\partial t'}$$

(Hint: You will need to make repeated use of the "chain rules"

$$\frac{\partial g}{\partial x} = \frac{\partial g}{\partial x'}\frac{\partial x'}{\partial x} + \frac{\partial g}{\partial t'}\frac{\partial t'}{\partial x}$$

and

$$\frac{\partial g}{\partial t} = \frac{\partial g}{\partial x'}\frac{\partial x'}{\partial t} + \frac{\partial g}{\partial t'}\frac{\partial t'}{\partial t}$$

where g is any function of x and t, or of x' and t'.) An exercise in Chapter 14 calls for a calculation which is similar to the one done here, except that the Lorentz equations for transforming position and time in the relativistic domain are applied to the wave equation for light propagation.

12-34. *The wave equation for a moving observer, II.* By direct substitution into the transformed wave equation found in Exercise 12-33, prove that the function

$$f(x', t') = A \cos\{B[x' - (v - V)t']\}$$

is a solution of the wave equation. Explain why the form of this wave function shows immediately that it describes a wave traveling with a velocity that observer O' measures to be $v - V$. How does the Galilean velocity transformation show immediately that this is just the velocity that you would expect observer O' to measure? In an exercise of Chap. 14 a result is obtained that is in striking contrast to the one obtained here.

12-35. *Spherical waves: the implications of geometry.* Modify the arguments leading to Eqs. (12-63), (12-64), and (12-65) so that they apply to uniform spherical waves. Show thereby that for such waves the average energy flux, average energy density, and amplitude obey the relations $\langle S \rangle = \langle P \rangle / 4\pi r^2$, $\langle \rho_E \rangle = \langle P \rangle / 4\pi v r^2$, and $A \propto 1/r$.

12-36. *Energy in a triangular wave pulse.* A triangular transverse wave pulse of amplitude A and wavelength $\lambda = 2l$, travels in the positive x direction along a string with tension F and linear density μ. It is described by the wave function $y = f(x, t)$, where $f(x, t) = A(1 - |x - vt|/l)$ for $|x - vt| < l$ and $f(x, t) = 0$ otherwise. Here $v = (F/\mu)^{1/2}$.

a. Find the total energy density $\rho_E(x, t)$ in this wave pulse. Express your result in terms of F, μ, A, and l.

b. Integrate the expression found in part *a* over all values of x to show that the total energy E carried by the traveling wave pulse is $2FA^2/l$.

c. Use the result of *a* to determine the energy flux $S(x, t) = \rho_E(x, t)\, v$.

d. For any fixed value of x, integrate $S(x, t)$ from $t = -\infty$ to $t = +\infty$ in order to verify that the wave pulse carries (past each point x) the total energy E found in part *b*.

12-37. *Wave energy in a nylon cord.* Use the results of Exercise 12-36 to evaluate the energy density ρ_E, the energy flux S, and the total energy E carried by a triangular wave with amplitude $A = 1.0$ cm and overall length $2l = 20$ cm traveling along the nylon cord described in Exercise 12-12. Express your results in SI units.

12-38. *Wave energy in a steel cable.* Use the results of Exercise 12-36 to evaluate the energy density ρ_E, the energy flux S, and the total energy E carried by a triangular wave with amplitude $A = 0.10$ cm and overall length $2l = 20$ cm traveling along the steel cable described in Exercise 12-25. Express your results in SI units.

12-39. *A second flight for the bumblebee?* The following quotation is due to British physicist Lord Rayleigh [*The Theory of Sound* 2d ed. (1896; reprinted by Dover, New York, 1945), **2**:154]:

"The pitch of a sound is liable to modification when the source and the recipient are in relative motion . . . if v be the velocity of the observer and a that of sound, the frequency is altered in the ratio $(a \pm v):a$, according as the motion is towards or from the source. [If the motion is from the source and] we could suppose v to be greater than a, a sound produced after the motion had begun would never reach the observer, but sounds previously excited would be gradually overtaken and heard in the reverse of the natural order. If $v = 2a$, the observer would hear a musical piece in correct time and tune, but *backwards.*"

a. Carefully explain why Lord Rayleigh's statement is true in principle.

b. At the conclusion of an outdoor performance of Rimsky-Korsakov's *The Flight of the Bumblebee,* a fanatical music lover (having read Lord Rayleigh's statement) speeds away at twice the speed of sound! How far from the bandstand will he be by the time he has heard the complete backward performance? (A typical performance time for the composition is 80 s.)

12-40. *Doppler effect in two dimensions: moving source.* Figure 12E-40 shows a wave source S of intrinsic frequency ν passing an observer O. Both the observer and the propagation medium are at rest. The source is moving with constant velocity $\mathbf{v}_S$. At the instant shown, the source is located at $\mathbf{r}_s = r_s\hat{\mathbf{r}}_s$ with respect to O. The wave speed in the medium is v.

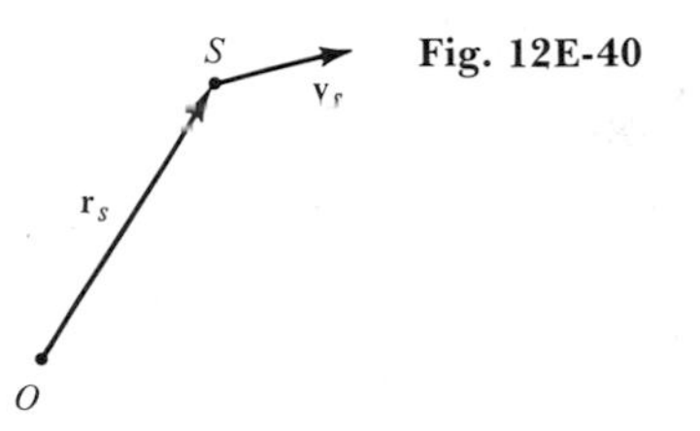

Fig. 12E-40

a. Show that when the waves emitted by S at the instant shown are later received at O, the received frequency ν' will be given by

$$\nu' = \nu \left(\frac{v}{v + \mathbf{v}_S \cdot \hat{\mathbf{r}}_s} \right)$$

b. Show that the result of part *a* reduces to Eq. (12-71) in the particular case that the source moves along a line through the observer.

12-41. *Doppler effect in two dimensions: moving observer.* Figure 12E-41 shows a wave source S of intrinsic frequency ν. It is immersed in a medium in which the wave speed is v. Both the source and the medium are at rest. An observer O moves past the source with constant velocity $\mathbf{v}_o$. At the instant shown, the observer is located at $\mathbf{r}_o = r_o\hat{\mathbf{r}}_o$ with respect to the source.

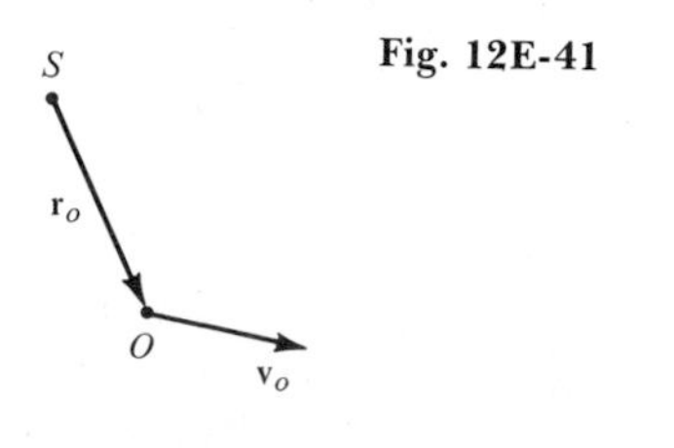

Fig. 12E-41

a. Show that the frequency ν' of the wave being received by O at the instant shown is given by

$$\nu' = \nu \frac{v - \mathbf{v}_o \cdot \hat{\mathbf{r}}_o}{v}$$

b. Show that the result of part *a* reduces to Eq. (12-72) when the observer moves along a line through the source.

12-42. *Waves from a supersonic source.*

a. Modify Fig. 12-20*a* so that it pertains to a case in which the speed $|v_s|$ of the source through the propagation medium is greater than the speed $|v|$ of the waves with respect to the medium, say twice as large.

b. Use the modified figure to show that all the waves emanating from the source are tangent to a cone of apex angle 2α, whose apex moves along with the source (assumed to be a point source). This conical envelope is the "shock wave" produced, for example, by a supersonic airplane. It is called the **Mach cone** after the Austrian philosopher and physicist Ernst Mach (1838-1916), who was the first to obtain photographs of the shock waves of supersonic projectiles. The half-angle α is called the **Mach angle.**

c. Find an equation which gives the Mach angle in terms of the ratio $M \equiv |v_s|/|v|$ of the two speeds. This ratio is the famous **Mach number.**

13

Superposition of Mechanical Waves

13-1 SUPERPOSITION OF WAVES

When a wave pulse or a wave train is passing through a certain region of space, it does not exclude other waves from that region. That is, as many waves as you like can pass simultaneously through the same location in a medium. This property of waves is diametrically opposite to a property we take for granted in studying the mechanics of rigid bodies, namely, that no more than one body can be present at a given point in space at a given moment.

This chapter is devoted to a study of what happens when two or more mechanical waves pass through the same region simultaneously. An understanding of many very important physical phenomena will emerge from this study.

Everyone has seen the complex pattern which is set up when two or more sets of ripples, generated on the surface of water, interpenetrate. A naturally occurring example and a carefully controlled laboratory demonstration of this phenomenon are illustrated in Fig. 13-1*a* and *b*, respectively. The primary question suggested by these pictures can be phrased in physical terms by returning to the example with which Chap. 12 began—a cork bobbing on the surface of a pond. The pond is covered with a circular pattern of ripples radiating from the spot where the pebble was dropped. If another pebble is dropped into the pond at a different point, a new set of ripples will come into being, centered on that point. How will the cork behave under the combined influence of the two sets of circular ripples? Or, more generally, how will any cork, located anywhere on the pond, behave when any number of pebbles are dropped into the pond at arbitrary locations and at arbitrary times?

This physical question must be rephrased in mathematical terms so that we can undertake an analysis. In these terms, the question becomes:

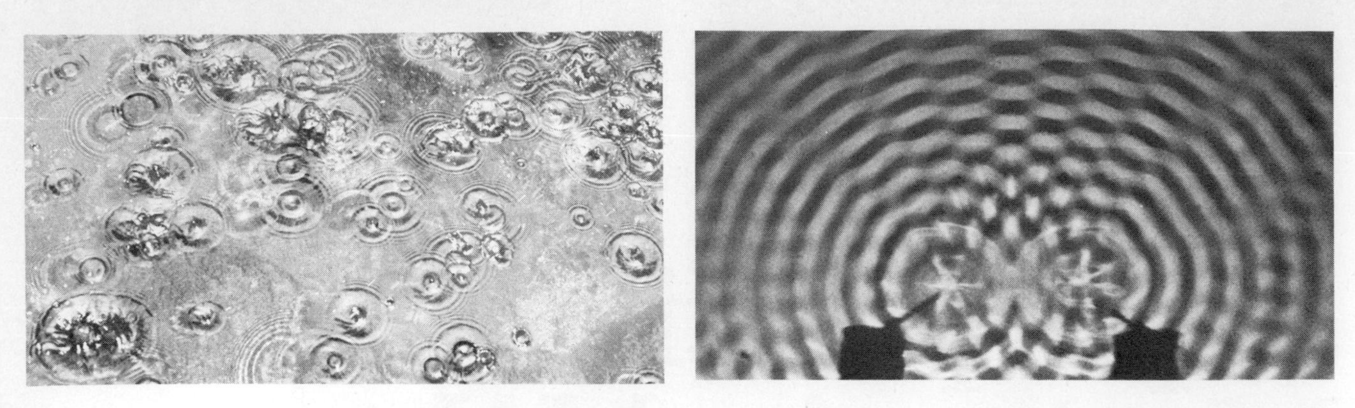

(a) (b)

Fig. 13-1 (a) Raindrops falling on a pond. (*Courtesy Elinor S. Beckwith.*) (b) A ripple-tank photo in which two oscillating rods are producing two interpenetrating circular wave trains. (*From PSSC Physics, 2d ed., Boston, D. C. Heath, 1965. Courtesy Education Development Corporation.*)

What is the form of the complete wave function produced by the *superposition* of two or more wave functions?

The answer to this question is the simplest conceivable one, provided the amplitude of the waves is not too great. (We will see what this restriction means later in this section.) If this condition is met, *the resultant wave function is simply the algebraic sum of the individual wave functions.* That is, at any particular location and at any particular moment, the displacement produced by the resultant wave is the algebraic sum of the displacements which would be produced by the individual waves if each were acting separately.

This statement is true of waves propagating in one dimension or in three dimensions, as well as in the two-dimensional situation exemplified by the ripples. In the most general case, n different wave trains propagate through three-dimensional space. An observer fixed in an inertial frame can specify the location of any point by means of a set of coordinates (x, y, z). The jth wave train is described by the wave function $f_j(x, y, z, t)$, where t denotes the time elapsed since a moment arbitrarily designated $t = 0$. In terms of these quantities, the rule for determining the resultant wave function, stated in the previous paragraph, can be expressed mathematically in the form

$$\begin{aligned} f(x, y, z, t) &= f_1(x, y, z, t) + f_2(x, y, z, t) + f_3(x, y, z, t) + \cdots + f_j(x, y, z, t) \\ &\quad + \cdots + f_n(x, y, z, t) \\ &= \sum_{j=1}^{n} f_j(x, y, z, t) \end{aligned} \tag{13-1}$$

This relation is called the **principle of superposition.** Note that it is based on two related but distinct assumptions:

1. Each individual wave, described by the wave function $f_j(x, y, z, t)$, is unaltered by the simultaneous presence of the other waves.

2. The resultant wave, described by the wave function $f(x, y, z, t)$, is the algebraic sum of the functions $f_j(x, y, z, t)$, as already stated.

The ultimate justification of the principle of superposition must be in experimental observation.

One practical consequence of this principle is quite familiar. In listening to an orchestra, your ears are immersed in the resultant wave train produced by the superposition of the wave trains whose sources are many instruments. The ear and the mind are responding to a very complex wave function. But it is not at all difficult, with a little experience, to pick out the sounds of the individual instruments. In doing this, the ear and the mind act together as a "filter" to single out the wave function typical of the particular instrument. This is possible because wave 1 (say that produced by the flute) is not altered by the simultaneous presence of wave 2 (that produced by the violin) or wave 3 (that produced by the harp).

In the case of waves propagating on a string, the superposition principle simplifies to the one-dimensional statement

$$f(x, t) = f_1(x, t) + f_2(x, t) + f_3(x, t) + \cdots + f_j(x, t) + \cdots + f_n(x, t)$$
$$= \sum_{j=1}^{n} f_j(x, t) \qquad (13\text{-}2)$$

Figure 13-2 illustrates the superposition of two one-dimensional wave pulses which arrive at the same point while moving in opposite directions. In Fig. 13-2*a*, both pulses displace the long spring from its equilibrium position in the same direction; that is, f has the same sign for both pulses. The two pulses, moving independently, come together, momentarily merge into a single large pulse, and then separate. Each one continues on its way *as though the other had not been there.*

In Fig. 13-2*b*, the pulses are of equal magnitude but opposite sign. The addition process is nonetheless identical to the previous one, if due attention is paid to signs in adding amplitudes. In this case, the result is a momentary "disappearance" of the pulses as they merge.

A natural question is: At the moment when the spring is straight, how does it "know" that it must not remain at rest, but must again distort as the pulses move apart? The answer is that while the displacement $f(x, t)$ is zero for all parts of the spring at this instant, the transverse velocity $\partial f(x, t)/\partial t$ is *not* zero.

While Fig. 13-2*a* and *b* illustrates two cases of the same phenomenon, their appearance is so different that they are often given the separate names **constructive interference** and **destructive interference,** respectively. In the more general case of repetitive wave trains superposing, however, both kinds of interference take place simultaneously. Therefore we will usually use the more general term **superposition** in discussing the subject, so as to avoid the confusion of unnecessary terminology.

Wave trains as well as pulses can be made to superpose so as to demonstrate constructive and destructive interference. Figure 13-3 illustrates the apparatus called the **acoustical interferometer.** It is a simple analogue of the *Michelson interferometer,* a very important instrument described in Sec. 14-2. A loudspeaker, which emits a pure sinusoidal sound wave of wavelength λ, is fitted to one end of the two-branched "trombone." The length of the trombone, measured via the upper arm, is fixed. In particular, the length of the path PRQ is fixed. The length of the lower arm PSQ can be adjusted by moving the close-fitting slide, as suggested by the double-headed arrow.

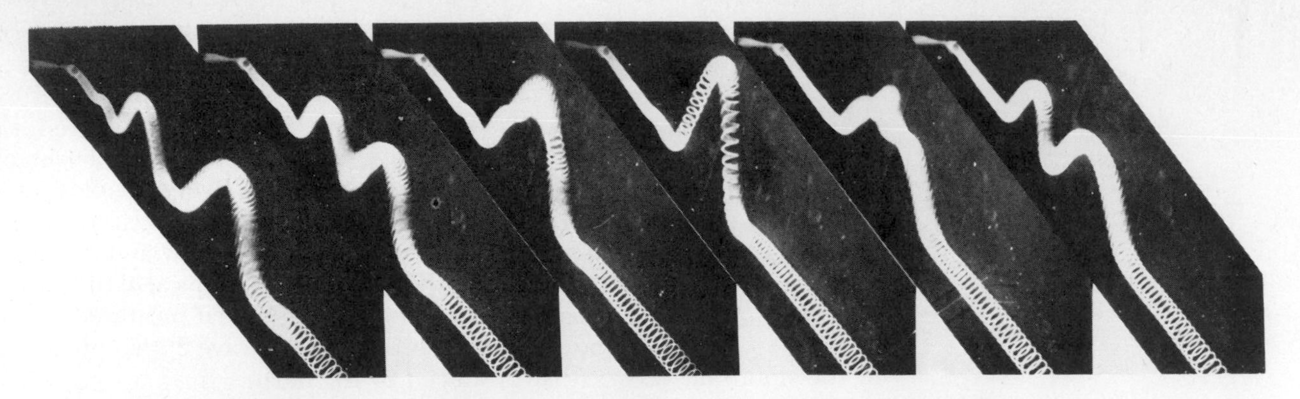

(*a*)

(*b*)

Fig. 13-2 (*a*) Two equal wave pulses of the same sign produce a large net displacement of the spring at the instant that they move through each other. (*b*) Two equal pulses of the opposite sign produce essentially no net displacement at the instant that they move through each other. (*Courtesy Education Development Corporation.*)

As the slide is gradually pulled out so as to lengthen the lower arm, the sound heard at the horn fades and disappears, then increases to a maximum, fades again, and so on, as far as the length of the slide permits the experiment to proceed. These observations can be explained in terms of the superposition principle.

The incoming sinusoidal sound wave may be described by the wave function of Eq. (12-21). With a slight change in notation, this can be written

$$f(x, t) = A\cos(kx - \omega t + \delta)$$

If we choose the origin at P, the sound wave arriving there is described by the space-independent wave function

$$f(t) = A\cos(-\omega t + \delta)$$

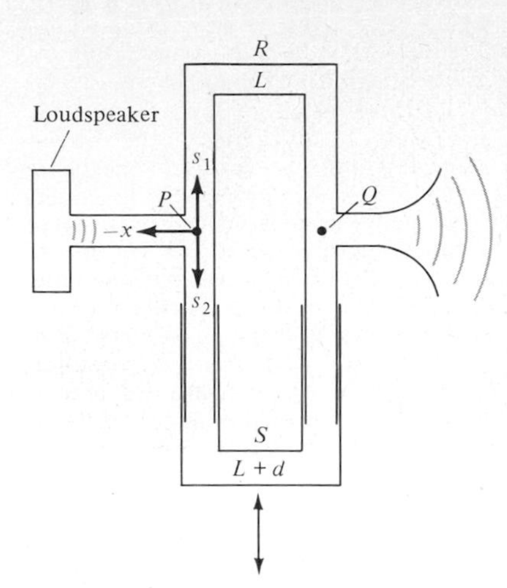

Fig. 13-3 The acoustical interferometer. A pure sinusoidal sound wave produced by the loudspeaker passes down a tube and splits into two parts at P. The two parts recombine at Q according to the superposition principle. The loudness of the sound heard at the horn depends on the relative phase of the two partial waves at Q, which in turn is determined by the difference in path length along the two arms PRQ, of length L, and PSQ, of length $L + d$.

The sound wave divides at P into two parts whose amplitudes are A_1 and A_2. The two separate wave trains pass through the arms PRQ and PSQ and superpose at Q. In order to describe the total wave function resulting from this superposition, we must describe the separate wave trains by the proper wave functions. The two sinusoidal waves may also be represented by wave functions of the form of Eq. (12-21). What is important here, however, is not the horizontal distance x but the distances s_1 and s_2 measured along the lengths of the fixed and movable arms, respectively. We choose the origin for both s_1 and s_2 at the branch point P. The wave functions can thus be written

$$f_1(s_1, t) = A_1 \cos(ks_1 - \omega t + \delta)$$

and

$$f_2(s_2, t) = A_2 \cos(ks_2 - \omega t + \delta)$$

(We assume for the sake of generality that the amplitudes A_1 and A_2 of the two wave trains are not equal.)

Since the two waves were obtained by "splitting" the original wave into two parts, the space-independent wave functions describing the two waves at point P must add to yield the space-independent wave function describing the original wave at that point. We thus have

$$f_1(t) + f_2(t) = f(t)$$

This gives

$$A_1 \cos(-\omega t + \delta) + A_2 \cos(-\omega t + \delta) = A \cos(-\omega t + \delta)$$

or

$$A_1 + A_2 = A$$

That is, the amplitude of the original space-independent wave function f at point P is the sum of the amplitudes of the space-independent wave functions f_1 and f_2 at the same point. (The equation $A_1 + A_2 = A$ can have meaning only at that point and at the symmetrical point Q, since these are the only locations at which all three wave functions are defined.)

Consider what happens when both arms of the apparatus have the same length L. The path length from P to Q measured along the fixed arm PRQ is $s_1 = L$. The path length from P to Q measured along the movable arm PSQ is $s_2 = L$. Because of the symmetrical geometry in the vicinities of points P and Q, the two waves combine at Q under the same circumstances as those under which they split at P. Thus at Q, the two waves superpose to yield the total space-independent wave function $g(t)$ given by the sum

$$g(t) = f_1(L, t) + f_2(L, t) = A_1 \cos(kL - \omega t + \delta) + A_2 \cos(kL - \omega t + \delta)$$

or

$$g(t) = (A_1 + A_2) \cos(kL - \omega t + \delta) = A \cos(kL - \omega t + \delta)$$

Rearranging terms to put together all the constants in the argument of the cosine gives

$$g(t) = A \cos[-\omega t + (\delta + kL)]$$

That is, the total space-independent wave function $g(t)$ at Q is identical to the original wave function at P, $f(t) = A \cos(-\omega t + \delta)$, except for an additional phase constant kL. The amplitude of $g(t)$, like that of $f(t)$, is equal to

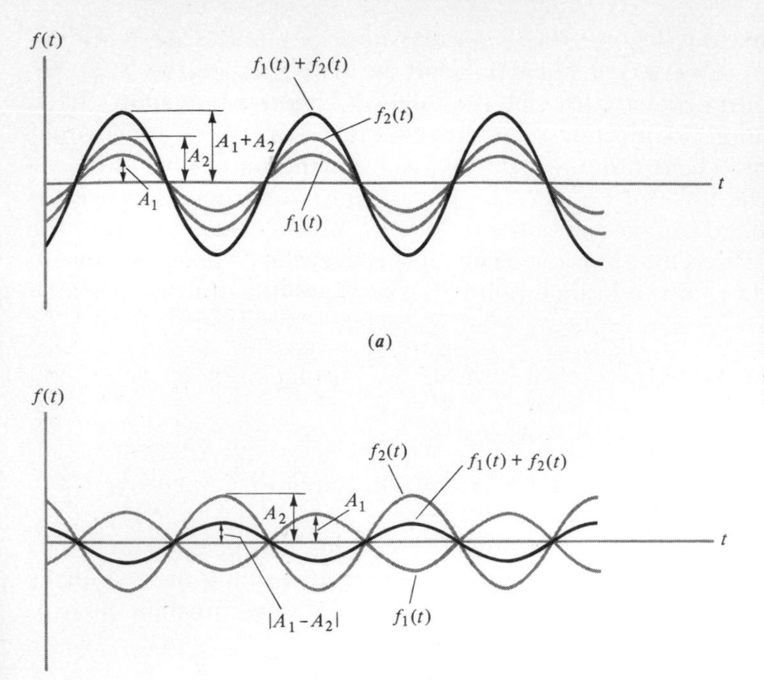

Fig. 13-4 (*a*) Two sinusoidal waves in phase. One is represented by the wave function $f_1(t) = A_1 \cos(-\omega t + \delta_1)$ and the other by the wave function $f_2(t) = A_2 \cos(-\omega t + \delta_2)$. The amplitudes A_1 and A_2 may be different, but the angular frequency ω is the same for both functions and the phase constants satisfy the equality $\delta_1 = \delta_2$. The two waves superpose constructively, as can be seen from the dark curve which represents the function $g(t) = f_1(t) + f_2(t)$, whose amplitude is $A_1 + A_2$. (*b*) Two sinusoidal waves out of phase. The only difference between the wave functions represented here and those in part *a* is that the phase constant δ_2 is given by $\delta_2 = \delta_1 + \pi$. The amplitudes A_1 and A_2, the angular frequency ω, and the phase constant δ_1 are the same as those in part *a*. The two waves superpose destructively, as can be seen from the dark curve which represents the function $g'(t) = f_1(t) + f_2(t)$, whose amplitude is $|A_1 - A_2|$.

the *sum* of the amplitudes of f_1 and f_2. Figure 13-4*a* depicts the superposition of the space-independent wave functions at point Q. The two functions f_1 and f_2 are said to be **in phase.**

Now consider what happens when the slide is pulled out so that the length $L + d$ of the movable arm is greater than the length L of the fixed arm by one-half the wavelength of the sound wave. That is, we have $d = \lambda/2$. Then the path lengths in the two arms are $s_1 = L$ and $s_2 = L + \lambda/2$, respectively. At Q, where the two waves superpose, the total space-independent wave function is given by the sum

$$g'(t) = f_1(L, t) + f_2(L + \lambda/2, t)$$

or

$$g'(t) = A_1 \cos(kL - \omega t + \delta) + A_2 \cos[k(L + \lambda/2) - \omega t + \delta]$$

But the wave number k is defined to be $k \equiv 2\pi/\lambda$, so that $k\lambda = 2\pi$ and we have $k(L + \lambda/2) = kL + \pi$. The second term on the right side of the last displayed equation thus becomes $A_2 \cos(kL - \omega t + \delta + \pi)$. But for any argument θ, we have $\cos(\theta + \pi) = -\cos\theta$, so $g'(t)$ can be written

$$g'(t) = A_1 \cos(kL - \omega t + \delta) - A_2 \cos(kL - \omega t + \delta)$$

or

$$g'(t) = (A_1 - A_2) \cos[-\omega t + (\delta + kL)]$$

That is, the total space-independent wave function $g'(t)$ has an amplitude equal to the magnitude of the *difference* of the amplitudes of f_1 and f_2. Figure 13-4*b* depicts the superposition of the space-independent wave functions at point Q. The two functions f_1 and f_2 are said to be **out of phase.**

Further increase in length of the movable arm, until d is equal to the wavelength λ of the sound wave, leads again to constructive superposition of waves which are in phase. Can you explain why? As the length of the movable arm is increased steadily, the two waves which superpose at Q shift gradually out of, into, and out of phase. The amplitude of the total outgoing wave therefore varies between $A_1 + A_2$ and $|A_1 - A_2|$. If it happens that $A_1 = A_2$, the sound will disappear completely when the waves are out of phase at Q.

The superposition principle can be justified mathematically, as well as on the physical grounds discussed earlier in this section. Consider two wave functions $f_1(x, t)$ and $f_2(x, t)$, each of which describes a possible wave propagating down a stretched string of linear density μ in which there is a tension F. If the function $f_1(x, t)$ represents a possible wave, then it must be a solution to Eq. (12-26), the wave equation for a stretched string. So we have

$$\frac{\partial^2 f_1(x, t)}{\partial x^2} = \frac{\mu}{F}\frac{\partial^2 f_1(x, t)}{\partial t^2} \tag{13-3}$$

Similarly, $f_2(x, t)$ must also be a solution to the same wave equation, so that we have

$$\frac{\partial^2 f_2(x, t)}{\partial x^2} = \frac{\mu}{F}\frac{\partial^2 f_2(x, t)}{\partial t^2} \tag{13-4}$$

What must be proved is that the form for $f(x, t)$ given by Eq. (13-2) is also a solution to the wave equation for the stretched string. That is, we must prove that the equation

$$\frac{\partial^2 f(x, t)}{\partial x^2} = \frac{\mu}{F}\frac{\partial^2 f(x, t)}{\partial t^2} \tag{13-5}$$

is valid when $f(x, t)$ is given by

$$f(x, t) = f_1(x, t) + f_2(x, t) \tag{13-6}$$

To do so, we substitute this expression for $f(x, t)$ into the equation it is supposed to solve, Eq. (13-5). We obtain

$$\frac{\partial^2}{\partial x^2}[f_1(x, t) + f_2(x, t)] = \frac{\mu}{F}\frac{\partial^2}{\partial t^2}[f_1(x, t) + f_2(x, t)]$$

Evaluating the derivatives of the sums on both sides of the last equation gives us

$$\frac{\partial^2 f_1(x, t)}{\partial x^2} + \frac{\partial^2 f_2(x, t)}{\partial x^2} = \frac{\mu}{F}\frac{\partial^2 f_1(x, t)}{\partial t^2} + \frac{\mu}{F}\frac{\partial^2 f_2(x, t)}{\partial t^2}$$

If this equation is satisfied, then Eq. (13-6) is a solution to Eq. (13-5). Applying Eqs. (13-3) and (13-4) shows immediately that it is satisfied, and so we have justified Eq. (13-6), the superposition principle.

The key feature used in our proof is the *linearity* of a derivative. That is, the derivative of the sum of two functions is equal to the sum of the derivatives of the functions. In different words, a derivative is linear because it is proportional to the first power of the dependent variable, which is the function being differentiated. Because each term in the wave equation is linear, it is said to be a *linear* partial differential equation. The superposition principle is a necessary and sufficient consequence of the linearity of the wave equation.

When does an actual *physical* system obey the superposition principle? To do so, it must be correctly described by a linear wave equation. This in turn will be true only if the system is itself linear—that is, if it obeys Hooke's law in the generalized sense discussed in criterion 1 toward the beginning of Sec. 12-6.

Hooke's law must ultimately fail to describe any real medium if the imposed distortions become too large. Waves will still propagate through the medium, but the superposition principle will not be obeyed. In this case, the behavior of the waves is much more complicated, since each wave train interacts with and is affected by the other wave trains which simultaneously pass through the same region of space. Such waves are called *nonlinear*. We discuss nonlinearity only qualitatively and briefly in connection with the behavior of the ear in Sec. 13-7.

Example 13-1 applies the superposition principle to the relatively simple case of two one-dimensional sinusoidal waves.

EXAMPLE 13-1

Show that the superposition principle holds true for the special case of two sinusoidal waves having the same wave number k and angular frequency ω and traveling along a stretched string in opposite directions. One of them has amplitude A and is moving in the positive x direction. It is described by the wave function

$$f_1(x, t) = A \cos(kx - \omega t)$$

This is Eq. (12-21) for the case of motion in the positive x direction, with the phase constant δ set equal to zero for simplicity. The other wave has amplitude B and is moving in the negative x direction. It is described by the wave function

$$f_2(x, t) = B \cos(kx + \omega t)$$

This is Eq. (12-21) for the case of motion in the negative x direction, with the phase constant δ again set equal to zero.

▪ You have already seen in Example 12-4 that the first of these two functions (there written in terms of the wavelength λ and period T instead of k and ω) satisfies the wave equation. You use much the same procedure as in Example 12-4 to show that the sum

$$f(x, t) = A \cos(kx - \omega t) + B \cos(kx + \omega t) \tag{13-7}$$

also satisfies the wave equation. First you evaluate the necessary partial derivatives, obtaining

$$\frac{\partial f(x, t)}{\partial x} = -kA \sin(kx - \omega t) - kB \sin(kx + \omega t)$$

$$\frac{\partial^2 f(x, t)}{\partial x^2} = -k^2 A \cos(kx - \omega t) - k^2 B \cos(kx + \omega t)$$

and

$$\frac{\partial f(x, t)}{\partial t} = -(-\omega)A \sin(kx - \omega t) - \omega B \sin(kx + \omega t)$$

$$\frac{\partial^2 f(x, t)}{\partial t^2} = -\omega^2 A \cos(kx - \omega t) - \omega^2 B \cos(kx + \omega t)$$

Inserting the values of the second partial derivatives into Eq. (13-5),

$$\frac{\partial^2 f(x, t)}{\partial x^2} = \frac{\mu}{F} \frac{\partial^2 f(x, t)}{\partial t^2}$$

you have

$$-k^2[A \cos(kx - \omega t) + B \cos(kx + \omega t)] = -\frac{\mu}{F} \omega^2 [A \cos(kx - \omega t) + B \cos(kx + \omega t)]$$

This equation is satisfied if $k^2 = \mu\omega^2/F$, or

$$\sqrt{\frac{F}{\mu}} = \frac{\omega}{k}$$

But according to the definitions of ω and k given by Eqs. (12-19) and (12-20), you have

$$\frac{\omega}{k} = \frac{2\pi\nu}{2\pi/\lambda} = \nu\lambda$$

where ν is the frequency of the wave. You know from Eq. (12-11*b*) that $\nu\lambda = |v|$, the speed of the wave. Furthermore, Eq. (12-35*b*) shows that $|v| = \sqrt{F/\mu}$. Consequently, ω/k is indeed equal to $\sqrt{F/\mu}$, and you have proved that the wave function $f(x, t)$ given by Eq. (13-7) does satisfy the wave equation, Eq. (13-5). Therefore the superposition principle is obeyed by the particular functions $f_1(x, t) = A\cos(kx - \omega t)$ and $f_2(x, t) = B\cos(kx + \omega t)$.

13-2 REFLECTION OF WAVES

When a pebble is dropped into a small pond, the wave pattern appears quite simple at first, as an array consisting of a number of concentric circular ripples moves outward from the original disturbance. As soon as the first ripples strike the bank, the pattern becomes more complicated since ripples are reflected. After a while the pattern appears quite confused as a result of multiple reflections. Figure 13-5 shows this effect in its simplest form in a ripple tank. A short-term disturbance has produced a sequence of a few circular ripples, which are being reflected by the plane surface at the bottom. Many and varied physical phenomena have their bases in the reflection of waves. A few examples are the vibration of a violin string, the possibility of radio communication between distant stations on the earth, and the operation of many optical instruments. We will now discuss briefly the interaction of waves with boundaries.

First consider the analogous situation in particle motion. When a particle strikes a barrier, what happens depends on the details of the collision. If the barrier and the particle are perfectly elastic, the particle bounces off without loss of mechanical energy.

Waves are reflected without loss of energy when the medium through which they propagate terminates abruptly at a barrier which cannot move. A three-dimensional example is a gas confined in a closed container, the walls of which will reflect sound waves produced inside. A two-dimensional

Fig. 13-5 Ripple-tank circular wave reflected by plane surface. (*Courtesy Education Development Corporation.*)

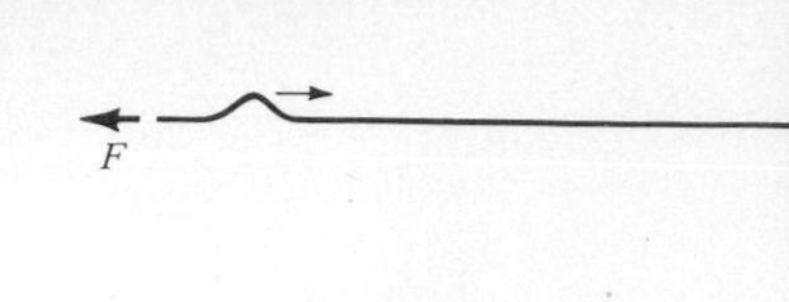

(*a*)

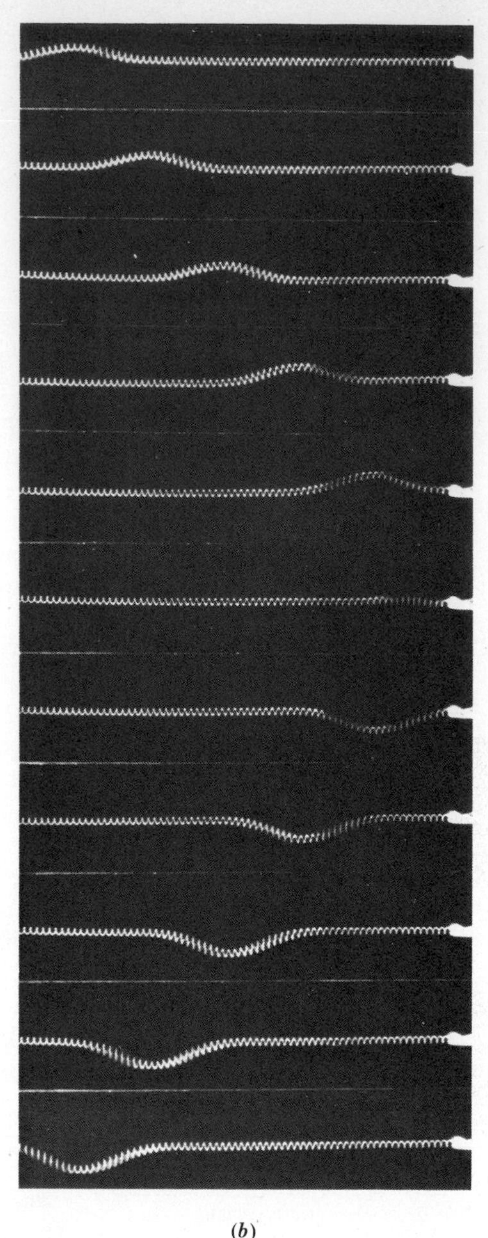

(*b*)

Fig. 13-6 (*a*) An idealized experiment in which a pulse propagates through a long string subjected to a tension F. The left end of the string is not shown. The right end is rigidly fixed. (*b*) A physical realization of the idealized experiment of part *a*, using a spring instead of a string. Note the inversion of the pulse upon reflection from the fixed end of the spring. (*From PSSC Physics, 2d ed., Boston, D. C. Heath, 1965. Courtesy Education Development Corporation.*)

example (shown in Fig. 13-5) is the surface of a pool of water in a container with vertical sides. The one-dimensional example on which we will concentrate is a stretched string.

We begin with the case shown in ideal fashion in Fig. 13-6*a*, where the right end of the string is tied down so that it cannot move. (This is not the same as, but is more stringent than, the condition that the barrier cannot move. In the case of water surface waves, for example, the sides of the container are rigid, but the water in contact with them can oscillate freely up and down. We will consider the analogous case for a stretched string later.)

Figure 13-6*b* is a series of photos of a pulse traveling through a stretched spring. (This spring behaves in essentially the same way as a stretched string. But it is easier to photograph because its flexibility makes practical a smaller tension; hence a large pulse amplitude is attainable and also the pulse travels relatively slowly.) The right end of the spring is rigidly fixed. A positive pulse—a pulse involving displacements only in the positive y direction—moves into the picture from the left. It reaches the fixed end and is reflected as a *negative* pulse—a pulse involving displacements only in the negative y direction—moving toward the left.

To see why this happens, compare the force exerted on an arbitrary element of the string by the element immediately to its right with the force exerted on the element at the extreme right end of the string by the object to which it is rigidly fixed. The arbitrary element experiences at its right end a force having a downward component during the first half of the pulse. (It nevertheless moves upward because it experiences a force at its left end having an upward component of greater magnitude.) When the peak of the pulse reaches the arbitrary element, the force on its right end is horizontal and has no vertical component. Subsequently, the force on the right end has an upward component, whose magnitude first increases and then decreases to zero as the pulse passes. The time average of the force over the entire pulse is zero, and the element is finally at rest in its undisturbed position.

But the element at the extreme right end of the spring experiences a downward force throughout the entire time during which the arriving pulse passes it. The result is a net downward acceleration which the element transmits to its neighbor to the left, leading to a negative pulse. The sixth photo of Fig. 13-6*b* shows the spring at the instant when the peak of the pulse has arrived at the right end of the spring. The first half of the nearly symmetric pulse has already been reflected to the left as a negative pulse. It has superposed with the still-arriving second half of the original pulse, leading to essentially zero displacement of the spring. But the spring is moving downward, as can be seen in subsequent photos.

It is a help to understanding what happens on reflection to mimic the process in the following way. Imagine that the spring, instead of terminating at the point where it is fixed, extends indefinitely to the right. And imagine that a negative "ghost pulse," or virtual pulse, identical to the actual pulse but inverted and moving to the left along the imaginary extension of the spring, arrives at the point where the spring actually ends at the same time as the actual pulse. The two pulses will then superpose on the actual spring, just as is shown in the photos of Fig. 13-6*b*. Just as in the actual situation, the endpoint (which is not fixed in the situation we are imagining) will never move. The actual pulse continues to the right, becoming a virtual pulse. And the negative virtual pulse, moving to the left, becomes the actual reflected pulse.

Let us now consider the opposite extreme case, in which the right end of a stretched string, instead of being fixed, is completely free to move in the y direction. In order for there to be tension in the string, the end of the string must still be constrained so that it cannot move to the left or to the right. The situation is shown in idealized form in Fig. 13-7*a*, where the string terminates in a massless ring which slides freely on a frictionless post. (This is the one-dimensional analogue of the surface of water confined in a container with vertical sides.)

It is not practical to attach a string to a ring sliding on a frictionless post. However, the situation can be reasonably well approximated by attaching the end of a stretched spring to a very light thread, as in Fig. 13-7*b*.

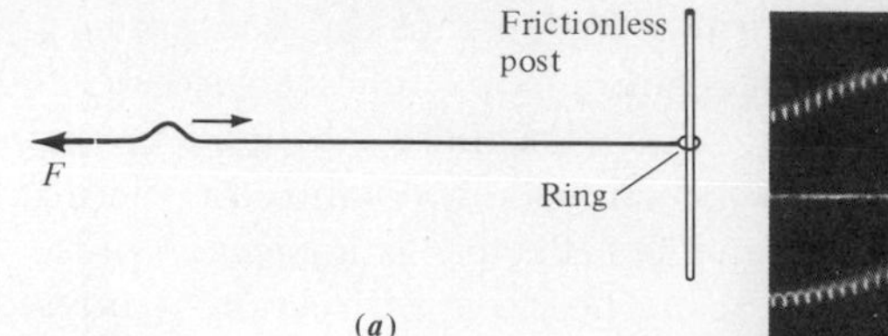

(a)

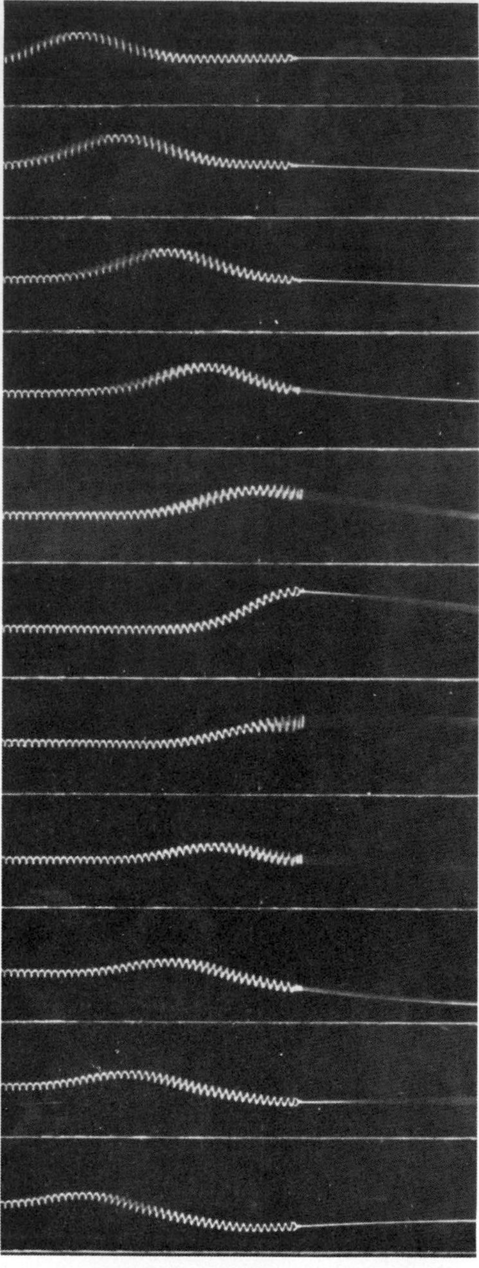

(b)

Fig. 13-7 (*a*) An idealized experiment in which a pulse propagates through a long string subjected to a tension F. The left end of the string is not shown. The right end, attached to a massless ring looped over a frictionless post, is free to move in the y direction. (*b*) A physical realization of the idealized experiment of part *a* using a spring whose end is attached to a very light thread. Note that the pulse is reflected from the end of the spring without change in sign, and the maximum displacement of the end of the spring is greater than the maximum displacement of any other part of the spring. (*From PSSC Physics, 2d ed., D. C. Heath, Boston, 1965. Courtesy Education Development Corporation.*)

The thread is strong enough to maintain the rather small tension in the spring. But its mass is negligible compared to that of the spring, and it offers negligible inertial resistance to transverse displacement.

Just as in Fig. 13-6*b*, a positive pulse arrives at the end of the spring from the left. In this case, however, the reflected pulse moving back toward the left is *not* inverted, but is identical to the original pulse except for its direction of motion. Note also that the maximum displacement of the end of the spring (where it is attached to the thread) is greater than the maximum displacement of other parts of the spring as the pulse passes through them.

These observations can again be understood by comparing the end element of the string to an arbitrarily located element. As noted in Sec.

12-3, the net force exerted in the y direction on an arbitrary element by its neighboring elements depends on its *curvature.* It is therefore accelerated upward during the arrival of the first quarter of the symmetrical positive pulse, downward during the second and third quarters, and upward during the fourth quarter. But the force exerted on the end element by the frictionless post cannot have a y component. It is therefore accelerated upward as long as the arriving pulse has a negative *slope,* that is, during the first half of the pulse. When the peak of the pulse arrives, the end element has a displacement equal to the pulse amplitude. Subsequently, the end element is accelerated in the negative y direction. But it has a positive velocity and therefore overshoots. The symmetry of the situation suggests that the end element comes to rest when its displacement is equal to twice the amplitude of the pulse. The slope of the string is now positive, and the end element is accelerated back toward its undisturbed position. The upward reaction force it exerts on its neighboring element to the left leads to the propagation of a positive pulse to the left. This is the reflected pulse.

In the preceding small-print section, the reflection of a pulse at the fixed end of a stretched string was interpreted in terms of an inverted virtual pulse traveling in the direction opposite to that of the actual pulse on an imaginary extension of the string. The same approach can be used in considering the reflection of a pulse from a string whose end is free to move in the y direction (the case represented in Fig. 13-7*b*). In this case, however, the virtual pulse is not inverted, but has the same sign as the actual pulse. What will be the maximum displacement of the point on the imaginary string of indefinite length, at which the two pulses come together?

13-3 STANDING WAVES

The behavior of musical instruments, lasers, electrons in atoms, and many other diverse systems depends on the phenomenon of *standing waves.* All the general characteristics of this special class of waves can be observed in a stretched string, the relatively simple system already studied.

Like all periodic mechanical waves, a standing wave involves a periodic oscillation of the elements comprising the medium. The elements of a stretched string, for example, oscillate in a direction transverse to the length of the string. But the most evident property of a standing wave is that there are certain fixed locations, called **nodes,** where there is never any motion. This is in contrast with the traveling-wave case, where every part of the medium sooner or later participates in the wave motion. But standing waves have an intimate connection with traveling waves. In the one-dimensional case typified by the string, for example, standing waves can be produced when traveling waves of equal amplitude, frequency, and wavelength move in opposite directions along the string.

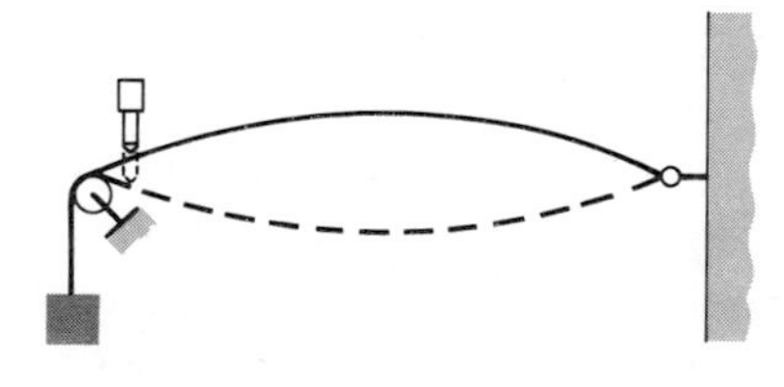

Fig. 13-8 A length of stretched rubber tubing is disturbed by a vibrator. A standing wave is set up in the fundamental mode, for which the wavelength is twice the length of the tubing. There must be nodes at the two ends, since they are fixed. There is a single antinode in the center.

In a popular lecture demonstration, a length of rubber tubing is stretched between two supports, as shown in Fig. 13-8. As in Fig. 12-11, the tension in the tubing is determined by the weight hanging at the left end. A vibrator, which acts as the source, is touched to the tubing at a point near the left end, and a wave immediately begins to travel to the right along the tubing. When the wave reaches the rigidly anchored right end, it is reflected. As we noted in Sec. 13-2, this reflection involves a sign inversion of the transverse displacement and a reversal of the direction of travel, but does not change the amplitude, frequency, or wavelength of the wave.

The reflected wave travels back to the left end of the tubing and is reflected a second time. There is a second sign inversion, so that the net change in sign of the transverse displacement after two reflections is zero. The wave then passes the vibrator again. If it happens that the length of the tubing is exactly half the wavelength of the wave, the twice-reflected wave will have traveled a total distance of one wavelength when it reaches the vibrator. It will therefore produce a transverse motion of the tubing at the vibrator which is exactly in synchronism with the motion which the vibrator is itself inducing in the tubing. The wave excited by the vibrator has the same frequency, wavelength, and propagation speed as the wave already moving down the tubing. Thus the displacement maxima, minima, and zeros of the two waves coincide everywhere, and the waves are in phase. This is because the wave functions describing the two waves have in their arguments the same phase constant δ. Figure 13-4*a* illustrates two sinusoidal waves which are in phase.

Since the wave already traveling down the tubing and the wave produced by the vibrator superpose constructively, there comes to be a wave of larger amplitude moving to the right. This process continues with each round trip, and a total wave is built up composed of two oppositely moving traveling waves of equal frequency and wavelength. A steady state is soon reached in which the two traveling waves have the same amplitude, whose value is such that the rate at which energy is lost to friction in the rubber tubing and to air resistance equals the rate at which energy is supplied by the vibrator.

The two oppositely moving traveling waves constitute a total wave which is a standing wave. The standing wave has, as it must, nodes at the two immovable ends of the tubing (see Fig. 13-8). The segment of the tubing located at the center executes transverse oscillations of the greatest amplitude. This amplitude is the sum of the amplitudes of the two traveling waves, that is, twice the amplitude of either. The point of greatest amplitude is called an **antinode.**

You can see from inspection of Fig. 13-8 that the condition for producing the standing wave illustrated is that half a wavelength of the traveling wave must be equal to the length of the tubing. This can be satisfied most conveniently by adjusting the frequency of the vibrator, which is also the frequency of the wave the vibrator induces. [For a given linear density and tension of the tubing, the speed $|v|$ of the traveling wave will be fixed according to Eq. (12-35*b*), $|v| = \sqrt{F/\mu}$. Thus Eq. (12-11*b*), $\nu\lambda = |v|$, specifies that the product of the frequency ν and the wavelength λ will be constant. Therefore, adjusting the frequency adjusts the wavelength.] The critical frequency for producing the standing wave of Fig. 13-8 is called the **fundamental frequency.** Since the standing wave is produced by superposition of two traveling waves moving in opposite directions, its frequency and wavelength are the same as those of the traveling waves. (This assertion is justified quantitatively later in this section.)

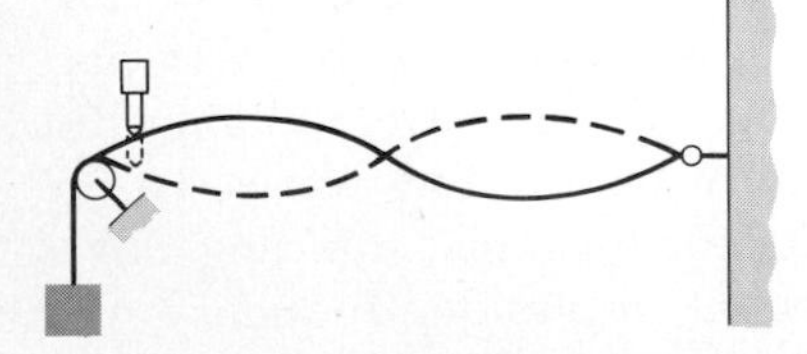

Fig. 13-9 The standing wave set up in the tubing has a wavelength equal to the length of the tubing, or half the wavelength of the fundamental mode. There is a third node at the center of the tubing, with two antinodes spaced evenly between the nodes.

If the frequency of the vibrator is doubled, the wavelength of the wave it produces will be halved, and a standing wave will be set up with two half-wavelengths contained in the length of the tubing. Such a standing wave has a node at the center of the tubing, in addition to the two mandatory nodes at its rigidly supported ends. See Fig. 13-9. For a frequency which is three times the fundamental, a standing wave will be formed as in Fig. 13-10, with two nodes in addition to those at the ends of the tubing. Thus

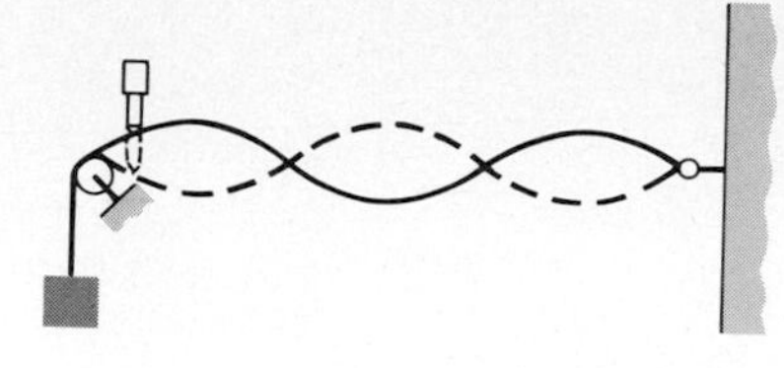

Fig. 13-10 Here the standing wave has a wavelength equal to two-thirds the length of the tubing, or one-third the wavelength of the fundamental mode. There are four nodes spaced evenly along the tubing and three antinodes between them.

it is divided into three equal parts, each half a wavelength long. In general, the condition for setting up a standing wave is that the frequency be an *integral multiple of the fundamental frequency*.

For frequencies which are not integral multiples of the fundamental frequency of the stretched tubing, it is not possible to form a standing wave. You can look at this from two different points of view. One is that at all frequencies except those for which an integral number of half-wavelengths of the standing wave fit exactly into the length of the tubing it is impossible to have nodes separated by half a wavelength and also to have nodes at the two ends. The other way to look at the situation is that the twice-reflected traveling wave passing the vibrator will not produce a transverse motion in the tubing which is in phase with that induced by the vibrator unless the frequency is such that an integral number of half wavelengths are contained in the length of the tubing. At all other frequencies, the superposition of the traveling wave being produced by the vibrator and the multiply reflected traveling waves already in the tubing sometimes increases the transverse displacements of the total wave and sometimes decreases them. The net result is confused and complicated but generally small transverse motions in the tubing.

Let us develop an equation that relates the particular frequencies ν, at which there can be a standing wave, to the linear density μ, the tension F, and the length L of the stretched rubber tubing. This will relate the standing-wave frequencies directly to the physical properties of the system under consideration. The discussion in the preceding paragraph, and Figs. 13-8 and 13-10, show that the condition for a standing wave is

$$n\frac{\lambda}{2} = L \tag{13-8a}$$

where λ is the wavelength of a wave traveling on the tubing and n is any of the integers

$$n = 1, 2, 3, \ldots \tag{13-8b}$$

That is, the condition for formation of a standing wave is that an integral number of half-wavelengths of the standing wave fit exactly into the length of the tubing.

The frequency of the standing wave is equal to ν, the frequency of the two traveling waves which superpose to produce it. Equation (12-11b) connects ν to the speed of the traveling waves $|v|$ and their wavelength λ (which is also the wavelength of the standing wave) by the relation $\nu = |v|/\lambda$. Inserting into this relation the value of $|v|$ given by Eq. (12-35b), $|v| = \sqrt{F/\mu}$, yields

$$\nu = \frac{1}{\lambda}\sqrt{\frac{F}{\mu}}$$

From Eqs. (13-8), we have

$$\lambda = \frac{2L}{n} \tag{13-8c}$$

So the equation determining the frequencies for standing waves in a stretched tubing, or string, is

$$\nu = \frac{n}{2L}\sqrt{\frac{F}{\mu}} \tag{13-9a}$$

where

$$n = 1, 2, 3, \ldots \tag{13-9b}$$

Example 13-2 applies this frequency condition to a particular system already considered in two examples in Chap. 12.

EXAMPLE 13-2

Predict whether a standing wave will be set up in the apparatus considered in Examples 12-5 and 12-8 for the vibrator frequency $\nu = 5.00$ Hz. If not, determine the nearest higher and lower frequencies for which there will be a standing wave, and describe the wave in each case.

■ Using the numerical values from Example 12-8 to evaluate the coefficient of n in Eq. (13-9a), you obtain

$$\frac{1}{2L}\sqrt{\frac{F}{\mu}} = \frac{24.6 \text{ m/s}}{2 \times 8.00 \text{ m}} = 1.54 \text{ Hz}$$

The frequencies for which standing waves will be produced in the tubing, in this particular case, are therefore given by $\nu = 1.54\, n$ Hz, with $n = 1, 2, 3, \ldots$. You thus have

$$\nu = 1.54, 3.08, 4.61, 6.15, \ldots \text{ Hz}$$

The vibrator frequency 5.00 Hz will not lead to a standing wave. But if the frequency is raised to 6.15 Hz, the standing wave corresponding to $n = 4$ will develop on the tubing. This wave has 3 nodes located at points on the tubing spaced equally between the 2 nodes at its ends. The 5 nodes divide the tubing into 4 parts, each a half-wavelength long. Alternatively, the vibrator frequency could be reduced to 4.61 Hz to produce a standing wave with one less node, corresponding to $n = 3$.

We are now ready to express the general physical ideas underlying standing waves in a precise mathematical way. We do so first for the simple and very important case of sinusoidal waves. In Sec. 13-4, we develop a general treatment which is valid for any wave function at all.

Suppose that a stretched string of length L is tied to rigid supports at both ends, and a sinusoidal wave train is excited in it by a vibrator placed near its left end, as in Figs. 13-8, 13-9, and 13-10. The wave train travels to the right, which we take to be the direction of positive x, the quantity x being measured from the extreme left end of the string. We have represented such a sinusoidal wave by means of Eq. (12-21) which, with a slight change in notation, can be written

$$f(x, t) = A\cos(kx - \omega t + \delta)$$

The minus sign specifies that the wave is traveling in the positive x direction, as desired. The value of the phase constant δ is determined by the displacement of the extreme left end of the string at the particular moment chosen as $t = 0$. It will turn out to be convenient to have this displacement be zero, that is, $f(0, 0) = 0$. (The advantage of making this choice will become evident shortly.) We thus wish to have

$$f(0, 0) = A\cos\delta = 0$$

One choice of δ which satisfies this equation is $\delta = -\pi/2$. So we can write $f(x, t)$ for arbitrary values of x and t in the form

$$f(x, t) = A\cos(kx - \omega t - \pi/2)$$

To save the trouble of carrying along the phase constant $-\pi/2$ in the calculations which follow, we note that for any argument θ we can write $\cos(\theta - \pi/2) = \sin\theta$. So the particular sinusoidal wave function we wish to use to represent the wave traveling along the string to the right can be written $f(x, t) = A\sin(kx - \omega t)$. Finally, using the shorthand notation $f_+ \equiv f(x, t)$ to emphasize the positive direction of motion, we have

$$f_+ = A\sin(kx - \omega t) \tag{13-10a}$$

You can satisfy yourself that this function is indeed a solution to the wave equation by substituting it into the general form given by Eq. (12-26), just as was done in Example 12-4 for a sinusoidal wave function written in terms of the cosine function.

When the wave train comes to the rigid support at the right end of the string, it is reflected as it arrives. The reflected wave is a sinusoidal wave train of equal amplitude traveling to the left, that is, in the negative x direction. Just as in Eq. (12-21), the reversal in direction can be represented in the wave function by changing the sign of the term ωt in the argument. Thus the wave function for the reflected wave is

$$f_- = A\sin(kx + \omega t + \delta) \tag{13-10b}$$

where the minus sign in the subscript refers to the direction of motion.

Why cannot the phase constant δ be set immediately to zero in Eq. (13-10*b*) as it was in Eq. (13-10*a*)? We wrote Eq. (13-10*a*) in such a way as to eliminate the need for a phase constant by choosing the location $x = 0$ and the instant $t = 0$ in a certain way. There are no more such free choices available, and f_- bears a fixed phase relationship—not yet known—to f_+. This phase relationship is specified by the phase constant δ.

The complete wave function $f \equiv f(x, t)$ is obtained by superposition. It is

$$f = f_+ + f_- = A\sin(kx - \omega t) + A\sin(kx + \omega t + \delta) \tag{13-11a}$$

What is the value of the phase constant δ? To find out, we apply the **boundary condition** at the left end of the string ($x = 0$), which is that the string at that point must *always* be motionless. When this condition is applied to Eq. (13-11*a*), the space-independent equation at $x = 0$ becomes

$$f(0, t) = A\sin(0 - \omega t) + A\sin(0 + \omega t + \delta) = 0$$

Since $\sin(-\omega t) = -\sin(\omega t)$, we can use the right-hand equality to obtain

$$\sin(\omega t) = \sin(\omega t + \delta)$$

This equation is satisfied if

$$\delta = 0 \tag{13-11b}$$

(It is also satisfied if $\delta = \pm 2\pi, \pm 4\pi, \pm 6\pi, \ldots$. But these values of δ yield nothing new. Why?)

The boundary condition used immediately above is the simplest example of a large class of such conditions encountered in the study of waves in physical systems. In general, *a boundary condition is a mathematical expression of some physical restriction placed on the motion of a system.* Such restrictions are most commonly imposed by the abrupt change in the nature of the physical conditions at the place where the system "ends." But other, more subtle, restrictions do occur in more complex systems, and they are not always imposed at an obvious physical boundary. They are nevertheless called boundary conditions because of their mathematical resemblance to the simple boundary condition leading to Eq. (13-11*b*). We encounter a variety of boundary conditions in Sec. 13-5 and later in this book.

The restriction on the value of δ given by Eq. (13-11*b*) means that the wave function given by Eq. (13-11*a*) is too general to describe the wave on the stretched string of length L unless $\delta = 0$. We apply this restriction so that Eq. (13-11*a*) becomes

$$f = A \sin(kx - \omega t) + A \sin(kx + \omega t) \tag{13-12}$$

In order to bring out the physical significance of Eq. (13-12), we perform some trigonometric manipulation. We use the identity

$$\sin F + \sin G = 2 \sin[\tfrac{1}{2}(F + G)] \cos[\tfrac{1}{2}(F - G)] \tag{13-13}$$

Setting $F = kx - \omega t$ and $G = kx + \omega t$, we obtain

$$f = 2A \sin(kx) \cos(-\omega t)$$

or, since $\cos(-\omega t) = \cos(\omega t)$,

$$f = 2A \sin(kx) \cos(\omega t) \tag{13-14}$$

Look at this equation with care. We started with two wave trains traveling in opposite directions. Each wave was described by a wave function which depended on both space (that is, position) and time through the argument $kx \mp \omega t$, as would be expected of traveling waves. The total wave function of Eq. (13-14) also depends on both space and time. But the dependences are separated, so that the sine function is *time-independent* and the cosine function is *space-independent.* This so-called *separation of variables* is a great convenience in understanding the behavior of the wave, as well as in mathematical manipulations.

We now apply the boundary condition at the right end of the string ($x = L$), which is that the string at that point must *always* be motionless, just as is the case at the left end of the string. It is particularly easy to apply this condition to the wave function in the form of Eq. (13-14). Whatever the value of $\cos(\omega t)$ may be in that equation, the value of f must be zero at the point $x = L$; that is, $f(L, t) = 0$. This can be true only if

$$\sin(kL) = 0$$

To satisfy this equation, we must have

$$kL = n\pi \qquad \text{where} \quad n = 1, 2, 3, \ldots$$

or

$$k = \frac{n\pi}{L} \qquad \text{where} \quad n = 1, 2, 3, \ldots \tag{13-15a}$$

Negative integral values of n, and the value $n = 0$, also satisfy the equation $\sin(kL) = 0$ mathematically. However, they are excluded on physical grounds. The wave number $k \equiv 2\pi/\lambda$ has as its physical meaning 2π times the number of waves per unit length of the string at a given instant, which cannot be negative. And L is the length of the string, which likewise cannot be negative. Hence the product kL cannot be negative. As for the value $n = 0$, this implies that $k = 0$. But $k = 0$ corresponds to an infinite wavelength—that is, no wave at all.

Equation (13-15*a*) can also be written in terms of the wavelength λ. Again using the definition $k = 2\pi/\lambda$ given by Eq. (12-20), we have $(2\pi/\lambda)L = n\pi$, or

$$\lambda = \frac{2L}{n} \qquad \text{where } n = 1, 2, 3, \ldots \tag{13-15b}$$

This is identical to Eq. (13-8*c*), which was derived by means of a more direct but less general and less quantitative argument. Its physical meaning bears reiteration: *Only those values of k or λ given by Eqs. (13-15a) or (13-15b) lead to standing waves. These are the values for which an integral number n of half-wavelengths fit exactly into the length of the string.* Whether they are propagated on a string or in a more complicated system, standing waves are invariably characterized by a restriction analogous to this one.

Consider a particular standing wave on a string. As is always true, the standing wave is described by the wave function of Eq. (13-14), $f = 2A \sin(kx) \cos(\omega t)$. But the particular wave is specified by a particular value of the integer n in Eq. (13-15*a*), so that the wave function is

$$f(x, t) = 2A \sin\left(\frac{n\pi}{L} x\right) \cos(\omega t) \qquad \text{where } n \text{ is a specific integer}$$

We now define an integer j such that j ranges over the values 0, 1, 2, 3, . . . , n. The locations along the string specified by

$$x = \frac{j}{n} L \qquad \text{where } j = 0, 1, 2, 3, \ldots, n$$

lie at the left end of the string and at fractions of the total length of the string equal to $1/n$, $2/n$, $3/n$, . . . , 1. Since the entire string contains n half-wavelengths, these points are thus spaced a half-wavelength apart. At these special points, the wave function has the value

$$f = 2A \sin\left(\frac{n\pi}{L} \frac{j}{n} L\right) \cos(\omega t) = 2A \sin(j\pi) \cos(\omega t)$$

$$\text{where } j = 0, 1, 2, 3, \ldots, n$$

But for these values of the argument of the sine function, the function itself always has the value zero. Thus at these points we always have

$$f = 0$$

regardless of the time t. These points are the *nodes*. Note that the boundary conditions *require* that the ends of the string, $x = 0$ and $x = L$, be nodes. They *permit* the existence of other nodes. For any particular value of n, there are $n + 1$ nodes on the string, including the two at the ends. Each value of n thus leads to a wave function which describes a particular way in which the string can vibrate. These particular ways are called **standing-wave modes,** or sometimes simply **modes** for short. The wave function for a particular mode is given by the equations

$$f = 2A \sin\left(n\pi \frac{x}{L}\right) \cos(\omega t) \qquad \text{where } n \text{ is a positive integer} \tag{13-16a}$$

or by using Eq. (13-15*b*), $\lambda = 2L/n$,

$$f = 2A \sin\left(2\pi \frac{x}{\lambda}\right) \cos(\omega t) \qquad \text{where } \lambda = \frac{2L}{n} \text{ and } n \text{ is a positive integer} \tag{13-16b}$$

Systems more complicated than the vibrating string also possess standing-wave modes, although they are not usually specified by conditions as simple as those given by Eqs. (13-15*a*), (13-15*b*), (13-16*a*), and (13-16*b*). But in cases where the modes *are* described by those equations, they are often given the special name **harmonic modes,** or (more commonly) **harmonics** for short. It was supposedly Pythagoras (in the sixth century B.C.) who first found the simple numerical relationship among the modes of a vibrating string and connected them with the pleasing sounds of musical harmony.

What happens between the nodes when a string is vibrating in a particular standing-wave mode? If we start at $x = 0$ and move along the string, the term $\sin(kx)$ increases gradually until it reaches the value $+1$, then decreases through 0 to -1, next increases through 0 to $+1$ again, and so on. As time passes, the term $\cos(\omega t)$ swings through all possible values between $+1$ and -1, but never passes beyond these limits. As a result, the point on the string located at some arbitrary value x oscillates between the extreme displacements $2A \sin(kx)$ and $-2A \sin(kx)$. Since t is the same for all points on the string at any particular moment, the value of $\cos(\omega t)$ is also the same for all points on the string, and the entire string oscillates together. All points on the string reach their extreme displacements at the same moment, and all pass through zero at the same moment. Figure 13-11 is a series of "snapshots" of the string, covering a half-cycle; that is, the time elapsed from the first sketch to the last is $T/2$, where the period T is defined by $T \equiv 2\pi/\omega$.

The points on the string which undergo the largest excursions from $y = 0$ are those for which $\sin(kx) = \pm 1$. They lie midway between the nodes, and are the antinodes. All points on the string other than the nodes oscillate between limits determined by the value of $\sin(kx)$ at their locations. The nature of the motion is suggested by Fig. 13-12. It should be clear why such waves are called standing waves.

Figure 13-13 illustrates the first three modes of oscillation of a stretched string. These are the oscillation modes shown separately in Figs. 13-8, 13-9, and 13-10.

So far we have considered the restrictions on the wave number k in the term $\sin(kx)$ of Eq. (13-14) if there are to be standing waves in a stretched string. As already noted at the beginning of this section, these restrictions lead to restrictions on the permissible values of the angular frequency ω (or the frequency $\nu \equiv \omega/2\pi$). This is because there is a fixed relation between k and ω (or between λ and ν) given by Eq. (12-11*b*) and by the discussion in Example 13-1. That relation is

$$\frac{\omega}{k} = |v| \qquad \text{or} \qquad \nu\lambda = |v|$$

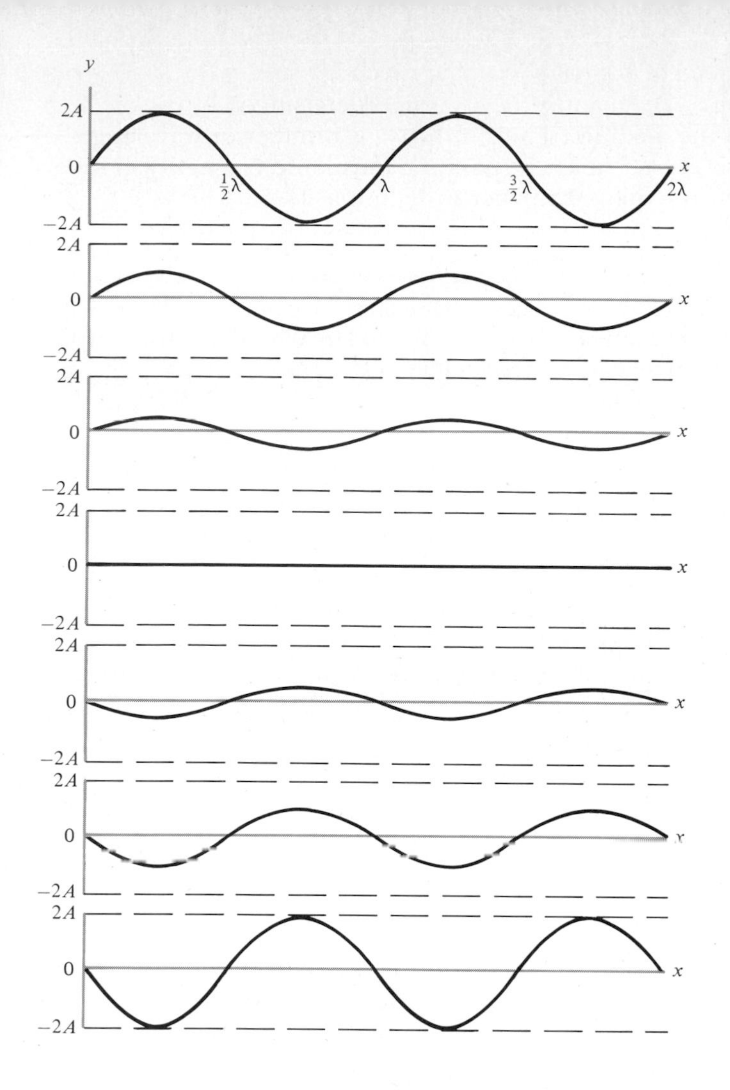

Fig. 13-11 A sequence of "snapshots" of a stretched string oscillating in a standing-wave mode. The wave does not move; that is why it is called "standing." The time elapsed from the first to the last of the snapshots is a half-period. The shape of the string is always sinusoidal, but the height of the sinusoidal varies from $2A$ to $-2A$, where A is the amplitude of each of the two sinusoidal traveling waves, moving in opposite directions, which may be considered to make up the standing wave.

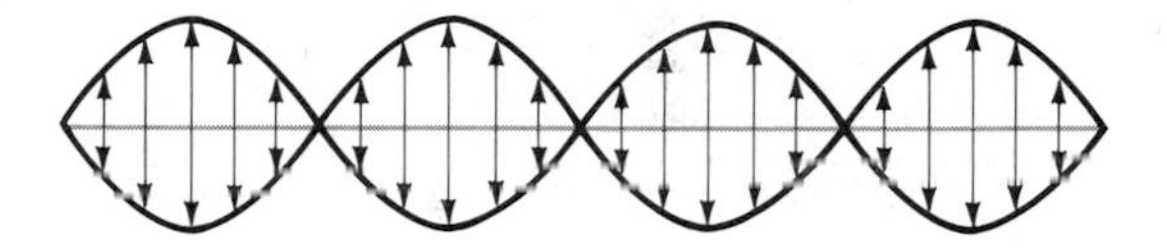

Fig. 13-12 Another way of looking at the motion of a stretched string oscillating in a standing-wave mode. Each point on the string oscillates up and down with harmonic motion, as indicated by the arrows. The amplitude of the oscillation of any point depends on its location along the string.

Fig. 13-13 The first three standing-wave modes (n = 1, 2, 3) for a stretched string. The modes are shown as having the same amplitude A. The vertical scale is grossly exaggerated for clarity. If a real string could actually oscillate with the amplitude shown, without breaking, it would almost certainly not constitute a linear system. That is, the restoring force exerted on an element of the string would not be proportional to its displacement, and the superposition principle would not be valid.

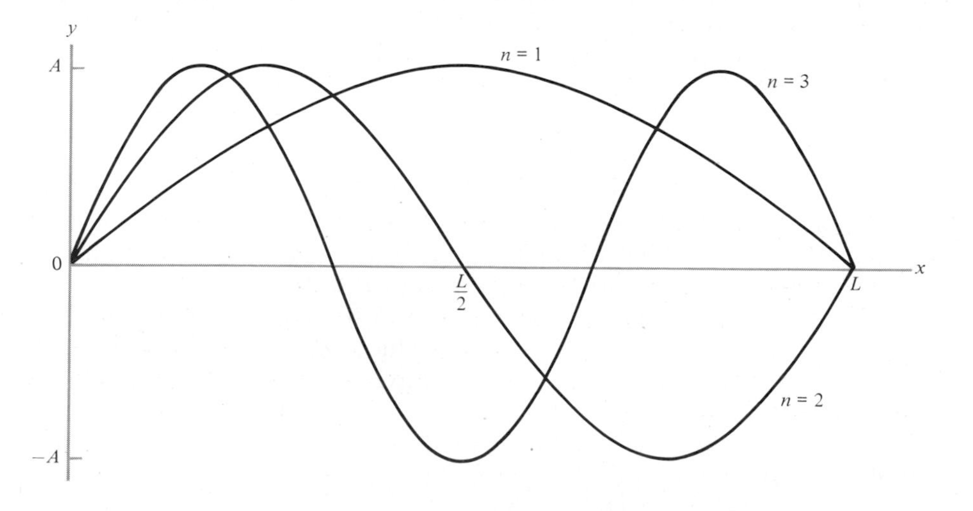

where $|v|$ is the speed of propagation of a traveling wave in the same string (or other medium). The quantity $|v|$, in turn, is determined by the mechanical properties of the medium. Thus, if the wave number or wavelength of a standing wave is given, the corresponding **harmonic frequency** is determined. This point is explored further in Example 13-3.

EXAMPLE 13-3

A string is fixed at both ends under tension. One of its harmonic frequencies is 360 Hz, and the next higher harmonic frequency is 420 Hz. What is the fundamental frequency, that is, the frequency corresponding to $n = 1$?
■ You can write Eqs. (13-9) in the form

$$\nu = n\frac{1}{2L}\sqrt{\frac{F}{\mu}} \qquad \text{where } n = 1, 2, 3, \ldots \tag{13-17}$$

The string tension F, the linear density μ, and the length L are all fixed for the particular string. Thus the quantity $(1/2L)\sqrt{F/\mu}$ on the right side of the equation is a constant. Since n is a dimensionless number, the dimensions of that quantity must be the same as those of the quantity ν on the left side of the equation, that is, hertz. (Can you verify this, beginning with the proper units for F, μ, and L?) Indeed, the quantity $(1/2L)\sqrt{F/\mu}$ is just ν_1, the fundamental frequency. Consequently, Eq. (13-17) can be written in the form

$$\nu = n\nu_1 \qquad \text{where } n = 1, 2, 3, \ldots$$

From this, you can see that the difference in frequency between two consecutive harmonics is equal to the fundamental frequency. So you have

$$\nu_1 = 420 \text{ Hz} - 360 \text{ Hz} = 60 \text{ Hz}$$

What would be wrong if the example stated that the consecutive harmonics were at 360 Hz and 440 Hz?

13-4 STANDING-WAVE SOLUTIONS TO THE WAVE EQUATION

Standing waves are encountered in physical situations of many different kinds. It is therefore important to study them in further depth. For the simplest case—that of one-dimensional sinusoidal standing waves in a stretched string—we have discussed the salient features in Sec. 13-3. These features include the form of the wave function, the permissible wavelengths or wave numbers, and the permissible frequencies. We obtained these results by combining several properties of traveling waves: (1) their superposition properties; (2) their reflection properties; and (3) the relations among the frequency and wavelength ν and λ, the traveling-wave speed $|v|$, the string tension F, and the mass per unit length μ of the string.

It is possible, however, to begin with the wave equation for a vibrating string in its most general form and to obtain *directly* the form of the wave function for standing waves, together with the restrictions on permissible wavelengths and the frequency condition $\nu = (n/2L)\sqrt{F/\mu}$. This will be done in this section. The general derivation will serve as a check on the results already obtained. More importantly, it will introduce techniques which will be used in Sec. 13-5 to study the vibration of a circular drumhead. It is not possible to obtain results for this more complicated system by simply applying features 1, 2, and 3 listed in the previous paragraph. The techniques developed in this section will also find application in the study of electromagnetic waves and in the quantum-mechanical systems considered in Chap. 31.

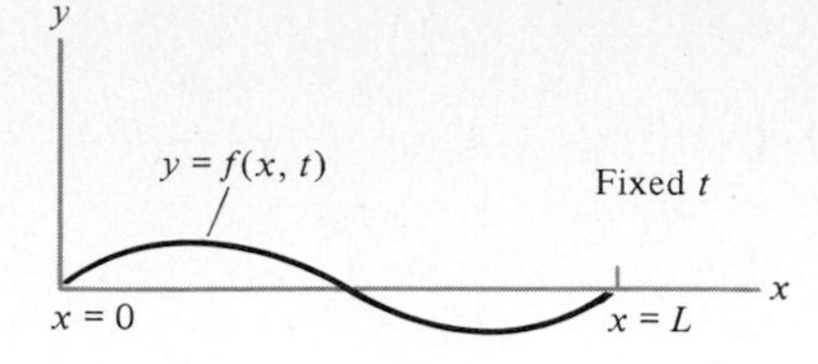

Fig. 13-14 Showing at a particular time t a wave function $f(x, t)$ which satisfies the definition of a standing wave.

In Fig. 13-14, the x axis lies along the undisturbed string, with the positive x direction chosen to the right. As before, we choose the origin $x = 0$ at the left end of the string, and the right end is at $x = L$. The transverse displacement of a segment of the string at a certain location and time is given by $y = f(x, t)$. In general, the function $f(x, t)$ is a solution to the wave equation for a stretched string. That is, it is capable of representing *any* wave in *any* part of *any* string. If it has the particular form $f(x, t) = f(x \mp |v|t)$, it represents a traveling wave. That is, the function represents a wave whose characteristic features, such as the locations where $f(x, t)$ is equal to zero, travel along the string in the positive or the negative direction at speed $|v|$. A different dependence on the variables x and t is required if $f(x, t)$ is to represent a standing wave—in other words, a wave in which the values $f(x, t) = 0$, and other characteristic features, occur at fixed locations.

Equation (13-14), $f(x, t) = 2A \sin(kx) \cos(\omega t)$, was obtained by considering the kinematics of standing waves. It suggests that for such waves $f(x, t)$ is a *product* of a function of x alone and a function of t alone. In general, $f(x, t)$ is not necessarily a *sinusoidal* standing wave like that described by Eq. (13-14). Nevertheless, there is a strong suggestion that a standing wave can be represented symbolically in the form

$$f(x, t) = g(x)h(t) \tag{13-18}$$

where $g(x)$ is some function of x and $h(t)$ is some function of t. At all locations where $g(x)$ happens to have the value zero, the displacement $f(x, t)$ of the string will be zero for *all values of t*. Thus the wave will have nodes at fixed locations. For instance, the form given in Eq. (13-18) makes it possible for $f(x, t)$ to have the property $f(x, t) = 0$ at both rigidly supported ends of the string, no matter what the value of t. All that is required is to have $g(x) = 0$ at both ends. These considerations would make it possible to guess that a standing wave must have the form of Eq. (13-18), even if Eq. (13-14) were not available to provide a hint.

Let us substitute Eq. (13-18) into the wave equation for a stretched string and see what happens. To do this, we must evaluate the second partial derivatives of $f(x, t)$ with respect to space and time, both of which appear in the wave equation. We have

$$\frac{\partial^2 f(x, t)}{\partial x^2} = \frac{\partial^2}{\partial x^2}[g(x)h(t)]$$

Since t is held constant in the partial differentiation, the function $h(t)$ is also held constant. Therefore we obtain

$$\frac{\partial^2 f(x, t)}{\partial x^2} = h(t)\frac{\partial^2 g(x)}{\partial x^2}$$

Since the partial derivative of a function of a single variable, such as $g(x)$, is not different from an ordinary derivative, there is no need to use partial-derivative notation on the right side of this equation. Thus we have $\partial^2 g(x)/\partial x^2 = d^2 g(x)/dx^2$, and

$$\frac{\partial^2 f(x, t)}{\partial x^2} = h(t)\frac{d^2 g(x)}{dx^2} \tag{13-19}$$

Similarly, we find

$$\frac{\partial^2 f(x, t)}{\partial t^2} = \frac{\partial^2}{\partial t^2}[g(x)h(t)]$$

$$= g(x)\frac{\partial^2 h(t)}{\partial t^2}$$

or

$$\frac{\partial^2 f(x, t)}{\partial t^2} = g(x)\frac{d^2 h(t)}{dt^2} \tag{13-20}$$

Now we substitute Eqs. (13-19) and (13-20) into Eq. (12-26), which is the general equation for waves on a stretched string. That equation is

$$\frac{\partial^2 f(x, t)}{\partial x^2} = \frac{\mu}{F}\frac{\partial^2 f(x, t)}{\partial t^2} \tag{13-21}$$

The substitution yields

$$h(t)\frac{d^2 g(x)}{dx^2} = \frac{\mu}{F}g(x)\frac{d^2 h(t)}{dt^2}$$

Dividing through by $g(x)h(t)$ isolates all functions of x on one side of the equation and all functions of t on the other:

$$\frac{1}{g(x)}\frac{d^2 g(x)}{dx^2} = \frac{\mu}{F}\frac{1}{h(t)}\frac{d^2 h(t)}{dt^2} \tag{13-22}$$

Equation (13-22) requires very careful analysis. First, we must realize that the symbol "=" really means that the quantities on the two sides of the symbol have a *common value*. It will be useful to show this explicitly by designating the common value as C and rewriting the equation in the form

$$\frac{1}{g(x)}\frac{d^2 g(x)}{dx^2} = C = \frac{\mu}{F}\frac{1}{h(t)}\frac{d^2 h(t)}{dt^2}$$

Next look at the pair of equations

$$\frac{1}{g(x)}\frac{d^2 g(x)}{dx^2} = C \tag{13-23a}$$

and

$$\frac{\mu}{F}\frac{1}{h(t)}\frac{d^2 h(t)}{dt^2} = C \tag{13-23b}$$

Since the left side of Eq. (13-23*a*) does not depend on the independent variable t, it is apparent that C cannot depend on t. And since the left side of Eq. (13-23*b*) does not depend on the independent variable x, it is also apparent that C cannot depend on x. Thus C depends on *neither* of the two independent variables, and so it must be a *constant*. Think about it!

In so doing, you may well ask yourself the following question: If C is a constant, it follows that the left side of Eq. (13-23*a*) is not a function of x either; how can this be? A similar question can be asked about Eq. (13-23*b*) with respect to the variable t. We will see very shortly that the answers are simple enough. The function $g(x)$ on the left side of Eq. (13-23*a*) will turn out to have the property that $d^2g(x)/dx^2$ is proportional to $g(x)$. Consequently, the x dependence of $1/g(x)$ will cancel the x dependence of $d^2g(x)/dx^2$. A similar thing will happen in Eq. (13-23*b*).

Transposing terms in Eqs. (13-23a) and (13-23b) yields

$$\frac{d^2g(x)}{dx^2} = Cg(x) \tag{13-24a}$$

and

$$\frac{d^2h(t)}{dt^2} = \frac{CF}{\mu} h(t) \tag{13-24b}$$

We started with a second-order partial differential equation involving the two independent variables x and t—the wave equation. We have separated it into two second-order ordinary differential equations, a time-independent one involving x and a space-independent one involving t. This was done by writing the solution to the partial differential equation as the product of the solutions to the ordinary differential equations. This is a significant accomplishment because we know how to find solutions to the ordinary differential equations, and we will therefore be able to solve the partial differential equation. The procedure we have followed is called **separation of variables,** and the constant C that arises in carrying out the procedure is called the **separation constant.** Separation of variables is one of the principal methods used for finding analytical solutions to partial differential equations of many kinds.

We will have no trouble in solving Eqs. (13-24a) and (13-24b). They both have the same basic structure as Eq. (6-16), the familiar ordinary differential equation for the harmonic oscillator, which is $d^2x/dt^2 = -\alpha x$. Different mathematical symbols (and physical meanings) are attached to the independent and dependent variables and the constant factors in each of these equations. But in each case, the second derivative of the dependent variable is proportional to the dependent variable itself. Equation (6-16) has a solution of the form given by Eq. (6-17), $x = A \cos(\omega t + \delta)$. In like manner, every such differential equation has a solution in which the dependent variable is equal to the product of some constant with a sinusoidal function. And the argument of the sinusoidal function is the independent variable multiplied by some other constant. The function can be a sine, or a cosine, or any expression of a sinusoidal. The only distinction among these three possibilities is the choice of the phase constant δ whose value specifies the zero point of the independent variable.

Applying these observations to the time-independent Eq. (13-24a), we write its solution in the form

$$g(x) = A \sin(kx) \tag{13-25a}$$

where k is a constant whose value has yet to be determined. (It will turn out to be the wave number.) The solution is taken to be a sine function rather than a cosine by making the proper choice of the phase constant δ. This is done because of the *boundary condition*

$$g(x) = 0 \qquad \text{at } x = 0 \tag{13-25b}$$

This condition states that there can be no displacement of the rigidly supported string at its left end. By choosing the sine function for the required sinusoidal, we satisfy this condition automatically with the convenient value $\delta = 0$ for the phase constant.

The constant k multiplying x in Eq. (13-25a) must have the value $k = 2\pi/\lambda$, where λ is the wavelength of the standing wave. The justification is that the time-independent function $g(x)$ controls the wavelength of the standing wave. Thus when x increases by one wavelength—by the amount λ—the function $g(x)$ must go through one oscillation. This requires that the argument of the sine function in Eq. (13-25a) increase by 2π when x increases by λ. That is precisely what happens if the argument is

$$kx = 2\pi \frac{x}{\lambda}$$

The function $g(x)$ thus assumes the specific form

$$g(x) = A \sin\left(\frac{2\pi x}{\lambda}\right) \tag{13-26}$$

The value of the wavelength λ is not yet known. It will be determined later by applying the proper boundary condition at the other end of the string, where $x = L$.

There is one more constant, A, in Eq. (13-26) whose value has not yet been determined. The quantity A is the amplitude of the wave, and its value is the magnitude of the maximum displacement of the string from its undisturbed position.

Let us verify that Eq. (13-26) actually does satisfy Eq. (13-24a). Differentiating both sides of Eq. (13-26), we find

$$\frac{dg(x)}{dx} = \frac{2\pi}{\lambda} A \cos\left(\frac{2\pi x}{\lambda}\right)$$

Differentiating a second time gives

$$\frac{d^2 g(x)}{dx^2} = -\left(\frac{2\pi}{\lambda}\right)^2 A \sin\left(\frac{2\pi x}{\lambda}\right)$$

Substitution of the values of $g(x)$ and its second derivative into Eq. (13-24a) produces

$$-\left(\frac{2\pi}{\lambda}\right)^2 A \sin\left(\frac{2\pi x}{\lambda}\right) = CA \sin\left(\frac{2\pi x}{\lambda}\right)$$

This is valid, and therefore Eq. (13-26) is a solution to Eq. (13-24a), *provided* that

$$C = -\left(\frac{2\pi}{\lambda}\right)^2 \tag{13-27}$$

Thus we have identified the separation constant C in Eqs. (13-24a) and (13-24b). It is a negative quantity whose value depends on the wavelength λ.

To determine the value of the wavelength, we apply the *boundary condition* at the right end of the string:

$$g(x) = 0 \qquad \text{at } x = L \tag{13-28}$$

The condition states that there can be no displacement of the string at that end either. Using Eq. (13-26), we have

$$A \sin\left(\frac{2\pi L}{\lambda}\right) = 0$$

Since the amplitude A is not zero if there *is* a standing wave in the string, we must satisfy this equation by having

$$\sin\left(\frac{2\pi L}{\lambda}\right) = 0$$

This can be achieved *only* if $2\pi L/\lambda$ has one of the values

$$\frac{2\pi L}{\lambda} = \pi,\ 2\pi,\ 3\pi,\ \ldots \qquad (13\text{-}29)$$

because $\sin \pi = \sin(2\pi) = \sin(3\pi) = \cdots = 0$. We do not list $2\pi L/\lambda = 0$, even though $\sin 0 = 0$, because it corresponds to the statement $\lambda = \infty$, and this means there is no standing wave.

A compact way to write Eq. (13-29) is

$$\frac{2\pi L}{\lambda} = n\pi \qquad \text{where } n = 1, 2, 3, \ldots \qquad (13\text{-}30a)$$

or

$$\frac{n\lambda}{2} = L \qquad \text{where } n = 1, 2, 3, \ldots \qquad (13\text{-}30b)$$

Compare this condition with Eqs. (13-8*a*) and (13-8*b*). The functions

$$g(x) = A \sin\left(\frac{2\pi x}{\lambda}\right)$$

are plotted in Fig. 13-15 for the first three values of the integer n. Each integer n produces a particular function $g(x)$. But all the functions $g(x)$ satisfy the boundary conditions $g(x) = 0$ at $x = 0$ and $x = L$, because for each function an integral number of half-wavelengths just fit into the length of the string. You can see this directly from Eq. (13-30*b*).

Now we attack the space-independent differential equation, whose solutions are the functions $h(t)$ which satisfy Eq. (13-24*b*):

$$\frac{d^2h(t)}{dt^2} = \frac{CF}{\mu}\,h(t)$$

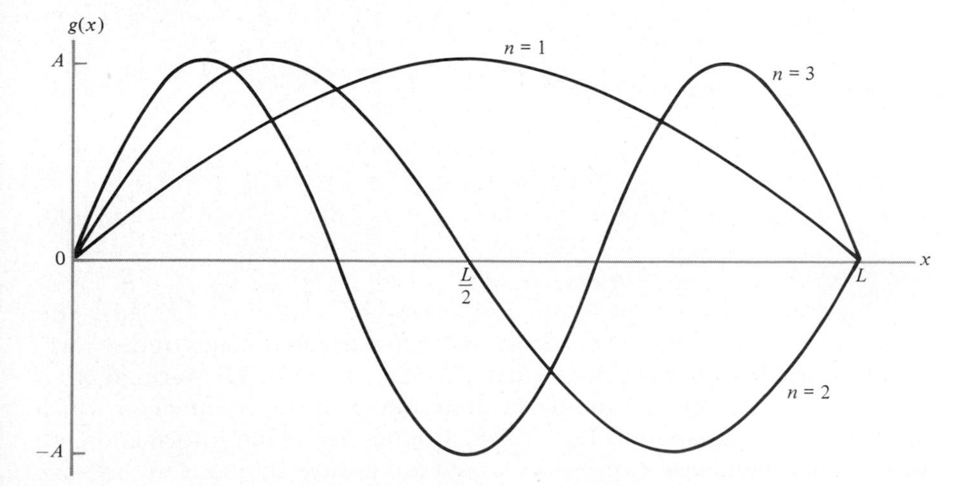

Fig. 13-15 Plots of the time-independent wave function $g(x) = A \sin(2\pi x/\lambda)$ with $\lambda - 2L/n$. The wave functions shown correspond to the values $n = 1, 2, 3$. Compare with Fig. 13-13.

From Eqs. (13-27) and (13-30a) we have

$$C = -\left(\frac{2\pi}{\lambda}\right)^2 = -\left(\frac{n\pi}{L}\right)^2$$

Using this value of C gives us

$$\frac{d^2h(t)}{dt^2} = -\left(\frac{n\pi}{L}\right)^2 \frac{F}{\mu} h(t) \tag{13-31}$$

Again, we have a differential equation which we know to have sinusoidal solutions. We choose to write them in the form

$$h(t) = \cos(2\pi\nu t) \tag{13-32}$$

Since $\cos 0 = 1$, this is tantamount to choosing the zero of time to be the instant when the displacements $f(x, t) = g(x)h(t)$ of all the segments of the string have their extreme values. The constant multiplying t has been written in terms of the frequency ν. Using the discussion preceding Eq. (13-26) as a guide, can you justify in detail the choice of the constant $2\pi\nu$ in Eq. (13-32)? What about the phase constant δ?

We do not multiply the cosine in Eq. (13-32) by an amplitude factor A. This is because the constant A which is part of the time-independent function $g(x)$ provides all the adjustability needed in the complete wave function $f(x, t) = g(x)h(t)$.

To verify Eq. (13-32), and also to determine the frequency ν, we differentiate once to obtain

$$\frac{dh(t)}{dt} = -2\pi\nu \sin(2\pi\nu t)$$

Differentiating a second time gives

$$\frac{d^2h(t)}{dt^2} = -(2\pi\nu)^2 \cos(2\pi\nu t)$$

Substitution of the values of $h(t)$ and its second derivative into Eq. (13-31) produces

$$-(2\pi\nu)^2 \cos(2\pi\nu t) = -\left(\frac{n\pi}{L}\right)^2 \frac{F}{\mu} \cos(2\pi\nu t)$$

This equality is valid, and therefore Eq. (13-32) is a solution to Eq. (13-24b), *provided* that

$$(2\pi\nu)^2 = \left(\frac{n\pi}{L}\right)^2 \frac{F}{\mu}$$

Thus the frequency must have one of the values

$$\nu = \frac{n}{2L}\sqrt{\frac{F}{\mu}} \qquad \text{where } n = 1, 2, 3, \ldots \tag{13-33}$$

We now have a complete solution to Eq. (13-18), $f(x, t) = g(x)h(t)$. The form of $g(x)$ is given by Eq. (13-26), and the permissible values of the wavelength λ which appears in that equation by Eq. (13-30b). The form of $h(t)$ is given by Eq. (13-32), and the permissible values of the frequency ν which appears in that equation by Eq. (13-33). Combining all this information, we have the mathematical expression for standing-wave solutions to the wave equation for a string with linear density μ, tension F, and length L:

$$f_n(x, t) = A_n \sin\left(\frac{2\pi x}{\lambda_n}\right) \cos(2\pi\nu_n t) \qquad \text{where } n = 1, 2, 3, \ldots \tag{13-34a}$$

with

$$\frac{1}{\lambda_n} = \frac{n}{2L} \tag{13-34b}$$

and

$$\nu_n = \frac{n}{2L}\sqrt{\frac{F}{\mu}} = \frac{1}{\lambda_n}\sqrt{\frac{F}{\mu}} \tag{13-34c}$$

We have used the subscript n as a label on the $f_n(x, t)$, and on amplitude A_n, the wavelength λ_n, and the frequency ν_n of each of these functions, to emphasize that for every value of the integer n there is a different standing wave. Every one represents, individually, a possible *mode* of the string. Note that in choosing a particular value of n to specify a solution of Eq. (13-34*a*), you must use the *same* value of n in evaluating $1/\lambda_n$ from Eq. (13-34*b*) and ν_n from Eq. (13-34*c*).

In the first mode, $n = 1$, the solution to the wave equation is

$$f_1(x, t) = A_1 \sin\left(\frac{2\pi x}{2L}\right) \cos\left(\frac{2\pi}{2L}\sqrt{\frac{F}{\mu}}\, t\right)$$

This describes the string oscillating at its fundamental frequency $\nu_1 = (1/2L)\sqrt{F/\mu}$ in a standing wave with wavelength $\lambda_1 = 2L$ and nodes at each end. At the start of every cycle of oscillation, it has the shape shown in Fig. 13-15 by the curve labeled $n = 1$ and an amplitude determined by whatever value A_1 happens to have. In the second mode, the solution is

$$f_2(x, t) = A_2 \sin\left(\frac{2\pi x}{L}\right) \cos\left(\frac{2\pi}{L}\sqrt{\frac{F}{\mu}}\, t\right)$$

Here the shape of the string at the beginning of every cycle is given by the $n = 2$ curve, for the appropriate value of the amplitude A_2. The wavelength is $\lambda_2 = L$, and the frequency is $\nu_2 = (1/L)\sqrt{F/\mu}$.

By using mathematical arguments to investigate standing waves in a string, we have obtained a description of these waves that agrees with the description we obtained earlier from simpler and more physical arguments. Compare Eqs. (13-9) and (13-34*c*). One advantage of the mathematical arguments is that they do not involve detailed assumptions about how the standing waves are excited, and thus they make it clear that the standing waves possible in a system are characteristic of the system, instead of the excitation process. Another advantage is that the mathematical arguments can be extended much more easily than the physical ones to the treatment of standing waves in systems of two or three dimensions. We make such an extension in Sec. 13-5.

Equation (13-34*c*) relates the frequency and wavelength of a standing wave on a string by means of the equation $\nu_n\lambda_n = \sqrt{F/\mu}$. According to Eq. (12-35*b*), $|v| = \sqrt{F/\mu}$, the quantity $\sqrt{F/\mu}$ occurring in Eq. (13-34*c*) is the speed $|v|$ of a *traveling* wave in the same string. We thus have the relation

$$\nu_n\lambda_n = |v|$$

between the *standing-wave frequency and wavelength* and the *traveling-wave speed.* This equation is identical in appearance to Eq. (12-11*b*), which relates the frequency and wavelength of a traveling wave to its speed; however, its meaning is different since a standing wave has zero speed.

While the equation $\nu_n \lambda_n = |v|$ has been derived for the special case of standing waves on a string, it is valid for all standing waves. Since the frequency and wavelength of standing waves can very often be measured with great accuracy, the relation is the basis for one of the most accurate methods for measuring the propagation speed of traveling waves. This method is exploited in Example 13-4.

EXAMPLE 13-4

A cellist tunes the "A string" of a cello by adjusting its tension until the fundamental frequency of the standing wave which he sets up in the string by bowing it is 220 Hz. The distance between the supports at the two ends of the string is 68.4 cm, and the mass of the string extending between the supports is 1.31 g.

a. What is the tension in the string?

■ Setting $n = 1$ in Eq. (13-34*c*) gives the fundamental frequency

$$\nu_1 = \frac{1}{2L}\sqrt{\frac{F}{\mu}}$$

Solving for the tension F, you find

$$F = 4L^2 \mu \nu_1^2$$

When you write the linear density μ in terms of the string's mass m and length L, the expression becomes

$$F = 4L^2 \frac{m}{L} \nu_1^2 = 4Lm\nu_1^2$$

Inserting the numerical values, you obtain a tension

$$F = 4 \times 0.684 \text{ m} \times 1.31 \times 10^{-3} \text{ kg} \times (220 \text{ s}^{-1})^2$$
$$= 173 \text{ N}$$

which pulls the supports together. To withstand this tension and that of the other three strings, the neck of the cello must be made reasonably strong. The total force exerted on the frame of a piano by the more than 200 strings stretched on it is very large; the frame must be made of heavy steel. ■

b. The cellist now plucks the string near one end. What is the speed of the traveling wave which propagates down the string?

■ You know the fundamental frequency; it is $\nu_1 = 220$ Hz. If you use Eq. (13-34*b*) to find the corresponding standing wavelength λ_1 from the known value of the length of the string $L = 68.4$ cm, you can use the equation $\nu_1 \lambda_1 = |v|$ to determine the traveling-wave speed. Setting $n = 1$, you have from Eq. (13-34*b*)

$$\lambda_1 = 2L$$

The speed is thus given by

$$|v| = \nu_1 \lambda_1 = 2\nu_1 L$$

Inserting the numerical values, you have

$$|v| = 2 \times 220 \text{ Hz} \times 0.684 \text{ m} = 301 \text{ m/s}$$

which is somewhat greater than the speed of a commercial jet plane.

13-5 STANDING WAVES ON A CIRCULAR MEMBRANE

For a one-dimensional system, such as a string, the frequencies of possible standing waves are related to each other in a very simple way. For a two-dimensional system, such as a drumhead, the relation is not so simple. In this section we extend the methods of Sec. 13-4 to study standing waves on a drumhead. In spite of some complications arising from the geometry of the two-dimensional case, the fundamental ideas carry over with little change. The results obtained for standing waves in both circular drumheads and strings are used in subsequent sections to understand some of the basic properties of musical instruments.

Just as for a stretched string, the properties of waves on a stretched membrane comprising a drumhead are determined by the relationship that Newton's second law imposes between the force exerted on each small segment of the membrane and its mass and acceleration. We will assume that the membrane is of uniform composition and is uniformly stretched across a circular rim, as a drumhead is. The force acting on any segment results from the tension in the membrane. We will again use the symbol F for the **tension.** However, here we have a two-dimensional membrane instead of a one-dimensional string, and F is now defined as the magnitude of the *force per unit length* acting across any line cutting the drumhead in any direction. This definition is illustrated in Fig. 13-16. The tension F is the same, no matter what the location or orientation of the line involved in defining its magnitude, because the membrane is stretched uniformly. The mass of each segment of the membrane is expressed in terms of its *mass per unit area,* called the **areal density** μ. We will assume that the membrane is of uniform composition, so that μ has the same value everywhere.

To recapitulate, we have redefined the quantities F and μ for the two-dimensional case now under consideration. You may feel that there is some awkwardness involved in using the same symbols for quantities slightly different from those used in discussion of the one-dimensional stretched string. There is a compensating advantage, however, in the similarity between the two-dimensional wave equation which we are about to develop and the one-dimensional equation which describes wave motion in

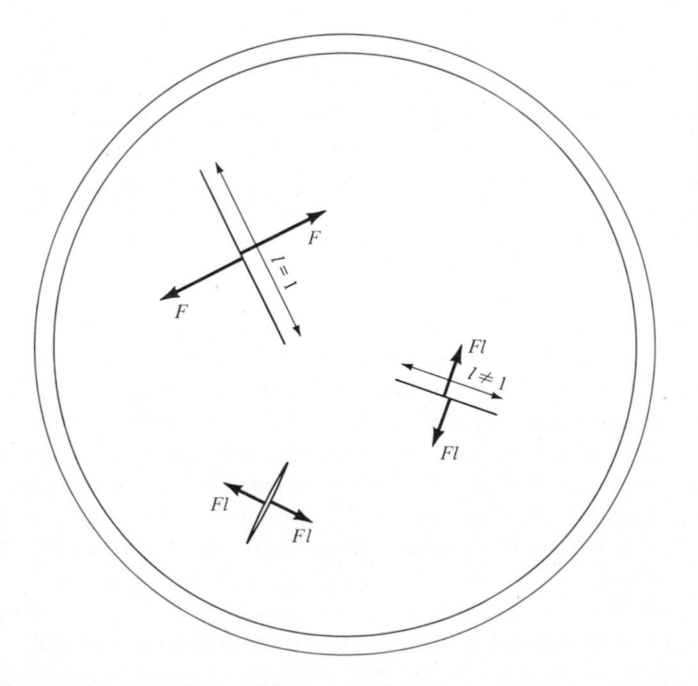

Fig. 13-16 Illustrating the tension F in a uniformly stretched membrane. If a line of unit length ($l = 1$) is constructed with any location and orientation, the total force exerted on the membrane at one side of the line by the membrane at the other side has magnitude F. If the length of the line is $l \neq 1$, each of these forces has magnitude Fl. If the membrane were actually cut along a line of length l, the two edges of the cut would pull apart because of the tension. To prevent this, a total force of magnitude Fl would have to be applied to each edge to balance the tension force.

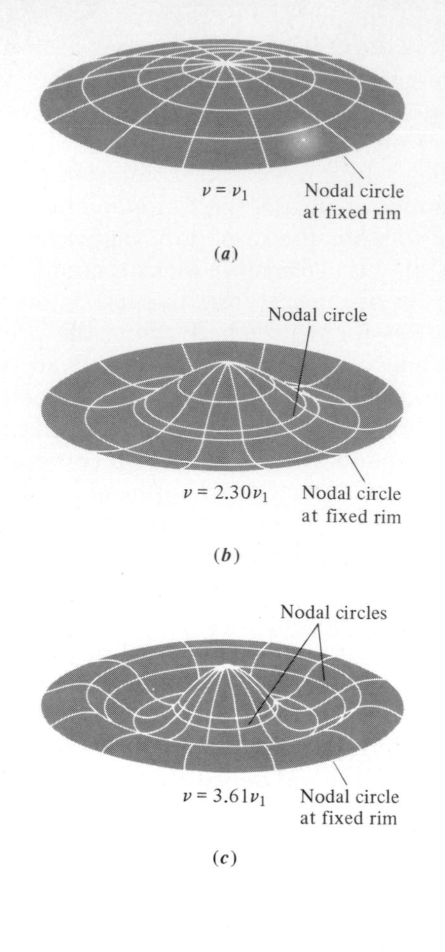

Fig. 13-17 "Snapshots" of a circular drumhead vibrating in the first three standing-wave modes possessing circular symmetry. A grid is drawn on the drumhead to facilitate visualization. As in the case of the vibrating string, each point on the drumhead oscillates in harmonic motion up and down with frequency ν and with an amplitude which depends on its position. But both the shape of the waveforms and the frequency ratios of the modes are more complicated than those for the vibrating string. (*a*) The fundamental mode, with frequency $\nu = \nu_1$. The value of ν_1 is determined by the radius, density, and tension of the drumhead, as discussed in the text. All points on the periphery of the drumhead are always nodes because the drumhead is fixed to the rim. In the fundamental mode they are the only nodes. Taken together, they comprise a *nodal circle.* (*b*) The second mode with circular symmetry. Its frequency is 2.30 times the fundamental frequency ν_1. There is a second nodal circle, located approximately 0.44 (less than half) of the distance from the center to the rim. At any instant, the displacement and direction of motion of points on the drumhead within this nodal circle are opposite to those of points outside it. The amplitude of the central antinode is greater than that of the second antinode. (*c*) The third mode with circular symmetry. Its frequency is 3.61 times the fundamental frequency ν_1. There are now two nodal circles in addition to the rim, located approximately 0.28 (less than one-third) and 0.64 (less than two-thirds) of the distance from the center of the rim, and the drumhead is divided into three regions which at any instant have displacements of alternate sign and are moving in alternate directions. The amplitudes of the three antinodes are successively smaller from the center outward.

a stretched string. Note that in going from one dimension to two dimensions, we give each of the quantities F and μ a corresponding dimensional change, dividing in each case by a length.

For simplicity, we will give a mathematical treatment only for standing waves which are symmetric about the center of the circular membrane. These are the ones that are excited in a drumhead if it is struck at its center. Three such standing waves are illustrated in Fig. 13-17. At the end of this section we will describe other types of standing waves that can exist on a drumhead.

As we did in studying waves on a stretched string, we study waves on a stretched drumhead by concentrating on a typical small segment of the drumhead. Because of the circular symmetry of the waves that we are considering, the appropriate shape of the segment is that of a thin ring concentric with the rim of the drumhead, as in Fig. 13-18. The ring extends from an inner radius r to an infinitesimally greater outer radius $r + dr$. The symmetry of the waves being treated implies that all points on the ring lying on a circle of a given radius have the same displacement z from the plane of the undisturbed drumhead. Note the strong analogy between the structure of Fig. 13-18 along any radial direction and the structure of Fig. 12-8, used in deriving the differential equation for any type of waves on a string. Here we derive the differential equation for symmetrical waves on a drumhead.

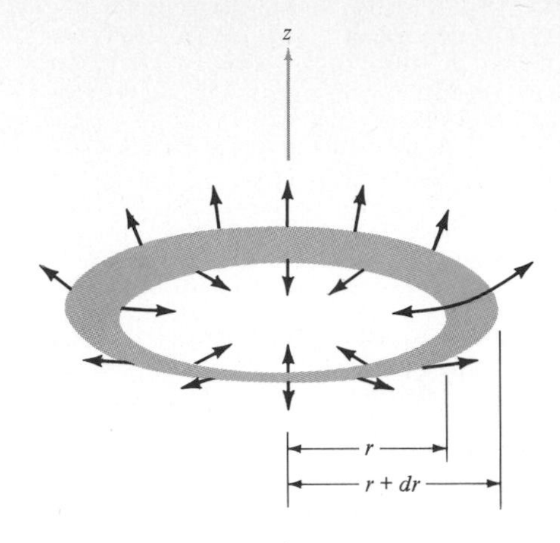

Fig. 13-18 Sketch for deriving the wave equation for the vibrating circular drumhead. Shown is a thin, ring-shaped element of the drumhead, at an instant when its outer periphery is distorted further in the positive vertical z direction than its inner periphery, and its shape is concave upward. The arrows schematically represent force vectors exerted on local regions of the ring by neighboring parts of the drumhead. They are everywhere tangent to the drumhead at their points of application, and their horizontal components (in the xy plane) lie along radii of the undisturbed drumhead. The tension F, or force per unit length, is everywhere the same in the drumhead. Consequently, more force vectors are shown along the outer periphery of the ring than along the inner periphery, because the outer periphery is longer.

At any point on the inner or outer periphery of the ring-shaped segment, the forces exerted on it by adjacent parts of the drumhead are locally tangent to the surface of the segment and locally perpendicular to its periphery. (What analogous statement can you make, based on Fig. 12-8, for the forces exerted on a segment of a stretched string?) Consider the inner periphery. Acting on it is the set of force vectors indicated in Fig. 13-18. Although they vary in direction, their z components (which are the ones associated with the z displacements of the segment) do not vary in magnitude. So the z component of the total force acting on a unit length of the inner periphery is just F times the slope, measured radially outward, of the drumhead at that periphery. (We are assuming here, as in Sec. 12-3, that the slope is small so that we can equate sines to tangents.) The slope is given by the partial derivative of $z = f(r, t)$ with respect to r, that is $\partial f(r, t)/\partial r$. Thus the z component of the total force acting on a unit length of the inner periphery is F, the tension force per unit length, times the slope $\partial f(r, t)/\partial r$. Multiplying this z component per unit length by $2\pi r$, the total length of the inner periphery, we find that the magnitude of the total force in the z direction acting on the inner periphery is

$$2\pi r F \frac{\partial f(r, t)}{\partial r}$$

The same expression, evaluated at $r + dr$, gives the magnitude of the total force in the z direction acting on the outer periphery of the segment, even though the forces in the z direction have opposite signs at the two peripheries. These two forces do not cancel completely because their magnitudes are not equal. In fact, the *net* force acting on the segment in the z direction is just the difference between the value of $2\pi r F\, \partial f(r, t)/\partial r$ at the outer periphery and its value at the inner periphery. This difference is given by the rate of change of the quantity with respect to r times dr, the change in r. So the net force acting on the segment in the z direction is

$$\frac{\partial}{\partial r}\left[2\pi r F \frac{\partial f(r, t)}{\partial r}\right] dr$$

Compare this with the equation just above Eq. (12-23), which gives the net force on a segment of a stretched string.

The mass of the ring-shaped segment is its area $2\pi r\, dr$ times the mass per unit area μ of the drumhead. So the mass is

$$2\pi \mu r\, dr$$

The acceleration in the z direction of the segment is

$$\frac{\partial^2 f(r, t)}{\partial t^2}$$

Equating force to mass times acceleration gives

$$\frac{\partial}{\partial r}\left[2\pi r F \frac{\partial f(r, t)}{\partial r}\right] dr = 2\pi \mu r \, dr \frac{\partial^2 f(r, t)}{\partial t^2}$$

Since the tension F is the same everywhere in the uniformly stretched drumhead it is a constant, just as is 2π. Thus we have

$$2\pi F \frac{\partial}{\partial r}\left[r \frac{\partial f(r, t)}{\partial r}\right] dr = 2\pi \mu r \, dr \frac{\partial^2 f(r, t)}{\partial t^2}$$

Canceling and transposing, we obtain

$$\frac{1}{r}\frac{\partial}{\partial r}\left[r \frac{\partial f(r, t)}{\partial r}\right] = \frac{\mu}{F}\frac{\partial^2 f(r, t)}{\partial t^2} \tag{13-35}$$

This is the **wave equation for circularly symmetrical waves** on the drumhead. [For the general case of waves that do not necessarily have this symmetry, the displacement of the drumhead must be written $z = f(r, \theta, t)$, and the left side of the wave equation contains partial derivatives with respect to θ.]

Because of the similarity of the right sides of the wave equations for a drumhead and a string—the sides containing the time derivative—it seems likely that for both the standing-wave solutions will have similar time dependences. With this in mind, we save ourselves some effort by assuming from the beginning that the standing-wave solutions to Eq. (13-35), $f(r, t) = g(r)h(t)$, can be written as

$$f(r, t) = g(r)\cos(2\pi\nu t) \tag{13-36}$$

Then we prepare to validate this assumption by calculating the partial derivatives. We obtain

$$\frac{1}{r}\frac{\partial}{\partial r}\left[r \frac{\partial f(r, t)}{\partial r}\right] = \cos(2\pi\nu t)\frac{1}{r}\frac{d}{dr}\left[r \frac{dg(r)}{dr}\right]$$

and

$$\frac{\partial^2 f(r, t)}{\partial t^2} = -g(r)(2\pi\nu)^2 \cos(2\pi\nu t)$$

Substituting these into the partial differential equation produces

$$\cos(2\pi\nu t)\frac{1}{r}\frac{d}{dr}\left[r \frac{dg(r)}{dr}\right] = -g(r)(2\pi\nu)^2 \frac{\mu}{F}\cos(2\pi\nu t)$$

This will be satisfied, and so Eq. (13-36) will be verified, if the following ordinary differential equation is satisfied:

$$\frac{1}{r}\frac{d}{dr}\left[r \frac{dg(r)}{dr}\right] = -\frac{4\pi^2\nu^2\mu}{F} g(r) \tag{13-37}$$

This is the time-independent equation corresponding to Eq. (13-35). The solutions $g(r)$ to Eq. (13-37) are the functions specifying the shapes of possible symmetrical standing waves on the drumhead. In finding these solutions, we will also determine for each of them the value of its frequency ν.

We will solve Eq. (13-37) numerically. (One reason for doing so is that it will introduce procedures of which we will make important use in Chap. 31 to solve the quantum-mechanical wave equation.) The first step is to work out the derivative of the term in brackets, so that the equation becomes

$$\frac{1}{r}\left[r\frac{d^2g(r)}{dr^2}+\frac{dg(r)}{dr}\right]=-\frac{4\pi^2\nu^2\mu}{F}g(r)$$

or

$$\frac{d^2g(r)}{dr^2}=-\frac{4\pi^2\nu^2\mu}{F}g(r)-\frac{1}{r}\frac{dg(r)}{dr} \tag{13-38}$$

Now the differential equation is in a form that allows us to apply the numerical method developed in Chap. 6. But before we do this, it is worthwhile to introduce the dimensionless variable u, defined as

$$u\equiv\frac{r}{a} \tag{13-39}$$

where a is the radius of the drumhead. The quantity u gives the distance from the center of the drumhead to some other location, expressed as a fraction ranging from 0 to 1 as that location ranges from the center to the rim. This will lead to an equation which can be applied to a drumhead of any radius. Employing the chain rule to convert the derivatives in Eq. (13-38) to the new variable, we have

$$\frac{dg(r)}{dr}=\frac{dg(u)}{du}\frac{du}{dr}=\frac{dg(u)}{du}\frac{1}{a}$$

and

$$\frac{d^2g(r)}{dr^2}=\frac{d}{du}\left[\frac{dg(r)}{dr}\right]\frac{du}{dr}=\frac{d}{du}\left[\frac{dg(u)}{du}\frac{1}{a}\right]\frac{1}{a}=\frac{d^2g(u)}{du^2}\frac{1}{a^2}$$

Substituting these derivatives into the differential equation and then multiplying through by a^2, we obtain

$$\frac{d^2g(u)}{du^2}=-\frac{4\pi^2\nu^2\mu a^2}{F}g(u)-\frac{1}{u}\frac{dg(u)}{du}$$

It is convenient to introduce the parameter α, defined as

$$\alpha\equiv\frac{4\pi^2\nu^2\mu a^2}{F} \tag{13-40}$$

If you insert the units for ν, μ, a, and F, you will see that α is a dimensionless number. In terms of α, the differential equation becomes

$$\frac{d^2g(u)}{du^2}=-\alpha g(u)-\frac{1}{u}\frac{dg(u)}{du} \tag{13-41}$$

The purpose of going from Eq. (13-38) to Eq. (13-41) is that the latter, since u and α are dimensionless quantities, has universal applicability. Once we find its solutions, they can be applied to *any* drumhead by simply using the appropriate value of a in evaluating the independent variable u and of μ, a, and F in evaluating the parameter α.

Writing Eq. (13-41) as

$$\frac{d^2g(u)}{du^2} = Q \tag{13-42a}$$

where

$$Q = -\alpha g(u) - \frac{1}{u}\frac{dg(u)}{du} \tag{13-42b}$$

we see that Eq. (13-42*a*) is mathematically identical to Eq. (6-14), $d^2x/dt^2 = Q$. This is the form we have used to solve second-order differential equations numerically. The numerical method for solving Eq. (13-42*a*) according to Eqs. (6-15) can be applied immediately. Doing this involves nothing more than adding to the harmonic oscillator program several steps which will generate the value of Q specified in Eq. (13-42*b*) and then running the calculator or computer with this new program. The vibrating drumhead program is listed in the Numerical Calculation Supplement.

As is always the case in numerical calculations, it is necessary here to specify the initial values of the quantity $g(u)$ and its first derivative $dg(u)/du$. The following points must be considered in determining these values:

1. In earlier numerical calculations, we have always begun by setting the initial value of the independent variable to zero, and we have then proceeded in small positive increments. But in the present case the quantity Q cannot be evaluated at $u = 0$ (the center of the drumhead) because at that point the factor $1/u$ in its second term goes to infinity. Fortunately, it is easy to circumvent this difficulty. The numerical calculation is started at $u = 1$ (the edge of the drumhead). *Negative* increments Δu are then taken and the calculation cycle is carried out repeatedly, each cycle yielding a value of $g(u)$ at a location closer to the center of the drumhead than its predecessor. This process is continued until $g(u)$ is evaluated very near, but not at, $u = 0$.

2. Since the drumhead is fixed at its outer edge, the "initial" condition on $g(u)$ at $u = 1$ is $[g(u)]_1 = 0$. In the terminology of Sec. 13-3, this condition is the boundary condition at one end of the range of the coordinate u.

3. The numerical method requires that we choose also an initial value for $dg(u)/du$ at $u = 1$. This quantity gives the slope of the drumhead immediately at the rim. But since the differential equation is linear in $g(u)$, the choice will affect only the vertical scale of a plot of the displacement $g(u)$ versus the coordinate u, and not the *shape* of $g(u)$. The vertical scale is of no real consequence, since it has to do only with the maximum displacement of the standing wave. Hence we can take any value we wish for $[dg(u)/du]_1$. To put it another way, we choose the units used to measure the displacement $g(u)$ so that its value at $u = 1$, in these units, is either -1 or $+1$.

4. The trickiest point is that the calculation of Q from Eq. (13-42*b*) requires a knowledge of the parameter α, which we do not know. Furthermore Eq. (13-40) shows that α depends on the value of the standing-wave frequency ν, and ν is what we are trying to determine. To deal with this difficulty, we try various "guess" values of α and use them to calculate $g(u)$ inward from $u = 1$, as outlined in point 1. Through this trial-and-error method, we search for those values of α which make it possible to satisfy the boundary conditions at the inner end of the range of the coordinate u, that is, at the center of the drumhead, where $u = 0$. Only those values of α can lead to standing waves. Therefore, only those values can lead to valid plots of the standing-wave functions $g(u)$. And once the values of α are known, Eq. (13-40) can be used, together with a knowledge of the constants μ, a, and F, to find the corresponding standing-wave frequencies ν.

What are the boundary conditions at $u = 0$? There is no restriction on the value of the displacement $g(u)$ at this point, because the drumhead is unconstrained at its center. But we must have $dg(u)/du = 0$ at $u = 0$. If this were not so, the value of Q given by Eq. (13-42b) would be infinite. The curvature of the drumhead given by Eq. (13-42a) would then be $d^2g(u)/du^2 = \pm\infty$ at $u = 0$, and the drumhead would have an infinitely sharp peak or depression at its center, as shown in Fig. 13-19. This is physically impossible, and the central boundary condition $[dg(u)/du]_0 = 0$ is thus justified.

In Examples 13-5 through 13-7, the vibrating drumhead program is used to find the values of the parameter α corresponding to the first three circularly symmetrical standing-wave modes of the drumhead.

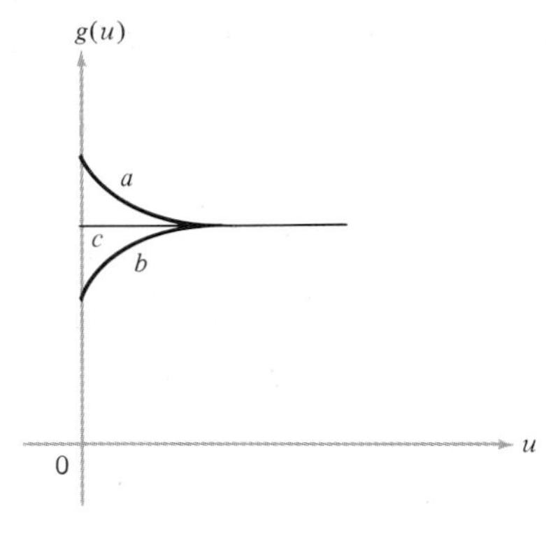

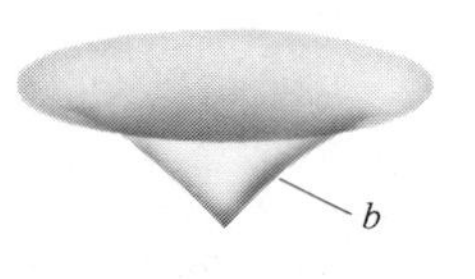

Fig. 13-19 Justification for the boundary condition $[dg(u)/du]_0 = 0$. The graph shows three possible behaviors of the wave function $g(u)$ in the immediate vicinity of the center of the drumhead, $u = 0$. We are considering only circularly symmetric wave functions. Symmetry therefore dictates that $g(u)$ must approach the vertical axis, at $u = 0$, in the same way from all directions. The curve representing $g(u)$ must approach the axis either with nonzero slope, as in the curve labeled a and b, or with zero slope, as shown by the curve labeled c. But curves a and b imply the correspondingly-labeled drumhead shapes shown in perspective. In each of these, the drumhead is distorted at its center into an infinitely sharp point, which is impossible for a real drumhead. Curve c in the graph implies the correspondingly labeled drumhead shape, which is flat and parallel to the horizontal plane in the immediate vicinity of the center. This is the only physical possibility for circular symmetry. The flatness, or zero slope, is specified by the boundary condition $[dg(u)/du]_0 = 0$.

EXAMPLE 13-5

Run the vibrating drumhead program with the following set of initial conditions and parameters:

$[g(u)]_1 = 0$; $[dg(u)/du]_1 = -1$; $u = 1$; $\Delta u = -0.01$; $\alpha = 5.750, 5.800, 5.825, 5.850, 5.900$.

■ The program displays the value of $g(u)$ at the end of every calculation cycle. Your concern centers on the slope $dg(u)/du$ of a plot of these values as u approaches zero, since the condition $[dg(u)/du]_0 = 0$ must be satisfied. So you begin by making a preliminary search using values of α more widely separated than those given above. You find that for values of α around 5.8, the slope $dg(u)/du$ is fairly small as u approaches zero. Then you use the values of α listed above to make a fine-grained search. In Fig. 13-20 the values of $g(u)$ are plotted versus u from the center of the drumhead to its periphery for each value of α.

The plot shows that when

$$\alpha \simeq 5.825 \tag{13-43}$$

the differential equation has a solution which satisfies the boundary condition $dg(u)/du = 0$ at $u = 0$, as well as the boundary condition $g(u) = 0$ at $u = 1$. The results for this value of α constitute a plot of the function $g(u)$ that solves the equation and satisfies the boundary conditions. The function describes the radial dependence of the standing-wave mode for which $\alpha = 5.825$. Physically, the plot shows a radial profile (or cross section) of the shape of the drumhead at a time when its displacements have a maximum positive value. The vertical scale in Fig. 13-20 is greatly exaggerated. The scale depends on the numerical value chosen for $[dg(u)/du]_1$. A value -1 was chosen simply to make the numerical values of $g(u)$ large enough to be plotted easily, and to lead to a positive displacement of the drumhead at the antinode $u = 0$. In reality, the displacements of the drumhead must be sufficiently small to satisfy the assumption that the slope of the displaced drumhead is everywhere small.

Remember that Fig. 13-20 is a "snapshot" of the profile of the drumhead from center to periphery at an instant when its displacement is maximum. It shows a scaled plot of $f(r, t) = g(r)\cos(2\pi\nu t)$ at an instant when $\cos(2\pi\nu t) = 1$. The center of the drumhead is displaced upward at this time. Shortly thereafter the center will be displaced downward, but with the same profile. Hence any point on the rim of the drumhead is a node (as it must be since the rim is fixed), and the center is an antinode.

Example 13-6 explores the properties of the second standing-wave mode of the vibrating drumhead.

EXAMPLE 13-6

Run the vibrating drumhead program with the following set of initial conditions and parameters:

$[g(u)]_1 = 0$; $[dg(u)/du]_1 = 1$; $u = 1$; $\Delta u = -0.01$; $\alpha = 30.85$.

■ A continued search for values of α satisfying the central boundary condition $[dg(u)/du]_0 = 0$—first with widely spaced values of α and then with more closely spaced values—yields

$$\alpha \simeq 30.85 \tag{13-44}$$

as the next value above $\alpha \simeq 5.825$ which does so. The resulting values of $g(u)$ are plotted versus u in Fig. 13-21. For this standing-wave mode, the drumhead has at a given instant a displacement of one sign at its center and a ring-shaped region near its edge with a displacement of the other sign. There is a node approximately—but not exactly—halfway out from the center at $u \simeq 0.44$, in addition to the node at the fixed rim. The antinodes lie at $u = 0$ and $u \simeq 0.70$. (Can you see why the value of $[dg(u)/du]_1$ was chosen to be $+1$ and not -1?)

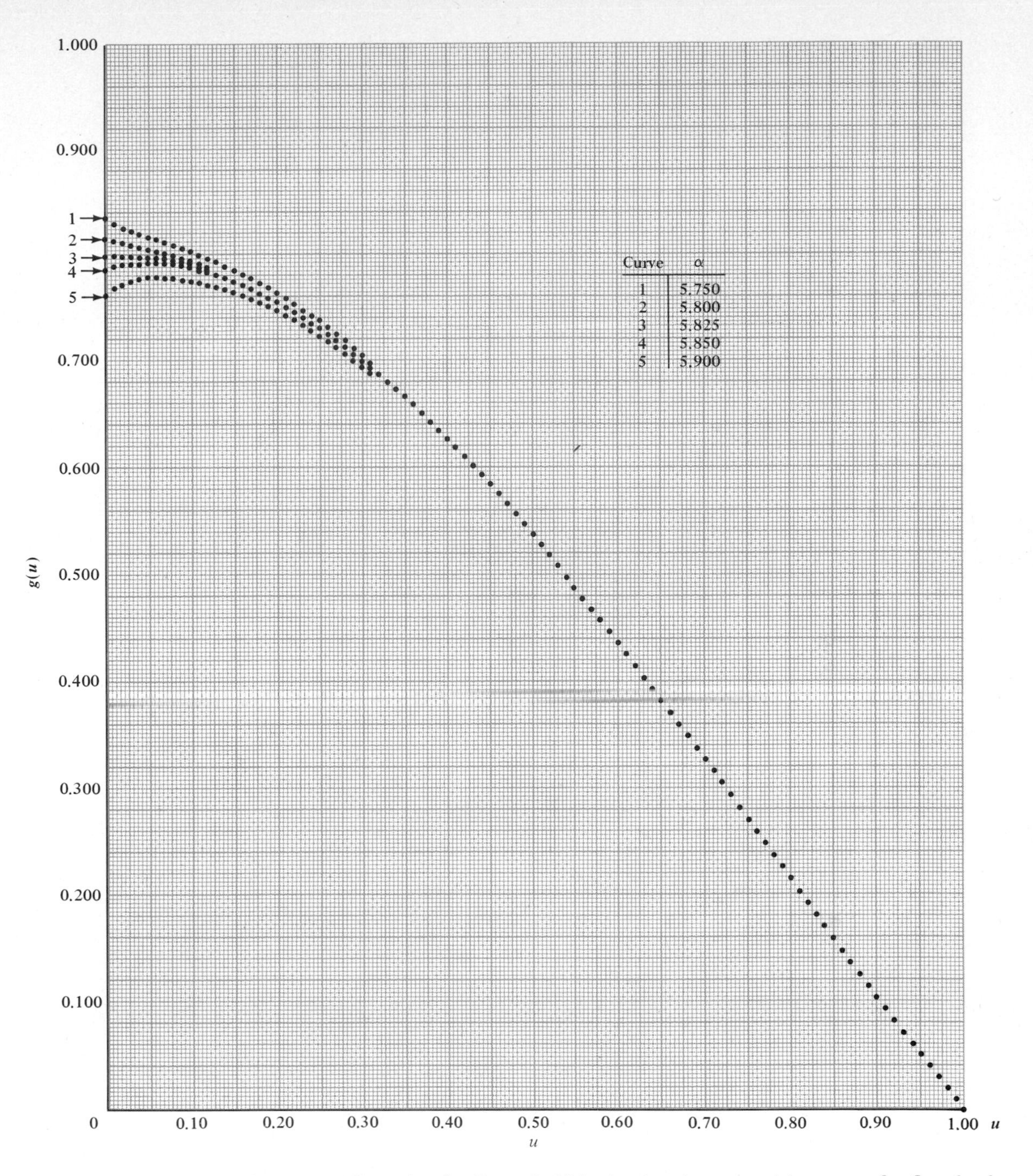

Fig. 13-20 Illustration for Example 13-5, showing the curve $g(u)$ versus u for five closely spaced trial values of the parameter α. Compare with Fig. 13-17*a*. Curve 3, representing the trial value $\alpha = 5.825$, comes close to approaching the $g(u)$ axis perpendicularly. It thus comes close to satisfying the boundary condition $[dg(u)/du]_0 = 0$. Beginning at the value $u = 1$, the curves representing the calculations for the five specified values of α nearly coincide at first, and have not been plotted separately. But as the calculation proceeds and the value of u approaches zero, the $g(u)$ curve becomes quite sensitive to the choice of α. The calculated curves therefore diverge and are plotted separately in the region of small u.

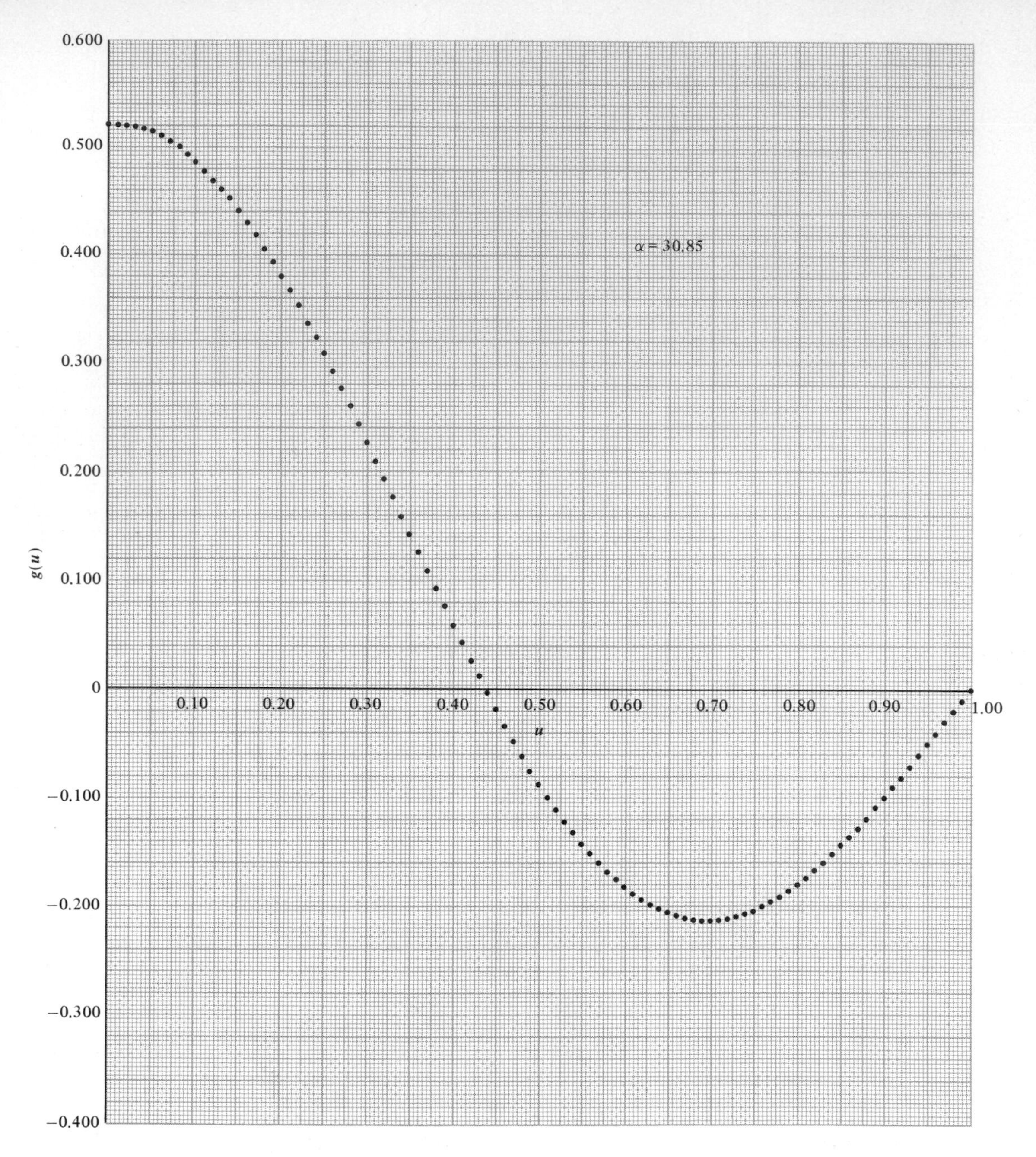

Fig. 13-21 Illustration for Example 13-6. The wave function $g(u)$ is plotted versus u for $\alpha = 30.85$, the second value of α for which the boundary condition $[dg(u)/du]_0 = 0$ is satisfied. Like Fig. 13-20, this graph may be regarded as a radial slice through the drumhead from center to periphery at an instant when every point on the drumhead achieves its maximum displacement from the undisturbed position represented by the u axis. Compare with Fig. 13-17*b*.

The standing wave is pictured in the perspective drawing of Fig. 13-17*b*. Note that the node at $u \simeq 0.44$ is actually a **nodal circle**—a circle having that radius, along which the drumhead remains at rest at all points.

Example 13-7 explores the properties of the third standing-wave mode of the vibrating drumhead.

EXAMPLE 13-7

Run the vibrating drumhead program with the following set of initial conditions and parameters:

$[g(u)]_1 = 0$; $[dg(u)/du]_1 = -1$; $u = 1$; $\Delta u = -0.01$; $\alpha = 75.9$.

■ The next value of α which satisfies the central boundary condition $[dg(u)/du]_0 = 0$ is

$$\alpha \simeq 75.9 \tag{13-45}$$

The corresponding values of $g(u)$ are plotted versus u in Fig. 13-22. Here there are two concentric ring-shaped regions surrounding the center region, and at any instant the sign of the displacement of the drumhead alternates in going from the center to the first ring and then to the second. The appearance of the standing wave is indicated by the drawing in Fig. 13-17*c*. There are now two nodal circles in addition to the fixed rim, located at $u \simeq 0.28$ and $u \simeq 0.64$.

The maximum displacement is smallest for the outer ring-shaped region, larger for the inner ring-shaped region, and largest for the central region. By considering the way the drumhead is supported, you should be able to give a physical explanation for this mathematical result.

Fig. 13-22 Illustration for Example 13-7. The wave function $g(u)$ is plotted versus u for $\alpha = 75.9$, the third value of α for which the boundary condition $[dg(u)/du]_0 = 0$ is satisfied. Compare with Fig. 13-17*c*.

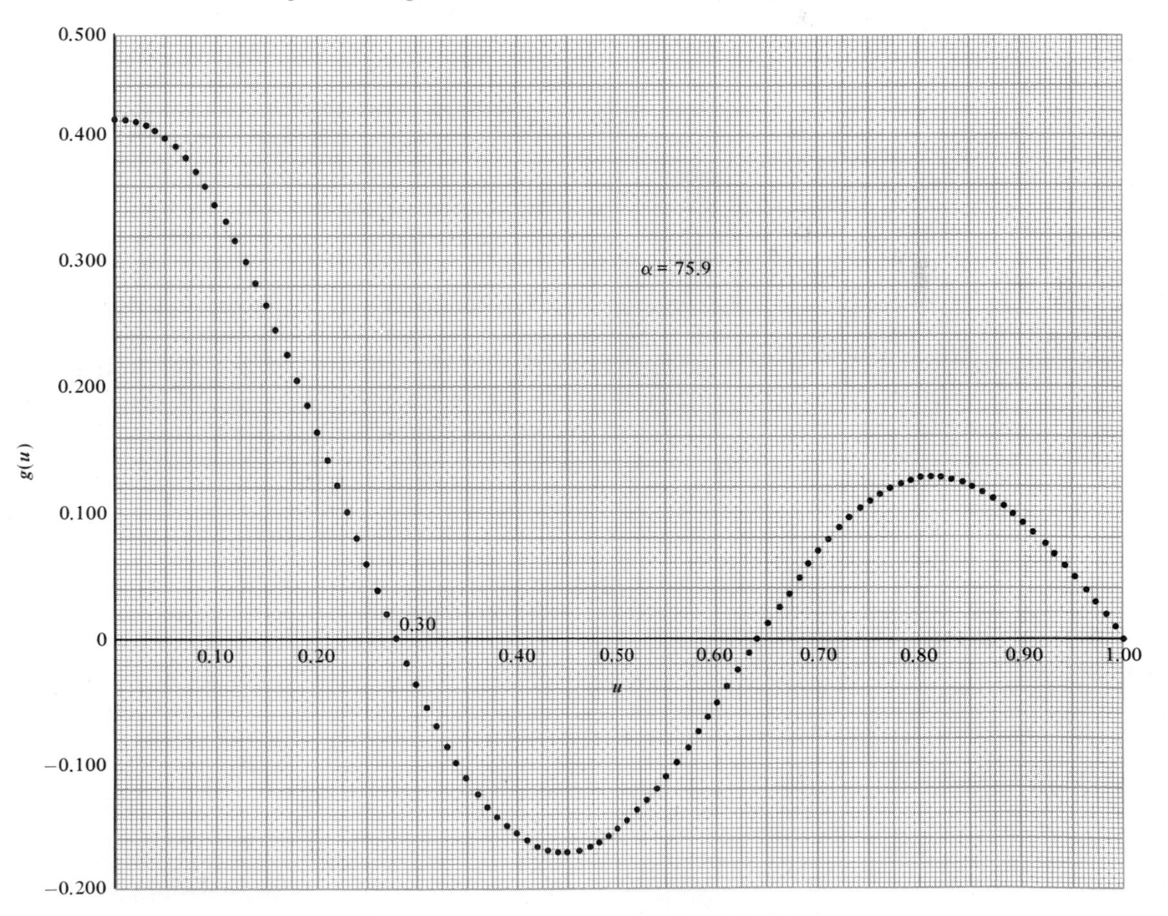

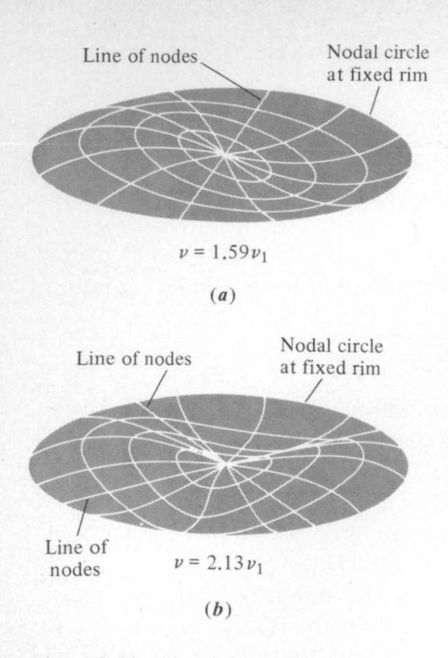

Fig. 13-23 "Snapshots" of a circular drumhead vibrating in two standing-wave modes which do not possess circular symmetry. The frequency ν of each mode is expressed in terms of ν_1, the frequency of the fundamental mode shown in Fig. 13-17*a*. (*a*) Here the drumhead vibrates symmetrically about a motionless **line of nodes** which is a diameter of the drumhead. The parts of the drumhead to the left and the right of the line of nodes have at any instant opposite displacements from the undisturbed position and oppositely directed velocities. (*b*) Here there are two lines of nodes which are perpendicular to each other. They divide the drumhead into a "four-leafed shamrock." At any instant, the "leaves" have displacements of alternate sign and move in alternate directions. In both of the modes shown in this figure, the center of the drumhead is a nodal point because it lies on at least one line of nodes.

In Examples 13-5 through 13-7 we have determined the frequencies (or at least the ratios of the frequencies) of the first three circularly symmetrical standing waves in a drumhead. According to Eq. (13-40), the frequency ν is

$$\nu = \frac{1}{2\pi a}\sqrt{\frac{F}{\mu}}\sqrt{\alpha} \tag{13-46}$$

The quantity $(1/2\pi a)\sqrt{F/\mu}$ is determined by the physical properties of the drumhead: its radius a, tension F, and density μ. So there is a certain fixed value of this quantity for a particular drumhead. Hence the value of ν for each standing wave on a particular drumhead is proportional to its value of $\sqrt{\alpha}$. Taking the values of α from Eqs. (13-43) through (13-45), we have

$$\nu \propto \sqrt{5.825}, \sqrt{30.85}, \sqrt{75.9}, \ldots$$

or

$$\nu \propto 2.41, 5.55, 8.71, \ldots$$

The lowest is the fundamental frequency for the drumhead. It is the frequency for the standing wave, illustrated in Fig. 13-17*a*, that has the simplest shape. If we write the fundamental frequency as ν_1, our results can be expressed as

$$\nu = \nu_1, \frac{5.55}{2.41}\nu_1, \frac{8.71}{2.41}\nu_1, \ldots$$

or

$$\nu = \nu_1, 2.30\,\nu_1, 3.61\,\nu_1, \ldots \tag{13-47}$$

In contrast to the situation for standing waves on a string, the frequencies of successively more complicated drumhead standing waves of circular symmetry are *not* equal to the fundamental frequency multiplied by the successive integers. That is, the standing-wave modes are *not* harmonics.

This is also true for the standing waves we have been ignoring—those in which the displacement of the drumhead at any instant does not have symmetry about its center. There are many such standing waves, each with its own characteristic frequency. Their mathematical treatment is beyond the level of this book. We indicate in Fig. 13-23 only the general character of two of them by showing the shape of each standing wave and its frequency in units of ν_1. These two are the lowest-frequency standing waves on a drumhead that do not have circular symmetry. In Sec. 13-8, you will see that there is a direct connection between the nonintegral relation among the frequencies of various standing-wave modes of a drumhead and the "nonharmonious" sound of a drum.

You may be interested to know that Eq. (13-35) is called *Bessel's equation of order zero*. It is a very important differential equation because it arises in many different fields of mathematical physics. Solutions to it can be obtained, with some difficulty, by analytical methods. The solutions, such as the ones plotted in Fig. 13-17, are called *Bessel functions of order zero*. They are evidently oscillatory, but not sinusoidal. Nor are they exponentially diminishing sinusoids of the sort encountered in Chap. 6.

So far we have dealt with the vibrations of drumheads only in a form reduced to dimensionless parameters. In Example 13-8 these results are applied to a specific drumhead having realistic size, tension, and mass per unit area.

EXAMPLE 13-8

Find the fundamental frequency for a drumhead of radius $a = 20.0$ cm, areal density $\mu = 500$ g/m², and tension $F = 1000$ N/m.

▪ Using Eq. (13-46), with $\alpha = 5.825$, you have

$$\nu = \frac{1}{2\pi a}\sqrt{\frac{F}{\mu}}\sqrt{\alpha}$$

$$= \frac{1}{2\pi \times 0.200 \text{ m}}\sqrt{\frac{1000 \text{ N/m} \times 5.825}{0.500 \text{ kg/m}^2}}$$

$$= 85.9 \text{ Hz}$$

13-6 ACOUSTICS

Sound waves in air and other media are quite typical of the longitudinal waves discussed in Sec. 12-6. They are usually small enough in amplitude, and the resistance to deformation of the propagation medium is usually linear enough—that is, obeys Hooke's law closely enough—that the superposition principle applies quite accurately. (There are notable exceptions, such as the crack of a whip and the sonic boom of high-speed airplanes.) However, the word "sound" used in the everyday sense embraces much more than the propagation of longitudinal waves through the air.

First, there must be some sort of vibrating source, and these are of many varieties. Second, there must be some way of coupling the vibrating source to the propagation medium. Third, there must be propagation through the medium. Except in unusual cases, the propagation process is strongly affected by reflection, bending of path, and absorption at intervening or nearby surfaces, including the ear canal itself. Fourth, the oscillating air sets the eardrum into vibration. Fifth, the eardrum sets into vibration the very complex solid and liquid mechanical system comprising the middle and inner ear. This system is very far from obeying Hooke's law for sounds of commonly encountered loudness and thus introduces complicated modifications of its own. Sixth, the mechanical vibration is translated into electric nerve impulses by an array of an enormous number of so-called hair cells whose detailed structure varies subtly from point to point. Seventh, the nerve impulses are organized and interpreted by the auditory nerve and the brain.

The first through the fourth of these processes lie in the domain of **physical acoustics,** to which we devote most of our attention in this section. The fourth through the seventh are the domain of **physiological acoustics,** and the seventh also falls into the domain of **psychological acoustics.** The remainder of this section provides an introduction to the physical bases of the acoustical impressions which we call *loudness, pitch,* and *quality* or *timbre.*

The ear, like most of the other sense organs, must respond in a useful manner to a tremendous range of stimulus intensity. To this task it is admirably adapted. The ratio of the energy flux in the loudest sound the ear can handle to that of the faintest sound it can detect is something like 10^{13} at frequencies of approximately 1 kHz. The normal ear is most sensitive at about 3.5 kHz, where the minimum audible energy flux is of order of magnitude 10^{-12} W/m². This flux involves a pressure fluctuation of the order of 10^{-10} times atmospheric pressure, and the oscillation amplitude of the air is something like 10^{-12} m, or about 1 percent of the diameter of a typical atom. (Individual atoms in the air move about randomly very much more than this, as you will see in Chap. 17. But this is the average displacement of the air taken as a whole, as it oscillates against the eardrum.)

In order for the enormous range of stimulus intensity to which the ear can respond to be compressed into a manageable range of perceived loudness, the sense of hearing, like that of sight and of touch, is highly compressed. That is, it takes much more than a doubling of the sound energy flux S to make a listener render a subjective judgment that a sound has become "twice as loud." And a redoubling of the subjective loudness requires a still greater increase in S. In fact the sensation of **loudness** is related roughly logarithmically to the energy flux incident on the ear. We can exploit this physiological-psychological fact to define a rough but convenient scale of *acoustic intensity*.

We call the energy flux associated with the faintest audible sound S_0 the **threshold of hearing.** The **acoustic intensity** α of a sound whose energy flux is S is defined to be

$$\alpha \equiv 10 \log \frac{S}{S_0} \tag{13-48a}$$

This equation says that a 10-fold increase in energy flux above S_0 leads to a 10-fold increase in α. But a 100-fold increase in energy flux above S_0 leads only to a 20-fold increase in α, and a 1000-fold increase leads to a 30-fold increase in α. You can see how this leads to the desired "compression" of the scale of α, which corresponds to what is perceived by the listener. The threshold of hearing varies substantially from individual to individual and rises significantly with increasing age. Nevertheless, it has been agreed by international convention to set

$$S_0 = 1 \times 10^{-12}\ \text{W/m}^2 \tag{13-48b}$$

The acoustic intensity α is evidently a dimensionless number; nevertheless, α is expressed in a unit called the **decibel** (dB) after Alexander Graham Bell (1847–1922), the Scottish-American teacher of the deaf, phonologist, acoustician, inventor, and painter. The decibel is of convenient size, since 1 dB represents, in the middle range of frequency and loudness, approximately the minimum change in loudness of a sound which is perceptible.

Table 13-1 lists the acoustic intensity in decibels of a number of familiar acoustic environments.

Table 13-1

Typical Acoustic Intensities

Acoustic environment	**Acoustic intensity α** (in dB)
Threshold of hearing	0 (by definition)
Intelligible whisper in quiet surroundings	20
Quiet street	40
Interior of jet plane in flight	50
Conversation	60–65
Busy downtown street	60–70
Heavy truck from side of road	90
Jackhammer at 5 m distance	95–105
Rock band in closed room	90–>120
Threshold of pain	120

EXAMPLE 13-9

By what factor must the acoustic energy flux increase if the change is to be barely perceptible? Assume that the initial acoustic intensity is 40 dB and the frequency range lies near 3 kHz, where the typical ear has normal response and maximum sensitivity.

■ Since the acoustic intensity and the frequency are both roughly in midrange for the ear, a change of 1 dB in α will be perceptible. Using Eq. (13-48*a*), calling the initial and final intensities α_i and α_f, and calling the corresponding energy fluxes S_i and S_f, you can therefore write

$$1 = \alpha_f - \alpha_i = 10\left(\log \frac{S_f}{S_0} - \log \frac{S_i}{S_0}\right) = 10 \log \frac{S_f}{S_i}$$

or

$$\log \frac{S_f}{S_i} = \frac{1}{10}$$

By the definition of the logarithm function, this means that

$$\frac{S_f}{S_i} = 10^{1/10}$$

or

$$\frac{S_f}{S_i} = 1.26$$

That is, the ear can just discern a 26 percent increase in acoustical energy flux.

Just as the ear responds to the physical stimulus of acoustic energy flux with the sensation of *loudness,* it responds to the physical stimulus of frequency with the sensation of **pitch.** The lowest frequency audible as a sound is somewhere between 16 Hz and 25 Hz for most normal ears. The upper limit depends very much on the individual and particularly on age. Many persons younger than 16 years can hear well above 18 kHz (say, to 20 kHz or so) while few persons older than 45 years can hear much above 12 kHz. (An interesting benchmark is the annoying whistle produced by poorly designed television receivers. Its frequency in North America and some other parts of the world is 15,750 Hz. Can you hear it?) The high-frequency response of the ear is particularly susceptible to damage from sustained loud sounds; rock musicians and persons who listen frequently to highly amplified music are often found to be substantially deaf at frequencies above 10 kHz or so.

The *sensation of pitch* is also logarithmic with respect to the *stimulus of frequency.* That is, a repeated additive increase in the sensation of pitch by a given amount requires a repeated increase of the frequency by the same multiplicative factor. To give the simplest and most familiar example, an increase in pitch of one *octave* requires a doubling of the frequency, and an increase in pitch of two octaves requires a quadrupling of the frequency, regardless of the value of the initial frequency. In these terms, the ear has a frequency range close to 10 octaves. This corresponds roughly to a 1000-fold range of frequency ($\simeq$20 Hz to $\simeq$20 kHz).

Besides its sensitivity to loudness and pitch, the ear possesses great sensitivity to *quality,* or *timbre,* as it is called in connection with musical instruments. It is this sensitivity which makes it possible for you to distinguish among the voices of your friends, even when they have been substantially distorted by the telephone. With a little practice you can distinguish not only among the various instruments in an orchestra, but also between two

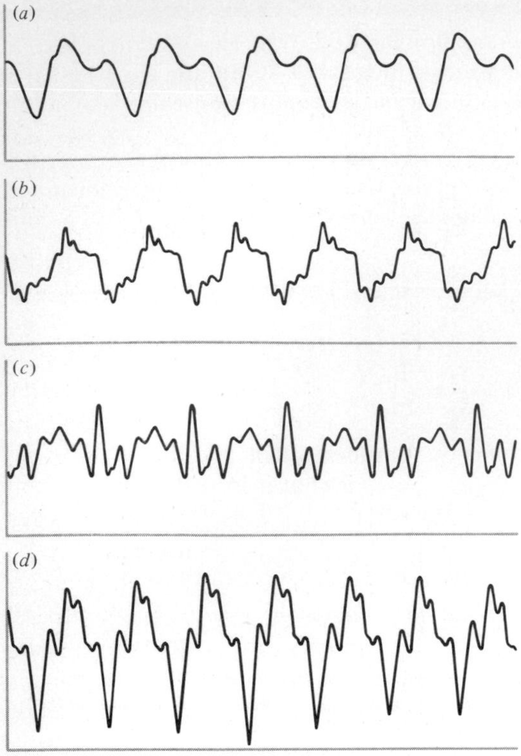

(a)

(b)

(c)

(d)

Fig. 13-24 Space-independent wave functions for the same musical note played on (a) a flute, (b) a clarinet, (c) an oboe, and (d) a saxophone. (*After D. C. Miller, Sound Waves, Their Shape and Speed, Macmillan, New York, 1937.*)

instruments of the same kind. That is one reason why musicians will go to great trouble and expense to obtain a particular violin, or guitar, or clarinet.

What is the physical stimulus corresponding to the subjective sensation of quality or timbre? The answer begins with Fig. 13-24, which depicts the space-independent wave function of a particular note (which happens to be middle C) played on several different musical instruments. A little study will convince you that the *periods* of the waves are all the same, but their *shapes* vary considerably. This consideration leads us directly to Fourier synthesis, which is the topic of Sec. 13-7.

13-7 FOURIER SYNTHESIS

It seems reasonable to ask: Can a wave of any shape whatsoever be *synthesized*—built up—by the proper superposition of simple sinusoidal waves? It was proved in 1822 by the French mathematician Joseph B. Fourier (1768–1830) that the answer is Yes. Fourier's theorem can be stated on several levels of generality, two of which concern us here:

1. Any perfectly periodic function $f(t)$—that is, any function which repeats itself identically and indefinitely—whose first derivative is well defined everywhere—that is, the function has no sharp corners—can be represented by a *finite* series of the form

$$f(t) = A_1 \sin[2\pi(\nu)t + \delta_1] + A_2 \sin[2\pi(2\nu)t + \delta_2] + A_3 \sin[2\pi(3\nu)t + \delta_3] + \cdots + A_n \sin[2\pi(n\nu)t + \delta_n] \quad (13\text{-}49a)$$

Expressed in more compact notation, this is identical to

$$f(t) = \sum_{j=1}^{n} A_j \sin[2\pi(j\nu)t + \delta_j] \tag{13-49b}$$

Written in either of these two ways, the series is called the **Fourier expansion** for a space-independent wave function $f(t)$. If this function represents the time dependence of the pressure oscillations in a sound wave arriving at the fixed position of the ear of a stationary listener, then it describes the situation most often analyzed in acoustics. However, the theorem is equally valid no matter what physical phenomenon is described by the mathematical function $f(t)$.

Each term in the summation of Eqs. (13-49*a*) and (13-49*b*) is called a **Fourier component** of the function $f(t)$. We have written these terms in such a way that they depend on the frequency $j\nu$ of the sinusoidal function giving the time dependence of the Fourier component, rather than on its angular frequency or its period. The reason is that it is customary to deal with frequencies in acoustics. In Eqs. (13-49*a*) and (13-49*b*), the amplitudes A_j of each of the components must be chosen properly to synthesize the desired function $f(t)$. That is, proper "amounts" of the various components must be used. Furthermore, the components must have the proper phases δ_j. But what is perhaps surprising is that the *only* components necessary have frequencies which are *integral multiples of the fundamental frequency* ν.

2. Even if a perfectly periodic function $f(t)$ does not have a well-defined first derivative everywhere—that is, if it has sharp corners like the "sawtooth" function illustrated in Fig. 13-25*a*—it can still be represented as

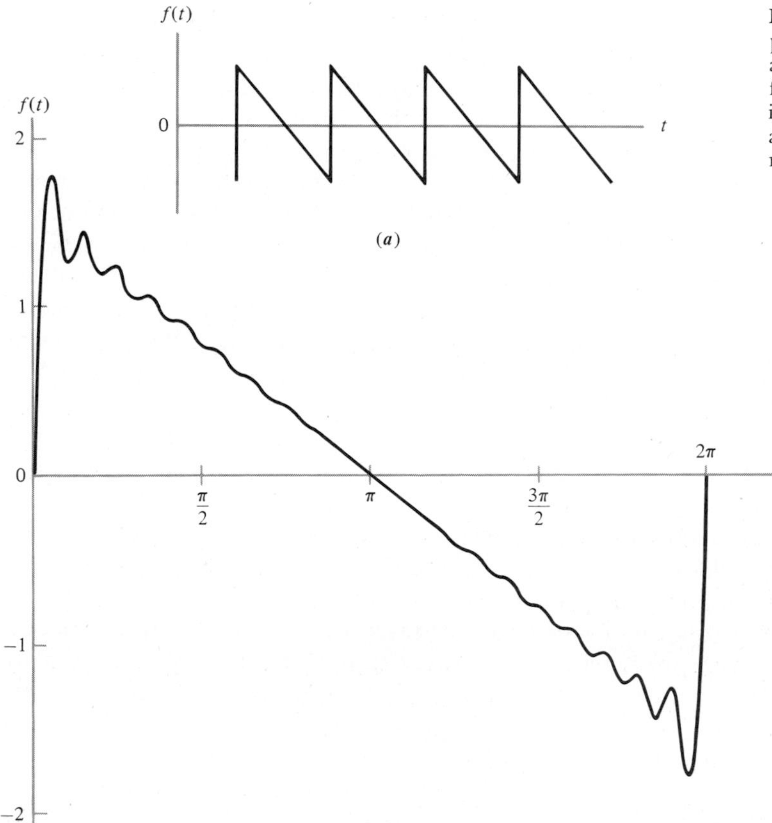

Fig. 13-25 (*a*) An idealized "sawtooth" function. The perfectly sharp corners shown are not realizable in an actual physical system. (*b*) The space-independent wave function obtained by superposition of the 20-term series in which the typical term has the form $\sin(2\pi j\nu t)/j$, and j runs from 1 to 20. The function bears strong resemblance to that in part *a*.

a sum of the form given in Eq. (13-49*a*), but with an infinite number of terms of frequency ν, 2ν, 3ν, ... For comparison, Fig. 13-25*b* represents the sum of 20 terms of the form

$$f(t) = 1 \sin[2\pi(\nu)t] + \tfrac{1}{2}\sin[2\pi(2\nu)t] + \tfrac{1}{3}\sin[2\pi(3\nu)t] + \cdots$$
$$= \sum_{j=1}^{20} \frac{1}{j} \sin[2\pi(j\nu)t]$$

It does not take a great stretch of the imagination to believe—and it can be rigorously proved—that Fig. 13-25*a* represents the sum of the infinite series

$$f(t) = \sum_{j=1}^{\infty} \frac{1}{j} \sin[2\pi(j\nu)t] \qquad (13\text{-}50)$$

In this particular case the **amplitude coefficients** A_j are given by $A_j = 1/j$. Consequently, the value of A_j decreases fairly rapidly as j increases. This is generally the case when series of the form of Eq. (13-49*b*) represent physical situations. Therefore, it is not usually necessary for practical purposes to use a very large number of terms in the summation.

The practical question now arises: Given a particular form $f(t)$ of a space-independent wave function, is it possible to determine the amplitudes A_j and the phases δ_j and thus to analyze the wave into its simple sinusoidal components? It is indeed possible. The process is the inverse of Fourier synthesis and is called **Fourier analysis.** Fourier analysis is an important tool both mathematically and physically. In acoustics, it is usually accomplished by the use of special instruments called Fourier analyzers or wave analyzers. A microphone picks up the sound and converts it into an electric waveform nearly identical to the waveform of the sound. (By **waveform** is meant the shape of an actual space-independent or time-independent wave. A distinction is thus made between a waveform and a wave function, which is a mathematical representation, either algebraic or graphical, of a wave.) The electric waveform is then fed into an array of devices called *filters,* or their equivalents, which separate the various Fourier components according to their frequencies. The individual amplitudes of the Fourier components are then measured as voltages and displayed in one of several convenient fashions.

The most useful form of display is usually the **Fourier spectrum,** which is simply a graphical plot of amplitude as a function of frequency. Figure 13-26 depicts the Fourier spectra of typical notes played on a variety of musical instruments. Such a Fourier spectrum, in which only specific, sharply defined frequencies contribute to the complete wave, is called a **discrete spectrum,** or **line spectrum.**

Of course, nothing in the real world is infinitely sharp, and the spectra of Fig. 13-26 are idealized. Figure 13-27 depicts Fourier spectra actually obtained from a flute and a bassoon, playing in various parts of their ranges, by a specialized wave analyzer called a sound spectrograph. In each case, the corresponding space-independent waveform also is shown.

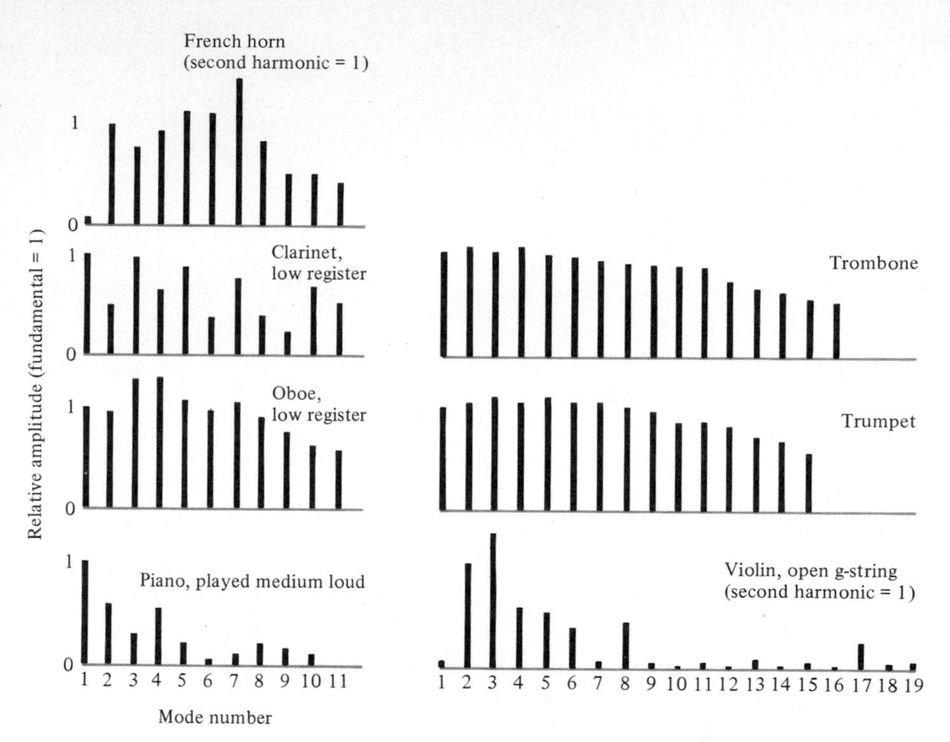

Fig. 13-26 Typical Fourier spectra of some common musical instruments. The Fourier spectrum of an instrument varies considerably over its range of pitch and depends to some degree on the loudness with which it is played and other factors. Nevertheless, the characteristic quality of the instrument is still clearly discernible. The reason for this is especially evident for the clarinet, where the odd harmonics ($j = 1, 3, 5, \ldots$) are considerably stronger than the even harmonics ($j = 2, 4, 6, \ldots$).

Fourier analysis can be applied to time-independent wave functions as well. Consider, for example, the vibrating string shown in the photo of Fig. 13-28. The shape of the string at any instant is quite complex. But it can be represented by a Fourier series of the form

$$f(x) = \sum_{j=1}^{n} A_j \sin\left[2\pi\left(\frac{j}{\lambda}\right)x + \delta_j\right]$$

This form is very similar to Eq. (13-49*b*). However, the independent variable is x rather than t. And the coefficient of x, that is $2\pi(j/\lambda)$, is the wave number, which depends on the fundamental wavelength λ corresponding to the fundamental frequency ν in Eq. (13-49*a*). In contrast, the quantity $2\pi(j\nu)$ which multiplies t in Eq. (13-49*b*) depends on the frequency ν and is the angular frequency.

For the particular case of the string in Fig. 13-28, the phase constants δ_j must all be equal to zero if the origin is chosen at the left end of the string. This restriction is imposed by the boundary conditions. Can you explain how?

A physical interpretation of the Fourier series describing the shape of the vibrating string is as follows. The string may be regarded as vibrating in several of its harmonic modes at the same time. Such vibration is the rule rather than the exception for stringed musical instruments. The modes are superposed. You may imagine the higher-j modes as "riding" on the lower-j modes, each with its own particular amplitude A_j. It is the superposition of varying proportions of harmonic modes, or harmonics, which determines in large measure the peculiar tone quality, or timbre, of each musical instrument. This point is considered in further detail in Sec. 13-8.

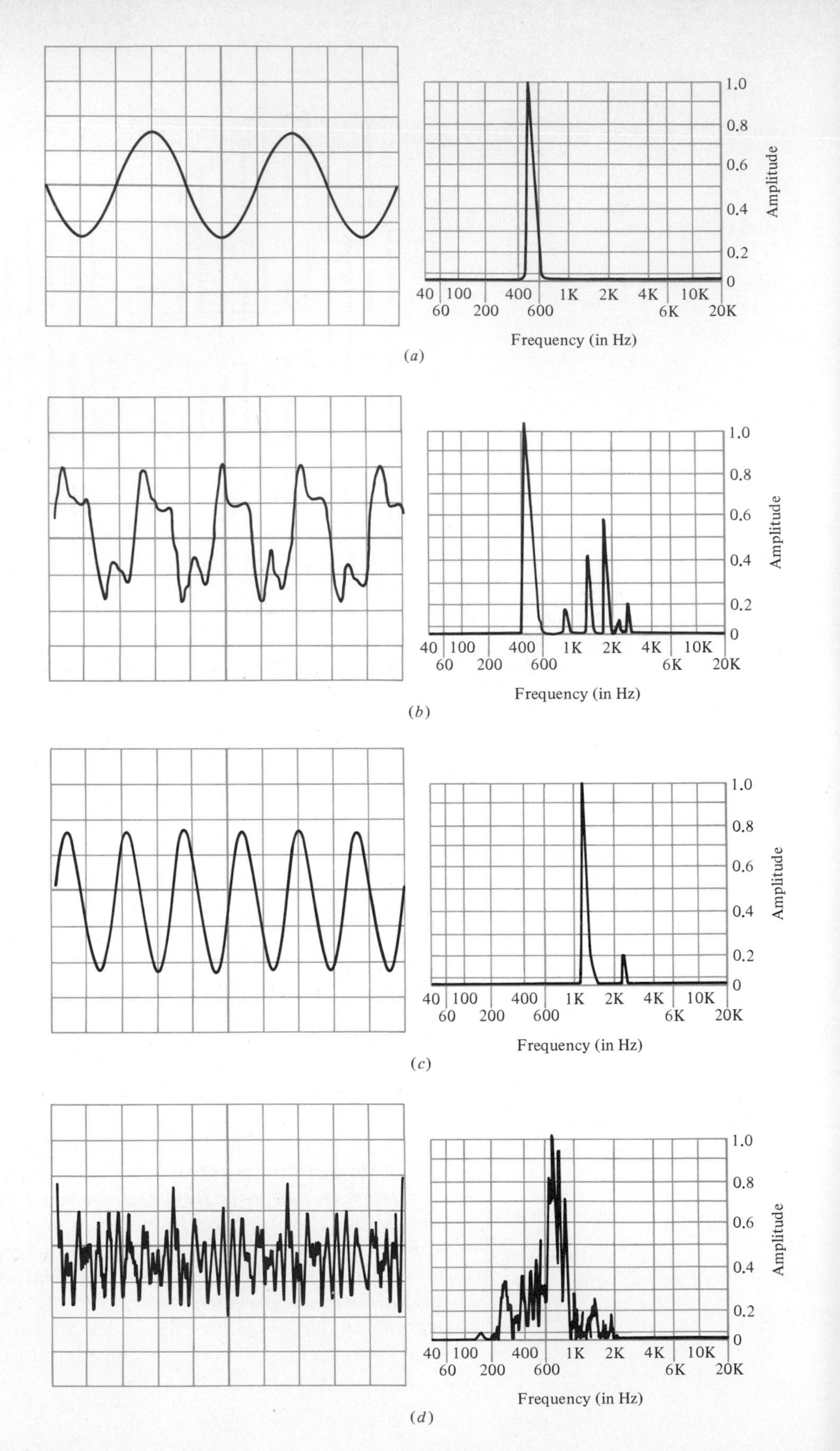
1.0
0.8
0.6
0.4
0.2
0
Amplitude
40
100
400
1K
2K
4K
10K
60
200
600
6K
20K
Frequency (in Hz)
(a)
(b)
(c)
(d)

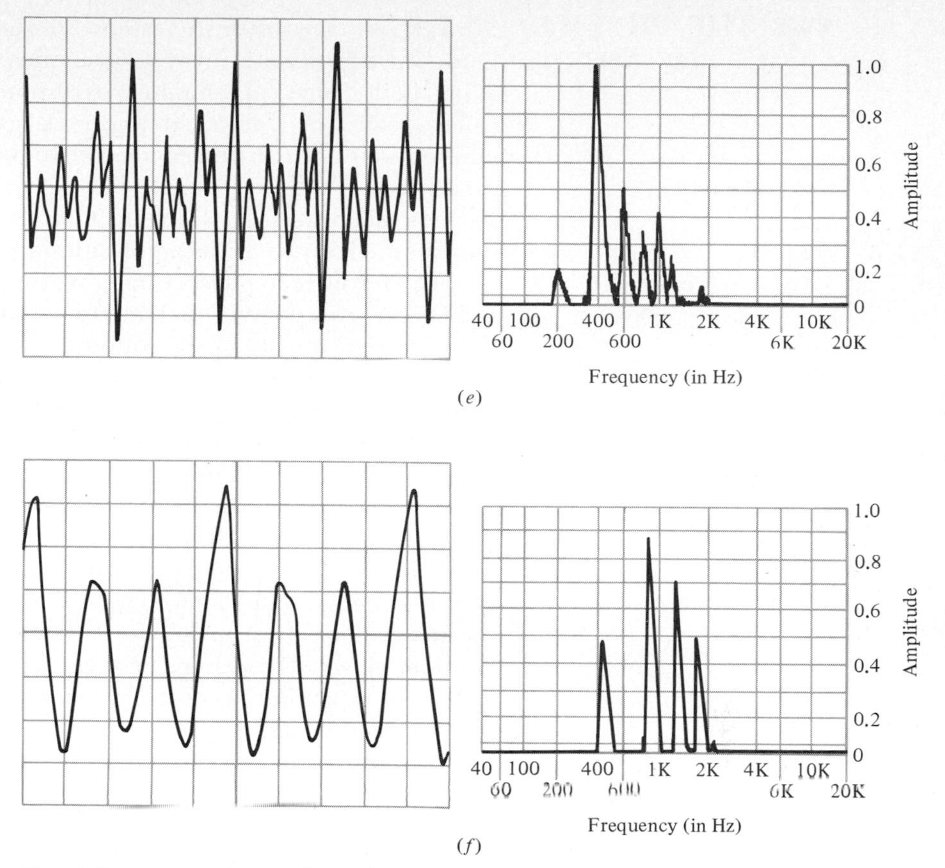

Fig. 13-27 (*a*) Waveform and sound spectrum of a pure (sinusoidal) tone generated electronically (500 Hz). (*b*) Waveform and sound spectrum for a flute playing near the lower end of its range (A = 440 Hz). The fundamental is strongest, but higher frequencies are well represented, too, especially the third and fourth harmonics. (*c*) Flute playing in the middle of its range (D = 1175 Hz). The tone has "cleaned up" to the typically pure quality we associate with this instrument. The second harmonic has only one-fifth the amplitude of the fundamental, and no other harmonics can be seen. The eye can barely distinguish the waveform from a pure sinusoidal. (*d*) Bassoon playing at the very bottom of its range (B-flat = 58 Hz). Note the complete absence of the fundamental and the great weakness of the lower harmonics. The maximum amplitude is in the twelfth harmonic. (*e*) Bassoon playing in the middle of its range (G = 196 Hz). The fundamental is now visible, but is weaker than five of the harmonics, especially the second. (*f*) Bassoon near the top of its range (A = 440 Hz). This is the same note being played by the flute in part *b*. The fundamental is now stronger, but still does not dominate the waveform. This sound is much smoother than the lower notes, but still quite characteristic of the bassoon.

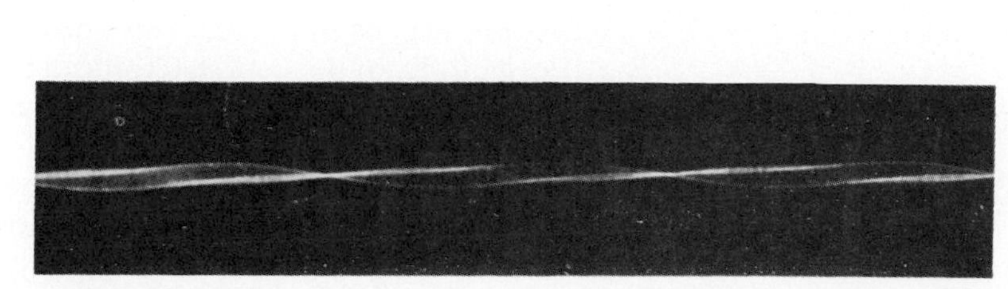

Fig. 13-28 Time-exposure photograph of a string vibrating in a complex fashion. The complicated shape is due to the simultaneous presence of several sinusoidal standing-wave modes, each with its own amplitude. The time-independent wave function representing the shape of the string at any moment can be represented as a Fourier series, as discussed in the text. (*From D. C. Miller, The Science of Musical Sounds, New York, Macmillan, 1916.*)

13-8 THE PHYSICS OF MUSIC

We can now return to the acoustical question, asked in Sec. 13-6, which launched the discussion of Fourier analysis and synthesis in Sec. 13-7: What is the physical stimulus corresponding to the subjective sensation called quality or timbre? It is very largely the Fourier spectrum of the emitted sound which determines the timbre of a musical instrument, or the quality of a sound in general. The ear (or rather the auditory sensory apparatus taken as a whole) is a quite sensitive, although nonquantitative, Fourier analyzer as far as amplitudes are concerned. (However, the ear is not at all sensitive to phase. Changing the values of the phase constants δ_j in the Fourier components making up a sound wave causes no change at all in the perceived quality of the sound.)

As far as the timbre of musical instruments is concerned, the Fourier spectrum of a more or less sustained note is not the whole story, because musical instruments do not usually play sustained notes. The *attack* and *decay* are also of great importance. The first term refers to the brief time during which the musical note "starts up," while the second refers to the time during which it "dies away." Not only does the amplitude of the total wave change during these intervals, but also the ratios of the Fourier components change in a way which contributes to the characteristic timbre of the instrument. In a piano, for instance, the note begins with a sudden hammer blow to the string (the attack) and dies away gradually (the decay). An organ pipe, on the contrary, takes a short but noticeable time to build up to a steady state. (Indeed, a tape recording of a piano played backward sounds vaguely like an organ.) The relatively highly damped violin string responds very sensitively to subtle variations in bowing technique. The singer imposes a *vibrato* (a slight oscillation of pitch at a frequency of approximately 5 Hz) on the sung note, either consciously or unconsciously. The violinist does the same by rocking the fingers on the fingerboard.

Most sources of sound do not produce a single frequency or a group of discrete frequencies in the way that most musical instruments do. Rather, there is sound energy at all frequencies within some range, although the amplitude may vary in a very complex way as a function of frequency. The human speaking voice is an example of such a source. There may well be a sense of generally "high" or "low" pitch—everyone with normal hearing can distinguish easily between normal female and male voices—but the sense of pitch is not specific. In such cases the Fourier spectrum is not discrete but *continuous*.

Discrete spectra are relatively easy to synthesize, and artificial musical instruments that do this have existed for some time. The electronic organ is the best known of these. It operates by reproducing the Fourier spectra characteristic of real organ pipes, using other sources which are cheaper and more compact. As a practical matter, all but the most expensive electronic organs produce a recognizable but very poor imitation of real pipe organs. Even the best electronic organs can be distinguished from the genuine article by a not very practiced ear. In part this has to do with the difficulty of reproducing the very large number of harmonics which contribute to the quality of an organ pipe. Sometimes there are more than 30 of these, while few electronic organs use more than 7 or 8 Fourier components in synthesizing their notes.

Usually, however, it is the attack and decay pattern characteristic of the original instrument that is most difficult to imitate with an electronic instrument. The problem can be solved by the expedient of using a mechanical system more or less similar to that in the parent instrument to originate the tone. This approach is obvious, and very successful, in the electric guitar.

The most successful electronic instruments are not intended to imitate the parent instrument as closely as possible. The best musical results are attained when an instrument is accepted on its own terms and a literature is devised for it. Such a trend is clearly discernible in the case of the electric guitar. When the instrument was first designed, about fifty years ago, the idea was to build something that would sound as much like a guitar as possible, but louder. Nowadays a vast variety of sound qualities (that is, a vast variety of Fourier spectra) are available. Except for the attack-decay characteristics and the technique of playing, many of these sounds bear little resemblance to those of the acoustic guitar.

A still more striking example of the trend away from imitation of traditional instruments is the synthesizer, which has no parent instrument at all. In the synthesizer, a wide variety of electronic tone generators are controlled by a standard keyboard. The musician also has very flexible control over the way in which the individual waveforms are superposed to produce the output waveform, as well as on the attack-decay pattern. In effect, the musician makes whatever Fourier synthesis best suits the music to be played. The nearest parallel among traditional instruments is the large pipe organ, in which various banks of pipes, called stops, can be turned on and off. However, the synthesizer has considerably greater flexibility. Its waveforms are not restricted to those typical of pipes, and the amplitudes of the individual component waveforms can be adjusted separately. Unfortunately, there is as yet little or no memorable literature written for the synthesizer.

What is it that underlies the "musicality" of musical instruments? Stringed instruments like the violin, for example, not only can play complex, sustained melodies, but also can be combined harmoniously to produce rich, pleasing musical textures. Percussion instruments such as the cymbals and most drums, on the other hand, are restricted to a "tzing-boom" which serves mainly to accentuate the music played by other instruments. And why is it that certain combinations of musical notes sound harmonious while others sound harsh?

The answer to these questions lies within the scope of the partially physical, partially physiological and psychological theory of harmony. The groundwork for this theory, of which we will consider mainly the physical aspects, was laid by the German physicist and physiologist Hermann von Helmholtz (1821–1894).

All harmony is founded on the fact that two or more musical notes sounded together are perceived as **consonant** (that is, pleasant-sounding or harmonious) or **dissonant** (that is, harsh or disharmonious). To some extent, this judgment is subjective and learned. But there can be no doubt that it has an objective basis. All listeners agree, for example, that two notes whose frequency ratio is 1:2 sound consonant when played together. Because of the logarithmic response of the ear to frequency, combinations of two notes sound similar, though not identical, to other combinations of two notes having the same frequency ratio. Musicians call the frequency ratio of two notes an **interval.** The interval with the 1:2 ratio, called the **octave,** is the simplest and most important.

An example of an octave is the combination of the note called middle C, whose frequency is normally 261.6 Hz, with the next-higher note given the name C, whose frequency is 2 × 261.6 Hz = 523.2 Hz.

On the other hand, all listeners agree that two notes whose frequency ratio is 1:1.06—an interval which musicians call a semitone—sound highly dissonant when played together.

An example of a semitone is the combination of middle C with the adjacent note, called C-sharp, whose frequency is 277.4 Hz.

The judgment of consonance and dissonance is ultimately based on the way the human ear hears *beats,* which will now be described. The first curve in Fig. 13-29 depicts a space-independent wave function $f_1(t) = \sin(2\pi\nu t)$, where ν is an arbitrary frequency. The second curve is the function $f_2(t) = \sin[2\pi(1.1\nu)t]$, which differs from the first only in that its frequency is 10 percent higher.

The third curve is the sum of the other two. It thus represents the sound wave which excites the ear when the two pure sinusoidal notes are played together. The salient feature of the superposition is the slow waning and waxing of the amplitude, as the two component waves repeatedly drift into and out of phase on account of their slightly different frequencies. The variations in perceived loudness are called **beats.**

Let us now consider the phenomenon of beats quantitatively. If two sinusoidal sound waves of frequency ν and $\nu + \epsilon$ are heard together, the

Fig. 13-29 The production of beats by the superposition of two sinusoidals of equal amplitude having slightly different frequencies. The resultant waveform is characterized by two frequencies. The higher of the two, ν_+, is the average of the frequencies of the two sinusoidals and is perceived as the pitch. The lower of the two frequencies, ν_-, is one-half the difference of the frequencies of the two sinusoidals. It is the frequency of the envelope of the resultant wave function. The quantity $2\nu_-$ is called the beat frequency, as explained in the text.

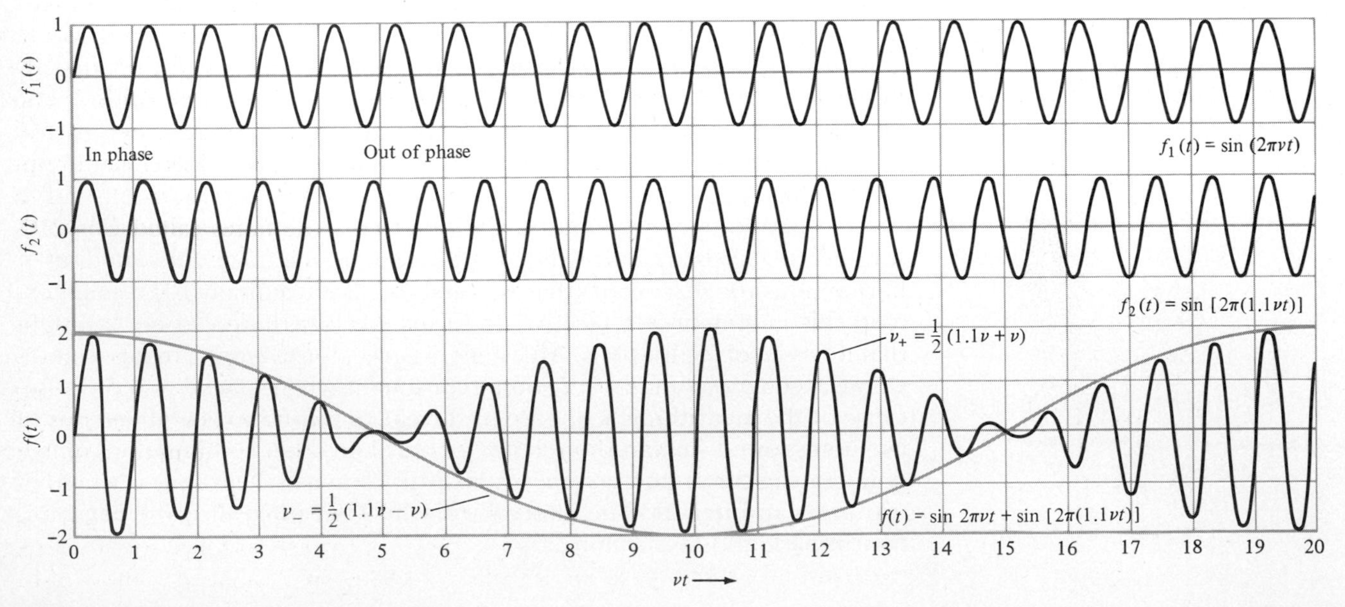

superposition principle gives the total wave function as

$$f(t) = \sin(2\pi\nu t) + \sin[2\pi(\nu + \epsilon)t] \tag{13-51}$$

We use the standard trigonometric identity of Eq. (13-13) to rewrite this expression in the form

$$f(t) = 2 \sin\{\tfrac{1}{2}[2\pi\nu t + 2\pi(\nu + \epsilon)t]\} \cos\{\tfrac{1}{2}[2\pi\nu t - 2\pi(\nu + \epsilon)t]\}$$

or

$$f(t) = 2 \sin\left[2\pi\left(\nu + \frac{\epsilon}{2}\right)t\right] \cos\left(2\pi \frac{\epsilon}{2} t\right) \tag{13-52}$$

While this equation is correct for any values of ν and ϵ, we are interested at present in the case where ϵ is very much smaller than ν, so that the two frequencies are very close together. When this is so, the two sinusoidal components are not heard as separate tones. Rather, the frequency heard is that of the sine term, $\nu + \epsilon/2$, which is the average of the frequencies of the two components. If ϵ is small, this frequency leads to a perceived pitch indistinguishable from ν or $\nu + \epsilon$.

This tone is **modulated**—that is, its amplitude is varied—by the cosine term, whose frequency is $\epsilon/2$. But the amplitude depends on the magnitude, not the sign, of the cosine function. Thus, each time the cosine passes through a complete cycle, the amplitude goes through two maxima and two zeros. The loudness of the sound thus varies with a frequency $2 \times \epsilon/2 = \epsilon$ times per second. If this frequency is low enough, the modulation produces perceptible pulsations in loudness.

This phenomenon is very useful in tuning two strings of a musical instrument (such as guitar strings or piano strings) to the same frequency. As the two strings are sounded together with frequencies ν and $\nu + \epsilon$, the tension in one of them is varied to make the beat frequency ϵ smaller and smaller, until the beats disappear. The strings are then tuned accurately to the same frequency. They are said to be in **unison,** from the Latin words meaning "one sound."

Beat frequencies greater than about 7 Hz can no longer be heard as countable pulsations in loudness, but are detected by the ear as a "roughness" or "harshness" which the listener interprets as a dissonance. The beat frequency corresponding to the most disagreeable dissonance depends on the pitch of the note heard. At low pitch it is about 15 Hz, in the middle range it is about 40 Hz, and at high pitch it increases to about 250 Hz. As a rough rule of thumb, any frequency ratio less than about 1:1.2—that is, any musical interval smaller than what musicians call a minor third (for example, the interval between the notes called C and E-flat)—is perceived as more or less dissonant.

Not all dissonances are produced by two notes very close together. There are two reasons for this. First, practically all musical tones carry a rich cargo of harmonics. Thus there is the possibility of dissonant beating between the fundamental of one of the notes and one of the harmonics of the other, or between one or more harmonics of one of the notes and one or more of the other.

Second, there is the intriguing role of the ear itself. The ear is very far from obeying Hooke's law. This is true even for the displacements of the eardrum and other parts of the ear produced by relatively faint sounds.

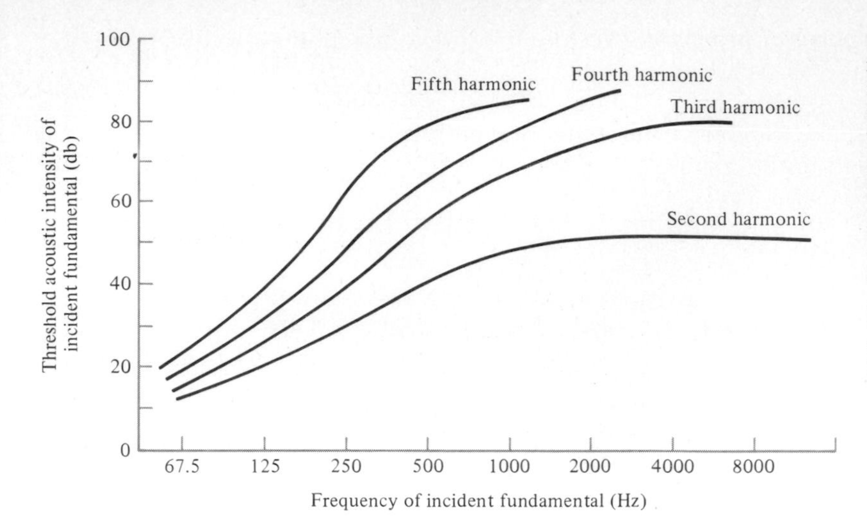

Fig. 13-30 The production of aural harmonics in the human ear. If a pure tone (a sinusoid) is to produce perceptible harmonics in the ear, it must have a minimum acoustic intensity whose value depends on its frequency. In general, the higher the frequency of the incident tone, the louder it must be to produce aural harmonics. For instance, the graph shows that a 1000-Hz incident tone must have an acoustic intensity of 40 dB if the listener is to peceive the second harmonic at 2000 Hz. But an incident 125-Hz tone at 18 dB will induce the listener to perceive the second harmonic at 250 Hz. Greater intensities are required to produce higher aural harmonics.

The effect becomes rapidly more important as the amplitude of the incident sound increases. Figure 13-30 shows the intensity level of an incident pure tone at which the ear begins to perceive **aural harmonics,** that is, harmonics which are not present in the incident sound but are generated in the ear. If you compare this figure with Table 13-1, you will see that aural harmonics make a substantial contribution to the sound you hear at almost all commonly encountered acoustic intensity levels.

In order to see how the nonlinearity of the ear leads to the production of aural harmonics, let us make the simplest possible (and crude) assumption as to the way the ear deviates from Hooke's law when its parts vibrate in response to the pressure oscillations of a sound wave. If a certain part of the ear obeyed Hooke's law, there would be a linear relation between its displacement y and the force F exerted on it by the sound wave. Such a relation can be written $y = aF$, where a is the reciprocal of the force constant defined in Chap. 6. But suppose that the ear part actually obeys the simplest possible nonlinear rule

$$y = aF + bF^2 \tag{13-53}$$

(Note that the displacement y would conform approximately to Hooke's law if the displacing force F were small enough or if the constant b were small enough, which is not the case here.)

Now suppose that a force is exerted on the nonlinear part as a result of the arrival at the ear of a purely sinusoidal sound wave, so that

$$F = \sin(2\pi\nu t)$$

(We take the amplitude $A = 1$ for simplicity.) The displacement of the nonlinear ear part is then found by substituting this expression for F into Eq. (13-53), which leads to the displacement

$$y(t) = a\sin(2\pi\nu t) + b\sin^2(2\pi\nu t) \tag{13-54}$$

As far as the first term on the right side of this equation is concerned, the ear part will oscillate so as to reproduce faithfully the incident frequency ν. But consider the second term. By using the trigonometric identity $\sin^2\theta = \frac{1}{2}[1 - \cos(2\theta)]$, it can be written

$$b\sin^2(2\pi\nu t) = \frac{b}{2}\{1 - \cos[2\pi(2\nu)t]\}$$

So the displacement $y(t)$ is given by

$$y(t) = \frac{b}{2} + a\sin[2\pi\nu t] - \frac{b}{2}\cos[2\pi(2\nu)t] \qquad (13\text{-}55)$$

That is, the function describing the displacement of the nonlinear part of the ear contains two terms that are Fourier components with frequencies ν and 2ν, respectively. The frequency 2ν is, so to speak, "manufactured" by the ear. But it is heard just as though such a frequency component were present in the incident sound wave.

In very similar manner, the *nonlinearity* of the ear (as its departure from Hooke's law is called) results in the creation of additional tones when the incident sound wave has two or more frequency components. Suppose, for example, that the displacing force is described by

$$F = \sin(2\pi\nu_1 t) + \sin(2\pi\nu_2 t) \qquad (13\text{-}56)$$

(Here we have again set the amplitudes at the values $A_1 = A_2 = 1$ for simplicity.) If the ear obeys Eq. (13-53), as above, we have

$$y(t) = a[\sin(2\pi\nu_1 t) + \sin(2\pi\nu_2 t)] + b[\sin^2(2\pi\nu_1 t) + \sin^2(2\pi\nu_2 t) + 2\sin(2\pi\nu_1 t)\sin(2\pi\nu_2 t)] \qquad (13\text{-}57)$$

As before, the term $b[\sin^2(2\pi\nu_1 t) + \sin^2(2\pi\nu_2 t)]$ leads to the "manufacture" of Fourier components of frequency $2\nu_1$ and $2\nu_2$. However, there is now a new effect produced by the mixed term (the last term on the right side of the equation). Using a trigonometric identity like Eq. (13-13), we can write

$$2b\sin(2\pi\nu_1 t)\sin(2\pi\nu_2 t) = b\cos[2\pi(\nu_1 - \nu_2)t] - b\cos[2\pi(\nu_1 + \nu_2)t]$$

Thus the ear adds to the perceived sound new components with frequencies equal to the difference and the sum of the incident frequencies. These are the **difference** and **sum tones**, or collectively simply the **combination tones.** The difference tones are far more important, since they tend to have much greater intensity.

The nonlinearity of the ear does not cease at the quadratic term in Eq. (13-53). Higher-order terms are present, though of smaller magnitude, and they add still further complications to the perceived sound in a manner similar to that described above. Not only do these tones exist in the combined sound perceived, but also they can be discerned separately by a listener who has a little practice.

If the amplitude is sufficiently great, *all* systems must ultimately fail to obey Hooke's law. In audio high-fidelity systems, the weak link in this respect is usually the loudspeakers. If the volume is set too high, they will manufacture the combination tones of Eq. (13-57). The result is **intermodulation distortion,** or IM distortion for short. However great the care taken in making a faithful recording of a musical performance, the effort will be wasted if the volume is turned so high as to produce significant IM distortion in the speaker. If the volume is set either too high or too low, it results in a level of IM distortion in the ear which is *different* from that experienced by the "live" or direct listener. Ideally, music should be heard at precisely the volume level at which it was recorded.

It is the presence of combination tones in the perceived sound which unravels an important musical paradox. You may have noted in Figs. 13-26 and 13-27 that the fundamental is by no means always the strongest component in the tone produced by a musical instrument. Nevertheless, we always perceive the pitch of the instrument as that of the fundamental. Indeed, it is only by concentrating that we can hear the harmonics, which may be much stronger, as separate entities. But *every pair of adjacent harmonics in a musical tone produces the fundamental as a difference tone.* As a result, the ear supplies a strong fundamental even when it is entirely or nearly missing in the incident waveform.

The satisfactory performance of compact and economical sound systems depends on the same effect in an essential way. It is not possible for a small loudspeaker to reproduce low frequencies at all. Such systems simply do not pass on to the ear the low-frequency part of the Fourier spectrum commonly present in music (and even in speech). Nevertheless, the ear reconstructs the missing frequencies from difference tones, at least well enough to make the music or speech recognizable. (In the light of the above discussion, how is it that small earphones of good quality can give truly excellent reproduction of low frequencies?)

We are now ready to consider the conditions for consonance and dissonance in a general way. When a musical tone is played on some instrument, what strikes the ear is a space-independent waveform which has as its Fourier components a fundamental of frequency ν_1 and higher harmonics of frequencies $2\nu_1$, $3\nu_1$, and so on. The relative amplitudes of these Fourier components depend on the particular instrument, the particular tone being played on it, and the particular way in which that tone is being played. If a second musical tone of fundamental frequency ν_2 is simultaneously played on some other instrument, what strikes the ear is a superposition of the waveforms of the two tones. Like the first tone, the second also comprises a set of Fourier components with characteristic amplitudes. To this complex superposed waveform, the ear adds further complexities. For each pair of Fourier components whose amplitudes are great enough, the ear manufactures aural harmonics.

There are consequently all sorts of possibilities for combination tones. If the fundamental frequencies of the two notes have a simple ratio, say $\nu_2/\nu_1 = 2$, the interval musicians call the octave, then the combination tones will all have frequencies which coincide with harmonics already present. But suppose the ratio is more complicated, for example, $\nu_2/\nu_1 = \frac{9}{8}$, the interval musicians call a second. Then some of the combination tones will fall into the range of beat frequencies—somewhere between 15 Hz and 250 Hz, depending on pitch—which the ear interprets as harsh. The more such combination tones that are present and the greater their amplitudes, the more dissonance will be perceived in the combination of the two musical notes.

In order to explore the possibilities for dissonance, the frequencies of the harmonics of two musical tones can be plotted along a horizontal axis whose scale is proportional to frequency. This is done in Fig. 13-31 for the fundamental frequency ratio $\nu_2/\nu_1 = \frac{3}{2}$, the interval which musicians call a fifth. In this case, many of the harmonics of the higher-pitch tone coincide with those of the lower-pitch tone. Of the others, only the harmonic of frequency $3\nu_2$ lies close enough to any other harmonic that their ratio is less than 1:1.2, which is the crude threshold criterion for dissonance. Taking the ratio of $3\nu_2$ to the frequency $4\nu_1$ of one of the neighboring harmonics of the lower-pitch tone, we have

$$\frac{3\nu_2}{4\nu_1} = \frac{\frac{9}{2}\nu_1}{4\nu_1} = 1.13$$

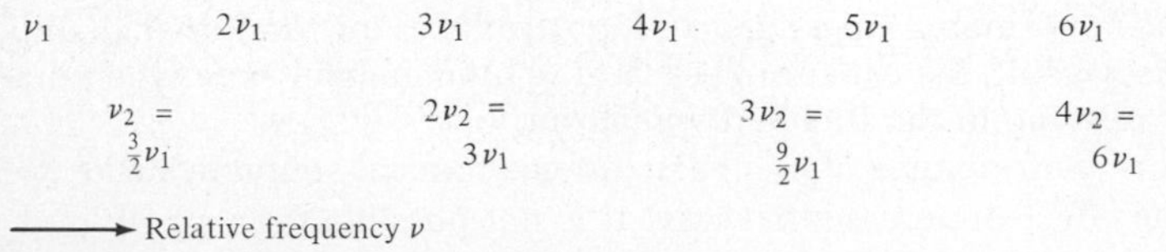

ν_1 $2\nu_1$ $3\nu_1$ $4\nu_1$ $5\nu_1$ $6\nu_1$

$\nu_2 = \frac{3}{2}\nu_1$ $2\nu_2 = 3\nu_1$ $3\nu_2 = \frac{9}{2}\nu_1$ $4\nu_2 = 6\nu_1$

Relative frequency ν

Fig. 13-31 Plot of the relative frequencies of a musical tone of fundamental frequency ν_1 and its harmonics, together with the relative frequencies of a musical tone of fundamental frequency $\nu_2 = 3\nu_1/2$ and its harmonics. This is the musical interval called the fifth.

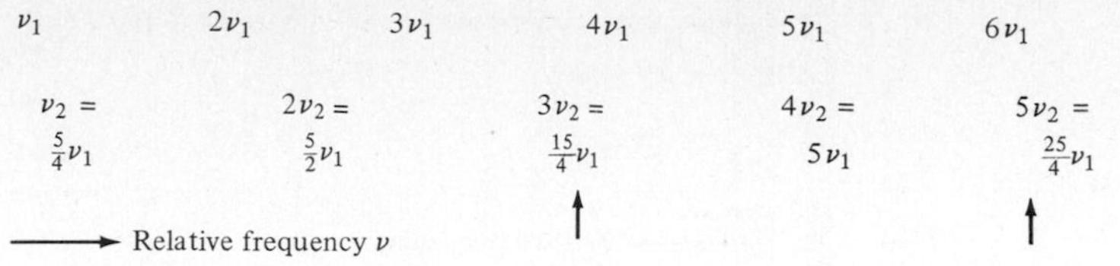

ν_1 $2\nu_1$ $3\nu_1$ $4\nu_1$ $5\nu_1$ $6\nu_1$

$\nu_2 = \frac{5}{4}\nu_1$ $2\nu_2 = \frac{5}{2}\nu_1$ $3\nu_2 = \frac{15}{4}\nu_1$ $4\nu_2 = 5\nu_1$ $5\nu_2 = \frac{25}{4}\nu_1$

↑ ↑

→ Relative frequency ν

Fig. 13-32 Relative frequency plot for Example 13-10, with $\nu_2 = 5\nu_1/4$. This is the musical interval called the major third.

The ratio of the frequency $5\nu_1$ of the other harmonic of the lower pitch tone to $3\nu_2$ is

$$\frac{5\nu_1}{3\nu_2} = \frac{5\nu_1}{\frac{9}{2}\nu_1} = 1.11$$

The interval called the fifth is perceived as relatively consonant, although less so than the octave. This is suggested by the fact that the frequency ratio for the fifth, 3:2, is "more complicated" than the frequency ratio for the octave, 2:1.

Examples 13-10 and 13-11 consider progressively more dissonant musical intervals.

EXAMPLE 13-10

The musical interval having the frequency ratio $\nu_2/\nu_1 = \frac{5}{4}$ is called by musicians a major third. Explore the possibilities for dissonance in the major third. Is it more or less dissonant than the fifth, discussed immediately above?

■ Using a relative frequency plot like that of Fig. 13-31, you plot the frequencies ν_1 and $\nu_2 = \frac{5}{4}\nu_1$ together with their associated higher harmonics, as in Fig. 13-32. Inspection suggests that the frequency pairs denoted by arrows may lead to beat frequencies small enough to produce dissonance. You find the frequency ratios of the pairs to be

$$\frac{4\nu_1}{3\nu_2} = \frac{4\nu_1}{\frac{15}{4}\nu_1} = 1.07$$

and

$$\frac{5\nu_2}{6\nu_1} = \frac{\frac{25}{4}\nu_1}{6\nu_1} = 1.04$$

You compare this with the rough criterion that any frequency ratio less than about 1.2 is likely to sound dissonant, and you conclude that the major third is somewhat dissonant. Comparison suggests that it is more dissonant than the fifth, both because there are more possibilities for beats and because the frequency ratios are smaller.

How dissonant the major third sounds depends on the intensity of the dissonant harmonics in the particular instruments used and the pitch of the fundamentals. If the notes are played loudly enough, there is also the possibility of combination tones which are dissonant with some of the harmonics.

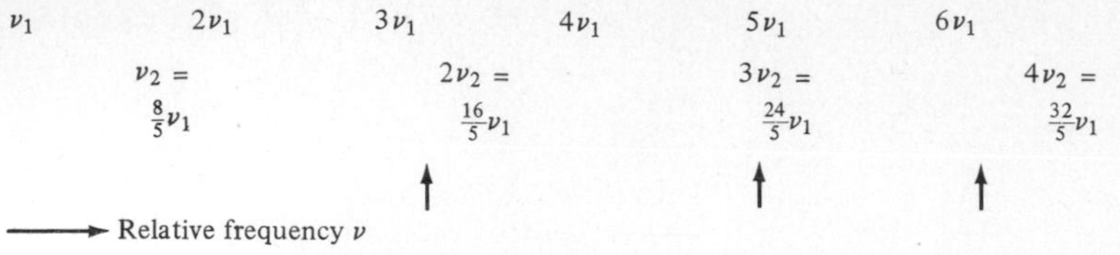

ν_1 $2\nu_1$ $3\nu_1$ $4\nu_1$ $5\nu_1$ $6\nu_1$

$\nu_2 = \frac{8}{5}\nu_1$ $2\nu_2 = \frac{16}{5}\nu_1$ $3\nu_2 = \frac{24}{5}\nu_1$ $4\nu_2 = \frac{32}{5}\nu_1$

Relative frequency ν

Fig. 13-33 Relative frequency plot for Example 13-11, with $\nu_2 = 8\nu_1/5$. This is the musical interval called a minor sixth.

EXAMPLE 13-11

The musical interval having the frequency ratio $\nu_2/\nu_1 = \frac{8}{5}$ is called by musicians a minor sixth. Explore the possibilities for dissonance in the minor sixth. Is it more or less dissonant than the major third, discussed in Example 13-10?

■ Using a relative frequency plot like that of Fig. 13-31, you plot the frequencies ν_1 and $\nu_2 = \frac{8}{5}\nu_1$ together with their associated higher harmonics, as in Fig. 13-33. Inspection suggests that the frequency pairs denoted by arrows may lead to beat frequencies small enough to produce dissonance. You find the frequency ratios of the pairs to be

$$\frac{2\nu_2}{3\nu_1} = \frac{\frac{16}{5}\nu_1}{3\nu_1} = 1.07 \qquad \frac{5\nu_1}{3\nu_2} = \frac{5\nu_1}{\frac{24}{5}\nu_1} = 1.04 \qquad \frac{4\nu_2}{6\nu_1} = \frac{\frac{32}{5}\nu_1}{6\nu_1} = 1.07$$

Here there are three dissonant intervals. There are, moreover, relatively few consonant intervals, and there are abundant possibilities for combination tones. So the minor sixth is relatively dissonant, compared to the major third and the fifth.

In Examples 13-10 and 13-11, why is it unnecessary to consider the possibility of dissonances between harmonics higher than the sixth?

In actual music, it is quite common to have harmonies involving three or more musical notes. The analysis of such situations from a physical point of view becomes rapidly more complicated, but the principles involved are the same. It should be remembered that dissonance is not universally undesirable in music. A piece consisting entirely of consonances would be unbearably tedious, and the resolution of dissonances into consonances is a very important aspect of Western music.

We conclude this section with a brief discussion of actual musical instruments. Every instrument must perform two functions. It must generate any of a number of fundamental frequencies (perhaps along with a series of harmonics) at the will of the musician. Then it must transmit the vibrational energy to the air in a reasonably efficient way.

In the stringed instruments, the frequency is set by the linear mass density, tension, and length of the string. The string is set into oscillation by striking it with a hammer (as in the piano), by plucking it (as in the guitar, the lute, the harpsichord, and the violin played pizzicato), or by bowing it (as in the viols and violins). The bowing mechanism is the only one whose function is not obvious. The string adheres to the bow by friction until the maximum static frictional force is exceeded. The string then slips back until it is again caught by the bow at the other end of its oscillation. (Such action is encountered in a wide variety of mechanical systems. It is called "stick-slip" action.)

The details of the excitation mechanism are very important in determining the quality of the instrumental sound, since they influence the amplitudes of the various harmonics. In the piano, for instance, the hammer strikes the string in the region between about one-eighth and one-sixth of the distance along its length. This strongly suppresses all the harmonics which possess nodes in that region. (Why?)

A string by itself is a very inefficient transmitter of energy to the air, because it is so thin. In stringed instruments, therefore, one or both ends of the string are stretched across a (usually) relatively light wooden **bridge,** which rests on a wooden plate called the **sounding board.** In most stringed instruments, the sounding board is part of a box, usually of complex shape. The sounding board (and the air it contains, if any) is set into vibration by the oscillations transmitted from the string via the bridge. Since the sounding board is large, it sets the surrounding air into motion rather efficiently, as required.

If the instrument is to have relatively uniform tone quality over its range of pitch, the entire system must be fairly highly damped (see Sec. 6-6) so that it will not "ring" at certain frequencies and sound "dead" at others. The complex shape also aids in "smoothing out" the response of the sounding board to string vibrations of different frequencies, for reasons whose discussion would require too much digression here.

In brass and woodwind instruments, and in the organ, it is the vibrating source which has no well-defined oscillation frequency. Contrary to the situation in the stringed instruments, it is the equivalent of the sounding board—in this case the air in the pipe which is the main body of the instrument—which vibrates in standing-wave modes of sharply defined frequency and thus controls the frequency of oscillation of the source.

In most of the woodwinds (and some organ pipes) the source is a **reed.** In the clarinet and the saxophone, for example, the reed is a thin, flexible plate of cane which fits over a hole in the mouthpiece. As it vibrates, at a frequency controlled by the body of the instrument, the reed opens and closes the hole, allowing puffs of air to enter the instrument. In the brasses, the same effect is achieved by the musician's lips, which are coupled to the instrument by a cup- or funnel-shaped mouthpiece. The flute and most organ pipes are driven by the oscillating turbulences produced (in a very complicated way) when air from a narrow slit (in the case of the flute, the player's lips) is directed against a sharp edge. In the recorder, the ocarina, and some organ pipes the same effect is achieved by a built-in whistle.

The resonant frequency of the air in the body of the instrument is determined by the length of the column and by whether the pipe is open at both ends or closed at one end (and to a lesser extent by its shape.) As usual, the physical situation sets the boundary conditions. A standing wave in a pipe open at both ends, in which the air is excited near one end, must have *pressure nodes* at both ends (or very near them) since nothing that happens inside the pipe can possibly change the pressure of the outside air. Figure 13-34*a* is a schematic picture of a standing pressure wave set up in such an open pipe. The fundamental frequency corresponds to the longest wavelength satisfying this condition, which has just one antinode between the nodes at the ends. The length L of the pipe is thus one-half the wavelength λ_1 of the fundamental mode; that is, $\lambda_1 = 2L$. The fundamental frequency ν_1 is related to λ_1 according to the rule $\nu_1 = |v|/\lambda_1$, where $|v|$ is the speed of sound in air. Hence the frequency is

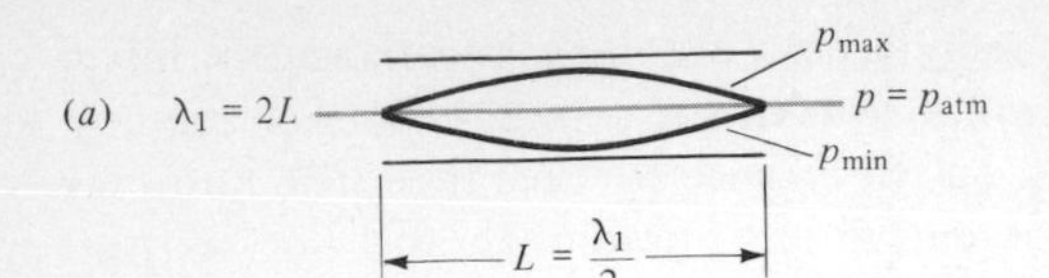

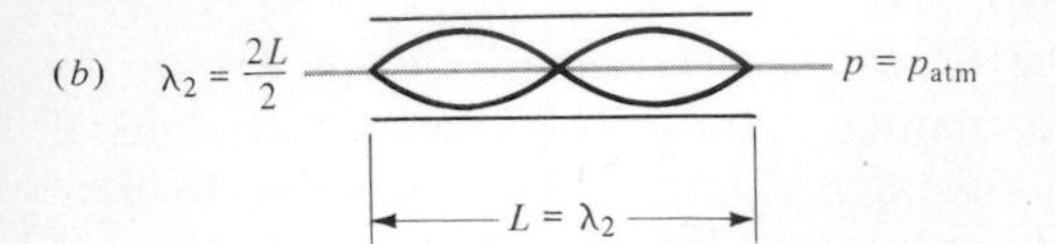

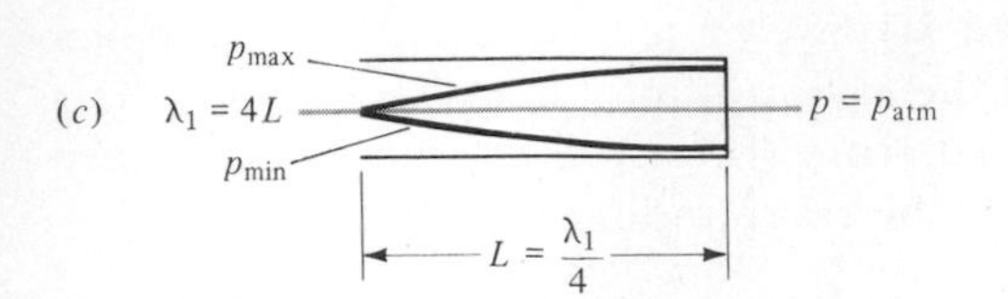

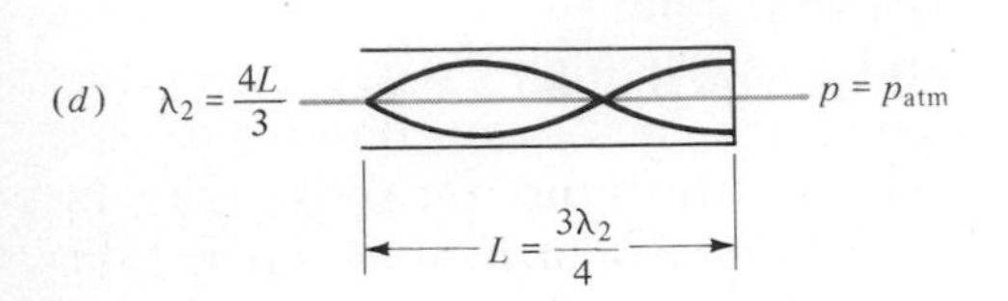

Fig. 13-34 Standing acoustic pressure waves produced in open pipes and in pipes closed at one end (called "closed pipes"). An open end must be a pressure node because the activity in the pipe cannot significantly influence the pressure of the open atmosphere. A closed end must be a displacement node because the air cannot move longitudinally past it. As discussed in the text, a displacement node must be a pressure antinode. (*a*) The fundamental standing pressure wave in an open pipe has nodes only at the two ends, and its wavelength is twice the length of the pipe. Compare with Fig. 13-8. (*b*) The second harmonic has an additional pressure node in the center of the pipe, and its wavelength is equal to the length of the pipe. Compare with Fig. 13-9. (*c*) The fundamental standing pressure wave in a closed pipe has a node at the open end and an antinode at the closed end. Its wavelength is four times the length of the pipe. (*d*) The next possible standing wave is the third harmonic, with an additional pressure node one-third of the distance from the closed end to the open end of the pipe. Only odd harmonics are possible in a closed pipe.

$$\nu_1 = \frac{|v|}{2L} \tag{13-58}$$

The next possible mode is the one in which there is a node between the two nodes at the ends. See Fig. 13-34*b*. The wavelength λ_2 is thus one-half that of the fundamental, λ_1, and the frequency ν_2 is twice that of the fundamental, ν_1. In like manner, harmonics can exist whose frequencies are 3, 4, 5, . . . times that of the fundamental, just as in a string.

If one end of the pipe is closed, there must be a *displacement node* at that point. Sound waves involve longitudinal displacement of the air, and this is prevented at the closed end. But we saw in Sec. 12-6 that a zero displacement corresponds to a pressure extreme (see Fig. 12-15). Thus there must be a *pressure antinode* at the closed end. The situation is that shown in Fig. 13-34*c*. Here the fundamental has a wavelength 4 times the length of the pipe, and its frequency is

$$\nu_1 = \frac{|v|}{\lambda_1} = \frac{|v|}{4L} \tag{13-59}$$

Hence a closed pipe has a fundamental frequency one-half that of an open pipe of the same length. But from the musical point of view there is a matter of greater interest. Note from Fig. 13-33*d* that the geometry of the situation requires that the next possible mode have a wavelength not one-half, but one-third that of the fundamental. Similarly, the next wavelength will be one-fifth that of the fundamental. In general, only the odd-numbered modes are possible, with frequencies 3, 5, 7, . . . times that of the fundamental. As you can see from the sound spectrum of Fig. 13-26, the clarinet approximates this situation, to which it owes its distinctive quality.

Example 13-12 applies the principles just developed to an organ pipe.

EXAMPLE 13-12

How long must an organ pipe be to produce the lowest audible musical note called C, whose frequency is 16.35 Hz, if it is

a. Open at both ends?
b. Closed at one end?

Take the speed of sound to be 344.0 m/s.

■ **a.** For the open pipe you have from Eq. (13-58)

$$L = \frac{|v|}{2\nu_1} = \frac{344.0 \text{ m/s}}{2 \times 16.35 \text{ s}^{-1}} = 10.52 \text{ m}$$

b. For the closed pipe, Eq. (13-59) gives

$$L = \frac{|v|}{4\nu_1} = \frac{344.0 \text{ m/s}}{4 \times 16.35 \text{ s}^{-1}} = 5.260 \text{ m}$$

Ranks of very large "organ pipes" are sometimes used as decorations in large churches. You can be sure that they are either nonfunctional or nonmusical if they are much more than 10 m long!

In the brass instruments, the pitch is altered by using a set of valves or a slide to vary the length of the tube and thus the resonant frequency of the air within it. In the woodwinds the same effect is achieved in a much more complicated fashion by opening and closing holes distributed along the tube.

The bell at the end of the instrument serves two related purposes. First it serves as an efficient way to couple the vibrating air inside the instrument to the outside air and thus to emit the sound. Second, it is not equally efficient in doing this at all frequencies. Thus the bell tends to suppress some harmonics relative to others and contributes to the timbre of the instrument as a whole.

Percussion instruments are a special class. In general, they are two-dimensional oscillators (like the drum and the cymbals) or even three-dimensional oscillators (like the bell and the wood block). The frequencies of their standing-wave modes do *not* constitute a simple arithmetic series, as we have seen for the drumhead in Sec. 13-5. The modes are thus *not* harmonic. Taken together, the nonharmonic modes and the combination tones constitute an irregular series in frequency, and it is not possible for the ear to identify or construct a well-defined fundamental frequency to which a pitch can be assigned. In some cases, such as the snare drum, the triangle, and the cymbals, this is what is desired since the instrument is essentially a "noisemaker." In other cases, the musical tone is produced by using one or more ingenious devices. In the xylophone, a musical tone is achieved by hollowing out the centers of the bars in such a way as to "tune" the second mode to an approximate integral multiple of the fundamental. A similar device is used in the bell. In the chimes, which consist of a set of tubular bars, the bars are suspended at the nodes of the desired modes, thus suppressing other, dissonant ones. In the marimba, tuned hollow containers called resonators, in which standing-wave oscillations can be set up very much as in the air column of a trumpet, are used to enhance the fundamental. Thus the dissonant higher modes are relatively weak.

The timpani (or kettledrums) is an interesting special case. The fundamental (see Fig. 13-17*a*), in which the entire drumhead moves together, transfers energy to the air so efficiently that it damps out very quickly and is not heard except as part of the initial "boom." The pitch heard is that of the mode having the next higher frequency, shown in Fig. 13-23*a*, while the higher modes contribute a somewhat unmusical quality to the instrument. Thus the pitch, while it can be discerned, is somewhat "fuzzy."

All percussion instruments have the property that elaborate chords cannot be played on them, even if the individual notes do have discernible pitches. It is not possible to obtain consonances with bells, for example, when more than two are played together. There are simply too many possibilities for dissonances among the various nonharmonic modes. As a consequence, there is a special musical literature written for the carillon, as a choir of bells is called. Rather than calling for the playing of many bells together, advantage is taken of the fact that while the sound of bells persists for some time, there is considerable variation among the decay rates of the modes. So ingenious and unexpected harmonies can be created by sounding one bell when only the desired modes of a previously sounded bell or bells still persist. Carillons are particularly common in Belgium and Holland. If you ever have the chance to listen to one, listen with the ear of a physicist as well as the ear of a music lover!

EXERCISES

Group A

13-1. *Superposition.* Two wave trains are traveling along a very long stretched string at the same speed $|v|$ but in opposite directions. The wave train traveling in the positive x direction is described by the wave function $f_1(x, t) = A_1 \cos[k_1(x - |v|t)]$ and the wave train traveling in the negative x direction by $f_2(x, t) = A_2 \cos[k_2(x + |v|t) + \delta_2]$.

a. Plot $f_1(x, t)$ over the range $-(4\pi/k_1) < x < (4\pi/k_1)$ for each of the following instants: (1) $t = 0$; (2) $t = \pi/2k_1|v|$.

b. On the same graphs, plot $f_2(x, t)$ for the positions and instants listed in part *a*. Assume that $A_2 = A_1$, $k_2 = 2k_1$, and $\delta_2 = \pi/6$.

c. On the same graphs, plot the wave function $f(x, t)$ that results from the superposition of the two traveling waves described in *a* and *b*.

13-2. *An experimental determination of the speed of sound in air.* The brass rod AB in Fig. 13E-2 is 0.50 m long and is clamped at its center E. When the rod is rubbed lengthwise with a rosined cloth, it is set into longitudinal vibration with E necessarily a node and A and B antinodes. At end B, a cardboard disk is fastened. The disk fits loosely into an open glass tube CD on the bottom of which a fine powder (such as talc) has been sprinkled. When AB is set into vibration, the powder gathers into little heaps at very regular intervals. Adjacent heaps of powder are separated by 4.8 cm. The speed of sound in brass has been measured; it is 3.5×10^3 m/s.

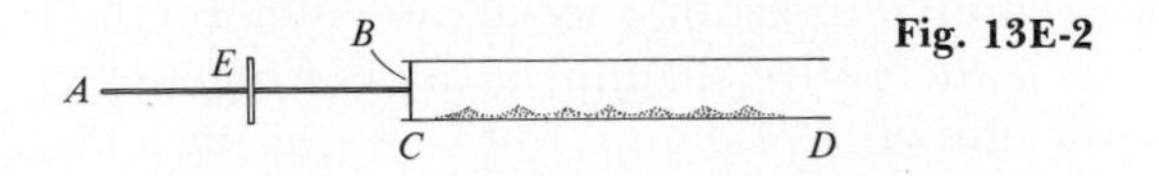

Fig. 13E-2

a. What is the fundamental frequency of longitudinal vibrations of the clamped rod AB?

b. Assuming that the vibrations described are vibrations at the fundamental frequency, what is the speed of sound in air?

13-3. *Piano wire.* The wire that is used to produce the tone "concert A" ($\nu = 440$ Hz) on a particular piano is made of a material with density 7.8 g/cm³. The wire is 0.80 mm in diameter and 60.0 cm long. What tension exists in the wire when its fundamental frequency is properly tuned to concert A?

13-4. *Standing wave I.* The equation for a standing wave in a string is given by $y = A \sin(kx) \cos(\omega t)$, where $A = 0.04$ m, $k = 4\pi$ m^{-1}, and $\omega = 800\pi$ s^{-1}.

a. What is the distance between nodes?

b. What is the wavelength of the traveling waves that superpose to produce the standing wave?

c. What is the frequency of the vibration?

d. With what speeds do the traveling waves propagate on the string?

e. What is the amplitude A' of each of the two traveling waves (of equal amplitude) that produce the standing wave?

13-5. *Standing wave, II.* Two waves each of wavelength 20 cm and of equal amplitude are traveling in opposite directions on a taut string fixed at both ends. If the string is 1.0 m long, how many nodes are there, counting the fixed ends, in the standing wave produced?

13-6. *Wave on a string.* A string 1.0 m long has a mass of 5.0×10^{-3} kg and is under a tension of 50 N.

a. What is the speed of a wave traveling on the string?

b. If the string is vibrating in four segments (with five nodes), what is the frequency of the sound it is producing?

c. Suppose that the standing wave has amplitude A. Write the wave function for the standing wave, taking the origin at one end of the string.

13-7. *The right location.* The first string on the violin is called the E string because it produces the tone E when its full length vibrates. The second string is called the A string for similar reasons. The ratio of the frequency of E to that of A is 3:2 (the major fifth). Where must a violinist place a finger on the A string so that it will produce the same tone as the E string?

13-8. *The fundamentals.* A number of strings of the same length but different diameters are made of the same material and are at the same tension. Prove that their fundamental frequencies are inversely proportional to their diameters.

13-9. *How much motion?* Points A, B, and C lie along a uniform stretched string of length L. The distances of A, B, and C from the left end of the string are exactly 0.1 L, 0.4 L, and 0.7 L, respectively. Find the amplitudes of the transverse motion of the string at points A, B, and C, if the string is vibrating with maximum amplitude h in its nth mode, for the following values of n:

a. $n = 1$ (fundamental mode); **b.** $n = 3$; **c.** $n = 10$

13-10. *Drums.*

a. Find the frequencies of the first three modes of oscillation of a drumhead of mass 50 g, radius 20 cm, and tension 600 N/m.

b. Three identical drumheads with the characteristics described in part *a* are mounted on three open cylinders. Find the lengths for the cylinders if each has one of the frequencies found in part *a* as its fundamental. (Notice that each assembled drum will constitute a pipe open only at the bottom end. Take the speed of sound to be 340 m/s.)

13-11. *Find the frequencies.* A glass tube 30 cm long is closed at one end. Take the speed of sound to be 340 m/s,

a. What is the frequency of the fundamental standing wave mode?

b. What is the frequency of the second standing wave mode?

13-12. *Low notes.* A pipe that is closed at one end has a fundamental frequency of 30 Hz. Assume a sound speed of 340 m/s.

a. What is the length of the pipe?

b. Suppose the cover is removed from the closed end. What is the fundamental frequency now?

13-13. *Listen to the beat.* A 256-Hz tuning fork produces four beats per second when sounded with another fork of unknown frequency. What are two possible values for the unknown frequency?

13-14. *Different pipes, same overtone.* For a particular pair of organ pipes, the first overtone (the second harmonic mode) of the closed pipe has the same frequency as the first overtone (the second harmonic mode) of the open pipe. What is the ratio of the pipe lengths?

Group B

13-15. *Constructive interference.* In Fig. 13E-15, source S produces a pure sound of a single frequency that can be varied. Observer O is situated so that OS is perpendicular to APB, a reflective wall. PS is equal to 2.0 m. What are the two lowest frequencies for which the sound heard at O has maximum loudness? The speed of sound in air is 340 m/s.

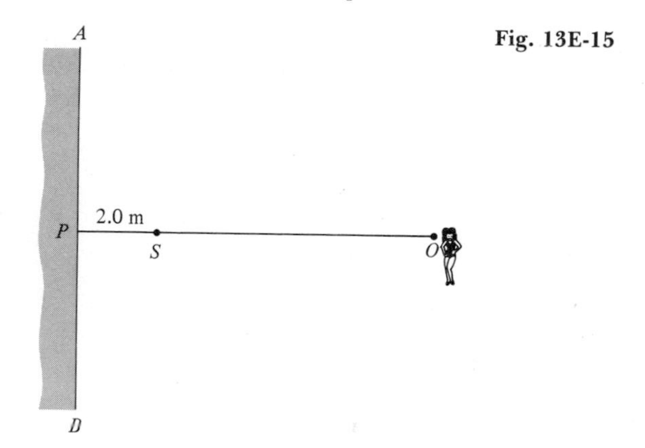

Fig. 13E-15

13-16. *Tune-up.* Some of the low keys of the piano have two strings. On a particular key one of the strings is tuned correctly to 100 Hz. When the two strings are sounded together, one beat per second is heard. By what percent must a piano tuner change the tension of the untuned string to make it match perfectly? The beating is due to superposition of the fundamental tones, which are by far the strongest.

13-17. *Energy in standing waves on a uniform string.* A uniform string (length L, linear density μ, and tension F) is vibrating with amplitude A_n in its nth mode. Show that its total energy of oscillation is given by $E = \pi^2 \nu_n^2 A_n^2 \mu L$.

13-18. *Acceleration in standing wave modes.* A 10-m length of nylon rope has a linear density of 0.10 kg/m and is under a tension of 1000 N.

a. Find the frequencies of the standing wave modes of this rope.

b. If the rope is vibrating in the nth mode, what is the required amplitude of vibration (at the antinodes) in order for the maximum acceleration to exceed the acceleration of gravity g?

13-19. *Can you explain it?* Inspection of Figs. 13-20 and 13-21 reveals that the node in the second mode of the circular membrane occurs at $u = 0.44$, while the nodes in the third mode occur at $u = 0.28$ and at $u = 0.64$. Equation 13-47 can be used to show that $\nu_1/\nu_2 = 0.44$, $\nu_1/\nu_3 = 0.28$, and $\nu_2/\nu_3 = 0.64$. Explain this agreement of node locations with frequency ratios.

13-20. *Working together.* S_1 and S_2 are identical sources of sound of a single frequency. They are equidistant from O. When S_1 alone is sounding, the amplitude reaching O is A.

a. If S_1 and S_2 are both turned on and are operated "in phase," what is now the amplitude of the sound at O?

b. How does the energy flow at O with both sources sounding compare with the flow when only S_1 is sounding?

c. What is the increase in acoustic intensity in decibels when S_2 is turned on?

13-21. *Fourier analysis.* Fig. 13E-21 represents a taut string fastened at A and C. It is initially raised a short distance in the center from B to B' and released. Show that the even-numbered harmonic modes are missing from the sound produced.

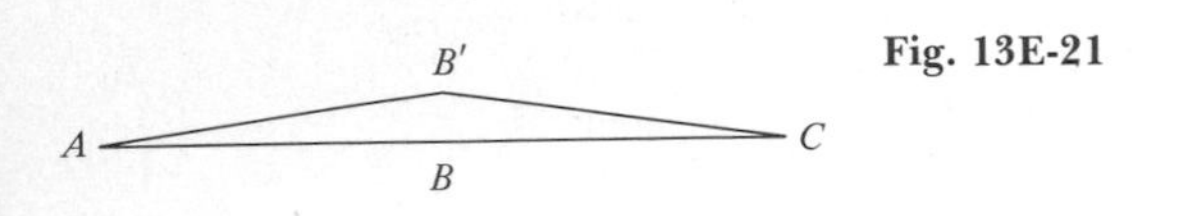

Fig. 13E-21

13-22. *Whistles and beats.* A boy is walking away from a wall at a speed of 1.0 m/s in a direction at right angles to the wall. As he walks, he blows a whistle steadily. An observer toward whom the boy is walking hears 4.0 beats per second. If the speed of sound is 340 m/s, what is the frequency of the whistle?

13-23. *Using a resonance to determine the speed of sound.* A sounding tuning fork whose frequency is 256 Hz is held over an empty measuring cylinder. See Fig. 13E-23. The sound is faint, but if just the right amount of water is poured into the cylinder, it becomes loud. When this occurs, the sound consists of the vibrations from the fork plus identical vibrations from the air column. If the length 0.31 m, what is the speed of sound in air to a first approximation? (For higher precision a correction must be made to allow for the fact that the pressure node of the air column occurs slightly outside the end of the column.)

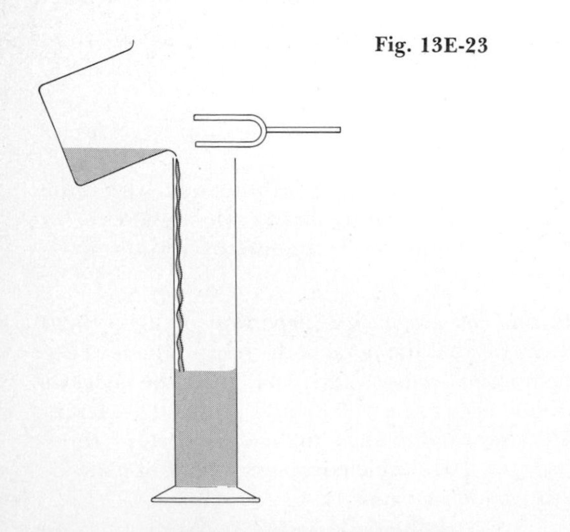

Fig. 13E-23

13-24. *What do you hear?* Suppose that the displacement $y(t)$ of a part of the human ear obeys the nonlinear equation, Eq. (13-53), and that the wave function describing the incident sound is given by Eq. (13-56), with $\nu_1 =$ 200 Hz and $\nu_2 =$ 260 Hz. If $y(t)$ is written as a sum of sinusoidal functions, what frequencies are involved?

13-25. *Superposing three waves.* Three wave sources, S_1, S_2, and S_3 are located in a large body of water and are equidistant from an observation point O. The wave sources have identical frequencies ν and power outputs. When only the jth source is operating, the surface displacement at O is $A\cos(2\pi\nu t + \delta_j)$, where $A > 0$ and $0 \leq \delta_j < 2\pi$.

a. Find the surface displacement $z(t)$ at O when all three sources are operating.

b. Combine the contributions from S_1 and S_2 to show that

$$z(t) = 2A\cos[\tfrac{1}{2}(\delta_1 - \delta_2)]\cos[2\pi\nu t + \tfrac{1}{2}(\delta_1 + \delta_2)] + A\cos(2\pi\nu t + \delta_3)$$

c. Suppose that A, δ_1, and δ_2 are given, and that $|\delta_1 - \delta_2| \leq \pi$, but that the value of δ_3 can be chosen at will. Find the value of δ_3 for which the overall surface displacement $z(t)$ has a maximum amplitude. What is the value of that maximum amplitude?

d. Evaluate the optimum value of δ_3 and the corresponding maximum amplitude for the following cases: (1) $\delta_1 = \delta_2 = 0°$; (2) $\delta_1 = 60°$, $\delta_2 = 0°$; (3) $\delta_1 = 90°$, $\delta_2 = 0°$; (4) $\delta_1 = 120°$, $\delta_2 = 0°$; (5) $\delta_1 = 180°$, $\delta_2 = 0°$.

e. For case (5) in part *d*, show that the amplitude does not depend on the particular choice made for δ_3. Explain why not.

f. Extend the results of part *c* to cover all possible values of $\delta_1 - \delta_2$.

13-26. *Steel guitar.* A steel guitar can readily be used to produce a *glissando* (Italian for "sliding") in which the pitch of a vibrating string is continuously varied while the string is sounding. This is accomplished by sliding a "stop" along the string to increase or decrease the effective length of the string. In a certain steel guitar, one particular string vibrates at a frequency of 440 Hz at its full length of 50 cm.

a. How far must the slide be moved to raise the pitch by one octave?

b. If the slide is moved at a constant speed of 20 cm/s to raise the pitch by an octave, find the duration of the glissando.

c. How many times does the string vibrate during the glissando in part *b*?

Group C

13-27. *Rocked in the cradle of the deep.* Two sources of water waves, S_1 and S_2, are separated by a distance d, as shown in Fig. 13E-27. The sources produce outgoing circular waves at the same angular frequency ω; the corresponding wave number k is given by $k = \omega/v$, where v is the wave speed. When only S_1 is operating, the instantaneous vertical displacement of the water at point O is given by

$$f_1(t) = \frac{C_1}{\sqrt{r_1}} \cos(kr_1 - kvt + \delta_1),$$

where r_1 is the distance from S_1 to O. When only S_2 is operating, the instantaneous displacement at O is given by

$$f_2(t) = \frac{C_2}{\sqrt{r_2}} \cos(kr_2 - kvt + \delta_2)$$

where r_2 is the distance from S_2 to O.

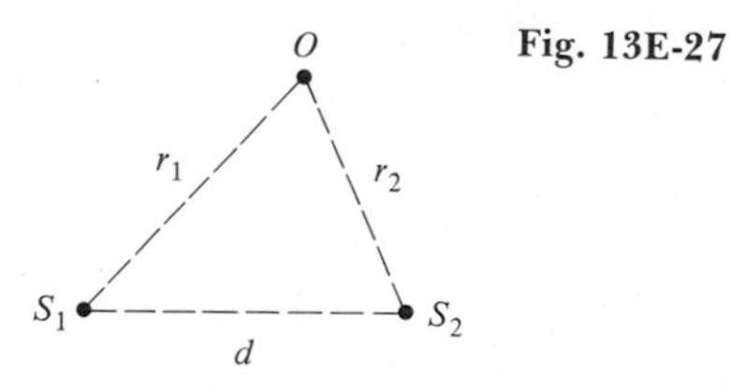

Fig. 13E-27

a. Find the instantaneous vertical displacement $f(t)$ at O when both sources are operating.

b. Using trigonometric identities, show that the result of part a can be rewritten as

$$f(t) = A_+ \cos\theta_+ \cos\theta_- - A_- \sin\theta_+ \sin\theta_-$$

where

$$A_\pm = \frac{C_1}{\sqrt{r_1}} \pm \frac{C_2}{\sqrt{r_2}}; \qquad \theta_+ = \frac{k(r_1 + r_2)}{2} - kvt + \frac{\delta_1 + \delta_2}{2}$$

and

$$\theta_- = \frac{k(r_1 - r_2)}{2} + \frac{\delta_1 - \delta_2}{2}$$

c. Find two restrictions relating r_1 and r_2 which, if both are satisfied, lead to the result $f(t) = 0$; that is, the point O is a node.

d. Suppose that $C_2 = C_1$ and $\delta_2 = \delta_1 - \pi$. That is, suppose that S_1 and S_2 are equal in strength but are oscillating 180° out of phase. Show that the water surface will remain motionless along the perpendicular bisector of the line joining S_1 and S_2.

13-28. *Reflection at a junction.* Two wires, made of different materials and/or having different thicknesses, have linear densities μ_1 and μ_2 respectively. The two wires are joined end to end. The combined wire is put under tension, which must be the same in both parts of the wire. A sinusoidal wave in the left part is traveling in the positive x direction and can therefore be represented by the wave function $y_i = A_i \cos(k_1x - \omega_1t)$. When this incident wave reaches the junction at $x = 0$, it is partially reflected and partially transmitted. The reflected wave is still in the left part of the wire, and can be represented by the leftward-traveling wave function $y_r = A_r \cos(k_1x + \omega_1t)$, where the amplitude A_r is yet to be determined. (Note: The omission of a phase constant from this wave function is a trick which leads to a simple method of determining the phase relationship between the incident wave and the reflected wave. As you will soon see, the omission leads to a value for the amplitude A_r of the reflected wave which may be either positive or negative, whereas, strictly speaking, a wave amplitude must be positive.) The transmitted wave is represented by the wave function $y_t = A_t \cos(k_2x - \omega_2t)$, with the value of A_t likewise to be determined. One boundary condition is that at the junction between the two wires, their instantaneous displacements $y_1 = y_i + y_r$ and $y_2 = y_t$ must be equal; that is, $y_1 = y_2$ at $x = 0$.

a. Show that this boundary condition leads to the results $\omega_1 = \omega_2$ and $A_i + A_r = A_t$. What is the physical meaning of these results?

b. The second boundary condition at the junction, $x = 0$, is that $\partial y_1/\partial x = \partial y_2/\partial x$; that is, there is no kink at the junction. This is necessary since the force exerted by the wires on each other at the junction must be equal and opposite. As in Chap. 12, assume that the displacements y are small. Use the boundary condition to show that $k_1A_i - k_1A_r = k_2A_t$.

c. Using the results of parts a and b, show that

$$\frac{A_r}{A_i} = \frac{k_1 - k_2}{k_1 + k_2} = \frac{|v_2| - |v_1|}{|v_2| + |v_1|}$$

where $|v_1|$ and $|v_2|$ are the wave speeds on the two parts of the wire.

d. If $\mu_1 < \mu_2$, what does the result of part d imply for the sign of A_r? (This result includes the case where μ_2 is infinite, which means physically that the left part of the wire is fixed rigidly at $x = 0$.) How can you rewrite the wave function y_r so that A_r will have a positive value? What is the resulting phase relationship between y_r and y_i?

e. If $\mu_1 > \mu_2$, what is the sign of A_r? (This includes the case where $\mu_2 = 0$, which means physically that the left part of the wire is completely free at $x = 0$.) What is the resulting phase relationship between y_r and y_i? Express your result in terms of a phase constant δ_r in the argument of the function y_r.

f. Show that the phase difference between y_t and y_i is always zero; that is, $\delta_t = 0$. One way is to show that at $x = 0$, $y_i + y_r = y_t$ regardless of the value of t.

g. Show that

$$\frac{A_t}{A_i} = \frac{2k_1}{k_1 + k_2} = \frac{2|v_2|}{|v_2| + |v_1|}$$

13-29. *Energy flow at a junction.* Equation (12-59) can be applied to an element dx of a sinusoidal traveling wave by replacing m by μdx and E by dE, giving $dE = \frac{1}{2}\omega^2A^2\mu dx$, where $\omega = 2\pi\nu$.

a. What is the expression for $\langle\rho_E\rangle$, the average energy per unit length?

b. Equations (12-55) and (12-56) relate energy flux and energy density. Use these equations and the result of part a to show that the energy arriving at a junction of two wires of different linear density joined together is equal to the energy carried away from the junction. It will be necessary to use the results of Exercise 13-28.

13-30. *The information is all there!* Consider carefully the relationship among the mode frequencies and node locations of successive symmetrical modes of a circular membrane. One aspect of this relationship is implicit in Exercise 13-19. Provide an explanation for the following assertion (which is correct): The frequencies and wave functions for modes 1 through $n - 1$ of the circular membrane can be determined directly by inspection of the results for the nth mode by itself, without any further numerical calculation.

Numerical

13-31. *Checking the boundary conditions, I.* Show that for the second standing-wave mode of the symmetrical vibrating drumhead $\alpha = 30.85$ gives a better fit to the boundary condition at $u = 0$ than that given by $\alpha = 30.80$ or $\alpha = 30.90$. Do this by running the vibrating drumhead program with the same initial conditions and parameters as in Example 13-7, except with α set equal to the lower value in one run and equal to the higher value in another run. Plot your data only in the region where $u < 0.10$. Add to your plot points taken from Fig. 13-22 in this region.

13-32. *Checking the boundary conditions, II.* Show that for the third standing-wave mode of the symmetrical vibrating drumhead $\alpha = 75.9$ gives a better fit to the boundary condition at $u = 0$ than that given by $\alpha = 75.8$ or $\alpha = 76.0$. Do this by running the vibrating drumhead program with the same initial conditions and parameters as in Example 13-8, except with α set equal to the lower value in one run and equal to the higher value in another run. Plot your data only in the region where $u < 0.10$. Add to your plot points taken from Fig. 13-23 in this region.

13-33. *The fourth standing wave mode.* Run the vibrating drumhead program to find the value of the parameter α for the fourth standing-wave mode of the symmetrical vibrating drumhead. Determine appropriate values for the initial conditions and the independent variable increment by considering the values employed in Examples 13-6, 13-7, and 13-8. Make a guess at a trial value to use for α by inspecting the values found for the first, second, and third modes in the three examples. Next look at Figs. 13-21, 13-22, and 13-23 in order to determine how many times $g(u)$ should change sign as u decreases from 1 to 0. Then carry out a run with your trial value of α. Do not plot, but just note the number of sign changes in $g(u)$, and its slope at $u = 0$. Search for a value of α which produces a $g(u)$ having the proper number of sign changes, and an acceptably small slope at $u = 0$. When you have found it, make a final run in which you plot $g(u)$ versus u. Compare your plot with the three figures. Finally, use your value of α to add one more frequency ν to the set displayed in Eq. (13-47).

13-34. *A loaded drumhead.* Describe how the procedure for determining the standing wave modes of the symmetrical vibrating drumhead would need to be modified if a point particle of mass m were attached to the center of the drumhead.

13-35. *Standing waves on a loaded drumhead.* Following the procedure you described in Exercise 13-34, perform numerical calculations to determine the frequencies and wave functions of the first three modes of a loaded drumhead for $m = \mu\pi a^2$.

13-36. *Fourier synthesis with $A_j = 1/j^2$.* Consider the Fourier series

$$f(t) = \sum_{j=1}^{\infty} A_j \sin(2\pi j\nu t + \delta_j)$$

where $A_j = 1/j^2$ and $\delta_j = 0$.

a. Let $f_n(t)$ represent the sum of the first n terms in the series; that is,

$$f_n(t) = \sum_{j=1}^{n} \frac{1}{j^2} \sin(2\pi j\nu t)$$

With the help of a programmable calculating device, plot $f_n(t)$ versus νt for $0 \leq \nu t \leq 1$, for each of the following values of n: (1) $n = 4$; (2) $n = 7$; (3) $n = 10$.

b. Sketch the form of the function $f(t)$ that would be obtained by summing the entire infinite series. Contrast the function $f(t)$ obtained here with the sawtooth function obtained in Sect. 13-7 by synthesis with $A_j = 1/j$.

13-37. *Fourier synthesis of a square wave.* Consider a space-independent wave function $f(t)$, given by $f(t) = A$ for $iT < t < (i + \frac{1}{2})T$ and $f(t) = -A$ for $(i + \frac{1}{2})T < t < iT$ for all integers $i = 0, \pm 1, \pm 2, \pm 3, \ldots$ The wave described by $f(t)$ is called a *square wave* of amplitude A and period T.

a. Construct a graph showing $f(t)$ for $-T \leq t \leq T$.

b. The square wave $f(t)$ can be written as a Fourier series in the form of Eq. (13-49*b*):

$$f(t) = \sum_{j=1}^{\infty} A_j \sin(2\pi j\nu t + \delta_j)$$

where $\nu = 1/T$, provided that the amplitudes A_j and phases δ_j are properly chosen. The correct choices are $A_j = 4A/\pi j$ for odd j; $A_j = 0$ for even j; $\delta_j = 0$ for all j. With the help of a programmable calculating device, plot the partial sums

$$f_n(t) = \sum_{\substack{\text{odd} \\ j=1}}^{n} \frac{4A}{\pi j} \sin\left(\frac{2\pi j}{T} t\right)$$

for each of the following values of n: (1) $n = 3$; (2) $n = 5$; (3) $n = 7$.

13-38. *Numerical Fourier analysis of a plucked string.* A stretched string of length L is plucked so that its midpoint is given a transverse displacement $L/10$. Define an x axis extending along the undisturbed string, and with an origin at one end. Then the shape of the string just before being released can be written as $f_{\text{exact}}(x) = x/5$, for

$x \leq L/2$; $f_{\text{exact}}(x) = (L - x)/5$, for $x > L/2$. Prepare a program for your calculating device which allows you to find approximate values of the first three nonzero amplitudes, A_1, A_3, and A_5, in the Fourier series for $f_{\text{exact}}(x)$. This is:

$$f_{\text{Fourier}}(x) = \sum_{j=1}^{n} A_j \sin\left(\frac{2\pi j}{\lambda} x\right)$$

where $\lambda = 2L$. The program should cause x to increase in 20 uniform steps from 0 to L. At each step the quantity $f_{\text{exact}}(x) - f_{\text{Fourier}}(x)$ should be evaluated, and then entered in a routine (accessible in a programmable calculator through a single key) which calculates the standard deviation, or variance, of a set of values. In running the program, first set $A_3 = 0$ and $A_5 = 0$ and then find by trial and error a value of A_1 which minimizes the standard deviation, or variance. Next, using the value previously found for A_1 and keeping $A_5 = 0$, find the value of A_3 which minimizes the standard deviation, or variance. Finally, using the values previously found for A_1 and A_3, find the value of A_5 which minimizes the standard deviation, or variance. Now run the program in such a way that you can make plots of $f_{\text{exact}}(x)$ and $f_{\text{Fourier}}(x)$. How is it known from the beginning that $A_2 = 0$ and $A_4 = 0$? See Exercise 13-21.

13.39. *Numerical Fourier analysis revisited.* Modify the procedure described in Exercise 13-38 so that you can use it to find approximate values of the first three nonzero amplitudes A_1, A_2, and A_3 in the Fourier expansion for a string plucked a third of the way from its end, at the point $x = L/3$. Why is $A_2 \neq 0$ in this case? Can you explain why $A_3 \simeq 0$?

14 Relativistic Kinematics

14-1 THE RELATIVISTIC DOMAIN

In the preceding chapters we have developed and used the mechanics of Isaac Newton to study the behavior of objects moving at speeds small compared to the speed of light. Now we turn our attention to objects moving at speeds comparable to the speed of light. In so doing we enter what is called the **relativistic domain,** temporarily leaving behind the newtonian domain. In the relativistic domain, motion must be treated in terms of Albert Einstein's **theory of relativity.**

The basic concepts of relativity occur in two forms: the **special theory of relativity** and the **general theory of relativity.** The special theory is used to treat the rapid motion of objects, of any size, as seen from inertial reference frames. It is the original form of the theory, developed by Einstein in 1905, and by far the simplest. The general theory has to do with noninertial (that is, accelerating) reference frames and their relation to gravity. This form of the theory, first set forth by Einstein in 1916, is used for the most part to treat the behavior of very massive systems. These are generally ones at the large end of the natural scale of sizes, that is, systems of astronomical size. Because of its mathematical complexity, we only mention the existence of the general theory. In this chapter and the next we concern ourselves with developing the special theory of relativity. And in subsequent chapters we make use of the special theory on a number of occasions.

The mechanics we will develop to analyze motion on the basis of the special theory of relativity is known as **relativistic mechanics.** It is used to study the high-speed motion of objects over the entire range of sizes. Relativistic mechanics applies over the entire range of speeds, too, its results merging smoothly into those of newtonian mechanics when the speed is low compared to the speed of light. It is the more broadly applicable mechanics, valid from zero speed to the speed of light. (We will find that the range of possible speeds for any object ends at the speed of light; higher speeds cannot be attained.) To put the matter another way, the rules of newtonian mechanics emerge as the low-speed special case of the more broadly applicable relativistic mechanics.

This close relation between newtonian and relativistic mechanics will be very helpful in our development of relativistic mechanics. It allows us to employ procedures analogous to those that worked so well in our development of newtonian mechanics. First, we concentrate our attention on accurately describing high-speed motion. This is the task of this chapter, which treats **relativistic kinematics.** In Chap. 15 we consider the question of what causes the motion, using arguments that involve the conservation of momentum to obtain relativistic mechanics.

We will take advantage of the close relation between newtonian and relativistic mechanics in another way to help us in our development of the latter. Since relativistic mechanics is supposed to be valid at all possible speeds, any relativistic result must agree with the familiar newtonian result if the speed is low. Thus we can check each new result of relativistic kinematics, or mechanics, against the newtonian result by considering the low-speed case of the relativistic result. If the check is satisfactory, we will gain confidence in the correctness of our arguments.

You may find the special theory of relativity to be one of the most interesting, challenging, and rewarding topics in this book. The arguments involve simple mathematics and simple physical systems. But you must pay very careful attention to the arguments. They often lead to conclusions that seem to be in serious conflict with your intuition. If you object to a conclusion on intuitive grounds, you should then feel obliged to find a fault in the argument *or* a fault in your intuition. After all, you should not expect "common sense" to take you very far in guessing about how systems should behave at speeds with which you have had no everyday experience at all. Be warned by Einstein's saying: "Common sense is that layer of prejudices acquired before the age of eighteen."

14-2 THE SPEED OF LIGHT

The speed of light plays a fundamental role in the theory of relativity. This speed enters into relativistic kinematics, and hence into relativistic mechanics, as the maximum speed at which information can be transmitted from one location to another. Before we can explain the connection between relativistic kinematics and the transmission of information, we must describe briefly how light travels. Then we must consider experiments which measure how fast it travels in various circumstances.

Light travels according to the laws of wave motion. Except for a *very* important difference, it travels in just the same way that mechanical waves do. Thus there is a strong analogy between light waves and transverse waves propagating along the connected particles of a stretched string, or pressure waves propagating through the interacting particles of the air. The very important difference is that *there is no mechanical medium involved in the propagation of a light wave.* This is made apparent by the fact that light travels from the sun to the earth through the essentially perfect vacuum of space.

In Chap. 27 we investigate what it is that is "waving" in a light wave, and we find that the answer is electric and magnetic fields. We also find that traveling light waves transport energy, just as traveling mechanical waves

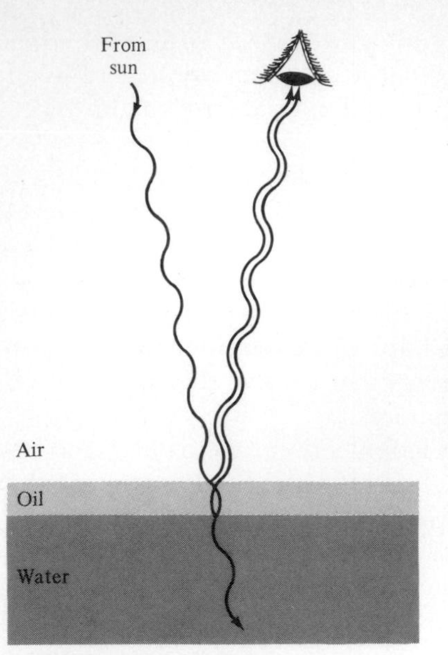

Fig. 14-1 The reinforcement in the reflection of light by a thin oil film floating on water.

do. This is the property of interest here. But you already know it and are reminded of it every time you stand under the hot sun on a clear day. Because light carries energy, it can be used to transmit signals that can actuate detectors at distant locations. In fact, light waves are commonly used for this purpose. Of course, this is just what happens in the phenomenon of vision.

In Chap. 28 we study a variety of experimental demonstrations proving that light travels as a wave. Here we describe two such demonstrations. Our first experimental proof of the wave nature of light is found in the multiple colors produced when white light is reflected from a very thin film of oil that has spread over a surface of water. You have surely seen this phenomenon at some time or other.

A very thin film of oil is itself quite transparent and colorless. It certainly does not have a multiplicity of intrinsic colors. The observed colors are formed because part of the illuminating light is reflected at the interface between the oil and the air above it, and the rest passes into the film where some of it is reflected at the interface between the oil and the water beneath it. An observer sees light reflected from both surfaces, the two beams of reflected light having traveled paths of different length to reach the eye. See Fig. 14-1. If the illuminating light is of a single wavelength, the two beams of reflected light will either reinforce or cancel each other when superposed in the observer's eye, depending on the way in which the wavelength of the light fitted into the difference in the two path lengths. If the additional length of path traveled by the light reflected from the lower surface is equal to an integral number of wavelengths (that is, λ, or 2λ, or 3λ, and so forth), then the two reflected light beams will be in phase when they combine, and they will reinforce each other. (For the case illustrated in the figure, the combining light beams reinforce each other because the length of the additional path is λ.) If the additional path length is equal to a half-integral number of wavelengths (that is, $\lambda/2$, or $3\lambda/2$, or $5\lambda/2$, and so forth), the beams will be out of phase when they combine and so will cancel each other.

White light is actually a mixture of wavelengths, the various wavelengths corresponding to what our eyes and brain perceive as the various colors. Also, an oil film is usually not of uniform thickness. So when white light is incident on such a film, regions of a particular thickness produce strong reflection of the particular wavelength which has the proper relation to the thickness, and we perceive the corresponding color. Regions of different thicknesses produce strong reflections of different wavelengths, perceived as different colors. The most satisfactory way to explain this familiar phenomenon (and the only way to explain many others which we discuss in the next paragraphs and in Chap. 28) is by saying that light travels as waves that superpose—that is, obey the superposition principle—just like mechanical waves involving small displacements.

A particularly striking demonstration and application of the superposition of light waves are provided by an instrument invented and refined in the 1870s by the U.S. physicist Albert A. Michelson (1852–1931). The instrument shown schematically in Fig. 14-2 is called a **Michelson interferometer.** Light of a single wavelength expanding from a point source S is rendered into a nondivergent beam by lens L. The beam falls on a glass plate P inclined at an angle of 45° to the beam. The plate is lightly silvered so that it

Fig. 14-2 Schematic diagram of a Michelson interferometer. Consideration of the paths followed by light at the two edges of the beam incident on plate P will show that the path difference is the same across the entire width of the recombining beams, providing that the surface of plate P is at precisely 45° to the direction of the beam incident on it and that the surfaces of mirrors M_1 and M_2 are at precisely 90° to the beams incident on them.

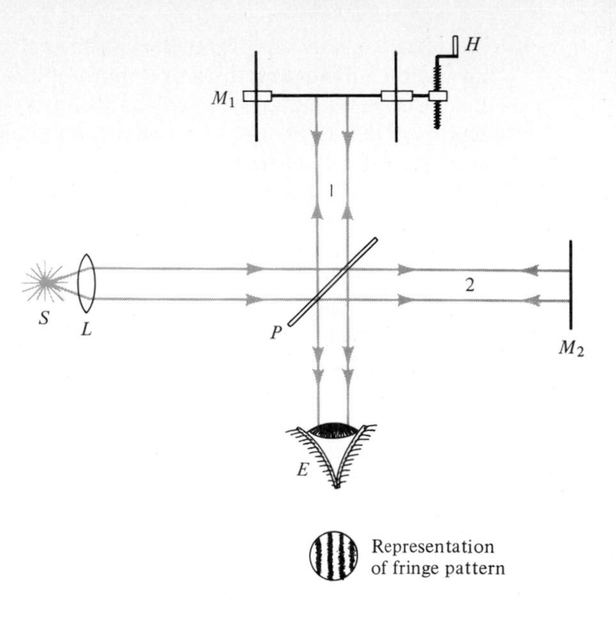

forms a partially reflecting mirror. Approximately half of the incident beam is reflected along path 1 to the fully silvered mirror M_1, and the remainder is transmitted through P to follow path 2 to the fully silvered mirror M_2. Mirrors M_1 and M_2 reflect the beams back on themselves toward P. On striking P, about half of the beam returning along path 2 is reflected into the observer's eye E, and about half of the beam returning along path 1 is transmitted through P into E. Thus the interferometer splits the beam of light from the source into two beams, which travel different paths until they are brought together to recombine in the eye of the observer.

The recombining beams will be in phase, and so will reinforce each other, if the round-trip distance along path 1 differs from the round-trip distance along path 2 by an integral number of wavelengths λ of the light. If these distances differ by a half-integral number of wavelengths, the recombining beams will be out of phase and will cancel each other. (Note the close analogy with the acoustical interferometer of Sec. 13-1.) In practice, M_1 and M_2 are not precisely perpendicular, so the path difference varies by a few wavelengths across the width of the recombining beams. The result is that the observer sees a pattern of a few alternating bright and dark bands, called **fringes,** as indicated in the insert below the eye in Fig. 14-2.

The mirror M_1 can be moved along a carriage by turning the handle H connected to a screw of fine pitch. When M_1 is moved through a distance $\lambda/4$, one-quarter of the wavelength of the light emitted by S, the round-trip distance traveled by the light beam in that arm of the interferometer changes by $\lambda/2$, one-half a wavelength. Thus all the path differences for the two beams change by half a wavelength. As a consequence, every region where the recombining beams had been in phase now becomes a region where they are out of phase, and every region where they had been out of phase now becomes a region where they are in phase. This makes all the regions that had contained bright fringes now shift to regions that contain dark fringes, and vice versa. The effect is striking. It can be understood only in terms of the superposition of waves.

A Michelson interferometer is the instrument used to establish the relation between the meter and the wavelength of light emitted by atoms of krypton-86. As discussed in Sec. 2-2, the standard meter is now defined to be 1,650,763.73 wavelengths of this light. The number was obtained by using a krypton-86 light source, stepping off a total distance with the movable mirror defined by the separation between the scratches at the ends of the meter bar that was the previous standard, and counting the total number of fringe shifts that are observed.

Interferometers using a light source of known wavelength are employed in science, engineering, and technology to measure lengths. The limit of accuracy is a fraction of the wavelength of the light used, that is, smaller than about 5×10^{-7} m. It is desirable for the light source to be a laser. The reason is that lasers emit an almost continuous train of light waves that will produce fringes in an interferometer even when the difference in the lengths of the paths followed by the two beams is very large. All other light sources emit light in a series of short-duration bursts, with the wave in each burst unrelated in phase to the wave in the others. Because the bursts of light from these sources are of short duration, they extend along the beam only a short distance—typically something like a few centimeters. When the difference in the total length of the two paths followed by the split beam through the interferometer exceeds the length of an individual burst of light—the **coherence length**—fringes are no longer observed. This is because a burst combines in the observer's eye no longer with itself, but instead with a preceding or following burst that is unrelated in phase. The accuracy of the krypton-86 interferometer definition of the meter suffers as a result of the necessity of measuring the length of a meter bar in a sequence of short steps, instead of in one single measurement. So it is likely that the standard meter will be redefined once more, using a laser interferometer. (Krypton does not appear to be usable in a laser, so a different material, producing light of a different wavelength, probably will have to be employed in such a redefinition of the meter.)

We turn now to a question of most basic importance to the theory of relativity: What is the speed of propagation of a light wave? The first reasonably accurate measurement of the speed of light was performed in 1849 by the French physicist A. H. L. Fizeau (1819–1896), using the apparatus sketched schematically in Fig. 14-3. Light from an intense source S was reflected by a mirror M_1 to a lens L_1 that focused it on a gap in a wheel W having 720 evenly spaced teeth. The light passed through the gap to a lens L_2, which converted it into a parallel beam that traveled a distance of 8.63 km to a mirror M_2. At the mirror the beam was reflected back on itself, finally passing through the gap in the wheel. Just missing M_1, the beam entered the observer's eye E. The wheel was then set into rotation, at gradually increasing speed. The brightness of the light seen by the observer decreased, reaching zero when the speed was 12.6 rotations per second, and then increased to a maximum when the rotation speed was 25.2 rotations per second. This is because the rotating toothed wheel chopped the light

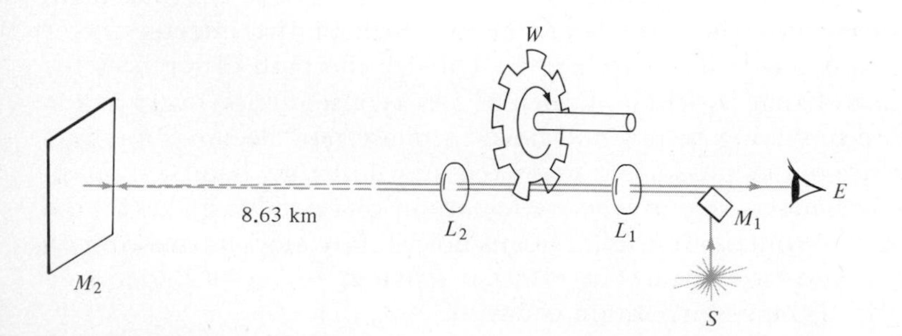

Fig. 14-3 Schematic diagram of Fizeau's apparatus for measuring the speed of light.

beam traveling toward the mirror into a series of pulses. At 12.6 rotations per second, the time required for a pulse to travel to the mirror and back equaled the time required for the wheel to rotate through such an angle that the tooth adjacent to the gap moved into the path of the beam, completely blocking the returning pulse. But when the rotation speed was 25.2 rotations per second, the next gap of the wheel had rotated into the light path, allowing all the returning pulse to pass into the observer's eye. These data are used in Example 14-1 to evaluate the speed of light.

EXAMPLE 14-1

Use Fizeau's data to determine the speed of light.

■ Since there were 720 teeth, and gaps, in the wheel, the time t required for one gap to rotate into the position of an adjacent gap is 1/720 of the time required for the wheel to make a complete rotation. At 25.2 rotations per second, the time for one complete rotation is 1 s/25.2. So

$$t = \frac{1}{720} \times \frac{1 \text{ s}}{25.2} = 5.51 \times 10^{-5} \text{ s}$$

In the time t the light pulse travels a distance d from the wheel to the mirror and back. This distance is

$$d = 2 \times 8.63 \times 10^{3} \text{ m} = 1.73 \times 10^{4} \text{ m}$$

Evaluating the speed $v = d/t$ of the light pulse, you have

$$v = \frac{d}{t} = \frac{1.73 \times 10^{4} \text{ m}}{5.51 \times 10^{-5} \text{ s}}$$
$$= 3.14 \times 10^{8} \text{ m/s}$$

This result is about 5 percent large, compared to more precise modern experiments, owing mainly to the uncertainty inherent in Fizeau's method of measuring the rotation rate of his toothed wheel.

In the intervening years many investigators have carried out progressively more accurate measurements of the speed of light. A central motivation for this difficult task is the very important role played by the speed of light in physics. These measurements have been made using not only visible light, but also other members of the family of electromagnetic radiation to which light belongs. All electromagnetic radiation travels at the same speed in vacuum. For technical reasons, it is possible to make especially precise measurements of the speed of the type of electromagnetic radiation called microwaves (very short-wavelength radio waves, with $\lambda \simeq 1$ cm). The speed of electromagnetic radiation, loosely called the **speed of light,** is so fundamental that it is given its own symbol c. The currently accepted value, obtained principally from microwave measurements in vacuum, is

$$c = 2.99793 \times 10^{8} \text{ m/s} \qquad (14\text{-}1)$$

with an accuracy of 1 in the last quoted decimal place. For most practical calculations the value used is $c = 3.00 \times 10^{8}$ m/s. Although measurements of the speed of light are sometimes made with the light traveling through air, the value given in Eq. (14-1) has been corrected so that it pertains to the speed of light *in vacuum.* The correction is very small; the speed in vacuum is higher than the speed in air by 0.00067×10^{8} m/s.

Consider a mechanical wave propagating through a medium that is at rest in an inertial reference frame, such as a sound wave propagating through still air. The wave travels at a speed that can be calculated from the wave equation (or looked up in a table reporting experimental measurements) in terms of the mechanical properties of the propagation medium. When we speak of the speed of this wave, it is perfectly unambiguous what we mean. We do not mean the speed measured with respect to the source of the wave. This would be inappropriate since the speed of the wave (in contrast to its frequency or wavelength) is *not* affected by any motion the source may have. Nor do we mean the speed of the wave measured with respect to the observer of the wave. It would not be appropriate to mean this because the speed measured with respect to the observer (in other words, the value that would be obtained by the observer in a measurement of the speed of the wave as seen by the observer) *is* affected by the motion of the observer. So there would be no unique value for the speed of a wave propagating through a particular mechanical medium if the speed were measured with respect to the observer. What we do mean when we speak of the speed of the wave is the speed *measured with respect to the mechanical medium through which it propagates.* For a particular medium this speed has a particular value, and it is the fundamental value characterizing the motion of the wave.

But *a light wave propagates through vacuum.* So there is no medium involved in its propagation. Yet *speed must always be measured with respect to something.* What then can we mean when we talk about the speed of light in vacuum? With no propagation medium with respect to which to measure the speed, it would seem that we must choose between measuring the speed with respect to the source of a light wave or measuring the speed with respect to the observer of the light wave. (What else is there?) We have no reason to expect that the motion of the source of a light wave is connected in any way to its speed—it certainly is not in the case of a mechanical wave. If we rule out the source, what remains is the observer. Then thinking carefully about what actually happens in an experimental procedure used to determine the speed of light, we soon come to the conclusion that if we say the speed of a light wave is to be measured with respect to the observer, we are completely consistent with the experimental procedure. In an experiment used to determine the speed of light, the light wave moves through some apparatus fixed in a reference frame that can be considered to be an inertial frame. Take as an example an experiment with the apparatus fixed to the surface of the earth. An observer stationed at the apparatus manipulates it and records the data. Thus the value obtained for the speed of light is a value measured with respect to the apparatus or, what amounts to the same thing, a value measured with respect to the observer. These considerations tell us what we mean by the speed of light by specifying the *operational definition* of the quantity—that is, the definition in terms of the experimental operations carried out in order to measure its value. What we mean is that *we measure the speed of light with respect to the observer.*

Say we are measuring the speed of a beam of light passing by us and that we are in the essentially inertial reference frame of the earth's surface. Another observer is in an essentially inertial reference frame of a rocket ship, moving uniformly with respect to us in the direction of the light beam at a very high speed. This observer uses an identical apparatus to measure the speed of the same light beam. Will we obtain the value quoted in Eq. (14-1), $c = 3.00 \times 10^8$ m/s, while the observer obtains some other value? If we answer Yes, we put ourselves in the following logical dilemma. We are on the earth, which moves at an appreciable speed with respect to the sun, while the sun moves with respect to the center of mass of our galaxy and our galaxy moves with respect to the center of mass of the universe. So it is not only the observer in the rocket ship who is moving with respect to the universe—we are, too. Who, then, will be measuring "*the*" speed of the light beam? If it is not the observer in the rocket ship, why should it be us? In the absence of a medium through which light waves propagate, there is no observer in the universe who has a privileged status with regard to a measurement of the speed of light.

For mechanical waves, there *is* a privileged observer—the one who is at rest with respect to the inertial frame of the propagation medium. This observer will obtain the fundamental value when measuring the speed of the mechanical waves. Other observers, in inertial frames moving with respect to that of the propagation medium, will find numerically different values when they measure the speed of the waves from their point of view. If an observer is moving through the medium in the same direction as that of the mechanical wave, the value he measures for the speed will be decreased by the value of his speed through the medium. If another observer is moving relative to the medium in opposition to the direction of the propagating mechanical wave, she will measure a value for its speed which is increased by the value of her speed. There is an abundance of accurate experimental proof that this is so. An example is found in any measurement of the Doppler effect for sound waves with a moving observer, discussed in Sec. 12-7. But it *cannot* work this way for light waves since light waves have *no* propagation medium.

Based, at least in part, on ideas such as these, in his landmark 1905 paper on relativity theory Einstein made this assertion: *The speed of light is measured by observers in all inertial reference frames to have the same value, despite the fact that such observers may be moving with respect to one another.*

In the nineteenth century, most physicists held strongly to quite a different idea. They were so impressed with the successful theory of the propagation of mechanical waves that they tried, as much as possible, to use it as a model for the theory of the propagation of light waves. Thus the generally held view at the time was that there is a privileged observer for light waves—the observer in a *special* inertial reference frame. It was believed that only that observer would obtain the fundamental value when measuring the speed of light waves. Other observers, in inertial frames moving with respect to the special one, would obtain different values for the speed. This usually was done by saying that the special frame is the inertial frame fixed with respect to the universe as a whole. Thus the idea was that an observer in a reference frame stationary with respect to the center of mass of the universe would measure the "real" value of the speed of any light beam. And an observer is an inertial frame moving with respect to the center of mass of the universe would measure an "apparent" value for the speed of the light beam which differed from the "real" value in a way that depends on the motion of the observer's inertial frame.

In 1881 Michelson began a series of experiments whose results, when finally they were understood, showed that this idea is incorrect. The experiments used a Michelson interferometer. In particular, a very careful series of measurements on an improved apparatus was performed by Michelson in collaboration with the chemist Edward W. Morley (1838–1923) in 1887.

The **Michelson-Morley experiment,** as it is called, takes advantage of the fact that any reference frame fixed to the surface of the earth is an essentially inertial frame that is moving at a high speed with respect to the inertial frame fixed to the center of mass of the universe. At the time no information existed concerning the motion of either our galaxy with respect to the universe or the sun with respect to our galaxy. But motion of the earth with respect to the sun was, of course, well known. Almost entirely because of its annual motion, the earth moves with respect to the sun at a speed of about 3×10^4 m/s.

In the experiment a large interferometer was mounted on a granite block. The block was floated in a pool of mercury, both to minimize bending of the block from any nonuniformity of support and to facilitate smooth rotation of the entire apparatus in the horizontal plane. One arm of the interferometer was aligned parallel to the direction of motion of the earth with respect to the sun, and the other arm was aligned perpendicular to this direction. In the parallel arm, the light first moved away from the observer, in the same direction as that of the observer's motion with respect to the sun. After reflection from the mirror at the end of the arm, the light moved in a direction opposite to that of the observer's motion with respect to the sun. In the perpendicular arm, the direction of motion of the light was at all times perpendicular to that of the observer. If the motion with respect to the sun of the observer stationed at the interferometer affected the speed of the light that he observed—because a motion with respect to the sun means a motion with respect to the universe as a whole—there would be very small but measurable differences between the speeds the observer would find for light traveling in the various directions. If these differences were present, they would show up as a change in the phase of one recombining light beam with respect to the other. This is because the total travel time for one beam would be affected differently from that of the other. In the experiment, the fringe pattern was observed as the interferometer was rotated through 90°. The 90° rotation interchanged the two arms of the interferometer, which would have reversed the sign of any phase change present and thereby led to a fringe shift.

Michelson and Morley made a calculation in which they assumed that the observed speed of light waves is affected by the motion of the observer's inertial frame with respect to one fixed to the center of mass of the universe in the same way that the observed speed of mechanical waves is affected by the motion of the observer's inertial frame with respect to one fixed to the mechanical propagation medium. The calculation predicted that they should have found a shift of about 0.4 fringe when they rotated the apparatus by 90°. (This does not take into account the motion of the sun with respect to the universe. When it is taken into account, the predicted fringe shift is considerably larger.) They knew that they could reliably detect a fringe shift as small as 0.01 fringe. But to this degree of accuracy they found *no evidence for a fringe shift.*

Thus the Michelson-Morley experiment is inconsistent with the idea that there is something special about an inertial frame fixed to the center of mass of the universe, as far as the speed of light is concerned. This is so because the experiment shows that rapid motion of an observer's inertial frame with respect to the "special" inertial frame does not change the speed of light measured by the observer. Michelson received the Nobel prize in physics in 1907 for the invention and utilization of the interferometer. He was the first U.S. citizen to win the prize.

In contrast, the Michelson-Morley experiment is consistent with Einstein's assertion that the speed of light is measured by observers in *all* inertial frames to have the same value.

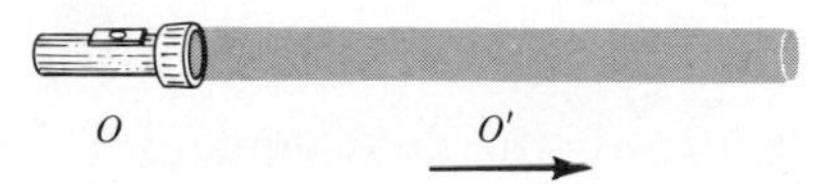

Fig. 14-4 Observer O is at rest in one inertial frame, and observer O' is at rest in another inertial frame. Observer O' is moving to the right relative to O, so the frames are in relative motion. But because all inertial frames are equivalent, the speed of light is the same in both frames. Thus O and O' will find that the speed of the light beam is the same.

What Einstein said is certainly not what Newton would have said. Einstein's assertion is completely contradictory to newtonian intuition. For instance, imagine the front of a light beam that is moving away from a flashlight just after it has been turned on, as in Fig. 14-4. Let there be two observers measuring the speed of the front of the beam. Observer O is stationed in the inertial frame containing the flashlight, and observer O', who is on a rocket ship, is stationed in an inertial frame moving with respect to the first one in the same direction as the light beam and at half the speed of light. If the speed of light is the same for both observers, then *both* O and O' will obtain the *same* numerical value, $c = 3.00 \times 10^8$ m/s, for their measurements of the speed at which the front of the beam moves away from them.

You cannot say that O is right while O' is wrong, just because O happens to be in the same inertial frame as the flashlight. The speed of light waves is not afected by the motion of the source any more than the speed of sound waves is affected by the motion of the source of sound. (Many experiments show this statement to be true. One of the earliest was a Michelson-Morley type of experiment in which the source of light was a star moving rapidly away from the observer making the measurements.) If both O and O' are in inertial frames relative to which the flashlight is in rapid motion, and the two observers are also moving rapidly with respect to each other, Einstein would still say that both obtain the same value $c = 3.00 \times 10^8$ m/s in their measurements of the speed of the front of the beam.

Before you use the argument of Fig. 14-4 to reject Einstein's assertion about the speed of light as an impossibility, remember that you most likely have no experience at all in the relativistic domain. Since you live in a world where obvious phenomena have to do with speeds that are extremely small compared to the speed of light, your intuition is based on evidence from the newtonian domain. It cannot legitimately be used to deny, or to confirm, a statement concerning the relativistic domain.

14-3 THE EQUIVALENCE OF INERTIAL FRAMES

The most important feature of Einstein's first paper on relativity theory is his postulate of the equivalence of inertial reference frames. Before stating the postulate, we will summarize very briefly the point of view concerning inertial reference frames that most physicists held around the turn of the twentieth century.

Different inertial reference frames move with respect to one another. But they do so with constant velocity and thus do not accelerate with respect to one another. For this reason, all inertial frames are completely equivalent with regard to all mechanical phenomena. A game of billiards played on a table in a railway car moving smoothly and uniformly along a

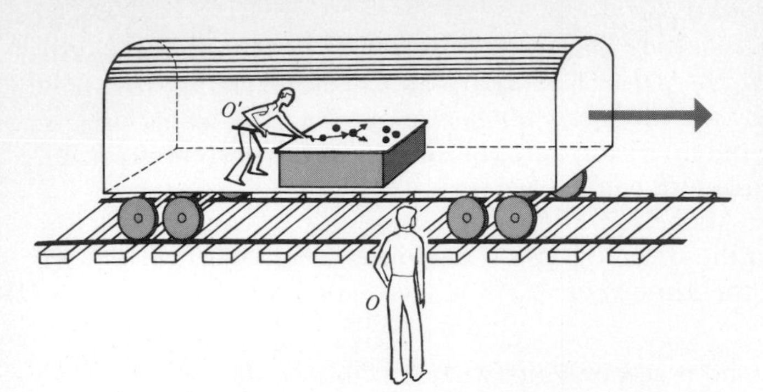

Fig. 14-5 Observer O is at rest in one inertial frame, and observer O' is at rest in another inertial frame. Observer O' is moving to the right relative to O, so the frames are in relative motion. But because all inertial frames are equivalent, the mechanical laws governing the motion of the billiard balls is the same in both frames. Thus O and O' will find that the mechanical laws governing the behavior of the billiard balls are the same.

straight track, as in Fig. 14-5, is governed by exactly the same equations of mechanics as when the game is played with the car standing in a station. We have seen other examples of this property on several occasions in earlier chapters, and it was certainly well known by physicists working before 1905. But, as we have said, physicists of that era thought that for optical phenomena (phenomena involving light) all inertial frames were not equivalent. They felt that there was one special inertial frame that was singled out by the properties of light propagation. In that frame—and in that frame alone—they believed the speed of light would be measured to have the numerical value $c = 3.00 \times 10^8$ m/s.

Einstein asserted that nature is *simpler* than others had believed to be the case because the same situation that applies for mechanical phenomena also applies for optical phenomena. All inertial reference frames are equivalent for *optical phenomena,* as well as for *mechanical phenomena,* since the speed of light will have the same measured value c in all such frames. In his 1905 paper Einstein generalized to make the statement include *all physical phenomena.* He did this in his **postulate of the equivalence of inertial frames:** *All inertial reference frames are completely equivalent for all physical phenomena.* This statement is often called the **principle of relativity.** A tremendous body of experimental data has verified Einstein's postulate. The experimental work started with Michelson and Morley and continues to this day. The experiments verifying the postulate of equivalence of inertial frames cover a wide variety of phenomena. No valid experiments have contradicted it. Relativistic kinematics is founded on this single postulate.

There has been controversy about the extent to which Einstein's work was influenced by the earlier experiments of Michelson. The logic of his 1905 paper is based on the explicitly electromagnetic properties of light, and he does not even mention the experiment in the paper. In fact, Einstein was silent on this point in all his writings. R. S. Shankland questioned Einstein about it in interviews given on several occasions not long before his death in 1955. According to Prof. Shankland,* Einstein said, "I am not sure when I first heard of the Michelson experiment. I was not conscious that it had influenced me directly during the seven years that relativity had been my life. I guess I just took it for granted that it was true." Shankland concludes by saying, "It should be noted that in 1905 it was not the practice to give specific references in published papers as it is today. Many important papers gave no references whatever, so the fact that Einstein's 1905 paper [which has no references] makes no explicit mention of the Michelson-Morley experiment is not in the least unusual. To what degree and at what stage of his activities he was directly influenced by the work of others, it is now impossible to determine. But I became convinced that his interest in the Michelson-Morley experiment . . . had existed before 1905."

* *American Journal of Physics,* vol. 41, 1973, p. 895; quoted by permission

14-4 SIMULTANEITY

It is an experimental fact that the measured speed of a beam of light is not affected by the measurer's own motion. But on first exposure the fact may seem hard to believe. It must have seriously bothered Einstein, too, at first. It certainly made him think more deeply than anyone had before about what speed really is, for it to have such perplexing properties.

Since speed involves the ratio of a space interval to a time interval, his thoughts led him in due course to consider separately the question of what is really meant by space and what is really meant by time. He told R. S. Shankland: "At last it came to me that time was suspect." So Einstein set out to think through, from scratch, how time should be defined *operationally* in physics—that is, what procedures should be used to *measure* time. To develop relativistic kinematics, we must do the same. In this we will follow closely many of Einstein's arguments.

Einstein argued that a measurement of time always involves a determination of **simultaneity.** He wrote, "If I say 'That train arrives here at 7 o'clock,' I mean something like this: 'The pointing of the small hand of my watch to 7 and the arrival of the train are simultaneous events.' " An **event** is anything that occurs at a particular time and place. **Simultaneous events** are events which occur at the *same time.*

If two events occur at the same place, as in Einstein's example, it is easy to determine whether they also occur at the same time and so determine whether they are simultaneous. Consider Fig. 14-6, which depicts the simultaneous events described by Einstein. An observer located where two events occur at the same place receives from each event, with essentially no time delay, the information he needs to judge their simultaneity. He sees both events at the instants when they occur because he is at the location where they both occur.

But simultaneous events do not necessarily occur at the same place. It is possible to determine whether two events are simultaneous in a case where the events occur at places separated by an appreciable distance. However, it is not as easy to do this as it is in a case where the events occur at the same place. The reason is that there are time delays between the instants at which the separated events occur and the instants at which the observer who is judging their simultaneity sees them occur. The information that they have happened is carried to the observer watching the events by light traveling at a speed which is not infinite. Hence an accurate determi-

Fig. 14-6 A measurement of the arrival time of a train. The arrival of the train and the pointing of the hour hand of the watch to 7 are simultaneous events.

Fig. 14-7 A measurement of the simultaneity of two spatially separated events. (The measurement requires the observer to look in opposite directions at the same time. One way to accomplish this is by using two mirrors, each mounted at 45° to the light beam incident on it.)

nation of the simultaneity of separated events requires that the time for light to travel to an observer not be neglected. (This crucial fact is ignored in newtonian kinematics.) Einstein said that the simplest way to take this time into consideration is to station the observer at the point *midway* between the separated events, as in Fig. 14-7. Then the time delays will certainly be equal because in all circumstances light moves from both events to the observer at the same speed. Thus the **simultaneity of separated events** can be determined by applying the following test. *If separated events are simultaneous, then light emitted from each event must reach an observer midway between them at the same time.*

Information that the events have occurred could be transmitted to the observer by means other than light signals. If the region between the locations of the events contained air, sound signals could be used. But it would be essential for the observer to measure *precisely* the relative motion of the air at *all* points along the paths of the signals, since any air motion would affect the apparent speed of sound and therefore the observed time delays. Einstein said light signals should be used, instead of sound signals, because light has the great simplifying property that its speed relative to the observer has the same value, regardless of the circumstances.

Given a way to determine the simultaneity of separated events, an observer in any reference frame can measure time at any location in that frame. First the observer obtains a number of clocks running at the same rate and places one of them near each point in the reference frame where time must be measured. The observer then moves to the point midway between a pair of clocks and directs a helper at one of these clocks to adjust the hands until the observer sees both clocks simultaneously reading the same time. Next another clock is paired with one of the first pair and is similarly set. The process is continued until all the clocks fixed in the observer's reference frame are set. The clocks are then said to be **synchronized.** The observer can use the synchronized clock at any location in the reference frame to measure time at that location and can make valid intercomparisons of these time measurements.

However, these clocks would *not* be considered synchronized by an observer who is moving relative to the reference frame containing the clocks. The reason is that two events which are simultaneous from the point of view of one reference frame are *not* simultaneous from the point of view of a second reference frame moving relative to the first, as we will prove immediately below. Thus the events used in synchronizing the clocks, although simultaneous as judged from the reference frame containing the clocks, are not simultaneous as judged from a reference frame moving with respect to the clocks. And so the observer moving with respect to the clocks does not judge them to be properly synchronized. Consequently, this observer will not consider comparisons of time measurements made with these clocks to be valid. (This crucial fact is also ignored in newtonian kinematics. Indeed, the assumption that there is a universal time scale, which is the same in all reference frames no matter what their state of motion, and the justification of this assumption, are considered in newtonian kinematics to be so self-evident that they are unvoiced. This incorrect assumption plays a vital role in deriving the Galilean transformations. See Sec. 3-8.)

We will prove that observers in relative motion disagree about the simultaneity of separated events, by thinking through what would happen if a certain experiment were carried out. Such a procedure is called analyzing a **thought experiment.** The experiment is illustrated in Fig. 14-8. It shows two successive views of a railroad train moving at a constant high velocity V along a straight track, from the viewpoint of an observer O fixed on the ground. Prior to the passage of the train, O has carefully measured off a distance in one direction down the track to a location B_1 and an equal distance in the opposite direction to a location B_2. At B_1 and B_2 she has placed two blasting caps (small explosive charges that are detonated electrically) near where the side of the train will pass. A modification of Einstein's test for determining the simultaneity of separated events implies that she can make them explode simultaneously in her reference frame by sending light signals from her midpoint location, which actuate detonators on their arrival at B_1 and B_2. She does this so that the explosions occur when, from her point of view, an observer O', stationed on the train at its center, is just passing her.

When the blasting caps explode, they leave marks B_1' and B_2' on the side of the train near its two ends. After the experiment is over, the observer O' on the train can measure the distances $O'B_1'$ and $O'B_2'$. As you would expect, he certainly finds the distances to be equal. (If this were not the case, space would have properties in one direction which differed from its properties in the opposite direction.)

The explosions also produce light flashes. The observer O on the ground finds that the flashes coming from the front and rear ends of the train reach her at the same time. This confirms that the blasting caps actually were detonated simultaneously in her frame of reference. She also finds that O' receives the flash coming from the front end *before* he receives the flash coming from the rear end. As the figure shows, O sees this to be a result of both the motion of O' toward the flash approaching him from the front and his motion away from the flash approaching him from the rear. So O explains the observations of O' by saying that the motion of O' causes him to receive the flash from the front of the train before the flash from the rear.

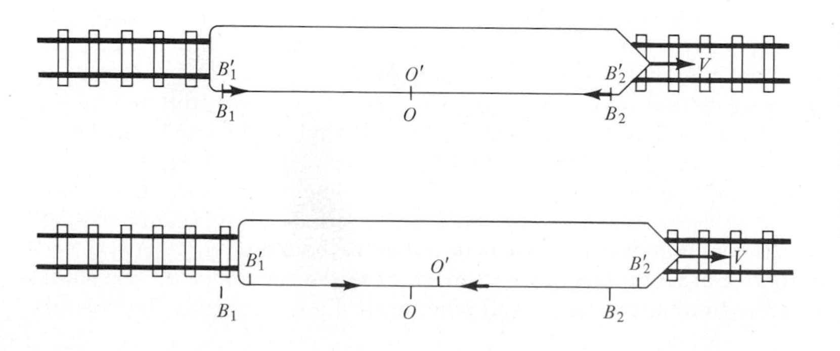

Fig. 14-8 A thought experiment concerning the simultaneity of separated events, which are measured in reference frames moving relative to each other. The two successive illustrations of the experiment both show it from the point of view of the ground-based observer O. The top one views the situation at the time of detonation, and the lower one at a slightly later time. The short arrows represent light flashes coming from the detonations.

Observer O' agrees that the flash from the front arrives at his location before the flash from the rear. (The observers cannot disagree about this since light detectors fixed at the center of the train can be used to make permanent records of when the flashes arrive, and anyone can inspect these records when the experiment is over.) But O' disagrees with O about the explanation. He perceives himself to be stationary in the frame of reference of the train and contends that the ground on which O stands is moving past him, going in the direction toward the rear of the train. At equal distances from his location are two points, B_1' at the rear of the train and B_2' at the front. Explosions occur at each of these points and produce light flashes. Since he does not receive these light signals from the equidistant events at the same time, when O' applies the test for determining the simultaneity of separated events, he concludes that the two explosions were *not* simultaneous.

Soon we will obtain a relation that allows us to make a quantitative statement about the departure from simultaneity of the two events, as judged by O'. But here the important result is the qualitative conclusion that according to O' the explosion at the front of the train occurred some time before the explosion at the rear, even though the explosions were simultaneous according to O. The reason for this disagreement is, again, just the fact that the observer on the train moves relative to the observer on the ground in the finite time it takes for the light signals to converge. The observers disagree about the simultaneity of the events because of their relative motion.

Who is "really" right, the observer on the ground or the observer on the train? It would not be surprising if you answer by saying that the ground-based observer is right and the train-based observer is suffering from some sort of optical illusion. But this is not so. Both observers have made valid measurements and have given them valid interpretations. The observer on the train is completely correct in saying that since he measures the events not to be simultaneous, they *are not* simultaneous. And the observer on the ground is just as correct in saying that they *are* simultaneous because she measures them to be so. Measurement gives the true reality in physics.

If you nevertheless feel that measurements made in the reference frame fixed to the ground have more significance than those made in the frame fixed to the train moving at constant velocity relative to the ground, you should remember that both are equally good approximations to inertial reference frames. Then remember that experiment shows that "All inertial reference frames are completely equivalent. . . ." The fact that you live on the ground gives you a natural bias toward the frame fixed to the ground. If you have ever taken a long train trip, you will realize that someone living permanently on a train moving at constant velocity over the ground would have the same bias toward the frame fixed to the train.

Another reason why it may be difficult to look at the experiment discussed above with an unbiased attitude is that the blasting caps, and the observer who caused them to be detonated, were on the ground. As a result, the explosions were simultaneous as seen from the frame of reference of the ground. You will find it instructive to repeat the analysis for a variation of the experiment in which the blasting caps were carried on the train and their detonation was initiated by the observer in the train, so that the explosions occurred simultaneously in the frame of reference of the train. It is also worthwhile contemplating an experiment in which two identical spaceships pass each other going in opposite directions. Identical-twin observers, one on each ship, both initiate simultaneity experiments. Each will conclude that the events initiated by the other were not simultaneous. Here there can be no bias in your reaction to the results of the measurements because the situation is completely symmetrical with regard to the two reference frames.

14-5 TIME DILATION AND LENGTH CONTRACTION

We continue our study of time and space by analyzing additional thought experiments. The experiments are designed to produce three quantitative results. One gives the relation between time intervals measured by observers moving relative to each other. It will be shown that there is a disagreement between the observer stationed in the reference frame in which clocks are distributed and the observer moving relative to that reference frame concerning the rate at which the clocks run (as well as their synchronization). The other two experiments give the relations between space intervals measured by observers in relative motion. They will show that the observers also disagree about the separation between the clocks in the direction of the observers relative motion. These results, along with the results of the simultaneity thought experiment considered in Sec. 14-4, are used in Sec. 14-6 as a starting point in an argument that leads to the transformation equations which in the relativistic domain replace the Galilean transformation equations.

Figure 14-9 shows an observer O in an inertial frame who is comparing a time interval measured on her clock C with a measurement made by an observer O' of the same time interval. Observer O' is in another inertial frame moving to the left at constant velocity relative to O, and so are his two clocks C'_1 and C'_2. We take the positive direction to the right and thus write the velocity of O' relative to O as $V = -|V|$, where $|V|$ is the speed of O' with respect to O. When the observers were not in motion, prior to the experiment, they verified that all their clocks were synchronized and ran at the same rate when at rest with respect to each other. When the observers are in relative motion, there is no complication at all in comparing the reading of the clock belonging to O with either of the clocks belonging to O' at the instant that the O and O' clocks pass by each other. If two clocks are at essentially the same location, any observer at that location can make a valid comparison of their readings, despite the relative motion, since there will be no appreciable time delay in receiving information from either clock. In the experiment, O sends a flash of light from C along a path which is perpendicular to a distant mirror. The light signal is reflected back on that path and subsequently returns to C. The beginning of the time interval to be compared is defined by the emission of the light signal at C, and the end of the time interval is defined by its reception at C. Observer O measures the duration of the time interval by the difference between the two readings of C.

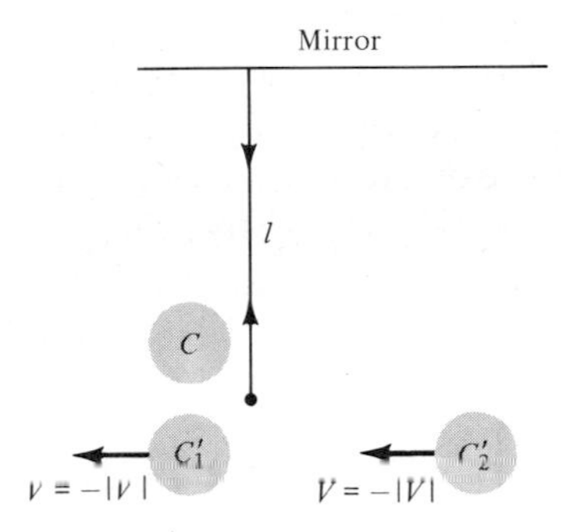

Fig. 14-9 A thought experiment concerning time dilation and length contraction, from the point of view of observer O.

The clock C'_1 belonging to O' is adjacent to the clock C when the light signal is emitted. This situation is illustrated from the point of view of O in Fig. 14-9. Because O' is moving to the left with respect to O, the clocks C'_1 and C'_2 move to the left while the light signal is traveling to the mirror and back. And at the instant when the light signal returns to C, the clock C'_2 has moved to the location adjacent to C. Thus observer O' can measure the beginning of the time interval with clock C'_1 and the end of the time interval with clock C'_2, since both clocks are in the right place at the right time.

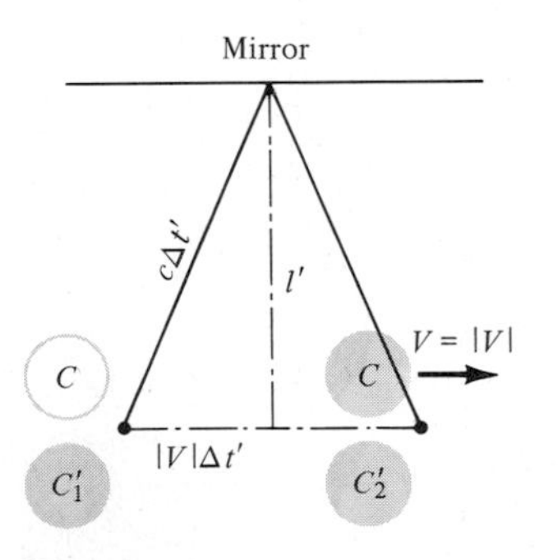

Fig. 14-10 The thought experiment of Fig. 14-9, from the point of view of observer O'.

Figure 14-10 illustrates the experiment, from the point of view of O', when the light signal returns to the permanent location of his clock C'_2 and the instantaneous location of the clock C belonging to O. Since O' is moving to the left relative to O with speed $|V|$, observer O must be moving to the right relative to O' with the same speed. If this were not the case, there would be an asymmetry between the reference frames of the two observers that is not allowed by Einstein's postulate of the equivalence of inertial frames. Thus the velocity of O relative to O' is $V = |V|$. Observer O' would describe the experiment by saying that O emitted a flash of light in the general direction of the mirror when her moving clock C was next to C'_1 and that the light returned to C when it was next to C'_2. Note the fundamentally important fact that according to O' the light does not follow the shortest path to the mirror and back, since it is not moving perpendicular to the mirror.

According to O, the duration of the time interval is $T = 2\Delta t$, where Δt is the time required for the light signal to travel to the mirror, or to travel back. Figure 14-9 shows that if l is the perpendicular distance to the mirror as measured by O and c is the speed of light, then

$$\Delta t = \frac{l}{c} \tag{14-2}$$

According to O', the time interval has a duration $T' = 2\Delta t'$, where $\Delta t'$ is the time for light to make the trip to the mirror, or the trip back. In Fig. 14-10, the quantity l' is the perpendicular distance to the mirror as measured by O'. The quantity $|V|\ \Delta t'$ is the distance traveled by clock C moving at speed $|V|$ in time $\Delta t'$. And the quantity $c\ \Delta t'$ is the distance traveled by the light signal moving at speed c in that time. From the figure and the pythagorean theorem, it is evident that

$$(c\ \Delta t')^2 = (|V|\ \Delta t')^2 + l'^2$$

To solve this equation for $\Delta t'$, we first write it as

$$c^2(\Delta t')^2 = |V|^2(\Delta t')^2 + l'^2$$

Now $|V|^2 = V^2$ because V^2 has a positive value no matter whether the value of the velocity V is positive or negative. Thus we have

$$c^2(\Delta t')^2 = V^2(\Delta t')^2 + l'^2$$

Then we gather the coefficients of $(\Delta t')^2$ to obtain

$$(c^2 - V^2)(\Delta t')^2 = l'^2$$

or

$$(\Delta t')^2 = \frac{l'^2}{c^2 - V^2} = \frac{l'^2}{c^2}\frac{1}{1 - V^2/c^2}$$

Taking square roots produces the result

$$\Delta t' = \frac{l'}{c}\frac{1}{\sqrt{1 - V^2/c^2}} \tag{14-3}$$

Two observers in relative motion cannot be in disagreement about distances measured *perpendicular* to the direction of motion. It is easy to prove this by the argument illustrated in Fig. 14-11 and explained in its caption. Thus the perpendicular distances to the mirror must be measured by both observers to have the same value $l = l'$. So Eq. (14-3) can be written

$$\Delta t' = \frac{l}{c}\frac{1}{\sqrt{1 - V^2/c^2}}$$

Substituting from Eq. (14-2), we obtain

$$\Delta t' = \Delta t\frac{1}{\sqrt{1 - V^2/c^2}}$$

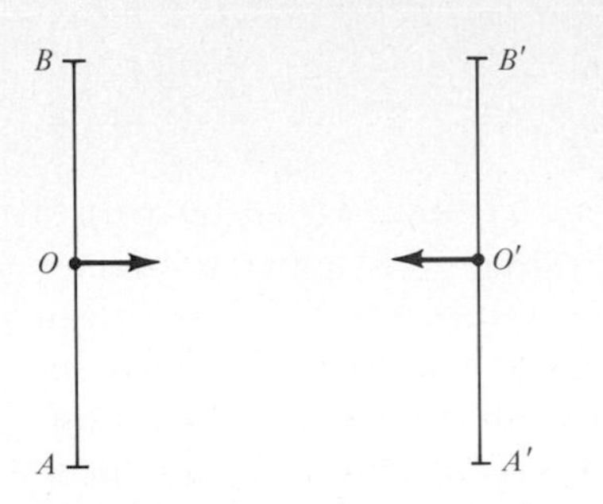

Fig. 14-11 A thought experiment proving that observers in different inertial frames are in agreement about lengths measured *perpendicular* to their direction of relative motion. Two rods AB and $A'B'$ are first measured to be of the same length when at rest with respect to each other. Then they are made to move by each other, in the manner illustrated. When they pass, observers at A and B mark on their rod the locations of the ends of rod $A'B'$ and also send light signals toward the center of the rods. Observers on the other rod perform similar operations. Since both observers at O and O' receive the signals at the same time, they both agree that the marks were made simultaneously, and so both accept the length comparisons. Now the procedure carried out in each reference frame is symmetrical to that carried out in the other. And the postulate of equivalence of inertial frames says that the properties of the reference frames themselves are symmetrical. Thus there is complete symmetry in the situation. Consequently the length comparisons must show that the rods have the same lengths when moving perpendicular to their orientation. This is the only result that could be agreed on by both observers and also be consistent with the complete symmetry.

Then multiplying through by the factor 2 produces the relation

$$T' = \frac{1}{\sqrt{1 - V^2/c^2}} T \tag{14-4}$$

Since the factor $1/\sqrt{1 - V^2/c^2}$ has a value greater than 1 if V^2/c^2 is greater than 0, this relation shows that O' measures the value T' for the time interval to be *longer* than the value T measured by O. The phenomenon is called **time dilation** (the word "dilation" means stretching). Recall that the time interval is defined by two events (the emission and reception of a light signal) which occur at the *same place* (C) from the point of view of O. She measures the shortest possible value T for the time interval, which is called the **proper time.** From the point of view of O', the events initiating and terminating the time interval occur at *different places* (C_1' and C_2'). He necessarily measures a longer value T' for the time interval, called the **dilated time.**

The reason for the time dilation found in the thought experiment is simply that the light connecting the mirror with the two events travels a *longer path* when the events are viewed from a frame of reference in which they occur at different locations. Since light travels at the *same speed* with respect to observers in both reference frames, it takes the light signal *more time* to follow its longer path in the frame of observer O'. So observer O' obtains a dilated value for his measurement of the time interval.

Now we develop a relation involving length measurements made by the two observers *parallel* to their direction of relative motion. We do this by reconsidering the same apparatus pictured in Figs. 14-9 and 14-10, except that a rigid rod extending from clock C_1' to clock C_2' has been fixed in the reference frame of O'. We use the symbol L' to represent the length of the rod as measured by O' and the symbol L to represent its length as measured by O. (Note that the conclusion obtained from the thought experiment in Fig. 14-11 does not apply here because O and O' are not comparing lengths measured perpendicular to their direction of relative motion.)

According to O, the quantity T is the time interval from when she observes the front end of the rod pass C to when she observes the rear end pass that location. See Fig. 14-9. Thus in time T the rod moves past her through its own length L. Since the rod is moving in her reference frame with speed $|V|$, its length L must be

$$L = |V|T \tag{14-5}$$

According to O', the quantity T' is the time interval during which the clock C moves the length L' of the rod. See Fig. 14-10. Since this clock is moving through his reference frame with a speed that also has the value $|V|$, the length L' is given by

$$L' = |V|T' \tag{14-6}$$

Dividing this into Eq. (14-5) gives

$$\frac{L}{L'} = \frac{T}{T'}$$

Then using the relation obtained from the time dilation argument, Eq. (14-4), to evaluate T/T', we have

$$\frac{L}{L'} = \sqrt{1 - V^2/c^2}$$

or

$$L = \sqrt{1 - V^2/c^2}\, L' \tag{14-7}$$

This result shows that observer O will measure the length L of the rod to be *shorter* than the length L' measured by observer O', since the factor $\sqrt{1 - V^2/c^2}$ has a value less than 1 if V^2/c^2 is greater than 0. The name given to this phenomenon is **length contraction.** Note that from the viewpoint of O' the length is measured over a space interval between *fixed* locations (along the fixed rod). He measures the largest possible value L' for the space interval, called its **proper length.** From the viewpoint of O the space interval is measured between *moving* locations (along the moving rod). She measures a shorter length, or **contracted length** L.

It is apparent from our argument that length contraction and time dilation are intimately related physical phenomena. The necessity for length contraction follows immediately from the necessity for time dilation, given the requirement that the speed of O' with respect to O equals the speed of O with respect to O', since all inertial frames are equivalent.

The phenomena of time dilation and length contraction predicted from the thought experiment that we have considered are not specific properties of this particular thought experiment. Rather, they are *general properties of time and space.* This is indicated by the fact that a variety of quite different thought experiments make the same predictions concerning time dilation and length contraction. We consider one of these in Sec. 14-6. Furthermore, there are many different real experiments which confirm these predictions. Several of these are discussed later in this section.

It should be pointed out that the two events occurring in the thought experiment considered here (the emission and reception of the light signal) are separated by a proper time in the O frame and separated by a proper length in the O' frame. So evaluating the dilated time defined by the events is a matter of calculating T' from T according to Eq. (14-4), and evaluating the contracted length involves calculating L from L' according to Eq. (14-7). One calculation goes from unprimed to primed quantities, and the other is reversed in that it goes from primed to unprimed quantities. The physical situation causes the equations to be used in reverse ways. But mathematically Eqs. (14-4) and (14-7) are completely analogous since the identical factor $1/\sqrt{1 - V^2/c^2}$ relates T' to T and L' to L.

The *result* given by Eq. (14-7) has the same mathematical form as an *assumption* made by G. F. Fitzgerald and later by H. A. Lorentz, two physicists who were attempting to explain the results of the Michelson-Morley experiment. Equation (14-7) is therefore often said to describe the **Fitzgerald-Lorentz contraction,** or the **Lorentz contraction.** The logical framework used by Fitzgerald and by Lorentz turned out to be inconsistent, even though some of their results are mathematically identical to those of Einstein. In contrast, the role of Eq. (14-7) in Einstein's theory is certainly not that of an assumption. It is an inevitable consequence of the postulate of the equivalence of inertial frames.

EXAMPLE 14-2

When a railroad train is stationary on the ground, it is measured to have a proper length of precisely 1 km. Predict its contracted length, measured from the ground, when its speed relative to the ground is 100 km/h.

■ You can make the prediction by evaluating Eq. (14-7),

$$L = \sqrt{1 - V^2/c^2}\, L'$$

The speed of the train is

$$|V| = \frac{100 \times 10^3 \text{ m}}{3.6 \times 10^3 \text{ s}} = 27.8 \text{ m/s}$$

The speed of light is

$$c = 3.00 \times 10^8 \text{ m/s}$$

So

$$\frac{|V|}{c} = \frac{27.8 \text{ m/s}}{3.00 \text{ m/s}} = 9.27 \times 10^{-8}$$

and

$$\frac{V^2}{c^2} = 8.59 \times 10^{-15}$$

Since $V^2/c^2 \ll 1$, the square root can be evaluated very accurately by using the binomial expansion approximation, Eq. (3-36), to give

$$\sqrt{1 - V^2/c^2} = 1 - \tfrac{1}{2}V^2/c^2$$

Thus you have

$$\sqrt{1 - V^2/c^2} = 1 - 4.30 \times 10^{-15}$$

Since the proper length of the train is

$$L' = 10^3 \text{ m}$$

its contracted length is

$$L = (1 - 4.30 \times 10^{-15})\ 10^3 \text{ m}$$

or

$$L = 10^3 \text{ m} - 4.30 \times 10^{-12} \text{ m}$$

The relativistic contraction in the length of the train is predicted to be 4.30×10^{-12} m. This is about 1 percent of the diameter of an atom. By the standards of the newtonian domain, the train is traveling at an appreciable speed and has an appreciable length. But the predicted total contraction is so small as to be completely unmeasurable.

Example 14-2 provides a typical demonstration of the fact that relativistic kinematics makes no predictions in disagreement with experiments carried out in the newtonian domain. In that domain objects move at speeds which are very small compared to the speed of light. Hence V^2/c^2 is extremely small compared to 1, so that $\sqrt{1 - V^2/c^2}$ and its reciprocal have values extremely close to 1. As a consequence, the predictions of relativistic kinematics merge smoothly into those of newtonian kinematics at speeds characteristic of the newtonian domain. But for objects moving at speeds comparable to the speed of light, relativistic kinematics predicts effects which are significant numerically and so should be measurable.

One of the first experimental confirmations of time dilation and length contraction was obtained in the 1940s from the study of **cosmic radiation.** A stream of very high-energy protons from cosmic sources is known to bombard the earth constantly. When these particles strike atomic nuclei in air at the top of the atmosphere, typically at an altitude of about 10,000 m, they produce particles called **pions.** The pions are unstable, decaying very quickly into particles called **muons.** Muons are also unstable. They soon decay into electrons, but this second decay is not as rapid as the first. Experiments have shown that a muon will live for an average time of 2.2×10^{-6} s, as seen in a reference frame in which the muon is at rest.

From the earth reference frame, the muons are moving downward at a speed about 99.9 percent of the speed of light. This is a result of the large downward momentum of the initially incident protons and is known from direct measurements of the muon momentum. If the "lifetime" of muons from birth to decay were 2.2×10^{-6} s, the distance the muons could travel when going at a speed of 3.0×10^8 m/s would be $3.0 \times 10^8 \text{ m/s} \times 2.2 \times 10^{-6} \text{ s} = 660$ m. The distance is short compared to the 10,000 m the muons must cover in order to reach ground level. It would seem that muons should not make it to ground level, and so none should be detected there. But muons are detected in abundance at ground level.

An explanation is found in the phenomenon of time dilation. The time 2.2×10^{-6} s is the average time a muon lives, measured from a frame of reference moving with the particle. (In other words, this lifetime is measured in a frame relative to which the muons are at rest.) The time is a proper time since it is a time interval measured between two events (the birth and death of the muon) which occur at the same place. To determine whether a typical muon can travel a distance of 10,000 m from the top of the atmosphere to the surface of the earth, the calculation should employ the dilated time evaluated from the frame of reference of the earth, since this is the frame from which the process is viewed. The calculation in Example 14-3 does so.

EXAMPLE 14-3

Evaluate the dilated time a muon lives before decaying when taveling at speed equal to $0.999c$ if the proper time it lives is 2.2×10^{-6} s. Then compare it to the time required to travel 10,000 m moving at that speed.

▪ If the proper time is T, then Eq. (14-4) says the dilated time will be

$$T' = \frac{1}{\sqrt{1 - V^2/c^2}} T$$

Here

$$|V| = 0.999c$$

or

$$\frac{|V|}{c} = 0.999$$

Squaring both sides, you obtain

$$\frac{V^2}{c^2} = 0.998$$

So

$$\sqrt{1 - V^2/c^2} = \sqrt{0.002} = 0.045$$

and

$$\frac{1}{\sqrt{1 - V^2/c^2}} = \frac{1}{0.045} = 22$$

Thus the dilated time is

$$T' = 22 \times 2.2 \times 10^{-6}\ \text{s} = 49 \times 10^{-6}\ \text{s}$$

The time t' required to travel 10,000 m going at 3.0×10^8 m/s is

$$t' = \frac{1.0 \times 10^4\ \text{m}}{3.0 \times 10^8\ \text{m/s}} = 33 \times 10^{-6}\ \text{s}$$

Since this is less than the available time $T' = 49 \times 10^{-6}$ s, there is no difficulty in explaining the fact that cosmic ray muons typically reach ground level before decaying.

Muons can be thought of as clocks that give one tick in a proper time equal to their average lifetime. Because they are microscopic, it is possible to build machines, like high-energy particle accelerators, to make them move at speeds high enough for relativistic effects to become large. And nature has provided such an accelerator in the form of the cosmic radiation. So muons are ideally suited to test time dilation. The test is completely successful. Time dilation provides a *qualitative explanation* of the seemingly impossible fact that any cosmic ray muons at all reach the ground. Furthermore, careful measurements of the exact fraction reaching the ground are in *quantitative agreement* with the expected dilation factor. As predicted by relativistic kinematics, the proper time interval defined by the rapidly moving muon clocks is observed in the reference frame of the earth to be a dilated time interval. Another way of stating this is to say that the *moving clocks run slow*.

In 1971 the prediction that moving clocks run slow was tested in an experiment using macroscopic clocks. Four extremely accurate cesium-133 clocks (see Sec. 2-3) were taken by commercial airlines on trips around the world. Two clocks traveled eastward in the same direction as the earth's rotation, and two clocks traveled westward in opposition to its rotation. Before and after their circumnavigations, the readings of the clocks were compared with the readings of reference clocks at the U.S. Naval Observatory. Prior to analyzing the results for the effect of time dilation, corrections had to be made for the fact that neither the flying clocks nor the ground-based clocks were in completely acceleration-free inertial reference frames. This was done by using Einstein's general theory of relativity, which treats relative accelerations. The general theory was also used to correct for the effects on the clocks, closely related to those of acceleration, produced by the earth's gravity. The results of the test were quite satisfactory. Time dilation was expected to cause a total time loss of about 3×10^{-7} s for each clock moving eastward. The corrected measured value agreed with expectation to within the experimental accuracy of about 0.2×10^{-7} s.

Length contraction provides a complementary explanation of the observations concerning cosmic ray muons. In the reference frame where the muons are at rest, the average time available before decay is the proper time 2.2×10^{-6} s. This frame is moving downward through the atmosphere relative to the earth at speed $|V| = 0.999c$ along with the muons. Thus from the point of view of an observer in this frame, the earth and atmosphere are moving upward at the same speed. Consequently, in the muon reference frame the length measured from the point at the top of the atmosphere, which passes them when they are born, to the point at the bottom of the atmosphere, which passes when they strike the ground, will be very much contracted. In Example 14-4 the numbers are worked out.

EXAMPLE 14-4

Evaluate the contracted length of the atmosphere, of proper length 10,000 m, moving by the muons at speed $|V| = 0.999c$. Then show that it takes less than the average muon lifetime, 2.2×10^{-6} s, for this length to move once along itself.

■ The contracted length L is related to the proper length L' by Eq. (14-7):

$$L = \sqrt{1 - V^2/c^2}\, L'$$

You already have evaluated the contraction factor for $|V| = 0.999c$ in Example 14-3. It is

$$\sqrt{1 - V^2/c^2} = 0.045$$

So the contracted length has the value

$$L = 0.045 \times 10{,}000 \text{ m} = 450 \text{ m}$$

The time t required for the atmosphere to move through its contracted length past the muon, going at essentially the speed of light, is

$$t = \frac{4.5 \times 10^2 \text{ m}}{3.0 \times 10^8 \text{ m/s}} = 1.5 \times 10^{-6} \text{ s}$$

Since this is less than the available time $T = 2.2 \times 10^{-6}$ s, there is no difficulty in understanding why the muon lives long enough for the atmosphere to move past it.

The two explanations of the muon measurements are summarized in Fig. 14-12*a* and *b*.

14-6 THE LORENTZ POSITION-TIME TRANSFORMATION

By analyzing appropriately chosen thought experiments, we have obtained four essential features of relativistic kinematics: (1) Observers in relative motion disagree about measurement of simultaneity. (2) They disagree about measurement of time intervals. (3) They disagree about measurement of length intervals parallel to the direction of motion. (4) They agree about measurement of length intervals perpendicular to that direction. We have quantitative relations involving the last three features, but only a qualitative description of the first one. In this section we analyze a thought experiment that will provide a quantitative relation concerning the simultaneity disagreement. It will also yield independent derivations of the second and third feature. All four features are contained in a set of four equations called the *Lorentz position-time transformation.*

As a preliminary to developing the Lorentz transformation of relativistic kinematics, it is worthwhile reviewing the Galilean transformation of newtonian kinematics. Figure 14-13 shows two observers O and O', with the primed observer moving relative to the unprimed observer at a constant velocity. The two observers have constructed mutually parallel sets of axes x, y, z and x', y', z' with origins at their own locations. For simplicity, they have aligned the x and x' axes parallel to the direction of their relative velocity. We continue to use the symbol V as a signed scalar to represent the velocity of O' with respect to O. Its numerical value is positive if O' moves

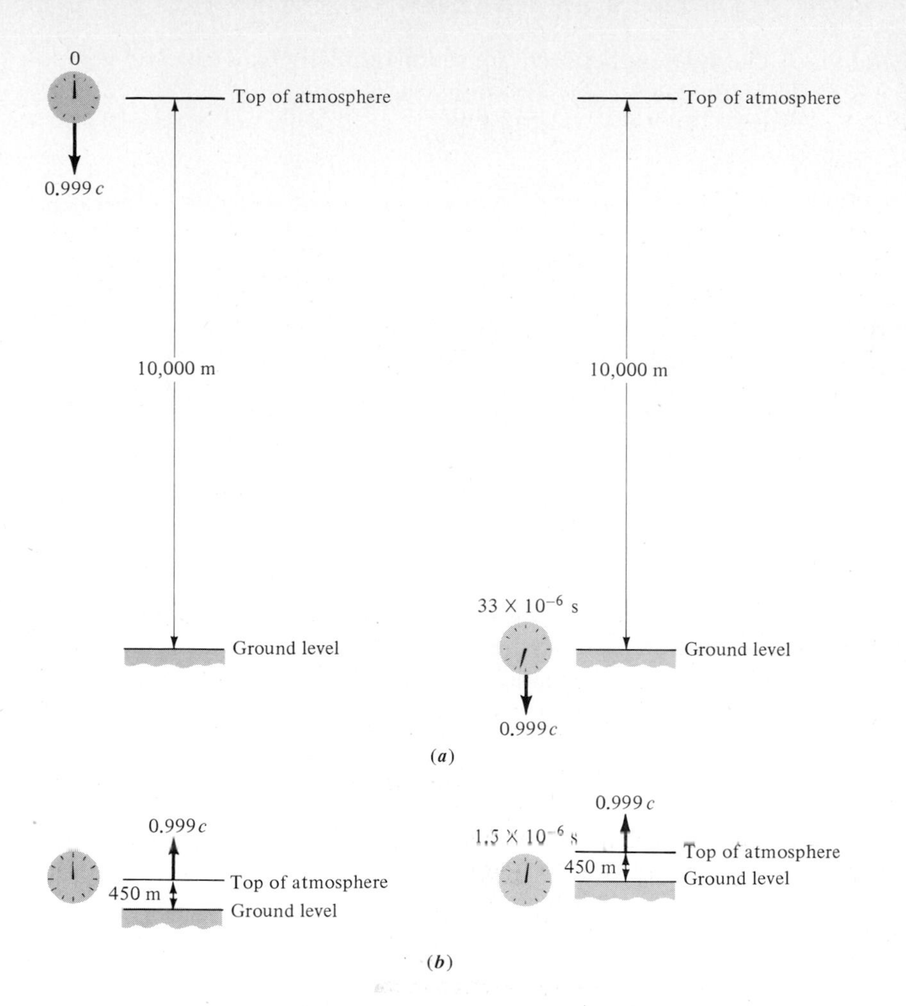

Fig. 14-12 (*a*) Measurement of the decay of cosmic ray muons from the point of view of an observer on the ground. (*b*) Measurement of the decay of cosmic ray muons from the point of view of an observer on a muon.

with respect to O in the positive direction of the x and x' axes, as in the case illustrated in Fig. 14-13. If O' moves with respect to O in the negative direction of the x and x' axes, the numerical value of V is negative.

Observer O specifies that an event occurs at a point P and at a certain time in terms of the values she measures for its coordinates x, y, z and the time t. Similarly, O' specifies the same event by stating the values x', y', z' which he measures for the coordinates and the value t' which he measures

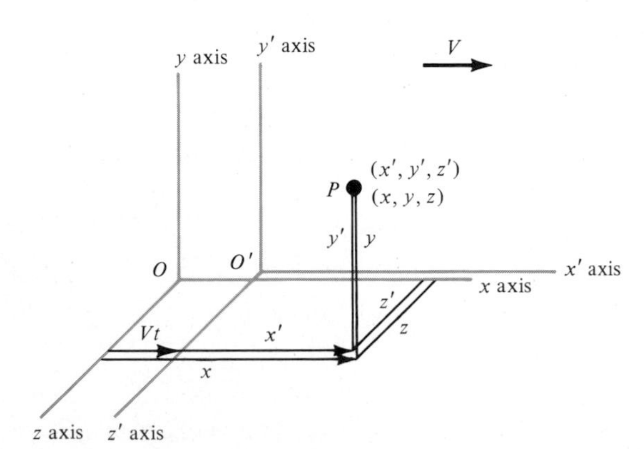

Fig. 14-13 Two frames of reference in uniform relative motion. The constant velocity of the primed frame with respect to the unprimed frame is V.

for the time. The relation between the set of numbers (x, y, z, t) and (x', y', z', t') is given by the *Galilean position-time transformation*

$$
\begin{aligned}
x' &= x - Vt \\
y' &= y \\
z' &= z \\
t' &= t
\end{aligned}
\tag{14-8}
$$

where O and O' measure t and t' from the instant they pass each other.

The first three of Eqs. (14-8) are exactly what would be obtained by taking components of both sides of Eq. (3-56), with the velocity of observer O' with respect to observer O directed along the x and x' axes. The fourth of Eqs. (14-8) is an implied, but unwritten, condition on Eq. (3-56). According to newtonian kinematics, the equation $t' = t$ is deemed to be so self-evident as not to warrant explicit mention. But here we have added it to the other three to emphasize that newtonian kinematics is based on the *assumption* that there is a universal time scale which is the same at all locations in all reference frames, independent of their state of motion. Because we now believe this assumption to be incorrect, we must investigate the correctness of the transformation equations to which it leads.

This will be done by arguing through another thought experiment. The argument will make tentative use of some ideas arising from earlier thought experiments. But you will see that its results provide complete justification of these ideas. Actually, the argument assumes only the validity of Einstein's postulate of the equivalence of inertial frames. This is because it is based directly and solely on one experimental fact which agrees with the postulate. This is the fact that the speed of light is independent of the motion of the observer and the motion of the source.

Two observers in inertial frames move past each other, with the constant velocity V of O' relative to O directed along the positive x and x' directions of the mutually parallel axes of their coordinate systems. Each observer is located at the origin of the observer's system. When they pass each other, they celebrate the event by igniting a flashbulb at the temporarily coincident origins of the coordinate systems. They also both set their clocks to zero, so that the two observers will measure time from the instant of ignition of the flashbulb.

The light produced by the flashbulb expands outward from its point of emission in all directions at speed c. According to O', at time t' the light is on a spherical shell of radius $r' = ct'$, centered on the origin of his coordinate system. Thus he finds the coordinates of a typical point on the shell to be related by the equation of a sphere of radius $r' = ct'$. The equation is

$$x'^2 + y'^2 + z'^2 = r'^2$$

or

$$x'^2 + y'^2 + z'^2 = c^2t'^2 \tag{14-9}$$

And according to O the light is on a spherical shell of radius $r = ct$ at time t, centered on the origin of her coordinate system. So she finds the coordinates of a point on the shell at that time to satisfy the equation of the sphere

$$x^2 + y^2 + z^2 = r^2$$

or

$$x^2 + y^2 + z^2 = c^2t^2 \tag{14-10}$$

Thus *both* O and O' find themselves to be at the *center* of an expanding spherical shell of light, as Fig. 14-14 illustrates. Is this possible? Yes, providing the relation between the sets of numbers (x', y', z', t') and (x, y, z, t) is such that both Eqs. (14-9) and (14-10) are satisfied. We will find a relation which satisfies them both by finding a set of equations which transforms Eq. (14-9) into Eq. (14-10). This set of equations will constitute the Lorentz position-time transformation.

To be specific, what we will do is let the qualitative ideas we have come across in studying the earlier thought experiments suggest how we might obtain the Lorentz position-time transformation by modifying the Galilean position-time transformation. Because of considerations explained below, it seems reasonable to try the following modified form for the transformation equations:

$$\begin{aligned} x' &= \gamma(x - Vt) \\ y' &= y \\ z' &= z \\ t' &= \gamma(t + \delta) \end{aligned} \tag{14-11}$$

Here γ is a dimensionless quantity, and δ is a quantity that must have the dimensions of time. If we are successful in making Eqs. (14-11) satisfy the requirement that they transform Eq. (14-9) into Eq. (14-10), it will be done by finding the necessary expressions for γ and δ. We will soon do just this. But even now we can say that γ and δ must involve the velocity V of relative motion of the two reference frames, and also the speed c of light, in such a way that γ should approach 1 and δ should approach 0 when $|V|$ becomes very small compared to c. This must be so in order that Eqs. (14-11) reduce to the Galilean transformation of Eqs. (14-8) in those circumstances. After all, we know from experiment that the Galilean transformation leads to correct predictions when the relative speed of the two reference frames is negligible compared to the speed of the light.

We also know that when the speed of relative motion is not small compared to the speed of light, there is a disagreement between the two reference frames concerning the synchronization of clocks, which arises from the disagreement about simultaneity. We have tried to allow for it by inserting the additive term δ in the fourth of Eqs. (14-11). This term is supposed to take into account the contention of O' that the clock used by O at

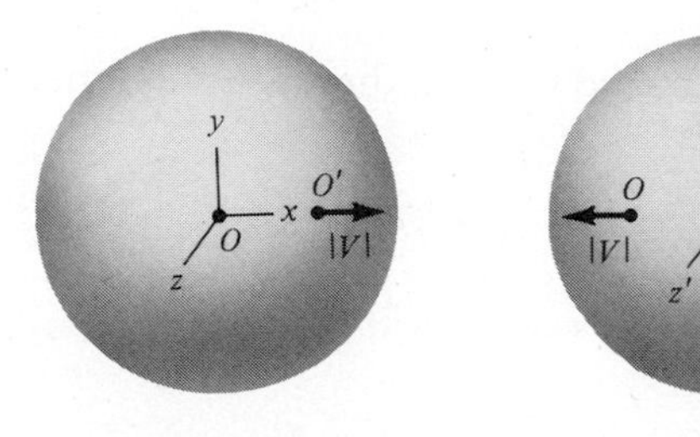

Fig. 14-14 (*a*) From the point of view of O, the light produced by a flashbulb is on an expanding spherical shell centered at the O origin of coordinates, and O' is moving to the right at speed $|V|$. (*b*) From the point of view of O', the light is on an expanding spherical shell centered at the O' origin of coordinates, and O is moving to the left at speed $|V|$.

the location of an event to time it is not synchronized to give the same reading as the clock located at the origin of the coordinate system of O. Having considered the synchronization disagreement, we have then attempted to take into account the disagreement between the two observers concerning the rates at which clocks located at their respective origins measure time intervals. This is done by inserting the multiplicative factor γ in the fourth equation. As discussed immediately before Example 14-2, the identical multiplicative factor should appear in the equation relating the unprimed and primed values of the coordinate that is measured along the direction of relative motion. So we have inserted the γ in the first transformation equation. Because the observers do not disagree about lengths measured perpendicular to the direction of motion, we have made no modification of the two equations relating the coordinates extending in the perpendicular directions.

Now let us confirm or deny the guesses we have made about the form of the Lorentz position-time transformation. We do this by seeing whether Eqs. (14-11) actually can transform Eq. (14-9) into Eq. (14-10), providing that δ and γ have the necessary forms. If we find the transformation works, then we will also find what these forms are. First, we use Eqs. (14-11) to write each of the primed variables of Eq. (14-9) in terms of the unprimed variables. If we make the substitutions, Eq. (14-9), $x'^2 + y'^2 + z'^2 = c^2t'^2$, becomes

$$\gamma^2(x^2 - 2Vxt + V^2t^2) + y^2 + z^2 = c^2\gamma^2(t^2 + 2\delta t + \delta^2) \qquad (14\text{-}12)$$

Our task is to obtain Eq. (14-10), $x^2 + y^2 + z^2 = c^2t^2$, from Eq. (14-12). Since the former has no terms in it that involve the product xt, there must be a cancellation of the second term in parentheses on the left side of Eq. (14-12) by some term, or terms, appearing on its right side. And since the term on the left side is proportional to the first power of t, the canceling terms on the right side must be also. Otherwise, the cancellation could not hold for all values of t. The only term on the right side having this property is the second term in parentheses. Therefore we must have

$$-\gamma^2 2Vxt = c^2\gamma^2 2\delta t$$

or

$$\delta = -\frac{Vx}{c^2} \qquad (14\text{-}13)$$

Next we substitute this result into Eq. (14-12) and work out the now simplified expressions in parentheses, producing directly

$$\gamma^2x^2 + \gamma^2V^2t^2 + y^2 + z^2 = c^2\gamma^2t^2 + \frac{\gamma^2V^2x^2}{c^2}$$

If we group the factors of x^2 on the left side and of t^2 on the right side, this becomes

$$\gamma^2(1 - V^2/c^2)x^2 + y^2 + z^2 = c^2\gamma^2(1 - V^2/c^2)t^2 \qquad (14\text{-}14)$$

Success! We see that Eq. (14-10), $x^2 + y^2 + z^2 = c^2t^2$, will be obtained immediately from Eq. (14-14), provided that

$$\gamma^2(1 - V^2/c^2) = 1$$

or

$$\gamma = \frac{1}{\sqrt{1 - V^2/c^2}} \tag{14-15}$$

Using the expressions given in Eqs. (14-13) and (14-15) for δ and γ in the form given by Eq. (14-11), we obtain the **Lorentz position-time transformation:**

$$\begin{aligned} x' &= \frac{1}{\sqrt{1 - V^2/c^2}}(x - Vt) \\ y' &= y \\ z' &= z \\ t' &= \frac{1}{\sqrt{1 - V^2/c^2}}(t - Vx/c^2) \end{aligned} \tag{14-16}$$

This transformation is given its name because H. A. Lorentz had obtained equations of the same mathematical form in 1895, while making an unsuccessful attempt to introduce consistency into the theory of moving electric charges.

Note that $\delta = -Vx/c^2$ does have the dimensions of time and does approach 0 when V/c approaches 0, as predicted initially. You should explain to yourself, in terms of the simultaneity thought experiment of Sec. 14-4, the physical reason why the quantity δ is proportional to the first powers of both V and x. The quantity V in the Lorentz transformation has the same significance as the V in the Galilean transformation of Eqs. (14-8). Its magnitude is the speed of one inertial reference frame with respect to the other. Its sign is positive if, as assumed in Figs. 14-13 and 14-14, O' moves in the positive x and x' directions relative to O. In such a case, the sign of the quantity $\delta = -Vx/c^2$ is opposite to the sign of the coordinate x. If the direction of relative motion is reversed, so that O' moves in the negative x and x' directions relative to O, the sign of V is negative. In these circumstances the sign of δ is the same as the sign of x. Use the thought experiment involving the train to explain to yourself the physical significance of the sign of δ and of its dependence on the signs of V and x.

Also note that the predictions initially made about $\gamma = 1/\sqrt{1 - V^2/c^2}$ have been borne out. It is dimensionless and approaches 1 as V/c approaches 0. Use the time dilation-length contraction thought experiment of Sec. 14-5 to explain physically the dependence of γ on V. In doing this, also give a physical reason why γ does not depend on the sign of V, as is expressed mathematically by the fact that V enters as a squared quantity.

Figure 14-15 is a plot versus $|V|/c$ of the factor $\gamma = 1/\sqrt{1 - V^2/c^2}$ which appears in the Lorentz transformation. Its value departs very little from 1 until $|V|/c$ becomes larger than about 0.5. Then it begins to increase very rapidly, going to infinity when $|V|/c = 1$. For $|V|/c > 1$, the factor becomes an imaginary number. Thus the condition $|V| = c$ acts as a limit to the range of validity of the Lorentz transformation. We will repeatedly come across c playing the role of a limiting speed as we continue our development of the theory of relativity.

All the features of relativistic kinematics discussed earlier are contained in the Lorentz position-time transformation. Example 14-5 uses the transformation to yield one of these features, length contraction.

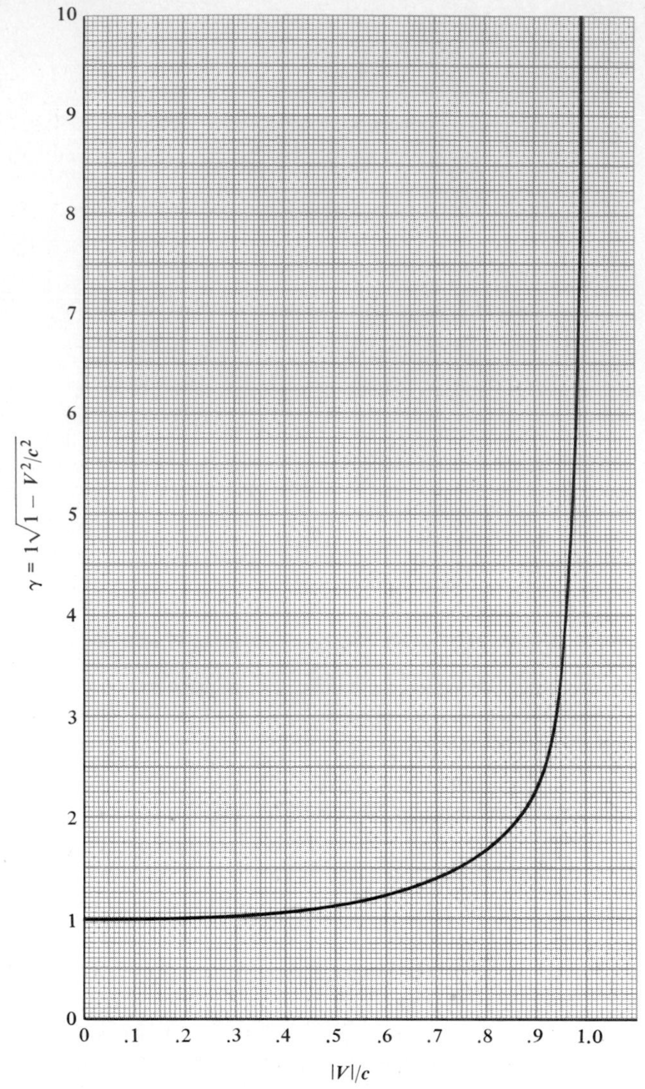

Fig. 14-15 A plot of the factor $\gamma = 1/\sqrt{1 - V^2/c^2}$ versus $|V|/c$.

EXAMPLE 14-5

Derive the length contraction formula from the Lorentz position-time transformation. To facilitate comparison with Eq. (14-7), make this derivation for a rod at rest in the primed reference frame.

■ Think of the rod lying along the x' axis of the primed frame, with one end always at coordinate x'_1 and the other end always at coordinate x'_2, where $x'_2 > x'_1$. Its proper length is

$$L' = x'_2 - x'_1$$

In the unprimed frame, which is moving relative to the primed one with velocity V, the ends of the rod have the changing coordinates x_1 and x_2. According to the first of Eqs. (14-16), the relations between the unprimed and primed coordinates are

$$x'_1 = \frac{x_1 - Vt_1}{\sqrt{1 - V^2/c^2}}$$

and

$$x'_2 = \frac{x_2 - Vt_2}{\sqrt{1 - V^2/c^2}}$$

So

$$x_2' - x_1' = \frac{x_2 - x_1 - V(t_2 - t_1)}{\sqrt{1 - V^2/c^2}}$$

Now the length of the rod moving through the unprimed frame is measured by the difference between the coordinates at which its endpoints lie simultaneously, as judged from that frame. In other words, the length is given by

$$L = x_2 - x_1$$

providing $t_2 = t_1$ so that the observer in the unprimed frame judges the measurements of x_2 and x_1 to be made simultaneously. Using this condition in the equation ending the last paragraph, you have

$$x_2' - x_1' = \frac{x_2 - x_1}{\sqrt{1 - V^2/c^2}}$$

or

$$L = \sqrt{1 - V^2/c^2}\, L'$$

The moving rod, of proper length L', is measured to have a contracted length L, and the contraction factor $\sqrt{1 - V^2/c^2}$ is in agreement with Eq. (14-7).

How would you use the Lorentz transformation to derive the time dilation formula?

Example 14-6 involves using the Lorentz position-time transformation to evaluate quantitatively the amount by which observers in relative motion disagree about simultaneity.

EXAMPLE 14-6

Starship Enterprise is overtaken by the new model, Starship Enterprise-prime, with E' passing E at a relative speed $|V| = c/2$. The captain of E salutes the captain of E' by blinking the bow and stern lights of E, simultaneously from the point of view of E. As measured by E, the distance between the lights is 100 m. By how much do the times of emission of the signals from the lights differ, as measured by E'?

■ In the reference frame of E the signals from the two lights are emitted at times t_1 and t_2 from locations x_1 and x_2. But the emission times are judged in the reference frame of E' to occur at times t_1' and t_2'. The last of Eqs. (14-16) shows you that

$$t_1' = \frac{t_1 - Vx_1/c^2}{\sqrt{1 - V^2/c^2}}$$

and

$$t_2' = \frac{t_2 - Vx_2/c^2}{\sqrt{1 - V^2/c^2}}$$

According to E', the difference between the emission times is

$$t_2' - t_1' = \frac{t_2 - t_1 - (V/c^2)(x_2 - x_1)}{\sqrt{1 - V^2/c^2}}$$

There is no difference in the times according to E, so $t_2 = t_1$. Thus the result simplifies to

$$t_2' - t_1' = -\frac{(V/c^2)(x_2 - x_1)}{\sqrt{1 - V^2/c^2}}$$

To aid interpretation, this is best written as

$$t_2' - t_1' = -\frac{1}{\sqrt{1 - V^2/c^2}} \frac{V(x_2 - x_1)}{c^2}$$

For numerical evaluation, you use $|V|/c = 0.500$ to calculate

$$\frac{1}{\sqrt{1 - V^2/c^2}} = 1.15$$

Then, choosing x_2 to be the coordinate of the bow light, with the forward direction of the ship being positive, you have $V = |V| = 0.500c$ and

$$\frac{V(x_2 - x_1)}{c^2} = \frac{0.500c\ (x_2 - x_1)}{c^2} = \frac{0.500 \times 100 \text{ m}}{3.00 \times 10^8 \text{ m/s}} = 1.67 \times 10^{-7} \text{ s}$$

So you obtain

$$t_2' - t_1' = -1.15 \times 1.67 \times 10^{-7} \text{ s} = -1.93 \times 10^{-7} \text{ s}$$

The minus sign means that the captain of E' judges the bow light of E to blink slightly earlier than the stern light. Explain from simple considerations why it is measured to be early. Also identify the origins of the two factors that combine to produce the measured time difference, $V(x_2 - x_1)/c^2$ and $1/\sqrt{1 - V^2/c^2}$.

14-7 THE LORENTZ VELOCITY TRANSFORMATION

Now we will use the Lorentz position-time transformation to obtain the transformation equations which show how to find the velocity of a particle as measured in a certain inertial frame, when given both its velocity as measured in some other inertial frame and the relative velocity of the two frames. This Lorentz velocity transformation will then be applied in two examples, which show how the speed of light c acts as a limiting speed.

In Fig. 14-16 the motion of a point P is viewed from two inertial reference frames x, y, z and x', y', z', which are in relative motion. The x and x' axes have been constructed on a common line, along which is the direction of relative motion. The constant velocity of the primed frame with respect to the unprimed frame is specified by the scalar V, whose sign is positive if this velocity is in the direction of the positive x and x' axes, as illustrated in the figure. The remaining axes of each frame have been constructed parallel to the corresponding axes of the other frame. In both frames of reference, time is measured from the instant at which the origins of the frames overlap. In the unprimed reference frame, the location of the point at a certain time is specified by the set of numbers (x, y, z, t). In the primed frame, the same information is given by the set (x', y', z', t').

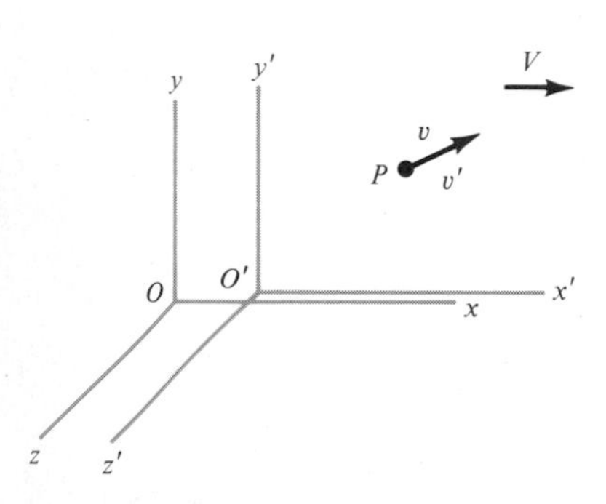

Fig. 14-16 A moving point P, as viewed from two inertial reference frames in relative motion. The constant velocity of the primed frame with respect to the unprimed frame is V. The speed of P measured from the unprimed frame is v, and its speed measured from the primed frame is v'.

Let us start by reviewing the velocity transformation of newtonian kinematics. A way to obtain this transformation is first to take the differentials of both sides of each of Eqs. (14-8), the Galilean transformation equations relating the two sets of numbers. Remembering that V is a constant while doing this, we obtain

$$dx' = dx - V\,dt$$

$$dy' = dy$$

$$dz' = dz$$

$$dt' = dt$$

Then the first three equations are each divided by the fourth, giving

$$\frac{dx'}{dt'} = \frac{dx}{dt} - V$$
$$\frac{dy'}{dt'} = \frac{dy}{dt} \tag{14-17}$$
$$\frac{dz'}{dt'} = \frac{dz}{dt}$$

These are the three components of a form of the velocity transformation equation we have used on a number occasions in newtonian kinematics. This is so since the three components of the velocity of P measured in the unprimed frame are

$$v_x = \frac{dx}{dt} \qquad v_y = \frac{dy}{dt} \qquad v_z = \frac{dz}{dt} \tag{14-18}$$

while, as measured in the primed frame, they are

$$v'_x = \frac{dx'}{dt'} \qquad v'_y = \frac{dy'}{dt'} \qquad v'_z = \frac{dz'}{dt'} \tag{14-19}$$

[Of course, in the newtonian domain no distinction is made between t' and t, so Eqs. (14-19) would conventionally be written with t replacing t' in each.] Thus Eqs. (14-17), (14-18), and (14-19) together give

$$v'_x = v_x - V$$
$$v'_y = v_y \tag{14-20}$$
$$v'_z = v_z$$

These are just the representation in components of the velocity transformation of Eq. (3-57) for a case in which the velocity of the primed frame with respect to the unprimed frame lies along the x and x' axes.

In relativistic kinematics the velocity transformation is obtained in a completely analogous manner, except that the Lorentz position-time transformation rather than the Galilean position-time transformation is used. First the differentials are taken of both sides of each equation of the Lorentz transformation, Eqs. (14-16), remembering that both V and c are constants. The results are

$$dx' = \frac{1}{\sqrt{1 - V^2/c^2}}(dx - V\,dt)$$
$$dy' = dy$$
$$dz' = dz$$
$$dt' = \frac{1}{\sqrt{1 - V^2/c^2}}(dt - V\,dx/c^2)$$

Then each of the first three of these equations is divided by the fourth, producing

$$\frac{dx'}{dt'} = \frac{\dfrac{1}{\sqrt{1 - V^2/c^2}}(dx - V\,dt)}{\dfrac{1}{\sqrt{1 - V^2/c^2}}(dt - V\,dx/c^2)} = \frac{\dfrac{dx}{dt} - V}{1 - \dfrac{V}{c^2}\dfrac{dx}{dt}}$$

$$\frac{dy'}{dt'} = \frac{dy}{\frac{1}{\sqrt{1 - V^2/c^2}}(dt - V\,dx/c^2)} = \frac{\frac{dy}{dt}}{\frac{1}{\sqrt{1 - V^2/c^2}}\left(1 - \frac{V}{c^2}\frac{dx}{dt}\right)}$$

$$\frac{dz'}{dt'} = \frac{dz}{\frac{1}{\sqrt{1 - V^2/c^2}}(dt - V\,dx/c^2)} = \frac{\frac{dz}{dt}}{\frac{1}{\sqrt{1 - V^2/c^2}}\left(1 - \frac{V}{c^2}\frac{dx}{dt}\right)}$$

Using Eqs. (14-18) and (14-19) to write these in terms of the velocity components, we have the **Lorentz velocity transformation:**

$$v'_x = \frac{v_x - V}{1 - Vv_x/c^2}$$

$$v'_y = \frac{\sqrt{1 - V^2/c^2}}{1 - Vv_x/c^2}\,v_y \qquad (14\text{-}21)$$

$$v'_z = \frac{\sqrt{1 - V^2/c^2}}{1 - Vv_x/c^2}\,v_z$$

These equations tell us how to transform a measured velocity from one inertial frame to another moving relative to the first—that is, how to use a velocity measured in one frame to predict the velocity that will be measured in the other.

Inspection will immediately show you that as $|V|/c$ approaches 0, the Lorentz velocity transformation equations approach Eqs. (14-20), the Galilean velocity transformation equations. Thus you see once more that relativistic predictions become indistinguishable from nonrelativistic predictions in the limit of speeds small compared to the speed of light. But the nonlinear Lorentz velocity transformation equations have properties very different from those of the Galilean velocity transformation for speeds comparable to the speed of light, as you will see in Examples 14-7 and 14-8.

EXAMPLE 14-7

Two particles are moving in opposite directions along the x axis of the x, y, z reference frame, as illustrated in Fig. 14-17a. The velocity of particle 1 is $v_1 = 0.90c$, and that of particle 2 is $v_2 = -0.80c$, relative to this frame. What is the velocity of particle 1 relative to particle 2?

■ Your newtonian intuition may tell you the answer is $1.70c$. But that intuition is wrong. The correct answer is obtained by using the first of Eqs. (14-21) to trans-

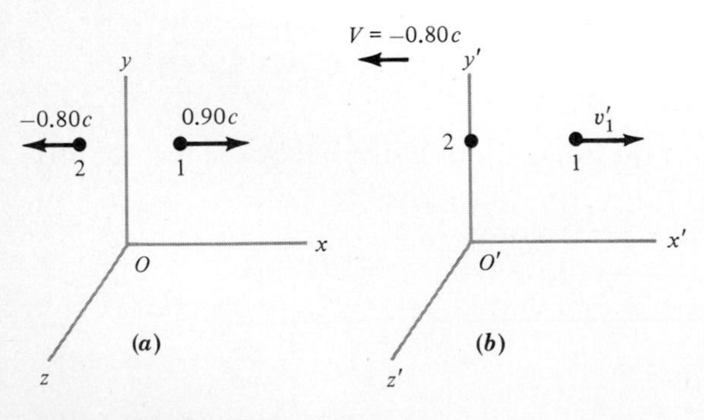

Fig. 14-17 (a) Particles 1 and 2 moving at high speeds in the opposite directions. (b) The situation as viewed in a primed reference frame in which particle 2 is stationary.

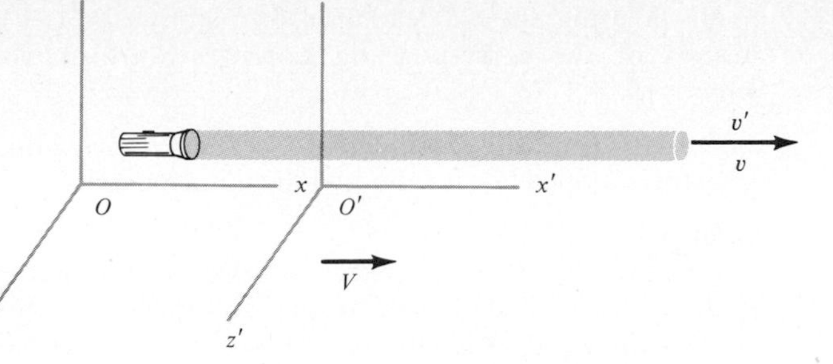

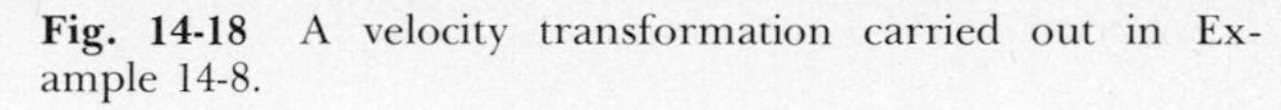
Fig. 14-18 A velocity transformation carried out in Example 14-8.

form the velocity of particle 1 to a reference frame moving along with particle 2, as in Fig. 14-17*b*. This is done by setting $V = v_2 = -0.80c$ and $v_x = v_1 = 0.90c$ in the transformation equation. You obtain

$$v'_x = \frac{0.90c - (-0.80c)}{1 - (-0.80c)(0.90c)/c^2} = \frac{1.70c}{1.72} = 0.99c$$

or, since $v'_x = v'_1$,

$$v'_1 = 0.99c$$

EXAMPLE 14-8

At the end of Sec. 14-2, we considered a thought experiment in which the speed of the same beam of light was being measured by two observers O and O'. Observer O' was moving relative to O in the same direction as the light beam and at half the speed of light. Yet Einstein asserts that both observers measure the same value c for the speed of the beam. Show that the Lorentz velocity transformation leads to an identical statement.

■ The experiment is illustrated in Fig. 14-18 from the point of view of observer O. The velocity to be transformed is that of the light beam, as measured by O, which is $v_x = c$. The velocity V of O' with respect to O is $V = c/2$. So the first of Eqs. (14-21) gives you

$$v'_x = \frac{c - c/2}{1 - (c/2)c/c^2} = \frac{c/2}{1 - 1/2} = \frac{c/2}{1/2}$$

or

$$v'_x = c$$

Thus both observers measure the same speed c for the beam of light.

Examples 14-7 and 14-8 show that a feature of the Lorentz velocity transformation equation for the x component of velocity is that relative velocities cannot be compounded in such a way as to produce a speed exceeding the speed of light. The y and z component equations have the same feature. In Chap. 15 Einstein's famous relation between mass and energy is obtained from arguments based on the Lorentz velocity transformation equation for the y component.

EXERCISES

Group A

14-1. *Ship time.* How fast must a space ship travel relative to the earth so that exactly 10 years of earth time correspond to exactly 1 year of space ship time?

14-2. *Strongly contracted.* A rectangle fixed in the O' system has one side 2.00 m long on the x' axis and one side 1.00 m long parallel to the y' axis. Observer O measures it as a square, 1.00 m on each side. What is the speed of O' relative to O along the common x axis?

14-3. *Muon watch.* The average lifetime of muons at rest is 2.2×10^{-6} s. A laboratory measurement made on muons produced at a high energy particle accelerator yields an average lifetime of 6.9×10^{-6} s. Determine the speed of the muons with respect to the laboratory.

14-4. *Blink-blink.* A spacecraft of length 10 m is moving at constant speed $|V| = 0.80c$ relative to an inertial frame O'. A light signal is sent from the front of the craft toward the tail.

a. In a frame O fixed to the spacecraft, how long does the light signal take to reach the tail?

b. How long does it take to reach the tail as viewed from the frame O'?

14-5. *A light in the distance.* When O' passes moving with respect to O at a speed $|V| = 0.80c$, they both set their clocks to zero time. One hour later according to his clock O' illuminates his dial, thus sending a light signal to O.

a. According to the clock of O, when was the signal sent?

b. According to O, how long does it take the signal to reach her? What does her clock register on the arrival of the signal?

c. According to O', how long does it take for the signal to reach O? What is the reading on his clock when the signal reaches O?

14-6. *Child of the galaxy.* It takes light about 10^5 years to travel from the most distant star in our galaxy to the earth. But in principle it is possible for a child born on a space ship as it leaves the earth to arrive at the star before dying of old age—if the space ship travels at a high enough speed. Explain why. Then estimate the required speed.

14-7. *A fast ship.*

a. What is the value of $1/\sqrt{1 - V^2/c^2}$ for a space ship that travels 1 light year as observed from the earth in one year of space ship time?

b. What is the value of $|V|/c$?

14-8. *Add 'em up?* A rocket ship approaches the earth with speed $0.80c$. Another ship approaches from the opposite direction with the same speed. What is the speed of the first ship relative to the second?

14-9. *Galileo versus Lorentz.* What percentage error is made in using the Galilean transformation $x' = x - Vt$ instead of the corresponding Lorentz transformation when $|v|/c = 1/7$?

14-10. *Lorentz in action.* Use the Lorentz position-time transformation to derive the time dilation formula.

Group B

14-11. *Tilt!, I.* A meter stick is at rest in the inertial frame O' with one end at the origin, and makes an angle of 60° with the x' axis. Observer O' sees observer O pass him by, moving in the positive x' direction with velocity $0.80\ c$.

a. If O fixes her x axis parallel to the x' axis, what angle does the meter stick make with the x axis according to her measurements?

b. What is the length of the meter stick as measured by O?

14-12. *Tilt!, II.* Just as O passes O' in Exercise 14-11, O' sets off a strobe light at his origin focused so that the light travels along the meter stick.

a. What are the x' and y' components of the velocity of the light flash?

b. Transform these components into the frame O.

c. What angle does the light path make with the x axis? Compare this result with the effect of relative motion on the tilted meter stick in Exercise 14-11.

d. Show that the speed of the light in the frame O is c.

14-13. *The relativity of shape.* Consider the two reference frames O and O' that are described in Sec. 14-6. A wooden slab is at rest in frame O' and is centered on the origin of O'. Observer O' finds that it is square, with sides exactly 1 m long. Find the coordinates of its corners and describe the shape and orientation of the slab, as measured in frame O at $t = 0$, when frames O and O' coincide, in each of the following cases.

a. The edges of the slab in O' are parallel to the x' and y' axes.

b. The edges of the slab in O' are inclined at 45° to the x' and y' axes.

14-14. *The relativity of confinement.* A rocket ship of proper length 100 m flies at a high speed into a straight tunnel bored through a mountain. The proper length of the tunnel is 50 m. To an observer stationed at the tunnel, the length of the ship is contracted so that he judges the ship to be slightly shorter than the tunnel. There are sliding doors at the entrance and exit of the tunnel, which operate independently. If a door is actuated, it slams shut and then immediately snaps open. When the observer finds the ship to be in the tunnel, he actuates both doors at the same time. He then describes what has happened by saying that the ship was momentarily confined between the closed doors. There is also an observer on the ship. For her it is the length of the tunnel that is contracted. She therefore judges the ship to be considerably longer than the tunnel. So she contends that the ship could not have been momentarily confined between the closed doors. Just how would she describe what has happened? Use the Lorentz position-time transformation to reconcile the two descriptions. That is, show that the relation between the descriptions is in quantitative agreement with the transformation. (Note: Persons who want to "prove Einstein wrong" often pose this problem as an "irrefutable inconsistency" in the theory of relativity.)

14-15. *Spatial separation versus length.* As seen by inertial observer O', a certain event 1 takes place at $x_1' = -L'/2$ at time $t_1' = L'/2c$. Another event 2 takes place at $x_2' = L'/2$ at time $t_2' = L'/2c$, so that for O' the two events are simultaneous.

a. Show that for another inertial observer O moving along the x' axis at velocity V with respect to O', the events are not simultaneous since $\Delta t = -\gamma L'V/c^2$, where $\gamma = 1/\sqrt{1 - V^2/c^2}$.

b. Determine the spatial separation of the two events, Δx, for observer O.

c. Does Δx represent the length of an object? Explain.

14-16. *Simultaneity: another view.* Repeat the analysis of the simultaneity thought experiment considered in Sect. 14-4, but have the observer stationed on the train cause the blasting caps to be detonated simultaneously from his point of view. Compare your analysis, and your results, with those of the text.

14-17. *A close look at delta.* Write a paragraph explaining, in terms of the simultaneity thought experiment considered in Sec. 14-4, the physical reason why the quantity $\delta = -Vx/c^2$ of Eq. (14-13) is proportional to the first powers of both V and x. Write a second paragraph explaining the physical significance of the sign of δ, and of its dependence on the signs of V and x.

14-18. *A close look at gamma.* Write a paragraph explaining, in terms of the time dilation-length contraction thought experiment considered in Sect. 14-5, the physical reason for the dependence of the quantity $\gamma = 1/\sqrt{1 - V^2/c^2}$ of Eq. (14-15) on V. Give a physical reason why γ does not depend on the sign of V.

14-19. *The inverse Lorentz transformations.* Since all inertial systems are equivalent, it follows that the laws of physics are the same for all inertial systems. The Lorentz transformation of Eqs. (14-16) is such a law.

a. Use this principle of equivalence to obtain the inverse Lorentz transformation, from the primed system to the unprimed one.

b. Obtain the same result by the more laborious method of algebraic manipulation.

14-20. *An alternative derivation of the Lorentz transformation.* When inertial reference frames O and O' coincide, let a flash of light be produced at the common origin. Each observer is justified in considering himself at the center of an expanding sphere of light. Experiment has revealed that each obtains the same value c for the speed of light. The Galilean transformation, $x' = x - Vt$, does not give this result. Therefore try a modification, $x' = \gamma(x - Vt)$ [Eq. (14E-1)], where γ is to be determined. The principle of equivalence (see Exercise 14-19) requires that Eq. (14E-1) hold for the inverse transformation, $x = \gamma(x' - V't') = \gamma(x' + Vt')$ [Eq. (14E-2)]. In this equation, we use the assertion that $V' = -V$. But for generality, the possibility has been allowed that t' may be different from t.

If x and x' are the intersections of the sphere with the axis at times t and t', respectively:

a. To what is x'/t' equal?

b. To what is x/t equal?

c. Use the results of parts a and b to eliminate x and x' in Eqs. (14E-1) and (14E-2) and thus to determine γ.

d. With this relation for γ, express t' in terms of t and x.

14-21. *Relativistic pursuit.* A spaceship passes the earth at $t = t' = 0$ with relative velocity v. At time t_1 on earth clocks, a super-spaceship leaves the earth with relative velocity $V > v$ to catch up with the first one. This will happen at earth time t_2, where $vt_2 = V(t_2 - t_1)$ or $t_2 = Vt_1/(V - v)$.

a. Is $(t_1 - 0)$ a proper or dilated time interval?

b. What does the clock on the "slow" spaceship read when the earth clock reads t_1?

c. How far from the spaceship is the earth then?

d. Is $(t_2 - 0)$ a proper or dilated time interval?

e. What does the clock on the slow spaceship register when the ship is overtaken?

f. In the frame of reference of the slow spaceship, how much time has elapsed since the pursuit started?

g. In the same reference frame, how large a distance was covered in the pursuit?

h. Divide the result of part g by the result of part f to obtain the velocity of the fast ship relative to the slow one. Compare this result with the one that could be obtained by using the formulas for the Lorentz velocity transformation.

14-22. *The speed of light in moving water.* The speed of light in still water is c/n, where $n \simeq 1.33$ and is called the *index of refraction.* If the water is moving with speed $|V| \ll c$ in the same direction as the light is traveling, show that the speed observed in the laboratory is equal to $c/n + |V|(1 - 1/n^2)$. This result was obtained by the French physicist Fizeau about the middle of the nineteenth century and can be explained only relativistically.

14-23. *Michelson-Morley fringe shift according to Galileo.* Let the distance to each mirror from the inclined lightly silvered mirror P in Fig. 14-2 be L. Assume that there is a stationary medium, the ether, through which the apparatus moves with velocity V in the direction of mirror M_2. Assume also that newtonian physics is valid, so that the Galilean transformations apply.

a. If the speed of light is c, what is the speed of light relative to M_2 as it approaches M_2 from P?

b. How long does it take the light to travel from P to M_2?

c. What is the speed of light relative to P on the return trip?

d. How long does it take the light to travel from M_2 to P? What is the total time for the round trip from P to M_2 and back to P?

e. Show that the speed of light relative to M_1 is $\sqrt{c^2 - V^2}$ and that this is also the speed relative to P on the return trip.

f. What is the total time for the round trip to the mirror M_1 and back to P?

g. If $|V|/c \ll 1$, show that the difference in time between the round trip to M_2 and the round trip to M_1 is equal to LV^2/c^3.

h. What distance does light travel in this time difference?

i. If the wavelength of the light is λ, to what fraction of a wavelength does this distance correspond?

j. Calculate this fraction if $L = 10$ m, $\lambda = 6 \times 10^{-7}$ m and $|V| = 3 \times 10^4$ m/s. Take c equal to 3×10^8 m/s.

k. If the apparatus is rotated through 90°, how much of a fringe shift is expected? (Michelson and Morley could have detected a shift of 0.01 fringe.)

14-24. *Squashing a circle.* A circular hoop of radius a' is at rest in the $x'y'$ plane of system O'. The inertial frame O' is moving as described in Sec. 14-6 at constant velocity V with respect to the inertial frame O.

a. Show that the measurements made in frame O will indicate that the hoop is elliptical in shape, with its semimajor axis parallel to the y axis and of length $a = a'$, and its semiminor axis of length $b = a'\sqrt{1 - V^2/c^2}$.

b. The *eccentricity* e of an ellipse is defined by $e \equiv \sqrt{1 - (b/a)^2}$. Derive a simple expression for the eccentricity of the ellipse in part *a*. Evaluate e for each of the following values of $|V|$, the relative speed: (1) $0.010c$; (2) $0.10c$; (3) $0.50c$; (4) $0.90c$; (5) $0.999c$.

Group C

14-25. *Relativistic Doppler effect.* A space ship traveling toward the earth with speed $|v_s|$ has a long rod sticking out at right angles to its direction of travel. When a light at the space ship end of the rod flashes, the light pulse which travels along the rod is reflected back to the space ship by a mirror at the end of the rod. The returning light pulse activates a very fast-acting mechanism which makes the light flash again. The length of the rod is such that observers on the space ship agree that the light flashes exactly once per microsecond. That is, its frequency is exactly $\nu = 1 \times 10^6$ Hz.

a. If the time dilation were the only effect to take into account, what would be the period of the flash on earth clocks?

Light from the flash also travels toward the earth. This period of flashing will be less than the period found in part *a* since the space ship moves toward the earth between flashes and the second flash has less far to travel than the first one.

b. How far toward the earth will an earth observer say the light has traveled in the time between flashes?

c. How far will this observer say the ship has traveled between flashes?

d. What is the distance between successive flashes for an earth observer?

e. Show that the time between flashes that reach the earth is equal to $(1/\nu)\sqrt{(1 - |v_s|/c)/(1 + |v_s|/c)}$.

f. If $|v_s| = \frac{3}{5}c$, what is the period T of the flashes as observed on the earth? What is their frequency ν'?

g. If the space ship were traveling *away* from the earth with the same speed, what would be the period between flashes observed on the earth?

h. Find an expression for the frequency ν' of the flashes as they are observed on the earth.

This change in the period and frequency of a signal emitted by a source moving toward or away from an observer is called the **relativistic Doppler effect.** Compare the result obtained in part *h* with Eq. (12-71). The latter equation gives the Doppler frequency shift for waves moving through a medium when the source is moving and the observer is at rest with respect to the medium.

14-26. *Doppler effect in action.* Light is emitted by a certain species of atoms in a star located in a distant galaxy. An identification of the species is made by studying the distribution pattern of the numerous sharply defined frequencies of the light arriving at the earth. (That is, the source is identified by studying the "spectrum" of the light it emits.) From this identification a certain component of the light is determined to have a frequency $\nu = 8.0 \times 10^{14}$ Hz, as would be measured by an observer stationed in the distant glalaxy. The frequency ν' of this component is measured by an observer stationed on the earth and found to have the value 5.0×10^{14} Hz. Use the relativistic Doppler effect formula obtained in Exercise 14-25 to determine the direction and speed of the motion of the galaxy with respect to the earth.

14-27. *Interstellar encounter.* A space ship coasting in interstellar space encounters an alien space probe which has a radio transmitter. As the probe approaches, the frequency initially received by the ship is 130 MHz = 130×10^6 Hz. As the probe recedes into the distance, the frequency eventually drops to 60 MHz. Consult Exercise 14-25 and then answer the following questions.

a. What is the relative speed of the ship and the probe?

b. What is the intrinsic frequency of the probe's transmitter?

c. Find the received frequency of the signals emitted by the probe at the time of its closest approach to the ship.

14-28. *Composition of two Lorentz transformations.* Consider three inertial reference frames O, O', and O''. Let O' move with velocity V with respect to O, and let O'' move with velocity V' with respect to O'. Both velocities are in the same direction.

a. Write the transformation equations relating x, y, z, t with x', y', z', t' and also those relating x', y', z', t' with x'', y'', z'', t''. Combine these equations to obtain the relations between x, y, z, t and x'', y'', z'', t''.

b. Show that these relations are equivalent to a direct transformation from O to O'' in which the relative velocity V'' of O'' with respect to O is given by

$$V'' = \frac{V + V'}{1 + VV'/c^2}$$

c. Show that this expression for V'' is in agreement with the first of the Lorentz velocity transformations, Eqs. (14-21).

d. Explain how the analysis you have gone through proves that two successive Lorentz position-time transformations are equivalent to one direct position-time transformation.

e. M, M', and M'' are meter sticks lying along the parallel x, x', and x'' axes. They are at rest in O, O', and O'', respectively. Construct a table that shows the lengths observers in each frame would assign to each meter stick.

14-29. *A more general form of the Lorentz transformation.* The coordinate origin of inertial frame O' moves with constant velocity V in the positive x direction, as measured from inertial frame O. The x' and x axes are parallel but not coincident. Axes y' and z' are parallel to axes y and z, respectively. At time t_0 (according to O) and time t_0' (according to O'), the point (x_0, y_0, z_0) in the frame O coincides with the point (x_0', y_0', z_0') in the frame O'. Construct an argument, based on the Lorentz transformation given in Eq. (14-16), which shows that the correct transformation of coordinates from O to O' must be:

$$x' - x_0' = \gamma[(x - x_0) - V(t - t_0)]$$

$$y' - y_0' = y - y_0$$

$$z' - z_0' = z - z_0$$

$$t' - t_0' = \gamma[(t - t_0) - V(x - x_0)/c^2]$$

where $\gamma \equiv 1/\sqrt{1 - V^2/c^2}$. (Hint: What is the transformation for two observers who are at rest with respect to one another, but who have simply chosen different origins for position and time coordinates?)

14-30. *Form-invariance of the wave equation.* The wave equation for a beam of light traveling in the x direction can be written

$$\frac{\partial^2 f(x, t)}{\partial x^2} = \frac{1}{c^2}\frac{\partial^2 f(x, t)}{\partial t^2}$$

Here c represents the speed of light and $f(x, t)$ represents the value of the "electric field," or of the "magnetic field," in the light beam. Together these fields form the linked pair of traveling transverse waves called the "electromagnetic field" that constitutes the light beam. The wave equation determines the properties of the light beam, as measured by an observer O in an inertial reference frame with position and time variables x and t. An observer O' is in an inertial reference frame moving with respect to the frame of observer O along the x axis at constant velocity V. The position and time variables used by O' are x' and t'. Their relations to the variables x and t used by O are given by the first and last of Eqs. (14-16), the Lorentz equations for transforming position and time. Use these relations to show that when transformed into the variables x' and t' the wave equation becomes

$$\frac{\partial^2 f(x', t')}{\partial x'^2} = \frac{1}{c^2}\frac{\partial^2 f(x', t')}{\partial t'^2}$$

That is, the wave equation is **form-invariant** under the Lorentz transformation. The variables are changed, but the form of the equation remains the same. (Hint: You will need to make repeated use of the "chain rules"

$$\frac{\partial g}{\partial x} = \frac{\partial g}{\partial x'}\frac{\partial x'}{\partial x} + \frac{\partial g}{\partial t'}\frac{\partial t'}{\partial x}$$

and

$$\frac{\partial g}{\partial t} = \frac{\partial g}{\partial x'}\frac{\partial x'}{\partial t} + \frac{\partial g}{\partial t'}\frac{\partial t'}{\partial t}$$

where g is any function of x and t, or of x' and t'.) Compare your results with the results of Exercise 12-33 which calls for a similar calculation, except that the Galilean equations for transforming position and time are applied to the wave equation for transverse waves in a stretched string. In particular, discuss the physical significance of the fact that the transformed wave equation for light has exactly the same mathematical form after it is transformed as it has before being transformed. What does the transformed wave equation predict for the speed of light, as measured by observer O'? Compare this prediction to the prediction obtained in Exercise 12-34 concerning the speed of the waves in the string, as measured by observer O'.

14-31. *Rocket astronomy.* An astronomer observes that a group of protons from the sun (part of the solar wind) passed the earth at time t_1. Later, she discovers that Jupiter has emitted a large burst of radio noise at time $t_2 = t_1 + \Delta t$. A second astronomer O' riding in a rocket traveling from earth to Jupiter at speed $|V|$, observes the same two events.

a. Assume that the earth is directly between the sun and Jupiter, 6.3×10^8 km from Jupiter. Let $|V| = 0.50c$ and $\Delta t = 900$ s. Calculate the time interval $\Delta t'$ measured by observer O' in the rocket. Could the protons from the sun have triggered the radio burst from Jupiter?

b. Is there a second rocket reference frame in which the two events were simultaneous? If so, what is its speed with respect to the earth? If not, why not?

c. Assume that a radio noise burst is triggered by a burst of protons. What limit can be placed on Δt?

d. Suppose that the two events were separated by $\Delta t = 60$ min. What was the speed of an observer who measured a proper time interval between these events? Calculate the time interval measured by the observer.

14-32. *Three equal speeds.* Frames O and O' are related as described in Section 14-6. There is a particle moving in the $x'y'$ plane of the frame O' whose speed v' equals V, the relative speed of O and O'. In frame O, the particle's speed v is also equal to V.

a. Find the angle θ between $\mathbf{v}$ and the x axis. Find the angle θ' between $\mathbf{v}'$ and the x' axis.

b. Evaluate the angles θ and θ' for the following values of V: (1) $0.10c$; (2) $0.30c$; (3) $0.90c$; (4) $0.99c$.

14-33. *Invariant spacetime interval.* As seen from inertial frame O, event 1 occurs at time t_1 at the location specified by the coordinates (x_1, y_1, z_1). Another event, event 2, occurs at time t_2 at the location specified by the coordinates (x_2, y_2, z_2). Observer O' in a different inertial frame observes the same two events, whose times of occurrence and locations he specifies by means of the corresponding primed quantities.

a. Use the Lorentz position-time transformation to express the quantity $c^2(t_2 - t_1)^2 - (x_2 - x_1)^2 - (y_2 - y_1)^2 - (z_2 - z_1)^2$ in terms of the primed quantities.

b. Show that the quantity given above reduces to the product of c^2 and the time interval between the two events if the events occur in the same place, and that the quantity is thus the product of c^2 and the proper time between the two events.

Note that $x_2 - x_1$ is not invariant using the Lorentz transformation, though it is invariant using the Galilean transformation. The same holds for $t_2 - t_1$. The combination of space and time quantities given in part *a* is **Lorentz invariant;** hence the expression "spacetime."

14-34. *The invariant speed.* A flash of light has components of velocity v_x, v_y, and v_z in the unprimed coordinate system. That is, $v_x^2 + v_y^2 + v_z^2 = c^2$. Using the Lorentz velocity transformation, calculate the speed of the light in the reference frame O' which is moving with speed V relative to O in the direction of the x and x' axes. The y and y' axes are parallel.

14-35. *The aberration of starlight.* Figure 14E-35 represents a star on the y axis sending light toward the earth,

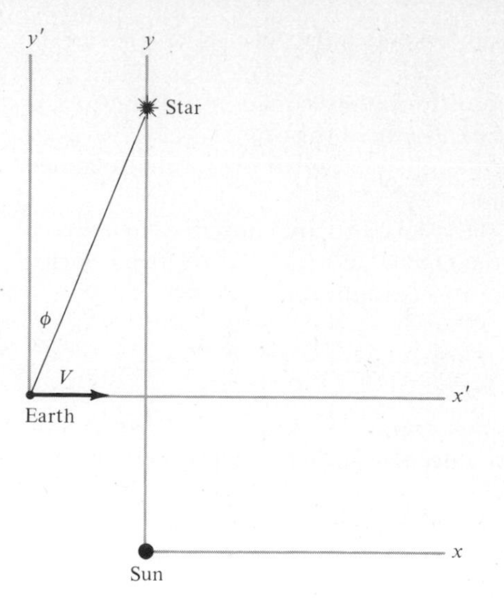

Fig. 14E-35

which is at the origin of a system of coordinates whose x' axis is in the direction of the earth's motion relative to the sun. Let ϕ be the angle between the y' axis and the direction of the star as observed from the earth or, what is the same, the observed path traveled by the light from the star to the earth.

a. Show that $\sin \phi = V/c$, where V is the speed of the earth relative to the sun.

b. Is the shift in the star's observed direction due to the motion of the earth toward or opposite the direction of the velocity of the earth relative to the sun?

c. Show the speed of the light in the coordinate system attached to the earth is still c.

d. Evaluate ϕ in arc seconds if $V = 3 \times 10^4$ m/s.

This phenomenon of shift of star position is called **stellar aberration.**

14-36. *The relativity of acceleration.* An accelerating object is observed from two inertial frames, O and O'. The x and x' axes are collinear with the path of the object and O' has a velocity V relative to O which lies along the x axis. Show that the accelerations as measured in the two frames are related by

$$a' = a(1 - V^2/c^2)^{3/2}/(1 - vV/c^2)^3$$

where v is the instantaneous velocity of the object as observed by O. [You will need to note that $dv/dt' = (dv/dt)(dt/dt')$.]

15

Relativistic Mechanics

15-1 THE BASIS OF RELATIVISTIC MECHANICS

Now that we have developed the most important properties of relativistic kinematics, we can use them as a foundation to build the structure of relativistic mechanics. The procedure we follow is completely parallel to one that served us well in our earlier work with newtonian mechanics. It starts from the law which states that the total momentum of an isolated system is conserved if the system is viewed from any inertial reference frame. There is an overwhelming amount of experimental evidence supporting this law in the newtonian domain; some is found in the strobe photos of puck collisions considered in Chap. 4. When Einstein was developing relativistic mechanics, there were no experiments in the relativistic domain to show that the total momentum of a system is constant. He nevertheless boldly assumed that this, the most fundamental law of newtonian mechanics, would also apply to relativistic mechanics. Then he analyzed the implications of the assumption by means of a momentum-conservation thought experiment.

We will analyze such a thought experiment. It will lead us to a relativistic definition of mass, just as our analysis in Chap. 4 of an experiment on momentum conservation for a puck collision led us to a newtonian definition of mass. We will then use the understanding of relativistic mass and momentum thus obtained to develop the relativistic concepts of force and energy.

Although our development of relativistic mechanics will follow our development of newtonian mechanics very closely in outline, it will be very significantly changed in detail. This is because of the fundamental differences between newtonian kinematics and the kinematics that must be used in the relativistic domain. These differences lead to equally fundamental differences between the predictions concerning rapidly moving objects which will be obtained from our study of relativistic mechanics and the predictions concerning slowly moving objects we obtained when studying newtonian mechanics. Particularly striking will be predictions concerning relativistic mass, relativistic energy, and the relativistic relation between mass and energy.

While Einstein in 1905 had to carry the law of momentum conservation over to relativity theory as an assumption, because experimental tests did not then exist, the situation now is quite different because of the advent in recent years of high-energy particle accelerators. Physicists working with these machines have made countless measurements confirming that momentum conservation holds for collisions between particles moving at speeds very close to the speed of light. And long before it was technically possible to obtain direct experimental confirmation of Einstein's assumption, there were experiments whose results agreed with predictions concerning relativistic mass and energy, which are logical consequences of relativistic momentum conservation. Nuclear fission certainly provides the most dramatic example. We discuss this and other experimental confirmations of Einstein's theory at appropriate places in this chapter.

15-2 RELATIVISTIC MASS AND MOMENTUM

We obtain the relativistic definition of mass by analyzing a thought experiment quite analogous to the experiment on the air-table puck collision analyzed in Sec. 4-3. Just as in Sec. 4-3, by *mass* we mean here and throughout this chapter the *inertial aspect of mass*—not the gravitational aspect of mass. In the thought experiment, observers O_1 and O_2 obtain two identical particles, say billiard balls B_1 and B_2. While the observers are at rest with respect to each other in an inertial reference frame, they carry out a preliminary experiment in which the balls collide while moving at speeds relative to the two observers that are very small compared to the speed of light. That is, they carry out an experiment exactly like one of those considered in Sec. 4-3. Then they analyze the experiment and verify that both balls have the same mass when the balls are observed to be moving very slowly or to be at rest. We use the symbol m_0 for the mass of either ball, measured when its speed relative to the observer making the measurement is zero or very small compared to the speed of light.

Next the observers repeat the collision experiment, but modify it so that each observer can compare the mass of one ball moving slowly with respect to himself/herself with that of the other ball moving rapidly with respect to himself/herself. To do this, each observer takes one ball, and the two of them separate. Then they start moving rapidly along parallel lines toward each other, so that they will pass each other at a constant relative speed comparable to the speed of light. The situation is illustrated in Fig. 15-1 from the point of view of an inertial reference frame with axes x, y, z. Observers O_1 and O_2 are moving in opposite directions parallel to the x axis at a constant speed that is an appreciable fraction of the speed of light. Just before they pass, O_1 throws ball B_1 in a direction that he judges to be perpendicular to the x axis and toward O_2. From the point of view of O_1, the speed with which he throws the ball has the value $v_\perp$. At the same instant, O_2 throws ball B_2 toward O_1 in a direction that she judges to be perpendicular to the x axis. From her point of view, O_2 throws B_2 at a speed which also has the value $v_\perp$. The two balls are thrown at such a time, and with such a speed, that they collide as indicated in the figure. The common speed $v_\perp$ at which the two observers throw their balls perpendicular to the x axis is very small compared to the speed of light c. But the speed with which the observers approach each other along the x axis is comparable to the speed of light. Thus the balls approach each other along trajectories that are seen in

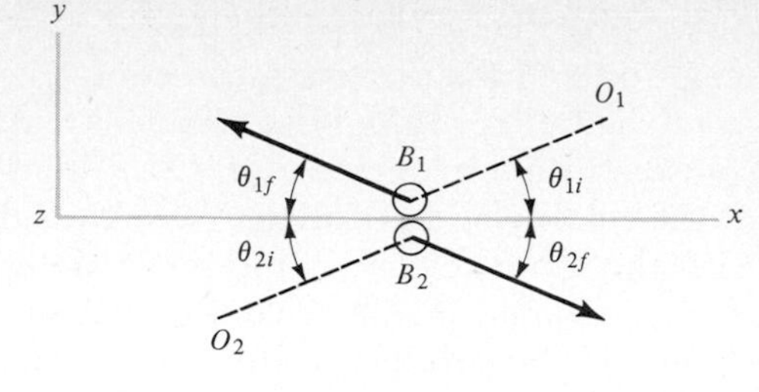

Fig. 15-1 A symmetrical collision between two identical billiard balls, as seen from an inertial reference frame. Since the speed of observers O_1 and O_2 with respect to this reference frame is supposed to be comparable to c, while the speed with respect to these observers of the balls B_1 and B_2 is supposed to be very small compared to c, all the labeled angles are really much smaller than shown.

the x, y, z frame to be inclined at very small angles to the x axis. (In the figure the angles have been exaggerated for the sake of clarity.)

From the point of view of the x, y, z reference frame, illustrated in Fig. 15-1, the balls B_1 and B_2 approach each other moving in opposite directions at equal speeds along paths that form equal angles $\theta_{1i} = \theta_{2i}$ with the x axis. This simplification is a consequence of the symmetrical way that we assume the observers throw the balls. After the collision, B_1 and B_2 recoil along paths inclined to the x axis at angles θ_{1f} and θ_{2f}.

We require that the total momentum of an isolated system be conserved in any inertial frame. This requirement will be satisfied by the collision we are studying, since the two balls form an isolated system after they are thrown, and since the x, y, z frame is inertial. The initial total momentum of the balls is zero. So momentum conservation demands that their final total momentum be zero also. This means that the balls must move apart after the collision with equal speeds along paths inclined at the equal angles $\theta_{1f} = \theta_{2f}$. The actual value of these angles depends, for given values of the speeds involved, on the precise timing of the throws made by O_1 and O_2. We assume for the sake of simplicity that this was done so as to make the balls collide as shown in the figure, recoiling from the collision at final angles equal to the initial angles at which they approach the collision. That is, we take $\theta_{1f} = \theta_{1i}$ and $\theta_{2f} = \theta_{2i}$. We further simplify the analysis by assuming that the collision is elastic, so that the balls move apart afterward at a relative speed equal to their relative speed of approach before the collision. This assumption will not restrict the validity of the results obtained by imposing the fundamental requirement that the total momentum of an isolated system be conserved in any inertial frame. Just as in the newtonian domain, momentum is conserved in a relativistic collision *whether or not* the collision is elastic. The consequence of these two final simplifying assumptions—that the final angles of recoil equal the initial angles of incidence and that the collision is elastic—is to make the collision *completely* symmetrical, as seen in the x, y, z reference frame.

The same collision is shown in Fig. 15-2 from the inertial frame of reference x_1, y_1, z_1 in which observer O_1 is stationed. As viewed from the x_1, y_1, z_1 frame, O_1 throws B_1 with an initial velocity of magnitude $v_\perp$ in the negative y_1 direction. After the collision it recoils with a final velocity of the same

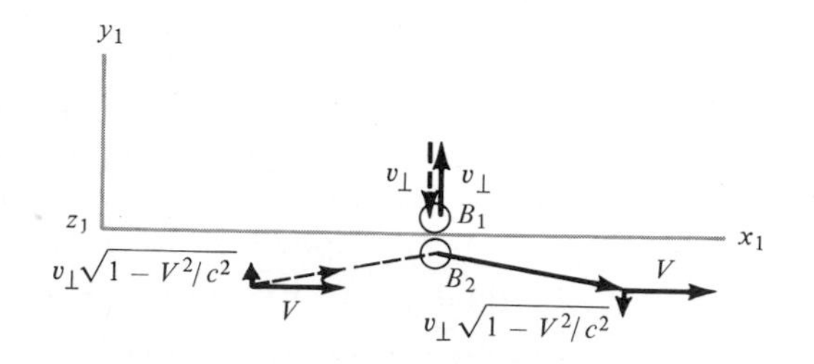

Fig. 15-2 The symmetrical collision between B_1 and B_2, from the point of view of O_1. The angles between the axis and the initial and final trajectories of B_2 again have been exaggerated for the sake of clarity. Actually, the angles are much smaller because V is comparable to c and $v_\perp$ is very small compared to c.

magnitude in the positive y_1 direction. So, from the viewpoint of O_1, ball B_1 never has an x_1 component of velocity. Also, O_1 observes B_2 move into the collision with a velocity that has an x_1 component equal to V, the magnitude of the velocity of O_2 relative to O_1. After the collision, O_1 sees B_2 move away with the same velocity component along the x_1 axis. Neither ball experiences a change in the x_1 component of its velocity, and there will be no change in the product of this quantity and the mass of the ball. Thus it is apparent that there can be no change in the component along the x_1 direction of the total momentum of the isolated system consisting of the two colliding balls, as measured from the inertial frame of O_1. The x_1 component of the total momentum of the system is conserved.

We investigate the conservation of the momentum component of the system along the y_1 direction by evaluating the velocity components of each ball along that direction, as measured by O_1. Now we know that O_2 measures the initial velocity of B_2 to have a component only along the y_2 direction of her x_2, y_2, z_2 reference frame, with a value equal to the quantity $v_\perp$. But the Lorentz velocity transformation shows that O_1 does *not* measure the y_1 component of the velocity of B_2 to have the same value. To find out what value he does measure, we apply the velocity transformation from the x_2, y_2, z_2 reference frame to the x_1, y_1, z_1 reference frame and thereby determine the initial y_1 component of the velocity of B_2 as seen by O_1. For this purpose, we use the second of Eqs. (14-21), with the notational change that we transform from the "sub-2" frame to the "sub-1" frame instead of from the "unprimed" frame to the "primed" frame. We write the equation with v_{y_2} replacing v_y, v_{y_1} replacing v_y', and v_{x_2} replacing v_x. Also, we set the signed scalar, used in Eqs. (14-21) to represent the velocity of the primed frame with respect to the unprimed frame, equal to $-V$. The reason is that the sub-1 frame is moving with respect to the sub-2 frame at speed V in the negative x direction. Then we have

$$v_{y_1} = \frac{\sqrt{1 - V^2/c^2}\, v_{y_2}}{1 + Vv_{x_2}/c^2}$$

Here v_{x_2} is the initial x_2 velocity component of B_2, as seen by O_2. Its value is $v_{x_2} = 0$. And v_{y_2}, the initial y_2 velocity component of B_2 according to O_2, has the value $v_{y_2} = v_\perp$. Thus we find that when O_1 measures the initial y_1 component of the velocity of B_2, he obtains the value

$$v_{y_1} = \sqrt{1 - V^2/c^2}\, v_\perp$$

After the collision O_1 sees B_2 to have a velocity with a y_1 component of the same magnitude but with a negative sign.

Now that we know all the y_1 components of the velocities of both balls, let us see if the y_1 component of the total momentum of the isolated system they form is constant, as viewed from the inertial reference frame of O_1. In the preliminary experiment, O_1 and O_2 were at rest with respect to each other and observed both balls moving at speeds small compared to the speed of light. They found that momentum was conserved if they used the newtonian definition that the momentum of each ball equals its mass m_0 times its velocity. If we apply the same definition in this experiment, where the balls are moving at speeds comparable to the speed of light, we find that the initial momentum of the system has a y_1 component given by

$$p_{y_{1i}} = m_0(-v_\perp) + m_0\sqrt{1 - V^2/c^2}\, v_\perp \qquad \text{(proves to be incorrect)}$$

The coefficient of m_0 in the first term is the initial velocity component in the y_1 direction of B_1, and the coefficient of m_0 in the second term is the initial velocity component of B_2 in the y_1 direction. (See Fig. 15-2.) We find also that the final momentum of the system has a y_1 component given by

$$p_{y_{1f}} = m_0 v_\perp + m_0\sqrt{1 - V^2/c^2}\,(-v_\perp) \qquad \text{(proves to be incorrect)}$$

Here the coefficient of m_0 in the first term is the final y_1 velocity component of B_1, and the coefficient of m_0 in the second term is the final y_1 velocity component of B_2.

These two expressions do *not* yield values of $p_{y_{1i}}$ and $p_{y_{1f}}$ which are equal. Indeed, they yield values which are the negative of each other because the right side of the first expression is just the negative of the right side of the second. Applying the newtonian definition that momentum equals the product of mass m_0 and velocity leads to the conclusion that the initial total momentum of the isolated system has a y_1 component which is different from that of its final total momentum. If so, the total momentum of the system will not be conserved in the x_1, y_1, z_1 inertial reference frame, even though it is conserved in the x, y, z inertial frame. This would contradict Einstein's postulate—verified experimentally in many different ways—that all inertial frames are equivalent. And it would invalidate in the relativistic domain the law of conservation of momentum, which we know to be the experimental foundation of mechanics in the newtonian domain. Thus we do not accept as correct the values of $p_{y_{1i}}$ and $p_{y_{1f}}$ displayed in the equations of the preceding paragraph.

How can we salvage Einstein's postulate and the fundamental law of momentum conservation and also retain the definition of momentum as the product of mass and velocity? By changing the properties of mass and/or velocity! Since we are already using properties of velocity which are different from those used in the newtonian domain and which are in agreement with experiments in the relativistic domain, we try changing the properties of mass. We do this by allowing the *relativistic mass* m of a particle to depend on the particle's speed, instead of taking the mass to be a constant, as in the newtonian domain. And we do it in such a way that no conflict can arise with experiments of the newtonian domain which show the mass of a particle to be independent of its speed when the speed is small compared to the speed of light. This is accomplished by requiring that the relativistic mass m of a particle approach the constant value m_0 when its speed becomes small compared to the speed of light.

We find an expression for the relativistic mass of a particle when its speed is comparable to that of light by equating the initial and final total momentum components along the y_1 direction of the system of two colliding balls. Abandoning the newtonian idea that the mass of each rapidly moving ball has the fixed value m_0, we insert into the expression for the initial total y_1 component of momentum the relativistic mass m. We have then

$$p_{y_{1i}} = m\,(-v_\perp) + m\sqrt{1 - V^2/c^2}\,v_\perp$$

The first term on the right side is the y_1 component of momentum of B_1, as measured by O_1. This observer sees B_1 to be moving at the speed $v_\perp$, which is small compared to c. So for O_1 the relativistic mass of B_1 is indistinguishable from its newtonian value m_0, and the equation becomes

$$p_{y_{1i}} = m_0\,(-v_\perp) + m\sqrt{1 - V^2/c^2}\,v_\perp \qquad (15\text{-}1a)$$

But O_1 sees B_2 to be moving at a speed that is comparable to c because it has the large x_1 component of velocity derived from the motion of O_2. Therefore in the second term the relativistic mass m cannot be replaced by the newtonian value m_0. In exactly the same way, we find the following equation for the final total y_1 component of momentum of the system

$$p_{y_{1_f}} = m_0 v_\perp + m\sqrt{1 - V^2/c^2}\,(-v_\perp) \tag{15-1b}$$

Now we *insist* that the y_1 component of momentum of the system be conserved by equating its initial and final values:

$$p_{y_{1_f}} = p_{y_{1_i}} \tag{15-2}$$

By using Eqs. (15-1a) and (15-1b), this gives us

$$m_0 v_\perp + m\sqrt{1 - V^2/c^2}\,(-v_\perp) = m_0(-v_\perp) + m\sqrt{1 - V^2/c^2}\,v_\perp$$

Dividing through by $v_\perp$ and then transposing, we have

$$2m\sqrt{1 - V^2/c^2} = 2m_0$$

or

$$m = \frac{m_0}{\sqrt{1 - V^2/c^2}} \tag{15-3}$$

We have found the expression for relativistic mass which allows the y_1 component of momentum of the system to be conserved. We will interpret the meaning of this expression shortly.

As was noted earlier, it is apparent from Fig. 15-2 that the initial and final total momentum components of the system along the x_1 direction cannot differ. This is a consequence of the system's symmetry. So we also have conservation of its x_1 component of total momentum:

$$p_{x_{1_f}} = p_{x_{1_i}} \tag{15-4}$$

Equations (15-2) and (15-4), taken together, give us the vector conservation equation

$$\mathbf{p}_f = \mathbf{p}_i \tag{15-5a}$$

A more complete expression of this important result reads

$$(\mathbf{p}_{\text{total}})_{\text{final}} = (\mathbf{p}_{\text{total}})_{\text{initial}} \tag{15-5b}$$

The total relativistic momentum of an isolated system is constant, as viewed from an inertial reference frame. In this equation the **relativistic momentum p** of a particle is defined as

$$\mathbf{p} = m\mathbf{v} \tag{15-6}$$

where m is its relativistic mass, given by Eq. (15-3), and $\mathbf{v}$ is its velocity.

To satisfy law of the momentum conservation, we have had to allow the mass of B_2 as measured by O_1 to be a function of the speed of B_2 as measured by O_1. From Fig. 15-2 and the pythagorean theorem, you can see that this speed v of B_2 relative to O_1 is given by

$$v = \sqrt{V^2 + v_\perp^2(1 - V^2/c^2)}$$

But since V is comparable to c, while $v_\perp$ is very small compared to c, the second term under the square root is completely negligible in comparison to the first term. Hence the speed v of B_2 relative to O_1 is essentially V. Thus we can rewrite Eq. (15-3), specifying the relativistic mass m of B_2 as measured by O_1, in terms of the ball's speed v as measured by O_1. We do this by setting $V = v$ in Eq. (15-3), to obtain

$$m = \frac{m_0}{\sqrt{1 - v^2/c^2}} \tag{15-7}$$

In this equation m is the **relativistic mass** of a particle which is observed to be moving at speed v. *The relativistic mass of a particle is its mass measured when it is observed to be moving at a speed comparable to the speed of light.* The mass m_0 appearing in Eq. (15-7) is called the **rest mass** of the particle. *The rest mass of a particle is its mass measured when it is observed to be either moving at a speed very small compared to the speed of light or at rest.*

What we have done is to define mass in relativistic mechanics in such a way that mass times velocity—that is, momentum—is conserved. We used exactly the same approach in Sec. 4-3 to define mass in newtonian mechanics. The difference is that velocity has relativistic properties which are not the same as its newtonian properties, and this leads to a change in the properties of mass.

The origin of this change in the properties of mass can be traced to time dilation. Observer O_2 measures the mass of B_2 to be m_0. She finds its initial velocity component in the perpendicular direction by dividing the length dy_2 it moves by the proper time interval dt_2 required for the motion. Observer O_1 accepts the measurement of the perpendicular length, but not of the time interval. In fact, O_1 measures a dilated time interval that is larger than dt_2 by the factor $1/\sqrt{1 - v^2/c^2}$, with v being equal to the relative speed of the observers. Momentum conservation requires they both agree as to the momentum component of B_2 in the perpendicular direction. According to O_2 it is $m_0\, dy_2/dt_2$. Since O_1 finds the time interval in the denominator to be increased by the factor $1/\sqrt{1 - v^2/c^2}$, there must be a compensating increase in his measurement of one of the terms in the numerator. Since there is no change in the length, the increase must be in the mass. So O_1 finds the relativistic mass of B_2 to be larger than m_0 by the factor $1/\sqrt{1 - v^2/c^2}$.

The cloud chamber photograph in Fig. 15-3 provides qualitative confirmation of the relativistic prediction that the mass of a particle is larger than its rest mass if its speed is an appreciable fraction of the speed of light. A set of dots shows the path of a high-speed electron, emitted from a radioactive source, that traveled from the left of the photograph to collide near the center with an electron which is part of a molecule of the gas in the chamber. The molecular electron is essentially free and stationary. That is, its speed and binding energy are negligible compared with the speed and energy of the electron from the source. Emerging from the point of collision are sets of dots showing the paths followed by two electrons, one being the recoiling incident electron and the other the struck electron. The way in which the cloud chamber makes the paths visible is explained in the caption to the figure.

The collision between the two electrons does not have the appearance of an elastic collision between two particles of equal mass, one of which is initially stationary. You can see this by comparing Fig. 15-3 with Fig. 15-4, which is reproduced from Figs. 4-14 and 8-9. It shows an elastic collision on an air table between an incident puck and a stationary puck of the same mass. As discussed in Chaps. 4 and 8, such a collision is characterized by the fact that the two particles leave the point of collision on paths that form a 90° angle. The angle in the electron-electron

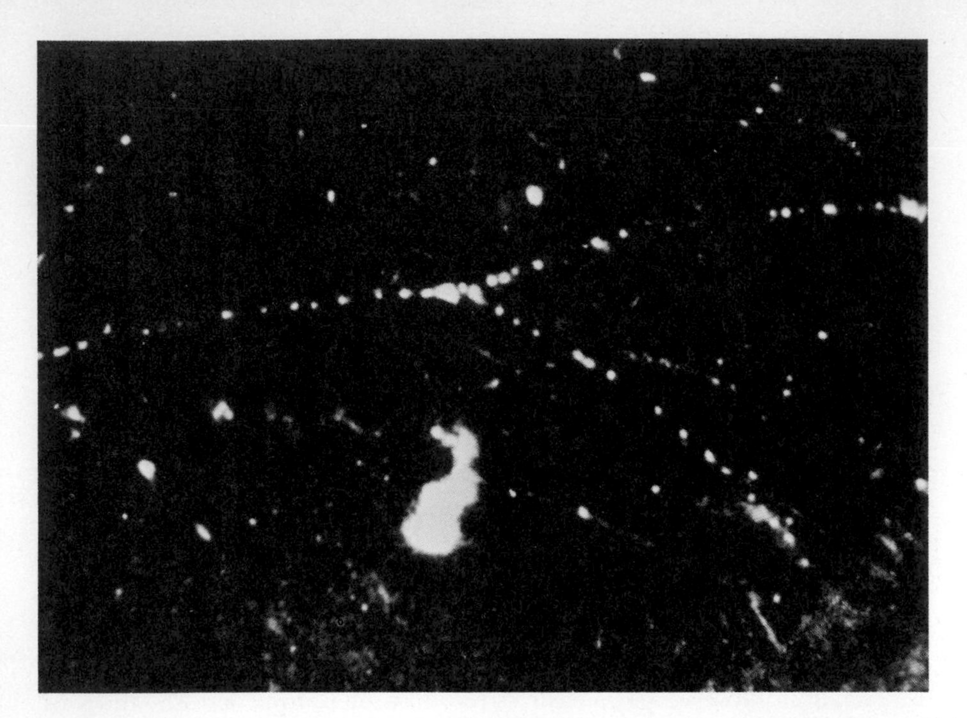

Fig. 15-3 Cloud chamber photograph of a collision between a high-speed electron and another electron which is essentially free and stationary. The incident electron, coming from the left, happens to approach the other electron located just above the center of the photograph. The two electrons interact through an electric force they exert on each other when they are close together. Momentum is thereby transferred from the incident to the struck electron, and the two move off at high speeds to the right. Their paths from the collision point form an angle of about 45°. The electron paths are recorded by droplets of liquid condensing from supersaturated vapor filling the cloud chamber. The droplets begin to condense on vapor molecules which the high-speed electrons have ionized (charged electrically by ejecting a molecular electron). The droplets occur at essentially random locations along the so-called tracks of the colliding electrons, and a single photograph is taken, not a stroboscopic photograph. Thus the separation between adjacent droplets does *not* provide direct information about the speeds of the electrons. Note that all three of the tracks are in good focus throughout their entire lengths. This shows that the plane in which they lie must be nearly perpendicular to the camera axis, so that the angle between the tracks emerging from the point of collision is not distorted in the photograph.

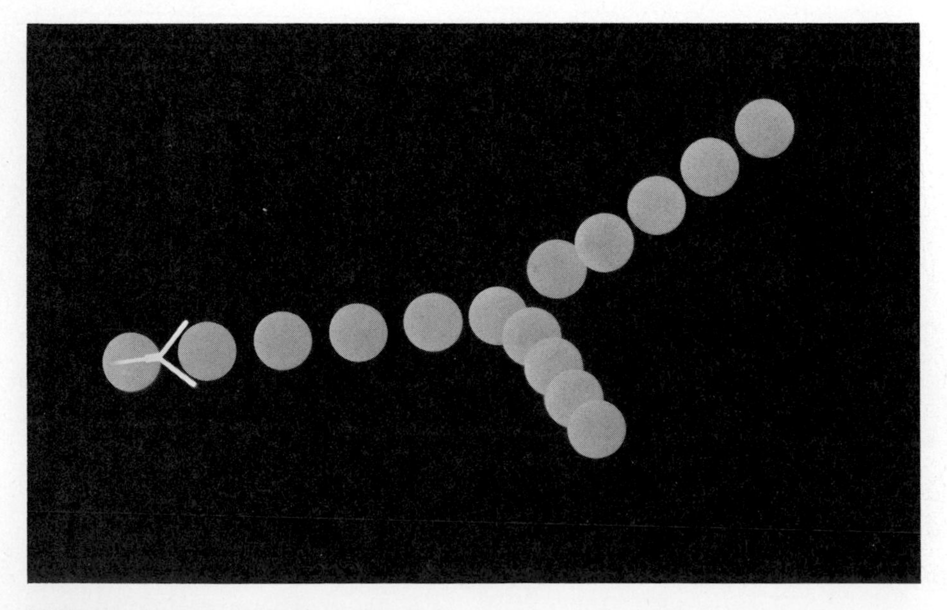

Fig. 15-4 A strobe photo of a collision on an air table between two magnetic pucks of equal mass. The incident puck comes from the left and makes a very close encounter with the initially stationary puck near the center of the table. While they are close together, the two pucks interact through a magnetic force. Note that the paths followed by the pucks after the collision form a 90° angle.

collision of Fig. 15-3 is clearly less than 90°. This makes it look like either an inelastic collision between two pucks of equal mass or an elastic collision in which the mass of the incident puck is significantly larger than the mass of the struck puck.

In the typical collision shown in Fig. 15-3, the electrons collide elastically. The physical reason is that it is difficult for electrons to get rid of mechanical energy when they collide. (To do so, they must emit an X ray.) Confirmation of the elastic nature of the collision is found in the photograph. Note that the electrons follow paths which are arcs of circles. This is because there is a "magnetic field" applied to the cloud chamber, perpendicular to the plane of the photograph. This results in a centripetal force on the charged electrons whose strength depends on their speeds. In Chap. 23 the details of this process are studied. Here it suffices to say that measurements of the radii of the arcs, when combined with results to be obtained in Sec. 15-3, can be used to determine the speeds of the electrons. In this way it can be shown that the collision between the two electrons actually is elastic.

The radius of curvature of the path followed by the incident electron shows that its speed is $v = 0.97c$. For such a speed, Eq. (15-7) predicts that the relativistic mass of the electron will be $m_0/\sqrt{1 - v^2/c^2} = m_0/\sqrt{1 - (0.97)^2} = 4.1m_0$, where m_0 is its rest mass. An elastic collision between two pucks, with the initially moving puck having a mass 4 times the mass of the initially stationary one, is shown in Fig. 15-5. The similarity is certainly evident between this photograph and the one showing the elastic collision between the electron of relativistic mass 4.1 times the rest mass of the initially stationary electron. It provides good qualitative confirmation of the relativistic prediction. (The confirmation is only qualitative because a newtonian collision cannot precisely model a relativistic collision. After a collision the ratio of the puck masses is still 4/1. But a collision reduces the ratio of the electron masses to a value smaller than 4.1/1, since the incident electron slows down and the struck electron speeds up.) Quantitative confirmation of the values of relativistic mass m, predicted by $m = m_0/\sqrt{1 - v^2/c^2}$, is presented in Sec. 15-3.

Examples 15-1 and 15-2 carry out numerical calculations using Eq. (15-6), $\mathbf{p} = m\mathbf{v}$, and Eq. (15-7), $m = m_0/\sqrt{1 - v^2/c^2}$.

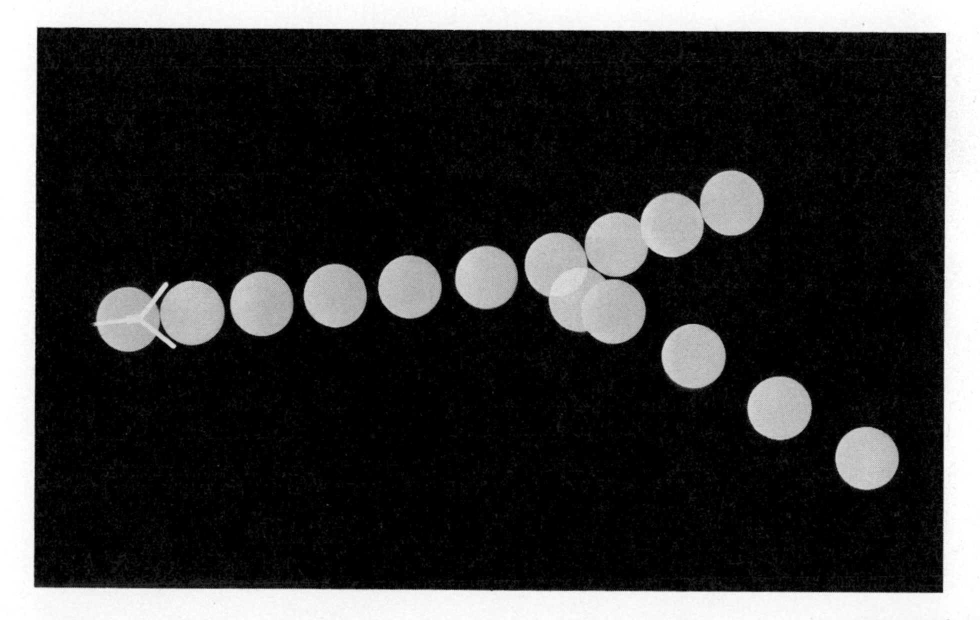

Fig. 15-5 A collision on an air table between magnetic pucks of unequal mass. The puck incident from the left has a mass which is 4 times as large as that of the struck puck. Here the angle between the paths followed by the pucks is about 45°. Compare this collision to the electron collision shown in Fig. 15-3.

EXAMPLE 15-1

Calculate the relativistic momentum and mass of a cosmic ray muon moving at speed $v = 0.999c$. The rest mass of a muon is 1.9×10^{-28} kg.

■ The relativistic mass is

$$m = \frac{m_0}{\sqrt{1 - v^2/c^2}}$$

In Example 14-3 you saw that $1/\sqrt{1 - v^2/c^2} = 22$ for $v/c = 0.999$. So you have

$$m = 22m_0 = 22 \times 1.9 \times 10^{-28}\text{ kg} = 4.2 \times 10^{-27}\text{ kg}$$

The relativistic momentum can then be calculated from

$$\mathbf{p} = m\mathbf{v}$$

The magnitude is

$$p = mv = 4.2 \times 10^{-27}\text{ kg} \times 3.0 \times 10^8\text{ m/s} = 1.3 \times 10^{-18}\text{ kg·m/s}$$

The direction of the momentum is the same as that of the muon velocity.

EXAMPLE 15-2

A high-energy particle accelerator produces a beam of protons (hydrogen atom nuclei) that have relativistic masses m which are 100 times their rest masses m_0. Determine the speed of the protons in the beam.

■ From the equation $m = m_0/\sqrt{1 - v^2/c^2}$, you have

$$\frac{m}{m_0} = \frac{1}{\sqrt{1 - v^2/c^2}}$$

Since the value of m/m_0 is given to be 100, you know that $1/\sqrt{1 - v^2/c^2} = 1.00 \times 10^2$, or

$$\sqrt{1 - v^2/c^2} = 1.00 \times 10^{-2}$$

You evaluate v^2/c^2 by carrying out the following sequence of calculations:

$$1 - v^2/c^2 = 1.00 \times 10^{-4}$$

$$v^2/c^2 = 1 - 1.00 \times 10^{-4}$$

$$v/c = (1 - 1.00 \times 10^{-4})^{1/2}$$

The binomial theorem approximation gives you

$$v/c \simeq 1 - \tfrac{1}{2} \times 1.00 \times 10^{-4} = 1 - 5.0 \times 10^{-5} = 0.999950$$

Thus you have

$$v \simeq 0.999950c$$

15-3 RELATIVISTIC FORCE AND ENERGY

Now that we have obtained an expression for momentum which satisfies the requirements of relativistic mechanics, we can use it to find a satisfactory expression for force. The procedure is exactly the same as in newtonian mechanics: force is obtained from momentum by definition. The **relativistic force F** acting on a particle is defined to be the rate of change of its relativistic momentum **p**. That is

$$\mathbf{F} \equiv \frac{d\mathbf{p}}{dt} \tag{15-8}$$

This definition of force has precisely the mathematical form of the second law of motion of newtonian mechanics. Thus the fundamental relation between force and motion is the same in both newtonian and relativistic mechanics—providing Newton's second law is written in terms of the fundamental mechanical quantity momentum, and not acceleration.

In newtonian mechanics the mass m_0 of a body on which a force acts is almost always a constant (for an exception, see Example 4-8). Thus it is almost always true in newtonian mechanics that

$$\mathbf{F} = \frac{d\mathbf{p}}{dt} = \frac{d(m_0\mathbf{v})}{dt} = m_0 \frac{d\mathbf{v}}{dt}$$

or

$$\mathbf{F} = m_0\mathbf{a}$$

The equation $\mathbf{F} = m_0\mathbf{a}$ is (with few exceptions) completely equivalent to the equation $\mathbf{F} = d\mathbf{p}/dt$ in newtonian mechanics. But it is *not* generally possible to write an equation analogous to $\mathbf{F} = m_0\mathbf{a}$ in relativistic mechanics, *not even* if the relativistic mass m is substituted for the rest mass m_0. The error in making this common mistake can be seen by repeating the calculation above using the variable relativistic mass m. It yields

$$\mathbf{F} = \frac{d\mathbf{p}}{dt} = \frac{d(m\mathbf{v})}{dt} = m \frac{d\mathbf{v}}{dt} + \mathbf{v} \frac{dm}{dt} \tag{15-9a}$$

or

$$\mathbf{F} = m\mathbf{a} + \mathbf{v} \frac{dm}{dt} \tag{15-9b}$$

In relativistic mechanics the mass m of a body on which a force acts is usually not constant, since the force will typically lead to a change in the speed of the body and therefore to a change in its relativistic mass. This means the term $\mathbf{v}\, dm/dt$ is generally not zero. Thus, in general, $\mathbf{F} \neq m\mathbf{a}$ for relativistic mechanics.

An exception is found in the special case of a body moving at a constant relativistic speed in a circular path under the application of a centripetal force. The electrons in Fig. 15-3, for example, are moving through circular arcs in a magnetic field. In such circumstances m = constant, $\mathbf{v}\, dm/dt = 0$, and Eq. (15-9b) simplifies to $\mathbf{F} = m\mathbf{a}$. This equation was used by Alfred H. Bucherer in 1909 to analyze an experiment that provided the first quantitative confirmation of the relativistic prediction $m = m_0/\sqrt{1 - v^2/c^2}$

Bucherer sent a beam of electrons from a radioactive source through a speed selector consisting of a region having both "electric fields" and "magnetic fields" in an arrangement that is explained in Chap. 23. For a particular adjustment of the fields, only electrons of a certain speed emerged from the selector. Those that did entered a region containing only a magnetic field, where they moved through a circular arc at constant speed. A measurement of the radius of curvature of the arc, combined with the known speed of the electrons, determined their centripetal acceleration $\mathbf{a}$. The centripetal force $\mathbf{F}$ acting on them could be determined from the measured strength of the magnetic field and the electron speed, in a way that also is explained in Chap. 23. Since the simplified form, $\mathbf{F} = m\mathbf{a}$, of Eq. (15-9b) applied to the motion of the electrons, the measured values of $\mathbf{F}$ and $\mathbf{a}$ could be used to measure the value of their mass m. Bucherer obtained data for several values of the electron speed v. His results are shown by the crosses in Fig. 15-6. The dots in the figure show some results obtained by others more recently. The solid

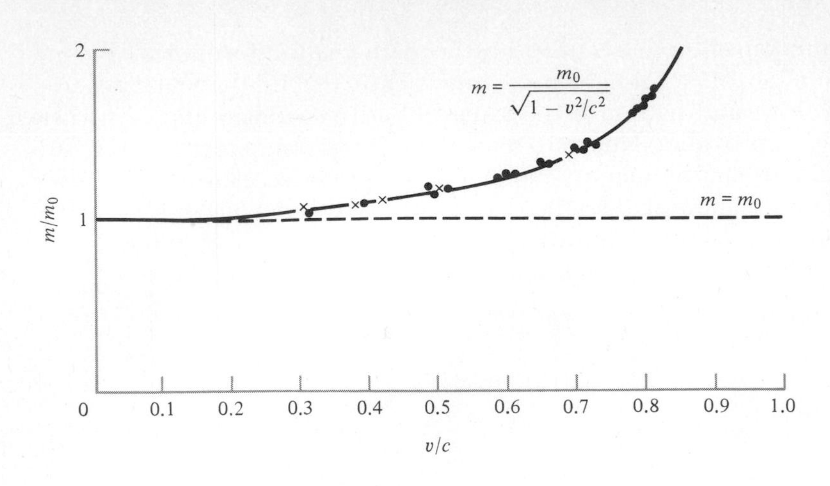

Fig. 15-6 Predicted and measured values of the mass of high-speed electrons. The solid curve is the prediction of relativistic mechanics for the ratio of the mass m of an electron to its rest mass m_0, versus the ratio of its speed v to the speed of light c. The dashed curve shows the prediction of newtonian mechanics. The crosses are values measured by Bucherer, and the dots are values obtained in more recent measurements.

curve represents the relativistic prediction for the mass, and the dashed curve gives the newtonian prediction. Agreement between the relativistic theory and experiment is certainly satisfactory.

The complications that typically arise from the presence of the term $\mathbf{v}\, dm/dt$ in Eq. (15-9*b*) make force a somewhat less useful concept in relativistic mechanics than in newtonian mechanics. But this is not a serious disadvantage because relativistic mechanics is most frequently used in connection with systems of microscopic size. At very small distances, forces are much more difficult to determine experimentally than energies. So even more emphasis is put on the advantages of using energy considerations with microscopic systems than with the macroscopic systems treated in newtonian mechanics—a fortunate match with the features of relativistic mechanics. Of the not-so-frequent uses to which force is put in relativistic mechanics, by far the most important is found in the following derivation of the relativistic expression for kinetic energy.

Fig. 15-7 A particle, initially at rest at x_i, which moves to x_f as a result of the application of a force F.

Imagine a particle of rest mass m_0 that is initially stationary at x_i. A force of magnitude F is applied to it in the direction of the positive x axis, as in Fig. 15-7. The force causes the particle to move with increasing speed along the axis. To evaluate the kinetic energy which the particle acquires, we calculate the work W done by the force while the particle moves to some final location x_f and then equate it to the particle's kinetic energy at x_f. The method is just the same as that used in Sec. 7-3 to evaluate kinetic energy in newtonian mechanics, but the calculation is more complicated because in the present case the mass of the particle increases as its speed increases.

First we carry over into relativistic mechanics the definition of the work W used in newtonian mechanics:

$$W \equiv \int_{x_i}^{x_f} F\, dx \tag{15-10}$$

Then we begin a sequence of manipulations which will lead to evaluation of the integral. As applied to this one-dimensional situation, the equations $\mathbf{F} = d\mathbf{p}/dt$ and $\mathbf{p} = m\mathbf{v}$ produce Eq. (15-9*a*) in the form

$$F = m\frac{dv}{dt} + v\frac{dm}{dt}$$

Hence

$$F\,dx = \left(m\frac{dv}{dt} + v\frac{dm}{dt}\right)dx$$

The variables m and v in the term in parentheses are related by the equation $m = m_0/\sqrt{1 - v^2/c^2}$. Multiplying through by the square root and then squaring both sides of the equality, we get

$$m^2(1 - v^2/c^2) = m_0^2$$

Multiplying through by c^2 converts this equation to

$$m^2c^2 - m^2v^2 = m_0^2c^2$$

We take the time derivative of each term, remembering that m_0 is a constant as well as c, and obtain

$$c^2\frac{d(m^2)}{dt} - \frac{d(m^2v^2)}{dt} = 0$$

Differentiation gives

$$2c^2m\frac{dm}{dt} - 2m^2v\frac{dv}{dt} - 2v^2m\frac{dm}{dt} = 0$$

Dividing through by $-2mv$ and transposing the first term produce

$$m\frac{dv}{dt} + v\frac{dm}{dt} = c^2\frac{dm}{dt}\frac{1}{v}$$

Using this relation in the expression for $F\,dx$, we have

$$F\,dx = c^2\frac{dm}{dt}\frac{1}{v}\,dx$$

Since $v = dx/dt$, it follows that $1/v = dt/dx$, so that

$$F\,dx = c^2\frac{dm}{dt}\frac{dt}{dx}\,dx$$

This simplifies to

$$F\,dx = c^2\,dm \qquad (15\text{-}11)$$

Now we use Eq. (15-11) in Eq. (15-10) to write the work done as

$$W = \int_{x_i}^{x_f} F\,dx = \int_{m_i}^{m_f} c^2\,dm = c^2\int_{m_i}^{m_f} dm$$

Here m_i and m_f are the masses of the particle when it is at x_i and x_f, respectively. The fundamental theorem of calculus allows us to evaluate the integral on the right side of the last equality immediately, to yield

$$W = c^2(m_f - m_i)$$

But m_i is the rest mass m_0 because the particle was stationary at x_i. And at x_f its mass m_f is the relativistic mass m. So we have the result

$$W = mc^2 - m_0c^2 \qquad (15\text{-}12)$$

The work done by the force in building up the speed of the particle, and thereby increasing its mass, equals its final relativistic mass times c^2 minus its initial rest mass times c^2.

We now relate the work W done in setting the particle into motion to the kinetic energy K that the particle has when moving. In newtonian mechanics this is done by defining the kinetic energy to be equal to the work:

$$K = W$$

The same definition is used in relativistic mechanics. So we have for the **relativistic kinetic energy** K of the moving particle

$$K = mc^2 - m_0c^2 \tag{15-13}$$

Confidence in the validity of Eq. (15-13) can be gained by using it to evaluate the kinetic energy of a particle moving at a speed v that is very small compared to c. We write $m = m_0/\sqrt{1 - v^2/c^2}$, and the equation becomes

$$K = \frac{m_0c^2}{\sqrt{1 - v^2/c^2}} - m_0c^2 \tag{15-14a}$$

or

$$K = m_0c^2[(1 - v^2/c^2)^{-1/2} - 1]$$

For v/c very small compared to 1, we can apply with accuracy the binomial expansion approximation

$$(1 - v^2/c^2)^{-1/2} = 1 + \tfrac{1}{2}v^2/c^2$$

From this we obtain

$$K = m_0c^2[1 + \tfrac{1}{2}v^2/c^2 - 1] = \tfrac{1}{2}m_0v^2 \qquad \text{for } v/c \ll 1 \tag{15-14b}$$

So we do find that the relativistic expression for kinetic energy reduces, as it must, to the newtonian expression when the speed of the particle is in the newtonian domain.

A particularly straightforward experimental verification of Eq. (15-14*a*) for v/c comparable to 1 was published in 1964 by Bertozzi. A beam of electrons having a controllable high speed was produced by a particle accelerator. For each setting used, the speed v was determined by measuring electronically the time required for the electrons in the beam to travel a distance of 8.40 m between two electron detectors. The kinetic energy of the electrons in the beam was found at each setting by stopping the beam in a thick aluminum absorber, where the kinetic energy was converted to heat energy. Measurements of the resulting increase in the absorber's temperature in a given amount of time were used to determine the total kinetic energy deposited in the absorber during that time, in a way that is explained in Chap. 17. Separate measurements of the electric charge collected by the absorber gave the number of electrons stopping in it, and so the kinetic energy K per electron could be determined. Results are shown in Fig. 15-8, which is a plot of K/m_0c^2 versus v/c. Experimental values are represented by the two dots. The solid curve is the prediction of relativistic mechanics, and the dashed curve is the prediction of newtonian mechanics. The experimental data agree quite well with the relativistic prediction.

We continue the interpretation of Eq. (15-13),

$$K = mc^2 - m_0c^2$$

Since the quantity K in this equation is an energy, the quantities mc^2 and m_0c^2 must also be energies. Specifically, mc^2 is an energy associated with the particle when it is moving and has mass m, and m_0c^2 is an energy associated with it when it is at rest and has mass m_0. To see their significance, we write the equation as

$$mc^2 = K + m_0c^2$$

and then consider what it means physically. We come to an unavoidable conclusion: The energy mc^2 is the total energy of the moving particle, since it is the sum of the energy K required to set it into motion and an energy m_0c^2 it has when at rest. The energy mc^2 is assigned the symbol E and the name **total relativistic energy.** The energy m_0c^2 is symbolized by E_0 and called the **rest-mass energy.**

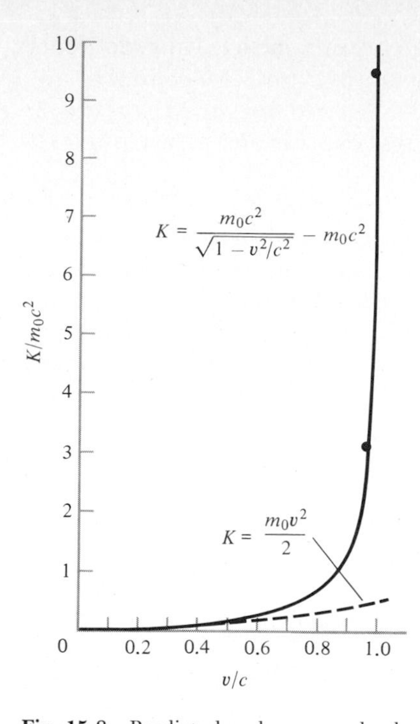

Fig. 15-8 Predicted and measured values of the kinetic energy of high-speed electrons. The solid curve is the prediction of relativistic mechanics for the ratio of the kinetic energy K of an electron to its rest mass energy m_0c^2, versus the ratio of its speed v to the speed of light c. The dashed curve shows the prediction of newtonian mechanics. The dots are values measured by Bertozzi.

We have obtained Einstein's famous relations between energy and mass: *The total relativistic energy of an object equals its relativistic mass times* c^2:

$$E = mc^2 \tag{15-15}$$

The rest-mass energy of an object equals its rest mass times c^2:

$$E_0 = m_0c^2 \tag{15-16}$$

In these equations, Einstein established the fundamental connection between energy and mass. The two mechanical quantities are inseparably related. *The energy content of an object can be measured by its mass, and vice versa, since energy is proportional to mass. And the square of the speed of light is the proportionality constant.*

Let us use the relations we have obtained to describe what a force does when it is applied to an initially stationary particle, as when one of the electrons is coming up to speed in the particle accelerator used by Bertozzi. It is always true that an increment of work done by the force leads to an increment in its total relativistic energy. You can see this from Eq. (15-11), $F\,dx = c^2\,dm$. But while the speed v of the particle has not yet become an appreciable fraction of the speed of light c, its kinetic energy K is small compared to its rest-mass energy m_0c^2. So while the particle is still in the newtonian domain, its total relativistic energy, which according to Eq. (15-13) is $mc^2 = K + m_0c^2$, is not significantly larger than m_0c^2. Thus the fractional increase in m is negligible. This is why the particle follows the newtonian prediction $K = m_0v^2/2$, as you can see from Fig. 15-8, and why the principal manifestation of an increment of work done by the force is an increment in the particle's speed.

But as the force continues to do work, v starts to approach c, and K starts to become significant in comparison to m_0c^2. This means the increase in m becomes important. It is as if the inertia of the particle increased because its mass increased, so that it becomes more difficult for the applied force to increase its speed. This is made clear in Fig. 15-8 by the way the value c acts as a limit to the values of v. In the highly relativistic domain where $v \simeq c$, the most apparent manifestation of an increment of work being done by the applied force is an increment in the particle's mass, not in its speed. The work done increases $K = mc^2 - m_0c^2$ by increasing m, and v hardly changes.

We now have a physical picture of why the speed of light is nature's speed limit, that is, why *no particle whose rest mass m_0 is greater than zero can move with a speed v exactly equal to c or greater than c.* For any such particle, $mc^2 = m_0c^2/\sqrt{1 - v^2/c^2}$ approaches infinity as v approaches c. Thus, the force applied to a particle would have to do an infinite amount of work to make it attain a speed equal to the speed of light. This would consume an infinite amount of energy, and that much energy is never available. If v is greater than c, then mc^2 is an imaginary quantity.

Example 15-3 applies Einstein's energy expressions to a cosmic ray muon.

EXAMPLE 15-3

Calculate the relativistic kinetic energy, total relativistic energy, and rest-mass energy of a cosmic ray muon moving at speed $0.999c$. The rest mass of a muon is 1.9×10^{-28} kg.

■ First you calculate the rest-mass energy

$$\begin{aligned} E_0 &= m_0c^2 \\ &= 1.9 \times 10^{-28}\text{ kg} \times (3.0 \times 10^8\text{ m/s})^2 \\ &= 1.7 \times 10^{-11}\text{ J} \end{aligned}$$

Then use the evaluation $1/\sqrt{1 - v^2/c^2} = 22$ for $v/c = 0.999$ from Example 14-3 to calculate the total relativistic energy

$$\begin{aligned} E = mc^2 &= \frac{m_0c^2}{\sqrt{1 - v^2/c^2}} = 22m_0c^2 \\ &= 22 \times 1.7 \times 10^{-11}\text{ J} \\ &= 3.8 \times 10^{-10}\text{ J} \end{aligned}$$

You can now obtain the relativistic kinetic energy from Eq. (15-13):

$$K = mc^2 - m_0c^2 = E - E_0$$

You have

$$\begin{aligned} K &= 3.8 \times 10^{-10}\text{ J} - 0.17 \times 10^{-10}\text{ J} \\ &= 3.6 \times 10^{-10}\text{ J} \end{aligned}$$

The total relativistic energy of the rapidly moving muon considered in Example 15-3 is very small by the standards of the everyday world, and its rest-mass energy is even smaller. But a muon is a microscopic particle. For a macroscopic body the total relativistic energy is enormous, even when the body is stationary, because it has so much rest-mass energy. For instance, the rest-mass energy of 1 kg of any material is $m_0c^2 = 1\text{ kg} \times (3 \times 10^8\text{ m/s})^2 \simeq 10^{17}$ J! It is fortunate, for safety's sake, that in bulk matter almost all this energy is permanently locked in. Even in uranium about 99.9 percent of the rest-mass energy is unavailable. But approximately 0.1 percent can be extracted by incorporating uranium-235 into a nuclear reactor. This amounts to a potentially available energy of about 10^{14} J in 1 kg of uranium-235.

The word "potentially" suggests that there is a relation between rest-mass energy and potential energy. The relation will be developed in the next section by considering yet another thought experiment.

15-4 RELATIVISTIC ENERGY RELATIONS

Figure 15-9 shows two stationary balls of rest mass m_{01} and m_{02}. A strong spring is compressed between the balls, but they are held together by a latch. For simplicity, we assume the rest mass of the spring and latch arrangement to be negligible. The rest-mass energy stored in the two balls is $m_{01}c^2 + m_{02}c^2$ and the potential energy stored in the spring is U. So the total energy content of the system is $m_{01}c^2 + m_{02}c^2 + U$.

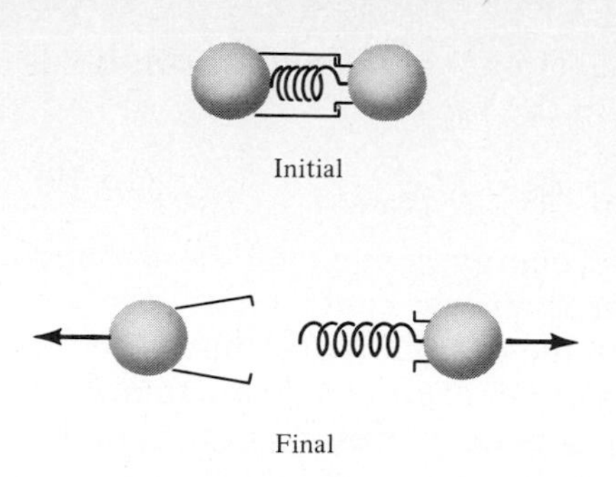

Fig. 15-9 A thought experiment used to investigate the relation between rest-mass energy and potential energy.

The latch is opened with an insignificant expenditure of energy, allowing the spring to revert to its normal length. This drives the balls apart at high speed, and they move away with relativistic masses m_1 and m_2. Since the spring has lost its potential energy U, the total energy content of the system is now the total relativistic energy $m_1c^2 + m_2c^2$ of the two balls. Equation (15-13) shows that the value of this quantity is $m_{01}c^2 + m_{02}c^2 + K_1 + K_2$, where K_1 and K_2 are the relativistic kinetic energies of the two balls.

In newtonian mechanics potential energy is defined so that the loss U in the potential energy of such a system is just equal to the gain $K_1 + K_2$ in its kinetic energy. We adopt this definition, and the resulting relation,

$$K_1 + K_2 = U$$

into relativistic mechanics. Then adding $m_{01}c^2 + m_{02}c^2$ to both sides, we have

$$m_{01}c^2 + m_{02}c^2 + K_1 + K_2 = m_{01}c^2 + m_{02}c^2 + U \qquad (15\text{-}17)$$

This energy conservation equation describes what happens in the isolated system. Its left side is the final value of the energy content of the system. It is comprised partly of the rest-mass energies of the two moving balls and partly of their relativistic kinetic energies. The right side of the equation is the initial value of the energy content. Part of this energy is contained in the rest-mass energies of the two balls, and part of it is in the potential energy of the spring.

Now let us reconsider what happens in a less detailed way. Initially there is a stationary object with some sort of internal structure that we do not describe. Nevertheless, we can still say what the value of its total energy content is because we can measure its mass and Einstein has shown us that the result, multiplied by c^2, gives this energy. Since the object is at rest, we will also say that this energy content is its rest-mass energy M_0c^2, having the numerical value

$$M_0c^2 = m_{01}c^2 + m_{02}c^2 + U \qquad (15\text{-}18)$$

What we have done is to incorporate into the rest-mass energy of the stationary object a contribution U, which is an energy arising from its internal properties.

If the stationary object had any other internal energy, we could do the same. For instance, if heated to a high temperature, the object would gain thermal energy because of the ensuing random motion of its internal constituents. We would add that heat energy to the rest-mass energy. If the constituents were given an organized internal motion, they would have kinetic energy. But we would lump this into the rest-mass energy of the object if it is considered as a *whole* and if as a whole it remains at *rest.* Providing that a body is at rest, overall, its total energy content *is* its rest-mass energy.

Of course, it is sometimes not convenient to ignore the internal structure of a composite object, if the structure is known. For instance, it might be more useful to continue to treat the potential energy U of the spring in the macroscopic object in Fig. 15-9 as such, and not use Eq. (15-18) to define M_0c^2. If so, then Eq. (15-17) would be used to describe what happens. But relativistic mechanics is frequently applied to microscopic particles which have an incompletely known internal structure and which are, therefore, best considered as a whole.

To take this point of view, we use Eq. (15-18) in the energy conservation equation, Eq. (15-17), converting it to the form

$$m_{01}c^2 + m_{02}c^2 + K_1 + K_2 = M_0c^2 \tag{15-19}$$

This says that since the initial rest-mass energy content of the isolated system M_0c^2 is greater than the final rest-mass energy content $m_{01}c^2 + m_{02}c^2$, the system gains kinetic energy $K_1 + K_2$ in the transition to compensate for the loss of rest-mass energy. An extension of our argument to a system containing any number of bodies leads to the general expression of the **energy conservation law of relativistic mechanics:**

$$(m_0c^2_{\text{total}} + K_{\text{total}})_{\text{final}} = (m_0c^2_{\text{total}} + K_{\text{total}})_{\text{initial}} \tag{15-20a}$$

For an isolated system the sum of the total rest-mass energy and total relativistic kinetic energy is conserved, as measured in any given inertial reference frame. Equation (15-13) shows that a completely equivalent form is

$$(mc^2_{\text{total}})_{\text{final}} = (mc^2_{\text{total}})_{\text{initial}} \tag{15-20b}$$

For an isolated system the total relativistic energy is conserved, as measured in any given inertial reference frame. In either form, the single law replaces the separate laws of conservation of mechanical energy and conservation of mass used in newtonian mechanics. Thus the two separate principles of conservation of energy and conservation of mass (which are, in particular, the foundation stones of the science of chemistry) are supplanted by a single more general **principle of the conservation of mass-energy.** In this view, mass is one manifestation of energy, in much the same sense that a compressed spring is a manifestation of energy. We have tried to make this principle seem plausible. But its real justification is found in an abundance of direct experimental confirmation. We present such evidence in later sections and give a precursor later in this section in Example 15-5.

First, however, we will obtain a set of very useful relations among the quantities that characterize the mechanical properties of a relativistic particle. For the purpose of relating its total relativistic energy mc^2, its rest-mass energy m_0c^2, and the magnitude $p = mv$ of its relativistic momentum, we evaluate

$$\begin{aligned} m^2c^4 - m_0^2c^4 &= m_0^2c^4\left(\frac{1}{1 - v^2/c^2} - 1\right) \\ &= m_0^2c^4\left(\frac{v^2/c^2}{1 - v^2/c^2}\right) = c^2\,\frac{m_0^2}{1 - v^2/c^2}\,v^2 \\ &= c^2m^2v^2 = c^2p^2 \end{aligned}$$

Thus

$$m^2c^4 = c^2p^2 + m_0^2c^4$$

and we obtain the result

$$(mc^2)^2 = (cp)^2 + (m_0c^2)^2 \tag{15-21a}$$

Equation (15-13) gives the relation between mc^2 and m_0c^2 involving the relativistic kinetic energy K, instead of p. Let us write it again in the form

$$mc^2 = K + m_0c^2 \tag{15-21b}$$

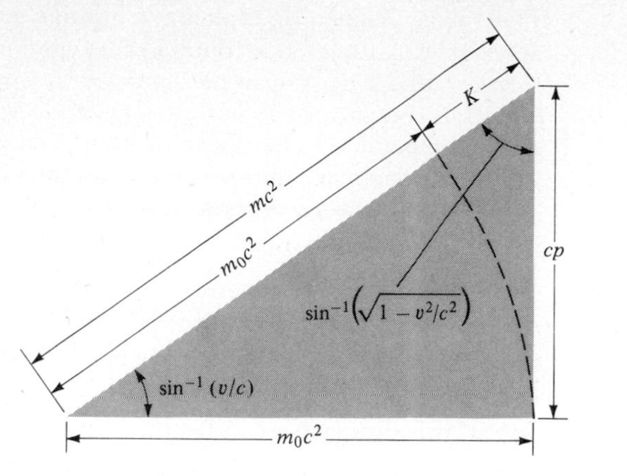

Fig. 15-10 A right triangle and circular arc which form a figure useful in remembering Eqs. (15-21). You should apply simple trigonometry to show that the relations between the lengths marked in the figure are in agreement with these equations.

Taking the square root of both sides of Eq. (15-21*a*) and substituting into Eq. (15-21*b*), we find immediately that the relation between K and p is

$$K = \sqrt{(cp)^2 + (m_0c^2)^2} - m_0c^2 \tag{15-21c}$$

A convenient way of remembering all three of Eqs. (15-21) is provided by the geometrical construction in Fig. 15-10. Justify to yourself that the construction correctly describes the three equations. Example 15-4 makes use of the third.

EXAMPLE 15-4

a. An electron has momentum of magnitude $p = 5.00 \times 10^{-22}$ kg·m/s. Evaluate its relativistic kinetic energy K. The rest mass of an electron is $m_0 = 9.11 \times 10^{-31}$ kg.

■ First you should calculate

$$cp = 3.00 \times 10^8 \text{ m/s} \times 5.00 \times 10^{-22} \text{ kg·m/s} = 1.50 \times 10^{-13} \text{ kg·m}^2/\text{s}^2$$
$$= 1.50 \times 10^{-13} \text{ J}$$

and

$$m_0c^2 = 9.11 \times 10^{-31} \text{ kg} \times (3.00 \times 10^8 \text{ m/s})^2 = 8.20 \times 10^{-14} \text{ kg·m}^2/\text{s}^2$$
$$= 8.20 \times 10^{-14} \text{ J}$$

Then you substitute these quantities into Eq. (15-21*c*) and evaluate

$$K = \sqrt{(15.0 \times 10^{-14} \text{ J})^2 + (8.20 \times 10^{-14} \text{ J})^2} - 8.20 \times 10^{-14} \text{ J}$$
$$= 17.1 \times 10^{-14} \text{ J} - 8.20 \times 10^{-14} \text{ J}$$
$$= 8.9 \times 10^{-14} \text{ J}$$

■

b. For the sake of comparison, use newtonian mechanics to evaluate the kinetic energy of the electron.

■ In newtonian mechanics at all speeds $m = m_0$, $p = m_0v$, and

$$K = \frac{m_0v^2}{2} = \frac{m_0^2v^2}{2m_0} = \frac{p^2}{2m_0}$$

Setting $p = 5.00 \times 10^{-22}$ kg·m/s and $m_0 = 9.11 \times 10^{-31}$ kg, you have

$$K = \frac{(5.00 \times 10^{-22} \text{ kg·m/s})^2}{2 \times 9.11 \times 10^{-31} \text{ kg}} = 1.37 \times 10^{-13} \text{ kg·m}^2/\text{s}^2$$
$$= 13.7 \times 10^{-14} \text{ J}$$

This value, obtained by applying newtonian mechanics to the electron, is considerably larger than the value found in part *a* by applying relativistic mechanics. It is *not* correct because newtonian mechanics is *not* applicable to an electron whose kinetic energy is comparable to its rest-mass energy, since such an electron moves at a speed comparable to the speed of light. Can you give a qualitative explanation of why newtonian mechanics overestimates the value of the electrons's kinetic energy?

The relation between the relativistic kinetic energy and momentum of a particle has a simple limiting behavior at both ends of the range of possible values of the particle's speed v. For v/c approaching 0, Eq. (15-14*b*) shows that K reduces to the expression $K = m_0v^2/2$. Also, p reduces to $p = m_0v$ in this limit where m becomes indistinguishable from m_0. We can combine these two expressions, just as in Example 15-4*b*, to obtain $K = p^2/2m_0$. Multiplying and dividing the right side of this expression by c^2, we have

$$K = \frac{(cp)^2}{2m_0c^2} \qquad \text{for } cp \ll m_0c^2 \tag{15-22}$$

The restriction is expressed in terms of quantities appearing in the equation. It is equivalent to $v/c \ll 1$. Why?

For v/c approaching 1, the quantity cp in Eq. (15-21*c*) increases very rapidly because $p = mv$ and m becomes ever larger. On the other hand, the quantity m_0c^2 is a constant. So in this limit cp becomes very large compared to m_0c^2, and Eq. (15-21*c*) reduces to the expression

$$K = cp \qquad \text{for } cp \gg m_0c^2 \tag{15-23}$$

In the region between these two limiting behaviors, Eq. (15-21*c*) must be used to connect K to cp. The overall relation, for different values of particle rest mass m_0, is pictured qualitatively in Fig. 15-11. What would a plot of K versus cp look like if K is evaluated for all cp by using the newtonian relation $K = (cp)^2/2m_0c^2$? How do the features of the correct relativistic relation between K and cp compare with those of the incorrect newtonian relation?

A particularly simple, although almost unique, case of Eqs. (15-21) is found in a very interesting particle called the **neutrino.** Several accurate, but indirect, experiments imply that the rest mass of a neutrino is zero! For the neutrino $m_0 = 0$, and Eqs. (15-21) reduce to

$$mc^2 = K = cp \qquad \text{for } m_0 = 0 \tag{15-24}$$

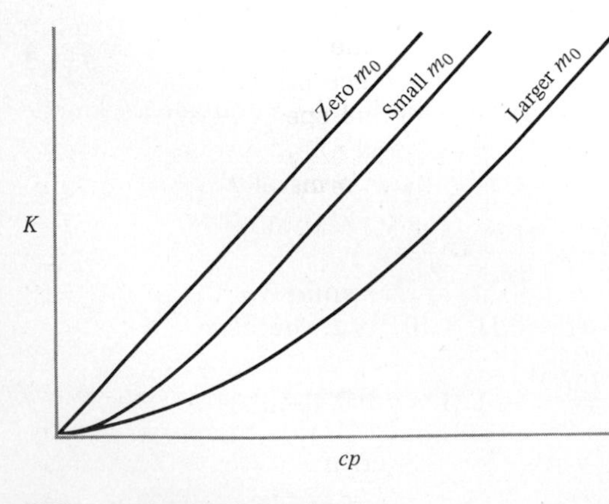

Fig. 15-11 The relation between a particle's kinetic energy K and the speed of light c times its momentum p. The form of the relation is indicated schematically for three different values of the particle's rest mass m_0.

Other experiments show that, even though they have no rest mass, neutrinos do carry relativistic kinetic energy K and momentum p (in amounts that vary depending on the circumstances). Therefore neutrinos do have relativistic mass. With this in mind, the equation

$$m = \frac{m_0}{\sqrt{1 - v^2/c^2}}$$

requires that *for a neutrino $v = c$ in all circumstances.* The denominator must always be zero since the numerator is always zero. Otherwise m, and also K and p, would always be zero. A neutrino's energy is entirely kinetic because it has no rest-mass energy. Furthermore, its momentum is directly proportional to its energy. A neutrino is a **completely relativistic particle** that is *always* in motion and moving with a speed that is *always* measured to have the value c. This is true, independent of the state of motion of the observer.

For a particle of zero rest mass, the concept of rest mass becomes more abstract. It is not possible to define zero rest mass operationally in terms of mass measurements made when the particle is moving relative to the observer at a speed small compared to the speed of light c. This cannot be done since a neutrino *always* moves at speed c. Instead, an observer can measure the particle's momentum p and kinetic energy K. If the results show that $K = cp$, then the particle is said to have zero rest mass.

Since neutrinos, having zero rest mass, move at the speed c under all circumstances, they could be used—just as well as light—to synchronize Einstein's clocks. This would not be a practical thing to do because it is much more difficult to detect neutrinos than to detect light. The reason is that neutrinos interact much more weakly with matter than light does. (Although neutrinos move at the same speed as light, it does not follow that their other properties must be related to those of light.) Nevertheless, the point being made is a very significant one. The speed c is not just the universal speed of light. Rather, *c is the universal limiting speed in nature.* This speed limit is reached by neutrinos because they have zero rest mass.

Another particle having zero rest mass is called the **graviton.** The existence of gravitons is predicted reliably by the theory of gravity. But their interaction with matter is so extremely weak that at the time of this writing they have not been detected experimentally. Although gravitons, like neutrinos, have nothing to do with light, they travel at the same speed c because they have zero rest mass.

There is one more type of zero-rest-mass particle, the **photon.** As we discuss at length in Chap. 30, the existence of photons is very well established experimentally. In contrast to neutrinos and gravitons, which have no relation to light, photons have the most intimate relation to light. Photons *are* light, when light is considered from the viewpoint of the quantum domain. But they travel at speed c because they have zero rest mass, not because of their relation to light.

Most kinds of particles have nonzero rest mass. So the typical particle cannot reach the limiting speed c. However, the very existence of this limit exerts a dominant influence on the particle's behavior when its speed becomes an appreciable fraction of c. This is why c plays a fundamental role in so many different phenomena that have no relation to light or to other forms of electromagnetic radiation.

The process involved in the formation of a cosmic ray muon is analyzed in Example 15-5.

EXAMPLE 15-5

A cosmic ray muon is formed from the decay of a particle called a pion. (See the discussion preceding Example 14-3.) In the decay a neutrino is also formed. The process can be expressed symbolically as

$$\pi \longrightarrow \mu + \nu$$

where the symbols represent a pion, a muon, and a neutrino. Measured values of the rest masses of these particles are $m_{0\pi} = 2.49 \times 10^{-28}$ kg, $m_{0\mu} = 1.89 \times 10^{-28}$ kg, and $m_{0\nu} = 0$. (These are 274, 207, and 0 times the rest mass of an electron.) Predict the kinetic energy K_μ of the muon, as measured in a reference frame moving along with the pion before it decays.

■ Since the pion is at rest in the stipulated reference frame, you can picture the decay by means of the initial-final diagrams of Fig. 15-12. Initially, the isolated system has zero total momentum. So the momentum conservation law, Eqs. (15-5*a*) and (15-5*b*), requires it to have zero total momentum finally. Therefore, you can conclude that the muon and neutrino are emitted "back to back," as shown, with momenta of equal magnitude:

$$p_\mu = p_\nu \tag{15-25}$$

Next you utilize the energy conservation law, Eqs. (15-20*a*) and (15-20*b*). It demands that

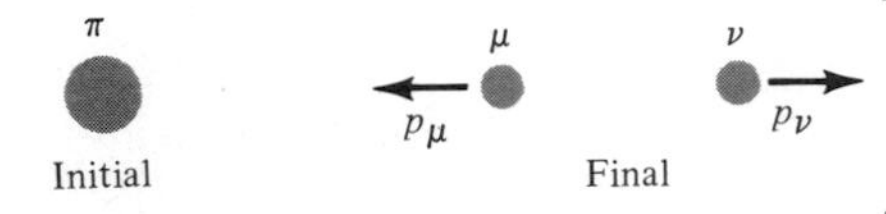

Fig. 15-12 The decay of a pion at rest into a muon and a neutrino.

$$m_{0\pi}c^2 = cp_\nu + \sqrt{(cp_\mu)^2 + (m_{0\mu}c^2)^2} \tag{15-26}$$

The term on the left side of this equation is the pion rest-mass energy, which is the initial total relativistic energy of the system. The system's final total relativistic energy is on the right side. In agreement with Eq. (15-24), the first term is the total relativistic energy of the zero-rest-mass neutrino. The second term is the total relativistic energy of the muon, evaluated by taking the square root of both sides of Eq. (15-21*a*).

Substituting Eq. (15-25) into Eq. (15-26) gives you

$$m_{0\pi}c^2 = cp_\mu + \sqrt{(cp_\mu)^2 + (m_{0\mu}c^2)^2} \tag{15-27}$$

Since $m_{0\pi}c^2$ and $m_{0\mu}c^2$ are known, you now have one equation in one unknown, cp_μ. Transposing and squaring produce

$$m_{0\pi}c^2 - cp_\mu = \sqrt{(cp_\mu)^2 + (m_{0\mu}c^2)^2}$$

and

$$(m_{0\pi}c^2)^2 - 2m_{0\pi}c^2cp_\mu + (cp_\mu)^2 = (cp_\mu)^2 + (m_{0\mu}c^2)^2$$

or

$$cp_\mu = \frac{(m_{0\pi}c^2)^2 - (m_{0\mu}c^2)^2}{2m_{0\pi}c^2}$$

To determine the corresponding value of K_μ, you can use Eq. (15-21*c*) to write Eq. (15-27) as

$$m_{0\pi}c^2 = cp_\mu + K_\mu + m_{0\mu}c^2$$

or

$$K_\mu = m_{0\pi}c^2 - m_{0\mu}c^2 - cp_\mu$$

Using the expression just obtained for cp_μ, you write

$$K_\mu = m_{0\pi}c^2 - m_{0\mu}c^2 - \frac{(m_{0\pi}c^2)^2 - (m_{0\mu}c^2)^2}{2m_{0\pi}c^2}$$

Then you can group the terms as follows:

$$K_\mu = \frac{2(m_{0\pi}c^2)^2 - 2m_{0\pi}c^2 m_{0\mu}c^2 - (m_{0\pi}c^2)^2 + (m_{0\mu}c^2)^2}{2m_{0\pi}c^2}$$

$$= \frac{(m_{0\pi}c^2)^2 - 2m_{0\pi}c^2 m_{0\mu}c^2 + (m_{0\mu}c^2)^2}{2m_{0\pi}c^2}$$

$$= \frac{(m_{0\pi}c^2 - m_{0\mu}c^2)^2}{2m_{0\pi}c^2}$$

or

$$K_\mu = \frac{(m_{0\pi} - m_{0\mu})^2 c^2}{2m_{0\pi}}$$

The numerical value of the kinetic energy of the muon is

$$K_\mu = \frac{(2.49 \times 10^{-28}\ \text{kg} - 1.89 \times 10^{-28}\ \text{kg})^2 \times (3.00 \times 10^8\ \text{m/s})^2}{2 \times 2.49 \times 10^{-28}\ \text{kg}}$$

$$= 6.51 \times 10^{-13}\ \text{J}$$

This value is in excellent agreement with experiments that measure the kinetic energy of the muon in a reference frame in which the pion is at rest when it decays. In these experiments the pions are produced in collisions between protons, emitted from a high-energy particle accelerator, and nuclei in a target. The pions are immediately stopped by an absorber, so that they decay while at rest.

Can you predict the kinetic energy of the neutrino?

Several aspects of Example 15-5 are worthy of comment. First, the solution of the equations arising in Example 15-5 might seem complicated, but it is actually simple compared to what is often involved in applying relativistic mechanics to a system in which two or more particles interact. The calculation was made easier by the zero initial momentum of the system and by the zero rest mass of one of the particles in the final system. In more typical cases, the algebraically complicated relations between momentum and energy that characterize relativistic mechanics can lead to equations which are quite difficult to handle.

Second, note should be taken of the strong analogy between the decay of a pion into a muon and a neutrino, pictured in Fig. 15-12, and the process, pictured in Fig. 15-9, in which a composite object decays by disintegrating into two pieces. If you wish, you can look closely enough at the macroscopic object to see the spring and latch arrangement that constitutes its internal structure before decay. This is very much harder to do for a microscopic particle. In fact, the internal structure of a pion is not completely understood, although it is known that what is called the **weak nuclear force** (see Sec. 1-2) is involved in its "spring and latch" arrangement. But in any case it is not necessary to know the internal structure of whatever it is that decays, if what you are interested in doing is predicting the energies of the decay products. All you need to know is the rest masses of the initial and final constituents of the system.

The third point to note is that in pion decay there is an appreciable decrease in the total rest mass of the isolated system. About 24 percent of the rest mass disappears. However, there is also a significant increase of the kinetic energy of the system. In describing such a process, it is often said that mass has been converted into energy. But it is more accurate to say that *rest-mass energy has been converted into kinetic energy.* The agreement between the measured and predicted values of the kinetic energy gain confirms the validity of the relativistic energy conservation law on which the prediction is based.

15-5 ENERGY AND REST MASS IN CHEMICAL AND NUCLEAR REACTIONS

Certainly the most important practical example of the conversion of rest-mass energy into kinetic energy is found in nuclear fission. In this section we discuss this nuclear reaction, ending up by evaluating the quite appreciable change in rest mass that occurs in the fission of a uranium-235 nucleus. We also consider a simple chemical reaction so that we can compare it to the nuclear reaction with regard to energy and rest-mass change. We start with the chemical reaction because Sec. 8-4 has already presented an essential feature of the explanation of what happens during that reaction.

The chemical reaction of interest is

$$\mathrm{Na} + \mathrm{Cl} \longrightarrow \mathrm{NaCl}$$

A sodium atom (Na) combines with a chlorine atom (Cl) to form a molecule of sodium chloride (NaCl). The process can be described most conveniently in terms of Fig. 15-13. This is a quantitative version of Fig. 8-18 that plots the potential energy U of the system as a function of the center-to-center separation r of the atoms Na and Cl, or of the ions Na^+ and Cl^-, that form the molecule. Consider the two electrically neutral atoms Na and Cl widely separated but approaching each other. When r decreases to the **dissociation separation** r_d, the atoms begin to overlap significantly. At this point an atomic electron will jump from Na to Cl, so that the two constituents of the system become the positive ion Na^+ (Na with a negatively charged electron missing) and the negative ion Cl^- (Cl with an electron added). As r continues to decrease, the system remains in the form of the two ions that together constitute the molecule.

To understand why the mutual ionization occurs, you should consider the fact that Na has 11 electrons and Cl has 17. When Na loses an electron, Na becomes the ion Na^+, which has the same number, 10, of electrons as the noble gas atom neon. At the same time Cl gains an electron to become Cl^-, having 18 electrons just as the noble gas atom argon does. The particular stability of these two noble gases arises from the energetically favorable arrangements of "closed shells" formed by 10 and 18 electrons. So the **ionization energy** E_i absorbed in the mutual ionization has the comparatively small value of about 2×10^{-19} J, because it leads to favorable electron arrangements in both ions. Furthermore, when the process occurs, energy becomes available as a result of the electric attraction between the oppo-

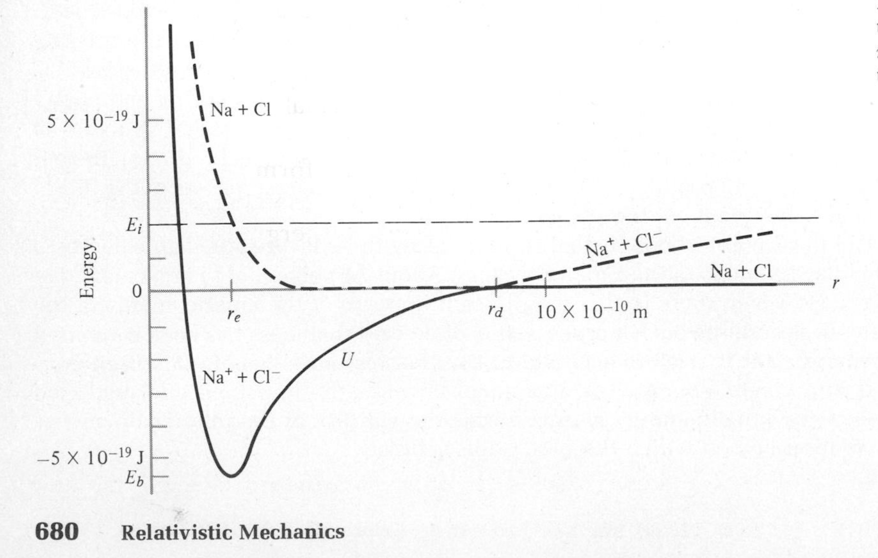

Fig. 15-13 The potential energy U as a function of the center-to-center atomic separation r in the chemical reaction leading to the formation of sodium chloride.

sitely charged ions. As you probably know (and will study in detail in Chap. 20), bodies of opposite electric charge attract one another with a force that obeys the same $-1/r^2$ law as does the gravitational force. Thus there is a negative electric potential energy, proportional to $-1/r$, for the ions Na^+ and Cl^-, just as there is a gravitational potential energy with the same sign and r dependence for a system of two gravitating masses. The electric force is extremely strong compared to the gravitational force, so the electric potential energy has a magnitude extremely large compared to the gravitational potential energy. Nevertheless, it is not large enough to supply the required energy E_i until r decreases to r_d.

As r continues to decrease below that value, the system consists of Na^+ and Cl^-, and its energy U becomes more negative in proportion to $-1/r$. But when r approaches the **equilibrium separation** r_e, the curve of U versus r departs from this behavior. It reaches a minimum at r_e and then turns up. This also arises from the energy associated with an electric force. But in this case it is the repulsive force exerted by the positively charged sodium nucleus on the positively charged chlorine nucleus (assisted by a quantum-mechanical effect called the "exclusion principle"). As you probably also know (and will study in Chap. 20), bodies of the same electric charge repel with a force law just like that for opposite charges except that the sign is positive. So the potential energy is positive in such a case. As r becomes smaller than r_e, there is less and less negative charge of the intervening electrons to "shield" the positive charge of one nucleus from that of the other. So as the highly charged nuclei come closer, the positive energy associated with their repulsion rapidly causes the energy U of the system to increase.

Now let us use Fig. 15-13 to describe what happens in the reaction. The two widely separated constituents of NaCl approach each other with an initial relative speed that is low enough for the corresponding kinetic energy K to be negligible. So the system has an initial total energy E equal to the constant value of U at large r, defined to be $U = 0$. As r decreases to a value less than r_d, U becomes negative and K becomes positive in such a way as to maintain the value $E = K + U = 0$. When r is near r_e, the system emits electromagnetic radiation that carries away an energy of about 5.8×10^{-19} J. This reduces the remaining total energy of the system to a value near the **binding energy** $E_b = -5.8 \times 10^{-19}$ J. With this value of E, the separation distance between the molecular constituents must be near the equilibrium value $r_e = 2.4 \times 10^{-10}$ m. Thus the molecule is formed with its constituents essentially fixed at that separation. In the environment of a typical chemical reaction (for instance, in a solution) the energy emitted in electromagnetic radiation is rapidly converted to thermal energy. This is the source of the heat produced by the reaction.

The amount of energy emitted in the reaction forming the molecule is measured by the binding energy of the molecule. Example 15-6 calculates the rest-mass change associated with this binding energy and the fractional rest-mass change.

EXAMPLE 15-6

a. Determine the difference between the combined rest mass of an Na atom and a Cl atom and the rest mass of an NaCl molecule.

b. Then compare this total rest-mass difference to the initial total rest mass. The rest mass of Na is very nearly equal to 23 u, where the **atomic mass unit** (u) is defined so that the mass of the carbon-12 atom is precisely 12 u. The numerical value of the unit is

$$u = 1.661 \times 10^{-27} \text{ kg} \tag{15-28}$$

There are two species (isotopes) of Cl. Assume that the more prevalent one, chlorine-35, is involved in the reaction. Its rest mass is very close to 35 u.

■ **a.** You consider a system consisting initially of the separated and almost stationary atoms Na and Cl, with a total energy content defined to be $E = 0$. The final state of the system can be considered to consist of the stationary molecule NaCl of total energy content $E = E_b = -5.8 \times 10^{-19}$ J, with the electromagnetic radiation having escaped the system. The energy has decreased since the system is not isolated. But the energy decrease in the system is just equal to the energy carried out of it by the radiation, so energy would be conserved in a larger system in which the radiation was considered to be a part. Both the initial total energy and final total energy of the system you are considering consist of rest-mass energies only, because in each case there is no appreciable kinetic energy. So the decrease ΔE in the total energy E means a decrease $\Delta m_0 c^2$ in the total rest-mass energy $m_0 c^2$, with

$$\Delta m_0 c^2 = \Delta E = -5.8 \times 10^{-19} \text{ J}$$

Thus

$$\Delta m_0 = \frac{\Delta E}{c^2} = \frac{-5.8 \times 10^{-19} \text{ J}}{(3.0 \times 10^8 \text{ m/s})^2}$$
$$= -6.4 \times 10^{-36} \text{ kg}$$

You should repeat the analysis, using the larger system that contains in its final state also the emitted radiation, and show that the same result is obtained.

b. The initial total rest mass m_0 equals the rest mass of Na plus the rest mass of Cl. Using the values given, you have

$$m_0 = 23 \text{ u} + 35 \text{ u} = 58 \text{ u} = 58 \times 1.66 \times 10^{-27} \text{ kg}$$
$$= 9.63 \times 10^{-26} \text{ kg}$$

To compare m_0 with Δm_0, you should evaluate the ratio

$$\frac{\Delta m_0}{m_0} = \frac{-6.4 \times 10^{-36} \text{ kg}}{9.63 \times 10^{-26} \text{ kg}} = -6.7 \times 10^{-11} \simeq -10^{-10}$$

This is the fractional decrease in rest mass during the chemical reaction. It is extremely small.

The results we have obtained from considering the formation of what is called an **ionically bonded** molecule are typical of all chemical reactions. In a **covalently bonded** molecule, mutual ionization does not occur, but the U versus r curve looks very much like Fig. 15-13. The reason for its behavior when r is greater than r_e is more difficult to explain for such a molecule, but it still involves electric attraction between charges of one sign (electrons shared by the molecule) and charges of the other sign (the nuclei). When r is less than r_e, the behavior of U is governed by just the same electric repulsion between the nuclei as in ionically bonded molecules. As a consequence of this similarity, the binding energy E_b for a typical covalently bonded diatomic molecule has about the same value as in NaCl, and the fractional decrease in rest mass also has a value $\Delta m_0/m_0 \simeq -10^{-10}$.

Because this fractional change is so minute in any chemical reaction, the difference between the final and initial rest masses cannot be observed by direct measurements of these masses. (If this were not the case, an experimental chemist probably would have discovered the relativistic mass-energy relations long before they were discovered by a theoretical physicist.) The unobservably small change in rest mass in any chemical reaction provides the practical justification for the law of conservation of mass in chemical reactions, which you have undoubtedly come across if you have studied chemistry.

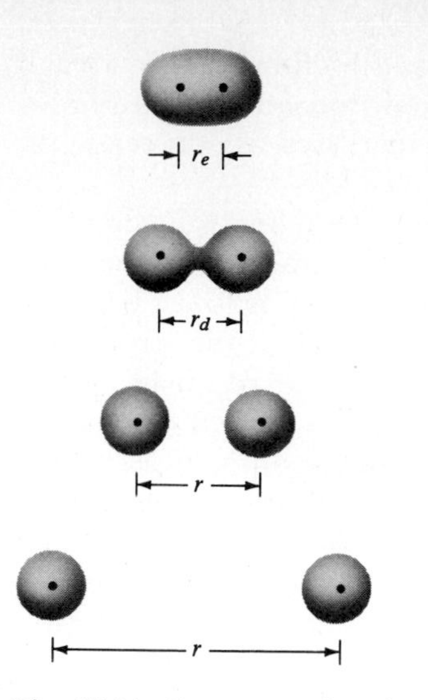

Fig. 15-14 Four successive views of a fissioning nucleus.

But in nuclear reactions $|\Delta m_0/m_0|$ can be larger than in chemical reactions by a factor of around 10^7. That is, in nuclear reactions the magnitude of the fractional change in rest mass is $|\Delta m_0/m_0| \simeq 10^{-3}$ in favorable cases. And in reactions in which a single microscopic particle decays, this ratio can be even larger. For the decay of the pion it is about $\frac{1}{4}$, as you saw in Example 15-4. Thus in nuclear and elementary-particle physics, rest-mass changes in reactions are much too large to be overlooked experimentally. The large values of $|\Delta m_0/m_0|$ encountered in nuclear and elementary-particle physics are a reflection of the interactions that take place between the constituents of nuclei and elementary particles. These interactions are much stronger than those involved in chemical reactions, as we explain below for the case of nuclear fission.

Fission is a nuclear reaction that is of great significance—in many different ways. Symbolically, it can be expressed as

$$F \longrightarrow f_1 + f_2$$

Here F stands for a nucleus of an atom, like uranium, that splits by fission into two smaller nuclei f_1 and f_2, called **fission fragments.** The reaction is depicted in Fig. 15-14. It is most conveniently described in terms of Fig. 15-15, which is a plot of the potential energy U of the system as a function of the center-to-center separation r of its two constituent parts. The behavior of U as r varies can be explained best by imagining that there are two positively charged atomic nuclei f_1 and f_2, each containing about half the rest mass and half the positive electric charge of the nucleus of a uranium

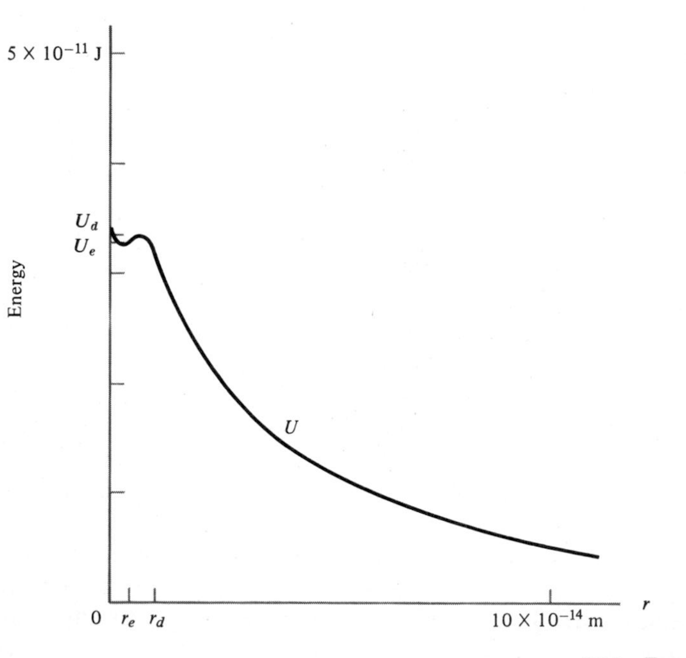

Fig. 15-15 The potential energy U as a function of the center-to-center separation r of the fission fragments in the nuclear reaction leading to the fission of uranium.

atom. They are widely separated but approaching each other. Since each exerts a repulsive electric force on the other proportional to $+1/r^2$, there is a positive potential energy U in the system which increases, as r decreases, in proportion to $+1/r$.

When r decreases to about 1×10^{-14} m, the nuclei f_1 and f_2 begin overlapping. At this point, but not before, they begin to attract each other as a result of the **strong nuclear force** (see Sec. 1-2) that begins to act between them. Like the weak nuclear force involved in pion decay, the strong nuclear force does not act over large distances. Both varieties of nuclear force cut off abruptly when objects between which the forces act cease being essentially in contact. They are said to be of **short range,** in contrast to the long-range electric and gravitational forces whose effects on objects interacting through them diminish only gradually as the objects are separated. The strong nuclear force is so named because it is somewhat stronger than the electric force, if both are operating. For example, two protons (hydrogen atom nuclei) separated by 2×10^{-15} m exert both forces on each other. The nuclear force is attractive, and the electric force is repulsive, with the magnitude of the former being about 10 times greater than that of the latter. But the nuclear force becomes repulsive, and even stronger, if the two particles approach closer than about 0.5×10^{-15} m. The same strong nuclear force acts between two neutrons (particles of almost the same rest mass as protons, but having no electric charge) or between a proton and a neutron—it has nothing to do with electric charge.

The strong nuclear force acting between the neutrons and protons in one fission fragment and those in the other commences when r decreases below $r_d \simeq 1 \times 10^{-14}$ m, the sum of the radii of the two fragments. This produces a decrease in U. But as r continues to decrease, repulsive effects set in and U increases again. So U goes through a minimum at the equilibrium value r_e.

There is a striking similarity between what happens in a nucleus undergoing fission and what happens in the system, pictured in Fig. 15-9, of two balls separated by the spring and latch. And the U versus r curves for each of these systems are qualitatively alike. Before fissioning, a nucleus has a shape specified by a point on Fig. 15-15 at the equilibrium separation r_e. That is, its two parts have separation r_e. The corresponding potential energy is U_e. Since there is no significant motion of the nucleus or its two constituent parts, its total energy content E equals its potential energy U_e. But if the nucleus is given enough extra energy, then it will be possible for it to elongate so that r passes by the value r_d where the potential energy curve has the maximum value U_d. That is, the nucleus must be given extra energy at least as great as $U_d - U_e$ to surmount what is called the **fission barrier.** If the nucleus at r_e receives this energy, it has a total energy greater than its potential energy. The difference can be in the kinetic energy of the two separating fission fragments. It is enough for the fragments to move apart, past r_d. They then continue to separate with increasing rapidity as the kinetic energy K of their motion increases while U decreases. This happens in such a way as to maintain a constant value of the total energy $E = K + U$. When they are widely separated, U has dropped essentially to zero and the total kinetic energy of the motion of the fission fragments has the value U_d. Since the fragments are about the same size, this energy is divided between them about equally. The net effect is that adding the energy $U_d - U_e$ to the nucleus leads to the liberation of the very much larger energy U_d.

You can describe in very similar terms what happens to the double-ball spring-latch system, when the bit of energy required to open the latch is supplied to the system. The analogy is apparent: The fission fragments play the role of the balls; their electric repulsion plays the role of the spring; their nuclear attraction plays the role of the latch.

The value of U_e is approximately 3.2×10^{-11} J, and the value of U_d is approximately 3.3×10^{-11} J. How does the fissioning nucleus get the energy, about 0.1×10^{-11} J, required to "open the latch"? For the uranium-235 nucleus, this energy can be supplied if the nucleus is hit by a slow-moving neutron. Since a neutron is uncharged, it experiences no electric repulsion and so can move freely up to a highly charged nucleus. At the nuclear surface, the strong attractive nuclear force starts to act on the neutron, accelerating it into the nucleus. The kinetic energy it thereby gains is distributed through the nucleus by collisions, and the value happens to be just the required amount, that is 0.1×10^{-11} J. So the nucleus, which is now actually uranium-236, can commence to fission. Uranium-238 will not fission upon capture of a neutron of low initial speed, since the energy provided is not quite enough to put uranium-239 over the top of its fission barrier. It takes a neutron of appreciable initial kinetic energy. Since such neutrons are not sufficiently abundant in a nuclear reactor, uranium-238 is not directly useful as a reactor fuel.

A **nuclear reactor** is a device in which fission takes place continuously, for the purpose of producing heat energy to run an electric power plant. What makes the process possible is the fact that in a typical fission reaction each fission fragment emits on the average about one neutron. This neutron emission takes away about 0.2×10^{-11} J from the total kinetic energy of the fragments. But it leaves approximately 3.1×10^{-11} J, which is degraded into heat energy as the fission fragments are brought to a stop in the reactor by a series of collisions. Of the two neutrons emitted on the average in each fission, one neutron typically is lost in some way or other. The other neutron undergoes collisions in the reactor material, which slow it down. But eventually the neutron hits a uranium-235 nucleus and triggers another fission. Thus the process perpetuates itself. This is the **chain reaction.**

The significant feature of the energetics of fission, and power production by reactors, is that somewhat more than 3×10^{-11} J of energy is produced in each fission reaction. Note that this is larger than the 6×10^{-19} J produced in a typical chemical reaction by a factor of 5×10^7. Comparison of Figs. 15-13 and 15-15 will show that the factor 5×10^7 is accounted for, in part, by the difference in size of a nucleus and of a molecule. In both cases the energy produced in the reaction depends on the value of U at the inner limit of the region where U is determined by the electric force. In this region U varies approximately like $1/r$. Since the value of r at this limit is smaller for nuclei than for molecules by the factor 0.5×10^{-4}, the value of U, and therefore the energy characterizing the reaction, will be larger by the factor 2×10^4 for fission. The rest of the factor 5×10^7 has its origin in the fact that the electric force between the two constituents of the molecule is exerted between single electric charges, while in fission it is exerted between the approximately 46 charges in one fragment (half the number in uranium) and a comparable number of charges in the other. Now, the strength of the electric force is proportional to the number of charges in one of the interacting bodies times the number in the other (just as the strength of the gravitational force is proportional to the product of the masses of the interacting bodies). So the electric force, and therefore the potential energy U, should be larger for the fissioning nucleus than for the molecule by a factor of about $(46)^2 \simeq 2 \times 10^3$. Since the product of the size factor and the charge factor is $2 \times 10^4 \times 2 \times 10^3 \simeq 5 \times 10^7$, we have obtained a very satisfactory explanation of why a fission reaction releases so much more energy than a chemical reaction. Because the point is frequently misunderstood, we emphasize once more that in nuclear fission the energy is supplied by the electric force, not by the nuclear force.

Example 15-7 evaluates the rest-mass change and the fractional rest-mass change in the fission of a uranium-235 nucleus.

EXAMPLE 15-7

a. Evaluate the decrease Δm_0 in the total rest mass of a system consisting initially of a uranium-235 nucleus plus a neutron of negligible kinetic energy that is about to hit it, and consisting finally of two widely separated fission fragments that have not yet emitted neutrons.

b. Then compare the decrease in total rest mass to the initial total rest mass.

■ **a.** Since there is no kinetic energy in the initial system, its total energy consists entirely of rest-mass energy. If you measure energies so that $E = 0$ when $U = 0$ in Fig. 15-15, then $E = U_d = 3.3 \times 10^{-11}$ J. In the final system the fission fragments share a total kinetic energy $K = U_d$. You know that this kinetic energy comes from a decrease in rest-mass energy of the same magnitude because the system is isolated. So you have

$$\Delta m_0 c^2 = -U_d = -3.3 \times 10^{-11} \text{ J}$$

and

$$\Delta m_0 = -\frac{3.3 \times 10^{-11} \text{ J}}{(3.0 \times 10^8 \text{ m/s})^2} = -3.7 \times 10^{-28} \text{ kg}$$

b. The initial total rest mass is approximately

$$\begin{aligned} m_0 &= 236 \text{ u} = 236 \times 1.66 \times 10^{-27} \text{ kg} \\ &= 3.9 \times 10^{-25} \text{ kg} \end{aligned}$$

Thus the comparison called for yields

$$\frac{\Delta m_0}{m_0} = -\frac{3.7 \times 10^{-28} \text{ kg}}{3.9 \times 10^{-25} \text{ kg}} = -9.5 \times 10^{-4} \simeq -10^{-3}$$

The ratio $\Delta m_0/m_0$ is a figure of merit for an energy-producing reaction. The value you obtained in Example 15-6 for a chemical reaction was $\Delta m_0/m_o \simeq -10^{-10}$. Much the same value would be obtained for any other chemical reaction, for instance the one involved in the production of energy by burning coal. On the basis of joules of energy produced per kilogram of fuel consumed, a nuclear power plant is 10^7 times more efficient than a power plant using coal or oil.

15-6 NUCLEAR REACTION Q VALUES

Most nuclear reactions involve nuclei and particles with electric charges appreciably smaller than the charges of a fissioning nucleus and its fission fragments. Such reactions are not dominated by the electric force, and therefore the nuclear force plays a crucial role in determining the energies involved in the reactions. The nuclear force is much more complicated than the electric force. As a consequence, it is difficult to predict the en-

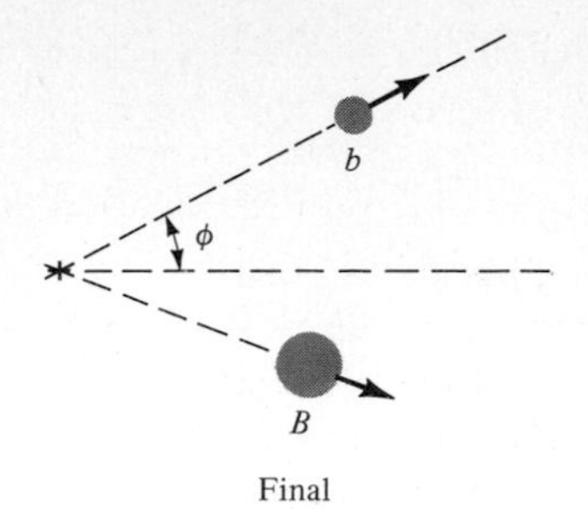

Fig. 15-16 Schematic illustration of a nuclear reaction.

ergies involved in a typical nuclear reaction from a knowledge of the forces that act during the reaction. But it is not at all necessary to do so if what is required is the relation between the initial and final values of the kinetic energies of the constituents of the system. The laws of conservation of momentum and energy show that this relation depends on only the values of the rest masses of the constituents before and after the reaction. And these rest masses can be obtained directly from appropriate experiments. The change in rest mass is a measure of the amount of binding produced by the nuclear force and hence provides a measure of the strength of the force.

A typical **nuclear reaction** is illustrated schematically in Fig. 15-16 and written symbolically as

$$a + A \longrightarrow b + B$$

Before the reaction, a **bombarding particle** a is incident on a **target nucleus** A. After the reaction, a **product particle** b is emitted at an angle ϕ with respect to the direction of the bombarding particle, and the **residual nucleus** B moves off at some other angle. A specific example is provided by the reaction in which a is an alpha particle, as the helium-4 nucleus is called; A is a nitrogen-14 nucleus; b is a proton, or hydrogen-1 nucleus; and B is an oxygen-17 nucleus. This "transmutation" of nitrogen into oxygen by alpha-particle bombardment was the first artificial nuclear reaction. It was first produced in 1919 by Ernest Rutherford, and collaborators, who obtained alpha particles from a radioactive source and used air to provide the nitrogen target nuclei.

In Rutherford's reaction, as in most other reactions studied in nuclear physics, each of the bodies involved has a kinetic energy quite small compared to its rest-mass energy. Thus their total relativistic energies are all only slightly larger than their rest-mass energies, and so they are moving at speeds small compared to the speed of light. Therefore it is a good approximation to use the newtonian relations

$$K_a = \frac{m_{0a}v_a^2}{2} \tag{15-29a}$$

and

$$p_a = m_{0a}v_a \tag{15-29b}$$

to evaluate the kinetic energy and momentum of a in terms of its rest mass m_{0a} and speed v_a, and similarly for b and B.

But relativistic mechanics still enters the reaction in a vital way because the total final kinetic energy $K_b + K_B$ of the isolated system is generally not equal to the total initial kinetic energy K_a. The kinetic energy difference ΔK is called the ***Q*** **value** of the reaction. That is,

$$Q = \Delta K = K_b + K_B - K_a \tag{15-30a}$$

The relativistic law of energy conservation shows that the Q value can also be written in terms of the rest-mass difference Δm_0 as

$$Q = -\Delta m_0 c^2 = -(m_{0b} + m_{0B} - m_{0a} - m_{0A})c^2 \tag{15-30b}$$

Some reactions have positive Q values. A positive Q value means that the total final kinetic energy is larger than the total initial kinetic energy, and thus the total final rest mass is smaller than the total initial rest mass. For other reactions where the Q value is negative, the total kinetic energy decreases in the reaction, and the total rest mass increases.

Using Eqs. (15-29a) and (15-29b), and the similar ones for b and B, in Eqs. (15-30a) and (15-30b), and employing momentum conservation to eliminate the difficult-to-measure quantity K_B, we obtain the **Q-value equation**

$$Q = \Delta K = K_b\left(1 + \frac{m_{0b}}{m_{0B}}\right) - K_a\left(1 - \frac{m_{0a}}{m_{0B}}\right) - 2\frac{\sqrt{m_{0a}m_{0b}}}{m_{0B}}\sqrt{K_a K_b}\cos\phi \tag{15-31}$$

The Q-value equation involves the rest masses of the bombarding particle m_{0a}, the product particle m_{0b}, and the residual nucleus m_{0B}. If approximate values are known for these quantities (and in practice they always are), then the equation allows a determination of the Q value from the measured kinetic energy K_a of the bombarding particle and the measured kinetic energy K_b and emission angle ϕ of the product particle. The measured Q value can then be used in Eq. (15-30b) to obtain an accurate evaluation of the difference between the initial and final total rest masses of the reacting bodies.

It is not worthwhile taking the space to derive Eq. (15-31) because the derivation is almost identical to the one given for Eq. (8-18b). You can see that this is so by comparing the two equations. The earlier one is just a special case of the present equation, pertaining to a type of collision called **scattering.** In scattering, a and b are the same body (called particle 1 in the earlier equation), and A and B also are the same body (called particle 2). That is, in scattering each of the two particles retains its separate identity. There are two kinds of scattering. In **inelastic scattering,** the Q value is negative; there is a decrease in the total kinetic energy of the system and a corresponding increase in the total rest mass. For isolated *macroscopic* bodies, the loss of kinetic energy appears as a gain in the thermal energy of the bodies; but the fractional rest-mass increase that results is so small as to be unmeasurable. In **elastic scattering,** the Q value is zero. An application of Eq. (15-31), reduced to the form for elastic scattering, was given in Example 8-10. An application of the equation in its general form to a nuclear reaction is given in Example 15-8.

EXAMPLE 15-8

In Rutherford's reaction, alpha particles of measured kinetic energy $K_a = 1.23 \times 10^{-12}$ J are used to bombard nitrogen-14 nuclei, producing protons and oxygen-17 nuclei. The protons emitted in the forward direction, where $\phi = 0$, are measured to have a kinetic energy $K_b = 9.53 \times 10^{-13}$ J. Determine the Q value of the reaction. Then evaluate Δm_0, the difference between the final total rest mass and the initial total rest mass. Express Δm_0 in atomic mass units.

■ The first thing you must do is to verify that Eq. (15-31) is applicable by showing that K_a and K_b are indeed small compared to $m_{0a}c^2$ and $m_{0b}c^2$, so that Eqs. (15-29*a*) and (15-29*b*) are usable. For this purpose it is certainly accurate enough to equate the rest masses of an alpha particle and a proton to the rest masses of the corresponding atoms helium-4 and hydrogen-1. (More than 99.9 percent of an atom's rest mass is in its nucleus.) The rest masses of the helium-4 atom and the hydrogen-1 atom are very close to 4 u and 1 u, where u = 1.66×10^{-27} kg. So you have for the nuclear rest-mass energies

$$m_{0a}c^2 = 4 \times 1.66 \times 10^{-27} \text{ kg} \times (3.00 \times 10^8 \text{ m/s})^2 = 5.98 \times 10^{-10} \text{ J}$$

and

$$m_{0b}c^2 = 1 \times 1.66 \times 10^{-27} \text{ kg} \times (3.00 \times 10^8 \text{ m/s})^2 = 1.49 \times 10^{-10} \text{ J}$$

Comparison with the kinetic energies given above justifies the use of Eqs. (15-29*a*) and (15-29*b*) from newtonian mechanics in deriving Eq. (15-31) for the Q value, as far as a and b are concerned. Can you explain why the same will be true for B?

Writing Eq. (15-31) with $\cos\phi = 1$ for $\phi = 0$, you have

$$Q = K_b\left(1 + \frac{m_{0b}}{m_{0B}}\right) - K_a\left(1 - \frac{m_{0a}}{m_{0B}}\right) - 2\frac{\sqrt{m_{0a}m_{0b}}}{m_{0B}}\sqrt{K_aK_b}$$

It is sufficiently accurate to use the values $m_{0a} = 4$ u and $m_{0b} = 1$ u in this equation, as justified above. The same justification shows that you can use $m_{0A} = 14$ u and $m_{0B} = 17$ u for the rest masses of the nuclei of nitrogen-14 and oxygen-17. With these numerical values you obtain

$$Q = 9.53 \times 10^{-13} \text{ J}\left(1 + \frac{1 \text{ u}}{17 \text{ u}}\right) - 1.23 \times 10^{-12} \text{ J}\left(1 - \frac{4 \text{ u}}{17 \text{ u}}\right)$$
$$-2\frac{\sqrt{4 \text{ u} \times 1 \text{ u}}}{17 \text{ u}}\sqrt{1.23 \times 10^{-12} \text{ J} \times 9.53 \times 10^{-13} \text{ J}}$$
$$= -1.89 \times 10^{-13} \text{ J}$$

Since $Q = \Delta K$, you have shown that 1.89×10^{-13} J of kinetic energy is lost in the reaction. The Q-value equation allows you to reach this conclusion even though K_B, the kinetic energy of the residual nucleus, is not measured.

The change in rest-mass energy is

$$\Delta m_0c^2 = -Q = 1.89 \times 10^{-13} \text{ J}$$

There is as much rest-mass energy gained as there is kinetic energy lost. The amount of rest mass gained is

$$\Delta m_0 = -\frac{Q}{c^2} = \frac{1.89 \times 10^{-13} \text{ J}}{(3.00 \times 10^8 \text{ m/s})^2} = 2.10 \times 10^{-30} \text{ kg}$$

To express this result in atomic mass units, you evaluate

$$\Delta m_0 = 2.10 \times 10^{-30} \text{ kg} \times \frac{1 \text{ u}}{1.66 \times 10^{-27} \text{ kg}} = 0.00127 \text{ u}$$

The result $\Delta m_0 = 0.00127$ u, obtained in Example 15-8, establishes a relation among the rest masses of the nuclei of the atoms helium-4, nitrogen-14, hydrogen-1, and oxygen-17. It also relates the rest masses of the atoms themselves. The helium atom has two electrons, and the nitrogen atom has seven. So the initial total nuclear rest mass in the reaction is approximately the rest masses of these atoms, less nine electron rest masses. The hydrogen atom has one electron, and the oxygen atom has eight. So the final total nuclear rest mass is also approximately the total atomic rest mass less nine electron rest masses. The word "approximately" is used because we are ignoring the mass equivalents of the binding energies of the atomic electrons. The error in so doing is appreciably smaller than the possible error in the measured value $\Delta m_0 = 0.00127$ u, which is about 1 in the fifth decimal place. Thus this value also equals the difference between the sum of the rest masses of the hydrogen-1 and oxygen-17 atoms and the sum of the rest masses of the helium-4 and nitrogen-14 atoms.

As indicated earlier, if very accurate measurements already have been made for three of the four rest masses, the value of Δm_0 obtained from the Q-value analysis of the nuclear reaction measurement can be used to determine accurately the fourth rest mass. This is a widely employed experimental procedure. In the particular case of the particpants in Rutherford's reaction, completely independent measurements have been made with great accuracy of *all* four of the atomic rest masses. The **mass spectroscopy** technique used in these measurements is very similar to that used by Bucherer and described in Sec. 15-3. A beam of atoms of the species of interest is sent through an electric discharge, which ionizes each atom by removing one electron. The charged ions then enter a region of electric and magnetic fields, where they follow a path that specifies their speed and their momentum. From these two quantities the mass of each ion can be evaluated. Because their speeds are quite low, the mass is the rest mass. Adding the rest mass of an electron to the rest mass of the ion gives the rest mass of the neutral atom. The measured atomic rest masses and their sums are

Helium-4:	4.002603 u
Nitrogen-14:	14.003074 u
Sum:	18.005677 u

Hydrogen-1:	1.007825 u
Oxygen-17:	16.999133 u
Sum:	18.006958 u

The rest-mass difference obtained from mass spectroscopy is therefore given by

Final m_0:	18.006958 u
Initial m_0:	18.005677 u
Difference:	0.001281 u

Thus the value obtained from this experimental technique is $\Delta m_0 = 0.001281$ u. The value found from the analysis of the completely independent nuclear reaction measurement is, according to Example 15-8, $\Delta m_0 = 0.00127$ u.

The two determinations agree to within the accuracy of the nuclear reaction measurement. Agreement to even greater precision is found in other cases. This provides the most accurate experimental verification of the relativistic energy conservation law, on which the Q-value equation used in the nuclear reaction analysis is based. Thus it verifies with the highest degree of accuracy one of the most basic predictions of Einstein's special theory of relativity.

We have completed our development and testing of the special theory of relativity. But we will make further use of relativistic kinematics and mechanics, where appropriate, as we proceed with our study of physics. For instance, relativistic kinematics is used to give a physical interpretation of the origin of magnetic fields and of their relation to electric fields. Several applications of relativistic mechanics are presented in our treatment of electromagnetism. And particularly important use is made of relativistic mechanics when we investigate the particlelike properties of electromagnetic radiation, in connection with the topic of quantum physics. On the other hand, we continue to employ newtonian mechanics whenever we can because it is simpler, and more familiar, than relativistic mechanics.

EXERCISES

Group A

15-1. *Muon watch.* Evaluate the relativistic mass, momentum, and kinetic energy of the muons considered in Exercise 14-3. These muons are produced at a high-energy particle accelerator with a speed such that a laboratory measurement of their average lifetime gave the value 6.9×10^{-6} s. The average lifetime of muons at rest is 2.2×10^{-6} s, and their rest mass is 1.89×10^{-28} kg.

15-2. *Nonrelativistic and extreme relativistic limits.*

a. What is the maximum value of v/c for which the relativistic kinetic energy of a particle can be expressed as $\frac{1}{2} m_0 v^2$ with an error of not more than 1 percent?

b. What is the minimum value of v/c for which the relativistic kinetic energy of a particle can be taken equal to the total energy with an error of not more than 1 percent?

15-3. *Survival distance.* The rest mass energy of a pion is 2.23×10^{-11} J. As seen by an observer with respect to whom the particle is essentially at rest, its half-life is 1.77×10^{-8} s. (This is the time required for half the particles to decay into something else.) Suppose that another observer measures the total relativistic energy of one of a group of such particles, finding $10 \times 2.23 \times 10^{-11}$ J. From the point of view of this observer, how far will the group of particles travel before half of them decay?

15-4. *Energy versus speed.*

a. If the total relativistic energy of a particle is E and its rest-mass energy is E_0, show that $v/c = [1 - (E_0/E)^2]^{1/2}$.

b. Calculate v/c for an electron whose kinetic energy equals its rest-mass energy.

15-5. *Rutherford's reaction.* Use the Q value obtained in Example 15-8 to determine the expected kinetic energy K_b of protons emitted at the scattering angle $\phi = 40°$.

15-6. *Mass loss in the freezing of water.* When 1.00 kg of water at 0°C freezes into ice at the same temperature, it liberates 3.34×10^5 J of heat energy. What is the fractional decrease of the mass of water when it freezes to ice?

15-7. *A colliding-beam experiment.* Find the minimum total relativistic energy and the speed of oppositely directed electron beams that will allow the production of electron-positron pairs by the reaction

$$\text{electron} + \text{electron} \rightarrow \text{electron} + \text{electron} + \text{positron} + \text{electron}$$

All the particles have rest-mass energy 8.2×10^{-14} J.

15-8. *Annihilation.* An electron and its antiparticle, a positron, annihilate one another to create photons, particles with zero rest mass.

a. If the electron and positron are at rest with respect to each other, why is it impossible for the annihilation to produce just one photon?

b. If two photons are produced, how are their motions related? What is the energy of each? The rest-mass energy of an electron or a positron is 8.2×10^{-14} J.

15-9. *Mass-energy equivalence.* Consider two identical objects, each of rest mass m_0, including the mass of a relaxed spring attached to it. See Fig. 15E-9*a*. An experimenter brings the two together, compressing the springs as shown in Fig. 15E-9*b*. He then releases them, and finds that the springs cause them to move quickly in opposite directions, each with speed v.

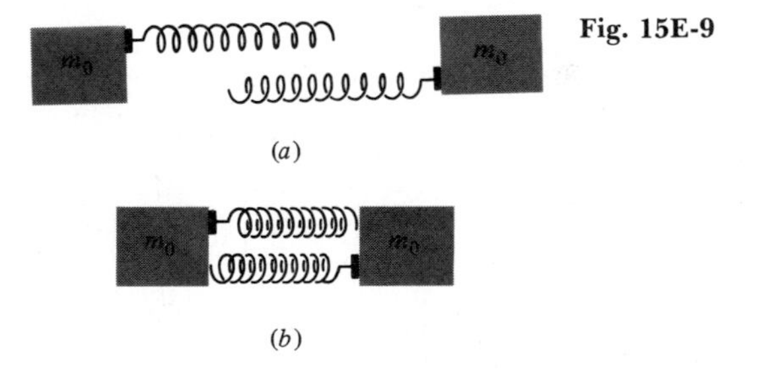

Fig. 15E-9

a. What is the relativistic expression for the total energy of the system when the objects are in motion?

b. On the basis of the conservation of total energy, what must have been the energy of the system when the objects were stationary?

c. What was the rest-mass energy of the masses alone when the objects were stationary? Account for the fact that the total rest-mass energy of the stationary system in Fig. 15E-9*b* is greater than the rest-mass energy of the separated masses in Fig. 15E-9*a*.

15-10. *Binding energy: the helium nucleus.* A helium nucleus consists of two protons each of mass 1.007277 u and two neutrons each of mass 1.008665 u. The mass of a helium nucleus is 4.001505 u. What is the binding energy of the helium nucleus per nucleon (nuclear particle)?

15-11. *Relativistic conservation law for mass-energy.* In the first nuclear disintegration experiment using an artificially accelerated particle, lithium nuclei were bombarded by swiftly moving protons (hydrogen nuclei), resulting in the reaction

$$\text{proton} + \text{lithium-7} \longrightarrow \text{helium-4} + \text{helium-4}$$

The kinetic energy of the proton was 0.80×10^{-13} J. Each helium nucleus had a kinetic energy of 14.24×10^{-13} J. The rest mass of the proton is 1.007277 u, that of the lithium-7 nucleus is 7.014358 u, and that of the helium-4 nucleus is 4.001505 u.

Show that the experimental data confirms the relativistic equation for the conservation relation of mass-energy.

15-12. *Pick your angle, choose your energy.* A nuclear reaction sometimes used to produce neutrons of uniform energy from a beam of protons of uniform energy is

$$\text{proton} + \text{lithium-7} \longrightarrow \text{neutron} + \text{beryllium-7}$$

The Q value of the reaction is -2.62×10^{-13} J. A target containing lithium-7 nuclei is bombarded by a beam of protons of kinetic energy 8.00×10^{-13} J. At what angle to the proton beam will neutrons of kinetic energy 5.00×10^{-13} J be emitted?

15-13. *Proton kinetic energies in Rutherford's reaction.* Use the Q value equation, together with information presented in Example 15-8, to calculate the kinetic energy of protons emitted in Rutherford's reaction at an angle of 20° to the beam of incident alpha particles.

15-14. *Threshold energy for a reaction of alpha particles with aluminum, I.* Radioactive phosphorus can be produced by bombarding aluminum with alpha particles (helium nuclei) according to the reaction

$$\text{helium-4} + \text{aluminum-27} \longrightarrow \text{neutron} + \text{phosphorus-30}$$

The masses of the atoms are: aluminum-27: 26.981532 u; helium-4: 4.002603 u; phosphorus-30: 29.978353 u; neutron: 1.0086653 u.

What is the minimum kinetic energy of the alpha particles required to bring about the reaction? Ignore the recoil energy of the products by assuming that the target nucleus is infinitely massive compared to the bombarding particle.

15-15. *Q value for a reaction producing carbon-14.* In the upper atmosphere, as a result of cosmic ray bombardment, the following reaction takes place

$$\text{neutron} + \text{nitrogen-14} \longrightarrow \text{proton} + \text{carbon-14}$$

The mass of the neutron is 1.0086653 u. A nitrogen-14 atom has a mass of 14.0030732 u; a carbon-14 atom has a mass of 14.003239 u; and for hydrogen-1 it is 1.0078252 u. What is the Q value of the reaction? (The nearly constant rate at which this reaction occurs over periods of millenia makes possible the technique of radiocarbon dating, which is of great importance in archaeology and related fields. The radioactive carbon-14 atoms become incorporated first into carbon dioxide molecules and thence through the process of photosynthesis into living plant matter. Since the radioactive decay rate of carbon-14 is known, the time elapsed since the death of the plant can be determined by measuring the amount of carbon-14 remaining in it.)

Group B

15-16. *Power play.* In newtonian mechanics the relation $dE/dt = \mathbf{F} \cdot \mathbf{v}$ is valid, where E is the total energy of a particle that is moving with velocity $\mathbf{v}$ and is acted on by a net force $\mathbf{F}$. Show that this relation is also valid in relativistic mechanics. (Note: You will need to express the squared magnitude of the momentum in vector form: $p^2 = \mathbf{p} \cdot \mathbf{p}$.)

15-17. *Velocity in terms of energy and momentum.* Show that the components of the velocity of a particle of energy E and momentum $\mathbf{p}$ are given by

$$v_x = \frac{\partial E}{\partial p_x} \qquad v_y = \frac{\partial E}{\partial p_y} \qquad v_z = \frac{\partial E}{\partial p_z}$$

These relations apply in both the relativistic and newtonian domains.

15-18. *Hit and stick.* A particle of rest mass m_0 traveling at a speed of v_i makes a completely inelastic collision with an identical particle initially at rest, and they stick together.

a. What is the rest mass of the composite particle?

b. What percentage of the kinetic energy was converted into rest-mass energy?

c. What happened to the remainder of the kinetic energy?

d. What is the speed of the composite particle?

15-19. *Decay of a neutral pion.* A moving π^0 particle decays into two photons, which have zero rest mass. One photon travels in the same direction as the π^0 was traveling, the other in the opposite direction. What is the energy of each photon if the total relativistic energy of the π^0, E_π, is twice its rest-mass energy of 2.16×10^{-11} J?

15-20. *Decay of a positive kaon.* A K^+ particle decays to a μ^+ particle and a neutrino. If the K^+ is at rest, what are the kinetic energies of the μ^+ and the neutrino? The rest mass energy of the K^+ is 7.90×10^{-11} J; that of the μ^+ is 1.71×10^{-11} J. The rest-mass of the neutrino is 0.

15-21. *Decay of a neutral kaon.* A moving K_1^0 particle decays into two π^0 particles. If one of these is observed at rest, what are the total relativistic energy, E_K, of the K_1^0 and the total relativistic energy, E_π, of the moving π^0?

The rest-mass energy of the K_1^0 is $m_{0K}c^2 = 7.96 \times 10^{-11}$ J; that of the π^0 is $m_{0\pi}c^2 = 2.16 \times 10^{-11}$ J.

15-22. *Decay of a pion into a muon and a neutrino.* Continue the calculation in Example 15-5 so as to obtain a prediction of the kinetic energy of the neutrino, measured in a reference frame moving along with the pion before it decays.

15-23. *The Q-value equation.* Derive the Q value equation, Eq. (15-31), by modifying the derivation in the text leading to Eq. (8-18b).

15-24. *Threshold energy for a reaction of alpha particles with aluminum, II.* In Exercise 15-14, you were asked to ignore the recoil energy of the products in calculating the reaction threshold.

a. Recalculate the threshold energy for the reaction

$$\text{helium-4} + \text{aluminum-27} \longrightarrow \text{neutron} + \text{phosphorus-30}$$

taking the finite mass of aluminum-27 into account.

b. What percentage error in the threshold is caused by assuming the target nucleus is infinitely massive as you did in Exercise 15-14?

15-25. *Decay of free neutrons.* Free neutrons decay by the process

$$\text{neutron} \longrightarrow \text{proton} + \text{electron} + \text{antineutrino}$$

The masses of the particles are as follows: neutron: 1.008665 u; proton: 1.007277 u; electron: 5.49×10^{-4} u; antineutrino: 0 u.

a. Find the Q value for this decay process.

b. In this decay, most of the released energy is carried by the electron and the antineutrino, because they are much less massive than the proton. This energy can be divided between the electron and the antineutrino in essentially any proportion. Suppose that in a particular decay the energy and momentum carried by the antineutrino are negligibly small. Find the energies and momenta of the proton and the electron.

c. Suppose that in another decay, the electron is emitted with negligible kinetic energy and momentum. Find the energies and momenta of the proton and the antineutrino.

15-26. *Threshold energy for a reaction of protons with beryllium.* The reaction

$$\text{proton} + \text{beryllium-9} \longrightarrow \text{neutron} + \text{boron-9}$$

has $Q = -2.97 \times 10^{-13}$ J. If a beam of protons is directed against a beryllium target, what is the minimum proton kinetic energy required for the reaction to occur?

Group C

15-27. *Relativistic dragster.* A particle of rest mass m_0 is acted on by a constant force of magnitude F directed along the x axis. The particle is at rest at the origin at time $t = 0$.

a. Find the particle's relativistic momentum as a function of the time t.

b. Find the particle's total relativistic energy as a function of its displacement x.

c. Use the relation $E = \sqrt{(cp)^2 + (m_0c^2)^2}$ to find the particle's displacement as a function of time t.

d. Use the result of part c to find the particle's velocity and acceleration as functions of time. Do your results agree with nonrelativistic mechanics for small values of t?

15-28. *Pair production.* The photon, a particle with zero rest mass, is sometimes transformed into two particles, an electron and a positron, each of which has rest-mass energy of 8.2×10^{-14} J. This process is called **pair production.** Suppose the new particles are emitted in the direction of travel of the original photon.

a. Show that the momentum of either particle is less than its energy divided by c.

b. Show from this that if the conservation of energy is satisfied by the transformation, the conservation of momentum is not. (This result means that the decay can occur only when some other particle is available to absorb the excess momentum of the photon.)

c. What is the threshold value (smallest possible value) of the photon energy for this transformation?

15-29. *An elastic collision.* A particle whose rest mass is m_0 and whose relativistic kinetic energy is K strikes an identical particle initially at rest. The collision is elastic, the particles remaining unchanged. The collision is also symmetrical, each particle moving off at an angle of $\theta/2$ with the original velocity.

a. Show that $\cos^2(\theta/2) = (K + 2m_0c^2)/(K + 4m_0c^2)$

b. From this show that $\cos\theta = K/(K + 4m_0c^2)$

c. Calculate θ if $K = m_0c^2$.

15-30. *The Lorentz invariance of $E^2 - (cp)^2$.* A particle with rest mass m_0 has speed v in the positive x direction in the "laboratory" frame of reference O.

a. Calculate its energy, kinetic energy, and momentum in the laboratory frame. Verify explicitly the result, taken from Eq. (15-21), that $E^2 - (cp)^2 = (m_0c^2)^2$.

b. Calculate the energy, kinetic energy, and momentum of the particle in a "rocket" frame of reference O' moving at speed V in the positive x direction with respect to the laboratory frame. Show that $(E')^2 - (cp')^2 = (m_0c^2)^2$. This is an example of an invariant quantity; $E^2 - (cp)^2 = E^2 - c^2(p_x^2 + p_y^2 + p_z^2)$ is the same in *every* inertial coordinate system. For another example, see Exercise 14-33.

15-31. *Lorentz momentum-energy transformation.* A particle of rest mass m_0 is moving at speed v in the positive direction of the x axis of inertial frame O. In that frame the components of its relativistic momentum are $p_x = m_0v/\sqrt{1 - v^2/c^2}$, $p_y = 0$, $p_z = 0$. Its total relativistic energy is $E = m_0c^2/\sqrt{1 - v^2/c^2}$. An observer O' is moving in the positive direction along the x axis at speed V, with $V < v$. In the inertial frame moving with O' the components of the relativistic momentum of the particle are $p'_x = m_0v'/\sqrt{1 - v'^2/c^2}$, $p'_y = 0$, $p'_z = 0$, and its total relativistic energy is $E' = m_0c^2/\sqrt{1 - v'^2/c^2}$. Here v' is the speed of the particle with respect to O'. Use the Lorentz velocity transformation, given by the first of Eqs. (14-21), to evaluate v'. Then use this value of v' in the right sides of the expressions for p'_x, p'_y, p'_z, and E' to derive the following set of equations:

$$p'_x = \frac{1}{\sqrt{1 - V^2/c^2}} (p_x - VE/c^2)$$

$$p'_y = p_y$$

$$p'_z = p_z$$

$$E' = \frac{1}{\sqrt{1 - V^2/c^2}} (E - Vp_x)$$

These equations constitute the **Lorentz momentum-energy transformation.** Compare them to Eqs. (14-16), the Lorentz position-time transformation. In the comparison show that the quantities p_x, p_y, p_z, E/c^2 transform in exactly the same ways as the quantities x, y, z, t, respectively. This circumstance is the starting point of a more advanced treatment of special relativity, using four dimensional vectors with components (x, y, z, t) and $(p_x, p_y, p_z, E/c^2)$. These vectors are usually called "four-vectors."

15-32. *Momentum-energy conservation.* Two particles with rest masses m_{01} and m_{02} are moving along the x axis in the inertial frame O with velocities v_1 and v_2. They collide head on; out of the collision emerge two different particles with rest masses m_{03} and m_{04} moving along the x axis with velocities v_3 and v_4. Conservation of momentum requires that the relativistic momenta of the four particles obey the relation $p_1 + p_2 - p_3 - p_4 = 0$.

a. Use the Lorentz momentum-energy transformation to obtain the relation that holds in a second inertial frame O' moving with velocity V relative to O along its x axis.

b. If the conservation of momentum holds for O', what other conservation law must hold simultaneously?

15-33. *Energy of a cosmic-ray muon.* In the formation process for cosmic-ray muons considered in Example 15-5, the pion is moving in the direction toward the earth's surface before it decays, and has a relativistic kinetic energy 30 times its rest-mass energy. The muon emitted when the pion decays happens to be emitted in the direction towards the earth's surface. Use the Lorentz momentum-energy transformation derived in Exercise 15-31 to evaluate the relativistic kinetic energy of the muon, as measured in a reference frame fixed to the earth's surface.

15-34. *Speed of the zero-momentum frame.* In the inertial frame of a laboratory, particle 1 is moving to the right with total relativistic energy E_1 and relativistic momentum p_1, and particle 2 is stationary with total relativistic energy E_2 equal to its rest-mass energy. Use the Lorentz momentum-energy transformation equations derived in Exercise 15-31 to show that in an inertial frame moving to the right with speed

$$V = \frac{cp_1}{E_1 + E_2} c$$

the system of two particles has zero total relativistic momentum.

15-35. *An elastic electron-electron collision.* An electron is moving through the inertial reference frame of the laboratory at a speed comparable to the speed of light. It experiences an elastic collision with another electron, which is free and initially stationary. Use the conservation of relativistic momentum and the conservation of total relativistic energy to derive an expression showing that the angle between the trajectories of the two electrons after the collision will be less than 90°. Then use this expression to show that the angle you measure between the electron trajectories in the cloud chamber photograph of Fig. 15-3 is consistent with a relativistic mass of the incident electron equal to 4.1 times its rest mass. (Hint: You may find it easier to use the equations displayed in Exercises 15-31 and 15-34 to perform the calculation as follows: transform to the inertial frame in which the total relativistic momentum of the electrons before the collision is zero; treat the collision; then transform back to the laboratory reference frame.)

15-36. *Threshold for production of a proton-antiproton pair.* An observer in a laboratory sees a proton of sufficient energy strike a stationary proton to produce a proton-antiproton pair in addition to the two original protons. (The antiproton is the antiparticle of the proton. It has the same mass as a proton but a negative charge. See Exercise 15-7 for an analogous process involving electrons and a positron.)

To determine the reaction threshold (the minimum kinetic energy of the proton moving in the laboratory reference frame to bring about the reaction), shift to a frame of reference in which initially the two protons have zero total relativistic momentum. In this frame, the two protons have equal and opposite speeds, equal masses, and therefore equal energies. Suppose the energies are such that after the collision the two original protons and the proton-antiproton pair are at rest in this frame of reference. The energy required to bring this about will be the reaction threshold energy.

a. In this reference frame, what is the minimum total relativistic energy needed to create a proton-antiproton pair? What is the minimum total relativistic energy of each original proton?

b. What is the speed of either original proton in this reference frame?

c. Suppose the speed of one of these protons is zero in the laboratory frame of reference. What is the speed of the zero-momentum frame of reference relative to the laboratory frame of reference?

d. What is the speed of the other proton in the laboratory frame of reference?

e. What is the total relativistic energy of this proton? What is its relativistic kinetic energy, the minimum required in the laboratory frame of reference?

f. Only one-third of the relativistic kinetic energy is converted to the rest mass energy of the proton-antiproton pair. What happens to the remainder of the relativistic kinetic energy?

15-37. *An inefficient process.* This problem illustrates the inefficiency of using a high-energy particle to collide with an identical initially stationary one to produce new particles. Let the initially moving particle have speed v_i and total relativistic energy $E = mc^2 = m_0c^2/\sqrt{1 - v_i^2/c^2}$ and let $E_0 = m_0c^2$ be the rest-mass energy of either particle. Let M_0 be the rest mass of the particle formed in the completely inelastic collision and v_f be its speed so that its total relativistic energy is $E = Mc^2 = M_0c^2/\sqrt{1 - v_f^2/c^2}$

a. From the conservation of relativistic momentum and total relativistic energy, show that $v_i/v_f = (m + m_0)/m$.

b. Use this result to show that $1/\sqrt{1 - v_f^2/c^2} = \sqrt{(m + m_0)/2m_0}$

c. From this, show that $M_0c^2 = \sqrt{2m_0\,(m + m_0)}\,c^2$.

d. Show that the energy which appears as new rest-mass energy, the useful energy E_u, is given by $E_u = 2m_0c^2[\sqrt{(m + m_0)/2m_0} - 1]$. Then show that the efficiency η of conversion to new rest-mass energy, defined as the ratio of E_u to the energy the accelerator must give to the incoming particle, is

$$\eta = \frac{2m_0}{m - m_0}\left(\sqrt{\frac{m + m_0}{2m_0}} - 1\right)$$

e. Accelerators have been built which accelerate protons so that their relativistic kinetic energy is 30 times their rest-mass energy. What is the efficiency of conversion of such accelerators in a proton-proton collision? Where is the remainder of the energy?

f. Apply the expression for E_u to the case of the formation of a proton-antiproton pair discussed in Exercise 15-36, and compare the results.

g. In some modern high-energy particle accelerators, this "inefficiency" problem is overcome by making the desired collisions occur between two beams of particles moving in opposite directions, instead of aiming a single beam of particles at a stationary target. Refer to Exercise 15-36, and explain why this approach is desirable.

Numerical

15-38. *Kinetic energy versus momentum.* Write a program that will make the calculating device you use evaluate Eq. (15-21*c*), $K = \sqrt{(cp)^2 + (m_0c^2)^2} - m_0c^2$. Check it against the result of Example 15-4*a* by taking $m_0c^2 = 8.20 \times 10^{-14}$ J, the electron rest-mass energy, and evaluating K for the value of cp employed in the example. Then use a number of other values of cp so as to obtain values of K from which you can produce a quantitative version of Fig. 15-11 for the case in which m_0 is the electron rest mass. Plot these values over a range of K extending from 0 to 5 m_0c^2.

Answers

1. Most answers are given to three significant figures. The departures from this convention occur where appropriate, based either on the problem statement or on the numerical details.

2. Unless otherwise indicated, g has been assigned the value 9.80 m/s^2.

3. For exercises which require estimates, the numerical answers are preceded by "Est."

CHAPTER 2

1. Est: (*a*) 0.1 m/s = 3.3×10^{-10} c; (*b*) 1 m/s = 3.3×10^{-9} c; (*c*) 6.7 m/s = 2.2×10^{-8} c; (*d*) 27 m/s = 9×10^{-8} c; (*e*) 291 m/s = 9.7×10^{-7} c; (*f*) 8.1×10^3 m/s = 2.7×10^{-5} c; (*g*) 1.01×10^3 m/s = 3.4×10^{-6} c; (*h*) 3×10^4 m/s = 10^{-4} c

3. 500 s = 8.33 min

5. (*a*) 9.47×10^{12}; (*b*) 4.33 light-years; (*c*) 8.66 yr

7. (*a*) 8.8×10^{-2} s; (*b*) 0.22

9. (*a*) 0.441 s

13. 8 m/s^2

15. (*a*) 4 m/s^2; (*b*) 4.55 m/s^2

17. Est: (*a*) 50 mi/h; (*b*) 2×10^3; (*c*) 0.82 h/day, 5%

19. $v = v_i + a(t - t_i)$, $x = x_i + v_i(t - t_i) + \frac{1}{2}a(t - t_i)^2$,

$$x = x_i + \frac{v^2 - v_i^2}{2a} \text{ (no change)},$$

$$x = x_i + \tfrac{1}{2}(v + v_i)(t - t_i)$$

21. (*a*) 0.5 m/s; (*b*) -0.5 m/s; (*c*) -0.167 m/s; (*d*) 0

23. 40.7 m

25. (*a*) $\pi/4$ s = 0.785 s; (*b*) 0; (*c*) $3\pi/4$ s = 2.36 s, 0.12 m/s^2

29. (*a*) 17.1 mi/(h · s) = 7.66 m/s^2 = 0.78 g; (*b*) 46.9 m; (*c*) 176 mi/h = 78.5 m/s, 88 mi/h = 39.3 m/s, 10.2 s

31. 11.2 m

33. (*a*) 29.4 m/s; (*b*) 44.1 m

35. (*c*) 1.78 s, 11.2 m

37. (*a*) 1.48 s; (*b*) 10.7 m

39. (*a*) 0.981 s, 5.39 m/s. (*b*) 0.563 s, 20.5 m/s. (*c*) Actual round-trip time is 1.54 s, while constant-speed time would be 1.33 s. (*d*) Balloon intended for Lou must be dropped 1.70 s before Hugh throws the ball; balloon intended for Hugh must be dropped 1.19 s before Hugh throws the ball. Yes. 0.05 s. 0.51 s. (*e*) 0.55 s. (*f*) Lou's tomato strikes first, 5.13 s after it was thrown (and 4.13 s after missing on the way up); it hits her at 20.3 m/s. Hugh's tomato hits 0.44 s later (5.57 s after it was thrown); it hits at 24.6 m/s.

CHAPTER 3

1. (*a*) 0.5 s; (*b*) 8.0 m/s^2

3. (*a*) 0.553 s, 0.553 s, 1.11 s; (*b*) v_x = 1.80 m/s, v_y = 5.42 m/s; (*c*) 71.6°, 5.71 m/s; (*d*) 5.71 m/s

5. 46.2 m

7. 6.14 m/s

9. (*b*) yes; (*c*) no

11. (*b*) $\mathbf{A} + \mathbf{B}$ = 5.0 cm at 0°, $\mathbf{A} - \mathbf{B}$ = 8.66 cm at 90°, $\mathbf{B} - \mathbf{A}$ = 8.66 cm at 270° = −90°

13. 247 km, 26.6° south of east

15. 20 cm at C, 45 cm at D

17. D

19. (*a*) 5.00 m/s; (*b*) 0; (*c*) 2.62 m/s^2 toward center of circle

21. (*a*) 189 km/h; (*b*) 5.6° east of north

23. 6.43 m/s = 14.3 mi/h downward, 7.84 m/s = 17.4 mi/h

27. (*a*) $t_1 = 1.41\, v_0/g$; (*b*) $t_2 = (1.04\, v_0/g) + \Delta t$; (*c*) t_1 = 5.04 s, t_2 = 4.20 s, relay method is quicker

29. (*a*) 0.344%; (*b*) 5.07×10^3 s = 84.5 min; (*c*) the two periods would be equal

31. (*c*) 53.1°, 5 m/s; (*d*) 133 m

33. (*a*) $R_x = (2v_0^2/g)(\sin\theta\cos\theta - \tan\alpha\cos^2\theta)$; (*b*) $\theta_{\max} = 45° + \alpha/2$; (*c*) 60°; (*d*) 41.2 m/s

35. (*a*)$\mathbf{w} = v_0 \cos\theta\, \hat{\mathbf{x}} + (u_0 - v_0 \sin\theta)\, \hat{\mathbf{y}}$, or $w = \sqrt{v_0^2 + u_0^2 - 2u_0 v_0 \sin\theta}$ at an angle of $\sin^{-1}(v_0 \cos\theta/w)$ with y axis;
(*b*) $w = (25.2 - 4.1 \sin\theta)^{1/2}$ at angle of $\sin^{-1}(0.41 \cos\theta/w)$, $w_{\max}$ = 5.4 m/s, the maximum angle between $\mathbf{w}$ and $\hat{\mathbf{y}}$ is $\sin^{-1}(v_0/u_0)$ = 4.7°, which occurs when θ = 4.7° and also when θ = 175.3°

CHAPTER 4

1. (*a*) $\frac{1}{32}$ slug; (*b*) 4.1 slug; (*c*) divide weight in pounds by g; (*d*) 32 lb; (*e*) $\frac{1}{32}$ slug

3. 0.5 m

5. (*a*) 6.25×10^3 N; (*b*) 213 times the weight

7. 1.86×10^2 N on moon, 1.13×10^3 N on earth

9. (*a*) 8.0×10^5 N; (*c*) net force is 4.1% of the weight

13. 6.67 m/s

15. (*a*) 1.1 m/s; (*b*) to minimize the recoil speed

17. The ground exerts forces on the ball during impact.

19. (*a*) 130.7 kg/s. (*b*) The thrust remains the same but the rocket's mass is smaller.

21. (*a*) They meet halfway between their initial positions. (*b*) a length L.

23. Block slides forward when stopping if $|a| > \mu_s g$. Block will not slide backward when starting if $|a| < \mu_s g$.

25. (*a*) $\mathbf{v}_{1i} = \mathbf{v}_{1f} + \mathbf{v}_{2f}$. (*b*) The three vectors form a closed triangle and therefore lie in a plane. (*d*) The vector $\mathbf{v}_{1i}$ is a diagonal of the parallelogram with sides $\mathbf{v}_{1f}$, $\mathbf{v}_{2f}$. (*e*) The vector difference $\mathbf{v}_{1f} - \mathbf{v}_{2f}$ is the other diagonal. (*f*) The parallelogram must then be a rectangle, and therefore $\mathbf{v}_{1f}$ must be perpendicular to $\mathbf{v}_{2f}$.

27. (*a*) 23 N; (*b*) 2.9×10^{-2} s; 0.58 m

29. (*a*) $\mathbf{v}_{1f} = -v_0\hat{\mathbf{x}}/2$, $\mathbf{v}_{3f} = v_0\hat{\mathbf{x}}/2$; (*b*) $\mathbf{v}_{1f} = -\frac{3}{2}v_0\hat{\mathbf{x}}$, $\mathbf{v}_{3f} = -\frac{1}{2}v_0\hat{\mathbf{x}}$

33. (b) $m_2/m_1 \simeq 2.9$. (c) The data are consistent with an elastic collision.

35. (a) 4.43 m/s; (b) 3.13 m/s; (c) 3.78×10^3 m/s^2; (d) 3.79×10^2 N; (e) 3.87×10^2

37. $k'' = k_1 + k_2$

39. (a) $x_M = \mu_s L$, $y_M = \mu_s^2 L/2$; (b) $x_M = 8m$, $y_M = 3.2m$

41. 4.2 m/s

43. (a) $-\frac{2}{3}$ m/s; (b) $\frac{4}{3}$ m/s; (c) 2 s; (d) $\frac{8}{3}$ m; (e) $-\frac{4}{3}$ m; (f) 0

45. (a) $|\mathbf{v}_{Bf} - \mathbf{v}_{Af}| = 4.29$ m/s, $|\mathbf{v}_{Bi} - \mathbf{v}_{Ai}| = 17.3$ m/s, no; (b) 2.48; (c) 15.9 m/s at 59.0°

47. (a) Taking the positive direction to be downward, the equation reads $Ma = Mg + T_l - T_u$.

49. Letting F_A and F_B represent the frictional forces at interfaces A and B, (a) $F_A = m_1 a$, $F_B = (m_1 + m_2)[a \cos \theta_B + g \sin \theta_B]$, $F_A/F_{A\max} = a/\mu_A g$, $F_B/F_{B\max} = \dfrac{1}{\mu_B}\left[\dfrac{a/g + \tan\theta_B}{1 - (a/g)\tan\theta_B}\right]$. (b) There is slippage at A if $a \geqslant \mu_A g$; there is slippage at B if $a \geqslant g\left(\dfrac{\mu_B - \tan\theta_B}{1 + \mu_B \tan\theta_B}\right)$ Interface A slips first if $\mu_A < (\mu_B - \tan\theta_B)/(1 + \mu_B \tan\theta_B)$. (c) The equations of part b apply with the substitution $\theta_B \to -\theta_B$. Interface A slips first if $\mu_A < (\mu_B + \tan\theta_B)/(1 - \mu_B \tan\theta_B)$. (d) $12.1° \leqslant \theta_B \leqslant 38.7°$.

51. (a) Trucker can stop without slippage if $v_0^2/2\mu_s g < S_0$. (b) $v_0^2/2\mu_s g = 91.8$ m $< S_0$, and so trucker can stop without slippage. (c) Trucker can stop without slippage if $v_0 \Delta t + v_0^2/2\mu_s g < S_0$. (d) $v_0 \Delta t + v_0^2/2u_s g = 107$ m$> S_0$, and so trucker cannot stop without slippage.

CHAPTER 5

1. $a = g$

3. 1.2 m/s^2

5. (a) 3.27 m/s^2 downward; (b) 4.9 m/s^2 downward; (c) 13.1 N; (d) 9.8 N

7. (a) 2.5 N directed opposite to motion; (b) 0.255

11. 0.857 m/s^2 forward

13. Est: fictitious force of 0.5 mg downward, 50% increase in perceived weight

15. (a) 2.25×10^5 N; (b) 7.5 m/s^2

17. (a) 128 N; (b) 64 N

19. (a) 6.53 m/s^2, 1 kg downward, 2 kg upward; (b) 13.1 m/s^2, 1 kg downward, 2 kg upward

23. (a) $\mu_s < \tan\theta$ for block to slide back; (b) $v_f/v_i = \left(\dfrac{\tan\theta - \mu_k}{\tan\theta + \mu_k}\right)^{1/2}$; (c) 18.6 m/s

27. 5.08×10^3 s = 84.6 min

29. (a) $\cos\theta = m/M$; (b) $L = \dfrac{l^2 gM}{4\pi^2 m}$; (c) $T = 2\pi\sqrt{h/g}$

31. $\sqrt{2}$

33. (a) $d_2 = 2d_1$; (b) $a_1 = 2.18$ m/s^2, $a_2 = 4.36$ m/s^2; (c) 0.545 N

35. The bananas rise with the same acceleration and speed as the monkey until they get stuck against the pulley.

37. (a) 9.8 N; (b) $a_2 = 4.9$ m/s^2 down, $a_A = 2.45$ m/s^2 down, $a_3 = 2.45$ m/s^2 up; (c) 19.6 N; (d) 1.6 kg

39. (a) no

CHAPTER 6

1. 0.25 kg

3. 1.58 Hz

5. 6.21 cm

7. (a) $x = (10.0 \cos 10\pi t - 10.0 \sin 10\pi t)$ cm; (b) $x = 14.1$ cm $\cos(10\pi t + \pi/4)$; (c) $x_{\max} = 14.1$ cm, $v_{\max} = 443$ cm/s, $a_{\max} = 1.39 \times 10^4$ cm/s^2

11. $2\pi\sqrt{d/g}$

13. 1.72 s

15. (b) $N = N_0 e^{Rt}$

19. (a) $\mu_s g/4\pi^2\nu^2$; (b) 1.65 cm

21. (a) The period $T = 2\pi\sqrt{\frac{m_1m_2}{(m_1+m_2)k}}$, the velocities of 1 and 2 are related by $v_2/v_1 = -m_1/m_2$, and the amplitudes A_1 and A_2 are related by $A_2/A_1 = m_1/m_2$; (c) $T = 2\pi\sqrt{m/2k}$, $v_2 = -v_1$, $A_2 = A_1$

23. (a) 112 s; (b) 30 s

25. (a) $2y$; (b) ρAl; (c) $F = -2\rho gAy$; (d) $2\pi\sqrt{l/2g}$

CHAPTER 7

1. (a) (1) 120 J, (2) 250 J, (3) 315 J, (4) 6.25×10^5 J, (5) 1.5×10^7 J; (b) (1) 0.171 m, (2) 0.357 m, (3) 0.45 m, (4) 893 m (5) 21.4 km

3. m/M

5. 42.9°

9. (a) $\frac{1}{2}m(v_f^2 - v_i^2)$; (b) $\frac{1}{2}m(v_f^2 - v_i^2)/s$; (c) $W = 225$ J, $F = 11.3$ N

11. (a) 9.8×10^7 J; (b) 4.9×10^7 J

13. (a) 866 J, (b) 86.6 N; (c) -866 J; (d) 0

15. (a) 6.26 m/s; (b) 49 N

19. $W = mgx$

21. (a) $\sqrt{\frac{2m_BgD}{m_A + m_B}}$; 7.67 m/s

23. (a) 783 J; (b) 695 J; (c) 2.82×10^3 J; (d) 371 J; (e) 2.62×10^3 J; (f) 92.4 J

25. (a)$V_f - V_i = |v_g'| \ln(m_f/m_i)$; (b) $m_f = 0.223\ m_i$

27. (a) 44 J; (b) 44 J; (c) 44 J; (d) yes

29. (a) 4.90 J; (b) 2.55 J; (c) 2.35 J

31. (a) 4.43 m/s; (b) 0.333

33. 20 m

35. (a) $\sqrt{gR}$; (b) $R/2$; (c) $\frac{K_{\text{actual}}}{K_{\text{ideal}}} = \frac{1}{1.3} = 0.769$

37. (a) $x_{\max} = \sqrt{(\mu_k mg/k)^2 + mv_0^2/k} - \mu_k mg/k$; (b) block returns if $kx_{\max} > \mu_s mg$; (c) $x_{\max} = 0.202$ m, block does not return

39. 2 mg

41. (a) 0.500 J; (b) 0.500 J; (c) 0.667 J; (d) no

43. (a) $\frac{1}{2}k(h + r - l_0)^2$; (b) $\frac{1}{2}k(|h - r| - l_0)^2$; (c) $2kr(h - l_0)$ if $h > r$, $2kh(r - l_0)$ if $r > h$; (d) $\frac{1}{2}k(\sqrt{h^2 + r^2 + 2hr\cos\theta} - l_0)^2$; (f) the answers for parts a, b, and d

45. (a) 0.791 m/s; (b) 0.601 m/s

CHAPTER 8

1. 206 W, 0.276 hp

3. (a) 16.3 W/m²; (b) 5.45×10^{-3}

5. (a) $2\pi r(T_1 - T_2)$; (b) $2\pi rn(T_1 - T_2)$; (c) 4.24×10^4 W = 56.9 hp

7. r_2/r_1

9. (a) $(R/r)^3$; (b) 125, 7.84 N

11. (a) 2; (b) 8; (c) 4

13. (a) NR/r; (b) 24

15. (a) MgD/d; (b) 490 N; (c) 1.63×10^6 N/m²; (d) 16.1 atm

17. $M/(M + m)$

19. (a) 96.1%; (b) 4.71 N; (c) 6.00×10^4 N/m² = 0.592 atm

21. (a) 1.98 m/s; (b) 1.01×10^{-3} s, 10^{-3} m, yes; (c) 2 mg; (d) 4×10^{-6} J, 4×10^{-3} W; (e) flea: 2×10^3 W/kg, climber: 4.93 W/kg

23. (a) m_1gh; (b) $\frac{1}{2}(m_1 + m_2)v^2 + m_2gh$; (c) $\sqrt{\frac{2gh(m_1 - m_2)}{m_1 + m_2}}$; (d) $\frac{(m_1 - m_2)h}{m_1 + m_2}$

25. (a) $\mathbf{v}_{1f} = \frac{m_1 - m_2}{m_1 + m_2}\mathbf{v}_{1i}$, $\mathbf{v}_{2f} = \frac{2m_1\mathbf{v}_{1i}}{m_1 + m_2}$; (b) $\frac{4m_1m_2}{(m_1 + m_2)^2}$; (c) for $m_2 = m_1$; (d) 0.284

27. (a) $K_{1i} = 3.13 \times 10^{-2}$ J, $K_{2i} = 0$, $K_{1f} = 1.25 \times 10^{-2}$ J, $K_{2f} = 6.29 \times 10^{-3}$ J, $|\Delta K| = 1.25 \times 10^{-2}$ J; (b) $|\Delta K| = 1.17 \times 10^{-2}$ J

31. Est: (a) 2.85 kg·m/s; (b) 4×10^{-3} s; (c) 710 N; 18 kW

33. (a) and (b): $m\omega^2A^2/4$

35. (a) elastic if $\epsilon = 1$, inelastic if $0 \leq \epsilon < 1$

37. (a) $14.5°$, $37.8°$; (b) $K_{pf} = 0.400K_{\alpha i} = 1.2 \times 10^6$ eV

39. (b) $\frac{1}{2}(m_1 + m_2)v_i^2 + \frac{1}{2}kd^2$;
(c) $u_f = \omega d$ at angle α with x axis;
(d) $v_{1f} = \left[v_i^2 + \left(\frac{m_2\omega d}{m_1 + m_2}\right)^2 - \frac{2v_i m_2\omega d \cos\alpha}{m_1 + m_2}\right]^{1/2}$,
$\beta_1 = \tan^{-1}\left[\frac{m_2\omega d \sin\alpha}{v_i(m_1 + m_2) - m_2\omega d \cos\alpha}\right]$,
$v_{2f} = \left[v_i^2 + \left(\frac{m_1\omega d}{m_1 + m_2}\right)^2 + \frac{2v_i m_1\omega d \cos\alpha}{m_1 + m_2}\right]^{1/2}$,
$\beta_2 = \tan^{-1}\left[\frac{m_1\omega d \sin\alpha}{v_i(m_1 + m_2) + m_1\omega d \cos\alpha}\right]$;
(e) $t_r = 7.85 \times 10^{-2}$ s, $K_f = 19.6$ J, $\mathbf{u}_f = 4.00$ m/s at $\alpha = 110°$, $\mathbf{v}_{1f} = 2.96$ m/s at $\beta_1 = 30.6°$, $\mathbf{v}_{2f} = 2.54$ m/s at $\beta_2 = 62.4°$

CHAPTER 9

1. 6.89 rad

5. (a) 1.57 rad/s^2; (b) 1800 revolutions

7. (a) v_0/r; (b) $v_0/(R - r)$; (c) 20 rad/s, 5 rad/s

9. The day would become longer.

13. 1700 km below the surface

15. $x_c = 0.5l$, $y_c = 0.3l$

17. $F_0 = mg(1 - r_1/r_2) + Mg/2$

21. $\mathbf{a} = -\alpha r\hat{\mathbf{v}} - \omega^2 r\hat{\mathbf{r}}$

25. 1.5 m

31. (a) $\mu_k h/2$; (b) $w/2\mu_k$; (c) 0.165 m, 0.758 m

35. (a) 5.88×10^3 N; (b) horizontal: 5.09×10^3 N; vertical: 1.96×10^3 N; (c) 2.29×10^4 Nm, balancing torque is due to forces acting underground

39. (a) $m_1\mathbf{v}_1/(m_1 + m_2)$; (b) $m_2\mathbf{v}_1/(m_1 + m_2)$; (c) $\frac{-m_1\mathbf{v}_1}{m_1 + m_2}$; (d) 0

45. $4MgR(1 - \cos\theta)/3\pi + U(0)$

47. (a) $26.6°$; (b) $45°$

49. (a) 225 N; (b) 298 N upward at an angle of $107.6°$ with top surface of obstacle; (c) $F_t' = 224$ N upward, $F_n' = 196$ N inward, $F_t'/F_n' = 1.14$

51. (a) $\theta_0 = \tan^{-1}\left(\frac{1 - \mu_G\mu_W}{2\mu_G}\right)$; (b) $46.4°$; (c) $51.3°$; (d) $90°$; (e) μ_G

53. (a) on the angle bisector, a distance $\frac{l\sqrt{2}}{4}$ from B; (b) $18.4°$

57. (a) $l/2$; (b) $l/4$, $3l/4$; (c) $l/2N$,
$\frac{l}{2}(1 + \frac{1}{2} + \frac{1}{3} + \cdots + 1/N) = \frac{l}{2}\sum_{n=1}^{N}\frac{1}{n}$

59. $5r/3$

CHAPTER 10

1. (a) hoop; MR^2; square: $\frac{4}{3}MR^2$

3. 19.7 km

7. $\frac{1}{3}MA^2$

11. (a) $3g\sin\theta/2l$; (b) $3g\sqrt{\frac{5}{4} - 2\cos\theta + \frac{3}{4}\cos^2\theta}$

13. (a) $2\pi\sqrt{2R/g}$; (b) $2\pi\sqrt{3R/2g}$; (c) part a, 15.5% longer

15. 2 rotations per second

17. (a) $g/51$; (b) $50\,Mg/51$

19. (a) $\frac{2g}{3}\sin\theta$; (b) $\frac{g}{2}\sin\theta$; (c) $3s_c/4$

25. $2R/5$ above center

29. $a = 0.134g$, $M/m = 3.05$

31. 1.1×10^3 J

33. (a) $\frac{1}{4}m_0 r_0^2\omega_s^2$, 1.97 J; (b) $\omega_p = \frac{2dg}{r_0^2\omega_s}$, 2.34 rad/s, $\omega_p/\omega_s = 3.72 \times 10^{-3}$, yes

35. (*a*) $(7R - 27r)/10$

37. (*a*) $3R_0/4$; (*b*) $R_i/R_0 = 0.823$, $V_c/V_0 = 0.557$

39. (*a*) $I_e/I_{sph} = 0.828$; (*b*) 5.86×10^{33} kg·m²/s; (*c*) 2.68×10^{40} kg·m²/s; (*d*) 2.19×10^{-7}; (*e*) 2.14×10^{29} J; (*f*) 2.66×10^{33} J (*g*) 8.06×10^{-5}

41. (*b*) $dM = \frac{3M}{R^3} r^2 dr$; (*c*) $dI = \frac{2M}{R^3} r^4 dr$; (*d*) $0 \leqslant r \leqslant R$

43. no

45. (*a*) $2l/3$ from A; (*b*) $2\pi\sqrt{2l/3g}$; (*c*) $2l/3$

47. (*a*) $\lambda^3 M_1$; (*b*) $\lambda^5 I_1$; (*c*) $\lambda = 1.2$: $M_2 = 1.73M_1$, $I_2 = 2.49I_1$; $\lambda = 2.0$: $M_2 = 8M_1$, $I_2 = 32\,I_1$; $\lambda = 5.0$: $M_2 = 125M_1$, $I_2 = 3125\,I_1$

49. (*a*) $I = \frac{2}{5}MR^2 + \frac{1}{2}mc^2$.
(*b*) $[(H + 5R/8)M + (H + R + l)m]/(M + m)$.
(*c*) $\dfrac{[(H/R + \frac{5}{8}) + (H/R + 1 + l/R)(m/M)]\,g}{[2/5 + mc^2/2MR^2]R\omega_s}$.
(*d*) For maximum precession rate, use $l = D - d/2$. For minimum precession rate, use $l = d/2$.
(*e*) 1.39.

51. Subscripts h, p, and c refer to hanging weight, pulley, and cylinder, respectively.
(*a*) $a_h = \frac{4g}{15}(2 - \sin\theta)$,
$\alpha_p = \frac{4g}{15R}(2 - \sin\theta)$, $a_c = \frac{2g}{15}(2 - \sin\theta)$,
$\alpha_c = \frac{2g}{15R}(2 - \sin\theta)$; (*b*) $T_h = \frac{Mg}{15}(7 + 4\sin\theta)$,
$T_c = \frac{Mg}{5}(1 + 2\sin\theta)$; (*c*) $\dfrac{1 + 7\sin\theta}{15\cos\theta}$

CHAPTER 11

(*Note:* Small discrepancies between your results and the answers given here may be due to use of differing values for G, the earth's mass and/or radius, and so on.)

1. 8.06×10^6 m, 1.68×10^6 m

3. 275 m/s²

5. 6.04

7. 3.26 s

9. $\frac{9}{10}$ of the way from the earth's center to the moon's center, also $\frac{9}{8}$ of the way from the earth's center to the moon's center, or 3.46×10^5 km and 4.32×10^5 km from earth's center; 8.6×10^4 km

13. Est: 30 km

15. (*a*) 2; (*b*) concave toward the sun

19. (*b*) 3.30×10^5

23. (*a*) $\frac{1}{2}m_1\omega^2 r_1^2$, $\frac{1}{2}m_2\omega^2 r_2^2$, $\frac{1}{2}(m_1 r_1^2 + m_2 r_2^2)\omega^2$

25. (*a*) $r_e/2h$; (*b*) $\frac{1}{2}$

27. (*a*) $\dfrac{2\pi R_1 R_2}{\sqrt{G\,M_e}}\sqrt{\dfrac{M_1R_1 + M_2R_2}{M_1R_2^2 + M_2R_1^2}}$;

(*b*) $\dfrac{G\,M_e M_1 M_2}{R_1^2}\left\{\dfrac{1}{M_1 + \left[M_2\left(1 + \frac{L}{R_1}\right)\right]}\right\} \times \left[\dfrac{3\left(\frac{L}{R_1}\right) + 3\left(\frac{L}{R_1}\right)^2 + \left(\frac{L}{R_1}\right)^3}{\left(1 + \frac{L}{R_1}\right)^2}\right]$ (*c*) 93.0 min, 2.66×10^{-3} N $= 3.88 \times 10^{-6}\,M_2 g$; (*d*) 95.0 N

29. (*c*) 1.57

31. (*a*) 7.14×10^3 km; (*b*) $\gamma = 1.006 \times 10^4$ km, $\delta = 9.45 \times 10^3$ km, $c = 3.44 \times 10^3$ km

33. (*a*) 8.19 km/s; (*b*) 0.64 km/s; (*c*) 5.73 km/s; (*d*) increase speed to 6.32 km/s (an increase of 0.59 km/s); (*e*) 141°

CHAPTER 12

(*Note:* In the answers to Chapter 12 exercises, the units of x and t are assumed to be m and s, respectively, unless otherwise specified.)

3. 0.75 s

5. 24.5 min

7. 3.0 m

9. (*a*) 10 Hz; (*b*) 0.4 m, 4 m/s, $y = 0.10$ m $\cos(20\pi t - 5\pi x)$

11. (*a*) $0.2\pi \sin(10\pi x - 40\pi t)\frac{\text{m}}{\text{s}}$; (*b*) 0.628 m/s, transverse displacement is zero.

13. (*a*) 4.0 m; (*b*) 80 N; (*c*) 0.987 J/m; (*d*) 19.7 W

15. (*a*) 1.13×10^{-4} W; (*b*) 0.113 J

17. 305.6 Hz, 294.6 Hz

23. $0.020 \text{ m} \sin(2\pi x - 10\pi t)$

25. 42.6 m/s

27. (*a*) 766 Hz; (*b*) 837 Hz

29. (*a*) 71.4 s; (*b*) 101.1 Hz, 98.9 Hz

31. (*a*) 10 Hz; (*b*) positive direction: $\lambda = \frac{0.4 \text{ m}}{8n + 1}$, $v = \frac{4 \text{ m/s}}{8n + 1}$ for n = 0, 1, 2, 3, and so on; negative direction: $\lambda = \frac{0.4 \text{ m}}{8n - 1}$, $v = \frac{4 \text{ m/s}}{8n - 1}$, for $n = 1, 2, 3, \ldots$ and so on.

37. $\rho_E = 5$ J/m and $S = 760$ W for $|x - vt| < 0.1$ m; ρ_E and S are both zero otherwise; $E = 1.0$ J

39. (*b*) 54.4 km

CHAPTER 13

3. 1.09×10^3 N

5. 11

7. $\frac{1}{3}$ of the distance from the end of the string (so that the vibrating portion has $\frac{2}{3}$ of its original length)

9. (*a*) $0.309h$, $0.951h$, $0.809h$; (*b*) $0.809h$, $0.588h$, $0.309h$; (*c*) 0, 0, 0

11. (*a*) 283 Hz; (*b*) 850 Hz

13. 252 Hz, 260 Hz

15. 42.5 Hz, 127.5 Hz

23. 317 m/s

25. (*a*) $A \cos(2\pi\nu t + \delta_1) + A \cos(2\pi\nu t + \delta_2) + A \cos(2\pi\nu t + \delta_3)$; (*c*) $\frac{\delta_1 + \delta_2}{2}$, $A[1 + 2 \cos\frac{1}{2}(\delta_1 - \delta_2)]$; (*d*) (1) 0°, $3A$, (2) 30°, $2.73A$, (3) 45°, $2.41A$, (4) 60°, $2A$, (5) 90°, A; (*f*) If $\cos\left(\frac{\delta_1 - \delta_2}{2}\right) < 0$ then use $\delta_3 = \left(\frac{\delta_1 + \delta_2}{2}\right) \pm \pi$.

27. (*a*) $\frac{C_1}{\sqrt{r_1}} \cos(kr_1 - kvt + \delta_1) + \frac{C_2}{\sqrt{r_2}} \cos(kr_2 - kvt + \delta_2)$; (*c*) $\frac{r_1}{r_2} = \left(\frac{C_1}{C_2}\right)^2$, $k(r_1 - r_2) = (2n + 1)\pi - (\delta_1 - \delta_2)$, where $n = 0, \pm 1, \pm 2, \ldots$.

29. (*a*) $\frac{1}{2}\mu\omega^2 A^2$

CHAPTER 14

1. $0.995c$

3. $0.948c$

5. (*a*) 1.67 h; (*b*) 1.33 h, 3 h; (*c*) 4 h, 5h

7. (*a*) $\sqrt{2}$; (*b*) $1/\sqrt{2}$

9. 1.0 percent

11. (*a*) 70.9°; (*b*) 0.917 m

13. (*a*) A': ($\frac{1}{2}$m, $\frac{1}{2}$m), B': ($-\frac{1}{2}$m, $\frac{1}{2}$m), C': ($-\frac{1}{2}$m, $-\frac{1}{2}$m), D': ($\frac{1}{2}$m, $-\frac{1}{2}$m), A: $\left(\frac{1}{2\gamma} \text{ m}, \frac{1}{2}\text{m}\right)$, B: $\left(-\frac{1}{2\gamma} \text{ m}, \frac{1}{2}\text{m}\right)$, C: $\left(-\frac{1}{2\gamma} \text{ m}, -\frac{1}{2}\text{m}\right)$, D: $\left(\frac{1}{2\gamma} \text{ m}, -\frac{1}{2}\text{m}\right)$; (*b*) E': $\left(\frac{1}{\sqrt{2}} \text{ m}, 0\right)$, F': $\left(0, \frac{1}{\sqrt{2}} \text{ m}\right)$, G': $\left(-\frac{1}{\sqrt{2}} \text{ m}, 0\right)$, H': $\left(0, -\frac{1}{\sqrt{2}} \text{ m}\right)$, E: $\left(\frac{1}{\gamma\sqrt{2}} \text{ m}, 0\right)$, F: $\left(0, \frac{1}{\sqrt{2}} \text{ m}\right)$, G: $\left(-\frac{1}{\gamma\sqrt{2}} \text{ m}, 0\right)$, H: $\left(0, -\frac{1}{\sqrt{2}} \text{ m}\right)$.

15. (*b*) $\gamma L'$; (*c*) no

21. (*a*) proper; (*b*) γt_1, where $\gamma = 1/\sqrt{1 - v^2/c^2}$; (*c*) $\gamma v t_1$; (*d*) dilated; (*e*) t_2/γ; (*f*) $(t_2/\gamma) - \gamma t_1$; (*g*) $\gamma v t_1$

23. (*a*) $c - V$; (*b*) $L/(c - V)$; (*c*) $c + V$; (*d*) $L/(c + V)$, $2Lc/(c^2 - V^2)$; (*f*) $2L/(c^2 - V^2)^{1/2}$; (*h*) LV^2/c^2; (*i*) $LV^2/\lambda c^2$ (*j*) $\frac{1}{6}$; (*k*) $\frac{1}{3}$ of a fringe

25. (*a*) $1 \mu s/\sqrt{1 - v_s^2/c^2}$;
(*b*) $(c/\nu)/\sqrt{1 - v_s^2/c^2} = \dfrac{300 \text{ m}}{\sqrt{1 - v_s^2/c^2}}$
(*c*) $(|v_s|/\nu)/\sqrt{1 - v_s^2/c^2} = \dfrac{(300 \text{ m})(|v_s|/c)}{\sqrt{1 - v_s^2/c^2}}$;
(*d*) $\dfrac{(c - |v_s|)}{\nu\sqrt{1 - v_s^2/c^2}} = 300 \text{ m} \sqrt{\dfrac{1 - |v_s|/c}{1 + |v_s|/c}}$; (*f*) 0.5 μs, 2 MHz; (*g*) 2 μs; (*h*) $\nu' = \nu\sqrt{\dfrac{1 \pm |v_s|/c}{1 \mp |v_s|/c}}$, with upper signs for motion towards the earth

27. (*a*) $0.368c$; (*b*) 88.3 MHz; (*c*) 82.1 MHz

31. (*a*) -173 s, no; (*b*) yes, $3\,c/7$; (*c*) $\Delta t > 2100$ s = 35 min; (*d*) $0.538c$, 2.92×10^3 s = 48.7 min

33. (*a*) $c^2(t_2' = t_1')^2 - (x_2' - x_1')^2 - (y_2' - y_1')^2 - (z_2' - z_1')^2$

35. (*b*) in the direction of the earth's motion; (*d*) 20.6 arc seconds

CHAPTER 15

1. $3.14m_0 = 5.93 = 10^{-28}$ kg, $2.97m_0c = 1.68 \times 10^{-19}$ kg·m/s, $2.14m_0c^2 = 3.64 \times 10^{-11}$ J

3. 52.8 m

5. 8.88×10^{-13} J

7. $E_{\text{min}} = 2m_ec^2 = 1.64 \times 10^{-13}$ J, $c\sqrt{3}/2$

9. (*a*) $2m_0c^2/\sqrt{1 - v^2/c^2}$; (*b*) $2m_0c^2/\sqrt{1 - v^2/c^2}$; (*c*) $2m_0c^2$

13. 9.34×10^{-13} J

15. $6.75 \times 10^{-4}\ \text{u}c^2 = 1.01 \times 10^{-13}$ J

19. $\left(1 + \dfrac{\sqrt{3}}{2}\right) m_{0\pi}c^2 = 4.03 \times 10^{-11}$ J,
$\left(1 - \dfrac{\sqrt{3}}{2}\right) m_{0\pi}c^2 = 0.29 \times 10^{-11}$ J

21. $E_K = 1.47 \times 10^{-10}$ J, $E_\pi = 1.25 \times 10^{-10}$ J

25. (*a*) 1.25×10^{-13} J; (*b*) $K_e = 1.25 \times 10^{-13}$ J, $K_p = 1.20 \times 10^{-16}$ J, $p_e = p_p = 6.33 \times 10^{-22}$ kg·m/s; (*c*) $E_{\bar{\nu}} = K_p = 1.25 \times 10^{-13}$ J, $K_p = 5.20 \times 10^{-17}$ J, $p_{\bar{\nu}} = p_p = 4.17 \times 10^{-22}$ kg·m/s

27. (*a*) Ft; (*b*) $Fx + m_0c^2$; (*c*) $x = \sqrt{c^2t^2 + \left(\dfrac{m_0c^2}{F}\right)^2} - \dfrac{m_0c^2}{F}$; (*d*) $v = (Ft/m_0)/\sqrt{1 + \left(\dfrac{Ft}{m_0c}\right)^2}$; $a = (F/m_0)/\left[1 + \left(\dfrac{Ft}{m_0c}\right)^2\right]^{3/2}$; yes.

29. (*c*) 78.5°

33. 6.95×10^{-10} J

37. (*e*) 20 percent

Index

Important Physical Constants*

Quantity	Symbol	Value
Universal gravitational constant	G	$6.67 \times 10^{-11}\ \mathrm{N \cdot m^2/kg^2}$
Speed of light	c	$3.00 \times 10^{8}\ \mathrm{m/s}$
Permeability of free space	μ_0	$4\pi \times 10^{-7}\ \mathrm{T \cdot m/A}$ (by definition)
	$\mu_0/4\pi$	$1 \times 10^{-7}\ \mathrm{T \cdot m/A}$
Permittivity of free space	$\epsilon_o\ (= 1/\mu_0 c^2)$	$8.85 \times 10^{-12}\ \mathrm{C^2/N \cdot m^2}$
	$1/4\pi\epsilon_0$	$8.99 \times 10^{9}\ \mathrm{N \cdot m^2/C^2}$
Elementary charge (magnitude of electron charge)	e	$1.60 \times 10^{-19}\ \mathrm{C}$
Boltzmann's constant	k	$1.38 \times 10^{-23}\ \mathrm{J/K}$
Avogadro's number	A	$6.02 \times 10^{26}\ \mathrm{kmol^{-1}}$
Universal gas constant	$R\ (= Ak)$	$8.31 \times 10^{3}\ \mathrm{J/kmol \cdot K}$
Faraday's constant	$\mathscr{F}\ (= Ae)$	$9.65 \times 10^{7}\ \mathrm{C/kmol}$
Electron rest mass	m_e	$9.11 \times 10^{-31}\ \mathrm{kg}$
Electron rest mass energy	$m_e c^2$	$5.11 \times 10^{5}\ \mathrm{eV}$
Electron charge/mass ratio	e/m_e	$1.76 \times 10^{11}\ \mathrm{C/kg}$
Proton rest mass	m_p	$1.67 \times 10^{-27}\ \mathrm{kg}$
Proton rest mass energy	$m_p c^2$	$9.38 \times 10^{8}\ \mathrm{eV}$
Planck's constant	h	$6.63 \times 10^{-34}\ \mathrm{J \cdot s}$
Bohr magneton	m_B	$9.27 \times 10^{-24}\ \mathrm{A \cdot m^2}$

* More precise values are cited in text; see index.

Mathematical Symbols

$\propto$ is proportional to
$=$ is equal to
$\neq$ is not equal to
$\simeq$ is approximately equal to
$\equiv$ is identical to
$>$ is greater than
$<$ is less than
$\geqslant$ is greater than or equal to
$\leqslant$ is less than or equal to
$|z|$ the absolute value of z
$\langle z \rangle$ the average value of z